Treatment Marshes for Runoff and Polishing

Treatment Marshes for Runoff and Polishing

Robert H. Kadlec

CRC Press
Taylor & Francis Group
Boca Raton London New York

CRC Press is an imprint of the
Taylor & Francis Group, an **informa** business

CRC Press
Taylor & Francis Group
6000 Broken Sound Parkway NW, Suite 300
Boca Raton, FL 33487-2742

CRC Press is an imprint of Taylor & Francis Group, an Informa business

Printed on acid-free paper

International Standard Book Number-13: 978-1-138-32262-2 (Hardback)

Library of Congress Cataloging-in-Publication Data

Names: Kadlec, Robert H., author.
Title: Treatment marshes for runoff and polishing / Robert H. Kadlec.
Description: First edition. | Boca Raton, FL : CRC Press/Taylor & Francis Group, 2019.
Identifiers: LCCN 2019008603 | ISBN 9781138322622 (hardback : acid-free paper)
Subjects: LCSH: Constructed wetlands. | Water–Purification–Biological treatment. | Sewage–Purification–Biological treatment. | Marsh ecology.
Classification: LCC TD756.5 .K333 2019 | DDC 628.1/682–dc23
LC record available at https://lccn.loc.gov/2019008603

Visit the Taylor & Francis Web site at
www.taylorandfrancis.com

and the CRC Press website at
www.crcpress.com

ISBN: 978-1-138-32262-2 (hbk)
ISBN: 978-0-429-45193-5 (ebk)

Contents

Preface

This book does not fit neatly into the compartmentalized world of publishing. It is the third edition of *Treatment Wetlands* but with significant deletions and a new emphasis on nonpoint source waters and ecological value. The focus is exclusively on surface flow wetlands. Subsurface flow wetlands are omitted because the sheer volume of new information does not allow adequate coverage of both surface and subsurface flow systems in a single volume. A third edition concerning subsurface flow wetlands and enhanced systems is in preparation elsewhere.

This work is further restricted to water quality improvement for low source pollutant concentrations, thus excluding the chemical performance of wetlands treating strong wastewater effluents. The subject source waters therefore include urban and agricultural runoff; streams, lakes, and rivers; and highly treated municipal effluents.

The first edition of *Treatment Wetlands* (Kadlec and Knight, 1996) was based upon treatment performance databases and the literature on general wetland science, both of which were deemed to be significant at the time. But a study by Zhang et al. (2010) documents the fact that, by the time the second edition of *Treatment Wetlands* (Kadlec and Wallace, 2009) was finalized, the number of wetlands papers had increased by a factor of 7.3. Similar publication measures show there are now many times more publications than available to Kadlec and Wallace. It is clear that the majority of information now available postdates the second edition, which in turn means that this third edition covers far more than small updates.

WHAT'S A TREATMENT MARSH?

The ecological meaning of the word *wetland* has been severely stretched in the broad treatment wetland literature. In a broad scientific sense, natural wetlands are defined and categorized according to principles set forth by Cowardin et al. (1979). They have plants, water and some variety of soil substrate; and they function under a variety of hydrologic conditions. *Treatment wetland* is a term that has extended the meaning to include planted gravel filters, thus encompassing all subsurface flow wetlands, which have no natural wetland analog. A classification system has been set forth by Fonder and Headley (2013), who proposed that treatment wetlands have three principal features: macrophytic vegetation; a waterlogged substrate; and inflow of contaminated water.

Because this book focuses on systems that have both ecological value and water quality improvement benefits, the term *treatment marsh* is employed. Among the many classification systems, a marsh in this book is a shallow wetland dominated by herbaceous rather than woody plant species. It may be continuously or intermittently flooded. This name is an attempt to bridge the gap between wetland science and lightly loaded treatment wetlands.

WHAT CONTAMINANTS ARE CONSIDERED?

Nutrients are the primary targets for most treatment marshes, including phosphorus, oxidized nitrogen, and ammonia. However, wetlands receiving high loads and concentrations of these substances often display undesirable ecological characteristics, and are here excluded. Trace metals, trace organics and pharmaceuticals are also considered, because of potential downstream aquatic effects, and because conventional treatment processes often cannot do the job.

ECOLOGICAL VALUES

There is growing understanding of the dichotomy between wastewater treatment wetlands of many types, and treatment marsh polishing systems. The realm of treatment marshes implies that they have considerable ecological value, and are part of the spectrum of restoration, creation, or enhancement of wetlands. Intense biological prejudice is often accorded to treatment wetlands by ecologists, in the mistaken belief that any system with less than pristine water quality has negligible value for diversity and habitat. The inverse position is often held by water quality treatment professionals (aka engineers), whereby only concentrations and flows have substantive meaning. Treatment marshes can provide significant benefits, ancillary to the primary purpose of water quality improvement, in the form of vegetative biodiversity, wildlife habitat, and human uses. Measuring such benefits is facilitated by biological indices, and by monetization of human uses of treatment marshes, with both new topics now included.

There is also a growing realization that humans have only limited control of the biological character of a newly built marsh. We can provide the basin, soils, initial plants, and water; but mother nature will decide the development of the wetland over the ensuing years. This means that the goals of mitigation wetlands are not often fully realized. Nor can the vegetation communities of treatment marshes be fully dictated by the designer. For the spectrum of wetlands between these extremes, we can only provide assistance to natural recruitment. Contrary to a widely held and slowly disappearing belief, mature treatment marshes need not be, and usually are not, monocultures of aggressive and invasive emergent species.

CORRECTING MISCONCEPTIONS

It is a purpose of this book to provide new and old knowledge about marsh functions, a good deal of which refutes overly simplistic, historical notions. Prominent among these are:

A. Black box descriptions of performance are sufficient. Everyone measures inlet and outlet concentrations and flows, and so percent removals and load reductions are easily calculated. Unfortunately, these metrics are insufficient for process understanding, and are inadequate for design.

B. The time scale of processes is dictated by water movement. The temptation is to ascribe removal from the water to concentrations and conditions in the water, which occur on a time scale of days. In reality, removal is accomplished by microbes (days), algae (weeks); macrophytes (months); and soil accretion (years).

C. Soils are a static matrix, with the primary function of storage. In addition to sorption and accretion, top soil layers contain plant roots, and undergo a large suite of hydrologic and microbial processes. In this book the topsoil layer is denoted as the "action zone."

D. Microbial conversion is the dominant treatment mechanism. A carry-over from wetlands treating strong wastewater, for BOD and ammonia, is a set of concepts that attribute removals to microbial processes. In contrast, removals in treatment marshes also involve vegetative cycles as a major pathway. Further, sorption and burial can dominate trace contaminant removal.

E. Treatment marshes don't work at cold temperatures. While microbial processes definitely do slow at cold temperatures, other marsh processes are not temperature sensitive. These include particulate settling and sorption. Vegetative cycles show largest uptakes during the cool spring season. Therefore, seasonality is often the modifier of marsh processes.

DESIGN TOOLS

Expectations for calculational detail continue to increase. It is no longer necessary to be given a single equation, arranged to be solved for the single variable of interest. This third edition is predicated on the extensive use of spreadsheets, and the large array of iterative and optimization tools that go with them. The scientific design team for a treatment marsh must include that capability.

Physical processes, such as hydraulics, sedimentation, and energy flows, are not sensitive to water quality. Therefore, the basis for their description includes surface flow wetlands in general, not just treatment marshes.

In the first edition, central tendency rate coefficients were presented, along with tables detailing the values for individual systems. Several investigators soon found that their results did not match the central tendencies, and incorrectly concluded that something must be wrong. In the second edition, and continuing in this third edition, the distributions of rate coefficients across numerous wetlands of all types are presented, so that new results may be placed in that spectrum, and designs may be selected with different positions across the intersystem landscape.

The scatter of wetland outlet concentrations around an often-seasonal trend is another type of variability to be accounted. This stochastic annual pattern is accounted separately, based on system performances, and percentile distributions are presented as a necessary part of a design basis.

TIME AND SIZE SCALES

Treatment marshes are in-the-ground, outdoor systems. In the absence of an available full-scale system, researchers sometimes consider marshes to be potted plants, and all manner of tubs, tanks, and troughs, sometimes located indoors. Such mini-systems do not experience wind, sun, birds, and animals. When the size is too small, the system is subject to severe edge effects. Although comparative results from small lab systems are useful, there is often the unstated assumption that they would represent the treatment performance of a full-scale wetland. Because that assumption is often unwarranted, this book avoids mesocosm data as much as possible.

There exist many studies based on infant or juvenile ecosystems, which have not had time to mature into the full suite of components that occur in fully developed wetlands. Long-term studies, involving several years of data, are much preferred, and are now fairly numerous.

WHAT'S NEW?

Of course, there is much more information available now than in 2008 when the second edition analyses were completed. The doubling time of the available data is on the order of two or three years, because old systems continue to return new information as more and more systems come on line in more and more application areas. As a consequence, much of the data used in support of this book was not available at the time of the second edition. It has been reassuring to find that most of the data and interpretations of the first and second editions have stood up well to the test of time, but not surprising to find that some numerical interpretations required updating.

Over the last decade, wetland technology has advanced considerably, as has ecological understanding. Treatment marsh applications have expanded worldwide, but with the majority in developed countries. Emphasis is shifting to treatment of nonpoint sources, to control agricultural and urban contamination. These are event-driven systems, which require modifications to the quantitative descriptions that are in common usage for continuous flow marshes.

The database for wildlife and human use of treatment marshes is also growing rapidly. Significant numbers of studies of bird and animal use of these wetlands have appeared in recent years, which permit evaluation of ecological merit. This growing knowledge base leads to an increased appreciation of the complexity of treatment marshes.

However, as much as things have changed, some things remain the same. The expectations of the first and second editions, forecasting rapid growth of treatment marshes, and treatment wetlands in general, have certainly proven true. Indications are for a continued increase.

THIS BOOK

This book has been updated to reflect the dramatic advances in wetland technology over the last 12 years. The authors of the second edition came from different backgrounds, and worked in different aspects of the treatment wetland field. By combining their knowledge and experience, they endeavored to present a broad range of information regarding the science, hydrology, hydraulics, reactor theory, applied design, implementation, cost, and operation and maintenance (O&M) of treatment wetland systems.

The format of this third edition reflects a multiple approach. The first six chapters allow the reader to explore marsh characteristics, and the internal mechanisms by which treatment wetlands operate. Hydrology, hydraulics, vegetation, and the soil action zone are discussed. Chapter 7 outlines the quantitative basis for interpreting performance.

The next three chapters explore processes and results for sediments, nitrogen and phosphorus. Existing projects and operating results from real-world treatment marshes are utilized extensively. Internal mechanisms, their influence on treatment performance, and their effect on system variability, are explored in detail.

Chapters 11–14 concern the specifics of event driven systems, and the treatment of other potential pollutants. For treatment marshes, some treatment wetland primary targets, such as BOD and pathogens, are relegated to ancillary target status. Trace metals and organics are discussed as a primary focus for treatment marshes.

Chapter 15 deals with biodiversity, including vegetation and all faunal trophic levels. It reports what has been found about inhabitants of marshes, and the biological interpretation of those results. The human use of treatment marshes is detailed.

Chapters 16 through 19 examine how performance data is analyzed and applied to the design process. Like previous editions, this book adopts a performance-based approach to design, in addition to presenting design tools such as loading charts and scaling factors. Continuing with the theme of practical implementation, these chapters summarize current knowledge that is key to getting wetland projects built, including construction methods, cost information, and O&M requirements.

All topics have been nearly totally rewritten, as required by the vastly increased data sources and understanding that have developed in the decade since the second edition.

Robert H. Kadlec
February 2019

Acknowledgments

The author wants to acknowledge the extreme patience of his wife Kelli, who put up with over a year's worth of working weekends and the virtual loss of a spouse.

This book expands upon many concepts advanced in the first edition, for which Robert Knight bears a full share of credit. He was a major architect of the foundation for this work. This book also reflects a foundation built with the shared efforts of Scott Wallace in the second edition of *Treatment Wetlands*. The sheer volume of information has demanded a bifurcation of the second edition, and Scott has elected to update the sections on subsurface flow and enhanced wetlands, which have been gratefully ceded to him. Nonetheless, Scott and his colleagues contributed greatly to the second edition in general.

My formal educational training was in engineering, which provided formidable analytical tools but zero biological and ecological understanding. However, my brother provided considerable ecological education (John A. Kadlec, wildlife management), and opportunities to learn during our joint project at the University of Michigan. I have also learned by association with numerous colleagues formally trained in ecology, including Chris Tanner (aquatic ecology), Hans Brix (aquatic botany), and of course Robert L. Knight (ecologist).

I have had many opportunities to learn different viewpoints from the dozens of colleagues with whom I worked over the past four decades. The experience gained from collaborating with other wetland scientists, engineering firms, and project owners has been a necessary and important component of my continuing wetland education.

The author wishes to acknowledge the efforts of the hundreds of engineers and scientists who have had the courage to create, innovate, and ultimately develop treatment wetlands as a viable technology to solve many environmental problems. The friendly and open communication between colleagues at international conferences has made this field a pleasure to work in, and the "lessons learned" greatly contributed to the rapid evolution of treatment wetlands, and now to the emergence of treatment marshes.

I am very appreciative of projects that have shared data. Without the data assembled from these diverse resources, this book could not exist. The list is long, and these hundreds of project owners are owed heartfelt thanks for their generosity.

Robert H. Kadlec
February 2019

Author Biography

Robert H. Kadlec is Professor Emeritus of Chemical Engineering at the University of Michigan He holds BS, MS, and PhD degrees in chemical engineering from the Universities of Wisconsin and Michigan, 1958–1962. That era saw the culmination of the "unit operations" approach to chemical processing, and the transition to the use of principles of transport phenomena to describe transfer and reaction rates in a wide variety of chemical and biochemical processes. Those techniques and analytical tools are also the foundation of today's environmental engineering. Bob began applying engineering analysis to wetland processes in 1970, with the goal of managing wetlands for water quality improvement. The result was the Houghton Lake, Michigan, wetland treatment system, which is still operating successfully.

Research on the Houghton Lake natural wetland, over 30 years of its operation for engineered treatment, formed the early framework for Dr. Kadlec's development of wetland process characterization. The technology has grown tremendously, and so has Bob's involvement in treatment wetland projects. In 1993, he retired from the University of Michigan, in order to devote full time attention to treatment wetlands as a private consultant, known as Wetland Management Services. He provides specialty consulting services to a wide range of governmental and private organizations.

He has participated in over 300 projects, ranging from simple feasibility studies to comprehensive university research projects. Early university studies focused primarily on wetland hydrology and water chemistry. In the course of many projects, a good deal of knowledge of practical ecology was imparted by his colleagues.

Bob has worked on treatment wetlands in many states and several other countries. He has participated in the design of over a hundred treatment wetlands. Major and long-running projects have included Houghton Lake, MI; Brighton, Ontario; Hillsdale, MI; Columbia, MO; and the Everglades Stormwater Treatment Areas. He is past chairman of the Macrophyte Specialist Working Group of the International Association on Water Quality. He has authored or coauthored over 150 publications on treatment wetlands, in addition to dozens of project reports. He was a proposer and developer of the USEPA North American Treatment Wetland Database.

This book is an effort to consolidate five decades of research and practical experience.

1 Introduction

Treatment wetlands in general are plant-soil-water systems utilized to promote water quality improvement. This book expands and modernizes information concerning a specific type of treatment wetland, the free water surface system, often called a marsh. It is further restricted to waters at relatively low contaminant concentrations. As a consequence, it is positioned in the gap between ecological desires for wetland restoration and preservation, and engineering concerns for reliable and economical methods for water quality improvement.

The focus of this book is upon constructed marshes rather than the use of natural marshes. Some treatment marshes will be built on former wetland sites, putting them in a category of restoration projects. Others will be built on an existing upland sites and are then in the category of newly constructed wetlands. In other cases, contaminated waters may be introduced into existing wetlands. The fundamental mechanisms of water quality improvement and ecological response are similar in all cases, and the principles of operation and performance forecasting are not different.

1.1 WHY THE WORD MARSH

Within the community of treatment wetland practitioners, systems with exposed water came to be known as free water surface wetlands, beginning somewhere in the late 1970s.They are part of the larger set of treatment wetlands in general (Fonder and Headley, 2013). Within the larger community of wetland scientists, that new terminology is not in common use, and the much older name of marsh has remained in primary usage. Because this book focuses upon systems that both ecological value and water quality improvement benefits, the term treatment marsh is employed. This name is an attempt to bridge the gap between wetland science and wastewater treatment technology, and it corresponds to the wetland systems that will be discussed.

This choice is also driven by the wish for simplicity of terminology. Marshes have single word designations in a variety of languages, whereas free water surface wetlands require three to seven word names (Table 1.1).

1.2 SCOPE AND CONTEXT

Relation to Restoration

Restoration is the manipulation of the physical, chemical, or biological characteristics of a site with the goal of returning the natural/historic functions of the wetland. Restoration is commonly divided into re-establishment, rehabilitation, and creation. Re-establishment refers to the goal of returning natural/historic functions to a former wetland Rehabilitation refers to the goal of repairing natural/historic functions of a degraded wetland. Creation is the construction of a wetland in an area that was not a wetland in the recent past. Implicit in this terminology is the expectation that all wetland functions are to be present, including water quality improvement, floodwater storage, fish and wildlife habitat, aesthetics, and biological productivity (U.S. EPA, 2002).

Mitigation is a term that frequently occurs in connection with restoration, and refers to "the restoration, creation, or enhancement of wetlands to compensate for permitted wetland

TABLE 1.1
Marshes and wetlands in several languages.

Language	Wetland	Marsh	FWS Wetland
English	wetland	marsh	free water surface wetland
Spanish	humedale	pantano	humedal libre de la superficie del agua
German	Feuchtgebiete	Sumpf	freie Wasseroberfläche Feuchtgebiet
French	zones humides	marais	surface humide de l'eau libre
Czech	mokřadů	bažina	volné vodní plochy mokřadů
Swedish	våtmark	kärr	fri vattenyta våtmark
Danish	vådområde	marsk	fri vandoverflade vådområde

losses" (Lewis, 1989). Under Section 404 of the Clean Water Act of the USA, wetlands may be legally destroyed, but their loss must be compensated for by the restoration, creation, or enhancement of other wetlands.

When the word treatment is added to the wetland category, it means that the water quality improvement function has primacy, but certainly not to the exclusion of other functions.

This book does not detail the various guidance publications concerning restoration and mitigation (e.g., Ruiz et al., 2013; Biebighauser, 2015; Craft, 2016). This subject is left to other publication venues.

Predecessor Books

Third Edition?

This work builds upon concepts set forth in two predecessor books:

Kadlec and Knight (1996). *Treatment Wetlands*, CRC Press; and

Kadlec and Wallace (2009). *Treatment Wetlands*, second edition, CRC Press.

Much of the material in these two predecessor volumes will not be repeated here, either because it is not relevant to the marsh context, or because it has receded to the level of historical interest only. Because of their functional utility, many of the mathematical diagnostics are retained.

These covered a wide range of wetland types and applications. It is no longer feasible to place the numerous modern science and technology developments in a single volume, hence this book is restricted to treatment marshes, as further defined later in this chapter. Nevertheless, the systems covered are the large majority of all treatment wetlands, based upon acreages and water volumes treated. The excluded treatment wetlands are more numerous, but of hugely smaller size.

Relevant Wetland Science Books

There exists a wealth of published information about general wetland science. The reader may consult any of several texts, prominently including:

- *Biogeochemistry of Wetlands: Science and Applications*, KR Reddy and R DeLaune, (2008). CRC Lewis Publishers; Boca Raton, Florida.
- *Wetland Ecology Principles and Conservation*, PA Keddy (2010). Cambridge University Press; Cambridge, United Kingdom.
- *The Biology of Freshwater Wetlands*, A van der Valk (2012). Oxford University Press, Oxford, England.

- *Wetlands*, 5th edition, WJ Mitsch and JG Gosselink, (2015). John Wiley and Sons, Hoboken, New Jersey, 456 pp.
- *The Ecology of Freshwater and Estuarine Wetlands*, DP Batzer and RA Sharitz (eds.) (2014). University of California Press, Berkeley, California.
- *Wetland Plants: Biology and Ecology*, JK Cronk and MS Fennessey (eds.) (2001). Lewis Publishers; Boca Raton, Florida, 462 pp.
- *Wetland Soils*, JL Richardson and MJ Vespraskas (eds.), (2001). Lewis Publishers, Boca Raton, Florida, 417 pp.
- *The Wetlands Handbook*, E Maltby and T Barker (eds.) (2009). John Wiley and Sons, Hoboken, New Jersey, 800 pp.

Additionally, there are many compilations of research results for specific regional situations. These are for diverse locations, including, for example:

Czech Republic:

Freshwater Wetlands and Their Sustainable Future: A Case Study of the Trebon Basin Biosphere Reserve, Czech Republic, J Kvet, J Jenik and L Soukupova (eds.) 2002. Parthenon Publishing, 495 pp.

Canada:

Prairie Wetland Ecology, HR Murkin, AG van der Valk and WR Clark (eds.) 2000. Iowa State University Press; Ames, Iowa, 413 pp.

Florida:

Phosphorus Biogeochemistry in Sub-Tropical Ecosystems, KR Reddy, GA O'Connor and CL Schelske (eds.) 1999. Lewis Publishers; Boca Raton, Florida, 707 pp.

The Everglades Experiments: Lessons for Ecosystem Restoration, CJ Richardson (ed.), 2008. Springer, New York, 698 pp.

When combined with periodicals focused on wetlands, such as the journal *Wetlands*, these form a formidable collection of scientific works that explore many facets of wetland behavior and character.

Exclusions

The information from marshes has been restricted in several ways in this work, so that the database is relevant and useful for data interpretation and design sizing. First, size matters, because of issues of scale-up. Microcosms and mesocosms are useful for investigating component processes, and in general are easily replicated to bolster statistical confidence. Because of edge effects, small systems do not have representative vegetation densities, nor are they typically exposed to a realistic set of meteorological conditions. They usually contain only a few of the components of a large system, such as soil without plants, litter, or animals. The components of the ecosystem, including plants, algae and microbes, are exposed to a time changing water chemistry, which is different from the spatially distributed patterns of these along the flow direction. Thus, there is a possibility that mesocosm studies produce replicated, well-controlled results of artifacts, rather than processes that actually occur in the field (Eriksson and Weisner, 1997; Komor and Fox, 2001; Ahn and Mitsch, 2002). Therefore, the database summarized here excludes studies of systems smaller than pilot-scale, generally defined as systems of several tens of square meters extent, located in the out-of-doors.

Also excluded are the many studies that are based on infant or juvenile ecosystems, which have not had time to mature into the full suite of components that occur in fully developed wetlands. There are many phenomena that take longer to develop than the tenure of a typical graduate student. Long-term viability, and maintenance requirements cannot be deduced from systems in the startup mode. Treatment marshes typically take a period of one or two years to outgrow the startup period.

It is also possible in some circumstances to augment the structure and/or functioning of treatment marshes, and such systems are termed intensified wetlands. For example, treatment marshes might include floatation structures for the plants, and operation might include plant harvest. In general, such modifications for marshes do not have long track records and raise questions of economic and physical feasibility. Thus, these will not receive great attention here.

1.3 TYPES OF TREATMENT WETLANDS

Fonder and Headley (2013) define several classes of treatment wetlands, split at the most fundamental level into surface flow and subsurface flow (SSF) categories. This book is only about the former category, free water surface systems (FWS), which in a broad context are also called marshes. Treatment marshes can be further classified according to vegetation growth forms, such as emergent macrophytes, submerged macrophytes, etc.; but in practice, marshes often contain more than one vegetation types found in different spatial distribution patterns.

Treatment marshes often contain areas of open water, floating vegetation, and emergent plants; either by design or as an unavoidable consequence of the design configuration. Berms, dikes, and water distribution features are used to control flow and depth. As water flows through the wetland, it is treated by the processes of plant uptake and release, sedimentation, filtration, oxidation, reduction, adsorption, and precipitation. The components in a typical FWS wetland are shown in Figure 1.1.

FWS Wetland Vegetation

The principal community types in FWS marsh systems are:

- Emergent aquatic vegetation (EAV)
- Submerged aquatic vegetation (SAV)
- Floating aquatic vegetation (FAV)
- Open unvegetated water (OW)

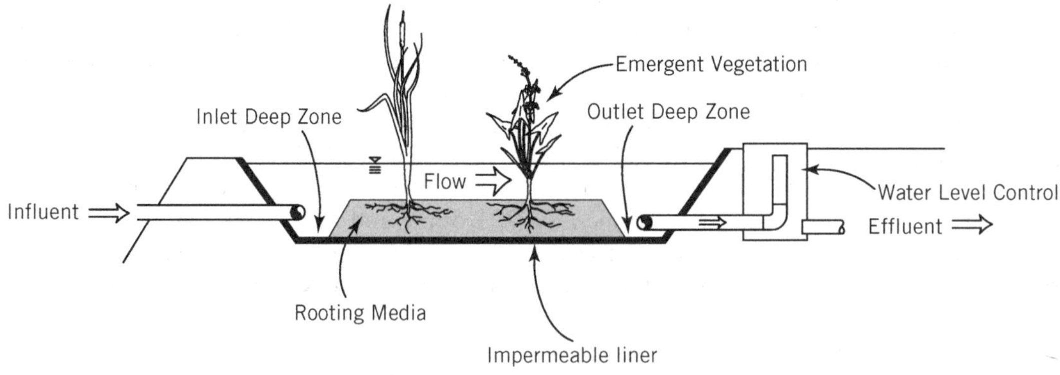

FIGURE 1.1 Basic elements of a FWS wetland. For treatment marshes for low concentrations, the liner is not a requirement. Structure designs vary.

Four groups of aquatic macrophytes (Figure 1.2) can be distinguished on a basis of morphology and physiology (Wetzel, 2001):

1. Emergent macrophytes grow on water saturated or submersed soils from where the water table is about 0.5 m below the soil surface to where the sediment is covered with approximately 1.5 m of water (e.g., *Acorus calamus, Carex rostrata, Phragmites australis, Schoenoplectus (Scirpus) lacustris, Typha latifolia*);
2. Floating-leaved macrophytes are rooted in submersed sediments in water depths of approximately 0.5 to 3 m and possess either floating or slightly aerial leaves (e.g., *Nymphaea odorata, Nuphar luteum*);
3. Submersed macrophytes occur at all depths within photic zone. Vascular angiosperms (e.g., *Myriophyllum spicatum, Ceratophyllum demersum*) occur only to about 10 m (1 atm hydrostatic pressure) of water depth;
4. Freely floating macrophytes are not rooted to the substratum; they float freely on or in the water and are usually restricted to nonturbulent, protected areas (e.g., *Eichhornia crassipes, Hydrocotyle spp., Lemna minor*).

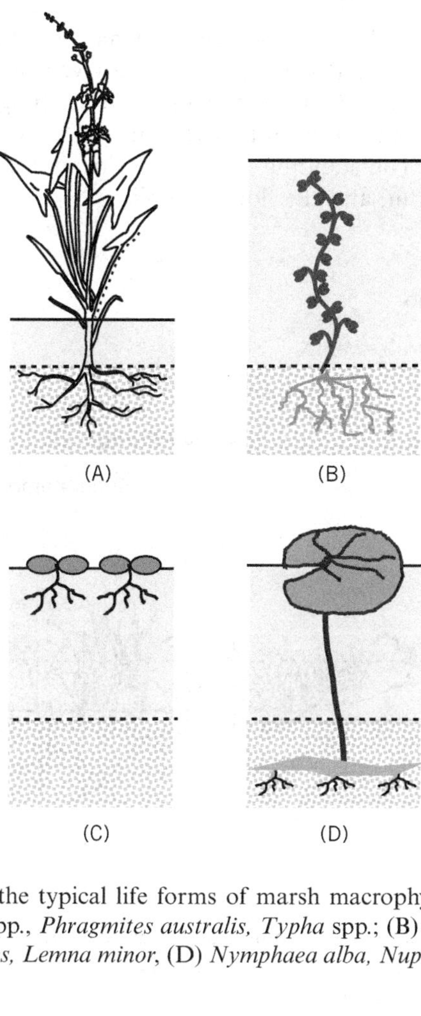

FIGURE 1.2 Sketch showing the typical life forms of marsh macrophytes. Examples of species illustrated are (A) *Schoenoplectus* spp., *Phragmites australis, Typha* spp.; (B) *Potamogeton* spp., *Ceratophyllum* spp. (C) *Eichhornia crassipes, Lemna minor*, (D) *Nymphaea alba, Nuphar lutea*.

In addition, a large number of the emergent macrophytes can be established in floating mats, either with or without a supporting structure. Some species have more than one of these growth forms; however, there is usually a dominant form that enables the plant species to be classified. In emergent plant species, most of the aboveground part of the plant emerges above the water line and into the air.

Floating species have leaves and stems buoyant enough to float on the water surface. Submerged species have buoyant stems and leaves that fill the vertical stratum between the sediment surface and the top of the water column. Floating and submerged species prefer deep aquatic habitats, but they may occur in wetlands when water depth exceeds the tolerance range for rooted, emergent species.

In natural lake margin systems, these community types are arranged along the water depth gradient from upland to the deeper parts of the lake (Figure 1.3).

Epiphytic and Floc Components

The macrophyte community in marshes (phytomass) usually coexists with a small-scale community consisting of algae, bacteria, and detritus. It usually exists as periphyton and floc. The US Environmental Protection Agency (USEPA) defines periphyton as a complex assemblage of algae, cyanobacteria, microinvertebrates, their secretions, and detritus attached to submerged surfaces (Figure 1.4).

Floc consists of an unconsolidated layer of biogenic, detrital material relatively rich in organic matter. It is mainly derived from the dominant vegetation at a particular location, and likely generated in situ (e.g., plant litter, cyanobacteria, fungi, etc.).

Floc and periphyton can be important factors in the nutrient and sediment cycling in marshes (Noe et al., 2003). The standing stock of nitrogen and phosphorus in floc is large compared to the water column, and the floc is mobile.

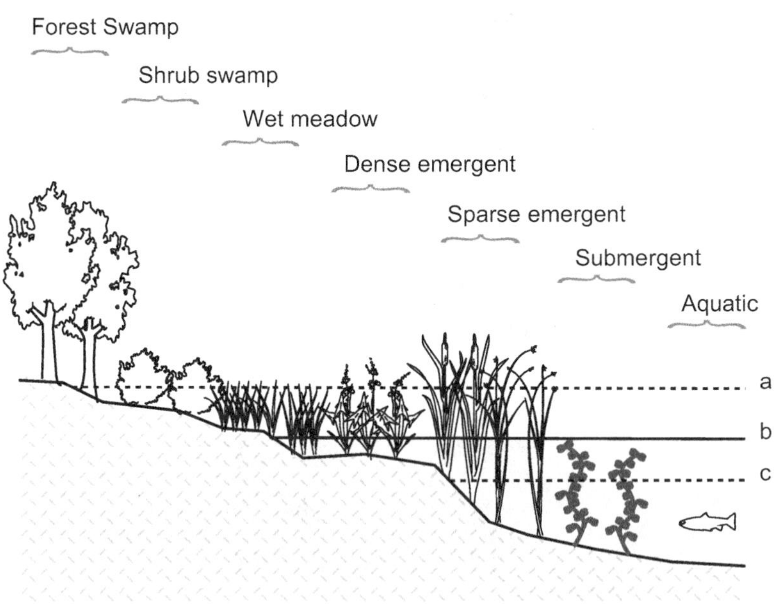

FIGURE 1.3 Wetland community types along a depth gradient from lake to upland. The marsh category includes the right five designations. Water levels may vary from high (a) to low (c), with mean at (b). (Based on information in Wilcox et al., 2007).

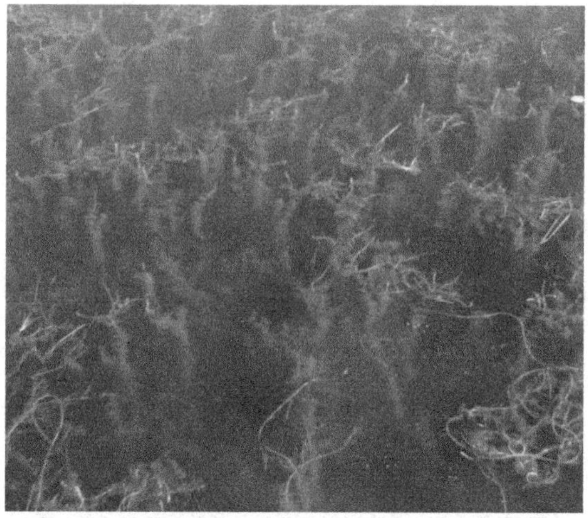

FIGURE 1.4 Submerged plant parts often have thick growths of attached periphyton.

1.4 HISTORY OF FWS TREATMENT WETLANDS

The Ijssel Lake Polder Authority in Flevoland, The Netherlands constructed the first European free water surface (FWS) wetland in 1967 (Veenstra, 1998). The wetland had a design depth of 0.4 m and the total area was 1 hectare. Longitudinal channels were implemented to facilitate mechanical biomass harvesting and system maintenance. The wetland design included channels of 3 m wide and 200 m long (Figure 1.5), separated by parallel stretches of 3 m. The system exhibited a very good treatment effect and by the early 1970s, about 20 FWS wetlands of this type, called a "planted sewage farm" (or Lelystad process) were in operation in the Netherlands (Greiner and de Jong, 1984; Veenstra, 1998). This system treated raw sewage.

In 1968, FWS constructed wetlands were created in Hungary near Keszthely in order to preserve the water quality of Lake Balaton and to treat municipal wastewater (Lakatos, 1998). The constructed wetland was established in place of an existing natural wetland. The system originally consisted of six ponds 40–60 cm deep with a surface area of 10 ha and the ponds were fed with 8000 m^3/d of mechanically pretreated wastewater. By 1985, the protection of Lake Balaton had grown to include the 1,800-ha Keszthely pond, which turned out to be a Submerged Aquatic Vegetation (SAV) system (Clement et al., 1998). This system was built to treat a polluted river.

In 1973, the Mt. View Sanitary District in Martinez, California constructed about 8.5-ha of FWS wetland marshes for wildlife habitat and wastewater discharge (James and Bogart, 1989). In 1976, the communities of Pinetop and Lakeside, Arizona, and in 1977 Show Low, Arizona, created a series of constructed lake/wetland areas for effluent evaporative disposal and wildlife production (Wilhelm et al., 1989). These were all to treat relatively strong domestic wastewater effluents.

History of Treatment Marshes

North America has a rich history of constructing large-scale FWS treatment marshes over the last 20 years. Florida has a number of the largest constructed wetland treatment areas in the world, including the Lakeland and Orlando constructed wetlands, both of which

FIGURE 1.5 The FWS wetland at Elburg, the Netherlands.

were started in 1987. Each wetland has about 500 ha for advanced treatment of municipal wastewater. Six treatment wetlands, called stormwater treatment areas, totaling over 23,000 ha, have been built in south Florida. More are under planning and construction. These are hoping to achieve a goal of 10 μg/L effluent phosphorus. Over 40 km of levees, and a comparable length of canals, are used to enclose each of these constructed treatment wetlands and convey water. The approximate cost, including supporting research, was $1 billion USD, of which about $800 million USD was for construction. These are to treat an estimated 30-year annual average flow of 4,400,000 m^3/d, and remove approximately 80 metric tons of phosphorus per year. One of these large stormwater treatment areas, STA2, is shown in Figure 1.6. Another smaller system located near Imperial, California is shown in Figure 1.7.

The basis for the first edition of *Treatment Wetlands* was largely the North American Database, version 1 (Knight et al., 1993). That was expanded to a North American Database, version 2 (NADB Database, 1998). The NADBv2 contained information on 351 wetlands, of which 30 were relevant to the treatment marsh category (source waters: 6 tertiary, 11 advanced secondary, 4 stormwater, 9 river and lake). At the time of the second edition of *Treatment Wetlands*, 2007, the number of marsh systems of all sizes reporting phosphorus data had grown to 295, of which 161 had source water of less than 2.0 mg/L. By early 2017, the number of marsh systems of all sizes reporting phosphorus data had grown to 366, of which 237 had source water of less than 2.0 mg/L. These totals contain systems not considered herein because of size or data quality, but nonetheless, it is apparent that treatment marsh technology has continued to grow.

In contrast to North America, FWS constructed wetland technology did not spread rapidly throughout Europe; and the main technology focus there has been on HSSF and VF

FIGURE 1.6 Part of STA2, a 6273 ha treatment marsh in south Florida.

FIGURE 1.7 A portion of the Imperial treatment marsh in southern California.

systems. However, FWS constructed wetlands are in operation in many European countries (e.g., Norway, Sweden, Denmark, Poland, Estonia, and Belgium).

In Sweden, FWS systems have been constructed with nitrogen removal as a primary goal but other aims, such as biodiversity and irrigation are also taken into consideration (Vymazal, 2006). Sometimes, the aim is to provide phosphorus polishing after chemical treatment

TABLE 1.2

Ranges of treatment marsh source water quality. Units are mg/L.

	Background	Lower Limit	Upper Limit
BOD	1–2	1	15
TSS	1–3	1	50
Ammonia nitrogen	<0.1	0.1	5
Oxidized nitrogen	<0.1	0.1	10
Total phosphorus	0.01–0.05	0.01	2

and a buffer in case of treatment failure in the conventional treatment plant (Sunblad, 1998). In total, approximately 7800 hectares of treatment marshes have been constructed in Sweden between 1990 and 2010 (Strand and Weisner, 2013).

Water Quality Boundaries Treatment Marshes

Approximate variable ranges under consideration here are shown in Table 1.2. These numbers are not hard and fast limits, as is case for regulatory limits; rather, they are rough guidelines.

Note that the lower limit under consideration for treatment marsh technology is here considered to be at or below the general levels of wetland background water quality. One implication is that incoming waters with below-background concentrations may be counter-treated; that is, they may increase in concentration Another implication is that there is no point in implementing a treatment marsh if water quality is already at or below wetland background.

Source waters with concentrations above the (approximate) upper limit can pose difficulties for a variety of reasons. High BOD can lead to anaerobiasis in the marsh. High TSS can block sunlight, and lead to excessive solids buildup in the system. High ammonia can lead to issues of toxicity for wetland biota. High nitrate generally will not threaten marsh biota, but there is human drinking water safety limit of 10 mg/L. High total phosphorus is a major factor in ecosystem eutrophication.

In later chapters, other water quality caveats will be discussed. These include some trace metals, such as selenium and mercury, as well as some non-trace metals, such as iron, aluminum, and calcium. And, the sulfur content of source waters can have large effects upon treatment.

1.5 APPLICATIONS OF TREATMENT MARSHES

The restrictions applied to define treatment marshes result in source waters coming from the following principal categories:

- Highly treated domestic wastewaters
- Rivers and lakes
- Agricultural runoff
- Urban runoff

Polishing

Polishing refers to the final municipal wastewater treatment stages before the wastewater is eventually discharged into natural water bodies. Polishing processes involve removal of remaining suspended solids and biological oxygen demand (BOD), as well as some degree of nutrient reduction (Tang et al., 2017). Modern municipal wastewater treatment plants produce better-than-secondary effluents. The effluent concentrations for the 91st Avenue treatment plant in Phoenix, feeding the Tres Rios wetland; and for the Orlando Easterly wetland, receiving effluent from the Iron Bridge treatment plant, are shown in Table 1.3. Tertiary and higher treatment usually produces compliant secondary discharges, but changing regulatory requirements can create the need for advanced treatment, which may be provided by constructed wetlands.

A high degree of treatment is also provided by some municipal wastewater lagoon systems, particularly those with long detention times. Table 1.3 shows effluent concentrations for Haines Junction, Yukon, and Houghton Lake, MI lagoons; both of which are polished in wetlands. The detention time in the Haines Junction lagoons is about a year and half, and about eight months at Houghton Lake. As add-ons to lagoons, the solids trapping properties of wetlands can compensate for the export of algal debris from facultative ponds and provide further nutrient removal.

Field Runoff

The amount of municipal effluent water that might be polished is usually small in comparison to agricultural runoff for the same watersheds. The runoff finds its way into rivers and lakes of the region, which may then be the target for marsh treatment. Table 1.3 contains two sample river source waters for wetlands, the Trinity River in Texas (Kadlec et al., 2011) and the New River in California (Kadlec et al., 2010a). There is also an example of lake source water for the Apopka, FL wetlands. Alternatively, agricultural runoff may be directly pumped into treatment marshes, such as those located in Imperial, CA, and in Norway (Table 1.3). Target contaminants from agricultural fields vary, depending upon the perceived threat to receiving ecosystems. The principal contaminants include suspended solids, nitrate, phosphorus and agricultural chemicals; but normally not all at the same time. Runoff from row crops and pasture areas may be low or high in mineral solids, depending on farming practices, rainfall intensity, soil types, and topography. Nutrient concentrations and loads from row crops and pastures depend on fertilization practices. As for urban runoff, there is usually no pretreatment prior to the wetland.

There is a potential role for wetlands in reducing solids loading coming from especially row crops. Such "dirt traps" are remarkably effective even at very short detention times (small systems) (U.S. Department of Agriculture, 1991; Braskerud, 2001a). However, short detention does not suffice to reduce dissolved nutrients, because those removals rely on the biogeochemical cycle, which operates at a much slower pace. Constructed wetlands also have the ability to abate the pulses of some pesticides, but not all, that are exported from fields in modern agriculture (Rodgers and Dunn, 1992). Despite such successes, the implementation of end-of-field wetlands has proceeded at a slow pace.

Wetlands are the only economically feasible means of controlling phosphorus in runoff reaching the Florida Everglades. The biggest constructed wetlands in the world are in operation there, with an aggregate area in excess of 20,000 ha. However, nitrogen pollution is a concern for the marine environments, and therefore wetlands are receiving considerable attention in connection with protection and restoration of the Baltic Sea and the Gulf of Mexico (Arheimer and Wittgren, 1994; Hey, 2002).

Urban Stormwater

Urban runoff carries modest amounts of pollutants and is therefore a logical application for treatment marshes. Table 1.2 lists the water quality entering two example systems,

TABLE 1.3

Source water quality for example treatment marshes.

	AWT				River		Lake	Agricultural		Urban		Various Stormwater
	Tres Rios	OEW	Haines	HLake	Tarrant	Brawley	Apopka	Imperial		Perth	Tahoe	
	AZ	FL	Yukon	MI	TX	CA	FL	CA	Norway	Australia	CA	NC
No. Wetlands	1	1	1	1	1	1	4	1	5	1	1	10
Reference	Data	Data	Data	Data	Data	Data	1,2	Data	3,4	5	6	7
Parameter												
BOD	2.0	1.8	3.8	21.1	-	12.5	-	7.2	-	-	-	-
TOC	8.9	9.0	-		-	8.8	-	7.9	-	-	-	-
TSS	4	1	6	34	75	34	48	20	50	-	34	40
NH_4-N	1.50	0.30	0.02	6.46	0.08	2.90	0.90	0.76	0.20	0.15	0.02	0.18
NO_3-N	3.00	0.70	0.00	0.54	2.90	0.49	0.05	2.45	2.15	0.40	0.59	0.37
TN	6.00	1.90	-		3.50	5.60	4.50	6.09	2.35	1.34	1.60	1.71
TP	3.000	0.185	0.170	3.510	0.900	1.400	0.104	0.812	0.300	0.140	0.279	0.270

References
Data refers to author's files.
1 Dunne et al. (2012)
2 Dunne et al. (2013)
3 Braskerud (2002a)
4 Braskerud (2002b)
5 Adyel et al. (2017)
6 Heyvaert et al. (2006)
7 Merriman et al. (2017)

located at Tahoe, CA, and Perth, Australia. Stormwater concentrations and loads are cyclic due to periods of dryfall and deposition, followed by the first flush of runoff after rain, followed by exponential decreases in runoff constituent concentrations as storages rinse from the landscape, and finally dry conditions and deposition until the next storm event. Pollutant concentrations and loads generally range from low levels from undeveloped and park lands, to low density residential and commercial, to higher density residential and commercial, and finally to high density commercial and industrial land uses. The use of constructed wetlands, usually with accompanying ponds, is now a routine best management practice (BMP) for controlling the quality of runoff. In the United States, the implementation of wetland stormwater BMPs has been very uneven, with numerous and early applications on both the east and west coasts, but much later and fewer systems elsewhere.

There is basically no pretreatment for urban stormwaters if the forebay settling basin is considered part of the wetland. At most, there may be a debris screen to catch major floating objects. Expectations are also typically lower than for many other application areas, with moderately good TSS reductions, but much lesser reductions in dissolved constituents. Although most urban stormwater wetlands are small, and do not acquire data, it is clear that their numbers are quite large.

1.6 NATURAL WETLANDS

The focus of this book is predominantly upon constructed wetlands rather than the use of natural wetlands. Although there are definitely circumstances in which it is logical and legal to utilize existing wetlands, it is far more common, at this point in history, that a treatment wetland will be built on an existing upland site. The principles of operation and performance forecasting are not different.

In 1972, the University of Florida began a research effort directed at assessing the effectiveness of natural cypress wetlands for municipal wastewater recycling. From March 1974 until September 1977, secondarily treated municipal wastewater from a trailer park north of Gainesville was discharged to two isolated cypress wetlands (domes), and two control wetlands were also monitored. Research studies measured nearly all aspects of the physical, chemical, and biological processes occurring in the wastewater and control cypress domes (Ewel and Odum, 1984).

Concurrently, the University of Michigan began the first in-depth study of using engineered wetlands for wastewater treatment in a cold climate region. Treatment lagoons provided the source water. Work began in 1971, with mesocosm plots in the peatland, continued with a three-year pilot study, and a finally a full-scale system that still operates (Figure 1.8) (Kadlec, 2009a).

Natural Wetlands in the USA

With wetlands being protected by federal law, the use of natural wetlands for wastewater treatment became limited at the end of the 20th century. Hammer and Bastian (1989) during the conference on constructed wetlands in Chattanooga, Tennessee, stated:

> Although some natural wetlands have been effectively used for water quality improvement, we do not wish to encourage additional use. We have recently become aware that natural wetlands are valuable resources that must not be wasted. Much remains to be learned about their many values and functions and the long-term consequences of wetland destruction. However, enough is known to conclude that it is not worth of risking the unnecessary loss of any remaining natural wetlands without a better understanding of their important role in biological productivity, fish and wildlife habitat, flood protection, groundwater discharge, base flow stabilization of rivers and water quality

FIGURE 1.8 The Houghton Lake, Michigan, system utilizes a pre-existing peatland for seasonal discharges.

improvement. On the other hand, constructed wetlands may provide a relatively simple and inexpensive solution for controlling many water pollution problems without detrimentally affecting our natural wetlands resources.

Natural Wetlands in Arctic Canada

The United States and Canadian north, including Alaska, the Yukon, Northwest Territories, and Nunavut (5,629,000 km^2), is a little larger than all of central and western Europe (4,595,000 km^2). The population of the United States and Canadian north is 856,000, while that of central and western Europe is 496,000,000. The north area is approximately 25% wetland, or about 1,400,000 km^2. Clearly, the amount of wetland per person (1.6 km^2) is huge. The use of wetlands for treatment is an option that does not threaten the resource. Treatment facilities that are commonly used in southern climates are complex and require operator skills that are typically not available in the north, and the economic means to build and operate high tech facilities does not exist. The only viable option for wastewater treatment is low-tech natural systems. As a consequence, there are over 40 treatment wetlands in the United States and Canadian north.

Those northern systems above the tree line typically treat strong lagoon waters. Tundra treatment systems often occupy, or discharge to, wet depressions. These have topographic features that influence plant communities and water retention and fit the general definition of wetlands. As a result they are referred to as treatment wetlands (Yates et al., 2012). Treatment wetlands of the low arctic boreal or taiga regions target a broader spectrum of source waters, including urban stormwaters and highly pretreated lagoon effluents.

1.7 GEOGRAPHIC SETTINGS

FWS wetlands are suitable in all climates, including the far north. However, ice formation can hydraulically preclude winter operation, and the rates of some removal processes are lower for cold water temperatures, notably nitrogen conversion processes. When ice covers the open water, the transfer of oxygen from the atmosphere is reduced, decreasing oxygen-dependent treatment processes. Other processes, such as TSS removal, are more effective under the ice than in summer conditions. It is generally more efficient to store water during winter and treat it during the warm part of the year.

Climate Zones and Hardiness

Wetlands exist in all climates, and on every continent except Antarctica. Wetlands in arctic climates may have temperatures as low –50°C, and wetlands in tropical climates may have temperatures as high as 50°C. In intermediate, temperate climates they experience warm summers and cold winters. The climate of a region, together with physical characteristics such as rainfall and elevation, determines its plant and animal life. As a consequence, the selection of initial plantings needs to be consistent with climate, by paying due attention to growth requirements. These are typically available from plant suppliers. The climate zones of North America are shown in Figure 1.9.

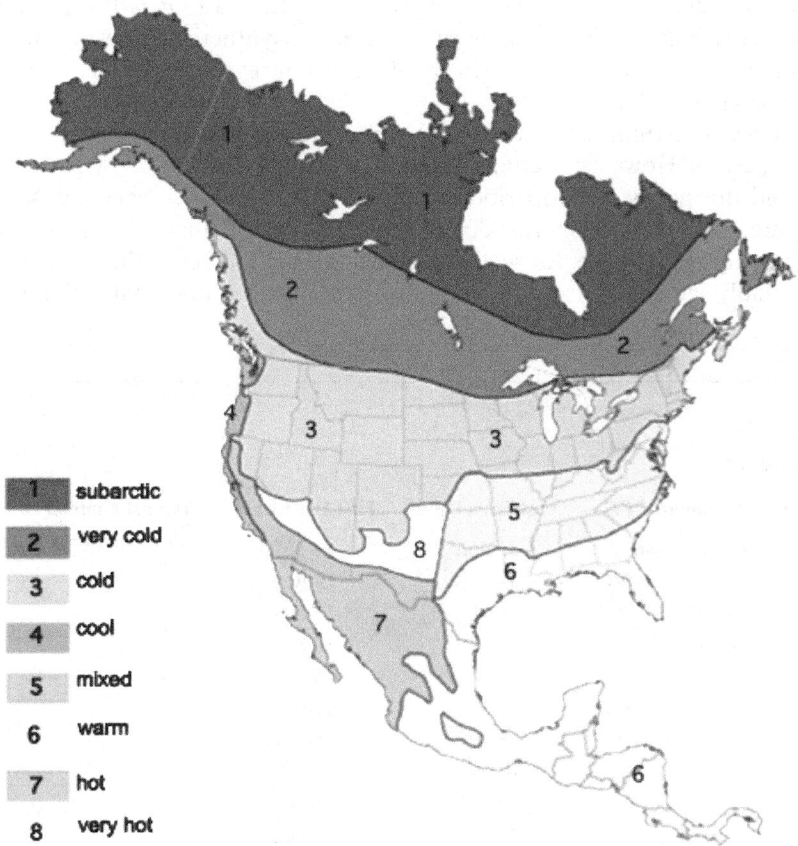

FIGURE 1.9 Climate zones of North America.

A hardiness zone is a geographically defined area in which a specific category of plant life is capable of growing, as defined by climatic conditions, including its ability to withstand the minimum temperatures of the zone. The main factors determining average minimum temperature are elevation, latitude and proximity to the coast.

A numerical scale may be set according to the minimum temperature to be expected for a particular locale (Daly et al., 2012). The zone scale was developed by the United States Department of Agriculture (USDA) as a rough guide to survival of plants. Zones for North America are listed in Table 1.4. Similar zones are applicable in other countries.

Interestingly, plants that are commonly abundant in treatment marshes have very large geographical distribution. Plants such as *Typha* spp., *Phragmites australis*, and *Shoenoplectus* spp. are found worldwide, and in all but the coldest plant hardiness zones. However, other less dominant components of the vegetative communities differ considerably across the hardiness zones. For example, arctic treatment wetland plant communities contain *Carex aquatilis, Ranunculus pygmeaus, Stellaria crassifolia, Poa artica*, and *Plantago juncoides* (Yates et al., 2012). At the another climatic extreme, EAV in the south Florida STAs was found to commonly contain, in addition to the dominant *Typha domingensis: Cladium jamaicense, Eleocharis interstincta, Panicum* spp., *Sagittaria lancifolia*, and *Ludwigia* spp. (Chimney, 2012).

Solar Radiation and Growing Seasons

A major driver of both wetland water temperature and vegetation growing season is the annual pattern of incoming solar radiation. In the former case, it is the primary factor in the wetland energy balance; in the later case, the photosynthetically active radiation (PAR) controls plant growth. The annual pattern of incoming extraterrestrial radiation depends only on latitude (Figure 1.10).

Plant hardiness determines the ability of a plant to survive and corresponds roughly to the frost-free period. However, wetland emergent plants often follow a pattern of growth, senescence, and dormancy. Photoperiod initiates winter dormancy, whereas soil, water and air temperature conditions primarily determine cessation of dormancy in the spring. This means the growing season days for wetland plants typically exceeds the number of frost-free days. Additionally, adaptation allows many, if not most, cold-climate plants to continue

TABLE 1.4
Plant hardiness zones.

Hardiness Zone	Minimum T (°C)	Last Frost Date	First Frost Date	Typical Number of Frost-Free Days
1	Below – 46°C	15-Jun	15-Jul	30
2	–46°C to – 40°C	15-May	15-Aug	90
3	–40°C to – 34°C	15-May	15-Sep	120
4	–34°C to – 29°C	10-May	15-Sep	125
5	–29°C to – 23°C	30-Apr	15-Oct	165
6	–23°C to – 18°C	15-Apr	15-Oct	180
7	–23°C to – 12°C	15-Apr	15-Oct	180
8	–12°C to – 7°C	10-Mar	15-Nov	245
9	–7°C to – 1°C	15-Feb	15-Dec	265
10	–1°C to 4°C	20-Jan	20-Dec	335
11	4°C and up	Frost-free		365

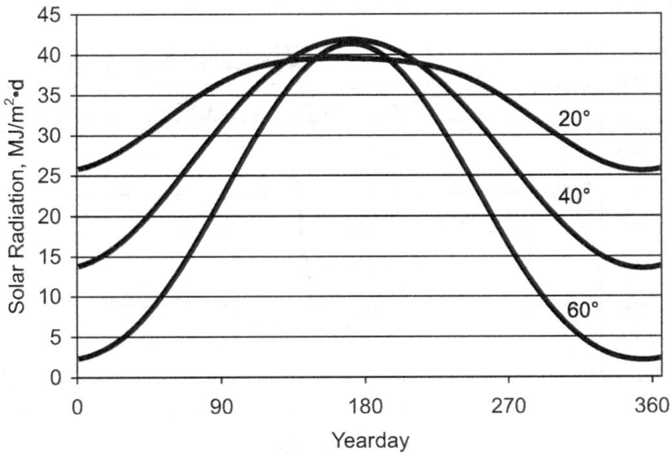

FIGURE 1.10 Undepleted solar radiation as a function of Julian day for various latitudes.

nondormant activity at air temperatures below 0°C. Growing seasons for wetland plants is discussed in Malone and Williams (2010).

1.8 HYDROLOGY

Hydrology is a primary determinant of the structure and functioning of a treatment marsh. There are a variety of water transfers to and from the wetland, to which the source water is added. An important ecosystem attribute is hydropattern, which describes the depth and duration of flooding in the system. Hydropattern is determined to a large extent by the volume and timing of source water deliveries.

Water Depths

Average marsh water depths typically range from about 10 to 90 cm. In the lower part of the range, rooted emergent macrophytes are favored, in the upper lower part of the range, conditions favor submerged and floating vegetation. Deeper zones may be intentionally included to aid in flow distribution. Depths may be both spatially and temporally distributed within a treatment marsh (Figure 1.11). Dryout sometimes occurs and will generally affect treatment negatively.

Inflow Patterns

Five distinct source inflow temporal patterns are considered here: (1) uniform, (2) large watershed runoff driven, (3) small watershed rainfall driven, (4) discretionary, and (5) unfrozen season. The differences among these create the need for performance analyses that properly account for the frequency and durations of the flow time series.

The uniform source category results from either an advanced wastewater treatment plant, or from an always-available water body, such as a large river or lake. Variations will usually be present, but they are small and usually unpredictable (Figure 1.12).

The large watershed runoff category results from agricultural landscapes at the regional scale. The storage capacity of the collection canals or streams is sufficient to ensure flows throughout an entire wet season (Figure 1.13).

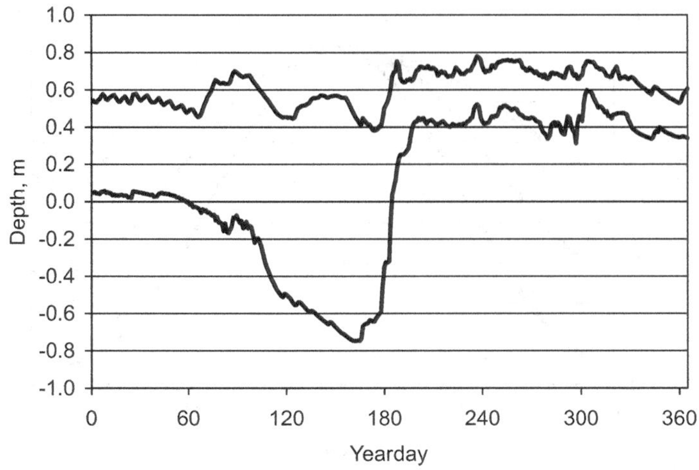

FIGURE 1.11 Annual pattern of water depth for the EAV front end (STA56C1A) and the SAV back end (STA56C1B) of a stormwater treatment area flow way.

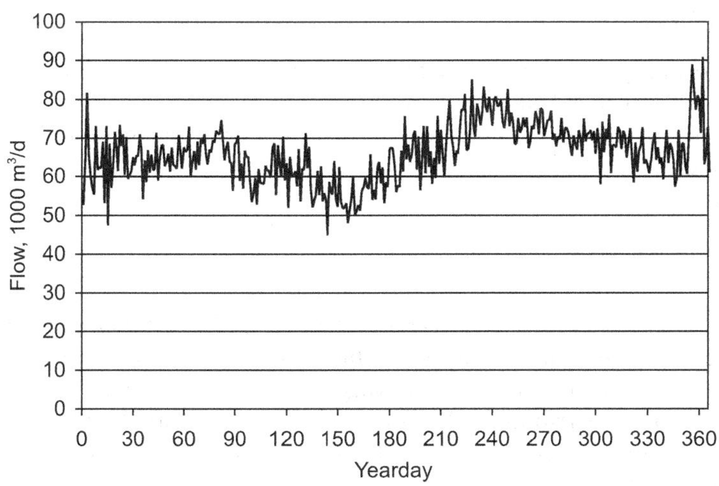

FIGURE 1.12 An example of uniform source flow for the Orlando Easterly Wetland (OEW) for the calendar year 2000. Water is pumped from an advanced municipal treatment plant.

Small watershed runoff occurs from farm fields and urban sources. Stormwater treatment wetlands often serve small catchments that respond rapidly to the onset and cessation of rain events (Figure 1.14).

Some source facilities have alternative destinations for the flows. Commonly, reuse for irrigation is an option, with water from the nonirrigation going to a treatment wetland. This is the case for the Blue Heron system at Titusville, FL, and at the Incline Village system near Carson City, NV. Another destination alternative is power plant cooling, as at Tres Rios, AZ, and Titusville, FL. The delivery window to the wetland does not necessarily correspond to wet seasons (Figure 1.15).

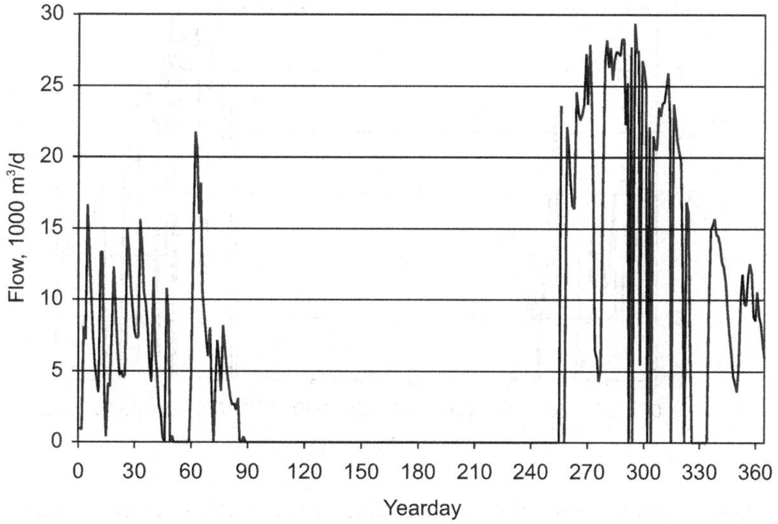

FIGURE 1.13 An example of wet season source flow for STA56 for the calendar year 2003. Water is pumped from a collection canal.

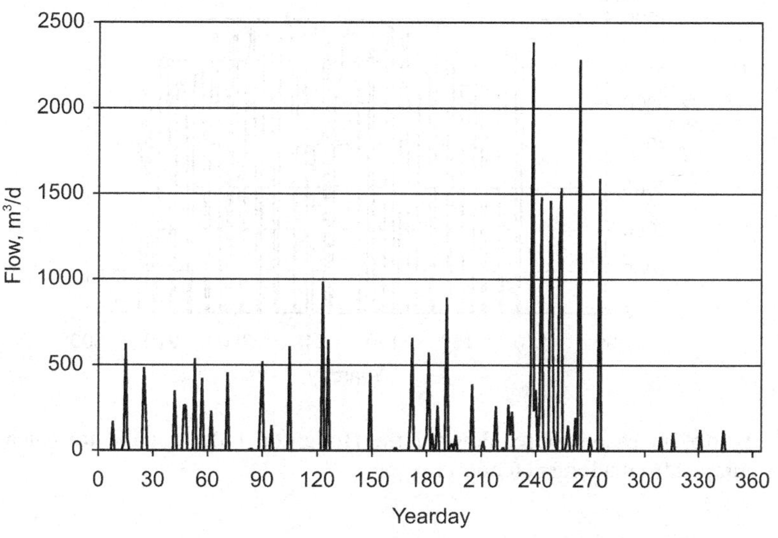

FIGURE 1.14 The rain event source flow for the Hidden River Wetland calendar year 1992. Water enters by gravity from a small impervious catchment.

In cold climates, marshes are frozen during the winter season, and incapable of providing treatment. Water to be treated may be generated year-round, and thus storage during the cold season is required. The marsh receives water from storage during the warm season, as dictated by marsh conditions or regulatory requirements. Water releases may not occur on every warm season day, depending upon remaining stored water, pump capacities, or the availability of operating personnel (Figure 1.16).

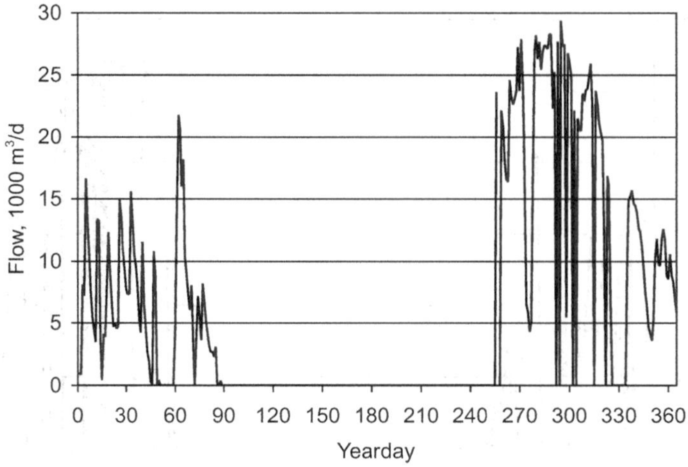

FIGURE 1.15 Discretionary source flow for the Blue Heron Wetland calendar year 1999. Water is pumped from an advanced municipal treatment plant at Titusville, FL.

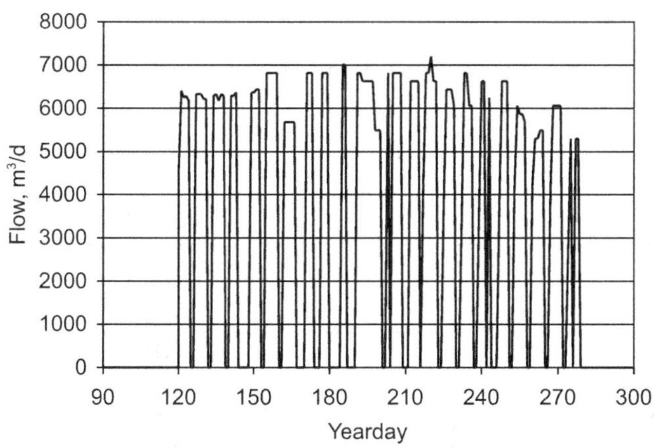

FIGURE 1.16 Unfrozen season source flow for the Houghton Lake wetland for calendar year 1997. Stored water is pumped from a long detention lagoon.

1.9 ECOLOGICAL BENEFITS OF TREATMENT MARSHES

Treatment marshes can provide significant benefits, ancillary to the primary purpose of water quality improvement, primarily in the form of vegetative biodiversity, wildlife habitat, and human uses.

Vegetative Diversity

The vegetation in young treatment marshes will evolve to a structure and diversity dictated by the water quality and quantity. This is true of both natural and constructed marshes.

TABLE 1.5
Number of plant species by plant group in the NADBv2 vegetation database for constructed wetlands.

Plant Group	Number of Species Recorded
Emergent Macrophyte	501
Floating Aquatic Plant	31
Submerged Aquatic Plant	10

Source: NADB Database (1998).

Initial plantings, if any, will in general not persist (Mitsch et al., 1998). Often, opportunistic species will develop to form the major cover in the wetland.

Even when treatment marshes are dominated by cattails or bulrush, dozens of other herbaceous and woody plant species are typically present (Table 1.5). Thus, the issue is not so much diversity as it is dominance.

Wildlife

Because they closely mimic natural wetlands, it should be no surprise that they attract a wide variety of wildlife, namely insects, mollusks, fish, amphibians, reptiles, birds, and mammals (Kadlec and Knight, 1996). FWS systems can provide significant ancillary benefits, primarily in the form of human uses and wildlife habitat.

A large number of studies of bird use in treatment marshes have been published since the summary presented in NADB Database (1998). The results of that early study still represent more modern research. For instance, Chimney and Gawlik (2007) report a total of 139 bird species representing 39 families in two of the south Florida stormwater treatment areas (STAs). Combined, the two STAs had 20 of the 21 families identified by Kadlec and Knight (1996). A later survey by Beck et al. (2013) found avian densities were 38 times greater than in the natural marsh. Local species richness in the STAs was 78% nearly four times greater than in the natural marsh.

Human Use

The human uses of treatment wetlands, including marshes, consist mainly of bird watching, waterfowl hunting, fishing, and recreational hiking. These uses are facilitated by design features such as vehicle parking, signage, observation points, and interpretive centers (Figure 1.17).

1.10 DATA SOURCES

This book is based upon data acquired over the past forty years. The growth of wetland literature has been exponential – literally. Zhang et al. (2010) compiled statistics on citation frequencies over an 18-year period (1991–2008). The exponential regression had a doubling time of 6.7 years, and an $R^2 = 0.99$. It is therefore undoubtedly the case that treatment wetland literature has experienced a similar exponential increase.

The information used herein comes from well over a hundred full-scale operating systems. Sources include published journal literature, published project reports, and project

FIGURE 1.17 The interpretive center at Green Cay wetland near Boynton Beach, FL.

base data files. Many of the base data files are public records provided by governmental owners; others were provided by generous owners, often to allow the author to interpret them. Referencing is done via standard literature citations where possible.

Many of the projects used herein are large and long-running. Those have had multiple research investigators, who have produced project reports and peer-reviewed publications.

A Floridian Bias

In particular, the Florida Everglades protection project has been producing information for over twenty years, by studying the more than twenty treatment marshes comprising that project. Over a hundred researchers, including those at the South Florida Water Management District (SFWMD) and contractors to that agency, have been involved in studies. Very large sums of money have supported this work.

As a result, each year the SFWMD, in collaboration with the Florida Department of Environmental Protection, issues a massive report, detailing many of the year's research results. There are twenty-five such reports, with citations as:

SFWMD (South Florida Water Management District) and FDEP (Florida Department of Environmental Protection), 1999–2004. Everglades Consolidated Reports, SFWMD, West Palm Beach, FL, USA.

SFWMD (South Florida Water Management District) and FDEP (Florida Department of Environmental Protection), 2005–2018. South Florida Environmental Report. SFWMD, West Palm Beach, FL, USA.

These are available at the SFWMD website:

www.sfwmd.gov/science-data/scientific-publications-sfer

Additionally, there are many other topic-specific reports available on the website. And, much of the raw data for Everglades protection wetlands is available on the public website DBHYDRO. DBHYDRO is the South Florida Water Management District's corporate environmental database that stores hydrologic, meteorologic, hydrogeologic, and water quality data. It can be accessed at:

www.sfwmd.gov/science-data/dbhydro

It is not feasible to cite particular SFWMD sources, or groups of such sources. The reader may assume that if information is given herein for a Florida stormwater treatment area (STA), it has come from one or more of these SFWMD sources.

Avoiding Bias

Given the huge amount of information available for the south Florida STAs, compared to other projects, it is unavoidable that there is a Floridian tilt to the overall treatment marsh database. For some purposes, this is not a serious difficulty. For instance, hydrology and hydraulics are not subjects with anticipated regional variability. However, for subjects that are temperature or climate sensitive, such as processing of oxidized nitrogen, regional variability can be a strong controlling factor. In such instances, there has been a conscious effort to address regional bias, by including information from multiple climate zones.

The Florida STAs typify only one of the four source categories addressed in this work. Municipal effluent polishing, urban and small agricultural runoff control, and lake and river improvement occur impressively large numbers. And, although their performance database is rapidly growing, they have not enjoyed the he financial resources available in south Florida.

Nomenclature

In this book, identification of specific treatment marsh units follows the general hierarchy given in Knight et al. (1993), which identifies sites, systems, and cells. One or more wetlands may exist at a particular geographic site. These are identified by information including site name, state, country, and community. Systems were defined as single or parallel wetland treatment areas that have separate outflow monitoring stations. Here, the term system means an independent marsh, receiving its own identified inflows and outflows, and hydrologically isolated from any other wetlands of the site. A single system may have multiple cells arranged in series and/or parallel. A cell is enclosed by berms and has identifiable inflow and outflow points.

A similar nomenclature has evolved for the Florida stormwater treatment areas. There, an STA is a geographical cluster of treatment marshes. The cluster is considered as a single entity in terms of regulatory permits. Compliance is determined for the combined outflow from the multiple systems that comprise the STA. Those component systems are locally known as flow ways and are composed of one or more cells.

1.11 SUMMARY

Natural wetlands have been used as receptors for wastewater since ancient times. The 20th century brought about the development of manmade wetland systems that are designed to emphasize specific characteristics of the natural wetland environment, aiming to improve the overall treatment capacity of the system. From this, three types of engineered wetlands evolved, with many variations of each: free water surface (FWS), horizontal subsurface flow (HSSF), and vertical flow (VF) systems. This work deals with unaugmented FWS systems that treat with mildly polluted waters.

Treatment marshes are mechanically simple systems that rely on passive treatment processes. Many systems are favorable for use in rural settings or areas of low population density because they may require large land areas. Systems in the urban setting are used to treat rainfall runoff and can be considerably smaller. Treatment marshes are perceived as a cost-effective and environmentally conscious treatment technology. They are demonstrably effective in providing multiple ecological benefits, in addition to water quality benefits.

FWS wetlands are similar to natural wetlands in that they contain areas of emergent vegetation, submerged vegetation, floating vegetation, and open water. They offer habitat benefits similar to natural wetlands, and invariably attract a variety of wildlife. These wetlands typically are used to polish effluent from advanced treatment processes such as long-detention lagoons and tertiary municipal plants, but here treatment of strong effluents is excluded: primary, secondary and industrial effluents are not considered for ecological reasons. FWS systems are virtually always the choice for stormwater treatment, including agricultural and urban runoff.

The information analyzed and summarized in this book represents the efforts of hundreds of scientists and engineers over the past decade. It builds upon the FWS sections of *Treatment Wetlands*, 2nd edition (Kadlec and Wallace, 2009). While a synthesis of this massive collection of information is necessary to carry the wetland treatment technology to a wider audience, it is still beneficial for the reader to refer to the original sources for more details and a regional perspective, and to examine the evolution of wetland engineering during this developmental period. The reader is encouraged to examine the references cited throughout the text and provided at the end of this book.

2 Hydrology

The success or failure of a treatment marsh is contingent upon creating and maintaining correct water depths and flows. In this chapter, the processes that add and subtract water from the wetland are discussed, together with the relationships between flow and depth. Internal water movement in wetlands is a related subject, which is critical to understanding of pollutant reductions.

The water status of a wetland defines its extent, and is the determinant of plant species composition in natural wetlands (Mitsch and Gosselink, 2015). Hydrologic conditions also influence the soils and nutrients, which in turn influence the character of the biota. Flow and storage volume determine the length of time that water spends in the wetland, and thus the opportunity for interactions between waterborne substances and the wetland ecosystem.

The ability to control water depths is critical to the operation of treatment wetlands. This operational flexibility is needed to maintain the hydraulic regime within the hydrologic needs of desired wetland plant species and may also be needed to tune the treatment efficiency of the treatment marsh. It is therefore necessary to understand the hydraulic factors that relate depth and flow rate, including vegetation density and aspect ratio. In FWS wetlands, this requires an understanding of stem drag effects on water surface profiles.

Water enters marshes via streamflow, runoff, groundwater discharge, and precipitation (Figure 2.1). These flows are extremely variable in most instances, and the variations are stochastic in character. Stormwater treatment marshes generally possess this same suite of inflows. Treatment wetlands dealing with continuous sources of wastewater may have these same inputs, although streamflow and groundwater inputs are typically absent. The steady inflow associated with continuous source treatment wetlands represents an important distinguishing feature. A dominant steady inflow drives the ecosystem toward an ecological condition that is somewhat different from a stochastically driven system.

Wetlands lose water via streamflow, groundwater recharge, and evapotranspiration (Figure 2.1). Stormwater systems also possess this suite of outflows. Evapotranspiration (ET) occurs with strong diurnal and seasonal cycles, because it is driven by solar radiation, which undergoes such cycles. Thus, ET can be an important water loss on a periodic basis.

Wetland water storage is determined by the inflows and outflows together with the characteristics of the wetland basin. Depth and storage in natural wetlands are likely to be modulated by landscape features, such as the depth of an adjoining water body or the conveyance capacity of an outlet stream. Large variations in storage are therefore possible, in response to the high variability in the inflows and outflows. Indeed, some natural wetlands are wet only a small fraction of the year, and others may be dry for interim periods of several years. Such periods of dry-out have strong implications for the vegetative structure of the ecosystem. Constructed treatment marshes, on the other hand, typically have some form of outlet water level control structure. Therefore, there is little or no variation in water level for continuous flow wetlands. Dry-out in continuous flow systems does not normally occur, and only the vegetation that can withstand continuous flooding will survive. Stormwater treatment wetlands, and those subject to discretionary inflows, may undergo drying.

The important features of wetland hydrology from the standpoint of treatment efficiency are those that determine the duration of water biota interactions, and the proximity of waterborne substances to the sites of biological and physical activity. There is a strong tendency in the wetland treatment literature to borrow the detention time concept from other

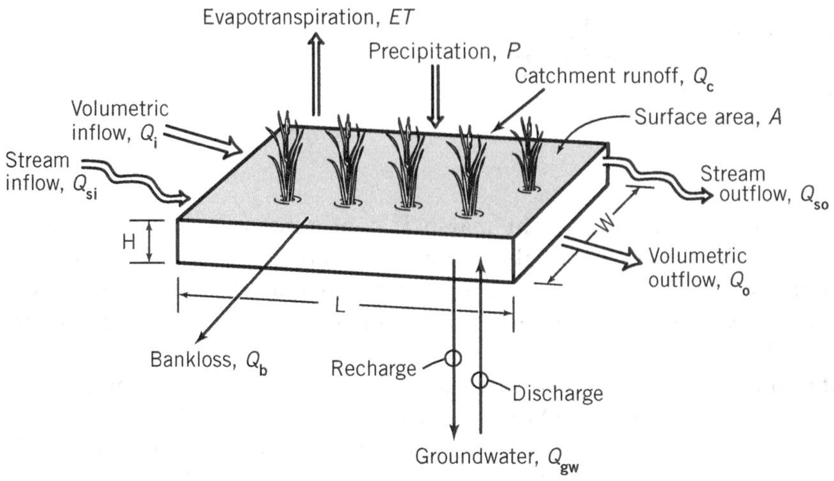

FIGURE 2.1 Components of the wetland water budget.

aquatic systems, such as "conventional" wastewater treatment processes. In purely aquatic environments, reactive organisms are distributed throughout the water, and there is often a clear understanding of the flow paths through the vessel or pond. However, wetland ecosystems are more complex, and therefore require more descriptors.

2.1 DESCRIPTORS

Literature terminology is somewhat ambiguous concerning the symbols used for hydrologic variables. Often, terms like depth and detention time are implied to be single numbers for a given system, when in fact they are temporally and spatially variable, and are in fact best represented by distribution functions. The notation and parent variables are illustrated in Figure 2.1.

Common Descriptors

Some hydrologic variables are used without regard for the underlying variability, and are of value only in a general sense, rather than quantitatively. These are here referred to as nominal descriptors.

Hydroperiod and Hydropattern

The term hydroperiod refers to the duration of time that the wetland is flooded and is therefore a single number fraction between zero and one. It gives a general idea of when the wetland has standing water. In turn, it reflects whether or not source water is being added. Because the goal of a treatment marsh is to improve source water quality, systems with hydroperiod much less than unity are not as efficient as those with hydroperiods near unity.

The timing, duration, and distribution of wetland water levels are, together, commonly referred to as the wetland hydropattern, which incorporates the duration and frequency of water level perturbations (Rasmussen, 2008). Hydropattern were developed as an evaluation metric for natural wetlands, but are currently also in use for treatment marshes.

Hydropattern metrics are used to constrain operations and to evaluate water quality processes. Two such metrics are the wetland hydrograph and the depth-duration distribution.

The hydrograph is a plot of water depth versus time. Because four of the five classes of source water flows considered here (see Chapter 1) are not continuous, treatment marsh water depths are not uniform in time. An example hydrograph of a system with discretionary deliveries is shown in Figure 2.2.

The wetland depth-duration can take different forms, principally depth-duration and depth-frequency (Figure 2.3). It is used to determine operating strategies that keep water depths in the range required by the desired vegetation.

Detention Time

Wetland detention time is defined as the wetland water volume involved in flow divided by the volumetric water flow:

$$\tau = \frac{V_{active}}{Q} = \frac{\varepsilon A_{active} h_{nominal}}{Q} \tag{2.1}$$

where: A_{active} = area of wetland containing water in active flow, m^2
h = wetland water depth, m
V_{active} = volume of wetland containing water in active flow, m^3
ε = fraction of active volume occupied by water, –
τ = detention time, d

It is sometimes convenient to work with the nominal parameters of a given wetland. To that end, a *nominal detention time* (τ_n) is defined:

$$\tau_n = \frac{V_{nominal}}{Q} = \frac{(LWh)_{nominal}}{Q} \tag{2.2}$$

A very common alternative designation for nominal detention time is HRT. Equation (2.2) is a rather innocuous relation, but has no less than four difficulties, which have led to

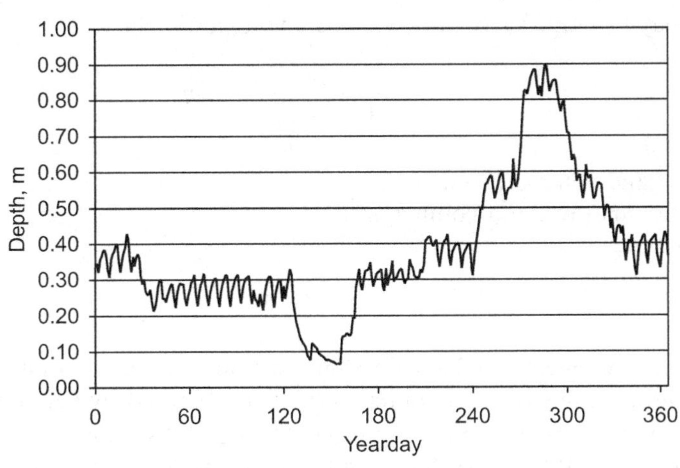

FIGURE 2.2 The 1978 mean depth hydrograph for the Boney Marsh treatment wetland located north of Okeechobee, FL. The source water delivery was from the Kissimee River. Pumps were not run on weekends, hence the weekly variability. There was a May shutdown and an autumn (wet season) spike.

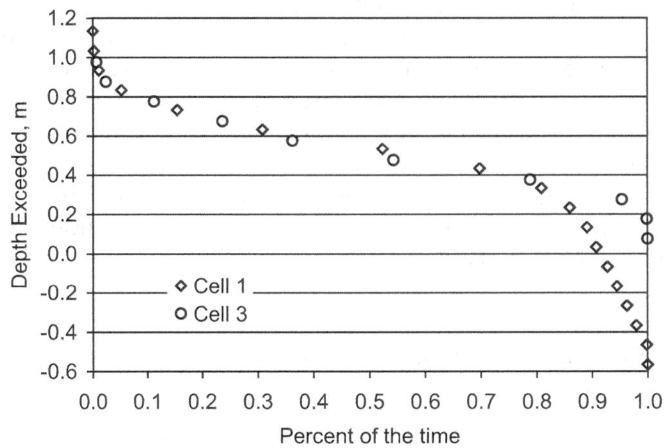

FIGURE 2.3 The depth-duration relationship for two independent wetlands in STA2, for daily and spatially averaged data. The period of record was ten years.

misunderstandings in the literature. First, there is ambiguity about the choice of the flow rate: should it be inlet, or outlet, or an average? This will be addressed in the next section.

Second, for FWS systems, some of the wetland volume is occupied by stems and litter, so that $\varepsilon \leq 1$. This quantity is difficult to measure, because of spatial inhomogeneity, both vertical and horizontal. It is known to be ca. 0.95 for cattails in a northern environment (Kadlec, 1998), and for submerged aquatic vegetation in the Everglades, and for an emergent community (Lagrace, 2000, as cited by U.S. EPA, 1999).

Third, not all the water in a wetland may be involved in active flow. Stagnant pockets sometimes exist, particularly in complex geometries. As a result, $A_{active} \leq A = L \cdot W$. A gross areal efficiency may be defined as $\eta = A_{active}/A$.

Fourth, the mean water depth (h) in a wetland is difficult to determine with a satisfactory degree of accuracy, especially for large wetlands. That variability translates directly to a comparable uncertainty in the water depths.

These effects may be empirically lumped, and a volumetric efficiency (e_V) defined as:

$$e_V = \frac{V_{active}}{(LWh)_{nominal}} = \varepsilon \eta \tag{2.3}$$

where: e_V = wetland volumetric efficiency, –
$LWh_{nominal}$ = nominal wetland volume, m^3
It is then clear that:

$$\tau_n = e_V \cdot \tau \tag{2.4}$$

Volumetric efficiency reflects ineffective volume within a wetland, compared to presumed nominal conditions. Portions of the nominal volume are blocked by submerged biomass (ε), bypassed (η), or do not exist because of poor bathymetry ($h/h_{nominal}$).

Confusion in nomenclature exists in the older literature, where e_V is sometimes identified as "wetland porosity." For dense emergent vegetation in FWS wetlands, it has presumptively been assigned a value in the range 0.65–0.75 (Reed et al., 1995; Crites and Tchobanoglous, 1998; Water Environment Federation, 2001), all of which use the symbol "n" in place of e_V). USEPA (1999, 2000a) presumptively assign the range 0.7–0.9 (both of which use the symbol "ε" in place of e_V).

It may be assumed that conservative tracer testing will provide a direct measure of the actual detention time in a wetland (Fogler, 1992; Levenspiel, 1995). Then, via Equation 2.4, there is a direct measure of e_V, although there is no knowledge gained about the three contributions to e_V by this process. At this point in the development of constructed wetland technology, there have been numerous such tracer tests. Results from 25 tests on FWS wetlands show an average of $e_V = 0.73$, with a range from 0.20–1.33. This is a strong indication that the term porosity is a misnomer, because e_V values are more strongly influenced by η and $h/h_{nominal}$ than by the voids among plants.

Wetlands routinely experience water gains (rainfall) and losses (evapotranspiration, seepage), so that outflows differ from inflows. If there is net gain, the water accelerates; if there is net loss, the water slows. A rigorously correct calculation procedure involves integration of transit times from inlet to outlet.

When there are local variations in total flow and water volume, the correct calculation procedure must involve integration of transit times from inlet to outlet. For steady flows, it may be shown that (Chazarenc et al., 2003):

$$\tau_{an} = \tau_{in} \cdot \left(\frac{\ln(R)}{R-1} \right) \tag{2.5}$$

where: R = water recovery fraction, $=Q_{out}/Q_{in}$

Q_{out} = outlet volumetric flow, m^3/d

τ_{an} = actual nominal detention time, d

τ_{in} = inlet flow-based nominal detention time, d

In terms of detention time alone, moderate amounts of atmospheric gains or losses (P – ET) are not usually of great importance, although there is ambiguity in the choice of flow rate (Q). Some authors base the calculation on the average flow rate (inlet plus outlet ÷ 2). This approximation is good to within 4% as long as the water recovery fraction is $0.5 < R < 2.0$.

Detention Time Distributions

A very important feature of marsh water flows is that they exhibit a distribution of detention times; i.e., detention time is not a single number. This fact cannot be ignored in the analysis of wetland water quality performance, as will be discussed in Chapter 7.

Hydraulic Loading Rate

The hydraulic loading rate (q or HLR) is the rainfall equivalent of whatever flow is under consideration. It does not imply the physical distribution of water uniformly over the wetland surface. The wetted area is usually known with good accuracy, because of berms or other confining features. The defining equation is:

$$q = \frac{Q}{A} \tag{2.6}$$

where: A = wetland area (wetted land area), m^2

Q = water flow rate, m^3/d

The definition is most often applied to the source water addition flow at the wetland inlet: $q_i = Q_i/A$. The subscript "i" is often omitted for simplicity.

The relation between nominal detention time and hydraulic loading rate is easily seen to be:

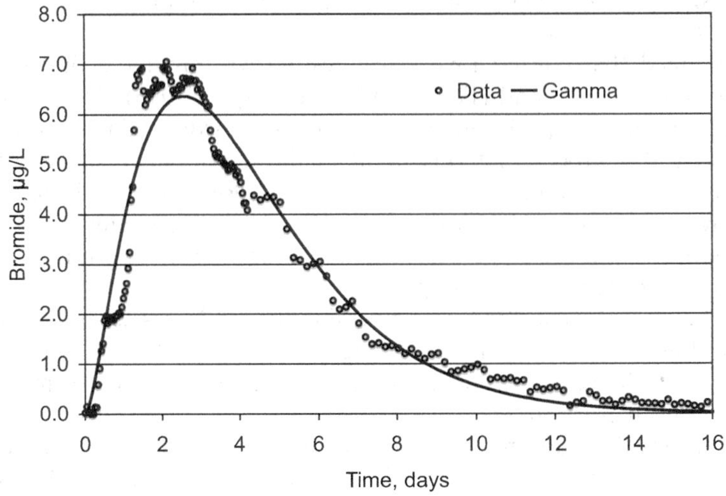

FIGURE 2.4 Results of a sodium bromide tracer test of a 1.33 ha marsh, Tres Rios Hayfield 1, receiving 2015 m^3/d. The nominal detention time was 4.22 days.

$$q = \frac{Q_i}{LW} = \frac{\varepsilon h_{nominal}}{\tau_n} \tag{2.7}$$

Thus, it is seen that hydraulic loading rate is inversely proportional to nominal detention time for a given wetland depth. Hydraulic loading rate therefore embodies the notion of contact duration, just as nominal detention time does.

Water Velocity

The nominal water velocity is the flow per unit water cross sectional area. In terms of the notation used here,

$$v_{nominal} = \frac{Q}{\varepsilon h_{nominal} W} \tag{2.8}$$

where: $\varepsilon h_{nominal}W$ = open area perpendicular to flow, m^2

The actual velocities (v) are those that would be measured with a probe in the wetland. It is noted that there is large spatial and temporal variation in "v," and hence individual spot measurements may be as much as a factor of ten different from the mean. Bias is due to high local measurements, because probes do not easily find small pockets of stagnant water.

The *superficial velocity* is the empty wetland velocity – again a spatial average. In terms of the notation used here,

$$u_{nominal} = \frac{Q}{h_{nominal} W} \tag{2.9}$$

where: $h_{nominal}W$ = total wetland area perpendicular to flow, m^2

The superficial velocity is also referred to as the cross-sectional hydraulic loading. For FWS wetlands, there is not much difference between "u" and "v," because FWS porosity is nearly unity.

The relation between superficial and actual velocities is:

$$u = \varepsilon v \tag{2.10}$$

Another term in common use is the flow per unit width, defined by:

$$\frac{Q}{W} = u_{nominal} h_{nominal} \tag{2.11}$$

Averaging Periods

Because wastewater treatment wetlands typically operate on a continuous basis, much of the performance literature for such systems deals with monthly or annual averages. But calendar definitions are not necessarily appropriate for systems with discontinuous source flows. For instance, one would not expect the same performance from a system running year-round as for a system that runs only six months of the year. Thus, for performance analyses and calculations, it is necessary to specify the time period involved in averaging, and whether or not it was receiving water. The common usage in the urban stormwater literature is to use event mean values. In the broader context, it is useful to consider the factors that determine whether a marsh is "running," or not.

In Service

There are times when routine, or emergency, maintenance activities indicate that water cannot be sent to the treatment marsh, and it is considered to be "offline." The list of reasons includes:

- Soil management. Activities in this group include soil removal or addition, and leveling.
- Vegetation management. The replacement or alteration of the vegetation may need to occur, to reestablish destroyed stands, or to change the plant communities, i.e., weed removal.
- Endangered species management. Snail kites, stilts, and burrowing owls have all been implicated in restricting or shutting down STA cells.
- Repair of water control structures.

During offline times, water is not sent to the marsh, and it is deemed to be out of service. Offline times are usually small. For instance, over a five-year POR, the five Florida STAs averaged 8% offline.

It is also possible that the listed activities will restrict, but not preclude, system operation. Such restrictions are usually on water depth; which in turn influences the allowable inflow. For instance, endangered species management cause restrictions on the five Florida STAs during about a quarter of the time during 2015–2016.

In Use

A treatment marsh is here considered to be *in use* if it is delivering treated water to downstream receiving systems. That means there is outflow through the discharge structures. Continuous flow systems are in use nearly 100% of the time, and annual averages can be

TABLE 2.1
Annual days that example STA paths are in use.

	STA56 North FW1	STA56 Center FW2	STA34 East FW1	STA34 Center FW2	STA2 Cell 1 FW1	STA2 Cell 3 FW3
POR, years	17	17	12	12	12	15
Max	205	211	300	284	342	343
Min	15	19	104	78	151	122
POR Mean	123	125	193	199	242	206
Percent	34	35	54	57	67	52

used. Cold climate wetlands are in use only during the unfrozen season, and season averages are normally used for analysis. Urban stormwater wetlands are in use only during rainfall events, leading to the utilization of event averages for analysis.

Discretionary flow systems have characteristics that are not totally seasonal, nor totally rain event driven. Both short and long periods of outflow can occur. The Florida STAs, treating agricultural runoff, are in use about half the time (Table 2.1).

2.2 WATER DEPTH

Mean water depth is here denoted by "h." The depth calculation requires a detailed survey of the wetland bottom topography, combined with a survey of the water surface elevation. The accuracy and precision must be better than normal, because of the small depths usually found in treatment wetlands. That means ground-level surveys are required, as opposed to aerial surveys. As-built surveys under dry conditions may not suffice for determination of ground levels, because of possible soil swelling and lift upon wetting. The ground and water surveys combine to give the local depth:

$$h = H - G \tag{2.12}$$

where: G = local ground elevation, m
h = water depth, m
H = local water elevation, m

In general for treatment marshes, there is not a well-defined soil-water interface. Surface water overlies an unconsolidated horizon, comprised of flocculent biological residuals and the pronate litter from senesced macrophytes. These are interspersed to varying degrees and are characterized by low bulk density. Below the floc-litter layer(s) is consolidated soil (Figure 2.5).

Marsh research has investigated the issue of ground level definition in topographic surveys (Choi and Harvey, 2014; Piccone et al., 2014). Standard surveyor's staffs, when "firmly placed," would typically penetrate the floc-litter layer, because of its low bulk density. Because of this ambiguity, recent Florida wetland surveys have opted to equip the surveyor's staff with a footplate; with an area of about 45 cm^2 for Choi and Harvey (2014) and 181 cm^2 for STA ground surveys (Piccone et al., 2014). There is no standard for foot plate size. Because the floc-litter layer is often in the range of 10–20 cm in thickness, there is a comparable differential between surveying techniques. For example, floc-litter thickness measurements for 13 STA cells in 2003 averaged 11 cm, with a range of 5–20 cm. The difference between ground elevation definitions is important, because treatment marsh water depths are often in the range of 20–70 cm.

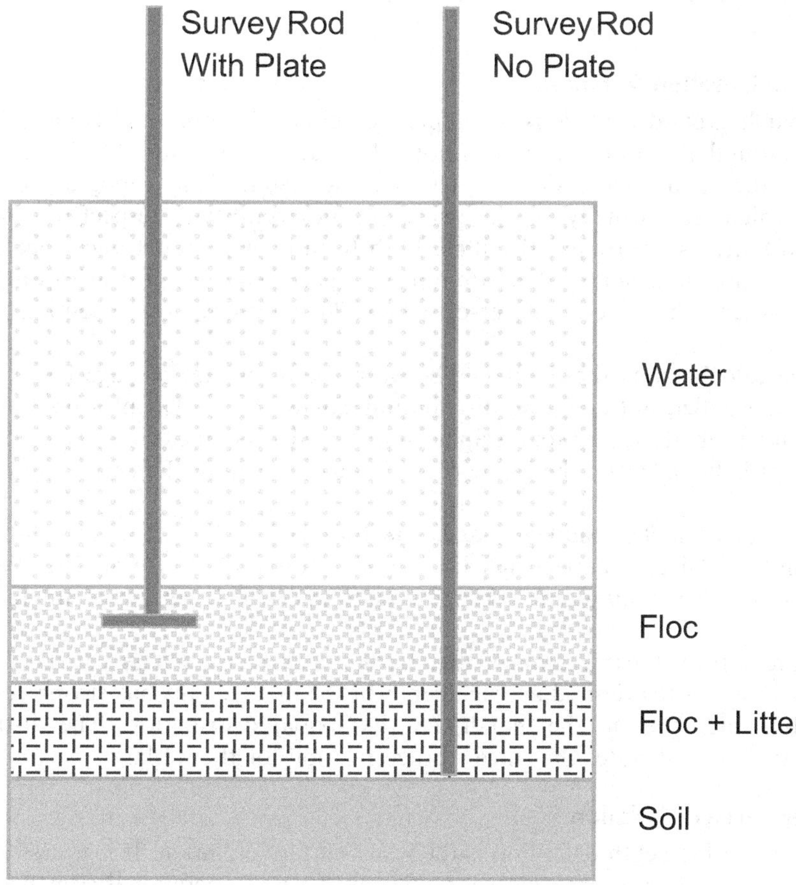

FIGURE 2.5 Vertical zones at the water-soil interface.

System Bathymetry

The cells that comprise a wetland treatment system contain deep areas of open water as well as shallow vegetated areas. The spatial arrangement of these can take several general forms:

- Mostly shallow marsh, possibly with an inlet spreader canal, and an outlet collector canal.
- Shallow marsh with one or more transverse deep zones.
- A deep pond inlet zone, followed by shallow marsh.
- A deep central channel, often sinuous, with fringing shallow marsh benches.
- Interspersed and irregular deep areas.

The overall treatment wetland system is comprised of one or more such cells. These may be terraced, to account for local topography.

The deep zones pose few difficulties in depth description, because the water surface is a level pool, and the bottom is contoured during construction. However, the shallow marsh sections have variability in both bottom elevation (G) and in water surface elevation (H). The average wetland water depth may be defined to include or exclude deep zones depth.

Marsh Ground Elevation Variation

Treatment marsh ground elevations are typically neither flat nor level (Figure 2.6). Treatment marsh ground elevations are not uniform but are spatially variable. There are several meanings for the term "water depth." Unevenness results from topographical hills and depressions, which are normally not excessive, yet important with respect to wetland water depth. Marsh bottoms are frequently tilted, both from inlet to outlet and from side to side across the flow direction. Survey data are normally specified on a grid, with one dimension parallel to flow and the other transverse to flow. That information encompasses both tilt and unevenness.

Tilt can be calculated as the difference between the transverse average levels at the inlet and those at the outlet, not counting distribution canals, if any. Tilt plus unevenness represents the total variability in the topographic survey and is characterized by the distribution of elevations. It is bounded by the range of survey elevations for the entire two-dimensional grid.

Unevenness for example treatment marshes in Florida is in the range of 0.3–1.3 m (Table 2.2). Slopes in these "flat" treatment marshes are of the order of 10^{-4} m/m. Thus for small systems, tilt is just a few centimeters, while for large systems it can range up to one meter (Table 2.2).

It is apparent from these examples that both tilt and unevenness can be of the same order of magnitude as the desired average depth in the marsh. This has strong consequences for where the water goes in the marsh, i.e., its spatial distribution. The distribution of ground elevations also dictates the stage storage relation for the wetland.

Marsh Water Surface Variation

The second part of the depth definition is the water surface elevation. It is necessary to distinguish between two circumstances: a level pool scenario and operational scenarios. The level pool arises when there is no flow, and the water surface is flat and horizontal. It is also termed a static water elevation. Operational scenarios occur under flow conditions, when the water surface is neither flat nor horizontal. Figure 2.7 depicts level pool and flowing situations.

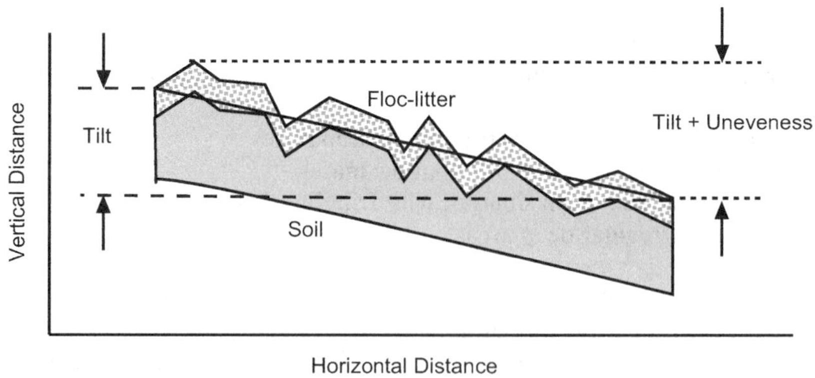

FIGURE 2.6 One-dimensional schematic of the longitudinal variability of ground elevations. The vertical scale has been exaggerated, and the horizontal scale compressed.

TABLE 2.2

Example tilt and unevenness in treatment marshes. The designation NTC refers to north test cells in STA1W.

	Cell	Flow Length m	Tilt + Unevenness cm	Tilt cm
STA2	1	5,273	98	21
STA2	2	4,758	91	9
STA2	3	4,596	91	12
STA56	1A	966	107	87
STA56	1B	966	85	22
STA56	2A	666	128	36
STA56	2B	992	70	60
STA1W	NTC9	84	47	3
STA1W	NTC10	84	29	3

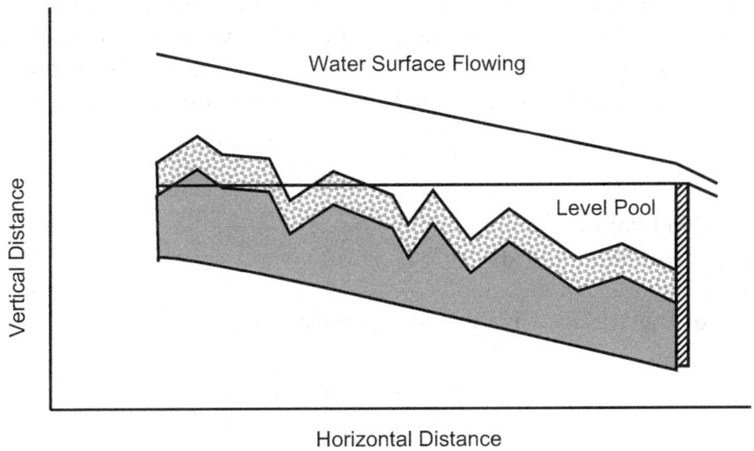

FIGURE 2.7 One-dimensional schematic of the level pool and flowing scenarios. The vertical scale has been exaggerated, and the horizontal scale compressed.

It is apparent that the mean water depth depends strongly on whether or not water is flowing, as does the volume of water in the wetland. In a later section, the water surface profiles will be examined in more detail. Those water surface profiles depend upon wetland size, wetland vegetation, and water velocity.

Alternate Water Depth Definitions

For non-flowing conditions, the water surface is a level pool. But because of ground tilt and irregularity, there are alternate interpretations of water depth.

1. Static water depth. This is defined as the difference between mean water stage under non-flowing conditions and the mean ground elevation. The mean ground elevation is taken to be the spatial average of surveyed elevations determined on a grid.

2. Depth of surface water. Because portions of the marsh may not have surface water under low stage conditions, the depth in those areas that do have surface water is different from the depth defined in 1. above. Indeed, when static water depth is zero, half of the marsh has surface water. So, an alternate measure represents the average depth of surface water where it exists. It is the volume of surface water divided by the area of surface water.
3. Flowing water depth. This is the volume of surface water in the marsh at a flow rate, divided by the total wetted marsh area. As indicated in Figure 2.7, flowing water depths are larger than static water depths. This is the depth to be used in computing detention times.
4. Water depth to consolidated soil. This is defined as the difference between water stage and the consolidated soil elevation. This definition has been used in some studies (U.S. FWS, 2007; Wasantha Lal et al., 2015).

Stage-Storage and Stage-Area Relations

The volume of water stored in a wetland under static conditions can be represented by the stage-storage relation for the wetland. It is a measure of the distribution of ground elevations. Related is the stage-area relation, which describes the wetted area as a function of level pool elevation. It is basically the areal distribution of ground elevations, ranging from the stage at which there is no standing water, to the stage at which the entire marsh surface is inundated.

Often, a polynomial fit of the stage-volume data is adequate, for instance a quadratic:

$$V = a + bG + cG^2 \tag{2.13}$$

where: V = wetland volume, m^3
G = wetland ground elevation, m
a,b,c = fitting constants
The derivative of this function is the water surface area:

$$A = \frac{dV}{dG} = b + 2cG \tag{2.14}$$

where: A = wetland area, m^2

Such a relation is only valid between the minimum and maximum ground elevations for the wetland.

As an example, STA56C1A was surveyed at 73 internal grid points, using the recognized standard datum (NAVD 88) (Piccone et al., 2014). That information was then processed to establish the stage-storage and stage-area relations (Figure 2.8).

The best fit was:
G_{min} = 3.51 m G_{max} = 4.49 m
a = 30.56 x 10^{-6} m^3 b = −17.41 x 10^{-6} m^2 c = 2.48 x 10^{-6} m
R^2 = 0.993

2.3 COMPONENTS OF THE MASS BALANCE

Transfers of water to and from the wetland follow the same pattern for surface and subsurface flow wetlands (refer to Figure 2.1). In treatment marshes, source water additions are normally the dominant flow, but under some circumstances, other transfers of water are also important. The dynamic overall water budget for a wetland is:

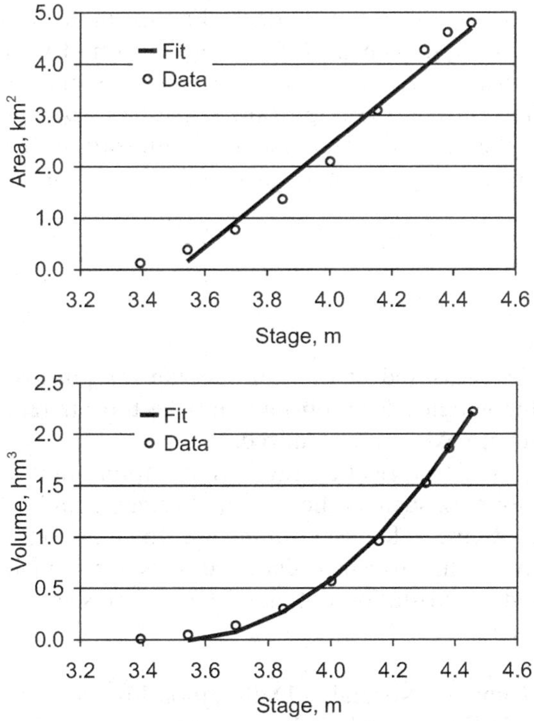

FIGURE 2.8 Stage-volume and stage-area relations for STA56C1A.

$$Q_i - Q_o + Q_c - Q_b - Q_{gw} + Q_{sm} + P \cdot A - ET \cdot A = \frac{dV}{dt} \tag{2.15}$$

where: A = wetland top surface area, m^2
ET = evapotranspiration rate, m/d
P = precipitation rate, m/d
Q_b = bank loss rate, m^3/d
Q_c = catchment runoff rate, m^3/d
Q_{gw} = infiltration to groundwater, m^3/d
Q_i = input source water flowrate, m^3/d
Q_o = output source water flowrate, m^3/d
Q_{sm} = snowmelt rate, m^3/d
t = time, d
V = water storage in wetland, m^3

Inflows and Outflows

Most moderate to large scale facilities will have input source flow measurement; a smaller number of facilities will have the capability of independently measuring outflows as well as inflows. Due a lack of outlet flow measurements, the overall water budget (Equation 2.15) is often used to calculate the estimated outflow rate. Usually, only rainfall is a significant addition, and only ET is a significant subtraction, to the inflow, simplifying the analysis. This calculation is most easily performed when there is no net change in storage.

The change in storage (ΔV) over an averaging period (Δt) can be a significant quantity compared to other terms in the water budget. For example, if the nominal detention time in the wetland is 10 days, then a 10% change in stored water represents one day's addition of water. Because water depths in treatment wetlands are typically not large, changes of a few centimeters may be important over short averaging periods. One of the difficult components of the surface water budget is infiltration. There are direct methods for measurement, but often infiltration is assumed to account for discrepancies in the water budget.

Precipitation

Rainfall

Rainfall amounts may be measured at or near the site for purposes of wetland design or monitoring. However, the gauging location must not be too far removed from the wetland, because some rain events are extremely localized.

For most design purposes, historical average precipitation amounts suffice. These may be obtained from archival sources, such as the National Oceanic and Atmospheric Administration (NOAA), National Centers for Environmental Information (NCEI). In the United States, a very large array of climatological data products are available online at: www.ncdc.noaa.gov/data-access/land-based-station-data/land-based-datasets.

Among their products are:

- 1981–2010 U.S. Climate Normals. Daily, monthly, and annual data includes temperature, precipitation, snowfall, heating and cooling degree days, frost/freeze dates, and growing degree days calculated from observations at approximately 9,800 USA stations.
- National Solar Radiation Database. The 1991–2010 database contains data for over 1,500 stations across the United States.
- Data Snapshot Maps. A variety of maps of normals for the United States, including temperature and precipitation, based on data from 1981 to 2010.

As an illustration of the map service, the June normal precipitation map is shown in Figure 2.9.

The total catchment area for a wetland is likely to be just the area enclosed by the containing berms and roads; and that area is easily computed from site characteristics. Rainfall on the catchment area will in part reach the wetland basin by overland flow, in an amount equal to the runoff factor times the rainfall amount and the catchment area (Figure 2.3). A very short travel time results in this flow being additive to the rainfall:

$$Q_c = \psi \cdot P \cdot A_c \tag{2.16}$$

where the units of each term are m^3/d, and where:

A_c = catchment surface area, m^2

ψ = catchment runoff coefficient, –

For small and medium sized wetlands, the catchment area will typically be about 10–20% of the wetland area, meaning that about 10–20% of a site will be taken up by berms and access roads which may drain to the wetland. Runoff coefficients are high, since the berms are impermeable; a range of 0.1–1.0 might be typical. The combined result of impermeable berms, their necessary area, coupled with quick runoff, is an addition, of about 10–20% to direct rainfall on the bed.

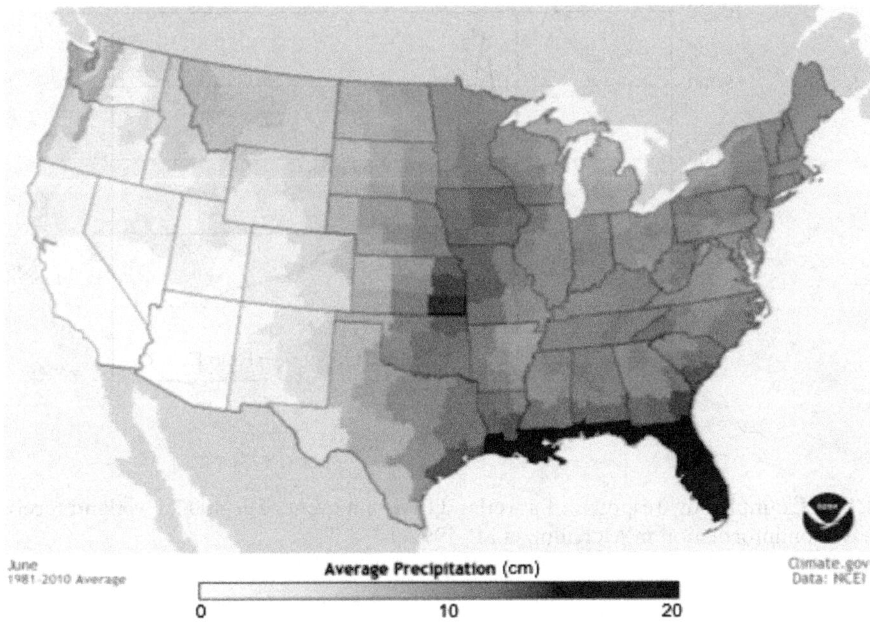

FIGURE 2.9 Normal precipitation map for the United States for June. (Source: NOAA, Climate.gov, www.climate.gov/maps-data, accessed 11/21/17).

For large wetlands, the catchment fraction of the wetted area is smaller. For example, for the 232 ha STA1EC3, levee runoff areas are 3.1%. Another 2.1% is included as power line peninsulas.

Dynamic Rainfall Response

For treatment marsh systems are fed a relatively constant flow of source water, there is a strong tendency to visualize a relatively constant set of system operating parameters – depths and outflows in particular. This is not the case in practice. There may be significant outflow response to rain events. A sudden rain event, such as a summer thunderstorm, will raise water levels in the wetland, and produce a pulse of outflow from the wetland (Figure 2.10). The changes may be magnified by increased catchment inflows.

The implications for water quality are not inconsequential. Samples taken during the period ensuing the rain event represent flows much greater than average. Water has been pushed through the marsh, and exits having been somewhat diluted. Velocity increases are great enough to move particulates that would otherwise remain anchored. Internal mixing patterns will blur the effects of the rain on water quality.

Sampling intervals are not normally small enough to define these rapid fluctuations. It is therefore important to realize that compliance samples may give the appearance of having been drawn from a population of large variance, despite the fact that the variability is in large part due to deterministic responses to atmospheric phenomena.

Snowmelt

In northern climates, snowmelt is a springtime component of the liquid water mass balance. The end of season snow pack is melted over time, in rough proportion to the temperature

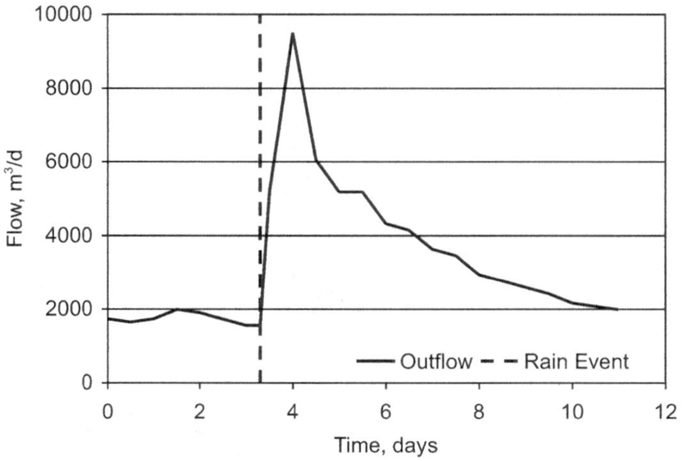

FIGURE 2.10 Example flow response of a wetland to a rain event. The 400 ha wetland received 5.1 cm of rain. (Based on information in McKillop et al., 1999).

excess above freezing. The amount of the snowpack is documented in weather records. The spring melt typically takes a few weeks.

Most, but not all, northern treatment marshes do not run during the snowmelt period. Notable exceptions are the arctic treatment wetlands, which often operate during the melt period. A more southerly example is the Brighton, Ontario, wetland, which operates year-round in a location that has snow cover four months of the year (Kadlec et al., 2012a).

There are many snowmelt runoff models, including SRM, widely used in both simulation and forecasting modes (Kustas et al., 1994; Lee et al., 2005; Butt and Bilal, 2011). The variables of the SRM model are temperature, precipitation and snow cover; and the input parameters are degree-days, runoff coefficient, temperature lapse rate, recession coefficient, critical temperature, time lag and rainfall contributing areas.

Evapotranspiration

Water loss to the atmosphere occurs from open water surfaces (evaporation), and through emergent plants (transpiration). This water loss is closely tied to wetland water temperature, and is discussed in detail in Chapter 5. Here the impacts of evapotranspiration (ET) on the wetland water budget are explored. The two simplest estimators are: (1) Large FWS wetland ET is roughly equal to lake evaporation, and (2) is roughly equal to 80% of pan evaporation. However, a more detailed estimation procedure is normally utilized, based on local meteorological conditions. The basis for these potential evapotranspiration calculations is a 100% hydroperiod, and a standardized vegetative cover is presumed. Figure 2.11 shows the distribution of annual evapotranspiration in different regions of the United States.

Evapotranspiration follows strong diel and strong monthly variability, because it is driven by solar radiation. This water transfer shuts off at night and reaches a maximum at mid-day. The low amount of incoming solar radiation in winter in northern latitudes causes a lower atmospheric loss than in summer, when there is higher solar radiation.

Marsh treatment systems frequently operate with small hydraulic loading rates, in the range of 1–5 cm/d. ET losses approach a daily average of 0.50 cm/d in summer in the southern United States; consequently, 10–50% the daily added water may be lost to ET under those circumstances. As an extreme example, the Incline Village treatment wetland near Carson City, NV, evaporated 83% of the pumping plus rainfall inflow (Kadlec et al., 1990).

In addition to the consumptive use of water, which may be critical in water-poor regions, ET acts to concentrate contaminants remaining in the water. This will be considered further in Chapter 6.

In mild temperate climates, annual rainfall typically slightly exceeds annual ET, and there is little effect of atmospheric gains and losses over the course of a year. But most climatic regions have a dry season and a wet season, which vary depending upon geographical setting (Figure 2.12). As a consequence evapotranspiration losses may have a seasonally variable impact. For example, ET losses are important in northern systems that are operated seasonally. In northern North America, about 80% of the annual ET loss occurs in the summer 6 months. Therefore, lightly loaded seasonal wetlands in cold arid climates are strongly influenced by net atmospheric water loss. Examples include the seasonal Roblin, Manitoba, treatment wetland system, which operates at zero discharge two summers out of every three (Mohr, 2000); and the seasonal lined Saginaw, Michigan, treatment wetland, which operated with 52% water loss over a ten-year POR (Kadlec and Zmarthie, 2010).

Transpiration Flow

Vertical flows of water in the upper soil horizon are also driven by rooted emergent plant water uptake to support transpiration. In aquatic and wetland systems with fully saturated soils or free surface water, the meteorological energy budget requires the vaporization of an amount of water sufficient to balance solar radiation and convective losses. Some of this

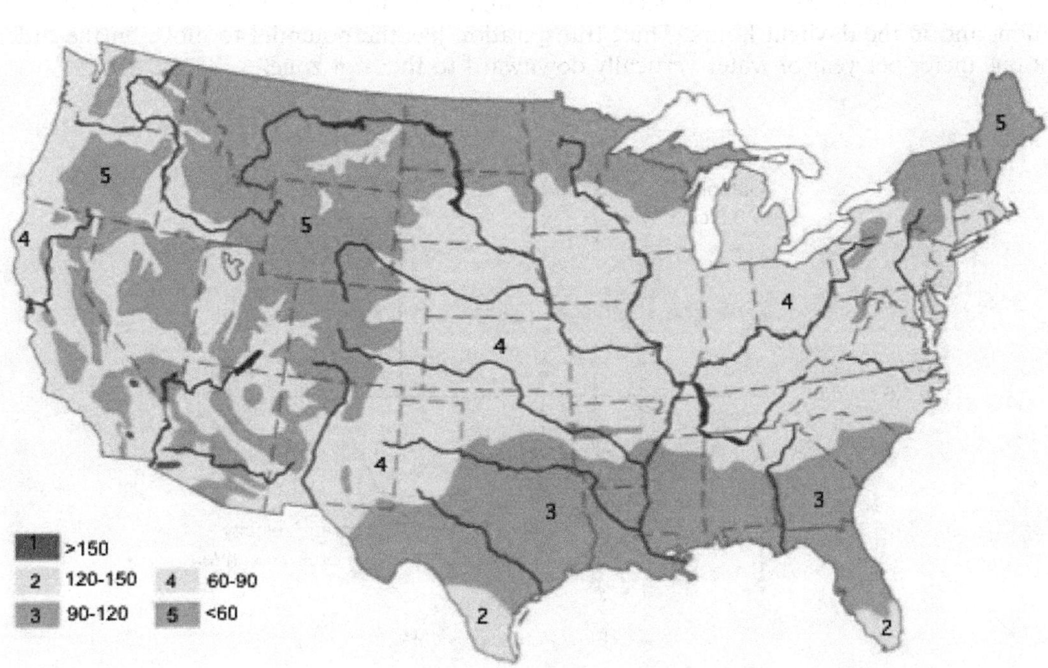

FIGURE 2.11 Potential annual evapotranspiration (cm) for the United States. (Adapted from Merritts, 2016).

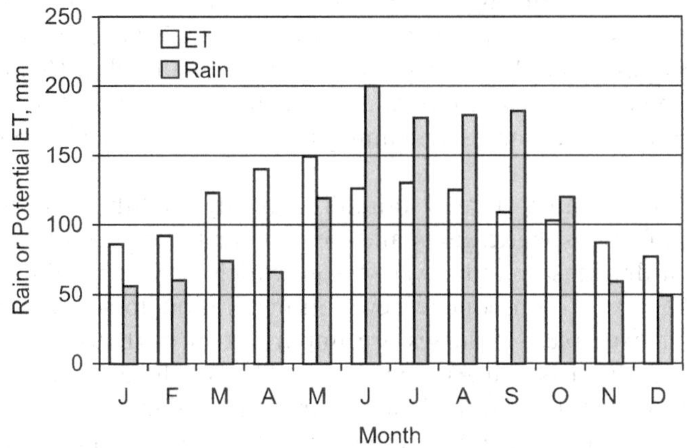

FIGURE 2.12 Monthly average rainfall and potential evapotranspiration for south Florida for a 15-year POR. (Based on information from Abtew et al., 2011).

vaporization is from the water surface (evaporation); some is from the emergent plants (transpiration). Emergent plants "pump" water from the root zone to the leaves, from which water evaporates through stomata, which constitutes the transpiration loss (Figure 2.13). In a densely vegetated wetland, transpiration dominates the combined process. Water for transpiration must move through the soil to the roots. That movement is vertically downward from overlying waters in most wetland situations. In warm climates, ET is about 100 cm/yr (Figure 2.10), but is concentrated in that part of the year with greatest solar radiation, and in the daylight hours. Thus, transpiration has the potential to move on the order of one meter per year of water vertically downward to the root zone.

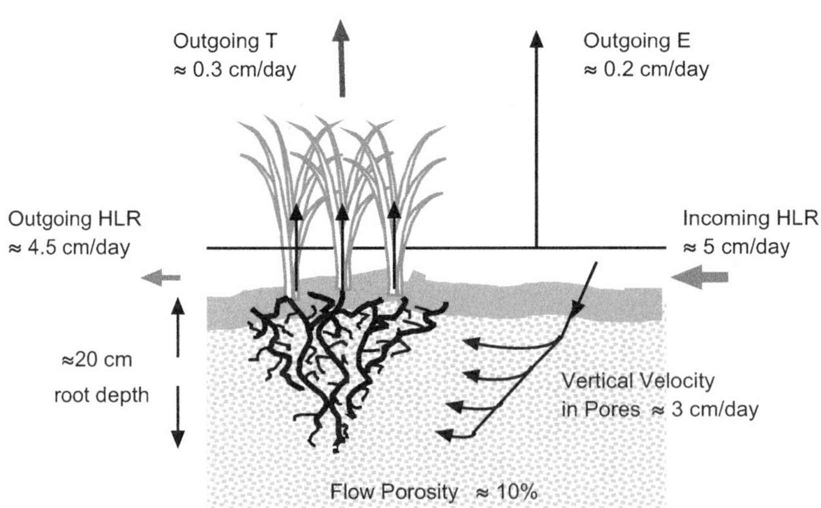

FIGURE 2.13 Example schematic of wetland flows, showing the routing of transpiration. Transpiration flows create a vertical flux of water that transports phosphorus from the litter-benthic mat zone down into the root zone. The vertical location of water extraction is dependent on the vertical position and density of the imbibing roots.

Trends in surface water quality data, and their interpretation, are potentially influenced by pore water hydrologic movement due to transpiration. The effects would be highest in wetlands of slow horizontal surface flow, dense vegetation, and large water residence times. The effect of transpiration on treatment efficiency in laboratory mesocosms has been reported by Wießner et al. (2013). They found a linear correlation between plant transpiration and ammonium removal. Bachand et al. (2014a) concluded that transpiration effects were important in full-scale marshes.

As will be discussed in Chapter 4, the root zone for emergent wetland macrophytes is typically about 20 cm in thickness, with root biomass decreasing with depth. It is therefore to be expected that underground transpiration flows will be active in that zone.

The vertical distribution of transpiration flows to roots has been studied and modeled in the agricultural literature (see, for example, Garrigues et al., 2006). Although such flows are fundamental for terrestrial crops, transpiration as a hydrologic transport pathway in wetlands has generally been ignored (Bachand et al., 2014b).

Beebe et al. (2014) conducted vertical tracer studies of mesocosms containing either plants (*Typha* spp.) and clipped plants. Tracer arrival time at a depth of 5 cm was 104 min for untrimmed cells, and 450 min in trimmed cells. The authors concluded that plant transpiration contributes significantly to vertical flow. The authors concluded that the vertical tracer tests demonstrated the importance of plant transpiration on vertical flow of water and dissolved constituents in treatment marshes.

Bachand et al. (2014a) determined transpiration flows for six wetlands in the Yolo Bypass area of California. The growing season there was approximately from early June to mid-September. They used hydrologic and mass budgets, and electrical conductivity as a natural conservative tracer, to quantify evaporation and transpiration during the irrigation season. The magnitudes of the underground transpiration flows were determined to be a large fraction of the inlet source water hydraulic loading rates (Table 2.3). The authors concluded: "These findings challenge the current science foundations" for analyzing solute transport in treatment marshes.

SEEPAGE

The surface water in a treatment wetland can exchange flows with groundwater underneath and adjacent to the wetland. Vertical water movement is termed infiltration, or deep seepage. If the water moves downward from the marsh, it is called recharge (of groundwater). If it moves upward, it is called discharge (of groundwater). Horizontal water movement to or from the marsh through, or just under, the confining levee, is termed seepage, or bank loss.

TABLE 2.3

Transpiration flows for Yolo Bypass wetlands.

Wetland	Area ha	Inlet Hydraulic Loading cm/d	Evapotranspiration cm/d	Transpiration cm/d
F20	42	1.38	0.65	0.44
F66	35	0.65	0.59	0.43
R31	63	0.98	0.64	0.50
R64	26	1.43	0.65	0.28
W32	30	1.16	0.70	0.51
W65	43	0.97	0.70	0.51

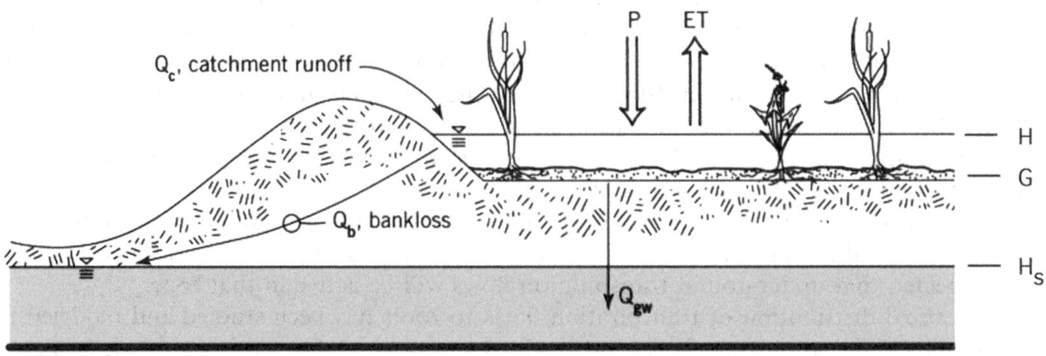

FIGURE 2.14 Water budget quantities.

Bank Seepage

Shallow seepage, or bank seepage, occurs if there is hydrologic communication between the wetland and adjacent aquifers. This is a nearly horizontal flow (Figure 2.14). If impermeable embankments or liners have been used, bank losses will be negligibly low. However, there are situations where this is not the case, notably for large wetlands where the only feasible embankment material is somewhat pervious. An empirical procedure may then be used in which the bank loss is calibrated to the head difference between the water inside and outside of the berm (Guardo, 1999) (Figure 2.14). A linear version of such a model is:

$$Q_b = \lambda L_b (H - H_s) \tag{2.17}$$

where: H = wetland water elevation, m
H_s = external water elevation, m
L_b = length of the berm, m
λ = empirical coefficient, m/d

Bank seepage can occur either out of or into the treatment wetland depending upon the relative elevations of the marsh surface water and the neighboring groundwater.

As an example, wetland levees in southern Florida are typically built from the peat and limestone soils native to the area. Leakage is therefore significant and has been studied extensively in connection with many canal, storage and treatment projects. Seepage collection canals are employed to collect losses (Figure 2.15).

A calibrated seepage coefficient of 0.0635 cubic meters per second per meter of head difference per kilometer of levee length ($(m^3/s)/m/km$) was used in analysis of STA water budgets (Zhao and Piccone, 2017).

Infiltration

Deep seepage, or infiltration occurs by vertical flow. Unless there is an impermeable barrier, wetland waters may pass downward to the regional piezometric surface (Figure 2.12). The soils under a treatment wetland may range in water condition from fully saturated, forming a water mound on the shallow regional aquifer; to unsaturated flow (trickling). Treatment marsh soils range in permeability, but two common types, peats and clays, have very small hydraulic conductivities, which limits vertical water movement. The hydraulic conductivity of clay is of the order of 1×10^{-7} m/s. Well decomposed peats have values of the order of 1×10^{-6}. These values, when used with typical marsh water depths, forecast small deep seepage rates. There are some unusual cases where the wetland is built on very porous

FIGURE 2.15 Aerial view of a portion of STA1E, with the flow direction from top to bottom. Four different canals are in use here: (1) collection canals at the outlet ends of cells 2 and 4N, (2) a spreader canal that redistributes water from cell 4N into cell 4S, (3) a discharge canal which receives water from cell 2, and (4) a seepage return canal. A seepage return pump transfers leaked water to the spreader canal for cell 4S. The purpose of the seepage return canal is to protect the neighboring urban area from flooding with leaked water.

substrates, such as the Tres Rios cobble site, which infiltrated 80% of the incoming water (Kadlec, 2008a).

Measured hydraulic conductivities in the ENRP soils were $0.7–2.1 \times 10^{-6}$ m/s (Harvey et al., 2004), which are in the range for clays and peats. The driving force for deep seepage is the hydraulic head differential between the wetland water surface and the regional groundwater table.

Infiltration can be measured locally by use of "seepage meters" (Harvey et al., 2000). The device is an inverted, cutoff drum inserted into the wetland soil. It is connected to a flexible collection bag that is partially filled with water. The change in the water in the bag over time represents the seepage under the drum.

Local infiltration rates may be either positive or negative (Table 2.4). The Everglades Nutrient Removal Project (ENRP) wetlands were bounded on the east side by the Loxahatchee National Wildlife Refuge, a marsh at higher stage. That stage differential caused groundwater flows into the ENRP on its east side. The ENRP was bounded on the west side by agricultural lands at lower water stage. Consequently, the ENRP lost water to seepage on its west side. Interior portions of the ENRP showed small seepage losses.

Total Seepage Flows

Under flowing conditions, it is usually not accurate to calculate total losses to groundwater from Equation 2.15, because it involves the difference between large numbers – inflow and outflow- with the associated uncertainty. However, during periods of no rainfall or inflow, the decrease in wetland stage is due to outflow, ET, and total seepage (Q_{gw} + Q_b), and some of the uncertainty is avoided. Equation 2.15 applied to this situation can be written:

TABLE 2.4

Vertical water fluxes measured at wetland sites in the treatment zone of WCA2A and the ENRP. Positive numbers mean groundwater discharge into wetland.

Wetland	Location	Distance	No. of Observations	Seep Rate, cm/d
WCA2A	S10C	Inlet+50m	1651	2.9
	E1	Inlet+2191m	1945	-0.03
	F1	Inlet+1968m	1885	0.04
	F4	Inlet+6906m	1616	-0.5
ENRP	Cell 1 East	East Side+50m	110	0.7
	Cell 1 Mid	East Side+867m	37	-0.04
	Cell 2 Mid	West Side-1027m	53	-0.03
	Cell 2 West	West Side-359m	43	-0.07
	Cell 2 West	West Side-50m	68	-3.3

Source: Harvey et al. (2004).

$$\frac{dh_{residual}}{dt} = \frac{Q_b + Q_{gw}}{A} = \frac{dh}{dt} + P - ET - \frac{Q_o}{A} \tag{2.18}$$

where: $h_{residual}$ = wetland water elevation adjusted for P, ET, and Q_o, m

The slope of the time progression for $h_{residual}$ is the total seepage loss. Figure 2.16 shows a two-week period of stage and flow data for the one of the south Florida marshes, identified as the A1 Flow Equalization Basin. The decrease in stage residual is strictly linear, and corresponds to a total seepage of 1.21 cm/d.

Such stage decrease analyses have been performed for STA34 flow way 3 (Polatel et al., 2014). A drawback to this method of seepage measurement is that periods of no rain–no inflow may be scarce. Thus, it is restricted to systems that undergo extended dry periods.

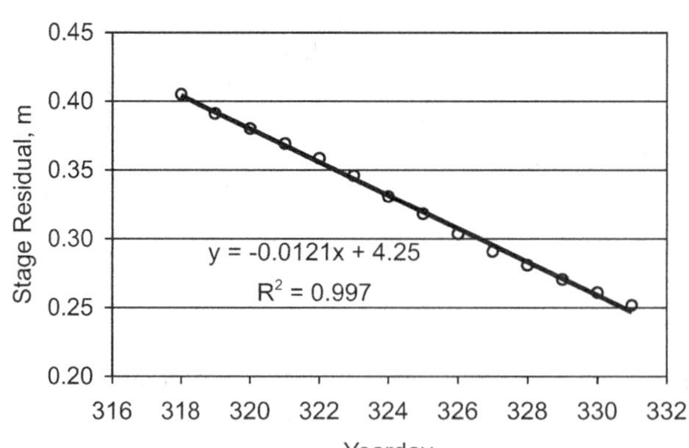

FIGURE 2.16 Decreasing adjusted stage for the A1 Flow Equalization Basin shows a decline of 1.21 cm/d due to seepage. The period is the second half of November 2016, during which the average depth was 34 cm.

In some instances, computer simulation methods may be warranted. There are several well-developed software packages that may be employed.

- MODFLOW is a computer program of the US Geological Survey (USGS) that numerically solves the three-dimensional ground-water flow equation for a porous medium by using a finite-difference method (Harbaugh et al., 2000; Harbaugh, 2005). MODFLOW was employed in the design of STA34 (Montgomery Watson Americas, Inc., 1999).
- The SEEP2D software was developed by the United States Army Engineer Waterways Experiment Station to model a variety of problems involving seepage (Jones, 1999). Sangoyomi et al. (2011) conducted a seepage analysis for STA-3/4 based on a SEEP2D model. They determined empirical bank loss coefficients in the range of 0.035–0.21 cubic meters per second per meter of head difference per kilometer of levee length (m^3/s)/m/km), depending upon position around the wetland perimeter.
- DRAINMOD is a hydrologic modeling package originally developed to simulate agricultural drainage systems in poorly drained soils like those of the Coastal Plain of North Carolina (Skaggs, 1978). It has been reported for simulating Carolina bays wetlands (Caldwell et al., 2007).

These illustrative examples are for large systems, but seepage may be of importance for all sizes of wetlands, and for all source waters. For instance, the Sacramento, CA, wetlands (7.5 ha) leaked about 40% of the added treated municipal water (Nolte and Associates, 1998a). The leakage was known to join a large river, which minimized risks to acceptable levels. The Champaign County, IL, wetlands (1.7 ha) leaked 41% of the added row crop runoff (Kovacic et al., 2000). The San Joaquin, CA, wetland (8.7 ha) leaked 71% of the added agricultural irrigation runoff (Brauer et al., 2015).

2.4 THE WETLAND WATER BUDGET

Equation 2.15, the wetland water balance, states that the change in storage in the wetland results from the difference between inflows and outflows. In theory, any one term may be calculated from Equation 2.15 if all the other terms are known. But in practice, none of the measurements are very precise, and large errors may result for such a calculation (Winter, 1981).

Temporal Patterns

An example of monthly variability of the wetland water budget is given in Table 2.5 for the Brighton treatment wetland (6.2 ha) (Kadlec et al., 2012a). The values are for a ten-year period of record and are folded and averaged for each month. Outflows were not measured and were calculated from inflows plus net additions. In general, rain snowmelt and evapotranspiration ore small components of the water budget. This wetland was built on dense clay soils, and hence seepage was negligible.

An example of annual variability of the wetland water budget is given in Table 2.6 for a large treatment marsh (STA2C1, 745 ha) (Zhao and Piccone, 2017). All flows were measured or estimated independently, and hence a closure error can be calculated. Importantly, the annual error in closure of the water budget ranged from –20% to +24%, with a root mean square (RMS) error of 12%. These percentages are based upon the total water inflow and outflow. For the entire POR, positive and negative errors tend to cancel, and the error for the 14-year POR was only 0.4%.

TABLE 2.5

Monthly flows into and out of the Brighton treatment wetland. These are averages for the ten-year POR. Units are ha·m.

	Inflow	Outflow	Rain	Snow	Snowmelt	ET
Jan	10.66	11.16	0.22	0.24	0.23	0.00
Feb	9.04	9.44	0.18	0.17	0.19	0.00
Mar	12.55	13.08	0.31	0.14	0.18	0.00
Apr	13.42	13.62	0.44	0.03	0.03	0.32
May	11.14	10.95	0.44	0.00	0.00	0.68
Jun	8.98	8.58	0.49	0.00	0.00	0.94
Jul	8.48	7.83	0.35	0.00	0.00	1.04
Aug	7.91	7.49	0.48	0.00	0.00	0.94
Sep	7.52	7.44	0.54	0.00	0.00	0.67
Oct	8.62	8.83	0.47	0.00	0.00	0.31
Nov	9.42	10.04	0.51	0.06	0.06	0.00
Dec	11.92	12.44	0.29	0.22	0.18	0.00
Annual	119.67	120.92	4.71	0.88	0.88	4.89

The RMS error increases with decreasing water budget period. For instance, for Boney Marsh, over an eight-year period, the daily, monthly, and annual RMS errors were 67%, 16%, and 7% respectively (Mierau and Trimble, 1988).

These are not extreme examples. Similar lack of closure has been reported for four wetlands at Sacramento, where all mass balance terms were measured independently, including infiltration measured by drawdown (Nolte and Associates, 1998b). The RMS monthly errors were 60%, 47%, 26%, and 19% for Cells 3, 5, 7, and 9, respectively. The annual percentage residuals were –56%, –44%, –15%, and –15%, respectively. The conclusion was that these apparent water losses were due to faulty inflow or outflow measurements.

These examples serve to alert the wetland designer or operator that care must be taken in water flow measurements and that water balance differencing is apt to provide estimates with large uncertainty. With great care, balance closure may be held to the ± 5 to 10% range (Martinez and Wise, 2001).

Water budgets also vary widely from wetland to wetland, which is termed inter-system variability. The magnitudes of the water gains and losses are compared to the source water inflow, for example, systems in Table 2.7. Rainfall ranges from 0.2 to 57% of source water, and evapotranspiration ranges from 2.8 to 113%. Seepage losses range from 0 to 115% of source water. Clearly, there are treatment marshes for which water gains and losses cannot be ignored.

Gains and losses are more likely to be important in treatment marshes than in free water surface wetlands treating strong effluents. That is partly because treatment marshes nearly always are built in unlined basins, whereas stronger effluents require groundwater protection, and hence include some form of impervious lining. As a consequence, infiltration may be significant in treatment marshes, and may in fact be the design goal of the system. It is also partly because source water loading rates are typically higher in more heavily loaded treatment wetlands, lessening the effects of rain and ET. Therefore, the literature on treatment wetlands in general pays scant attention to water balance effects on pollutant removal.

TABLE 2.6

Annual water budgets for STA2C1 for a 14-year POR. Units are ha·m.

	Inflow	In Seep	Rain	Ouflow	Out Seep	ET	Δ Storage	% Error
2003	5,424	314	957	3,786	341	974	399	-20
2004	7,335	2	805	6,092	519	985	−198	−9
2005	7,452	11	661	6,977	520	931	44	4
2006	7,387	18	860	6,800	487	1,005	−20	0
2007	6,876	11	776	5,125	653	959	−225	−16
2008	7,394	51	1,031	5,617	623	967	206	−14
2009	5,450	18	928	5,097	423	997	−201	−1
2010	6,652	19	1,033	7,839	593	925	386	24
2011	3,022	31	720	3,401	540	1,028	−229	22
2012	6,549	2	987	5,995	586	1,010	179	3
2013	7,071	35	926	6,550	354	995	−77	−3
2014	6,455	48	938	7,061	374	995	23	13
2015	18,709	4	598	17,785	554	961	119	1
2016	4,705	39	1,123	5,773	314	968	−191	16
POR	7,177	43	882	6,707	492	979	15	0.4

TABLE 2.7

Example water budgets for treatment marshes, as percentages of added inflow.

		Surface Inflow	Rain	Seepage In	ET	Seepage Out	Surface Outflow
Florida	STA34C2B	100.0	9.8	0.0	10.2	0.0	112.1
Venice, Italy	Novissimo	100.0	2.4	15.0	11.2	6.2	107.3
Texas	Shannon	100.0	5.3	0.0	7.5	0.2	97.8
Florida	STA2C3	100.0	7.7	0.0	9.5	8.9	93.8
Florida	STA34C2A	100.0	7.9	0.0	8.2	0.0	92.9
Florida	STA2C1	100.0	12.2	0.1	13.1	0.6	83.7
Arizona	Tres Rios Hayfield	100.0	0.4	0.0	5.4	15.5	79.5
Michigan	Houghton Lake, MI	100.0	56.6	0.0	77.9	0.0	78.2
California	Imperial	100.0	0.2	0.0	2.5	35.2	64.6
California	Sacramento C7	100.0	3.8	0.0	5.9	18.2	64.2
California	Brawley	100.0	0.4	0.0	5.2	43.6	54.1
Arizona	Tres Rios Cobble	100.0	0.2	0.0	2.8	80.6	16.9
Nevada	Incline	100.0	36.0	0.0	112.8	21.3	0.0
Florida	Kanapaha	100.0	9.1	0.0	7.3	114.9	0.0

Effects on Concentrations

Water losses and gains, which are augmentations of the source flow, fall in four general categories: precipitation (P), evaporation (E), transpiration (T), and infiltration (seepage, I). The first two are transfers of water only, but the second two carry solutes as well as water. It is instructive to consider the concentration of a conservative substance (tracer), as well as the flow increases and reductions that may occur. Table 2.8 illustrates water mass balance effects for an idealized situation, an averaged steady state condition in which infiltration is uniformly distributed along the length. Rainfall is assumed not to contain tracer. Each of the four possible augmentations is explored individually in a range that is representative of reported performances (Table 2.7). Of course, the four generally occur in some combination, so Table 2.8 is to be regarded as showing the boundaries of the augmentation flows.

Rainfall adds to the source flow and dilutes the tracer. There is therefore a concentration reduction, but no tracer mass flow reduction, since the rain does not contain tracer. Evaporation reduces source flow during passage through the marsh and acts to concentrate the tracer. There is no effect on the mass flow of tracer along the wetland length. Both of these are water-only effects.

In contrast, transpiration removes both water and tracer from the surface water. Transpiration flows are drawn from the root zone and replenished by downward movement of surface water and its solutes. There is no effect upon the concentration of tracer remaining in the surface water, but the tracer mass flow with surface water is reduced. The effect of gravity-driven infiltration is the same as for transpiration.

Detention times computed using the average of the inflow and outflow are not significantly different form detention times computed from the distance-variable behavior of the flow (maximum 104%). Thus, it will rarely be necessary to use Equation 2.5.

The results in Table 2.8 are not directly useable in design, but do alert us to the fact that flow augmentations can have large effects upon pollutant concentrations and apparent

TABLE 2.8

The effects of water gains and losses on flow and the concentration of an inert tracer. The nominal detention time is computed from the average of inlet and outlet flows; actual detention time is calculated from Equation 2.5.

	Outflow % of Inflow	Detention % of Nominal	Concentration Reduction %	Mass Reduction %
Precipitation				
0	100	100	0	0
10	110	100	9	0
30	130	101	23	0
50	150	101	33	0
Evaporation				
0	100	100	0	0
10	90	100	−11	0
30	70	101	−43	0
50	50	104	−100	0
Transpiration				
0	100	100	0	0
10	90	100	0	10
30	70	101	0	30
50	50	104	0	50
Infiltration				
0	100	100	0	0
10	90	100	0	10
30	70	101	0	30
50	50	104	0	50

removals. Further, while water loss can be described via ET, concentration effects require that we consider E and T separately: evaporation increases concentrations with no mass effect, while transpiration has no effect on concentrations, but causes a mass reduction. It is possible and practicable to account for these in the description of contaminant removal, as will be discussed in Chapter 7.

2.5 SUMMARY

This chapter has provided details on marsh water movement and internal water conditions. Quantitative descriptors – detention time, velocity, and hydraulic loading – are defined, and found to be more complicated than might first appear. Importantly, these all require knowledge of the treatment marsh operational schedule, because averages need to be carefully specified. Source water delivery timing often includes no-flow periods, which need not be included for proper quantitative description.

Wetland water depth is another deceptively simple concept, founded on not-so-simple considerations of the wetland topography and the operational conditions. In reality, wetland bottoms are neither flat nor level; they are lumpy and tilted. Further, as flows increase, water depths increase. Stage-storage and stage-area relations, which pertain to level pool conditions, and the depths so implied, do not substitute for flowing water depth. Thus, simple single-number depth characterization is generally inadequate for quantitative purposes.

Wetland water budgets are typically dominated by surface inflows and outflows, evapotranspiration, and precipitation. Groundwater interactions can also be important, especially for treatment marshes, which are not commonly lined. Surface flows are generally measurable with sufficient accuracy. Precipitation may be projected from historical weather data, with the possibility of some error due to changing climatic conditions. The resulting water budgets vary seasonally, because precipitation and evapotranspiration vary seasonally. Stochastic variability is present, and may be large, on several times scales of interest. Atmospheric additions and losses are predictable by several techniques, briefly described in this chapter.

In lightly loaded treatment marshes, the contributions of atmospheric augmentation and infiltration may be very important in design calculations. Mass balance considerations dictate that evaporation and transpiration be considered separately, because they have different effects on solute concentrations.

3 Surface Water Movement

Early treatment wetland researchers often conceptualized water movement in free water surface wetlands to be plug flow, meaning that water moved through the system in lockstep from inlet to outlet (e.g., Water Environment Federation, 1990; Reed et al., 1995). This allowed the use of (nominal) detention time in the quantification of pollutant reduction.

As is well known today, nothing could be further from the truth (Figure 3.1). Surface water movement is three-dimensional, with vertical velocity profiles, infiltration, and lateral inhomogeneities in addition to the dominant longitudinal flow from inlet to outlet.

From an overall geometric perspective, the surface water sheet is quite thin: less than a meter, in contrast to wetland lengths and widths of hundreds or thousands of meters. It is therefore very tempting to consider vertically averaged conditions for water flow. This almost universally done, because of the complexity of using three-dimensional computer fluid mechanics calculations. However, such vertical averaging implies that water depth becomes a significant factor in process descriptions.

Surface water depth simulations, regardless of dimensionality, are almost always possible for situations of complete inundation. But that is not necessarily always the case for treatment marshes.

3.1 GEOMETRICAL SETTING

Topographical Effects

The surface irregularity associated with wetland microtopography plays an important role in surface water movement, especially at low water depths. Choi and Harvey (2014) identified four threshold stages for the inlet treatment zone of Water Conservation Area 2A (WCA2A), based upon the measured distribution of ground elevations. The topographic distributions in Figure 3.2 represent the fraction of wetland cross-section that has an elevation equal to or less than a given elevation in the treatment region of the WCA2A gradient. This function characterizes the wetland cross-section available for surface flow, and the depth distribution for a given surface-water stage.

The WCA2A cattail marsh is a natural wetland that has been receiving agricultural runoff for over fifty years. Those thresholds characterized abrupt changes in the friction equations that were fit to water surface profile data (Figure 3.2). Above stage threshold (d), the entire marsh is covered by water, i.e., there was complete inundation. Between thresholds (c) and (d), higher areas of the wetland are at or above water. Between thresholds (b) and (c), flow is only in deep channels. Between thresholds (a) and (b), surface water exists only in isolated, non-connected pools. Below threshold (a), water is all subsurface.

When water stage drops below the maximum elevations of ridges and hummocks, threshold (d), the surface-flow paths tend to become more tortuous as water is forced around and between the exposed ground surfaces. Thus, microtopography can contribute to flow resistance when surface water levels are low.

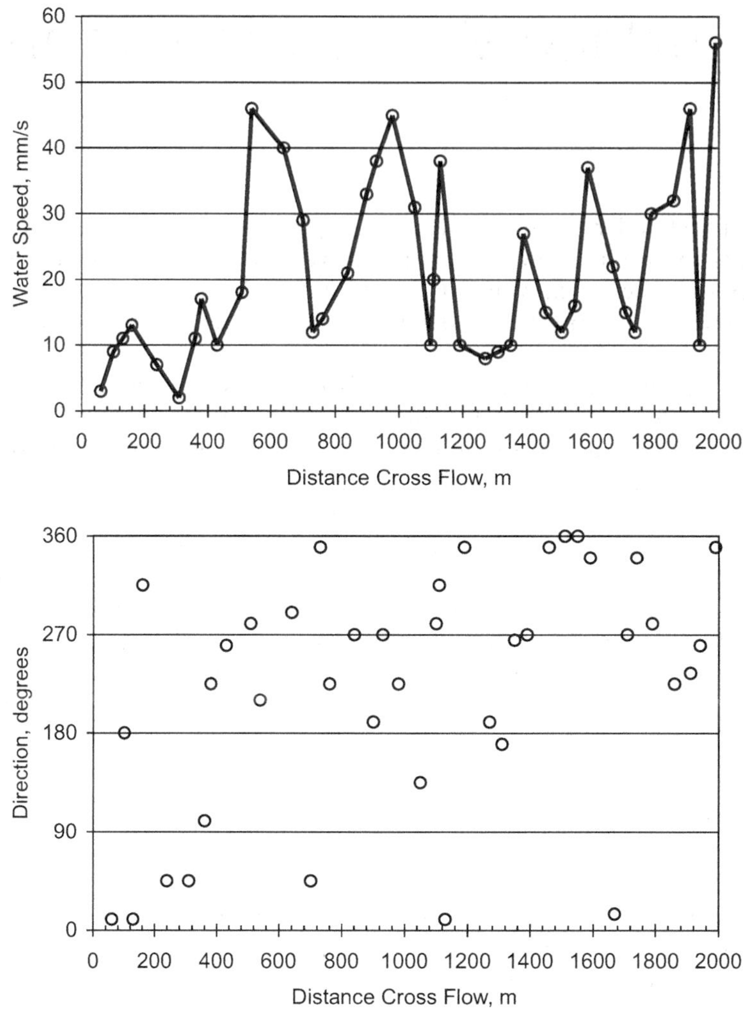

FIGURE 3.1 Flow speed (upper panel) and direction (lower panel) near the inlet of STA2C3 on a transect transverse to the nominal flow direction. Flows are every which way. The average nominal flow is ≈ 6 mm/d, and the nominal flow direction is 180°. (Based on data from Dessu et al., 2017).

Dimensionality

Averaging water movements across the width transverse to the principal flow direction is more problematic, especially for wetlands with irregular shapes. Two-dimensional flow nets are not as daunting as 3-D descriptions and have seen considerable use for treatment marshes. Figure 3.3 shows the 2D analysis of a less-than-ideal system geometry, which cannot be accurately portrayed as one-dimensional. In contrast, Figure 3.4 shows the nearly one-dimensional flow pattern of a long slender marsh cell.

One-dimensional representations have value as a framework from which to understand overland flow depth profiles and their dependence on wetland geometry and flow, and structure settings. They also form the basis for interpreting the effects of detention time distributions on pollutant removal.

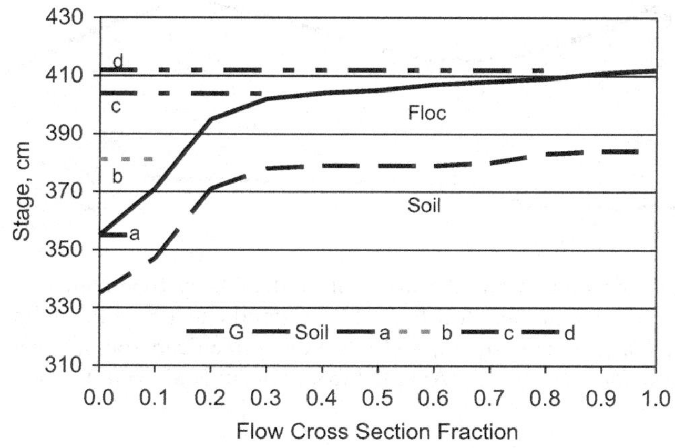

FIGURE 3.2 Distribution of ground elevations and threshold stages for the northern end of WAC2A. (Redrawn from Choi and Harvey, 2014)

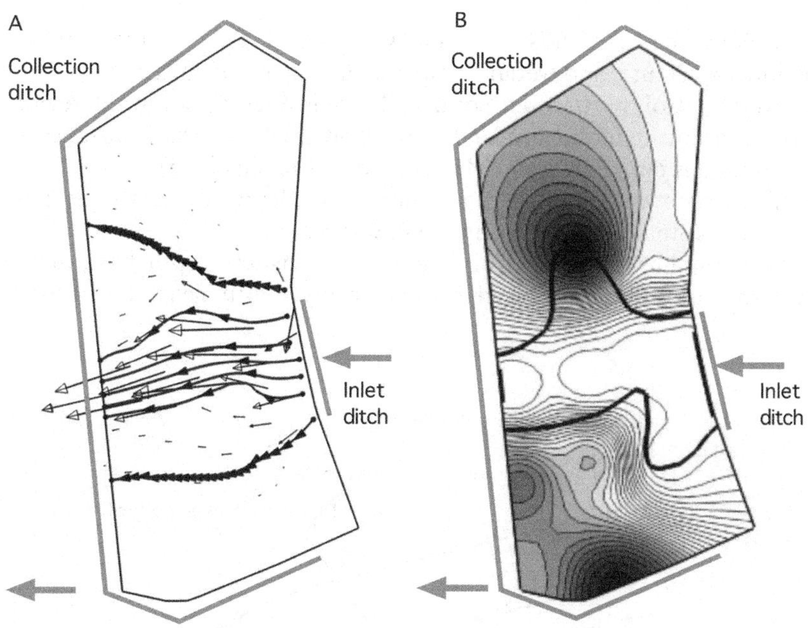

FIGURE 3.3 Surface flow simulations and tracer response of the Kompsasuo treatment wetland in northern Finland. This 2.2 ha shallow sedge marsh received peat mining wastewater that fit the criteria in Chapter 1. (A) The 2-D MODFLOW simulation of flow velocities. (B) The distribution of added tracer after 21 days. (Adapted from Ronkanen and Kløve, 2008)

3.2 ONE-DIMENSIONAL FRAMEWORK

Wetland water depths and flow rates are controlled by two major wetland features; the outlet structure and resistance to flow within the wetland. In general, it is very desirable to have control at the outlet structure, because then the operator has control over water depth.

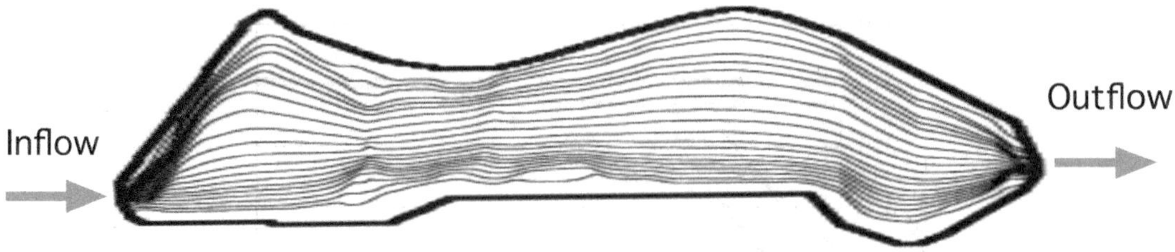

FIGURE 3.4 Surface flow simulations the first cell of the Ekeby treatment wetland near Eskilstuna, Sweden. The 2-D simulation identified flow lines. The wetland had 1.8 ha of shallow zones, vegetated by *Glyceria maxima* and *Phragmites australis*, and 0.8 ha in three deep zones, vegetated by *Potamogeton pectinatus* and *Ceratophyllum demersum*. (Adapted from Kjellin et al., 2007)

Under complete outlet control, a level pool of water exists upstream of the outlet structure, regardless of what is growing there. However, that is not always possible, particularly for large or densely vegetated wetlands. Water may be held up by the vegetation, at a depth that is independent of the outlet structure setting.

Four different situations may occur, and are easily visualized (Figure 3.5):

1. Very low flow, small wetlands; complete weir control. There is a level pool upstream of the outlet structure, and wetland water stage is spatially invariant.
2. Partial weir control, partial weir control. There is a level pool in the region near the weir, but a gradient in stage near the wetland inlet. This is a distance thickening sheet flow, and is designated an "M1" profile by French (1985).
3. Normal depth flow. Vegetation drag controls the depth to exactly the stage created by the weir. Depth is uniform along the flow direction.
4. Large flow, minimal weir control. There is a constant depth flow, at the normal depth, near the inflow, followed by decreasing depth near the outflow. This is

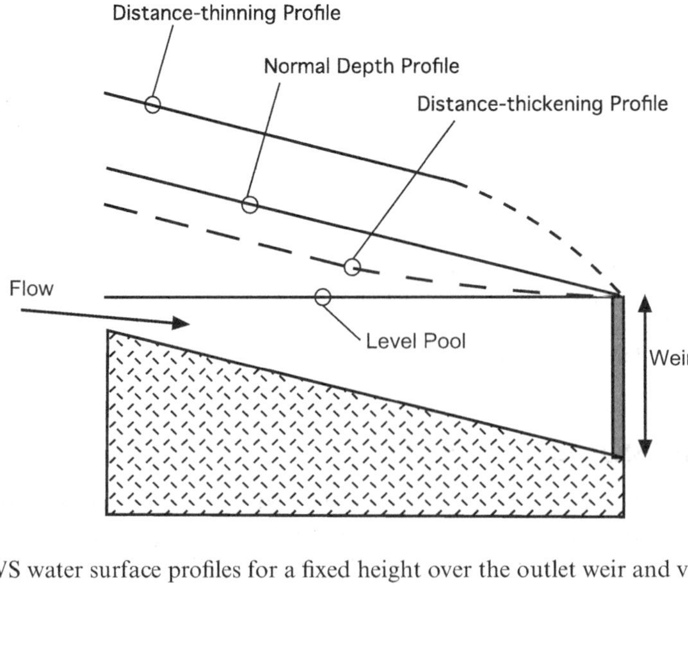

FIGURE 3.5 FWS water surface profiles for a fixed height over the outlet weir and various inlet flows.

a distance thinning sheet flow, and is designated an "M2" profile by French (1985).

These various possibilities are covered by a backwater calculation. Because wetlands nearly always meet the criterion for gradually varied flow with a small Froude number (French, 1985), the water flow momentum balance can be simplified to contain only gravitational and friction terms. The component pieces are the spatial water mass balance, the friction equation, and specification of inflows, geometry and outlet depth setting. Notation is given in Figure 3.6.

For a one-dimensional (rectangular) system, the spatially variable water mass balance is:

$$\frac{\partial(\phi h)}{\partial t} + \frac{\partial(hu)}{\partial x} = P - ET - I \tag{3.1}$$

where ET = evapotranspiration rate, m/d
h = water depth, m
I = infiltration rate, m/d
P = rainfall rate, m/d
u = superficial velocity, m/d
x = distance in flow direction, m
t = time, d
ϕ = storage porosity, –

A Simplified Exploratory Model

The steady state version of Equation 3.1 is amenable to computer spreadsheet calculations. Input information includes time-averaged values of P, ET, and I, as well as a friction equation that provides velocity u as a function of water depth, and also possibly a function of

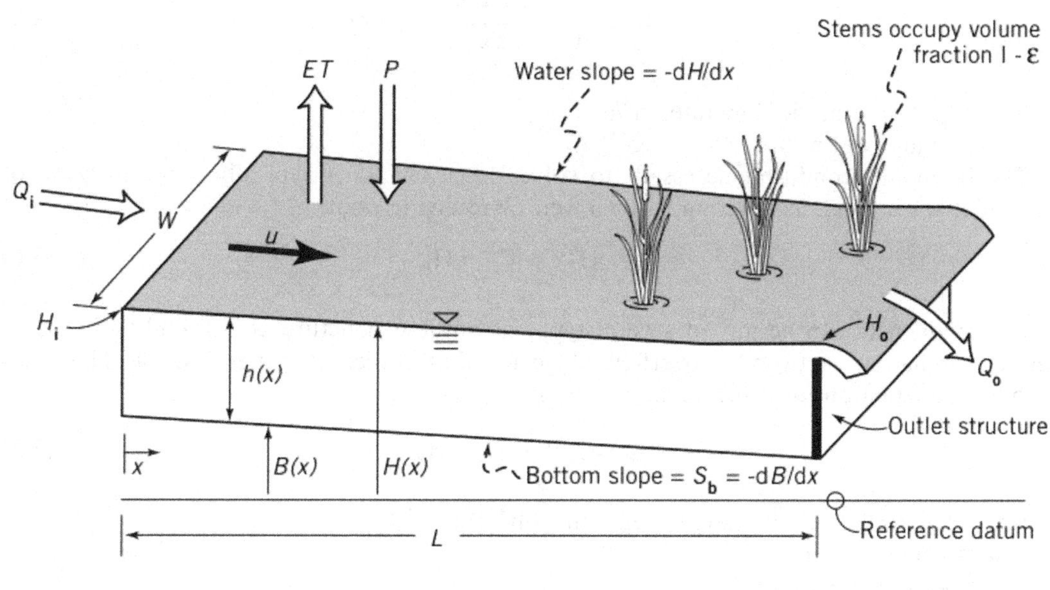

FIGURE 3.6 Notation for marsh water sheet hydraulics.

distance for variable vegetation patterns. The calculations are driven by an outlet water level specification and the incoming flow rate.

Here, a more simplified version is used to illustrate hydraulic features of marshes. For steady flow, one-dimensional (rectangular) systems, with uniform vegetation, in the absence of rain, infiltration or ET effects, the flow situation can be simplified as indicated in Equation 3.2. The spatial water mass balance, water depth (h) and superficial flow velocity (u) are distance-variable:

$$\frac{d(hu)}{dx} = 0 \tag{3.2}$$

One choice for frictional losses is a general power law relationship. This is discussed in further detail in the next section of this chapter:

$$u = a\, h^{b-1}\, S^{c} \tag{3.3}$$

where a,b,c = constants
u = superficial velocity, m/d
S = – dH/dx = negative of the water surface slope, m/m

The superficial water velocity is the volumetric flow divided by the total cross section normal to flow. Part of that cross section is occupied by vegetation, so the actual velocities are higher. As will be seen in Chapter 5, that fraction is typically less than 5% of the area, and hence there is not a great deal of difference between actual and superficial velocities in treatment marshes.

Water elevation is water depth plus bed bottom elevation profile (Figure 3.6):

$$H = B + h \tag{3.4}$$

where B = elevation of the bed bottom above datum, m
H = elevation of the water surface

Equations 3.2, 3.3 and 3.4 combine to give:

$$aW\, h^{b}\left(-\frac{d(h+B)}{dx}\right)^{c} = Q \tag{3.5}$$

where Q = volumetric flow rate, m^3/d
W = wetland width, m

The boundary condition necessary to solve Equation 3.5 is typically a specification of the outlet water level, as determined by a weir or receiving pool:

$$H(x = L) = H_L \tag{3.6}$$

Although there can be any of several types of outflow structure, it is useful to illustrate the determination of the weir overflow stage for that choice of outlet control. The commonly used equation for a rectangular weir is:

$$Q = C_E W_W (H_L - H_W)^{1.5} \tag{3.7}$$

where C_E = effective discharge coefficient, $(m^3/d)/(m^{2.5})$
W_W = width of weir, m
H_W = weir crest elevation, m

H_L = water surface elevation at wetland outlet, m

Of course, if $H_L < H_W$, then there is no outflow.

Equation 3.5 cannot be solved analytically to a closed-form answer, but numerical solution is easy via any one of a number of methods. Required input parameters are the wetland width and length, bottom slope profile, flow rate and the height over the weir, together with the friction parameters a, b, c.

An example spreadsheet is shown in Table 3.1, and the accompanying output is graphed in Figure 3.6. This example is for a very large wetland, approximately the size of STA2C1.

TABLE 3.1

Spreadsheet for backwater calculation for STA2C1 at high flow.

Input data and derived quantities. Bold are input data.

L=	**5818**	m	A =	**7,450,000**	m^2
W =	1281	m	Aspect =	4.5	
Q =	**1,000,000**	m^3/d	Q =	11.574	m^3/s
a=	**1.94E+05**	m^(1-b)/d	q =	0.134	m/d
b+1 =	**3.73**		HLR =	13.42	cm/d
c=	**0.476**		Δx =	291	m
dB/dx =	**0.000069**	m/m	ΔhB =	-0.40	m
Outlet Depth =	**0.600**	m	ho/L =	0.000103	m/m
Apparent Gradient =	0.000099	m/m	Inlet h =	0.773	m
Apparent n =	0.714	1/m^.33/s	Mean h =	0.746	m
Detention Time =	5.56	days	Normal h =	0.774	m

Output elevations and depth

Distance Back	Distance	Bottom Elev	Water Elev	Depth
m	m	m	m	m
0	5818	0.000	0.600	0.600
291	5527	0.020	0.693	0.673
582	5236	0.040	0.744	0.703
873	4945	0.061	0.782	0.722
1164	4654	0.081	0.815	0.734
1455	4364	0.101	0.844	0.743
1745	4073	0.121	0.871	0.750
2036	3782	0.141	0.896	0.755
2327	3491	0.162	0.920	0.759
2618	3200	0.182	0.943	0.762
2909	2909	0.202	0.966	0.764
3200	2618	0.222	0.988	0.766
3491	2327	0.242	1.010	0.767
3782	2036	0.262	1.031	0.769
4073	1745	0.283	1.052	0.770
4364	1455	0.303	1.073	0.770
4654	1164	0.323	1.094	0.771
4945	873	0.343	1.115	0.772
5236	582	0.363	1.135	0.772
5527	291	0.384	1.156	0.772
5818	0	0.404	1.176	0.773

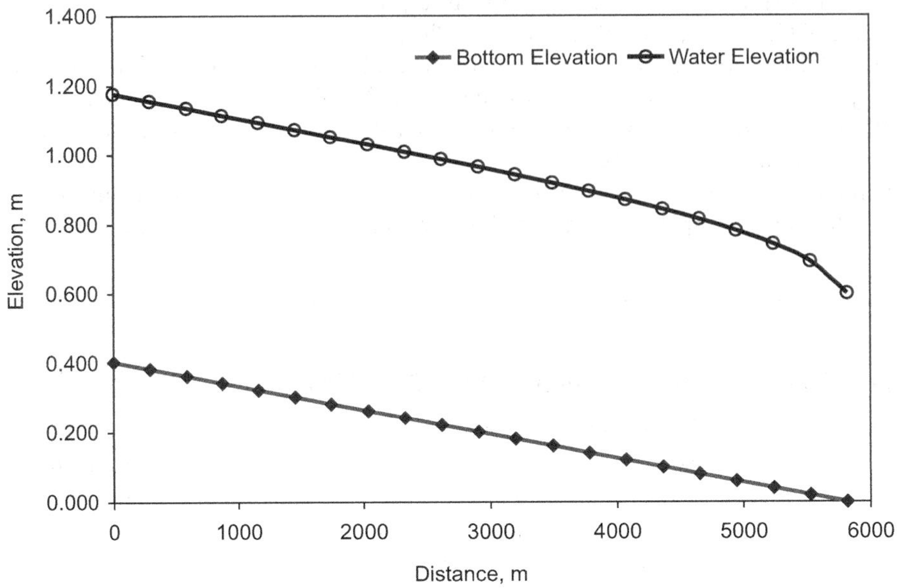

FIGURE 3.7 Output water surface profile for the Table 3.1 example.

This is a case of a distance thinning water sheet. Most of the length of the wetland operates at the normal depth determined by vegetation resistance. Only the last 20% of the flow length shows any effect of the outlet weir setting.

This highly simplified model may be used to explore the dependence of the water surface elevation of a treatment marsh in response to variations in the imposed flows, size, geometry and vegetation density (via a friction equation). For example, Figure 3.8 illustrates the effect of size, with other parameters being held constant. It is presumed that the hydraulic loading has been set to achieve the desired contaminant removal. However, the more water there is to treat, the larger the required area of marsh. While the simple model results are not precise, they do point to potentially significant profile effects. For this example, for a large wetland, water depths at the system inlet are 80 cm, versus the outlet depth set at 40 cm. That difference is enough to threaten the survival of some EAV species. Secondly, the detention time for the large wetland is 66% higher than that determined for the outlet depth, which may affect treatment performance. Such profile effects are not present for the smaller wetland in this example.

As a second exploratory example, the dependence of the water surface elevation upon aspect ratio (AR, length to width ratio). As will be discussed in Chapter 13, long slender wetlands have a tendency to be more hydraulically efficient than short, wide systems. However, the long slender wetlands also have a greater resistance to flow, causing water to pile up at the inlet. Figure 3.9 illustrates the effect of aspect ratio, using the simple model, with other parameters being held constant. It is again presumed that the area, and hence the hydraulic loading, has been set to achieve the desired contaminant removal. For AR = 25, the inlet depth is about double the outlet depth. Large AR have been used, for instance, AR = 50 for a 2 ha pilot project in Florida (CH2M Hill, 2003), and AR = 34 for the Imperial wetland in California (Kadlec et al., 2010a). The later experienced overflow at the front end for high flows.

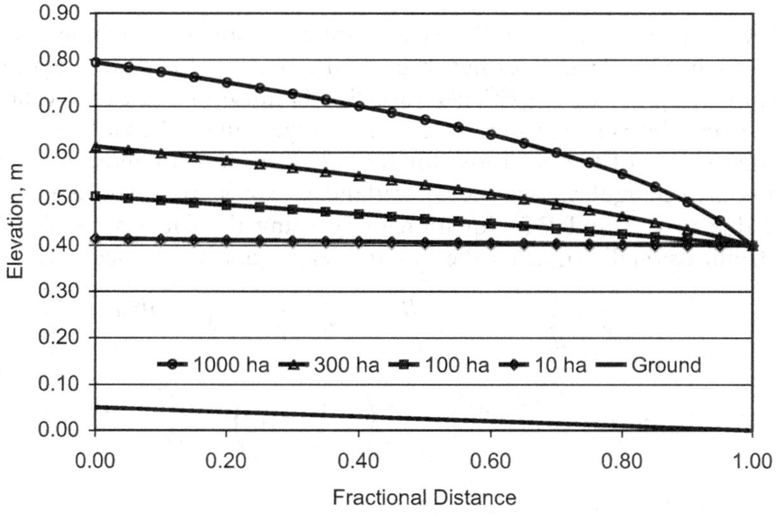

FIGURE 3.8 The dependence of water surface elevation on wetland size according to a 1-D simplified model. The hydraulic loading is fixed at 5 cm/d, and the aspect ratio at 5.0, and friction is for a moderately dense EAV. The outlet water level is fixed at 40 cm, and there is a 5 cm drop in the bottom elevation. Nominal detention times increase with size, from 7.5 to 12.5 days.

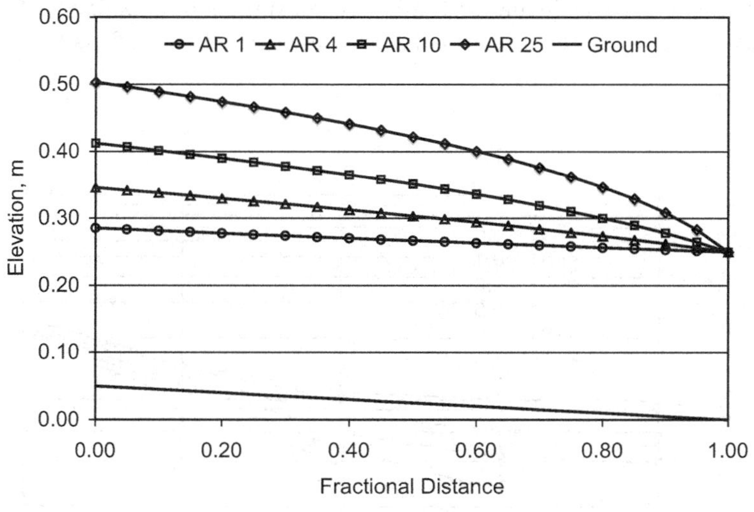

FIGURE 3.9 The dependence of water surface elevation on aspect ratio (AR) according to a 1-D simplified model. The hydraulic loading is fixed at 5 cm/d, and the size at 10 ha at 5.0, and friction is for dense EAV. The outlet water level is fixed at 25 cm, and there is a 5 cm drop in the bottom elevation. Nominal detention times increase with AR, from 4.8 to 7.7 days.

Head Loss Estimation

The spreadsheet implementation of the steady state version of equation (3.1) allows for exploration of a number of variables but in general requires numerical integration, which is inconvenient in conceptual design calculations. But because of the extreme nonlinearity of the equations, it can be very inaccurate to use average values. Accordingly, it is better to use precalculated values of the head loss for the intended design conditions. To accomplish this, the case of a rectangular constructed wetland is considered, with a negligible loss or gain of water due to P, I and ET. Equation (3.5), using the friction coefficients a, b = 3, and c =1, is dedimensionalized, using the wetland length and the outlet water depth:

$$y = \frac{h}{h_o} \qquad y_B = \frac{B}{h_o} \qquad z = \frac{x}{L} \qquad S_1 = \frac{dy_B}{dz} \tag{3.8}$$

$$y^3\left(-\frac{d(y+y_B)}{dz}\right) = y^3\left(-\frac{dy}{dz} - S_1\right) = \frac{qL^2}{ah_o^4} = M_1 \tag{3.9}$$

where h_o = water depth at outlet, m
L = wetland length, m
q = hydraulic loading rate, m/d
The rest of the new variables are defined in (3.8) and (3.9). It is presumed that the outlet water depth is fixed. Analytical integration of (3.9) is possible, and yields the inlet water depth, and hence the head loss for a given wetland. Solutions depend on two dimensionless groups: S_1, which represents the bed slope, and M_1, which contains the friction coefficient, the hydraulic loading rate, outlet depth and wetland length.

Figure 3.10 presents the solution of Equation 3.9 for different parameter values. It may be used to provide preliminary estimates of head losses in surface flow wetlands. The outlet water level is often set by an independent relation for the outlet structure, such as Equation 3.7.

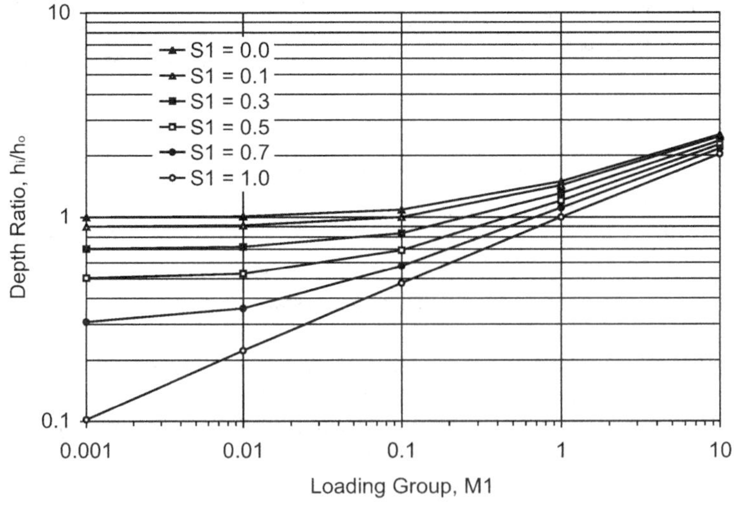

FIGURE 3.10 Inlet/outlet depth ratio for FWS wetlands of different slops and different loading rates. The friction power law is used, with b = 3 and c = 1. See text for definitions of the dimensionless groups M_1 and S_1.

A general feature of backwater calculations is that the surface water elevation profiles for large, long wetlands tend to be dominated by vegetation resistance. That in turn means that most of the profile, excepting the region near the outlet, is at or close to the normal depth. Figures 3.5 and 3.6 and Table 3.1 depict this phenomenon.

In contrast, small short wetlands have nearly level pool surface water elevation profiles. Those elevations reflect the depth control exerted by the outflow structure.

Uniform Depth

If the water depth is constant along the flow direction, then dh/dx = 0. Equation 3.5 then becomes a simple algebraic relation:

$$aW\,h_N^b\left(-\frac{dB}{dx}\right)^c = Q \tag{3.10}$$

where h_N = normal depth, m

This may be rearranged to calculate the normal depth:

$$h_N^b = \frac{Q}{aW\left(-\frac{dB}{dx}\right)^c} = \frac{Lq}{a\left(-\frac{dB}{dx}\right)^c} \tag{3.11}$$

The normal depth is indeterminate unless there is a bottom slope to the wetland. Given that there is bottom slope, Equation (3.11) forecasts that depth increases with volumetric flow rate; and that depth is proportional to the (1/b) power of the flow. Table 3.3 indicates b ≈ 2, which means that depth is proportional to the square root of the flow rate.

$$h_N = Q^{\frac{1}{b}}\left(aW\left(-\frac{dB}{dx}\right)^c\right)^{-\frac{1}{b}} = \alpha Q^{\frac{1}{b}} \tag{3.12}$$

where α = constant grouping defined by (3.12)

This dependency has an interesting effect on the calculation of nominal detention time, because the wetland water volume depends on flow rate. Equation 2.2 becomes:

$$\tau_n = \frac{V_{nominal}}{Q} = \frac{(Ah)_{nominal}}{Q} = A\alpha Q^{\left(\frac{1}{b}-1\right)} \tag{3.13}$$

This leads to the rather startling result that for b = 1, the wetland detention time is independent of the flow rate, and hence independent of the hydraulic loading rate.

At the other extreme, for small, flat bottom wetlands with outlet weirs, there is also a flow dependence for detention time. For the case of an overflowing rectangular weir, $H_L = H_B + h$, and Equation 3.7 becomes:

$$\mathrm{h} = (\mathrm{H_W} - \mathrm{H_B})\left(\frac{\mathrm{Q}}{\mathrm{C_E W_W}}\right)^{\left(\frac{1}{1.5}\right)} = \beta Q^{\left(\frac{1}{1.5}\right)} \tag{3.14}$$

where β = constant grouping defined by (3.14)

This dependency also shows an effect on the calculation of nominal detention time, because the wetland water volume depends on flow rate. Equation 2.2 becomes:

$$\tau_n = \frac{V_{nominal}}{Q} = \frac{(Ah)_{nominal}}{Q} = A\beta Q^{\left(\frac{1}{1.5}-1\right)} \tag{3.15}$$

These dependencies have important consequences for the mathematical description of pollutant removal, which will be discussed in Chapter 7.

TRANSIENTS

Unsteady flow causes changes in the water surface profile with time. In most circumstances, it is not feasible to acquire stage profile data during periods of unsteady flow, because it would require water surface elevations as a function of both distance and time. There are exceptions, and a few will be discussed here.

Response to a Flow Pulse

Several of the flow categories under consideration are not steady flow, i.e., runoff, seasonal and discretionary systems. These wetlands receive flow in pulses, and the flows and water levels rise and fall in response.

When the wetland is small, the water surface is a nearly level pool, and the outflow and level are controlled by the outlet structure. Such is the case for a 60-hr pump event at the Des Plaines, IL, site (Figure 3.11). For that example, S1 = 0 and M1 = 0.003, which corresponds to a level pool according to Figure 3.9. The nominal detention time was 2.4 d (Kadlec, 2010). The outflow response time was approximately a day and a half. During the transient, the water level in the wetland rose by 16 cm.

When the wetland is large, the water surface is not a level pool, and depth is mostly under vegetation control. That was the situation for ENRPC1, a 522-ha wetland that received a 14-day flow pulse in 1998 (Figure 3.12). The stages in the inlet end responded very quickly to the increased inflow, and the outlet stages also increased, but with a lag time of several days. As a result, the tilt of the water surface changed during the flow event. Similarly, outlet flows lagged the inlet flows, also by a few days.

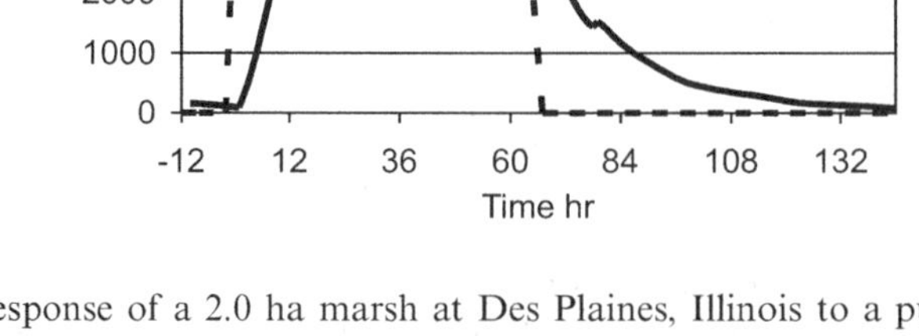

FIGURE 3.11 Flow response of a 2.0 ha marsh at Des Plaines, Illinois to a pumped flow pulse addition. The volume added to wetland EW3 was equivalent to the volume of water in the system. (Redrawn from Kadlec, 2010)

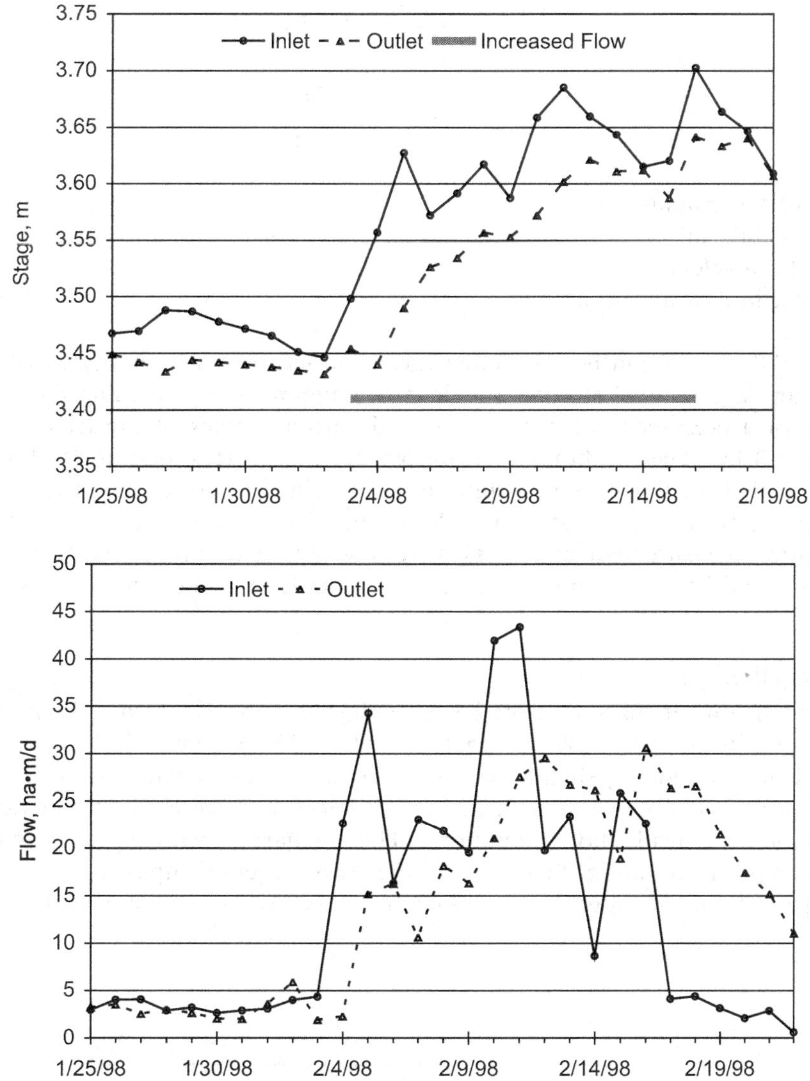

FIGURE 3.12 The response of inlet and outlet water stages to an increase in flow rate. Data are from the ENRP, Cell 1. Lower panel: Flows, with an average increase from 34,000 to 249,000 m^3/d for the 522-ha wetland. Upper panel: Inlet and outlet stages, with an average increase of 14 cm.

Radial Flow Startup

A study of historical interest is the radial flow pilot project at Houghton Lake, MI, because it is the first attempt to quantitatively describe overland flow in marshes (Kadlec et al., 1981). Source water was discharged at a single point interior to a sedge marsh and spread radially over a flat marsh bottom. Nonetheless, this site did possess considerable ground elevation variability as described in Kadlec (1990). Stage data were acquired at multiple points on concentric rings, utilizing a surveyor's level, a several times during the onset of pumped additions. The describing equations analogous to equations 3.1 and 3.3 are:

$$\frac{\partial h}{\partial t} + \frac{1}{r}\frac{\partial (rhu)}{\partial r} = P - ET - I \tag{3.8}$$

$$u = a\,h^{b-1}\left[-\frac{\partial h}{\partial r}\right]^{c} \tag{3.9}$$

where a,b,c = constants
h = water depth, m
u = superficial velocity, m/d
r = distance in flow direction, m
t = time, d

Equations 3.8 and 3.9 can be solved analytically, in terms of a similarity variable, $s = r^2/t$. Results for an August 1977 startup are shown in Figure 3.13, including data and model calculations for a presumed value of c = 1, and various values of the friction parameter b (β in Figure 3.13). The best fit is clearly for b = 2, with a corresponding $a = 9 \times 10^6$ 1/m·d. It is also clear that for this wetland, Manning's formula does not fit data well. A feature of this radial geometry is that the stage profile is quite distinct and of a magnitude that minimizes elevation measurement errors. This work was in a wetland area of approximately 2 ha, which would have very minimal stage differentials in a one-dimensional rectangular geometry.

Wave Propagation

Field tests in treatment marshes have been conducted using sinusoidal discharge disturbances that create water waves (Wasantha Lal et al., 2015; Wasantha Lal, 2017). Field tests were carried out by sending shallow water waves across the wetland using discharge perturbations created at control structures. The water waves were superimposed on near-equilibrium or near-steady flow conditions. The discharges, wave speeds, and the wave attenuation rates and resulting stage data were used to develop empirical functions expressing discharge in terms of water depth and energy slope. The results showed that Manning's

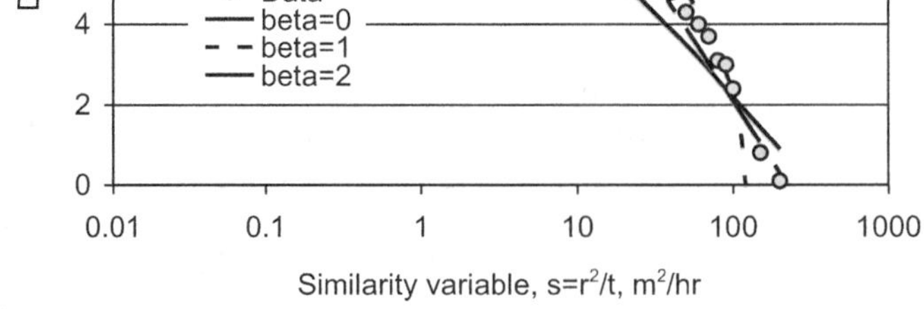

FIGURE 3.13 Startup depth profiles for a pilot-scale radial flow wetland. (Based on information from Kadlec et al., 1981.)

equation is not applicable for wetlands with thick emergent vegetation. The coefficients for power law expressions for vegetation resistance varied with depth range.

3.3 TWO-DIMENSIONAL ANALYSES

Treatment marsh vegetation patterns are often spatially variable in two dimensions, as well as the wetland bottom topography. Accordingly, flow calculations can be implemented on a 2-D mesh that accounts for these planar distributions. For instance, Paudel et al. (2013) used a finite-element mesh of 1135 unstructured triangular elements and 632 nodes to represent Cell 2 of STA2. This wetland had a triangular shape that was not conducive to a one-dimensional analysis (Figure 3.14), and had large scale vegetation patches. Each cell was assigned a bottom elevation based on survey data, and each cell was assigned a depth-dependent Manning's coefficient based upon vegetation type. Four vegetation types were considered: cattail, sawgrass, open water, and a transition zone that was converting from cattail to SAV. The Hydrologic Simulation Engine (HSE) of the Regional Simulation Model (RSM) software was used (Wasantha Lal et al., 2005). The authors concluded that the water level profiles observed in the field were in very good agreement with simulations. Further, the triangular shape of the wetland created a backwater effect at the outlet end of the system (Figure 3.15).

3.4 FRICTION EQUATIONS

Water flow through the wetland is associated with a local frictional head loss, given by equation (3.3). This is a generic power law representation of the fact that the water velocity

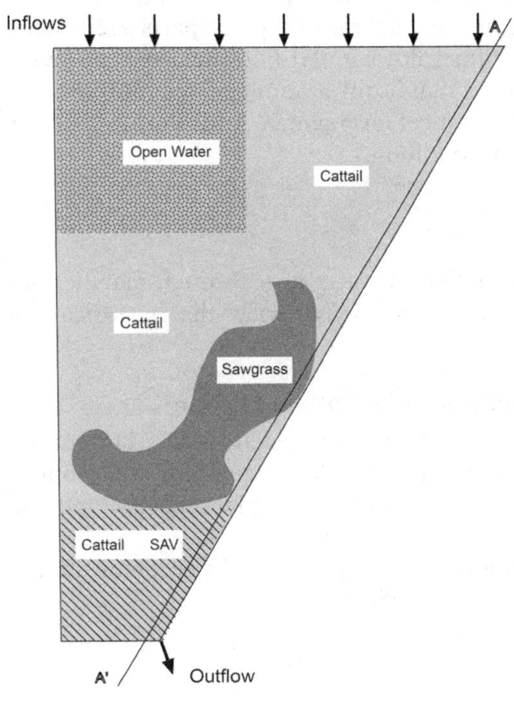

FIGURE 3.14 Cover types and 2D grid for STA2C2. (Based on information from Paudel et al., 2013.)

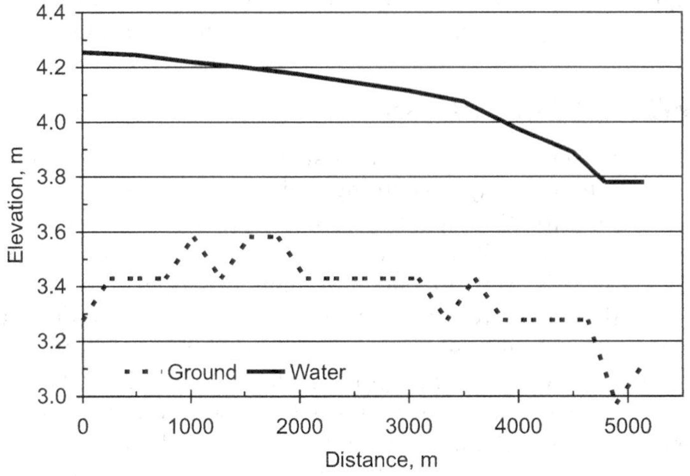

FIGURE 3.15 Water and ground elevations for a high flow condition in STA2C2. The 960 ha wetland received 327 ha·m/d. The transect A-A' (Figure 3.10) is represented. Note that the water depth is nearly uniform along the flow direction, with a mean ± sd = 64 ± 10 cm. (Based on information from Paudel et al., 2013.)

is related to the water surface slope (S = -dH/dx) and to the depth of the water (h). This generalized form (3.3) was first suggested by Horton (1938). It encompasses several different hydraulic situations.

The friction equation is a vertically depth-averaged, steady state result, based upon a reluctance to go to three-dimensional computational fluid mechanics. This results in three difficulties in the wetland environment. First, there is a vertical profile of vegetation resistance in many cases, because the submerged plant parts are often stratified. Second, a good deal of the literature presumes flows in flat-bottomed systems, which is not the case for wetlands. It is usual to have a significant amount of microtopographical relief in the wetland, which also factors into vertical averaging. And third, many treatment marshes are in a continual unsteady flow condition.

Limiting Condition Relations

A sizeable literature on friction in water flow through marshes has developed over the past two decades. It is informative to briefly review that historical framework that has guided these modern efforts.

Open Channel Flows Controlled by Bottom Friction

The framework that is very often borrowed from the literature is adaptation of constant depth, flat-bottomed, open channel flow equations. Of course, wetlands are not flat-bottomed, and they are full of submersed plant parts. Nonetheless, it is informative to briefly consider some of these limiting cases.

When a = 1/K, b = 3 and c = 1; equation (3.3) becomes the equation for laminar flow in an open channel (Straub et al., 1958):

$$u = \frac{1}{K} h^2 S \tag{3.10}$$

where K = laminar flow friction coefficient, s·m

Note that a unit conversion is necessary to convert to the mass balance unit of days. This equation is based upon the presumption that drag exerted by the bed bottom is the only retarding force for water movement. The limit of this formulation for a channel devoid of vegetation is the (Re_h) less than 2,500:

$$Re_h = \frac{h\rho u}{\mu} < 2500 \tag{3.11}$$

where μ = water viscosity, kg/m·s

ρ = water density, kg/m^3

For average warm water properties and a depth of 30 cm, this means velocities less than about 700 m/d (0.8 cm/s), a range that includes most FWS wetlands, except for the very largest.

When a = 1/n, b = 5/3 and c = 1/2; equation (3.3) becomes Manning's equation (Manning, R., 1891, as referenced by Powell, 1968):

$$u = \frac{1}{n} h^{2/3} S^{1/2} \tag{3.12}$$

where n = Manning's coefficient, s/m$^{1/3}$

This equation is also based upon the presumption of bed drag being the only retarding force for water movement.

Therefore, equations 3.10 and 3.12 are both inappropriate for vegetated wetlands, but modifications of Manning's equation are still in use for marshes, as discussed later.

Flows Controlled by Stem Drag

Another extreme or limiting situation is that of vegetated drag uniformly distributed over the entire vertical profile. That is essentially the case of flow in a loosely packed bed of resistive elements. Historically, the most commonly accepted friction relationship for porous media states that superficial velocity is proportional to the slope of the water surface:

$$u = -k\frac{dH}{dx} = kS \tag{3.13}$$

where k = hydraulic conductivity, m/d

u = superficial water velocity (Q/Wh), m/d

This is the one-dimensional version of Darcy's law (Darcy, 1856; as referenced by Freeze, 1994). It is restricted to the laminar flow regime. Laminar flow is defined in terms of a stem Reynolds number (Re_d):

$$Re_d = \frac{d\rho u}{\mu} \leq (\approx 1) \tag{3.14}$$

where d = particle diameter, m

ρ = density of water, kg/m^3

μ = viscosity of water, kg/m·d

Wetland flows are outside the conditions ranges for Darcy's law and its accompanying laminar flow criterion, but adaptations have been put forth that extend the range to include both laminar and turbulent flow (Ergun, 1952, as referenced by Bennett and Myers, 1982). The laminar term, Equation (3.13) is preserved, and a turbulent term is added:

$$S = -\frac{dH}{dx} = \frac{1}{k}u + \omega u^2 \tag{3.15}$$

where ω = turbulence factor, d^2/m^2

Both k and ω are predictable from properties of the submerged matrix (Bennett and Myers, 1982). The authors based their work on granular materials, with porosities on the order of 40%. Wetland vegetation resembles fiber beds, with porosities on the order of 95%. The predicted values of k and ω are two to three orders of magnitude different from Ergun's basis. This disparity has been reviewed for fiber beds of high porosity (Åström et al., 1992), who found it to be much less accurate.

Therefore, the physical dependencies of the Ergun parameters for Equation 3.15 are inappropriate for vegetated wetlands.

The nature of flow around immersed cylinders is better visualized in terms of the wakes behind the cylinders. As flow increases, wakes begin to appear at about $Re_d = 5$. For $40 < Re_d < 150$, there are laminar vortices behind the cylinder. And for $Re_d > 300$, the vortices are turbulent.

THE DRAG COEFFICIENT: SMALL SCALE CONTROLLED STUDIES

Many recent wetland channel and mesocosm studies utilize a drag relationship (see for instance, James et al., 2004a; James et al., 2008a). These typically presume or use a flat-bottomed system, and uniform arrays of submerged elements, either plants or simulated plants. In its simplest form, for uniform vertical distribution, flow through the submersed matrix is described by:

$$S = C_D \cdot a \cdot \left(\frac{u^2}{2g}\right) \tag{3.16}$$

where C_D = drag coefficient, –

g = acceleration of gravity, m/s^2

The presence of submerged stems, leaves, and litter creates an underwater environment dominated by drag on those surfaces, rather than the channel bottom. The common measures of vegetation density involve the presumption of cylindrical stems, the number of stems per square meter, and their diameter:

$$a = n_s d \tag{3.17}$$

$$(1 - \varepsilon) = n_s \frac{\pi d^2}{4} \tag{3.18}$$

where a = projected plant area normal to flow per unit volume, m^2/m^3

d = cylinder diameter of vegetation, m

$(1-\varepsilon)$ = fraction of volume occupied by plants, m^3/m^3

n_s = number of stems per unit area, $\#/m^2$

In a more general sense, there are leaves as well as stems, thus invoking the traditional measure of vegetative surface area, the Plant Area Index (PAI), which is discussed in Chapter 4. In the context of immersed surfaces and drag, it is the fraction of the total PAI below water, and its vertical distribution that are of interest. While PAI and area normal to flow are not identical, a direct, monotonic relation between them would be expected (Jalonen et al., 2012).

If the stem Reynolds number (Re_d) within the array is less than 200, flows proceed with interfering laminar wakes (Nepf, 1999).

$$Re_d = \frac{d\rho u}{\mu} < 200 \tag{3.19}$$

As a point of reference, stems of one cm diameter in a flow of 1000 m/d would produce Re = 116, still within the laminar range. At most wetland velocities, this implies that Stem densities are such that drag is determined by obstruction of flow (form drag).

If flow is wake-shedding laminar within the submerged canopy,

$$C_D = \frac{K_1}{Re} \tag{3.20}$$

where K_1 = constant, –

These circumstances lead to parameters in Equation 3.3 of b = 1 and c = 1:

$$u = \frac{K_{stem}}{n_s} S \tag{3.21}$$

where n_s = number of stems per unit area, $\#/m^2$

K_{stem} = conveyance coefficient, $m^{-1} \cdot s^{-1}$

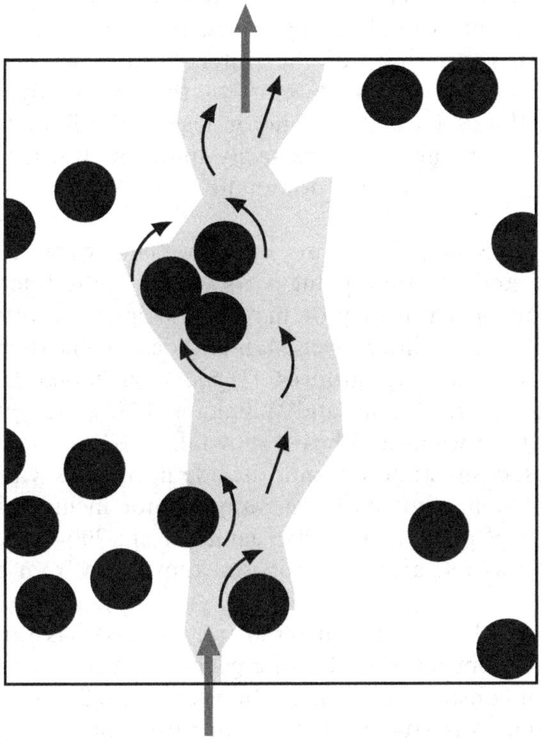

FIGURE 3.16 Overhead view of dye moving (top to bottom) in an array of lab stem elements (large circles) of porosity 0.01. The stem Reynolds number was 113 for this point injection. (Based on information from Tanino and Nepf, 2008.)

There is no depth effect in this formulation, which is in effect just Darcy's law for uniform porous media – a bed of submerged vegetation in this case. Data from channels with vertical rods indeed support this analysis (Nepf, 1999; Schmid et al., 2004; James et al., 2008a). Hall and Freeman (1994) confirmed the direct proportionality of resistance to stem density for bulrushes, which are quite comparable to rod geometry.

At high Re_d, in the range of 2,000–20,000, C_D is constant for submerged plants (Armanini et al., 2005). These circumstances lead to parameters in Equation 3.3 of b = 1 and c = 0.5:

$$u = \sqrt{\frac{2g}{aC_D}}\sqrt{S} \tag{3.22}$$

These idealized relations for vegetated systems provide insights into friction law parameters but are generally insufficient for real ecosystem calculations. There are, several other important features of wetland flows that must be taken into account to gain a description of a real, full-scale system. There are vertical and spatial profiles of stem/leaf density, a litter layer on the wetland bottom, wetland spatial patchiness, wind forces that can move water (Jenter and Duff, 1999), and a wetland bottom that is not flat (Kadlec, 1990).

Profiles of Drag Surfaces

The vertical location of plant stems and leaves varies with the type of vegetation under consideration (see Figure 1.2). One limiting case is floating plants, such as water hyacinths (*Eichhornia crassipes*), which populate only the top-most stratum of the water column. Rooted plants with floating leaves, such as water lilies (*Nymphaea* spp.), also place most drag in the vicinity of the water surface, with a lesser amount in the water column due to stems. In contrast, most of the commonly used emergent macrophytes in treatment wetlands have stems and/or leaves distributed throughout the water column, but the distributions are not necessarily uniform. Bulrushes continue with stem morphology exclusively, but leaves are dominant at mid-depths for cattails, sedges and reeds.

Both live and dead plant material occurs in the wetland water column. A bottom layer normally contains dead and prostrate plant parts, which is the litter layer. As discussed in Chapter 2, the definition of depth may or may not include the litter layer as part of the water column. Above the prone litter layer, standing dead erect stems or culms may be an important contributor to the drag surfaces. *Typha angustifolia* L. shoots often remain standing for as long as 3 years (Jordan and Whigham, 1988). However, dead shoots can be bent and flattened by strong winds and heavy snowfall.

Almost all flume mesocosm studies do not include necromass. Channel drag studies may involve artificial or real stems, but both generally do not include either standing dead or litter (see for instance Nepf and Vivoni, 2000; James et al., 2004a; Tanino and Nepf, 2008). A summary of current theoretical and mesocosm knowledge from channel studies can be found in Nepf (2012a).

The stratification of total vegetative material (phytomass) has rarely been quantitatively studied in emergent treatment marshes, but the general structure of marsh vegetation often follows that measured in constructed research flumes (Figure 3.17).

Vegetative stratification was studied at the Houghton Lake, MI, cattail treatment wetland, for two varieties of cattail (*Typha latifolia* and *Typha angustifolia*) (Table 3.2). Porosities (ε) were very high, 0.91–0.99, and stem counts (n_S) were in the range of 37 and 110 per m^2 respectively, during mid-summer. Frontal areas (a) decreased with increasing

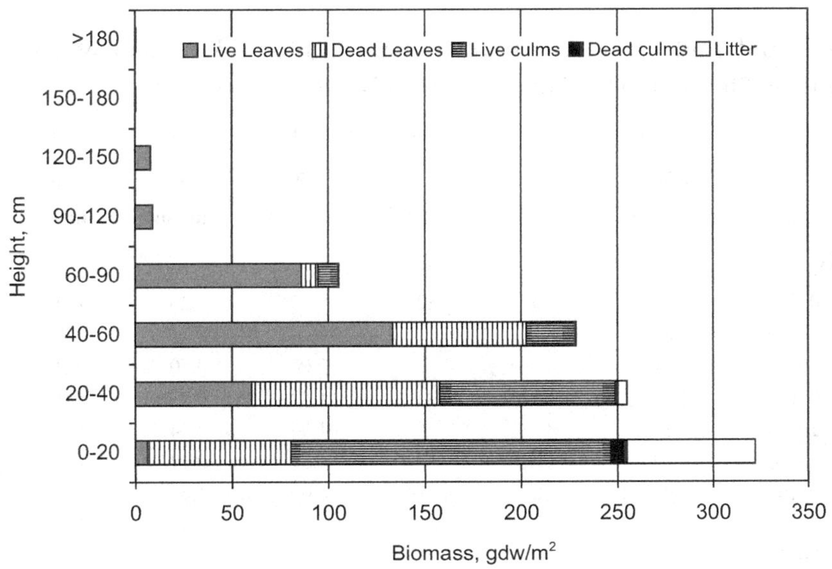

FIGURE 3.17 Biomass by category and vertical layer for sawgrass (*Cladium jamaicense*). Data from October, when the plants were 34 months old. (Based on information in Rybicki et al., 2000.)

depth. Accordingly, the values of b in Equations 3.21 and 3.22 would be somewhat higher than 1.0 due to the stem density decrease with height.

Spatial Patchiness

Most marshes are not spatially uniform but display patches of vegetation interspersed with open water zones. Thus, we are faced with the problem of describing friction in the context of spatial distributions of vegetation density. One alternative is to conduct two-dimensional calculations, which can accommodate differing resistances in the various vegetation patches, as described above.

A second, simpler alternative is to seek the dependence of the friction coefficient for the entire wetland upon the extent and distribution of the patches. This concept has been explored by Luhar et al. (2008). There are three ways to view the blockage of flow by macrophytes: (1) the fraction of the flow cross section blocked, (2) the fraction of the volume in a vegetation patch that is blocked (its porosity), and (3) the fraction of the wetland surface area that contains blocking vegetation, $B = A_V/A$. The last choice (B) has been utilized by Green (2006) and Luhar et al. (2008). Green (2006) examined data from 38 shallow rivers containing SAV, and found an exponentially increasing Manning's n with the blockage (B), with an $R^2 = 0.74$. Luhar reinterpreted that data, and determined a slightly different relation:

$$n = n_{ow}\left(\frac{1}{1-B}\right) \tag{3.23}$$

where B = blockage, –

n = Manning's coefficient, $(m^{1/3}/s)$

The blockage range extended from B = 0 to about B = 0.85. The n-value for full vegetation was not determined, but presumably would have been about 1.0 for SAV (see Table 3.4). That value corresponds to a blockage of 0.95 in Equation 3.23. Luhar also

TABLE 3.2

Stem densities and frontal areas for cattails at the Houghton Lake treatment marsh. Litter comprised about 30% of the total phytomass but is not included here.

		d	n_S	n_S	a	a	a	ε
	Stratum	**Diameter**	**Live**	**Dead**	**Live**	**Dead**	**Total**	**Porosity**
	cm	cm	#/m^2	#/m^2	m^2/m^3	m^2/m^3	m^2/m^3	
Typha latifolia								
May	0–10	1.48	0	82	0.00	1.20	1.20	0.94
	10–20	0.87			0.00	0.71	0.71	0.98
	20–30	0.61			0.00	0.50	0.50	0.99
July	0–10	1.48	92	37	1.42	0.58	2.00	0.91
	10–20	1.10			1.06	0.44	1.50	0.95
	20–30	0.92			0.89	0.37	1.26	0.96
October	0–10	0.99	0	174	0.00	1.69	1.69	0.95
	10–20	0.77			0.00	1.31	1.31	0.97
	20–30	0.66			0.00	1.13	1.13	0.98
Typha angustifolia								
May	0–10	1.02	117	112	1.20	1.15	2.35	0.93
	10–20	0.68			0.80	0.77	1.58	0.97
	20–30	0.52			0.59	0.57	1.16	0.98
July	0–10	0.94	178	110	1.57	0.92	2.49	0.93
	10–20	0.75			1.26	0.74	2.01	0.95
	20–30	0.68			1.14	0.66	1.80	0.96
October	0–10	0.83	0	260	0.00	2.15	2.15	0.94
	10–20	0.67			0.00	1.72	1.72	0.96
	20–30	0.69			0.00	1.77	1.77	0.96

determined that Equation 3.23 did not depend upon the spatial arrangement of the blockage.

Wetland Data

Small-scale and theoretical studies contribute to our theoretical understanding of wetland flows, but it is not possible to construct overall wetland flow dependencies from these fundamentals. Typically, the information on bathymetry and vertical and lateral spatial patterns of vegetation density is not available. Accordingly, for practical purposes, it is necessary to interpret data from full-scale operating systems.

Major advances in formulating correct and improved approaches to overland flow in wetlands have been made in the past two decades (e.g., Nepf, 1999; Oldham and Sturman, 2001; Choi et al., 2003; Wasantha Lal et al., 2005; Cheng and Nguyen, 2010; Marjoribanks et al., 2014). This section utilizes the emerging knowledge and calibration database to provide simple estimation methods to predict depths and velocities in surface flow wetlands. Florida emergent marsh studies comprise a large fraction of the available wetland friction information. These serve to provide general guidelines for site-specific factors.

Generalized Friction Parameters

It would be desirable to have predictive methods for the parameters a, b, and c in friction equations such as 3.3. Data exist for a few types of wetlands (Table 3.3). As discussed above, site-specific factors are known to be very important, and it is very dangerous to extrapolate from non-wetland information. Manning's "constant" is clearly not constant for the wetland environment, and it is preferable to utilize a model which describes the depth variability, namely Equation 3.3. The exponent "c" is 0.5 in the turbulent open channel formulation. However, many wetland investigations indicate a higher value (Table 3.3). As a limiting value, laminar flow around a uniform array of submerged objects over a flat bottom is theoretically described by $c = 1.0$. Specific marsh situations display a range from 0.46 to 1.00. The exponent "b" is 1.67 in the turbulent open channel formulation. But the

TABLE 3.3

Power law coefficients for example marshes.

		Depth m	Conveyance Coefficient a (m/d)/m^b-1	Depth Exponent b	Slope Exponent c	Reference
Treatment						
Houghton Lake	Sedge	0.05–0.25	2.00E+08	2.00	0.71	1
Houghton Lake	Sedge	0.08–0.30	5.00E+07	2.50	1.00	2
Houghton Lake	Cattail	0.05–0.21	6.00E+07	3.00	1.00	3
WCA2A	Cattails	0.30–0.85	1.80E+07	1.60	1.00	4
WCA2A	Sawgrass	0.30–0.85	4.70E+07	1.64	1.00	4
ENRPC1	55% EAV	0.27–0.70	2.85E+08	3.11	0.46	5
ENRPC4	94% SAV	0.36–0.81	3.79E+09	2.43	0.85	5
STA34C3A	Cattail	0.41	8.60E+04	0.86	0.49	6
STA34C3A	Cattail	0.26	4.50E+05	1.43	0.58	6
STA34C3A	Cattail	0.10	3.32E+06	0.95	0.88	6
STA34C2A	Cattail	0.45	1.07E+06	2.55	0.59	6
STA34C2A	Cattail	0.26	1.16E+06	1.67	0.66	6
Natural						
Shark Slough	Emergents	0.20–0.80	6.20E+06	1.44	1.00	7
WCA3A	Slough	0.10–0.70	6.90E+05	1.67	0.50	8
Shark Slough	Emergents	0.20–0.70	1.24E+06	1.12	0.66	9
Mean				1.98	0.77	

References:
1 Kadlec et al. (1981)
2 Kadlec (1990)
3 Hammer and Kadlec (1986)
4 Choi and Harvey (2014)
5 Unpublished Data SFWMD
6 Wasantha Lal (2017)
7 Bolster and Saiers (2002)
8 Min et al. (2010)
9 He et al. (2010)

depth variability measured for wetlands increases this value, due to bottom irregularity and other factors. Example marshes yield values from 0.86 to 3.11. The coefficient "a" is a function of vegetation and litter density, as well as the selected values of "b" and "c." Methods for forecasting the three parameters do not currently exist.

Manning's Coefficients

Early in the history of research and development related to overland flow in wetlands, mathematical descriptions were often adaptations of turbulent open channel flow formulae. These are discussed in detail in a number of texts, for example the work of French (1985). The general approach is utilization of mass, energy and momentum conservation equations, coupled with an equation for frictional resistance. Historically, the most common friction equation for open channel flow was and is Manning's equation, which became firmly entrenched in computer software and widely used in engineering practice. Suppose that hydraulics textbook open channel information was to be used to estimate Manning's n for a wetland. Guidance may be found in estimation procedures in the hydraulics literature, for instance French (1985). The value of n may be estimated from information on the channel character, type of vegetation, changes in cross section, surface irregularity, obstructions, and channel alignment. Using the highest value of every contributing factor, the maximum open channel n value is 0.29 $s/m^{1/3}$ French (1985). This is approximately one order of magnitude less than values determined from actual wetland data. Clearly, hydraulics textbook information is inadequate to describe the densely vegetated, low-flow wetland environment.

As noted by Piercy, 2010:

> Models to estimate hydraulic resistance due to vegetation in emergent wetlands are crucial to wetland design and management. Hydraulic models that consider vegetation rely on an accurate determination of a resistance parameter such as a friction factor or a bulk drag coefficient. At low Reynolds numbers typical of flows in wetlands, hydraulic resistance is orders of magnitude higher than fully turbulent flows and resistance parameters are functions of the flow regime as well as the vegetation density and structure.

The fundamental problem with the utilization of Manning's equation for marsh surface water flows is that Manning's equation is a correlation for turbulent flows in open channels; while marsh flows are nearly always controlled by stem drag. Under these conditions, Manning's n is not constant, but is strongly depth dependent (Wilson, 2007), and also velocity dependent (Hall and Freeman, 1994). There is also a difficulty with the extension of open channel flow concepts to densely vegetated channels (James et al., 2004a). The frictional effects that retard flow in open channels are associated primarily with drag exerted by the channel bottom and sides. Wetland friction in dense macrophyte stands is dominated by drag exerted by the stems and litter, and microtopography, with bottom drag playing a very minor role (Kadlec, 1990). As a consequence, overland flow parameters determined from open channel theory are not applicable to wetlands. In particular, Manning's coefficient is no longer a constant; it depends upon depth as well as stem density. Predictions from previous information on nonwetland vegetated channels are seriously in error.

Much of the existing information on wetland surface flow has been, and to some extent still is, interpreted and reported via Manning's equation. Old software is often constrained to its use. Much-used programs such as the Soil and Water Assessment Tool (SWAT) (Nietsch et al., 2011), HEC-RAS (Brunner, 2010), and MIKE-SHE (Danish Hydraulic Institute, 2011) continue to utilize Manning's equation for wetland applications. In contrast, some of the older software has been updated to include flow equations more appropriate for marshes, such as MODFLOW (Restrepo et al., 1998; Harbaugh et al., 2000). Perhaps unfortunately, some wetland authors continue to base interpretations on Manning's equation (Su et al., 2009; Bal et al., 2011).

Although not appropriate for surface flow wetlands, Manning's equation (3.12) has nevertheless been widely used and calibrated for some treatment marshes. Table 3.4 gives the general range of Manning's n for a sampling of wetlands.

TABLE 3.4

Manning's coefficients for example marshes. These are generalities, because values depend on depth, flow, and vegetation densities.

Project	Vegetation	Depth	Velocity	Manning's n	Reference
		m	m/d	$s/m^{1/3}$	
Treatment					
Open Water	None			0.03–0.05	1
Sacramento Cell 3	Dense Bulrush	0.45–0.60	50–60	5.9–6.7	2
Sacramento Cell 7	Dense Bulrush	0.45–0.60	40–75	2.1–7.6	2
Houghton Lake	Dense Cattail	0.15	67	4.7	3
Benton Cell 2	Woolgrass	0.12–0.42	110–358	3.3	4
Boney Marsh	Mix	0.30–0.70	35–135	1–4	5
ENR Cell 1	Cattails + SAV	0.36–0.79	30–867	0.43–2.50	6
STA2C2	Cattail	0.40–1.00	173–690	0.72	7
STA2C2	Sawgrass	0.40–1.00	173–690	0.55	7
STA2C2	Open Water	0.40–1.00	173–690	0.09	7
STA34C3A	Cattail	0.10–0.41	980–1100	0.18–0.45	8
ENR Cell 4	SAV	0.36–0.81	277–1562	0.42–1.33	6
STA2C3	SAV	0.88	490	0.99	9
STA2C3	SAV	0.76	406	1.42	9
Natural					
WCA1	Sawgrass	0.15–1.50		0.33–1.20	10
WCA2A	Sawgrass	0.15–1.50		0.32–1.20	10
Chandler Slough	Water Hyacinth	0.40–0.70		0.20–0.55	11
WCA3A	Sawgrass	0.10–0.70	340–1000	0.34	12
Shark Slough	Sawgrass	0.20–0.70	242–1640	0.46	13

References:
1 French (1985)
2 Dombeck et al. (1998b)
3 Unpublished Data Kadlec
4 Unpublished Data TVA
5 Mierau and Trimble (1988)
6 Unpublished Data SFWMD
7 Paudel et al. (2013)
8 Wasantha Lal et al. (2015)
9 Wasantha Lal (2017)
10 Shih et al. (1979)
11 Shih and Rahi (1982)
12 Min et al. (2010)
13 He et al. (2010)

Generally, Manning's n is strongly depth dependent for FWS systems, decreasing as depth increases. The nature of this dependence is illustrated in Figure 3.18 for the ENRP marsh study. Over a depth range of 30–80 cm, n values decrease by a factor of five for this emergent-SAV wetland. For this system, the n-values decrease with the square of the depth. Other studies have utilized $n = Ah^{-0.77}$, for instance, Min et al. (2010) and Paudel et al. (2013). Swain et al. (2004) utilized $n = Ah^{-2.00}$.

Likewise, n-values are dependent on vegetation density, because stems and litter provide the dominant drag surfaces, which has been shown above to depend upon the amounts of immersed live and dead stems and leaves, and litter. Since both litter and live and dead stems are involved, the relation is not easily predictable. It depends on litterfall events, and the transition of standing dead to prone dead (litter).

The Influence of Unsteady Flow

The calibration of a friction law to wetland data should theoretically be for periods of steady flow, because of the variability of depths and water slopes during transients (see, for instance, Figure 3.11). But many treatment marsh applications have few, if any, such steady flow periods. If inlet and outlet stage and flow data is available at some frequency, perhaps as high as daily, then the question arises about how to extract friction law coefficients from it.

Here, three methods are compared: (1) use daily data, (2) use monthly average data, (3) select steady operating periods. Because Manning's formulation specifies both the depth and slope exponents, direct calculations of n may be made for each selected period. For the power law, the three parameters (a,b,c) are selected to minimize the squared error.

Table 3.5 shows an example of the results for ENRPC1, for which daily data were available. Daily and monthly information display high coefficients of variability (CV) for depth water slope and flow. By selecting steady flow periods, the CVs are much reduced. The resulting Manning's n values depend heavily upon the selected data period, but the power law coefficients much less so.

The Influence of Bathymetric Variability

As discussed above, the bottom elevation of many wetlands is irregular, with local depressions and hummocks. On a large scale, these are quantified by depth-area-volume relations

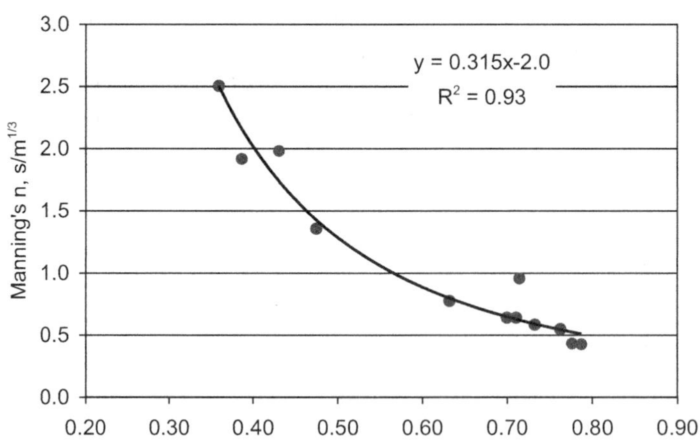

FIGURE 3.18 Manning's n versus depth for ENRP Cell 1, an emergent-SAV mix. Data span four years, and include only stable periods of constant flow.

TABLE 3.5

Variability of friction law coefficients with selection of data period. These are calibrations to four years' data from ENRP Cell1.

		Daily	Monthly	Steady
Depth, m	mean	0.55	0.55	various
	CV	0.23	0.17	0.03
Flow, m^3/s	mean	2.2	2.2	various
	CV	0.99	0.75	0.16
Water Slope, m/m	mean	1.60E-05	1.60E-05	various
	CV	0.66	0.50	0.19
Power Law	a	2.07E+08	2.61E+08	1.94E+08
	b	3.07	2.69	3.73
	c	0.49	0.52	0.48
	R^2	0.77	0.64	0.99
Manning's	n	0.30	0.88	1.06

(see Figure 2.8). On a small scale, these features define the micro-topography of the wetland bottom and are represented by a soil surface elevation distribution. Small constructed wetlands are typically graded to a specified tolerance, such as ± 5 cm. In practice, these tolerances often either are not achieved, or are not carried over into the operational regime. Interestingly, some natural wetlands have about the same fine-scale distributions of soil elevations as do constructed wetlands.

The effect of such uneven bottoms upon the friction model depends upon the orientation and shape of the high spots and depressions (Stothoff and Mitchell-Bruker, 2003). Ridge features may either be parallel to flow, and act as flow-straighteners, or be perpendicular to flow and act as "speed bumps." To illustrate the potential effects, assume the bottom elevation distribution represents the flow cross-section (Kadlec, 1990; Choi and Harvey, 2014). In order that water depth remain positive, depth is measured with respect to the lowest soil elevation. A purely geometric effect prevails: there is not much cross-section available for flows at low water stage. This effect is not lost until water depths are well above the point of complete inundation. Application of equation 3.21 to a linear soil elevation distribution produces a depth effect on conveyance capacity. For example, a straight line approximation for immersed cattail area with depth, applied to flows at depths up to 60 cm, introduces depth dependence represented by b = 1.94 in the general Equation 3.3, with c = 1.0.

Stage-Discharge Relations

Equations 3.12 and 3.14 suggest that for limiting circumstances, there should be a relation between flow and depth only. The limiting circumstances are vegetation-controlled normal depth flow, and structure-controlled flow. It is therefore tempting to describe all marsh flows using a stage-discharge formulation:

$$Q = C_V h^b \tag{3.24}$$

where b = constant, e.g., power law b or weir formula exponent
C_V = generalized discharge coefficient, $(m^3/s)/(m^b)$
Q = flow, m3/s

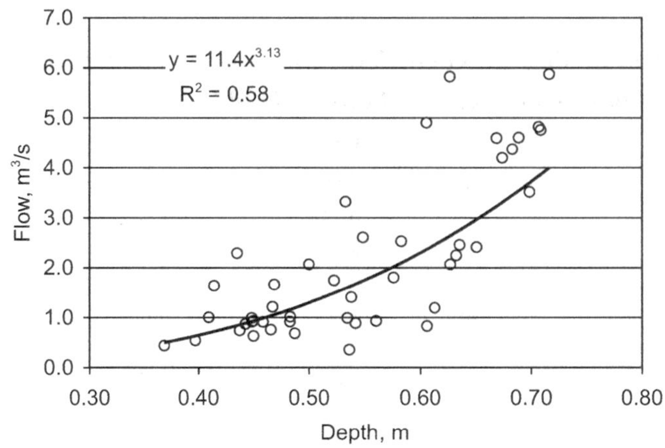

FIGURE 3.19 Stage-discharge curve for four years of monthly data from ENRPC1.

h = depth, m

The generalized discharge coefficient (C_V) is dependent upon conditions in a specific wetland, including ground slope, vegetation density, and outlet structure characteristics. It is therefore not possible to determine the value *a priori*, but such a relation can give a reasonable representation of wetland data (Figure 3.19).

The utility of a stage-discharge relation resides in the fact that does not attempt to describe the longitudinal depth profiles but rather presumes those depths are uniform.

3.5 VERTICAL PROFILES

Marsh surface flows are three-dimensional. The principal flow direction is logically from inlet to outlet and displays a vertical distribution of speeds determined by the type of vegetation and possibly wind effects. But there are also vertical water movements that can occur, although these would normally be small in the context of the entire wetland. Lateral flows will occur, transverse to the principal flow direction, driven by wind, bathymetry, and possibly transpiration. In practice, the complexity of the combination of these effects requires simpler procedures for pollutant removal forecasting.

Vertical Velocity Profiles

Although it is common practice to use vertically averaged longitudinal water velocities, there are many different vertical velocity profiles that occur. These are caused by the different growth forms of the wetland vegetation (Figure 3.20).

Unvegetated systems display profiles that have maxima at the water surface, and are concave upward. The laminar profile is parabolic; the turbulent profile is logarithmic (Figure 3.20A). Wetlands with emergent vegetation that extends above water also have profiles that maximize at the surface and are concave downward if the vegetation frontal area decreases with elevation above the bottom (Figure 3.20B). However, if the frontal area increases with elevation, then the upper layers encounter more resistance, and the profile may become concave upward. If the vegetation is overtopped, then the lower layers encounter more resistance than the overtopping layers, and the velocities above the vegetation are higher than within it (Figure 3.20C).

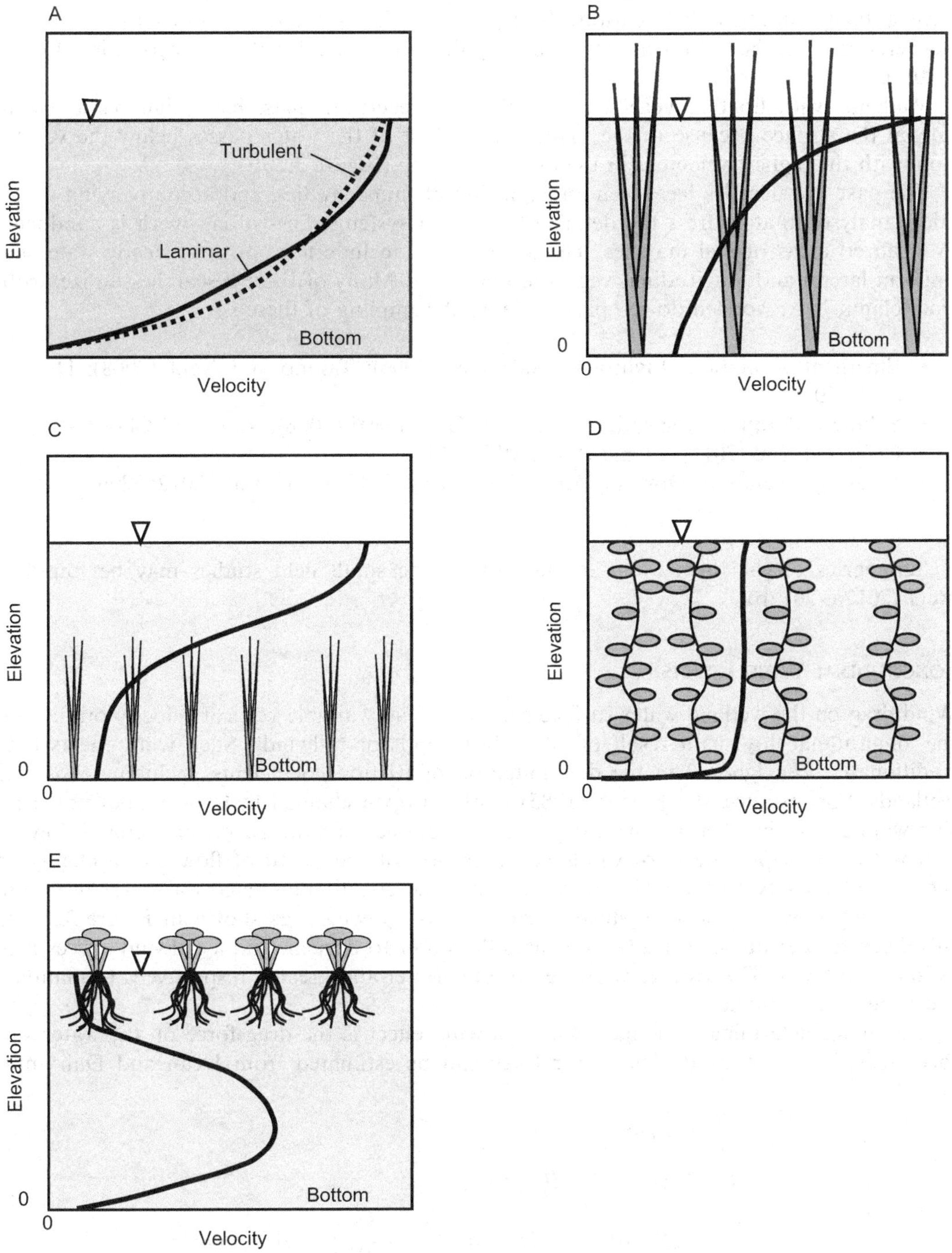

FIGURE 3.20 Longitudinal velocity profiles in various vegetation types. These are the result of throughflow of source water. (A) Unvegetated channels. (B) Emergent vegetation. (C) Overtopped vegetation. (D) Uniform submerged vegetation. (E) Floating vegetation.

Submerged aquatic vegetation often occupies the entire vertical profile and may have a fairly uniform distribution. In that case, the wetland functions as a uniform (loosely) packed bed, and there is no significant vertical velocity variation (Figure 3.20D).

Wetlands with floating vegetation, or floating vegetated mats, have high resistance to flow at the surface because of the plant cover. Most of the water passes below the vegetation, with the highest velocities in the underflow zone (Figure 3.20E).

The past two decades have seen an explosion of flume studies, and accompanying theoretical analyses, that address the details of idealized systems. Less often, work is conducted in localized areas of real marshes. The idealizations include flat bottoms, steady state, and uniform lateral and longitudinal vegetation densities. Many of these researches utilize artificial "plants," i.e., wooden dowel, plastic plants. A sampling of these is:

- Emergent vegetation: Lightbody and Nepf (2006); Tanino and Nepf (2008); Huai et al. (2015)
- Submerged aquatic vegetation: Nepf and Ghisalberti (2008); Yang and Choi (2010); Chen and Kao (2011); Zhang et al. (2013b)
- Floating vegetation: Downing-Kunz and Stacey, 2012; Huai et al., 2012; Zhao et al., 2012

Summaries of the theory and results from flume small field studies may be found in Nepf (2012a, 2012b).

Longitudinal Wind Effects

Wind drag on the wetland water surface can also create variable vertical velocity profiles for the longitudinal flow, as a result of either headwinds or tailwinds. Such wind effects have traditionally been ignored in the determination of friction coefficients, including those for wetlands. For instance, the French (1985) book on open channel hydraulics does not mention wind effects. Neither are there full-scale field studies of wind effects on wetland flows.

The two principal effects of winds are alteration of the depth of flow and a change in the vertical velocity profiles. The former can be theoretically illustrated for a system in uniform steady flow, experiencing steady headwinds or tailwinds, as shown in Figure 3.21. In this theoretical example, a headwind causes the water to deepen, and a tailwind causes it to become shallower. The average velocities become lesser and greater, respectively, to maintain the same volumetric flow.

The major determinant of the extent of a wind effect is the drag force on the water surface created by that wind. This shear force can be estimated from Dean and Dalrymple (1991):

$$\begin{aligned} \tau_S &= \rho k W^2 \\ k &= 1.2 \times 10^{-6} \quad W < 5.6 m/s \\ k &= 1.2 \times 10^{-6} + 2.25 \times 10^{-6}\left(1 - \frac{5.6}{W}\right)^2 \quad W > 5.6 m/s \end{aligned} \tag{3.25}$$

where k = coefficient, –
ρ = density of water, kg/m^3
τ_S = shear stress, Pa
W = wind speed, m/s

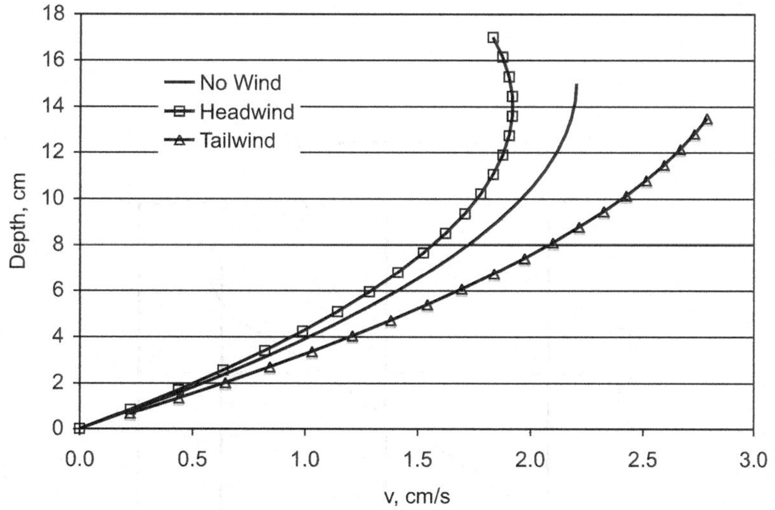

FIGURE 3.21 The effects of headwinds and tailwinds on theoretical vertical velocity profiles in a sparsely vegetated channel. The friction law coefficients are a = 5.6E+10 1/m·d, b = 3, and S = 1.0. The results are for a fixed volumetric throughput for the system, i.e., Q = constant.

Flume studies have been conducted for shallow vegetated systems (Jenter and Duff, 1999; Oldham and Sturman, 2001; Banerjee et al., 2015; Pang et al., 2015). These serve to confirm theoretical expectations, but do not provide information that is immediately transferable to full-scale wetlands. As an example, Figure 3.22 shows the vertical velocity profiles for EAV and SAV measured by Banerjee et al. (2015).

Two major factors prohibit the use of theoretical and flume-scale data in full-scale systems: the wind does not blow all the time, and the direction is variable. Winds typically follow a diel pattern, with maxima during the daytime and near-zero minima at night. In some geographic regions, there is also a seasonal pattern to wind speeds. Thus, wind usually would affect flow only for part of the day, or part of the year. While there is normally a prevailing wind direction, it would not normally align with the wetland flow direction. Thus, wind driven flows are just as likely to be transverse to the nominal flow direction as they are to be parallel to it. As a result, it is difficult to determine wind impacts on wetland friction equations. But, because wind driven flows are not trivial, they do play a role in determination of cross flow and mixing and are therefore of importance for the flow efficiency of the wetland. This is further discussed in Chapter 7.

Vertical Velocities

It might seem that there are no consequential vertical velocities excepting those driven by hydraulic head differences (seepage). In reality, there are other forces that create vertical flows, including transpiration, secondary stem flows and thermal destratification.

Vertical transpiration flows occur because plants draw water to satisfy their functionality, and this need would most often be supplied from the water above the root zone. This subject has been introduced in Chapter 2 and will be examined in more detail in Chapter 6. These flows are on the order of millimeters per day and occur in all wetlands with rooted emergent vegetation. Despite their small magnitude, such flows can carry contaminants to the root zone, where they can interact with soils and plant roots.

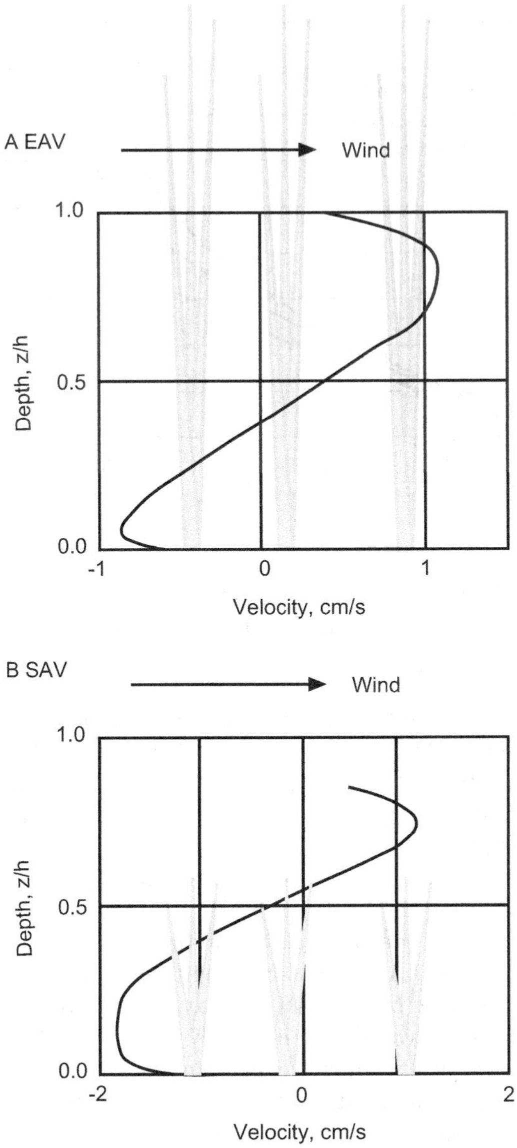

FIGURE 3.22 Vertical velocity profiles measured in a non-throughflow flume "vegetated" with plastic plant elements. The water surface is at z/h = 1.0. The wind tunnel flume was operated at a 2.8 m/s wind (2.25 mph). Panel A represents EAV, panel B represents SAV. (Based on data from Banerjee et al., 2015.)

Vertical secondary flows can occur behind stems in the wetland. Such flows form through the interaction of the protruding structure with the vertical gradient of longitudinal velocity in the benthic region. These move water upward from the wetland bottom into the main longitudinal flow. Nepf and Koch (1994) conducted flume studies that showed that vertical secondary flows were as high as 15% of the local longitudinal velocity can develop at the base of individual stems (Figure 3.23). Such flows can carry nutrients up from the sediments, where they have been regenerated by microbial processes, to areas higher in the canopy, where they can be used by epiphytes. As yet, there is no full-scale wetland data that reports this effect.

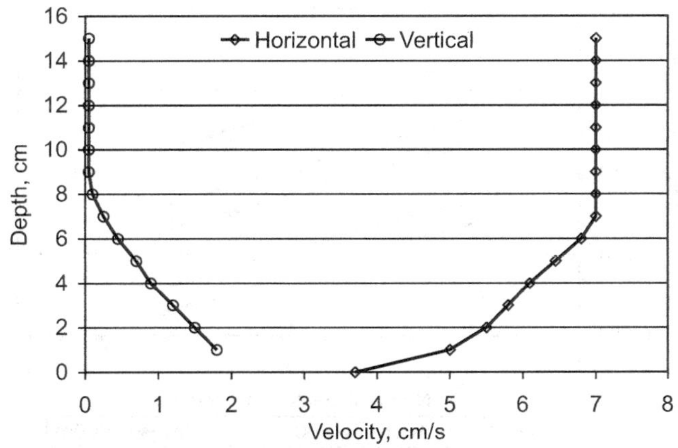

FIGURE 3.23 Vertical and horizontal velocities in a flume with emergent stems (dowels) at a density of 216 per m^2. (Based on data from Nepf and Koch, 1999.)

Thermal destratification occurs in wetlands, a phenomenon caused by the diel cycle of heating and cooling. During the day, solar radiation creates a vertical gradient of temperature in the water column, with the warmest water at the top. This gradient is stable, because the low-density warm water is on top of cooler, higher density water. During the nighttime hours, the top layer cools, and becomes denser than the underlying waters. The result is that the top waters sink through the less dense waters below. Jenter et al. (2003) measured the magnitude of the process in a marsh, and found local plunge velocities up to 5 cm/s, occurring in the middle of the night. This phenomenon can be regarded either as a flow or as mixing. More discussion of stratification is given in Chapter 6.

3.6 WIND DRIVEN FLOWS

OPEN WATER WIND EFFECTS

As discussed earlier, wind can either assist or impede longitudinal water flows, when aligned with the nominal flow direction. But wind driven flows occur irrespective of the nominal direction and are often present in cross-flow directions. In this section, more detail is given on the expected magnitude of such flows. Because wind currents have been studied in lakes for many decades, a brief review of these is appropriate. There are important implications of those results for wetlands.

There are two wind factors that contribute to currents in a water body, wind-generated setup, and wind stress.

Wind Setup

Under extreme circumstances, water surface setup in the direction of the wind stress can occur. The amount of setup is dependent on a number of factors but it is interesting to note that the underflow generated by the setup is in the direction opposite to the wind stress (Figure 3.24).

The method used by the U.S. Army Corps of Engineers (U.S. ACE) to calculate wind set-up, based on Bodine and Hebler (1978), can be approximated by:

$$S = \left(2.01 \times 10^{-4}\right)\frac{W^2 F_U}{h} \tag{3.26}$$

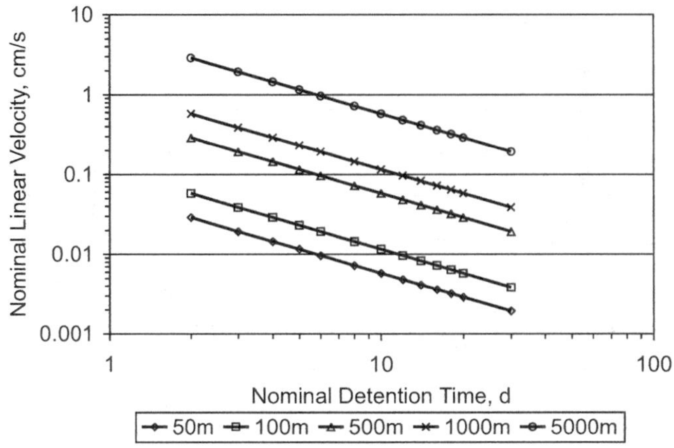

FIGURE 3.24 Ranges of nominal linear velocities for various wetland lengths and nominal detention times.

where h = water depth, m
F_U = fetch, km
S = setup, m
W = wind speed, m/s

For a marsh of depth 60 cm and fetch 1.6 km, a 4.5 m/s wind will cause about one centimeter of set-up. Thus, only hurricane winds will cause significant set-up; e.g., one meter for 45 m/s (100 mph).

Although there are sections of open water in some treatment marshes, most of their area is occupied by either EAV or SAV. The presence of vegetation alters the effects of wind, both velocities and setup. Set-up is negligible under non-hurricane conditions, and wind-driven velocities are modified compared to open water conditions.

Wind Driven Surface Velocities

Wind creates a surface stress on water that is proportional to the wind velocity squared, which in turn creates surface water movement in the wind direction. Empirical relations derived for Lake Mendota in Wisconsin give a rough estimate of the magnitude of wind driven currents (Haines and Bryson, 1961).

$$u = 5.3 + 0.013W \tag{3.27}$$

where u = water velocity (top 60 cm), cm/s
W = wind velocity at 5m, cm/s

This regression allows for a surface lake current driven by thermal and other forces, in addition to a wind effect. The speed of wind-generated surface currents is about 1.3% of the speed of the forming wind, at low wind velocities. According to this rule of thumb, a 10 mph wind creates a surface velocity of 5.9 cm/s.

Wind current speed increases as wind speed increases up to a critical point where surface disturbance and the conversion of wind energy to wave motion starts to reduce flow. Researchers have found winds over about 6 m/s (13 mph) decrease surface current speed.

Such wind-driven currents are by no means trivial in large bodies of unprotected, unvegetated water. For instance, Jin and Sun (2007) measured vertical velocity profiles in Lake

Okeechobee under a variety of nonhurricane conditions. They measured horizontal velocities in the upper meter of the water column in the range of 12–18 cm/s. Deeper flows were smaller, generally less than 5 cm/s, and often in the reverse direction.

The surface currents are compensated by flows in other directions at depth below the surface. Measured vertical profiles of the horizontal velocities in Lake Okeechobee suggest a thin surface layer with relatively strong currents in the wind direction compensated for by slow but broad return flow deeper in the water column (Jin and Sun, 2007).

For open water systems, these velocities are one or more orders of magnitude greater than the linear nominal velocities due to added flows to the treatment marshes. The range of source water flows in such marshes depends upon the desired detention time, depth, and size of the wetland (Figure 3.24). Thus, surface wind-driven currents in marshes may not be ignored.

Wind Effects in Vegetated Wetlands

There are but few studies of wind-driven currents in full-scale marshes.

Vegetated systems occur in categories that depend upon the water depth h and the vegetation height above the bottom h_V. When $h > h_V$, the wetland is an SAV system; conversely when $h < h_V$, the wetland is an EAV system. SAV may occupy the entire water column, $h = h_V$. For flow-through systems, wind-driven velocity profiles change markedly as the wetland character changes from over-topped SAV to EAV.

SAV Systems

The presence of submerged macrophytes creates the possibility of an upper zone, free of vegetation; and a lower zone, occupied by SAV. The upper zone is frequently a feature of beds of SAV in lakes but may or may not be present in treatment marshes.

For SAV beds that are overtopped by water, wind-driven flows may be present in the top layer, but the effects are dampened out with depth into the SAV. Researchers therefore have focused upon the penetration depth of the wind effect (Nepf and Vivoni, 2000; Herb and Stefan, 2005).

There can be little doubt that the top layer of an overtopped SAV system behaves like an open water body with respect to wind-driven flows. The wind flume study by Banerjee et al. (2015) bears this out, with surface velocities above the SAV moving in the wind direction and compensatory return flows in the lower layers in the SAV (Figure 3.22B).

The magnitude of wind currents in large SAV systems is illustrated by the results of Li et al. (2017). Their studies were conducted in a shallow bay of Lake Taihu in China. The coverage was 74% SAV and 18% FAV, and the depths of measurement were 40, 80, and 120 cm. Thus although named a "lake," this system was clearly a shallow SAV bed.

Li et al. (2017) found that compensation currents occurred in the lower middle and bottom layers of the vertical water profile in Lake Taihu in China. They found downwind water velocities in the range of 1–10 cm/s in response to winds of approximately 1–4 m/s. Flow reversals were identified in 60–100% of the data, and were most prevalent for moderate, sustained winds (1–2 m/s).

EAV Systems

The presence of emergent macrophytes creates an above-water zone in which the effects of wind are presumably lessened. However, laboratory and field studies have shown that the water column still receives wind impacts (Poindexter, 2014; Banerjee et al., 2015; Tse et al., 2016).

TABLE 3.6

Wind and water velocities in a non-advective EAV wetland at two locations. In-canopy wind was measured at 1.35 m above the water. The wind speed at 5 m above the water surface was 5 m/s.

	In-canopy Wind, m/s			Water, cm/s		
	Downwind	Crosswind	Vertical	Downwind	Crosswind	Vertical
Location 1						
Surface	0.81	-0.21	0.16	0.37	0.03	0.35
4 cm deep	0.72	-0.13	0.22	0.11	0.02	0.46
Location 2						
Surface	0.80	-0.21	0.18	-0.15	-0.07	0.40
4 cm deep	0.73	-0.14	0.20	-0.05	0.01	0.26

Source: Tse et al. (2016)

Tse et al. (2016) and Poindexter (2014) reported on wind effects on water flow in an emergent marsh. The study site was a 3-ha marsh located on Twitchell Island in the Sacramento-San Joaquin Delta of Northern California. Wetland projects in this region have the purpose of reducing organic soil loss and to improve the quality of agricultural runoff (Diaz et al., 2012). The study was conducted on a single day with moderately strong winds, and thus cannot be regarded as conclusive. Nonetheless, this appears to be the first report of field measurement of wind-driven momentum transfer across the air-water interface within a wetland with emergent vegetation. The water depth was about 0.25 m, and the *Typha* spp. and *Schoenoplectus acutus* plants reached a height of hc = 1.75±0.25 meters above the water surface.

The results in Table 3.6 confirm that the canopy is effective in reducing the wind speed, 0.8 m/s vs. 5 m/s. Further, the rule of thumb for open water (water speed = 1.3% of air speed) is probably an overestimate for this wetland (0.45% of in-canopy wind speed). But, the wind-driven water speeds at the surface are comparable to those for advective flow in treatment marshes. Further, there appear to be return flows at relatively small depths below the surface, indicating that the wind is causing exchange flows.

Interestingly, a concept of linear winds creating linear water speed responses is probably not correct. For the Twitchell Island study, under strong, persistent wind forcing, surface water velocities oscillated, with a distinct peak at the same frequency as periodicity of the canopy-scale wind billows, known as Kelvin-Helmholtz billows, or honami[1] (Finnigan, 2000; de Langre, 2008). These wind vortices were advected downwind as they formed and penetrated the entire depth of the vegetation canopy. They caused wind-driven water motions that carried momentum deeper into the water column than viscous transport alone and were directly responsible for the motions measured in the water.

Tse et al. (2016) concluded that wind effects near the water surface were (a) strong enough to influence biogeochemical budgets and (b) caused primarily by transfer of momentum from above the canopy, through the air in the canopy, and across the surface via shear.

3.7 LATERAL WATER MOVEMENT

Horizontal water flows frequently occur in directions other than the nominal flow direction, driven by wind, density differences, and evapotranspiration. Flows transverse to the nominal flow direction (lateral) are very important, because they can move water and associated contaminants into regions not aligned with the nominal flow direction, and not in the main flowpath. Wind and density driven currents are typically exchange flows, meaning flows into side regions at one depth are compensated by return flows at a different depth.

Density Driven Lateral Flows

Temperature differences can occur between the shallow, vegetated regions of a wetland and adjacent, deeper regions of open water. During daylight hours, solar radiation is uniform over the area, but there is a greater heating effect in the shallow zones. In contrast, nighttime cooling is more rapid in the shallow waters. This diurnal cycle of temperature difference has been observed to generate diurnally varying exchange flows between the littoral zone and the main body of a lake, and by extension would also occur between similar zones in wetlands (Zhang and Nepf, 2009). This potential effect has not been well studied in treatment marsh environments, but was found to be dominant in the summer for the flushing of littoral wetlands in field studies (James and Barko, 1991; Andradottir and Nepf, 2001). These are exchange flows, meaning that water enters the vegetated stand at one depth, and water returns from the vegetation at a different depth.

Evapotranspiration Driven Lateral Flows

Sanchez et al. (2016) studied lateral flows in the Tres Rios full-scale wetlands near Phoenix. They determined that water moved transverse to the central wetland flow channel into the vegetated fringe areas (Figure 3.25). They hypothesized that lateral movement of water into

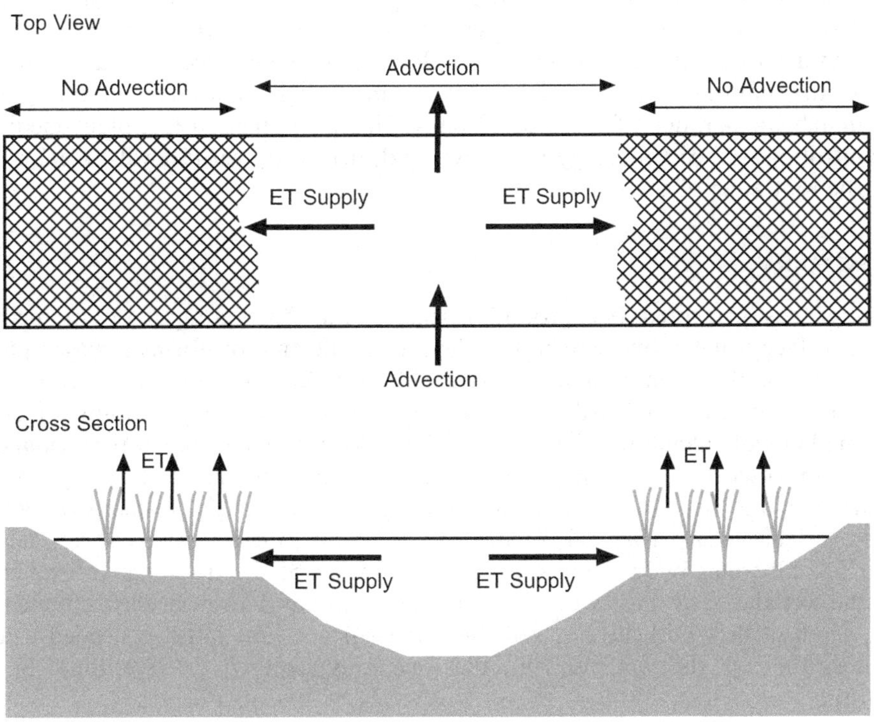

FIGURE 3.25 Plant evapotranspiration can draw water from a main flow channel, which moves vertically into the soil to satisfy the loss through the plants.

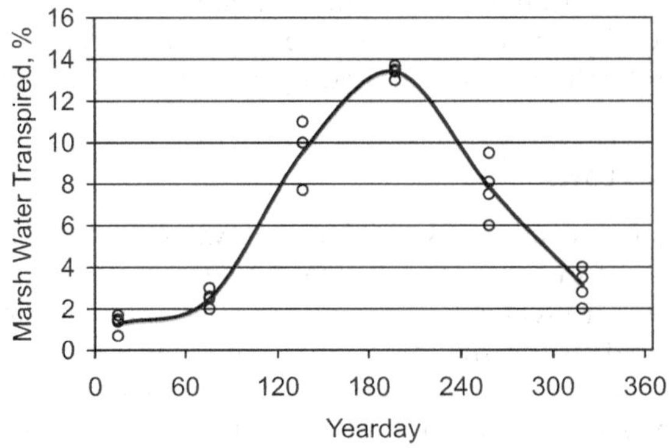

FIGURE 3.26 Percent of the marsh water evapotranspired daily at Tres Rios. (Adapted from Sanchez et al., 2016; Bois et al., 2017.)

the marsh was to replace losses via plant transpiration in the marsh, particularly in the summer when the plants were most productive and transpiration is high. They termed this lateral movement of water into the vegetated marsh as a "biological tide." It may be speculated that the driving force is more properly evapotranspiration, because both evaporation and transpiration cause water loss from the fringing, non-advective zones. This presumptive lateral flow was deemed to be enough to replace the marsh surface water in about seven days in mid-summer (Figure 3.26).

Bois et al. (2017) continued the Sanchez studies, and used more than six years of data on plant community composition and biomass production, transpiration and evaporation rates, and water quality to verify the lateral flow at the Tres Rios marsh. Tracer studies were used to determine the ET-driven lateral flows. They concluded that lateral flows bring water from the main flow channel into the vegetated marsh. Estimates of the replacement time for marsh water ranged from 4 to 13 days. This phenomenon is a plant-mediated biological control of surface hydrology in the wetland, driven by withdrawals of marsh surface water to the root zone of the marsh.

3.8 SUMMARY

This chapter has presented background research and some computational methods for describing surface water movement in marshes. This information forms a major part of the basis for understanding contaminant removal. Much is known about the various contributory processes, but it is not feasible to incorporate that entire huge amount of detail into practical methods of calculating the hydraulic basis for treatment marsh performance.

The first sections deal with the measurement and forecasting of the shape of the free water surface, and the depths that accompany it. The basic framework rests upon a spatially variable, and perhaps also temporally variable, water mass balance. The required pieces include all water budget items, plus a description of the frictional resistance to flow through the wetland. Two pieces of design information can be retrieved from such an analysis: the depth of flow and the slope of the water surface. The latter is needed to calculate the headloss through the wetland, and the former is required to determine the wetland detention time.

Wetland water surface profiles in the flow direction are determined by the slope and irregularity of the land, the density and distribution of vegetation, the incoming flow rate, and the type and setting of the outflow structure. For wetlands that have intermittent deliveries of source water, the steady state desired by model calculations is not often achieved. Many treatment marshes are irregular in shape and in vegetation coverages and must therefore be considered in both longitudinal and lateral dimensions.

The key ingredient in the mass balance analysis of flows and depths is the resistance to flow created by friction with the immersed plant parts. Several versions of these friction equations are in use, including power law empirical correlations, drag relations for immersed objects, and modifications to Manning's equation for channel flow. The use of these to estimate hydraulic resistance due to vegetation in emergent wetlands is a crucial part of wetland design and management.

Although vertically averaged water flow forms the basis for many of the available mathematical descriptions of flow, there is strong vertical variation in water velocities. In general, water in layers near the wetland bottom moves more slowly than the uppermost layers. Such stratification has implications for the wetland detention time distribution.

Gravity-driven water flows are not the only water movements. There are additional flows driven by wind, density gradients and evapotranspiration, all of which possess annual and diel cycles. While a good deal is known about the mechanisms creating the various flows, it is not (yet?) feasible to synthesize that knowledge into full-scale system behavior. Stochastic variability is superimposed upon all of the flows: e.g., which way the wind blows. Stochastic variability in depth and flow is then propagated into the contaminant processing descriptions. In Chapter 7, methods of quantifying the complex behavior of surface flows will be presented.

Note

1 Honami are the regular marching waves seen in cereal crops on windy days.

4 The Marsh Underground

The subsurface regions of treatment marshes are an active ecosystem that interacts with the free water ecosystem above. This subsurface zone consists of three major components: (1) phytomass, consisting of living (biomass) and dead (necromass) plant roots and rhizomes; (2) soils and sediments; and (3) water. As the knowledgeable reader will know, this subsurface region has traditionally been called "wetland soils." This popular designation conjures an image of an inert nonbiological substrate, and obscures the fact that the subsurface zones are an active ecosystem, in which physical, chemical and biological processes all occur. Those underground processes are all important in the functioning of the total wetland in pollutant removal.

There has been increasing attention to the wetland underground, resulting in books on just this subject, such as Richardson and Vepraskas (eds.) (2001), and Vepraskas and Craft (2016).

In a broad sense, the top layers of substrate are the wetland action zone, comprised of the sediments and soils and housing the roots of vegetation. This action zone is broadly describable as a layer of loose floc and litter resting upon a consolidated soil layer in which the plant roots are situated. These two subzones are often of like thickness, but differ greatly in mobility, bulk density and chemical content. Both layers are opaque: it is not possible to see what maybe happening in either. Both layers are typically regarded as stationary, but in fact both are moving vertically, albeit at speeds that normally go unnoticed. Changes in the action zone storages and processes proceed slowly, with change times measured in months, or more likely, in years.

It is fully recognized that in some special instances, litter, floc and roots may be absent; in which case the reader will interpret accordingly. Several decades of scientific research have shown conclusively that much of the processing of nutrients occurs within the "brown stuff," the soils and sediments that reside below the vegetative canopy and below the water (Figure 4.1). The theme of this chapter is to describe the processes occurring in and near this action zone.

The action zone interacts with emergent plant parts, overlying water, and soils beneath. There is no intent to slight other zones of the wetland, or other processes outside the action zone. Overlying water carries dissolved constituents, and is subject to flow patterns determined by bathymetry, stem density and wind. Indeed, most of the data acquired from P treatment wetlands has to do with flow rates and water chemistry. Emergent plant parts provide functions of transpiration, shading and shelter from wind; and contribute to P uptake. Deep soils may be of importance in wetland treatment where there are sizable infiltration or exfiltration flows.

Plant roots are contained in the action zone, usually in the soil, but sometimes in the floc layer as adventitious roots. Although not viewable, roots undergo the same life cycle as the emergent and submergent plant parts that they support: they grow, die, and their litter decomposes. Importantly, roots extract nutrient and other chemicals from porewater, and may pass transpiration water flows to emergent plant parts.

4.1 WETLAND SOILS IN GENERAL

Treatment marsh soils are hydric soils that fall into two general categories: (1) organic soils or histosols which typically contain more than 20 to 35% organic matter (12–20 % carbon) on a dry-weight basis, and (2) mineral soils that typically have less than 20 to 35% organic

FIGURE 4.1 The action zone for the Houghton Lake wetland. The thickness is approximately 30 cm. The white circles are the cuts through rhizomes. Live roots are the white tendrils. A cut through a dead rhizome is visible at the left center. Fragments of dead leaf litter have fallen across the section just below the live stem bases. There is no floc, because this section is from a floating mat.

matter. However, many natural and constructed wetlands have soil/sediment profiles that include organic and mineral horizons or layers, and also have spatial heterogeneity, so that some physical, chemical, and biological properties of each general soil type may be present.

Organic soils typically have lower bulk density (less than 1 g/cm^3) and have a higher porosity and higher water holding capacity than mineral soils; and consequently, flooded organic soils often hold more water than flooded mineral soils. In spite of higher porosity, organic soils frequently have low hydraulic conductivities indicating that they may impede horizontal or vertical groundwater flows. Thus, organic soils can act as an aquitard to seal a wetland and increase hydroperiod over geologic time.

A floc layer may occur above the first soil horizon in treatment marshes. This non-soil layer consists of newly deposited or partially decomposed plant material and epiphytic debris, and can be easily removed from the soil surface. This floc layer is not considered to be part of the wetland soils but is a source of new soil material after decomposition.

New wetland soils form in response to external and internal loadings of carbon and mineral sediments to the wetland. In closed systems without inputs and outputs, new sediment formation is dependent upon the net difference between carbon fixation by plants and the rate of carbon degradation. In treatment marshes, the carbon cycle generation of new sediments and soils is augmented by sedimentation of incoming mineral solids and carbon accretion.

Wetland Mineral Soils

The definition of hydric soils indicates that any upland soil utilized for construction of a wetland treatment system will become a hydric soil following a short to long period of flooding and continuous anaerobiosis. Since many treatment marshes are constructed in areas that were formerly uplands, they begin existence dominated by mineral soils. As constructed wetland treatment systems mature, the percent of organic matter in the soil generally increases.

Mineral soils are classified by particle size distributions, color, depth, and a number of other factors. The three major mineral soil classes are clays, silts, and sands. Clays are soils with very fine particles packed closely together. In some clay soils, these fine minerals become organized into crystalline lattices. Because of their very fine texture and low hydraulic conductivity, clays may function as aquitards.

Like clays, silts may also reduce the flow of water due to low hydraulic conductivities. However, silts have larger particle sizes than clays, and do not typically exhibit the crystalline properties of some clays. Loamy soils consist of mixtures of clays, silts, and sands and typically have excellent plant growth characteristics because of adequate nutrient holding capacity and higher hydraulic conductivities.

Sandy soils are made of larger mineral particles, typically with a preponderance of quartz crystals that are chemically inert compared to clays. Thus sandy soils are less likely to be able to bind chemical nutrients important for plant nutrition. The hydraulic conductivity of sandy soils may be quite high, depending on the percentage of silts and clays in the soil.

After startup of a new marsh on mineral soil, the soils can undergo a process of gleying. Gleying is the process of waterlogging and creation of reducing conditions in soils. In waterlogged soils, water replaces air in pores, oxygen is quickly used up by microbes feeding on soil organic matter, which creates anaerobic conditions. Organic deposits can speed the development of anaerobic conditions, and waterlogging slows organic decay processes, aiding the survival of organic matter.

Calcitic Muds and Marl

Wetland soils that form in high-calcium environments can have relatively little carbon in relation to calcium. Such wetlands develop with conditions of short hydroperiod (the annual period of inundation). Dry periods promote the oxidation of organic detritus, and thus inhibit the formation of peat. Under flooded conditions, an ecosystem may develop that contains dense algal populations that can quickly utilize nutrients. Given a supply of dissolved calcium, the algae can create calcium carbonate precipitates. In combination with silt, these solids become calcitic muds (Gleason and Stone, 1994).

If the calcium carbonate precipitates are mixed with sands and clay, the result is marl, a substrate in the alkaline fens of northern climates, as well as the Everglades marshes of south Florida. These are fine-grained tan or whitish materials, which may be hard or soft depending on exact composition and conditions of deposition. Such biogeochemistry is also a feature of SAV treatment marshes, which contain a complex mixture of algae, bacteria, microbes, and detritus that is attached to submerged surfaces. Thus calcitic additions to SAV marsh sediments occur despite long hydroperiods.

Wetland Organic Soils

Organic soils, known as histosols, occur in wetland environments where the rate of organic matter formation is greater than the rate of decomposition. This condition is most likely to occur in areas with relatively low dissolved oxygen and high carbon fixation rates by herbaceous emergent macrophytes and/or algae.

Organic soils may be classified by their extent of decomposition, and may be called peat, muck, or mucky peat. Organic soils that have the least amount of decomposition are called fibrists (peat). Fibric peats have more than two thirds of the plant fibers still identifiable. Saprists or mucks have greater than two-thirds of the original plant materials decomposed, and hemists are in between saprists and fibrists.

Bulk density of organic soils tends to increase with decomposition. Slightly decomposed organic soils have larger pore spaces and higher rates of saturated water movement compared to well-decomposed soils, which may have hydraulic conductivity rates lower than clay soils. Even a slight degree of unsaturation creates an extremely large capillary suction pressure in the soil micropores. This means that organic soils and sediments are virtually undrainable; they retain a very high percentage of water.

Drying of organic soils promotes oxidation and gasifies carbon, but not the mineral nutrients associated with those soils. There are large amounts of nitrogen, phosphorus, sulfur and other mineral constituents organically bound in unavailable forms. Oxidation destroys these recalcitrant organics, and releases the associated substances. Upon reflooding, those substances can dissolve, and provide relatively high concentrations of nutrients and other dissolved minerals.

4.2 COMPONENTS AND STRATIFICATION

All three components of the marsh underground are interspersed in the horizontal plane, but with differing vertical distributions. Naming of these vertical strata differs depending upon which branch of wetland science is reporting (Figure 4.2). Soil scientists consider various soil *horizons*, as follows (Reddy and Delaune, 2008):

- O Horizon. The surface organic layer, dominated by fresh or partly decomposed organic material, including plant detrital matter. The mineral content is small.
- A Horizon. This layer is characterized by an accumulation of humified organic matter, perhaps with a significant mineral content.
- B Horizon. Soil layer that has accreted minerals and humus from above, by water percolation or vertical soil movement.
- C Horizon. Consolidated material that is unaffected by the layers above.

Peatland scientists are conditioned to think in terms of two general zones, the acrotelm (meaning highest portion of the marsh), and the catotelm (meaning lower portion of the marsh). This terminology, of Greek etymology, was specifically defined by Ingram (1978):

The acrotelm:

- Contains the oscillating water table (if dryout occurs) and variable water content (if dryout occurs).
- Has high hydraulic conductivity.
- Is rich in peat-forming aerobic bacteria and other microorganisms.
- Has a live matrix of growing plant material.

The catotelm

- Has a water content invariable with time.
- Has a negligibly small hydraulic conductivity.
- Is anoxic or anaerobic.
- Is devoid of peat-forming aerobic microorganisms and poor in microbes in general.

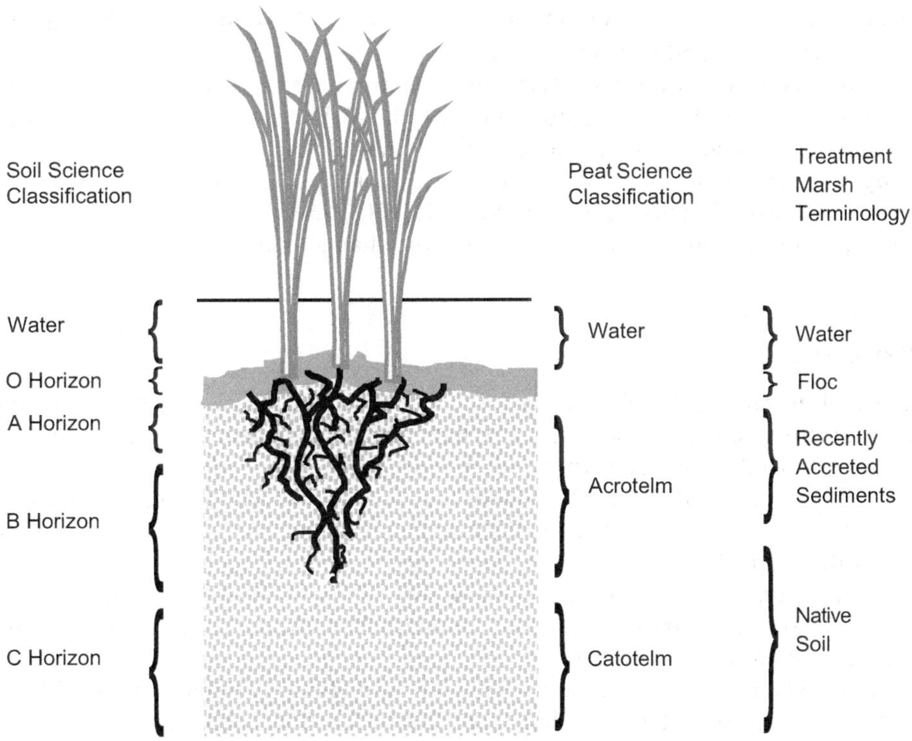

FIGURE 4.2 Three stratification classifications for underground wetland regions.

The boundary between the two layers is defined by the transition from peat containing living plants to peat containing dead plant material.

The von Post scale is a semi-quantitative scale that is used world-wide to describe the degree of humification of peats. Treatment marshes vertical sequences follow much the same pattern as in natural peatlands, prior to or subsequent to operational stability. This scale ranges from H1 to H10, described in part by these definitions (Verry et al., 2011):

- H1. Completely undecomposed peat which, when squeezed, releases almost clear water. Plant remains easily identifiable. No amorphous material present. This category is generally the fibric class.
- H5. Moderately decomposed peat, which, when squeezed, releases very "muddy" water with a very small amount of amorphous granular peat escaping between the fingers. The structure of the plant remains is quite indistinct although it is still possible to recognize certain features. The residue is very pasty. This category is generally the hemic class.
- H10. Completely decomposed peat with no discernible plant structure. When squeezed, all the wet peat escapes between the fingers. This category is generally the sapric class.

Wetland science practitioners in Florida have suggested a layer terminology that is based upon perceived timing or origin and physical characteristics:

- Floc is the uppermost layer, and is comprised of decaying macrophyte tissue, soil particles, algae, and microbes (DeBusk et al., 2001; Corstanje et al., 2006a). It has low solids content, and is therefore "pourable."
- Recently accreted sediments (RAS) are the topmost consolidated layer, that is, are not pourable. The lower boundary of the RAS is defined both visually, by different color and texture; and chemically via determination of change points along the vertical profile. Bhomia et al. (2015) reported change points for four variables: soil bulk density, TP, δ15N, and δ13C.
- Native, or antecedent, soils are those below the change point depth.

Root Phytomass

Roots form a nonnegligible fraction of the action zone. Root phytomass is typically comprised of three component parts: (1) live rhizomes (biomass), (2) live roots (biomass), and (3) dead roots and rhizomes (necromass). The entire belowground marsh phytomass often weighs 1,000–5,000 gdw/m^2 (grams dry weight per square meter). That total is usually in the top 25 cm of soil. Water content is high, around 90% of the wet weight is water and 10% is dry solids. Thus, the volume occupied by the root mass is up to 50,000 grams wet per meter squared, or 20% of the action zone volume. The action zone is therefore mostly wet soil. Nonetheless, those roots perform important functions in contaminant processing.

Roots and rhizomes are continually growing and dying. Something like 25% of the total root mass is dead material, which upon decomposition is converted to soil (Figures 4.3 and 4.4).

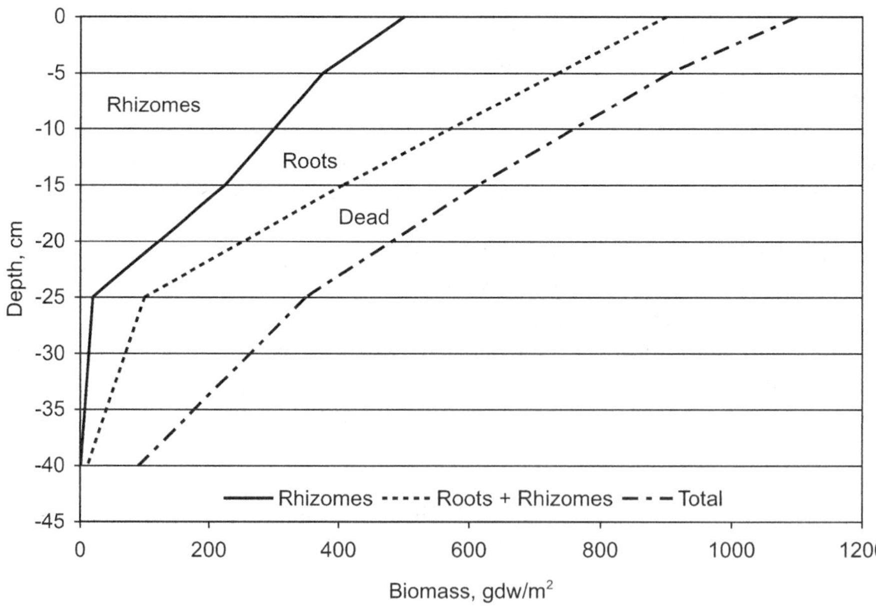

FIGURE 4.3 Vertical distribution of belowground biomass of *Typha angustifolia*. Total includes rhizomes, roots, and dead roots and rhizomes. (Based on data from Gross et al., 1993.)

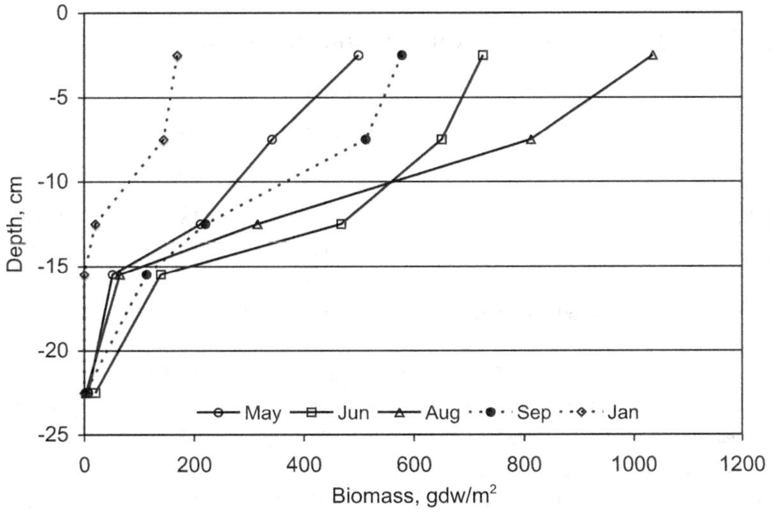

FIGURE 4.4 Depth profiles of live roots plus rhizomes for *Sparganium erectum* versus season. The site was in the southern United Kingdom. (Based on data from Liffen et al., 2013.)

Vertical Distributions

Roots in marshes generally display an exponentially downward decreasing density. This may be described by:

$$M = M_T\left(1 - e^{-z/b}\right) \tag{4.1}$$

where: M = root mass above depth z, gdw/m^2
M_T = total root mass, gdw/m^2
b = depth at which 63.2% is above, cm
z = depth below soil surface, cm

Examples of the fitting parameters are shown in Table 4.1. It is noteworthy that the exponential model has a very high R^2, averaging 0.98. On average for these examples, 80% of the root mass is located at depths less than 17 cm. As a consequence, the effects of roots on underground chemistry are most prevalent in the uppermost underground zones.

Floc and Litter

Treatment marsh investigators have often observed a detrital layer, consisting of fibrous dead plant parts, or "litter"; and flocculent material or "floc", overlaying the soil surface. However, there is no standard (e.g., bulk density or particle size fraction) for defining the layer, or for distinguishing the boundary between underlying organic soils and the litter and floc.

Plant Litter

Plants in a treatment marsh undergo a cycle of growth and death, thus producing standing dead necromass, which eventually falls to the bottom of the water column. This cycle will be more fully described in Chapter 5. The breakdown of dead plant material forms a gradient between larger, partially decomposed plant fragments above the soil surface and

TABLE 4.1

Root mass and depth distribution for several treatment wetland species.

		Approximate Root Mass gdw/m²	Depth, cm 80% Above	R^2 Exponential	Reference
Phragmites australis			18	0.996	1
Phragmites australis		1,300	14	0.993	2
Typha angustifolia		1,500	15	0.998	3
Typha Latifolia	High Loading	2,500	21	-	4
Typha spp., *Scirpus* spp.		875	17.8	0.982	5
Schoenoplectus tabernaemontani	Light Loading	1,040	20	0.984	6
Schoenoplectus tabernaemontani	Moderate Loading	1,030	16	0.987	6
Schoenoplectus tabernaemontani	High Loading	1,110	15	0.998	6
Scirpus californicus	High Loading	5,455	13	0.996	7
Baumea articulata		3,960	22	0.972	8
Cladium jamaicense	Light Loading		17	-	9
Carex spp., *Juncus* spp.		2,820	15.5	0.917	5
Sparganium		2,050	15	0.994	10
Ceratophyllum spp., *Potamogeton* spp.		285	16.8	0.989	5
Mean			**16.9**	**0.984**	

References
1 Daniels and Parr (1990)
2 Kuusemets et al. (2002)
3 Gross et al. (1993)
4 Unpublished data Houghton Lake, MI
5 Unpublished data Minnesota
6 Tanner (2001)
7 Nolte and Associates (1998b)
8 Adcock and Ganf (1994)
9 Pumo et al. (2010) 10 Liffen et al. (2013)

smaller highly decomposed fragments that have become incorporated into the organic soil below. Under continuous flooding, plant fragments can persist for many months or years. Figure 4.5 shows this litter layer for a Typha treatment marsh (the Houghton Lake system).

The mass of the litter ranges from near-absence in SAV systems, to amounts comparable to the aboveground biomass. For example, at the end of the growing season at Houghton Lake, erect biomass was 2,600 gdw/m^2, while the litter layer was 2,900 gdw/m^2. On a mass basis, the litter is mostly water (e.g., 90% at Houghton Lake). It occupies about 10–20 cm of vertical depth. Most of this necromass is ultimately lost to decomposition, but residuals add to the buildup of new soils.

Floc

The fine material in the floc layer is a watery matrix of dissolved and particulate organic matter. In areas of high nutrient enrichment, a soupy floc layer may be present, containing

a
b

FIGURE 4.5 The litter layer at the Houghton Lake treatment marsh. The frame marks out a 0.25 m^2 area, from which erect live and dead plants have been removed. (A) before litter removal, (B) after litter removal.

plant fragments in a matrix of decomposing algae and other decayed organic matter, that grades to a more "consolidated" soil with no clear boundary between floc and soil.

The effects of elevated nutrient waters on existing wetland sediment/soil profiles has been observed to be development of a low density, mushy material occupying the "water" column (e.g., Kadlec and Bevis, 1990).

The particulate material in the floc is situated in and on the pronate litter, and is considerably greater than the suspended material above it in the water column. The accumulation of several years' sedimentation, less decomposition, form this unconsolidated layer. There is not an accepted common terminology for the floc. Nolte and Associates (1997) called it the "A layer," and described it as follows:

> The A layer consists of a slurry of dark, decomposing, loosely structured detrital material that pours out when the sampler is tipped. The material in the A layer has settled to the bottom, but has not been integrated into the matrix of the basin floor.

Ad hoc methods have been developed for measuring the amount of the floc. At Houghton Lake, replicate plots were isolated and harvested for suspendable material. A stainless steel cylinder, with a sharpened bottom edge and with handles attached to the top edge, was twisted down into the underlying soil, thus isolating 0.111 m^2 of surface water, vegetation and suspendable material. Emergent vegetation was clipped and discarded along with plant litter so that the sediments could be stirred. The water depth was measured at least three locations. The water was stirred gently, to suspend solids, including phytoplankton, but not violently enough to detach periphyton. A water sample was then immediately taken. A subsample was returned to the laboratory, filtered, oven dried at 80°C and weighed. The amount of floc was then scaled to the entire cylinder area. The average floc amount in the discharge zone was 8,000 gdw/m^2. The amounts suspended by this stirring technique were not expected to be the same as those that are susceptible to entrainment by natural processes, however, it was an indicator of the relative amounts of suspendable material.

Studies of the Florida STAs have employed a visual technique for distinguishing the floc, which takes advantage of the difference in coloration of floc and soil. This visual distinction was aided by "pourability" of the floc material, which could be decanted from the underlying soil. Despite the low bulk density of the floc, substantial amounts of solids material are involved (Table 4.2). That is because the floc depth in those treatment marshes is typically about 10–12 cm. In those marshes, dry floc solids exceed the amount of dry material to be found in phytomass.

There can also be a vertical gradient in the density of the floc material, with increasing solids with depth. Data were acquired at the Houghton Lake treatment marsh by freezing the water column around a pipe filed with a cryogenic mixture of dry ice and acetone (T = –80°C). The ice annulus was then sliced into sections, and analyzed for solids content. A linear profile was found:

$$\begin{aligned} \rho_B &= 0.0155 + 0.0019z \\ R^2 &= 0.84 \end{aligned} \tag{4.2}$$

where: ρ_B = bulk density, gdw/cm^3
z = depth below water surface, cm

TABLE 4.2
Floc amounts found in various treatment marshes.

		Bulk Density	Floc Depth	Floc Mass
		gdw/cm3	cm	gdw/m2
STA1W	EAV	0.085	18.4	15,489
STA2	EAV	0.054	5.2	2,821
STA2	SAV	0.113	5.1	5,700
STA5	EAV	0.071	8.9	6,085
STA6	EAV	0.045	8.3	3,718
ENRP	EAV	0.082	18.9	15,588
ENRP	SAV	0.092	16.7	15,364
Sacramento	EAV	0.069	14.3	9,761
Apopka	EAV	0.051	33.0	16,830
Houghton Lake	EAV	0.030	20.0	8,109
Mean		0.069	14.9	9,947

Source: Florida STAs and ENRP, unpublished data; Sacramento, Nolte and Associates (1997); Apopka, Coveney et al. (2002); Houghton Lake, unpublished data.

The floc solids bulk density was about tripled as the depth increased downward to 17 cm, the total water depth at that location.

In contrast to the other components of the wetland underground, floc is mobile. It can be suspended and transported with water, due to stirring events. Suspension can occur due to temporarily high water velocities or bioturbation.

Antecedent Organic Soils in Agriculture

The soil substrates for treatment marshes are often altered peatlands that have been in agriculture. Dawson et al. (2010) note:

> Drainage and tillage operations for any form of agricultural activity increase soil aeration and exacerbate degradation by providing optimal environmental conditions for aerobic microbial mineralization of soil organic matter. Indeed, the initial biochemical oxidation of peat soil for agricultural exploitation is a prerequisite for successful agricultural activity, as it facilitates the release of soil organic nutrients that are essential for crop growth. Such continued degradation does, however, significantly alter the soil texture and structure causing dramatic changes in soil hydrology.

Treatment marsh substrates are often drained peats, which have undergone subsidence and oxidation. Such soils are generally referred to as mucks in the United States, and they generally fit into a modified sapric category. Across the United States, expansive areas that once supported the high-fiber "sawgrass peats" in the early-1900s have since decomposed into low-fiber "mucks." The upper layers are most altered by agricultural drainage, and soil subsidence can be considerable. Due to drainage and subsequent cropping, organic decomposition exceeds production, resulting in soil subsidence and lowering of the surface elevations. Losses of one to two meters of organic soil are not uncommon. Such subsidence makes the lands difficult to drain, thus impairing their agricultural usefulness.

Agricultural organic soils are classified as Histosols, suborder Saprists (from Greek *saprose*, rotten; implying a high degree of decomposition). Saprists are predominately black in color, with bulk densities generally exceeding 0.20 gm/cm^3 and saturated water contents on the order of 80%. These soils are *Haplosaprists* (from Greek haplous, simple; implying minimal evidence of true soil horizon development).

The taxonomy of organic agricultural soils is quite different in Europe and Scandinavia. For instance, 90% of Poland's peatlands have been drained, which inspired their peat scientists to develop classifications for agriculturally drained peats, The European term for agricultural alteration of peatlands is *moorshing*. This term is explained by Gnatowski et al. (2010) as:

> Moorshing of peat soils comprises biological, chemical and physical changes caused by a decrease in the water content and, consequently, by aeration of the soil profile. These processes lead to the formation in the top layers of a new material called moorsh.

Treatment Marsh Soils Development

Treatment marsh soils follow a time sequence of alteration from antecedent soils. Antecedent soils are those prior to marsh establishment, if a constructed system; or those prior to the inception of source water additions, if an existing wetland. They are replaced by soil materials generated within the wetland by treatment operations. This soil alteration proceeds in two ways: (1) alteration of existing soils, via overflooding and the action of plant roots on their environment, and (2) replacement, by accretion of new material within and on top of the existing soils, and. Accretion is addressed in a later section of this chapter. Replacement typically takes many years to accomplish for the upper 20 cm of the soil column.

However, changes in the action zone can happen on a much shorter time scale. Obviously the flooding of previous upland soils quickly changes the oxygen availability in the topsoil layers. The long hydroperiods of treatment wetlands are conducive to the buildup of organics: first litter and micro-detritus, then the sediments formed from their decomposition, and finally the organic soils generated from those sediments and deposited mineral solids.

If the treatment marsh is established on former upland, then there will be a period of root development for the aquatic plants. This occurs for the situation of planted systems and also for systems that are allowed to develop via natural ingrowth. This is a relatively rapid process, because wetland plant root systems grow at high rates. Studies suggest that roots and rhizomes will fill their desired niche in well under one year's time (see for instance Hussner et al., 2009; Miller and Fujii, 2010; Headley and Tanner, 2012; Leto et al., 2013). Once roots are established, they function to alter the chemistry of the soil column, primarily by alteration of the redox potential, as will be discussed subsequently.

In short, the wetland rearranges itself to accommodate the environment created by the addition of source waters. The functioning of the wetland after such adaptation is no longer dependent upon the previous condition and type of soils, hydrology, and biota. It is totally dependent upon the new soils, hydrology, and biota. It is this new sustainable mode of wetland operation that is the target of most designs.

Thus the final state of a treatment wetland, and the accompanying suite of water quality functions, are largely independent of the initial condition of the real estate upon which it is built. During the interim period of adaptation, antecedent conditions are important, because they dictate the short-term performance of the wetland. That period of adaptation appears to extend for up to two years for newly constructed wetlands, and for alteration of natural wetlands to a treatment function.

Soils Physical Properties

The quantification of soil components and their vertical distributions requires a few definitions.

Porosity or Water Content

$$\theta = \frac{V_W}{V_T} \tag{4.3}$$

where: V_S = solids volume, cm^3

V_W = water volume, cm^3

V_T = total volume of wet soil, = $V_S + V_W$, cm^3

θ = porosity, cm^3/cm^3

Porosity is an ambiguous quantity for organic treatment marsh soils, because there is no crisp physical boundary for the soil particles. Rather, water is bound to and interspersed with organics with ill-defined particle boundaries. Further, the tightness of the bond between water and the organic solids varies considerably. For mineral solids, water is freely drainable, except for a residual held by surface tension in interparticle crevices. For organic solids, there are binding forces ranging from chemical bonds to surface tension. Binding strengths range from simple porewater retention in large pores to chemically bound water, attached to hydration sites on the molecules. As a result, organic marsh soils are not gravity drainable.

This water-holding capacity of organic sedimentary materials and soils results from the hydration of the complex organic molecules that form these solids. A single monolayer of water on humic or fulvic acid components of an organic soil requires a water content of about 5% by weight (Schnitzer, 1986). As a result, organic treatment marsh soils cannot be completely air-dried, because the more tightly bound water cannot evaporate. An air-dry organic soil still retains about 30% water by weight.

As a consequence, air-dried organic soils contain up to 50% water, and still are not totally devoid of water if dried at 80°C. Further, the volume occupied by solids is not an easily measurable quantity. As a result, researchers most often turn to bulk density, the mass of "dry" solids per unit wet volume.

Bulk Density

This is an easier laboratory measure, derived from the dimensions of a core slice and its dry weight of solids (usually taken to be done at 80°C):

$$\rho_B = \frac{M_s}{V_T} \tag{4.4}$$

where: M_S = mass of soil, gdw
M_W = mass of water, g
M_T = total mass, = $M_S + M_W$, g
ρ_B = bulk density, gdw/cm^3

Mass Fraction Water

This is determined from the initial and final weights of a core slice after drying.

$$F_w = \frac{M_w}{M_T} \tag{4.5}$$

Mineral (Ash) Content

To remove the organics from a sample, it is heated at 550°C to a constant weight (ignition), which theoretically burns off all organic constituents.

$$F_{ash} = \frac{M_A}{M_T} = 1 - LOI \tag{4.6}$$

where: M_A = mass of mineral soil (ash), gdw
F_{ash} = fraction of solids that are mineral, –
LOI = fractional loss on ignition, –

In the field setting, decomposition processes also remove organic material, converting it to carbon dioxide and methane. However, those biochemical processes are less destructive than ignition, and typically leave a residual of ash plus refractory organics. The refractory organics have origins as the lignin component of the biomass from which the soil formed (Figure 4.6).

Marsh soils are comprised of layers, with differing physical properties, and consequently there are vertical profiles of bulk density, water and ash content. These result partly from the loss of light organic material (cellulose) along the vertical profile, and partly from the accretion of materials of composition dictated by the new source waters. The examples in Table 4.3 illustrate the variability due to ecosystem type and operational longevity.

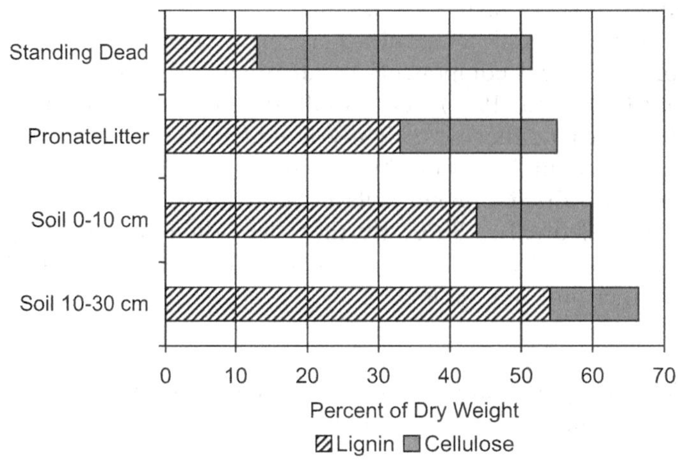

FIGURE 4.6 Lignin and cellulose content of action zone layers for the WCA2A treatment marsh. (Based on data from DeBusk and Reddy, 1998.)

TABLE 4.3

Example vertical profiles of the constituents of treatment marsh soils. WCA2A had been in operation for over thirty years, while the ENRP cells were in their first and second years. Shown for comparison are properties of the marl soils of Everglades National Park (ENP) marshes.

	Depth	Bulk Density	Water	Carbon	Ash	Ca	Al	Fe
	cm	g/cm^3	% Volume	% dw	% dw	g/kg	g/kg	g/kg
WCA2A	0–10	0.064	90.0	42.4	15.4	37.6	0.80	0.43
EAV	10–20	0.075	88.1	45.8	13.5	29.5	1.00	0.28
	20–30	0.067	89.7	47.6	13.1	27.4	0.80	0.37
	30–40	0.053	91.4	49.9	9.3	22.1	0.57	0.29
ENRPC1	0–5	0.126	88.9	49.8	13.9	30.7	1.67	1.24
EAV	5–10	0.203	80.7	51.0	12.3	37.0	1.61	1.06
	10–30	0.240	76.8	51.8	11.2	28.9	1.52	1.53
ENRPC4	0–5	0.213	79.8	51.3	11.6	36.3	1.50	2.35
SAV	5–10	0.265	74.8	52.1	10.4	33.5	1.50	2.38
	10–30	0.258	75.5	53.8	10.2	33.0	1.50	2.48
ENP	Floc	0.120	92.0	23.0	64.1	241	3.48	5.87
EAV	0–10	0.250	83.3	27.3	52.3	162	10.10	10.20
	10–20	0.300	80.0	28.9	49.3	137	12.40	10.40

Source: WCA2A, Reddy et al. (1993); ENRP, unpublished data SFWMD; ENP, Osborne et al. (2011).

4.3 WATER MOVEMENT

The water phase of the action zone is not static but rather exhibits movement, both vertically and horizontally. In contrast to the free water above, water movement in the wetland soil is driven by suction pressures as well as gravity (Freeze and Cherry, 1979).

Saturated Hydraulic Conductivity

For fully saturated soils (i.e., with standing surface water above), underground gravity-driven flows are presumed to follow Darcy's law; for which water velocity is proportional to the hydraulic head:

$$u = k\,S \tag{4.7}$$

where k = hydraulic conductivity, m/d
u = superficial velocity, m/d
S = – dH/dx = negative of the piezometric water slope, m/m

Hydraulic conductivities in treatment marsh soils have received only scant attention, and thus it is necessary to infer information from peat and soil science, which have produced large amounts of information. In general terms, the hydraulic conductivity of the floc layer is very high, but rapidly diminishes with depth as the soils become better decomposed. When a bulk density of 0.20 gdw/cm^3 is reached, the organic soil is essentially impervious

Table 4.4 illustrates this strong vertical gradient in hydraulic conductivity for organic soils, and compares a few classes of mineral soils.

TABLE 4.4

Saturated hydraulic conductivities of soils. Porosity is also the saturated water content.

		ρ_B	θ_P	θ_{res}	θ-θ_{res}	k	
	von Post	Bulk Density	Porosity	Undrainable	Drainable	Conductivity	Lateral Drainage Extent
	Scale	gdw/cm^3	%	%	%	cm/d	m
Peat							
Fibric	1	0.04	0.94	0.34	0.60	18,317	105
Fibric	2	0.06	0.93	0.59	0.34	7,690	82
Fibric	3	0.08	0.91	0.62	0.29	4,170	65
Hemic	4	0.09	0.90	0.67	0.23	2,160	52
Hemic	4.5	0.10	0.90	0.70	0.20	1,296	44
Hemic	5	0.11	0.90	0.72	0.18	788	36
Hemic	5.5	0.12	0.89	0.73	0.16	409	27
Hemic	6	0.13	0.89	0.76	0.13	215	22
Hemic	6.5	0.14	0.88	0.76	0.12	86	14
Hemic	7	0.15	0.87	0.75	0.12	35	9
Hemic	7.5	0.16	0.86	0.75	0.11	26	8
Hemic	8	0.17	0.86	0.76	0.10	17	7
Sapric	9	0.19	0.85	0.76	0.09	11	6
Sapric	10	0.20	0.82	0.74	0.08	8	6
Mineral Soil							
Sand		1.53	0.38	0.05	0.32	646	
Silt		1.33	0.49	0.05	0.44	44	
Clay		1.39	0.46	0.10	0.36	15	
Loam		1.37	0.40	0.06	0.34	12	

Source: Peat, Verry et al. (2011); Mineral Soil, Twarakavi et al. (2009).

Unsaturated Suction Flows

Several source water addition strategies call for periods of no inflow, during which water levels will recede. When water levels in a treatment marsh recede enough to expose areas of higher ground, underground water flows come into play. This phenomenon is often called "dryout," but marsh soils are not dry at such low stages. The typical meaning of dryout in the marsh context is that the mean water stage recedes to the level of the mean ground elevation. However, the concept of a flat bottom, with uniform depth water standing above it, is not at all accurate. In that erroneous context, "dryout" occurs when the (uniformly flat) water level recedes down to the (uniformly flat) ground surface. This concept is flawed for three very important reasons: (1) the wetland "soils" are not dry when the water table recedes down to the surface, (2) the upper substrate horizons are vertically layered, with large variations in water holding and conveyance properties, and (3) the topography is not uniform or flat. When the water table is below ground, the driving force for both vertical and horizontal water movement is the height of water plus the (negative) suction head (Freeze and Cherry, 1979).

Soils in treatment marshes are often peats of some description; and sometimes sands or clays. Peats are well known to exhibit very large capillary suctions if dried even a slight amount. Sands are also susceptible to large capillary suctions, but have a lesser water content.

The fractional saturation of a porous medium is a normalized water content, and is given by:

$$f = \frac{(\theta - \theta_{res})}{(\theta_P - \theta_{res})} \tag{4.8}$$

where θ = fraction of total volume occupied by water

θ_{res} = residual fraction of total volume occupied by water

θ_p = porosity of the substrate

f = fractional saturation

The residual water is so tightly bound that it does not move.

Only a slight amount of drying causes organic substrates and sands to exert considerable suction. The van Genuchten equation for capillary pressure is often used to compute the pressure exerted by a substrate at partial saturation (van Genuchten, 1980; Twarakavi et al., 2009):

$$\psi = \frac{1}{\alpha}\left(f^{\left(-\frac{1}{m}\right)} - 1\right)^{\left(\frac{1}{n}\right)} \tag{4.9}$$

where α = scaling parameter, cm^{-1}

m= parameter, =1–1/n

n= parameter

ψ = suction head, cm

Equation (4.9) has been parameterized for drained agricultural peatlands in Poland (Gnatowski et al., 2010) and in the United Kingdom (Kechavarzi et al., 2010). Typical values are: fibric: α = 0.08, n = 1.9, hemic: α = 0.02, n = 1.7, sapric: α = 0.003, n = 1.6, moorsh: α = 0.02, n = 2.17, sand: α = 0.125, n = 1.3. There has been no parameterization of equation (4.9) for treatment marshes.

Only a slight amount of drying causes organic substrates and sands to exert considerable suction (Grover and Baldock, 2013). For instance, a well-decomposed organic soil (sapric) at half saturation (f = 0.5) exerts one meter of suction head (Figure 4.7).

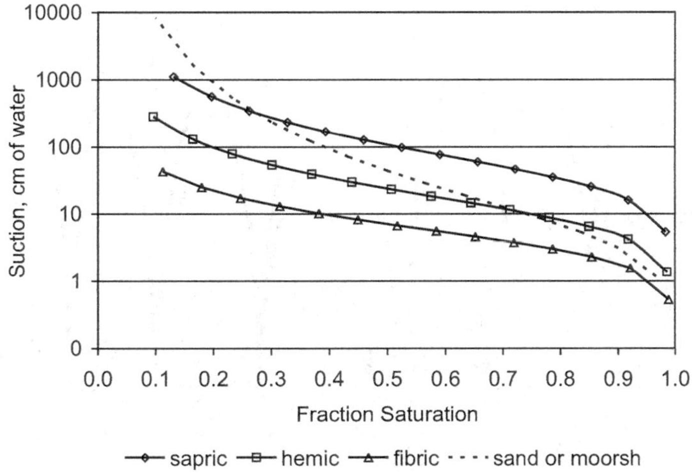

FIGURE 4.7 Suction heads created by partial drying of organic and sandy soils. For a treatment marsh, the top of the action zone corresponds to the fibric category, while the bottom of the action zone corresponds to the sapric category.

The vertical structure of the substrates in treatment marshes includes a litter floc layer overlying a more consolidated substrate, usually peat, and occasionally sand. The peats are of the same character as the better decomposed organic soils studied elsewhere, and termed sapric. The litter floc layer is likely quite closely related to the fibric layer in other situations. Therefore, the firm underlayer in the STAs is expected to exert considerably more suction than the litter floc layer. The later will exert only about ten centimeters suction under medium dry conditions, while the former will exert more than a meter of suction head.

The elevation differentials in typical treatment marshes are on the order of 30 cm, partly due to longitudinal gradient and partly due to local irregularities. Water suctions in the firm base soils would counteract the hydraulic tendency of the water table to recede. However, suctions in the floc litter layer are not likely to be able to maintain wetness under drying conditions. Therefore, local regions of floc dryout are expected for high spots in the elevation, but those high spots would still retain considerable moisture in the base soils. In effect, there is a drainable surface layer over a nondrainable substrate.

Thus the action zone behaves like a thin sponge, covering the entire marsh surface, and resisting drainage. If water is removed via evapotranspiration, the sponge blanket will draw in replacement water, either from any standing water pools or from groundwater below the action zone (Figure 4.8). The result of this suction water movement is that the root zone continues to have access to water, even if the surrounding water table is low.

Peatland drainage has been extensively studied, in connection with water removal for peat harvesting (Verry et al., 2011). Those reaches have shown the lateral extent of drainage that may be expected in organic soils of different degrees of decomposition, due to the construction of drainage ditches. Such drainage ditches are analogous to the deep zones that often occur in treatment marshes, at inlets, outlets and possibly intermediate locations. Table 4.4 lists the lateral extent of drainage to a ditch expected for different degrees of decomposition.

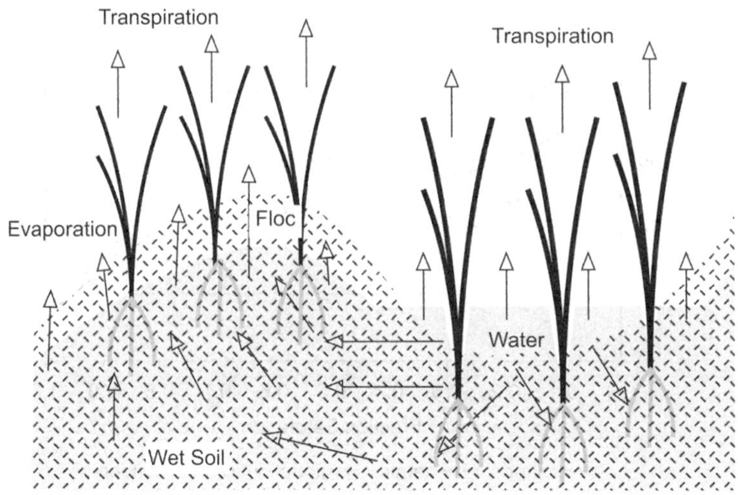

FIGURE 4.8 Suction water movement in the action zone, to satisfy evapotranspiration losses. Note that the suction from wetland soils causes water to rise to levels above the nearby free water stage.

Flows Driven by Evapotranspiration

Systems with Standing Water

When there is surface water present, as there would be during periods of source water addition, evaporation occurs from that surface water. Additionally, water is transpired from the plants, and that water is drawn from the root zone (see Chapter 2, Figure 2.13). Plants transpire water through their above-water parts, but that water is drawn from the root zone. The root water uptake at any depth is generally considered directly proportional to the root density and the maximum potential evapotranspiration (Pumo et al., 2010). Thus, about 60% of the transpired water moves through the upper 10 cm of the soil root profile. Only a fraction of the soil volume is comprised of movable water, perhaps 80%. In a warm climate, transpiration in a fully vegetated wetland is about 0.4 cm/d. Consequently, the upper 10 cm of the soil profile is completely flushed about every 20 days. In temperate climates, the rate is comparable, but transpiration only functions in the warm growing season, during which the porewater is flushed at the same rate as in warm climates.

Detailed flow models, describing the vertical transpiration-driven flow of porewater, have been developed for terrestrial agricultural crops (Schoups and Hopmans, 2002; Zhu et al., 2010b). These utilize porous media flow equations with spatially dependent withdrawals of water to satisfy transpiration. Withdrawals are proportioned to the root density at a particular vertical position. This same concept has been carried over to the wetland environment. A downward-decreasing wetland root density is presumed to result in a downward-decreasing water uptake to supply transpiration (Laio et al., 2009; Šimůnek and Hopmans, 2009; Pumo et al., 2010). Therefore, an exponentially decreasing root mass with depth results in an exponentially decreasing lateral removal of transpiration water to the roots. More detailed models of root water uptake have been proposed for terrestrial systems (e.g., Vrugt et al., 2001), but have not been adopted for wetlands.

Root-proportional transpiration flows are employed in the HYDRUS model commonly used for wetlands (Šimůnek et al., 2012). HYDRUS presumes that transpiration is a root uptake function, which operates throughout the rooting depth, with variability in the root density. Thus, withdrawals are determined as function of depth, driven by the overall areal transpiration rate.

Floating mat and floating plant systems differ somewhat from emergent marshes with respect to replenishment of ET losses from the action zone, because the root/soil action zone is replenished from below rather than from above. These systems are distinguished by the fact that replacement water travels through the root zone with ever-increasing transpiration withdrawals as it moves upward. This is in contrast to emergent systems, which show a decreasing movement as the water moves downward.

"Dry" Systems

The upper portion of the firm layer is the site of the roots for an emergent system. Plants take up water from this zone of soil water. In general, transpiration is not impaired until about half the available water is depleted in organic soils. That situation corresponds to a period of extreme water shortage in a treatment marsh.

Evapotranspiration does not shut off when surface water disappears. Capillary suction pressure keeps the surface wet, and solar energy continues to drive both transpiration and some amount of evaporation.

A study of ET in a Florida peatland with a subsurface water table was conducted by Dolan et al., 1984. The wetland contained primarily emergent aquatic macrophytes. The marsh substrate consisted of an ~1.5-m thick layer of highly organic (88%) Brighton peat soil. This peat layer is underlain by a zone of saturated coarse sand. During this study, the water table remained at an average depth of 0.2 m below the surface of the peat soil throughout. Over the 13-month study, the ratio of ET/PET was 1.00.

The current practice in computer modeling of ET for the condition of no standing water centers on capillary driving forces with root extraction. When the water table is deeper than the root zone, water extracted by the roots from the unsaturated zone is replenished by upward flow from the water table. This flow is modeled using the van Genuchten parameters. Water is sucked into the root zone, and lost in response to the energy balance. Shah et al. (2007) further examined the HYDRUS model, in terms of the effect of depth to the water table. The depth over which there is no effect on ET/PET ranged from 16–39 cm, depending upon vegetation cover and soil type. The lower values are for no vegetation and sandy soil.

4.4 CHEMISTRY

The action zone of a treatment marsh is the location of numerous chemical reactions, which have differing effects on the wetland's treatment potential. Most of these involve the soil organic material, and are microbially mediated. Wetland plants also have an active role in the subsurface chemistry. Wetland biogeochemistry is the subject of entire books, notably Reddy and Delaune (2008) and DeLaune et al. (Eds.) (2013). While these and other references provide a wealth of information about wetland soil processes, it has not yet been possible to completely bring that knowledge to bear directly upon the quantification of treatment marsh pollutant reduction. A brief summary is presented here, to dispel the notion that the wetland action zone is merely a physical support for surface processes.

Chemical Composition of Marsh Soils

The principal components of interest here are carbon, nitrogen and phosphorus. These solids constituents of treatment marsh soils vary along the treatment gradient from inlet to outlet, and with depth. For instance, the inlet end of WCA2A experienced elevated phosphorus in source water for several decades, creating new soil accretions with elevated P content (Figure 4.9).

The decomposition of these solids contributes the nutrients to porewaters. Near the inlet, radioisotope studies indicated that the top 30 cm of soil were deposited under the influence of the elevated P source water (Reddy et al., 1991). However, at the downgradient site, radioisotope studies indicated that only the top 14 cm were deposited under the influence of the elevated P source water. Soils below the 14 cm horizon were presumed to have been created under lower nutrient, antecedent conditions.

The nitrogen content of soils along the treatment gradient in WCA2A substantiates the results in Figure 4.9, with little change as treatment progresses (DeBusk and Reddy, 1998).

The general shape of the profiles in Figure 4.9 have also been observed at the Houghton Lake treatment marsh, ten years into its treatment history. Data (unpublished) show relatively constant vertical profiles of carbon, with 414±38 mg/gdw at the discharge and 411±65 mg/gdw at a distance of 300 m from the discharge, with no vertical trend. There were also relatively constant vertical profiles of nitrogen, with 33.8±2.5 mg/gdw at the discharge and 30.0±3.8 mg/gdw at a distance of 300 m from the discharge, with no vertical trend. Soil phosphorus decreased downward over the top 15 cm, by about a factor of two.

Profiles in general reflect the nutrient ratios of the biomass from which they were formed. These have much more carbon than nitrogen, and much more nitrogen than phosphorus. The historical measure of these proportions is the Redfield ratio, first suggested for marine plankton, of C:N:P = 106:16:1. Some literature suggests that the marine proportions might apply to wetland soils (e.g., Kadlec and Knight, 1996; Reddy and Delaune, 2008), but those proportions do not hold for freshwater organisms (They et al., 2017). The ratios for organic wetland soils are much different. For instance, the newly accreted soils in Figure 4.9 have C:N:P = 353:24:1, while the underlying antecedent organic soils have C:N:P = 1190:88:1

Soil Oxidation-Reduction Potential

Oxidation-reduction is a chemical reaction in which electrons are transferred from a donor to an acceptor. The electron donor loses electron and increases its oxidation number or is

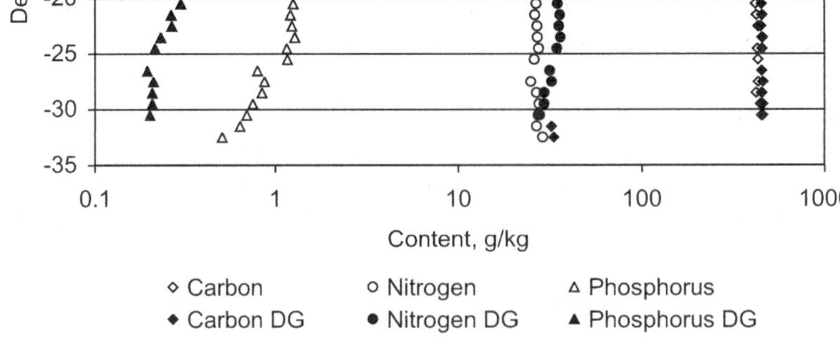

FIGURE 4.9 Depth profiles for carbon, nitrogen and phosphorus in soil solids for a treatment marsh (WCA2A). The open symbols are for a site 300 m from the discharge; the solid symbols are for a site 1920 m downgradient (DG) from the discharge. (Based on data from Reddy et al., 1991.)

oxidized; the acceptor gains electrons and decreases its oxidation number or is reduced. The driving force of a chemical reaction is the tendency of the free energy of the system to decrease until, at equilibrium, the sum of the free energies of the products equals that of the remaining reactants. In a reversible oxidation-reduction reaction, this driving force can be measured in Joules or in (milli)volts. Consider a reaction in which n electrons are transferred:

$$Ox + ne^{-} \ddot{A}\ \mathrm{Re}\, d \tag{4.10}$$

If the free energy change, represented in voltage, is measured against the standard hydrogen electrode, it is denoted by Eh. The equilibrium relation is then:

$$Eh = E_o + \frac{RT}{nF}\ln\left(\frac{[Ox]}{[Red]}\right) \tag{4.11}$$

where E_o = reference potential, mV, zero for the standard hydrogen electrode
Eh = oxidation reduction potential, mV
F = Faraday's constant, 96.4 J/mol·mV
n = number of electrons transferred
R = gas constant, 8.314 J/mol·K
T = temperature, K

The brackets denote concentrations. The interested reader may find more details in chemistry references, such as Ponnamperuma (1972); Pankow (1991), or Morel and Hering (1993).

Eh is a quantitative measure of the tendency of a given system to oxidize or reduce susceptible substances. Eh is positive and high in strongly oxidizing systems; it is negative and low in strongly reducing systems.

Oxidation-reduction conditions affect chemical and microbial processes, and have a very large effect on the biological availability of major and trace nutrients in soils in general (Patrick et al., 1985; Gambrell et al., 1987).

In submerged sediments and soils, redox potential ranges from around -400 mV (strongly reduced) to +700 mV (well oxidized). The oxidation of organic matter yields energy; the amount of energy depends on the nature of oxidant, or electron acceptor. Energetically, the most favorable oxidant is oxygen; after oxygen is depleted there follows a succession of organisms capable of reducing NO_3^-, MnO_2, FeOOH, SO_4^{2-} and CO_2 with each oxidant yielding successively less energy for the organism mediating the reaction (Westall and Stumm, 1980). This succession leads to zonation, either in the vertical direction with depth into sediments in FWS wetlands, or in the radial direction around roots. The former case is illustrated in Figure 4.9, in which upper layers of the wetland bed display the more energetic reaction zones (Reddy and D'Angelo, 1994). It should be noted that the intermediate zones of Figure 4.10, in which the transition from oxic to anaerobic conditions occurs, are thin in FWS wetlands, typically comprising no more than one or two centimeters. Depending on the magnitude of the vertical transpiration flow, this zone thickness is controlled by downward advection of surface water and its redox potential, with much lesser contributions from diffusion. However, the zonation around wetland plant roots is much smaller still, with typical zone thicknesses of a millimeter or two.

The chemistry of these thin transition layers maybe summarized in a number of equivalent ways (Reddy and D'Angelo, 1994; Mitsch and Gosselink, 2015); here a simple version is chosen with organic matter represented by CH_2O. Oxygen is the terminal electron acceptor in aerobic zones, and is reduced while electron donors are being oxidized, notably organic

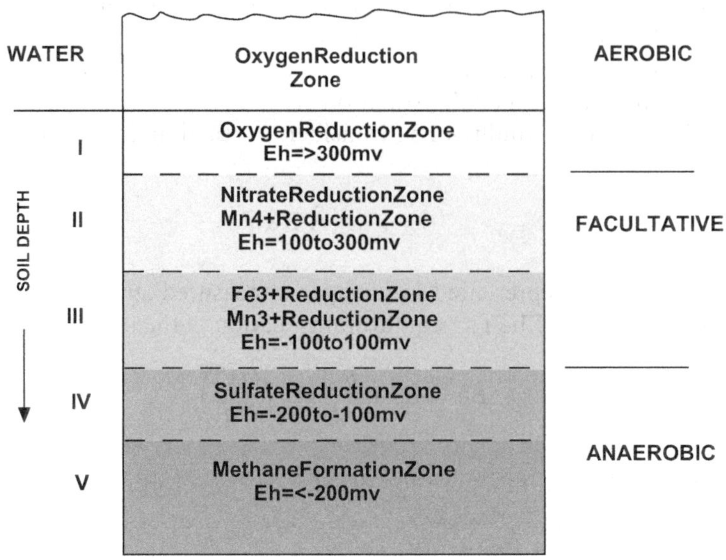

FIGURE 4.10 Hypothetical vertical redox zonation in the soils under a FWS wetland.

substances and ammonia. This reduction of O_2 to H_2O is carried out by true aerobic microorganisms, and CO_2 is evolved as a waste product:

$$CH_2O + O_2 \rightarrow CO_2 + H_2O \tag{4.12}$$

As O_2 is depleted, nitrate will be used as electron acceptor followed by oxidized manganese compounds and then followed by ferric iron compounds. The order of these reductions is the same as that indicated by thermodynamic considerations (Reddy et al., 1986).

Nitrate is the next oxidant to be reduced following oxygen depletion. Many microorganisms can utilize NO_3^- as terminal hydrogen acceptor instead of O_2, which is the denitrification process (refer to Chapter 9):

$$5CH_2O + 4NO_3^- \rightarrow 2N_2 + 4HCO_3^- + CO_2 + 3H_2O \tag{4.13}$$

As the redox potential continues to decrease, manganese is transformed from manganic to manganous compounds at about +200 mV (Laanbroek, 1990).

$$CH_2O + 3CO_2 + H_2O + 2MnO_2 \rightarrow 2Mn^{2+} + 4HCO_3^- \tag{4.14}$$

When the reduction of nitrate stops by depletion of this electron acceptor, the reduction of ferric oxide starts. A wide range of anaerobic bacteria are able to conserve energy through the reduction of Fe^{3+} to Fe^{2+} (Younger et al., 2002). Many of these microorganisms also have the ability to grow through the reduction of Mn_4^+ to Mn_2^+.

$$CH_2O + 7CO_2 + 4Fe(OH)_3 \rightarrow 4Fe^{2+} + 8HCO_3^- + 3H_2O \tag{4.15}$$

Sulfate reduction occurs when the redox potential drops below −100 mV. Only a small amount of reduced sulfur is assimilated by the organisms and virtually all is released into the external environment as sulfide (Wake et al., 1977).

$$CH_2O + SO_4^{2-} \rightarrow H_2S + 2HCO_3^- \quad (4.16)$$

Sulfate reduction is promoted by design in wetlands built to remove metals with insoluble sulfides (Younger, 2000).

Methane production requires extremely reduced conditions, with a redox potential below -200 mV, after other terminal electron acceptors have been reduced.

$$4H_2 + CO_2 \rightarrow CH_4 + 2H_2O \quad (4.17)$$

$$4H_2 + CH_3COOH \rightarrow 2CH_4 + 2H_2O \quad (4.18)$$

Methanogenic bacteria utilize hydrogen as an electron source, but can also use formate ($HCOO^-$) or acetate (CH_3COO^-) (Equation 4.18). Methane is either released to the atmosphere or is oxidized to CO_2 by methanotrophic bacteria as soon as it enters the oxic zone.

Redox Potentials in Treatment Wetlands

The redox potential of wetland soils decreases with vertical depth into the sediments because the principal source of free oxygen is from atmospheric diffusion at the top of the sediment layer. The typical oxygen gradient in wetland sediments includes a thin (less than a few centimeters) oxidized surface horizon at the sediment/water interface; underlain by increasingly reduced conditions with depth based on the amount of biological and chemical reducing activity in the sediments (Figure 4.10). Vertical redox gradients in treatment wetland soils will vary in response to distance from the point of wastewater loading (Figure 4.11).

Northern wetlands are sometimes sealed by an ice cap in winter, which prevents the supply of oxygen to the water and/or soils. This then shifts the redox profile to much lower eH values, causing sulfate reduction and methanogenesis to dominate even the upper soil horizons. Gaseous sulfur compounds cannot escape, and may reach lethal levels for wetland biota.

Porewater

Wetland soils typically contain considerable amounts of porewater, and solute concentrations in that porewater are typically quite different from solute concentrations in the overlying water. Here the focus will be upon phosphorus and ammonia, the two nutrients that are often the target for treatment marshes. However, profiles of other chemicals, such as iron and manganese, often follow similar vertical patterns in the action zone (Mayer et al., 1999). Free oxygen and nitrate nitrogen rarely appear in porewaters.

The origins of chemicals in porewater are principally from the decomposition of soil materials (Carpenter et al., 2009).

There are several techniques for measuring porewater solute concentrations, including sectioning and centrifugation, vertical arrays of sampling ports, and porewater equilibrators. The last has emerged as the technique of preference for wetlands.

Peepers are in situ dialysis samplers, which make use of semi-permeable membranes to extract solution from the sediment and associated porewater (Teasdale et al., 1995). A peeper unit has a vertical array of small chambers, typically spaced one or two cm in height. Dissolved constituents diffuse across the membrane, and at equilibrium it is presumed that the water in the chamber has the same composition as the adjacent porewater. The chambers are initially filled with deoxygenated deionized water. The required time for equilibration depends upon the membrane selected, but is typically on the order of two

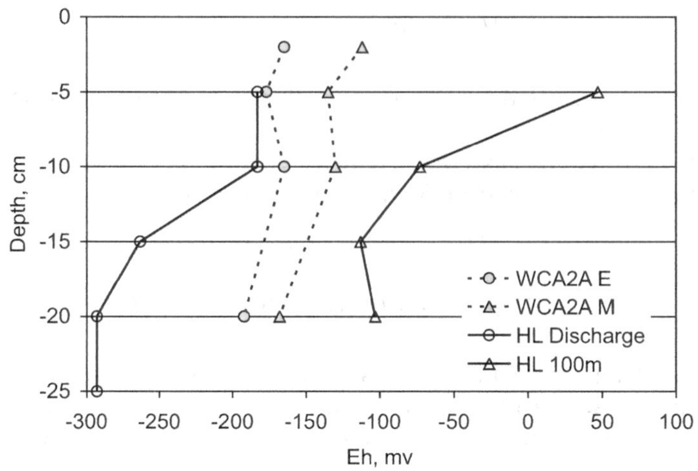

FIGURE 4.11 Example depth profiles of redox potential in treatment marsh soils. WCA2A: E = enriched, M = moderately enriched; data from Thomas et al., 2009. HL: Houghton Lake, discharge = source water input point, 100 m = 100 m downgradient; unpublished data.

weeks. The peeper is then removed from the sediment, and each chamber is sampled with a micropipette.

Phosphorus in Porewater

There are vertical gradients of porewater phosphorus in wetland soils; but these are not necessarily unidirectional at all depths. The action zone (top 10 cm) often has been observed to possess upward decreasing gradients (Novak et al., 2004, 2007; Bhadha et al., 2011; Palmer-Felgate et al., 2011; Dierberg et al., 2012; Hill and Robinson, 2012).

Examples of such gradients are shown in Figure 4.12. In most circumstances, the vertical porewater P gradients are strongly decreasing upward in the top half of the action zone. But if the profile is extended to depths below the action zone, there is a decreasing downward trend. These profiles of dissolved phosphorus are examined in more detail in Chapter 10.

It is important to note that porewater phosphorus is many times higher than the overlying surface water. These high concentrations are due to soil decomposition processes. This has strong implications for phosphorus removal in the overall ecosystem; namely, that mass transfer or diffusion will not transfer P from the overlying water to the action zone. Instead, there is an implied upward leakage of porewater phosphorus to surface water.

Ammonia in Porewater

There are vertical gradients of porewater ammonia nitrogen in wetland soils; but these are not necessarily uni-directional at all depths. The action zone (top 10 cm) often has been observed to possess upward decreasing gradients (Figure 4.13). In lower soil horizons, ammonia decreases with increasing depth. These profiles of dissolved phosphorus are examined in more detail in Chapter 9.

The general profiles in Figure 4.13 apply more widely. Payne et al. (2009) studied eight wetland sites in southern Michigan, using peepers. Shallow samples at 2.2 cm intervals were aggregated into the 0–20 cm depth range and deep samples were aggregated over the

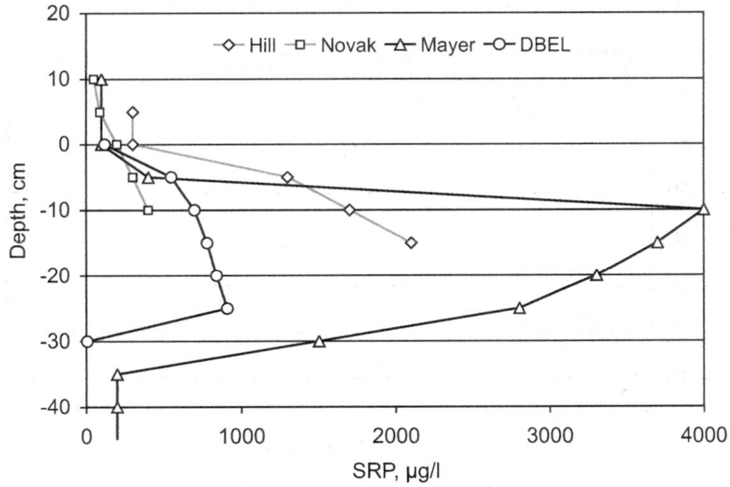

FIGURE 4.12 Example depth profiles for porewater phosphorus in marshes. The Hill and Robinson (2012) data are for a system in the United Kingdom treating agricultural runoff, vegetated by *Sparganium erectum*. The Novak et al. (2004) data are for a system treating agricultural runoff in North Carolina vegetated by *Leersia oryzoidesi*. The Mayer et al. (1999) data are for a septage-contaminated, groundwater-fed system in Ontario, vegetated by *Typha*. The DBEL (unpublished) data are for a SAV system (STA2C3) treating agricultural runoff.

20–50 cm depth range. Shallow samples had hundreds of times greater ammonia than surface waters, and about double the ammonia as the deep samples.

It is important to note that porewater ammonia is very much higher than in the overlying surface water. As for phosphorus, these high concentrations are due to soil decomposition processes. This has strong implications for ammonia removal in the overall ecosystem; namely, that mass transfer or diffusion will not transfer ammonia nitrogen from the overlying water to the action zone. Instead, there is an implied upward leakage of porewater ammonia to surface water.

Respiration – Gas Flows

The generic chemical transformations represented by Equations 4.12 through 4.18 indicate the consumption and formation of several gases. These include:

- Utilization of oxygen and carbon dioxide
- Production of carbon dioxide, nitrogen, nitrous oxide, and methane

There are three potential transfer routes for gases entering and leaving the action zone: (1) by diffusion (mass transfer) through the water column, (2) by bubbling from the sediment (ebullition), and (3) by gas flows through emergent rooted wetland plants (convective flux). In most cases, plant transport is the dominant release mechanism in vegetated wetlands (Brix, 1990; Brix et al., 2001).

Wetland emergents contain aerenchyma, a spongy tissue that forms spaces or air channels in the leaves, stems and roots, which allows exchange of gases between the shoot and the root. The channels of air-filled cavities provide a low-resistance internal pathway for the exchange of gases between the plant above the water and the submerged tissues. The driving

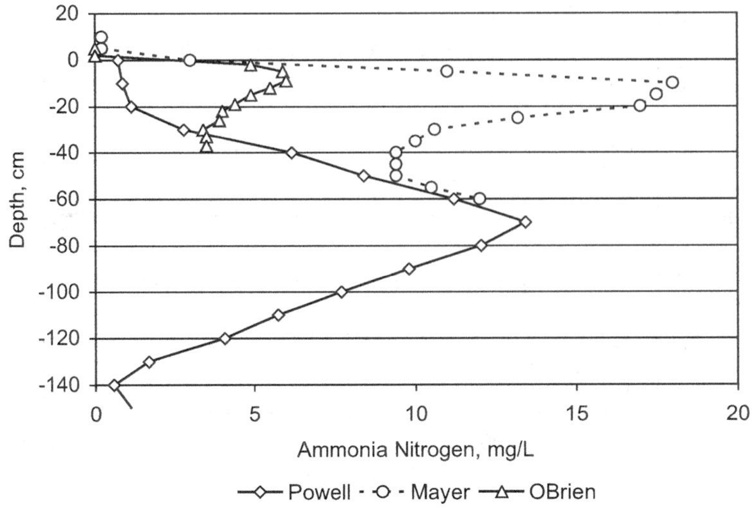

FIGURE 4.13 Example depth profiles for porewater ammonia nitrogen in marshes. The Powell (2010) data are for a groundwater-fed system in Ohio, vegetated by *Scirpus* and *Carex*. The Mayer et al. (1999) data are for a septage-contaminated, groundwater-fed system in Ontario, vegetated by *Typha*. The O'Brien et al. (2012) data are for a SAV system treating stream water in Michigan, vegetated by *Potamogeton* and *Ceratophyllum*.

forces for plant gas flow are gradients in temperature and water vapor between the internal gas spaces of the plants and the ambient atmosphere (thermal transpiration and humidity-induced pressurization). Brix and Orr (1992) found significant plant gas flows in *Baumea articulata, Eleocharis sphacelata, and Schoenoplectus validus*; but much higher flows in *Typha domingensis, Typha orientalis, and Phragmites australis.* A clear diel variation in pressurization and convective flow was observed.

The presence of hypoxic soils in wetlands requires that plants aerate their root zones in order to survive. Oxygen is transported downward through the aerenchyma, most of which is consumed in oxidation of soil organics near the root surfaces, but some of which leaks through root pores into the surrounding soil.

Within a stand of emergent vegetation, some plants (culms) act as influx conduits, and others as efflux conduits. Sorrell and Boon (1992) found that *Eleocharis* influx culms were erect and undamaged, while older efflux culms were often partially decomposed or had collapsed near the water surface.

The dominant gas-forming process in soil is the decomposition of soil organics, also called soil respiration, which can occur aerobically near roots exuding oxygen or anaerobically at sites distant from roots. Aerobic decomposition rates are much higher than anaerobic rates (DeBusk and Reddy, 1998).

Aerobic processes consume oxygen and produce carbon dioxide. The reduction of nitrate consumes nitrate and produces CO_2, N_2O, and N_2; and hence is a major contributor to outflow of these three gases. These processes occur in the upper soil horizons (Figure 4.10).

In lower soil horizons, the redox potentials are low enough to support the anaerobic decomposition, which produces methane. There are clear diel patterns of methane flows in plants in wetlands, with higher methane outflows during the day (Brix et al., 2001; Minke et al., 2014). There are also seasonal variations in methane emission, with higher rates in the summer months (Sha et al., 2011; Jitka et al., 2017).

Figure 4.14 shows the conceptual set of processes for carbon gas formation and usage in the action zone. Plants utilize carbon dioxide in their photosynthetic metabolism, and pass carbon into the root zone via surface litterfall and root death. A significant portion contributes to soil accretion, which is discussed in the next section. Another portion produces dissolved organic carbon, which may either be oxidized to CO_2 (majority) or converted to methane (minority). The product methane and carbon dioxide pass to the atmosphere, either through plants (majority) or through diffusion and ebullition (minority). The numerical values in Figure 4.14 are those reported by Brix et al. (2001), for a Phragmites wetland in Denmark.

There is a large literature concerning the emissions of gases from marshes and other treatment wetlands, including the reviews of Søvik et al. (2006) and Mander et al. (2014). The motivation is the modern concern about greenhouse gases, which notably include carbon dioxide, methane and nitrous oxide. The concern is whether wetlands, including treatment marshes, are net contributors to greenhouse gas production, or function to reduce greenhouse gases. Flooded wetlands generally sequester carbon dioxide from and release methane to the atmosphere. The combination of these two factors determines whether these offsetting processes make a wetland system an overall contributor to the greenhouse effect (Kayranli et al., 2010). Despite numerous studies, there is no clear answer to this question, because of the variability of treatment marsh performances (Kayranli et al., 2010).

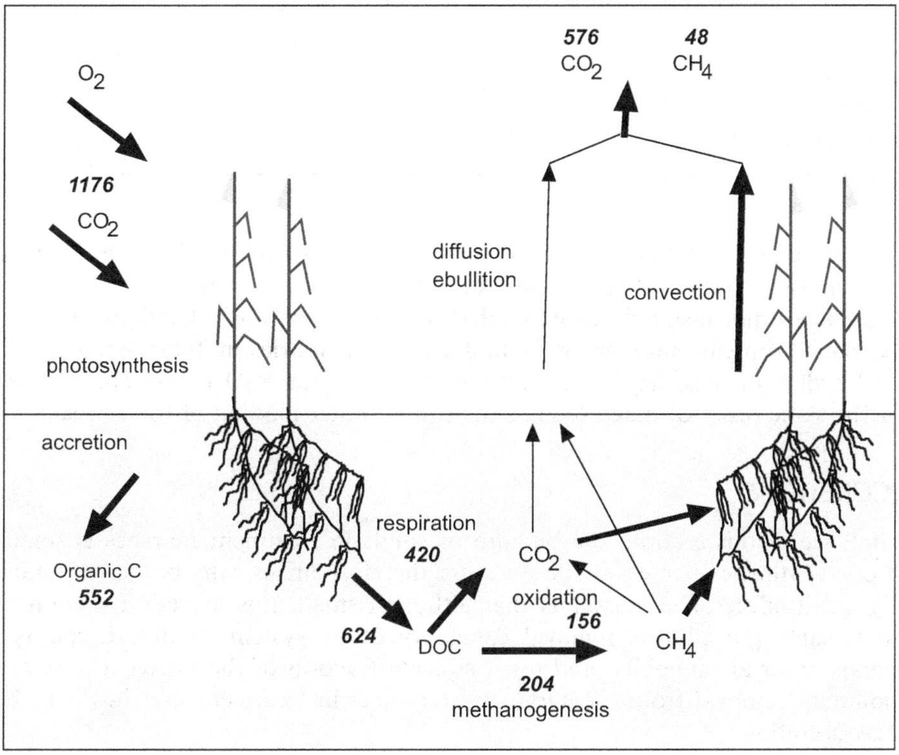

FIGURE 4.14 Mechanisms of carbon conversions and gas exchanges in marshes. The numerical values are for the Vejlerne system in Denmark.

TABLE 4.5

Gaseous emissions from example treatment marshes.

Country	Wetland	TOC In $g/m^2 \cdot d$	CO_2-C $g/m^2 \cdot d$	CH_4-C $mg/m^2 \cdot d$	N_2O-N $mg/m^2 \cdot d$	Reference
Sweden	Magle					1
	Juncus	0.62	7.12	367	2.4	
	Typha	0.62	6.90	169	3.0	
	Phragmites	0.62	6.85	250	4.7	
Finland	Hovi					2
	Summer	1.01	1.20	29	0.4	
	Winter	1.01	0.21	46	0.1	
Norway	Skjonnhaug					2
	Summer	4.00	1.40	72	4.0	
	Winter	4.00	0.98	330	1.5	
Denmark	Vejlerne					3
	Annual		2.28	132		
Nova Scotia	Truro Dairy					4
	Aug-Oct	0.73	5.30	214	21.0	
	Nov-Mar	0.73	0.50	29	1.0	
	Apr-Jun	0.73	4.10	172	12.0	
	Jul-Sep	0.73	7.10	479	33.0	

References
1 Ström et al. (2007)
2 Søvik et al. (2006)
3 Brix et al. (2001)
4 VanderZaag et al. (2010a)

The answer to the greenhouse gas question does not directly affect the descriptions of pollutant removal processes that are the focus of this book. However, the carbon emissions from wetlands are not insignificant compared to the surface water loadings of carbon. The general variability in gas emissions is exemplified by the entries in Table 4.5. The CO_2 flows are several orders of magnitude greater than the CH_4 and N_2O flows. The vented CO_2-C flux is of the same order of magnitude as the source water loading of total organic carbon.

4.5 ACCRETION

As we shall see in this section, the buildup of solids in treatment marshes is small, of the order of one centimeter per year. The question therefore arises, why bother to quantify this seemingly small effect? The answer is that although small, this accretion accounts for the entire sustainable phosphorus removal capability of he system. It also accounts for the entire metals removal capability, and for a sizeable fraction of the nitrogen removal. Much of the pollutant removal from the surface water sheet in treatment marshes is to be found in this new accretion.

Solids accumulate in wetlands from two principal sources: (1) the sedimentation of incoming TSS entering with water inflows, and (2) production of solids within the wetland due to biological processes. Biological production includes debris formation from bacteria, algae, and above and below ground plant parts. Sedimentation produces buildups preferentially in the inlet zone of the wetland, while bioaccretion is spread across the entire wetland, although internal spatial gradients may be present. Material accumulating in wetlands ranges from largely mineral sediments transported into the wetland to largely organic matter produced by primary productivity in the wetland, depending on sediment supply and wetland hydrology. Models and mass balances that incorporate a complete separation of incoming sedimentation and internal bioaccretion have been set forth (Reinhardt et al., 2005).

Sedimentation may often be distinguished from bioaccretion based upon several soil chemistry indicators. Particulate matter from outside the wetland often is characterized by high proportions of clay, silt and sand, whereas internally produced materials tend to have a high organic content. As a consequence, the bulk density of allochthonous sediments is typically higher than 0.4 gdw/cm^3, whereas in situ production tends to be less than 0.2 gdw/cm^3.

Measurement of Solids Deposition and Elevation Change

The action zone of a treatment marsh rests upon a subsoil base, which itself may be in a condition of vertical movement. The subsidence of this basal material is of great concern to coastal wetlands, which often display measurable rates of sinking. Thus, wetland science has focused on two independent measures of vertical soil movement: the elevation change in the top of the soil column, and the accretion of new soil materials. The rates of both processes are likely to be quite small, thus requiring either long-term studies or specialized techniques.

Direct Methods

Direct measurement can sometimes be accomplished. For instance, Kadlec (2009b) performed repeated surveys of the marsh surface elevations over a span of 35 years, and was able to document an elevation change of 30 cm near the source water entry. Such long-term studies are seldom feasible, and procedures that function over shorter time frames are desirable. Another circumstance that favors direct measurement is the existence of an antecedent basal soil that differs from the sediments that are being deposited. Braskerud (2003) researched marshes that were established on clay, and collected sediments that were mostly sand and silt, thus the boundary between sediment and the original bottom was easy to detect. Further, buildups were significant, ranging from 1–3 cm/yr over the six-year study. In neither of these two studies was there any process of basal subsidence.

Marker Methods

The influence of vertical accretion and belowground processes on marsh elevation may be investigated by use of surface elevation tables (SETs, Cahoon et al., 2000). SETs measure changes in marsh elevation by comparison of a time series of marsh surface elevations with replicated stable established benchmarks. Changes in marsh elevation can thus be tracked over time. Vertical accretion can be measured by use of horizon markers, typically using feldspar (Callaway et al., 2013). When both methods are applied simultaneously, the components of surface elevation change attributable to accretion or to compaction/subsidence may be differentiated: shallow subsidence = accretion − elevation increase. Both techniques are described in detail in Callaway et al. (2013).

Feldspar or other marker horizons can provide measures of the short-term accretion above them (Braskerud, 2001b; Harter and Mitsch, 2003). In the case of compaction and decomposition occurring underground, the surface elevation increase will be less than the accretion of new material. Surficial accretion is also potentially lessened by resuspension of the topmost layers.

Marker methodology is simple in concept: place a recognizable material on the soil surface, and come back later to see how much new material has been deposited on top. But care must be taken in measurement. One method is the use of cryogenic sampling, in which the soil is frozen so as to not disturb the profile. Cryogenic cores are obtained using a straight copper tube with a sharp, closed tip pushed into the sediment far enough to extend through the feldspar marker horizon, and then filled with liquid nitrogen to freeze the sediment adjacent to the tube. The frozen cores were then pulled out and the distance from the sediment surface to the feldspar marker is measured in several places around the circumference of the core. The top is determined by simply observing the interface between brown ice and white ice.

In practice, things are not quite so simple. For example, 41 feldspar markers were installed in the ENRP (SFWMD, unpublished data). By year three, 16 had been destroyed, for 13 no marker could be found, leaving 12 measurable stations. Similarly, Mitsch and coworkers lost 30% of their feldspar markers in one year (Mitsch et al., 2014a). These losses were attributed to erosion, resuspension, bioturbation or mixing. They attempted to quantify bioturbation by installing exclosures to keep out large biota. Exclosed sites experienced approximately twice the amount of sedimentation as did the open sites. But, 50% of the exclosures were damaged by animals, and 25% lost the feldspar marker. Mitsch et al., 2014a) therefore concluded that horizon markers were not appropriate for their treatment marshes.

Inference Methods

Other quantitative studies have relied upon atmospheric deposition markers such as radioactive cesium (^{137}Cs) or radioactive lead (^{210}Pb). The ^{137}Cs procedure is described in detail in Ritchie and McHenry (1990); and the ^{210}Pb procedure is described in detail in Brenner and Kenney (2013). These techniques require several years of continued deposition for maximum accuracy, and pose difficult analytical challenges. These dating methods enable the development of age–depth relations for sediment cores that extend from the date of core collection, to several tens of years prior. Use of these dating methods is widespread and routine. Bhomia et al. (2015) lists 27 wetland studies employing ^{137}Cs and/or ^{210}Pb procedures.

Radioisotope dating is usually based upon a set of assumptions regarding the nature of the deposition process, including:

- There is a known time sequence of the additions of the radioisotope.
- New accretions originate from deposition on the top of the soil, forming layers of new material over the course of time, meaning roots do not contribute to the construction of the vertical profile of the radioisotope.
- The layers are noninteractive, meaning there is no mixing.
- The radioisotopes partition exclusively to soil particles, meaning there are no transport processes in porewaters or plants.

The first of these requires some care, because the time sequence can depend upon when and where the system is located. The input of 1950–1960 bomb-derived ^{137}Cs fallout has ceased and inventories are decreasing due to radioactive decay. In the future, the use of ^{137}Cs may no longer be possible, especially in southern hemisphere. However, the Chernobyl accident of 1986 will continue to provide some areas of Europe and eastern Asia with a cesium time marker. Since the fallout of ^{210}Pb is effectively continuous, the use of this radionuclide does not face these problems. However the fallout of excess ^{210}Pb is low in the western areas of the continents, where the air masses have traveled over the oceans.

The second of these assumptions, the "layer cake" idea, will be further explored in a subsequent section.

The third assumption is questionable, especially in the aquatic and wetland environments. While the solids do not mix of their own accord, the biological activity in lakes and wetlands can cause stirring of surficial sediments (see, for example, Robbins, 1986; Robbins et al., 1990). The range of animals that can cause such mixing ranges from worms and clams, up to fish, waterfowl and water-loving rodents.

The last assumption is probably the most firmly founded, because the radioisotopes are nonexchangeable, and on delivery to the land surface as fallout they are rapidly and strongly fixed by the surface soil or small sediment particles (Mabit et al., 2008).

Stable Isotopes

Bhomia et al. (2015) utilized vertical chemical profiles to identify recently accreted material. Variables included the stable isotope ratios δ15N (‰) and δ13C (‰), as well as bulk density and phosphorus content. Linear regressions of the profiles in the top zone and bottom zone were found to intersect at a change point along the depth profile.

Bhomia et al. (2015) presumed that the change point reflected the age of the soil above. However, their study did not include roots or their effects. It is known that there is preferential plant transport of 13 C, causing enrichment in the rhizosphere (Chanton, 2005), and Saintilan et al. (2013) concluded that core profiles were in major part due to root activity. Thus there is a possibility that the Bhomia change points were in part due to the rooting depth of plants.

Results

Sedimentation

If large concentrations of suspended material enter a quiescent wetland environment, there can be high amounts of sediment trapping compared to generation and resuspension, and buildup typically occurs preferentially in the inlet section of the wetland. Therefore, a "delta" of accreted sediments may build in the inlet region of the wetland (Walker, 2001; Braskerud, 2003).

A small constructed wetland, treating mixed sources from an agricultural catchment in southeast Sweden, was investigated for TSS retention (Johannesson et al., 2011). A small inlet zone (6.4% of total area) experienced a buildup of 4 cm/yr, with a mass accretion of 22 kgdw/m·yr. This material was comprised of clay particles, with a bulk density of 0.55 gdw/cm^3. The authors speculate that particles generated within the wetland were exported. A second such small Swedish P removal wetland removed 36 kgdw/m·yr of TSS, primarily clay particles, and 6.9 gP/m·yr (Kynkäänniemi et al., 2013). In this system, an inlet sedimentation basin or deep zone was included.

Highly loaded wetlands treating mineral solids were observed to accrete 2–8 cm/yr (Braskerud, 2001b, 2003). These Norwegian end-of-field systems captured solids from runoff events, at hydraulic loading rates of 66–180 cm/d. Much of this sediment accretion was in the inlet of the system (Figure 4.15)

Two small wetlands were operated on the Ohio State University campus for ten years, treating river water for the removal of sediments and nutrients (Anderson and Mitsch, 2006; Mitsch et al., 2014a). Sediments from pumped river water accumulated in the wetlands at rates of about 5 kgdw/m·yr, of material of bulk density about 0.65 g cm^{-3}. The deposits were largest in the inlet zone of one wetland, but more uniformly distributed in the other.

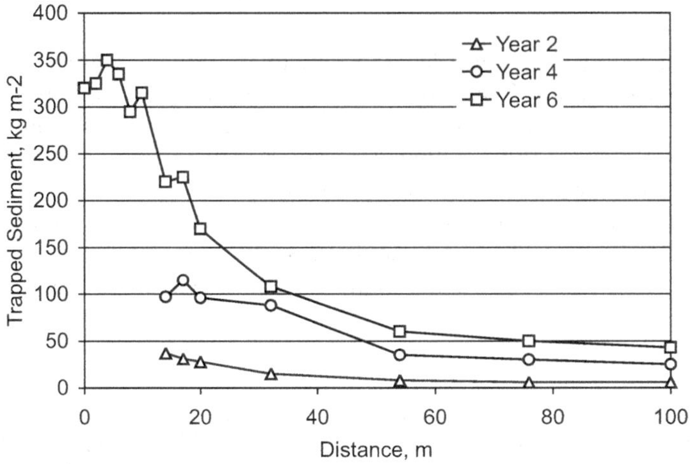

FIGURE 4.15 Sediment accumulation in a constructed wetland treating field runoff. The first 26 m were occupied by a sedimentation basin (14 m) and an inlet delta zone (12 m), and the balance was emergent marsh. The wetland accumulated about 15 cm of a mostly mineral material over the six-year study. (Based on data in Braskerud, 2003.)

The Apopka constructed wetlands remove algal phytomass (biomass plus necromass) from incoming lake waters (Dunne et al., 2012). Over a four-year period, these constructed marshes removed an average of 1.4 kgdw/m·yr of incoming solids.

Bioaccretion

Marshes create new soils and sediments. Not all of the dead plant material undergoes decomposition. Some small portions of both above ground and below ground necromass resist decay, and form stable new accretions. The origins of new sediments may be from remnant macrophyte stem and leaf debris, remnants of dead roots and rhizomes, and from undecomposable fractions of dead microflora and microfauna (algae, fungi, invertebrates, bacteria). The accumulation of internally produced solids is here termed bioaccretion.

The manner of accretion has mostly been presumed to be sequential vertical layering (Kadlec and Walker, 1999; Rybczyk et al., 2002b), but that view is likely to be overly simplified. At least two factors argue against simple layering: vertical mixing of the topsoils and sediments (Robbins et al., 1990), and the injection of accreted root and rhizome residuals at several vertical positions in the root zone. Nonetheless, new residuals are deposited on the wetland soil surface, from various sources. The most easily visualized is the litterfall of macrophyte leaves, which results in top deposits of accreted material after decomposition. However, algal and bacterial processing, which occurs on submersed leaves and stems, results in litterfall and accretion of microdetrital residuals.

Bioaccretion has been quantified for a small fraction of the total treatment marshes (Table 4.6). Bioaccretion may be spatially non-uniform, due to gradients in deposition and productivity. This has been found to be true even in wetlands of very low nutrient status (Reddy et al., 1993). Inlet zones may therefore be subject to solids accumulations that are double the wetland average.

TABLE 4.6

Examples of soil bioaccretion measured via different methods. Ag = agricultural runoff.

Location	Reference	Method	Source Water	Bioaccretion cm/yr
Andersen Marsh, CA	1	210Pb	Lake	0.45
Everglades WCA2A, FL	2	210Pb	Ag	0.58
Upper Klamath, CA	3	210Pb	Ag	0.37
Everglades WCA2A, FL	4	137Cs	Ag	0.53
Everglades WCA2A, FL	5	137Cs	Ag	1.13
Upper Klamath, CA	3	137Cs	Ag	0.38
Twitchell Island East, CA	6	Feldspar	Ag	1.92
Twitchell Island West, CA	6	Feldspar	Ag	2.10
ENRP Typha, FL	7	Feldspar	Ag	2.15
ENRP Najas, FL	7	Feldspar	Ag	2.88
ENRP Ceratophyllum, FL	7	Feldspar	Ag	2.39
ENRP Pistia, FL	7	Feldspar	Ag	0.80
ENRP Open Water, FL	7	Feldspar	Ag	3.95
STA1W 1–4, FL	8	13C, 15N	Ag	0.90
STA1W 5, FL	8	13C, 15N	Ag	1.20
STA2, FL	8	13C, 15N	Ag	1.10
STA34 FW1,2, FL	8	13C, 15N	Ag	1.70
Mean				**1.44**

References
1 Kim (2003)
2 Craft and Richardson (1998)
3 Graham et al. (2005)
4 Craft and Richardson (1993)
5 Reddy et al. (1993)
6 Miller et al. (2008)
7 Unpublished data SFWMD
8 Bhomia et al. (2015)

Root and Rhizome Death

Soil elevation changes are the result of material additions to the top of the soil/sediment column (accretion) accompanied by changes within that column. Underground processes associated with roots can be grouped into four categories in addition to the basal process of soil organic matter decomposition (Kuzyakov, 2006). They are root respiration, rhizomicrobial respiration, decomposition of root residues, and root exudation. Root contributions are typically ignored (Rogers et al., 2006). Rasse et al. (2005) concluded:

> The relative contributions to soil C pools of roots vs. shoots is one aspect that has been mostly overlooked, although it appears a key factor that drives the fate of plant tissue C either as mineralized CO_2 or as stabilized soil organic matter (SOM).

In the absence of subsidence of the basal soils, the below-ground soil processes include root necromass generation, and decomposition. Root necromass is identifiable, but integrated within the general soil matrix. Therein lies a fundamental difference between top accretion and root accretion: fallen aboveground macrophyte litter is not typically considered part of the soil column, while dead root material is. That root/rhizome necromass may be considered separately from the general soil matrix, although it undergoes similar decomposition processes.

It is nearly impossible to directly measure the death rate of live underground plant material, and thus current wetland science uses informed guesses for the formation of underground dead plant material. The basis for such estimates can be the rate of root/rhizome decomposition, which is amenable to field measurement, coupled with a mass balance.

A simple conceptual model for underground necromass involves creation from live underground roots and rhizomes (death), and the decomposition of the necromass (respiration):

$$\frac{dM_{RN}}{dt} = k_{DL}M_{RL} - k_{DR}M_{RN} \tag{4.19}$$

where M_{RL} = mass of root/rhizome live biomass, gdw/m^2
M_{RN} = mass of root/rhizome necromass, gdw/m^2
k_{DL} = death rate coefficient for root/rhizome biomass, yr^{-1}
k_{DR} = decomposition rate coefficient for root/rhizome necromass, yr^{-1}
t = time, yr

This concept has been employed in very detailed models of underground carbon and nutrient cycling in wetlands (Rybczyk et al., 1998; Van der Peijl and Verhoeven, 1999, 2000). If the system is at a stationary condition, meaning that there are no long-term trends in biomass or necromass, then:

$$k_{DL} = k_{DR}\frac{M_{RN}}{M_{RL}} \tag{4.20}$$

Thus, the production rate coefficient for root/rhizome necromass may be estimated from the standing crops and the decomposition rate coefficient, if there is available data. But root necromass information for marshes is very rare. Neill (1992) found dead roots of *Scholochloa* were three times the amount of live roots, but Gross et al. (1993) found dead *Typha* roots to be only half of the live. Bernard and Fiala (1986) found dead *Carex* roots were 84% of the total underground phytomass for three species; and Silvan et al. (2004) found similar high amounts of dead roots in a bog. These later two studies would not be representative of treatment marshes.

Based on this meager information, it appears that the root death rate coefficient is comparable to the root decomposition rate coefficient, or about 0.7 yr^{-1} (Table 4.7). This is supportive of estimates of the root death rate coefficients (turnover rates), which are as follows:

- Gustafson, 1976, as cited by Prentki et al. (1978) *Typha* 0.57 yr^{-1}
- Fiala, 1973, as cited by Prentki et al. (1978) *Phragmites* 0.33 yr^{-1}
- Rybczyk et al. (1998) 0.30 yr^{-1}
- Van der Peijl and Verhoeven, 1999 0.52 yr^{-1}

Thus it appears that in marshes underground plant parts grow, die and are replaced on the time scale of one to two years. This is a relatively large turnover rate, which implies that root death is an important factor in accretion.

TABLE 4.7
Root decomposition rates in wetlands.

Species	One-year loss %	Rate coefficient yr^{-1}	Reference
Carex aquatilus	60	0.51	1
Salix planifolia	20	1.61	1
Scolochloa festucacea	63	0.46	2
Phragmites aystralis	67	0.40	2
Scirpus lacustris	48	0.73	2
Typha glauca	51	0.67	2
Typha latifolia	72	0.33	3
Sparganium eurycarpum	58	0.54	3
Scirpus fluviatilis	40	0.92	3
Eleocharis root	48	0.73	4
Eleocharis rhizome	50	0.69	4
Typha root	58	0.55	4
Typha rhizome	22	1.53	4
Mean	**51**	**0.74**	
sd	**16**	**0.40**	

References
1 Thormann et al. (2001)
2 Wrubleski et al. (1997)
3 Puriveth (1980)
4 Snyder and Rejmánková (2015)

A Speculative Accretion Model

It is useful to explore simple models of accretion, for the purpose of estimating the potential effects of root death additions to the accumulation of soil material. Here a modification of the model set forth by Rybcyzk et al. (1998) is utilized. It considers both top inputs from litter and floc, and internal inputs from root death. The concept is top down layering accompanied by side injection of carbonaceous material.

Modified Layer Cake Model

The choice is for an annual time step, and no vertical mixing. It is assumed that each year a layer is deposited on top of the existing soils, which is termed a cohort. That layer originates as a "rain" of decomposition products from aboveground biogeochemical activity. In subsequent years, the cohort of material that accrued during year one will gain no new material from above. It will, however, accrue dead root material (Figure 4.16).

Top Layer

The new top accretion is assumed to be comprised of a refractory component, that is very resistant to decomposition, and a decomposable fraction, that undergoes further continual decomposition. Roughly, the resistant fraction is lignin-rich, and the decomposable fraction is cellulose-rich. It is presumed that no decomposition occurs during the top accretion period of one year. Of course, there is such decomposition loss, but it will be subsumed in the net accretion during that first year.

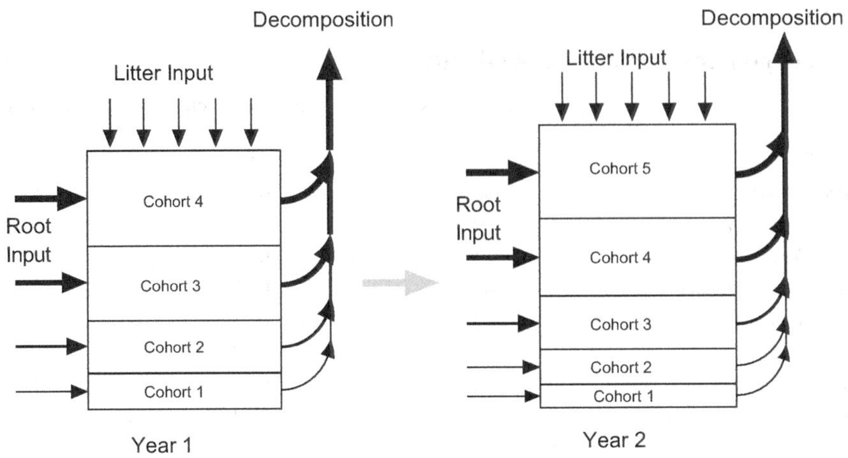

FIGURE 4.16 Schematic of the modified layer cake model of bioaccretion.

It is further assumed that there are no roots in this new layer during its first year of existence. However, roots will populate the layer as time progresses.

As a benchmark, 1.0 cm/yr of top accretion at a bulk density of 0.20 gdw/cm^3 would produce 2,000 $gdw/m^2 \cdot yr$ of root soil mass.

Lower Layers

In subsequent years, the cohort of material that accrued during year one will accrue dead root material. That soil addition from root death is presumed to consist of a refractory component and a decomposable fraction, as for the biological rain onto the top layer.

The remaining decomposable material from the top rain is combined with the decomposable material from root death, and undergoes a first order decomposition process. The refractory material from the top rain is combined with the refractory material from root death, and continues to pass to lower depths without further alteration.

Bulk Density

It is known that bulk density of the soil increases as the degree of decomposition increases. Here a linear relation is proposed:

$$BD = BD_{refrac} f_{refrac} + BD_{decomp} f_{decomp} \tag{4.21}$$

where BD = bulk density, gdw/m^3

f = mass fraction

The volume of a cohort of soil material will be given by its mass divided by its bulk density, and on an areal basis will give the cohort thickness:

$$T = \frac{M}{BD} \tag{4.22}$$

where M = soil mass, gdw/m^2

T = cohort thickness, m

Root Soil Creation

Roots experience a life cycle of growth, death, and decomposition. Here it is assumed that dead root material is a component of the wetland soil. In the absence of models for wetland plant root death, it will be presumed that a first-order process prevails:

$$G_{dr} = k_{dr} M_r \tag{4.23}$$

where G_{dr} = root soil generation, $gdw/m^2 \cdot yr$
k_{dr} = root soil generation rate coefficient, yr^{-1}
M_r = root mass, gdw/m^2

The total generation of root soil, over the whole root profile, is therefore:

$$G_{Tdr} = k_{dr} M_{Tr} \tag{4.24}$$

where G_{Tdr} = total root soil generation, $gdw/m^2 \cdot yr$
M_{Tr} = total root mass, gdw/m^2

As a benchmark, 3,000 gdw/m^2 of root mass would produce 1,500 gdw/m^2 of root soil mass for a turnover rate coefficient of 0.5 yr^{-1}. Thus, the amount of new soil accretion from roots is about the same as that from the biogeochemical rain on top.

Root Biomass

Roots are distributed over the depth of the root zone, with an exponentially decreasing profile. As a result, the mass of roots in a vertical slice between z_1 and z_2 is:

$$M_r = M_{Tr}\left(e^{-z1/b} - e^{-z2/b}\right) \tag{4.25}$$

where: M_r = root mass between depths z_1 and z_2, gdw/m^2
M_{Tr} = total root mass, gdw/m^2
b = depth at which 63.2% is above, cm
z1 = top depth below soil surface, cm
z2 = bottom depth below soil surface, cm

Illustrative Model Results

The model was exercised for conditions that are representative of treatment marshes. The parameters were:

Two cases were calculated; one with root death input and the second without. The annual cohorts are tracked through time, with respect to both the cohort layer thickness and the cohort mass remaining (Figure 4.17).

Litter necromass rain	2,000	$gdw/m^2 \cdot yr$
Root death input	2,000	$gdw/m^2 \cdot yr$
Fraction decomposable	0.70	
Decomposable loss rate	0.20	yr^{-1}
Bulk density decomposable	0.10	gdw/cm^3
Fraction refractory	0.30	
Refractory loss rate	0.00	yr^{-1}
Bulk density refractory	0.20	gdw/cm^3

Several features of accretion become apparent. Firstly, if decomposition and root input are ignored, accretion is quite large. For the presumed conditions, an annual accretion of about 3.1 cm would occur, leading to a new soil depth of 62 cm in twenty years. The assumed decomposition processes reduce that to 24 cm for a top-rain system, and to 20 cm for a roots+top rain system.

Secondly, as to be expected, the addition of root necromass increases the amount of material that collects over time. Perhaps less obviously, the root necromass additions slow the rate of contraction of individual cohorts. That in turn has the effect of evening out the accretion thickness increase over time.

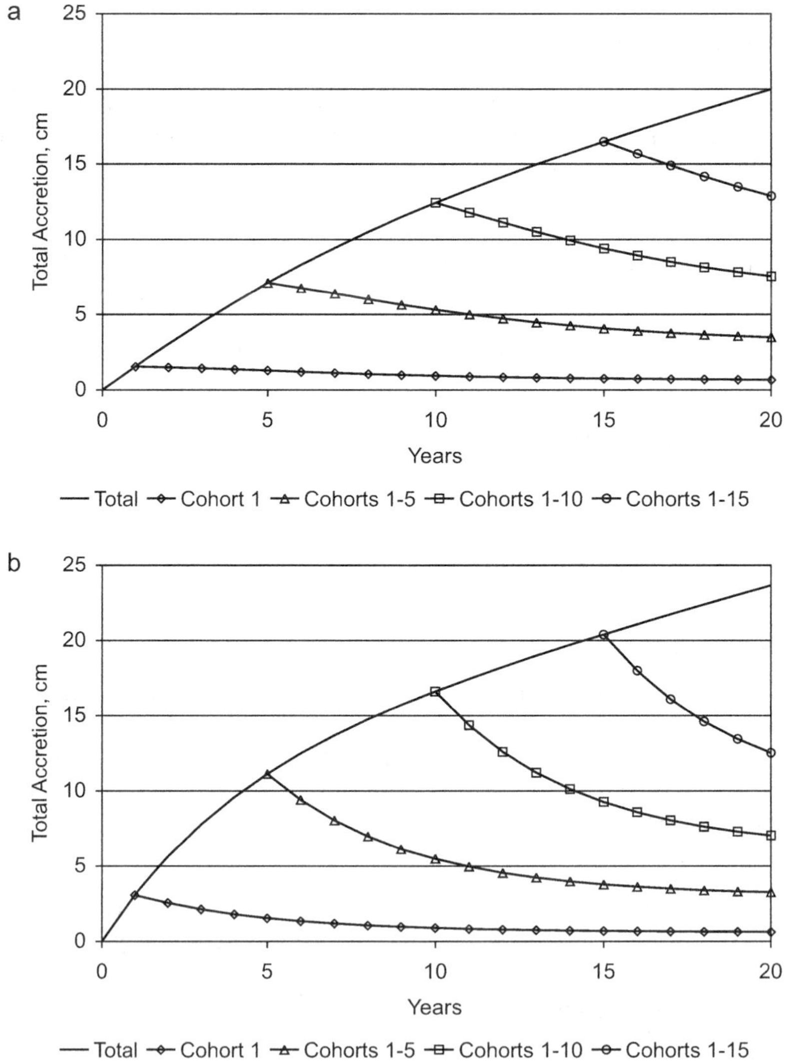

FIGURE 4.17 Accretion model for cases of roots plus top rain (A) and top rain only (B). The total necromass addition in is 4,000 gdw/$m^2 \cdot$yr.

The age of the soil at various elevations is strongly influenced by the assumed manner of accretion (Figure 4.18). The 20-year-old cohort is 4 cm deeper if one assumes top additions only.

Although these model results are not supported by specific calibration results, they are an aid to understanding the general importance of root death in the overall accretion process. The interpretation of vertical profiles in soil cores depends in significant measure upon the presence or absence of roots and rhizomes, and their rate of death and decomposition.

4.6 SUMMARY

This chapter has attempted to refute the image of the wetland soil compartment as a mere passive support for wetland plants and surface water. To that end, it has been named the action zone, or the wetland underground. Wetland soils support plant and microbial growth that provide the substrates necessary for water quality enhancement in treatment marshes. The action zone is importantly involved in many pollutant removal processes; indeed, it is the most important zone for storage of phosphorus and metals. Depending upon the scientific discipline that is giving consideration to wetland processes, there are many nomenclatures that are in common use. Here the focus has been upon the development of treatment marsh soils, and their properties that link them to pollutant removal in the overall ecosystem. Many treatment marshes function to produce organic soils, but there are also many treatment marshes that accrete mineral material.

The three action zone components, phytomass, water, and soils and sediments, are all active rater than passive. Plant roots grow and die, and transpire water, thus moving dissolved substances into the action zone. Water movement is present and important in the hydrology of the marsh and attendant functions. Soils and sediments are biologically active, and change characteristics with time and vertical position.

The action zone is vertically stratified with respect to many of its characteristics. Soil bulk density and mineral content vary vertically. Plant roots are exponentially distributed, and therefore so too are the root-driven processes. Vertical gradients are the result of a chain of oxidation/reduction reactions that provide many wetlands with their typical

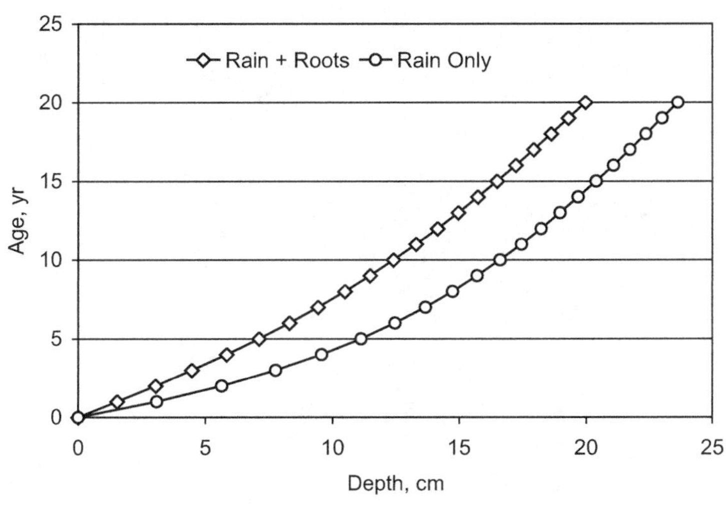

FIGURE 4.18 Depth distribution of soil ages under the assumptions of root participation (rain + roots) or not (rain only).

profile of declining redox potentials with depth. Redox in turn affects the microbial processes that are important in most aspects of wetland use for water quality improvement. Porewater chemistry is drastically different from that of the marsh surface waters, with nitrogen and phosphorus concentrations that are one to two orders of magnitude greater than surface water.

Saturated conditions prevail continually in the soil zone. Water flows in the action zone are driven primarily by evapotranspiration and capillary suction, with exceptions being infiltrating systems. The typical organic soils are capable of exerting up to a meter of suction pressure, and as a consequence, the action zone is a "wet blanket" draped over surface irregularities.

Gas flows are also part of the underground activity. The action zone imports oxygen (air) via the root and rhizome system, to support oxic conditions in the vicinity of the roots. Several underground chemical reactions create gasses that are vented through plants and from the soil surface. Thus the wetland underground "breathes."

Accretion of new soil materials is a common feature of all treatment marshes. Accumulations are due to either or both of incoming solids deposition and internal solids generation. The decomposition of deposited organics leads to carbon dioxide and methane emissions, and creates vertical shrinkage that accompanies deposition. New accretions can eventually affect hydrology, in addition to participating in removal of various pollutants.

5 Microbial and Plant Communities

This chapter is about the aboveground plant communities and their associated microbial populations. These plant assemblages are the dominant visual components treatment marshes. The characterization of the wetland vegetative communities depends strongly upon the background discipline of the researcher. A taxonomist or phytosociologist is focused on the listing of the individual species that are present. An ecologist may well be concerned with the biological diversity and richness of the system, and the productivity and community dynamics. An aquatic botanist is likely to emphasize the functional details of plant physiology, and their adaptations to environmental conditions. An engineer may use only broad concepts of vegetation, to tie together data analyses and predictive modeling.

The purpose of this book is to better understand the role of vegetative communities in the pollutant removal performance of the marsh. Those removals are virtually all mediated by the microbial and plant communities, which operate at different speeds. To that end, all disciplinary points of view can contribute. But, a somewhat different method of dividing and compartmentalizing the subject matter is useful. Three groupings are considered here: macrophytes (large plants), mesophytes (small but visible), and the microbiota (need a microscope). It should be noted that the term mesophyte has another common meaning; it refers to plants that survive on a modest amount of water. Herein, it means aquatic plant assemblages of intermediate size. There is not a spatial distinction, because these groups are interspersed in the aboveground marsh structure. And this is not just a size categorization, because these three groups differ in functions and the speed of processes.

5.1 GENERAL CATEGORIZATION

MACROPHYTES

Table 5.1 lists plant families and genera that are typically dominant in treatment marshes. There are very large numbers of other macrophytes that will also be found in a marsh. For example, at the Houghton Lake marsh, there were over 20 species represented in addition to the dominant *Typha* (Kadlec and Bevis, 2009). Schwartz et al. (1994) found 185 species in the Orange County wetland, in which 21 species were planted.

Macrophytes serve several important functions in treatment systems (Brix, 1997), with five being dominant in treatment marshes: (1) they provide a large surface area for microbes, (2) they assist in filtration of suspended material, (3) they control the energy flows that support evapotranspiration and modulate water temperature, (4) they control water depth and flow, and (5) they utilize nutrients and other substances, and convert them to new wetland residuals.

MESOPHYTES

This grouping is actually comprised of both flora and fauna. The parent assemblage is typically dominated by floristic components, but also contains invertebrate fauna and microbes. A further subdivision is typically employed (Figure 5.1).

TABLE 5.1

Classification of major macrophytic plants occurring in treatment marshes. All are angiosperms in the group monocots.

Family	Typical Genera	Typical Growth Habit
Araceae	*Lemna, Pistia*	Floating
Ceratophyllaceae	*Ceratophyllum*	Submergent
Cyperaceae	*Cyperus, Carex, Cladium, Eleocharis, Scirpus, Bolboschoenus*	Emergent
Haloagaceae	*Myriophyllum*	Emergent
Hydrocharitaceae	*Hydrilla, Elodea, Najas*	Rooted, submergent to floating
Juncaceae	*Juncus*	Emergent
Nymphaeaceae	*Nuphar, Nymphaea*	Floating leaved
Poaceae	*Panicum, Phragmites, Arundo, Phalaris, Glyceria,*	Emergent (occasionally rooted with floating leaves)
Pontederiaceae	*Eichhornia, Pontederia*	Floating and emergent
Potamogetonaceae	*Potamogeton*	Submergent
Typhaceae	*Typha*	Emergent

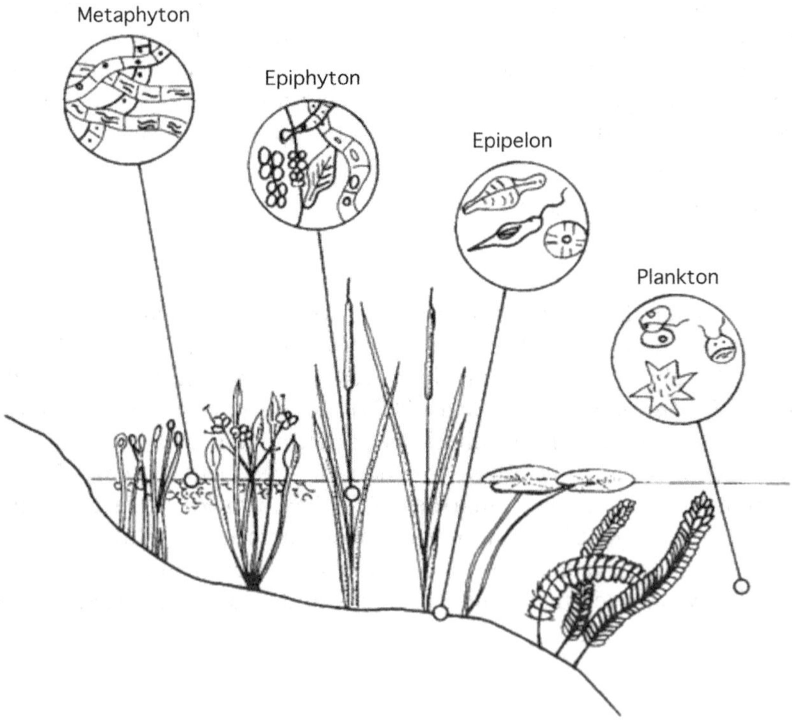

FIGURE 5.1 Algal assemblages found in freshwater marshes.

In the literature, there is a lot of confusion and controversy about terminology (Vymazal, 1995). Although here the choice is for the categories depicted in Figure 5.1, there are other common usages. The subdivisions used here are those proposed and used by Stevenson et al. (1996) and van der Valk (2012).

Periphyton refers to microfloral growth upon substrata in fresh waters. It is a mixture of algae, cyanobacteria, microbes, and detritus that is attached to submerged surfaces in most aquatic ecosystems. A more explicit manner of expression is to refer to the organisms with appropriate modifiers descriptive of the substrata upon which they grow in natural habitats. These algal communities can be classified as epipelic, algae growing on sediments, called *epipelon*; or epiphytic algae, growing on macrophytic surfaces, called *epiphyton*. Periphyton can also mean floating or anchored cyanobacterial mats, which is the common usage in south Florida.

Phytoplankton consist of the assemblage of small plants having no or very limited powers of locomotion; they are therefore more or less subject to distribution by water movements. Planktonic algae move by means of flagella, or possess various mechanisms that alter their buoyancy. However, most algae are slightly denser than water, and sink, or sediment from, the water.

Metaphyton is a group of aggregated algae that is neither strictly attached to substrata nor truly suspended. It can be clouds of filamentous algae, such as *Spirogyra*. In the Florida Everglades, metaphyton consists of mats of blue-green algae, laden with calcareous depositions. At times, these mats may reside on the bottom.

Because mesophytes are located within the wetland water column, they can utilize and process substances directly from the water. While this ability is also present to some degree in submerged macrophytes, it is the sole source for mesophytes. These collections of small organisms are also capable of rapid growth and death cycles, and the attendant rapid uptake of nutrients. The cycle speeds are many times faster than the cycle speeds of the macrophytes. The biomass may be fairly low, but the nutrient storage capacity can rival that of the macrophytes.

Microbes

Diverse microbial populations are ubiquitous on the solid surfaces afforded by macrophytes and mesoflora. The numbers of free moving bacteria in the water is very small by comparison. Of course, microbes also pervade the wetland underground, as discussed in the previous chapter. Microbial metabolism not only depends on the presence of appropriate enzymes but also on environmental conditions such as temperature, dissolved oxygen (DO), and hydrogen ion concentration (pH). The concentration of the chemical substrate undergoing the transformation is of primary importance in determining reaction rate. In later chapters, the role of various types of bacteria in removal of specific pollutants will be explored further. They are a major factor in processing of nitrate and ammonia nitrogen, phosphorus, and carbon compounds

Emergent wetland plant species are the primary concern of this chapter because they provide surface area for epiphytic and microbial growth, which is important in many of the chemical assimilation processes in wetland treatment systems. The associated cycle of growth, senescence, death and decomposition is also a major factor for the light nutrient loadings that define treatment marshes.

5.2 MICROBES AND MESOPHYTES

Microbial Specifics

Almost all nutrient cycling processes in wetlands are mediated by bacteria (Batzer and Sharitz, 2014). Bacteria are found free-floating in water, growing on plants, litter and sediments, and growing in the soil.

Microbes can be classified by their metabolic requirements. Two categories are commonly assigned to wetlands, autotrophs and heterotrophs.

Autotrophs are organisms that produce organic compounds, such as carbohydrates, from simple substances present in their surroundings, usually carbon dioxide. They use energy from light (photosynthesis) or inorganic chemical reactions (chemosynthesis). Phototrophs use light as an energy source, while chemotrophs use electron donors as a source of energy, whether from organic or inorganic sources; however in the case of autotrophs, these electron donors come from inorganic chemical sources.

An important group of autotrophs are the cyanobacteria, blue-green algae that can occur in wetlands. Nitrogen fixation is a metabolic pathway found in these bacteria. Many groups of anaerobic bacteria can fix nitrogen, and some aerobic organisms possess special units that function anaerobically to fix nitrogen (van der Valk, 2012).

Chemoautotrophs generally fall into several groups: methanogens, halophiles, sulfur oxidizers and reducers, nitrifiers, anammox bacteria. They do not have photosynthetic pigment and hence utilize chemical energy to reduce CO_2 to organic food. The chemical energy is obtained from the oxidation of certain chemicals such as ammonia, nitrites, methane, carbon monoxide, molecular hydrogen, iron salts, sulfur, and sulfur compounds. These reactions are the suite that defines most of the chemical removals that occur in treatment marshes. The microorganisms involved include nitrifying bacteria, some denitrifying bacteria, iron bacteria, and sulfur bacteria.

Chemoheterotrophs derive energy from organic compounds and also use the same or other organic compounds for cell synthesis. Heterotrophic bacteria conduct the decomposition of wetland plant, sediment and soil material, mostly via aerobic respiration. However, heterotrophic bacteria can use a wide range of substances as oxidants or electron acceptors (Batzer and Sharitz, 2014).

The majority of denitrifying bacteria are facultative aerobic heterotrophs that switch from aerobic respiration to denitrification when oxygen as an available terminal electron acceptor runs out. This forces the organism to use nitrate to be used as a terminal electron acceptor. The diversity of denitrifying bacteria is large, and this group can exist in a wide range of habitats.

Phosphorus is involved in microbial metabolism. It comprises something like 1.5–2.5% of microbial tissue (Reddy and Delaune, 2008). In the wetland environment, dominant forms of organic phosphorus include monoesters, diesters, and nucleic acids (Turner and Newman, 2005; Turner et al., 2006). Monoesters include inositol phosphates, sugar phosphates, and mononucleotides. Much of the organic phosphate in plant and microbial residues is phosphate diesters such as phospholipids and nucleic acids.

Microbial Biomass Model

Microbes exist at very high numerical density in wetlands, with reported number densities of 10^6–10^{10} per liter. But, because of their tiny size (0.5–2.0 μm), they are less dominant on a mass basis. The inventory of microbial biomass reflects both growth-death processes, and a dependence upon the substrate that subsidizes growth. In simplest form, the generally accepted formulation is (Metcalf and Eddy, 1991; Maier et al., 2000):

$$\frac{dX}{dt} = \mu_{\max}\frac{S}{S+K_S}X - k_d X \tag{5.1}$$

where X = microbial biomass, gm/m^2

S = substrate concentration, gm/m^2

k_d = death rate coefficient, d^{-1}

K_S = substrate half saturation constant, gm/m^2

μ_{max} = specific growth rate coefficient, d^{-1}
t = time, d

Equation 5.1 describes the batch behavior of a collection of microbes and must have flow terms for a flow-through system. The substrate concentration also requires a mass balance, so the treatment wetland situation requires considerable modification. Wetland applications are rare, but have been set forth by McBride and Tanner (2000) and Pálfy and Langergraber (2014) for gravel bed wetlands.

Bacteria have very high metabolic rates, with generation times on the order of one hour (van der Valk, 2012). A key feature of this model and analogs is that the rate coefficients, μ_{max} and k_d, are quite large; on the order of 1.0 d^{-1}. This means that microbial communities rapidly respond to changing conditions, within a day or so. The rate coefficients are typically quite temperature sensitive. These high-speed processes make it very difficult to include details in a removal model that also has to deal with seasonal and annual behavior.

The Mesophytes: Algae

The assemblage of primitive plants that are collectively referred to as algae includes a tremendously diverse array of organisms. Algae may size from single cells as small as one micrometer to large seaweeds that may grow to over 50 meters. Many of the unicellular forms are motile. Algae are ubiquitous, they occur in every kind of water habitat (freshwater, brackish and marine). Algae are often an important component of a treatment wetland, either as an early colonizing community or as the intended dominant design community. The reader is referred to Vymazal (1995) and Stevenson et al. (1996) for a more complete description of algae and element cycling in wetlands.

For the most part, algae depend on light for their metabolism and growth and serve as the basis for an autochthonous foodchain in aquatic and wetland habitats. Organic compounds created by algal photosynthesis contain stored energy which is used for respiration or which enters the aquatic foodchain and provides food to a variety of microbes and other heterotrophs. Alternatively, this reduced carbon may be directly deposited as detritus to form organic peat sediments in wetlands and lakes.

Algae also depend on an ample supply of the building blocks of growth including carbon, typically extracted from dissolved carbon dioxide in the water column, and on macro and micronutrients essential to all plant life. When light and nutrients are plentiful, algae can create massive populations and contribute significantly to the overall food web and nutrient cycling of an aquatic or wetland ecosystem. In many treatment marsh waters, high algal productivity can pull CO_2 out of the water, shift the whole carbonate equilibrium, and drive the pH *as* high as 9 or 10 on a diurnal basis. Under these conditions, co-precipitation of phosphorus with calcium phosphate can be significant.

Plankton

Plankton is composed of small organisms suspended in the water column. These organisms are often carried by the surface water currents. They are often denser than water, and require those currents to keep them in suspension (van der Valk, 2012). Phytoplankton encompasses a surprising range of cell size and cell volume from the largest forms visible to the naked eye, for example, *Volvox* (500–1500 μm), to the algae as small as 1 μm in diameter (Vymazal, 1995). Phytoplankton algae are mainly unicellular, though many colonial and filamentous forms occur, especially in fresh waters.

Planktonic or free-floating algae are generally not important in marsh ecosystems unless open or deep-water areas are present. In the case of treatment marshes, there is an exception, which is the possibility of high abundance of phytoplankton in the source waters. For instance, the treatment marshes north of Lake Apopka in central Florida receive waters with high concentrations of planktonic TSS, and the accompanying nitrogen and phosphorus (Bachmann et al., 1999; Coveney et al., 2002; Dunne et al., 2012, 2013). In this situation, the processing of incoming planktonic biomass and necromass is of central interest for understanding performance.

Epipelon

Benthic algal communities are not commonly an important feature of treatment marshes, because there is rarely any bare bottom for attachment. Exceptions are the deep zones in the marsh, if any, and the intentional exclusion of macrophytes. Examples of the later are the open water constructed wetlands designed to maximize photolytic processes (Jasper et al., 2013). Such systems are shallow basins that contain macrophyte soil barriers and are maintained open by removal of floating vegetation. What remains is an aquatic system with a benthic mat, which is a consortium of organisms dominated by algae, aerobic bacteria.

Epiphyton

Epiphytic algae generally show little substrate specificity. In spite of this relative indifference of epiphytic algae to their substrate, the epiphytic habitat has several distinctive attributes. The surface itself has a definite life span. New leaves are colonized as they develop during the growing season resulting in a summer and autumn peak in epiphytic biomass and productivity. The canopy of aquatic macrophytes often creates light-limiting conditions for epiphytic algae. On the other hand, decreases in growth and photosynthetic rates, as well as abundance and occurrence of submersed macrophytes, have been attributed to light attenuation by the periphyton complex.

Several methods have been reported for determination of the biomass and chemical composition of epiphytes (Goldsborough, 2001). The epiphytic material is removed from leaves of SAV by: (1) shaking the plants in water, (2) scraping or brushing their surfaces, or (3) chemical dissolution in dilute acid. The last of these relies upon the calcitic nature of the epiphyton formed in hard water (Ostrofsky and Miller, 2017). Epiphytic encrustations can be up to 20% of the weight of the leaves upon which it resides, and up to 3 cm thick (Scinto and Reddy, 2003).

Epiphytic algae generally show little substrate specificity; many epiphytic species are encountered in natural epilithic communities and on artificial substrates. In spite of seeming relative indifference of epiphytic algae to their substrate, the epiphytic habitat has several distinctive attributes. The surface itself has a definite life span. New leaves are colonized as they develop during the growing season resulting in a summer and autumn peak in epiphytic biomass and productivity. The canopy of aquatic macrophytes often creates light-limiting conditions for epiphytic algae (Darley, 1982). On the other hand, decreases in growth and photosynthetic rates, as well as abundance and occurrence of submersed macrophytes, have been attributed to light attenuation by the periphyton complex (Vymazal, 1995).

Periphyton

Periphyton in its broad definition includes all aquatic organisms (microflora) growing on submergent substrates. Although periphyton usually begin colonization of new plant surfaces by attached algal growth of filamentous and unicellular species, this functional component also includes a variety of free-living algae (not attached to the surface), fungi, bacteria, and protozoans following a period of maturation. Periphyton growing on plants is here called epiphyton. These surfaces frequently include living vascular plants as well as dead plants, leaves, and exposed substrates. Benthic algae are here called epipelon.

Metaphyton

Filamentous algae typically occur in marshes as periphyton or mats may dominate the overall primary productivity of the wetland, controlling dissolved oxygen and carbon dioxide concentrations within the wetland water column. They are opportunistic, because they can grow very rapidly compared to macrophytes. Therefore, the early period of constructed wetland life may create ideal conditions for algal establishment. However, macrophytes can later easily shade out the algae.

The calcareous algal mats that form in subtropical climates, under conditions of sparse macrophyte densities, are a second type of metaphyton (Rejmánková and Komárková, 2000; McCormick et al., 2006). Figure 5.2 shows that both floating (metaphyton) and benthic (epipelon) mats can coexist.

Algal Biomass Models

The literature is rich with models of algal communities, dating to the stream periphyton models of the 1970s (McIntire, 1973; McIntire and Colby, 1978). These are exemplified by the water quality module of the Environmental Fluid Dynamics Code (EFDC) (Tetra Tech, Inc., 2007; U.S. EPA, 2016b). EFDC is a multifunctional surface water modeling system, which includes hydrodynamic, sediment-contaminant, and eutrophication components.

The inventory of algal biomass reflects growth, death (metabolism and respiration in EFDC), predation and settling processes. Additionally, there is potential import and export in flow through systems. With notation consistent with other use herein, the model is:

$$\frac{\partial B}{\partial t} = kB - k_D B - k_P B - \frac{1}{h} k_S B + \frac{1}{A}(Q_i B_i - Q_o B) \tag{5.2}$$

where A = wetland area, m^2

B = algal biomass, gm/m^2

FIGURE 5.2 A metaphyton-dominated wetland contains sparse emergents, for protection an anchoring. The algal mats are both floating and resting on the bottom.

B_i = incoming algal biomass, gm/m^2
h = water depth, m
k = growth rate coefficient, d^{-1}
k_D = death rate coefficient, d^{-1}
k_P = predation rate coefficient, d^{-1}
k_S = settling rate coefficient, m/d
Q_i = inlet flow rate, m^3/d
Q_o = outlet flow rate, m^3/d
t = time, d

The production rate coefficient (k_P) is dependent upon light intensity, temperature and nutrient availability, which may be written for an extreme of simplicity as:

$$k = k_{20}\left(\frac{I}{I + K_I}\right)\left(\frac{C}{C + K_C}\right)\left(\theta^{(T-20)}\right) \tag{5.3}$$

where k_{20} = growth rate coefficient at 20°C, d^{-1}
C = nutrient concentration, g/m^3
I = light intensity, langleys/d
K_C = half saturation nutrient concentration, g/m^3
K_I = half saturation light intensity, langleys/d
T = water temperature, °C
θ = temperature coefficient, –

Of course, there is usually more than one nutrient involved, and light penetration can be much more complicated. EFDC also speculates upon dependence of the death rate and predation rate upon temperature. However, the code does not track predator populations, and assumes they consume in proportion to availability of algae. Equations 5.1 and 5.2, in their respective expanded forms, are applied in EFDC to four kinds of algae: cyanobacteria, diatoms, greens, and stationary. The stationary category is intended to apply to macroalgae, such as *Chara* spp., or to epipelon or periphyton.

The concepts of the EFDC algal model were incorporated into a specific attempt to use it to describe the ENRP treatment marsh (Moustafa and Hamrick, 2000), called the Wetland Water Quality Model. It was exceedingly computationally cumbersome and required inordinate amounts of computer time. Most of the over 200 coefficients had to be estimated, because of insufficient calibration data. As a result, this model has not been put to use.

CE-QUAL-W2 is a two-dimensional, longitudinal/vertical, hydrodynamic and water quality model (Cole and Wells, 2016). The algal portion of the model is essentially the same as the EFDC model, but considers the "death rate" in Equation 5.2 to be comprised of mortality, respiration and excretion. It also considers that for epiphyton, there is no transport or settling, but rather there can be direct burial.

This model structure, specifically tailored for epiphytic algae, was calibrated to a SAV-algal system (Zhang et al., 2015). The algal portion of the model retains several of the terms of Equation 5.2, but includes the modifications that account for the fact that the community is attached:

$$\frac{\partial B}{\partial t} = kB - k_D B - k_E B \tag{5.4}$$

where k_E = rate coefficient for sloughing, d^{-1}

A key feature of these models is that the rate coefficients are quite large. Hence turnover times for algae are of the order of 100 yr^{-1}. This means that the algal communities rapidly respond to changing conditions, within a week or so.

Water Chemistry Effects

Because of the use of nutrients from the overlying water, epiphyton and metaphyton can play an important role in nutrient cycling (Scinto and Reddy, 2003; McCormick et al., 2006; Ostrofsky and Miller, 2017). Algae store and transform essential growth nutrients in wetlands and aquatic habitats. Because of their relatively low contribution to the overall fixed carbon in wetlands, algae do not constitute a major storage reservoir for these elements in wetlands. However, because of their high turnover rates in some aquatic habitats, algae may be important for short-term nutrient fixation and immobilization with subsequent gradual release and recycling. The functional result of this nutrient cycling is that intermittent high inflow concentrations of pollutants used by algae for growth may be immobilized and transformed more effectively than would be possible without these components, thereby reducing the amplitude of wetland constituent outflow concentrations.

Dissolved oxygen (DO) profiles in wetlands with substantial populations of algae and submerged plants undergo major diel changes. Wetland water column DO can fluctuate from near zero during the early morning following a night of high respiration to well over saturation (> 20 mg/L) in high algal growth areas during a sunny day. Dissolved carbon dioxide and consequently the pH of the water vary proportionally to DO because of the corresponding use of CO_2 by plants during photosynthesis and release at night during respiration. As CO_2 is stripped from the water column by algae during the day, pH may rise by 2 to 3 pH units (a 100- to 1000-fold decrease in H+ concentration). These daytime pH changes are reversible, and the production of CO_2 at night by algal respiration frequently returns the pH to the previous day's value by early morning.

The structure of algal populations is sensitive to water quality (McCormick et al., 1997a, 1997b). Consequently, algae respond to the gradients in water quality along the flow direction in treatment marshes (McCormick et al., 1996). In very general terms, high nutrient status causes light limitations that reduce algal density. Additionally, high nutrients favor assemblages of green algae, whereas low nutrient status can favor blue-green (cyanobacterial) populations.

5.3 VEGETATION TYPES AND COMMUNITIES

There are many general functions of macrophytes in wetlands. Physical functions include transpiration, flow resistance, particulate trapping, all of which are related to vegetation type and density. Ecological functions include wildlife habitat and human use values. The focus here is water quality, and in particular, the processing of potential pollutants.

In turn, there are many influences of vegetation on chemical processing and removal in treatment wetlands. Several authors have presented lists of the roles of plants in treatment wetlands (Brix, 1997; Merz, 2000; Tanner, 2001b; Rejmankova, 2011)

These may include:

1. Plants and their density may regulate the water depth and flow rate.
2. The arrangement of vegetative community types influences the internal flow patterns, including short-circuiting.
3. Emergent plant transpiration drives water flows to underground regions.
4. Emergent plants influence the water temperature, via shading and the balance between evaporation and transpiration.
5. Submerged plant parts influence particulate processes, such as mixing and re-suspension, particle interception and filtration, and flocculation.
6. The plant growth cycle seasonally stores and releases nutrients, thus providing a "flywheel" effect for a nutrient removal time series.

7. The creation of new, stable residuals that accrete in the wetland. These residuals contain chemicals as part of their structure, or in absorbed form, and hence accretion represents a burial process for nitrogen.
8. Submersed litter and stems provide surfaces on which microbes and mesophyton reside. These include nitrifiers and denitrifiers, and other microbes that contribute to chemical processing.
9. The presence of vegetation influences the supply of oxygen to the water. Submerged vegetation may provide photosynthetic oxygen supply directly in the water. Emergent plants may impede surface aeration, but to some limited extent, stem flow plant oxygen flux supplies protective nitrification in the immediate vicinity of plant roots.
10. The carbon content of plant litter supplies the energy need for heterotrophic denitrifiers.
11. Aerial plant parts cause light attenuation, thus reducing the potential for algal growth, and diminishing the role of evaporation.
12. In marshes, plants provide habitat for diverse animal populations, which in turn influence chemical processing via herbivory and bioturbation.

Many of these mechanisms are structural in nature. Consequently, selecting the "perfect" species is not nearly as important as establishing a functional plant canopy. Since microbiological transformations within the wetland are a function of area available for biofilm growth, the creation of surface area by emergent aquatic plants and associated leaf litter is an important contribution to the treatment process. Plant species that provide structure year-round generally perform better than species that die below the water line after the onset of cold temperatures. For these reasons, fast-growing emergent species that have high lignin contents and that are adapted to variable water depths are the best suited for FWS systems.

Vegetation types in FWS wetlands exhibit small performance differences, but these differences are often masked by other unavoidable differences in comparable wetlands. At the time of this writing, the case for superiority of a particular plant species has not been proven or disproven. The evidence points toward minimal net differences among plant species. A more diverse mix of plant species will be better able to accommodate changes in water quality and flow. In other words, a polyculture is preferable to a monoculture.

The importance of these effects depends upon the type and purpose of the treatment wetland under consideration. Emphasis varies greatly amongst the broad group of treatment wetlands, including surface and subsurface flow, and light and heavy loadings, and different target pollutants. For instance, oxygen supply (number 9 earlier) implications depend upon the need for oxygen and the internal location of that need.

Vegetation and Oxygen

A great deal is known about oxygenation processes in marshes, which include surface reaeration of the water, photosynthetic production of oxygen, and delivery of oxygen to the root zone. It is established that the presence of an emergent canopy, whether EAV or FAV, can deplete oxygen with respect to open water (e.g., Hamilton et al., 1995). But, SAV fosters underwater oxygen production (e.g., Drapcho and Brune, 2000; Colmer and Pedersen, 2008). Therefore, when the various plant communities in a treatment marsh are studied to determine water column oxygen, large differences are observed. Chimney et al. (2006a) documented these differences for the ENRP treatment marsh (Figure 5.3).

Thus, if there is a need for oxygen in the water column, EAV and FAV are not favored, but SAV is better than open water.

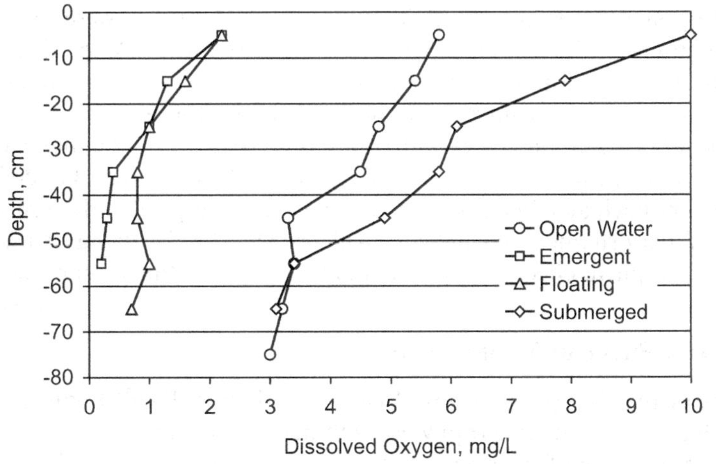

FIGURE 5.3 Vertical profiles of dissolved oxygen in the various vegetation types in the Everglades Nutrient Removal Project FWS wetlands, Florida. Data are from 141 profiles collected over a 2.5-year period. (Based on data from Chimney et al., 2006a)

Alternatively, the need for oxygen may be in the root zone. That would be the case for a marsh with dominant transpiration flows to the root zone, which accompany the low hydraulic loadings that are characteristic of the treatment marshes under consideration here. It would also apply to subsurface flow wetlands (Huang et al., 2010). There are large numbers of studies on the mechanism of oxygen transfer by plants to their root zone, usually termed radial oxygen loss (ROL) (e.g., Inoue and Tsuchiya, 2008; Lai et al., 2012; Cheng et al., 2014; Mei et al., 2014). These studies also suggest that pollutant removal hinges upon ROL.

Thus, if there is a need for oxygen in the root zone, EAV and FAV are favored.

Carbon compounds (BOD) require oxygen, and ammonia (NH_3-N) may require oxygen for microbial conversion to by-products (CO_2, N_2, N_2O). Ammonia can also be removed via plant uptake and growth in treatment marshes (light loadings). The removal of high loadings of BOD and ammonia in the water column of free water surface (FWS) wetlands would speculatively be enhanced by delivery of oxygen to the water, which in turn implies the absence of EAV or FAV. This idea led to the publication of a USEPA design manual that specified a central open water area, containing SAV, in FWS wetlands (U.S. EPA, 2000a; Kreissl and Brown, 2000). This USEPA Manual focuses on constructed wetlands designed to be the central process in a wastewater treatment system. Clearly, these concepts do not fit the range of water chemistry in a treatment marsh, as considered herein, as was recognized by the authors:

> Unlike most constructed wetlands reported in the literature, which are lightly loaded tertiary systems used to polish wastewater, these "blue collar" passive systems perform most of the treatment.

Based primarily upon oxygen transfer, the U.S. EPA (2000a) manual argues for a three-zone system design: shallow emergent marsh, followed by an open water/SAV component, ending with a shallow emergent marsh. Others have speculatively selected different sequences, such as the four zone concept of Merz (2000). His four zones were: deep, open water and SAV (>1 m); deep marsh (0.4–0.6 m), emergents; marsh (0.2–0.4 m), emergents;

shallow marsh (0.1–0.2 m), emergents. Merz (2000) specified emergents-only in the marsh sections, based upon the perceptions that "submerged plants ... were vulnerable to being overgrown by biofilms ... ", and that floating plants "present a risk to effective treatment."

Yet other studies have found essentially no effect of deep, open water/SAV zones on either oxygen concentrations or nitrogen removal (Kadlec, 2007).

This example of oxygen processes illustrates that treatment marshes can differ greatly from FWS systems in general, in terms of the effects of vegetation. Guidelines and design criteria based on municipal effluent treatment cannot be applied indiscriminately, although some elements of performance and analysis will transfer. In following chapters, methods of accounting for macrophyte effects on specific pollutants in treatment marshes will be sought.

Classification and Physical Arrangement

The four groups of marsh aquatic macrophytes (see Chapter 1 and Figure 1.2) are emergent macrophytes, floating-leaved macrophytes, submersed macrophytes, and freely floating macrophytes. The category of unvegetated open water also needs consideration. In addition, a large number of the emergent macrophytes can exist in floating mats, either with or without a supporting structure. Floating mats are discussed in the next section.

Treatment marshes are often thought of as aquatic ecosystems containing emergent plants, thus focusing upon one of these four categories. Vymazal (2013) has reviewed the usage of various species of emergents in FWS treatment wetlands of all kinds. Worldwide, he found 643 FWS systems and a total of 150 species of emergents. In Europe and Asia, *Phragmites* is the most used, while in the Americas and Oceania, *Typha* and *Scirpus* are the most common (Table 5.2).

Botanical classifications provide an accurate concept of the types of macrophytes that may appear in treatment marshes but are generally insufficient to characterize a full-scale operating system. It is necessary to know *where* the vegetation is located, in addition to *what* is growing there. An overall treatment marsh system is comprised of one or more cells, which may or may not have varying vegetation types, intentionally or unintentionally. Polycellular systems are quite common, containing up to 30 cells in series and parallel in one system.

Coverage Fractions

Coverage fraction is defined to be the area of a given community type divided by the total cell area (Table 5.3). The community types may be either species specific, or macrophyte

TABLE 5.2

Percentages of commonly used macrophyte genera in FWS constructed wetlands.

	Asia	Europe	North America
Typha	33%	35%	54%
Scirpus	7%	13%	28%
Phragmites	54%	46%	4%
Juncus	3%	5%	7%
Eleocharis	4%	1%	7%

Based upon Vymazal (2013).

TABLE 5.3
Vegetation community percentages in the cells of the ENRP. These are four-year means.

Cell	SAV/OW	FAV	Typha	Other EAV	Total
Buffer	0.063	0.386	0.430	0.121	1.000
1	0.416	0.060	0.354	0.170	1.000
2	0.228	0.073	0.678	0.022	1.000
3	0.158	0.009	0.387	0.446	1.000
4	0.946	0.009	0.032	0.013	1.000

type specific. If the treatment marsh is compartmentalized, then fractions may differ from cell to cell. This measure serves to specify the relative proportions of species or types within the mosaic of a wetland cell.

For example, the ENRP system consisted of five cells, of varying community types (Table 5.3). Cell 4 was established and managed with minimal EAV, and the other cells developed via natural recruitment. No single macrophyte designation can describe the vegetation of this system as a whole.

Constructed treatment marsh systems treating runoff often contain deep cells near the inlet. These are usually devoid of vegetation, and function as settling basins for removing suspended sediment. Such sedimentation basins may or may not be considered to be part of the overall treatment marsh. For purposes of performance analysis and description, it is preferable to consider them separately.

Treatment marsh cells may or may not contain monotypic stands that dominate the entire system. Commonly, there exist areas of vegetation that are termed patches. Vegetation influence factors also include patch fractions and patch arrangement.

Fringing, Banded and Clumped Vegetation

Many natural wetlands are riverine, meaning that the wetland contains a central stream channel, which may meander through a zone of emergent vegetation. Indeed, this wetland form is so common that it is mistakenly presumed to be a good prototype for a treatment wetland. Nothing could be further from the truth. A central channel will convey most of the water, and the fringing vegetation will effectively receive flow only on rising flood conditions and discharge it on the ebbing stage. Simply stated, if the water needs to go through the plants, don't plant them off to the side. Jenkins and Greenway (2005) provide numerical calculation results for several aspect ratios and cover percentages, which show that hydraulic efficiencies are low for fringing vegetation.

The Jenkins and Greenway (2005) computational research demonstrated that cross-banded vegetation is a better option for hydraulic efficiency than fringing vegetation. Therefore, any open water that may be included in the cell should be oriented across the flow direction, whether or not it is deeper than the average. However, adequate conveyance may be difficult to maintain if large aspect ratios are combined with significant banded or uniform vegetation.

These concepts were utilized in the Imperial, California treatment wetland (Figure 5.4). The first cell was built with a sinuous path and fringing vegetation; the second cell was built with banded vegetation. Only overall system water chemistry data were taken, thus no information on relative efficiencies is available.

FIGURE 5.4 Fringing (A) and banded (B) vegetation patterns in the Imperial, CA treatment marsh.

A third concept for the vegetation pattern is the use of small islands or hummocks, either emergent or submerged (Thullen et al., 2005). At the present time, there is no information on the hydraulic efficiency of this alternative, nor any pollutant removal comparison.

Deep Zones

Unvegetated deep zones in surface flow constructed wetlands are supposed to serve several purposes (Knight and Iverson, 1990). These deeper areas extend below the bottom of the vegetated basin areas by at least 1 m to discourage the development of rooted emergent macrophytes (Figure 5.5). When situated at the wetland inlet, deep zones serve to distribute water across the width of the wetland. When situated at the wetland outlet, deep zones serve to collect water evenly across the width of the wetland.

Internal to a marsh cell, such unvegetated cross ditches provide a low resistance path for water to move laterally and provide a nearly constant head across the wetland. They also provide for extra detention time, but in a deep-water zone. Such ditches often become covered with duckweed (*Lemna* spp.) or can contain SAV. These redistribution ditches change the potential for short-circuiting within the wetland, because high-speed rivulets are intercepted and mixed with quiescent water in the deep zone. However, the redistribution ditch adds a potential for wind mixing that compensates the reduced short-circuiting. But water is more effectively distributed over the wetland, improving the gross areal efficiency to some extent. Lightbody et al. (2007, 2008) provided evidence from laboratory models and

FIGURE 5.5 The Saginaw, Michigan treatment wetland system contains one intermediate deep zone, in addition to inlet and outlet deep zones. *Lemna* spp. covers this zone.

computations that indicated improvement due to deep zones at high degrees of removal. However, that improvement was contingent on the assumption that serious short-circuiting occurred in the vegetated sections of the system.

Most marshes treating urban stormwater contain an unvegetated deep pond or forebay at the inlet. This is intended to serve as a repository for settled solids, preventing them from clogging the vegetation in the remainder of the system. This deep-water area also provides storage for captured events, and is typically sized as a fraction of the overall footprint. For instance, Schueler (1992) show options with 10–45% of the area allocated to the deep inlet pool. The balance of the footprint is usually devoted to shallow and deep emergent marsh, and possibly to deep cross-zones. This patterning of the bathymetry is completely in agreement with the principles elucidated for good hydraulic efficiency for continuous flow systems.

Examples

The spectrum of vegetation and community structures in treatment marshes is very large. It is of interest to consider a potpourri of examples that demonstrate that diversity among systems.

Tucush

A water treatment facility was constructed in 2005/2006 to manage drainage from the expanding Tucush Valley waste rock dump at Antamina Cu-Zn-Mo mine, located in the rugged high Andes in Peru, 4,200 m above sea level. Although situated in a tropical latitude (9°31′ S, 77°03′ W), the climate is cold, with monthly mean temperatures in a narrow annual range of 7–11 °C. The wetlands cover 6 ha and are designed to treat up to 115 L/s for removal of nitrates (ca. 8 mg/L), plus trace ammonia and metals (Manrique et al., 2012).

The project is the highest operating wetland in the world and provides an important precedent for water treatment at other operating and abandoned high altitude mines, particularly in the Andes. Monitoring to date indicates the wetland is functioning as designed with effective removal of nitrate and zinc. However, all water parameter concentrations measured at the treatment system discharge point are compliant with established water quality requirements.

The wetlands were planted with *Scirpus californicus* and *Juncus arcticus*, but volunteer species also occur (Table 5.4). Each of the 30 wetland cells displays a spatial pattern of vegetation types (Figure 5.6).

Onaway

The City of Onaway, Michigan has implemented a wastewater treatment system consisting of aerated lagoons, infiltration beds, ferric chloride addition and UV disinfection. This

TABLE 5.4

Coverage fractions for the dominant species of macrophytes in the Tucush treatment marsh. (Note: percentages need not add to 100 because of understory and overstory accounting.)

Genus	Species	Inlet	Inlet	Outlet
		Parallel 1	Parallel 1	
Rosaceae	*Alchemilla pinnata*	47	0	0
Poaceae	*Calamagrostis stiligulata*	4	1	0
Juncaceae	*Juncus arcticus*	1	17	12
Juncaceae	*Juncus bufonius*	23	24	19
Brassicaceae	*Rorippa nasturtium-aquaticum*	10	9	31
Cyperaceae	*Scirpus californicus*	13	57	40
	Open water	9	8	8
	Total	96	108	102

FIGURE 5.6 The vegetation mosaic in a cell of the Tucush treatment marsh.

combination produces an effluent with very little BOD or TSS, but with 0.68 mg/L phosphorus. The discharge from the system is further polished in a three cell, 3.13 ha constructed treatment marsh. This system operates year-round in fairly cold climate, with a mean annual air temperature of 7°C, and four months of the years averaging below freezing (Kadlec, 2008b).

The vegetation in the marsh is dominated by *Typha* spp. and *Lemna* spp., which occur in a randomly patchy spatial pattern (Figure 5.7).

Wakodahatchee

The Wakodahatchee wetlands are a nine-cell, 16-ha constructed treatment natural treatment system, built in 1996. The incoming water is high in both nitrogen and phosphorus (Bays et al., 2000a). The system is located in Palm Beach County, Florida. There is large public use of this wetland (Ghermandi, 2016).

Emergent marsh zones were planted with, and are composed of, bulrush (*Scirpus validus, S. californicus*), duck-potato (*Sagittaria lancifolia*), arrowhead (*Sagittaria latifolia*), spikerush (*Eleocharis cellulosa*), fireflag (*Thalia geniculata*), and pickerelweed (*Pontederia cordata*) (Figure 5.8). Duckweed (*Lemna* sp.) has become naturally ubiquitous throughout the wetland, and there are other volunteer species as well. Cattails (*Typha latifolia*) were not planted and their anticipated colonization is controlled.

Roblin

The community of Roblin is located in west central Manitoba. It is an agricultural community of about 1,000 inhabitants. Wastewater is treated in a four-cell lagoon system prior to introduction to, during the summer months, the four-cell wetlands (Mohr, 2000). The water entering the wetlands is of high quality (BOD < 6, TSS = 9 mg/L; total N = 2 mg/L), except for phosphorus (1.3 mg/L). The 24-ha constructed wetland is a zero discharge system for the excess water during four years out of five. During flow-through operation in wet years, nutrient removal will be nearly complete because of long detention time, on the order of two months in a very wet year. The wetlands have been in operation since summer 1999.

The wetland cells were designed with islets, a range of depths from deep pools to very shallow areas, and a number of macrophyte species. The most prominent species seeded for

FIGURE 5.7 Cattails and duckweed intersperse in the Onaway treatment marsh.

FIGURE 5.8 The Wakodahatchee wetlands possess intentional and unintentional vegetative diversity, and spatial patterning. In this cell zone, duck-potato (*Sagittaria lancifolia*) is thriving next to pickerelweed (*Pontederia cordata*).

the wetland includes the common cattail (*Typha latifolia*), great bulrush (*Scirpus validus*), and reed grass (*Phragmites australis*). The deeper areas of the system developed dense growths of SAV (Figure 5.9).

FIGURE 5.9 SAV occupies the open water portions of the Roblin, Manitoba treatment marsh. Cattails dominate the shallower zones.

Richland Chambers

The George W. Shannon Wetlands, located near Corsicana, Texas, cleans water on its way to the Richland-Chambers drinking reservoir. The Trinity River is the source. The 102-ha system, established in 2003, contains four marsh cells. The target contaminants were suspended solids, nitrogen, and phosphorus (APAI, 2008; Mokry et al., 2008; Kadlec et al., 2011).

All cells were planted with giant bulrush (*Schoenoplectus californicus*), grassy arrowhead (*Sagittaria graminea*), crowfoot sedge (*Carex crus-corvi*), squarestem spikerush (*Eleocharis quadrangulata*), and coontail (*Ceratophyllum demersum*). About 60 species have subsequently been identified in cells. Dominant species include the plantings except for spikerush, and volunteer species: water primrose (*Ludwigia peploides),* muskgrass (*Chara* spp.), duckweed (*Lemna* spp.), barnyard grass (*Echinochloa crus-galli)*, and smartweed (*Polygonum* spp.). However, these occur in selected small areas, while open water and floating aquatics dominated within the marsh zones, specifically containing duckweed and algae (Figure 5.10).

Musselwhite

The Musselwhite underground gold mine is located in northwestern Ontario, Canada, at approximately 53° North Latitude (Bishay and Kadlec, 2005). The mean annual air temperature is –0.9°C, and the record low was –51.1°C. There is an average of 90 days frost-free. Water discharge from pond storage usually begins in mid-May and can continue to November 30. The 2.5-ha wetland operates, at a depth of about 30 cm and a hydraulic loading rate of 50 cm/d. The site was a former forested peatland, with the trees cut down, and logs and brush left in the wetland. This wetland provides exceptional treatment of nitrogen and phosphorus (Bishay and Kadlec, 2005).

The original ecosystem was a spruce-sphagnum wetland that also contained sedge meadows. A road now functions as a berm to retain water. A portion of that original ecosystem remains in the inlet region of the treatment wetland, comprising about 10% of the total area. The remainder of the treatment wetland is vegetated by a mixture of sedges (*Carex rostrata* and *Carex aquatilis*), horsetail (*Equisetum* spp.) and cattails (*Typha*

FIGURE 5.10 The vegetation mosaic in a cell of the Shannon treatment marsh.

latifolia). *Equisetum* is dominant overall, with sedges forming boundary bands and cattails dominating the outflow region (Figure 5.11).

Five of these six examples are not included in the Vymazal (2013) summary, which simply illustrates the near-impossibility of producing complete inventories of marsh vegetation. What is apparent is the amazing intersystem diversity of treatment marsh vegetation. It presents a daunting task of generalizing the performance of these systems, yet some intersystem generalities for performance can be elucidated, as will be shown in subsequent chapters.

SUCCESSION

Vegetative succession is the process of change in the species structure of a community over time. In general, communities in early succession will be dominated by fast-growing, well-dispersed species (opportunist). As succession proceeds, these species will tend to be replaced by more competitive species. Allogenic succession is caused by external environmental influences and not by the vegetation or its inhabitants, that is, the nutrient status of the introduced water. Autogenic succession is caused by the vegetation or its inhabitants, that is, herbivory by wetland animals. The later have been termed "ecosystem engineers" (Jones et al., 1997; Mitsch and Gosselink, 2015). In general, one may think of three successional phases: pioneer, transitional, and mature (Burrows, 2012).

Wetland macrophyte communities in treatment systems are often not mature (static), and they are often not determined by the discretionary choices of the wetland designer. The development of the vegetative community is dependent upon the antecedent landform, whether prior upland or prior wetland. These are differentiated by terminology, with prior upland systems called constructed wetlands, and prior wetland systems called natural wetland treatment systems.

For new treatment marshes on previous upland, there are generally two phases of change: startup, during which grow in takes place; and drift, during which slow changes in

FIGURE 5.11 Equisetum with sparse cattails characterize the Musselwhite treatment marsh.

the communities occur. A constructed wetland begins its existence with the vegetation placed by the constructors, or by the process of natural recruitment from the seed bank associated with the selected soils. Vegetation may be initially selected by design, by introducing purchased or harvested propagules. Such wetlands have been termed "designer wetlands." The alternative of allowing natural recruitment is termed a "self designing wetland" (Mitsch et al., 2012, 2014b). Both types of systems undergo changes driven by the imposed water quality. The continual addition of species propagules, conveyed by water, wind or animals, drives development of these ecosystems, rather than the introduced species. As a consequence, planted and unplanted wetlands often develop along similar paths (Mitsch et al., 2012). This observation is termed convergence, meaning that no matter what you start with, the mature treatment marsh will be the same.

For systems starting from existing wetlands, there is a process of community replacement, which has a variable duration. A natural wetland will have evolved over time to contain a mix of vegetation commensurate with the hydropattern and water quality conditions prior to wastewater addition. That wetland vegetation is likely to change over the course of time, as local adaptations to the treatment hydropattern and quality occur. The plant community that develops over time is a function of organic loading, hydrology, and climate.

FWS wetlands that are heavily loaded with organic matter and nutrients will typically develop a less diverse plant community since fewer plant species are able to tolerate the reducing conditions that develop under these circumstances. In treatment marshes with higher water quality, a diverse species composition has a better chance of developing.

Alterations of Natural Marsh Vegetation

Historically, there have been a few long-term treatment marsh studies that focus upon an impact zone situated within a larger natural marsh. These prominently include the Houghton Lake marsh (Kadlec and Bevis, 2009) and WCA2A (Walker, 1995; Kadlec and Walker, 1999). Nutrient loadings create zonation within the receiving wetland, with stronger effects near the point of discharge, diminishing in the direction of water flow as nutrients are stripped from the water (Figure 5.12). In general terms, the zone nearest the new discharge may undergo species alteration; zones further away may retain their species under nutrient enrichment, and at long distances the background ecosystem will continue to prevail (Lowe and Keenan, 1997). The inlet zone in which nutrients are reduced to a background concentration has been called a number of different names, such as the impacted zone, or the zone of total containment. There is extensive literature on what grows in such impact zones in phosphorus treatment marshes, and the time progressions of size (e.g., Kadlec and Bevis, 1990, 2009; Rutchey et al., 2008).

If the original natural wetland is of large areal extent, then the impact zone can grow to occupy an area determined by the nutrient loading. This is exemplified by the time progression of the impact area at the Houghton Lake treatment marsh (Kadlec and Bevis, 2009). The peatland irrigation site originally contained sedges (*Carex* spp.), willows (*Salix* spp.) and bog birch (*Betula pumila*); and leatherleaf (*Chamaedaphnae calyculata*) and bog birch (*Betula pumila*). These accounted for 87% of the whole peatland. Isolated small patches of cattail were also present, accounting for 1.8% of the cover.

The addition of treated wastewater created an impacted region. This zone, of about 94 ha extent in the fortieth year of the project, became strongly dominated by Typha spp. Aerial photography was used to delineate the area of greatest impact in 25 of the first 30 of 40 years, and satellite imagery was used in 8 of the last 25 of 40 years. Regions of dense cattail were able to be clearly demarcated on color and false-color photos due to elevated reflectance in wavelengths corresponding to elevated chlorophyll levels. Satellite imagery offers less discrimination, and was therefore adjusted to match photography in the 15 overlap years. The visually affected area grew at a decreasing rate throughout the 40-year period

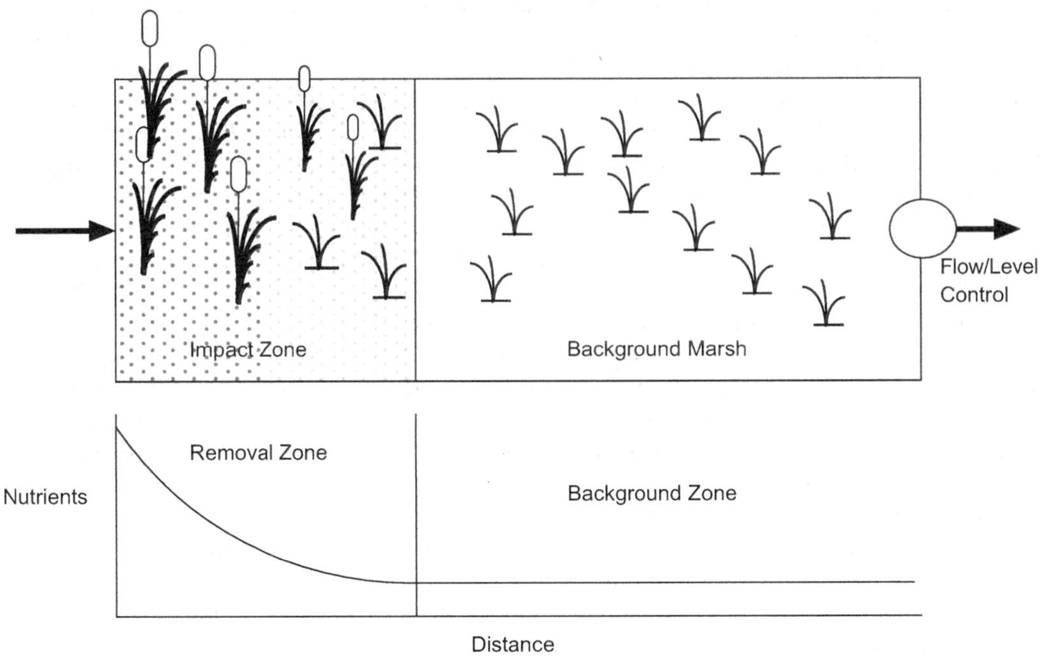

FIGURE 5.12 Nutrient inputs create zonation in the vicinity of source water addition to a natural wetland.

of record (POR) of the project (Figure 5.13). An exponential model describes the time trend in the affected area:

$$A = A_{\max} \cdot (1 - \exp(-t/b)) \tag{5.5}$$

where: A = affected wetland area, ha
A_{max} = maximum affected wetland area, ha
b = time constant for area expansion constant, yr^{-1}
t = time since discharge began, yr

This model fits the time trend (Figure 5.13) with an $R^2 = 0.91$. The maximum area was projected to be A_{max} = 87 ha. The model time constant for areal expansion, during which 63.2% of the expansion has taken place, was b = 10.3 years.

If the original natural wetland is of limited areal extent, then the impact zone can grow to occupy the entire original footprint. That occurred at Kinross, Michigan, where treated wastewater was introduced into the upstream end of a 320-ha natural wetland dominated by sedge (*Carex* spp.), leatherleaf (*Chamaedaphne calyculata*) and sparse spruce (*Picea mariana*). Source water additions began in the 1940s, emanating from a secondary WWTP built in conjunction with the U.S. air base at the site. The 100-ha wetland water track was bounded at its downstream end by a highway. Historical areal photography clearly showed the area occupied by cattail (Kadlec and Bevis, 1990). Over the course of three decades, cattails replaced the original vegetation, with the impact area [progressing from the inlet until the highway outlet was reached (Figure 5.14)].

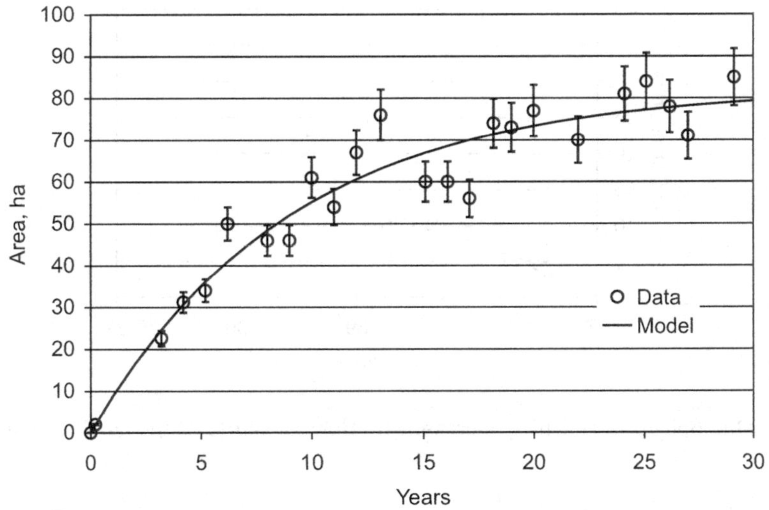

FIGURE 5.13 The visually affected area at the Houghton Lake marsh as a function of time. Circle data is from color aerial photography. Square data is from satellite imagery, adjusted to the same interpretation as photography. The fit is to and exponential rise, which calibrates to a final area of 87 ha, at the current irrigation rate. Error bars are estimated from repeated photo interpretations.

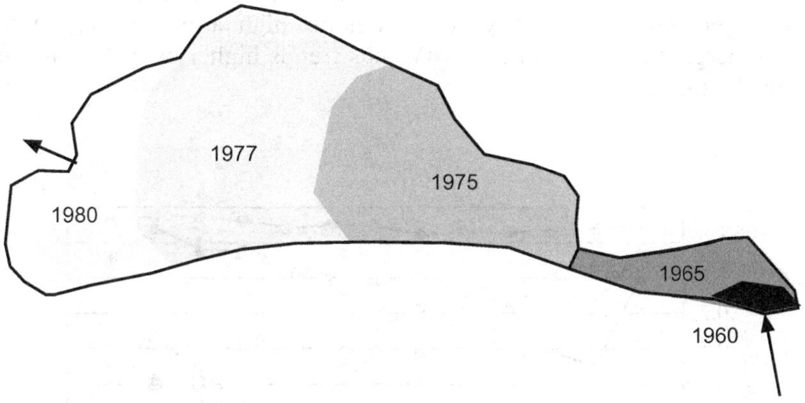

FIGURE 5.14 Cattail area expansion into a natural wetland at Kinross, Michigan, due to source water addition. (Based on data from Kadlec and Bevis, 1990.)

Startup of Constructed Marshes

There is not a large database on the length of time required for a newly constructed marsh to attain full vegetative coverage, which is the pioneer period of succession. Macrophytes vary considerably in their ability to spread, but the time scale for fill-in is surely more than weeks. One to two years seems reasonable for planted systems (Figure 5.15).

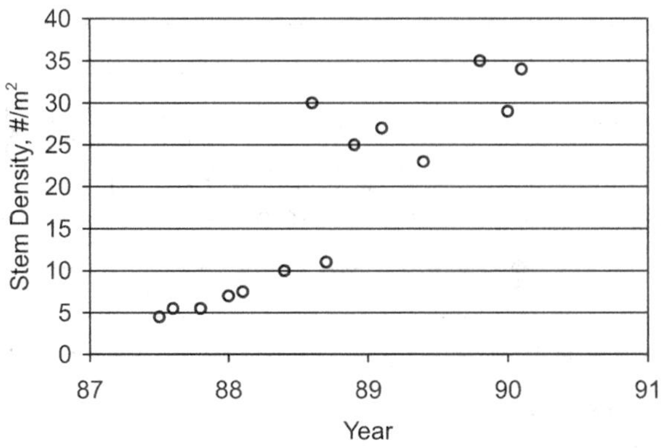

FIGURE 5.15 Macrophyte stem density increase with time for cell 11 of the OEW system.

The biofilms that conduct a good share of the pollutant removal develop in shorter times, more likely measured in weeks or months (Jones et al., 2018). And, it is very likely that microbial communities develop at even faster rates.

Transitions of Marsh Vegetation

In the absence of allogenic effects or events, vegetation types show differing degrees of stability in a treatment marsh environment. Observations of the percentages of EAV and SAV in the Florida STAs indicate that EAV is more stable over time than EAV, since the later shows a tendency to decrease over time. This is illustrated in Figure 5.16 for cells in STA34. In general, emergent coverage in EAV cells remains high and relatively stable over the 14-year POR. Emergent coverage in two SAV cells trends higher over the 14-year POR, but trends lower in the third.

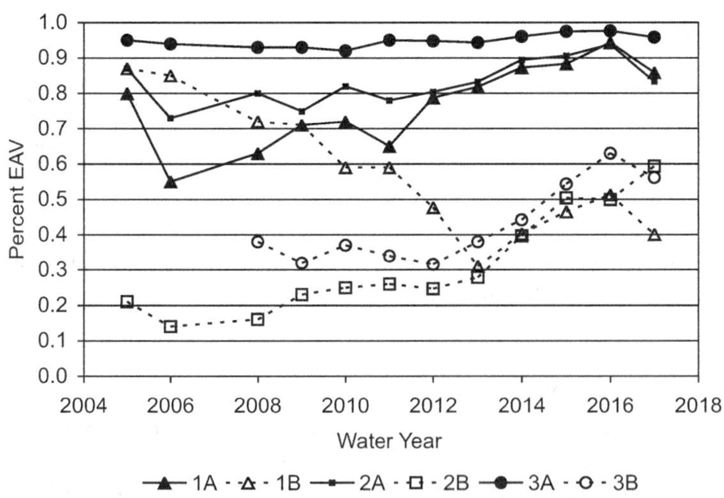

FIGURE 5.16 EAV coverages with time for six cells of STA34. Solid symbols represent the generic EAV classification; open symbols represent the generic SAV classification.

Severe allogenic and autogenic effects are not uncommon in treatment marshes. For instance, hurricanes can literally blow all the SAV out of a Florida marsh. Waterfowl have decimated SAV in the Florida STAs. Unknown external influences totally killed the emergent vegetation in the Tres Rios demonstration wetlands, which had been healthy for the preceding two years (Kadlec, 2008a). Muskrats have eaten out all the cattails in many northern wetlands (Kadlec et al., 2006). Usually these catastrophic vegetation losses impair treatment, but not always.

5.4 FLOATING MATS

Marsh ecosystems can contain zones of floating mats. These differ from zones of floating plants because they contain tall emergent macrophytes (Figure 5.17). Floating mats occur with some regularity in treatment marshes. Here we are concerned about two aspects of floating mat systems: their unintended development in treatment wetlands, and the intentional design of floating plant mats for wastewater quality improvement. All floating marshes have low bulk density, organic substrates, and nearly all are fresh marshes (Sasser et al., 1996)

Typical floating mats vary in thickness from about 25 to 45 cm. The mat consists of a layer of densely intertwined live, dead and decaying roots with some litter collection on the surface (Figure 5.17). That litter layer is aerobic, but thin, with a typical thickness of about 5 cm. Beneath the mat a zone of relatively clear free-water exists, that varies in depth. The depth of the free-water zone (from the underside of the floating mat to the basin surface) in natural shallow lake systems is typically between 1 and 2 m on average. In treatment marshes, the clear water zone is thinner, in the range of 30–70 cm.

Floating mat thickness tends to be stable. There is apparently a balance reached between accretion and the loss of peat-like organic material that detaches and falls from the bottom of the floating substrate. The thickness of a mat tends to be related to the maximum

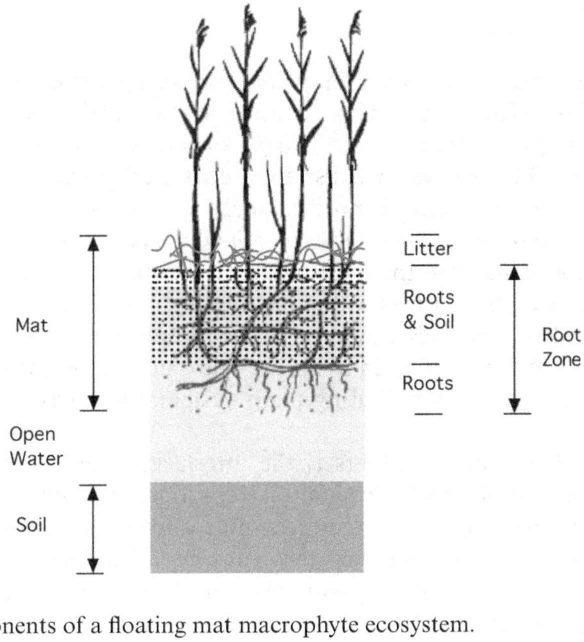

FIGURE 5.17 Components of a floating mat macrophyte ecosystem.

rooting depth of the vegetation, as the roots play an important role in binding the organic material together.

Floating islands or mats are widespread vegetation formations that occur in all climatic regions of the globe (van Duzer, 2004). These range in character from the floating sedge fens of Alaska (Racine and Walters, 1991) to the papyrus swamps of equatorial Africa (Gaudet, 1977; Kansiime and Nalubega, 1999). For example, very large areas of the Mississippi River delta wetlands are floating mat systems (Sasser et al., 1996), comprising over 70% of the western Terrebonne Basin.

Terminology varies considerably across the globe. Headley and Tanner (2006) identified the following.

- Floating Island: general term used to describe any floating piece of land, commonly consisting of peaty soil and/or buoyant aquatic vegetation.
- "Flotant": natural floating marshes in Louisiana, USA.
- Floating Marsh: technical term used to describe "Flotants" listed above.
- Floating Typha Mats: floating wetland dominated by Typha spp. in Canada.
- "Tussocks": terminology used in Florida, USA, to describe, often undesirable, floating marsh islands.
- "Plav": extensive floating mats dominated by Phragmites australis covering approximately 100 000 ha of the Danube River Delta in Romania.
- "Sudd": floating wetlands in central and east Africa.
- "Embalsados" – (floating soils): floating peatlands in Argentina.
- Floating Peat- term used predominantly in the Netherlands in reference to peat that becomes buoyant when cut-over peat lands are rewetted.

It is certain that floating mats provide removal of contaminants from water. What is not known at this point in our understanding is whether they provide greater or lesser removals than rooted marshes of equivalent size. Removal capabilities of floating systems will be addressed further in Chapter 7. Floating mat systems may be more akin to floating plant systems, such as water hyacinths, but no direct comparisons have been done.

Mat Stability and Formation

Many common aquatic macrophytes are buoyant, that is, they float as individuals, because of the air content of the plant parts. Those with a low vertical profile, such as water hyacinths (*Eichhornia crassipes*), water lettuce (*Pistia stratiotes*), and duckweed (*Lemna* spp.), float stably in their upright position, because their center of gravity is below their center of buoyancy. Those with a tall vertical profile, such as cattails (*Typha* spp.), phragmites (*Phragmites australis*), and panicgrass (*Panicum hemitomon*), do not float stably in their upright position as individuals – they tip over. However, a connected community with enough areal extent can easily exist as a floating system. Small communities are sometimes termed "tussocks," while larger ones would be termed floating islands or floating mats.

If the plants can normally float as individuals, without any support from a substrate or their neighbors, then the system is a floating plant system. Otherwise, it is termed a floating mat system.

Often, the linkage that connects plants is the interlaced root system. All common wetland plants, including cattails and phragmites, reproduce by sending out runners from which new aerial pant parts emerge. Large areas may appear to be comprised of many different individuals of the same species, but they may be all the same plant. If the runners become detached or die off, individual plants remain. The collection is called a genet, and the apparent individuals of the genet are called ramets.

In other situations, the connectivity is the result of a cohesive organic substrate, or peat blanket. Floating mats must be almost entirely organic in order to be buoyant enough to float. They derive their buoyancy from gas spaces in rhizomes (Hogg and Wein, 1987, 1988; Krusi and Wein, 1988), and also from gases generated by decomposition processes.

There are at least three natural formation mechanisms (Clark and Reddy, 1998):

1. The delamination and floating of unvegetated organic substrates from deeper sediment. Germination of plants occurs after emergence. This is a peat float-up process.
2. The rhizomes of aquatic plants colonize the water surface from a nucleus of aquatic vegetation that is either unattached or expanding from the shore. This is the grow-over process.
3. Units of rooted vegetation and substrate split simultaneously from the bed, and float to the wetland surface. This is a mat floating process.

In treatment marshes, it is often the third mechanism that is relevant. It has been observed that the soil strength below the action zone of marshes is lower than elsewhere on the vertical profile (Ameen et al., 2017). Wetland roots increase soil cohesion by physically binding together sediment grains. A result of adding nutrient-rich source water is a lowering of the root biomass (root-shoot ratio) and consequently its vertical extent (Chen et al., 2010; Ameen et al., 2017). Changes in water chemistry can weaken cohesion. For example, organic soils have a fulvic component, defined by solubility in basic waters. High pH water is capable of liquefying that fulvic component, thus weakening the soil.

An increase in water depth creates an increased upward buoyant effect. That, in combination with weakened soils, is apparently sufficient to cause the root mat to detach from the underlying soils and become floating (Kadlec and Bevis, 2009; Chen et al., 2010).

Mats do not tear loose from clay soils because of the high strength of those soils. Therefore, floating mats are found where the antecedent soils are organic, which would be the case for treatment marshes on previous or existing wetland soils.

Mat Physics

There are several physical characteristics of floating mats that distinguish them from EAV and SAV systems (refer to Figure 5.17). First, the water flow is under the mat, and at least partially through the lower portions of the root zone. Therefore, the plants have direct access to the nutrients and chemicals in the flowing water. However, if there is a large separation of the mat from the wetland bottom, most water will bypass underneath the action zone.

Second, transpiration is the dominant form of water loss to the atmosphere. That transpiration flow is supplied from underneath the mat, with water being drawn up into the roots and upward to the leaf surfaces. Solutes will be drawn into the action zone from below by the transpiration stream, where root absorption and microbial activity can decrease their concentrations.

Third, the dissipation of solar energy is almost entirely in the above-mat leaf canopy, because there is not free-standing water on top of the mat under most circumstances. In turn, this implies that there is little evaporative cooling of the water.

Fourth, the mat totally shades the water column from photosynthetic radiation. That means that algal activity is limited to organisms that might be below the mat by advection from sunlit areas outside the mat area. The biofilms that exist in the mat root zone will be composed predominantly of nonphotosynthetic bacterial communities (Headley and Tanner, 2006).

Fifth, litterfall from the aboveground plant parts deposits on the top of the mat, which is typically not submerged. Litter decomposition is therefore aerobic, and contingent on internal and rainfall nutrient inputs, rather than on flowing water chemistry. There is not a component of litterfall due to mesophyton.

Sixth, the process of accretion builds from inside the mat. Mat growth occurs mainly from within through deposition of detritus of belowground roots and rhizomes, rather than by accumulation of material at the mat surface. Unlike rooted marsh systems, the floating mat does not act as a repository for settling particulate matter contained in the water column. Particulate matter detaches from the underside of the floating mat (Gantes and Torremorell, 2005), and is added to the underflowing water.

Additionally, isolated floating mat islands in open water are mobile, and move with the wind. If the open water zone under the mat is small or close to nonexistent, then such motion stirs the soil surface, creating turbidity (Chimney et al., 2000).

Examples of Floating Mats in Treatment Marshes

In several instances, treatment wetlands have developed floating mats of vegetation, which were unplanned and unexpected. A few illustrative examples are given here. Treatment has sometimes continued to be effective; but has been impaired in other cases.

Kis-Balaton, Hungary

The constructed shallow treatment impoundments (wetlands) on the Zala River, as it enters Lake Balaton, function for nutrient removal (Hatvani et al., 2011, 2014). The original vegetation of the second unit consisted mainly of reed beds (*Phragmites*). As a consequence of routing the river, the reed-beds were damaged. Dead rhizomes produced gas, which buoyed fragments of the reed-bed to the surface as a floating mat. These mats formed a matrix for secondary succession. These floating islands are partly attached to the still-living fragments of reed beds rooted in the sediment, and have an approximate diameter of about 15 m. The approximate rhizome mat thickness of the floating islands was 0.5m in 2001. The islands were characterized by willows (*Salix cinerea*), sedges (*Carex* spp.), cattail (*Typha angustifolia*), and ruderal species (e.g., *Bidens cernuus*) (Somodi and Botta-Dukat, 2004).

Kinross and Houghton Lake, Michigan

These two natural wetlands developed near-monocultures of cattails on preexisting peatlands. Over the course of time, these *Typha* communities became floating mats (Kadlec and Bevis, 1990, 2009). A possible cause was the retreat of the root zone to a smaller biomass located high in the soil profile, compared to prior conditions. This physical effect, coupled with possible partial peat dissolution into the less acidic added wastewater, led to a 50 cm soil-free water zone topped by the floating mat. The mats are closely woven beds of roots, rhizomes, and sediments (see Figure 4.1). These had enough strength to permit foot and small all-terrain vehicle travel on the mat. Treatment continued to be generally effective under the mat at Houghton Lake. The Kinross system received enormous phosphorus loads in its early history, and did not recover, but ammonia removal was quite good (Kadlec and Bevis, 1990).

Lake Apopka, Florida

The constructed marshes at Lake Apopka, Florida, developed into floating mats (Stenberg et al., 1998; Coveney et al., 2002). Different vegetation strategies were employed, and all underwent significant conversion to floating mats, over the period 1990–1995. In 1995, 73% of natural succession areas contained floating vegetation mats, while 55% of planted sites were floating. Belowground biomass declined as roots and rhizomes shifted to floating vegetation mats.

STA1W, Florida

Cell 2 of stormwater treatment area 1W (STA1W) was constructed on peatlands formerly in agriculture (Chimney et al., 2000). Cattails developed early in the project life, comprising about 50% cover of the 413-ha wetland. Changes in the mode of operation of this cell after six years of operation caused greater water depths, and the cattail areas separated from the base substrate and became floating mats. Because there was less than 100% cover, these floating islands moved with the wind. The water was shallow enough that portions of the mat bottom scraped the base substrate below, creating suspension of the base soils. The water in the cell was therefore very turbid, and the associated nutrients were exported from the wetland. Performance was severely impacted. The cell was later reconfigured, and partially converted to submerged aquatic vegetation.

Floating Rafts

Floating plant mats may also be artificially fostered in aquatic systems, by use of flotation rafts of one sort or another (Figure 5.18). A raft is tethered in place, and vegetation is established in the raft medium. Floating raft systems have been intentionally employed at many sites, and in great variety. In general, they employ floating matrices for plant support.

The float method has potential advantages, such as:

1. It directly takes up nutrients from water column, and does not remobilize the nutrients in sediment.
2. It enables the use of a diversity of aquatic plants.
3. It could be used in any waterbody regardless of its depth and bottom characteristics.

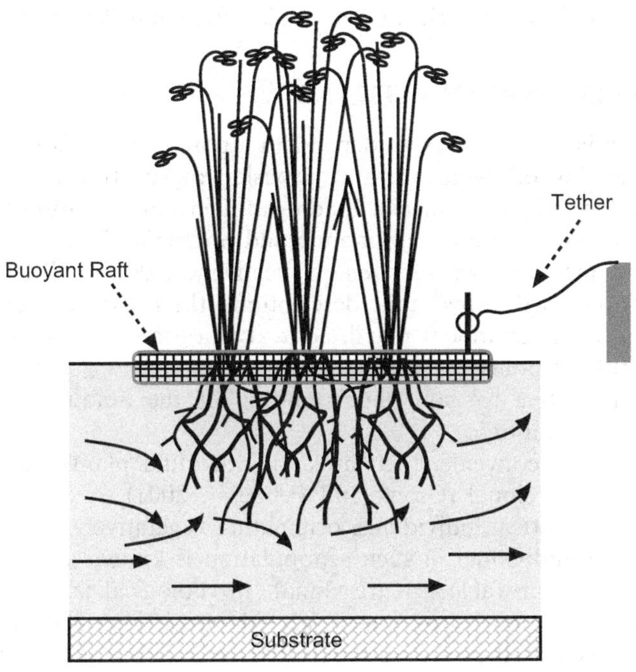

FIGURE 5.18 Basic components of a raft macrophyte system.

The principal drawback for raft systems is cost: the frames are expensive. Support structures for the mat are quite varied in design. Dozens of ideas have been patented (for instance: Balogh, 1982; Ishikawa and Mizuno, 1998; Hondulas, 1994; Schuck et al., 2007; Kania and Stewart, 2011,2012). Complete units are commercially available (van Duzer, 2004).

This concept has been utilized for over twenty years (see for example, Hiley, 1990; Boutwell, 2001; Smith and Kalin, 2001; Hefty, 2002; Wen and Recknagel, 2002). A very large category of plants may be established in floating systems in which supporting media and neighbors are required. For instance, 67 different plants tested in Hungary (Lakatos, 1998).

Applications of floating rafts have included (Headley and Tanner, 2012):

- Combined stormwater-sewer overflow
- Sewage
- Mine drainage
- Piggery effluent
- Poultry processing wastewater
- Eutrophic lake water
- Water supply reservoirs
- Stormwater

Existing floating raft systems fail to meet most of the criteria set forth in Chapter 1 for inclusion in this book. Many are of mesocosm scale, and almost all fall in the category of juvenile systems. They all contain augmentation of a natural system, and many are pond retrofits, not marshes. Many applications are for strong effluents, which are also excluded herein. We are left with a small fraction of the existing systems that are under consideration, and they are in the last three Headley and Tanner (2012) categories. Nonetheless, performance data for some raft systems is of use in further defining the capabilities of marshes that develop floating mats, which is topic for later Chapters.

For readers wishing more information on floating rafts, there are several review papers to consult: Headley and Tanner (2012), Chen et al. (2015) and Pavlineri et al. (2017).

5.5 BIOMASS AND GROWTH PATTERNS

The treatment potential of marshes is dependent upon the vegetation in two important ways: storage and cycling of chemicals, and provision of substrates for microbial and epiphytic activity. The low range of nutrient loadings under consideration in this book fosters both, meaning the systems are both agronomic and microbial. These are designated as the *cycling function* and the *substrate function*, respectively. Additionally, Chapter 3 has discussed a third function, used in hydraulic descriptions, the *resistance function*. The meaning of the term function is the treatment marsh context is a group of related actions contributing to the larger action of pollutant removal.

It is therefore imperative to gain understanding of the amounts, types and seasonal dynamics of wetland vegetation.

While it common and convenient to think of individual plants, wetland vegetation is, almost without exception, clonal (Cronk and Fennessy, 2001). A clonal colony or genet is a group of genetically identical individuals, originating vegetatively, not sexually, that grows in a given location. An individual in such a population is known as a ramet. All the treatment marsh plants listed in Table 5.1 are clonal. In ecological jargon, treatment marshes are often deemed to be monocultures, that is, growth consisting of a single plant type. That is not quite correct, because they are not collections of a single plant type; rather, they are a single plant (genet), or at most a collection of a few genets. The entire marsh community of roots and live and dead shoots may therefore be comprised of a few individual genets.

Yet, most wetland plants are also capable of sexual reproduction, and many produce propagules that can separate from the parent genet and form a new colony (Cronk and Fennessy, 2001). Thus it is possible to assist marsh establishment via seeding or transplanting. This is the common practice when beginning with bare soils in a new system. In many instances, natural seed banks may exist in the antecedent soils.

Growth Cycles

Biomass is commonly divided into belowground (roots, rhizomes, tubers, etc.) and aboveground biomass (all vegetative and reproductive parts above the ground level). The term standing crop includes live parts and dead parts of live plants that are still attached. These dead parts of plants together with still standing dead plants are called standing dead. The term litter refers to those dead parts of the plant that have fallen on the ground or sediment, but in some cases also includes standing dead. These compartments exchange material, but not uniformly over the course of the year (Figure 5.19).

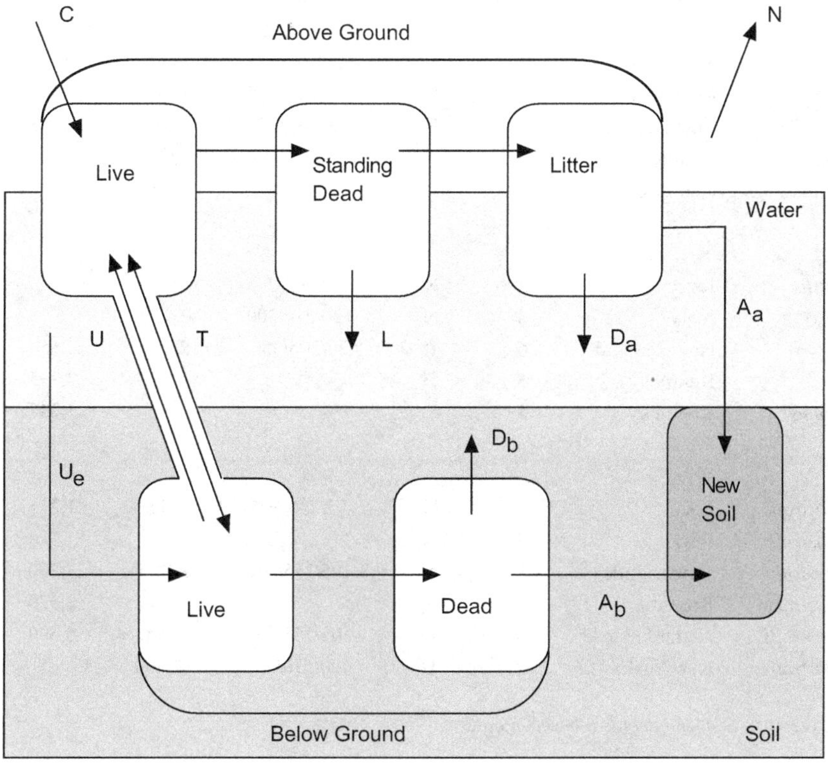

FIGURE 5.19 Transfers of plant materials in the biosphere of wetlands. Biomass consists of living, above and below ground components. Necromass consists of dead roots and rhizomes, plus aboveground standing dead and litter. Phytomass is the combination of biomass and necromass. Transfer to the phytomass occurs by external plant uptake from water (U_e) or air (C). Uptake (U) moves material from the root zone to aboveground parts. Translocation (T) moves material to or from below ground to aboveground. Transfer back to surface water and porewater occurs via leaching (L) and decomposition (D_a and D_b). Necromass residuals lose their identity, and accrete as new soils and sediments (A_a and A_b). Gases may be lost from the entire system (N).

Peak standing crop is defined as the single largest value of plant material present during a year's growth (Richardson and Vymazal, 2001). In tropical communities, with an almost constant biomass, it is not profitable to search for an annual maximum (Westlake, 1969). However, in all other climatic regions the biomass fluctuates widely throughout the year,

TABLE 5.5

End of season plant biomass in wetlands. Water type is N = no wastewater, S = nutrients at secondary treatment levels, P = nutrients at primary treatment levels, L = landfill leachate at around 300 gN/m^3. S/P/E refers to the start, peak and end yeardays of the growing season (add 182 days for southern hemisphere).

Species	Location	Ref.	Water	S/P/E	Live Above gm/m^2	Total Above gm/m^2	Roots & Rhizomes gm/m^2
Cattails							
Typha latifolia	Wisconsin	1	N	105/245/290		1,400	450
Typha latifolia	Texas	2	N	60/240/345		2,500	2,200
Typha glauca	Iowa	3	N	120/265/290	2,000		1,340
Typha latifolia	Michigan	4	N	120/245/275	490	890	6,200
Typha latifolia	Michigan	4	S	120/245/275	1,240	2,310	2,900
Typha angustifolia	Michigan	4	S	120/245/275	1,886	3,615	
Typha latifolia	Kentucky	5	P		5,602		3,817
Typha angustifolia	Kentucky	5	P		5,538		4,860
Bulrushes							
Scirpus fluviatilis	Iowa	3	N	130/265/285	790		1,370
*Scirpus validus**	Iowa	3	N	120/210/300	2,100		1,520
Scirpus validus	New Zealand	6	P	30/205/350	2,100	2,650	1,200
Scirpus validus	Kentucky	5	P			2,355	7,376
Scirpus cyperinus	Kentucky	5	P			3,247	12,495
Phragmites							
Phragmites australis	UK	7	N	75/220/305	942	1,275	
Phragmites australis	Iowa	3	N			1,110	1,260
Phragmites australis	Netherlands	8	N	105/255/350	2,900	3,200	7,150
Phragmites australis	Brisbane	9	S		1,460	2,520	1,180
Phragmites australis	Netherlands	8	P	105/255/355	5,000	5,500	3,890
Phragmites australis	New York	10	L	100/270/330	10,800		8,700

* Currently known as *Schoenoplectus tabernaemontani*

References:
1 Smith et al. (1988)
2 Hill (1987)
3 van der Valk and Davis (1978)
4 Unpublished data from Houghton Lake, MI
5 Pullin and Hammer (1989)
6 Tanner (2001a)
7 Mason and Bryant (1975)
8 Mueleman et al. (2002)
9 Greenway (2002)
10 Peverly et al. (1993)

with periods of growth, senescence and dormancy. The range of peak standing crop of wetland plants is quite large (Table 5.5).

Gross Primary Production (also called Gross Primary Productivity, or GPP) is normally defined as the assimilation of organic matter by a plant community during a specified period, including the amount used by plant respiration. Net primary production, or NPP, is defined as the biomass that is incorporated into a plant community during a specified time interval, less that respired. This is the quantity that is measured by harvest methods.

NPP of freshwater marshes is estimated most frequently through harvest of annual peak standing stocks of live and dead plant biomass. When root biomass is measured, it is usually an important part of net annual plant production. Some researchers consider net primary productivity estimates by peak standing stock to be underestimates because they do not account for biomass turnover during the growing season (Pickett et al., 1989). Kvet (1982) estimates turnover rates (productivity/biomass) in the range of 1.1–1.5 yr^{-1} for submerged species, 1.05–1.5 for short emergent species, 1.05–1.3 for tall emergent species and 1.15 for tall graminoids.

Aboveground parts of plants in northern environments have a life span dictated by the photoperiod and frost conditions of the region. They live through one growing season, and new plants emerge the next year, from root stock or from seed. However, in warm climates, individual plants may persist for more than one year. Davis (1989) tagged individual leaves of 43 individual shoots of *Typha domingensis*, and followed their growth over their entire life history in a Florida wetland (Figure 5.20). He found that leaf growth and mortality continued throughout the life span of each tagged plant. New leaves emerged and grew, even while total biomass declined. Older leaves senesced, broke or died even while total biomass increased. This continual growth and mortality resulted in an annual turnover of 4.4 ± 0.7 times the mean standing crop (Davis, 1989).

The frequency and timing of formation of new cohorts of vegetation can be variable. If propagation is vegetative, new ramets form in response to temperature and solar radiation. In temperate climates, it is often the case that new shoots appear in autumn, remain dormant over winter, and begin growth in early spring. However, Dickerman and Wetzel (1985)

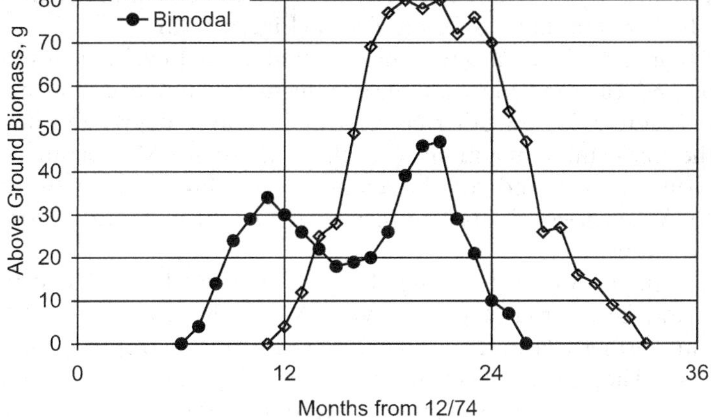

FIGURE 5.20 Growth of single shoots of *Typha domingensis* in a subtropical wetland. (Based on information in Davis, 1989.)

found that cohorts of *Typha latifolia* can also originate in midsummer and late summer. Overlapping cohorts combine to form the biomass of the overall population.

The concept of individual plant life history becomes important when, as is a common case, an entire wetland is planted at one time, creating a cohort of plants that will all live about the same length of time. Clearly, without regeneration this wetland will be devoid of plants after a few years. Therefore, the key to a self-sustaining wetland plant community is not only the survivability of plants in the treatment environment, but also the ability to regenerate.

Seasonal Patterns

The growth and senescence of the soft tissue macrophytes commonly found in treatment marshes all follow a common seasonal pattern in temperate climates. In north temperate climates, growth begins at the time of frost disappearance (around April), and senescence begins in early autumn (around September). As one moves north, the growing season shortens, until at 75°N it is only about eight weeks long (mid-June to mid-August) (Arndal et al., 2009). This autumnal decline creates standing dead aboveground plant material, which subsequently in part decomposes, and in part falls to the soil surface.

Examples for *Typha domingensis, Typha latifolia* and *Phragmites australis* are shown in Figure 5.21, which are representative of other emergent macrophytes as well. New growth proceeds from small shoots that may be initiated as early as late summer of the preceding year for *Typha* (Bernard, 1999), but remain tiny and dormant over the winter season. Aboveground biomass increases rapidly in early spring, typically commencing from late February to Late April, depending on climate. Growth tapers off, causing aboveground biomass to peak in late summer to early autumn. The size of the peak standing crop varies considerably with plant species and degree of nutrient availability (see Table 5.5). Typically, there is some degree of senescence that accompanies the later portions of the growth period, so that the total peak standing crop exceeds the live peak standing crop. During autumn, more rapid senescence occurs, leaving only a residual of standing and/or prostrate aboveground dead material.

In tropical or subtropical climates, seasonality still exists, despite the more uniform seasonal temperatures and insolation (Figure 5.21). There may be periods of dormancy and of regrowth, but there is typically not complete senescence and death of all aboveground plant parts.

The peak standing crop is a strong function of both the nutrient status of the water, and the species under consideration. For instance, *Phragmites* growing in high ammonia waters displays a much higher standing crop than *Typha* (Figure 5.22).

Belowground biomass follows a much more muted annual cycle. In some cases, available methods of root and rhizome biomass measurement are not accurate enough to clearly define a pattern (Figure 5.21). In other cases, a mid-summer depression has been found, to about 50% of the mid-winter maximum (Smith et al., 1988; Mueleman et al., 2002). But mid-summer maxima were found for Sparganium and Phragmites in Iowa (van der Valk and Davis, 1978). When root biomass is measured, it is usually an important part of net annual plant production.

Two factors are important in assessing the growth of wetland plants: the length of the growing season, and belowground productivity. All of the growth for the year occurs in about 100 days in northern latitudes, whereas systems in the tropics grow year-round (see Tables 5.3 and 5.4). Therefore, the instantaneous growing season rate is much higher than the annualized rate for northern systems. Belowground biomass is typically comparable to aboveground biomass, although the root-to-shoot ratio is sensitive to nutrient status and other variables. The ratio of below to aboveground biomass is generally less in a fertilized environment than in a lower nutrient (natural) environment (Mueleman et al., 2002).

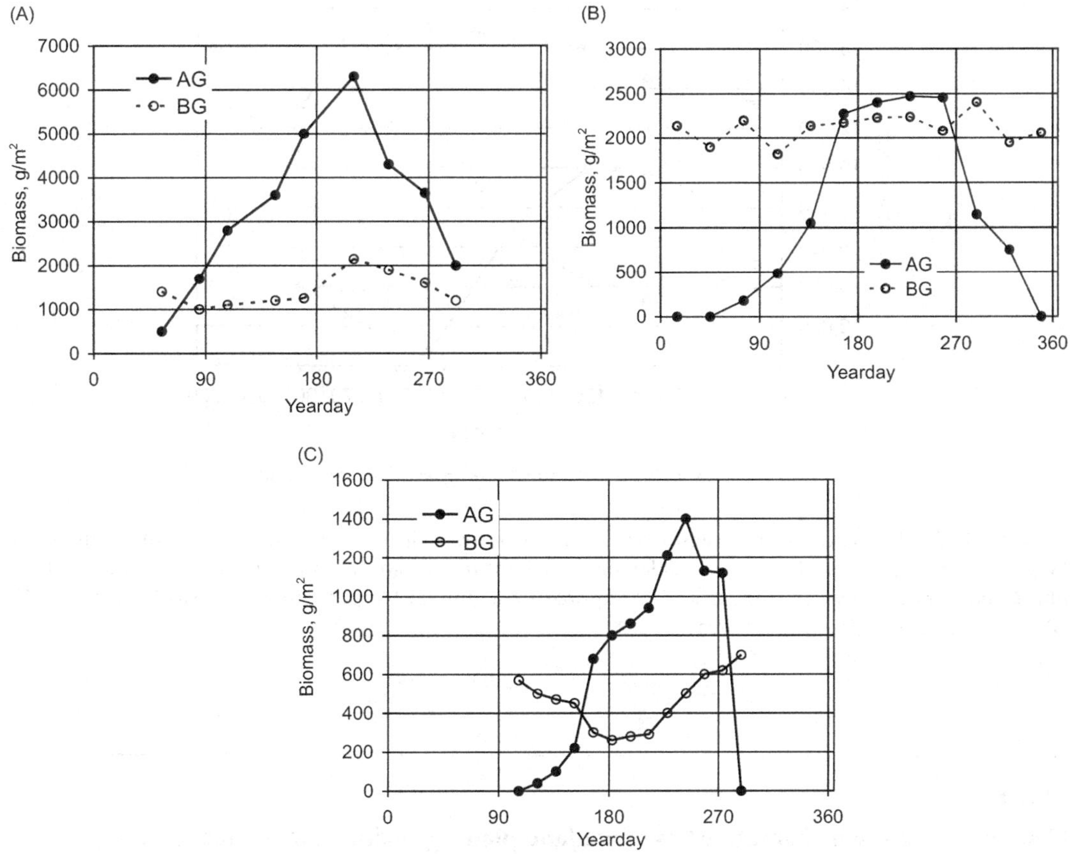

FIGURE 5.21 Seasonal patterns of above and below ground cattail biomass. (A) *T. domingensis*, Lake Burullus 31°N, based on Eid et al. (2012a, 2012b). (B) *T. angustifolia*, Texas 33°N, based on Hill (1987). (C) *T. latifolia*, Wisconsin 43°N, based on Smith et al. (1988).

Kadlec and Alvord (1989) indicated that belowground biomass responded to fertilization differently from aboveground biomass. The initial vegetation showed greatly reduced root biomass in response to the added nutrients; 1,500 gdw/m^2 versus 4,000 gdw/m^2 at the end of the growing season. There are some reports that root growth and activity continues much longer than for aboveground plant parts (Prentki et al., 1978).

Roots and rhizomes persist over winter in northern climates, and therefore standing crop alone is not a measure of productivity. Estimates of turnover times are on the order of two to three years for herbaceous wetland plants. For example, Tanner (2001a) estimated a lifetime of 18 to 24 months for *Schoenoplectus* rhizomes, and Prentki et al. (1978) reported 1.5 to 2 years for *Typha* rhizomes and at least three years for *Phragmites* rhizomes. Therefore, the total growth rate for wetland plants is much higher than for aboveground parts alone.

These factors lead to the conclusion that plant growth is much higher than one standing (aboveground) crop per year. Table 5.6 presents a hypothetical illustration of factors for two climate zones. The growth of plant biomass during the respective growing seasons is about the same, but the growing season is much attenuated in northern climates. As a result, the annualized growth is higher in the warmer environment.

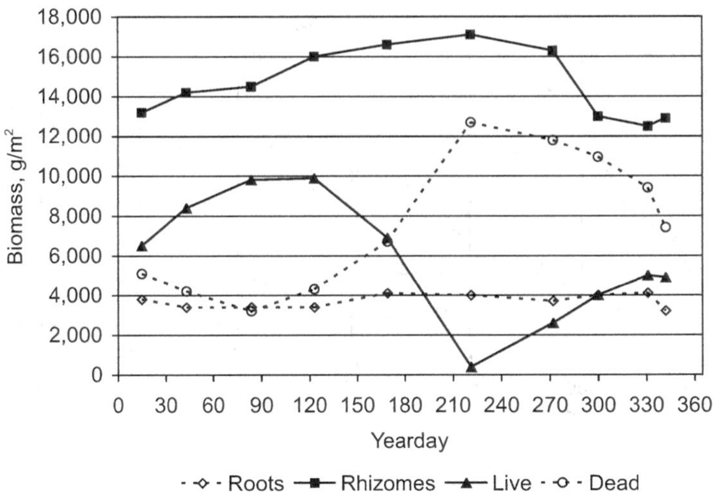

FIGURE 5.22 The pattern of growth of *Phragmites australis* in the warm dry continental climate of Griffith, Australia (34°S). Note the 182-day displacement for the southern hemisphere. The water phosphorus concentration was 12 mg/L, and the approximate annual temperature range was from 10–23°C. (Based on data from Hocking, 1989a)

TABLE 5.6

Hypothetical growth characteristics of wetland plants growing in temperate and subtropical conditions. These both grow at about the same rate during their respective growing seasons, which are year-round in the warm climate.

		Temperate		Subtropical
		Growing Season M-J-J-A	Annual	Annual
Peak Standing Crop Aboveground	gdw/m^2	2,000	2,000	2,000
Growth (GPP/NPP = 1.3)	gdw/m^2	2,600	2,600	
Growth (4 turnovers per year)	gdw/m^2			8,000
Growing Season	days	120	365	365
Growth Rate Above	$gdw/m^2 \cdot d$	21.7	7.1	21.9
Belowground Crop (root/shoot = 1.0)	gdw/m^2	2,000	2,000	2,000
Growth (0.5 turnovers per year)	gdw/m^2	1,000	1,000	1,000
Growing Season	days	240	365	365
Growth Rate Below	$gdw/m^2 \cdot d$	4.2	2.7	2.7
Total Growth Rate	$gdw/m^2 \cdot d$	25.8	9.9	24.7
Annualized Instaneous Growth Rate	$gdw/m^2 \cdot yr$	9,429	3,600	9,000
Undepleted Solar Radiation	$MJ/m^2 \cdot d$	38	24	31

Nutrient Effects

Primary productivity of wetland plants is increased by the availability of water, light, and nutrients. Adding nutrient-rich water to wetlands generally results in the stimulation of gross and net primary productivity of these ecosystems (Guntenspergen and Stearns, 1981; Nixon and Lee, 1986).

The growth of wetland plants, like that of terrestrial plants, is stimulated by fertilization (Boyd, 1971; Jordan et al., 1999; Mueleman et al., 2002). When a wetland becomes the recipient of waters with higher nutrient content than those it has been experiencing, there is a response of the vegetation, both in species composition and in total biomass. This response has been detailed for the Houghton Lake wetland by Kadlec and Alvord (1989). The increased availability of nutrients produces more vegetation during the growing season, which in turn means more litter during the nongrowing season. This litter requires several years to decay, and hence the total pool of living and dead material grows slowly over several years to a new and higher value. A significant quantity of structural components are thus retained in the wetland.

Primary productivity of wetland plants is increased by the availability of water, light, and nutrients. Adding source waters to wetlands generally increases the availability of both water and nutrients and consequently results in the stimulation of gross and net primary productivity of these ecosystems. The maximum rate of plant growth is attained as nutrient levels are initially increased. However, at higher nutrient levels, plant growth levels off, and at very high nutrient concentrations, phytotoxic responses may be observed. Phytotoxic nutrient levels are not observed in treatment marshes as defined here.

Figure 5.23 gives an example of this fertilizer response, for cattails grown in treated wastewater. A tenfold increase in water phosphorus concentrations produced a doubling of live shoots, tripling of standing dead, and an eightfold increase in litter (see Table 5.7).

Increased nitrogen concentrations also stimulate plant growth. Tanner (1994) found that as water nitrogen concentration was increased, both above and belowground biomass

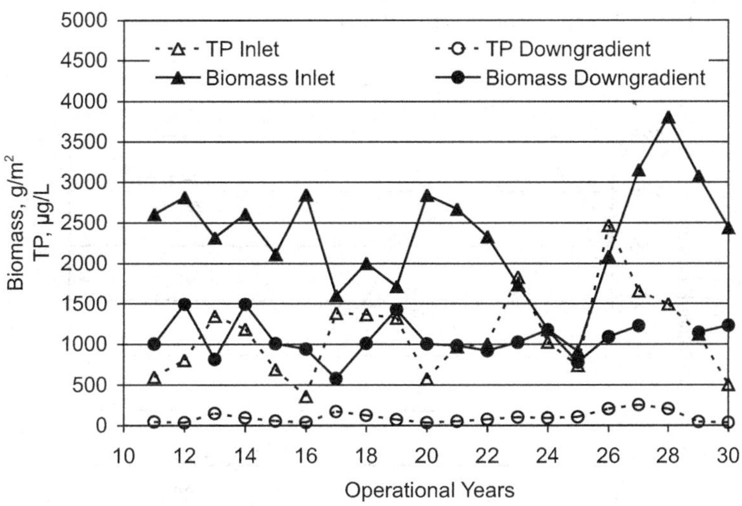

FIGURE 5.23 Biomass of live aboveground (live plus standing dead) *Typha latifolia* at the Houghton Lake treatment marsh, and the associated phosphorus concentrations. Locations are the source water inlet, and a downgradient location about 400 m from the inlet. The first 10 years were a startup mode. The means were 1,124 µg/L, 2,343 gm/m^2 at the inlet, and 101 µg/L, 1,071 gm/m^2 downgradient.

TABLE 5.7

Biomass (gdw/m²) of aboveground vegetation groups for *Typha latifolia* at the Houghton Lake (HL) treatment marsh, and for *T. angustifolia* and *T. glauca* at Lake Ontario fringe marshes in New York (NY).

	Live	Standing Dead	Litter	Total Above	Total Dead	Ratio Dead:Live
NY Snake	995	0	438	1,433	438	0.44
HL Downgradient	665	363	250	1,278	614	0.92
NY Deer	1,031	98	883	2,012	981	0.95
NY Sodus	1,172	589	1,276	3,037	1,865	1.59
HL Inlet	1,263	1,030	2,182	4,476	3,212	2.54
NY Mudge	910	961	1,635	3,506	2,596	2.85
NY South	415	52	1,200	1,667	1,252	3.02
NY Beaver	816	911	1,620	3,347	2,531	3.10
Mean	**908**	**501**	**1,185**	**2,595**	**1,686**	**1.93**

Source: NY, Vaccaro et al. (2009).

increased. Other studies have also established similar effects for other treatment wetland plants. Ho (1979) found *Phragmites* growth to be stimulated by nutrients in Scottish systems. Hill et al. (1997) found dry matter production of *Schoenoplectus (Scirpus) acutus* to be maximized in the 30–50 mg/L range, and then to rapidly fall off above 60 mg/L.

Because most treatment marshes have nutrient gradients in the flow direction, they may also display vegetation biomass gradient as well. This is illustrated by data from the Houghton Lake treatment marsh (Kadlec and Bevis, 2009). Relative biomass was calculated as the product of the number of plants per meter squared, the height of the plants, and the number of leaves per plant. In very low nutrient systems, virtually no effect has been found (Miao and Sklar, 1997; Weisner and Miao, 2004).

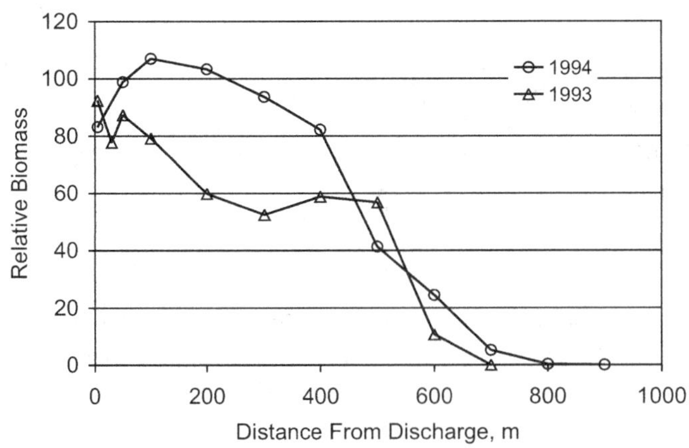

FIGURE 5.24 Relative cattail biomass along the flow direction at the Houghton Lake treatment marsh.

5.6 STANDING DEAD AND LITTER

Wetland plants lose biomass via shedding of dead leaves during the growing season, and by the end-of-season death of living erect shoots. The first process creates fragments, termed litter, throughout the growing season, which fall to the marsh surface. The second process creates erect dead stems and leaves, which also shed fragments, but which also can fall to the marsh surface in their entirety. The transfer of standing dead to pronate material on the marsh surface is also litterfall.

Wetland plant tissues fall at variable rates depending on the survival strategy of the individual plant species. Herbaceous plant species typically recycle the entire aboveground portion of the plant annually in temperate environments. The growth season may vary from 10 or more months in subtropical regions to less than three months in colder climates. Also, most herbaceous species lose a fraction of living leaf and stem material as litter throughout the growing season, so there is a continuous rain of dead plant tissues throughout the year with seasonal highs and lows of litterfall. In the northern hemisphere, large amounts of flowers are shed during the spring, and leaves and fruiting bodies are lost during the fall.

It is important to note that a newly established system will not possess a complete set of stable compartments. Constructed emergent marshes typically begin life with live above and belowground material, but do not possess standing dead or litter. Those necromass compartments establish over a period of one to three years, due to slow decomposition, as detailed later in this chapter.

The necromass of standing dead and litter in emergent marshes often exceeds the biomass of live plant material (Table 5.7). It is an important storage for nutrients and other chemicals.

Standing Dead

In a temperate climate in autumn, the green expanse of emergent marsh vegetation changes to a brown expanse of standing dead in a few weeks. Most herbaceous wetland plants do not directly fall to the wetland floor after senescence and death. Instead, plants remain in an upright stance until meteorological conditions cause them to topple. Wind, rain, decompositional weakening, and especially weight of snow, cause the standing dead material to fall. Senesced cattail plants typically remain upright for 6–18 months before fragmenting, toppling over, and forming a litter layer (Davis and van der Valk, 1978). Terminology varies, and so dead material is sometimes called litter, regardless of whether it is upright or not. At other times, and in this book, a distinction is drawn between standing dead and prone material called litter.

Because live material transitions to standing dead, the relative quantities of these materials are similar, but peak at different times of the year The transition from live to dead aboveground material is illustrated for Phragmites in Figure 5.22. For a detailed description of the transfers of carbon and nutrients amongst the live, standing dead and litter compartments, the reader is referred to the work of Davis and van der Valk (1978) for *Typha* and *Scirpus.*

In many emergent wetland macrophytes, collapse of shoot material to the sediment surface does not occur until long after senescence. Consequently, large amounts of standing dead plant matter can accumulate in freshwater wetlands, and it remains for well over one year. Decay of this material is not rapid. For instance, Kuehn and Suberkropp (1998) measured exponential decay in *Juncus effuses* with $k = 0.40\ yr^{-1}$, $R^2 = 0.99$, which corresponds to a half-life of 1.7 years. Welsch and Yavitt (2003) measured $k = 0.54\ yr^{-1}$ for *Typha latifolia*. Kuehn et al. (2011) measured $k = 0.75\ yr^{-1}$ for *Typha angustifolia*.

In cold climates the presence of standing dead material provides an excellent adjunct to insulation. Firstly, the standing material protects the wetland soil or water surface from direct exposure to the wind. This windbreak function is probably secondary to the function of catching snow, often to a depth of a meter or more. Snow is held up on dead plants, creating a zone of air spaces interlaced with plant material and captured snow (Figure 5.25). This composite is an excellent thermal insulator, and often prevents freezing in vegetated natural wetlands at times when open water is deeply frozen.

Thatch

In warm and arid climates, treatment wetlands can accumulate excessive quantities of dead plant biomass, because of the lack of snow and rain to cause collapse. Regionally, this is referred to as thatch. This accumulation results from the low decomposition rates occasioned by lack of water to support decomposer organisms, plus the upright orientation of the necromass, which keeps the material in the air rather than in the water. The high productivity of the litter, coupled with slow decomposition, leads to very large standing crops of standing dead thatch (Figure 5.26). Mechanical harvesting may be used to remove standing dead aboveground material (combing or thinning), or removal of both dead above and belowground (thatching) (Nolte and Associates, 1998b; Thullen et al., 2002). Controlled burning is one alternative to remove excess plant biomass in wetland treatment systems, although ash produced by burning will re-introduce nutrients into the water column. This can potentially cause a short-term decrease in treatment efficiency. Burning has been

FIGURE 5.25 Standing dead wetland plants capture snow and provide thermal insulation.

FIGURE 5.26 Thatch at the Tres Rios Hayfield wetland near Phoenix, Arizona. This standing dead material is over two meters in height, and has totally blocked light penetration to the wetland water.

implemented at sites that permit such activities. Since accumulated plant necromass can regenerate, the benefits of removal are only temporary (Thullen et al., 2002).

Litter

Litter consists of plant fragments, generated from live plants or from standing dead, or from epiphytes. There is no standardized procedure for harvesting this material. It is common practice to cut and remove all erect live and dead plants before harvesting the litter. One technique is to remove material by gentle raking with the fingers, without any tearing of roots or dead leaves. Litter typically occupies a vertical thickness of about 10–20 cm.

Increased nutrients create a greatly increased amount of litter (Figure 5.27). Because of its longevity, it can take several years to build a more or less stable litter necromass pool.

Litterfall includes both macrophytic and mesophytic materials. In some instances, litterfall traps have been used to measure the fall of macrophyte detritus, but there are no studies of the mesophytic "rain." Murkin et al. (2000) estimated about a ten to one ratio for macrophyte: mesophyte litterfall from emergent marshes in Manitoba, but that ratio is very likely much lower for southern climates, and for SAV wetlands, that is, micro-litter is relatively more important.

Decomposition

Decomposition generally refers to the disintegration of dead organisms into particulate form (or detritus), and the further breakdown of large particles to smaller and smaller particles, until the structure can no longer be recognized. In wetland studies, the term decomposition is mostly confined to the breakdown and subsequent decay of dominant

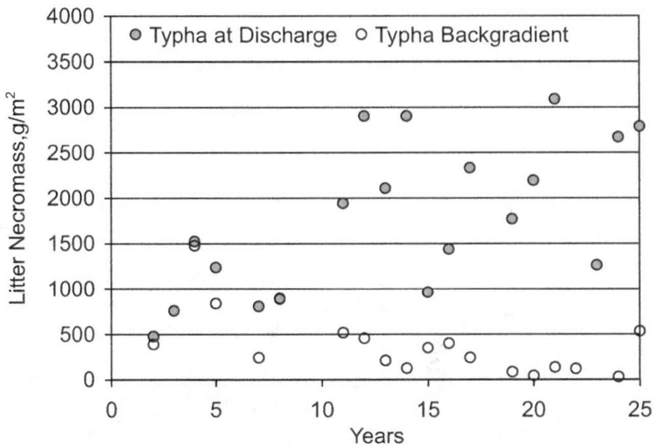

FIGURE 5.27 The time progression of cattail litter at the Houghton Lake treatment marsh.

macrophytes, which leads to the production of detritus. Rates of litter decomposition vary in wetlands and the fate of materials released and adsorbed during decomposition depends on the physical and chemical composition of material as well as environmental conditions at the site of decomposition.

Studies of litter decomposition are very numerous in the literature. Techniques for such studies have been compiled in books (Barlocher et al., 2005). Most of these studies have been concerned with aboveground plant parts.

Flooding in wetlands has been found to increase the litter decomposition rate through physical leaching of inorganic and organic compounds from the plant tissues (Whigham et al., 1989) and by providing habitat for aquatic microbes and invertebrates, which are important mediators in this process. However, if floodwaters are anaerobic, biological activity is greatly reduced (Tupacz and Day, 1990) and only the leaching mechanisms and anaerobic respiration will occur.

Patterns of Weight Loss

Chemical analysis of plant material reveals different rates of decomposition for different components of the plant material (soluble components, cellulose, hemicellulose, and lignin). Rates of decomposition of each component change over time, such that the specific rate of decay for each fraction decreases as decomposition proceeds (Moran et al., 1989).

The decomposition of litter and resultant release of nutrients can therefore be conceived to involve three phases (Godshalk and Wetzel, 1978a, 1978b, 1978c). An initial loss of soluble materials is attributed to abiotic leaching. This process is quite rapid and accounts for the majority of mass reduction during the early stages of decomposition. Leaching occurs very quickly under both aerobic and anaerobic conditions with most of the water-soluble organic substances being released within a few days. The initial drop is typically of the order of 10–20% for soft-tissue emergent macrophytes in treatment marshes (Kadlec and Wallace, 2009).

The initial sharp drop in necromass is followed by an exponential decline to an undecomposed residual, which lasts for months or years in EAV tissues, but only weeks in SAV tissues.

Because there are undecomposable residuals, there is a terminal phase three, in which very slow or negligible necromass loss is observed. This residual is difficult, if not impossible, to measure, because it consists of fine detrital material that is not retained in typical mesh bags. The time required is also several years. The residual of recalcitrant substances is estimated to be on the order of 5–20%, as inferred from long-term accretion studies.

Models

Individual Cohorts

The oldest and simplest model that accounts for a decreasing amount of a cohort of litter with time is an exponential decline. However, the observed three-phase behavior of the decomposition process argues for a better representation. For instance, Godshalk and Wetzel (1978c) argued for a double exponential. These two approaches are represented by:

$$\frac{M}{M_0} = e^{-kt} \tag{5.6}$$

$$\ln\left(\frac{M}{M_0}\right) = \frac{a}{b}\left(e^{-bt} - 1\right) \tag{5.7}$$

where: a,b = constants
k = decay coefficient, 1/d
M = litter necromass remaining, gdw/m^2 M_0 = starting litter necromass, gdw/m^2
t = time, d
Another way to account for the three phases of decomposition is:

$$\left(\frac{M - M*}{M_0 - M*}\right) = Ae^{-kt} \tag{5.8}$$

where: A = fraction of initial necromass lost rapidly
M* = residual litter necromass remaining, gdw/m^2

In the vast majority of literature studies, the value of M* is chosen to be zero; and the value of A is selected to be unity. There is then only one parameter to consider, the lumped mass loss rate coefficient. Chimney and Pietro (2006) provide rates of litter decomposition of 140 different wetland plant varieties, some of which are shown in Table 5.8. Considering the full set reported by Chimney and Pietro (2006), mean first order rate coefficients (k) for emergent macrophyte leaf litter decomposition averaged 1.4 yr^{-1} for 32 studies of Phragmites australis, 1.7 yr^{-1} for 23 studies of 10 Scirpus species, and 1.4 ± 0.9 for 72 studies of 8 Typha species.

Litter decomposition is largely mediated by vertebrates, invertebrates, and microbes living in wetlands. New litter is typically conditioned by fungi and bacteria before it is shred into smaller particles by aquatic macroinvertebrates (Merritt and Lawson, 1979). The activity of these organisms is conditioned by temperature, and therefore a temperature effect on decomposition is to be expected. Studies by Alvarez and Becares (2006) confirmed this effect for *Typha* in a treatment wetland. They found k = 0.73 yr^{-1} for 5°C, and 1.57 yr^{-1} for 20°C water. Inter-system comparisons show that warmer climates have higher rates of litter decomposition on an annual basis. However, the effect of frozen winter conditions typically interrupts the decay processes, which effectively come to a halt in frozen water and soils (Figure 5.28). Therefore, part of the variability across data sets has to do with this winter-season shutdown, which is typically not accounted.

TABLE 5.8

Summary of lumped loss rate coefficients for herbaceous plants in various wetlands.

	Type	Datasets	Median k	Half Life
		N	yr^{-1}	days
Juncus effusus	EAV	4	0.77	338
Phragmites australis	EAV	32	0.66	179
Typha angustifolia	EAV	6	0.69	306
Typha domingensis	EAV	6	1.57	154
Typha latifolia	EAV	21	1.13	96
Scirpus acutus	EAV	4	1.64	126
Scirpus californicus	EAV	1	2.41	105
Scirpus fluviatilis	EAV	3	0.51	341
Scirpus lacustris	EAV	4	0.66	237
Scirpus validus	EAV	1	2.04	124
			1.21	**201**
Eichhornia crassipes	FAV	27	7.48	18
Pistia stratiotes	FAV	1	5.58	45
Lemna minor	FAV	2	14.24	18
			9.10	**27**
Ceratophyllum demersum	SAV	4	7.15	36
Myriophyllum spicatum	SAV	6	9.53	23
Najas flexilis	SAV	5	9.49	23
Potamogeton crispus	SAV	8	36.24	7
			15.60	**22**
Nuphar lutea	FLAV	4	4.42	42
Nymphaea alba	FLAV	6	16.97	14
Nelumbo lutea	FLAV	2	2.59	98
			7.99	**51**

Source: Chimney and Pietro (2006).

Combined Effects of Successive Cohorts

Research has in general focused on the fate of a particular cohort of necromass, placed in a porous bag and isolated from other materials in the wetland. However, the litter layer in the wetland is the result of many such cohorts that accrue over the years, and the decomposition processes that reduce each of them over time. A conceptual model of this successive accrual and decomposition has been described in Chapter 4, for underground materials. However, it has only recently been quantitatively applied to a wetland litter compartment (Christensen et al., 2009).

As a simple example, consider litter that has a half-life of one year, being deposited once per year in cold temperate climate. At the end of year one, a fresh "crop" of litter of mass W_{01} is present. At the end of year two, half that remains, and another crop of W_{02} is added, with the total now being 1.5 W_{01}. A bit of arithmetic shows that, after a period of some years, this process will lead to an end-of-season litter crop that is twice the annual litterfall. It will take five years to build the litter to 97% of the final value. Of course, events

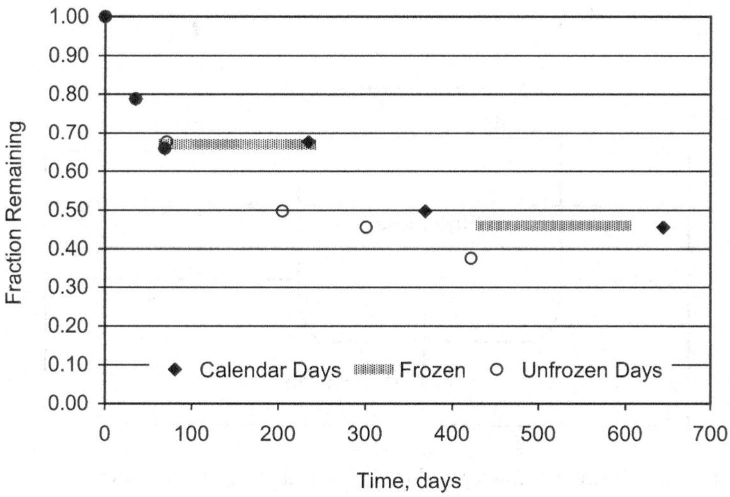

FIGURE 5.28 Decomposition of cattail (*Typha* spp.) litter in the inlet area of the Houghton Lake, Michigan, wetland. Material was placed on September 2. Freeze-up occurs around November 1, and thaw around May 1. No weight loss occurred during frozen conditions. Open symbols represent modified time, which runs only during unfrozen conditions.

are not so simple in a real situation, but this conceptual model serves to illustrate that a wetland has considerable "memory" via the process of litter accumulation and decomposition. Figure 5.27 illustrates this long-term development of the litter compartment necromass for the Houghton Lake wetland.

5.7 THE CYCLING FUNCTION OF PHYTOMASS

The preceding sections have detailed the complexities of the various vegetative compartments. The synthesis of those components into comprehensive descriptions has been accomplished, notably by Asaeda and coworkers (Asaeda and Karunaratne, 2000; Karunaratne and Asaeda, 2000; Asaeda et al., 2002, 2005; Eid et al., 2012a). As expected, these models involve many compartments and contain dozens of rate parameters that must be estimated. They are reasonably successful in describing seasonal behavior of biomass, but unfortunately have not been extended to consider necromass. They consider only rhizomes, roots, newly formed rhizomes, shoots, and panicles, and do not include standing dead above- and belowground plant parts.

To gain a more tractable description, it is possible to consider lumping phytomass into one compartment. The network shown in Figure 5.19 is amenable to such grouping, as illustrated in Figure 5.29. This lumping internalizes intercompartment transfers, but retains the external transfers of uptake, release, and accretion.

Phytomass undergoes seasonal variation in temperate wetlands, but those patterns are less pronounced in tropical and subtropical systems. For instance, the phytomass of Phragmites in the northern Netherlands displays a peak in midsummer that is about double that of spring and autumn (Figure 5.30). The amounts of nitrogen and phosphorus involved in the uptakes and returns can be large compared to light loadings in treatment marshes and must be considered in the quantification of nutrient removal. This marsh cycling function will be addressed in Chapters 9 and 10.

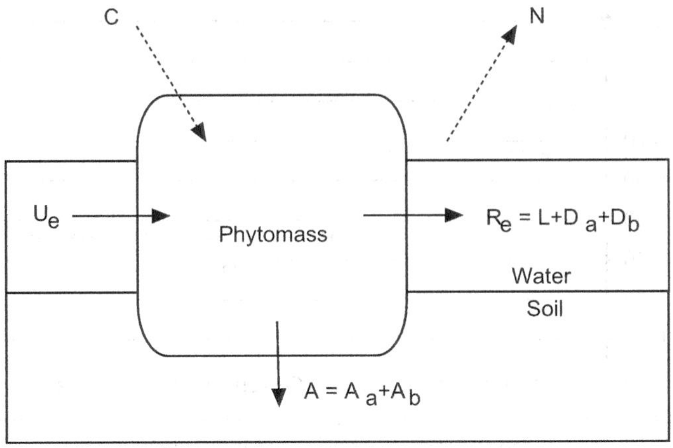

FIGURE 5.29 Lumped phytomass conceptual model. Nomenclature is shown on Figure 5.19.

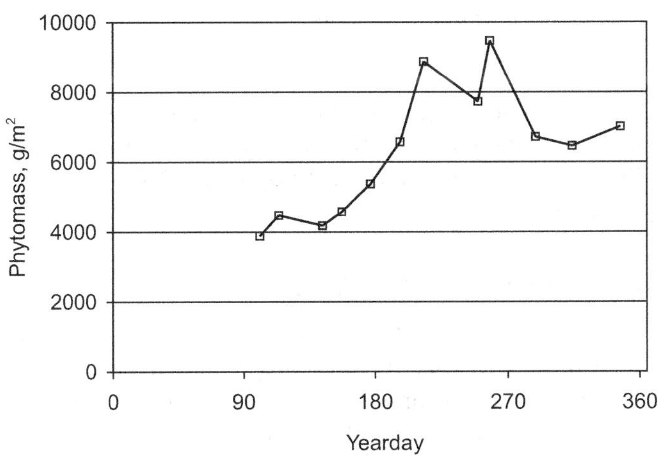

FIGURE 5.30 Seasonal variability of phytomass of *Phragmites australis* in a treatment wetland in Northern Netherlands. Pronate litter not included. (Prepared from data in Meuleman et al., 2002)

5.8 PHYSICAL MEASURES AND THE SUBSTRATE FUNCTION

Apart from the cycling function, vegetation provides large surface areas, which are involved in several wetland processes. Both stems and leaves contribute to the total. The four major surface area groupings, as depicted in Figure 5.31, are:

- Live (green) canopy; above water
- Dead (brown) canopy; above water
- Live and dead erect plant parts, under water
- Pronate litter, under water

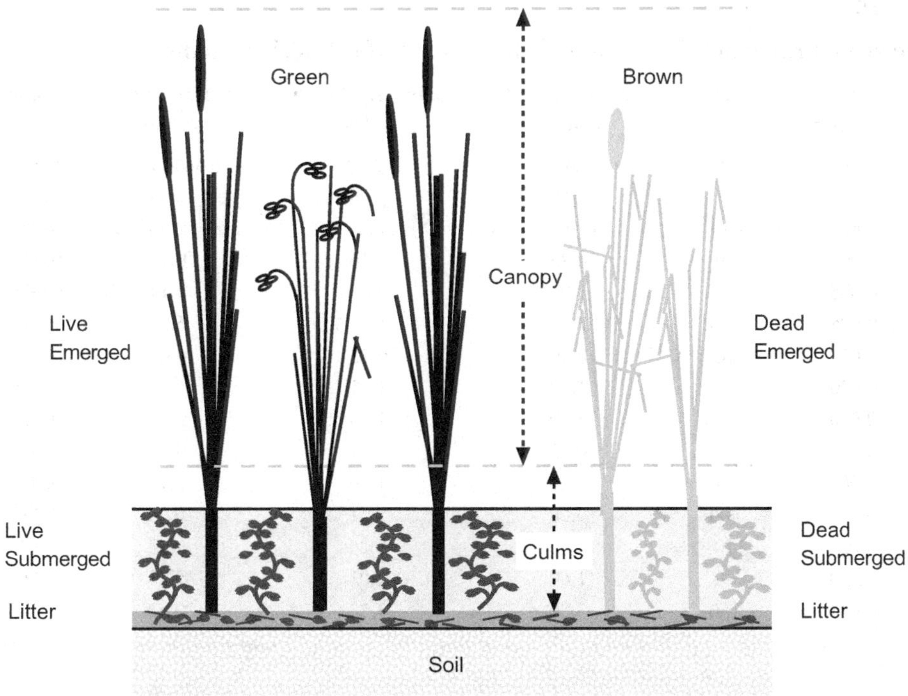

FIGURE 5.31 Categories of plant surface areas.

The canopy area affects solar radiation, convective heat and mass losses, and transpiration. The under-water areas affect the extent of biofilms and resistance to flow. There are several measures of vegetative structure that can be used to quantify effects on treatment potential (Table 5.9). In general, structural measures are functions of elevation, either depth in the water or vertical position in the canopy. Here, we first consider elevation or depth-averaged values, and then discuss the vertical distributions.

Under Water Structure

As a baseline for considering the voluminous information on under water structure, the relevant measures can be tabulated for the highly idealized situation of cylindrical stems (Table 5.10). The hydraulic resistance function of vegetation hinges upon the frontal submerged area that impedes flow. However, the substrate function of vegetation depends upon the

TABLE 5.9
Structural measures of wetland vegetation.

Whole Plants	Submerged portion	Canopy portion
Stem density	Frontal area	Green Leaf Area Index (GLAI)
Height	Surface area	Plant Area Index (PAI)
Biomass	Volume fraction	

TABLE 5.10

Baseline structural quantities for the idealized case of cylindrical stems.

Depth	Stem diameter	Stem density	Frontal Area/ Volume	Submerged Area Index	Surface Area/ Volume	Volume fraction
h	D	n	$a_F/Ah =$ nD	$a_S/A =$ nπDh	$a_S/Ah =$ nπD	1-e = $(n\pi D^2/4)$
m	m	$\#/m^2$	m^2/m^3	m^2/m^2	m^2/m^3	m^3/m^3
0.2	0.005	50	0.25	0.16	0.79	0.001
0.2	0.005	200	1.00	0.63	3.14	0.004
0.2	0.005	400	2.00	1.26	6.28	0.008
0.2	0.020	10	0.20	0.13	0.63	0.003
0.2	0.020	80	1.60	1.01	5.03	0.025
0.2	0.020	150	3.00	1.88	9.42	0.047
0.4	0.005	50	0.25	0.31	0.79	0.001
0.4	0.005	200	1.00	1.26	3.14	0.004
0.4	0.005	400	2.00	2.51	6.28	0.008
0.4	0.020	10	0.20	0.25	0.63	0.003
0.4	0.020	80	1.60	2.01	5.03	0.025
0.4	0.020	150	3.00	3.77	9.42	0.047

surface area of immersed vegetation, which is here termed the Submerged Area Index (SAI). Literature terminology is confusing, because authors concerned with hydraulics often refer to frontal area as leaf area index (LAI) (for instance, Rybicki et al., 2000, 2002).

Stem Density

The stem density of wetland plants is important because the resistance to water flow is determined in part by stem density. Only a small fraction of the ultimate plant density is planted in a new wetland. Planting densities range from 1,000–40,000 plants per hectare (0.1–4.0 plants per m^2), depending on the rate of spread of the selected plant species and the acceptable timeframe for plant establishment. Through vegetative reproduction, these plants will eventually spread to much greater densities.

Tanner reported 1,400–1,500 stems per m^2 for *Schoenoplectus tabernaemontani* growing in dairy wastewater (Tanner, 2001a), and over 2,000 stems per m^2 for *Schoenoplectus validus* (Tanner, 1994). In contrast, stem counts for *Scirpus acutus* in the Sacramento, California project were typically only 150 per m^2 in secondary effluent, although accompanied in some cases by 15–30 per m^2 *Typha latifolia* plants (Nolte and Associates, 1997, 1998a).

Cattails generally have many fewer stems per unit area than bulrushes. For instance, the discharge area at Houghton Lake, Michigan had 71 ± 23 per m^2 for *Typha latifolia*, and 89 ± 22 per m^2 for *Typha angustifolia*. A nutrient poor location at the same wetland had only 35 ± 22 per m^2 for Typha latifolia. Glenn et al. (1995) measured 140 per m^2 for *Typha domingensis* in northern Mexico. *Phragmites australis* has comparable numbers in secondary reedbeds, 70–100 per m^2 in the United Kingdom (Daniels and Parr, 1990; as referenced by Cooper et al., 1996). However, *Phragmites australis* grows to higher densities in warm climates, around 250 per m^2 in Australia (Hocking, 1989a).

Hydraulic modeling has therefore adopted similar stem density numbers. For instance, Nepf et al. (1997) used stem (cylinders) densities of 200–2,000 per m^2 in constructed flume experiments, to represent *Juncus roemerianus*. Hall and Freeman (1994) studied hydraulics in constructed flumes, with bulrush plants, at densities of 400 and 800 per m^2. In laboratory flumes, Schmid et al. (2004) used 12.8 stems (cylinders) per m^2 as representative of *Typha latifolia*.

Plant Height

The height of an EAV canopy is a major determinant of the convective heat and mass transfer that control evapotranspiration and the wetland energy balance. Those relations will be explored in Chapter 6. The live parts of common treatment marsh plants ultimately grow to about two meters tall (Table 5.11 and Figure 5.32). At the end of the growing season, they then become standing dead, for which there are no reported seasonal progressions through the nongrowing season.

Frontal Area

The role of frontal area in surface flow has been discussed in Chapter 3. Several hydraulic studies document its use in hydraulic calculations (e.g., Lightbody and Nepf, 2006; Huai et al., 2015). Studies by the U.S. Geological Survey focused upon frontal area as a controlling hydraulic parameter in the cattail and sawgrass environments of south Florida (Rybicki et al., 2000, 2002). Measured values for the Houghton Lake marsh are shown in Table 3.2.

Submerged Area

Since microbial transformations within a marsh are largely a function of area available for biofilm growth, the creation of underwater surface area by emergent and submergent aquatic plants and associated leaf litter is an important contribution to the treatment process. One method to assess the relative contribution of the plants is to measure the amount of submerged plant surface area available per area of wetland, which will be referred to as the Submerged Area Index (SAI). For instance, a waste stabilization pond would have a specific surface area of 1.0 m^2/m^2 since the only wetted surface area is the bottom of the pond. Specific surface areas for wetlands are higher, averaging 2.8 m^2/m^2 at depth 30 cm for various species (Table 5.12). Submerged aquatic vegetation offers an even greater surface area for epiphytes than does EAV The examples in Table 5.12 average 3.7 m^2/m^2, and none of those are nutrient-rich treatment systems.

The whole aboveground plant provides area from the culms (CPAI) and from the leaves (LPAI). For EAV, it is very often the case that only plant culms (stems) contribute to under water surface area, because leaves do not occur until an elevation in excess of the typical water depth.

For purposes of flow resistance or microbial substrate, it does not much matter whether the plant material is alive or dead. No distinction need be made for under water descriptions

Plant culms typically occupy only about 5% of the area of the wetland bottom. Therefore, bottom planar area adds to the total under water area, and provides surface area for biofilms, that is, epipelon. Often, that bottom area is covered by a litter layer. When this bottom area is added to the immersed plant surfaces, the result is the WSAI, the wetland submerged area index. WSAI is very nearly SAI plus 1.0.

Historically, one of the first attempts at quantifying pollutant removals in marshes involved extending the concepts of sewage treatment trickling filters to wetlands. To that end, the wetland submerged area was estimated at 15.7 m^2/m^3 (Reed et al., 1988; WPCF, 1990), based upon cylindrical stems in a cattail stand. That value was derived from a guess of 400 stems per square meter, of diameter 1.27 cm, corresponding to a porosity of 95%. As we now know, that estimate was about five times too high (Tables 5.10 and 5.12). Perhaps

TABLE 5.11
Examples of heights of emergent species.

Location		Species	Canopy Height m	Dead Height m	Reference
Czech Rep.	Nesyt	*Phragmites australis*	3.9		1
Japan	Kasumigaura	*Phragmites australis*	2.9		1
Scotland	Forfar	*Phragmites australis*	2.7		2
Scotland	Balgavies	*Phragmites australis*	2.4		2
Scotland	Lowes	*Phragmites australis*	2.2		2
Poland	Swarzewo	*Phragmites australis*	1.9		3
Nebraska	Ballards	*Phragmites australis*	2.9		4
Nebraska	Platte	*Phragmites australis*	2.5		5
Florida	ENRP	*Typha domingensis*	1.5		6
Arkansas	Fayetteville	*Typha domingensis*	2.8		7
Florida	WCA2A	*Typha domingensis*	2.5		8
Sweden	Bysjön	*Typha latifolia*	2.4		9
Arkansas	Fayetteville	*Typha latifolia*	2.3		7
Michigan	Houghton Lake	*Typha latifolia*	2.1		10
Michigan	Kinross	*Typha latifolia*	1.8		11
Sweden	Bysjön	*Typha angustifolia*	3.0		9
Canada	Mer Bleue	*Typha angustifolia*	2.5		12
Michigan	Houghton Lake	*Typha angustifolia*	2.3		10
Michigan	Houghton Lake	*Typha angustifolia*	2.4		10
Iowa	Eagle Lake	*Typha glauca*	2.7		13
Nebraska	Ballards	*Scirpus acutus*	2.0	0.2	4
Iowa	Eagle Lake	*Scirpus fluviatilis*	1.6		14
Iowa	Eagle Lake	*Sparganium eurycarpum*	1.5		14
Florida	Evgl	*Cladium jamaicense*	2.1	0.75	15
Florida	WCA2A	*Cladium jamaicense*	2.2		8
Michigan	Houghton Lake	*Carex* spp.	0.5		10
California	Mayberry	*Typha* spp.	3.0	1.4	16
California	West Pond	*Typha* spp.	3.2	1.9	16
Mean			**2.4**		

References:
1 Asaeda and Karunaratne (2000)
2 Ho (1979)
3 Hardej and Ozimek (2002)
4 Burba et al. (1999b)
5 Irmak et al. (2013)
6 Abtew and Obeysekera (1995)
7 Grace (1989)
8 Weisner and Miao (2004)
9 Weisner (1993)
10 Kadlec unpublished data
11 Kadlec and Bevis (1990)
12 Bonneville et al. (2008)
13 van der Valk and Davis (1980)
14 Burba et al. (1999b)
15 Rybicki et al. (2000)
16 Dronova and Taddeo (2016)

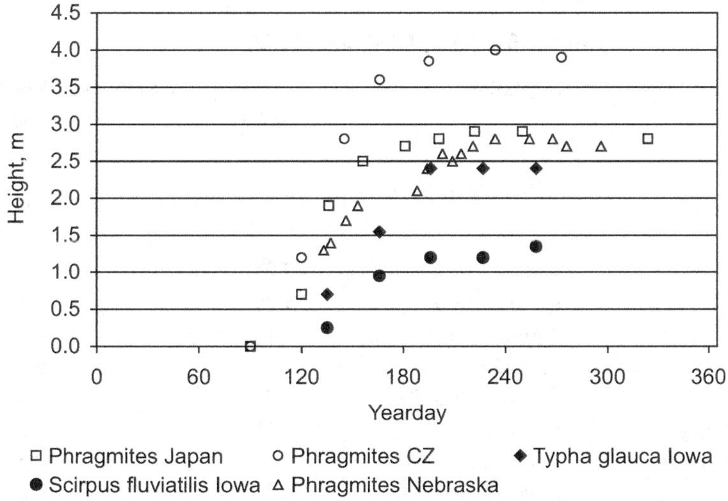

FIGURE 5.32 Examples of the seasonal progression of live plant height. Data ends with autumn senescence. Data sources: Japan, CZ: Asaeda and Karunaratne, 2000;Iowa: van der Valk and Davis, 1980; Nebraska: Irmak et al., 2013.

in recognition of this fact, the authors dropped the trickling filter concept in favor of empirical formulations (Reed et al., 1995)

Underwater Porosity

The nominal detention time in an FWS wetland is the wetland water volume divided by the volumetric flow rate. In turn, the actual water volume is less than the bathymetric value, because submerged stems take up space. The literature contains pronouncements of appropriate estimates ranging from 0.65 (Reed et al., 1995) to 0.95 (Kadlec and Knight, 1996). In fact, vegetative bed porosity depends upon stem density and stem size. For cylindrical stems, the relationship is given in Table 5.10.

For instance, at the Houghton Lake, Michigan, site, there were 96 *Typha latifolia* stems per m^2, and the mean stem diameter in the 30 cm depth was 1.2 cm. The cylinder porosity was therefore 99%. It is only when there are very large numbers of stems of large diameter that porosity drops below 95%, for example more than 100 per m^2 at diameter 2.5 cm. Such extreme sizes and densities are uncommon, but may be encountered in warm climates. For instance, Hocking (1989a, 1989b) reports stem densities of 250 per m^2, and basal diameters of one cm may be inferred from his data for *Phragmites australis* in a nutrient-rich warm climate. The corresponding cylindrical porosity is 96%.

Vertical Distributions of SAI

The depth dependence of specific surface is nearly linear for EAV, as illustrated in the examples in Figure 5.33. The meaning of this is simple: each new increment of depth immerses a proportional increment of underwater surface area. Data for the depth distribution of SAV are lacking, but anecdotal observations argue for similar behavior: a nearly uniform depth distribution.

TABLE 5.12

Submerged area index examples for FAV, EAV, and SAV.

Location	Vegetation	Submerged Area Index, m^2/m^2	Water Depth, m	Reference
FAV				
Thailand	*Eichhornia crassipes*	2.2	1.0	1
EAV				
Arcata, California	*Scirpus acutus*	3.95	0.3	2
Arcata, California	*Typha latifolia*	1.26	0.3	2
Benton, Kentucky	*Scirpus cyperinus*	3.1	0.3	2
Benton, Kentucky	*Typha latifolia*	2.1	0.3	2
Pembroke, Kentucky	*Typha angustifolia*	3.2	0.3	2
Houghton Lake, Michigan	*Typha latifolia*	6.61	0.3	2
Houghton Lake, Michigan	*Typha angustifolia*	5.37	0.3	2
WCA2A, Florida	*Typha domingensis*	2.3	0.3	3
Mean EAV		**2.8**	**0.3**	
SAV				
Trans-Ural Russia	*Elodea canadensis*	2.7	0.6	4
Trans-Ural Russia	*Myriophyllum verticillatum*	3.1	0.5	4
Trans-Ural Russia	*Potamogeton crispus*	2.0		4
Trans-Ural Russia	*Ceratophyllum demersum*	3.6	0.4	4
Denmark, Dystrup	*Ceratophyllum submersum*	11.0	1.0	5
Denmark, Stigsholm	*Potamogeton pectinatus*	4.2	0.8	5
China, Lake Taihu	*Potamogeton malainus*	1.6	1.0	6
New York, Lake Memphremagog	*Myriophyllurn spicaturn*	5.6	0.6	7
New York, Lake Memphremagog	*Vallisneria americana*	2.0	0.3	7
New York, Lake Memphremagog	*Vallisneria americana*	1.4	0.4	7
Mean SAV		**3.7**	**0.6**	

References:
1 Khatiwada and Polprasert (1999a)
2 U.S. EPA (1999)
3 Rybiki et al. 2002
4 Korlyakov and Nokhrin (2011)
5 Theil-Nielsen and Søndergaard (1999)
6 Zhao et al. (2012)
7 Petticrew and Kalff (1992)

Vegetation Canopy Structure

The canopy is comprised of the above-water components of the plants, living and dead. The canopy is a major influence on transpiration (green part only), and upon heat and mass transfer from the water to and from the air above (green and brown parts). The canopy density dictates where the incoming solar radiation heat is absorbed, and how far photosynthetically active radiation (PAR) penetrates. In turn, PAR penetration determines whether submerged biofilms are phototrophic or chemotrophic.

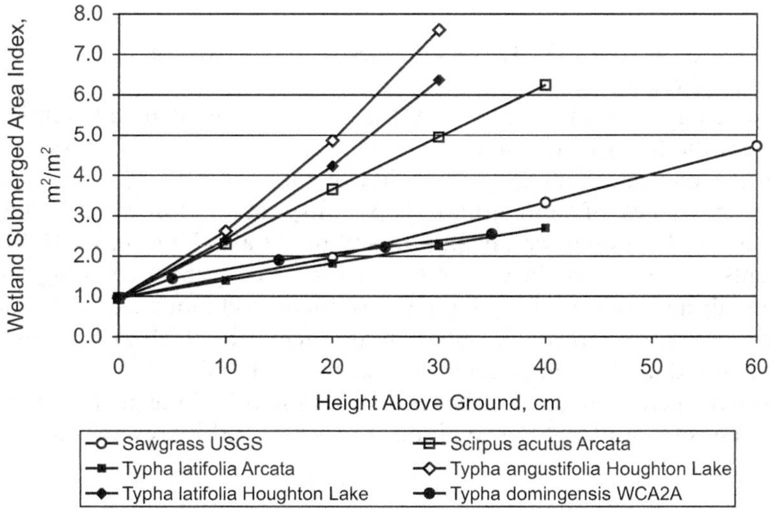

FIGURE 5.33 Submerged area indices (SAI) for various plants in treatment wetlands.

The physical attributes of the canopy are quantified by two categoric variables, canopy height and leaf area index (LAI). However, LAI means a number of different things, depending on the focus of the investigation, and its vertical distribution is also important. Consequently, it is necessary to be more specific when describing wetlands.

Leaf and Plant Area Indices

The leaf area of above-water plant parts is a major determinant of the wetland energy balance, and evaporation and transpiration. Under most normal depths of operation, the large majority of leaf area will be above water, on the order of double or triple the submerged area (Figure 5.31).

The presence of a canopy of above-water plant material influences the thermal condition of the wetland water below. The canopy shades the water, preventing some or all of the incoming solar radiation from reaching the water. The transpiration component of wetland ET is emitted from the canopy, and the loss of latent heat is therefore situated in the above-water plant material. The canopy also is a strong determinant of the convective transfer of heat and water vapor from the plants and water into the atmosphere above.

Canopies of emergent vegetation in treatment wetlands are characterized by the type, height and biomass of the vegetation, and by the density of stems and leaves. The density of above-water plant material has been the primary focus of studies of wetland energy balances, and thus water temperatures.

The EAV canopy is typically comprised almost entirely of leaves, because culms are often approximately confined to the underwater zone. Leaf Area Index (LAI) is a measure of the amount of leaf material, which is a key parameter in many of the models describing vegetation-atmosphere water and energy interactions. LAI was first defined as the total one-sided area of photosynthetic tissue per unit ground surface area. However, arguments can be made that aerodynamic resistance depends on the area of both sides of the leaf. We are thus faced with designating the index as either "1X," meaning one-sided, or "2X," meaning two-sided.

The term Plant Area Index (PAI) is used when standing dead is included, which is usual when measuring light transmission.

The term Green Leaf Area Index (GLAI) is sometimes used to differentiate from PAI. Green material is the focus of transpiration studies and the reader is cautioned that LAI is often used in that context, without specification of the exclusion of standing dead. This is especially true of studies of agricultural crop transpiration, for which standing dead is absent. GLAI is usually a one-sided definition and might also be termed GLAI-1X.

Measurements may be made in two general ways: directly, by harvesting and measuring the plant material; and indirectly, by light transmission techniques. In the indirect procedure, all canopy elements intercepting radiation are included, which in the case of treatment wetlands means standing dead material as well as live material.

LAI is strongly correlated to the aboveground biomass of the plants, which is in turn related to their density and height. The biomass relation used by several authors is:

$$GLAI = A \cdot Bio^B \tag{5.9}$$

where A,B = constants

Bio = Biomass, gdw/m^2

GLAI = Leaf area index, m^2/m^2

Although not explicitly stated, the biomass is live aboveground, and the index is GLAI-1X. Examples of coefficients A and B are given in Table 5.13.

Seasonality

Wetland EAV GLAI shows a bell-shaped annual pattern (Wilson et al., 2007) (Figure 5.34). The appropriate model equation is an exponential:

$$GLAI = GLAI_{max} \exp\left[-0.5\left(\frac{D - D_{max}}{b}\right)^2\right] \tag{5.10}$$

where b = regression parameter, /d

D = day of the year, d

D_{max} = day of the year for maximum LAI, d

GLAI = green leaf area index, m^2/m^2

$GLAI_{max}$ = green leaf area index, m^2/m^2

TABLE 5.13

Regression coefficients for biomass and GLAI.

	A	B	R^2	Reference
Phragmites	0.0136	1.02	0.91	Asaeda and Karunaratne (2000)
Phragmites	0.00277	1.00	0.925	Ho (1979)
Typha latifolia	0.0254	1.00	0.965	Weisner (1993)
Typha angustifolia	0.0122	1.00	0.999	Weisner (1993)
Arundo donax	0.0293	0.853	0.912	Watts and Moore (2011)

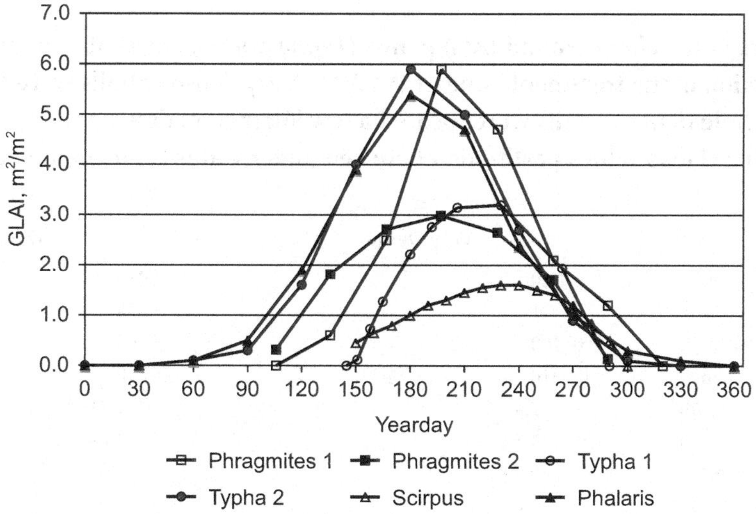

FIGURE 5.34 Seasonal patterns of green leaf area indices for several wetland vegetation types.

Vertical Distributions

Plant attributes are vertically variable, and it is therefore useful to define distribution functions. These have the general form shown below for biomass:

b(x) = biomass per unit wetland area contained in a vertical zone of thickness dx

The vertical total of biomass is then given by:

$$B = \int_0^{H_P} b(x)dx \tag{5.11}$$

where: B = biomass, gdw/m^2

H_P = height of the plant canopy, m

x = elevation aboveground surface

Because we are separately interested in hydraulic effects and canopy effects, the overall distribution may be broken into an underwater portion and an above water (canopy) portion.

The acquisition of field data from which the various indices may be calculated is requires a very large effort. Vertically stratified clip plots produce material that must be then sorted into live, dead, leaf and culm portions. Each of those portions is then measured for length and width. Small wonder that this has rarely been done, but results from those few studies provide an excellent quantification.

Studies by the U.S. Geological Survey were conducted in the treatment zone of WCA2A in Florida, at several points along the flow direction (Rybicki et al., 2002). Table 5.14 summarizes results from two locations, one near the inlet, where very dense cattails were found; and one near the outlet end of the phosphorus gradient, where dense cattails were found. Interestingly, the perceived density does not result from stem density, which was nearly identical at the two locations; but from the height and canopy leaf density. The plants were 50% taller at the upgradient location, 3.1 m vs. 2.1 m; and the plant area index (PAI-1X) was about 2.5 times greater, 10.3 m^2/m^2 vs. 3.9 m^2/m^2. Aboveground biomass was more

TABLE 5.14

Characteristics of the aboveground (AG) plants (*Typha domingensis*) at two locations along the flow direction in the treatment zone of WCA2A. Very dense cattails were located in the inlet region, while dense cattails were located a few kilometers downgradient. The lower part of the table shows values partitioned between under water (<40 cm) and above water (>40 cm).

		Very Dense			Dense		
H plant	cm	310			210		
Culm density	#/m^2	31			32		
Culm Zone	cm	0–100			0–60		
Leaf Zone	cm	40–310			40–210		
CPAI	m^2/m^2	3.45			1.81		
GLAI-1X	m^2/m^2	4.83			1.04		
LPAI-1X	m^2/m^2	6.84			2.06		
PAI-1X	m^2/m^2	10.29			3.87		
LPAI-2X	m^2/m^2	13.68			4.12		
PAI-2X	m^2/m^2	17.12			5.93		
SAI@40cm	m^2/m^2	1.61			1.57		
WSAI@40cm	m^2/m^2	2.55			2.53		
		Total	<40cm	>40cm	Total	<40cm	>40cm
Live leaves	gdw/m^2	898	0	898	155	0	155
Live culms	gdw/m^2	1,085	659	426	214	185	28
Live AG	gdw/m^2	1,983	659	1,324.3	369	185	183
Dead leaves	gdw/m^2	454	0	454	169	0	169
Dead culms	gdw/m^2	216	41	175	177	177	0
Dead litter	gdw/m^2	431	324	107	141	46	95
Dead AG	gdw/m^2	1,101	365	736	487	223	264
Total AG	gdw/m^2	3,084	1,024	2,061	856	408	447

Source: Rybicki et al. (2002).
Acronyms:
CPAI Culm plant area index
GLAI-1X Green leaf area index based on one side of the leaf
LPAI-1X Leaf area index, live plus dead, based on one side of the leaf
PAI-1X Plant (leaves plus culms) area index based on one side of the leaf
LPAI-2X Leaf area index, live plus dead, based on two sides of the leaf
PAI-2X Plant (leaves plus culms) area index based on two sides of the leaf
SAI@40cm Submerged plant area index
WSAI@40cm Wetland submerged area index (plants plus litter)

than triple at the upgradient location, 3,084 vs. 856 gdw/m^2. Water depths were approximately 40 cm, meaning that culms constituted almost the entirety of submerged area, about 1.6 m^2/m^2. The leaf zone was almost entirely above water.

The vertical profiles of area and biomass from the Rybicki et al. (2002) at the same two locations are shown in Figure 5.35. Both sets of profiles display upward decreasing densities, with much larger biomass and area at the low elevations. Both also show two-piece profiles, in which there is an upward decrease in the culm zone, followed by a jump to higher values at the bottom of the leaf zone, followed by an upward decrease in the leaf zone.

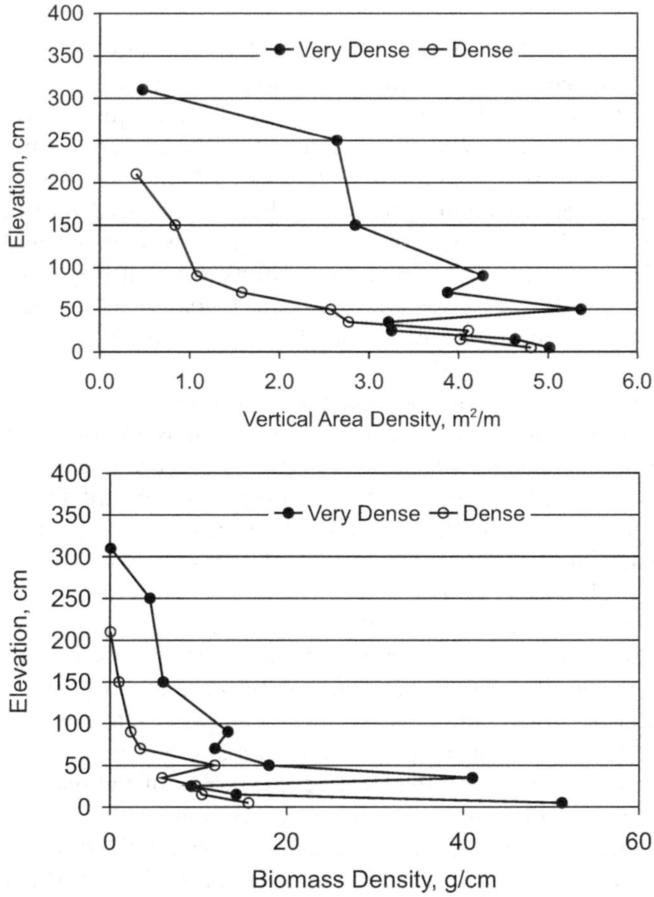

FIGURE 5.35 Vertical distributions of (A) plant area index and (B) total biomass for an inlet (very dense) and a downgradient (dense) location in the WCA2A treatment marsh. (Based on data in Rybicki et al., 2002)

5.9 SUMMARY

Wetland environments support a wide variety of bacteria, fungi, algae, and macrophytes. Treatment wetlands have been implemented that use periphyton, algae, submerged macrophytes, floating vegetation, and woody plants, although emergent macrophytes remain the most common choice.

The mesophytic community is the assemblage of microbes and algae that exist within and upon the macrophyte "forest." The mesophytes contribute in major ways to phosphorus processing in marshes. It is well established that the mesophytes are the first responders for nutrient removal from water, because they are situated within the water column. However, their storage capability is limited. The detritus from the mesophytic growth cycle is the dominant source of wetland particulate matter. Mesophytic detritus is first deposited in, and is a major contributor to, the floc bed.

Macrophytes serve several important functions in treatment systems. They provide a large surface area for microbes, which is the substrate function. They control the energy flows that support evapotranspiration and modulate water temperature. They control water depth and flow, and they utilize nutrients and other substances, and convert them to new wetland residuals.

The variety of wetland plants is very large, and they occur in a number of different growth forms. The most common are emergent, rooted vegetation and submerged aquatics. However, floating plants and floating leaved plants are frequently found in treatment marshes and are sometimes intentionally fostered. Emergents can form floating mats, and there is increasing use of created floating mats, especially for small stormwater treatment applications. A wide variety of plant species have been introduced into new treatment wetlands. Initial plant selection is a function of hydropattern, climate, and cultural choices. Regardless of the initial planting, the plant community will self-organize over time as additional plant species invade the system. If the project goal is to maintain a specific plant community, human intervention will be required to remove plants that are viewed as being undesirable.

The growth, death, and decay of plant biomass form an important biogeochemical cycle in treatment wetlands and imposes a seasonal cycle on many internal processes. During the growing season, nutrients such as nitrogen and phosphorus are taken up by the plants, and temporarily stored in the plant canopy. This uptake is significant for juvenile ecosystems where the plant canopy being established, and for periods of peak plant growth. At the end of the growing season, nutrients are returned to the system after the emergent portion of the plants die back. The decay of plant biomass imposes nonzero background concentrations for many constituents in treatment wetlands and is important in some treatment processes, such as denitrification. Some portion of the phytomass is resistant to degradation, leading to a net accretion of refractory organic matter in treatment wetlands.

The quantification of the amounts and structure of plant materials can involve many different metrics, including stem density, height, biomass and leaf area indices. These measures form part of the basis for describing the treatment function of the marsh. The various treatment functions rely upon different aspects of the vegetation structure. Growth of live material provides water and chemical uptake. Energy and water balances require knowledge of the entire live and dead canopy. Hydraulic functions depend upon the submerged portion of the vegetation.

The body of knowledge on marsh vegetation is so large that consolidation is necessary to support descriptions of system-wide performance. That sorting will be done in subsequent chapters.

6 Energy Flows and Temperature

Water temperatures in treatment wetlands are driven by energy flows (gains and losses) that act on the system. During warm conditions, the largest energy gain is solar radiation, and the largest energy loss is evapotranspiration. Energy flows are cyclical and act on both daily (diurnal) and seasonal time scales.

As water flows through the wetland, energy gains and losses drive the water temperature towards a balance point temperature, at which energy gains equal energy losses. This results in a longitudinal change in water temperatures as the system trends towards the balance point. The balance point temperature may be warmer or cooler than the influent water temperature, depending on the relative magnitude of the energy flows.

Because temperature exerts a strong influence on some chemical and biological processes, it is important to wetland design. In cold climates, freezing of the wetland will be an operational concern. Successful design requires that forecasts be made for expected or worst-case operating conditions, which implies prediction rules and equations. This chapter reviews the data on treatment wetland water temperatures and explores the tools available to wetland designers to predict water temperatures that result from energy flows within treatment wetlands.

The water temperature in treatment marshes is of interest for several reasons:

1. Temperature modifies the rates of several key biological processes.
2. Temperature is sometimes a regulated water quality parameter.
3. Water temperature is a prime determinant of evaporative water loss.

In the first instance, there is extensive literature supporting the strong effect of temperature on microbial nitrogen processing, with doubling of rates over a temperature range of about 10°C. In the second case, cold-water fishes, such as salmonids, are sensitive to water temperature, and cannot survive or breed in warm environments. In the third case, net water loss (and associated increases in total dissolved solids) is a detriment in arid climates, where water rights and water return credits are of increasing importance. Additionally, water temperature is strongly connected to evapotranspiration, which in turn is a major factor in the water budget for the wetland. Finally, freezing of the wetland can create operational problems in mildly cold climates, and preclude operations in deeply cold conditions.

6.1 THE UNDERPINNINGS: WETLAND ENERGY FLOWS

The energy flows that determine water temperature and the associated evaporative losses are shown in Figure 6.1. These processes are driven and dominated by solar radiation. Incoming solar radiation is partially reflected, with the remainder intercepted by the vegetative canopy and water column. Solar radiation intercepted by the vegetative canopy drives transpiration in plants. The remaining solar radiation is absorbed by the wetland water and drives evaporation. The combined water loss is termed evapotranspiration and is commonly abbreviated as ET.

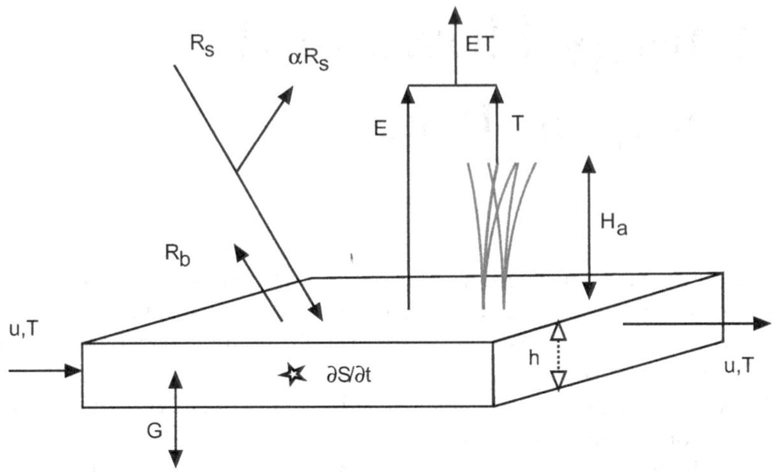

FIGURE 6.1 Components of the wetland energy balance. See text for explanation of notation.

Convection and diffusion carry water away from the surface, and transfer heat from the air to the wetland. The driving force for convective and diffusive heat transfer is the temperature difference between the wetland and the air above. For water vapor transport, the driving force is the water partial pressure difference between the wetland and the air above. Additionally, heat is radiated from the wetland. Heat may also be transferred from soils to the wetland, but that contribution is usually very small. The net effect of these processes will be a difference between the sensible heats of incoming and outgoing water flows, as well as a change in the energy stored in the wetland.

In a simplified plug flow case, energy additions and losses combine to change storage and advective changes. The system for the energy balance is here taken to be the wetland water body and the associated phytomass. Over a given balance period, the net incoming radiation is dissipated to heat and evaporated water:

$$\frac{\partial(S)}{\partial t} + \frac{\partial(uS)}{\partial x} = R_N - \rho\lambda_m \cdot ET - H_a - G \tag{6.1}$$

where c_P = heat capacity of water, MJ/kg·°C
$S = \rho\ c_P\ hT$ = sensible heat of the water, MJ/m^2
G = conduction heat loss to soil, $MJ/m^2 \cdot d$ ==
h = water depth, m ==
H_a = convective transfer loss to air, $MJ/m^2 \cdot d$
ET = water lost to ET, m/d
R_N = net radiation reaching the water, $MJ/m^2 \cdot d$
t = time, d
T = water temperature, °C
u = water velocity, m/d
λ_m = latent heat of vaporization of water, MJ/kg (2.453 at 20°C)
ρ = density of water, kg/m^3

It is informative to examine these terms, with a view to understanding the magnitude of the various heat fluxes.

NET RADIATION (R_N)

The net radiation that reaches the wetland is the difference between incoming radiation and reflected plus back heat radiation:

$$R_N = R_S(1 - \alpha) - R_b \tag{6.2}$$

where R_b = net outgoing long wave radiation, MJ/m^2·d
R_S = incoming solar radiation, MJ/m^2·d
α = fraction incoming radiation reflected (wetland albedo), –

Incoming Radiation (R_S)

The incoming solar radiation (R_S) is often available from local weather station data, because of the large current interest in harvesting solar energy. If not, it may be estimated from incoming extraterrestrial radiation, R_A (Allen et al., 2005). In turn, R_A can be calculated by any of several algorithm ms available in the literature, for instance that presented by Nikolov and Zeller (1992).

The amount of radiation, which makes it through the outer atmosphere, is solar radiation, which has traditionally been estimated from the Angstrom equation (Angstrom, 1924). Other modifications are reviewed by Muzathik et al. (2011).

$$R_S = (0.25 + 0.5\frac{n}{N}) \cdot R_A \tag{6.3}$$

where n = duration of sunshine, hr/d
N = maximum possible duration of sunshine, hr/d
R_A = extraterrestrial radiation, MJ/m^2·d
When n = N, the clear-sky solar radiation, R_{SO}, is 0.75 times the extraterrestrial radiation.

Reflected Radiation (αR_S)

A fraction α, the wetland albedo, of incoming solar radiation is reflected by the wetland. The albedo is variable for different surfaces and vegetation conditions. A terrestrial green vegetation cover has an albedo of about 0.20–0.25, and a value of α = 0.23 is commonly used as standard (Allen et al., 2005). Marsh data indicates lower values, in the range of 0.15–0.22 (Figure 6.2). This might be the result of the brown standing dead, which is not found in terrestrial crops.

Radiative Heat Loss (Back Radiation)

Net outgoing long wave (heat) radiation, also called back radiation, can be computed based on atmospheric characteristics of cloud cover, absolute temperature and moisture content. Various formulations exist, and a common choice is:

$$R_b = \left[1.35\frac{R_S}{R_{SO}} - 0.35\right]\left(0.34 - 0.139\sqrt{P_w^{sat}(T_{dp})}\right)\,\sigma(T + 273)^4 \tag{6.4}$$

where R_b = net outgoing long wave radiation, MJ/m^2·d
$P_w^{sat}(T_{dp})$ = water vapor pressure at the dew point, kPa
T = air temperature, °C
T_{dp} = dew point temperature, °C
σ = Boltzman's constant, = 4.903x10^{-9} MJ/m^2·d
In combination, the net incoming radiation is given by equation 6.2.

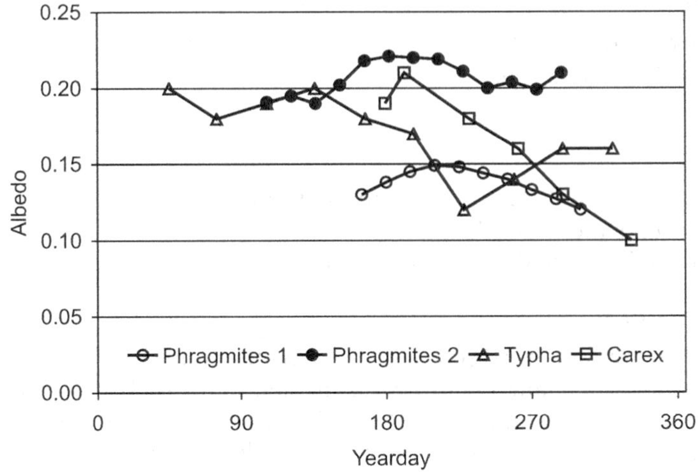

FIGURE 6.2 Albedo of example marshes. Data from Phragmites 1: Burba et al. (1999a); Phragmites 2: Lenters et al. (2011); Typha: Abtew and Obeysekara (1995); Carex: Priban et al. (1992).

Convective Losses and Gains to Air

Although lumped together in Equation 6.1, there are two major and distinct components of heat exchange with air. Wind blows through the wetland plant canopy, and either warms or cools the leaves. In the process, it removes the water transpired through the leaves. Secondarily, this air also may heat or cool the water underlying the canopy. The relative proportions depend upon the extent of vegetative cover, and the relative areas of leaves and water. The effect in the canopy is to control transpiration, while the effect in the wetland below is to control evaporation and water temperature.

Accompanying the heat transfer in the canopy, there will be a corresponding mass transfer of water vapor from the leaves to the air passing through. Additionally, there will be a corresponding mass transfer of water vapor from the water surface to the air. Calculations utilize the known relations between the transfer rates and wind speed. Both heat and water transfer are the result of aerodynamic processes, in which eddy circulation above the canopy moves energy and water (Figure 6.3).

The common representation is a transport coefficient times a temperature or water vapor concentration difference from canopy to overlying air (Allen et al., 2005). The concentration of water in air is typically measured as the partial pressure of water. Water vapor flow is evapotranspiration (ET) and is calculated as a mass transfer coefficient times the water vapor pressure difference between the water or leaf surface and the ambient air above the wetland.

$$ET = K_e[P_w^{sat}(T_w) - P_{wa}] = K_e \Delta P_w = \frac{1}{r}\Delta P_w \tag{6.5}$$

where ET = evapotranspiration, m/d
K_e = water vapor mass transfer coefficient, (m/d)/kPa
P_{wa} = ambient water vapor partial pressure, kPa
$P_w^{sat}(T_w)$ = saturation water vapor partial pressure at T_w, kPa
r = water vapor transfer resistance, kPa/(m/d)
T_w = water temperature, °C

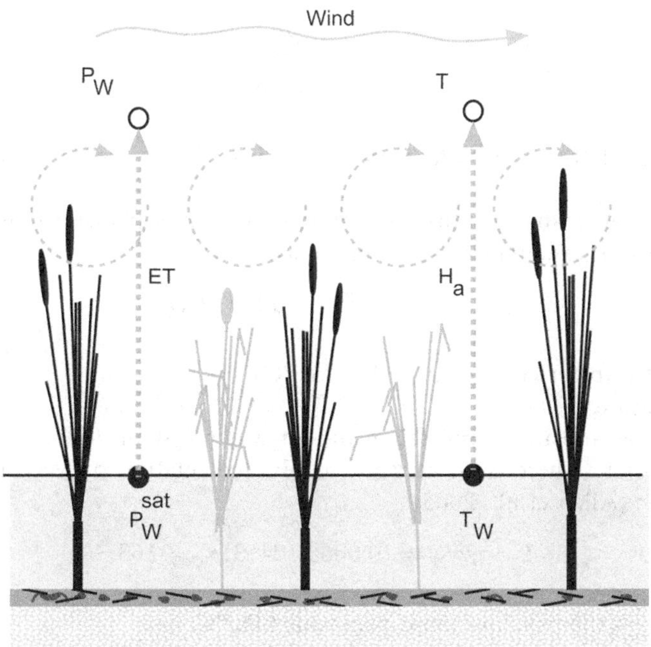

FIGURE 6.3 Driving forces for heat and water vapor transfer in the marsh canopy.

The partial pressure driving force between water and the atmosphere (ΔP_W) is sometimes termed the vapor pressure deficit. Either a resistance (r) or a conductance (K_e) may be used to represent the transfer coefficient. Confusingly, the driving force may be considered to be either the vapor pressure deficit or the corresponding difference in water concentrations in the air. The recent evapotranspiration literature often chooses a resistance version of the mass transfer equation.

$$\rho_L ET = K[C_w(T_w) - C_{wa}] = K\Delta C_w = \frac{1}{r}\Delta C_w \tag{6.6}$$

where ET = water lost to ET, m/d
K = water vapor mass transfer coefficient, m/d
C_{wa} = water concentration in the atmosphere, g/m^3
$C_w(T_w)$ = water concentration in the atmosphere at water surface, g/m^3
T_w = water temperature, °C
r = water vapor mass transfer resistance, d/m (often reported as s/m)
ρ_L = water liquid density, g/m^3

The relation between water concentration and partial pressure may be calculated from the ideal gas law:

$$C_w = \frac{(\overline{MW}_w)}{R_g T_a} P_w \tag{6.7}$$

where $\overline{MW}_w$ = molecular weight of water, g/mol
T_a = air temperature, °K = °C + 273
R_g = gas constant, 8.314 x 10^{-3} kPa·m^3/mol·°K

From equations 6.5 and 6.6:

$$K_e = \frac{(\overline{MW}_w)}{R_g T_a \rho_L} K \tag{6.8}$$

It should be noted that most literature uses units of m/s for the mass transfer coefficient K.

The convective heat transfer from the water to the air is likewise represented as a heat transfer coefficient times the temperature difference:

$$H_a = U[T_w - T] = U\Delta T \tag{6.9}$$

where U = heat transfer coefficient, MJ/(m^2·d·°C)

T = air temperature, °C

Because both heat and mass transfer in the air-water system derive from the same aerodynamic considerations, there is an accurate, calibrated relation between the heat and mass transfer coefficients (Allen et al., 2005):

$$U = \gamma\lambda K_e = (0.0666)(2453)K_e = 163.3K_e \tag{6.10}$$

where $\gamma = \frac{c_p P}{[0.622]\lambda}$ = the psychrometric constant, kPa/°C

(γ = 0.0666 at 20°C and 1 kPa)

(0.622 = 18/29 = molecular weight ratio, water/air)

c_p = heat capacity of air, MJ/kg·°C

P = ambient air pressure, kPa

The amount of water in the ambient air is a known quantity; calculated as the relative humidity times the saturation pressure of water at the ambient air temperature:

$$P_{wa} = RH \cdot P_w^{sat}(T) \tag{6.11}$$

where RH = relative humidity, –

The saturation pressure of water (vapor pressure) is a function of temperature. It is well represented (R^2 = 0.99999) by:

$$\ln\left[P_w^{sat}(T)\right] = 19.0971 - \frac{5,349.93}{T + 273.16} \tag{6.12}$$

This vapor pressure is exerted at the air-water interface.

The convective coefficients U and K_e are functions of wind speed and properties of the vegetative canopy, as will be subsequently detailed.

Conduction Heat Loss from Water to Ground

Heat may enter or leave the wetland surface water from the underlying soils, via the process of heat conduction. The vertical energy gains and losses from soils below the water are typically negligible compared to radiation and ET during summer, but are of considerable importance in winter, when they are the only gains. This heat flux is calculated as the product of a thermal conductivity and the soil temperature gradient:

$$G = -k\frac{dT}{dz} \tag{6.13}$$

where k = soil thermal conductivity, (MJ/d·m$^{2)}$/(°C/m)
T = soil temperature, °C
z = depth downward into soil, m

The thermal conductivities of soils vary with type and water content, with a typical range of 30–100 kJ/m·°C·d. Most marsh soils are organic and contain a large fraction water. These have thermal conductivities of about 50 kJ/m·°C·d. Dissanayaka et al. (2013) report thermal conductivities from three different peaty soil profiles at Bibai marsh, Hokkaido in Japan. They determined k = 44.1θ+7.8 50 kJ/m·°C·d, R^2 = 0.91, where θ is the water content in m^3/m^3.

When the soil T decreases downward, dT/dz < 0, and there is a positive heat loss (G > 0) into the soil.

Vertical energy transport in wetland soils has been modeled as transient heat conduction, and fits data quite well for other wetlands (Priban et al., 1992; Kadlec, 2001a, Kadlec and Wallace, 2009). The temperature profiles T(z,t) in the soils below a wetland are governed by the unsteady state heat conduction equation together with the boundary condition of a fixed temperature mean annual temperature, a constant at deep locations:

$$\frac{\partial^2 T}{\partial z^2} = \frac{1}{\alpha}\frac{\partial T}{\partial t} \tag{6.14}$$

$$T(^\circ, t) = T_s \tag{6.15}$$

For a sinusoidal surface temperature, the solution to this periodic, dynamic heat balance is (Priban et al., 1992):

$$T(z,t) = T_s + A\exp\left(-\frac{z}{H_P}\right)\cos\left[\omega(t - t_{\max}) - \frac{z}{H_P}\right] \tag{6.16}$$

where

$$H_P = \sqrt{\frac{2\alpha}{\omega}} \tag{6.17}$$

and

$$\alpha = \frac{k}{\rho_s c_s} \tag{6.18}$$

and where A = amplitude of surface temperature cycle, °C
c_s = soil heat capacity, MJ/kg·°C
H_P = penetration depth, m
k = soil thermal conductivity, MJ/m·d·°C
t = time, Julian day
t_{max} = time of maximum surface temperature, Julian day
T = temperature, °C
T_s = mean annual temperature of the soil surface, °C
z = vertical depth, m

α = thermal diffusivity of soil, m^2/d
ρ_s = soil density, kg/m^3
ω = annual cycle frequency = $2\pi/365$ = 0.0172 d^{-1}.

The penetration depth (H) is the depth at which the mean annual temperature swing is 63.2% of that at the soil surface (A). The heat flux (G) into (or out of) the water from the soil is then:

$$G = \left[\frac{kA}{H}\right] \cdot \{\cos[\omega(t - t_{\max})] - \sin[\omega(t - t_{\max})]\} \quad (6.19)$$

The subsurface temperature profiles under wetlands have been the subject of some few studies (Kadlec, 2009c; Worrall et al., 2015). The soils under the wetland act to store energy during the warm seasons, and release it back to the surface in cold seasons. Such energy storage, and the energy conduction processes that cause it, can be important in cool weather operation of treatment wetlands (Kadlec and Wallace, 2009). Worrall found the median soil heat flux was positive from April to August with a maximum in June and the flux was negative from September through March with its minimum in December.

Data on soil temperatures for the Houghton Lake treatment marsh were obtained over a five year period, at five depths at seven locations along the direction of flow (Kadlec, 2009c). Equation 6.8 fit that data well, with R^2 = 0.84 ± 0.05. The mean annual surface temperature was 8°C, and the penetration depth was 1.0 m. For a presumed thermal conductivity of 51.8 kJ/m·°C·d, the maximum downward heat flux was 0.77 $mJ/m^2 \cdot d$, occurring on July 15. This is on the order of 8–10% of the net incident solar radiation. The monthly progression of soil heat flux is shone in Figure 6.4.

6.2 TRANSFER RESISTANCES

The early Penman (1948) formulations utilized a single wind function as the principal factor for determination of the conductance of water vapor away from a grass surface. Other investigators then added the complexities of canopies of variable height and density (e.g., Montieth, 1981). Virtually all of the early papers on evapotranspiration were focused upon agricultural crops and their water usage.

The principles emerging from that work have since been adapted to wetland conditions. The principal difference from agriculture is the presence of surface water, distributed in and

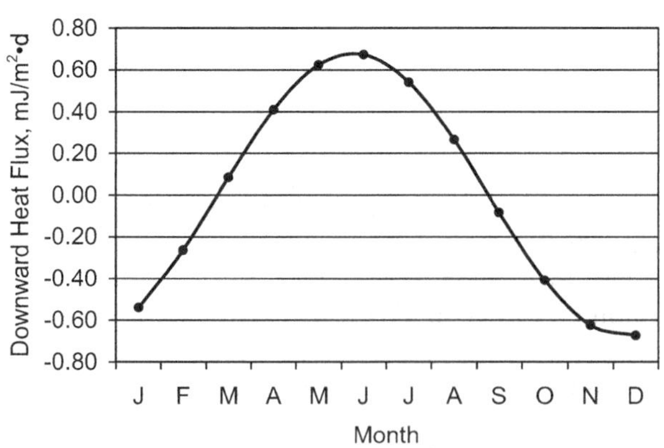

FIGURE 6.4 Heat flux from surface to soil at the Houghton Lake treatment marsh.

amongst aquatic plant canopies. Another difference is desire to deal with the temperature of that surface water, as a determinant of wetland water chemistry.

Wetland water temperatures are determined by the wetland energy balance, which also contains a strong dependence on the latent heat of the evaporating water. Thus, an understanding of wetland water temperature relies upon the concepts that determine evapotranspiration. The long history of research on evapotranspiration is a very strong platform on which to build descriptions of wetland water temperatures.

Wind Functions

The early Penman (1948) approach made use of a mass transfer coefficient, termed a wind function.

$$ET = K_e \Delta P_w = W(u) \Delta P_w \tag{6.20}$$

The water transport coefficient was expressed as a linear function of the wind velocity, (ASCE, 1990):

$$W(u) = (a + bu) \tag{6.21}$$

where a = empirical constant, (mm/d)/kPa
b = empirical constant, (mm/d)/kPa/(m/s)
ET = water lost to evaporation, mm/d
ΔP_w = vapor pressure deficit, kPa
u = wind speed at two meters elevation, m/s
W(u) = mass transfer wind function, (mm/d)/kPa

The empirical parameters a and b depend upon vegetation characteristics in an unspecified manner.

Resistances

The Penman-Monteith energy balance calculations utilize resistance instead of conductance (Montieth, 1981). The currently accepted version of Penman-Monteith assigns two parts to the mass transfer resistance for an emergent marsh (ASCE, 1990; Price, 1994; Abtew, 1996; Allen et al., 1998; Gavin and Agnew, 2004; Peacock and Hess, 2004). There is proposed to be a mass transfer resistance due to the canopy in series with a resistance due to aerodynamic transfer. The canopy resistance is related to stomatal openings in the leaves. (There is not such a dual control proposed for heat transfer.)

$$r = r_c + r_a \tag{6.22}$$

where r_c = canopy resistance, s/m
r_a = aerodynamic resistance, s/m

The aerodynamic resistance is often supposed to depend inversely upon the wind velocity, measured at two meters height. For instance, Montieth (1981) asserted that r_c was leaf area index dependent, and r_a was wind dependent:

$$r_a = \frac{200}{u} \tag{6.23}$$

where u = wind speed at two meters elevation, m/s

$$r_c = \frac{100}{GLAI} \tag{6.24}$$

where GLAI = green leaf area index, m^2/m^2

Atmospheric Resistance

The atmospheric, or aerodynamic, resistance is generally represented in terms of a multi-parameter aerodynamic model (Montieth, 1981; Jensen et al., 1990; Allen et al., 1998). Eddy diffusion theory was been used by those investigators to develop canopy height-dependent relations for r_a. This relation is complex, and contains three parameters not commonly measured (Allen et al., 1998):

$$r_a = \frac{\ln\left[\frac{z_m - d}{z_{om}}\right] \ln\left[\frac{z_h - d}{z_{oh}}\right]}{k^2 u_z} \tag{6.25}$$

where d = zero plane displacement height, m
k = von Karman's constant, 0.41
r_a = aerodynamic resistance, s/m
u_z = wind speed at height z, m/s
z_h = height of humidity measurements, m
z_m = height of wind measurements, m
z_{oh} = roughness height for heat and vapor transfer, m
z_{om} = roughness height for momentum transfer, m

This formulation was developed several decades ago for estimating crop evapotranspiration (Thom, 1971; Jacobs and Van Boxel, 1988; Allen et al., 2005). Equation (6.25) has been adopted for use in wetlands (see, for instance, Abtew and Obeysekera, 1995; Gavin and Agnew, 2004; Chin, 2011; Mohamed et al., 2012).

The wind velocity at 2.0 meters above ground is a commonly used agricultural reference. However, there are obvious complications for emergent wetland vegetation, which can have heights exceeding two meters. Accordingly, research measurements are often made well above the canopy (Tolk, 1992; Jacobs et al., 2002; Goulden et al., 2007; Irmak et al., 2013).

The three parameters d, z_{om}, and z_{oh} characterize the wind profile and the resulting heat and mass transport. For instance, at a height above water d + z_{om}, the wind velocity in the canopy goes to zero. All three are dependent on vegetation height.

Reported sets of wetland values are similar to those in common use for agricultural crops as set forth in Allen et al. (1998), and similar to those reported for maize (Jacobs and Van Boxel, 1988; Tolk, 1992). Interestingly, all of these sets of values lead to a linear relation for the atmospheric conductance $C_a = 1/r_a$ as a function of vegetation height (Table 6.1). This linear height dependence, with the proportionality to wind, is shown for a wide variety of canopies in Dingman (2015), which produce regressions with a correlation coefficient $R^2 = 0.96$.

The general range of wetland atmospheric resistances is illustrated in Table 6.2.

Because atmospheric resistance depends upon the seasonally variable vegetation height, and the seasonally variable wind speed, an annual cycle in resistance is expected. This will be muted in tropical climates, because of a small annual vegetation pattern. Vegetation height in treatment wetlands is also strongly dependent upon the nutrient status of the introduced water. This is in addition to species-dependence and season-dependence. Coupled with the gradients that are often present in treatment wetlands, these effects create complex and situation-specific heights. Atmospheric resistance is determined in large part by the vegetation, and thus so are evapotranspiration and water temperature.

TABLE 6.1

Examples of atmospheric resistance parameters.

		d	z_{om}	z_{oh}	Regression	R^2
Abtew and Obeysekera (1995)	Wetland	0.68h	0.13(h-d)	0.013(h-d)	$C_a = (0.0064h + 0.0024)u$	0.99
Jacobs et al. (2002)	Wetland	0.67h	0.10h	0.01h	$C_a = (0.067h + 0.0028)u$	0.99
Allen et al. (1998)	General Crop	0.67h	0.123h	0.0123h	$C_a = (0.079h + 0.0029)u$	0.98
Jacobs and Van Boxel (1988)	Maize	0.75h	0.25(h-d)	0.025(h-d)	$C_a = (0.051h + 0.0026)u$	0.99

TABLE 6.2

Reported atmospheric resistances for wetlands. These representative values are mostly averages for the growing season.

Site	System or Condition	Vegetation	Atmospheric Resistance s/m	Literature Reference
Ft. McMurray, Alberta	Fen	Carex, *Scirpus*	80	Nicholls (2015)
Wolfe Island, Ontario	Marsh	*Typha*	25	Price (1994)
ENRP, Florida	Marsh	*Typha*	88	Abtew and Obeysekera (1995)
Sudd, Africa	Marsh	Mixed	62	Mohamed et al. (2012b)
Tadham, UK	Wet Meadow	Grasses	83	Acreman et al. (2003)
Ham Wall, UK	Marsh	*Phragmites*	41	Acreman et al. (2003)
Paynes Prairie, Florida	u =2m/s	*Panicum*	66	Jacobs et al. (2002)
Nebraska	Agricultural	Maize	28	Irmak and Mutiibawa (2010)
ET Reference	u =2m/s	Grass 12 cm	104	Allen et al. (1998)
Open Water	u =2m/s	Open Water	121	Chin (2011)

A major difference between agricultural crops and wetland vegetation is the absence of harvest in wetlands. Crops such as maize are cut at the end of the growing season, which is typically about 110 days to maturity. Harvest may occur after various elapsed times from maturity. Consequently, the vegetation height is essentially nil for half the year.

In contrast, treatment marshes are almost never harvested, and there exists a standing dead canopy after autumn senescence. That standing dead can persist for more than one year, as the plant material undergoes slow decomposition. The result is a more uniform canopy height throughout the calendar year. Of course, in tropical or subtropical wetlands, the growing season is longer and the period of standing dead dominance is short.

The winter vegetation height is often a significant fraction of the height of the live maximum standing crop. Jordan and Whigham (1988) measured the standing dead height of Typha angustifolia in Maryland, four times during winter, and found that the height remained nearly constant at 2.0 meters. League et al. (2006) measured standing dead heights of 2.90 and 3.09 m for two haplotypes of *Phragmites* in Delaware. These were not far different from the maximum live shoot heights of 2.50 and 3.25 m.

Canopy Resistance

The canopy resistance is typically argued to be inversely proportional to the green leaf area index (e.g., Allen et al., 1998):

$$r_C = \frac{\alpha}{GLAI} \tag{6.26}$$

$$C_C = \frac{GLAI}{\alpha} \tag{6.27}$$

where α = a constant, s/m

GLAI = green leaf area index, m2/m2

The apparent simplicity of equations (6.26) and (6.27) disappears when site-specific values of α are found to cover a large range for wetlands, and when the interannual variability of LAI is considered. Allen et al. (1998) put forth that only the top half of the canopy is active in transpiration. Allen et al. (1998) make the approximation α = 200 s/m, which is common practice to calculate agricultural reference evapotranspiration.

The GLAI depends upon the type and density of vegetation. There are generally strong correlations between GLAI and canopy height, and between GLAI and canopy biomass, so the practitioner has a choice of how to parameterize GLAI. (See the section on LAI in Chapter 5).

Table 6.3 illustrates the general range of values for r_C for wetland situations and the generally recognized reference crops. The review of Mohamed et al. (2012) explores the effects of the wetland range of 50–400 s/m upon ET.

Because canopy resistance depends on the status of the wetland vegetation, it follows a seasonal pattern dictated by the growth and senescence of the plants. In temperate climates, that pattern is conditioned by the timing of the warm growing and cold winter season, but in tropical climates, the pattern is determined by the occurrence of the wet and

TABLE 6.3

Reported canopy resistances for wetlands. Seasonality and plant size and density effects are not included.

Site	Location	Vegetation	Canopy Resistance s/m	Literature Reference
Ballards Marsh	Nebraska	*Phragmites, Scirpus*	75–150	Burba et al. (1999b)
Paynes Prairie	Florida	*Panicum, Polygonum*	50	Jacobs et al. (2002)
ENRP	Florida	*Typha domingensis*	25–90	Abtew and Obeysekera (1995)
WCA2A	Florida	*Cladium jamaicense*	170	Koch and Rawlik (1993)
		Typha domingensis	50	(estimated LAI = 4)
Everglades	Florida	*Cladium jamaicense*	55	Douglas et al. (2009)
Sudd Wetland	Sudan	*Phragmites, Typha*	86	Mohamed et al. (2012)
Lake margin	Ontario	*Typha*	74	Price (1994)
Hudson Bay	Ontario	*Carex*	39–87	Lafleur (1990)
Sanjiang Plain	China	*Carex, Glyceria*	72	Sun and Song (2008)
ET Reference	General	Grass 12 cm	70	Allen et al. (1998)
ET Reference	General	Alfalfa 50 cm	45	Allen et al. (2005)

dry seasons. In the case of temperate climate, canopy resistance is essentially infinite during the (frozen) winter months, and is at minimum values during the growing season pattern (Burba et al., 1999a). The reciprocal of resistance, the canopy conductance, also follows an annual (Figure 6.5). The winter season in temperate zones is accompanied by a complete absence of transpiration. In tropical climates, canopy resistance is at a minimum during the wet season, but attains higher values during the dry season (Mohamed et al., 2012).

6.3 SYNTHESIS: EVAPOTRANSPIRATION

Evapotranspiration is the primary energy loss mechanism for the wetland and serves to dissipate the majority of the incoming solar energy. In this context, evapotranspiration can be thought of as the cooling system for the treatment wetland. Without the attendant energy loss through the latent heat of vaporization of water, the "wetland" temperature would increase to a hot, desert-like condition since incoming solar radiation could not be effectively dissipated. Some estimation methods rely on energy balance calculations, while others rely on surrogate measurements While evapotranspiration is best thought of in terms of the wetland energy balance, sometimes only the water volume lost through ET is of concern, and the attendant energy flows associated with ET can be ignored. As a result, there are a variety of methods to estimate ET.

Empirical Estimation Methods

There are several related measurements of lake and wetland water losses. These measurements are not interchangeable, and indiscriminate use can lead to confusion. Information that can be used to estimate ET includes the following procedures.

Lake Evaporation

Lake evaporation is the water loss from large, unvegetated water bodies (E). It is roughly the same as wetland evapotranspiration (Parkhurst et al., 1998), but differences will be

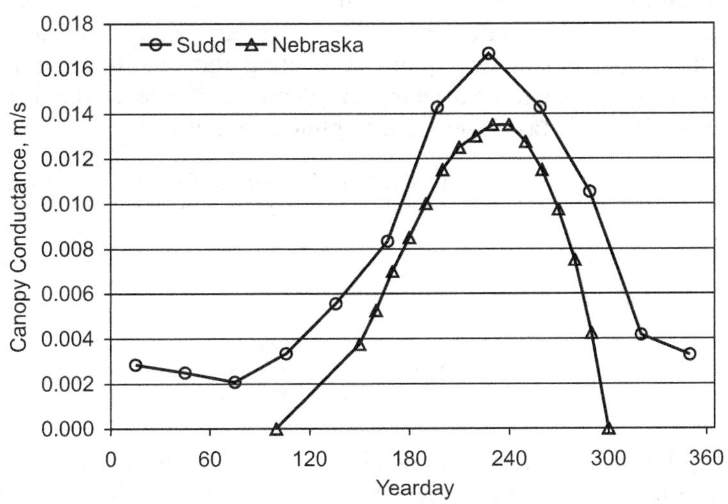

FIGURE 6.5 The seasonal course of wetland canopy resistance for a temperate *Phragmites* wetland in Nebraska, and for a tropical wetland in Sudan, which contains areas of *Phragmites, Echinochloa, Cyperus*, and *Typha*. (Based on data in Burba et al., 1999a; Mohamed et al., 2012).

discussed in a subsequent section of this chapter. This information is typically widely available. For example, for the State of Texas, historical records may be downloaded from: www.twdb.texas.gov/surfacewater/conditions/evaporation/. A worldwide summary is presented by van der Leeden et al. (1990).

Pan Evaporation

There are two types of lysimeter data available in the literature, site-specific studies and standardized pan data. The former can take several forms. One group includes evaporation from closed-bottom lysimeters (pans) of varying design (EP), containing only water (see for example, Abtew, 1996). These may be place in stands of emergent vegetation (EPV) or in areas of open water, with or without submergent or floating plants (EPO). A second type measures evapotranspiration from closed-bottom lysimeters (pans) of varying design, which contain soil, plants and water (ETP). These are placed in stands of comparable vegetation.

Class A pan evaporation (EA) is the water loss from a shallow pan of specific design, situated on a specified platform. The pan is placed on a platform above ground, and therefore evaporates more water than a lake or large wetland. The Class A pan integrates the effects of many of the meteorological variables, with the notable exception of advective effects. Wetland evapotranspiration, ET, over at least the growing season, can be approximated as about 0.70–0.85 times Class A pan evaporation, EA, from an adjacent open site.

Class A pan data is collected at weather service data stations. The number of pan stations is limited in most geographical regions, for example, 22 for all of Canada (NRC, 1978), and approximately 100 stations across the southern and midwestern tier of the United States. Data are reported in Climatological Data, a publication of the National Oceanic and Atmospheric Administration (NOAA), National Climatic Data Center, and available at (www.ncdc.noaa.gov). Historical summaries for the United States appear in Farnsworth and Thompson (1982).

Energy Balance Methods

For large wetlands, the principal driving force for ET is solar radiation. A good share of that radiation is converted to the latent heat of vaporization. About half the net incoming solar radiation is converted to water loss on an annual basis. Equations based upon equation 6.1, with no storage or advection, and its variants are widely used in the literature to predict ET. Its use is dependent on equations relating the quantities in Equation 6.1 to meteorological and environmental variables. A common choice is the Penman-Montieth equation (Montieth, 1981). It takes the form (Abtew et al., 2011):

$$ET = \frac{1}{\lambda}\frac{\Delta_{VP}(R_N - G) + \rho c_P(\Delta P_W)\frac{1}{r_a}}{\Delta_{VP} + \gamma(1 + \frac{r_c}{r_a})} \tag{6.28}$$

where

$$\Delta_{VP} = \frac{4098 P_W^{sat}}{(237.3 + T)^2} \tag{6.29}$$

The Montieth (1981) derivation is very complex, and only two aspects are noted here. First, it sought an explicit equation, in which one plugged in numbers on the right side and produced ET as result on the left side. To do that, it was necessary to linearize the vapor pressure equation (6.12), and thus the parameter Δ_{VP} was introduced. Second, the surface (water) temperature was considered to be of no interest for agricultural applications, and hence unnecessary. The main agricultural focus is upon the water requirements of crops, vis-à-vis water losses. Thus, the derivation used the relation between heat and mass transfer to eliminate the surface temperature.

In contrast, wetland applications are centered upon water temperature, as needed for performance calculations. We will therefore want to retain water temperature when Equation 6.1 is used for wetland calculations. Further, in this modern era of computer spreadsheets, it is no longer necessary to obtain explicit equations such as 6.26. Available spreadsheet goal-seeking tools make it just as easy to deal with the parent nonlinear relations. So, in later sections we will not need the Montieth (1981) derivation.

Nonetheless, there has been, and continues to be, a very large body of results from Equation 6.26 and its use to predict ET. These are widely available in the form of reference ETo information

Potential Evapotranspiration as a Reference

For large wetlands, a common assumption is that ET may be represented by the reference crop ETo computation, perhaps with modifications to represent the difference in vegetation between wetland and reference crop. The Environmental and Water Resources Institute (EWRI) of the American Society of Civil Engineers (ASCE) established a benchmark reference evapotranspiration equation that standardizes the calculation of reference evapotranspiration (Allen et al., 2000); (www.kimberly.uidaho.edu/water/asceewri/). The intent was to produce consistent calculations for reference evapotranspiration for use in agriculture. A spreadsheet program, PMday.xls, is available (Snyder and Eching, 2000; Snyder, 2001). Inputs include the daily solar radiation ($MJ/m^2 \cdot d$), air temperature (°C), wind speed (m/s), and humidity (e.g., dew point temperature (°C) or relative humidity (%)). The program calculates ETo using the Penman-Monteith equation (Montieth, 1981) as presented in the United Nations FAO Irrigation and Drainage Paper by Allen et al. (1998).

This procedure has been calibrated and verified for a green alfalfa crop, with a fetch of at least 100 meters. Other cover types may vary, due to changes in albedo and convective transport and other factors. It is critical to recognize that small wetlands will have significantly greater convective heat transfer, and consequently ET is amplified in small wetlands.

Reference Crop ETo from Reporting Services

In the United States, arid states provide extensive documentation of ETo in support of agricultural irrigation, such as the California Irrigation Management Information System (CIMIS, wwwcimis.water.ca.gov/cimis/welcome.jsp), the Arizona Meterological Service (AZMET), and the Washington State University Public Agricultural Weather System (PAWS) (http://paws.prosser.wsu.edu/). A comparable system in the United Kingdom is the Meteorological Office Rainfall and Evaporation Calculating System (MORECS) (Fermor et al., 1999). MORECS is based on a variation of the Penman-Monteith Equation and uses solar radiation, air temperature, vapor pressure, and wind speed to estimate both potential and actual evapotranspiration. Ten northeastern states in the United States use MORECS to provide localized estimates of ETo (DeGaetano et al., 1994). These services provide the results of energy balance model calculations, usually on a daily time step, for current and recent weather conditions.

Crop Coefficients

Reference evapotranspiration is a rough first approximation for ET for large marshes, meaning ET = ETo. For more accuracy, crop coefficients are required:

$$ET = K_C \cdot ETo \tag{6.30}$$

where K_C = crop coefficient, –

In agriculture, this approach leads to crop coefficients that influence ET at a specific site. This approach has the advantage of retaining the energy balance used in other ecosystems but modifying it slightly for site-specific circumstances. Allen et al. (1998) list crop coefficients for over one hundred agricultural crops, and also for a few marsh types. A seasonal variability is identified, with lesser values in the early growing season, and again at the end of the growing season. Cattails, bulrushes and *Phragmites* are all listed with an average $K_C = 1.20$; but *Phragmites* has a lesser seasonal variability.

Subsequent to the Allen et al. (1998) results, several investigators have reported K_C values for marshes (Table 6.4). These show that seasonality of marsh K_C is very minimal, but a considerable range of values is apparent.

Size is a very important factor in the determination of crop coefficients. Reference evapotranspiration is regionally derived and is a poor representation of the localized climatological variables in the energy balance. Whereas large wetland ET displays variability from ETo on the order of ± 20%, small systems can have ET that is two or three times ETo. Thus microcosms, mesocosms, and systems of less than about one hectare can have huge variability. For instance, Fermor et al. (1999) found $0.7 < K_C < 3.8$ for a 0.05 ha *Phragmites* system; Towler et al. (2004) found $2.5 < K_C < 4.0$ for *Typha* mesocosms; and Beebe et al. (2014) found $K_C = 2.5$ for *Typha* microcosms.

Size Effects on ET

Very small wetlands will display strong reactions to the surrounding microclimate. Linacre (1976) refers to this as the "clothesline effect" and cites several studies that show enhanced evapotranspiration for what amount to potted plants. Since some constructed treatment marshes tend to be small, it is reasonable to enquire at what size this effect becomes important, but there is very little information available on the size effect.

The Koerselman and Beltman (1988) study was on a wetland of "less than one hectare" and displayed no large differences from similar studies on larger wetlands. However, as size is decreased, the advective air energy terms in the energy balance become important at

TABLE 6.4

Crop coefficients for emergent marshes. The reference ETo is Penman-Monteith.

	Typha		Phragmites		Open Water
	Drexler et al., 2008	Mao et al., 2002	Anda et al., 2014	Irmak et al., 2013	Mao et al., 2002
January		0.51			0.68
February		0.61			0.74
March		0.64			0.78
April		0.73	1.03		0.77
May	0.80	0.87	1.24	0.65	0.85
June	0.92	0.87	1.24	0.57	0.85
July	1.02	0.78	1.46	0.57	0.76
August	1.09	0.76	1.22	0.61	0.79
September	1.01	0.86	1.03	0.80	0.70
October	0.90	0.78		0.99	0.79
November		0.65			0.57
December		0.56			0.53
Mean	0.96	0.72	1.20	0.70	0.73

some point, and regional methods are no longer adequate. Ratios to pan and lake evaporation, and to radiation, would not be expected to hold.

The use of energy balance information to estimate regional wetland ET is predicated on the assumptions of uniform, equilibrated water temperature and negligible effects of energy contributions from the air passing through the canopy. There are consequently two factors that may increase water losses from treatment marshes, in comparison to large-scale wetlands in the same locality. The first is the potential for incoming warm water to evaporate to a greater extent than regional waters at ambient conditions. This enhancement is greatest at the point of entry, and diminishes along the flow direction.

The second factor has to do with the microclimate created by the wetland. Small wetlands are subject to the "clothesline" and "oasis" effects, in which warm dry air can contribute to heat input and to water loss, well in excess of the loss driven by radiation alone. Indeed, this is the entire basis for the Danish willow systems, which are zero-discharge SSF wetlands (Brix and Gregersen, 2002). Estimation of the magnitude and distance scale of this effect may be done by considering the energy balance on the air passing through the canopy of the wetland. If the prevailing wind broadsides the wetland, there is convective transfer of heat to the canopy until the air has lost its heat excess over the regional wetland ambient air. Factors such as Leaf Area Index (LAI), canopy height, and air temperature and humidity influence the energy balance on the air as it moves through the wetland vegetation. Typical wetland widths for the dissipation of the incoming temperature excess and humidity deficit are on the order of 50 to 100 meters (Brix, 2006).

The clothesline effect is expected to be important in small urban stormwater systems. Unfortunately, there are not yet guidelines for the ET losses from these systems.

6.4 WATER TEMPERATURES

The energy flows that determine water temperature and the associated evaporative losses are shown in Figure 6.1 for an emergent marsh. A treatment wetland may contain one or two thermal regions, depending on water loading (detention time). For long detention times, there is an inlet region in which water temperatures adjust to the prevailing meteorological conditions, and an outlet region in which that adjustment is complete (Figure 6.6). After adjustment, temperature does not change further with distance, or detention time. The value reached is determined by the balance of energy flows, and is termed the balance temperature. For short detention times, near the wetland inlet, the adjustment may not be completed, and the balance temperature is not reached. In this adjustment or accommodation region, there will be a difference between the sensible heats of incoming and outgoing water flows; in contrast, they are equal in the balance region. In the balance region, sensible heat of the flowing water is therefore not a factor in the energy budget.

OBSERVED MARSH WATER TEMPERATURES

Annual Cycles

The annual cycle of wetland water temperatures in mild to warm climates follows a sinusoidal pattern, with a summer maximum and a winter minimum. In northern climates, the onset of frozen conditions typically is accompanied by under-ice water temperatures of 1–2°C. The sinusoidal model, truncated for frozen conditions, is:

For the unfrozen season ($t_1 < t < t_2$):

$$T = T_{avg}(1 + A \cdot \cos[\omega(t - t_{\max})]) \tag{6.31}$$

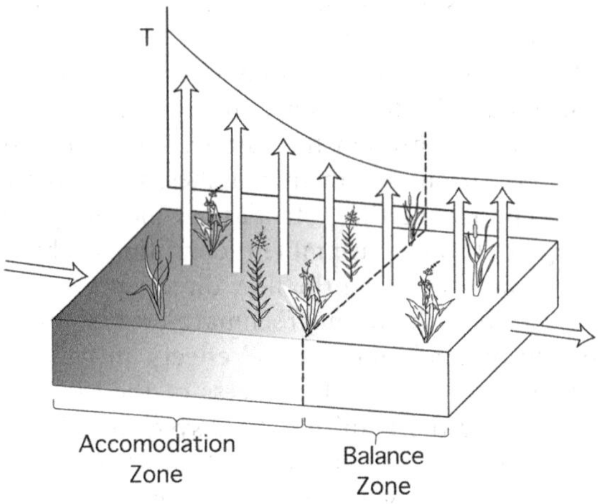

FIGURE 6.6 Gradients in temperature and evapotranspiration in a wetland.

For the frozen season ($t_2 < t < t_1$):

$$T = T_o \tag{6.32}$$

where: A = fractional amplitude of the sinusoid, –
ω = yearly cycle frequency = $2\pi/365 = 0.0172\ d^{-1}$
t = time, Julian day
t_1 = ice-out time, Julian day
t_2 = freeze up time, Julian day
t_{max} = time of annual maximum temperature, Julian day
T = water temperature, ° C
T_{avg} = annual average water temperature, °C
T_o = under-ice water temperature, °C

The various quantities associated with this time series model are illustrated in Figure 6.7. The Brighton, Ontario cycle requires truncation, and the weekly data fit has $R^2 = 0.91$. Fitting parameters are:

T_{avg}	9.8	°C	t_1	70	days
A	1.24	–	t_2	340	days
T_o	1.50	°C	t_{max}	206	days

Five parameters are required for Equation 6.31: T_{avg}, A, t_1, t_2, and t_{max}. Also required for 6.30 is T_o. Examples of regressions of data from ten treatment marshes to this truncated, sinusoidal time series model are shown in Table 6.5. Data from four to fourteen years at each site were folded into a composite annual pattern. From this information, it is seen that the time of maximum wetland water temperature is essentially fixed at $t_{max} = 200 \pm 4$ days (mean ± std. dev., N = 10).

Important: T_{avg} and A refer to the sinusoidal portion of the annual time series, and are not the overall annual means for truncated times series sinusoids.

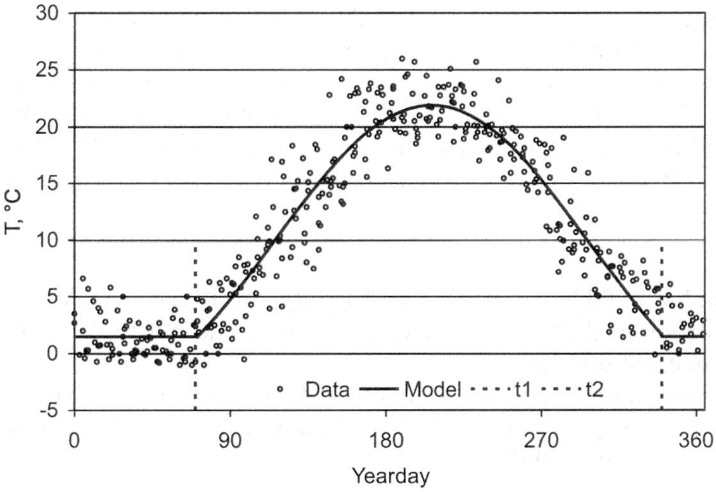

FIGURE 6.7 Annual pattern of wetland water temperatures at the Brighton, Ontario facility. Data span ten operating years.

TABLE 6.5

FWS water temperature regression parameters.

Site	Location		T_{avg} °C	A	$A \cdot T_{avg}$ °C	R^2	T_{max} °C	T_{min} °C	t_{max} day	t_1 day	t_2 day	Years
Musselwhite	Ontario	Wetland	1.9	8.43	16.0	0.72	18.1	2.0	206	112	300	4
Musselwhite	Ontario	Air	–1.0	–19.96	20.0	0.99	18.0	–20.1	205	–	–	
Listowel	Ontario	Wetland	6.3	1.80	11.3	0.95	17.7	1.5	204	88	320	4
Listowel	Ontario	Air	7.2	2.65	19.1	0.99	18.9	–8.5	206	–	–	
Brighton	Ontario	Wetland	8.1	1.63	13.2	0.91	21.2	1.5	203	82	325	10
Brighton	Ontario	Air	7.2	1.95	13.9	0.99	21.1	–6.8	205	–	–	
Hillsdale	Michigan	Wetland	13.5	0.88	11.9	0.98	25.4	2.0	196	28	365	7
Hillsdale	Michigan	Air	8.7	1.51	13.1	0.95	21.9	–4.5	209	–	–	
Commerce	Michigan	Wetland	11.8	0.89	10.5	0.96	22.3	2.0	204	44	365	4
Commerce	Michigan	Air	10.6	1.19	12.6	0.99	23.3	–2.1	202	–	–	
Columbia	Missouri	Wetland	14.3	0.68	9.7	0.99	24.1	4.6	201	–	–	14
Columbia	Missouri	Air	13.0	1.04	13.6	0.99	26.6	–0.5	201	–	–	
Imperial	California	Wetland	20.2	0.44	11.3	0.97	29.2	11.3	201	–	–	4
Imperial	California	Air	20.3	0.56	11.4	0.95	31.7	8.9	204	–	–	
Tres Rios	Arizona	Wetland	16.5	0.55	9.1	0.95	25.7	7.3	194	–	–	4
Tres Rios	Arizona	Air	21.3	0.53	11.3	0.87	32.6	10.0	202	–	–	
Shannon	Texas	Wetland	20.8	0.46	18.9	0.81	30.3	11.3	198	–	–	4
Shannon	Texas	Air	21.0	0.39	16.5	0.54	29.3	12.8	196	–	–	
ENRP	Florida	Wetland	24.4	0.23	5.6	0.77	30.2	18.7	196	–	–	4
ENRP	Florida	Air	24.3	0.18	4.4	0.98	28.7	19.9	207	–	–	

The remaining four parameters are site-specific. The treatment wetland designer will be able to find or estimate the times of freeze-up and thaw for the site in question. However, there is not a lot of variability for the time of freeze-up for north temperate climates, $t_2 = 335 \pm 29$ days (Table 6.5 mean ± std. dev., N = 5). There is more variation in the time of spring thaw, with $28 < t_1 < 112$ days. In general terms, A, T_{avg}, depend upon site climatic conditions, and T_{avg} increases and A decreases as the site moves to warmer latitudes. As a constraint, because of the symmetry of the sinusoid around t_{max}, there is a necessary relation between t_1 and t_2:

$$t_{\max} - t_1 = t_2 - t_{\max} \tag{6.33}$$

For the examples in Table 6.5, $t_{max} = 201 \pm 4$ (July 20). Parameters A and T_{avg} pertain to the sinusoidal portion of the temperature time sequence, and not to the entire annual profile in the case of truncated profiles

This cyclic model allows quantification of existing data sets, so that information from a variety of wetlands may be compared. It is, however, not predictive, because parameters depend upon site climatic conditions.

Water Temperature Variability

The deterministic trend expressed in Equations 6.31 and 6.32 represents the central tendency of water temperatures, but there are also stochastic variations. Daily meteorological variations in air temperature, cloudiness, windiness and relative humidity cause responses in water temperatures, as do changes in incoming water temperatures. Together, these factors create the need to add variability to the trend:

$$T = T_{avg}(1 + A \cdot \cos[\omega(t - t_{\max})]) + E \tag{6.34}$$

where E = stochastic contribution to water temperature, °C

The values of E follow a distribution that is nearly normal (Figure 6.8). The breadth of the scatter does not change materially during the course of the year, so that E does not depend upon time (t). However, the breadth of the E distribution does depend upon sampling frequency. The standard deviation of the daily Brighton distribution (Figure 6.8) is 2.5°C, whereas for monthly means it is 1.7°C.

PREDICTING WATER TEMPERATURES

Edinger et al. (1968) introduced the concept of an equilibrium water temperature, here termed the balance temperature, defining it as the water temperature at which the sum of the heat fluxes through the surface water is zero. That sum of fluxes is the right-hand side of Equation 6.1:

$$0 = R_N - \rho\lambda_m \cdot ET - H_a - G \tag{6.35}$$

The individual terms in Equation 6.35 have been examined earlier in this chapter and are characterized by Equations 6.2–6.27.

These equations may be used to estimate the balance water temperature and evaporative loss simultaneously. Edinger (1970) outlines the ideas. The procedure is to find the water temperature and corresponding vapor pressure that satisfy the energy conservation Equation (6.35). The required information includes the air temperature, humidity, sunshine, and wind speed; together with the extraterrestrial incoming radiation. These calculations may be easily done via an iterative (goal seeking) spreadsheet, which searches for the water temperatures (and corresponding vapor pressures, via equation 6.12) that balances the energy flows.

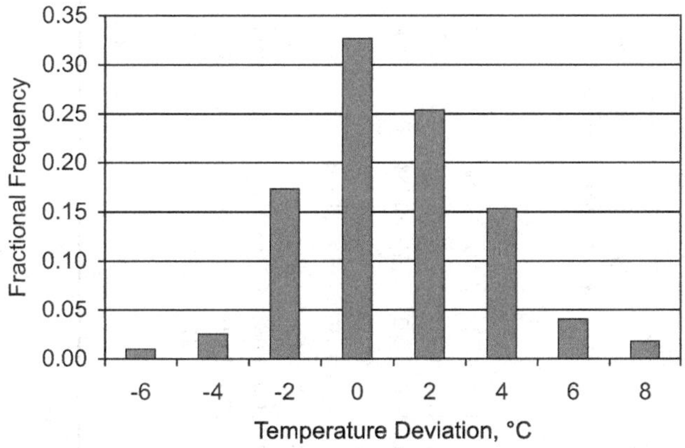

FIGURE 6.8 Deviations of daily temperatures from the sinusoidal trend for the Brighton, Ontario treatment marsh.

Table 6.6 illustrates such a spreadsheet for monthly T and ET calculations. The winter months, when there is snow and ice, are outside the bounds of this simplified energy balance procedure, and hence are removed from consideration.

This version of the energy balance utilizes a wind function, which is site-specific. This may be selected from knowledge of other wetlands, or from a calibration for the site in question. The fact that we have retained water temperature in the description provides the means by which to calibrate the wind function. The parameters "a" and "b" are selected so that calculated water temperatures match the observations from the site. This is a second goal-seeking procedure, which minimizes the mean square error between calculated temperature and temperature data.

Adjustment of the parameters "a" and "b" for the Brighton example results in the fit shown in Figure 6.9. Best fit was achieved using a = 4.80 $(MJ/m^2{\cdot}d)/kPa$ and b = 6.40 $(MJ/m^2{\cdot}d{\cdot}kPa)/(m/s)$.

The water vapor driven off by solar radiation must be convected into the air according to a water partial pressure difference from the water or leaf surface to the ambient air. The water content of the air is determined by both the air temperature and the relative humidity. At high humidity, water temperatures must be high to sustain the mass transfer gradient; conversely, at low humidity, water temperatures are lower. Example calculations show that balance water temperatures are approximately equal to air temperatures for relative humidities of about 50%. But, in arid regions water may experience significant evaporative cooling upon transit through the wetland (Kadlec, 2006).

This spreadsheet analysis may be applied whenever meteorological data are available, and there are wetland balance condition data to be fit. Twelve examples are summarized in Table 6.7, in terms of wind parameters and temperature maxima. There is considerable variability in wind parameters, as expected from different site conditions. Importantly, the maximum water balance temperature is not necessarily close to the maximum air temperature. The range, from −5.9 °C to +5.1 °C, indicates that air temperatures are not an acceptable estimator of balance temperatures.

Short Term Cycles

In recent years, high-frequency water temperature measurements have become possible due to the availability of small inexpensive temperature loggers, which can record high

TABLE 6.6

Energy balance spreadsheet for the Brighton, Ontario treatment marsh

Month	Ta °C	RH %	W m/s	S %	ex terr Rsmax	Rs MJ/ m²•d	Rso	Rs/Rso	albedo	Rin MJ/ m²•d	Psat (Ta) kPa	P kPa	Rback MJ/ m²•d	Rnet MJ/ m²•d	(neg = up) SoilFlux MJ/m2•d	Mean Water T°C (Select)	Water Psat kPa	λk	U	Heat Gain from Air MJ/ m²•d	Heat Loss from ET MJ/ m²•d	Net Gain MJ/ m²•d	Sq Err	ET mm/ d
Jan	-7.6	78	2.70	38	12.77	5.62	9.58	0.59							-0.50									
Feb	-6.5	77	2.60	40	18.06	8.12	13.54	0.60							-0.23									
Mar	-1.0	74	2.35	44	25.79	12.12	19.35	0.63	0.16	10.18	0.571	0.423	3.33	6.85	0.10	**0.7**	0.647	19.8	1.321	-2.301	4.454	*0.00*	0.00	1.82
Aspr	6.4	69	2.00	46	33.71	16.18	25.28	0.64	0.16	13.59	0.961	0.663	3.49	10.10	0.41	**8.3**	1.090	17.6	1.172	-2.176	7.519	*0.00*	0.00	3.07
May	12.5	68	1.60	50	39.42	19.71	29.57	0.67	0.16	16.56	1.446	0.983	3.63	12.93	0.61	**14.6**	1.661	15.0	1.002	-2.131	10.188	*0.00*	0.00	4.15
Jun	17.8	70	1.50	55	41.75	21.92	31.31	0.70	0.16	18.41	2.034	1.424	3.64	14.77	0.65	**19.6**	2.282	14.4	0.959	-1.760	12.363	*0.00*	0.00	5.04
Jul	20.6	68	1.50	63	40.33	22.79	30.25	0.75	0.16	19.14	2.423	1.648	3.93	15.21	0.52	**21.7**	2.594	14.4	0.959	-1.059	13.629	*0.00*	0.00	5.56
Aug	19.7	72	1.55	66	35.35	20.51	26.52	0.77	0.16	17.22	2.292	1.650	4.04	13.18	0.25	**20.8**	2.455	14.7	0.980	-1.084	11.845	*0.00*	0.00	4.83
Sep	15.3	74	1.75	62	27.92	15.64	20.94	0.75	0.16	13.13	1.734	1.283	4.08	9.06	-0.08	**16.0**	1.810	16.0	1.066	-0.709	8.425	*0.00*	0.00	3.43
Oct	9.2	74	2.05	55	19.90	10.45	14.92	0.70	0.16	8.77	1.162	0.860	3.91	4.86	-0.39	**9.1**	1.157	17.9	1.193	0.077	5.323	*0.00*	0.00	2.17
Nov	3.2	78	2.25	45	13.75	6.53	10.31	0.63	0.16	5.49	0.770	0.601	3.35	2.13	-0.59	**3.0**	0.758	19.2	1.279	0.296	3.014	*0.00*	0.00	1.23
Dec	-4.5	78	2.50	40	11.22	5.05	8.42	0.60							-0.63									
																						SSQE = 0.00		

Wind Function
a = 4.80 MJ/m2/d/kPa
b = 6.40 (MJ/m2/d/kPa)/(m/s)

Mass Transfer Coefficient
λk = a+b*W MJ/m2/d/kPa

Heat Transfer Coefficient
U = γλk MJ/m2/d/°C
γ = 0.0666

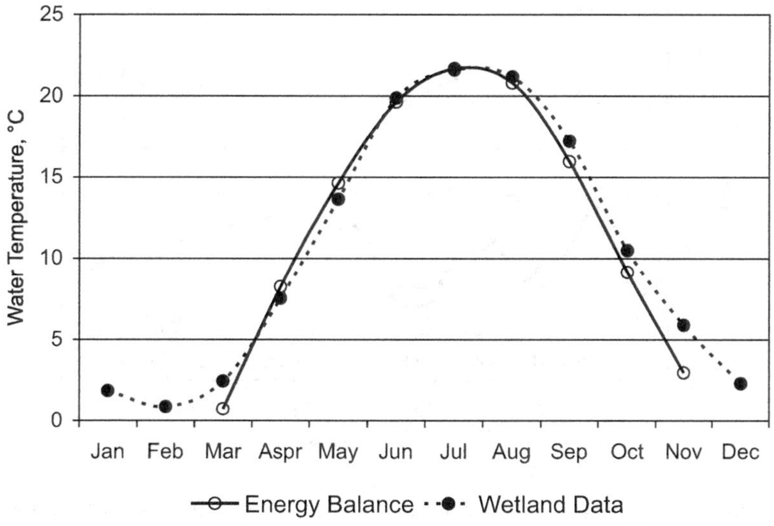

FIGURE 6.9 Energy balance fit for water temperatures at the Brighton, Ontario treatment marsh.

TABLE 6.7

Wind function parameters and forecast balance temperatures for 12 example treatment marshes. Units are a = (MJ/m^2·d/kPa and b = (MJ/m^2·d)/kPa)/(m/s).

	a	b	Ta, max °C	Tmax °C	Delta °C
Salem, OR	5.00	7.00	19.2	20.4	1.2
Tres Rios, AZ	3.30	2.90	32.5	26.6	–5.9
Sacramento, CA	1.50	5.00	26.1	23.2	–2.9
ENRP, FL	1.80	1.80	27.0	29.2	2.2
Houghton Lake, MI	4.82	7.00	18.9	20.0	1.1
Des Plaines, IL	5.00	1.00	20.3	25.4	5.1
Imperial, CA	1.00	4.00	32.1	29.5	–2.6
Shannon, TX	0.50	2.00	27.8	30.7	2.9
Columbia, MO	3.00	1.00	25.0	27.4	2.4
OEW, FL	1.00	3.00	28.2	27.2	–1.0
Brighton, Ontario	4.80	6.40	20.6	21.7	1.1
Commerce, MI	11.50	4.10	23.1	22.3	–0.8
Median	3.15	3.50			
Mean	3.60	3.77			
sd	3.02	2.20			

frequency data and have extensive memory. (See, for instance, the Onset HOBO® data logger at www.onsetcomp.com/products/data-loggers/mx2201)

Marsh waters are shallow and have only small thermal inertia. Therefore, the diel swings in solar radiation drive diel variations in wetland water temperature. Minima are observed in early morning and maxima in late afternoon. The general pattern is a marked diurnal swing in water temperature, which can be as large as 8 to 10°C in the warm months (Figure 6.10).

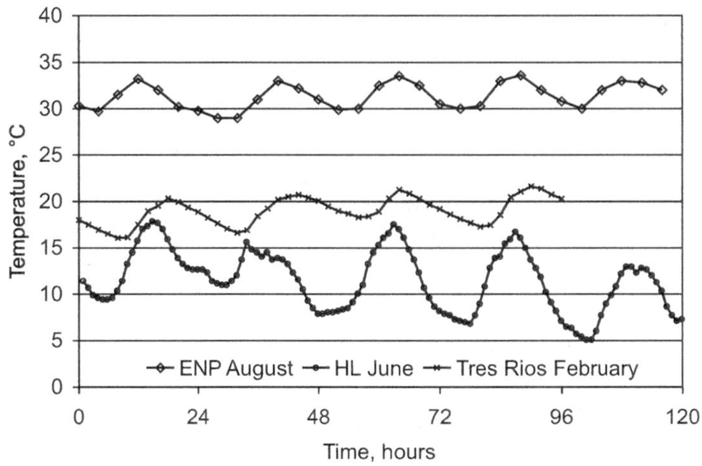

FIGURE 6.10 Diurnal temperature fluctuations in marshes. ENP = Everglades National Park (July 1 – August 4) (Swain and Decker, 2010); HL treatment marsh (June 1–8) (Kadlec, 1999a); Tres Rios treatment marsh (February 26–29).

Ordinarily, these daily cycles may be averaged to interpret wetland performance, but there are some exceptions. For instance, daily monitoring at the Tres Rios demonstration project was routinely conducted in the early daylight hours, because of the extreme heat later in the day in southern Arizona (Wass, 1997). Interpretation of the diurnal variation indicated that those morning values were about 2°C lower than the daily average. Determination of the temperature coefficients for microbial processes was therefore based upon adjusted temperatures (Kadlec, 2006).

The amplitude of the daily water temperature swing depends upon site meteorology, and type and density of vegetation. The general pattern is a marked diurnal swing in water temperature, which can be as large as 8 to 10°C in the warm months.

Diel water temperature variation is not unique to wetlands. Lakes, ponds and streams also possess such cycles. Woolway et al. (2016) quantified the pattern and magnitude of diel variability using data from 100 lakes and found average summer diel ranges in near-surface temperatures of between 4 and 7°C for small lakes. They concluded such variations have the potential to influence important temperature-dependent ecological and biogeochemical processes in lakes. Similarly, Maheu et al. (2015) studied diel variability in stream water temperature at 135 streams across the United States. They found the mean summer daily range varied between 0.4 and 9.6 °C with a mean value of 3.7°C.

Calculating Diel Cycles

The marsh energy balance (Equation 6.1) may be written, for the case of no gradients and changeable heat storage, as:

$$\frac{\partial(\rho c_P h T)}{\partial t} = R_N - \rho\lambda_m \cdot ET - H_a - G \tag{6.36}$$

The time scale for the short-term cycles is hourly. Simulation software has been developed to carry out diel calculations, such as FTLOADDS (Swain and Decker, 2010) and the Heat Source Wetlands (HSW) model (Smesrud et al., 2014). Their methods are essentially refinements of that presented above as a spreadsheet (Table 6.6), but on an hourly time scale and including storage.

On an hourly time scale, incoming solar radiation peaks in the early afternoon, and is zero during the nighttime. Back radiation continues for all 24 hours, and thus net radiation is negative at night and positive during the daylight hours. The energy for evapotranspiration stays positive for all 24 hours, but with a marked peak in early afternoon. Heat transfer to and from air is a small contribution, and changes direction at night. These trends are illustrated for the Houghton Lake marsh in Figure 6.11. The combined effect is a net energy loss at night, and a net energy gain during the day. This net effect causes the water to heat during the day and cool during the night. In the Houghton Lake example, there was a 15 cm deep water column that responded to the energy fluxes.

The energy balance spreadsheet is used to achieve the hourly integration of the energy balance Equation 6.36. In contrast to the monthly implementation of the energy balance, there is usually not the opportunity to fit model parameters to match datasets. In the construction of Figure 6.11, the starting and ending cycle temperatures were assumed equal, and the wind function was adjusted to achieve a diel profile match. The wind diel parameters so determined were a = 3 and b = 6, compared to the values a = 4.8 and b = 7.0 determined from the monthly fit for this system.

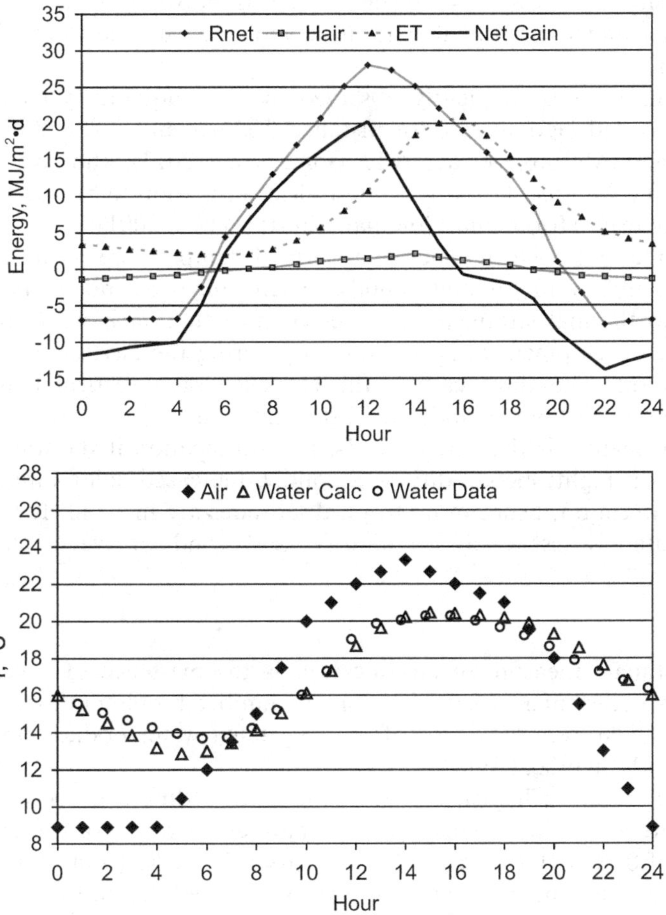

FIGURE 6.11 Measured and calculated energy fluxes and water temperatures for the Houghton Lake treatment marsh for August. Upper panel: progression of estimated energy fluxes. Lower panel: progression of air and water temperatures.

Notably, the mean daily water temperature of 17.0°C is not the same as the mean daily air temperature of 15.8°C. A single water temperature measurement at 8:00 in the morning would be 14.0°C, whereas a single value measured in the afternoon would be 20.0°C. Further, the temperature progression is dictated almost entirely by radiation and ET, with the convective heat flux being nearly negligible.

Stratification

Because wetlands are very shallow systems, they are almost invariably considered to be vertically uniform. That may be a carry-over from studies of deeper lakes, which would suggest stratification to be improbable. As a consequence, wetland water characteristics are nearly always vertically averaged for analysis. However, a review of information on vertical profiles in treatment wetlands and other shallow water bodies suggests that most are vertically stratified for a portion of the diel cycle or for a portion of the year.

Water density is a function of temperature; with the unusual property that the maximum density is achieved at 4°C (Lide, 1992). Changes in water temperature may result in layers of water with different densities and partition the water column into discrete density/temperature layers. In the summer, solar radiation raises the temperature of the surface water, reducing its density. The less-dense surface water is buoyant relative to the cooler (and denser) water layer underneath. Daily stratification can occur due to the diurnal fluctuation in solar radiation.

Thermal stratification is frequently observed in most aquatic systems, including waste stabilization ponds and lagoons (see for instance, Kellner and Pires, 2002; Abis and Mara, 2006), urban ponds (McEnroe et al., 2013; Song et al., 2013), and rivers (Bormans et al., 1997; Bormans and Webster, 1998). Lake stratification is thoroughly described in the literature on limnology (e.g., Horne and Goldman, 2001; Wetzel, 2001).

There are three potential regimes for vertical temperature profiles that have been observed in wetlands and shallow ponds. There may be no vertical profile at all, a condition of no thermal stratification. The second situation is no vertical profile during the night, but the development of surface heating during the daytime hours. This is termed diurnal stratification. The third case is the existence of a vertical temperature gradient throughout the entire 24-hour period, called stratification.

One important feature is that there will be no stratification if the water body is cooling, which is typically a nighttime condition. Secondly, increased solar radiation increases the likelihood of stratification, hence more tropical latitudes are more likely to exhibit stratification. Deeper water is less likely to stratify, and wind or water currents can defeat stratification.

Metrics

Although the ultimate measure of stratification is the existence of a vertical temperature gradient, it is also convenient to think of single-number representations of such gradients. The metrics listed here were developed for unvegetated water bodies, but are applicable to wetlands, some with qualifications.

Perhaps the simplest such metric is the temperature difference between the upper and lower layers of the water. Researchers differ somewhat on what constitutes a stratified delta T, varying from 1.0 °C (McEnroe et al., 2013) to 1.5–2.0 °C (Vallentyne, 1957). It is also possible to calculate the linearized gradient over a specified depth, for example, °C/m.

The relative thermal resistance to mixing (RTRM) is a normalized density difference from top to bottom of a water column. The normalizing factor is the density difference for one degree Celsius at the 4°C density maximum (0.008 kg/m^3). This index originated in the early 20th century (Birge, 1916, as referenced by Vallentyne, 1957), and has been in use since that time (see for instance Chimney et al., 2006 or Song et al., 2013).

$$RTRM = \frac{(\rho_{BOT} - \rho_{TOP})}{(\rho_4 - \rho_5)} \tag{6.37}$$

where ρ = density of water, kg/m^3
BOT = at bottom of water column
TOP = at top of water column
4 = at 4°C
5 = at 5°C

An index value of 50 is often taken to represent the conditions for stratification, with larger values indicating stronger stratification. The corresponding temperature difference is 1.5–2.0 °C. Alternatively, a temperature difference of 0.6–1.0 °C has also been used as a stratification indicator.

Condie and Webster (2001) present an energy ratio criterion for stratification based on pond/wetland models and data from a shallow unvegetated Australian billabong. Their work extended the concepts set forth by Holloway (1980) for lakes. The ratio of the input of kinetic energy by the wind to the input of potential energy by solar radiation scales with a nondimensional parameter:

$$S = \frac{\rho c_P w^3}{g \alpha h Q} \tag{6.38}$$

where α = thermal expansion coefficient of water, K^{-1} (T-dependent)
c_P = specific heat of water, kJ/kg·K (4.187)
g = gravitational acceleration, m/s^2 (9.81)
h = water depth, m
ρ = density of water, kg/m^3 (1,000)
Q = incoming shortwave radiation flux, kJ/s·m^2=W/m^2
w = wind speed, m/s

The stratification conditions were found, for two different models and the data, to separate as follows:

$S > 10^8$ no stratification
$10^8 > S > 10^7$ diurnally mixed
$10^7 > S$ always stratified

Condie and Webster (2001) recognized several limitations on the general application of this criterion to wetlands.

1. There is a very strong dependence on the wind speed. Sheltering and fetch can alter the effect on the water surface.
2. If there an externally driven flow-through, as exists in treatment wetlands, the turbulence generated by such a flow would significantly modify the stratification behavior of the water body.
3. The depth of light penetration, and hence the depth of solar heating, can vary; and this alters the calculation basis for the criterion. There are, however, some important features of all the various theoretical criteria. Firstly, there will be no stratification if the water body is cooling ($Q < 0$), which is typically a nighttime condition.

Condie and Webster (2001) did not consider vegetated systems, and there are additional considerations for wetlands. The presence of EAV can reduce the speed of the wind at the water surface. An EAV canopy can intercept radiation before it reaches the water (due to absorption and reflectance), and the presence of SAV can limit the depth of penetration.

These caveats and considerations combine to make the uncertainty in stratification criterion numbers too large to be useful.

Wetland Data

Stratification is exemplified by a measured diel vertical pattern for the Sacramento, CA treatment wetland (Figure 6.12). That marsh contained two predominant species: *Schoenoplectus acutus* and *Typha latifolia*. Shallow areas were densely vegetated with average densities on the order of 320 live and dead plants per square meter (see Dombeck et al., 1998a; Dombeck et al., 1998b). Temperature profiles exhibited a vertical temperature gradient in the water column that was approximately 6°C over 46 cm of depth in mid-afternoon, when air temperatures were about 6°C warmer than the water surface. At the end of the afternoon, air temperatures dropped by about 8°C, and the gradients disappeared until about 8:00 AM the next day (Figure 6.12). Temperatures within the water column remain uniform through the remainder of the night. This mixed layer continues to cool and deepen throughout the night, not warming until just after sunrise the next day.

Chimney et al., (2006) reported vertical temperature gradients in the Florida ENRP treatment marsh project. The intensity of thermal stratification differed among habitat types. The water column at EAV and OW sites was close to being isothermal on most dates. In contrast, SAV sites often had markedly higher surface water temperatures, especially during the summer months. Diel variation in hourly mean water temperature also differed markedly between the open-water and *Typha*-dominated EAV sites, with marked thermal differences at the EAV site. The amplitude of diel variation in bottom temperatures at this location was noticeably dampened compared to variation at the surface. Periods when surface water was cooler than at the bottom were noted from March through August and September to October. Similar vegetative effects were reported by; Crepeau and Miller (2014), for marshes in the San Joaquin Delta of California.

Vertical stratification was studied in an emergent (*Typha orientalis*) wetland near Sydney, Australia (Waters, 1997; Waters and Luketina, 1998). Large stratifications (5–7 °C/m) were observed year-round. Stratifications were strongest in summer and persisted for up to weeks at a time.

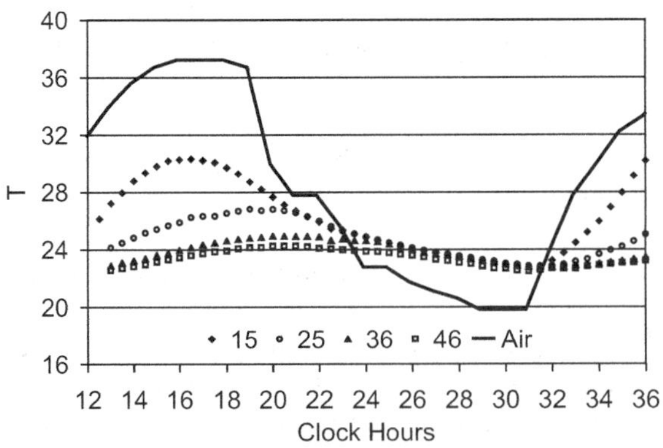

FIGURE 6.12 Diel temperature patterns versus depth for the Sacramento treatment marsh. Thermal stratification occurred between 800 and 2400 on August 11. The maximum vertical gradient of 8.6°C over 30.5 cm occurs at 13:00 hours. (Source: Nolte and Associates, unpublished data)

Jenter et al. (2003) found that in the slowly flowing wetlands of the Everglades, thermally driven convection was the dominant vertical mixing mechanism. Data was collected within an EAV area of medium-dense spike rush (*Eleocharis cellulosa*) on the edge of a large area of dense sawgrass (*Cladium jamaicense*) in lower Shark River Slough within Everglades National Park They measured both temperature profiles and velocity fluctuations in the water column. The mean water velocity at their site was about 0.8–1.3 cm/s. Results were very similar to those at Sacramento. Temperature profiles exhibited a vertical temperature gradient in the water column that was approximately 3°C over 60 cm of depth in mid-afternoon, when air temperatures were about 6°C warmer than the water surface. At the end of the afternoon, air temperatures dropped by about 10°C, and the gradients disappeared until about 8:00 AM the next day. This pattern was repeated nearly every day.

Consequences of Stratification/Destratification

Relatively little attention has been paid to marsh thermal mixing. Wind energy and heat transfer interact at the marsh water surface, resulting diurnal vertical mixing. Wind shear provides kinetic energy for vertical mixing at the water surface, which is transmitted down via mean flow and turbulent diffusion to establish the extent of the mixed layer. During periods of positive surface heat flux (surface heating), stable density gradients (stratification) act as a potential energy barrier to turbulent diffusion and reduce the mixed layer depth. During periods of negative heat flux (surface cooling), an unstable density stratification is produced in the water column, resulting in penetrative convection (plunging thermals). The mixing consequences of stratification are explored in more detail in Chapter 7.

Wetland Inlet and Outlet Temperature Comparison

The desire to mitigate thermal pollution of receiving waters has led to investigations of the effects of marshes on the temperatures of inflows and outflows (Jones and Hunt, 2010; Sun et al., 2012; Apfelbaum et al., 2013; Smesrud et al., 2014). These investigations have commonly proceeded on the hypothesis that marshes cool the water passing through them. In fact, wetlands may either cool or heat flow-through waters, depending upon the relation of inflow temperature to the balance temperature and the travel time. The balance temperature varies considerably with climatological conditions, and thus no general conclusion is possible. Three example situations are explored in more detail.

Wetland cooling occurred at The Tres Rios, Arizona, H1 demonstration marsh, where water temperatures were collected daily on weekdays. Water temperatures coming from the advanced treatment plant were quite warm year-round, varying from 21°C to 34°C. The water cooled on passage through wetlands in both winter and summer (Figure 6.13).

Wetland heating was observed at the Imperial, California demonstration marsh, which operated with weekly temperature monitoring at inlet, outlet and interior locations. Water temperatures coming from field runoff were moderate to warm year-round, varying from 10°C to 27°C. The water heated on passage through wetlands in all months of the year (Figure 6.14).

Both wetland cooling and heating occurred at the Commerce, Michigan, project. It operated with daily temperature monitoring at inlet, and outlet locations. Water temperatures coming from the advanced treatment plant were moderate to warm year-round, varying from 10°C to 21°C. The water cooled on passage through wetlands in during winter, spring and autumn, but heated during summer (Figure 6.15).

The most obvious thermal adjustment indicator is that if source waters enter cooler than the balance temperature, they will heat, and if source waters enter warmer than the balance temperature, they will cool. Source waters from treatment plants typically have a more muted annual cycle than air of balance temperatures and are warmer in winter. Therefore,

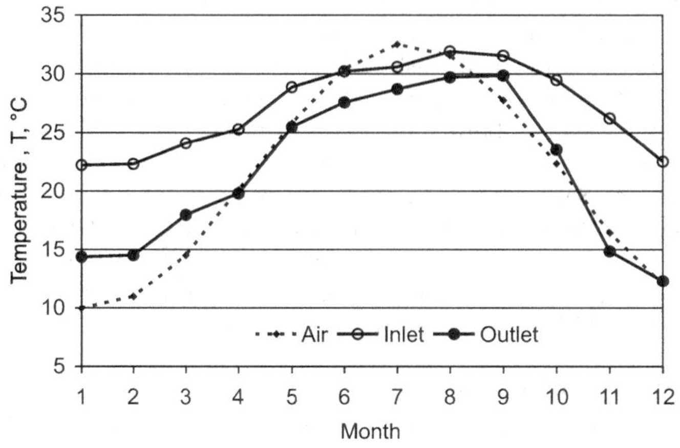

FIGURE 6.13 Annual temperature patterns for the H1 Tres Rios treatment marsh. Four years' data folded are represented.

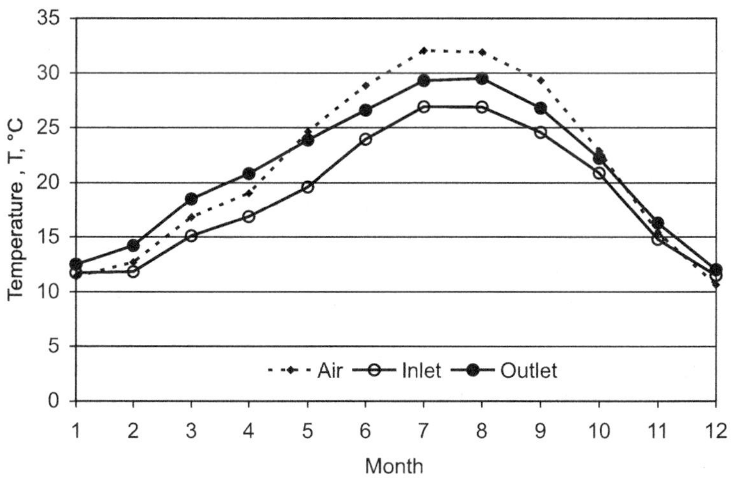

FIGURE 6.14 Annual temperature patterns for the Imperial treatment marsh. Four years' data folded are represented.

a wetland is likely to cause winter cooling, but summer effects reflect other variables, and present no common heating/cooling direction. Source waters from runoff, rivers and lakes are likely already at or close to the balance temperature, and further wetland temperature effects will be minimal.

Temperature Accommodation

The inlet zone of a treatment wetland may exhibit temperature changes, as the water approaches the balance temperature (Figure 6.6). If the source water originates from runoff, ponds or rivers, then incoming water temperatures are likely to be close to the balance

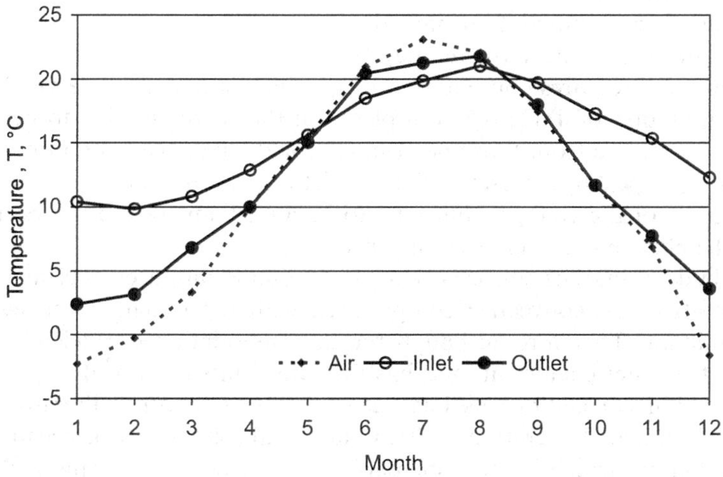

FIGURE 6.15 Annual temperature patterns for the Commerce treatment marsh. Four years' data folded are represented.

temperature, and therefore changes along the flow direction will be minimal (Gregory, 2009). But if the source water originates from an indoor treatment facility, the inflow temperature to the marsh can be quite different from the balance temperature.

The energy budget (Equation 6.1) is capable of providing a good representation of temperature variation with distance, or its surrogate detention time. For a steady flow, this energy budget is:

$$\frac{d(uS)}{dx} = R_N - \rho\lambda_m \cdot ET - H_a - G = E_{net} \tag{6.39}$$

where E_{net} = net energy gain or loss at location x, $MJ/m^2 \cdot d$

Using the definition of sensible heat, and recognizing that E_{net} is a function of water temperature,

$$\frac{d(u\rho c_P hT)}{dx} = E_{net}(T) \tag{6.40}$$

The net energy gain is a complicated function of water temperature, as shown in Equations 6.2–6.29 and the various terms of Table 6.6. Chaudhry et al. (1983) set forth a complex algorithm for this determination of water temperature However, E_{net} (T) may be linearized in the region of the balance temperature Tb, resulting in:

$$\rho c_P hu \frac{d(T - T_b)}{dx} = \left[\frac{\partial E_{net}}{\partial T}\right]_{Tb} (T - T_b) = \eta(T - T_b) \tag{6.41}$$

where η = accommodation coefficient, $MJ/m^2 \cdot d \cdot °C$

This balance may be integrated to give:

$$T = T_b + (T_i - T_b) \cdot \exp(-\eta\,\tau/\rho\, c_P\, h) = T_b + (T_i - T_b) \cdot \exp(-\,\tau/\tau_A) \tag{6.42}$$

where $\tau = x/u$ = nominal detention time, d
τ_A = characteristic accommodation time, d
T_i = source water temperature (incoming), °C

This approximation requires computation of η, which is readily done via Table 6.6. Once the balance computation of Table 6.6 is in place, all that is required is to perturb the water temperature by one degree from those determined for the balance condition.

The quantity $\tau_A = \rho c_P h/\eta$ represents characteristic accommodation time for the wetland, during which 63.2% of the change from inlet to balance temperature has been achieved. At $3\tau_A$, 95.0% of the change has been accomplished.

This energy budget analysis suggests that the accommodation coefficient is comprised of radiative, evaporative and convective components, with the radiative and evaporative portions being dominant. Therefore, although the accommodation coefficient is analogous to a convective heat transfer coefficient, and has the same units ($MJ/m^2 \cdot d \cdot °C$), it is not predictable from convection correlations as has been presumed in other literature (Reed et al., 1995; Water Environment Federation, 2001; Crites et al., 2006, 2014), because those correlations ignore radiation, which is the principal heat input in summer, and soil heat retrieval, which is the major energy source in winter, and evapotranspiration, which is the principal heat loss.

Curiously, while a significant amount of the previous wetland thermal literature emphasizes the dependence of removal processes upon wetland water temperature, it does not provide any suggestions about how to estimate or predict those water temperatures (Reed et al., 1988; Water Environment Federation, 1990; Crites and Tchonanoglous, 1998; U.S. EPA, 2000a; Reddy and DeLaune, 2008; Dotro et al., 2017). Another sector of previous wetland thermal literature suggests that the balance temperature is the air temperature (Reed et al., 1995; U.S. EPA, 1999; Water Environment Federation, 2001; Crites et al., 2006, 2014; Nietsch et al., 2011), which is clearly not the case except in summer when the relative humidity is approximately 50%. It is further not the case in winter, when water temperatures are driven to within a degree or two of the freezing point, and not lower. The reader is cautioned that major simulation packages can contain highly unsatisfactory representations of wetland water temperatures. For instance, the Surface Water Assessment Tool (SWAT) utilizes a modified air temperature, stating that:

> In addition to air temperature, water temperature is influenced by solar radiation, relative humidity, wind speed, water depth, ground water inflow, artificial heat inputs, thermal conductivity of the sediments and the presence of impoundments along the stream network. SWAT assumes that the impact of these other variables on water temperature is not significant.

Calculation Procedures

The most direct method of estimating the size of the accommodation zone is via longitudinal profile data. Equation 6.40 suggests an exponential decline in temperature with distance, or the equivalent nominal travel time. There are very few such studies, but those support an exponential decline (Figure 6.16). The Tres Rios, Arizona, demonstration project operated twelve research wetlands (0.12 ha) and four pilot scale wetlands (about 1.0 ha). The research wetlands were operated at three detention times, approximately quadruplicated. Transects were monitored along the flow direction in the pilot wetlands. Consequently, on any given transect day, data were available for both distance and loading variations of detention time. Water temperatures coming from the advanced treatment plant were warm year-round, varying from 21°C to 34°C. The water cooled on passage through wetlands in both winter and summer (Figure 6.16).

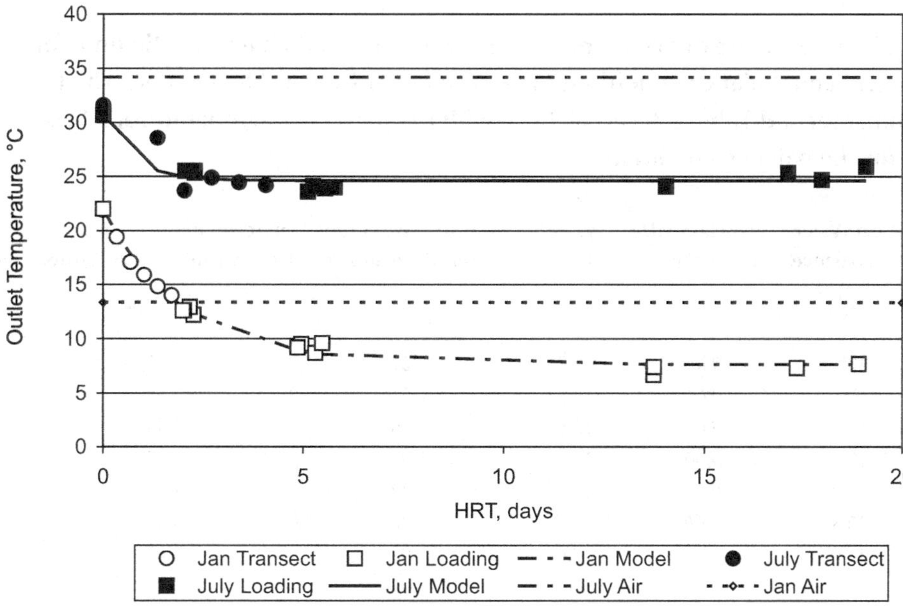

FIGURE 6.16 Wetland water temperature profiles through various Tres Rios, Arizona, wetlands. Squares represent a transect in wetland H1. Circles represent research wetlands operated at difference detention times (loadings). The upper data and curves are for July 2, with $R^2 = 0.81$; the bottom data and curves are for January 27, with $R^2 = 0.99$.

Energy Balance Estimate of Accommodation Zones

Equation 6.39 may be used to estimate accommodation coefficients. This calculation proceeds from the same meteorological data as the calculation of the wetland balance temperature, which finds the water temperature and corresponding vapor pressure that satisfy the energy conservation Equation (6.33). The required information includes the air temperature, humidity, sunshine, and wind speed; together with the extraterrestrial incoming radiation. All that is necessary is to (a) duplicate the balance temperature spreadsheet, and (b) change the water temperatures by one degree. (Because the E_{net} function is nearly linear, one could also change by anything up to about 5°C.) The results for Brighton example are shown in Table 6.8.

How Large Is the Adaptation Zone?

The wetland designer or data interpreter needs to know whether there is an adaptation zone, and if so, how much of the wetland it may occupy. This may be assessed either through estimates of η, the accommodation coefficient, or through τ_A, the time for 63.2% accommodation (see Equation 6.40). Data for treatment marshes indicates that adaptation takes on the order of 1–3 days detention (Table 6.9). This implies that many treatment wetlands will totally contain the temperature adaptation gradient if the incoming water is colder or warmer than the balance point. However, very short detention wetlands may never reach the balance temperature (see Figure 6.16).

6.5 VEGETATION EFFECTS

In agricultural applications, it is not crucial whether water losses proceed from plants as transpiration, or form water as evaporation. In treatment wetlands, transpiration drives

TABLE 6.8

Energy balance estimate of accommodation coefficients and characteristic time for the Brighton wetland. Balance conditions are carried over from Table 6.6. Perturbed values come from a second Table 6.6 calculation, with the perturbed temperature, all else the same as for the balance condition.

Month	Mean Water T °C balance	Net Gain MJ/m^2·d balance	Mean Water T °C	$\delta E_{net}/\delta x$ = Net Gain MJ/m^2·d perturbed	η Accommodation Coefficient MJ/m^2·°C·d perturbed	$\tau_A = \rho c_P h/\eta$ days
Jan	ice					
Feb	ice					
Mar	0.7	*0.00*	1.7	–2.27	2.27	0.80
Apr	8.3	*0.00*	9.3	–2.51	2.51	0.72
May	14.6	*0.00*	15.6	–2.66	2.66	0.68
Jun	19.6	*0.00*	20.6	–3.07	3.07	0.59
Jul	21.7	*0.00*	22.7	–3.32	3.32	0.54
Aug	20.8	*0.00*	21.8	–3.28	3.28	0.55
Sep	16.0	*0.00*	17.0	–2.97	2.97	0.61
Oct	9.1	*0.00*	10.1	–2.63	2.63	0.69
Nov	3.0	*0.00*	4.0	–2.33	2.33	0.77
Dec	ice					
Mean					**2.78**	**0.66**

Note: h = 0.43 m; ρC = 4.2 MJ/m3·°C; perturbation = T – T_b = 1.0 °C

TABLE 6.9

Accommodation and 63% change detention times (| $_A$) for treatment marshes. Estimated from energy balance methods, except for three fits to profile data.

Site	Location	η MJ/m^2·°C·d	τ_A days
Houghton Lake	Michigan	2.91	0.44
Commerce	Michigan	3.56	0.73
Des Plaines	Illinois	0.89	2.86
Shannon	Texas	1.99	0.96
ENRP	Florida	4.13	0.61
Salem	Oregon	2.31	1.12
Brighton	Ontario	2.78	0.66
Imperial	California	2.15	0.93
Tres Rios	Arizona	1.95	0.71
Tres Rios Fit	Arizona	0.57	1.67
OEW	Florida	2.79	0.92
OEW Fit	Florida	0.61	3.07
Sacramento	California	1.76	1.08
Sacramento Fit	California	1.48	1.89
Mean		**2.14**	**1.26**
St. Dev.		**1.05**	**0.82**

underground water flows, and removes energy from the plant canopy rather than the water below. Consequently, the apportionment is of great interest for treatment wetlands.

VEGETATION COVERAGE

A treatment wetland nearly always will contain patches of open water, as well as patches of emergent vegetation (EAV). Consequently, the water loss is partitioned accordingly:

$$ET_W = F_V ET_V + (1 - F_V)E \tag{6.43}$$

where F_V = fraction vegetated, –
ET_W = entire wetland evapotranspiration, mm/d
ET_V = evapotranspiration from the vegetated part, mm/d
E = evaporation, mm/d

This patch equation indicates that total ET is the areal average of water losses from vegetated areas (ET_V) and from open water areas (E). Secondarily, part of the vegetated ET_V occurs from the vegetation (canopy transpiration), part from the water below the vegetation (subcanopy evaporation). Thus, vegetated ET_V combines with open water patch evaporation to yield wetland ET (Kustas and Norman, 1997). This means that the fraction transpiration depends firstly upon the fraction open water, and secondly upon the vegetation density within vegetated patches. Sanchez-Carrillo et al. (2004) documented this concept for a marsh that went from fully vegetated to open water over a span of six years. More open water resulted in a greater fraction of evaporation.

Effect of Plant Cover on Wetland Evapotranspiration

A separate question is the magnitude of ET_V versus E, that is, does a plant canopy enhance or suppress total water loss? Answers vary. Allen et al. (2016) found that presence or absence of a cover of *Salvinia minima, Lemna* spp., and *Eichhornia crassipes* had no effect on ET_W. Gavin and Agnew (2004) asserted ET_W > E, while van der Weert and Kamerling found ET_W < E. The review of Abtew (2005) found that studies of the past 20 years had $0.7 < ET_W/E < 1.1$. Mohamed et al. (2012) suggested that the ration E/ET_W depended upon latitude and wind speed, and that the concept of ET_W = E does not hold.

Open Water Evaporation and Temperature

Portions of all treatment wetlands are open water. These may be design elements, such as deep zones or collection/distribution canals, or they may occur because of topographic lows in the wetland bottom. It is common for a treatment wetland to contain 20–80% open water, meaning water with no rooted emergent vegetation (EAV), such as *Typha* spp., *Schoenoplectus* spp., or *Phragmites* spp. However, it should be noted that open water may contain submerged plants (SAV), and may in part be covered by floating vegetation (FAV), such as *Lemna* spp., *Eichhornia crassipes, Hydrocotyle* spp., or *Pistia stratiotes.*

The fraction of open water is not insignificant in the large majority of cases. For instance, forty of the treatment wetland cells that comprise the Everglades protection project contain an average of 39% open water. It is uncommon for a treatment wetland to be completely open water (such as the Commerce wetland in Michigan – Kadlec (2011)); or to be completely covered by emergent vegetation (such as STA2 Cell1 – SFWMD (2017)). Constructed wetlands often have deep zones, which preclude most of the rooted emergents. It has been suggested that deep zones should be included in treatment wetland design, to improve oxygenation (Hammer and Knight, 1994; U.S. EPA, 2000a) and to ameliorate the potential effects of short-circuiting (Lightbody et al., 2009). However, evidence of such advantages is not present in some performance datasets (Kadlec, 2007). It remains a fact that deep zones are commonly part of design.

As a consequence of these varied ecologies, the energy balance for shallow open water may an important component of the overall energy balance that determines wetland water temperature,

Estimation Procedures

Typically, a mass transfer equation or some version of the Penman energy balance approach is used to estimate evaporation from open water bodies. By extension, either may be used to estimate the water surface temperature, which is part of the driving force for evaporation. The water surface temperature will adjust so that the heat supply from radiation and convection is compensated by the latent heat of vaporization. The spreadsheet method of solving the energy balance applies equally well to open water systems. The review of Finch and Hall (2001) concludes:

> The equilibrium temperature is essentially a simple model and therefore could be programmed relatively easily using a spreadsheet.

In so doing, it should be recognized that the albedo of a water surface is 0.06–0.12, considerably lower than for vegetated systems.

Mass Transfer Coefficients

The mass transfer calculation proceeds from a water vapor flux equation:

$$E = K_e[P_w^{sat}(T_w) - P_{wa}] = C_m u \Delta P_w \tag{6.44}$$

where C_m = velocity coefficient, (m/d·kPa)/(m/s)
E = water lost to evaporation, m/d
K_e = water vapor mass transfer coefficient, m/d·kPa
P_{wa} = ambient water vapor pressure, kPa
$P_w^{sat}(T_w)$ = saturation water vapor pressure at T_w, kPa
T_w = water temperature, °C
u = wind velocity, m/s

Over the years, meteorological data have been inconsistently measured using a variety of different standards, resulting, according to Singh and Xu (1997), in over 100 such evaporation formulae. It is therefore not possible to find a value of C that is applicable to all water bodies.

Wind Functions

The open water vapor flux coefficient has been postulated to be a linear function of the wind velocity by several researchers. Penman (1948) proposed:

$$E = W(u)\Delta P_w \tag{6.45}$$

where E = water lost to evaporation, mm/d
ΔP_w = vapor pressure deficit, kPa
u = wind speed at two meters elevation, m/s
W(u) = mass transfer wind function, (mm/d)/kPa

There have been many such wind function proposals. Sweers (1976) reviewed 22 of these, and concluded that the best choice was:

$$W(u) = (3.67 + 2.70u) \tag{6.46}$$

It is noteworthy that these open water wind function parameters are very near to the central tendency of parameters for treatment marshes (Table 6.7).

A size effect has been reported in more recent reviews. Shuttleworth (1993) proposed the transfer equation:

$$E = 2.909A^{-0.05}u_2\Delta P_w \tag{6.47}$$

where A = area of the water surface, m^2

E = evaporation, mm/d

ΔP_w = vapor pressure deficit, kPa

u_2 = wind velocity at two meters height, m/s

The area dependence has been presumed to represent a fetch dependence. This is suggested suitable for lakes in the range of 2500 m^2 < A < 10000 km^2 that are not in a humid environment. In effect, this area factor represents the "clothesline" effect for lakes.

Recent studies have recommended a reduced wind dependence (McJannet et al., 2012 (19 systems); McMahon et al., 2013 (27 studies)):

$$E = (2.36 + 1.67u_2)A^{-0.05}\Delta P_w \tag{6.48}$$

The Split Between Evaporation and Transpiration

As noted earlier, canopy transpiration occurs from the vegetation, and subcanopy evaporation comes from the water below the vegetation. Therefore:

$$ET_V = T + E \tag{6.49}$$

where E = $(1\text{-}\Psi)ET_V$ = transpiration from the vegetated area, mm/d

T = ΨET_V =evaporation from the vegetated area, mm/d

ET_V = evapotranspiration from the vegetated part, mm/d

¬ = fraction transpiration

This layer equation indicates that total ET_V is comprised of water losses from vegetation and from open water underneath the canopy. This situation is well described in the literature (Nobel, 1999), by considering the canopy and water as separate components of the wetland ecosystem for energy budget purposes.

Wetland Data

Only a few wetland studies have attempted to separate transpiration and evaporation. Those include measurements of the two components of ET in the very early studies of the Houghton Lake marsh, which showed that shading reduces surface water evaporation, while transpiration continues from the canopy (Kadlec et al., 1987). Nontreatment marshes have provided some data. Herbst and Kappen (1999) reported on four years of studies at a *Phragmites* system in northern Germany, and Burba et al. (1999a) studied a *Phragmites* system in Nebraska. Sanchez-Carillo et al. (2004) report transpiration and evaporation from a *Typha* marsh in Spain. More recent studies include treatment marshes (Kadlec, 2006; Bachand et al., 2014a; Sanchez et al., 2016). Results are summarized in Table 6.10. During the course of an entire year, transpiration accounts for 0.32–0.74 of total ET.

Transpiration is present when emergent green plants are present, and is therefore confined to the growing season. Evaporation occurs whenever there is an ice-free water surface under the canopy. These two seasons do not necessarily coincide, especially in colder climates, where there is normally an ice-free period in spring, before the new live shoots

TABLE 6.10

Evaporation and transpiration for example treatment marshes. Transpiration season length varies.

Location	Vegetation	Years	Season Months	E mm/d	T mm/d	ET mm/d	T/ET	Ref.
Germany	*Phragmites*	4	12	1.06	1.97	3.03	0.64	1
Spain	*Phragmites*	6	6	9.88	7.52	17.40	0.43	2
Nebraska	*Phragmites*	1	4	2.04	2.24	4.28	0.52	3
Spain	*Typha*	6	6	9.88	4.72	14.61	0.32	2
California	*Oryza sativa*	1	9	1.54	4.48	6.02	0.74	4
Arizona FS	*Schoenoplectus*	2	12	6.61	9.54	16.15	0.59	5
Arizona Demo	*Schoenoplectus*	1	12	1.83	4.24	6.07	0.70	6

References:
1 Herbst and Kappen (1999)
2 Sanchez-Carillo et al. (2004)
3 Burba et al. (1999a)
4 Bachand et al. (2014a)
5 Sanchez et al. (2016)
6 Kadlec (2006)

appear, and again in autumn, after complete senescence. Thus, although both transpiration and evaporation peak in summer, the T-season is typically shorter than the E-season. The proportion of transpiration is highest in midsummer, when it reaches 70–80% of ET.

Figure 6.17 shows the annual progression of E and T for a warm climate system in Arizona, the full-scale Tres Rios wetland (Table 6.10, reference 5). That system was put in place based upon a demonstration project (Table 6.10, reference 6). The full-scale project was configured differently, with different vegetation patterns and densities; and received

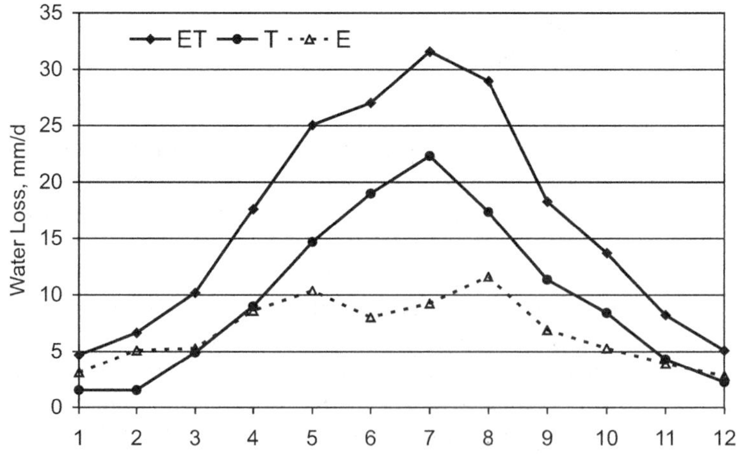

FIGURE 6.17 Evaporation and transpiration for the Tres Rios full-scale wetland. (Based on data from Sanchez et al., 2016)

a much higher hydraulic loading. As a result of the last factor, water temperatures in the full-scale system were considerably higher than in the demonstration project, which drove higher ET rates.

Estimating Procedures and Models

One of the first approaches for estimating the T/E partition was advanced by Shuttleworth and Wallace (1985). They used modeling to relate the T/ET_V ratio to leaf area index, and in general their results can be represented by:

$$\Psi = 1 - \exp(-(0.31 + 0.64GLAI)) \tag{6.50}$$

where GLAI = green leaf area index, m^2/m^2

Leaf area index to estimate wetland transpiration remains in use in water loss modeling calculations (Neitsch et al., 2011).

Shuttleworth and Wallace (1985) also advocated a two-layer approach, in which the canopy and the underlying substrate were considered separately. They used the Penman Monteith calculations for each layer, with appropriate modifications for plant structure. That approach was subsequently adopted by a number of authors (Kustas, 1990; Massman, 1992; Kustas and Norman, 1997, 1999; Best, 1998; Guan and Wilson, 2009). Notably, these Penman Monteith calculations invoke the two (unnecessary) Penman simplifications, linearization of the vapor pressure equation and elimination of surface temperature.

Goulden et al. (2007) investigated the San Joaquin freshwater marsh, a remnant of La Cienega de las Ranas (The swamp of the frogs), a wetland that once covered 2100 ha along the San Diego Creek in southern California. It comprises dense stands of *Typha latifolia* and *Scirpus californicus* and a large amount of standing litter. Goulden et al. (2007) considered transpiration from emergent vegetation and evaporation from standing water beneath vegetation, and evaporation from open water. They found about 20% of the total annual ET originated from beneath vegetation and 80% was transpiration.

Treatment Marshes

Kadlec (2006) postulated that the energy lost via transpiration is not abstracted from the flowing water, but from the plant canopy of the wetland, which is little affected by water temperature. As a simplified and approximate correction, latent heat losses were split between evaporation and transpiration. Effectively, the net energy loss from the system has been partitioned between the canopy, functioning at the balance temperature, and the surface water, functioning at the local water temperature.

A more complete characterization involves partitioning of radiative and convective heat in addition to latent heat. Smesrud et al. (2014) performed simulations for the Salem, Oregon treatment marshes, based on such a scheme, as depicted in Figure 6.18. A factor β partitioned incoming radiation. Radiative and convective energy exchanges between atmosphere, canopy and water were partitioned according to the GLAI. The resulting model was named Heat Source Wetlands (HSW). HSW provided good representations of seasonal and diurnal changes in wetland water temperatures over a range of hydraulic and heat loading, water depths and meteorological conditions.

6.6 COLD CLIMATES

Treatment wetlands that operate in cold (subfreezing) environments face several unique design challenges. Once an ice layer forms on the wetland, water temperatures will be 1 to 2°C, and the thickness of the ice layer becomes a design consideration. The formation of an ice layer will reduce the depth of the water column, reducing detention times, unless the

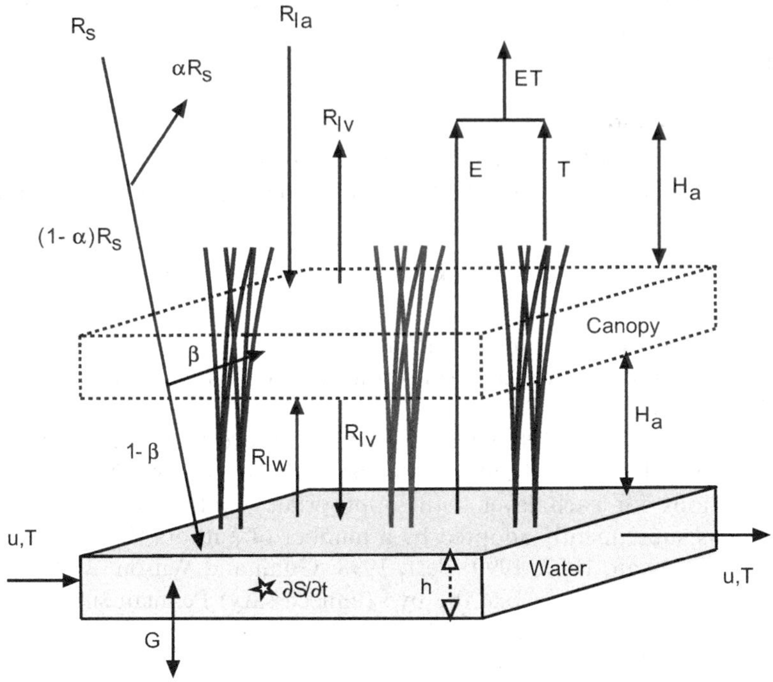

FIGURE 6.18 The energy balance terms for an emergent marsh, partitioned between a canopy section and a water section.

This depiction follows that described by Smesrud et al. (2014). Terminology is that of Figure 6.1, plus additionally: β = fraction solar radiation absorbed by the canopy R_{la} = longwave (heat) radiation from atmosphere R_{lv} = longwave (heat) radiation from vegetation in canopy R_{lw} = longwave (heat) radiation from water

water level is increased in the fall to accommodate the anticipated thickness of the ice layer. Energy balance calculations are required to determine the extent of ice formation.

Ice thickness can vary significantly from year to year due to variations in snowfall and temperature. The principal factor is the insulation provided by the snow layer. In fact, water in natural northern swamps and marshes may not freeze in winter, due to the presence of an insulating layer of snow. The standing dead vegetation is an effective snow trap that collects drifting and falling snow. Thus, the marsh snow depth is frequently greater than the accumulated snowfall (see Figure 5.25). If snow accumulates before a significant ice layer forms, subsequent freezing is strongly inhibited. Consequently, ice thicknesses in vegetated wetlands are often very much less than lake ice thicknesses because of the insulating value of captured snow.

Operational options for FWS treatment wetlands in cold climates are typically either full year-round discharge, allowing for ice formation, or pond storage over the frozen season, with discharge through treatment marshes during the unfrozen season.

In this section, methods for estimating the extent of ice formation are presented. Ice cover in wetlands causes the energy balance to split into a balance on the canopy and a separate balance on the water and ice below. It is the latter that is of interest in understanding the degree of ice formation. Radiation and vaporization are no longer factors for the water-side balance, because the ice layer blocks these processes from the underlying water.

Spatial Extent and Distribution of Ice in Treatment Marshes

The point discharge introduction of warm water into a treatment wetland causes an unfrozen, un-snow-covered inlet area to persist even in the event of extremely cold air temperatures. As the water moves out into the wetland, the incoming exotherm is dissipated, and a snow and ice cover becomes possible. This cover may consist of snow, ice, or a combination, depending on the vegetation density. If the discharge is into an unvegetated inlet zone, snow trapping is not possible, and ice covers the inlet pond in areas away from the discharge point. Flow then proceeds away under or over the ice. If the inlet zone is densely vegetated, snow may be held up above the water by standing dead vegetation or by a floating litter layer. In that event, flow from the unfrozen, uncovered inlet area proceeds away under a snow blanket.

At the Hillsdale, Michigan treatment marsh, the unfrozen zone advanced and retreated during freezing temperatures, depending upon short-term meteorological conditions (Kadlec et al., 2012a) (Figure 6.19). A complete ice cover formed under very cold conditions, only to have open water appear during less-cold, but freezing conditions. The incoming water was venting groundwater, with an annual temperature cycle from 4 to 21°C, and a mean of 12.5°C. It is this sensible heat that is capable of overcoming freezing for a short period of travel time.

This adaptation zone is not easily amenable to model calculations, because all of the terms in the energy balance (see Equation 6.1) are operative. A rough approximation may be obtained from Equation 6.40, with an assumption of the accommodation coefficient. It is also necessary to assume that the balance temperature is approximately equal to the air temperature for the subfreezing conditions. For conditions that approximate those for Figure 6.19, a small zone of open water at the inlet is correctly predicted.

The overall ice cover for the wetlands responds to the air temperature (Figure 6.20). When the mean air temperature drops below zero, ice formation quickly covers most of the

FIGURE 6.19 The inlet zone of this FWS wetland at Hillsdale, Michigan remains unfrozen in this February 1, 2001, photo. Downstream portions of the wetland are fully ice and snow covered. Note the preferential channel extending outward to the muskrat mound.

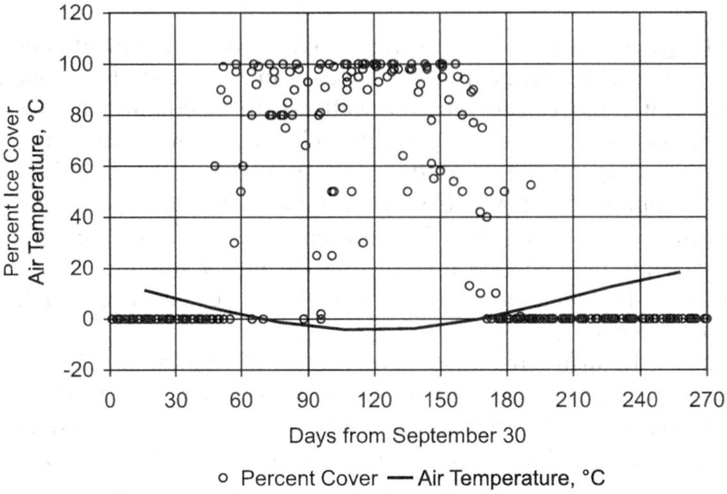

FIGURE 6.20 Extent of the ice cover at the Hillsdale, Michigan FWS treatment wetlands. Data span four winters, for four wetland cells. The average monthly air temperature swing was from –4 to 22°C, with an annual mean of 9°C. Note that the proverbial "January thaw" was indeed experienced, with nearly complete loss of the ice cover during January.

wetland, excepting the inlet zone. In most years, this cover remains intact until the mean air temperature rises above freezing, whereupon the ice disappears rapidly, over a week or two.

Ice in Quiescent Ponds

An approximate estimating method for determination of ice thicknesses is based upon the length and intensity of subzero conditions. Ice formation in still water at a constant cold surface temperature proceeds according to the well-known Stefan model (Ingersoll et al., 1948):

$$h_{ice} = a\sqrt{-T_s \cdot t} \tag{6.51}$$

where a = constant, $m/(°C{\cdot}d)^{0.5}$
h_{ice} = ice thickness, m
t = time, d
T_s = ice top surface temperature, °C

Crites et al. (2006) suggest that the value of a = 0.027 $m/(°C{\cdot}d)^{0.5}$ for open water, 0.018 for open water with snow, and 0.010 for dense vegetation and litter; however, no calibration data or source are offered.

However, it must be noted that Equation 6.51 is for still open waters, with no incoming sensible heat, and a fixed below-zero temperature at the top ice surface, none of which prevail in treatment wetland projects. Snow insulation, heat gain from soils, and moving wind and water all combine to slow ice formation. Consequently, the predicted values of ice thicknesses for wetlands are very conservative, often by factors of two or three. For instance, winter operation has been proven feasible at FWS projects located in Northern Michigan and Ontario, where the water would be predicted to freeze to the bottom according to Equation 6.51. Consequently, a more detailed energy balance is required for treatment wetland estimates.

The Balance Condition for Under-Ice Flow

The system chosen for energy balancing is selected to be the water under the ice (Figure 6.21). Any storages of sensible heat in the ice layer, or in the wetland water body, are ignored. Any net heat loss is compensated by ice formation and cooling:

$$\rho\lambda_f \frac{dh_{ice}}{dt} = H_a - G = U(T_w - T_a) - G \tag{6.52}$$

where G = heat gain from soil, $MJ/m^2 \cdot d$
H_a = heat loss to air, $MJ/m^2 \cdot d$
h_{ice} = ice thickness, m
t = time, d
T_a = air temperature, °C
T_w = water temperature, °C
U = overall heat transfer coefficient, $MJ/m^2 \cdot °C$
ρ = density of water, kg/m^3
λ_f = heat of fusion of water, MJ/kg
The overall heat transfer coefficient is comprised of the several layer components:

$$\frac{1}{U} = \frac{1}{U_{water}} + \frac{h_{ice}}{k_i} + \frac{h_{snow}}{k_s} + \frac{1}{U_{air}} \tag{6.53}$$

In terms of heat transfer resistances:

$$\frac{1}{U} = R = R_{water} + R_{ice} + R_{snow} + R_{air} \tag{6.54}$$

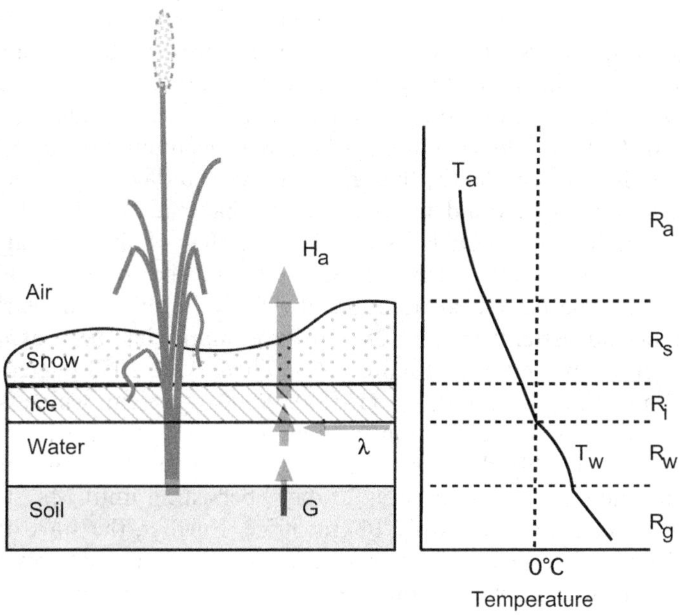

FIGURE 6.21 Temperature gradients through the vertical profile of a FWS wetland in winter. Partially frozen, balance conditions are assumed.

where h_{ice} = ice thickness, m
h_{snow} = snow thickness, m
k_i = thermal conductivity of ice, MJ/m·d·°C
k_s = thermal conductivity of snow, MJ/m·d·°C
R = overall heat transfer resistance, (m^2·°C)/MJ
R_{water} = overall heat transfer resistance, (m^2·°C)/MJ
R_{ice} = overall heat transfer resistance, (m^2·°C)/MJ
R_{snow} = overall heat transfer resistance, (m^2·°C)/MJ
R_{air} = overall heat transfer resistance, (m^2·°C)/MJ
U_{air} = snow to air heat transfer coefficient, MJ/m^2·d·°C
U_{water} = water to ice heat transfer coefficient, MJ/m^2·d·°C

Most of these resistances can be estimated with some accuracy, such as thermal conductivities. However, the under-ice, water-side heat transfer coefficient has not been the subject of research. It is tempting to estimate the value based upon known relationships simple geometries. There are relationships for laminar and turbulent flow near flat plates, but these do not deal with the tortuous flow path and under-water obstructions. Based on these other geometries, the water-side heat transfer coefficient is expected to be higher than the air-side coefficient.

The purpose of energy balance calculations for the frozen season is the estimation of either the amount of ice formed. The conservative assumption is that there is zero resistance to heat transfer on the water-side of the air-water interface (R_{water} = 0).

Example Detailed FWS Ice Calculation

Because of the time series progressions of air temperature and other variables, calculations require spreadsheet techniques for the solution of Equation 6.52, on a daily basis. There is an accommodation zone, in which incoming warm water cools to the winter balance point of 1–2°C. This example focuses upon energy fluxes in a downstream zone, in which that balance condition prevails. The driving force for ice formation is determined by the air temperature, which is presumed to follow a sinusoidal time progression fitted to data from 1997–1999 from Duluth, Minnesota. The annual swing in mean daily temperature is –11 to 19°C. This leads to 150 days of freezing air temperatures, with a mean of -7 °C, from November 8 to April 7. The water temperature in the balance zone is assumed to be 2°C for the frozen season. The heat loss transfer coefficients for the air-side of the ice is assumed to be 0.15 MJ/m^2·d·°C. the assumed pattern of snow is accumulation over the first 20 days, persisting until March at 15 cm depth, then melting over 20 days.

Heat from the underlying ground tends to counterbalance heat loss to the atmosphere. For this example, this heat addition is assumed to be that of the cyclical model described earlier in this chapter. Importantly, energy is returned from the soil to the water during most of the freezing period. The peak heat addition of 0.66 MJ/m^2·d occurs early in the freezing period, on November 30. After February 28, soil heat supply has been exhausted. For these conditions, 16 cm of ice are forecast (Figure 6.22). There were 1058 degree-days below zero, resulting in a Stefan prediction of (0.027 × (1058)0.5)(100) = 88 cm, which is much larger.

A common occurrence is the collection of snow on top of the ice layer. If snow is considered, there is a considerably different result for the modeled ice thickness. A presumed pattern of snow accumulation over the first 20 days, persisting until March at 15 cm depth, then melting over 20 days, produces only 16 cm of ice. Further, the start of ice formation is delayed until mid-December due to heat from the soil. We note that this early winter phenomenon, of snow in the canopy and unfrozen water beneath, is common in north temperate wetlands (see Figure 6.22).

No attempt has been made to describe the melting portion of the annual ice cycle.

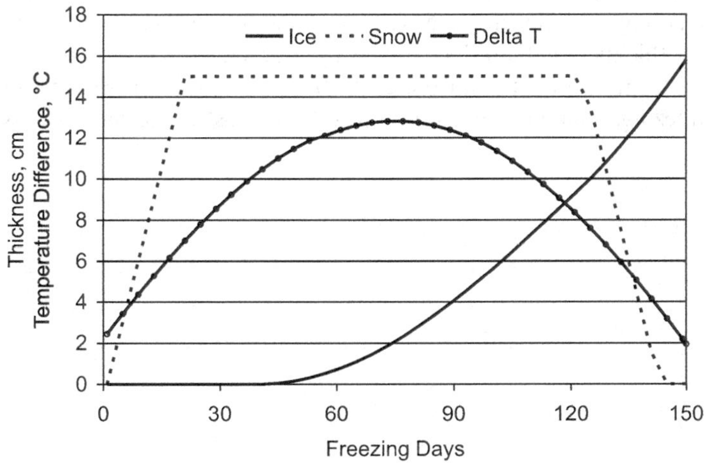

FIGURE 6.22 Progression of snow and ice thickness for an example marsh. See text for presumed conditions. Delta T represents the temperature difference between water and air.

6.7 SUMMARY

Water temperature exerts a strong influence on many chemical and biological processes. The energy flows that determine water temperature and the associated evaporative losses are radiation, convective heat transfer, and evaporative heat loss. The entire suite of processes is well understood, in part because of precursor work in agriculture. Wind is a primary factor in both convective heat and water vapor mass transfer, but the driving forces are the temperature difference and vapor pressure difference between water and air. The vegetative canopy modulates those convective processes, and intercepts and reflects incoming radiation.

Solar radiation is the primary driving force for evapotranspiration, which displays a strong annual cycle. In moderate temperate climates, ET losses are on the order of half a meter per year but can easily be double that in hot arid climates. There now exist a large number of wetland studies of ET, which indicate a sizeable variability due to site-specific conditions. As a result, simple generalities are not adequate for a particular wetland situation, but spreadsheets and more complex estimation routines are widely available.

As water flows through the wetland, energy gains and losses drive the water temperature toward a balance point temperature. Hot dry conditions can produce cooling, while humid conditions can produce heating. The equilibrium wetland water temperature represents a balance between the dominant transfers, which are incoming solar energy gains and evaporative energy losses. The adjustment of the incoming water temperature to this balance occurs at a modest pace, with acclimation complete in about a few days' detention. Both simple and complex methods now exist to estimate the change of temperature from incoming source to wetland balance conditions.

The distribution and density of wetland vegetation are very important to both temperature and water loss calculations. Emergent vegetation is the conduit for transpiration losses, and can shade the water below, thus inhibiting evaporation. Transpiration flows draw water into the root zone, and are thus important in pollutant processing. The wetland transpiration database has recently grown and permits estimates of that water flow.

In the winter, the insulation provided by snow and ice, plus heat returns from summer-warmed soils, are enough to prevent water from entirely freezing under moderately cold climate conditions. The ice thickness is then determined by losses upward through the insulating layers as well as gains vertically upward from the earth by conduction. However, the under-ice water stays just above freezing, which is low enough to inhibit most microbial processes.

7 Representing Treatment Performance

This chapter examines the available means of collecting and analyzing the large amount of pollutant removal performance data that now exists for treatment marshes. These systems are "open" systems heavily influenced by environmental factors, and are often dominated by biomass storage compartments that are very large relative to pollutant mass in the water column. These biomass storage compartments are affected by seasonal cycles, as well as by temperature cycles.

Treatment marshes, as defined herein, are not wastewater treatment systems in the environmental engineering sense. They do not receive strong influents, although some source waters may originate from multicomponent advanced wastewater treatment trains. Wastewater treatment systems, including those that employ surface or subsurface flow wetland components, are typically dominated by microbial processes, whereas treatment marshes rely upon biogeochemical cycles, which do include, but are not dominated by, microbial processing. Thus, only a very few concepts from the engineering science of biological wastewater treatment reactors have any relevance. However, concepts from the chemical engineering of multiphase and non-ideal reactors are relevant and will be adapted to the wetland environment.

Wetland biogeochemical processes are numerous, and consequently a single contaminant is exposed to a number of routes of removal or alteration. Plant uptake, sedimentation, sorption, and biochemical conversion contribute to varying extents, depending upon the constituent under consideration.

Treatment performance is represented by two components: the central treatment tendency for a wetland (or group of wetlands) and the anticipated variability away from that central tendency. Central tendencies are driven by flows and concentrations, in concert with environmental factors. Random events within the wetland will produce stochastic variations in effluent performance. Both must be assessed to describe treatment performance in constructed wetlands.

Different types of wetlands (wetland configuration, vegetative community) function differently. Therefore, a set of "universal" parameters for describing treatment performance in wetlands is not to be expected. The purpose here is to provide the wetland designer with a quantitative spectrum of anticipated performance.

7.1 FUNDAMENTALS

There are many measures and models for pollutant reductions in treatment wetlands. In this section, various definitions and options for constituent description are explored, as a necessary precursor to the discussions of individual pollutants that follow in ensuing sections and chapters.

CONCENTRATIONS

Concentrations of pollutant constituents are determined by a variety of methods, and at varying frequency. Common techniques involve grab samples, but autosamplers are also often used, either time-proportional or flow-proportional.

Individual concentration measurements are very often averaged, to eliminate some of the variability inherent in wetlands. The time-average concentration, temporarily denoted by an overbar, is defined as:

$$\bar{C} = \frac{1}{t_m} \int_0^{t_m} C dt \tag{7.1}$$

where: C = chemical concentration, mg/L
t = time, d
t_m = averaging period, d
Such average concentrations may be acquired from time-proportional autosamplers or computed from a time series.

An average mass flow of a chemical, $\overline{QC}$, is the product of the average flow and the flow weighted (or mass average) concentration, defined by:

$$\hat{C} = \frac{\frac{1}{t_m}\int_0^{t_m} QCdt}{\frac{1}{t_m}\int_0^{t_m} Qdt} = \frac{1}{\bar{Q}t_m}\int_0^{t_m} QCdt \tag{7.2}$$

where the hat notation indicates a flow-weighted average.

$$\overline{QC} = \bar{Q}\,\hat{C} \tag{7.3}$$

Percent concentration reduction is often used in the literature:

$$\%\ Concentration\ \text{Re}duction = 100\frac{C_i - C_o}{C_i} \tag{7.4}$$

This term is quite ambiguous, because it usually refers to the average of one or more synchronous samples for selected stream flows. Such contemporaneous measures do not properly reflect the internal chemical dynamics of the wetland, such as production of the chemical. Further, dilution or concentration due to rain, ET or other unaccounted flows render this an imperfect measure of true removal. Nevertheless, this terminology is frequently used in the literature.

Removal and Loading

It is important to distinguish among the various measures of global wetland chemical removal. Definitions used in this book are specified here.

Inlet Mass Loading Rates

$$M_i = Q_i C_i \tag{7.5}$$

$$m_i = \frac{M_i}{A} = \frac{Q_i C_i}{A} = q_i C_i \tag{7.6}$$

where: M_i = inlet mass loading, g/d
m_i = (specific) inlet mass loading, g/m^2·d

Acronyms are also often used for designating the chemical; PLR would denote *P*hosphorus *L*oading *R*ate, for instance. A chemical loading rate is a measure of the distributed, "rainfall" equivalent of a chemical mass flow. It does not imply the physical distribution of water uniformly over the wetland.

Mass Removal Rate

$$\tilde{J} \cdot A = (Q_i C_i - Q_o C_o) \tag{7.7}$$

where: $\tilde{J}$ = specific mass removal rate, $g/m^2 \cdot d$

This represents the areal average amount of a chemical that gets stored, destroyed or transformed. This single number measure of wetland performance can be misleading in the common event of strong concentration gradients and removal gradients.

Percent Mass Removal

This quantity links water losses and gains to chemical losses and gains.

$$\%\,Mass\ \mathrm{Re}moval = 100\frac{(Q_i C_i - Q_o C_o)}{Q_i C_i} = 100\frac{m_i - m_o}{m_i} \tag{7.8}$$

$$\left(1 - \frac{\%M}{100}\right) = \left(1 - \frac{\%W}{100}\right)\left(1 - \frac{\%C}{100}\right) \tag{7.9}$$

where: %C = percent concentration reduction, –

%M = percent mass removal, –

%C = percent flow reduction, –

This term is less ambiguous than concentration reduction, because it traces the chemical of interest and accounts for the effect of the quantity of water in which that chemical is located. However, the difficulties of contemporaneous measurement remain.

The Utility of Reduction Numbers

It is very easy to compare the amounts of a pollutant in the inlet and outlet streams of a wetland, and to compute the percentage difference. Unfortunately, this information is of very limited use in design, or in performance predictions, because it reflects none of the features of the ecosystem that are the target of design. By implication, it would be necessary to replicate the wetland that produced the percentage data, and to replicate the operating and environmental conditions that prevailed during data acquisition. The second is clearly impossible, and past experience has given strong indications that the first is also difficult.

The literature is replete with review papers that tabulate removals for a selected spectrum of wetlands (for example, Strecker et al., 1992; Cueto, 1993; Johnston, 1993). The implication is that wetlands of a similar type will achieve a similar reduction. While such groups of data begin to elucidate the bounds of performance, the effects of size, loading, flow patterns, depth and other design variables cannot be deduced from efficiency values alone.

In some instances, the incoming concentration of a particular chemical may be small for some period of time. Then, due to measurement errors or small transfers from wetland storages and productions, outflow concentrations may be greater than the incoming values. A one-time calculation of a “reduction efficiency” will properly reflect that condition as a (large) negative percent reduction. At other times, a larger inflow concentration may be reduced by the wetland, leading to a positive percentage removal. If the removal percentages are then averaged, the large negative value improperly dominates the calculation.

As a result of these considerations, great care must be exercised in interpretation of percentage reduction values.

Chemical Mass Balances

Measurements of chemical composition of wetland inflows and outflows are the most obvious method of characterizing water quality functions. However, such measurements by themselves can be very misleading. A much better characterization is achieved by computing the mass balance, or budget, for an individual chemical constituent.

A proper mass balance must satisfy several conditions:

1. The system for the mass balance must be defined carefully. A system in this context means a defined volume in space; and this is often taken to be the surface water in the wetland in the case of a treatment marsh. A precise definition is needed to compute the change in storage. The mass balance is termed global when the entire wetland water body is chosen as the system. In later chapters, it will be useful to compute the internal mass balance, which is based on an internal element or subdivision of the water body.
2. The time period for totaling the inputs and outputs must be specified. It may be desirable to express inflows and outflows in terms of rates, but these must then be averaged over the time period chosen.
3. All inputs and outputs to the chosen system must be included. The concept of mass conservation may be invoked to calculate one or a group of material flows. A partial listing of some of the inflows and some of the outflows does not constitute a mass balance.
4. Compounds undergo chemical reactions within a wetland ecosystem. Any production or destruction reactions that occur within the boundaries of the chosen system are to be included in the mass balance. Reactions outside the boundary are not counted, because an outflow must occur to transport the chemical to the external reaction site, and that is accounted as an outflow.
5. Water borne chemical flows are determined by separate measurements of water flows and concentrations within those waters. Therefore, an accurate water mass balance is a prerequisite to an accurate chemical mass balance.
6. If at all possible, it is desirable to demonstrate closure of the mass balance. This is achieved by independently measuring every component of the mass balance. The degree of closure is often expressed as a percentage of total inflows. Unfortunately, closure has rarely been demonstrated for any chemical in any wetland.

The foundation for chemical mass balances is the wetland water mass balance (see Chapter 2). Transfers of water to and from the wetland follow the same pattern for both surface and subsurface flow wetlands. In treatment wetlands, source water additions are normally the dominant flow, but under some circumstances, other transfers of water are also important. The dynamic overall water budget for a wetland is:

$$Q_i - Q_o + Q_c - Q_b - Q_{gw} + Q_{sm} + A(P - ET) = \frac{dV}{dt} \tag{7.10}$$

where: A = wetland top surface area, m^2

ET = evapotranspiration rate, m/d

P = precipitation rate, m/d

Q_b = bank loss rate, m^3/d

Q_c = catchment runoff rate, m^3/d

Q_{gw} = infiltration to groundwater, m^3/d

Q_i = input wastewater flowrate, m^3/d

Q_o = output wastewater flowrate, m^3/d
Q_{sm} = snowmelt rate, m^3/d
t = time, d
V = water storage in wetland, m^3

It is difficult to establish detailed chemical mass balances over the wetland surface water because of the number and complexity of the possible transfers to and from the water, and their nonsteady character. It is common practice to measure only the principal inflows and outflows, and to ascribe the difference to "removal," which may be positive or negative. This lumping of all transfers to and from the water body is often unavoidable due to economic constraints. It is possible to write a general mass balance equation for a generic chemical species:

$$Q_iC_i - Q_oC_o + Q_cC_c \pm Q_{gw}C_{gw} + A\left(PC_p - \tilde{J}\right) = \frac{d(VC_s)}{dt} \qquad (7.11)$$

where: C_c = concentration in catchment runoff, g/m^3
C_{gw} = concentration in groundwater recharge or discharge, g/m^3
C_i = concentration in inflow, g/m^3
C_o = concentration in outflow, g/m^3
C_p = concentration in precipitation, g/m^3
C_s = concentration in wetland surface water, g/m^3
$\tilde{J}$ = spatially averaged removal rate, $g/m^2{\cdot}d$

In Equation (7.11), bank losses and snowmelt have been omitted for the sake of simplicity. All of the transfers have been lumped into one removal rate. Flow rates are instantaneous. The removal rate is the average over the entire wetland area, and the system concentration is averaged over the entire water volume.

The time period for the global mass balance is of critical importance, because of the time scale of interior phenomena. Many wetlands, whether treatment or pristine natural, have long nominal detention times, which usually reflect long actual detention times. A two-week detention is not uncommon. If the wetland were in plug flow, an entering cohort of water would exit two weeks later. Clearly, same day samples taken from inlet and outlet should not be used to compute "removals." In fact, wetland flow patterns are always more complex than plug flow; the entering cohort breaks up, and pieces depart at various times after entry, some earlier and some later than the implied two-week detention. This difficulty of synchronous sampling may be alleviated in the mass balancing process, by selecting a mass balance period that spans several detention times.

The removal term is the result of transfers to and from the soils and biomass compartments in the wetland, as well as of transfers to and from the atmosphere, and chemical conversions. Those biomass and soils compartments dominate the overall wetland storage and transformations for most chemicals. Therefore, the water body mass balance is very sensitive to small changes in transfers, reactions and storages in biomass and soils. The removal rate depends very strongly on events in these solids compartments, and hence is determined in major part by the changing ecological state of the wetland. Because wetland biological processes are more or less repetitive on an annual cycle, the long-term performance of the wetland is best characterized by global mass balances, which span an integer number of years. Seasonal effects require a time period of three months, which is usually long enough to avoid storage errors and detention time offset.

Removal in Equation (7.11) is an areal average. However, in most flow through wetlands, there is a strong gradient in the un-averaged removal, in the direction of flow. As the downstream wetland system "boundary" is moved successively further from the inlet, the areal average removal rate decreases. The average removal rate depends on the size of the portion of the overall wetland that is chosen for the global mass balance. This weakness of the global mass balance can be corrected by using the internal mass balance, which reflects distance effects.

Data Folding

Performance data variability includes seasonal, stochastic, and year-to-year changes. Year-to-year changes occur because of changing vegetative communities, hydraulic or organic loadings, or weather conditions. Seasonal patterns frequently exist, which implies that short-term (a few months) studies can be misleading.

Seasonal changes in treatment performance can often be represented by cosine trends (Kadlec, 1999a). For a relatively stable treatment marsh, stochastic variability will report as a "cloud" around the seasonal trend line.

$$C = C_{avg}[1 + A \cdot \cos(\omega(t - t_{\max}))] + E \tag{7.12}$$

where: A = fractional amplitude of the seasonal cycle, –
C = instantaneous outlet concentration, mg/L
C_{avg} = average (trend) outlet concentration, mg/L
E = random portion of the outlet concentration, mg/L
t = time of the year, Julian day
t_{max} = time of the year for the maximum outlet concentration, Julian day

Data from several years may be "folded" to create an annualized grouping, distributed across the year according to Julian day. This use of many annual cycles has the advantage of including year-to-year variations in climate, flow, and ecosystem condition.

An example of data folding is shown in Figure 7.1, for the Brighton, Ontario treatment wetland. It is difficult to identify any annual pattern in the times series of concentrations, but when folded, a pattern of lower concentrations in summer and higher in winter is very apparent.

The deterministic portion of this representation may be modeled by the k-rate technique with appropriate rate constants and background concentrations, both of which may respond to temperature and season, as will be further discussed later in this chapter.

The existence of the error term (E) means that sampling must either be at high frequency, or cover many annual cycles before meaningful trend averages can be determined.

Graphical Representations

There exist a large number of datasets for some of the more common pollutants, such as TSS, BOD, phosphorus and nitrogen species. Several types of graphs may be used to compare performances within and across systems, and these have been used in prior treatment wetland literature:

A Output concentration vs. input concentration.
B Output concentration vs. input areal loading.
C Output loading vs. input loading.
D Load removed vs. input areal loading.

The first two of these are useful representations, but the last two very often lead to spurious relationships that serve no useful purpose. Many important variables are lost in these plots, because of their restrictive two-dimensional nature.

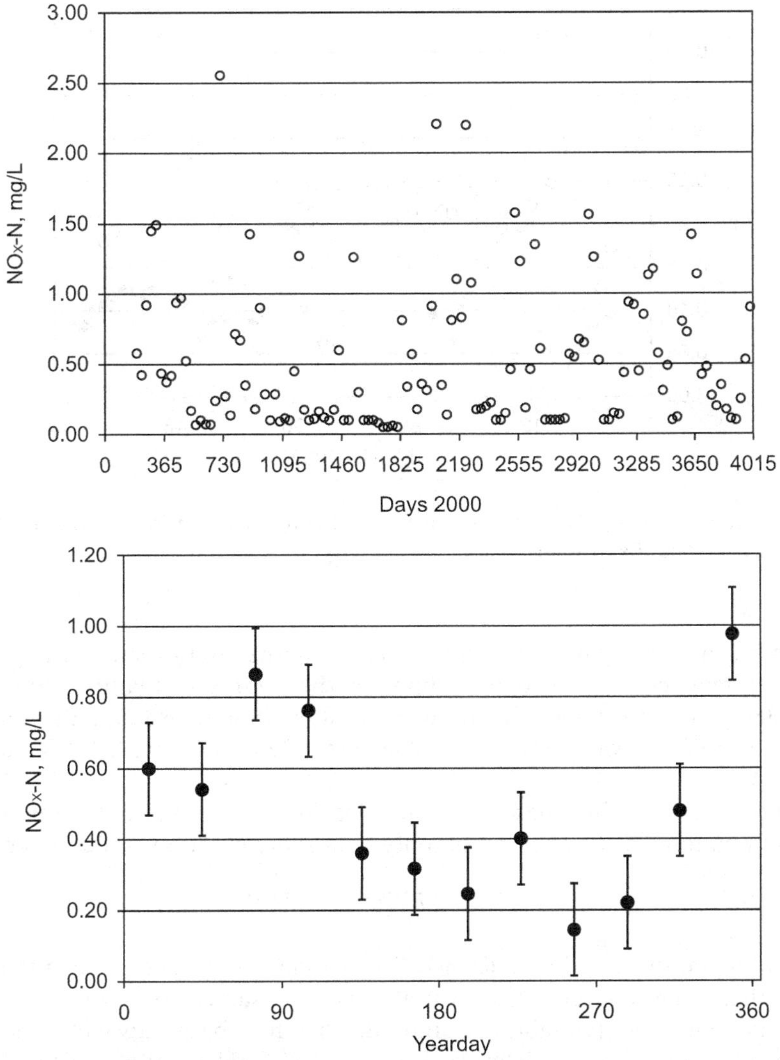

FIGURE 7.1 Annual nitrate nitrogen (NO_X-N) concentrations leaving the Brighton, Ontario treatment marsh over a ten-year period of record. Upper panel: Time series of monthly means. Lower panel: Ten-year means for each month.

Outputs vs. Inputs

The input/output concentration graph essentially extends the idea of percent removal to a time series of measurements from a single wetland, or to a group of wetlands (Figure 7.2). It is useful for obtaining first estimates of the potential of a class of treatment wetlands to reduce a particular contaminant. But, such plots are of no value in sizing the wetland, because they do not contain any information on the detention time or hydraulic loading.

Output Concentration vs. Input Areal Loading

Outlet concentrations produced in treatment wetlands depend upon three primary variables (area, water flow and inlet concentration), as well as numerous secondary variables

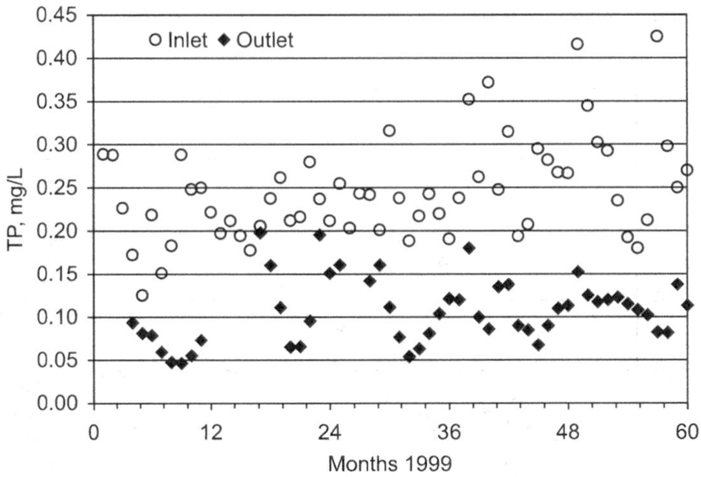

FIGURE 7.2 Inlet and outlet TP concentrations for the Ekeby treatment marsh at Eskilstuna, Sweden. (Source: unpublished data, Eskilstuna Energi och Miljö.)

(vegetation type, internal hydraulics, depth, event patterns, and others). It is presumed that the area effect may be combined with flow as the hydraulic loading rate (q = HLR) since two side-by-side wetlands with double the flow should produce the same result as one wetland. Therefore, two primary variables are often considered, HLR and inlet concentration (C_i).

It is tempting to combine these two variables into the input pollutant mass loading according to Equation 7.5, in order to simplify data representation or design criteria.

$$C_{out} = f(q_{in} C_{in}) = f(m_i) \tag{7.13}$$

In graphical form, this relation is termed the loading graph. The representation is bivariate, meaning that outlet concentration (variable 1) is presumed to depend uniquely on pollutant mass loading rate (variable 2). Indeed, this has been advanced as a basis for treatment marsh design (e.g., Richardson et al., 1997; U.S. EPA, 2000a; Wallace and Knight, 2006). The implication of Equation 7.13 is that the effect of doubling the hydraulic loading is the same as the effect of doubling the inlet concentration. Unfortunately, that is not the case for treatment marshes. Trivariate data regressions show that those effects differ (Kadlec and Knight, 1996). The behavior of concentrations along the flow path also shows that flow and concentration must be considered separately. For example, concentrations often show an exponential decline along the flow path, which can be described by a first order areal rate coefficient (k_A).

$$C_{out} = C_{in} \exp\left(-\frac{k_A}{q_{in}}\right) \tag{7.14}$$

An illustrative calculation for a 5,000 m^2 wetland with k_A = 10 m/yr shows:

- For Q_{in} = 250 m^3/d, C_{in} = 20 g/m^3: m_i = 1.00 g/m^2·yr C_{out} = 11.6 g/m^3
- For Q_{in} = 500 m^3/d, C_{in} = 10 g/m^3: m_i = 1.00 g/m^2·yr C_{out} = 7.6 g/m^3

The use of a loading relation has therefore gained us no simplification, because outlet concentration remains trivariate, requiring both inlet loading and inlet concentration. Nonetheless, the loading graph remains in use and has utility in exploring intersystem and intrasystem behavior. In those applications, parametric curves for different inlet concentrations are required. The outlet concentration load graph is a useful addition to the design toolkit for treatment wetlands. However, it cannot be used in isolation as a design sizing basis, because it does not separate the individual effects of inlet concentration and hydraulic loading.

Pitfalls of Graphical Representations

The purpose here is to illustrate the fallacy of graphical data representations and associated regressions between variables that contain the same multiplier, such as the water flow rate in graph types C and D above. This subject has been elucidated for natural treatment systems by von Sperling (1999). As a hypothetical example, consider concentrations entering the wetland presumed to vary randomly between 0.2 and 1.2 (gm/m^3). Likewise, the concentrations leaving are also random, ranging from 0.1 to 0.3 (gm/m^3). Therefore, the mean inlet concentration is 0.7 gm/m^3 and the mean outlet concentration is 0.2 g/m^3, and the resulting average concentration reduction is 70%.

A set of fifty "experiments" is run, in which the hydraulic loading is varied linearly between 1 and 50 m/yr. For any experiment, the inlet and outlet concentrations are independently random within the ranges selected (Figure 7.3). Not surprisingly, linear regression of the input-output concentrations explains virtually none of the variability. There is a 70 ± 18 % (mean ± sd) concentration reduction, and that is all that may be determined.

Next, the correlation between pollutant load reduction and inlet pollutant loading is examined. Pollutant loading is defined as hydraulic loading times concentration, for both the inlet and outlet. Pollutant load reduction is the difference between inlet and outlet pollutant loadings. An excellent correlation is obtained, with an R^2 = 0.93, which makes the data look great, and makes us feel that we can use this for design (Figure 7.4). Unfortunately, there is no connection of performance to inlet loading, no matter how much this load graph appeals to us. The hydraulic loading appears in both the ordinate and the abcissa, thus causing a stretching of a random two-dimensional cloud along a diagonal axis. Many examples of this inappropriate representation and analysis are to be found in

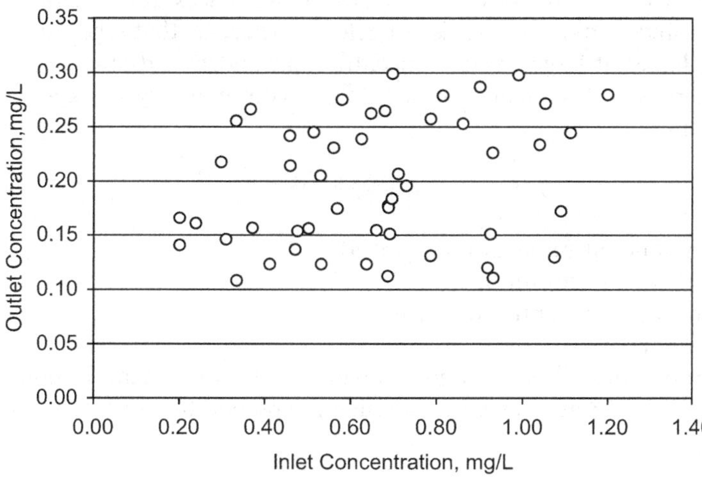

FIGURE 7.3 Scatter plot of input and output concentrations for a hypothetical dataset for 70% reduction.

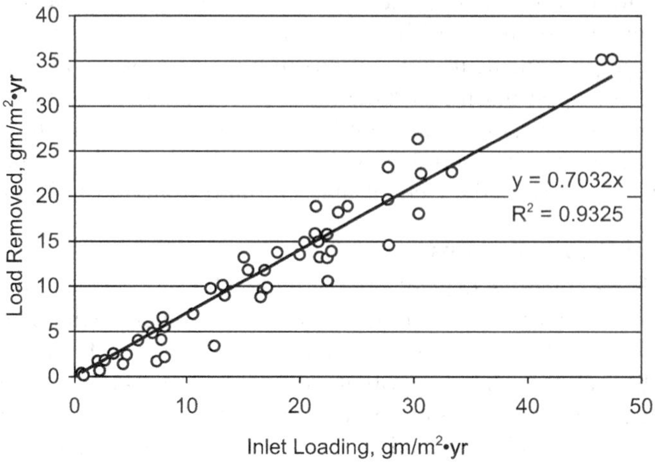

FIGURE 7.4 Load reduction versus incoming load for the hypothetical, random dataset of Figure 7.3.

the treatment wetland literature (Knight et al., 1993; Hammer and Knight, 1994; Vymazal, 2001a), but they are of no value in design.

7.2 CONCEPTS FROM REACTION ENGINEERING

The pollutant removal rate (J) in Equation 7.11 is the key to describing removals. That removal rate depends upon ecosystem attributes and concentrations, which must be quantified to complete the mathematical description. The marsh ecosystem chemically converts many substances into other chemical forms. There exists a rich history of chemistry and reaction engineering modeling from which descriptive relations may be drawn (for example, Levenspiel, 1999; Fogler, 2016).

Homogeneous reactions have the reactants and products in the same phase, while heterogeneous reactions have reactants in two phases. Pollutant removals in wetlands are nearly always heterogeneous reactions, involving dissolved substances and wetland solids, both biological materials and sediments. It is therefore expected that the rate of removal of a contaminant is larger at larger water concentrations, and that it is larger at larger amounts of biological materials and sediments. A simple and common way to express this is a second order rate expression:

$$J = kC_cC_s \tag{7.15}$$

where: J = contaminant removal rate, $g/m^3 \cdot d$
C_c = contaminant concentration, g/m^3
C_s = ecosystem solid concentration, g/m^3
k = rate coefficient, $(m^3/g)/d$

If the concentration of ecosystem solids remains relatively constant, its concentration can be absorbed in the expressed rate constant, obtaining the pseudo first order reaction constant.

$$J = k'C_c \tag{7.16}$$

where: k' = pseudo first-order rate coefficient, $(m^3/g)/d$

Bioreactors

A major sector of our economy depends upon the implementation of biological systems to produce a wide spectrum of products, ranging from pharmaceuticals to agricultural products. Consequently, a large literature exists concerning the quantitative description of bioreactors (for instance, Bailey and Ollis, 1986; Katoh et al., 2015). A commonly used rate expression is the Monod formulation:

$$J = k\frac{C_c C_s}{K_s + C_s} \tag{7.17}$$

where: K_s = half saturation constant, g/m^3

This expression is second order, involving the biomass (C_s) and the contaminant (C_s). As described in chapter 5, both microbial and algal biomass is typically represented by the Monod formulation.

This expression is pseudo first order at high values of C_s, and second order at low values of C_s.

Any kinetic relation (J) must be incorporated into the wetland mass balances for the contaminant, and possibly the wetland solids, whether inanimate or biologically active.

Autocatalysis

An autocatalytic reaction is defined as a chemical reaction that is catalyzed by its own products. The simplest autocatalysis is obtained by the chemical equation:

$$A + B \leftrightarrow B + B \tag{7.18}$$

for which the rate expression may be:

$$J = k_1 C_A C_B - k_2 C_B^2 \tag{7.19}$$

An intriguing facet of wetland nutrient processing is the fact that removals feed the biological cycle, creating a greater capacity for nutrient removal. Biomass (B) consumes nutrient (A), and thus creates more biomass (B). This aspect caused Kadlec (1997a), and Kadlec and Walker (1999), to describe the wetland process models to be "autobiotic," and the biota as the "biomachine."

Recently, this concept has been advanced as the central principle that forms the backbone of the organization in systems ecology, analogous to autocatalytic loops in systems chemistry (Veldhuis et al., 2018). Veldhuis et al. (2018) have coined (or at least greatly emphasized) the term ecological autocatalysis. They suggest that preemptive resource competition by consumers and plant resource diversity drive the emergence of autocatalytic loops at the ecosystem level. They go on to describe how interactions between such autocatalytic loops can explain pattern and processes observed at the ecosystem scale, and summarize efforts to model different aspect of the phenomenon.

Chemical Networks

Several wetland chemical removal processes involve more than one reaction, and more than one chemical species. Many removal reactions create products that are themselves contaminants of interest. An important example is the (microbial) sequential conversion network for nitrogenous species:

$$organic\ N \rightarrow ammonia\ N \rightarrow oxidized\ N \rightarrow gaseous\ N_2 \tag{7.20}$$

Each of the first three species is important by itself as a potential contaminant. Both consumption and production can occur, and it is therefore misleading to isolate one species and compute its "removal."

Another example is the reductive dechlorination of a chlorinated organic compound. Trichloroethylene has daughter products that are sequentially formed in a wetland environment:

$$trichloroethylene \rightarrow dichloroethylene \rightarrow vinyl\ chloride \rightarrow CO_2 + H_2O + Cl^- \tag{7.21}$$

In cases such as these, it is essential to utilize reaction models that account for both production and destruction. Each step may individually have a simple model, such as first order, but in combination, removal is quantitatively more complex.

7.3 PROCESSES THAT CONTRIBUTE TO POLLUTANT REMOVAL

A large number of wetland processes may contribute to the removal or reduction of any given pollutant. Here, some of the most important are described and the commonly used rules for quantification are presented. More details are presented in the following chapters for the most common chemicals of interest. The discussion here relates to localized phenomena. Removal processes must also be quantitatively placed in the context of internal wetland hydraulics and the topography and vegetative structure of the wetland.

Microbially Mediated Processes

Many wetland reactions are microbially mediated, which means that they are the result of the activity of bacteria or other microorganisms, and also the activity of algal species. Very few of such organisms are found free-floating; rather, the great preponderance are attached to solid surfaces. Often, the numbers are sufficient to form relatively thick coatings on immersed surfaces. This component of the marsh ecosystem has been described in Chapter 5 and has been termed the mesophytic component.

Transfer of a chemical from the water to immersed solid surfaces is the first step in the overall microbial removal mechanism. Those surfaces contain the biofilms responsible for microbial processing, as well the binding sites for sorption processes. The sediment-water interface is one such active surface; but the floc, litter, leaves and stems within the water column comprise the dominant wetted area. The following discussion analyses the transport of dissolved constituents to reaction sites located within the biofilms that coat all wetland surfaces. Mass transfer takes place both within the biofilm, and in the bulk water phase.

Dissolved materials must move from the bulk of the water to the vicinity of the solid surface, then diffuse through a stagnant water layer to the surface, and then penetrate the biofilm while undergoing chemical transformation (Figure 7.5). This sequence of events has been described and modeled in the text of Bailey and Ollis (1986), and is outlined here. The case of zero wetland background concentration will be described here; but extension to the case of nonzero background is possible.

The rate of transfer across the two films is:

$$J_{mt} = \frac{D_w}{\delta_w}(C - C_i) \tag{7.22}$$

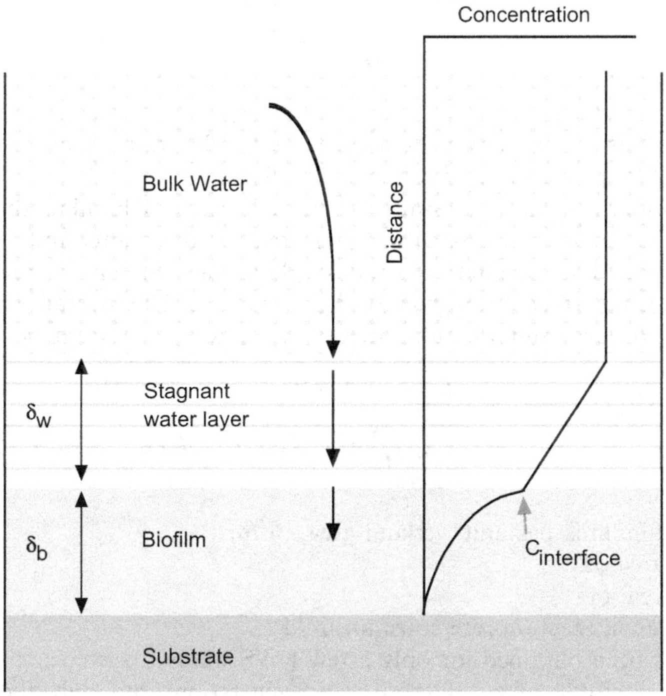

FIGURE 7.5 Path way for movement of a pollutant from the water across a diffusion layer and into a reactive biofilm. The solid may be sediment, a litter fragment, or a submerged portion of a live plant.

$$J_{mt} = E\, k_b\, \delta_b\, C_i \tag{7.23}$$

where C = concentration in the bulk water, mg/L = g/m^3
C_i = concentration at the biofilm surface, mg/L = g/m^3
D_w = diffusion coefficient in water, m^2/d
D_b = diffusion coefficient in biofilm, m^2/d
δ_b = thickness of the biofilm, m
δ_w = thickness of the stagnant boundary layer, m
E = tanh(ϕ)/ϕ, biofilm effectiveness factor, –
J_{mt} = mass transfer rate, g/m^2·d
k_b = reaction rate constant inside biofilm, d^{-1}
and where

$$\phi = \delta_b \sqrt{\frac{k_b}{D_b}} \tag{7.24}$$

Combining equations (the rate of transport of the pollutant from the bulk water to the biofilm is then):

$$J_{mt} = \left[\frac{E\, k_b\, \delta_b}{1 + M}\right] C = k_i\, C \tag{7.25}$$

where k_i = intrinsic first order areal reaction rate constant, m/d

$$M = \frac{E k_b \delta_b \delta_w}{D_w} \tag{7.26}$$

It is seen that this theory produces a local first-order rate of overall reaction, which depends upon biofilm properties and diffusion coefficients.

In a field situation, it is also necessary to know the area of biofilms that occupy a given area of wetland. In Chapter 5, this submerged area for the plants and bottom was called the WSAI, the wetland submerged area index (plants plus litter). Presuming that biofilms cover all submerged surfaces, this means A_b = WSAI. The overall removal rate from a wetland area A_w occurs from a biofilm area of A_b, and hence the rate of removal is:

$$\begin{aligned} J\, A_w &= k_i\, A_b\, C \\ J &= k_i\, \frac{A_b}{A_w}\, C = k_i\, a_s\, C = k_A\, C \end{aligned} \tag{7.27}$$

where a_s = biofilm area per unit wetland area, m^2/m

A_w = wetland area, m^2

A_b = biofilm area, m^2

k = first order areal reaction rate constant, m/d

Data for a_s has been obtained for only a few FWS wetland systems, giving only a rough estimate of the magnitude of a_s. If there is no vegetation, and only the wetland bottom serves as the potential location of biofilms, the value of $a_s \leq 1.00$. If the emergent vegetation is considered as additional biofilm area, a dense stand of plants can yield $a_s \approx 5$. Inclusion of the litter can further increase the value to $a_s \approx 10$.

This theory has been calibrated for treatment wetlands (Polprasert et al., 1998; Khatiwada and Polprasert, 1999b), who determined a_s in the range 2.2–2.9. Measurements of immersed vegetation surface area were made at Arcata, CA, and Houghton Lake, MI, and produced a_s in the range 1.0–9.0 (USEPA, 1999).

Microbially mediated reactions are affected by temperature. Response is much greater to changes at lower end of the temperature scale, (<15°C) than the warmer range (20–35°C) (Kadlec and Reddy, 2001). Processes regulating organic matter decomposition are affected by temperature, and all N cycling reactions (mineralization, nitrification, and denitrification) are affected by temperature. Phosphorus reactions are least affected by temperature. Temperature dependencies are discussed in connection with each contaminant in subsequent chapters.

Volatilization

The marsh environment provides considerable opportunity for losses of volatile compounds from the water to the atmosphere. As discussed in Chapter 5, some gases, principally oxygen, nitrogen oxides, methane and carbon dioxide, vent mainly through the plants. Other gases, such as ammonia, hydrogen sulfide and light hydrocarbons, volatilize from free water surfaces. The large areal extent of the wetland, coupled with relatively long detention times and shallow water depths, are conditions that foster convective and diffusional transport to the air-water interface, upward to bulk air and laterally off-site under the influence of winds. Because the definition of a treatment marsh precludes high concentrations of volatile substances, attention is directed to trace amounts.

There is typically equilibrium between air-phase and water-phase concentrations at the interface, which separates two vertical transport zones. Henry's law expresses the equilibrium ratio of the air-phase concentration to the water-phase concentration of a given soluble chemical:

$$P_i = H \cdot C_i \tag{7.28}$$

where C_i = interfacial water phase concentration, mol/m^3
= C_w (mg/L) ÷ Molecular Weight (MW, g/mol)
H = Henry's Law Constant, atm·m^3/mol
P_i = interfacial partial pressure in air, atm

Transport in both the air and water phases may involve convective currents as well as molecular diffusion, and therefore the transport flux (flow per unit area) is commonly modeled with mass transfer coefficients (Welty et al., 1984):

$$J = k_w(C - C_i) = k_a(P_i - P) \tag{7.29}$$

where C = water phase concentration, mol/m^3
J = loss flux, mol/m^2 hr
k_a = air-side mass transfer coefficient, (m/hr)(mol/m^3)/atm
k_w = water-side mass transfer coefficient, m/hr
P = partial pressure in air, atm

Air-side mass transfer coefficients are quite large, which places nearly all the mass transfer resistance on the liquid side. For instance, Mackay and Leinonen (1975) found over 80% of the transfer resistance in the water when $H > 10^{-4}$ atm·m^3/mol. In many instances, there is a zero bulk air concentration, and the transfer model reduces to:

$$J = k_w C \tag{7.30}$$

It is noteworthy that this theory leads to a first order areal removal rate.

Values of k_w depend upon the degree of convective mixing, as well as on the size of the molecule being transported. A large body of knowledge concerning oxygen and other gases in ponds was reviewed by Ro et al. (2007). They determined a general correlation, scaled to oxygen, from data concerning several gases:

$$k_w = 1.706 \cdot Sc^{-\frac{1}{2}} U_{10}^{1.81} \left(\frac{\rho_a}{\rho_w}\right)^{\frac{1}{2}} \tag{7.31}$$

where Sc = Schmidt number, = ν/D, –
D = diffusivity of gas, m^2/s
U_{10} = wind speed at 10 m height, m/s
ρ_a = density of air, kg/m^3
ρ_w = density of water vapor, kg/m^3
ν = kinematic viscosity of gas, m^2/s

According to 7.31, as wind speeds range from 0.5 to 5.0 m/s, the transfer coefficients range from 0.003 to 1.30 m/hr. Open water experimental studies of Peng et al. (1995) verified the strong effect of mixing in the water phase, and established a diffusion-only value of $k_w \approx 0.03$ m/hr for benzene, toluene, TCE and PCE. Open water studies of Crusius and Wanninkhof (2003) for the tracer gas SF_6 propose a form very similar to Equation 7.31. Their experimental range was $0.64 < k_w < 9.0$ m/hr.

Wetland studies differ in the effects of vegetation upon wind-driven water currents. Variano et al. (2009) measured $0.27 < k_w < 1.2$ cm/hr. Poindexter and Variano (2013) discuss the effects of thermal mixing upon k_w in marsh environments.

In the context of treatment wetlands, these rate constants are in the range of 20 to 20,000 m/yr. Therefore, light molecules are very likely to be effectively stripped in wetlands that are designed to remove other constituents with equal or lower rate constants.

Sedimentation

Chapter 8 deals with the removal of suspended solids. Here it is noted that a first order areal removal model is the outcome of theory and practice. Many pollutants partition to suspended solids, and thus removal of those sorbed substances also is expected to follow that model:

$$J = k_{TSS} \cdot C_{TSS} \cdot K_p \cdot C \tag{7.32}$$

where C_{TSS} = suspended matter concentration, mg/L=g/m^3
k_{TSS} = TSS removal rate constant, m/d
K_p = partition coefficient, m^3/g
The necessary connection to solids removal is the quantification of sorption.

Wetlands may create chemical precipitates, such as iron oxyhydroxides and calcium carbonate, which may be removed by sedimentation. Further, wetlands produce particulate matter as the result of biological processes.

Sorption

Partition coefficients relate the amount of sorbed pollutant to the concentration in the water under equilibrium conditions. Three types of sorption isotherms are in common use in wetland technology:

$$\textit{Linear} \quad C_s = K_p C \tag{7.33}$$

$$\textit{Freundlich} \quad C_s = K_f C^n \tag{7.34}$$

$$\textit{Langmuir} \quad C_s = K_L \frac{C}{C+b} \tag{7.35}$$

The sorption potentials for the principal contaminants of interest are discussed in the chapters pertaining to those contaminants. Here, a few generalities are noted:

- Sorption is important for phosphorus during the startup period for a treatment wetland. If initially absent in the sediments, phosphorus will be stored until the existing soils and sediments reach equilibrium with the overlying water. If initially present, phosphorus may be released.
- Sorption is important for ammonia nitrogen in intermittently dosed or operated wetlands. Short-term storage may be oxidized during drawdown periods.
- Sorption is important for hydrophilic organic chemicals, which partition strongly to the carbonaceous content of wetland sediments.
- The water-phase concentration that is experienced by wetland sediments and soils is porewater, which can have very different concentrations than the bulk water overlying those sediments and soils.

- Sorption sites are a partially renewable resource, because they may be added from the accumulation of newly formed sediments.
- Sorption may be partially irreversible, due to mineralization of sorbed materials, or to the formation of very strong chemical bonds.
- Linear sorption, Equation (7.33), results in a theoretical first order removal process at the local level.

Photodegradation

Sunlight can degrade or convert many waterborne substances, including pathogens and trace organic compounds (Nguyen et al., 2015; Bear et al., 2017). Photolysis of trace organic contaminants can occur via direct and indirect mechanisms. Direct photolysis occurs when a contaminant absorbs sunlight and undergoes a transformation reaction. Indirect photolysis involves other chemicals, Including nitrate and organic matter, which absorb light and produce reactive intermediates that subsequently react with contaminants (Jasper and Sedlak, 2013).

Photodegradation is usually unimportant in marshes because emergent macrophytes and floating vegetation shade of the water surface, and SAV reduces light penetration. Pathogen reduction effectiveness is presumptively determined by the radiation dose rate as well as the concentration of organisms. Although this is theoretically a second-order process, the sunlight dose in the wetland is relatively constant in the long term, and the elimination rate is therefore pseudo first order in the organism concentration.

Photodegradation has recently received attention in treatment wetland research and development, due to research activities in unvegetated wetlands, which are essentially shallow ponds (Jasper et al., 2013).

Plant Uptake

Plants take up nutrients to sustain their metabolism. They also may take up trace chemicals found in the root zone, which may then be stored, or in some cases, expelled as gases. Submerged plants may absorb nutrients and metals from the water column into stems and leaves, but EAV uptake is via roots. Roots are most often located in the wetland soils, although adventitious roots may sometimes be found in the water column. Roots access solutes from porewater, and associated sorption storages, which may originate from soil decomposition. Dissolved substances are also moved into and out of the root zone by diffusion and water flows, as discussed in Chapter 5.

In the treatment marsh environment, plant uptake is an important mechanism. Nutrient removal displays a pattern of seasonality, keyed to the vegetative cycle. Temperature is not always an acceptable surrogate for seasonality for nitrogen removal. Vegetative uptake in temperate climates is maximum during the spring period, at moderate temperatures; but release via decomposition is maximum during fall, also at moderate temperatures. Plants utilize phosphorus, nitrate and ammonium, and decomposition processes release nitrogen and phosphorus back to the water (Kadlec, 2005a).

Transpiration Flux

Vertical flows of water in the upper soil horizon are driven by plant water uptake to support transpiration. Emergent plants "pump" water from the root zone to the leaves, from which water evaporates through stomata, which constitutes the transpiration loss (Chapter 5). In a densely vegetated wetland during the growing season, transpiration dominates evapotranspiration. Water for transpiration must move through the soil to the roots. That movement is vertically downward from overlying waters in most wetland situations.

Thus, transpiration has the potential to move on the order of one meter per year of water vertically downward to the root zone. That water carries with it the contaminant concentrations associated with the bottom layer of overlying water, which is the litter/benthic zone of the wetland. This flow is termed the transpiration stream (TS), and it draws from porewater that is typically at a concentration different from that of the bulk surface water. In turn, the plant may block a portion of the dissolved pollutant, and take up a concentration less than that of porewater. These factors combine to determine the amount of plant uptake (Seeger et al., 2011; Reid and Jaffe, 2012; San Miguel et al., 2013):

$$J_U = TS \cdot TSCF \cdot C_{pw} \tag{7.36}$$

where C_{pw} = porewater concentration, g/m^3
J_U = uptake flux, $g/m^2 \cdot d$
TS = transpiration flux, m/d
TSCF = transpiration stream concentration factor, –

Predictive equations for the transpiration stream concentration factor have been set forth by a number of authors, however their results are not specific to wetland plants (Trapp, 2004; Dettenmaier et al., 2009). Nonetheless, wetland researchers have accepted those relations.

In a moderately dense emergent FWS wetland, the transpiration flux is far greater than the estimated diffusion fluxes.

Accretion

One of the least studied aspects of pollutant transfer in wetlands is in the creation of new soils and sediments, with their attendant chemical content. Not all of the dead plant material undergoes decomposition. Some small portions of both above ground and below ground necromass resist decay, and form stable new accretions. Such new stores of chemicals are presumed to be resistant to decomposition. The origins of new sediments may be from remnant macrophyte stem and leaf debris, remnants of dead roots and rhizomes, and from undecomposable fractions of dead microflora and microfauna (algae, fungi, invertebrates, bacteria).

As noted in Chapter 4, the amount of such accretion has been quantified for some treatment marshes (Reddy et al., 1991; Craft and Richardson, 1993; Rybczyk et al., 2002; Kadlec, 2009b; Bhomia et al., 2015). Historical studies have relied upon either atmospheric deposition markers (radioactive cesium or radioactive lead), or introduced horizon markers, such as feldspar or plaster. More recent reports rely upon elevation surveys and change points on the vertical soil profiles. All of these techniques require several years of continued deposition for accuracy.

7.4 INTERNAL HYDRAULICS AND MIXING

The removal of pollutants within a constructed wetland occurs through the diverse range of interactions between the sediments, substrate, microorganisms, litter, plants, the atmosphere, and the wastewater as it moves through the system. The dynamics of water movement through the wetland has a significant influence on the efficiency and extent of these interactions. Many of the important biogeochemical reactions rely on contact time between wastewater constituents and micro-organisms and the associated substrate, while wastewater velocity can be an important determining factor for other pollutant removal processes, such as mass transfer. Any short-circuiting or dead zones that occur within a wetland will consequently have an effect on contact time and flow velocities, and therefore impact on treatment efficiency. Nonideal flow patterns can have very large effects upon the removal of pollutants in wetland treatment systems (Kadlec et al., 1993; Carleton, 2002). It is therefore necessary to consider flow pattern effects, and the related mixing, in the design of wetland treatment systems.

Wetlands can be thought of as intermediate between two theoretically ideal reactors: the plug flow reactor and continuously stirred tank reactor (CSTR). The plug flow (PF) reactor represents a situation where there is no internal mixing within the reactor; water parcels move in unison from the inlet to the outlet. The CSTR represents the ideal of perfect mixing; water entering the system is instantaneously mixed throughout the reactor. Both are extensively described in the reaction engineering literature (e.g., Bailey and Ollis, 1986; Levenspiel, 1999; Fogler, 2016). Based upon a now very large literature on mixing behavior of treatment marshes, neither of these ideal extremes occurs very often, if at all. Nevertheless, major literature sources continue to utilize the extreme concepts (e.g., Crites et al., 2014 for plug flow, Marois and Mitsch, 2016 for CSTR).

The distinction between the two extremes is very important for high fractional removals, but of little importance for small percent reductions (Figure 7.6). For the simple example in Figure 7.6, if the desired pollutant reduction is only 10%, then it essentially does not matter what the extent of mixing may be. The effect is significant for reductions greater than about 50% (18% size penalty for the CSTR), and is extremely important for removals greater than about 90% (200% size penalty for the CSTR).

The extent of mixing and nonideality in the marsh flow environment depends upon several factors, which include:

- Shape
- Length-to-width ratio
- Turbulence, dictated by depth and flow rate, and thermal mixing
- Bathymetry
- Vegetation type, density, and spatial distribution
- Inlet and outlet water distribution, created by deep zones and types and location of hydraulic structures
- Compartmentalization
- Wind

At this point in the evolution of treatment marsh technology, all of these have received detailed investigations. Studies fall in two general groups, those that operate at a local, small spatial scale, and those that operate at the total system scale.

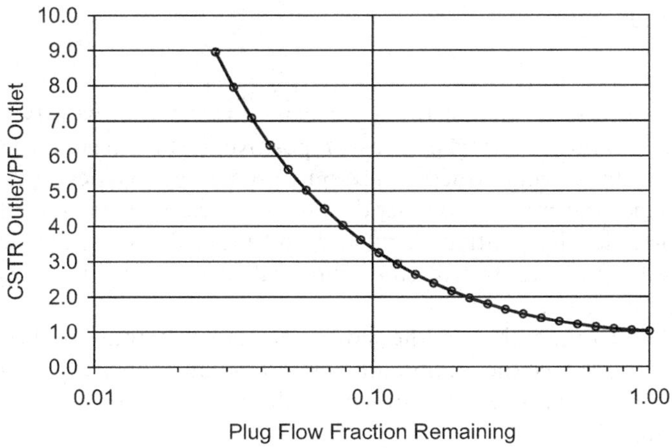

FIGURE 7.6 Outlet concentration ratio for a CSTR and plug flow reactors for a first order reaction rate as a function of the reactant remaining.

Localized Phenomena

In general, this group of studies furthers our understanding of marsh processes, but cannot yet be synthesized into an overall system description. Laboratory mesocosm platforms have been utilized, in conjunction with field studies. The one-dimensional chemical mass balance equation that embodies localized mixing is the reaction with dispersion equation (Fogler, 2016):

$$\frac{\partial C}{\partial t} + \frac{\partial (uC)}{\partial x} - D\frac{\partial^2 (C)}{\partial x^2} = -J \tag{7.37}$$

where C = contaminant concentration, g/m^3
D = dispersion coefficient, m^2/d
J = contaminant removal rate, $g/m^3{\cdot}d$
t = time, d
u = water velocity, m/d
x = longitudinal distance, m

Turbulent Mixing

The passage of water through wetland vegetation is accompanied by mixing, in all three directions; parallel to flow, transverse to flow and vertical. Nepf (2012b) reviews results for dispersion in submerged EAV and SAV wetland canopies, and concludes that mixing is due to turbulent diffusion at low to moderate stem densities, but shifts to mechanical dispersion at higher stem densities. The shift occurs at a submerged stem vegetation volume fraction of about 10%. Based mostly on laboratory flume studies, Tanino and Nepf (2008) suggest that for moderately high flow rates,

$$\frac{D}{ud} \simeq 0.2 \tag{7.38}$$

where d = stem diameter, m

Entire marsh systems are not uniform in either depth or vegetation density, and it is therefore not surprising that whole-marsh determinations do not generally correspond to the factor 0.2 in Equation 7.38 (Kadlec, 1994; Keefe et al., 2004). In the absence of stem diameter values, whole-marsh dispersion is more often represented by the system dispersion number, D/uL, where L is the flow path length. Values have been reported in the general range of $0.05 < D/uL < 0.5$.

Thermal Mixing

Heat transfer losses at the surface of the water can cause vertical density differences, resulting diurnal vertical mixing. During periods of positive surface heat flux (surface heating), stable density gradients (warm and less dense on top) act as a barrier to turbulent diffusion. In a lake, this reduces the mixed layer depth. During periods of negative heat flux (surface cooling), an unstable density stratification (cold and more dense on top) is produced in the water column, resulting in penetrative convection called plunging thermals. Those plunging thermals generate turbulence and mixing.

Jenter et al. (2003) found that in the slowly flowing wetlands of the Everglades, thermally-driven convection was the dominant vertical mixing mechanism. Measurements of vertical temperature gradients were combined with turbulence measurements. Jenter et al. (2003) concluded that:

> Diel coincidence of the thermal destratification and pronounced velocity fluctuations indicates the onset of mixing in the water column driven by thermal convection. The mixing process begins

> abruptly within approximately the upper 40 cm of the water column. Temperatures within this layer coincide through the remainder of the night. This mixed layer continues to cool and deepen throughout the night, not warming until just after sunrise the next day. A thin layer at the water surface remains slightly cooler than the mixed layer through the night, providing a constant source of negatively buoyant water that drives the mixing process. The mixing process continues until the air temperature rises above the temperature of the water at the surface.

Jenter et al. (2003) also concluded that the presence of vegetation was the primary factor causing thermally-driven convection to dominate the vertical mixing process. The intensity of thermal stratification differed among vegetation types. The water column at open water sites was close to being isothermal on most dates. In contrast, SAV sites often had markedly higher surface water temperatures, especially during the summer months, thus promoting thermal mixing

Wind Mixing

Wind can strongly influence the flow structure in both emergent and submergent treatment wetlands. A field study in a California marsh concluded that stirring near the water surface was strong enough to cause transfer of momentum from above the canopy, through the air in the canopy, and across the surface via shear, and influence biogeochemical budgets (Tse et al., 2016). Herb and Stefan (2005) formulated an analytic model that relates wind-driven vertical mixed layer depths in macrophyte beds of shallow lakes to several measurable physical parameters ~water depth, plant height, plant biomass density, and plant surface area per volume, as well as wind and heat input on the water surface.

Zeng et al. (2012) proposed a rather complicated model of the effects of wind on contaminant dispersion in a marsh environment. They concluded that for the fully developed steady wetland flow, wind dominated the extent of dispersion. Their model includes vegetation and friction effects, the direction of the wind relative to the water flow direction, and the strength of the wind.

Bentzen et al. (2008) performed field and numerical studies on a shallow runoff pond, and showed that there were very large effects of wind strength and direction on the system-wide circulation patterns. Guganessharajah (2001) performed field studies with drogues, and employed computational fluid dynamics (CFD) modeling, which showed that surface velocities may be 30-fold greater than for the deep layers, and that the velocity profile is most affected in the top 10–15 cm.

In laboratory flume experiments, with wooden dowels representing plants, Stephan et al. (2004) showed that mixing performance varied depending on whether the wind was with or against the flow. Stairs (1993) found that a tail wind of 5.7 m/s caused early water departure.

Despite these computational and experimental results on vegetated operating systems, the variability of wind speed and direction remains a barrier to formulation of quantitative long-term descriptions.

Whole System Behavior

The main method by which wetland scientists and engineers have gained information about internal hydraulic processes is through the use of inert tracers, which provide a means of tracking the movement of water through a wetland. The theory and practice behind hydraulic investigations have predominantly evolved out of the field of chemical reaction engineering (Levenspiel, 1999; Fogler, 2016). The basic concept is that reactions proceed as detention time increases, and consequently, if the distribution of detention times for a nonideal wetland is known, then the pollutant reduction can be calculated.

The factors that affect flow patterns and the associated detention time distribution are generally known, as tabulated earlier. They combine in complicated ways, and it is therefore necessary to measure the combined effect upon system behavior. The commonly utilized whole-system test is the inert tracer test. Most often, a tracer test is conducted by introducing an impulse of an inert substance into the wetland inlet at time zero. If water moved through the wetland in lock step, then such a tracer impulse would also exit as an impulse (a sharp spike of concentration). This result has never been observed in a wetland tracer test; the exit tracer is always a blurred, skewed bell-shaped curve. This is the result of mixing processes on a number of different distance scales. Expanses of open water permit surface wind-driven currents to develop, which are matched by return flows in lower water layers. Deeper parallel zones in the wetland carry more flow because of the depth effect on hydraulic resistance. These preferential channels may also be due to a lower vegetation density along some flow paths. The tracer impulse added to the incoming water provides a way to find such preferential paths, since the tracer will later be found preferentially in those wetland zones. Both natural and constructed marshes display such flow variability (Figure 7.7). In particular the results for constructed wetlands indicate that it is not possible to avoid such flow irregularities even with extreme care in construction.

For details of how tracer tests are conducted, the reader is referred to Headley and Kadlec (2007).

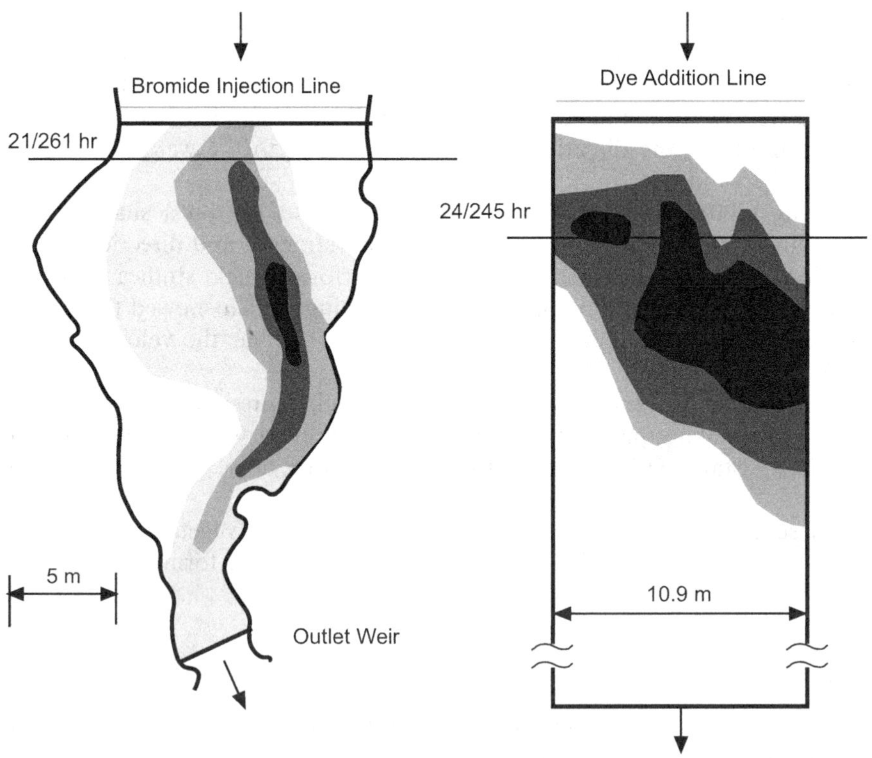

FIGURE 7.7 Tracer isopleths in a natural (A) and constructed (B) wetland. In both cases, tracer was added uniformly across the inlet width. The theoretical location of the pulse centroid is shown by the horizontal line, labeled with elapsed time and nominal detention time. (A) *Typha orientalis* natural wetland in New Zealand (Data source: Cooper, 1992). (B) *Typha latifolia* constructed marsh in Ontario (Data source: Herskowitz, 1986).

Interpretation of Data

Visual studies, like those of Figure 7.7, are informative but not quantitative. Numerical interpretation follows very well established procedures. The results of an impulse tracer test provide the volumetric efficiency of the wetland, together with information on the distribution of detention times in the system. The first requirement of a valid tracer test is that the tracer should be recovered nearly in its entirety at the wetland outlet. To that end, a simple check is made by summing the tracer at the wetland outfall:

$$M_e = \int_0^\infty Q_e C dt = ? = M_0 \tag{7.39}$$

where C = outlet tracer concentration, mg/L
M_0 = mass of tracer charged, gm
M_e = mass of tracer recovered, gm
Q_e = exit water flow, m^3/d
t = time, days

The detention time distribution (DTD) represents the time various fractions of incoming water spend in the wetland, and hence is the contact time distribution for the system. In a broader context, the DTD is the probability density function for residence times in the wetland. This time function is defined by:

$$f(t)\Delta t = \text{fraction of the incoming water which stays in the wetland for a length of time between } t \text{ and } t + \Delta t \tag{7.40}$$

where f = DTD function, 1/d

For an impulse input of tracer into a steadily flowing system, the function [f(t)] is:

$$f(t) = \frac{Q_e C}{\int_0^\infty Q_e C dt} = \frac{C}{\int_0^\infty C dt} \tag{7.41}$$

The first numerator is the mass flow of tracer in the wetland effluent at any time t after the time of the impulse addition. The first denominator is the sum of all tracer collected and thus should equal the total mass of tracer injected.

The mean *tracer detention time* is presumed to be the actual mean detention time, and is calculated from:

$$\tau = \frac{1}{M_0} \int_0^\infty tQC dt \tag{7.42}$$

where τ = tracer detention time, days

A wetland may have internal excluded zones, which do not interact with flow, such as the volume occupied by plant materials. In a steady state system without excluded zones, the tracer detention time (τ) equals the nominal residence time (τ_n). This is true whether the flow patterns are ideal (plug flow or well mixed), or nonideal (intermediate degree of mixing). An adsorbing tracer will produce an artificially short detention time, which may then be erroneously presumed to result from a large excluded zone. An incorrect topography may create either positive or negative differences between τ and τ_n. The ratio of tracer to nominal detention time is the *volumetric efficiency*:

$$e_V = \frac{\tau}{\tau_n} \tag{7.43}$$

where t_n = V/Q = nominal detention time, days
V = wetland water volume, m^3

There are a variety of reasons why the value of e_V is different from unity. Perhaps the most frequent is the existence of areas of the wetland that are not in the flow path. An extreme example is shown in Figure 7.8, for cell 7 of the Orlando easterly wetland (Min and Wise, 2009). Topographic data for this wetland indicated a channel, 0.3–0.8 m deeper than surroundings, leading from wetland inlet to outlet. As a result, only 24% of the wetland area was involved in flow (e_V = 0.24). Tracer data were fit using the MIKE 21 model (DHI, 2017b).

Jenkins and Greenway (2005) provide a numerical investigation, based upon 2D modeling, that confirms the low efficiency of wide short wetlands with opposed point inlets and outlets.

A second parameter, which can be determined directly from the detention time distribution, is the variance (σ^2), which characterizes the spread of the tracer response curve about the mean of the distribution, which is t:

$$\int_0^\infty (t-\tau)^2 f(t)\, dt = \sigma^2 \tag{7.44}$$

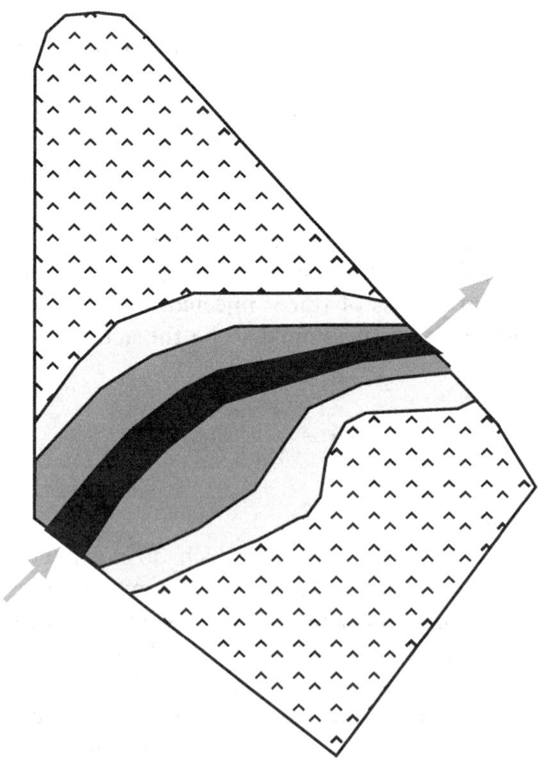

FIGURE 7.8 The modeled flow path of tracer through the 11.8 ha OEW Cell 7. (Based on information in Min and Wise, 2009.)

where σ^2 = DTD variance, d^2

The variance of the DTD is created by mixing of water during passage, or equivalently by a distribution of velocities of passage. This can be lateral, longitudinal or vertical mixing, or parallel flows of different velocities. An adsorbing tracer will lead to a narrowed response pulse, and hence to an erroneously low degree of mixing. This measure of dispersive processes may be rendered dimensionless by dividing by the square of the tracer detention time:

$$\sigma_\theta^2 = \frac{\sigma^2}{\tau^2} = \frac{1}{N} \tag{7.45}$$

The new parameter is σ_θ^2, the *dimensionless variance* of the tracer pulse. The dimensionless variance may be used to determine parameters in DTD models, such as the tanks-in-series model, for which it determines the parameter "N". Or, it may be used to compute the *DTD efficiency* of the wetland (Persson et al., 1999):

$$e_{DTD} = \left(1 - \sigma_\theta^2\right) = 1 - \frac{1}{N} \tag{7.46}$$

This efficiency is unity for plug flow, and zero for perfect mixing.

Persson et al. (1999) go on to suggest a combined measure, the *hydraulic efficiency* (λ) that reflects the excluded volume and the mixing pattern of constructed wetlands. This measure is defined as the product of the volumetric efficiency and the DTD efficiency:

$$\lambda = e_V \cdot e_{DTD} \tag{7.47}$$

This is useful for ranking wetlands for their combined efficiency but is not directly useful for parameter estimation. The two pieces are needed separately for quantitative estimates of performance. For the TIS model described here, the hydraulic efficiency is given by a very simple result:

$$\lambda = \frac{\tau_p}{\tau_n} \tag{7.48}$$

where: τ_p = tracer peak time, d

Persson et al. (1999) have explored both real wetland systems, and, via calibrated two-dimensional dynamic modeling, simulated situations. They found $0.11 < \lambda < 0.90$. The same parameter has been evaluated for other hypothetical wetland situations (Jenkins and Greenway, 2005).

The Gamma Distribution

Several decades of research on hundreds of wetland tracer tests have shown that most wetland DTDs can be represented by the gamma distribution function:

$$f(t) = \frac{N^N t^{N-1}}{\tau^N \, \Gamma(N)} \exp\left(\frac{-Nt}{\tau}\right) \tag{7.49}$$

where the gamma function $\Gamma(N)$ is defined by

$$\Gamma(N) = \int_0^\infty \exp(-x) x^{N-1} dx \tag{7.50}$$

The parameter "N" represents the number of well-mixed units in series for the TIS model.

While it may seem daunting to deal with Equations 7.49 and 7.50, it is in fact not difficult. The GAMMADIST function is available in Microsoft Excel™ and returns values of f for the time t and the parameters N and τ. The SOLVER application in Microsoft Excel™ allows the modeler to simultaneously solve for the variables N and τ. The procedure involves minimizing the residual error between data and the fitted DTD (Equation 7.49). The sum of the squared errors (SSQE) is minimized using SOLVER. An example of the results of this approach is shown in Figures 7.9 and 7.10. Figure 7.9 displays the two-dimensional pattern of tracer at two times during the test, while Figure 7.10 shows the fit of the gamma distribution to the outflow concentration data.

In the limit as N becomes very large, the gamma distribution becomes the plug flow (PF) distribution, with all water departing after exactly one detention time. When N = 1.0, the result is the DTD for a s CSTR.

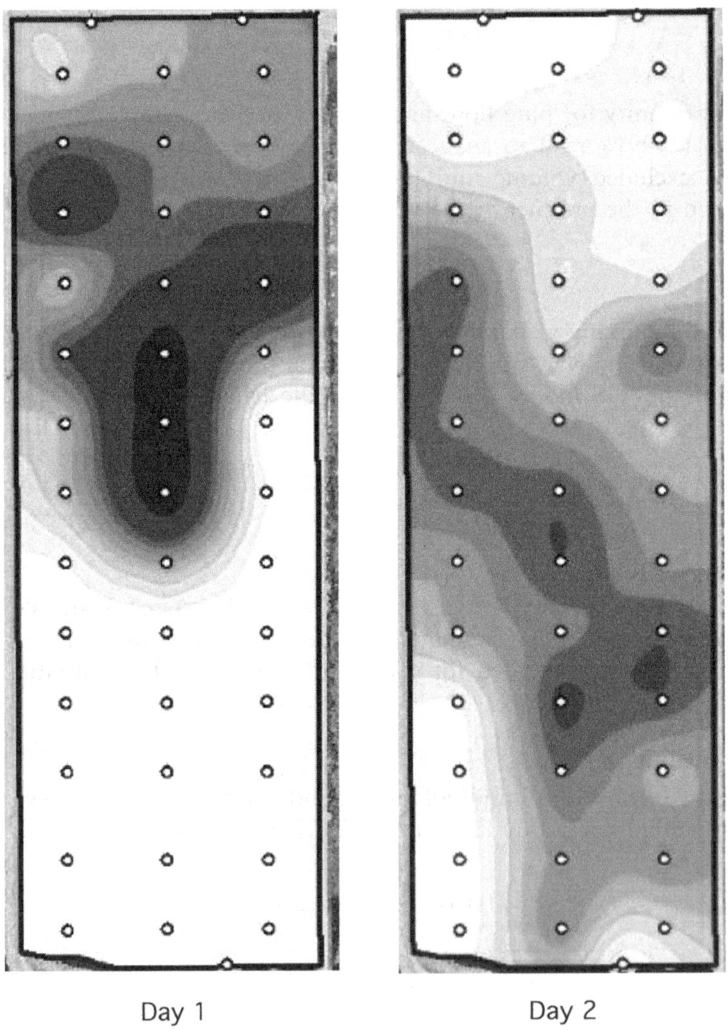

FIGURE 7.9 Progress of a RWT tracer through the 40 ha periphyton cell of STA34 in Florida. The nominal detention time was 3.85 days. The resultant outlet detention time distribution calibrated to 6 stirred tanks. Source: DB Environmental, Inc. unpublished data.

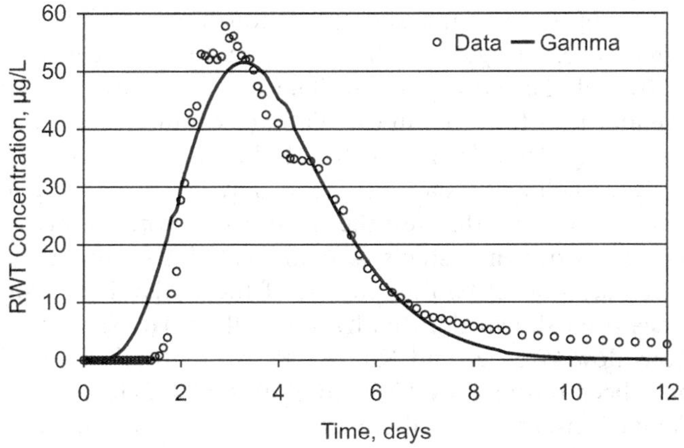

FIGURE 7.10 The gamma distribution representation of the tracer test of Figure 7.9.

The Method of Moments

Virtually the entire early literature on tracer testing of wetlands and ponds utilized parameter estimation methods, which reflected the computational tools available when they were developed circa 45 years ago. The most common method involves computation of the first and second moments of the experimental outlet concentration distribution, via numerical integration. A serious failing of the moment method of parameter estimation is that it emphasizes the "tail" of the response much more than the central portion of the response – that is, the peak area. Minor concentration anomalies on the "tail" of the concentration response curve may yield spurious parameter values. The better procedure is to utilize a robust parameter determination routine, such as a search to minimize the sum of the squared errors between the selected DTD function and the data. The interested reader can find details of the moment method either in reaction engineering textbooks (e.g., Fogler, 2016), or in the treatment wetland literature (e.g., Kadlec and Knight, 1996; Kadlec and Wallace, 2009).

Factors that Affect Tracer Results

A wide variety of substances have been used to trace the progress of water through treatment wetlands. These have included salt ions: lithium, bromide, chloride, iodide, fluoride; fluorescent dyes: rhodamine RWT and B, fluorescein; and tritiated water.

Although it is tempting to think that the entire wetland water volume is swept by the flow in a moderately uniform manner, giving rise to a gamma DTD, deviations from that pattern are commonly found. Two pathological situations are the existence of short-circuits from inlet to outlet, and/or the existence of deadwater zones that are not swept by flow.

The loss or gain of water to or from the atmosphere does not carry tracer in or out of the wetland. However, the transpiration component causes downward flow of water and tracer, onto the litter and soil. And, evapotranspiration (ET) does cause the water to slow as it passes through the system, and rain causes it to accelerate. Because there is normally level control at the outlet, the depth remains unchanged, but the linear velocity of the remaining water is altered. Accordingly, for ET, the detention time is increased and dissolved constituents become more concentrated. Conversely, for rain, the detention time is decreased, and dissolved constituents become diluted.

Recovery of tracer is an extremely important quality check for the measured DTD. Failure to recover 100% of the tracer may mean that the tracer adsorbed or was degraded during passage through the wetland. Organic compounds, including the fluorimetric dyes often used in water tracing in mineral systems, are notorious in this respect: they disappear in wetlands. Noninteractive, inorganic substances are to be preferred; lithium ions and bromide ions being two of the more popular. The use of a sorbing tracer can distort the tracer response curve and lead to errors in calculating hydraulic characteristics. It is possible that molecular tracers may degrade during travel through the wetland. This is especially true for the organic dyes, such as rhodamine. These may be subject to photolytic decomposition, or to the action of microbes that use the tracer as a carbon source.

Tracer analysis of event-driven systems is complicated by two factors: the flow is not steady, and the tracer may not all be flushed out of the wetland by a single event. Despite these difficulties, it has been shown that the hydraulic flow patterns are similar under event-driven and continuous flow (Werner and Kadlec, 1996). Methods for dealing with unsteady flows and depths have been outlined by Holland et al. (2004, 2005).

A much more detailed discussion of these effects can be found in Appendix B of Kadlec and Wallace (2009).

Models for Internal Hydraulics

Tracer testing is not an end in itself; rather, it is conducted to support the modeling and calculation of contaminant removals in the wetland system. Understanding flow behavior in constructed wetlands is necessary because it provides the framework for a pollutant transport and removal model, which in turn and directly affects the treatment efficiency. Accordingly, the tracer information is combined with the local, or intrinsic, removal rate to produce the wetland outlet concentration.

There are many candidate models that may be used, which typically involve series and parallel combinations of two idealized flow elements: perfectly mixed units and plug flow sections (Figure 7.11). It is clear from numerous studies that treatment wetlands are neither plug flow nor well-mixed. As the model networks increase in complexity, such as the parallel path and finite stage models, they are able to resolve the last bits of detail in the responses, but do so at the expense of adding more calibration parameters. In the extreme, it is possible to use complex computer codes to model wetland tracer responses (Martinez and Wise, 2003b; Keefe et al., 2004). In this book, the TIS model is utilized, as a spreadsheet compromise between too many parameters and too little detail.

Because all of the concepts in Figure 7.11 have been or are being used in treatment marsh studies and literature, each will be described in a bit more detail with wetland references. All are one-dimensional representations.

Figure 7.11A Plug Flow

This model presumes no mixing occurs as the water passes through the system. The simplicity of the model caused it to be used extensively in early wetland designs (often assumed, rarely observed). See for example Reed et al. (1995), U.S. EPA (1999), Crites et al. (2006). It remains the basis for wetland design in some literature sources (e.g., Crites et al., 2014).

Figure 7.11B Single CSTR

There is a long history of modeling nutrient and sediment dynamics in lakes, which typically involved the assumption of the lake as a single well-mixed unit (Vollenweider, 1975; Chapra and Canale, 1991). But because marshes at all scales display spatial variability in the flow direction, as single CSTR is rarely assumed. Isolated instances of the single CSTR assumption include, for example, Moustafa (1998); Christensen et al. (1994); Wang and Mitsch (2000), and Marois and Mitsch (2016).

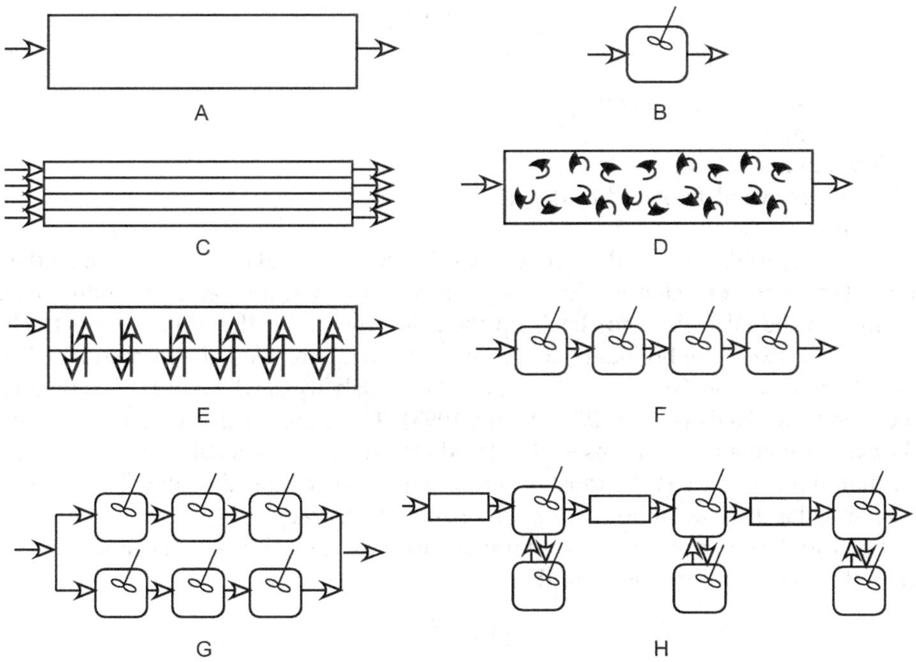

FIGURE 7.11 A sampler of various models to represent wetland tracer responses. The plug flow model (A) produces an impulse output at one detention time. The well-mixed model (B) produces an exponential decline. The rest of the models produce skewed bell shaped responses. See text for more details. (11A) Plug Flow, (11B) Single CSTR, (11C) Independent Stream Tubes, (11D) Plug Flow with Dispersion, (11E) Plug Flow with Side Storage Zones, (11F) The Tanks in Series Model, (11G) Parallel Path TIS, (11H) Finite Stage Model.

Figure 7.11C Independent Stream Tubes

Levenspiel (1999) terms this the *pure convection model.* This assumes that each element of fluid slides past its neighbor with no interaction by diffusion. Thus the spread in residence times is caused only by velocity variations. This concept supposes that a wetland has multiple "parallel" noninteracting paths for water movement from inlet to outlet. Each of these flow paths is considered to be plug flow. This model is a *segregated flow model,* because the fluid elements that enter together always stay together and are surrounded at all times by the fluid elements of the same age. At the wetland outlet, each independent flow mixes with the others, in proportions dictated by the detention time distribution.

Levenspiel (1999) posits that for the dispersion or tanks-in-series models the variance of the DTD grows linearly with distance: $\sigma^2 \propto L$. The consequences of this model assumption are discussed in Kadlec (2000), Carleton (2002), and Jenkins and Greenway (2005).

Figure 7.11D. Plug Flow with Dispersion

Another model uses of a dispersion process superimposed on a plug flow model (PFD). Mixing is presumed to follow a convective diffusion equation. A one-dimensional spatial model is chosen, because analytical expressions are available for computation of pollutant removal for the one-dimensional case (Levenspiel, 1999). The tracer mass balance equation includes both spatial and temporal variability:

$$D\frac{\partial^2 C}{\partial x^2} - \frac{\partial(uC)}{\partial x} = \frac{\partial C}{\partial t} \tag{7.51}$$

where C = tracer concentration, g/m^3
u = velocity, m/d
D = dispersion constant, m^2/d
x = distance from inlet toward outlet, m
t = time, d

The appropriate wetland boundary conditions for this mass balance are known as the *closed-closed* boundary conditions (Levenspiel, 1999). These imply that no tracer can diffuse back from the wetland into the inlet pipe, nor back up the exit structure at the wetland outlet. These are different from the *open-open* boundary conditions that are appropriate for river studies. There are analytical, closed form solutions to the later case, the availability of which has led to their repeated misapplication to wetlands (Bavor, 1988; Stairs, 1993). There are no closed form solutions to the wetland case, but numerical solutions to the closed-closed tracer mass balance have been available for more than three decades (Levenspiel, 1999). It is possible to calculate the dispersion constant that fits a particular data set, although there are issues of accuracy.

The dimensionless parameter that characterizes equation (7.51) is the Peclet number, or its inverse, the wetland dispersion number:

$$\mathbf{D} = \frac{D}{uL} = \frac{1}{Pe} \tag{7.52}$$

where Pe = Peclet number, dimensionless
D = wetland dispersion number, dimensionless
L = distance from inlet to outlet, m

The two results of interest from modeling of the pulse test are the tracer detention time and the dimensionless variance:

$$\tau = L/u \tag{7.53}$$

$$\sigma_\theta^2 = 2D \cdot 2D^2\left(1 - e^{-1/D}\right) \tag{7.54}$$

A potential problem with application of the PFD model to wetlands has to do with meeting the assumptions implicit in the model. Levenspiel notes:

> Because of all this we should be very careful in using the dispersion model where backmixing is large…

The dispersion model is commonly part of the algorithms embedded in aquatic simulation packages, such as MIKE 21 (DHI, 2017b), or TARSE (Transport and Reaction Simulation Engine) (James et al., 2009). Alternatively, many authors have chosen to code their own PFD wetland simulations (e.g., Musner et al., 2014).

Figure 7.11E Plug Flow with Side Storage Zones

This option derives from the literature on mixing in streams, for which there is a main channel exhibiting dispersion, and exchanges with side storages in the porous bed.

$$\begin{aligned}\frac{\partial C}{\partial t} + \frac{\partial(uC)}{\partial x} - D\frac{\partial^2 C}{\partial x^2} &= \alpha(C_S - C)\\ \frac{\partial C_S}{\partial t} &= \alpha\frac{A}{A_S}(C_S - C)\end{aligned} \tag{7.55}$$

where A = main channel cross section, m^2
C = tracer concentration, g/m^3
α = exchange coefficient, 1/d
Note that two additional parameters are required, α and A/A_S.

This model has been implemented by several studies. The OTIS (One Dimensional Transport with Inflow and Storage) software is available from the U.S. Geological Survey at https://water.usgs.gov/software/OTIS/ (Runkel, 1998). OTIS has been implemented for wetlands by several researchers, including Martinez and Wise (2003) and Keefe et al. (2004, 2010).

The commercial model Mike 21 (Danish Hydraulic Institute (DHI), 2017b) was used by Mike 21 model was used to simulate field and laboratory data by Nepf et al. (1997), Serra et al. (2004), and Burke and Wadzuk (2009).

Werner and Kadlec (2000b) termed this the zones of diminished mixing (ZDM) model, which contains four parameters: the detention time, the fraction dead zones, the fraction bypassing and a dispersion number for the main flow path. Zahraeifard and Deng (2011) called their version the VART (Variable Residence Time) model.

These PFD/Side storage models do not lend themselves to an easy representation for computing pollutant removal, because they require the use of programmed finite difference algorithms. Thus, although they often can do an excellent job of representing the tracer response, they are of limited use in wetland evaluation and design.

Figure 7.11F The Tanks in Series Model

The tanks in series (TIS) model captures the important features of wetland DTDs that produce the skewed bell-shaped response. The TIS model requires two parameters: the number of "tanks" (N), and the mean (tracer) detention time (τ). It corresponds to the gamma distributions described above. It may be implemented on spreadsheets in the process of treatment marsh sizing and performance descriptions. Therefore, the TIS model is selected for the design procedures outlined in this book.

Figure 7.11G Parallel Path TIS

The parallel paths model has been described earlier as an example of a method for dealing with the long tails evidenced in some tracer. This model may also be easily coded in a spreadsheet, but there are five parameters: the flow split, and the detention time and the number of tanks along both paths. Some authors recommend the two-path response model (Keller and Bays, 2001; Wang et al., 2006). However, if a two-path model is warranted, then there is either a real parallel compartmentalization, or something seriously wrong with the wetland. In the latter case, it is the wetland that needs improvement, not the model.

Figure 7.11H Finite Stage Model

This model is small number of TIS with side storage zones and a plug flow element. The concept of the finite stage model was advanced for ponds by early authors (Mangelson and Watters, 1972; Watters, 1972). A single stage of this model consists of a plug flow unit (of volume V_{PF}), followed by a well-mixed unit (of volume V_a) that exchanges material with a side-storage unit (of volume V_d). There are five parameters in this model: the number of stages, the detention time, the fraction plug flow, the fraction active well-mixed and the exchange flow ratio. The exchange flow between V_a and V_d is f, and the ratio to the through-flow is f/Q. The finite stage model is comprised of a small number (2–5) of these stages in series. The finite stage model can provide an excellent fit of data from both ponds and wetlands (Kadlec, 1994; Alvord and Kadlec, 1996; Nolte and Associates, 1998b).

There are of course many other series and parallel networks of CSTRs and PFRs that may be appropriate for a given situation. A network approach is appropriate for interconnected wetland cells in a larger overall system, when that connection is not linear (see for example Wang and Jawitz, 2006).

Computational Fluid Dynamics (CFD)

Many treatment wetlands are not linear in their design, but rather assume irregular shapes with irregular bathymetry. Flow distribution may be at a point along the inlet boundary, rather than distributed across the width of the wetland. Additionally, vegetation density may not be spatially uniform. Under such circumstances, the internal flow patterns are variable in at least two dimensions, lateral and longitudinal. As with all other models, vertical averaging will typically be utilized, although that assumption is questionable under many

TABLE 7.1

A sampling of tracer study results from various sizes of treatment marshes.

State or Country	Project	Vegetation	Size m^2	Tracer	Recovery	Tracer HRT days	Volumetric Efficiency	NTIS	Ref.
Sweden	Halmstad	Algae, Typha	29	Lithium	87%	3.24	74%	2.0	1
Sweden	Halmstad	Phragmites	29	Lithium	68%	2.19	52%	4.3	1
China	Jiangxi	Juncus	91	RWT	67%	0.084	92%	3.1	2
China	Jiangxi	Iris	91	RWT	84%	0.061	67%	5.5	2
FL	Everglades	SAV	2,000	RWT	83%	5.9	105%	5.9	3
FL	Everglades	Cattails	2,457	Lithium	95%	2.1	60%	4.3	3
CA	Sacramento	Bulrushes	2,926	Lithium	76%	4.7	98%	4.2	4
CA	Sacramento	Bulrushes	2,926	Lithium	78%	6.4	133%	5.2	4
AZ	Trés Rios	Bulrushes	9,100	Bromide	95%	2.8	60%	5.8	5
AZ	Trés Rios	Bulrushes	9,200	Bromide	80%	3.3	77%	8.6	5
IL	Des Plaines	Cattails	20,000	Lithium	98%	8.4	68%	2.7	6
FL	OEW	Typha, Scirpus	117,409	Bromide	96%	2.38	24%	2.6	7
Sweden	Alhagen	Typha, Phragmites	149,000	Iodide		6.19	62%	3.4	8
FL	OEW	Mixed Emergent	230,800	Bromide	123%	5.67	88%	2.1	9
FL	Lakeland	Cattails	300,000	Lithium	84%	5.7	29%	2.8	10
FL	Lakeland	Cattails	770,000	Lithium	67%	0.7	4%	0.3	10
FL	Lakeland	Mixed Emergent	810,000	Lithium	75%	5.8	33%	1.5	10
FL	Everglades	SAV	1,470,000	RWT	83%	4.3		3.5	3
FL	Everglades	Mixed Emergent	4,140,000	Lithium		11.3		5.6	3
FL	Everglades	Mixed Emergent	5,780,000	Lithium		13.4		7.9	3
FL	Everglades	SAV	6,119,907	Lithium		4.35	105%	12.0	3
FL	Everglades	SAV	8,984,021	Lithium		10.6	98%	10.3	3
Mean					**85%**	**5.0**	**0.70**	**4.7**	
Median					**83%**	**4.5**	**0.68**	**4.3**	

References
1 Bodin et al. (2012)
2 Guo et al. (2017)
3 Unpublished data, South Florida Water Management District
4 Nolte and Associates (1998b)
5 Unpublished data, City of Phoenix
6 Kadlec (1994)
7 Min and Wise (2009)
8 Wörmann and Krönnas (2005)
9 Wang et al. (2006)
10 Unpublished data, City of Lakeland

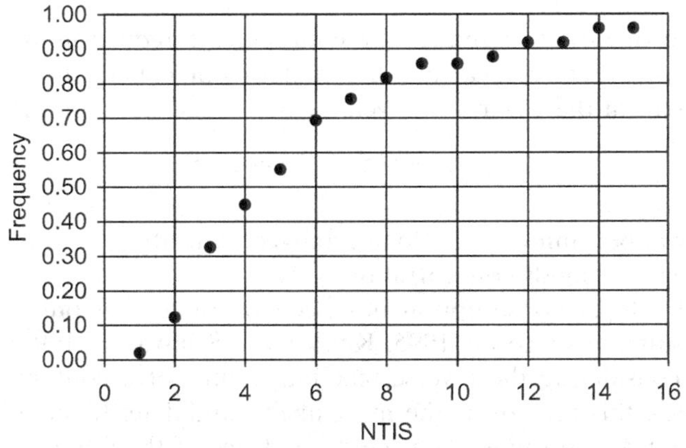

FIGURE 7.12 Frequency distribution of the number of tanks in series determined for marshes.

wetland circumstances. Flow modeling involves solution of mass and momentum balances on a two dimensional grid, together with a tracer mass balance. This procedure has been utilized, for example, by Walker (1998), Persson et al. (1999), Koskiaho (2003), Jenkins and Greenway (2005), Kjellin et al. (2007); Min and Wise (2010); and Sabokrouhiyeh et al. (2017). Several software packages are available for this purpose, including RMA4 and MIKE-21. The advantage of CFD modeling is the ability to explore the consequences of alterations in the basin morphology on the volumetric and DTD efficiencies of a system.

Wetland TIS Model Parameters

A cross section of reported literature values yields N = 4.6 ± 0.4 (median ± standard error) for182 tracer tests of 133 wetlands of varying sizes. Those tests had a median tracer recovery of 76%, and a median volumetric efficiency of 73%. Examples of these reported values are shown in Table 7.1. The distribution of N-values for these marshes is shown in Figure 7.12.

The range of values is quite large, and depends strongly on wetland configuration, inlet and outlet structures, and vegetation density. These influences are discussed in Chapter 14.

7.5 REACTION RATE MODELS

In this section, the concepts of local pollutant reduction are blended with wetland hydraulic considerations, and environmental and ecosystems features, to develop pollutant removal models.

Intrinsic Chemistry

The removal of a contaminant may depend upon the local concentrations of that contaminant in any of a number of ways, depending upon the mechanism(s) or pathways involved. Often, the rates of reaction are strongly dependent upon the water phase concentration of a specific contaminant. Additionally, other substances are usually involved in the conversion process, which in the case of treatment marshes are biomass (plants and mesophyton), necromass (floc and litter) and soils. Thus, although almost the entire treatment wetland literature presumes reaction rates that depend upon water phase concentrations (i.e., first order or similar), those rates are in fact also dependent on ecosystem reactive solids. This implies that most removal processes are second order, and when the solid dependence is ignored, we are dealing with pseudo first order (or other) models.

Zero Order

The most simplistic quantitative model for contaminant reduction is a constant rate of removal, termed zero order removal because it does not depend upon how much of the contaminant is present in the water at a given location. The local load removal is given by:

$$J = f(S) = "constant" \tag{7.56}$$

where: J = removal per unit area, or load removed, $g/m^2 \cdot d$
S = ecosystem reactive solids concentration, g/m^3
Such a constant rate of consumption has been postulated in only a few instances for treatment wetland situations (Horne, 1995; Kadlec and Srinivasan, 1995).

However, when considering the reverse reaction, return of a contaminant from the ecosystem reactive solids, this relation is the most likely candidate. If there is a gradient in the flow direction of the reactive solids, then the dependence of the uptake or return rates upon those reactive solids.

First Order

As seen earlier in this chapter, many individual wetland processes are basically first order, such as mass transport, volatilization, sedimentation, and sorption. It is therefore not unreasonable to presume that these behave similarly in combination, at least over some range of pollutant concentration. The local removal rate equation is:

$$J = k' f(S) C = kC \tag{7.57}$$

where: C = concentration, g/m^3
J = removal per unit area, or load removed, $g/m^2 \cdot d$
k' = intrinsic rate constant, m/d
k = pseudo first-order rate constant, m/d
This rate equation is the most prevalent in treatment wetland literature, although in many instances it is only presumptively advocated.

Saturation: Monod

Many biologically mediated reactions are first order only for concentrations lower than a saturation value. The premise is based upon the limited ability of the biological community to respond to increases in chemical availability, and this concept is implemented in models of other wastewater treatment technologies (Metcalf and Eddy, 2003). Such a model interpolates between zero- and first-order limits:

$$J = k' \frac{Cgf(S)}{K + C} = k' \frac{CS}{K + C} = k \frac{C}{K + C} \tag{7.58}$$

where: K = half-saturation constant, g/m^3
For low values of $C \ll K$, this is a first-order model. For high values of $C \gg K$, it is a zero-order model. This model has been frequently been implemented for microbial processes in treatment wetlands. For example, this Monod model was applied to nitrate removal in a California marsh (Karpuzcu and Stringfellow, 2012; Stringfellow et al., 2012). The half saturation constant was 48 mg/L, thus implying a first order model. The Monod model may also be applicable to the macrophytic community in a treatment marsh. Kadlec

(1997a) reported that this model was appropriate for phosphorus removal in marshes that depend upon the vegetation for removal, with a half saturation concentration of 0.8 mg/L.

Limiting Reactants

Some removal processes require a second reactant to achieve the transformation or removal process. Nitrification requires oxygen, and denitrification requires a carbon source. In such cases, removal rates may be limited by the supply of the second reactant, in addition to the concentration of the contaminant in question. Other treatment technologies utilize removal models that incorporate such supply limitations. For instance, for microbial nitrification:

$$k = k_{\max} \frac{C_{DO}}{K_{DO} + C_{DO}} \tag{7.59}$$

where: K_{DO} = dissolved oxygen half-saturation constant, g/m^3

The value of K_{DO} is suggested to be about 1.0 mg/L (U.S. EPA, 1993b). In a similar fashion, the carbon limitation for denitrification may be expressed as:

$$k = k_{\max} \frac{C_{OrgC}}{K_{OrgC} + C_{OrgC}} \tag{7.60}$$

where: K_{OrgC} = organic carbon half-saturation constant, g/m^3

The value of K_{OrgC} is suggested to be about 0.1 mg/L (U.S. EPA, 1993b). As a result, the effect of the carbon supply is forecast to be not large, unless that supply is very low.

In virtually all of the treatment wetland literature, supply limitations are implicit in the overall rate constants that are reported.

Return Fluxes and Background Concentrations

For many chemicals, the return rate to the water from the static compartments of the ecosystem – the soils and biomass – can be a significant (negative) contribution to the net rate. There are presently very few scientific studies to provide guidance on modeling this transfer. Therefore, the simplest option is used here: a constant (zero-order) return rate. The lumped rate equation for the net reduction of a chemical with no precursors is therefore written as:

$$J = kC - r^* = k(C - C^*) \tag{7.61}$$

where C^* = background concentration, g/m^3

k = removal rate constant, m/d (or with unit conversion, m/yr)

r^* = return rate of chemical, $g/m^2 \cdot d$

In the terminology of reaction engineering, the model is first order in the forward direction, and zero order in the reverse direction. The concentration C^* is achieved when there is no net uptake or conversion of the chemical in question, and is therefore termed the "background" concentration. When inlet waters have $C > C^*$, there will be a decrease with travel or time to this background concentration. When inlet waters have $C < C^*$, there will be an increase up to this background concentration.

There are several possible reasons for the existence of a real or apparent nonzero background concentration for a specific chemical constituent. First, there may be some portion of the incoming chemical that is resistant to storage or conversion in the wetland environment. This is particularly possible when the concentration measures a lumped set of species, one or more of which may be resistant to degradation in the wetland. For instance, total phosphorus in water may exist as particulate and dissolved, organic and inorganic forms. Some portion of the organic P may be highly resistant to uptake by the biogeochemical cycle. An extreme example would be the organophosphate pesticide diazinon, which is not efficiently degraded in wetlands. However, more benign sources may contain a biologically unavailable fraction, by virtue of the size and character of the molecules embodying the phosphorus (Proctor et al., 1999). Such P fractions may pass through the system untouched.

A second reason for a nonzero background concentration is the association of the chemical with particulates. Because the chemical is associated with (sorbed or incorporated in the structure) suspended particulate matter, a nonzero background level of TSS entails a nonzero background level of the chemical. For instance, at the Des Plaines, IL wetlands, the export of 8 mg/L of TSS carried 16 µg/L of phosphorus, which is due to a P content of 0.2% in the exported TSS. Although TSS is notoriously difficult to measure inside wetlands, background levels of 5–10 mg/L are commonly found in densely vegetated systems.

A third reason for a nonzero background concentration is a set of wetland processes that provide inputs distributed across the entire areal extent of the system. Groundwater discharge and rainfall may bring a specific compound into all portions of a wetland (Raisin et al., 1999). The chemical may be utilized in the biogeochemical cycle, which is also distributed across the entire wetland area. That same cycle can produce return of the substance to the water column, usually by the processes of decomposition and leaching (Kadlec, 1999a).

A fourth reason for nonzero background is seasonality. Dry seasons may be accompanied by loss of surface water, and dryout of the surface sediments in the wetland. The organics in the surface sediments may then be oxidized, resulting in the mineralization of previously organic-bound substances. Upon subsequent rewetting, these mineralized materials may dissolve and contribute to surface water concentrations.

Another factor influencing concentration gradients, and the possibility of plateaus, is hydraulic bypass of the reactive wetland environment. Bypassed water carries with it the inlet substances, which may reblend with treated water at downstream wetland locations (Kadlec, 1999a). This process cannot create a true plateau or background, but may easily lead to an inferred background concentration, derived from extrapolation of gradients in the upstream portion of the system. For some chemicals, very few treatment wetlands extend beyond the zone of total containment, and such extrapolation, via curve fitting, is the norm rather than the exception. As shown in Kadlec (2000), the nature of internal flow patterns leads to a data-fitted background concentration, which varies strongly with the hydraulic loading rate to the wetland. Higher loading rates lead to higher background concentrations for flow-through wetlands. In this chapter, methods will be set forth to minimize this effect.

Continuous Flow Wetlands

There are two prominent candidates for describing the longitudinal profiles of contaminant concentrations: the PFD and TIS models. Although the PFD model has sometimes been advocated for wetlands (e.g., Pardue et al., 2000; Wang and Chen, 2017), it is doubtful that it is the most appropriate model of comparable complexity. The DTDs for FWS wetland systems are characterized by a large amount of apparent dispersion, with $0.07 \leq D/uL \leq 0.35$ (Kadlec, 1994). The dispersion coefficient describes eddy transport of water elements both upstream and downstream. In marsh flows, such mixing may not occur, because flow is often predominantly laminar.

PFD longitudinal concentration profiles are available in the literature (Danckwerts, 1953), but have rarely been examined for wetland situations. Kadlec and Wallace (2009) provide discussion in connection with wetlands. In a treatment wetland, the PFD model

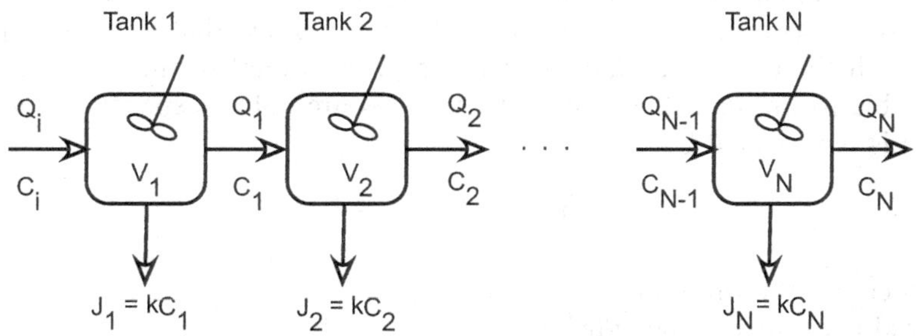

FIGURE 7.13 The tanks-in-series model for wetland hydraulics and first-order removal.

conceives of swirls, which cannot move back into the inlet distribution works, nor move forward into the outlet collection works. As a consequence, the longitudinal concentration profile is predicted to display an instantaneous drop at the wetland inlet. For D/uL = 0.2 and k/q = 3, the decrease at the inlet is 30%. This unrealistically large concentration drop has not been observed in practice, and hence the PFD model is not considered further here.

The TIS Model

In its simplest form, the TIS model accounts for the first order removal of a contaminant. Water passes through N tanks in series and loses contaminant in each (Figure 7.13). For the case of no water losses or gains, the steady flow contaminant mass balance for the jth tank is:

$$(QC_{j-1} - QC_j) = kAC_j \tag{7.62}$$

where A = area of a tank, m^2
k = reaction rate coefficient, m/day
C_j = concentration leaving tank "j", and entering tank "j+1", g/m^3
Q = flow, m^3/d
For the entire sequence of tanks, these mass balances combine to:

$$\frac{C_{out}}{C_i} = \left(1 + \frac{k}{Nq}\right)^{-N} \tag{7.63}$$

where q = system hydraulic loading rate, m/day
C_{out} = concentration leaving tank N, g/m^3
C_i = inlet concentration, g/m^3
Note that there are two reaction parameters in this model: the rate constant "k" and the hydraulic parameter "N." In the limit of a very large number of TIS, the plug flow limiting form is attained:

$$\frac{C_{out}}{C_i} = \exp(-k/q) \tag{7.64}$$

Equation 7.63 forms the basis for the models that are used in following chapters, but there are many necessary modifications to describe the marsh environment.

7.6 AREA OR VOLUME SPECIFICITY

In general terms, chemical reactions may be homogeneous, meaning they take place throughout the water column, or heterogeneous, meaning they take place on solid surfaces within the water column. In the first instance, they are volume-specific; and in the second instance they are area-specific. Thus far, it has been presumed that the marsh processes are area-specific, with the reactive surface area being proportional to the wetland surface area. It is also possible to presume that removal processes are volume-specific, which for a first-order reaction is represented by:

$$J_V = k'_V f(S)C = k_V C \tag{7.65}$$

where: C = concentration, g/m^3
J_V = removal per unit volume, $g/m^3 \cdot d$
k'_V = intrinsic rate constant, 1/d
k_V = pseudo first-order rate constant, 1/d
The relation between the two different rate coefficients is:

$$k = hk_V \tag{7.66}$$

where h = water depth, m
The fundamental difference between the use of k and k_V arises from the assumption of constancy of k-values or k_A-values, as follows:

- If k_V is assumed constant, then deeper water (greater h) provides more detention time, without any penalty of reduction of pollutant removal.
- If k is assumed constant, then deeper water (greater h) provides more detention time, but with a penalty of reduction of pollutant removal, which compensates for, and removes the advantage of, the greater detention.

This is a very significant difference that has serious implications for performance modeling and design. The fundamental consideration is the nature of the vertical profiles within the water column, as they relate to removal processes. The computational specter of considering a third dimension (vertical) is daunting, and simplicity demands that we not abandon vertical averaging in the water column.

Removal Processes

The various removal processes have been briefly discussed above, and their dependence upon vertical profiles can be summarized.

- Plant uptake is area-specific.
- Accretion is area-specific.
- Sedimentation removes particulates and sorbed chemicals. It is area-specific.
- Volatilization is area-specific.
- Sorption is area-specific with respect to the wetland soils.
- Photodegradation depends upon the vertical distribution of photosynthetically active radiation (PAR) within the water column.
- Microbially mediated processes, or more generally, mesophytic processes, depend upon the vertical distribution of epiphytes within the water column.
- Mesophytic processes depend upon dissolved oxygen (DO) and pH, as well as PAR, which can all possess vertical distributions.

Thus, it is seen that both area-specific and distributed volume-specific processes may be involved. The last three items require a further examination of vertical profiles.

As described in Chapter 5, the distribution of solid surfaces in marsh water column is generally fairly uniform. Therefore, increasing the water depth increases the immersed surface area for attachment of mesophyton (biofilms), and consequently more removal potential. Therefore, it might seem logical that if mesophyte uptake is the principal factor in treatment, then the volumetric rate equation is indicated. But, the *functioning* of the mesophyte community depends strongly upon other factors, including DO, PAR, temperature and pH. Thus, the depth-averaged removal rates may not increase with water depth, as the immersed area availability might indicate.

The vertical profiles of these determinants of mesophytic uptake have been studied in the ENRP treatment marsh (Chimney et al., 2006). The four major community types, EAV, SAV, FAV and open water (OW), displayed differing vertical patterns (Figure 7.15). Temperature profiles were nearly vertically uniform at this site, but other mesophytic drivers had vertical trends, whose strength depended on community type.

Dissolved oxygen was close to vertically uniform for EAV and FAV, with surface saturation values of about 25% (Figure 7.14A). But SAV and OW had large DO decreases with depth over the top 30 cm of the water column, with surface saturation values of about 120% and 70% respectively. Microbial processes that rely upon DO, such as biochemical oxygen demand (BOD) reduction or ammonia reduction, would be greatly emphasized in this top layer for SAV and OW. Conversely, microbial processes that rely upon low DO, such as denitrification, would be emphasized in the bottom layer of water.

Water pH was close to vertically uniform for EAV and FAV, with values of about 7.0 ± 0.2 (Figure 7.14B). OW pH values were also vertically uniform, but higher at 7.6 ± 0.1. SAV showed a steep gradient, from 8.3 at the surface to 7.3 at 35 cm depth. This affects the processing of calcium by periphytic biofilms, which in turn influence the co-precipitation of phosphorus.

Photosynthetically active radiation has the strongest dependence on community type and depth. The incident PAR at the water surface is unabated from incoming solar vales for SAV and OW, and in this Floridian setting is in the range of 120–180 $mol/m^2 \cdot d$. The presence of dense EAV reduces this by about 95% (Chimney et al., 2006; Crepeau and Miller, 2014), and a complete cover of FAV would do the same. Attenuation in the water column, which is exponential, reduces PAR by 80–95% in the first 40 cm of depth (Figure 7.14C). Therefore, photodegradation would be confined to that upper layer. Similarly, phototropic biofilm activity (algal) would also be confined to that layer.

These layering effects point to the concept that mesotrophic processes generally occur in a water layer of about 40 cm. Extra depth, although providing more attachment sites, will not materially contribute to their function. Taken together, these factors support a strong argument for use of an area-specific removal model.

Supporting Wetland Data

Depth studies have not been conducted for full-scale treatment marshes, so it is necessary to explore the rate coefficients via mesocosms and test cells, and by sequential depth variability. Side-by-side studies were conducted at Arcata, CA, in 12 372-m2 EAV test cells, for a period of two years (Gearheart et al., 1983). Analysis of their data showed that k_V decreased with increasing depth. According to Equation 7.66, if k is constant, then k_V values should be inversely proportional to depth. In any case, we can test the Arcata data for depth dependence using a power law:

$$k_V = ah^{-n} \tag{7.67}$$

where a = volumetric k_V at one meter depth, 1/d
n = exponent for depth

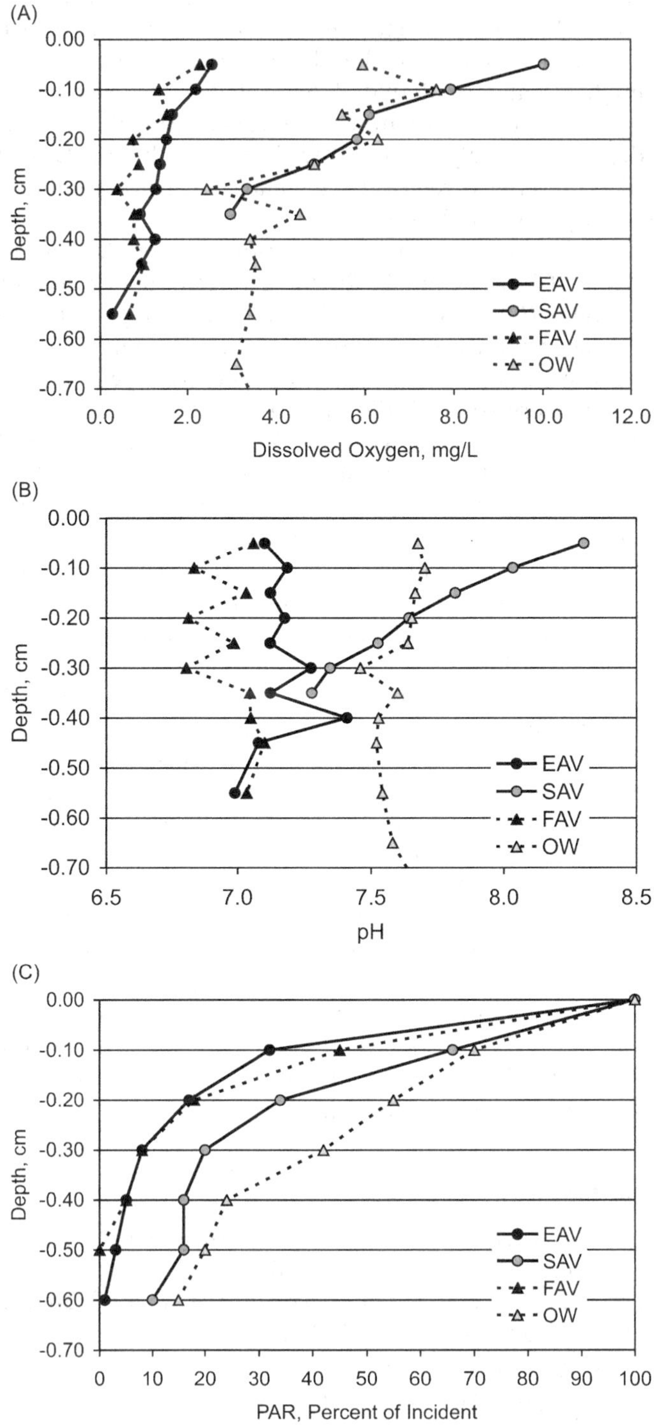

FIGURE 7.14 Vertical profiles of environmental factors in the Everglades Nutrient Removal Project. Data points are averages over a three-year period. Data courtesy of M.A. Chimney, as summarized in Chimney et al. (2006).

The calculated k_V values for four typical contaminants decrease rapidly with depth, indicating poorer treatment as depth increases (Table 7.2).

A side-by-side depth study of phosphorus removal in SAV mesocosms over a one-year period in Florida demonstrated similar k_V depth effects (DBE, 1999). As well, the depth dependencies of DO and pH were similar to those of Figure 7.15. The mesocosms were planted with *Najas* and *Ceratophyllum*, and were tracer-tested to establish the NTIS. This data is shown in Figure 7.15, along with the data fit to Equation 7.67, with the exponent n set to 1.00. The depth dependence of k_V is striking confirmation of the preference for an areal rate coefficient, even for SAV systems with surface area distributed in the water column.

Another side-by-side depth study of phosphorus removal in EAV (*Typha domingensis*) test cells over a one-year period also demonstrated similar k_V depth effects (SFWMD, 2003). The test cells were planted with *Najas* and *Ceratophyllum*, and were tracer-tested to establish the NTIS. This data is shown in Figure 7.16, along with the data fit to Equation 7.67, with the exponent n set to 1.00. The depth dependence of k_V is striking confirmation

TABLE 7.2

Volumetric removal rate constants in side-by-side pilot-scale wetlands operated at different depths.

		Concentration		k_V at one meter		
		Inlet	Outlet	depth, 1/d	Exponent n	R^2
TSS	mg/L	33.8	5.8	0.66	1.38	0.36
BOD	mg/L	24.5	13.3	0.081	1.58	0.46
Ammonia	mg/L	14.1	11.7	0.021	0.87	0.43
Fecal Coliforms	#/100ml	3,049	708	0.229	1.00	0.56

Source: Gearheart et al. (1983).

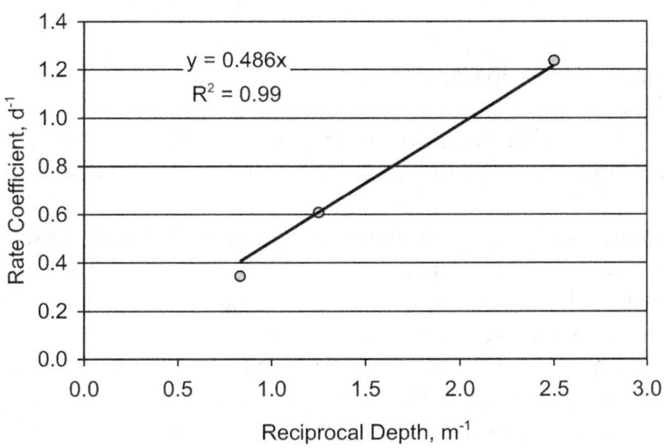

FIGURE 7.15 Depth variation of the phosphorus volumetric rate coefficient k_V for SAV mesocosms. Data from DBE (1999).

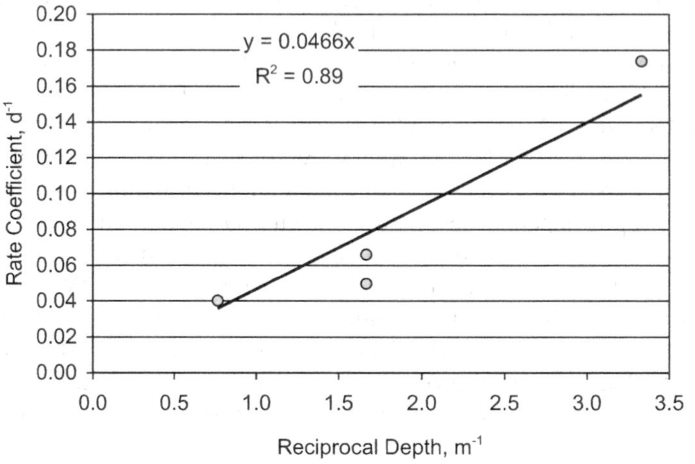

FIGURE 7.16 Depth variation of the phosphorus volumetric rate coefficient k_V for EAV test cells. Data from SFWMD (2003).

of the preference for an areal rate coefficient, for these EAV systems with surface area mostly attributable to stems (culms) in the water column.

Although these wetland studies are not exhaustive or conclusive, they do provide very strong indications that the better choice for a removal rate equation is the area-specific version. That choice is utilized in this book.

Consequences

Either k or k_V can be used to represent a data set or be used in design. However, the use of k_V requires the accompanying information on water depth (h) in the sizing equations. Equation 7.63 may be written for either area- or volume-specific rate coefficients:

$$\frac{C_{out}}{C_i} = \left(1 + \frac{k}{Nq}\right)^{-N} = \left(1 + \frac{k_V \tau}{N}\right)^{-N} \tag{7.68}$$

where q = system hydraulic loading rate, m/day
τ = detention time, days

There are choices for which detention time is used in the volumetric version, but all of these require the wetland water depth (h). In contrast, the areal version does not depend on the water depth.

In the design mode, the wetland designer using the volumetric version may be tempted to select a smaller footprint with deeper water, when in fact he should recognize that the k_V must then be reduced. The k_V depth dependence usually means that more detention time created by deeper water is counteracted by a decrease in the volumetric rate constant.

Wetland water depth is by no means a simple, easily determined parameter. It depends upon accurate bathymetry and upon the shape of the flowing water surface. Often, as-built (dry) bathymetry, or pre-project elevation surveys, are not good predictors of the actual operational conditions. For example, 10 test cells of the ENRP were resurveyed after being in operation for a few years (see Kadlec and Wallace, 2009, chapter 2). These were operating at nominal depths of 15 to 63 cm, but had actual depths ranging from 12 to 96 cm, and the RMS depth error was 32%. This is a surprisingly high variability for these small (0.2 ha) wetlands.

The water surface elevation is not a simple, easily determined parameter either. Weir settings give level, stable water surfaces for small, steady flow systems, but many other situations cause tilted, nonlevel surfaces. This has been identified as a major factor in the determination of wetland detention times (Jadhav and Buchberger, 1995) and has been discussed in detail in Chapter 3. Vegetation type and density, together with patch structure, combine with bathymetry to create a fairly detailed but well-understood calculation procedure. In the case of event driven systems, the added complexity of system dynamics becomes important.

Nonetheless, data analysis and design require knowledge of the water depth, because of the implied changes in the type of vegetation that may be fostered and maintained. In fact, it is often not possible to deepen the water and retain the same ecology, because of the hydropattern requirements of vegetation.

7.7 ADAPTING THE TIS MODEL TO WETLANDS

In this section, the steady state, or temporally averaged, condition is considered. The purpose is to incorporate as many of the known modifiers of the NTIS model as can be easily accommodated.

RETURN FLUXES

Because some contaminants display a background concentration, we first broaden the TIS concept to include such background concentrations. The appropriate rate expression is Equation 7.61. It can easily be shown that the result analogous to Equation 7.63 is:

$$\frac{(C - C^*)}{(C_0 - C^*)} = \left(1 + \frac{k\tau}{N}\right)^{-N} \tag{7.69}$$

It is noted that we could retain the return flux r* instead of C* = r*/k.

SEGREGATED OR MIXED FLOW?

There are two extreme cases that can yield the same DTD at the wetland outlet, as given in Equation 7.63: completely segregated flow and completely mixed flow. Segregated flow embodies the idea of numerous parallel, noninteracting stream tubes, and is represented by Figure 7.11C. If we imagine the wetland cut into a number of sections (N) along the flow path, the segregated flow (stream tube) concept means that each of those sections has a DTD represented by N TIS, and a detention time of τ/N.

Mixed flow is based upon the idea of transverse and vertical mixing at all locations along the flow direction within the wetland. If the wetland is cut into N sections, each will have a DTD represented by one tank, and a detention time of τ/N.

Although we cannot distinguish these two situations based on the outlet system DTD, there are major differences in the internal concentration profiles.

In segregated flow, the longitudinal concentration profile is given by:

$$\frac{(C(y) - C^*)}{(C_0 - C^*)} = \left(1 + \frac{ky}{Nq}\right)^{-N} \tag{7.70}$$

where y = fractional distance along the flow path, –

This profile is a smoothly decreasing concentration that starts at the inlet concentration and levels off at a plateau value of C*. In this case, all segments of the flow path behave individually as N stirred tanks. This situation could physically correspond to noninteracting parallel flow paths through the wetland.

In mixed flow, the longitudinal concentration profile is given by:

$$\frac{(C(y)-C^*)}{(C_0-C^*)}=\left(1+\frac{k}{Nq}\right)^{-yN} \tag{7.71}$$

where yN = number of tank along the flow path, –

This profile is also a smoothly decreasing concentration that starts at the inlet concentration and levels off at a plateau value of C*. In this case, all N segments of the flow path all behave individually as one stirred tank. This situation could physically correspond to high levels of vertical and lateral cross mixing along the flow path.

It is easily seen that for y = 1, Equations 7.70 and 7.71 give the same result

Both cases reduce to the same exponential form in the limit of very large N:

$$\frac{(C(y)-C^*)}{(C_0-C^*)}=\exp(-k\tau y) \tag{7.72}$$

The internal tracer responses along the flow path can distinguish these two situations, because for mixed flow, N increases with distance, while for segregated flow, N is constant.

Data sets for internal tracer DTDs are rare, but those available favor the mixed flow idea. The Hemet, California treatment wetland displayed a linear increase in NTIS with distance for two different flow paths (Figure 7.17) (Crohn et al., 2005). Those wetlands were structured with banded vegetation. A SAV treatment marsh in Florida, with relatively uniform vegetation, also displayed this linear increase in NTIS along the flow path (DB Environmental, unpublished data).

If there is channelization, or fringing vegetation, then there will be a small number of macroscopic preferential flow paths. That situation is considered separately.

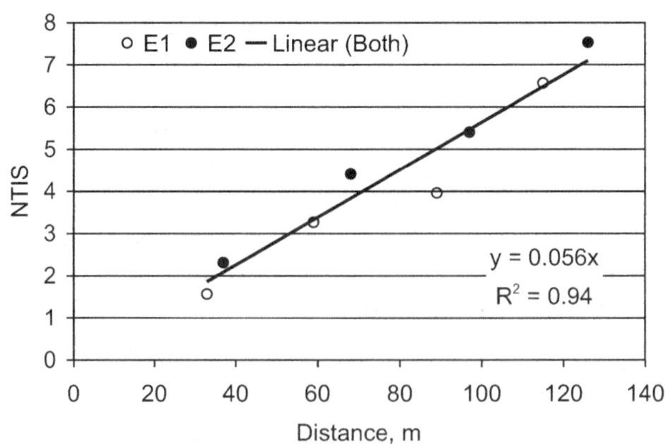

FIGURE 7.17 Tracer response versus distance for two flow paths at the Hemet, California treatment wetland. Data from Crohn et al. (2005).

More on Longitudinal Profiles

Equations (7.70) and (7.71) represent smoothly decreasing longitudinal concentration profiles that start at the inlet concentration and level off at a plateau value of C*. The parameters of the model can be determined from an analysis of longitudinal transect data for C and y. In practice, there are difficulties, which arise from three principal reasons. First, if there is time variability in the inlet concentration, then profiles reflect that effect, together with a transport delay. In such a dynamic situation, profile data must be averaged over a sufficiently long period to determine mean behavior. Second, it is typically impossible to determine where to take a sample across the width of the wetland so that it is spatially representative. There is usually a bias toward open water areas, because of the ease of obtaining the sample. However, such open water is often a preferential flow path (short-circuit). Therefore, multiple parallel paths should be sampled. This effect has been clearly elucidated by Dierberg et al. (2005). Thirdly, it is not enough to collect a set of spatially uniform samples in an effort to gain access to all flow paths. Spatially uniform sampling across the flow direction will produce an average that is substantially different from the mean (flow-weighted) concentration at that distance (Levenspiel and Turner, 1970; Kadlec, 1999b). Despite these obstacles, transect studies have been conducted, because there is a very large knowledge gain potential.

The size of large wetlands invites examination of the internal concentration profiles. Three distinctly different situations can be present. If the wetland is compartmentalized, with structures conveying flow from one cell to the next, then those intercell flows can be sampled for pollutant concentrations. If flow rates are also measured, then concentrations in the intercell flow can be flow-weighted. There are numerous examples of cell-wise transect monitoring of treatment marshes (e.g., the Floridian STAs, the Texas Shannon wetland, the Imperial California wetland).

In other instances, the wetland may contain transverse deep zones, which promote some degree of mixing of the varying flow elements. Internal flow measurements are not available in the case of deep zones. Several treatment marshes have been so sampled (e.g., the Tres Rios demonstration wetlands, the Sacramento, California, wetlands).

If the wetland is not compartmentalized, and does not contain deep zones, then a grid of interior points may be sampled for TP and other parameters. Often, a spatially uniform grid is sampled, with cross-flow points representing a specific distance from the inlet. In general, it is not possible to measure the flow rate that may be representative of a given sample point. Flow weighting is not possible. Studies at STA2 Cell 3 of the Everglades protection wetlands provide an example of internal transect measurements in large wetlands (Figure 7.18). As described in Juston et al. (2013), measurements were done along five parallel rows in the flow direction, at eleven distances along that flow direction. Internal tracer studies show that the flow was transversely mixed, with NTIS = 6 (see Equation 7.71). Nonetheless, considerable variability was present from row to row, with an average coefficient of variation of 25%.

Incorporating Volumetric Efficiency

As seen in the discussion of tracer testing in this chapter, there may be considerable difference between the nominal detention time and the mean tracer (actual) detention time, exemplified by the volumetric efficiency, e_V. Table 7.1 shows that for a sampling of wetlands, on average, $e_V = 0.70$. If we knew in advance what the volumetric efficiency would be, then we could use the intrinsic rate coefficient k'. Often e_V is not known, and an option exists to lump volumetric efficiency with the intrinsic rate coefficient.

$$k = e_V h k_V = e_V k' \tag{7.73}$$

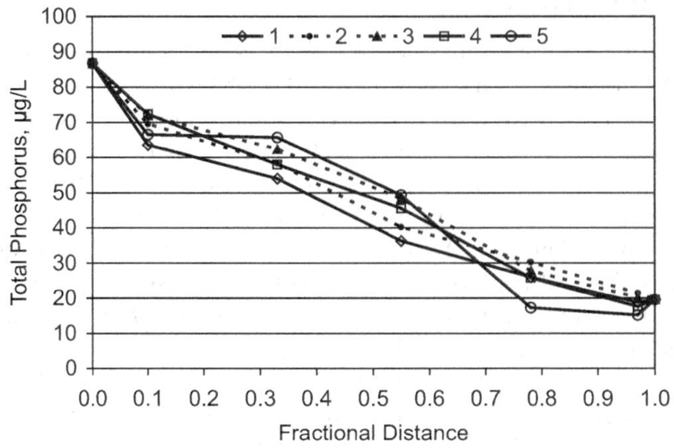

FIGURE 7.18 Average total phosphorus on five transects parallel to flow in STA2C3. Twenty transects taken over five years are represented. These are grab samples taken within the wetland. (Data courtesy of DB Environmental),

where k' = intrinsic rate coefficient, m/yr

k = nominal rate coefficient, m/yr

During data analysis, either may be used, if a tracer test is available. Equation 7.73 indicates the resultant calibrated values of k will differ markedly, depending upon which alternative is chosen. In this book, the nominal rate coefficient is chosen, because tracer testing may not be available to provide the actual detention time. This implies that the volumetric efficiency is absorbed into the rate constant.

Previous treatment wetland literature has sometimes elected to retain e_V as an independent design variable (e.g., Crites et al., 2014). These sources identify e_V as the wetland porosity, meaning the fraction of the water column unblocked by plant materials, and suggest that $0.7 < e_V < 0.9$. However, the data presented in Chapter 5 indicate that marsh porosities are usually greater than 95%. And, results discussed earlier in this Chapter indicate that e_V is mostly related to internal flow patterns, not porosity. The dependences of e_V on wetland structure are discussed in detail in Chapter 14.

A good portion of the treatment wetland literature (Crites and Tchobanoglous, 1998; U.S. EPA, 1999; WEF, 2001) also utilizes a volumetric rate coefficient. That choice also necessitates an estimate of the wetland water depth. Removal rate calculations under the volumetric assumptions require the designer to choose three parameter values: depth, volumetric efficiency and the intrinsic rate coefficient.

Water Losses and Gains

The several components of the water budget can be accommodated in the TIS model framework. Flows other than the source water are likely to be important for treatment.

Water Budget

The annual water budget forms the basis for a first approach to understanding pollutant reductions and the area required. Under the assumption of a constant water level, the flows out of the wetland may be computed from the inflow and the meteorological data. Some infiltration is assumed here, if leakage is present in the wetland. The relevant equations are:

$$Q_{out} = Q_{in} + A \cdot (P - ET - I) \tag{7.74}$$

or

$$q_{out} = q_{in} + (P - ET - I) \tag{7.75}$$

where A = wetland area, m^2
ET = evapotranspiration rate, m/d
I = infiltration rate, m/d
P = precipitation rate, m/d
q = hydraulic loading rate, m/d
Q = flow rate, m^3/d

The inlet hydraulic loading will be increased by rainfall (ca. 0.5–1.5 m/yr), and decreased by evapotranspiration (ca. 0.5–1.5 m/yr). However, there may be seasonal imbalances. These amounts are important if the wetland is to have a very low hydraulic loading, or correspondingly, a long detention time. The ratio $\alpha = (P\text{-}ET)/q_i$ is the atmospheric augmentation.

Evapotranspiration (rain) has two effects: lengthening (shortening) of detention time, and concentration (dilution) of dissolved constituents. The use of an average flow rate compensates for altered detention time, but not for dilution or concentration. The fractional error in a first order model prediction of concentration, due to flow averaging, is approximately equal to α, for $\alpha > -0.5$. Thus, if 25% of the inflow evaporates, use of a first order model with average flow predicts concentrations 25% lower than required by the mass balance. If rain adds 25% to the flow, use of a first order model predicts concentrations 25% higher.

A possibly important feature of ET is that transpiration carries water into the root zone (see Chapter 6). Therefore, if the pollutant mass balances are done on surface water, then the transpiration component is the same as infiltration: it carries materials into the soil. For a fully vegetated wetland, about half of ET is attributable to transpiration on an annual basis, and transpiration may be nearly the entire loss during the growing season. Here we represent transpiration as a fraction of ET:

$$T = ET - E = \alpha ET \tag{7.76}$$

where: E = evaporation, m/d
T = transpiration, m/d
α = transpiration fraction of ET, –

The pollutant mass balances will be conducted on a cells-in-series basis (see Figure 7.19). Results of the overall water mass balance are apportioned to the cells according the chosen number of TIS. For the first unit in the series:

$$Q_1 = Q_{in} + A_1(P - ET - I) \tag{7.77}$$

where: A_1 = area of tank number 1, m^2
Q_1 = flow rate out of tank 1, m^3/d

Flows are thus computed sequentially, from inlet to outlet, for the number of tanks chosen (NTIS).

The input data requirements for the tank water mass balances are:

1. Inflow (Q_{in});
2. Rain (P);
3. Evapotranspiration (ET);

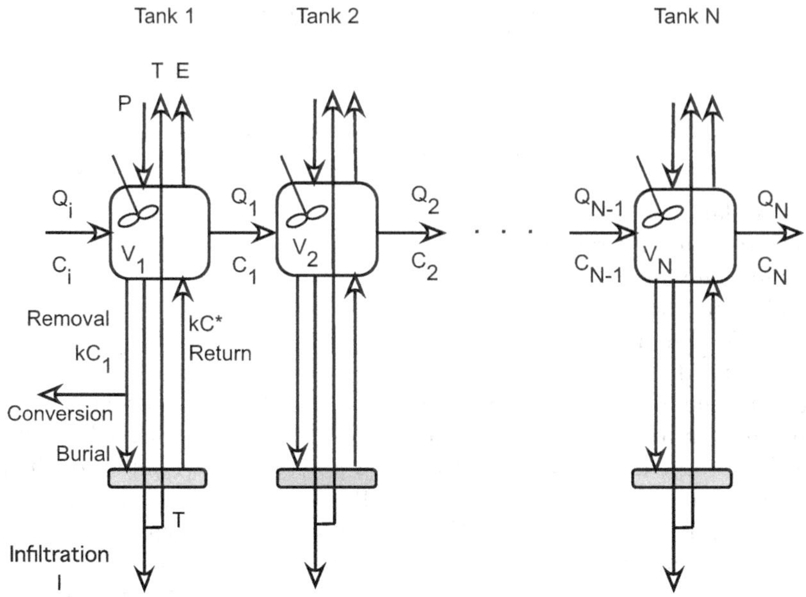

FIGURE 7.19 Conceptual TIS model for pollutant reduction.

4. Infiltration (I);
5. Area (A);
6. Number of tanks (N).

An example is detailed in Table 7.3. This hypothetical example is configured to include small rainfall, considerable evapotranspiration, and some infiltration; in other words, a leaky arid region system. The net loss of incoming water is 41%, of which 44% is ET and 56% is infiltration.

Pollutant Mass Balance

The TIS model is then carried forward via a sequential calculation of pollutant concentrations for each "tank" in the chosen hydraulic model.

A first order areal model with rate constant k is selected, with a wetland background concentration C* as necessary. A volumetric first order model may also be chosen, for which $k = \varepsilon h k_V$. The pollutant mass balance for the first of the wetland segments, designated by subscript "1", for steady state, nonuniform flow is:

$$Q_1 C_1 = Q_{in} C_{in} - I \cdot A_1 C_1 - \alpha ETA_1 C_1 - k \cdot A_1 \cdot (C_1 - C^*) \tag{7.78}$$

In this simple version, rainfall has been assumed to have a zero pollutant concentration, but it is easy to add an atmospheric input of the pollutant if it exists. Infiltration is assumed to occur at the tank outlet concentration. Transpiration flow of the contaminant has been included. Combining (7.77) with (7.78) gives the concentration exiting hypothetical segment number one:

$$C_1 = \frac{Q_{in} C_{in} + k \cdot A_1 \cdot C^*}{Q_1 + (\alpha ET) \cdot A_1 + I \cdot A_1 + k \cdot A_1} \tag{7.79}$$

TABLE 7.3

Water budgets for a treatment marsh design example. Sequential tank-to-tank calculations are based on Equation 7.77. This example has a combined detention time of 19.2 days, whereas averaging the flows leads to 17.2 days. The required input data is shown in bold.

Parameters

Flow Rate	**5,000**	m^3/d
Rain	**0.05**	cm/d
ET =	**0.4**	cm/d
Infiltration =	**0.5**	cm/d
NTIS =	**3**	
Area =	**24**	ha
Area/Tank =	8	ha
Depth =	0.3	m
Porosity =	0.95	
Volume/Tank =	22,800	m^3

Water Budget

		Inflow	Tank 1	Tank 2	Tank 3 = Outflow	Total
Flow Rate	m^3/d	5,000	4,320	3,640	2,960	
Rain	m^3/d		40	40	40	120
ET	m^3/d		320	320	320	960
Infiltration	m^3/d		400	400	400	1,200
Nominal Detention	days	13.7	5.3	6.3	7.7	19.2
HLR	cm/d	2.08	5.4	4.55	3.7	

or

$$C_1 = \frac{q_{in}C_{in} + k \cdot C^*}{q_1 + (\alpha ET) + I + k} \tag{7.80}$$

Note that the hydraulic loading rates in (7.80) are the tank loading rates, not the system loadings. This computation is then repeated sequentially for the remaining segments, in each case using the outlet concentrations and flows from the preceding unit. The wetland outlet concentration is that exiting from the final hypothetical segment.

Outgoing pollutant loads are calculated as the product of the volumetric outflow (m^3/d) and the outflow concentration (kg/m^3).

The additional input data requirements for the pollutant mass balances are:

7. Input concentration (C_{in});
8. Background concentration (C^*);
9. Rate coefficient (k);
10. Transpiration fraction (α);

TABLE 7.4

Contaminant budget for a treatment marsh design example. This example is based upon the water budget shown in Table 7.3. Sequential tank-to-tank calculations are based on Equation 7.80. The required input data is shown in bold.

Parameters

Ci =	**2.00**	mg/L
C* =	**0.01**	mg/L
k =	**10**	m/yr
k =	0.0274	m/d
Area/Tank =	80,000	m^2
T factor (α) =	**0.5**	

Contaminant Mass Balance

		Inflow	Tank 1	Tank 2	Tank 3 = Outflow	Removed	Reduction
Concentration	mg/L	2	1.42	0.96	0.62		69%
Load In Surface Water	kg/yr	3,650			666	2,984	82%
Load Infiltrated	kg/yr		207	140	90	437	15%
Load Stored	kg/yr					2,547	70%

The hypothetical example for the water budget of Table 7.3 is extended in Table 7.4 to illustrate these pollutant removal calculations, continuing the choice of P = 3 TIS. A hypothetical contaminant is chosen, entering the wetland at 2.00 mg/L. The value of C* is chosen to be low, 0.01 mg/L. The rate constant is selected to be k = 10 m/yr. Because the transpiration flux is presumed to draw contaminant into the root zone, the fraction of ET that is transpiration must be selected, and in the example is picked as $\alpha = 0.5$.

The computed concentration in the surface outflow water is $C_O = 0.62$ mg/L, or a concentration reduction of 69% (Table 7.4). However, the existence of ET losses and infiltration create a different result for mass removal: 82% of the contaminant entering does not leave with surface water. If infiltration water departs vertically downward without further treatment, then 15% of the contaminant mass removal is due to infiltration. Thus, it is seen that load reduction and concentration reduction are two different goals, which may lead to different designs. Both of these types of design sizing require the mass balance computations for water, and rate calculations for pollutants.

Mixture Weathering

Equation (7.69) represents the reduction of a single compound on transit through a treatment wetland. However, many contaminants are in fact mixtures. In almost all instances, water quality parameters are measured by procedures that lump individual chemical compounds into an overall, or total concentration for that class of materials. Biochemical oxygen demand (BOD) and total suspended solids (TSS) are examples of such lumping. It is clear that the individual components of such mixtures may be degraded or removed at different rates, and that there is a corresponding difference in removal rate constants (Crites and Tchobanoglous,

1998; Tchobanoglous et al., 2000; Shepherd et al., 2001; Kadlec, 2003a). There is therefore a distribution of rate constants across the various mass fractions of the mixture. Such a distribution may be discrete, for the case of a countable and very small number of individual compounds; or it may be continuous, for the case of a very large and possibly uncountable number of constituents. Combinations of both types of distribution are also common, such as for total nitrogen (TN). TN consists of a few separately identifiable compounds (nitrate, ammonia) plus lumped classes of compounds (particulate N, organic N). Total phosphorus (TP) is comprised of particulate (PP), and dissolved organic (DOP) and soluble reactive (SRP) forms. As water containing such a mixture passes through the wetland, its composition changes because different fractions of the mixture are reduced at different rates. The mixture becomes "weathered," a term coined to describe the selective stripping of light volatile materials upon exposure to outdoor environments.

Each fraction of the lumped material will in general possess its own k-value. Therefore, there is a distribution of k-values, designated by f (k):

$$\begin{aligned} f(k)dk &= \text{massfractionofmaterial} \\ &\text{withrateconstantintherangektok} + dk \end{aligned} \quad (7.81)$$

This k-value frequency distribution across the mass fractions of the lumped material is termed the kVD. It may also be shown for gamma distributions of k-values (Kadlec, 2003a) that the average k-value at any time during the reduction process is:

$$k = \frac{k_o}{(1 + \beta t)^n} \quad (7.82)$$

where: k = rate constant during the weathering process, m/yr

k_o = inlet rate constant, m/yr

n = mixture k-value distribution breadth parameter, –

t = length of time the mixture has weathered in the wetland, d

β = mixture k-value distribution weathering parameter, d^{-1}

In the wetland environment, the DVD and the kVD interact to produce the overall observed reduction in a lumped category of pollutants. However, batch testing eliminates the DVD effect, since there is no distribution function for batch time. The DVD effect is also removed in theory for the (unachievable) ideal of true plug flow.

It has been noted that observed weathering behavior in real wetland situations may be represented by the TIS model equation (7.69), wherein the parameter values are relaxed to become fitting parameters (Kadlec, 2003a). The *relaxed TIS concentration model* is therefore here defined to be:

$$\frac{C - C^*}{C_i - C^*} = \frac{1}{(1 + k_m/Pq)^P} \quad (7.83)$$

where: k_m = modified first-order areal constant, m/d

P = apparent number of TIS

A gamma DTD/kVD model was used to generate concentration profiles for an incoming $k_o = 0.2\ d^{-1}$, and ranges $1 \leq n \leq \infty$ and $1.5 \leq N \leq \infty$ (Table 7.5). That value of k_o is appropriate for nutrients, such as total phosphorus (TP) or total nitrogen (TN). That computer data was then fit with equation (7.83), generating k_m and P values for each profile. The value of the modified rate constant was constant across all distributions, $k_m = 0.198 \pm 0.006$ (mean ± sd). The apparent number of TIS varied systematically, but always $P \leq N$ and $P \leq n$

TABLE 7.5

Apparent TIS numbers (P-value) for hypothetical double continuous distributions. The relaxed TIS model has been fit to doubly distributed behavior, with "N" stirred tanks and a kVD parameter of "n".

	n = 1	n = 4	n = 8	n = ∞
N = 1.5	0.54	0.99	1.13	1.50
N = 2	0.68	1.37	1.62	2.00
N = 4	0.86	2.19	2.86	4.00
N = 8	0.95	2.90	4.29	8.00
N = ∞	1.00	4.00	8.00	

(Table 7.5). In other words, the presence of a weathering mixture will cause a reduction in the N-value determined from an inert tracer experiment.

To illustrate the effect of model choice, the progression of concentration reduction along a flow direction is compared for plug flow, NTIS and PTIS models (Figure 7.20). Plug flow produces a linear decline on the semi-logarithmic plot for first order disappearance kinetics. If the hydraulics are represented by four TIS in a tracer test, and the contaminant experiences first-order decay, the decline is no longer linear, because rates slow at longer detention time because of the early exit of some fraction of the material. If in addition the material is a mixture that undergoes weathering, the apparent number of tanks is reduced, to PTIS = 2, for instance. There is an even greater slowing of the decline, because the rate constant decreases as travel time progresses.

As seen in Figure 7.20, there is not much difference in concentration reduction among the various models for low removals, for example, when removal is less than 50%. However, there is a very large difference when removals are in the high range, and there can be as high as a factor of ten difference in concentrations when reductions are above 99%. This

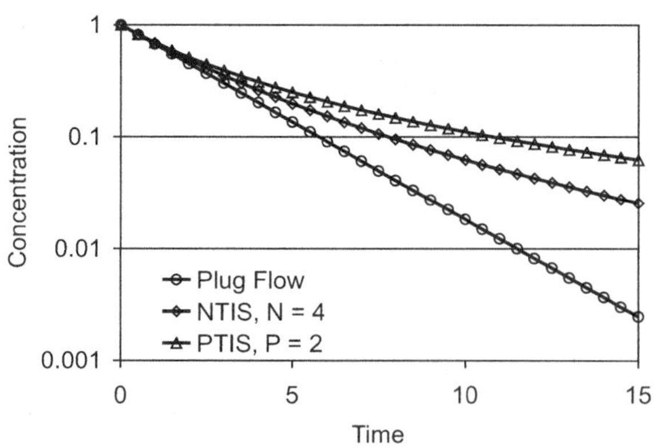

FIGURE 7.20 Comparison of time progressions along a flow path for three models: plug flow, *N*TIS, and *P*TIS.

large sensitivity to the P-value implies that high pollutant reductions cannot be achieved unless the wetland is hydraulically very efficient.

7.8 MODEL UTILIZATION

Temperature and Season

Temperature effects on k or k_V have often been summarized by use of the modified Arrhenius equation:

$$k_T = k_{20}\,\theta^{(T-20)} \tag{7.84}$$

where k_T is the rate constant at temperature T (°C) and k_{20} is the rate constant at 20°C. Values of the temperature correction factor (θ) may be estimated for data sets with adequate operational temperature data. However, it should be noted that incorrect historical presumptions still pervade the treatment wetland literature.

Microbial processes have been studied in several types of "conventional" wastewater treatment devices, including activated sludge plants and trickling filters. The Arrhenius factors for some contaminant reductions have been documented for those situations (see Metcalf and Eddy Inc, 1991, for instance). This prior information for other technologies led early wetland workers to assume that the temperature coefficients would be the same for wetlands, which they viewed as microbial reactors. For instance, a BOD temperature coefficient of $\theta = 1.06$ was assumed by Reed et al. (1995), and has been propagated in other subsequent publications (Crites and Tchobanoglous, 1998; Water Environment Federation, 2001; Crites et al., 2006, 2014). However, examination of treatment wetland data (and other natural system data) does not yield such a value for BOD (Kadlec and Reddy, 2001). For 23 FWS wetlands they found an average $\theta = 0.983$. The situation in wetlands is apparently more complicated than just microbial processing.

The situation for nutrients is even more complex, because light nutrient loadings are strongly influenced by plant and algal uptake on a seasonal basis. Microbial activity follows an annual pattern with a peak in mid-summer, but uptake peaks earlier, in spring. To the extent that growth has a claim on nutrients, removal will be out of phase with the annual temperature cycle. The loading region sensitive to growth uptake has been postulated to be up to 120 g/m^2·yr for nitrogen (Kadlec, 2005b). Phosphorus removal would be expected to be in phase with growth, since microbial conversions do not remove total phosphorus. When growth effects are dominant, the k-values change seasonally:

$$k = k_j j \;=\; 1,\ 2,\ 3, \ldots 12 \tag{7.85}$$

where j is the month number. Although higher frequency might be used for the seasonal change pattern, monthly values will be sufficient in cases where detention time lags are on the order of a week.

Rate Constant Distributions

The data from any one wetland may be used to calibrate the PTIS model. Then, the calibrations from a number of such wetlands may be used to determine the frequency distribution of those k-values. Here, the reduction of phosphorus is used as an example. In Chapter 10, this aspect of phosphorus removal is dealt with in much more detail. Often, the N-value is not known, because a tracer test has not been run. So, for illustration purposes, it will be assumed that P = N = 4, which is a representative N-value for FWS wetlands. The value of

C^* is presumptively taken to be 0.006 mg/L in this analysis. Equation 7.69 may be used to calculate a long-term average k-value, from the hydraulic loading and the inlet and outlet concentrations averaged over the period of record. By this procedure, temperature effects are lumped into the variability, which is expected to be small for phosphorus. Each wetland is accorded one value for its entire period of record, thus averaging over a number of annual periods, which differ for each wetland.

The probability distribution of these k-values for a set of 313 marshes is quite broad (Figure 7.21). The mean is k = 20.4 m/yr, but the distribution contains some very high rate constants. Accordingly, the median k = 11.2 m/yr. However, the range is 0 to 282 m/yr, and the standard error of the mean is 1.7 m/yr (standard deviation = 30 m/yr). Clearly this distribution is too broad to give much confidence in design for a mean k-value. It is obviously necessary to understand the components of the wetland environments and layouts that contribute to either high or low k-values. And, very importantly, it would be exceedingly dangerous to place great trust in any one wetland as a prototype for all others. In addition, it would be pure coincidence if any new wetland were to behave as the mean of the distribution.

As the result of these considerations, recommendations for design must go beyond the concept of a universal, or average, k-value that may be used for any wetland.

Random Variability

Many factors contribute to random variability in the outlet concentrations from treatment wetlands. This variability is typically not small, with coefficients of variation of 20–60% being common. The deterministic models presented above reproduce the central tendency of performance, but not the random variability. Whether there is microbial or vegetative control, seasonal patterns of wetland variables are the rule, accompanied by a random variable term (Kadlec, 1997b). When analyzing performance data, there are two ways to establish the seasonal deterministic pattern (trend): via use of a model, or via use of a presumptive trend. The model basis relies upon Equations 7.79 and 7.83, and is most

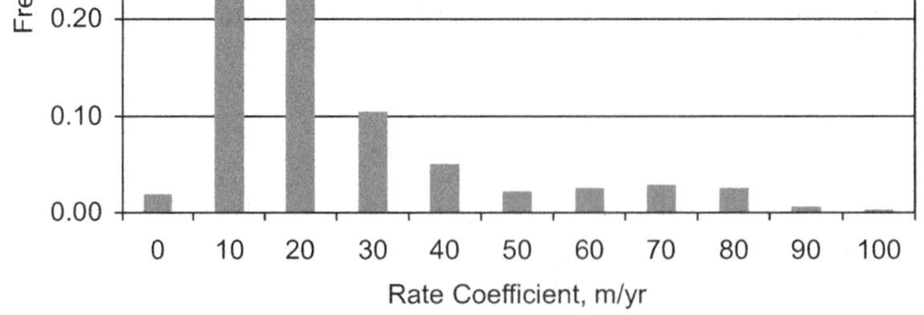

FIGURE 7.21 Frequency distribution of areal rate coefficients for phosphorus removal. The period of record for each of 313 marshes is represented.

readily applied to microbial, temperature-driven removals. The presumptive trend is very often an annual sinusoidal pattern.

Model based

$$C_{out} = C_{\text{model}} + E \tag{7.86}$$

Trend based

$$C_{out} = C_{avg}[1 + A \cdot \cos(\omega(t - t_{\max}))] + E \tag{7.87}$$

where: A = fractional amplitude of the seasonal cycle, –
C_{out} = instantaneous outlet concentration, mg/L
C_{model} = model outlet concentration, mg/L
C_{avg} = average (trend) outlet concentration, mg/L
E = random portion of the outlet concentration, mg/L
t = time of the year, Julian day
t_{max} = time of the year for the maximum outlet concentration, Julian day ω = annual frequency, 0.172 d^{-1}

The existence of the error term (E) means that sampling must either be at high frequency, or cover many annual cycles before meaningful trend averages can be determined. Data from several years may be "folded" to create an annualized grouping, distributed across the year according to Julian day. This use of many annual cycles has the advantage of including year-to-year variations in climate, flow, and ecosystem condition.

As an example, the Imperial, California agricultural runoff marsh received modest nitrate loadings (Kadlec et al., 2010a). The deterministic portion of this representation was modeled by the k-rate technique with a temperature-dependent rate constant. That fit had R^2 = 0.84 for the four-year period of record. Monthly data and model results are shown in Figure 7.22. The stochastic portion of the response is representable by the relative error (E/C_{model}). For this example, the 20th and 80th percentiles of relative error are 22% lower and 38% higher than the trend.

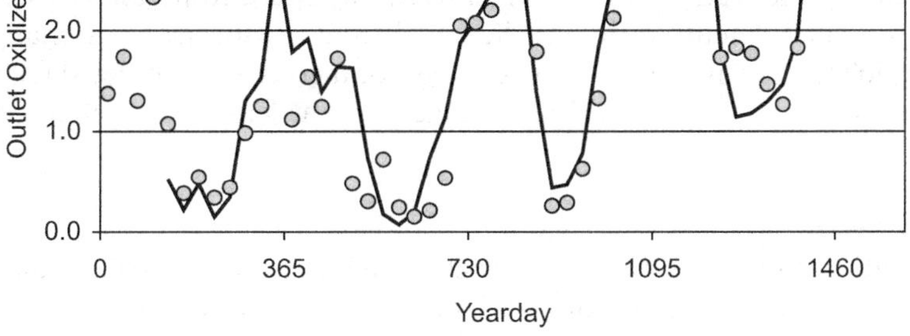

FIGURE 7.22 Variability of the nitrate nitrogen outlet concentrations for the Imperial, California treatment marsh.

It is important to note that the probabilistic portion of the wetland performance model is not a "safety factor," as utilized in some wetland design procedures (Water Environment Federation, 2001). In the usual sense of the term, a safety factor provides extra capacity in design to accommodate unforeseen events and phenomena and is a judgment call made by the designer. The stochastic variability that exists in all treatment wetland outlet concentration data is not unforeseeable. In fact, these probabilistic variations are just as quantifiable as the deterministic variations caused by changes in detention time, and must be accounted in design apart from any considerations of safety factors. It is simply unacceptable to ignore this "half" of wetland behavior in design, as is the case in U.S. EPA (2000a) and other sources.

In subsequent chapters, the percentile multipliers on the seasonal trend are presented, for each of the common contaminants and the various wetland types. The random variation, E, can be captured through the use of a multiplier on the trend value, C_{trend}:

$$E = C_{trend}\Psi \tag{7.88}$$

$$C_{out} = C_{trend}\,(1 + \Psi) \tag{7.89}$$

where: Ψ = stochastic portion of C/C_{trend}, –

Temporal Considerations

Synoptic Error

In the course of data analysis, there will typically be a set of contemporaneous values of wetland inlet and outlet concentrations, together with accompanying flows and other information. Calibration consists of selection of rate parameters that minimizes the error between those field observations and the calculated model values.

It is clear that paired contemporaneous measurements of input and output flows and concentrations have little chance of providing a quantitatively accurate description of removals, because of the transport delay in the wetland. Typical hydraulic detention times are of the order of several days to more than a week, which implies a significant shift in event timing if the wetland is in plug flow. However, tracer testing of hundreds of treatment wetlands has shown conclusively that no treatment wetland exhibits plug flow; rather, the detention time distribution extends to three or more nominal detention times (Kadlec, 1994). Therefore, there are remnant effects of inlet events at the outlet after three or more nominal detention times. Because of the complexity of modeling both the time and distance effects, the only chance of avoiding a transport delay artifact (synoptic error) is to compare inlet and outlet measurements averaged over more than those three nominal detention times.

The water leaving the wetland may have entered anywhere from about a tenth to three or four times the nominal detention time earlier. If there is a time series of changing inlet concentrations or flows, then instantaneous, contemporaneous inlet/outlet data should not be used to calibrate the model. Rather, time averages over at least three detention times should be used.

Solids Storage Effects

The water detention time does not appropriately reflect transport delays due to storage in or on stationary compartments of the wetland – the substrate, sediments and biota. The contaminant residence time will vary markedly depending on how many times it has been "parked" and recycled in various active or passive storage compartments during its passage through the wetland.

If the contaminant in question merely partitions to the wetland solids and sediments, adsorption-desorption phenomena will create a transport delay, known in the literature as retardation. Studies of this phenomenon in treatment wetlands have been conducted (McCarter and Price, 2017).

The correct concept for the longitudinal transport of interactive substances, such as nitrogen and phosphorus, includes major interactions with wetland solid compartments. Transport with water is rapid, but the exiting compounds also originate from storage, and not from the water that entered only one or a few detention times earlier. Pulses of incoming nutrients in part work their way through solid storages. Consequently, there is a second type of synoptic error that must be accounted, which results from the wetland storage delay.

Source Flow Time Sequences

Source waters are delivered to treatment marshes on different time schedules, ranging from discretely episodic in urban stormwater systems to continuous flow from advanced treatment plants, and also discharge on similar time schedules. The tanks in series model can be applied to any such sequence, even event-driven systems (Wong et al., 2006; Merriman et al., 2017). However, care must be taken in the selection of the time period for model calibration or forecasting. As noted in Chapter 2, a treatment marsh is here considered to be *in use* if it is delivering treated water to downstream receiving systems. The TIS model should be applied to in use periods and should not include times when the system is not in use. In other words, it is not reasonable to average over periods that include down time.

For an event-driven system, the selected period(s) would be the times when the system is receiving and discharging water, and the concentrations would be the event mean concentrations. Wong and Geiger (1997) therefore suggested that event mean detention times (or equivalent event mean hydraulic loadings) should be used in conjunction with event mean concentrations (EMCs). Wong and Geiger (1997) went on to suggest that multiple events over the course of time could be averaged, a procedure that has been detailed by McCarthy et al. (2018). These authors utilized Site Mean Concentrations (SMCs), which were the volume-weighted average of the EMCs for each event.

$$EMC = \frac{\sum_1^n C_i V_i}{\sum_1^n V_i} \tag{7.90}$$

$$SMC = \frac{\sum_1^p EMC_j V_j}{\sum_1^p V_j} \tag{7.91}$$

where: n = number of samples per event
C_i = concentration of sample "i" during event
V_i = volume represented by sample "i" during event
p = number of measured events
EMC_j = event mean concentration of event "j"
V_j = volume of event "j"
The flows involved can be similarly defined.

$$EMQ = \frac{\sum_1^n V_i}{t_i} \tag{7.92}$$

$$SMQ = \frac{\sum_1^p EMQ_j V_j}{\sum_1^p V_j} \tag{7.93}$$

where: EMQ = event mean flow, m^3/d
SMQ = site mean flow, m^3/d
t_i = duration of event "I", d

For continuous flow systems, the common selections for the period of model calibration or forecasting include annual, seasonal, or monthly. Northern systems may only be in use during the unfrozen season, and that would be a logical choice of period.

Startup

The models discussed earlier pertain to fully developed treatment wetlands. Plants are introduced into some constructed treatment wetlands as seeds or propagules at biomass densities far less than those that will ultimately develop. This process of grow-in typically takes a period of one to three years, depending upon climate. The amounts of nutrients taken up during the initial grow-in are substantial in many cases and represent a removal process that is not sustainable in the long term. Development of a litter layer takes even longer. Because submerged plant material creates attachment substrate for microbes, and a carbon source for metabolism for some of them, the wetland may not be fully functional during startup. Nonnutrient chemicals are likely to be less affected, but processes such as sorption still require the full ecosystem to reach their potential.

These large ecosystem start-up effects can dominate nutrient removal for many months. Therefore, calibration of steady-state models during such a period will not produce rate constants that are representative of long-term sustainable operation.

Dangers of Extrapolation

The models discussed above are all quite simplistic, in that they purport to describe general features of complicated processes. Many ecosystem compartments are lumped together, and the models are therefore termed "highly aggregated." The decision to use highly aggregated design models (all of those under consideration fit that description) carries the implied penalty of great risk in extrapolation beyond the calibration conditions. The safest criterion for extrapolation is to not do it; only interpolation on the calibration sets should be allowed. A less conservative approach would be to avoid extrapolation to conditions that are known to threaten the integrity of the vegetative community that typified the calibration. In any case, extrapolation should not be to water chemistries known to be outside the range of the calibration.

Hydraulic variables present difficulties in two ways: they may affect the performance of a specific vegetative community, and they may affect that community itself. For instance, removal performance may not differ too much from steady to pulsed operation. As long as the vegetation is not altered, we are faced only with the issue of how that ecosystem averages the event driven environment. Depth variation may not a large factor within certain ranges if the variations are not of long duration, but average depth is likely to be an important factor. The larger risk is that the ecological communities become affected and change to communities that do not match the calibration. For example, cattails find it difficult to survive at long hydroperiods in deep water. SAV may suffer if hydroperiods are too short. Prolonged dryout may pose problems under some design scenarios, or at least affect the subsequent treatment performance.

Internal flow patterns are controlled by topography, depth and compartmentalization. Pollutant removal known to be sensitive to those flow patterns. Therefore, the applicability of a specific TIS model is conditioned on the anticipated TIS for the extrapolation. There are very large implications, especially for high removals.

Extrapolation from a wetland of one type to another is clearly not a reasonable step. The microbial communities, and the character and magnitude of the biogeochemical cycles, may differ markedly. Forested ecosystems will not necessarily perform with the same rate constants as emergent marshes, even if the hydrologies are comparable.

As a consequence of these considerations, it is not a good practice to use model parameters (P, k, C*, θ) in situations outside the ranges of operating values from which they were derived. There are not "universal" values of rate constants, as offered in many literature sources (e.g., Reed et al., 1995; Water Environment Federation, 2001; Crites et al., 2006, 2014). Nor should the reported central tendencies of parameters, such as rate constants (Kadlec and Knight, 1996; Kadlec and Wallace, 2009), be interpreted as "universal" values. The parameter values obtained from various operating wetland systems vary widely, depending upon factors described in preceding sections. It is therefore prudent to examine the origin of a particular calibration set (P, k, C*, θ) before using it in design calculations. The questions to be addressed include:

1. Do the inlet and outlet concentration ranges of the calibration set include the ranges to be considered in design?
2. Do the ranges of detention times and hydraulic loadings of the calibration set include the ranges to be considered in design? Are the intended water depths in the calibration range?
3. Is the configuration of the system under design comparable to that of the calibration wetland(s)? This includes aspect ratio, number of physical compartments, and size. Consideration should be given to the N-values (or P-values) of calibration and design systems. And beware: small mesocosms may not be representative of a large system.
4. Are the climatic conditions for the calibration set similar to those of the intended design? Extremes of rainfall, evapotranspiration and temperature may affect the calibration values. The seasonal freeze up of temperate wetlands separates them from subtropical systems, for instance.
5. Is the ecology of the calibration wetland(s) comparable to that of the intended wetland under design? At a minimum, wetland types must correspond, across the variants of emergent marshes, floating plants, submerged aquatic vegetation, and possibly open water fraction.

Insofar as the intended design departs from calibration conditions and information, an increasing degree of risk is engendered. Intersystem data is a valuable aid to quantifying that risk. However, many systems do not possess enough information to gain firm estimates of rate model parameters.

7.9 SUMMARY

The various measures of marsh pollutant removal performance include concentration and mass reductions, as well as graphical representations. These ultimately rely upon the water and component mass balances for the wetland. The long history of development of chemical reaction engineering has much to offer for wetland pollutant removal reactions. Concepts of reaction order, reversible reactions, solid-fluid reactions, biochemical reactors and autocatalysis all find analogs in the marsh environment.

Many different processes cause pollutant reduction in marshes, including microbially mediated conversions, volatilization, sedimentation, sorption, photodegradation, and – very importantly – macrophyte and mesophyte uptake and cycling.

To add to this complexity, the flow patterns in marshes are very complicated. The dynamics of water movement through the wetland has a significant influence on the efficiency and extent of contaminant removal. Turbulent mixing caused by flow through vegetation is accompanied by wind and thermal mixing. As a descriptive technique, tracer testing has grown to be a larger contributor to understanding than its chemical reactor predecessor. A very large body of wetland tracer testing knowledge is now available to support quantitative models of pollutant removal. Simple removal rate rules combine with descriptions of internal hydraulics to form process models. The tanks-in-series (TIS) hydraulic model can describe most wetland situations.

Simplifying circumstances often apply to treatment wetland data analysis: steady flow, negligible water gains and losses, and concentration averages. The central tendency of wetland outlet concentrations are then often well-represented by the *PkC** model:

$$\frac{C - C^*}{C_i - C^*} = \frac{1}{(1 + k/Pq)^P} \tag{7.94}$$

for microbial control

$$k_T = k_{20}\,\theta^{(T-20)} \tag{7.95}$$

for vegetative control

$$k = k_j \quad j = 1, 2, 3, \ldots 12 \tag{7.96}$$

This is the tanks-in-series (TIS) model, with a modified number of tanks $P \leq N$, where N = the tracer-determined number of tanks.

Treatment marshes often are subject to water losses and gains, due to rainfall, evapotranspiration and infiltration. The incorporation of those effects means that the simple Equation (7.94) can no longer be written. Instead, tank-to-tank calculations must be employed. It is also necessary to exercise care in the selection of the time period for a data fitting or design application, so as to avoid synoptic error and inclusion of inoperational periods.

The k-values for a group of similar wetlands will display a frequency distribution. Choice of a high or low value from such a distribution for design purposes requires assessment of the desired degree of risk and other wetland factors.

There are quantifiable excursions around this central tendency, which can be included as a stochastic addition. This variability is quite large for treatment marshes. Variability information may also be configured as a multiplier on the deterministic portion of the prediction.

8 Suspended Solids

Marsh ecosystems interact with suspended sediments in water moving through the wetland. Removals and additions are the end result of a complicated set of internal processes, including the production of transportable solids by wetland biota.

Low water velocities, coupled with the presence of plant stems, leaves and litter, promote settling and interception of solid materials. This transfer of suspended solids from the water to the wetland sediment bed has important consequences for the quality of the water, as well as the properties and function of the wetland ecosystem. Many pollutants are associated with the incoming suspended matter, such as metals and organic chemicals, which partition strongly to suspended matter. In marshes used to polish municipal wastewater, the accretion of solids contributes to a gradual increase in the bottom elevation of the wetland. However, wetlands used to treat urban or agricultural stormwater, or those exposed to periodic ancillary flooding, may have rapid accretions in the inlet zone.

8.1 SOURCE CHARACTERIZATION

Laboratory Methods

Total suspended solids (TSS) are measured gravimetrically after filtration and drying (APHA, 2017) (Method 2540D), and reported in mg/L. Two filter sizes are in common use: 0.7 μm and 1.5 μm. The organic content is characterized as volatile suspended solids (VSS), determined from the weight loss on ignition at 550°C. The TSS method has been subjected to considerable criticism by Gray et al. (2000) for use on "natural" waters, and these authors recommend a suspended sediment concentration (SSC) analysis as a replacement (Method D 3977–97, ASTM 2000). One fundamental difficulty is the representativeness of aliquots, especially if they contain sand particles. A second difficulty is the wide variability of the TSS method in low concentration ranges. Gray et al. (2000) quote the Standard Methods precision as a 33% coefficient of variation at 15 mg/L. TSS measurements are likely to be biased low compared to SSC measurements.

Turbidity in water is caused primarily by suspended matter, although soluble colored organic compounds can contribute. Therefore, turbidity is sometimes used as a surrogate for gravimetric measurement of suspended matter. The measurement technique involves light scattering. The instrument is the turbidimeter, consisting of a nephelometer, light source, and photodetector. The standard unit is the nephelometric turbidity unit (NTU). The correlation between TSS and NTU is often good for a specific wetland system, but care must be taken in the extrapolation from one site to another (Table 8.1). From these results, it may be concluded that the NTU/TSS relationships for FWS wetland effluents differ substantially from those for activated sludge effluents and vary somewhat between natural systems.

Organic suspended solids in wetlands often contain a major fraction of phytoplankton (Mulling et al., 2013). The chlorophyll concentration in the water is an indicator of that fraction of TSS. Although there is not a precise numerical correlation between biomass and chlorophyll, the ratio has been reported in the range of 1.5–3.7% of the carbon content (Riemann et al., 1989). Since carbon comprises about 40% of biomass, the ratio is in the range of 4–9% of the planktonic biomass.

TABLE 8.1

Regressions between total suspended solids and turbidity for wetlands, forced through the origin (TSS = 0, NTU = 0).

	NTU/TSS	R^2	TSS Range mg/L	Turbidity Range NTU	Number	Ref.
Everglades	0.25	0.80	1–18	0.4–3.4	126	1
River water	0.83	0.77	0–145	0–125	64	2
River water	0.66	0.95	50–1400	100–1000	23	3
Agricultural runoff	0.75	0.52	0–143	0–121	1013	4
Submerged vegetation	0.74	0.93	0–215	0–150	>100	5
Water hyacinths	1.39	0.54	4–18	6–21	12	6
Oxidation pond	0.47	0.06	1–15	1–27	96	7

References:
1 South Florida Water Management District, unpublished data
2 Des Plaines River project unpublished data
3 Harter and Mitsch (2003)
4 Everglades Nutrient Removal project unpublished data
5 James et al. (2002)
6 Crites and Tchobanoglous (1998)
7 Gearheart et al (1983)

Organic Content

The suspended solids entering a treatment wetland may display widely varying characteristics, according to the source water involved. Domestic wastewaters at all pretreatment stages contain suspended materials that are primarily organic. Runoff waters, both urban and agricultural, may contain high proportions of mineral matter. Other source waters may involve highly specific characteristics, such as the colloidal materials that discharge from milking parlors. The two principal ways of describing solids are the mineral content and the size distribution.

Soil definitions of are often also applied to suspended matter. The generic fractions are organic, clay, silt and sand. TSS originating from urban or agricultural runoff is likely to be mostly mineral, while TSS originating from ponds and advanced treatment plants is likely to be mostly organic.

The volatile suspended solids (VSS) fraction of the solids is usually taken to be a measure of the organic fraction (Table 8.2), and the remaining nonvolatile suspended solids (NVSS) are assumed to be the mineral fraction of the overall TSS. For incoming waters derived from runoff from mineral soils, the fraction organic may be rather low. At the Des Plaines site, river water entering averaged 11–16% organic, while water leaving the treatment wetlands averaged 16–26% organic. Harter and Mitsch (2003) reported 9% organic for both entering and leaving waters from the Olentangy River wetlands. However, the Houghton Lake natural peatland showed 77% organic, and after lagoon wastewater addition showed 56% organic (unpublished data). As an extreme example, the fraction VSS in a potato wastewater treatment wetland was 94% (unpublished data). Obviously, no generalizations may be made across the spectrum of treatment wetlands and source waters, but it should be noted that organic materials can be generated within the wetland and may be subject to decomposition after deposition.

TABLE 8.2
Organic content of various source waters entering, and interior to, treatment wetlands.

System	Source	TSS Inlet mg/L	% NVSS
Houghton Lake, MI	Lagoon	25	56
Houghton Lake, MI	Natural Marsh	-	23
Houghton Lake, MI	Impacted Marsh	20	44
Estevan, SAS	Lagoon	27	40
Des Plaines, IL	River	80	87
Des Plaines, IL	EW3 Marsh	32	75
Des Plaines, IL	EW4 Marsh	35	85
Tarrant, TX	Sed Basin Full Scale	75	65
Tarrant, TX	Sed Basin Pilot	37	81
Onaway, MI	Lagoon + Sand Filter	3.1	26

Particle Sizes

Both mineral and organic TSS vary widely in size (Table 8.3). Coarse mineral particles, such as sands, are usually not present in treatment marsh source waters, but clays and silts

TABLE 8.3
General benchmark values for TSS characteristics.

	D µm	Density kg/m^3	Settling Velocity m/hr	Settling Time one meter	
Mineral					
Coarse sand	500	2700	209	17	s
Medium sand	200	2681	72	50	s
Fine sand	100	2665	25.2	2	min
Coarse silt	50	2650	6.8	9	min
Medium silt	20	2633	1.044	57	min
Fine silt	10	2617	0.263	3.8	hr
Very fine silt	5	2600	0.065	15.4	hr
Clay	2	2325	0.006	6.4	d
Organic					
Everglades floc	1000	1010	6.2	9.7	min
Everglades floc	100	1110	0.62	96.8	min
STA2	100		6.41	9.4	min
DP River	61	1019	1.19	50.4	min
HL Lagoon	131	1004	0.86	69.8	min
HL Original Marsh	80	1014	0.95	63.2	min
DP EW3	146	1002	0.55	109.1	min
Plankton	40	1.02	0.07	14.3	hr

Source: Mineral, MAPAQ, 2013.

are common in agricultural runoff. Particles entering wetlands are likely to be either clays or silts, ranging in size from 2–50 µm. Organic materials can range from bacteria to aggregated flocs. Plankton can be classified as macroplankton (200–2000µm), microplankton (20–200µm), or nanoplankton (2–20µm) (Makoto and Ikeda, 1984). Treatment marshes show typical particle sizes of 50–200 µm.

Neither the particles entering the wetland, nor those leaving, are of a single size. Frequency distributions of particle sizes are always present (Figure 8.1). As a result, particle processing also becomes distributed, with large particles behaving differently from small.

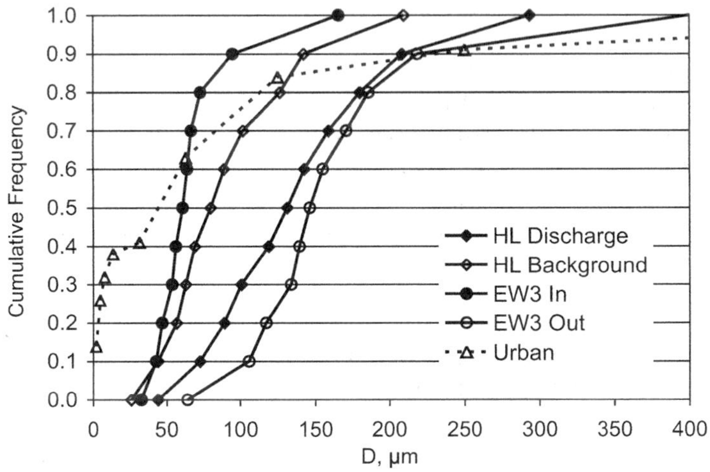

FIGURE 8.1 Particle size distributions for two FWS wetlands. At Des Plaines (EW3), the outlet particles are larger than those entering. At Houghton Lake (HL), the discharge area particles are larger than those in wetland background areas. Data source HL and EW: Unpublished data. The information for urban runoff is from Selbig and Bannerman (2011).

Source: Water Time Series.

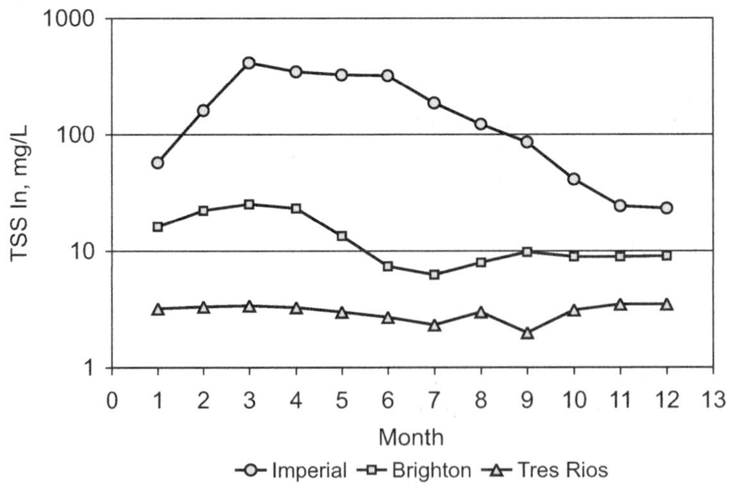

FIGURE 8.2 Time series of in source water TSS for three example wetlands. The periods of record were 10 years for Brighton, and 4 years for Tres Rios and Imperial. All had nearly constant flow.

The TSS concentrations incoming to treatment marshes can be variable during the course of the year (Figure 8.2). Treatment plants, such as that delivering water to the Tres Rios systems, are nearly constant through the year, but both agricultural and urban runoff may be quite variable during the annual cycle, such as that reaching the Imperial system to the Tres Rios. Treatment lagoons can also behave quite differently during the seasons, as was the Brighton lagoon. The peaks of TSS, if any, can occur at different times of the year. Figure 8.2 also illustrates the large differences in concentrations entering the marshes, which for the examples vary by two orders of magnitude.

8.2 PARTICULATE PROCESSES IN MARSHES

The FWS wetland processes sediments and suspended solids in a number of ways (Figure 8.3). After the suspended material reaches the wetland, it joins large amounts of internally generated suspendable materials, and both are transported across the wetland. Sedimentation and trapping, and resuspension, occur en route, as does "generation" of suspended material by activities both above and below the water surface. Chemical reactions may cause the formation of solids that can deposit. Algal and plant debris may form at one location and deposit downgradient in the wetland.

The removal of TSS requires inclusion of the size distribution and the nonideal flow hydraulic model. However, to gain understanding of the component processes, a useful conceptual framework is offered by a one-dimensional, vertically averaged, lumped mass balance model of TSS removal.

$$\frac{\partial(hC)}{\partial t} + \frac{\partial(uhC)}{\partial x} = G - S - F + R \tag{8.1}$$

where: C = concentration, g/m^3 = mg/L

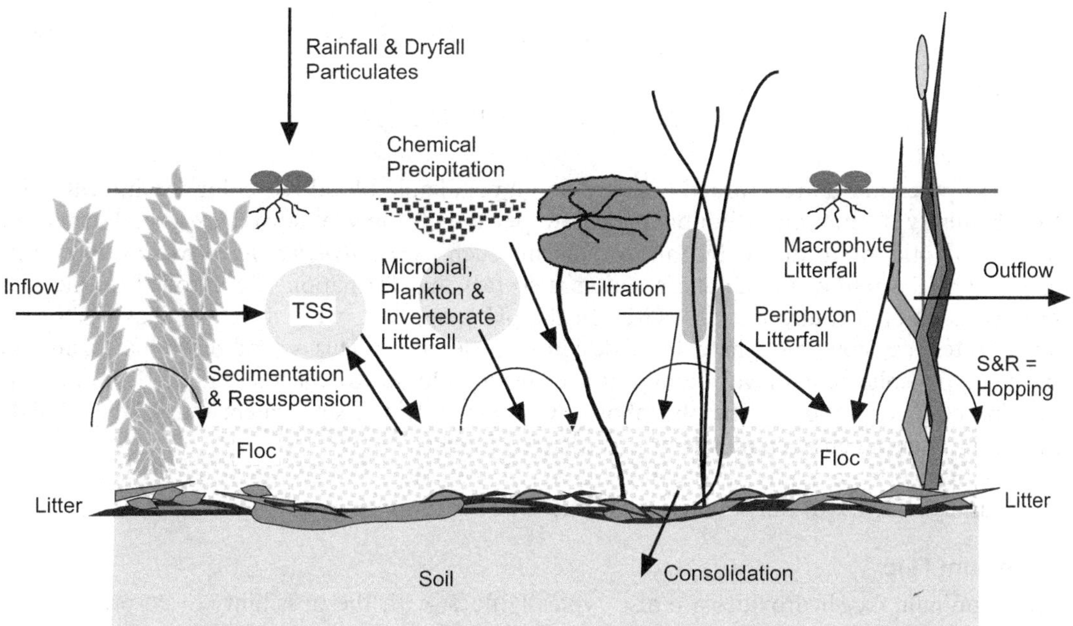

FIGURE 8.3 Processes affecting particulate matter removal and generation in FWS wetlands.

F = filtration rate, g/m^2·d
G = generation rate, g/m^2·d
h = water depth, m
R = resuspension rate, g/m^2·d
S = settling rate, g/m^2·d
t = time, d
u = superficial water velocity, m/d
x = distance, m

It is informative to examine these terms, with a view to understanding the magnitude of the various TSS fluxes. The first observation is that accumulation in the water column, the first term on the left-hand side of Equation 8.1, is virtually never of consequence in performance.

Generation by Precipitation

Two main types of particle generation may be manifest in a treatment marsh, chemical precipitation and biological generation. Several chemical reactions can produce wetland sediments under the proper circumstances. Some of the more important are the oxy-hydroxides of iron, calcium carbonate, and divalent metal sulfides. And as conditions of chemical composition, pH and redox change in the wetland, these and other compounds may undergo dissolution and be removed from the sediment bed.

Iron Flocs

The iron oxy-hydroxides are typically flocs, with the possibility of co-precipitates. They may form under conditions of elevated dissolved ferric iron and oxygen-rich water. The processes may be represented as (Younger et al., 2002):

$$Fe^{+2}+\frac{1}{4}O_2+H^{+}\ e^{+3}+\frac{1}{2}H_2O \tag{8.2}$$

$$Fe^{+3}+2H_2O\rightarrow FeOOH(sus)+3H^{+} \tag{8.3}$$

$$FeOOH(sus)\rightarrow FeOOH(sed) \tag{8.4}$$

These precipitates are characterized by an unmistakable blood-red color. As indicated by the chemistry, formation is inhibited by low pH and by low dissolved oxygen. Formation may be abiotic, or mediated by microorganisms such *as Thiobacillus ferrooxidans*. However, at pH > 9, the rate of the abiotic reaction is so fast that formation is controlled by the rate of oxygen supply (Younger et al., 2002). In the pH range 6 < pH < 8 that typifies FWS wetlands, rates are slow enough to be a design consideration. This set of reactions forms the basis for phosphorus removal by addition of ferric chloride to wastewaters, and the accompanying co-precipitation of the phosphorus. Consequently, the subsequent fate of these solids in polishing treatment wetlands is of considerable interest.

Although not generally in use for treatment wetlands, models such as CE-QUAL-W2 (Cole and Wells, 2016) purport to model the redox reaction 8.2.

Aluminum Flocs

The aluminum oxy-hydroxides are also typically flocs, with the possibility of co-precipitates. They may form under circumneutral pH conditions, and do not require oxygen. The processes may be represented as (Sobolewski, 1999):

$$Al^{+3} + 3H_2O \rightarrow Al(OH)_3 \downarrow +3H^+ \tag{8.5}$$

These precipitates are characterized by their formation of a "pin floc" material that does not readily settle (Bachand et al., 2000). This set of reactions also forms the basis for phosphorus removal by addition of alum to wastewaters, and the accompanying co-precipitation of the phosphorus. Consequently, the subsequent fate of these solids in polishing treatment wetlands is of considerable interest (Stumpner et al., 2018).

Calcium Carbonate

Calcium carbonates may be formed in wetlands, under conditions of elevated pH and dissolved calcium. The operative chemistry may be summarized as:

$$Ca^{+2} + HCO_3^{-} + H_2O \rightarrow CaCO_3 \downarrow +H^+ \tag{8.6}$$

This reaction may occur abiotically, but perhaps more importantly it may be mediated by algae. Algal activity can drive up pH, and create conditions that foster creation of calcium-rich solids (Vymazal, 1995). Indeed, this process has contributed to the formation of marl prairies as a form of natural wetlands. New sediments in Everglades protection treatment wetlands contain a significant fraction of calcium compounds (Dierberg et al., 2002).

Metal Sulfides

Many metals form very insoluble sulfides, including mercury, lead, cadmium, and zinc. These precipitates are important in the processes of metal removal in mine water wetlands, and follow the general chemistry (Sobolewski, 1999):

$$CH_3COO^- + SO_4^{-2} \rightarrow 2HCO_3^- + HS^- \tag{8.7}$$

$$M^{+2} + HS^- \rightarrow MS + H^+ \tag{8.8}$$

However, for many treatment marsh applications, metals are present at only very low concentrations. Consequently, the formation of insoluble sulfides does not create measurable additions to the sediments of the wetlands.

Biological Generation

This mechanism is the principal determinant of TSS performance of marshes, because they contribute to an irreducible background level of TSS in the wetland.

The generation of sedimentary material is a very important internal process in nutrient-rich treatment wetlands. The typically generous supply of nutrients assures a large production of a wide variety of transportable organisms and associated dead organic material. Particulate matter is formed via processes of death, litter fall, and litter attrition. These occur for biota at a number of different size scales, ranging from macrophytes on down to bacteria. Planktonic productivity can be a major generator of suspended solids in systems that do not create mats of periphyton; but periphyton is also a generator of detritus. Another portion of TSS or floc originates from leaf, stem and seed litter from macrophytes. Litter can be reduced to particulates by invertebrate shredder organisms. Further, particulates can be produced by the action of vertebrates (e.g., mosquito fish (*Gambusia holbrooki*) or carp (*Cyprinus carpio*)), which feed by grazing the epiphytic community. The proportions of microbial/algal debris versus macrophyte debris is variable, with the former dominating in oligotrophic situations, and the later in more eutrophic conditions (Noe and Childers, 2007). In either case, the particulates produced are organic in character.

Many treatment wetlands are characterized by high water chlorophyll content and high sediment accumulation. Bacterial and algal growth is promoted, and decomposition products form a new pool of suspendable material. A host of wetland invertebrates, such as *Daphnia*, also die and contribute to the sediments. Zooplankton feed on phytoplankton and detritus and are primarily found in surface waters where food resources (phytoplankton) are abundant.

These processes are difficult to predict and quantify, but available simulation engines attempt to do so. Both EFDC (Tetra Tech, Inc., 2007) and CE-QUAL-W2 (Cole and Wells, 2016) consider four categories of mesophyton: blue-green algae, diatoms, green algae and epiphyton. The models simulate the biomass of the various mesophytic components and assign particulate generation as a first order process that results entirely in settling. Importantly, these models presume that the amount of mesophytic biomass, and hence the amount of TSS production, is directly dependent upon the nutrient status of the water. Those dependencies are assigned a Monod formulation, as expressed in Chapter 5, Equations 5.2 and 5.4. In functional terms:

$$G = G(I, C_N, C_P) \tag{8.9}$$

where: C_N = nitrogen concentration, g/m^3

C_P = phosphorus concentration, g/m^3

I = light intensity, mol/m^2·d

As a consequence, it is to be expected that the wetland background concentration will depend upon nutrient status. Also, because there are longitudinal gradients in nutrient concentrations, longitudinal gradients in G are to be expected.

Indirect, but very strong, evidence of the production of particulate matter is in the large amounts of particulates that accumulate in treatment marshes. Large amounts of floc have developed in Everglades wetlands of all kinds, including the STAs. For example, the mean floc depth in 13 STA cells in 2003 was 10.9 ± 5.5 cm, weighing 8.1 ± 5.5 kg/m^2. The P content was 6.4 ± 3.8 g/m^2. These accumulations are very much larger than the sum of all TSS inputs, which were approximately 0.1 kg m^{-2} yr^{-1}. This leads to the inescapable conclusion that the STAs, and other P treatment wetlands, manufacture particulate matter and floc. Using cup collectors, Kadlec (2009b) determined that internal production and resuspension of sediments (about 13 g/m^2·d), in a wetland receiving lagoon water, was about forty times higher than the incoming sediment load. Similarly, Fennessy et al. (1992) measured about 30 g/m^2·d in cup collectors in a river water treatment wetland, which was about six times the incoming sediment load.

Thus, it is highly likely that in most instances, the effluent TSS from a treatment wetland is determined by internal biological processes, and not by a "removal efficiency" for incoming TSS. As a corollary, the solids leaving the wetland will very often not be related to the solids entering, but rather to the macro and micro detrital fragments originating internal to the system.

Settling of Particulates

In general, the vertically averaged settling rate may be written as first order in the suspended solids concentration:

$$S = w\,C \tag{8.10}$$

where: C = TSS concentration, gm^3
S = settling rate, $g/m^2 \cdot d$
w = solids settling velocity, m/d

Single Particles

The slow moving waters in the SF wetland environment often permit time for physical settling of TSS. The settling velocity of the incoming particulates, combined with the depth of the wetland, gives an estimate of the time and travel distance for those solids. Solids sink in water due to the density difference between the particle and water. For single, isolated spherical particles, the terminal velocity is reached quickly:

$$w^2 = \frac{4}{3}\frac{gd}{C_D}\left(\frac{\rho_s - \rho}{\rho}\right) \tag{8.11}$$

where: d = particle diameter, m
C_D = drag coefficient, –
g = acceleration of gravity, m/s^2
w = terminal velocity, m/s
ρ = density of water, kg/m^3
ρ_s = density of solids, kg/m^3

In turn, the drag coefficient is a function of the particle Reynolds number:

$$C_d = \left(\frac{24}{\mathrm{Re}_p}\right)\left(1 + 0.15\mathrm{Re}_p^{\ 0.687}\right) \tag{8.12}$$

where the Reynolds number is:

$$\mathrm{Re}_p = \frac{d\rho w}{\mu} \tag{8.13}$$

where: $\mathrm{Re_p}$ = particle Reynolds number, –
μ = viscosity of water, kg/m s (= 0.001 × μ in centipoise)

If all physical properties are known, Equations 8.11, 8.12, and 8.13 combine to determine the settling velocity. This iterative calculation is easily automated on a spreadsheet, with the results shown in Figure 8.4.

In the laminar flow region, $\mathrm{Re_p} < 1.0$, the drag coefficient is inversely proportional to the particle Reynolds number, and the settling velocity of the particle is then calculable from *Stokes law*:

$$w = \frac{gd^2}{18\mu}(\rho_s - \rho) \tag{8.14}$$

In the wetland environment, neither the density nor the particle diameter are known, and the particles are not spheres or discs. While it is possible to correct for nonspherical shapes (Dietrich, 1982), there is not a convenient method for determination of the particle density. Further, particles may agglomerate to larger size, or be subject to interference from neighboring particles.

Settling of Mixtures: The Vertical Dimension

Settling of particulate matter may be described by a first-order model, with a settling rate coefficient w, for each size fraction. In general, settling velocities are proportional to the

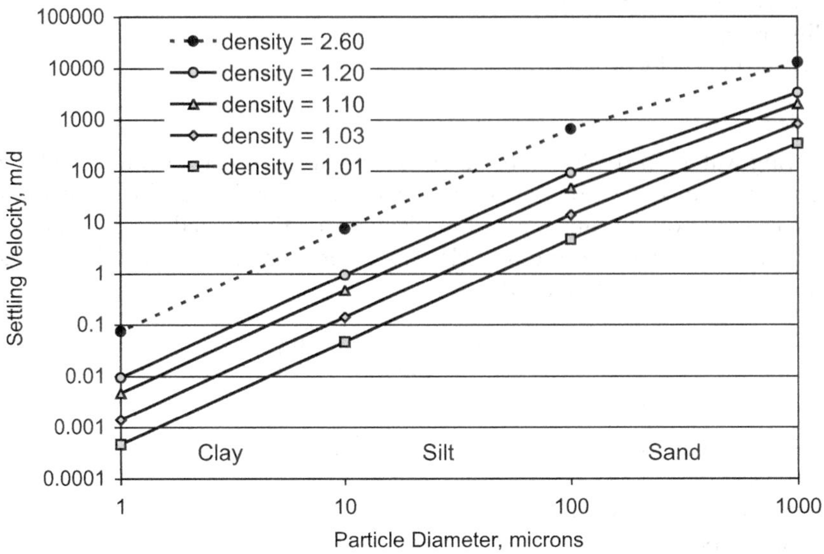

FIGURE 8.4 Settling velocity of spherical particles in water at 20°C, for different particle densities.

square of particle size, with variation including shape factors and particle density. Particle mass may be estimated to be roughly proportional to the cube of size. The time of fall of a particle through a vertical distance (h) is determined from its settling velocity:

$$t_{fall} = \frac{h}{w} \tag{8.15}$$

where: h = water depth, m

t_{fall} = time to fall, s

If the water is moving through the wetland length (L) at velocity (u), then the time of travel is:

$$t_{travel} = \frac{L}{u} \tag{8.16}$$

where: L = wetland length, m

t_{travel} = time to traverse wetland, s

u = water velocity, m/s

Theoretically, all particles of a size corresponding to a given fall velocity will be removed by settling if the travel time exceeds the settling time from the top of the water:

$$\begin{aligned} \frac{L}{u} &> \frac{h}{w} \\ N_{fall} &= \frac{Lw}{uh} > 1 \end{aligned} \tag{8.17}$$

where: N_{fall} = particle fall number, –

These concepts were applied to mixtures in shallow overland flow in grass (Deletic, 1999), with mean particle diameter used to determine the settling velocity (w). Values of

N_{fall} were found to be above 10 for complete removal, reflecting the difficulty of settling of the small end of the particle size distribution.

These innocuous representations are in fact somewhat misleading, because both w and u have distributions, the size distribution for w, and the velocity distribution for u. The settling rate distribution is in turn dependent on the particle size distribution, according to Equation 8.15. Figure 8.1 shows example size distributions. There is also a wealth of information available for velocity distributions, in the form of the detention time distributions. In a one-dimensional concept of a treatment marsh, the water velocities are simply the wetland length divided by the detention times. Conversion of the DTD to the velocity distribution leads to the example results shown in Figure 8.5, for a variety of TIS. When the flow is close to plug flow, as assumed in channel flow (N = 20), the velocity distribution is narrow, and centered on the mean velocity. However, when a wetland situation is considered (N = 4), a significant fraction of the velocities are 2–10 times greater than the mean. And, for a hypothetical settling basin (N = 2), most of the velocities are greater than the mean.

Stormwater wetland empirical data (Fletcher et al., 2004; Li and Fletcher, 2007; Yi et al., 2010) indicate that the settling efficiency (fractional removal), is representable by:

$$E_S = \frac{N_f^a}{N_f^a + b} \tag{8.18}$$

where: a,b = regression parameters, –

E_S = settling efficiency (fractional removal), –

Example values are a = 0.43 and b = 1.42 (Li and Fletcher, 2007).

For large treatment marshes, the water detention times (L/u) are often approximately 10 d, depths (h) 0.5 m, resulting in a hydraulic loading rate of 0.05 m/d. For an organic settling velocity of 10 m/d (see Table 8.3), the particle fall number would be 200, indicating that settling of incoming particulates would occur in the extreme front end of the wetland.

Mineral sediments in aquatic systems can possess settling velocities in the range of 100–500 m/d (Hamilton and Mitchell, 1996). The sediments entering a stormwater wetland are often mineral in character, and have high density, and therefore possess high settling rates. For instance, Li and Fletcher (2007) found w ≈ 100 m/d for an urban stormwater

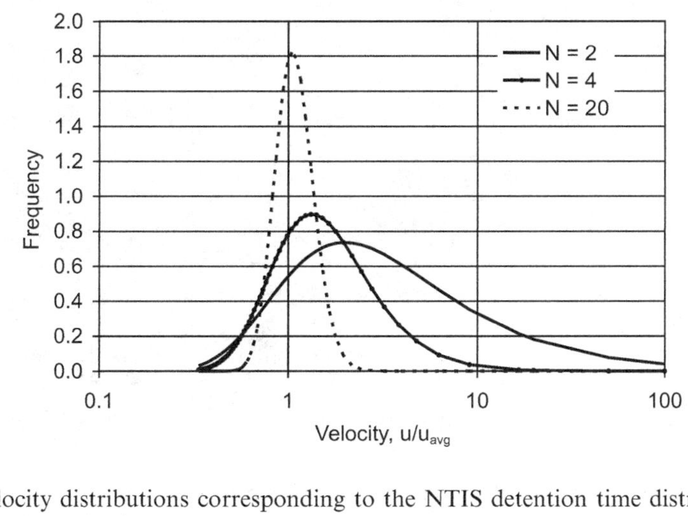

FIGURE 8.5 Velocity distributions corresponding to the NTIS detention time distributions. Note the log scale for the horizontal axis.

wetland. A mineral settling velocity of 100 m/d, and a hydraulic loading rate of 1.0 m/d, would apply to small urban or small field runoff control wetland. The particle fall number would be 100, again indicating that settling would occur in the extreme front end of the wetland. Experimental evidence of this was found by Braskerud (2001a).

As a result, system input/output data are of essentially no use in determining particulate removal rates (settling rates) in large or small wetlands. "Black box" removals do not provide a sufficient basis for interpretation or prediction of TSS removal processes. However, the available information indicates that incoming particulates are rapidly removed in the wetland environment, with essentially complete removal in the inlet section of treatment wetlands. This concept was initiated 25 years ago (USEPA, 1988), and has been substantiated by an enormous amount of data acquired since that time. However, the background TSS levels are also of interest, and thus performance information is still of interest.

Aggregation: Floc Particulates

Particles in aquatic systems are known to aggregate, both mineral and organic materials (Droppo, 2001; Lick, 2009). The aggregates that form are very irregular in shape (Figure 8.6). Because flocs are very porous, they exhibit effective densities that are lower than those of the component individual particles. The size of the floc is obviously greater than the size of the component individual particles. In combination these two effects in validate Stokes law. Typically, the settling velocity is higher than for individual particles.

Mineral flocs in source waters have been found to importantly affect TSS retention in wetlands (Braskerud et al., 2000, 2005a; Sveistrup et al., 2008; Johannesson et al., 2015).

Wetland floc particles can be much different from incoming mineral suspended solids, because they are composed of primarily organic materials. They are very irregular in shape, and agglomerate to relatively large size, for which Stokes law is unlikely to apply.

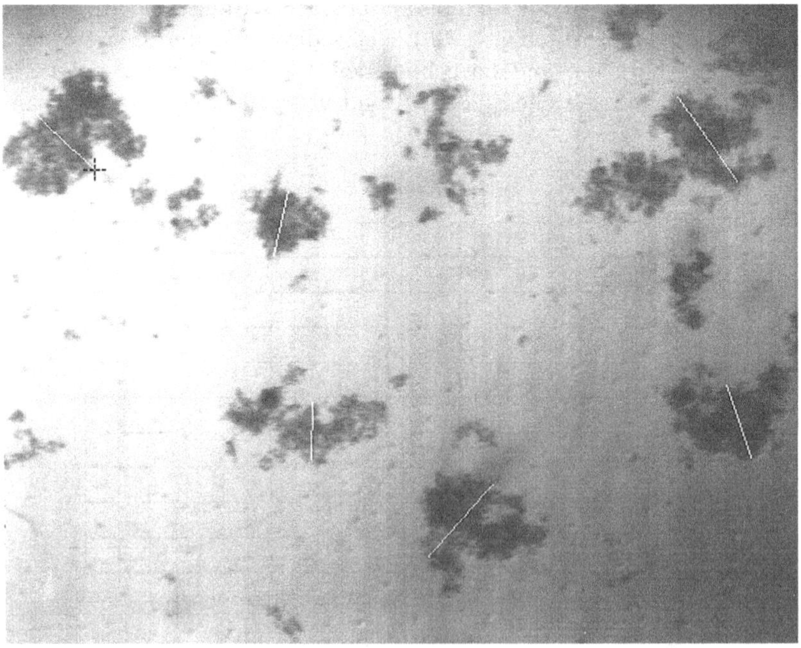

FIGURE 8.6 Photomicrograph of flocculent suspended particulate matter in the effluent from Des Plaines wetland EW3.

Nonetheless, bigger, denser particles fall faster, at a rate determined by the balance between gravitational and drag forces. Empirical data for flocs in freshwater show a lesser dependence on diameter:

$$w = ad^m \tag{8.19}$$

where: a = regression coefficient, $(m/d)/\mu m^m$

d = particle "diameter," μm

w = particle settling velocity, m/d

Regressed values reported for mineral flocs are a = 0.023 m/d and m = 1.56 (Lick, 2009), and for organic flocs are a = 0.32 m/d and m = 0.88 (Larsen et al., 2009a); where d is in μm.

Larsen et al. (2009a) performed tests of the settling properties of floc particles in laboratory columns and in field flumes located in WCA3A. They determined that particles aggregated into very large, fluffy conglomerates, with equivalent diameters in the range of 150–450 μm in slow moving waters. Aggregation was ascribed to the presence of extracellular polymeric substances, which consist of polysaccharides and proteins, wrapping around multiple constituent particles and binding them. The corresponding settling velocities were 10–100 m/d. For a water depth of 0.6 m, the settling times would be 10–100 minutes.

Aggregation has been studied, and methods of quantification developed over many decades of research (Lick, 2009). The basis is population balance modeling, which accounts for collisions that remove small particles from the size distribution and add larger particles to it. The data requirements are onerous, because the size distributions must be known (measured), which is not commonly the case for wetlands. Second, the mathematics are also onerous, because integro-differential equations are involved (Thomas et al., 1999; Kryven et al., 2014; Oyegbile et al., 2016). An integrated summary appears in Elimelech et al. (2013). At present, model quantification of the aggregation process in wetlands is not available.

Filtration

In addition to settling, particles may also be removed from the flowing water via collisions with underwater objects, which in marshes are dominated by underwater plant parts. Macrophytes and their litter form a nonhomogeneous "fiber bed filter" in the wetland context. A typical wetland "fiber" is a bulrush stem, of about one centimeter diameter. Stem densities are often too low to provide much opportunity for collision. A typical particle might be on the order of 1–100 μm. A typical water velocity is on the order of 10–100 m/d. Under these conditions, the collection efficiencies of mechanisms of diffusional or inertial impact deposition are predicted to be vanishingly small (Palmer et al., 2004).

However, there is evidence that direct interception and biofilm sticking is operative and significant for small particles. Lloyd (1997) (as referenced by Breen and Lawrence, 1998) examined the submerged surfaces of bulrushes (*Schoenoplectus validus*) and found particles as small as 0.5–2.5 μm sticking to biofilms.

In general, the direct interception filtration rate may be written as first order in the suspended solids concentration:

$$F = w_f\, C \tag{8.20}$$

where: C = TSS concentration, g/m^3

F = filtration rate, $g/m^2{\cdot}d$

w_f = solids filtration rate coefficient, m/d

The filtration rate coefficient is generally considered to be dependent on water velocity, and upon the frontal area facing the flow:

$$w_f = \eta u a_f \tag{8.21}$$

where: a_f = frontal area per unit wetland area, m^2/m^2
u = water velocity, m/d
η = filtration efficiency, –

Palmer et al. (2004) calibrated mechanistic interception and sticking filtration equations to flume experiments that simulated wetland conditions. Filtration efficiencies were in general less than 1.0 %. Their results also showed that this mode of particle removal was comparable to gravitational settling, for 50 and 300 μm particles, for water velocities of 3 cm s^{-1}. The travel distances for 50% removal through sparse Spartina were estimated to be 50–100 m at 5 cm/s and 300 μm particles; and 1–3 m at 5 cm/s and 300 μm particles.

The frontal area increases with the water depth, as more vegetative surfaces become immersed. For instance, if the immersed culms are cylindrical, then

$$a_f = (Dh) \cdot n \tag{8.22}$$

where: D = stem diameter, m
h = water depth, m
n = stem density, $\#/m^2$

The filtration efficiency depends upon the nature of the water flow, that is, whether creeping or with vortices; and upon the nature of the submerged surfaces, that is, whether rough or smooth, and their stickiness. The frontal area depends upon vegetation morphology and density. Consequently, the filtration rate coefficient is volumetric, in addition to being dependent upon water velocity. In combination, several authors have written:

$$F = \lambda h C \tag{8.23}$$

where: λ = volumetric filtration rate coefficient, 1/day

Thick sticky periphyton encrustations may be present, which enhance collision efficiency. Saiers et al. (2003) performed studies in a surface-water flume facility constructed within Shark River Slough Everglades wetlands, in which they dosed titanium dioxide particles average diameter of 0.3 μm. They found:

> The TiO2 particles were too small to be removed by settling, and the particles did not diffuse to the channel bottom, where they would be susceptible to removal by adsorption to the sediments. Interception of particles by aquatic vegetation represented the primary mechanism of particle removal within the surface-water channel.

The Saiers et al. (2003) data analysis concluded that a single stem was capable of scavenging 29% of the particles that approached its projected cross-sectional area from the upstream direction, or a first order volumetric removal rate constant of 3.55 hr^{-1}. Under the prevailing natural flow conditions of an *Eleocharis* spp. community in the Everglades, the travel distance for particles would be approximately 10 meters for 63% removal. However, Huang et al. (2008) found a much lesser filtration of 1.1 μm particles in a sawgrass (*Cladium jamaicense*) community. The mean first order volumetric removal rate constant was 0.58 hr^{-1} (Noe et al., 2010), leading to travel distances before removal of over 100 m.

Filtration requires water movement, and thus does not occur in stagnant water. However, nonflow conditions in wetlands are not stagnant. Wind mixing creates water flows, which then invoke the filtration mechanisms. Such wind-driven currents are by no means trivial in large bodies of unprotected EAV-free water, as documented in Chapter 4. Therefore, wind-driven currents can provide filtration in the absence of the addition of source water.

Resuspension

Particles that settle in the shallow water environment of wetlands are more susceptible to resuspension than those that settle in deep sedimentation basins. Settled particles may not "stay put" for a number of reasons:

1. Hydrodynamic shear forces may tear particles loose from the sediment bed, which is a dominant mechanism in streams and rivers. The sloughing of epiphytic solids may also be considered as resuspension. Shear resuspension is associated with high water velocities.
2. Wind and wave action are major drivers of resuspension in lakes and are also operative in open water areas of marshes. Wind resuspension is nonexistent in dense EAV marshes, but can be dominant in SAV systems with large fetch.
3. Biological activity (bioturbation) may result in the movement of particles from the sediments to overlying water. Benthivorous and zooplanktivorous fish, waterfowl, and mammals have all been implicated in sediment resuspension.

Hydrodynamic Resuspension

Much is known about the resuspension of particulates from flat surfaces (ASCE, 1975). Most interpretations are made in terms of the force per unit area (shear stress) required to tear a particle loose from the sediment surface. The concepts involve purely physical forces and apply most readily to mineral substrates and river systems. Most theoretical results are for planar sediment bed bottoms with no extraneous objects. Vegetated wetland bottoms do not fit these conditions. Nevertheless, planar resuspension serves as a basis for terminology and definition of the relevant controlling factors.

The shear stress on a flat bed of particles is written as:

$$\tau_b = \rho u_*^2 = \rho g h S \tag{8.24}$$

where: τ_b = shear stress at the bed surface, Pa
g = acceleration of gravity, m/s^2
h = water depth, m
ρ = density of water, kg/m^3
S = slope of the water surface, m/m
$u_\star$ = friction velocity, m/s

The slope of the water surface is given by a friction equation, which for open channel flow is either laminar or turbulent:

Laminar

$$S = \left(\frac{3\mu u}{\rho g h^2}\right) \tag{8.25}$$

Turbulent

$$S = \left(\frac{nu}{h^{2/3}}\right)^2 \tag{8.26}$$

where: n = Manning's coefficient, $s/m^{1/3}$
u = water velocity, m/s
μ = viscosity of water, kg/m·s

When these Equations 8.24, 8.25, and 8.26 are exercised, it is found that the friction velocity is on the order of 10% of the vertically averaged water velocity. In any case, the bed shear stress may be computed from knowledge of the flow condition.

No particle entrainment occurs until the bed shear stress is greater than a critical value, τ_{crit}. Critical shear stresses for wetlands have been reported in the range of 0.01 Pa (Larsen et al., 2009a) to 0.14 Pa (James et al., 2004b). If the critical shear stress is exceeded, then marsh TSS entrainment rates have been found to be represented by:

$$J_r = M\left(\frac{\tau_b - \tau_c}{\tau_c}\right)^n \qquad \tau_b > \tau_c \tag{8.27}$$

where: J_r = resuspension flux, $gdw/m^2\cdot d$
M = regression coefficient (resuspension flux at twice τ_c), $gdw/m^2\cdot d$
n = regression coefficient, –
τ_b = shear stress at bed surface, Pa
τ_c = critical shear stress for commencement of resuspension, Pa

The parameters in Equation 8.27 were determined for Everglades floc to be M = 19 gdw m^{-2} d^{-1}, n = 1.6 and 0.01 Pa (Larsen et al., 2009a). This model has also been applied to resuspension in shallow lakes with SAV (Hamilton and Mitchell, 1996).

Tsanis et al. (1998) concluded that a critical shear stress of 0.01 Pa applied to the Cootes Paradise treatment wetland. However, Harvey et al. (2011), in field studies in the Everglades, found significant resuspension at bed shears well below the 0.01 Pa level; that is, 40–60 $gdw/m^2\cdot d$ at τ_b= 0.008 Pa. Part of the entrained material originated from epiphyton. Wetlands differ from open channel situations, because submerged plant parts may be coated with epiphytic encrustations. Those biofilms grow and shed particles and are exposed to shear forces different from those on the bottom bed. Harvey et al. (2011) found floc entrainment at velocities below those that would cause bed resuspension. They speculated that P-rich, coarse particles were episodically dislodged from epiphytic biofilms under quiescent conditions, due to the activities of fish and macroinvertebrates. At higher velocities, they concluded that particles would originate from both the epiphytic films and from the bottom bed.

In the treatment marsh environment, nominal water velocities (wetland length divided by nominal detention time) are often not high enough to cause planar resuspension. But, as shown in Figure 8.5, there is a distribution of velocities to be expected, with some elements of water moving through the wetland at speeds considerably greater than the nominal average. Part of such DTD velocity variation is due to patchiness of vegetation. Kleeberg et al. (2010) noted that patchy mosaics lowered the velocity inside SAV patches, providing a low energy environment promoting particle trapping, but between the patches were open water areas of higher velocities, which retained only coarse-grained sediments.

Wind Resuspension

Wind creates waves and is a primary source of bottom shear stress in a lake, because the presence of waves may stir sediments. Lake studies typically involve zero flow-through velocities, but large wind fetches with few or no emergent macrophytes (James et al., 2004a). In large wetlands, wind fetch may be as much as several kilometers. This particularly affects wetlands without emergent vegetation, such as SAV cells. Resuspension occurs when deep water waves enter water shallower than one-half of the wavelength. Fetch and water depth are controlling factors.

Waves of such magnitude are unlikely in treatment marshes, but in open water areas, wind-driven currents cause surface flow in the wind direction and return flows along the bottom in the opposite direction. These recirculation velocities can far exceed the net velocity from inlet to outlet. Lake studies suggest both processes are wind-dependent. For instance, Malmaeus and Hakanson (2003) suggest resuspension is proportional to the square of the wind speed.

A study of surface (wind) shear on resuspension for EAV, SAV and open water was conducted by Wu and Hua (2014). This study investigated the effect of vegetation on sediment resuspension using a surface-driven raceway to simulate wind-driven current. Two common types of vegetation, *Vallisneria natans* (SAV) and *Acorus calamus* (EAV) of were planted in raceways. The results indicated that bed shear stress in the presence of vegetation was much less than unvegetated. In the absence of vegetation, large TSS concentrations were observed for high wind speeds, and much lower concentrations in SAV and still lower in EAV. The TSS release mechanism was mainly related to wind speed and began at about 5 m/s (11 mph).

However, many wetlands differ from this situation of bed resuspension, because submerged plant parts may be coated with epiphytic encrustations. Those biofilms grow and shed particles and are exposed to shear forces different from those on the bottom bed. Harvey et al. (2011) found floc entrainment at velocities below those that would cause bed resuspension. They speculated that coarse particles were episodically dislodged from epiphytic biofilms under quiescent conditions, due to the activities of fish and macroinvertebrates. At higher velocities, they concluded that particles would originate from both the epiphytic films and from the bottom bed.

The presence of submerged macrophytes inhibits particle resuspension from bottom sediments, by damping the velocities created by wave action (Li et al., 2008). It is known that vegetation increases the retention of particulates in both lake and stream environments. For instance, Horppila and Nurminen (2003) found that beds of *Ceratophyllum demersum* and *Potamogeton obtusifolius* in a lake environment effectively reduced resuspension, which they attributed to a reduction in wind and wave action. Similar results were found by James et al. (2002).

Bioturbation

The activities of several types of animals can create resuspension in wetlands, including different types of fish (such as common carp, *Cyprinus carpio;* bream (*Abramis brama*), rodents (nutria, *Myocastor coypus*); waterfowl and others, such as alligators (*Alligator mississippiensis)*). Nutria have been found to severely impact the TSS performance of treatment marshes (Kuehn and Moore, 1995; Mokry, 2010). Human activities in the interior of treatment wetlands may also result in locally elevated concentrations of suspended solids. For instance, the passage of a drifting boat can cause extreme resuspension (Figure 8.7). Carp effects on TSS performance have received some attention, and some approximate quantitative relations developed.

There have been only a few attempts to quantify bioturbation as a wetland sediment resuspension process. Breukelaar et al. (1994) used replicated, shallow, sparsely vegetated ponds stocked with carp and bream at different densities to determine the effect of fish on resuspension. Solids were captured in sediment traps. The relationship, determined for carp densities up to 600 kg/ha, was:

$$R = 0.236B_C \tag{8.28}$$

FIGURE 8.7 Passage of a drifting boat can stir up a cloud of floc. This site is in the interior of the A. R. Marshall Loxahatchee National Wildlife Refuge. The water was about 45 cm deep, and the vegetation was sparse.

where: B_C = carp biomass, kg/ha
J = resuspension rate, gdw/m^2·d

The corresponding levels of TSS also varied linearly with B_C, corresponding to a settling velocity of about w = 3 m/d. At 600 kg/ha, TSS ≈ 40 mg/L, and the resuspension rate was 140 gdw/m^2·d.

Studies of carp effects were conducted at Delta marshes in Manitoba (Badiou and Goldsborough, 2010, 2015). In systems comparable to Delta Marsh, biomasses typically range between 450 kg/ha and 800 kg/ha (Badiou and Goldsborough, 2015). Therefore, this mesocosm enclosure investigation utilized stocked densities of 650 and 1400kg/ha. For the low density mesocosms, TSS rose rapidly from about 10 mg/L to about 50 mg/L. The resuspension rates were measure at 45–85 gdw/m^2·d. Mesocosms with carp plus added nutrients showed much larger TSS and resuspension.

The Cootes Paradise marsh in Ontario contains large numbers of carp. Quantitative studies of their effect on TSS were conducted by Lougheed et al. (1998). Twelve 50-m^2 in situ enclosures were built in a shallow (40–70 cm) nonvegetated area of the marsh. These contained 0–2100 kg/ha carp. They determined the relation TSS = $0.013B_C + 54$ mg/L, of which 80–95% were inorganic solids. At a fish density of 500 kg/ha, the resuspension rate was 118 gdw/m^2·d. This formulation was applied by Tsanis et al. (1998) in a TSS model of the Coots Paradise wetland.

Resuspension Data

There are two common methods of quantifying resuspension using field measurements, chemical tagging and comparative collectors. The tagging method usually relies upon differences in the organic content of the solids, expressed as the volatile fraction (Bloesch, 1994). New material settling from the water column is presumed to differ in organic content from the resuspended material coming from the upper surface of the deposited sediments, and in combination result in the organic content of the material collected in a trap.

$$R = S\left[\frac{f_S - f_T}{f_R - f_T}\right] \tag{8.29}$$

where: f_S = organic fraction of settled(trapped) material

f_R = organic fraction of surficial bed material

f_T = organic fraction of TSS (seston and tripton)

Horppila and Nurminem (2001) used the tagging method in marshes, and reported resuspension in a *Typha angustifolia* stand at 10.4 gdw/m^2·d. In an adjacent zone of *Nymphaea alba* and *Nuphar lutea* (water lilies), resuspension was 26.1 gdw/m^2·d. In both cases, resuspension was wind-driven. In another study, Horppila and Nurminem (2003) measured resuspension in zones containing SAV (*Ceratophyllum demersum* and *Potamogeton obtusifolius*), where resuspension was 9.5 gdw/m^2·d. In adjacent open water zones, resuspension was 20.2 gdw/m^2·d. Variation during the warm seasons is shown in Figure 8.8.

The second method is the tandem use of "cups and plates." The cup collector has high aspect ratio (height: diameter), typically greater than 5. The plate collector typically has aspect ratio zero and may be either a solid plate or another identifiable horizon marker, as discussed in Chapter 4. In principle, the cup sidewalls prevent resuspension. Thus, the cup material represents settled material (S). The plate collector presumably represents an amount S-R, and by difference the resuspension amount is determined.

Sedimentation fluxes are not trivial in wetlands, as indicated by cup measurements. For example Dieter (1990) measured fluxes of 42–118 gdw/m^2·d in areas of dense emergent vegetation, compared to 105–147 gdw/m^2·d in areas of sparse emergent vegetation, both with a fetch of more than one kilometer.

Large sedimentation and resuspension fluxes also occur in wetlands receiving mineral TSS. Geranmayeh et al. (2018) measured both fluxes in four treatment marshes receiving field runoff mineral solids in Sweden, over a three-year period, using the cup and plate technique. Annual sedimentation ranged from 11–521 gdw/m^2·d, and resuspension from 8–458 gdw/m^2·d. Resuspension was 3–7 times greater than the net accumulation of particulates.

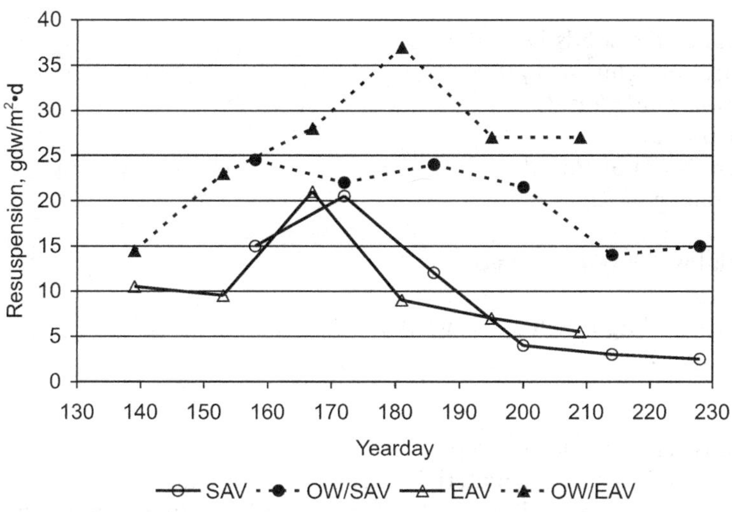

FIGURE 8.8 Summer resuspension rates in SAV, EAV, and open water adjacent to each. Data from Horppila and Nurminen, 2001, 2003.

And, resuspension was a nearly constant fraction of sedimentation, 77–87%. Similar results have been reported for the Olentangy, Ohio marshes receiving mostly mineral solids in river water (Mitsch et al., 2014a). And, similar results have been reported by WRI (1992), for the Des Plaines, Illinois marshes, also receiving mostly mineral solids in river water. Wetland EW3 was heavily loaded when the pump was operating and contained relatively sparse emergent vegetation. Settling columns data and cup measurements gave R = 46.0 gdw/m^2·d.

8.3 PROCESS QUANTIFICATION

From the process studies discussed above, it is clear that internal cycles dominate TSS processing in marshes. Therefore, the utilization of input and output data to compute percent removals is an inadequate representation of the processes that lead to those removals. A more detailed computational framework is necessary.

Spatial Mass Balances

Equation 8.1 describes one-dimensional plug flow movement of TSS in wetland waters, and includes terms for the various processes. In nearly all cases, changes in water column storage are negligible compared to other inputs and outputs. And, depth variations can usually be averaged. The water column TSS mass balance then becomes a steady state model. There is an accompanying sediment bed balance, in which the change in storage is the dominant feature. In long-term calculations, accreted solids undergo decomposition, but that process occurs on a time scale of years and is not a factor on the time scale of days or months, which is appropriate for the water phase calculations. The long-term, time-average profiles calculated from the vertically averaged mass balances for TSS in a linear flow wetland are:

$$uh\frac{\partial C}{\partial x} = G + R - S \tag{8.30}$$

$$\frac{\partial B}{\partial t} = S - R \tag{8.31}$$

where: B = mass of the solids bed, g/m^2

C = concentration, g/m^3 = mg/L

G = generation rate, g/m^2·d

h = water depth, m

R = resuspension rate, g/m^2·d

S = settling rate, g/m^2·d

t = time, d

u = superficial water velocity, m/d

x = distance, m

In general, the settling rate may be written as:

$$S = w\ C \tag{8.32}$$

where: w = solids settling velocity, m/d

The return flux of suspended material originates from generation (G) and resuspension (R), which may be distinguished by the tagging method described above. In a spatially uniform wetland, as may occur after inlet settling effects no longer prevail, there will be little or no concentration gradient, and:

$$wC^* = G + R \tag{8.33}$$

where: C* = background concentration, g/m^3 = mg/L

Because TSS generation is nutrient-dependent, there may be reductions in G along the flow direction after inlet settling is complete, which are the result of nutrient reductions along the gradient.

The Inlet Settling Zone

Equations 8.30 and 8.32 combine to give a representation of the change in TTS along the flow direction, which may then be written, for the plug flow assumption, as:

$$uh\frac{dC}{dx} = w(C^* - C) \tag{8.34}$$

Integration gives:

$$\frac{(C - C^*)}{(C_i - C^*)} = \exp\left(-\frac{wx}{uh}\right) = \exp\left(-\frac{w\tau_x}{h}\right) = \exp\left(-\frac{\tau}{\tau_A}\right) \tag{8.35}$$

where: C = TSS concentration, g/m^3

C_i = inlet TSS concentration, g/m^3

x = distance along the flow path, m

τ = nominal detention time along the flow path, d

τ_A = accommodation time (for 63% reduction), d

The quantity wx/uh = N_f, the fall number at distance x. It is probably easier to visualize the reduction along the flow path in terms of detention time. A useful concept is the accommodation time (τ_A), defined as the time for 63% of the excess incoming TSS above background to settle. For systems dominated by resuspension and settling, such accommodation times are often quite short. However, if generation is the dominant mechanism, accommodation times will be dictated by the speed of growth and death of organisms, which are typically longer than for settling and resuspension. And, instead of an inlet settling zone, the wetland displays a TSS growth zone, in which TSS concentrations increase with distance.

Particle Size Distributions

Equation 8.35 may be applied to mixtures of particles, of different sizes, shapes and densities. When vertically averaged, settling is described by the volumetric coefficient k = w/h. Solids properties combine to produce a settling rate coefficient distribution, f (k), defined by:

f(k)dk = mass fraction of particles with settling velocities w to w + dw

The first-order settling is then applied across the distribution to yield the TSS mass remaining

$$\frac{\bar{C}}{C_0} = \int_0^\infty e^{-kt} f(k)dk \tag{8.36}$$

where: $\bar{C}$ = vertically averaged concentration, g/m^3

C_0 = starting concentration, g/m^3

k = volumetric settling rate coefficient, d

t = elapsed time, d^{-1}

This approach was utilized to describe the batch (plug flow) removal of TSS for the Houghton Lake treatment marsh solids (Kadlec, 2003a). The settling rate distribution was found to be uniform over 0 < k < b, leading to:

$$\frac{\bar{C}}{C_0} = \int_0^b e^{-k\tau}\frac{1}{b}dk = \frac{(1-e^{-bt})}{bt} \tag{8.37}$$

Settling rate distributions are not typically available, and thus the utility of this procedure is limited. Additionally, the settling rate distribution would need to be concatenated with the detention time distribution in a nonideal flow wetland.

Fall Number Distributions

It is also possible to retain the concept of a fall number, while acknowledging that there will be a distribution of N_f corresponding to the size distribution of particles. This approach has been proposed by Fletcher et al. (2004) and Li and Fletcher (2007), as shown in Equation 8.18. The result is a system removal described by the fall number N_f:

$$E_S = 1 - \frac{(C_o - C*)}{(C_i - C*)} = \frac{N_f^a}{N_f^a + b} \tag{8.38}$$

where: a,b = regression parameters, –

E_S = settling efficiency (fractional removal), –

Li and Fletcher (2007) determined values of a = 0.45 and b = 1.4 for urban runoff solids from Melbourne, Australia, with a mean diameter of 40 μm.

Accommodation

Because accommodation zones are small (short), they are very often contained inside the wetland. In that circumstance, calibration of Equation 8.35 requires internal transect data for TSS, which is rarely available. Examples of internal transect profiles of TSS are given in Figure 8.9. For these examples, the accommodation times are one day or less. In general,

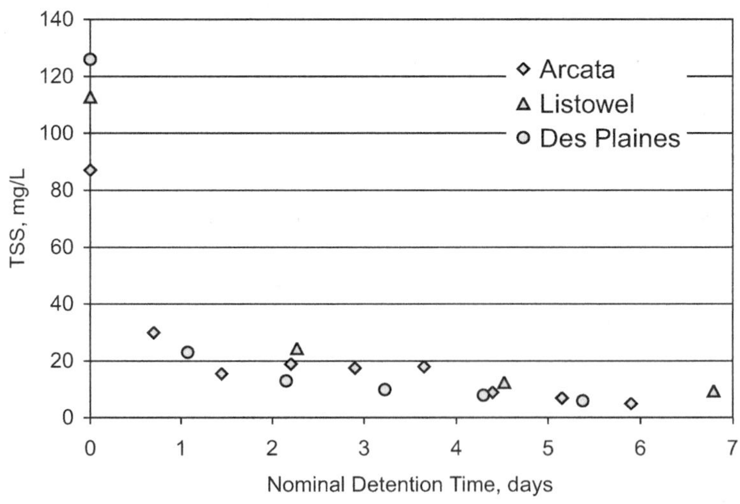

FIGURE 8.9 Gradients in suspended solids along the flow direction in treatment wetlands. Data source for Arcata, California: Gearheart et al. (1989); Data source for Listowel, Ontario: Herskowitz (1986); Data for Des Plaines, Illinois: unpublished data.

there are not enough points along a transect to provide a good estimate of the k-value, because most of the data are in the region near C*. The parameter "w" therefore cannot be determined accurately for nearly all FWS datasets.

Further, because most of the treatment marsh operates at a background concentration, there is not a necessity for the tanks in series model.

A detailed gradient study to provide calibration of the kC* model (Equation 8.34) was done at the Hallam Valley wetlands in Melbourne, Australia (Wong et al., 2006). Twenty-micron particles of either clay or sand were dosed into 60 m^2 channels, at high concentrations (200–600 mg/L). Exceedingly high water flows (nominal HRT < 3 hours) were required to detail the rapid decrease of TSS. Model fits were excellent, with w-values in the range of 16 to 21 m/day, for both vegetated and unvegetated channels, for both clay and sand particles. The corresponding accommodation times were 20–40 minutes. The C* value for the unvegetated channel was about double that for that for the vegetated channel (60 vs. 33 mg/L). This is consistent with resuspension being greater in the open channel. Wong et al. (2006) also fit the kC* model to data from vegetated filter strips and swales, with excellent results.

System Mass Balances

Equation 8.30 may be integrated over the entire length of the wetland system, under the presumption that the right-hand terms do not vary much in the wetland.

$$(M_o - M_i) = \frac{1}{A}(Q_o C_o - Q_i C_i) = G + R - S \tag{8.39}$$

where: A = wetland area, m^2

M_i = mass flux of the solids entering, $g/m^2{\cdot}d$

M_o = mass flux of the solids leaving, $g/m^2{\cdot}d$

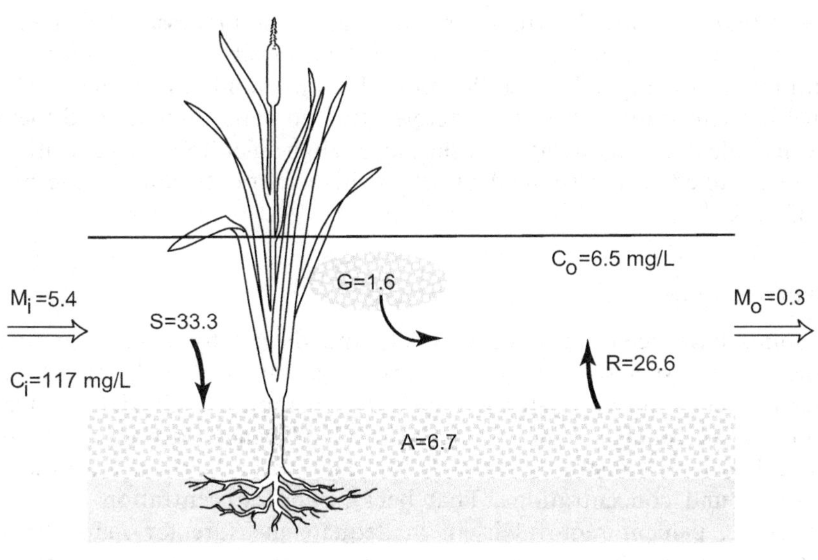

FIGURE 8.10 Components of the sediment mass balance for wetland EW3 at Des Plaines, Illinois. Accumulation A = S – R. The balance period is the 23-week pumping period in 1991. Data source: WRI (1992).

TABLE 8.4

Solids mass balances for example treatment marshes in Sweden and the United States.

	HL	TSS In	TSS Out	Load In	Settled	Accumulated	Generated or Resuspended	R
	m/yr	mg/L	mg/L	gdw/m²·d				(% of S)
Nyb	120	137	131	66	521	63	458	87
Ber	70	108	65	27	82	16	66	83
Ski	60	143	19	41	27	5	22	84
Wig	400	n.d.			11	3	8	77
EW3	20.7	33	6.5	5.4	33.3	6.7	26.6	80
HL	1.5	34.3	7.0	0.15	3.20	0.12	3.08	96

Sources: Sweden, Geranmayeh et al. (2018); EW3 and HL, unpublished data.
Key: Nyb = Nybble, Ber = Bergaholm, Ski = Skilleby, Wig = Wiggeby.

Q_i = volumetric inflow rate, m^3/d
Q_o = volumetric outflow rate, m^3/d

The magnitudes of these mass balance terms are illustrated in Figure 8.10. System inputs and outputs do not adequately describe the performance of this wetland, because they do not reflect the large internal cycling of suspended material.

This example is not an anomaly, because other investigators have found similar large internal cycles (e.g., Gerenmayeh et al., 2018; see Table 8.4). When there is no flow through a marsh, the same large internal cycles are observed, as evidenced in Figure 8.8.

8.4 OBSERVED PERFORMANCE

Many if not most treatment marshes remove TSS down to background levels in an inlet region. Background levels are determined by internal marsh processes. This means the k-C* model is dominated by C*, and other factors must responsible for the differences in outlet TSS concentrations at comparable inlet loadings. Likely candidates are inlet TSS concentration and inlet nutrient status. These two factors often go hand-in-hand, and there is not yet a study that has identified the relative importance. High inlet TSS concentrations could be partially short-circuited to the wetland outlet, or high nutrients could cause more internal generation of TSS.

Input-Output Relations

Suspended solids have been measured at inlets and outlets for a large number of treatment marshes. It is instructive to examine this large inter-wetland dataset, to ascertain the existence of trends among systems. A popular method of TSS data representation is the quotation of percentage removal, or removal efficiency. However, the presence of a background TSS level constrains removal efficiency to be below a level dictated by the inlet and background concentrations. That background concentration is often achieved. As a consequence, percent removal is an inadequate measure for many treatment wetlands. Indeed, some efficiencies are negative, in situations where pretreatment includes removal of TSS prior to the wetland, because influent TSS concentrations are below the wetland background concentrations.

First consider treatment marshes without preceding settling basins, receiving moderately low TSS source waters. The source waters are from lakes, rivers and well-treated municipal effluents. Inter-system outlet concentrations are quite low and increase slightly with the areal loading of TSS to the wetland (Figure 8.11). At any given loading rate, the data cloud spans about ±2.0 mg/L outlet concentrations. These outlet TSS values are strongly left censored, because of the lower detection limit of 1–2 mg/L that accompanies the measurement. Within this group are seven wetlands that have negative removal rates, because TSS is very low in the source water, and increases upon passage through the wetland. The wetlands considered exclude those with known issues of bioturbation or lack of vegetation.

Wetlands that target waters leaving row crop fields typically receive much higher loadings and concentrations, and the solids are of mineral origin. An example set of performances for these is shown in Figure 8.12. These span a large range of TSS loadings, spanning four orders of magnitude. The outlet TSS values average 40 mg/L, and have a very large scatter.

There is also a sizeable database of performances of urban stormwater wetlands. In the United States, this database originated with the work of Strecker et al. (1992) and Schueler (1992), and grew through continued efforts of Godrej et al. (1999). In contrast to continuous flow system data, event systems have much greater variability in design. Loadings are extremely variable, as are the geometries and arrangements of deep pools, shallow marsh zones and interconnecting channels. Therefore, the intra-system database gives an impression of what has been accomplished, rather than what might be accomplished by "optimal" designs. Percent removals are high in many cases, with 60% of the U.S. data showing greater than 60% removal.

Duncan (1998) (as referenced and reported by Wong et al., 1999) performed an empirical correlation of performance of 76 stormwater wetlands, including those in the United States, and complemented by systems in Australia. That relationship (R^2 = 0.80) is:

$$TSS_{out} = 0.78q^{0.33}TSS_{in}^{0.51} \tag{8.40}$$

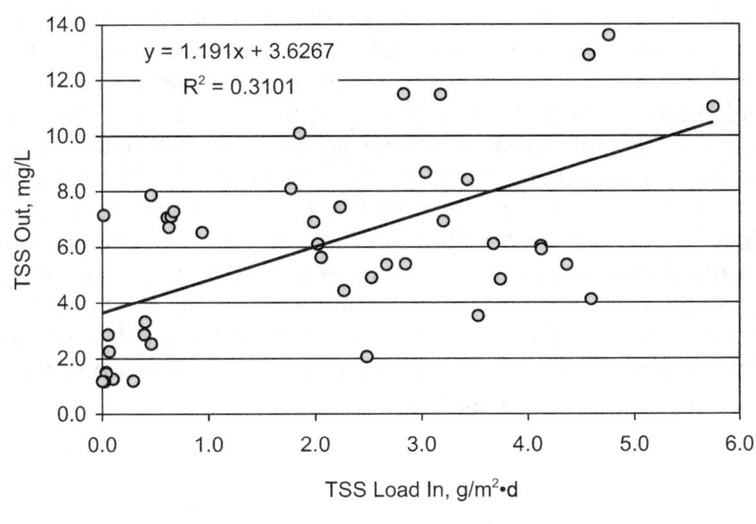

FIGURE 8.11 Load response for TSS in treatment marshes without or after settling basins. Data represent the POR for one wetland, N = 45 wetlands. The mean entering TSS = 33 mg/L (range 1–111), and the hydraulic loadings averaged 6.8 cm/d (range 0.4–30).

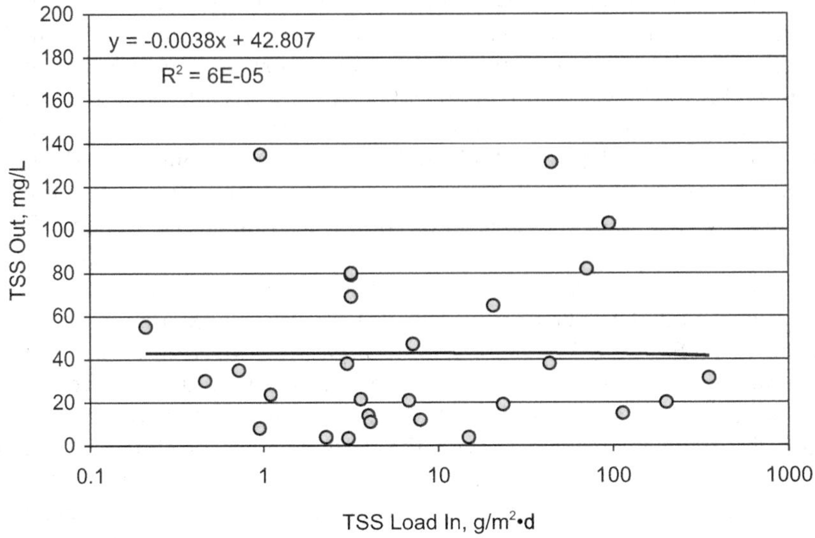

FIGURE 8.12 Load response for TSS in marshes treating agricultural row crop runoff. Data represent the POR for one wetland, N = 28 wetlands. The central line is a linear regression, with R2 = 0. The mean entering TSS = 153 mg/L (range 6–611), and the hydraulic loadings averaged 20 cm/d (range 1–100).

where: q = wetland hydraulic loading, m/yr
TSS = concentration, g/m^3

This regression is entirely consistent with the inter-system continuous flow wetland data (Figures 8.11 and 8.12), and may also be used to represent continuous flow systems.

Input-Output Data and the kC* Model

The PkC* model may be used to describe TSS reductions, but only if due care is taken to acknowledge the possibility of a steep internal drop in concentration, followed by levels at the C* plateau. Input/output data typically do not contain that information. It is known that k-value determinations are strongly influenced by those internal plateaus (see Kadlec and Knight, 1996), with calibrations producing serious under-estimates of the appropriate k. With that caveat in mind, there are reported calibrations for TSS for treatment marshes (Table 8.5). The stormwater systems were analyzed based on flow-proportional sampling, and EMC values. The assumed C* values for Table 8.5 are low, thus it is not representative of biologically disturbed or open water systems. The k-values so estimated are high, in the vicinity of 100 m/yr. For treatment marshes with depth about 50 cm, the detention time to the plateau would be about 1–2 days. That range matches the data shown in Figure 8.9. Thus, for forecasting TSS performance, the focus is upon C* whenever the detention times are greater than a couple days.

Variability

TSS removal is variable along the flow direction. And, temporal variability in performance may be comprised of three contributions: (1) event-to-event inflow and concentration behavior, (2) annual seasonal patterns, and (3) stochastic scatter in removal efficiencies.

TABLE 8.5
PkC* model parameters for TSS removal in marshes.

	P = 4, C* = 3 Continuous Flow	C* = 2 Event Driven	
Count	**33**	**11**	
	k, m/yr	**P**	**k, m/yr**
20th%tile	33	2.4	25
50th%tile	62	3.0	84
80th%tile	119	4.0	171
Min	12	1.6	10
Mean	78	3.5	125
Max	266	7.6	421

Source: Event Driven, Merriman et al. (2017).

Spatial Patterns

Removals of TSS along the flow direction lead to decreasing concentration profiles, which plateau at C*. The speed of decline varies but is often quite sudden (Figure 8.9). High rates of decline are the result of a dominant mechanism of particulate settling. When biological processes are involved, as in the growth of algae, the decline along the flow direction can be much slower, or there may be an increasing profile of concentrations if the source water is devoid of TSS. Data analysis via the TIS model requires the internal gradient of TSS concentrations, which is typically not available. One instance is the Shannon wetland near Corsicana, Texas (Figure 8.13).

Annual Patterns

For marshes receiving fairly steady flows, monthly trends in marsh TSS are variable and weak. Settling processes are not strongly temperature dependent, because the controlling

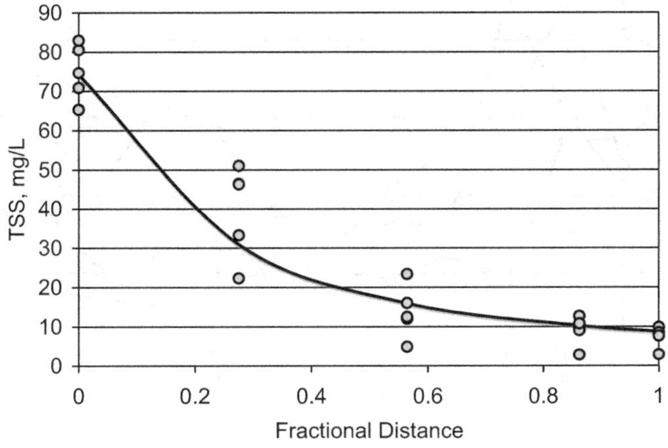

FIGURE 8.13 Fit of profile data for TSS in the field-scale wetland train at the Shannon wetland near Corsicana, Texas. Annual averages for five years are shown. The model line is from the PkC* model, with P = 4, k = 79 m/yr and C* = 6 mg/L.

factors of water density and viscosity are not strongly variable. The filtration capacity relies upon the presence of vegetation, but winter standing dead is arguably as effective as summer live submersed vegetation.

Annual patterns of TSS concentrations exiting from marshes treating runoff and river water show higher TSS in winter than in summer (Figure 8.14). The reasons for this (small) effect are not apparent but may have to do with wind and solar driven mixing. It is not likely an inlet TSS effect, because the systems in Figure 8.14 received varying levels of incoming TSS: 74 mg/L for Shannon, 32 mg/L for Brawley, and 19 mg/L for Imperial.

Systems receiving lagoon-pretreated water display the opposite pattern (Figures 8.15 and 8.16). During the warm summer months, these wetlands show elevated TSS, probably the result of algal growth and the associated particulates. Again, it is not likely an inlet TSS effect, because the Listowel system in Figure 8.15 received incoming TSS = 111 mg/L, while the Onaway wetland received 3.1 mg/L. TSS processes under the ice in winter have not been investigated, and the differences seen in Figures 8.15 and 8.16 are quite large. Onaway displays elevated winter values, but Listowel does not.

Given the lack of predictable seasonal behavior across systems, it is necessary to include seasonality in the anticipated stochastic variability.

Event Driven Systems

Performance data acquisition is difficult because flows occur during storm conditions, and close-interval sampling is needed to define the course of a single event. As discussed in Chapter 7, the appropriate analysis for event-driven wetlands is the use of event mean concentrations, or in-use time periods. Both water and solids dynamics influence the performance of event-driven wetlands. If the event period is long compared to the detention time of the wetland under the event flow, then the first part of the event completely flushes the antecedent water, and the outflow concentration ramps up toward the level dictated by the sustained event flow. After the cessation of event flow, the wetland outflows and outlet concentrations ramp back down to nonflow conditions. The majority of the treatment occurs

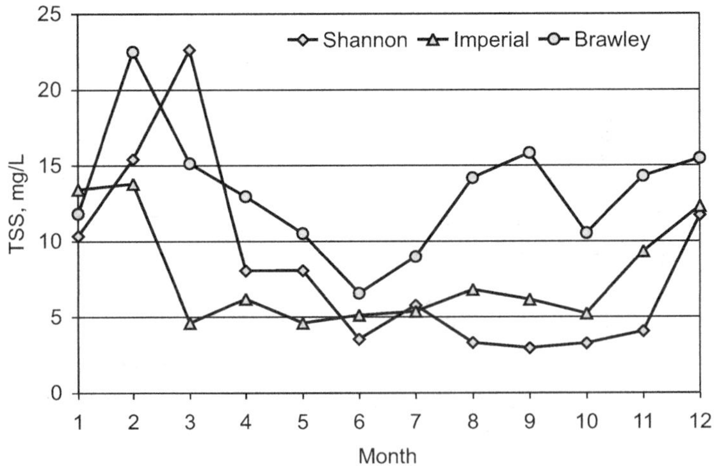

FIGURE 8.14 Monthly values of TSS in the Shannon, Texas and Imperial and Brawley, California treatment marshes. Annual averages for four years are shown. These southern systems have no ice cover. They receive runoff (Imperial) and river water (Shannon, Brawley).

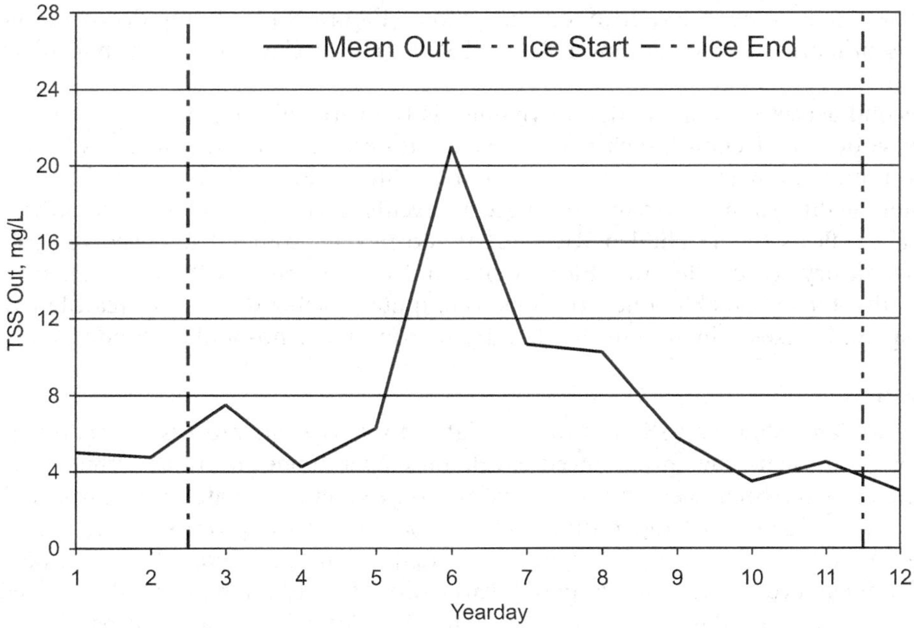

FIGURE 8.15 Monthly values of TSS in the Listowel, Ontario Unit 4 treatment marsh. Annual averages for four years are shown. This northern system experienced ice cover of 5–15 cm during winter.

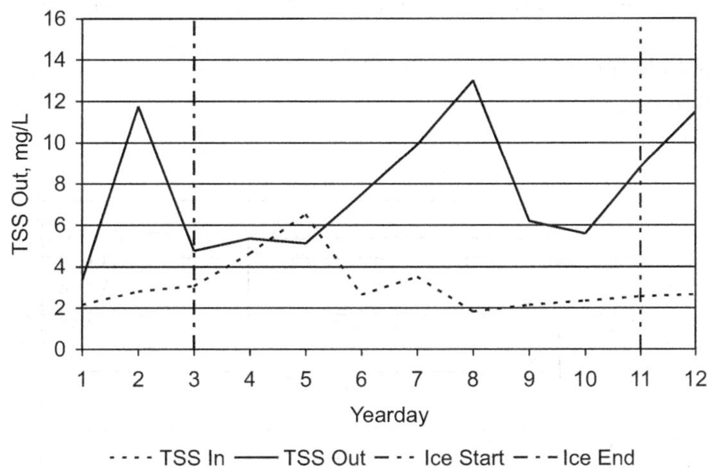

FIGURE 8.16 Monthly values of TSS in the Onaway, Michigan treatment marsh. Annual averages for five years are shown. This northern system experiences ice cover beginning about the end of November, and lasting until about the end of March.

during the period of high flow, and the expulsion of the highly treated antecedent water contributes little to overall TSS mass removal during the event.

In contrast, if the event period is short compared to the detention time of the wetland under the event flow, then the event does not completely flush the (highly treated) antecedent water. Ramp-up of the exported solids load starts to occur but stops far short of a new steady state determined by the inflow (Figure 8.17). Ramp-down of the mass export is primarily due to the lower flows; TSS remained elevated in the immediate after-even period.

A second aspect of event wetland dynamics is the variability from event to event during the course of time. Defining such patterns is a data-intensive effort, which has been accomplished in only a few instances (e.g., Carr and Rushton, 1995). Those researchers examined the water quality from a system for 83 storm events over a period of 30 months (Figure 8.18). The inflows to the Hidden River urban stormwater wetland followed the pattern of the wet and dry seasons for this Florida site, and it functioned with zero discharge in the dry months. Incoming TSS concentrations were quite variable during the year, but the outflow displayed spikes primarily during the dry season, under no-outflow conditions.

Stochastic

Because wetland effluent TSS is strongly related to ecosystem processes, random physical and biological events have pronounced effects on effluent concentrations. Principal bioturbation effects have been seen due to muskrats, beaver, nutria, waterfowl, carp, and shad. Additionally, season and temperature may be modifiers of the processes that generate and cycle solids. These effects may be separated by de-trending the data, which typically follow a mild annual cycle with superimposed variability. The trend may be determined most accurately if there are data spanning many annual cycles, which may then be "folded" into one multiyear display and averaged.

TSS data time series sometimes display some degree of sinusoidal behavior through the course of a calendar year. Therefore, detrending may be accomplished by fitting the (folded) time series to:

$$C = C_{trend} + E = C_{mean}\{1 + A\cos[\omega(t - t_{\max})]\} + E \tag{8.41}$$

where: A = amplitude fraction
C = concentration, g/m^3

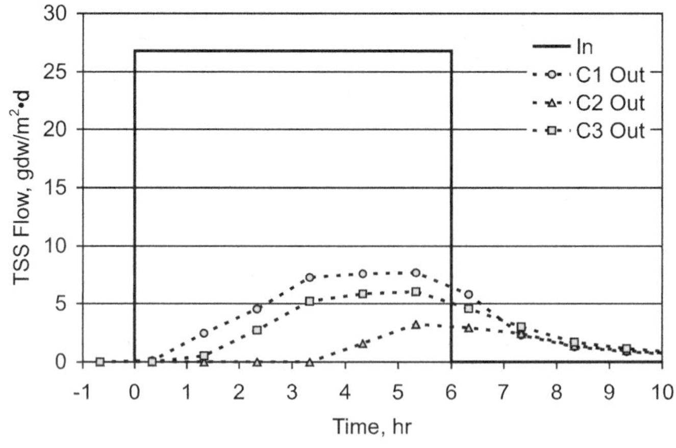

FIGURE 8.17 Mass inputs and outputs for three wetlands at Des Plaines, IL, driven by a six-hour pumping event. Nominal detention times at the event flow rate were 7.2, 15.4, and 8.2 hours for C1, C2, and C3, respectively. The wetlands were therefore not completely flushed of initial water. Mass removals were 73, 90, and 79% respectively.

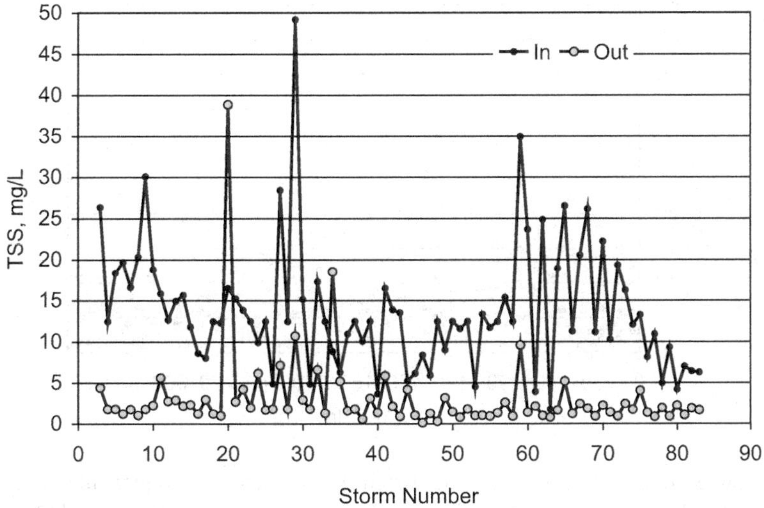

FIGURE 8.18 Results of a 30-month study of the performance of the Hidden River, FL, urban stormwater treatment wetland. TSS and flows were monitored for 83 rain events. Data source: Carr and Rushton, 199.

C_{trend} = trend concentration, g/m^3
C_{mean} = mean annual concentration, g/m^3
E = stochastic departure (error) of an individual measurement, gm/m^3
t = Julian time, d
t_{max} = Julian time of TSS maximum, d
ω = annual frequency, $2\pi/365$, radians/d

The scatter of TSS data is large, and the trend typically accounts for less than 50% of the variability. An example of this model fit to data from the Brighton, Ontario treatment marsh is given in Figure 8.19, for which $R^2 = 0.06$, implying that only 6% of the variability is accounted by the trend. The amplitude of the annual cycle was 0.27 times the mean. Examples may be found of both weaker and stronger annual trends, but typically there is only slight variability across the seasons.

It is useful to examine the multiplier factors, Equations 7.86 and 7.87, associated with the various (higher) percentiles of the effluent distributions, because these may be involved in permitting or licensing of the treatment wetland. Examples of these outlet multipliers are shown in Table 8.6, for a sampling of wetlands spanning a range of inlet concentrations from 1 to 100 mg/L. It may be seen that in several instances, excursions of outlet concentrations exceed the average inlet concentration, despite long-term average concentration reductions. It is only when the inlet TSS reaches about 25 mg/L that not more than 10% exceedances of the inlet concentration occur.

Because wetland effluent TSS distributions are only weakly seasonal, it is possible to ignore these trends, and to consider the total variability. This is sometimes done in the treatment wetland literature (e.g., USEPA, 1999). The frequency distributions of the inlet and outlet TSS measurements are displayed graphically. Figure 8.20 shows an example of this procedure, derived from data from the Arcata Allen enhancement marsh. Note that the 50th percentile represents the median of the data, not the mean. Further note that these are not paired point graphs, so that reductions cannot be computed at any specified frequency level.

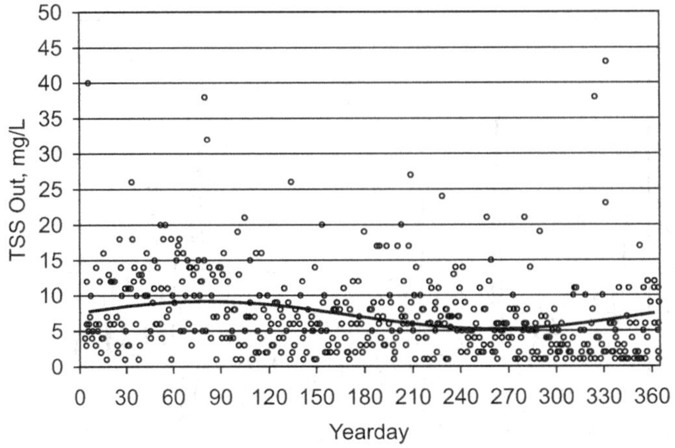

FIGURE 8.19 Suspended solids leaving the Brighton, Ontario treatment marshes versus day of the year. The mean outlet TSS = 7.2 mg/L. Ten years of weekly data are represented (N = 500).

TABLE 8.6

Trend multipliers for a sampling of FWS wetland effluents. Note that the 50th percentile is the median, not the mean. Frequencies are weekly.

		Inlet	Outlet	Outlet Multipliers (1+ψ)		
		mg/L	mg/L			
Percentile		50	50	80	90	95
Commerce Township	MI	1	9	1.6	2.0	2.4
Tres Rios H1	AZ	3	3	2.0	2.3	3.6
Brighton	ONT	13	7	1.8	2.3	2.9
Estevan	SAS	11	6	2.2	3.3	4.3
Columbia	MO	12	6	1.6	2.3	3.7
Brawley	CA	18	8	1.7	1.8	1.8
Listowel 3	ONT	18	6	1.9	2.7	3.7
Arcata Enhancement	CA	26	3	1.2	1.3	1.3
Imperial	CA	32	10	1.3	1.4	1.4
Tarrant WC1	TX	39	5	1.7	2.5	2.8
Cannon Beach	OR	58	6	2.0	2.4	2.9
Des Plaines EW3	IL	83	7	1.6	1.8	2.1
Listowel 4	ONT	100	5	1.6	3.1	3.9
Mean				**1.7**	**2.2**	**2.8**

Unvegetated Areas

The subject of inclusion of open water areas in treatment marsh systems has been *de facto* separated into three categories:

1. Ponds preceding wetlands
2. Deep zones inside the wetland
3. Entire systems without plants

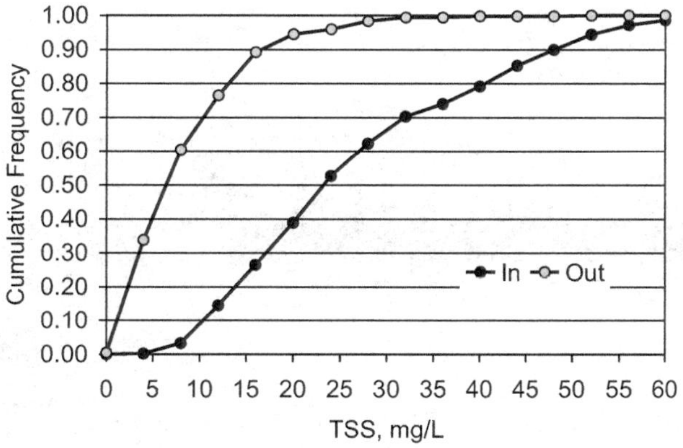

FIGURE 8.20 Probability distributions for inlet and outlet TSS for the Arcata Allen marsh. The median inlet TSS was 24 mg/L; the median outlet TSS was 7 mg/L. Data were weekly for 8 years. (Data source: Treatment Wetland Database, Finney, 2000.)

The first grouping includes settling basins, usually separated from the marsh by a structure; and inlet pools in stormwater treatment systems, usually not separated from the marsh by a structure. Ponds function to settle incoming TSS but are conducive to the production of TSS via algal cycling, as discussed in Chapter 3. Internal and outlet open water areas are settling zones, but are subject to wind resuspension and algal growth.

The primary function of internal deep zones is to create a transversely uniform distribution of flows. An inlet deep zone inside the wetland is essentially a wide, very short pond located inside the wetland boundary. U.S. EPA (2000a) suggests that open zones be incorporated into treatment wetlands as a means of enhancing TSS removal, along with other purposes. The reason given is that these "... can provide conditioning and transformation processes which may improve overall removal of TSS ...". Open zones may contain submerged vegetation, or be devoid of plants. This differentiation is very important, because open zones with submerged aquatic vegetation (SAV) will provide TSS reduction benefits, whereas unvegetated open zones may not, and may in fact increase TSS.

There are some few treatment marshes that contain essentially no vegetation. These may be intentionally fostered, for the purpose of maximizing light penetration (Jasper et al., 2013). In other instances, vegetation has been lost for unknown reasons (Kadlec, 2008a). And deep water created by design can preclude emergent vegetation, and foster waterfowl that destroy SAV (Kadlec, 2011).

Settling Basins

Because incoming TSS is rapidly settled and filtered in the wetland environment, it is possible and desirable to provide a first element of the treatment wetland complex that traps the fastest settling fraction of the suspended material. A pond provides for that pre-settling and is more easily cleaned than an emergent or submergent macrophyte bed. It is further desirable to collect solids and their partitioned metals and chemicals in a location that is not foraged by sediment-feeding vertebrates. This pre-settling pond may require infrequent dredging to remove the accumulated deposits.

Data from the Shannon wetland, Texas illustrates the mean performance of three parallel marsh wetland cell trains of three cells each, following two parallel unvegetated settling ponds (see Kadlec et al., 2011) (Figure 8.21). The settling ponds occupied 3.5% of the area but accounted for 67% of the solids removal (Figure 8.22). The wetland cells complete the

FIGURE 8.21 The settling basin at the Shannon wetland in Texas. The 3.6-ha pond was designed to be 2.2 m deep.

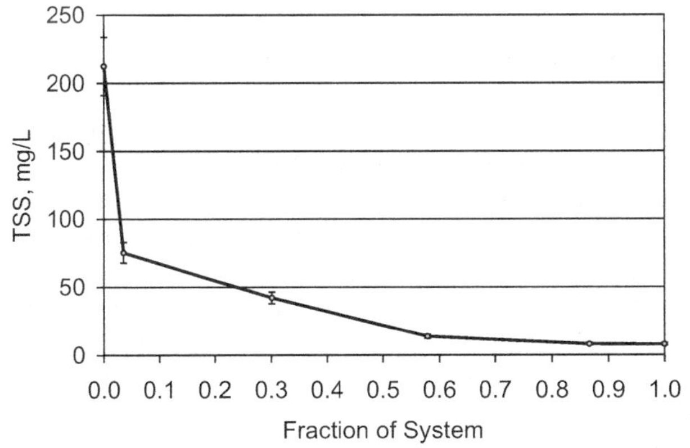

FIGURE 8.22 Profiles of TSS along the flow direction in the Shannon, Texas treatment train over a four-year period. Error bars represent the standard error of about 10%. Note that the wetlands exhibit a plateau, or background TSS, at about 10 mg/L. Data courtesy of the Tarrant Regional Water District.

solids removal. The last wetland cells were, however, needed for phosphorus removal. The performance of other pond-wetland systems is comparable. The large majority of TSS in all systems was retained in a minority fraction of the total footprint.

The placement of a pond as the final element in a wetland treatment system is generally not desirable from the standpoint of TSS reduction. The planktonic production in such a pond is typically quite high, leading to the reintroduction of high-chlorophyll microdetritus, much of which remains in suspension. An example of this phenomenon was the Lakeland, Florida system. Entering TSS was reduced in the first marsh cells but was regenerated in later, open water cells because of planktonic activity (Bays et al., 1993).

Deep Zones

A treatment wetland may include an inlet deep zone, within the footprint of the wetland. Such a feature is in effect a pre-treatment pond element, as discussed in the previous section. In contrast, the wetland may also contain internal deep zones, ranging from narrow ditches to large expanses of open water. Both the number and size of such zones has been variable across a number of projects. Here we examine the efficacy of such zones in TSS reduction.

Knight et al. (1994) operated a set of six treatment wetlands receiving treated paper pulp mill effluent for two years. Two internal deep zones comprising 25, 35, and 45% of the area were included in three of the six. There was variability in flow (different HLRs) and aspect ratios (2.5, 5 and 10). If the pairs of cells with the same aspect are compared on the basis of deep zones, the concentrations produced were lower, and the load removed was higher, in wetlands with deep zones in two of the three pairs.

Eidson et al. (2005) studied side-by-side wetlands at Augusta, Georgia. Six 12-ha cells were studied, two of a marsh-pond-marsh configuration (60% marsh), and four of a marsh-ditch configuration (82% marsh). Incoming water was very low in TSS, averaging 3.3 mg/L over a one-year period. The marsh-pond-marsh cells averaged 12.2 mg/L at their outlets, while the marsh-ditch cells averaged 4.1 mg/L. The effect of the large open water areas in the marsh-pond-march cells was to raise the TSS concentration, presumably due to generation and resuspension. The marsh-ditch systems created a negligible increase.

The Tres Rios, Arizona, project studied the effect of deep zones in a triplicated side-by-side study (Kadlec, 2007). The research site contained a set of twelve 0.12-ha (24 m wide × 50 m long) wetlands, built in a triplicate design with 2, 3, 4, and 5 deep zones, including one at the inlet and one at the outlet. No significant difference could be found in outlet TSS concentrations ($\alpha = 0.05$).

Unvegetated Zones

Further insights can be gained from systems that lost their vegetation over the course of time. A FWS treatment wetland in Commerce Township, Michigan, was "eaten out" by muskrats and waterfowl, leaving virtually no emergent or submergent vegetation (Kadlec, 2011). Prior to the loss, the effluent TSS was 5 mg/L; after, it was 13 mg/L.

The Tres Rios Hayfield wetlands are a second example Kadlec (2008a). During the spring and summer of the third year after startup, the vegetation essentially all died, for reasons that have not been resolved, and regrowth did not occur. Incoming TSS was low (3 mg/L) and remained relatively low during the vegetative period (2 mg/L) (Figure 8.23). After loss of vegetation, effluent TSS climbed to 27 mg/L. This wetland had unvegetated deep zones. In both cases it is apparent that the effluent solids could not have derived from incoming TSS, but rather were the result of internal generation and resuspension.

Accretion

Removed TSS accumulates on the wetland soil surface, and the organic portion may undergo decomposition. These processes, and measurement techniques, have been discussed in Chapter 4.

For high amounts of sediment trapping compared to generation and resuspension, buildup typically occurs preferentially in the inlet section of the wetland. Therefore, a "delta" of accreted sediments builds in the inlet region of the wetland. In contrast, lighter loadings and open water areas may foster the redistribution of suspendable material. For instance, Brueske

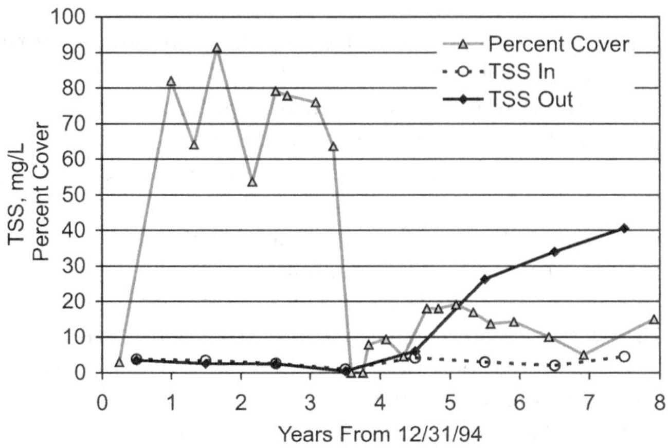

FIGURE 8.23 The transition of Tres Rios wetland H1 from a vegetated to an unvegetated state. Outlet TSS averaged 2 mg/L for the first four years, then climbed to 40 mg/L in year eight.

and Barrett (1994) found a "delta" in a highly loaded wetland (around 3.6 $g/m^2 \cdot d$ TSS), but little or no "delta" for a lower loading (around 0.8 $g/m^2 \cdot d$ TSS). Both Harter and Mitsch (2003) and Brueske and Barrett (1994) found greater sediment accretion in open water areas, which may have been attributable to most of the flow traveling through such areas, or to bioturbation. In contrast, Benoy and Kalff (1999) found a linear relation between sediment accumulation and biomass for submerged species *Myriophyllum spicatum, Potamogeton* spp., *Ceratophyllum demersum*, and *Elodea canadensis* beds in Lake Memphremagog between Québec and Vermont. It is apparent that the processes involved in sediment accumulation in wetlands are too complicated to permit generalities.

In the long term, solids accretion may raise the elevation of the wetland bottom, and thus impact system hydraulics and treatment. U.S. EPA (2000a) estimates treatment wetland accretion of 2–4 cm/yr, which would apply to source waters of moderately high nutrient content. That is excessively high. For example, the removal of 30 mg/L of TSS at a hydraulic loading rate of 10 cm/d results in solids storage of 1,095 $g/m^2 \cdot yr$. At a density of 0.2 g/cm^3, this gives 0.55 cm/yr if there is no decomposition. However, municipal TSS is about half mineral, and half decomposable solids and hence long-term external accretion would be about 0.27 cm/yr. Highly loaded wetlands treating mineral solids have been observed to accrete more than 2.0 cm/yr (Braskerud, 2001a).

Accretion is typically spatially nonuniform, due to gradients in deposition and productivity. This has been found to be true even in wetlands of very low nutrient status (Reddy et al., 1993). Inlet zones may therefore be subject to solids accumulations that are double the wetland average. However, some wetlands appear to redistribute solids fairly evenly from inlet to outlet.

To the authors' knowledge, only one municipal wastewater polishing FWS wetland has been serviced for solids removal, the Orlando, Florida, Easterly Wetland inlet cells (White et al., 2004). Removal of accumulations restored good hydraulic patterns and restored original water quality performance. It was suspected that uneven accumulations of new sediments were affecting flow patterns, and reducing efficiency (Sees, 2005). The inlet 9 % of the wetland was excavated 45 cm, after 15 years of operation. This overexcavation restored more than the original freeboard, and resulted in a great improvement in hydraulic efficiency, from 34% to 74%.

8.5 SUMMARY

The suspended solids entering treatment marshes contain varying fractions of mineral and organic materials, depending upon source water. Mineral materials are composed of varying fractions of clays, sands and silts. These are typically characteristic of urban runoff and agricultural row crop runoff. Organic suspended solids are characteristic of polishing lagoon effluents and lake and river waters. Particles vary greatly in size, density and shape. Fine clay particles may be only a few microns in size, while plankton and aggregates of organic particles can reach 1000 μm. Invariably, there is a broad size distribution. Particulate matter can be produced interior to the wetland. Redox reactions can produce precipitates driven by the wetland water chemistry. Processes such as phytoplankton growth and death lead to TSS generation.

Settling and filtration are the dominant removal mechanisms. High density mineral particles settle rapidly, in an hour or two, and are thus removed whether or not there is vegetation present. Organic materials are of lower density, and settle less rapidly, but still are settled in a matter of several hours. Large particles settle more quickly than fine material. Formation of aggregates increases size and speeds settling. Filtration occurs because of the sticky biofilms that coat under water plant parts act as collectors.

Resuspension moves particles from deposits on the wetland bottom or the immersed vegetation surfaces. This remobilization occurs due to hydrodynamic shear forces, which result from the bulk movement of water from inlet to outlet, and from wind and wave action. Episodic or sustained bioturbation can be caused by fish, waterfowl, and mammals; and is known to affect several operational treatment marshes.

The mass balance for TSS along the flow direction indicates that removal is the difference between settling and filtration, and generation and resuspension. Several studies have shown that these are very large quantities and result in a large internal cycle of suspended solids in the wetland. The net removal is therefore the difference between very large rates of settling and resuspension. This is evidenced by the dominance of background concentrations along much of the flow direction distance. For source waters with sizeable amounts of TSS, longitudinal TSS profiles are characterized by a rapid decrease to a background concentration, which is dictated by internal water chemistry. For source waters devoid of TSS, internal wetland processes will cause an increase along the flow direction. Consequently, when applying a removal model, care must be taken to acknowledge the probability of declines completed well inside the wetland.

Removal of TSS causes accretion of solids within the wetland. If large amounts of incoming TSS are present, a settling basin is implemented as the first element in the system.

Treatment wetlands are consistently effective at reducing elevated concentrations of suspended solids. Since most treatment wetlands are overdesigned in terms of TSS reduction, existing data are primarily useful for estimating background TSS outflow concentration variability and are not helpful in estimating an area-based reaction rate constant for this parameter. Considerable stochastic variability in outlet TSS is present. Design for TSS removal must therefore often rely upon very general intersystem data trends.

9 Nitrogen

Nitrogen compounds are among the principal constituents of concern in aquatic systems because of their role in eutrophication, and their impacts on aquatic invertebrate and vertebrate species. Nitrogen compounds also augment plant growth, which in turn stimulates the biogeochemical cycles of the wetland. The wetland nitrogen cycle is very complex, and control of even the most basic chemical transformations of this element is a challenge in ecological engineering. This chapter describes the wetland nitrogen cycle, summarizes current knowledge about environmental factors that control nitrogen transformations, and provides approaches that can be used to design wetland treatment systems to treat nitrogen.

Treatment marshes, by virtue of the defining ranges of nitrogen species concentrations, are agronomic systems (Kadlec and Wallace, 2009). That means that plant uptake and return fluxes of nitrogen are of importance in all cases. The proportions of plant and microbial processes are variable form system to system, and from one source water to another, but performance is contingent on both. The removal of nitrogen is very likely to be strongly mediated by the growth and decay of biomass. As a rough guideline, wetland vegetation requires about 20–40 $gN/m^2 \cdot yr$ to support biomass uptake (Kadlec, 2005a), which at a nitrogen loading of 120 $gN/m^2 \cdot yr$ is 25% of incoming nitrogen. This loading refers to dissolved inorganic nitrogen (DIN, nitrate plus ammonia), which are generally considered to be the forms utilized by vegetation.

There two very important general consequences of the low N-loading regime of treatment marshes: atmospheric aeration is adequate to support nitrification, and marsh carbon production is adequate to support denitrification.

9.1 NITROGEN FORMS, STORAGES AND PROCESSES IN MARSHES

Nitrogen Forms in Wetland Waters

The most important inorganic forms of nitrogen in treatment marshes are ammonia (NH_4-N) and oxidized nitrogen (NO_X-N), the latter being composed of nitrite (NO_2-N) and nitrate (NO_3-N). Nitrogen is also invariably present in FWS wetlands in organic forms (ON). Both dissolved and particulate forms may be present, but in most cases there is little particulate nitrogen in settled wetland surface waters.

Common analytical methods include procedures for determination of total or dissolved forms (APHA, 2005). These include:

- Nitrate
- Nitrite
- Ammonia
- Kjehldahl nitrogen (TKN if total)

From these basic measures, several derived concentrations may be computed:

- Oxidized nitrogen = nitrate + nitrite
- Inorganic nitrogen = oxidized nitrogen + ammonia
- Dissolved inorganic nitrogen (DIN) = dissolved oxidized nitrogen + ammonia

- Organic nitrogen (ON) = TKN – ammonia
- Total nitrogen (TN) = TKN + oxidized nitrogen

Each may represent an important feature of wetland water quality, depending upon the nature of source waters. Organic nitrogen in wetland waters is generally not a concern for the treatment marsh context, because elevated levels are primarily associated with human and animal waste treatment source waters. Consequently, this chapter deals principally with oxidized nitrogen and low levels of ammonia.

Ammonia

Ammonia exists in water solution as either as un-ionized ammonia (NH_3) or ionized ammonia (NH_4^+, ammonium ion), depending on water temperature and pH:

$$NH_3 + H_2O \leftrightarrow NH_4^+ + OH^- \tag{9.1}$$

Total ammonia is equal to the sum of the un-ionized and the ionized ammonia, and is designated as ammonia nitrogen in this book. The fraction of un-ionized ammonia in water may be estimated from equilibrium conditions, given by:

$$\log_{10} K_d = \log_{10} \frac{C_{IA}}{C_{UA}} = 0.09018 - \frac{272,992}{T + 273.16} - pH \tag{9.2}$$

where C_{IA} = ionized ammonia concentration, mg/L
C_{UA} = unionized ammonia concentration, mg/L
K_d = dissociation constant, –
T = water temperature, °C

The ionized form is predominant in most wetland systems because of moderate pH and temperature and is designated as ammonium nitrogen in this book. For a typical "average" environmental condition of 25°C and a pH of 7, un-ionized ammonia is only 0.6% of the total ammonia present. At a pH of 9.5 and a temperature of 30°C, the percentage of total ammonia present in the un-ionized form increases to 72%. At lower pH and temperature values, this percentage is insignificant. Un-ionized ammonia is toxic to fish and other forms of aquatic life at low concentrations typically at concentrations greater than 0.2 mg/L. U.S. EPA promulgates acute and chronic criteria for toxicity, and the reader is encouraged to consult the latest publication of such limits. Wetlands are useful for modulation of unionized ammonia, because they create circumneutral pH, and may lower water temperatures for warm effluents.

Ammonia typically comprises more than half of the total nitrogen in a variety of municipal and domestic effluents, where concentrations often are in the range of 20–60 mg/L. Ammonia concentrations in food processing wastewaters treated in wetlands can exceed 100 mg/L (Kadlec et al., 1997). Landfill leachates, particularly from recently closed and capped landfills, can contain hundreds of mg/L (Kadlec and Zmarthie, 2010). The restriction on ammonia under consideration in this book is to low levels of ammonia, below about 5 mg/L. Thus unionized ammonia toxicity is not of great concern.

Oxidized Nitrogen

Nitrite (NO_2^-) is an intermediate oxidation state of nitrogen (oxidation state of +3) between ammonia (–3) and nitrate (+5). Because of this intermediate energetic condition, nitrite is not chemically stable in most wetlands and is generally found only at very low concentrations. Nitrate (NO_3^-) is the most highly oxidized form of nitrogen (oxidation state of +5) found in wetlands. Because of this oxidation state, nitrate is chemically stable and would persist unchanged if not for several energy-consuming biological nitrogen transformation processes that occur. Nitrate can serve as an essential nutrient for plant growth, and in excess, nitrate leads to eutrophication of surface water. Nitrate is often the dominant form in nitrified secondary effluents. Nitrate is present seasonally in agricultural runoff, due to the oxidation of ammonia fertilizers in the vadose zone of farm fields and may reach 40 mg/L in some cases. Therefore, nitrate is the principal target of many treatment marshes.

Wetland Nitrogen Storages

Organic nitrogen compounds are a significant fraction of the dry weight of wetland plants, detritus, microbes, wildlife, and soils. The mass of these nitrogen storages varies in different wetland types. A general idea of the sizes of these different storage compartments is necessary to understand the nitrogen fluxes discussed below (Figure 9.1).

Porewater

As noted in Chapter 4, most of the marsh soil action zone is water, existing as porewater. Porewater is in intimate contact with soils, and therefore nitrogen concentrations are dictated by that interaction, and not by the surface water conditions immediately above. Porewater ammonia concentrations are several gN/m^3, even in the presence of surface waters of much lower concentrations (Figure 9.2). Ammonia concentrations in the surface waters in the wetland represented in Figure 9.2 were in the range 0.04–0.14 gN/m^3, yet the porewaters below reach levels 10–50 times higher. Porewater is mobile, and can move in response to flows driven either by hydraulic head differentials or transpiration demands. Plants draw their nitrogen requirements from this porewater, not the surface water above (with the possible exception of a portion of SAV N requirements).

Soils and Sediments

The total of newly accreted organic materials at the Sacramento, California FWS site had about 1.5% nitrogen (Nolte and Associates, 1998b). At the Houghton Lake, Michigan and

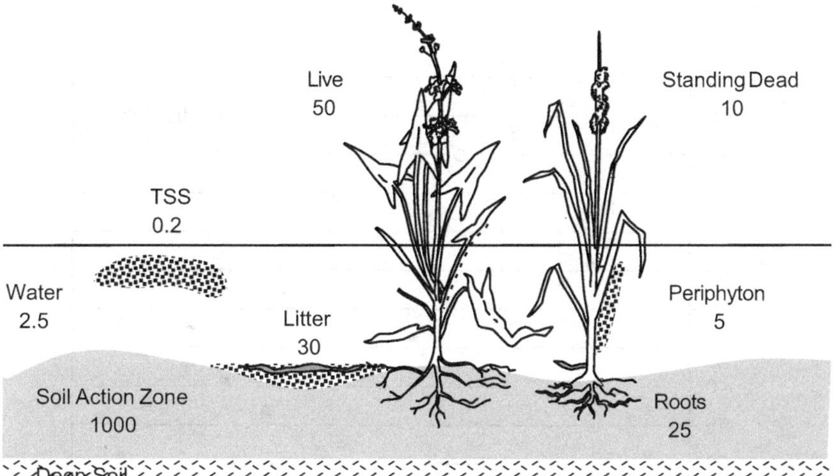

FIGURE 9.1 Nitrogen storages in a densely vegetated hypothetical treatment marsh. Most of the stored nitrogen is in soils and sediments (≈ 1,000 gN/m2), second most is in plant materials (≈ 100 gN/m2), then litter and floc (≈ 30 gN/m2), and least is in mobile forms in the water column (≈ 3 gN/m2).

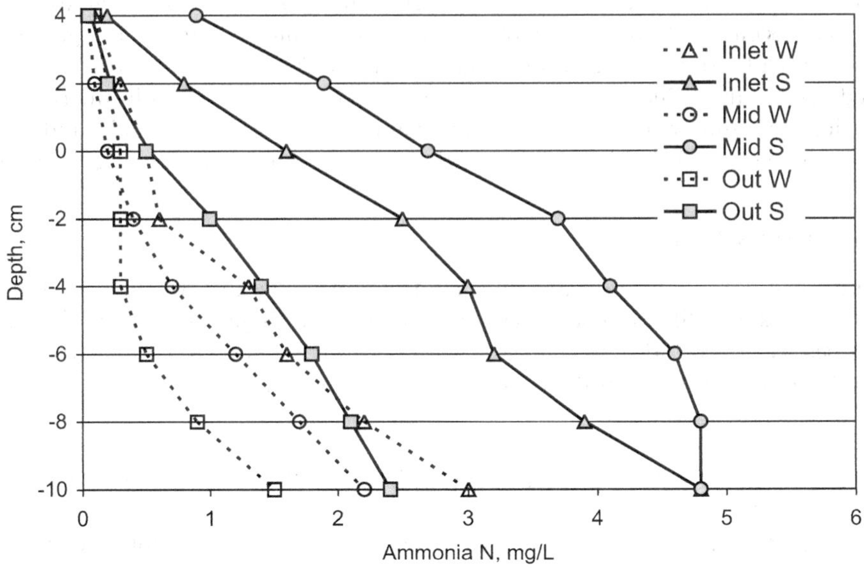

FIGURE 9.2 Porewater ammonia concentrations in STA2C1 in Florida. Data source: Dierberg et al. (2012).

WCA2A, Florida FWS sites, the organic sediments and soils averaged 3.13 ± 0.26 and 2.97 ± 0.37% nitrogen by dry weight, respectively. At both these sites, there was essentially no vertical profile in mass nitrogen percentage, but there was an increase in soil bulk density with depth for both. As a result, the volumetric storage of nitrogen increased with depth (Figure 9.3). The resulting nitrogen storage is about 500 to 2,000 gN/m^2 in the upper 30 cm of organic wetland sediments, which comprises the action zone.

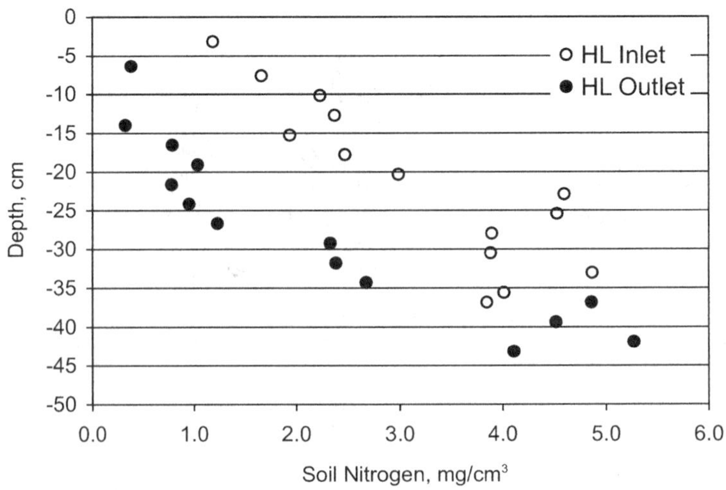

FIGURE 9.3 Volumetric soil nitrogen concentrations in the Houghton Lake treatment marsh.

It is not common for the new sediments and soils in a treatment wetland to be inorganic in character. However, systems treating runoff may receive considerable quantities of inorganic solids from soil erosion in the watershed, which then combine with organic materials generated within the wetland. An example is Chiricahueto marsh in Mexico (Soto-Jimenez et al., 2003). Agricultural runoff brought water at about 15 mg/L of total nitrogen to the marsh for over 50 years. The soil column is now mostly inorganic, with less than 5% carbon. Mineral matter typically has a low nitrogen content, and consequently the nitrogen percentages were low, less than 0.4% dry weight. Both carbon and nitrogen decreased together as depth increased, indicating that most of the soil nitrogen was associated with the organic content. The nitrogen content of the upper 30 cm at Chiricahueto was 330 gN/m^2.

Biomass

The total nitrogen content of living biomass in marsh wetlands varies considerably among species, among plant parts, and among wetland sites. There is little variation from location to location within a homogeneous stand (Boyd, 1978). Example ranges of dry weight nitrogen percentages in natural wetlands are: 0.9–2.6% for emergent plants; 1.96–3.8% for floating leaved plants; and 2.4–2.9% for submersed plants (Boyd, 1978).

Treatment wetlands are often nutrient-enriched and display higher values of tissue nutrient concentrations than natural wetlands. For instance, live cattail leaves in the discharge area of the Houghton Lake, Michigan FWS wetland averaged 2.0% N; those in nutrient-poor control areas averaged 1.1% N, dead leaves showed 1.6 versus 0.7% N, and litter leaves showed 3.6 versus 1.5% N, respectively (Table 9.1). Total biomass is enhanced by fertilization with effluent, and this compounds the effect of increased nutrient content, to produce large storages in treatment areas compared to unfertilized natural wetlands.

Different plant parts may show large differences in nitrogen content, and the seasonal variability may be very large. Biomass collected at the end of the growing season displays much lower nitrogen content than in spring, by factors of three to four. The decline of above ground tissue nutrient content is a common phenomenon in both treatment and natural wetlands and results in a markedly lower tissue nitrogen concentration at the end of the growing season. This is partly due to translocation to below ground rhizomes, which is discussed in a following section. Kühl and Kohl (1993) documented the extent of this variability for *Phragmites australis*, in the margin of Templiner See, a heavily loaded eutrophic shallow lake in Germany (see also Kadlec and Wallace, 2009). Klopatek (1978) has shown trends of the same magnitude for cattail roots and shoots. It is apparent that the timing and location of vegetation samples can greatly affect subsequent calculations of nitrogen storage in biomass.

These seasonal storages reflect the growth cycle of the plant in question. The processes of growth, death, litterfall and decomposition operate year-round, and with different speed and seasonality depending on climatic conditions and genotypical habit. Even in northern climates, the total annual growth is slightly larger than the end-of-season standing crop, by about 20% (Whigham et al., 1978). In southern climates, measurements show 3.5 to 10 turnovers of the live aboveground standing crop in the course of a year (Davis, 1994). Decay and translocation processes release most of the nitrogen uptake, with the residual accreting as new sediments and soils.

A time often used to assay biomass nitrogen is the end of the growing season. The compartments most often analyzed are live aboveground plant tissues, standing dead and litter, and below ground roots and rhizomes. The considerable fraction of the biomass is below ground, which is particularly troublesome from the standpoint of sampling, and hence often omitted. For these reasons, end-of-season, aboveground biomass nitrogen surveys do not represent the storage or seasonal cycling of biomass nitrogen.

TABLE 9.1

Nitrogen content (gN/m^2) of vegetation in treatment and natural areas at the Houghton Lake, Michigan treatment wetland site. DIN = Dissolved Inorganic Nitrogen = Oxidized plus Ammonia Nitrogen.

	Control, DIN ≤ 0.1 mg/L			Discharge, DIN ≈ 15 mg/L		
	Biomass gdw/m^2	Content %	Crop gN/m^2	Biomass gdw/m^2	Content %	Crop gN/m^2
Live						
1995	368	1.08	4	1,086	1.98	18.9
1996	773	1.08	8.2	1,323	2.37	30.3
1997	504	1	5.1	1,200	2.11	25.4
1998	311	1.11	3.4	1,333	1.65	22.5
4-year mean	489	1.07	5.2	1,235	2.03	24.3
Standing Dead						
1995	642	0.69	4.8	917	1.07	10.1
1996	390	0.58	2.3	392	1.54	5.8
1997	190	0.77	1.4	1,642	2.02	32.6
1998	401	0.61	2.1	1,336	1.59	22.3
4-year mean	406	0.66	2.7	1,072	1.55	17.7
Litter						
1996	84	1.6	1.4	1,769	3.6	62.2
1997	42	1.75	0.8	2,193	3.63	79.3
1998	87	1.7	2.3	3,090	3.55	111.1
3-year mean			1.5	2,351	3.59	84.2
Total Above	**982**	**1.08**	**9.4**	**4,658**	**2.37**	**126**

Nitrogen Transformations in Wetlands

Figure 9.4 shows the principal components of the nitrogen cycle in wetlands. The various forms of nitrogen are continually involved in chemical transformations from inorganic to organic compounds and back from organic to inorganic. Some of these processes require energy (typically derived from an organic carbon source), and others release energy, which is used by organisms for growth and survival. Most of the chemical changes are controlled through the production of enzymes and catalysts by the living organisms they benefit.

The several nitrogenous chemical species are inter-related by a reaction sequence. Nitrogen is speciated in several forms in wetlands, and partitioned into water, sediment and biomass phases. A marsh is also stratified vertically into zones that promote different nitrogen reactions. And, as a further complicating factor, micro-environments around individual plant roots may differ from the bulk surroundings (Reddy and D'Angelo, 1994). The detailed processes are well known but require quantification as an integrated network for the wetland environment.

A number of processes transfer nitrogen compounds from one point to another in wetlands without resulting in a molecular transformation. These physical and biological transfer processes include but are not limited to the following: (1) particulate settling and resuspension, (2) diffusion of dissolved forms, (3) plant translocation, (4) litterfall, (5) ammonia volatilization, (6) atmospheric deposition, and (7) sorption of soluble nitrogen on substrates. In addition to the

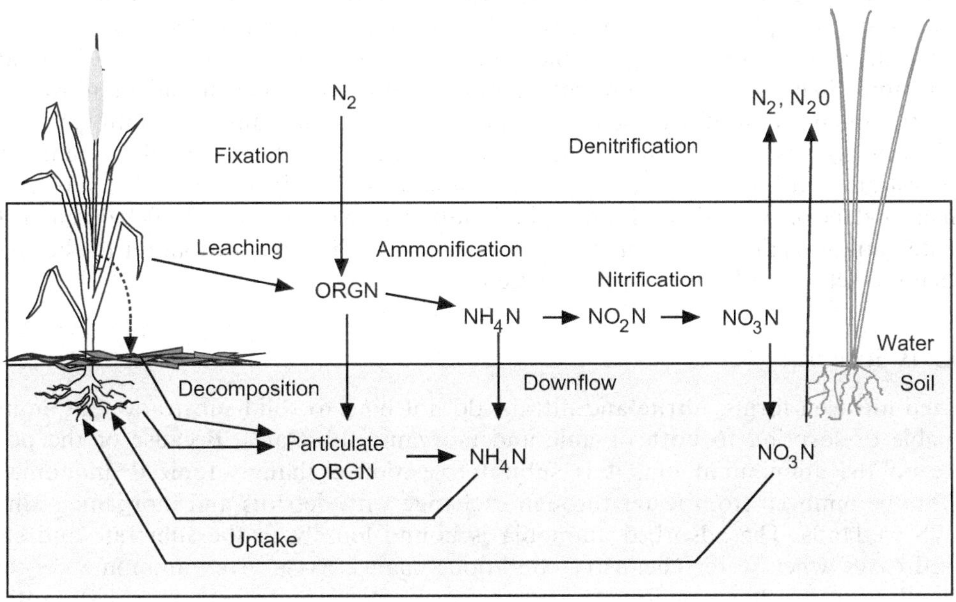

FIGURE 9.4 Simplified nitrogen cycle for a treatment marsh.

physical translocation of nitrogen compounds in wetlands, five principal processes transform nitrogen from one form to another: (1) ammonification (mineralization), (2) nitrification, (3) denitrification, (4) assimilation, and (5) decomposition. A detailed understanding of these nitrogen transfer and transformation processes is important for understanding wetland treatment systems. The sections below describe these processes and the environmental factors that regulate the transformations. Later in this chapter, empirical and theoretical design methods are presented for predicting the treatment wetland area necessary to accomplish the given nitrogen transformations.

9.2 WETLAND NITROGEN PHYSICAL PROCESSES

The wetland nitrogen cycle includes a number of relatively minor physical processes, including atmospheric nitrogen inputs, ammonia adsorption, ammonia volatilization and diffusion. In the context of treatment marshes, these processes are likely to be insignificant. For that reason, they have been omitted on Figure 9.2. Nonetheless, there are methods of quantification that can be used to confirm their relative importance.

Significant physical processes include sedimentation and infiltration or seepage. Sedimentation may also remove particulate nitrogen from the water, either as a structural component of the TSS, or as sorbed ammonia (see Chapter 8). Removal of incoming mineral TSS and its associated nitrogen is an important mechanism in marshes treating runoff. Infiltration flows also cause a term to be added to the nitrogen mass balances.

Atmospheric Deposition

Atmospheric deposition of nitrogen contributes measurable quantities of nitrogen to receiving land areas (Vet et al., 2014). All forms are involved: particulate and dissolved, inorganic and organic (Table 9.2). Wetfall contributes more than dryfall, and rain contributes more than snow (Zhang et al., 2012). The nitrogen concentration of rainfall is highly variable depending on atmospheric conditions, air pollution, and geographical location.

A typical range of total nitrogen concentrations associated with rainfall is 0.3 to 1.0 mg/L with more than half of this present as nitrate nitrogen. Some dryfall of nitrogen also occurs, from deposition of organic dust containing organic and ammonia nitrogen, particularly in dry regions. Typical dryfall nitrogen inputs are less than wetfall amounts. These concentrations can be used with local rainfall amounts to estimate rainfall inputs in nitrogen mass balances. Annual total atmospheric nitrogen loadings are 10–20 kgN/ha·yr. Consequently, atmospheric deposition is almost always a negligible contribution to the wetland nitrogen budget for all but the lowest source nitrogen.

Ammonia Sorption

Oxidized nitrogen forms, nitrite and nitrate, do not bind to solid substrates, but ammonia is capable of sorption to both organic and inorganic substrates. Because of the positive charge on the ammonium ion, it is subject to cation exchange. Ionized ammonia may therefore be removed from water through exchange with detritus and inorganic sediments in FWS wetlands. The adsorbed ammonia is bound loosely to the substrate and can be released easily when water chemistry conditions change. At a given ammonia concentration in the water column, a fixed amount of ammonia is adsorbed to and saturates the available attachment sites.

The character of the substrate is an important determinant of the amount of sorption or exchange. Organic sediments and peats in treatment marshes have moderate sorption capacities. Partition coefficients are defined by the relation:

$$K_{exch} = \frac{C_S}{C_L} \tag{9.3}$$

where C_L = ammonia concentration in water, mg/L

C_S = ammonia concentration on solid, mg/kg

K_{exch} = partition coefficient, L/kg

Partition coefficients for organic-based treatment marshes are on the order of 10 L/kg (see Kadlec and Wallace, 2009). The mass of sorbed ammonia nitrogen on detritus and

TABLE 9.2

Atmospheric deposition of nitrogen.

	Central USA	Western USA	Central Europe
Annual flux, kgN/ha·yr			
Total	4–8	1–4	10–20
Wet	1–5	< 1	6–10
Oxidized	2–4	< 1	2–6
Reduced	2–4	< 1	2–6
Concentrations, mg/L			
Total	0.5–0.9	< 0.3	0.7–2.0
Oxidized	0.3–0.6		
Reduced	0.2–0.4		

Sources: Zhang et al. (2012), Vet et al. (2014).

sediment in a wetland is not large and is very labile. The top 10 cm of the wetland substrate may contain up to 2.0 gN/m^2 in exchangeable form, for a peat exposed to 10 mg/L ammonium nitrogen. Because ammonia concentrations are seasonal, some of the stored ammonia is released during the summer, low concentration period, and replaced during the winter high concentration period. Such amounts are forecast to be small compared to wetland source loadings and removals.

Ammonia Volatilization

Un-ionized ammonia is relatively volatile and can be removed from solution to the atmosphere through diffusion through water upward to the surface, and mass transfer from the water surface to the atmosphere. The overall process has been quantified (see for example Kadlec and Wallace, 2009). The result of consideration of ammonia solubility, and gas and liquid mass transfer resistances, allows the resulting loss rate of ammonia to be expressed as:

$$J = K_L \frac{C_{ATL}}{1 + K_d} = kC_{ATL} \tag{9.4}$$

where: C_{AL} = concentration of free ammonia in the bulk water, gN/m^3
C_{ATL} = concentration of total ammonia in the bulk water, gN/m^3
K_d = dissociation constant, –
K_L = overall mass transfer coefficient, m/d
k = first order volatilization rate constant based on total ammonia, m/d

Many factors influence component processes, most of which will not be known or measured for field situations involving treatment wetlands. Solubility depends on temperature, and degree of ionization depends on temperature and pH. Additionally, both temperature and pH undergo large diurnal swings in some treatment wetlands, of up to 8°C and 2 pH units. In some few situations, there may be vertical stratification of the water column, leading to interfacial temperature and pH conditions that deviate from those in the bulk water (Jenter et al., 2003).

The water-side mass transfer coefficient depends upon the degree of turbulence (mixing) in the water, which in turn depends on depth, velocity and the amount of submersed plant and litter material (Serra et al., 2004), together with the wind speed (Liang et al., 2002). The air-side mass transfer coefficient depends upon the degree of turbulence (mixing) in the air, which in turn depends on wind speed and amount of emergent plant biomass. Therefore, ammonia loss rates should depend not only upon temperature and pH, but also on site-specific conditions.

Several studies of ammonia volatilization from free water surface wetlands provide data from which first order rate constants may be calculated (Table 9.3). The ammonia levels in water are higher than those encountered in treatment marshes, and thus evaluations in those systems requires extrapolation to much lower ammonia levels. A modified Arrhenius temperature factor developed from the data of Stratton (Stratton, 1969) is $\theta = 1.094$. This was used to adjust rate constants to 20°C in Table 9.3. The k_{20} values so computed for wetland systems span the range 0.28–0.68 m/yr, with mean ± sd = 0.47 ± 0.14. Table 9.3 Rate constants for ammonia volatilization for ponds and wetlands. Values based on total ammonia are shown.

Volatilization rate constants for vegetated wetlands are quite small compared to rate constants for other mechanisms, as will be discussed below. However, the same is not necessarily true for open water components.

TABLE 9.3

Rate constants for ammonia volatilization for wetlands. Values based on total ammonia are shown.

	T °C	pH	Total NH_3-N gN/m^3	Unionized NH_3-N gN/m^3	Loss Rate $gN/m^2 \cdot yr$	Total NH_3-N k, m/yr	Total NH_3-N k_{20}, m/yr	Ref.
Duplin Co., North Carolina								1
Grass	23	7.2	55	0.51	30	0.46	0.35	
Bulrush	24	7.2	46	0.42	19	0.4	0.28	
Cattail	22	7.2	46	0.36	20	0.57	0.48	
Greensboro, North Carolina								2
Bulrush & Cattail	23	7	60	0.54	46	0.7	0.53	
Ujjain, India								3
Cattail	35		14		37	2.6	0.68	
Griffith, Australia								4
Rice	20	8	73	2.78	35	0.48	0.48	

References:
1 Poach et al., 2002; Poach et al., 2003 Field: large-scale chambers
2 Poach et al., 2004 Field: large-scale chambers
3 Billore et al., 1994 Field: small-scale chambers
4 Freney et al., 1985 Field: air-side measurements

DIFFUSION

There are typically large concentration differences between porewater and surface water. The presence of the soil matrix prevents convection currents, and therefore the diffusive process is restricted to molecular diffusion in the porespace. The model for this process is the diffusion equation:

$$J_D = -\frac{\phi D}{\theta^2}\frac{dC}{dz} = -D_{EFF}\frac{dC}{dz} \tag{9.5}$$

where C = concentration, g/m^3
D = open water diffusion coefficient, m^2/d
D_{EFF} = effective diffusion coefficient, m^2/d
J_D = vertical diffusion flux, $g/m^2 \cdot d$
z = vertical distance, m
ϕ = matrix blockage coefficient, –
θ = tortuosity factor

The value of the effective diffusion coefficient in wetland sediments is of the order of 0.8 x 10^{-4} m^2/d at 25°C, for ammonia (Krom and Berner, 1980), and 0.4 x 10^{-4} m^2/d for nitrate (Reddy and Patrick, 1984).

Some idea of the expected magnitude of nitrogen diffusion may be gained by examining the situation of mildly eutrophic surface waters overlying a fully saturated peatland. Measured upward-decreasing ammonia gradients have been measured in the Florida STAs (Dierberg et al., 2012) The gradients were about 30 $gm/m^3/m$ in the top 10 cm of soil. Under these circumstances, the diffusion flux predicted by equation (9.5) is:

$$J_D = -\left(0.8x10^{-4}\right)(365)(30) = 0.88\,gN/m^2 \cdot yr \tag{9.6}$$

This estimated flux is irrelevant for plant growth and does not defeat the transpiration flux. It is concluded that the vertical movement of N into the root zone must be driven by vertical flow from the water column into the root zone, in order to overcome negative (upward) driving forces.

9.3 WETLAND NITROGEN MICROBIAL PROCESSES

Wetlands are a rich environment for a large suite of microbes that mediate or conduct numerous chemical reactions involving nitrogen. Heterotrophic bacteria derive carbon from preformed organic compounds, whereas autotrophs acquire energy and carbon from inorganic sources. Denitrification is often, but not always, accomplished by heterotrophs in wetlands, while nitrification is carried out autotrophically. Microbes also produce enzymes that can break down complex molecules, both inside and outside the cell. Microbes are preferentially associated with solid surfaces, rather than as free-floating organisms. The principal nitrogen microbial wetland processes are therefore carried out in biofilms located on soils, sediments, and submerged plant parts.

In the following sections, the principal nitrogen conversions are discussed in more detail (see Figure 9.2).

Ammonification of Organic Nitrogen

Ammonification is the biological transformation of organic nitrogen to ammonia and is the first step in mineralization of organic nitrogen (Reddy and Patrick, 1984). This process occurs both aerobically and anaerobically, and releases ammonia from dead and decaying cells and tissues. Heterotrophic microorganisms are considered to be the group involved (U.S. EPA, 1993b). The reactions can take place intracellularly or extracellularly, via the action of enzymes acting upon proteins, nucleic acids and urea (Maier et al., 2000). The sources of nitrogenous organics are plant and animal tissues, and direct excretion of urea. Organic nitrogen found in wetlands is 80–90% in the form of amides and peptides (Watanabe et al., 2014).

A typical ammonification reaction is breakdown of amides and peptides, causing the excretion of ammonia:

$$Amide/Peptide \rightarrow Organic + NH_3 + CO_2 \tag{9.7}$$

The ammonifying bacteria that accomplish it include *Bacillus, Clostridium, Proteus, Pseudomonas*, and *Streptomyces*.

Wastewater treatment units experience fairly large quantities of organic nitrogen, originating from human or animal waste. Treatment marshes exclude those applications, and thus only small residual levels of organic N are involved. Kinetically, ammonification proceeds more rapidly than nitrification. Thus large source concentrations create the potential for increasing ammonia concentrations along the flow-path of a treatment wetland and requiring design for nitrogen removal to include both ammonification and the slower nitrification process (Kadlec and Wallace, 2009). The strong influents that cause this phenomenon are not under consideration here.

Nonetheless, treatment marshes contain and create soluble organic nitrogen. There is an organic nitrogen background concentration, which may consist of irreducible residuals, or be due to return fluxes of organic nitrogen from decomposing solids. These background concentrations are on the order of 1.0 mg/L. For many source waters, it is probable that background organic compounds are mostly recalcitrant, being comprised of lignin (Lusk and Toor, 2016). Organic nitrogen accompanies the smaller, labile fraction of the total of dissolved organics.

Oxidation of Ammonia

Nitrification is the microbially mediated conversion of ammonium to nitrate. It is the principal transformation mechanism that reduces the concentration of ammonia nitrogen in many wetland treatment systems, by converting ammonia nitrogen to oxidized nitrogen. Nitrification has been typically associated with the chemoautotrophic bacteria, although it is now recognized that heterotrophic nitrification occurs and can be of significance (Keeney, 1973).

Processes and Organisms

In older conventional wastewater treatment literature, overall nitrification is considered to be a two-step, microbially mediated process (USEPA, 1993b). These are considered to be mediated primarily by autotrophic bacteria in the genus *Nitrosomonas* and the second step, nitrification, by bacteria in the genus *Nitrobacter*. Both steps can proceed only if oxygen is present, and thus the actual nitrification rate may be controlled by the flux of dissolved oxygen into the system.

$$\text{Nitritation} \qquad 2NH_4^+ + 3O_2 \rightarrow 2NO_2^- + 2H_2O + 4H^+ \tag{9.8}$$

$$\text{Nitrification} \qquad 2NO_2^- + O_2 \rightarrow 2NO_3^- \tag{9.9}$$

Based on this stoichiometric relationship, the theoretical oxygen consumption by the nitritation reaction is about 3.43 gram of O_2 per gram of NH_3-N oxidized, and 1.14 by the nitrification reaction, for a total of 4.57. Recent studies of WWTPs have confirmed nearly the same values (Liu and Wang, 2012).

Research over the past two decades has determined that the processes involved are often more complicated and variable than Equations 9.8 and 9.9 (Daigger, 2014). For instance, overall nitrification can be considered to be a three-step, microbially mediated process (Bothe et al., 2000; Maier et al., 2000; Stein et al., 2012; Ni et at., 2014):

$$NH_4^+ + O_2 + 2H^+ \rightarrow NH_2OH + H_2O \tag{9.10}$$

$$NH_2OH + H_2O \rightarrow NO_2^- + 4H^+ \tag{9.11}$$

$$2NO_2^- + O_2 \rightarrow 2NO_3^- \tag{9.12}$$

The situation is further complicated by the now-established routes that convert hydroxylamine (NH_2OH) and nitrite (NO_2^-) to nitrous oxide (N_2O). Furthermore, the dissolved oxygen requirement for Equations 9.10 and 9.11 is 2.29 g O_2 per g ammonia nitrogen, rather than the 3.43 suggested by Equation 9.8.

Natural environments are considerably more complex than the situations in biological nutrient removal systems in conventional wastewater treatment plants (WWTPs). There are now enough wetland data to understand some differences, and to appreciate that WWTP results may not apply to wetlands.

There are more genera potentially involved in natural systems than those associated with WWTPs. Ammonia oxidizing bacteria (AOB) include *Nitrosospira* and *Nitrosococcus* in addition to *Nitrosomonas* (Bothe et al., 2000). Austin et al. (2003) found *Nitrosospira* just as abundant as *Nitrosomonas* in a treatment wetland, with lesser numbers of *Nitrosococcus*. Likewise, nitrite is oxidized by *Nitrospira* and *Nitrococcus* as well as *Nitrobacter*, and the former was found to be much more prevalent in a treatment wetland (Austin et al., 2003). Furthermore, heterotrophic bacteria are capable of nitrification, such as *Paracoccus denitrificans* and *Pseudomonas putida* (Bothe et al., 2000). Nevertheless, *Nitrosomonas* is found in treatment wetlands (Silyn-Roberts and Lewis, 2001).

Nitrite Oxidizing Bacteria (NOB) were found not to include *Nitrobacter* in two FWS treatment wetlands (Flood et al., 1999). Similarly, Austin et al. (2003) found *Nitrospira* (4% of total) to be much more abundant than *Nitrobacter* (0.1% of total) in a treatment wetland. Importantly, nitrite may also be converted to nitrous oxide, in competition with its conversion to nitrate.

Oxygen Requirements

The first step in the removal of ammonia requires oxygen, and in some proposed pathways later steps also require it. Therefore, O_2 needs vary with the presumed removal pathways. The traditional WWTP stoichiometric factor of 4.57 g O_2 per g NH_4-N oxidized has been used in many treatment wetland publications as a means of inferring the maximum amount of oxygen transferred into the water to support nitrification. However, alternative pathways have the potential to substantially reduce the oxygen fluxes required, in two ways. First, if hydroxylamine is an intermediate, then the stoichiometric factor is 3.43 g O_2 per g NH_4-N oxidized to nitrate. Second, the portion of the intermediate nitrite or hydroxylamine that is converted to nitrous oxide (N_2O) requires no additional oxygen. Thus, the overall removal of ammonia has a correspondingly reduced oxygen requirement. These possible alternative scenarios produce a range of 2.29–4.57 g O_2 per g NH_4-N.

Consequently, in wetlands treating high levels of ammonia, oxygen supply limits the rate of ammonia removal. Oxygen supply requirements are also present for the oxidation of carbonaceous biochemical oxygen demand (CBOD), which can add to the limitation. The treatment marshes under consideration here do not contemplate source waters with high BOD or high ammonia, and generally oxygen supply is not rate limiting. Respiration in the subsurface action zone of marshes requires considerable oxygen. Oxidative processes in the litter layer may receive a good share of the required oxygen from the overlying water column, but deeper zones are oxygenated by transfer via plant roots. This process is discussed in the following section of this chapter.

Oxygen may also be introduced into the water column by plants. Plankton and SAV produce O_2, and there is oxygen loss from the roots of floating aquatic plants and rafts. Water column production of O_2 is discussed in Chapter 11.

Oxygen Transfer from Air

Oxygen transfer from the atmosphere occurs because of DO levels that are below saturation in the water. Oxygen must be transported to the air-water interface, and then into the bulk of the wetland water column. Resistance to mass transfer in the air is negligible, and the concentration adjacent to the interface is approximately 21% O_2. This results in a saturation DO concentration in an extremely thin layer of water at the surface. That concentration is primarily determined by temperature, and may be calculated (Elmore and Hayes, 1960):

$$C_{DO}^{sat} = 14.652 - 0.41022T + 0.007991T^2 - 0.00007777T^3 \quad (9.13)$$

where C_{DO}^{sat} = equilibrium DO concentration at 1.0 atmosphere, mg/L

T = water temperature, °C

The applicable mass transfer equation is:

$$J_{O_2} = K_L\left(C_{DO}^{sat} - C_{DO}\right) \tag{9.14}$$

where C_{DO}^{sat} = saturation DO concentration at water surface, mg/L = gO/m^3

C_{DO} = DO concentration in the bulk of the water, mg/L = gO/m^3

K_L = mass transfer coefficient, m/d

J_{O_2} = oxygen flux from air to water, gmO/m^2·d

Relevant system studies have been conducted for rivers and streams, ponds, and EAV wetlands. The simplest of these is the O'Connor and Dobbins (1958) correlation for streams, which is driven by water current speed:

$$\text{Streams} \qquad K_L = \sqrt{\frac{Du}{h}} \tag{9.15}$$

where D = molecular diffusivity of oxygen in water, m^2/d

(D = 1.76x10-4 m^2/d @ 20°C)

h = water depth, m

u = water speed, m/d

In more quiescent waters, wind is the most important factor influencing mass transfer coefficients. A review by Ro and Hunt (2006) produced an estimating equation for ponds, which for 20°C may be written:

$$\text{Ponds} \qquad K_L = 0.064W^{1.81} \tag{9.16}$$

where W = wind speed, m/s @ 10 m height

For lakes, K_L values are higher, because of the increased fetch (MacIntyre et al., 2010). Wind speeds at other heights vary as approximately the 1/7th power of the measurement height (Ro and Hunt, 2006; Ro et al., 2006).

Bavor et al. (1988) operated an open water, unvegetated wetland at Richmond, NSW. This system maintained high DO, ranging from 4.3 to 14.6 mg/L over the seasons. The values of K_L calculated from Bavor's open water system were 0.2–1.0 m/d.

Wind has a lesser effect in EAV wetlands, but thermal stratification/destratification also drives the mass transfer process. Poindexter and Variano (2013) studied DO transfer for EAV, and determined the associated mass transfer coefficients. A two-piece linear fit represents their results, given by:

EAV Wetland

$$\begin{aligned} K_L &= 0.21 & W &< 0.6m/s@40cm\ height \\ K_L &= 0.21 + 1.24(W - 0.6) & W &> 0.6m/s@40cm\ height \end{aligned} \tag{9.17}$$

These formulae are representative of some wetland situations, but cannot be regarded as definitive. When treatment marsh values of wind and water speeds, depths and temperatures are employed, ranges in the expected transfer flux to low DO waters are in the following ranges:

Streams K_L = 0.05–1.0 m/d J_{O_2} = 0.4–8.0 gO/m^2·d

Ponds K_L = 1–15 m/d J_{O_2} = 8.0–120 gO/m^2·d

EAV Marsh K_L = 0.21–0.95 m/d J_{O_2} = 1.7–7.6 gO/m^2·d

Belanger and Korzun (1990), working in sparse *Cladium* and moderately dense *Typha* wetlands, found no effect of wind up to about 3.2 m/s (measured at ten meter height), followed by a direct proportionality to the excess of wind speed above that threshold. Thus, for light winds, up to 3.2 m/s, K_L = 0.2 m/d, whereas K_L increased dramatically to ten times that value at wind of 5.5 m/s. The presence of sparse emergent macrophytes therefore does not block physical oxygen transfer. Low values of K_L in wetlands are due in large measure to low flow rates, and the attendant low degree of water mixing. In addition to the effect of wind, rain also creates surficial mixing and increases the mass transfer coefficient. Belanger and Korzun (1990) measured a linear dependence of K_L on rainfall intensity, with K_L = 1.2 m/d at a rainfall rate of 5 mm/hr.

As a further point of reference, the ammonia loading associated with a 5 cm/d hydraulic load and ammonia at 5 mg/L is 0.25 gN/m^2·d. Thus, in the treatment marsh context, atmospheric replenishment is highly likely to provide the oxygen necessary for nitrification.

Denitrification

The removal of oxidized nitrogen in marshes is the result of a complex set of processes that is often termed "denitrification." Nitrate disappearance occurs via many pathways that potentially remove nitrate (Burgin and Hamilton, 2007). The list includes:

- Denitrification is respiratory denitrification, in which organic matter is oxidized by nitrate in sediments and biofilms. Typically, most of the nitrate is converted to N_2, but a variable fraction is converted to nitrous oxide (N_2O).
- Dissimilatory nitrate reduction to ammonium (DNRA) is a heterotrophic process that utilizes organic matter as the energy source to reduce nitrate via fermentation.
- Sulfur-driven autotrophic denitrification couples the reduction of nitrate to the oxidation of reduced sulfur forms, including free sulfide (H_2S and S_2^-) and elemental sulfur.
- Anammox (anaerobic ammonium oxidation) is an autotrophic process by which ammonium is combined with nitrite under anaerobic conditions, producing N_2. The nitrite is derived from the reduction of nitrate.

These alternative reduction routes have been documented to comprise from 1 to 34% of the total nitrate loss (Bartlett et al., 1979; Stengel et al., 1987; Cooke, 1994; Van Oostrom and Russell, 1994). Nonetheless, denitrification often dominates the disappearance of oxidized nitrogen.

Denitrification is most commonly defined as the process in which nitrate is converted into dinitrogen via intermediates nitrite, nitric oxide and nitrous oxide (Hauck, 1984; Paul and Clark, 1996; Jetten et al., 1997).

Denitrification (nitrate dissimilation) is carried out by facultative heterotrophs, organisms that can use either oxygen or nitrate as terminal electron acceptors. Starting from nitrate via nitrite, there is sequential production of nitric oxide (NO), nitrous oxide (N2O), and nitrogen gas (N_2) (e.g., Cox and Payne, 1973; Koike and Hattori, 1978):

$$2NO_3^- \rightarrow 2NO_2^- \rightarrow 2NO \rightarrow N_2O \rightarrow N_2 \quad (9.18)$$

Diverse organisms are capable of denitrification. In the array are organotrophs (e.g., *Pseudomonas, Alcaligenes, Bacillus, Agrobacterium, Flavobacterium, Propionibacterium, Vibrio*), chemolithotrophs (e.g., *Thiobacillus, Thiomicrospira, Nitrosomonas*), photolithotrophs (e.g., *Rhodopseudomonas*), diazotrophs (e.g., *Rhizobium, Azospirillum*), archaea (e.g.,

Halobacterium) and other such as *Paracoccus* or *Neisseria* (Focht and Verstraete, 1977; Knowles, 1982; Killham, 1994; Paul and Clark, 1996).

The overall stoichiometric nitrate dissimilation reaction based on methanol (CH3OH) as a carbon source is summarized by the following (U.S. EPA, 1993b):

$$NO_3^- + 0.833CH_3OH \rightarrow 0.5N_2 + 0.833CO_2 + 1.167H_2O + OH^- \qquad (9.19)$$

The carbon (energy) requirement is 1.90 g methanol per gram of nitrate nitrogen. Some nitrate and carbon are also used by denitrifying bacteria for cell synthesis. For instance, another 0.57 g methanol is required for bacterial growth, bringing the total to 2.47 g of methanol to support the denitrification of one gram of nitrate nitrogen. In the absence of this or another equivalent carbon source, denitrification is inhibited. Denitrification produces alkalinity, with an observed yield of about 3.0 grams alkalinity as $CaCO_3$ per gram of NO_3-N reduced. This increase in alkalinity is accompanied by an increase in the pH of the wetland surface water.

Carbon Sources

The carbon source in wetlands is not methanol, but rather organic matter that is sometimes characterized by the Redfield ratio C:N:P = 106:16:1 (Davidsson and Stahl, 2000). The denitrification reaction is then written:

$$\begin{aligned} 70.7NO_3^- + (CH_2O)_{106}(NH_3)_{16}(H_3PO_4) \rightarrow\ & 35.3.4N_2 + 106CO_2 + 16NH_3 \\ & +106H_2O + H_3PO_4 \end{aligned} \qquad (9.20)$$

The carbon (energy) requirement is 3.58 grams organic matter per gram of nitrate nitrogen. On an elemental basis, the requirement is 1.29 gC/gN. If the product is N_2O, then the carbon requirement is 1.07 gC/gN, or 3.02 grams organic matter per gram of nitrate nitrogen. Some ammonia is theoretically liberated, which can support growth or add to the overall wetland ammonia pool.

As most denitrification is accomplished by heterotrophic bacteria, the process is strongly dependent on carbon availability. There is a general correlation between total soluble organic matter content and denitrification potential, but much better correlation occurs with the supply of easily decomposable organic matter or water-extractable organic carbon (Bremner and Shaw, 1958; Broadbent and Clark, 1965; Paul and Clark, 1996). Organic substances able to act as sources of energy and as hydrogen donors may be present in sediments and soils through the decomposition of tissues or be provided by living roots exudates (Stefanson, 1973; Bailey, 1976).

A number of treatment wetland studies have investigated the use of carbon supplements in the form of added plant biomass (Gersberg et al., 1983, 1984; Burchell et al., 2002; Hume et al., 2002a, 2002b). Another study added methanol (Gersberg et al., 1983), with good effect. Burgoon (2001) provided carbon by feed-forward of un-nitrified influent to wetlands receiving nitrified potato processing waters. All such studies have shown that carbon can be limiting in wetlands at high nitrate loadings. The amount of total carbon in dead and decomposing biomass is on the order of 40% of the dry biomass (Ingersoll and Baker, 1998; Baker, 1998; Hume et al., 2002b). Not all of the total carbon produced is available for denitrifiers. Baker (1998) has suggested that the C:N loading ratio be at least 5:1 so that carbon does not become limiting, which in his work translated to 20% availability. Hume et al. (2002b) suggest 8% availability. Presuming a carbon content of 40%, the required productivities are at the lower end of the range for emergent marshes (Kadlec and Knight, 1996). However, realization of higher nitrate removal rates, corresponding to higher inlet concentrations, may stress the ability of the wetland to generate the required carbon energy source. If carbon is limiting, then the rate of denitrification will depend strongly on the rate of carbon supply (Hume et al., 2002a).

It should be noted that the most labile form of organic carbon in wetland environments is the influent BOD, which is likely used preferentially (when available) to reduce oxidized forms of nitrogen.

Oxygen Inhibition

Theoretically, denitrification does not occur in the presence of dissolved oxygen. Yet, denitrification has been observed in numerous wetland treatment systems that have considerable dissolved oxygen in their surface waters (Phipps and Crumpton, 1994; Van Oostrom and Russell, 1994). This apparent anomaly is due to the complicated spatial zonation in a wetland. Oxygen gradients occur between surface waters and bottom sediments in wetlands, allowing both aerobic and anoxic reactions to proceed in close vertical proximity (millimeters) near the sediment-water interface (Figure 9.5). Thus, nitrate formed by nitrification in surface waters may diffuse into top anoxic soil layers where it is effectively denitrified (Reddy and Patrick, 1984).

Significant quantities of oxygen pass down through the airways to the roots (Brix and Schierup, 1990; Brix, 1993); and significant quantities of other gasses, such as carbon dioxide and methane, pass upward from the root zone. Some – perhaps most – of the oxygen passing down the plant into the root zone is used in plant respiration (Brix, 1990). However, there is a great deal of chemical action in the micro-zones near the roots of wetland plants. Figure 9.6 shows that the oxygenated microzone around a rootlet can conduct nitrification reactions, while denitrification reactions can be occurring only microns away in the anaerobic bulk soil. Diffusion easily connects these zones because of their close proximity.

Bacteria attached to surfaces are usually more numerous than free-living (planktonic) bacteria (Bastviken et al., 2003, 2005). Attached bacteria form microbial communities that

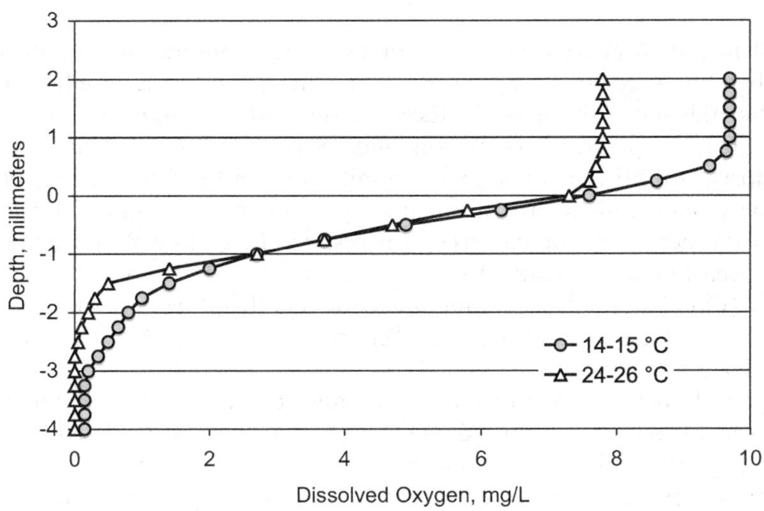

FIGURE 9.5 Oxygen distribution above and below the sediment water interface at two different temperatures. (Based on data in Crumpton and Phipps, 1992)

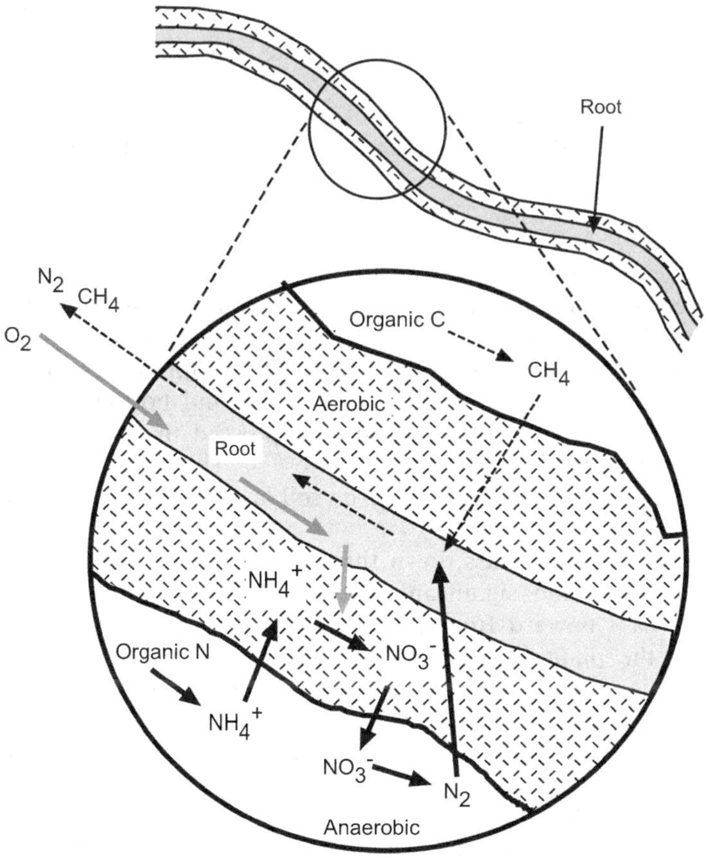

FIGURE 9.6 Pathways of nitrogen transformations in the immediate vicinity of a plant root. Adapted from Reddy and Graetz (1988).

are embedded in polysaccharide matrixes, for example, biofilms, and the bacterial activity within these biofilms is regulated by diffusion of nutrients into the biofilm and by internal processes within this layer. In wetlands, these surfaces are as important as the sediment for the nitrogen turnover processes (Eriksson and Weisner, 1997; Eriksson, 2001). Biofilms therefore comprise a third type of spatial nonuniformity in the wetland environment. Diffusion within the biofilm controls the internal supplies of oxygen, nitrate and ammonia, and thus regulates the net effects of bacterial conversions. In surface flow treatment wetlands, biofilms have been found to contain 10^8 to 10^9 organisms per cm^2, mostly beta and gamma Proteobacteria (Flood et al., 1999). Ammonia oxidizers (beta) were more prevalent near the inlet; denitrifiers (gamma) were more prevalent near the outlet. Alum addition was found to totally eliminate these bacteria.

Another type of spatial nonuniformity exists due to the presence of longitudinal gradients in dissolved oxygen in the flow direction. Oxygen may be depleted by heterotrophic activity, as well as nitrification; but atmospheric reaeration also occurs.

Clearly, wetland oxygen environments are much more complex than either the complete-mix situation that dominates activated sludge processing or the attached growth environment of trickling filters. Results from those technologies should not be extrapolated to treatment wetlands.

Alternative Microbial Processes

Anaerobic Ammonia Oxidation (Anammox)

There is now solid evidence for anaerobic elimination of by ammonia by nitrite, also called anaerobic ammonia oxidation (anammox) in a number of wastewater treatment environments (van Loosdrecht and Jetten, 1998). In an environment with nitrite and ammonia present, a reaction to dinitrogen has been demonstrated commercially:

$$NH_4^+ + NO_2^- \rightarrow N_2 + 2H_2O \tag{9.21}$$

This reaction is microbially mediated by five *Candidatus* anammox genera: "*Ca. Kuenenia*", "*Ca. Brocadia*", "*Ca. Anammoxoglobus*", "*Ca. Jettenia*", and "*Ca. Scalindua*". These genera make up a clade inside the phylum Planctomycetes.

The overall chemistry, including nitrite formation and bacterial growth requirements, has been proposed to be (Furukawa et al., 2001):

$$NH_3 + 0.85O_2 \rightarrow 0.44N_2 + 0.11NO_3^- + 1.43H_2O + 0.14H^+ \tag{9.22}$$

The process proceeds through nitrite, formed according to Equations (9.10) and (9.11), and carries an oxygen requirement of only 1.94 gO per g of NH_4-N. It is autotrophic, and has no organic carbon requirement.

Various commercial wastewater treatment processes are now available which capitalize on the advantages of this alternative route for nitrogen removal (van Dongen et al., 2001; Third et al., 2005).

Given advances in the ability to search for and detect nitrogen-processing organisms, it is not surprising that anammox bacteria have also been found in natural treatment systems. The organisms and processes have now been identified in several treatment marshes (Shipin et al., 2004; Erler et al., 2008; Zhu et al., 2010a; Tao et al., 2012; Song et al., 2014; Chu et al., 2015; Ligi et al., 2015; Waki et al., 2015; Jones et al., 2017). In general, anammox bacteria are present in marshes, but the anammox reaction represents only a minority of the total loss of ammonia. An exception was the treatment section of WCA2A in Florida, where relative anammox rates were higher than denitrification rates in the water column (Kim et al., 2017). The opposite relationship was observed in sediment samples. The importance of this alternative pathway for ammonia and oxidized nitrogen removal for treatment wetland analysis lies in the reduced carbon and oxygen requirements: less than half the oxygen and no carbon, compared to more conventional concepts.

Dissimilatory Nitrate Reduction to Ammonium Nitrogen

Dissimilatory reduction to ammonium nitrogen (DNRA) is a studied candidate mechanism in wetlands for reduction of oxidized nitrogen. An organic substance reduces nitrate (Reddy and DeLaune, 2008:)

$$C_6H_{12}O_6 + 3NO_3^- + 3H^+ \rightarrow 6CO_2 + 3NH_3 + 3H_2O \tag{9.23}$$

Nitrate ammonification is found in facultative anaerobic bacteria belonging to the genera *Bacillus, Citrobacter* and *Aeromonas*, or in the members of Enterobacteriaceae (Cole and Brown,

1980; Grant and Long, 1985; Maier et al., 2000). However, strictly anaerobic bacteria belonging to the genus *Clostridium* are also able to reduce nitrate to ammonia (Caskey and Tiedje, 1979). For many of the bacteria responsible for dissimilation to ammonium, formate is a major electron donor both for nitrate and nitrite, although most of the research on the nitrate reductase activity has been restricted to enteric bacteria such as *Escherichia coli* (Killham, 1994).

Conversion of NO_3-N to NH_4-N and ON increases markedly with decreasing redox potential, high pH, and large quantities of readily oxidizable organic matter (Buresh and Patrick, 1978, 1981). DNRA typically occurs at Eh values of less than -100 mV. Plants do not appreciably affect the process (Zhang et al., 2017).

Bartlett et al. (1979) measured production of ammonium, dinitrogen and nitrous oxide for microcosms with soils from a treatment wetland, but with no plants. From 1–6% of the product was ammonium nitrogen, the balance was measured as dinitrogen, with only trace amounts of nitrous oxide. Cooke (1994) measured 15N-labelled nitrate, ammonium and organic nitrogen in unvegetated microcosms in a treatment wetland. He found 34%, 6%, and 60% of $K^{15}NO_3$ converted by dissimilatory processes, microbial assimilatory processes, and denitrification, respectively; at one site; and 25%, 5%, and 70% at a second site. van Oostrom and Russel (1994) measured 16% dissimilatory nitrate reduction in microcosms containing *Glyceria maxima* mats. Scott et al. (2008) studied DNRA in the Lake Waco constructed wetland in Texas, and found that DNRA was operative, although small compared to denitrification. Sgouridis et al. (2011) found denitrification greater than DNRA by a factor of 13–20 in floodplain wetlands near Coleshill, UK.

The relative importance of denitrification and dissimilatory reduction of nitrate to ammonium in the marsh environment is far from certain. Denitrification may be the dominant process in environments rich in nitrate but poor in carbon, whereas the dissimilatory reduction of nitrate and nitrite to ammonium increases in importance in carbon-rich environments, which are preferably colonized by fermentative bacteria. Thus, nitrate-ammonifying bacteria are favored by nitrate-limited conditions (Laanbroek, 1990).

Sulfur-Driven Autotrophic Denitrification

Sulfur-driven autotrophic denitrification, as an alternate to carbon-driven, heterotrophic denitrification, is well known (Koenig and Liu, 2001; Soares, 2002). The bacterium *Thiobacillus denitrificans* can reduce nitrate to nitrogen gas while oxidizing elemental sulfur, or reduced sulfur compounds including sulfide ($S^=$), thiosulfate ($S_2O_3^=$) and sulfite ($SO_3^=$). For example, the chemistry proposed for utilization of elemental sulfur is (Batchelor and Lawrence, 1978):

$$NO_3^- + 1.1S + 0.40CO_2 + 0.76H_2O + 0.08NH_4^+ \rightarrow 0.5N_2 + 0.08C_5H_7O_2N + 1.1SO_4^= + 1.2H^+ \quad (9.24)$$

If sulfide is the primary species of reduced sulfur, the proposed chemistry is (Komor and Fox, 2002):

$$NO_3^- + 0.74S^{-2} + 0.1886CO_2 \rightarrow 0.48N_2 + 0.74SO_4^= + 0.037C_5H_7O_2N + 0.1H^+ + 0.37H_2O \quad (9.25)$$

This reaction requires 1.69 g sulfide sulfur per g of nitrate N. Other postulated reactions also exist. For instance, iron pyrite may be oxidized (Pauwels and Talbo, 2004):

$$14NO_3^- + 5FeS_2 + 4H^+ \rightarrow 7N_2 + 5Fe^{+2} + 10SO_4^= + 2H_2O \quad (9.26)$$

Treatment wetlands can have many forms of sulfur in sediments, arising from the introduction of sulfate in the incoming water. Reducing conditions can form sulfides and elemental sulfur in the sediments. Those sediments also contain carbon compounds, and consequently both heterotrophic, carbon-driven and sulfur-driven denitrification have been observed to occur simultaneously in wetland sediments (Nahar et al., 2000; Komor and Fox, 2001, 2002; Wass, 2003).

Nitrogen Fixation

Biological nitrogen fixation is the process by which nitrogen gas in the atmosphere diffuses into solution and is reduced to ammonia nitrogen by autotrophic and heterotrophic bacteria, cyanobacteria (blue-green algae), and higher plants. The reduction of gaseous nitrogen (N_2) to ammonia (NH_3) takes place very rapidly and for this reason the individual steps in the reaction have not been investigated in detail. It is supposed that the whole reaction is a 3-step, 2-electrons-per-step mechanism (Winter and Burris, 1976):

$$N{\equiv}N \rightarrow \underset{\text{diimide}}{HN = NH} \rightarrow \underset{\text{hydrazine}}{H_2N - N_2H} \rightarrow NH_3 \tag{9–27}$$

Fixation can be an important contribution to N supply to wetland macrophytes and can be accomplished by either by symbiotic N-fixers (e.g., *Frankia* sp.) or N-fixing bacteria; collectively termed diazotrophs. Bacteria are either autotrophic cyanobacteria, which occur in benthic, floating or periphytic metaphyton assemblages, or heterotrophic N-fixers that reside in the rhizosphere or sediments (Šantrůčková et al., 2010). N-fixing organisms include facultative anaerobes such as *Bacillus* spp. *and Klebsiella* spp., and strict anaerobes such as *Enterobacter* spp., *Vibrio* spp., *Desulfobacter* spp., *Desulfovibrio* spp., and *Clostridium* spp. In wetland systems, free-living bacteria, cyanobacteria (blue-green algae), N_2-fixing bacteria loosely associated with the roots of certain plants and probably *Frankia* are the most important N_2-fixing organisms.

Under anaerobic conditions, microbial assemblages in the root zone of *Typha* spp. and *Glyceria borealis* were shown to fix considerable quantities of atmospheric nitrogen (Bristow, 1974). The measured rates of nitrogen fixation were estimated to be about 6 $gN/m^2 \cdot yr$, and thus able to supply 10–20% of the growth requirement for *Typha*, and 100% for *Glyceria*. Under aerobic conditions, fixation dropped by an order of magnitude.

The Waco, Texas inlet wetland cell had measured rates of nitrogen fixation averaging 6.3 $gN/m^2 \cdot yr$ (range: 1.3–14.6)(Scott, 2006; Scott et al., 2007, 2008). The amount of denitrification, for comparison, was 23 $gN/m^2 \cdot yr$ (range: 12–34). This marsh was dominated by *Typha, Schoenplectus, Pontederia*, and *Sagittaria*. Deep areas of the marsh contained *Najas, Chara, Nuphar*, and *Nymphaea*.

Research at the calcitic marshes of Belize and south Florida contributes to the understanding of N-fixation in such oligotrophic systems. Inglett et al. (2011) measured N fixation along the nutrient gradient in the WCA2A Florida treatment marsh. The inlet, impacted *Typha domingensis* zone displayed fixation dominated by bacteria in the detritus, soil and floc at 12 $gN/m^2 \cdot yr$, while in the downstream unimpacted *Cladium* and slough zone, it was dominated by cyanobacteria in mat form at 10 $gN/m^2 \cdot yr$. Šantrůčková et al. (2010) found similar results for the calcitic marshes of Belize. In high phosphorus environments, for both *Typha domingensis* and *Eleocharis* marshes, heterotrophic fixation prevailed, at an average of 3.0 $gN/m^2 \cdot yr$, compared to 0.04 gN/m2·yr from cyanobacterial mats. At low phosphorus concentrations, the reverse was found, with an average of 4.5 $gN/m^2 \cdot yr$ from cyanobacterial mats, compared to 0.04 $gN/m^2 \cdot yr$ due to heterotrophic fixation. The undosed water concentrations were in the range of 0.2–1.1 mg/L for NH4-N, and less than

10 μgP/L; and phosphorus was added to test plots at the rate of 10 $gP/m^2 \cdot yr$ to create a high P environment.

These results were obtained for low nutrient wetlands, and do not permit extrapolation to other treatment wetlands, but do indicate the ability of wetland plants and soils to fix appreciable nitrogen. It appears that fixed N_2 may supply a sufficient quantity of N to offset N limitation in low-N systems. Speculatively, it is unlikely that the rates of fixation in nutrient-rich treatment marshes contribute materially to overall nitrogen cycling.

9.4 VEGETATION EFFECTS ON NITROGEN PROCESSING

Plants are an important component of nutrient cycling in treatment marshes. They utilize nitrate and ammonium, and decomposition processes release nitrogen back to the water. There are two direct effects of vegetation on nitrogen processing and removal in treatment wetlands:

- The plant growth cycle seasonally stores and releases nitrogen, thus providing a "flywheel" effect for a nitrogen removal time series.
- The creation of new, stable residuals that accrete in the wetland. These residuals contain nitrogen as part of their structure, and hence accretion represents a burial process for nitrogen.

On an instantaneous basis, plant uptake is important for many treatment marshes. A benchmark instantaneous growing season rate is suggested to be 120 $gN/m^2 \cdot yr$ (Kadlec, 2005a). The majority of the assimilated nitrogen is subsequently released during death and decay, but a small amount is permanently stored as new soil and sediment. The net removal of ammonia to accretion, via the vegetative cycle, is on the order of 10 $gN/m^2 \cdot yr$. This amount is of great importance for very lightly loaded wetlands, but of no importance for heavily loaded systems.

The two forms of nitrogen generally assimilated by plants are ammonia and nitrate nitrogen. Research has shown that many wetland plants are either relatively indifferent to which form of nitrogen is available or exhibit a preference for ammonia nitrogen. The hydroponic studies represented in Table 9.4 include only one instance of a nitrate preference. But in nitrate rich waters, nitrate may become a more important source of nutrient nitrogen. Aquatic macrophytes utilize enzymes (nitrate reductase and nitrite reductase) to convert oxidized nitrogen to useable forms. The production of these enzymes decreases when ammonium nitrogen is present (Melzer and Exler, 1982). Plants such as cattails (*Typha latifolia*) are very able to utilize either nitrate or ammonia (Brix et al., 2002b), and so are algae (Naldi and Wheeler, 2002) and cultivated rice (Kronzucker et al., 2000). Dhondt et al. (2003) found that about half of the applied nitrate in a riparian wetland was utilized by plants, while half was denitrified.

Conditions in a marsh are different from hydroponic situations, because of the greater abundance of microbial organisms and the availability of soil carbon to support denitrification. Further understanding of the potential field situation can be gained from the study of Messer et al. (2017a), who followed the fate of $^{15}NO_3$-N in recirculated batch bulrush mesocosms. Both high and low doses of nitrate (ca. 4 and 12 mg/L) were eliminated from the water over the span of about a week. The corresponding nitrate loss rate coefficients were in the range of 22–42 m/yr. At low nitrate loading, plant uptake (48%) dominated, while denitrification was minimal (4%). At high nitrate loading, plant uptake (25%) was reduced, while denitrification was dominant (32%). In both cases, the plant uptake ^{15}N was predominantly found in the plant roots.

TABLE 9.4

Examples of the preferences of treatment wetland plants for ammonia or nitrate. Groupings are subjective, and parent references should be consulted for details.

Not much effect	
Arundo donax	Tho et al., 2017
Acorus calamus	Chang et al., 2010
Iris pseudacorus	Chang et al., 2010
Coix lacryma-jobi	Chen et al., 2015
Reineckia carnea	Chen et al., 2015
Phragmites australis	Tylová et al., 2008
Glyceria maxima	Tylová et al., 2008
Sesbania sesban	Dan and Brix, 2009
Prefers Nitrate	
Coix lacryma-jobi	Jampeetong et al., 2013
Prefers ammonia	
Cyperus laevigatus	Piwpuan et al., 2013
Phormium tenax	Piwpuan et al., 2013
Canna indica	Konnerup and Brix, 2010
Salvinia natans	Jampeetong and Brix, 2009
Typha latifolia	Brix et al., 2002b

These considerations imply that plants are unlikely to preferentially deplete ammonia from wetland surface waters. Therefore, in this book, it is assumed that plant uptake depletes NO_3-N and NH_4-N in proportion to the quantities in surface water. Others have adopted this same presumption (Martin and Reddy, 1997).

The Effects of Vegetation Growth and Cycling

The removal of nitrogen from water by wetland plants has been the subject of many studies (e.g., Reddy and DeBusk, 1985; Rogers et al., 1991; Weller et al., 2016). Many such studies have been characterized by measurements of gross nitrogen uptake, with no deduction for subsequent fluxes due to plant death and decomposition, with the attendant leaching and resolubilization of nitrogen. From the standpoint of nitrogen removal from wetland water, it is the net effect of the macroflora on water phase concentrations that is of interest. Some package simulation models deal with plant uptake, but only in a superficial way. CE-QUAL-W2 (Cole and Wells, 2016) lumps all phytomass compartments, and considers only ammonium uptake. MIKE-SHE (DHI, 2017a) considers only uptake proportional to the transpiration flow and assigns arbitrary transpiration flow concentration factors.

Although it is possible to identify five separate phytomass compartments (see Figure 5.19), there is not sufficient information available to quantify changes in all of them. Consequently, some lumping is required, which can be total phytomass (live above and below ground, standing dead, above- and belowground litter) (see Figure 5.29), or biomass (living vegetative material) and necromass (dead vegetative material). The latter is illustrated in Figure 9.7.

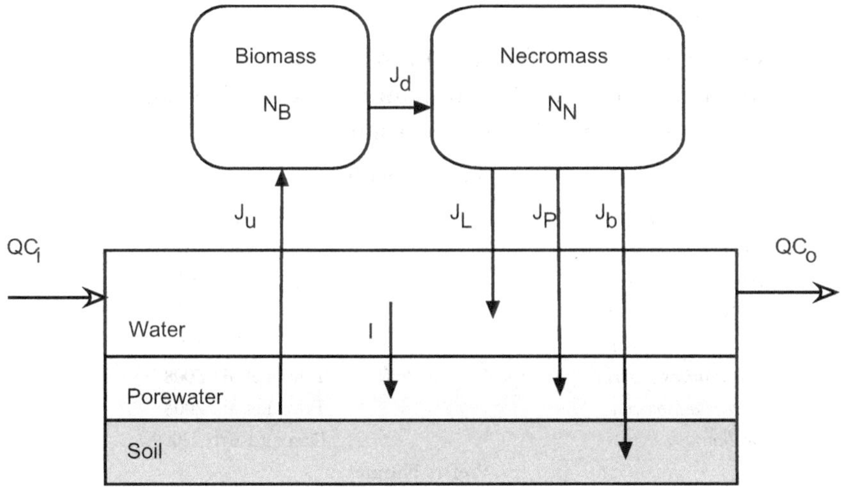

FIGURE 9.7 Conceptual routes of nitrogen transfer between biomass and necromass and wetland soils and water.

The seasonal patterns of vegetation growth and nitrogen storage embody complex patterns of biomass allocation among plant parts, as well as the nitrogen content of those various portions of living and dead material. However, from the point of view of the annual ecosystem removal of nitrogen, uptake and return from the combination of biomass and necromass are the principal features of concern. On an annual average basis, the only concern is net removal to permanent storage. However, during the course of the year, uptake and return may occur at different times, thus influencing removals differently in different seasons. For these reasons, it is necessary to examine the transfers to and from the collective parts of the macrophytes, which is here defined as phytomass. During the course of the year, especially in temperate climates, phytomass increases during the growing season, and shrinks during the senescence season. The same pattern is followed by phytomass nitrogen.

A Mass Balance Framework

The purpose here is to make order-of-magnitude assessments of the role of vegetation in the overall set of ammonia nitrogen processes. This choice has the effect of establishing a "green and brown box," which interacts with the balance of the wetland ecosystem (Figure 9.7). The nitrogen mass balances for biomass and necromass are (instantaneously):

$$\frac{dN}{dt} = J_U - J_r - J_b \tag{9.28}$$

where: $N = N_B + N_N$ = phytomass N content, gN/m^2
J_u = uptake of nitrogen by phytomass, gN/m^2·d
$J_r = J_P + J_L$, release of nitrogen from phytomass, gN/m^2·d
J_b = burial of nitrogen from phytomass, gN/m^2·d
dN/dt = storage change rate of nitrogen in phytomass, gN/m^2·d
or for a fixed time period Δt:

$$\Delta N = (J_U - J_r - J_b)\Delta t \tag{9.29}$$

In these mass balances, biomass and necromass have been lumped, as have surface water and porewater. This avoids the need for knowledge of the senescence and death processes, and the infiltration and transpiration flux.

The uptake of nitrogen is via the root system, which is usually below ground. Nitrogen must therefore be transported into the rhizosphere, by processes of diffusion (minor) and vertical movement driven by transpiration flux (major) (Tinker and Nye, 2000; Reddy et al., 2005). Some of the new plant growth nutrient requirement is supplied by translocation from stores in the rhizomes, and some from uptake from porewater. Nitrogen is returned to surface waters and porewaters by leaching and decomposition. It is likely that the majority of nitrogen in the necromass is returned, with lesser amounts transferred to permanent burial in the form of new soils and sediment.

Data to support the use of Equation 9.29 are limited in two ways. Firstly, virtually all research studies have measured the nitrogen stock in biomass (N_B), but not in necromass (N_N). Secondly, neither the N-flux due to biomass death (J_d) nor the return fluxes (J_P, J_L) are measured. The burial flux (J_b) has been measured in some instances, as the accretion of nitrogen in the newly formed soils. As a consequence, only the biomass piece of the network in Figure 9.7 is typically measured, and is represented by:

$$\frac{dN_B}{dt} = J_U - J_d = J_{NET} \tag{9.30}$$

where: J_d = biomass death transfer of nitrogen to necromass, $gN/m^2 \cdot d$
J_{NET} = net uptake of nitrogen by biomass, $gN/m^2 \cdot d$

An example of the annual pattern of biomass nitrogen in a natural marsh is shown in Figure 9.8. Above and below ground nitrogen storages follow different annual patterns, and

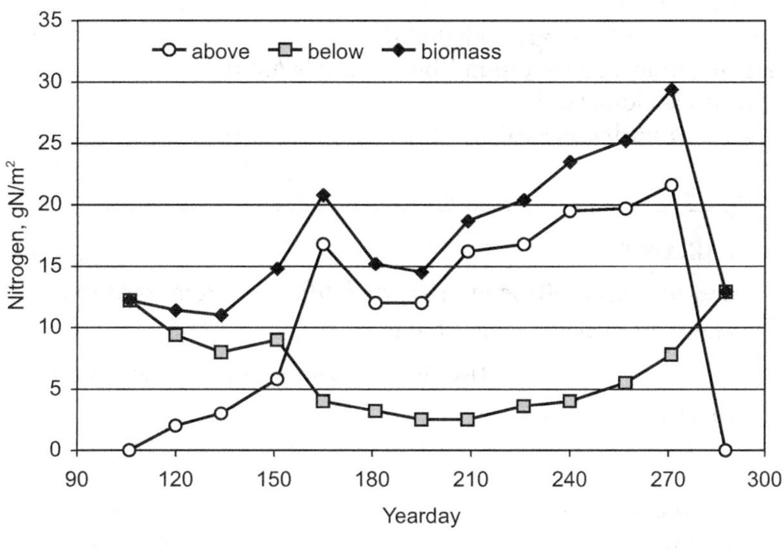

FIGURE 9.8 Seasonal pattern of nitrogen in above and below ground biomass for a *Typha latifolia* stand, (Data from Smith et al., 1988)

typically the live below ground biomass nitrogen concentrations are not much different from those for live above ground (Smith et al., 1988; Tanner, 2001b; Weller et al., 2016).

The majority of the standing phytomass nitrogen is in biomass at low ambient nitrogen concentrations in the surface water, but in high nitrogen environments, the necromass compartments become dominant (Table 9.5). Typical live biomass standing stocks are in the range 20–60 $gN{\cdot}m^2$ (Kadlec and Wallace, 2009).

Over the course of a full calendar year, for a repetitively stable ecosystem, there is no change in the total phytomass, and $\Delta N = 0$. For that annual period, plant uptake is either returned (more) or buried (less). But, as can be seen from Figure 9.8, the total phytomass nitrogen grows in spring and early summer, and recedes in autumn. This annual cycle is more pronounced in northern climates, in response to the more pronounced seasonal conditions.

At this point in the development of knowledge about wetland plant nitrogen cycling, there is some good idea of the change in storage (ΔN) for a given time interval, but less about the three individual fluxes that lead to the storage (J_U, J_r, J_b).

An Example Numerical Assessment

Weller et al. (2016) measured macrophyte biomass at the Tres Rios, AZ treatment wetland over a three-year period, and determined its nitrogen content and role in nitrogen removal in the wetland. Plant uptake was calculated as the foliar nitrogen content multiplied by the difference between maximum and minimum biomass. This uptake was found to be 19% of the marsh inorganic N removal over the study period. This calculation determines J_U for the growing season period but does not consider the subsequent transfers of nitrogen to necromass and the return fluxes (J_d, J_r, J_b), which occur over the entire year, and possibly persist for more than one year. Therefore, it is of interest to consider this data in terms of both uptake and return.

Biomass can be represented by a Gaussian equation (Wilson et al., 2007):

$$B_{LA} = B_{LA\max} \exp\left[-0.5\left(\frac{t - t_{\max}}{b}\right)^2\right] \tag{9.31}$$

where: b = fitting parameter, d
B_{LA} = live above ground biomass, $gdw/m^2{\cdot}d$
B_{LAmax} = maximum live above ground biomass, $gdw/m^2{\cdot}d$
t = yearday of peak biomass, d
t_{max} = yearday of peak biomass, d

TABLE 9.5

End-of-season nitrogen in *Typha latifolia* phytomass at the Houghton Lake treatment marsh.

	Live Above	Standing Dead	Litter	Total
Nitrogen Content, %dw				
Discharge	2.03	1.55	3.59	
Control	1.07	0.66	1.70	
Nitrogen Stock, gN/m^2				
Discharge	24.3	17.7	84.2	126.2
Control	5.2	2.7	1.5	9.4

This model fits the Tres Rios data well, with $R^2 = 0.89$ (Figure 9.9A). The tissue nitrogen content was measured to be nearly constant at 1.64%dw, and hence the live nitrogen storage can be represented by:

$$x_N B_{LA} = N_{LA} = N_{LA\max} \exp\left[-0.5\left(\frac{t - t_{\max}}{b}\right)^2\right] \tag{9.32}$$

where: N_{LA} = nitrogen in live above ground biomass, gN/m^2

N_{LAmax} = maximum nitrogen in live above ground biomass, gN/m^2

x_N = fraction nitrogen in live above ground biomass, gN/gdw

The rate of uptake (or release) of nitrogen from live biomass is the slope of the storage time series, dN_{LA}/dt. This is a positive quantity during plant growth, and a negative quantity during senescence and death (Figure 9.9B). The swing in net uptake is from about 100 to -100 $gN/m^2{\cdot}yr$. During the period of maximum growth, the rate of plant nitrogen uptake is large, and for this example slightly exceeded the average annual nitrogen removal in the

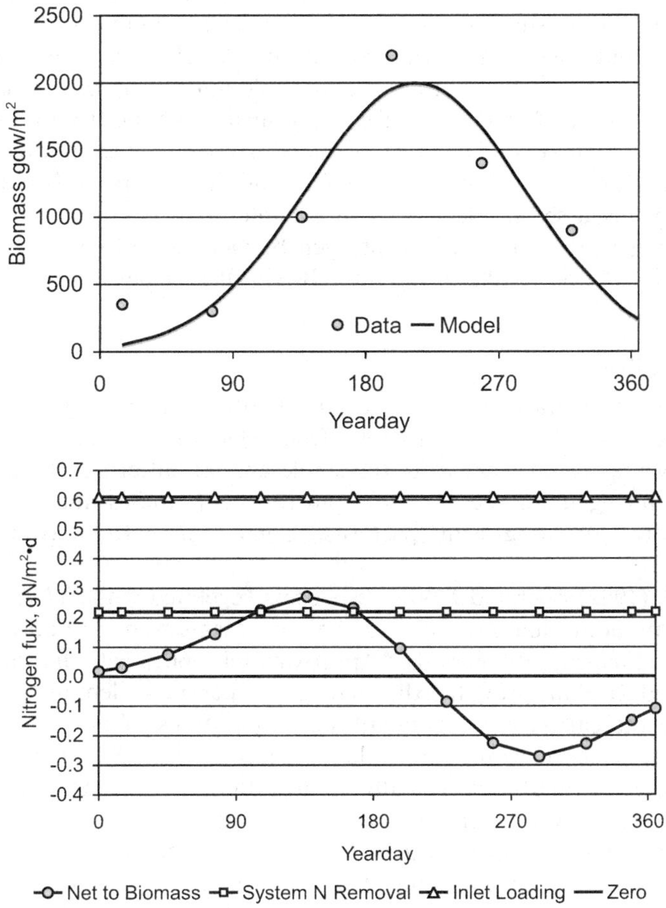

FIGURE 9.9 Seasonal pattern of nitrogen uptake to biomass for a mixed *Typha-Schoenoplectus* stand. Panel (A) shows live above ground biomass, and panel (B) shows the rates of N loading, removal. (Data from Weller et al., 2016)

marsh. But during the ensuing period of senescence and death, the live plant material returns its nitrogen, first to necromass, and eventually to either burial or nitrogen return to water and porewater, presumably in the form of organic nitrogen. Thus only a (small) burial fraction of the cycled plant nitrogen is removed from the active zones of the wetland, comprised of phytomass, water and soils in the action zone

Treatment marsh data provides estimates of growing season vegetative uptakes of 20–100 gN/m^2, which occurs during a four- to six-month period in temperate climates, but over a longer period in warm climates. This uptake is driven by the sequential creation and death of standing stocks of live macrophyte N that are typically in that range. Performance data shows total nitrogen removals averaging 83 gN/m^2yr, with a median removal of 47 gN/m^2yr (N = 110 marshes). The total N loadings to those marshes average 220 gN/m^2yr, median = 93 gN/m^2yr. The range of removals is from 1.6 gN/m^2yr (10th percentile) to 127 gN/m^2yr (90th percentile). Therefore, plant uptake is usually of importance in the annual cycle of N removal.

Accretion of Nitrogenous Residuals

The least studied aspect of nitrogen transfer in wetlands is in the creation of new soils and sediments, with their attendant nitrogen content. Not all of the dead plant material undergoes decomposition. Some small portions of both above ground and below ground necromass resist decay, and form stable new accretions, which correspond to the flux J_b in Equation 9.29 and as shown in Figure 9.7 and Table 9.6. Such new stores of nitrogen are resistant to decomposition. The origins of new sediments may be from remnant macrophyte stem and leaf debris, remnants of dead roots and rhizomes, and from undecomposable fractions of dead microflora and microfauna (algae, fungi, invertebrates, bacteria). The manner of such accretion is discussed in more detail in Chapter 4.

In this book, when preparing annual nitrogen budgets, it will generally be assumed that accreted nitrogen is 10% of uptake to biomass. It will further be assumed that stable operating conditions prevail.

Short-Term Anomalies

In addition to the considerations of long-term repetitive annual vegetation effects on wetland nitrogen processing, there are transient effects related to startup of treatment wetlands. These transient events are different from the stable annual pattern of swelling and shrinking of the phytomass nitrogen storage. Results from transient studies must not be construed as being representative of long-term patterns. Some case study transient results are informative.

Busnardo et al. (1992) operated FWS mesocosms vegetated with *Schoenoplectus californicus*. The startup ammonia loading rates to the mesocosms were 330 and 670 $gN/m^2 \cdot yr$ for two consecutive seven-month periods. Approximately 60% of the ammonia nitrogen removed was found in plant growth. Although this experiment demonstrated that emergent macrophytes have the capacity to assimilate large quantities of ammonia, Busnardo et al. (1992) speculated that plants would have a lesser effect in mature wetlands.

Newly constructed wetlands are typically planted sparsely compared to the ultimate grow out of vegetation. The development of the new vegetation creates a nitrogen demand that persists only during that grow-in period. For example, Sartoris et al. (2000) reported on the first two years of ammonia removal and plant coverage for a 9.9-ha FWS constructed wetland at Hemet, California. As the plant coverage went from near zero (planted clumps on 1.2 m spacing) to about 80% of *Schoenoplectus* spp., and the vegetation density increased by 67%, the ammonia load removed went from 98 down to 15 $gN/m^2 \cdot yr$. Sartoris et al. (2000) concluded that plant uptake was most likely the primary sink for nitrogen during the

TABLE 9.6

Estimated nitrogen accretion rates in treatment marshes.

Location	Ref.	Method	Water NH_4-N (typical), mg/L	Accretion cm/yr	Nitrogen Burial $gN/m^2 \cdot yr$
Louisiana	1	Feldspar	0.05	0.14	4
Michigan	2	Lead 210	0.1	0.2	6
Florida WCA2A	3	Cesium 137	0.3	0.5	9
Florida WCA2A	4	Cesium 137	0.3	0.4	12
Florida WCA3	4	Cesium 137	0.1	0.3	11
Florida	5	Lead 210	0.3	0.5	11
Florida ENRP	6	Feldspar	0.1	0.85	35
Manitoba	7				6
Australia	8				8
Klopatek	9				5

References:
1. Rybczyk et al. (2002)
2. Kadlec and Robbins (1984)
3. Reddy et al. (1991)
4. Craft and Richardson (1993)
5. Robbins et al. (1999)
6. Chimney, M.J., unpublished data
7. Murkin et al. (2000)
8. Hocking (1989b)
9. Klopatek (1978)

two-year study. In this case of a treatment marsh, the increase in coverage by plants reduced the fraction of open water, and hence created a lesser potential for atmospheric reaeration to support nitrification.

Harvest to Remove Nitrogen

Nitrogen removal is theoretically possible via the harvest of plants and their associated nitrogen content. However, above ground standing crops do not display a large potential for removal of nitrogen, even under the assumption that the entire crop could be recovered. Direct harvesting experience has shown that only a small fraction of the applied nitrogen can be recovered in harvested biomass. However, systems operating in tropical climates may be capable of greater sustained annual vegetative removals, which are enhanceable by repeated harvest. Amounts of nitrogen that can be removed by harvest were determined in a University of Florida study (Zhou et al., 2017). For plants of various kinds in low-nutrient waters, they found about 2–10 $gN/m^2 \cdot yr$ could be harvested in a single harvest (Table 9.7). In warm climates that foster regrowth, multiple harvests are possible, and enhance these amounts (Verhofstad et al., 2017).

Harvest of EAV is very disruptive to a marsh ecosystem. Harvest of floating plant (FAV) biomass is more practical and less destructive and has been practiced at fairly large scale.

Harvest may involve complete removal in the case of floating plants *(Lemna minor, Eichhornia crassipes*), or cutting of aboveground parts of rooted plants such as *Typha, Schoenoplectus, Phragmites.* Harvesting typically requires expensive mechanical equipment, and is labor intensive for large systems. The problem of harvested biomass disposal is often not easily resolved. Harvested *Phragmites* can be used for roof thatching in the European

TABLE 9.7

Harvestable nitrogen for different plant types in a single 70% removal of harvestable material.

Plant	Form	Total Biomass gdw/m^2	Harvestable Biomass gdw/m^2	Harvestable Biomass N gN/m^2	70% Harvest Biomass N gN/m^2
Hydrilla verticillata	Submerged	579	579	12.60	8.90
Potamogeton spp.	Submerged	152	152	3.37	2.36
Typha orientalis	Emergent	2,515	901	8.40	5.90
Panicum hemitomon	Emergent	298	254	4.70	3.29
Pontederia cordata	Emergent	1,039	237	3.58	2.51
Salvinia spp.	Free-floating	118	118	2.64	1.85
Pistia stratiotes	Free-floating	719	719	14.60	10.22
Nuphar advena	Floating-leaved	258	130	3.39	2.37

Source: Zhou et al. (2017).

context (Verhofstad et al., 2017), but its use as fuel is more often contemplated. Harvested biomass may either be composted or digested to form a biogas product. Composting requires transportation costs, and dedicated land area. Biogas generation from water hyacinths has been shown to be feasible (Biljetina et al., 1987), however sludge disposal remains a problem. The capital cost of harvesting for gas generation has been found to be prohibitively expensive (Chynoweth, 1987).

Wichmann (2017) provides an economic analyses of several uses of *Phragmites* biomass, including biogas production, direct combustion, and thatching. The measure employed was retrievable revenue minus costs of harvesting but did not include transportation costs. Biogas was unprofitable, combustion fuel was slightly profitable, and thatching was profitable.

As a consequence of these difficulties, plant harvesting is typically not employed for nitrogen removal.

9.5 ANNUAL NITROGEN MASS BALANCES

The foregoing discussion highlights the complexity of nitrogen processing in marshes. Multiple interconversions and internal loadings due to vegetation cycling indicate that inlet-outlet percent removals for individual nitrogen species are virtually meaningless. However, the individual process considerations discussed above may be combined via mass balances to form an integrated concept of nitrogen fluxes in treatment wetlands. This is a powerful diagnostic tool with which understanding of marsh performance can be furthered. This interpretive step is very important, because it:

1. Identifies the true rates of ammonification, ammonia oxidation and denitrification;
2. Places the role of the vegetative nitrogen cycle in the context of the microbial processes; and
3. Allocates the fate of added nitrogen to storage, leakage and gasification.

Both microbial processes and vegetative processes are involved. The (conventional) sequential nitrogen conversions of mineralization of organic nitrogen, nitrification of ammonia, and denitrification of oxidized nitrogen are presumed. These processes lead to nitrogen loss of the product dinitrogen gas along with its companion nitrous oxide. These processes are known to be temperature-sensitive, but available data generally only allow consideration of annual averages. The mass balance system includes water, porewater and phytomass.

Phytomass is presumed to take up dissolved inorganic nitrogen (NO_3-N and NH_4-N), and to return most of it as organic nitrogen, while burying a small fraction in accretion. The quantification of vegetation in the nitrogen cycle is somewhat speculative, because it depends upon estimates of biomass and tissue nitrogen content. Nonetheless, much is known about standing stocks and turnover rates, as well as the (narrow) bounds on nitrogen percentages in that biomass. Because data are typically not available to track the seasonal pattern of vegetation and its nitrogen content, these mass balances must of necessity consider long-term annual performance. Plant N species uptake is assumed to be in proportion to the concentrations of nitrate and ammonia in the water.

A stable annual cycle is considered, for which there are no accumulations of nitrogen except in the newly formed sediments. Data requirements are inlet and outlet flow rates, inlet and outlet concentrations of all N species. The uptake and burial of plant tissue nitrogen must be estimated. This information then determines the inter-species N transfers that are present in the system, via N species mass balances.

Mass Balance Equations

For the conventional conversion sequence, four mass balances are involved:

1. Mass balance for nitrogen in phytomass. Nitrogen uptake for growth is primarily returned as organic nitrogen to the water and porewater, but a small specified fraction is accreted as undecomposable residuals.

The basis is Equation 9.28, but it requires expansion to account for the three forms of nitrogen.

$$\frac{dN_B}{dt} = J_{UA} + J_{UN} - J_{rO} - J_b \tag{9.33}$$

where: N_b = phytomass nitrogen content, gN/m^2
J_b = burial transfer of nitrogen from phytomass, $gN/m^2 \cdot yr$
J_{rO} = return transfer of organic nitrogen to water, $gN/m^2 \cdot yr$
J_{UN} = uptake of oxidized nitrogen by biomass, $gN/m^2 \cdot yr$
J_{UA} = uptake of ammonia nitrogen by biomass, $gN/m^2 \cdot yr$

Over an annual cycle, the accumulation term (left-hand side) will be zero for a stable, repetitive ecosystem. Over shorter time periods, for example, monthly, there will be nonzero values for biomass storage and necromass storage,

2. Mass balance for organic nitrogen in water and porewater. The observed removal from water, plus internal loading from decomposing phytomass, must be accounted by mineralization (ammonification) to ammonia.

$$J_{outO} = J_{inO} + J_{rO} - J_{ammonification} \tag{9.34}$$

where: J_{inO} = incoming organic nitrogen, $gN/m^2 \cdot yr$
J_{outO} = outgoing organic nitrogen, $gN/m^2 \cdot yr$
J_{rO} = return transfer of organic nitrogen to water, $gN/m^2 \cdot yr$
J_{UA} = uptake of ammonia nitrogen by biomass, $gN/m^2 \cdot yr$
$J_{ammonification}$ = conversion of organic nitrogen to ammonia nitrogen, $gN/m^2 \cdot yr$

Accumulation within the water column is presumed to be negligible for either annual or monthly time scales.

3. Mass balance for ammonia nitrogen in water and porewater. The excess ammonia resulting from removal from water, plus ammonification, must be accounted by nitrification to oxidized nitrogen.

$$J_{outA} = J_{inA} - J_{UA} + J_{ammonification} - J_{nitrification} \tag{9.35}$$

where: J_{inA} = incoming ammonia nitrogen, $gN/m^2 \cdot yr$
J_{outA} = outgoing ammonia nitrogen, $gN/m^2 \cdot yr$
$J_{nitrification}$ = conversion of ammonia nitrogen to oxidized nitrogen, $gN/m^2 \cdot yr$
Accumulation within the water column is presumed to be negligible for either annual or monthly time scales.

4. Mass balance for oxidized nitrogen in water and porewater. The excess NO_X-N resulting from removal from water, plus nitrification, must be accounted by denitrification.

$$J_{outN} = J_{inN} - J_{UN} + J_{nitrification} - J_{denitrification} \tag{9.36}$$

where: J_{inN} = incoming oxidized nitrogen, $gN/m^2 \cdot yr$
J_{outN} = outgoing oxidized nitrogen, $gN/m^2 \cdot yr$
J_{UN} = uptake of oxidized nitrogen by biomass, $gN/m^2 \cdot yr$
$J_{denitrification}$ = conversion of oxidized nitrogen to gaseous forms, $gN/m^2 \cdot yr$
Accumulation within the water column is presumed to be negligible for either annual or monthly time scales. In some instances, data may be available that allow an intermediate mass balance for nitrite.

Inlet N flux to the marsh may include settleable particulate nitrogen, which is not likely to directly participate in wetland cycling, and hence can be accounted separately. If source water contains large amounts of particulates (and its N content), the first process unit would be a settling basin, which is not included in marsh performance. Outlet fluxes of the various N species may include seepage losses. It is usually not known where along the flow path seepage occurs, and the assumption is made that seepage transports N species at their mean concentration along that flow path.

The identification of the amounts of nitrification and denitrification allows determination of the presumed sufficiency of internal and external supplies of oxygen and carbon. The absence of such sufficient supplies is indicative of the presence of alternative nitrogen processing mechanisms.

Two examples of treatment marsh nitrogen mass balances will be explored, for (1) a lightly loaded treatment plant effluent polishing marsh, and (2) a leaky marsh treating contaminated river water.

Orlando Easterly Florida Treatment Marsh

This wetland has been in operation since 1987, and is described in general terms in U.S. EPA (1993a). It is a 494-ha constructed free water surface wetland with 17 compartments in a series and parallel arrangement, which receives about 60,000 m^3/d of highly treated municipal effluent. The cells were vegetated with soft-tissue emergent plants, and the vegetative communities evolved over time to a mixed marsh condition. In addition to annual and specialty project reports, there have been several published papers (Jackson and Sees, 2001; Martinez and Wise, 2003; Wang et al., 2006). Data used here are from the ten-year period 1993 to 2002.

Nitrogen totals less than 3 mg/L entering the system, and less than 1.4 mg/L in the effluent from the wetland. Atmospheric contributions are not negligible under these circumstances, and are estimated at 2.0 mg/L based upon other Florida data. The inlet hydraulic loading was 1.2 cm/d, and rainfall averaged about 0.4 cm/d (Table 9.8A). Particulate nitrogen is not a factor, because the TSS content of the incoming water is very low (1.2 mg/L). The data combine to produce a total nitrogen inlet loading of 11.3 gN/m^2·yr, apportioned across the species as indicated in Figure 9.10. This is much less than the required nitrogen for even modest plant growth, indicating that the vegetative cycle must draw upon internal sources of nitrogen. There was net removal of all forms of nitrogen, summing to a 70% reduction in the load of total nitrogen. The inlet-outlet concentration reduction was less, 55%, because it does not include the contribution of rainfall nitrogen.

Since measurements were not made of vegetative nitrogen processes, assumptions must be made. The wetland was moderately well vegetated, with some open water, leading to the assumption of an annual productivity of 1,000 gdw/m^2·yr with an assumed nitrogen content of 2%. Of this, 10% was assumed to be buried as new sediments (Table 9.8B). Both nitrate and ammonia were presumed to be used to support growth, in proportion to their availability in the water. Average concentrations were used to determine the uptake ratio, although selective spatial utilization may have occurred.

This information is adequate to calculate all the average annual transfers within the wetland via mass balances. The pattern of nitrogen transfers is dominated by the vegetative cycle (Figure 9.10). Production of ammonia from decomposition of biomass is eight times higher (20.65 gN/m^2·yr) than the reduction in organic N in the water from inlet to outlet (2.64 gN/m^2·yr). Nitritation is seven times higher than the reduction in the flowing ammonia load (11.32 versus 1.57 gN/m^2·yr), and that high internal load of nitrite is subsequently nitrified to nitrate. Some nitrate is lost through denitrification, but more is used to support plant growth. However, denitrification amounts to 52% of the net nitrogen input, whereas accretion of new sediments represents only 18%.

The required supplies of ancillary chemicals were present in the wetland (Tables 9.8A and 9.8B). Dissolved oxygen is present to support ammonia oxidation and the observed reaeration, which is calculated to need 0.16 gO/m2·d, well within the range of expected atmospheric reaeration discussed above. The required alkalinity is also available to support

TABLE 9.8A

Average inlet and outlet concentrations for the Orlando Easterly, Florida FWS wetland for 1993–2002.

Parameter	Inlet mg/L	Outlet mg/L	Mean mg/L	Fraction	Assumed Rain mg/L
HLR, cm/d	1.17	1.15	–	–	0.41
Organic N	1.67	0.98	1.32	–	1.0
Ammonia N	0.33	0.14	0.23	0.545	0.5
Nitrite N	0.60	0.04	0.32	–	0.5
Nitrate N	0.26	0.13	0.19	0.455	–
TSS	1.2	4.0	–	–	–
$CBOD_5$	2.0	2.6	–	–	–
DO	6.1	8.9	–	–	–
Alkalinity	94	92	–	–	–

Source: City of Orlando operating data.

TABLE 9.8B

Assumptions for the Orlando Easterly wetland carbon and oxygen supplies. Biomass is the assumed source of carbon, and oxygen requirements are determined from Figure 9.10 fluxes.

Assumption			Notes
Biomass Produced =	1,000	gdw/m^2·yr	
Carbon Content =	0.5	–	50%
Useable Carbon Fraction =	0.3	–	30%
Carbon Available =	150	gC/m^2·yr	
Denitrification Carbon Requirement =	140	gC/m^2·yr	1.07xN
Biomass N Uptake =	20	gN/m^2·yr	2%N
Biomass Buried =	100	gdw/m^2·yr	10%
Nitrogen Buried =	2.0	gN/m^2·yr	2%N
Oxygen Needed =	36	gO/m^2·yr	3.43NO_2N+1.14NO_3N
Daily Oxygen Needed =	0.16	gO/m^2·d	

ammonia oxidation. There is no carbon in the inlet water to support denitrification ($CBOD_5$ = 2 mg/L), but the biomass cycle produces enough available carbon to fuel heterotrophic denitrifiers.

Imperial, California Treatment Marsh

This wetland system began operation in 2000, and the data used here are from the four-year period 2001–2004. It consists of a 3.88 ha sedimentation basin, followed by 4.72 ha in four marsh cells in series. The system received 16,600 m^3/d of agricultural runoff. The cells are about 75% open water and 25% vegetated with bulrushes. Data were summarized in Kadlec et al. (2010a). The total nitrogen areal loading was over forty times that at the Orlando Easterly Wetland.

The hydraulic loading to the system is high (19.3 cm/d), and 35% infiltrates. The incoming water has high TSS (179 mg/L, Table 9.9A), which is effectively removed in the sedimentation basin and wetland cells. However, particulate nitrogen entering the marsh cells was low, and was not reduced in the system. Oxidized and dissolved organic nitrogen dominate the inflow, which had total nitrogen of 6.8 mg/L; the outflow had 3.8 mg/L TN (44% concentration reduction) (Table 9.9A). About 25% of the nitrogen load is infiltrated (Figure 9.11). In contrast to the Orlando system, the vegetative cycle at Imperial has almost no effect on the nitrogen budget. Vegetation was sparse, and gross uptake was estimated to be only 2% of the incoming nitrogen load.

Ammonification primarily reduces the load of dissolved organic nitrogen. Nitrification and denitrification dominate the processing matrix (Figure 9.11). The required supply of oxygen, in excess of the observed depletion of the water column DO, was 1.45 gO/m^2·d, which is reasonably within the range of expected atmospheric reaeration (Table 9.9B). Sufficient alkalinity was present to support nitrification. However, there was estimated to be not enough carbon available from the decomposition of the sparse vegetation, or incoming $CBOD_5$, to support denitrification. A possible candidate mechanism was sulfur-driven autotrophic denitrification. The incoming water contained over 600 mg/L of sulfate. If only a small fraction, less than one percent, of this were reduced to sulfide in the wetland

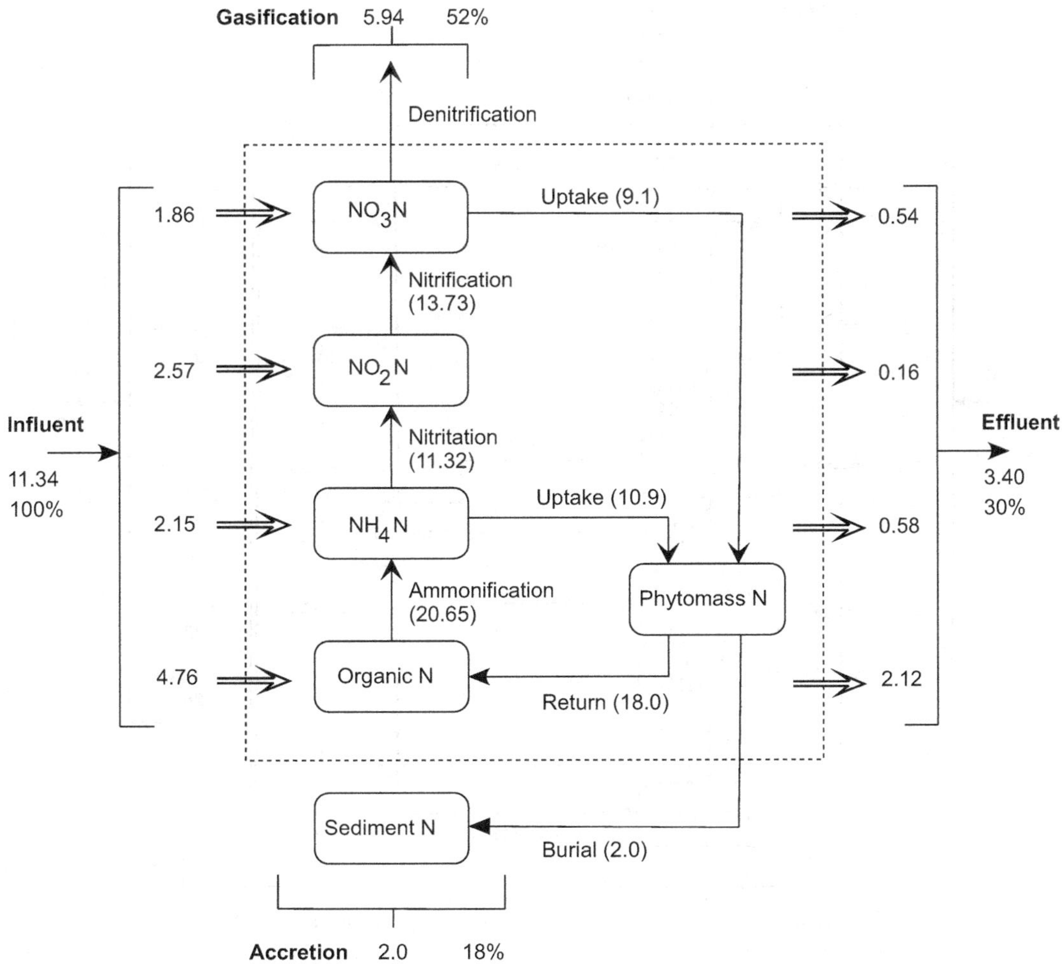

FIGURE 9.10 Estimated annual nitrogen fluxes in the Orlando Easterly treatment wetland ($gN/m^2 \cdot yr$). The vegetation cycle dominates this lightly loaded system.

sediments, then that sulfide could have supported the balance of the observed denitrification over carbon-driven, heterotrophic denitrification.

Variability and Implications

A few important points emerge from this integrated view of nitrogen processing. First, the magnitude of the vegetative nitrogen cycle is by no means trivial for treatment marshes, because uptake represents a good portion of the net removal for these lightly loaded systems. However, net burial is only a fraction of plant uptake. Secondly, the influence of the biomass decay causes the true amount of ammonification to exceed the apparent rate based only on water analyses. Third, the true amount of nitrification greatly exceeds the amount based only on input/output (I/O) water analyses. A sequential nitrogen kinetic model corrects for the production of ammonium from organic nitrogen, and calibrates to have higher rate coefficients accordingly. Finally,

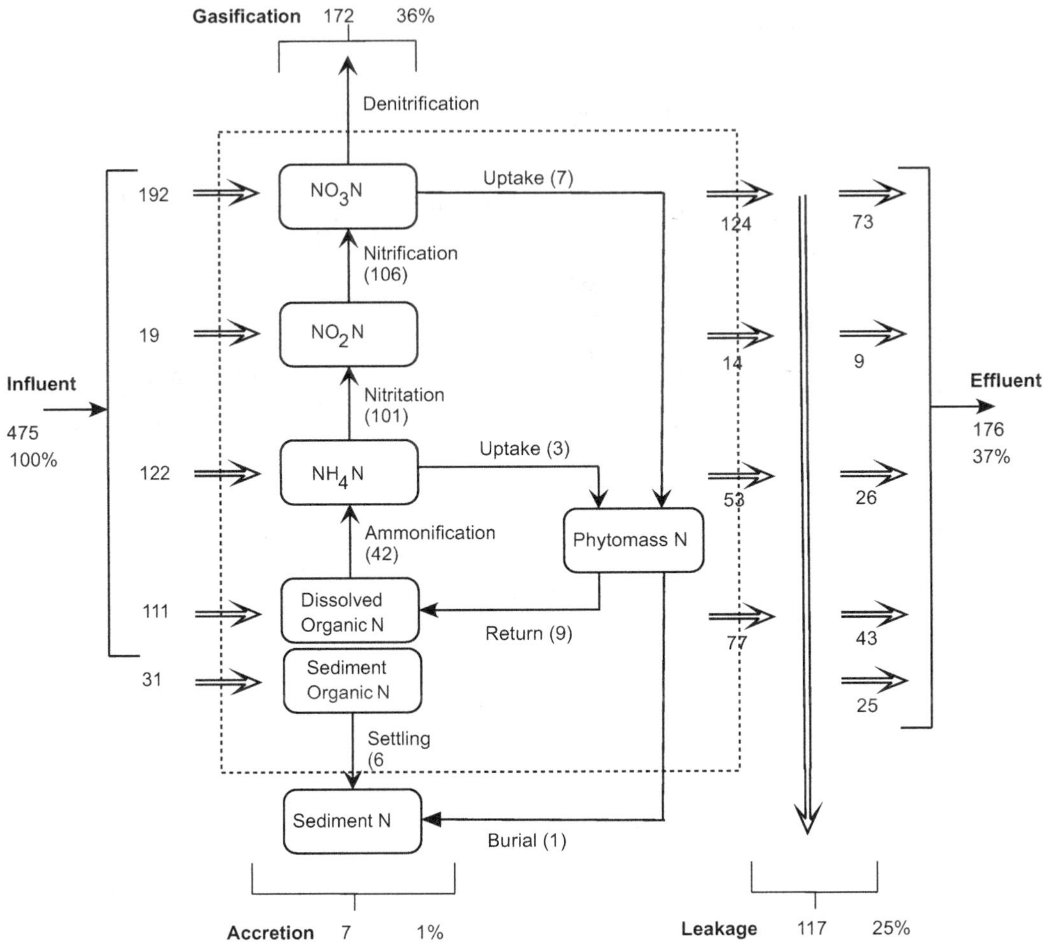

FIGURE 9.11 Estimated annual nitrogen fluxes in the Imperial, California demonstration wetland ($gN/m^2 \cdot yr$). The vegetation cycle is of little importance in this system.

the rate of denitrification far exceeds the rate based only on input/output (I/O) water analyses. The contribution of nitrification means that (I/O) denitrification is much smaller than the true value.

The OEW and Imperial systems treat different source waters, and display different nitrogen mass balances. This variation among marshes is typical, and further illustrated by the mass balance results in Table 9.10. Source waters vary in concentration by about a factor of ten for organic, ammonia and oxidized N, and hydraulic loadings vary by a similar amount. It is therefore not surprising that the inherent processes of ammonification, nitrification and denitrification also vary widely. Phytomass cycling, which is dependent on phytomass density, also displays a large degree of variability among the example systems.

There is no guarantee that the conventional sequential nitrogen conversion processes are in fact operative. The mass balances can identify anomalies, by identifying features that do not fit that conceptual model. Table 9.10 contains examples of such disparities.

TABLE 9.9A

Average inlet and outlet concentrations for the Imperial, California FWS wetland for 2001–2004.

Parameter	Inlet mg/L	Outlet mg/L	Leakage mg/L	Fraction
HLR, cm/d	19.3	12.5	6.8	–
Dissolved Organic N	1.77	1.01	1.39	–
Particulate Organic N	0.46	0.60	0.53	–
Ammonia N	1.64	0.53	1.09	0.344
Nitrite N	0.25	0.18	0.22	–
Nitrate N	2.65	1.49	2.07	0.656
TSS	179	11	–	–
$CBOD_5$	7.2	5.2	–	–
DO	8.2	7.2	–	–
Sulfate S	661	618	640	–
Alkalinity	239	203	–	–

Source: Imperial Irrigation District operating data.

TABLE 9.9B

Assumptions for the Imperial, California FWS wetland carbon and oxygen supplies. Biomass is the assumed source of carbon, and oxygen requirements are determined from Figure 9.11 fluxes.

Assumption			Notes
Biomass Produced =	500	$gdw/m^2{\cdot}yr$	–
Carbon Content =	0.5	–	50%
Useable Carbon Fraction =	0.3	–	30%
Carbon Available =	75	$gC/m^2{\cdot}yr$	–
Denitrification Carbon Requirement =	179	$gC/m^2{\cdot}yr$	1.07 × N
Denitrification Sulfide Requirement =	164	$gS/m^2{\cdot}yr$	1.69 × N excess
Sulfate Incoming =	47,000	$gS/m^2{\cdot}yr$	–
Biomass N Uptake =	10	$gN/m^2{\cdot}yr$	2% N
Biomass Buried =	0.1	$gdw/m^2{\cdot}yr$	10%
Nitrogen Buried =	1.0	$gN/m^2{\cdot}yr$	2% N
Oxygen Needed =	529	$gO/m^2{\cdot}yr$	$3.43NO_2N+1.14NO_3N$
Daily Oxygen Needed =	1.45	$gO/m^2{\cdot}d$	–

Excessive Carbon and Oxygen Requirements

The Musselwhite, Ontario system has been described in Bishay and Kadlec (2005), and appeared to be unable to provide the carbon and oxygen needed to support the conventional N conversion sequence (Table 9.10). The carbon requirement was 1,051 $gC/m^2{\cdot}yr$. There was essentially no organic carbon in the water entering the Musselwhite wetland, because that water originated from inorganic chemical processing. There was unusually high nitrite, 0.85 mg/L. Therefore, carbon supplies had to be generated by wetland processes, principally decomposition of dead biological material (necromass). The vegetation in the Musselwhite wetland was sparse, partly because of climate and partly because of severe phosphorus limitation, less than 20 μg P/L in the water column. Based on experience at other treatment wetlands, the annual production of biomass was estimated at 600 gdw/m^2·yr, which might yield about 90 gC/m^2·yr of useable carbon – only a small fraction of that needed to support the observed nitrogen loss in the wetland.

The conventional oxygen requirement was 14.8 gO/m^2·d This is much higher than the 0.5–4.0 g O/m^2·d estimated for reaeration under Musselwhite wetland conditions. Thus there does not appear to be enough oxygen transfer to support the conventional nitrification concept.

Bishay and Kadlec (2005) therefore examined the possibility that autotrophic nitrification and denitrification (anammox) was the conversion route for ammonia. Most of the processing was assumed to follow equation (9.21). There were no anomalies in these nitrogen fluxes, and more satisfying consequences in the ancillary chemistry that accompanies these conversions. The dissolved oxygen flux requirement reaeration would be required to supply 7.24 g O/m^2·d. Although still high, this is half the requirement of the traditional assumption. The carbon supply requirement under this scenario was 72 gC/m^2·yr, a value commensurate with the ability of the wetland to produce it. The unconventional anammox mass balances are shown in Figure 9.12.

Negative Nitrification

The Titoki, New Zealand wetland produced ammonia while reducing nitrate (Table 9.10) (Tanner and Sukias, 2011). When the conventional nitrogen conversion sequence is invoked, a negative nitrification is produced. This is possible if there is reduction of nitrate to ammonia, a process identified above as dissimilatory nitrate reduction to ammonia (DNRA). It is also possible to invoke a very high biomass production, but it would need to be about triple that seen in similar wetlands.

Loss of Vegetation

Two columns in Table 9.10 refer to the Tres Rios, AZ demonstration wetland H1, one computed for a fully vegetated healthy condition (before) and the other for the same wetland after it had lost most of its vegetation due to unknown causes (after). More details can be found in Kadlec (2008a). Vegetation growth amounts were estimated for purposes of the mass balances, at 2,000 gdw/m^2 before loss, and 600 gdw/m^2 after, both at 2% nitrogen. Of course, N uptake was in those same proportions.

Total nitrogen concentration reduction dropped from 52% to 32% due to vegetation loss. What is not obvious is the shift in other nitrogen conversion processes that accompany such vegetation loss. Ammonification and nitrification are virtually absent after vegetation loss, and denitrification was reduced (Table 9.10).

Limitations

Although mass balances are a good tool for understanding, there are several aspects of wetland nitrogen performance that are not elucidated.

- The distinction between porewater and surface water. Emergent plants extract nitrogen from porewater, and return some of the decomposition product organic N to porewater. Communication between surface water and porewater is difficult to quantify. Mass balances of necessity lump these two waters.

TABLE 9.10

Example annual nitrogen mass balance components.

		Brighton	ENRP	STA2	Brawley	Listowel 4	Shannon	Musselwhite	Titoki	Tres Rios H1 Before	Tres Rios H1 After
Years		10	4	4	4	4	4	6	4	2	2
HLR In	cm/d	5.32	2.66	3.78	8.87	1.95	5.45	52	14.54	12.39	11.51
HLR Out	cm/d	5.32	2.66	4.27	4.83	1.95	5.45	52	14.38	10.19	8.08
ON In	mg/L	1.70	2.37	2.12	1.71	9.97	0.54	1.18	2.68	1.45	1.13
ON Out	mg/L	1.85	2.20	2.16	0.89	2.66	0.63	0.75	2.64	0.91	1.45
NH_4-N In	mg/L	10.85	0.33	0.33	4.05	8.58	0.08	11.61	1.01	1.57	1.46
NH_4-N Out	mg/L	8.71	0.04	0.04	0.14	6.10	0.13	3.18	2.51	0.39	1.59
NO_X-N In	mg/L	0.87	0.52	0.60	1.06	0.40	2.94	6.64	8.81	2.57	3.95
NO_X-N Out	mg/L	0.52	0.05	0.11	0.58	0.20	0.70	10.29	2.76	1.37	1.46
ON In	$gN/m^2 \cdot yr$	32.9	21.60	33.35	55	71	10.6	224	142	66	47
ON Out	$gN/m^2 \cdot yr$	35.9	19.31	34.47	16	19	12.5	142	139	34	42
NH_4-N In	$gN/m^2 \cdot yr$	210.6	3.84	6.72	131	61	1.6	2,204	54	71	61
NH_4-N Out	$gN/m^2 \cdot yr$	169.2	0.39	0.95	2	43	2.5	604	132	15	47
NO_X-N In	$gN/m^2 \cdot yr$	16.9	5.67	10.59	34	3	58.5	**1,260**	468	116	166
NO_X-N Out	$gN/m^2 \cdot yr$	10.1	0.42	2.47	10	1	13.9	**1,953**	145	51	43
TN In	$gN/m^2 \cdot yr$	260.40	31.11	50.66	220	135	70.7	3,688	664	253	274
TN out	$gN/m^2 \cdot yr$	215.20	20.12	37.89	28	63	28.9	2,699	416	100	132
Ammonification	$gN/m^2 \cdot yr$	64.5	11.28	7.88	40	88	8.7	92	40	**61**	**5**
Nitrification	$gN/m^2 \cdot yr$	35.8	10.49	10.24	133	65	7.1	1,682	**-47**	**100**	**0**
Denitrification	$gN/m^2 \cdot yr$	37.7	9.98	11.77	146	65	39.8	983	245	155	100
Biomass N Uptake	$gN/m^2 \cdot yr$	75	10.00	10.00	20	45	12.5	12	40	**40**	**12**
Uptake NO_3-N	$gN/m^2 \cdot yr$	5	5.75	6.59	7	2	11.8	6	31	28	7
Uptake NH_4-N	$gN/m^2 \cdot yr$	70	4.25	3.41	13	38	0.7	6	9	12	5
Accretion	$gN/m^2 \cdot yr$	7.5	1.00	1.00	2	4	2	1	4	4	1
Leakage	$gN/m^2 \cdot yr$	0	0	0.00	45	0	0	0	0	22	49
Carbon Needed	$gC/m^2 \cdot yr$	40.3	12.4	12.6	166	70	42.6	**1,051**	262.2	166	107
Oxygen Needed	$gO/m^2 \cdot d$	0.42	0.15	0.13	1.73	0.54	0.08	**14.76**	0.00	1.27	0.15

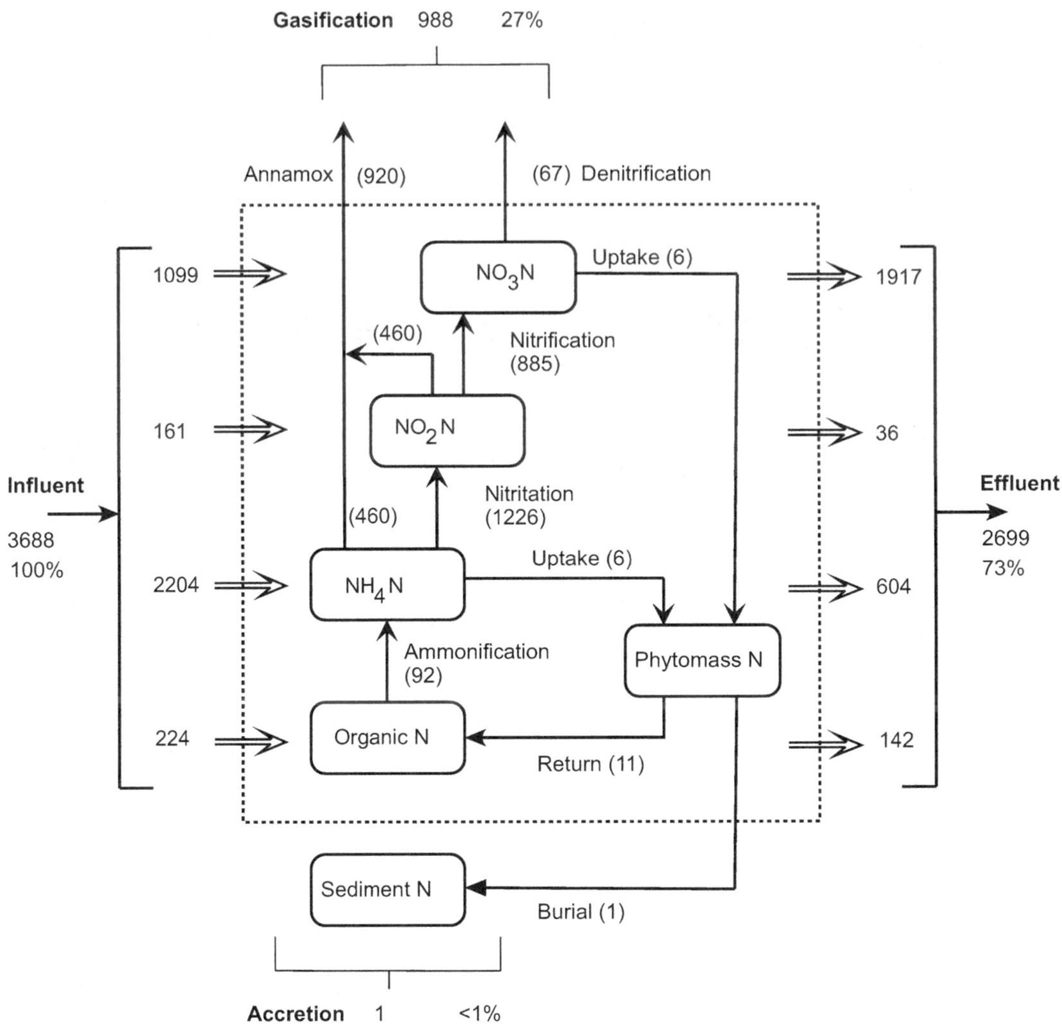

FIGURE 9.12 Mass balance for nitrogen flows in the Musselwhite, Ontario FWS wetland (gN/m^2·yr), for an autotrophic nitrification/denitrification assumption. Base data were means of six years' measurements. Adapted from Bishay and Kadlec (2005).

- The distinction between mesophytic and macrophytic phytomass. Mesophytic phytomass is located in the surface water, and cycles nitrogen therein. Rooted macrophytes cycle nitrogen from below ground and return some compounds to surface water via "macrophyte mining."
- Effects of temperature and season. Annualization precludes determination of seasonal growth effects or microbial temperature effects.
- Effects of macrophytic growth form. Floating plants cycle nitrogen directly to and from surface water, not porewater.

Accordingly, it is necessary to examine nitrogen removals by other methods of performance analysis, in addition to annual mass balances.

9.6 MONTHLY NITROGEN MASS BALANCES

Monthly nitrogen mass balances are speculative, because the required phytomass cycling data are not available. However, such monthly nitrogen budgets can give some idea of the complex seasonal behavior of nitrogen in treatment marshes. The Brighton, Ontario system will be used to illustrate seasonality (see Kadlec et al., 2012b for more information). The annual nitrogen mass balances for the 10-year POR for this wetland are shown in Table 9.10.

The POR total nitrogen loading to the wetland was 260 $gN/m^2 \cdot yr$, and 45 $gN/m^2 \cdot yr$ were removed. There were only slight inter-annual trends. This load reduction places the wetlands in a category in which the vegetative uptake and return are important, as well as microbial processes. The annual fluxes are based on the ten-year mass balances, together with assumptions, which are: (1) plant uptake is proportioned to ammonia and oxidized nitrogen according to their concentrations; (2) plant uptake was 37.5 $gN/m^2 \cdot yr$, incorporated into 1,500 gdw/m^2 biomass production at 2.5% N dry weight; (3) 90% of the annual uptake is recycled as organic N, and 10% buried in new sediments; (4) ammonification converts organic N to ammonia; (5) nitrification converts ammonia to oxidized N; and (6) denitrification converts oxidized N to gaseous N products. Plant uptake and release are of the same magnitude as the microbial processes (Table 9.10). Therefore, nitrogen removal was amplified in the spring growing season, as well as in the warmth of midsummer. Equation (7.85) is expected to apply to microbial processes, and Equation (7.86) to plant uptake.

It was not feasible to attempt to directly measure the nitrogen content of the biomass and necromass, because of spatial variability and the large level of effort required (Reinhardt et al., 2005). However, the annual patterns of growth, senescence and burial can be estimated based on information from other marshes (e.g., Kadlec and Hammer, 1988; Kadlec, 1999a), which allows the month-by-month accounting for nitrogen in the biomass and necromass (Kadlec and Wallace, 2009). The phytomass nitrogen mass balance for a fixed time period is given by Equation 9.29. The seasonal patterns for the 12 months were assumed to be:

Month	1	2	3	4	5	6	7	8	9	10	11	12
Uptake:	0.00	0.00	0.04	0.20	0.47	0.20	0.07	0.02	0.00	0.00	0.00	0.00
Return:	0.03	0.03	0.03	0.03	0.04	0.07	0.1	0.15	0.23	0.21	0.05	0.03

Burial: uniform at 0.083 each month

These patterns were used to apportion the annual totals of 37.5 $gN/(m^2 \cdot yr)$ uptake, 90% returned and 10% buried. The resulting monthly fluxes of nitrogen are shown in Figure 9.13A. Uptake and storage occur in spring, and release occurs in autumn. During the spring growth period, more than the entire external nitrogen load removal was consumed to create the standing crop. The balance was assumed to be withdrawn from soil storage.

The vegetative uptake and return, in conjunction with mass balances for the nitrogen species, allows computation of the monthly amounts of ammonification, nitrification and denitrification (Figure 9.13B, Table 9.10). Ammonification is highest in the autumn months, corresponding to the arrival of organic nitrogen from senescing and decomposing phytomass. Nitrification is favored in the warm months.

Denitrification is generally supposed to be the ultimate annual removal process for total nitrogen in excess of the net burial. The amounts of inferred denitrification, computed from mass balances, are higher in the warm months than in the cold months (Figure 9.13B). Because there are only small amounts of oxidized nitrogen in the wetland water at all times

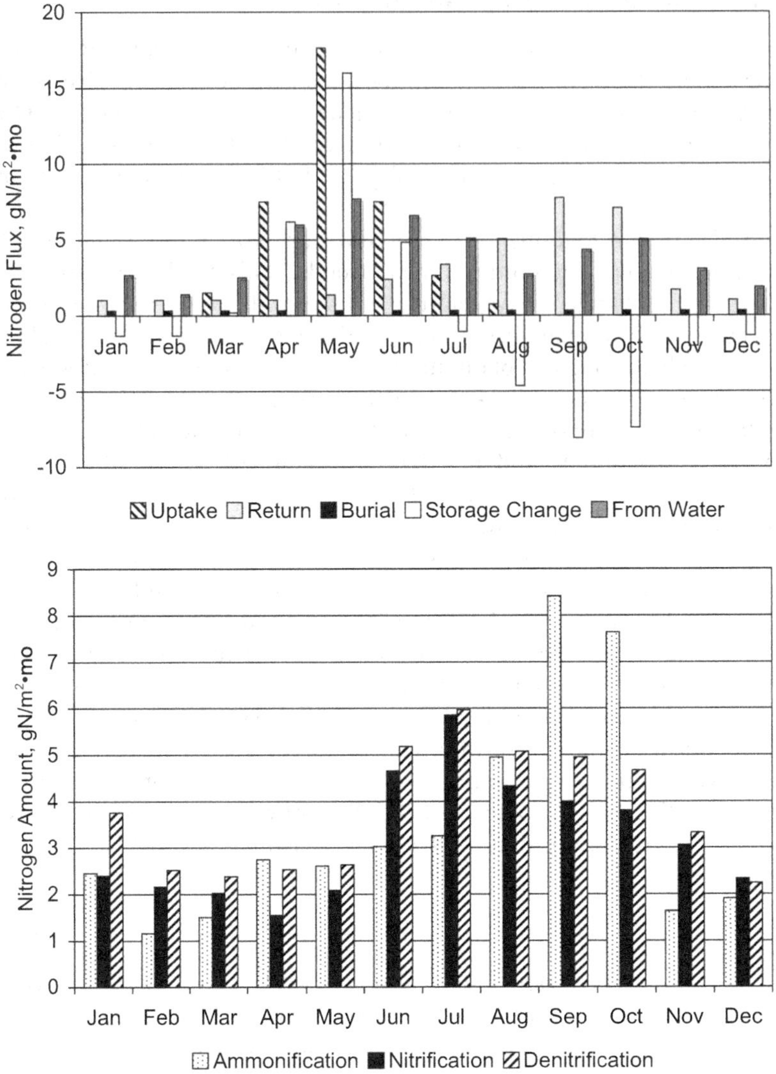

FIGURE 9.13 (A) Estimated nitrogen utilization and return from the biomass in the Brighton wetlands. Uptake is during the growing season, and return is during autumnal senescence. (B) Estimated microbial nitrogen processing.

(0.85 mg/L in; 0.46 mg/L out), it is likely that oxidized nitrogen is rapidly denitrified as soon as it is formed. The apparent removal of oxidized nitrogen, based on input and output, was 7.0 $gN/m^2{\cdot}yr$, but mass balances indicate that the total denitrification was many times higher, at 45.2 $gN/m^2{\cdot}yr$. The difference is due to production of oxidized nitrogen from nitrification. Total nitrogen reduction is the sum of burial and denitrification (Figure 9.8), and the POR average was 49 ± 3 $gN/m^2{\cdot}yr$ (mean ± interannual s.e.). The rate coefficient was restricted because of low nitrification, to 3.9 m/yr. This value is low compared to other wetlands, at the 10th percentile (Kadlec and Wallace, 2009).

The constructed marsh was successfully operated year-round in a moderately cold climate. Some measure of treatment was achieved in the cold season, but the highest removals were observed in the spring growing season. The estimated phytomass cycle was strongly involved in nitrogen processing.

9.7 ORGANIC NITROGEN PERFORMANCE

Organic nitrogen (ON) is present in domestic and municipal effluents, at concentrations up to 10–20 mg/L. Wastewater treatment wetlands typically receive these wastewaters after partial treatment, and the wetland influent then contains varying amounts of the original organic nitrogen, depending upon the type of pretreatment. But, the restriction to treatment marshes, as considered herein, means that source waters are very much lower in organic nitrogen, most often in the range of 1–5 mg/L.

Wetlands are themselves organic-rich sites, with considerable internal production of nitrogenous compounds. Incoming organic nitrogen is reduced, but not below the background concentration created by residuals and wetland return fluxes. Organic nitrogen is rarely if ever a targeted or regulated water quality parameter.

Despite the active ammonification that takes place in phytomass decomposition, organic nitrogen enters and remains at low concentrations in treatment marsh surface waters. This is partly due to the fact that most ammonification takes place in the wetland underground. As a result, organic nitrogen in treatment marshes is at background concentrations, and displays little or no reduction upon passage through the marsh. The median net period-of-record removal rate for 103 treatment marshes is 4.6 $gN/m^2 \cdot yr$ (s.d. = 56.9 $gN/m^2 \cdot yr$), which means essentially zero reduction.

Organic Nitrogen Loading

As detailed in Chapter 7, it is possible to represent annual wetland performance as the effluent concentration produced (C_o) at a given inlet loading rate and concentration (C_i) (LRI = HLR × C_i). For most contaminants, multiple datasets are represented by a trend that shows increasing C_o with increasing LRI. However, in the case of organic nitrogen, there is essentially no load response, and the outlet organic N remains essentially unaffected by load over three orders of magnitude change in inlet loadings (Figure 9.14). The outlet ON averages 1.36 ± 0.66 mg/L. Most of the variability in Figure 9.14 is due to hydraulic loading rates, because inlet ON is in a narrow range (1.7 ± 1.2 mg/L). However, these differences may be due to phytomass decay processes. Higher nutrient loadings spur higher biomass.

Variability

In general, there are four types of variability that may be manifested: inter-system, temporal, spatial and stochastic. The observed organic nitrogen leaving marshes depends to a small degree upon the character of the source water. Treated municipal wastewaters and agricultural runoff enter the wetland at marginally higher levels (ca, 1.8 and 1.5 mg/L respectively) than river waters or urban runoff (0.8 and 1.1 mg/L respectively). The marsh responds slightly to these levels, by producing higher levels from treated municipal wastewaters and agricultural runoff (ca, 1.5 and 1.5 mg/L respectively) than river waters or urban runoff (0.7 and 0.8 mg/L respectively).

Temporal variability can be present both in inter-annual trends and in intra-annual trends, or seasonality. The year-to-year data from long-running systems display slight but inconsistent trends. For instance, the OEW system showed a slight increasing trend over ten years, from 0.6 to 0.8 mg/L, while the Estevan system showed a decreasing trend over ten years, from 2.4 to 1.0 mg/L. The Brighton system showed an increasing trend over ten years, from 1.5 to 2.0 mg/L.

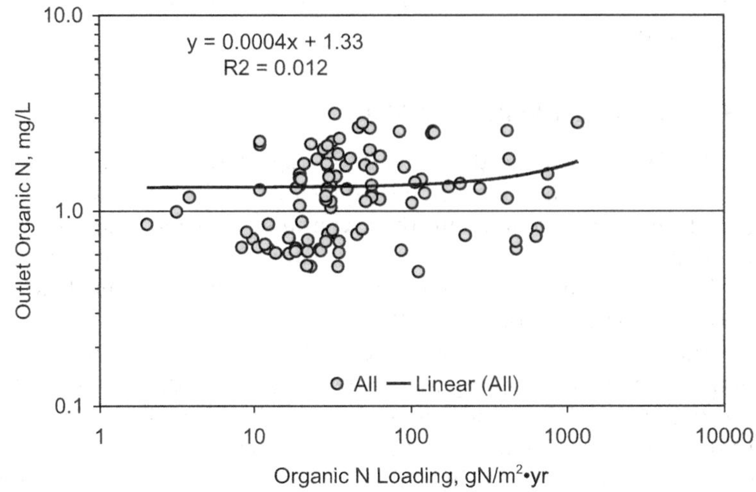

FIGURE 9.14 Load – concentration plot for organic nitrogen in treatment marshes.

Seasonality

Typically, there is essentially no seasonality in marsh effluent organic nitrogen. Data is only stochastically variable (Figure 9.15). Likely causes include the belowground location of ammonification, where temperatures are lower and steadier than in surface water (see Chapter 6). The return flux of organic nitrogen, which produces the observed near-background surface water concentrations, results from necromass decomposition, which does not depend strongly upon the above ground biomass growth cycle. The apparent result is a - steady year-round return flux of ON.

Longitudinal Profiles

Because organic nitrogen is near marsh background levels in the selected categories of entering source water, there is very little alteration along the flow path. There are but few studies that document the ON profile through the wetland. Figure 9.16 shows three profiles, from systems treating river water (Imperial, CA), lagoon effluent (Listowel 3, Ontario), and tertiary WWTP effluent (OEW, FL).

Rate Coefficients

Because organic nitrogen concentrations are close to background at most locations in treatment marshes, data fitting procedures for the PkC* model cannot determine rate coefficients with any reasonable degree of accuracy. Two other alternatives can be used: extrapolation of the rate coefficients from FWS treatment wetland that receive larger incoming concentrations of ON, and determination from the annual mass balance calculated rates of ammonification. As will be seen below, these yield very similar results for annual rate coefficients.

Surface Water Disappearance Rates

This method focuses upon the concentrations changes that occur in the surface water as it passes through the wetland. The return flux of ON is viewed as internal loading, which is described by the C* parameter. The area-based first order removal rate is utilized here:

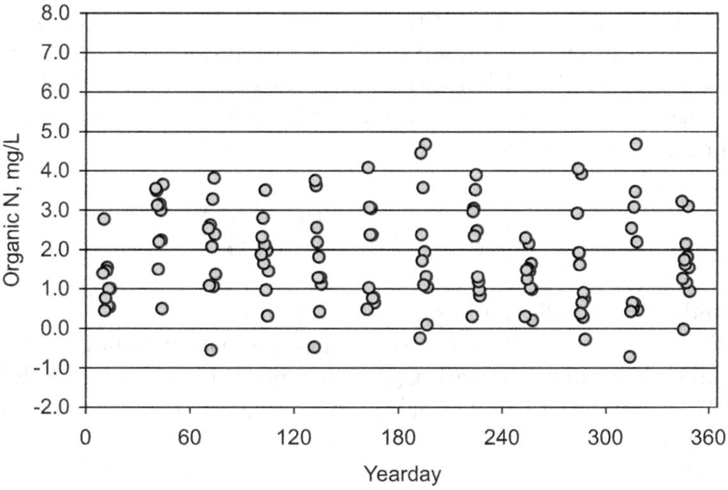

FIGURE 9.15 Organic nitrogen in the effluent from the Brighton, Ontario treatment marsh. Data are monthly during the ten-year period 2001–2010, with an arithmetic mean of 1.85 ± 0.53 mg/L.

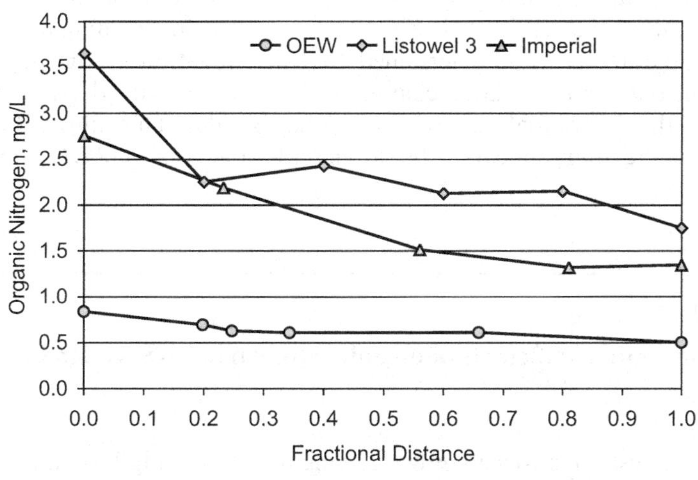

FIGURE 9.16 Organic nitrogen profiles along the flow direction in three treatment marshes. Data represent average values over 4 years for Listowel and Imperial, and ten years for OEW.

$$J_{ON} = k_{ON}\left(C_{ON} - C^*_{ON}\right) \tag{9.37}$$

where: C_{ON} = wetland organic nitrogen concentration, mg/L
C^*_{ON} = background wetland organic nitrogen concentration, mg/L
J_{ON} = removal rate of organic nitrogen, gm/m^2·yr
k_{ON} = removal rate constant for organic nitrogen, m/yr

The wetland environment may have actual hydraulics ranging from a few tanks in series (TIS) up to a large number, approximating plug flow, depending on design. However, organic nitrogen is expected to show weathering effects due to its complex speciation, thus reducing the effective number of TIS (refer to Chapter 7). Accordingly, the PkC* model is chosen, with P < N. In order to compare results across systems that in general do not have known N-values, the value P = 3 is chosen here. Further, there is a fairly narrow band of C* values, and therefore C* = 1.5 mg/L is chosen here to allow comparisons. The remaining model parameter is the k-value, selected to fit the model:

$$\frac{C_{ON,out} - 1.5}{C_{ON,in} - 1.5} = \left(1 + \frac{k_{ON}}{3q}\right)^{-3} \quad (9.38)$$

where: q = wetland hydraulic loading, m/yr

Because of the selection of C* = 1.5, parameter estimation is not reliable for low inlet concentrations, and those wetlands with $C_{ON,in} < 5$ mg/L have been excluded from calibration. Out of 147 wetlands with data for organic nitrogen, 60 systems met this criterion (Kadlec and Wallace, 2009).

Results of calibration of k-values for entire periods of record for the qualifying FWS wetland are summarized in Table 9.11. The median k-value for organic nitrogen is 17.3 m/yr, but the range is wide. The 10th to 90th percentile range is 5.0–61.9 m/yr. Accordingly, there is a large design window that encompasses varying degrees of risk.

Ammonification Rate Coefficients

This method focuses upon ammonification as determined from annual mass balances. The internal loading due to the return flux of ON is accounted in the annual mass balance, and therefore the C* parameter is predetermined by the estimation of the phytomass nitrogen transfers. Further, there is not a large change in the ON concentrations through the marsh, and consequently the differential reactor assumption is valid. This allows the rate coefficient to be calculated as the ammonification flux divided by the mean ORGN concentration:

TABLE 9.11

Annual loss rate coefficients of organic nitrogen in FWS wetlands.

Stipulations

1. Data restricted to wetlands receiving inlet C > 5 mg/L organic nitrogen.
2. Period of record averages are used in calculations.
3. For k-value calculations, the following PkC* parameters are selected:

   a. C* = 1.5 mg/L
   b. P = 3 TIS

4. Ranges of variables:

	HLR cm/d	ON In mg/L	ON Out mg/L
mean	6.3	18.2	8.1
median	3.9	10.8	5.7
max	49.9	69.5	29.6
min	1.0	5.7	1.6

Results (N = 60 wetlands)	
Percentile	Rate Coefficient m/yr
0.1	5.0
0.2	7.9
0.3	10.7
0.4	14.6
0.5	17.3
0.6	19.7
0.7	27.4
0.8	36.2
0.9	61.9

$$J_{ammonification} = k_{ON}\left(\frac{C_{ONin} + C_{ONout}}{2}\right) \tag{9.39}$$

where: C_{ONin} = wetland organic nitrogen concentration entering, mg/L
C_{ONout} = wetland organic nitrogen concentration leaving, mg/L
$J_{ammonification}$ = ammonification of organic nitrogen, $gN/m^2 \cdot yr$
k_{ON} = ammonification rate coefficient, m/yr

Table 9.12 summarizes ammonification rate coefficients determined from annual mass balances and concentrations. The median k-value for ammonification is 19.8 m/yr, but the range is wide. The 10th to 90th percentile range is 5.2–45.7 m/yr.

The value of C^*_{ON} may also be calculated, as the phytomass return flux divided by the rate coefficient. Results in Table 9.12 show a median of C^*_{ON} = 1.39 mg/L, with a 10th to 90th percentile range of 0.45–2.38.

The average (± sd) is 1.40 ± 0.98 mg/L. For the larger dataset represented in Figure 9.14, outlet ON averages 1.36 ± 0.66 mg/L. Thus the C* determined from mass balances does not significantly differ from the observed outlet ON concentrations.

TABLE 9.12

Annual ammonification rate coefficients for treatment marshes.

Stipulations

1. Period of record averages are used in calculations.
2. For k-value calculations, a differential reactor is selected.
3. Ranges of variables:

	HLR cm/d	ON In mg/L	ON Out mg/L
mean	16.15	1.84	1.50
median	8.87	1.50	1.45
max	110	4.79	2.82
min	2.01	0.25	0.49

Results (N = 31 wetlands)			
Percentile	Rate Coefficient m/yr	Percentile	Background C* mg/L
0.1	5.2	0.1	0.45
0.2	7.0	0.2	0.58
0.3	14.9	0.3	0.77
0.4	15.2	0.4	1.02
0.5	19.8	0.5	1.39
0.6	26.3	0.6	1.41
0.7	30.8	0.7	1.69
0.8	37.2	0.8	1.74
0.9	45.7	0.9	2.38

Seasonal Dependence

Ammonification in marshes displays seasonal dependence, but not water temperature dependence. This ecosystem result is at variance with the experiences in other situations. In conventional WWTP system design, ammonification is assumed to pertain to soluble organic nitrogen, and is modeled as a second-order process, first order in each of soluble organic nitrogen and first order in the biomass of heterotrophic microorganisms (U.S. EPA, 1993b). The ammonification rate increases with a doubling of the rate constant for a temperature increase of 10°C ($\theta = 1.07$) (U.S. EPA, 1993b). The rate of ammonification in flooded soils also depends on temperature and pH (Reddy and Patrick, 1984). The ammonification rate increases with a doubling of the rate constant for a temperature increase of 10° C ($\theta = 1.07$).

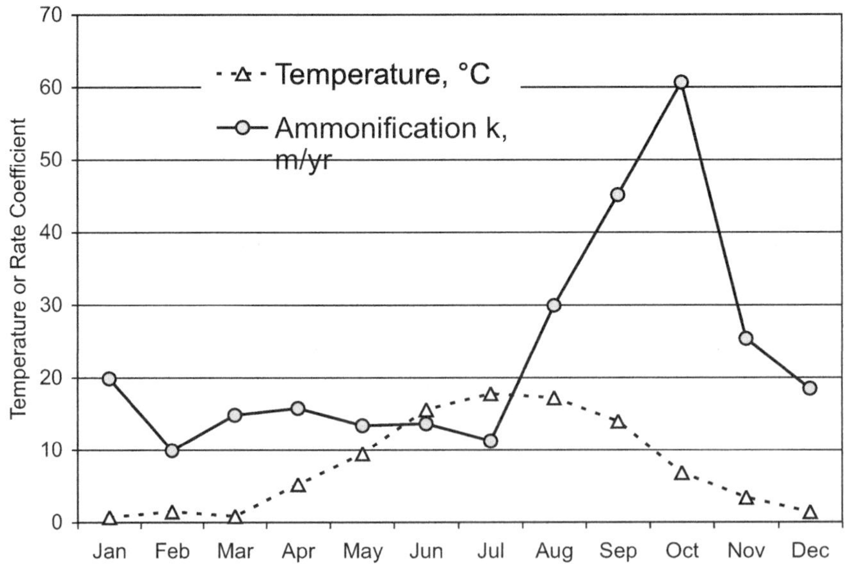

FIGURE 9.17 Ammonification rate coefficients and water temperatures for the Listowel 3 marsh in Ontario. Data are averages for four years.

However, the seasonality of ammonification in marshes is likely driven by the autumnal increase in microbial numbers that attends the (dramatic) increase in the amount of necromass. ON does not display increase, but presumptively microbial activity soars. Therefore, ammonification rate coefficients display maxima in the autumn months, not in midsummer at the peak temperatures. The seasonal progression of ammonification rate coefficients for the Listowel 3 system, as determined from mass balances, is presented in Figure 9.17. Biomass growth and senescence have been estimated month by month for the conditions of this northern wetland. Because ammonification is low in spring and summer, at moderate temperatures, and high in autumn, also at moderate temperatures, there is not a correlation of the rate coefficient with temperature. For instance, $R^2 = 0.02$ for the determination of a theta factor for the Listowel 3 system. A similar analysis for ten years' data for the Brighton, Ontario marsh produces, $R^2 = 0.03$.

9.8 AMMONIA NITROGEN PERFORMANCE

Ammonia is an intermediate in the sequential processing of nitrogen in treatment wetlands. It is produced by ammonification of organic nitrogen, and oxidized by aerobic processes. Because of toxicity of unionized ammonia in receiving aquatic ecosystems, this nitrogen species is often singled out for regulation. The fraction of unionized ammonia depends upon water temperature as well as total dissolved ammonia, and hence regulation may be seasonal, with lower concentration limits in summer months. Treatment marshes are by definition restricted to source waters with low to moderate concentrations of ammonia nitrogen. They therefore do not present a threat of ammonia toxicity to receiving waters.

As described in the preceding section, there is a large internal loading of ammonia due to ammonification of phytomass organic nitrogen. This process is obscured when the marsh is considered only as a device that lowers incoming ammonia concentrations. Ammonia processing is strongly tied to internal marsh production of ammonia by decomposing phytomass. The general observation is that marsh processes are capable of lowering ammonia concentrations essentially to zero, meaning there is a zero background concentration.

AMMONIA LOADINGS AND OUTLET CONCENTRATIONS

Reductions of ammonia in treatment wetlands systems can be represented by a loading analysis. The total ammonia loading to a treatment marsh is comprised of two major pieces: the external load imposed by source water, and the internal load imposed by ammonification. Because there are not direct measurements of ammonification, the magnitude of the internal loading must be estimated. That may be done by assuming the internal source organic nitrogen is generated by decomposition of phytomass. Here it is presumed that 90% of the biomass nitrogen produced during the growing season returns as organic nitrogen. Biomass nitrogen is considered to be approximately 2% of the biomass. Biomass is considered to be produced as a function of the available nutrients, following the concepts outlined by Tilman (1982), Thomann and Mueller (1987) and Kadlec (1997a); and as function of wetland design. In the absence of wetland data on biomass crops and their N-content, a default biomass production of 2,000 gdw/m^2 is used as an alternative. This leads to a growth requirement of 40 gN/m^2·yr, and the return of most as organic nitrogen, which is then mostly ammonified. Here, a default ammonification of 90% of the growth requirement is assumed, or 36 gN/m^2·yr.

There are alternatives for the graphical representation of the load response information. The procedure of displaying outlet ammonia as a function of total Kjehldahl nitrogen (TKN), was used by Kadlec and Wallace (2009). Here the choice is made to display outlet ammonia N concentrations as a function of the source water ammonia loading. The internal ammonia load adds to this, and so the marsh actually processes a greater total ammonia loading.

The greater the external ammonia loading, the greater the outlet ammonia concentration (Figure 9.18). The data display considerable scatter, but can be represented by a log-linear regression:

$$C_{Aout} = A(J_{inA})^m \tag{9.40}$$

where: A = fitting parameter
m = fitting exponent
C_{Aout} = outlet ammonia concentration, mg/L
J_{inA} = incoming ammonia nitrogen, $gN/m^2 \cdot yr$
The values shown on Figure 9.18 are A = 0.067 and m = 0.458, with $R^2 = 0.54$, N = 88 systems for their POR.

Profiles and Background Concentrations of Ammonia

There is not an apparent lower limit to exit ammonia concentrations in Figure 9.18 indicating a near-zero background for this species. This result has also been observed in treatment wetlands. For instance, transect water chemistry was acquired at the Houghton Lake treatment wetland over a 28-year period, including measurements of ammonium nitrogen. These data show a near-exponential decrease in NH_4-N concentrations with distance from the discharge, thus supporting a first order model. This system operated seasonally in the unfrozen months, and showed very low background concentrations (0.01–0.05 mg/L), both for the wetland prior to wastewater discharges and for the unaffected zones of the present-day wetland. Wastewater concentrations were much higher (up to 10 mg/L), therefore it is accurate to represent the data with a first-order model with a very small background. The detection limit for ammonia is typically 0.05 mg/L, and thus the presumptive background represents some value less than the detection limit. The assumption of C* = 0 mg/L appears to approximate operating data from all types of treatment marshes.

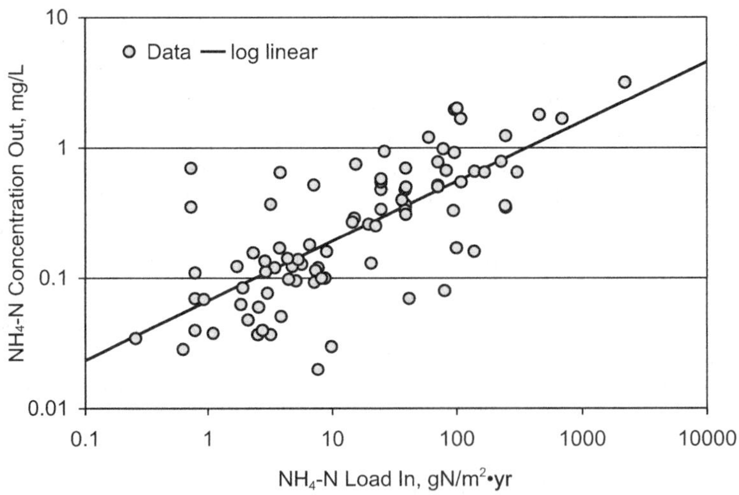

FIGURE 9.18 Load – concentration plot for ammonia nitrogen. Inlet ammonia loading from source water is used. The log-linear regression line has an R2 = 0.54 (N = 88).

AMMONIA LOSS RATE COEFFICIENTS

There are two different rate coefficients that may be calculated for ammonia processing: an ammonia loss rate coefficient, and a nitrification rate coefficient.

The loss rate concept is that ammonia flows through the wetland, and is reduced by an unspecified set of internal processes. Under the PkC* presumption, with C* = 0, as set forth in Chapter 7:

$$\frac{C_{ANout}}{C_{ANin}} = (1 + k'_{AN}/Pq)^{-P} \tag{9.41}$$

where: C_{ANin} = wetland ammonia nitrogen concentration entering, mg/L
C_{ANout} = wetland ammonia nitrogen concentration leaving, mg/L
k'_{AN} = first order areal removal rate coefficient, m/yr
P = apparent number of TIS
q = hydraulic loading rate, m/yr

This is a particularly simple formula to implement, because we only need to know the HLR and the inlet and outlet concentrations for a given internal flow pattern (P value). However, internal processes of plant uptake and ammonification easily lead to confusing results, such as negative rate coefficients (k'_{AN}). On a practical level, Equation 9.41 is best used only when ammonia dominates the inflow nitrogen, and when phytomass cycling is negligible. Those conditions are sometimes met for treatment marshes. In those circumstances, microbial processes are dominant, and oxygen may become limiting. Temperature coefficients are applicable.

An example of the data fit of Equation 9.41 is shown in Figure 9.19.

Annual ammonia loss rate coefficients for a set of treatment marshes are shown in Table 9.13. The median value for agronomic wetlands (11 m/yr) (DINLI < 120 $gN/m^2 \cdot yr$) is lower than for those that rely more upon microbial processes and for which the internal fluxes are less important (23 m/yr) (DINLI > 120 $gN/m^2 \cdot yr$).

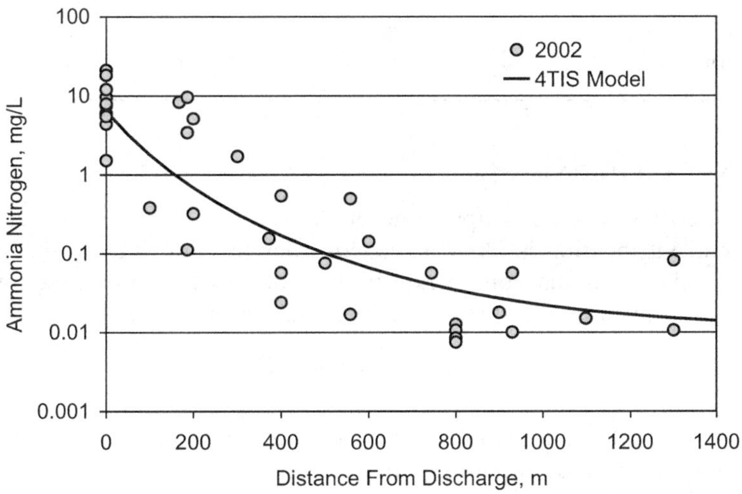

FIGURE 9.19 Decrease in ammonia nitrogen along the flow direction for the Houghton Lake treatment marsh in 2002. The model line is for C* = 0.01 mg/L. The nominal detention time to reach 1,000 m was about 30 days.

TABLE 9.13

Annual ammonia loss rate coefficients for treatment marshes.

Stipulations

1. Period of record averages are used in calculations.
2. For k-value calculations, the following PkC* parameters are selected:
   a. C* = 0.0 mg/L
   b. P = 3 TIS
3. Ranges of variables for (N = 115):

	HLR cm/d	NH_4-N In mg/L	NH_4-N Out mg/L
mean	10.3	3.0	1.4
median	4.9	1.7	0.5
max	110.0	19.3	11.0
min	0.2	0.1	0.0

Results (N = 128 wetlands)

	Rate Coefficient, m/yr	
	DINLI < 120	DINLI > 120
Percentile	N = 73	N = 42
0.1	-2	-10
0.2	1	3
0.3	4	11
0.4	8	15
0.5	11	23
0.6	15	35
0.7	23	49
0.8	30	59
0.9	37	96

Nitrification Rate Coefficients

Nitrification rate coefficients focus upon one of the several ammonia fluxes, its removal to oxidized nitrogen. Nitrification is likely a second-order reaction, involving the DO concentration as well as the ammonia concentration. Typically there is not data for wetland DO concentrations, and thus a pseudo-first order rate equation is utilized:

$$J_{nitrification} = k_{AN} C_{AN} \tag{9.42}$$

where: C_{AN} = wetland ammonia nitrogen concentration, mg/L
$J_{nitrification}$ = removal rate of ammonia nitrogen, $gN/m^2{\cdot}yr$
k_{AN} = pseudo-first order nitrification rate coefficient, m/yr

The total nitrification flux can be computed from the mass balance web described above, easily on an annual basis, and with difficulty and approximations on a monthly scale.

The ammonia concentration varies along the flow path, presumptively according to a TIS model. The appropriate wetland average ammonia concentration may then be calculated from:

$$C_{ANavg} = \frac{C_{ANin} - C_{ANout}}{P\left(\left(C_{ANin}/C_{ANout}\right)^{1/P} - 1\right)} \tag{9.43}$$

where: C_{ANin} = wetland ammonia nitrogen concentration entering, mg/L
C_{ANout} = wetland ammonia nitrogen concentration leaving, mg/L
P = apparent number of TIS

Further, often there is not a large change in the NH_4-N concentrations through the marsh, and consequently the differential reactor assumption is valid. This allows the rate coefficient to be calculated as the nitrification flux divided by the mean NH_4-N concentration:

$$J_{nitrification} = k_{AN}\left(\frac{C_{ANin} + C_{ANout}}{2}\right) \tag{9.44}$$

Table 9.13 summarizes nitrification rate coefficients determined from annual mass balances and concentrations. The median k-value for ammonification is 78 m/yr for agronomic conditions, but only 22 m/yr when microbial processes are more important. The 10th to 90th percentile ranges are 10–300 m/yr and 5–146 m/yr respectively.

Oxygen Supply

Because all mechanisms of ammonia reduction require oxygen to varying degrees, it is useful to estimate the amount consumed in treatment marshes. Most studies of ammonia removal in constructed wetlands assume the occurrence of the "classical" sequence of autotrophic nitrification followed by respiratory denitrification. The nitrification step requires 4.3 gO/gN, as well as an additional 0.3 gO/gN to supply the organisms.

If an upper limit to the necessary oxygen is sought, then CBOD reduction via aerobic degradation must be added to the requirement, and included in the overall oxygen demand. However, treatment marshes are not typically the recipient of high-BOD source waters.

TABLE 9.14

Annual nitrification rate coefficients for treatment marshes.

Stipulations

1. Period of record averages are used in calculations.
2. For k-value calculations, a TIS average NH_4-N concentration is used.
3. Ranges of variables for DINLI<120 (N = 73):

	HLR cm/d	NH_4-N In mg/L	NH_4-N Out mg/L
mean	4.8	2.1	0.9
median	3.9	1.0	0.3
max	20.0	17.2	7.9
min	0.2	0.1	0.0

	Results (N = 128 wetlands)			
	Rate Coefficient, m/yr m/yr		Oxygen Required $gO/m^2 \cdot d$	
	DINLI < 120	DINLI > 120	DINLI < 120	
Percentile	N = 73	N = 42	Percentile	N = 73
0.1	9	11	0.1	0.20
0.2	26	28	0.2	0.32
0.3	45	42	0.3	0.36
0.4	56	46	0.4	0.39
0.5	73	61	0.5	0.43
0.6	99	83	0.6	0.45
0.7	181	102	0.7	0.50
0.8	245	146	0.8	0.61
0.9	299	257	0.9	0.81

The use of "alternate" nitrogen chemistry, coupled with different ranges of BOD degradation (aerobic or anaerobic) will result in estimates of the implied oxygen supply considerably lower than that associated with "classical" nitrification and denitrification. Table 9.14 shows the distribution of implied oxygen supply to a sampling of agronomic treatment marshes. All values are well within the expected range of atmospheric reaeration discussed earlier in this chapter.

Temperature and Seasonal Dependence

Because treatment marshes function with a combination of microbial and vegetative processes, they display features of both. Microbial processes are traditionally considered to be temperature dependent, and vegetative processes are seasonal in nature. Consequently, annual ammonia patterns show a mixture of these.

Agronomic Marshes

Plant uptake rates do not correspond to the annual cycle of water temperatures, and hence ammonia removal in agronomic wetlands cannot be characterized by modified Arrhenius θ-factors. The contrast between water temperature sequences and ammonia loss rate coefficients is illustrated for Estevan, SAS in Figure 9.20. The DIN loading to this wetland was 34 $gN/m^2 \cdot yr$, placing it well within the agronomic range. The ammonia loading to this wetland was 28 $gm/m^2 \cdot yr$, average during the growing season. No correspondence between monthly rate constants and temperature is present. For this agronomic wetland, it would be appropriate to utilize monthly k-values:

j	= month # =	5	6	7	8	9	10	11
$k'_{AN,\,j,\,m/yr}$	=	17.1	12.5	11.7	9	12	14.6	9.1

Similar patterns occur for other agronomic systems, such as Listowel 1, 2, and 3; and Arcata enhancement wetlands.

Merriman et al. (2017) calibrated ammonia loss temperature coefficients to data from eleven stormwater wetlands in North Carolina. They report $\theta = 1.003 \pm 0.030$ for the ammonia loss rate coefficient for these agronomic wetlands. This near-unity value indicates

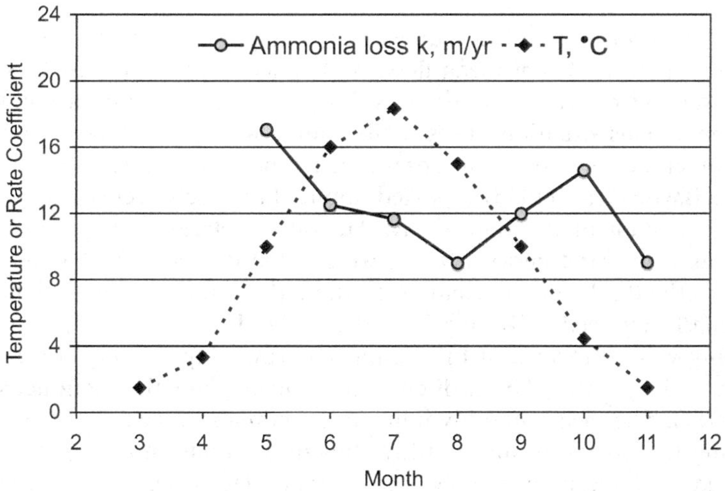

FIGURE 9.20 Seasonal ammonia rate constants for the Estevan, Saskatchewan treatment wetland. This system was operated seasonally (May – November) for nine years.

the lack of any water temperature effect for ammonia loss, as would be expected for a vegetation-uptake dominated wetland.

Nitrification

Treatment marshes that have high ammonia loadings show ammonia losses that are strongly dependent on nitrification, because there is a large excess of ammonia beyond that needed for plant growth. These are in the category of microbially dominated marshes with respect to ammonia. If the seasonality of the phytomass cycle is estimated, the nitrification process can be separated and analyzed for agronomic wetlands as well. Either situation allows estimates of the temperature dependence of nitrification. There are strong seasonal effects on nitrification, which for microbially mediated systems is in synchrony with the water temperature. An Arrhenius temperature coefficient (θ) is capable of accounting for these effects (see Chapter 7).

Temperature coefficients for nitrification have been the subject of study in both microbiology and wastewater treatment technology. For instance, Knowles et al. (1965) measured the activity of the two consortia involved, and reported $\theta = 1.100$ for *Nitrosomonas*, and $\theta = 1.06$ for *Nitrobacter*. Metcalf and Eddy (1991) list $\theta = 1.052$ for nitrification in general. Crites and Tchobanoglous (1998) list $\theta = 1.103$ for suspended growth wastewater treatment devices, thus promulgating the value in U.S. EPA (1993b). Based in part on these WWTP estimates, Crites et al. (2014) advance a value of list $\theta = 1.048$ for free water surface wetlands.

Results of calibration of k'_{AN}-values for entire periods of record for representative microbial wetlands are summarized in Kadlec and Wallace (2009), and slightly updated here. Monthly averages were used to avoid synoptic error (transit time offset). Calibrations were performed for best estimates of the internal hydraulics for each wetland. The median k'_{AN20}-value for ammonia loss is 13 m/yr, but the range is wide. Temperature coefficients had a mean value of 1.051 ± 0.028 ($N = 14$), indicating a relatively strong thermal effect on the suite of microbial processes that contribute to ammonia reduction.

Ammonia Concentration Pulses

Stormwater wetlands operate with a sequence of flow pulses, and these are included in the datasets represented above. Event mean flows and concentrations are used. A separate question arises if a wetland receives a continuous flow, which contains occasional concentration pulses of ammonia. This situation arises if the source water is a WWTP conducting nitrification, which sometimes undergoes an upset that temporarily causes nitrification to fail.

For instance, Bavor et al. (1995) reported about three such events per year over a four-year period of operation of the Byron Bay, Australia pretreatment system, some of which reached 20–45 mg/L for brief periods. These were reduced to less than 3 mg/L by the Byron Bay treatment wetlands. Under normal operation, the ammonia in wetland source water was about 5.5 mg/L, reduced to 1.0 mg/L by the wetland.

Daily sampling was performed at the Commerce Township, Michigan treatment wetland, which was preceded by an oxidation ditch pretreatment plant that conducted full nitrification (Kadlec, 2011). The mean monthly influent ammonia nitrogen to the treatment wetland averaged 0.42 mg/L, and the mean monthly effluent ammonia nitrogen from the treatment wetland averaged 0.35 mg/L over a five-year period. However, these averages were influenced by two WWTP upsets that sent high ammonia to the treatment wetland, over 68 days in 2000 and nine days in 2002. These peaked at approximately 13 mg/L and 20 mg/L, respectively. If these months are excluded, the mean monthly influent ammonia nitrogen to the treatment wetland averaged 0.085 mg/L, and the mean monthly effluent ammonia nitrogen from the treatment wetland averaged 0.086 mg/L.

The average k-value during the period 1999–2003 was 12.0 m/yr, which is close to the median expected value based on other wetlands. The year 2000 event had an event mean wetland inlet ammonia concentration of 4.1 mg/L, which was reduced to 2.3 mg/L. In the year 2002 event, 5.6 mg/L was reduced to 3.3 mg/L (Figure 9.21). The corresponding ammonia removal rate constant, for 3TIS at the prevailing water temperature of 20°C, was 40 m/yr. This high value, together with the synchrony between the inlet and outlet concentrations, signals a rapid response of the ammonia removal mechanisms within the wetland. Removal may be attributed to microbial processes, or to algal uptake. In either case, water-phase storages are adequate for calculations.

Open Water Zones

U.S. EPA (2000a) speculates that oxygenation is better provided by zones of open water, containing submerged vegetation if possible. Such zones are then incorporated as a necessary part of the wetland system design for ammonia removal. We here observe that such reasoning would apply only to those treatment marshes that are under microbial control of ammonia removal, and within that category, only to those marshes that do not have an adequate amount of dissolved oxygen. As defined herein, treatment marshes do not fall in that category, because atmospheric reaeration is adequate without the help of open water zones.

Variability

Two time scales are of interest: year-to-year, which reflects sustainability; and month-to-month, which reflects random events in the wetland.

Long Term Trends

Ammonia loss is expected to be stable over long periods of record, as long as the wetland vegetative community is stable. Additional microbial losses are expected to be stable

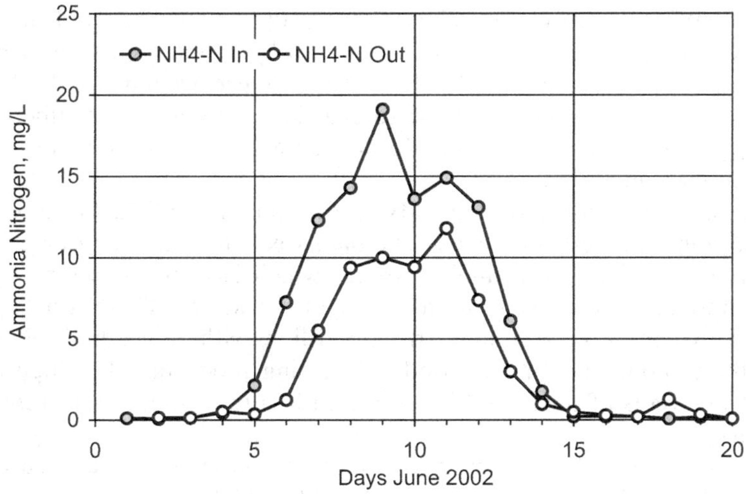

FIGURE 9.21 Response of the Commerce Township, Michigan treatment wetland to an ammonia pulse from the wastewater treatment plant in June 2002. Data source: Commerce Township unpublished daily monitoring data.

because of the pervasive character of the nitrifier populations. Ammonia loss was studied at the Houghton Lake, MI site for three decades (Figure 9.22), and displayed a slight decreasing trend.

Stochastic Variability

Seasonal trends in wetland effluent ammonia follow water temperatures for microbially controlled systems, but seasonal patterns are more complicated for agronomic wetlands, because agronomic (lightly loaded) systems will be heavily influenced by plant biomass

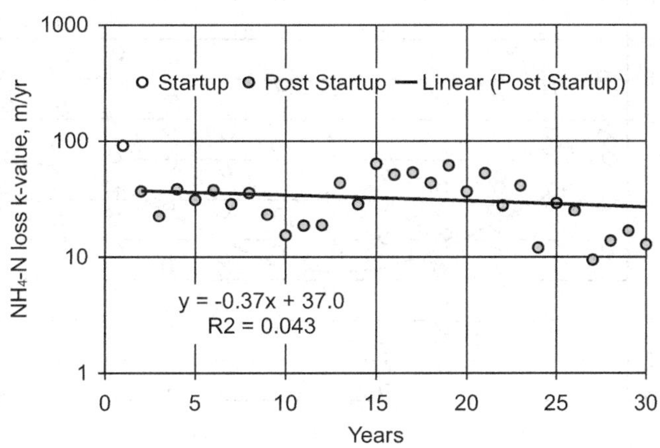

FIGURE 9.22 Time series of ammonia loss coefficients for the Houghton Lake treatment marsh. Each annual value is derived from 3–6 transects during the warm operating season. The PkC* model is represented, using C* = 0 mg/L and P = 3.

cycling. Ammonia nitrogen is very frequently the target of regulatory limits for wetlands treating strong WWTP source waters. Those limits properly seek to avoid the consequences of eutrophication and possible aquatic toxicity caused by ammonia, and may be imposed at varying frequencies of measurement. Wetland design must account for both the seasonal variability in the ammonia trends, and the stochastic variation that is superimposed upon it. In contrast, treatment marshes usually do not target ammonia reduction, and are rarely subject to ammonia regulations.

An example of such trends and variability is given in Figure 9.23 for ammonia nitrogen at the Brighton, Ontario wetland facility. The data frequency was monthly, rolled up from weekly measurements. There is an annual trend, which is nearly obscured by the scatter. The highest ammonia trend concentrations occur in the winter. There is a large amount of scatter of the daily concentrations about the trend line, with values both above and below the trend, as there should be for any good data fitting procedure. The high concentration end of the distribution is of regulatory interest, and it is therefore useful to look at the high percentile points.

It is useful to examine the variability of monthly average outlet concentrations, represented by a fractional addition to the trend value (Ψ), which corresponds to a multiplier on the trend value of $(1+\Psi)$ (see Chapter 7). Table 9.15 lists percentiles of the monthly ammonia Ψ-distributions for some representative treatment marshes, separated according to whether they are agronomic or microbial. These data indicate that the median of the 95th percentile is an additional 170% above the trend for agronomic systems, and that the median of the 95th percentile is an additional 87% above the trend for microbial systems. This difference reflects the fact that low ammonia loadings lead to low ammonia concentrations, from which infrequent but large excursions occur.

To incorporate this variability into design, the wetland designer must oversize the wetland, according to the steps outlined in the section on total nitrogen. Because deterministic equations (k-rate or other) represent the central estimate of performance, failure to incorporate variability means that 50% of the anticipated concentrations will be above the calculated values. If the design calculation is set to target the central tendency, exceedance

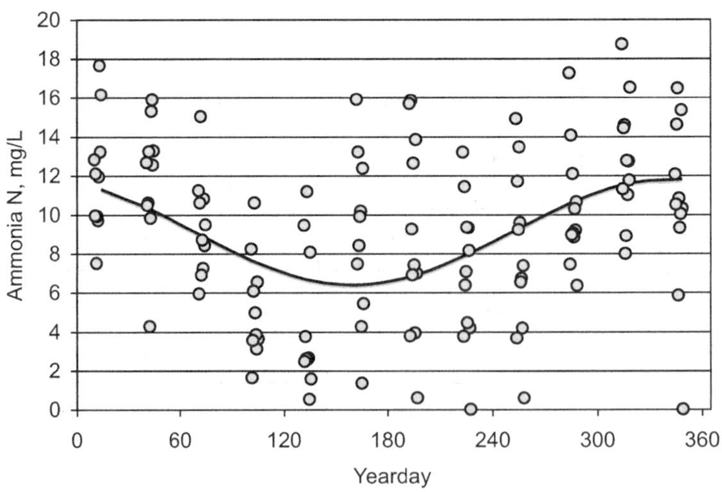

FIGURE 9.23 Distribution of monthly ammonia concentrations leaving the Brighton wetland over a 10 year POR.

TABLE 9.15

Trend multipliers for effluent ammonia concentrations for agronomic and microbial treatment marshes. The trend multiplier is (1+ψ), see Equation 7.89.

		Trend Mulitpliers			
	Years	50%	80%	90%	95%
Agronomic					
Estevan	9	0.81	1.27	3.56	6.11
Titusville	7	0.86	1.51	2.30	2.66
Imperial	4	0.90	1.83	2.91	3.29
Shannon	4	0.40	1.08	1.94	4.03
OEW	10	0.36	0.98	3.90	4.84
Listowel 1	4	0.95	1.53	1.91	2.10
Listowel 2	4	0.85	1.57	1.99	2.68
Listowel 3	4	0.96	1.56	1.91	2.73
Listowel 4	4	0.96	1.60	1.98	2.08
Listowel 5	4	1.11	1.45	1.65	1.80
Median		0.88	1.52	1.99	2.70
Mean		0.81	1.44	2.41	3.23
Microbial					
Columbia	3	0.97	1.37	1.70	1.80
Duplin	7	0.29	1.40	1.71	2.58
Oxelösund	5	1.00	1.29	1.38	1.41
Hassleholm	5	0.91	1.40	1.54	1.67
Brighton	4	0.95	1.30	1.46	1.54
New Hanover	4	0.94	1.18	1.34	1.68
Linköping	3	0.88	1.49	1.73	1.93
Musselwhite	6	0.91	1.53	2.44	2.91
Augusta	6	0.72	1.33	2.35	2.59
Brawley	4	0.83	1.27	1.87	2.80
Median		0.91	1.35	1.70	1.87
Mean		0.84	1.36	1.75	2.09

frequencies of 50% are expected to result. The risk of exceedance is lowered by increasing wetland size to the selected percentile of the Ψ-distribution.

9.9 OXIDIZED NITROGEN PERFORMANCE

The combination of nitrite (NO_2-N) and nitrate (NO_3-N) is oxidized nitrogen (NO_X-N), often referred to simply as nitrate, because nitrite is usually a small fraction of the total. Nitrate can serve as a source of nitrogen for plant growth, but may not be the preferred form in the presence of ammonia nitrogen because plants must reduce nitrate prior to further use. Nitrate and nitrite are also important in water quality control because, when present in drinking water, they may result in a potentially fatal condition known as methylglobanemia, or "blue baby" syndrome. The current regulatory criteria for nitrate in groundwater and drinking water supplies in the United States is 10 mg/L. This limit has on occasion been exceeded in water supplies in the Midwestern United States due to agricultural impacts, leading to deliveries of bottled water for babies and pregnant women. Eutrophication of marine environments is also a nitrate concern. The nitrogen content of the streams and rivers of the Midwestern United States is of particular importance at this point in history, because of hypoxia in the Gulf of Mexico, together with the associated ecological and economic consequences (U.S. EPA, 2017a). Nitrogen pollution is of similar concern for the Baltic Sea. Treatment marshes for agricultural runoff are an efficient and integral part of the suite of remedies (Hansen et al., 2018).

Curiously, nitrate removal has been discussed in the more recent treatment wetland literature, but often with widely diverging viewpoints. U.S. EPA (1999) states that " ... essentially no relationship exists between nitrate loading and effluent quality ... " U.S. EPA (2000a) states that the idea that "Constructed wetlands can remove significant amounts of nitrogen" is a "misconception." U.S. EPA (2000a) focuses on the removal of ammonia, but indicts nitrate removal by association. In contrast, Crites and Tchobanoglous (1998) conclude "When nitrogen is present in the nitrate form, nitrogen removal is generally rapid and complete." Crites et al. (2006) state " ... nitrate will be denitrified within a few days of detention."

The terminology of nitrate removal in wetlands can be somewhat confusing. A fraction of the nitrate may be taken up by algae and plants, and a small portion of this nitrate ultimately is converted to sediment organic nitrogen and stored. Another portion may be denitrified to dinitrogen gas and dissipated. In combination, these mechanisms are sometimes called "retention," although very little nitrogen is actually retained. Wetland processes reduce concentrations and masses of nitrate, and so the total set of removal processes also has been termed nitrate "reduction." Denitrification is used elsewhere as an umbrella term for all removal processes, but more than denitrification is involved. Here, the term "loss rate" is used, to indicate the entire suite of mechanisms that lead to less nitrate nitrogen departing the wetland.

Treatment wetlands have been proven effective in removal of nitrate (Kadlec and Wallace, 2009; Kadlec, 2012). They may be utilized at several general locations in the watershed, including end-of-field, at treatment plants, on low-order streams, and on major rivers. It is the purpose here to summarize existing performance data, and interpret it into rational design criteria. The attendant economics, ancillary benefits, and institutional impediments are discussed.

Loading Considerations

Reductions of nitrate in treatment marshes systems can be represented by a loading analysis. The total nitrate nitrogen loading to a treatment marsh is comprised of two major pieces: the external load imposed by source water, and the internal load imposed by nitrification. The external load is simply the inflow rate multiplied by the incoming NO_X-N concentration. Because there are not direct measurements of nitrification, the magnitude of the internal loading due to nitrification must be estimated. That may be done via the nitrogen mass balance, as detailed previously in Section 9.8.

The selected alternative for the graphical representation of the load response information is to display outlet nitrate nitrogen concentrations as a function of the source water nitrate nitrogen loading. The internal nitrate load adds to this, and so the marsh actually processes a greater total nitrate loading.

The greater the external nitrate loading, the greater the outlet nitrate concentration (Figure 9.24). The data display considerable scatter, but can be represented by a log-linear regression:

$$C_{NNout} = A(J_{inNN})^m \tag{9.45}$$

where: A = fitting parameter
m = fitting exponent
C_{NNout} = outlet ammonia concentration, mg/L
J_{inNN} = incoming ammonia nitrogen, gN/m^2·yr
The values shown on Figure 9.18 are A = 0.020 and m = 0.885, with R^2 = 0.76,and N = 167 systems for their POR.

Profiles and Background Concentrations of Nitrate

Nitrate is entirely consumable in treatment wetlands. The presumed value for C* is zero for all types of treatment wetlands, because no investigation has shown a lower limit to the reduction of nitrate. Minimum reported concentrations appear to be bounded only by the analytical detection limits.

There is not an apparent lower limit to exit nitrate concentrations in Figure 9.24 indicating a near-zero background for this species. For example, transect nitrate chemistry was acquired at the Moscow, Idaho treatment wetland over a two-year period, including measurements of nitrate nitrogen. These data show a near-exponential decrease in NO_X-N concentrations with distance from the discharge, thus supporting a first order model (Figure 9.25). This system operated year-round, with temperatures near freezing in December and January. The assumption of C* = 0 mg/L appears to approximate operating data from all types of treatment marshes.

Nitrate Loss Rate Coefficients

There are two different rate coefficients that may be calculated for nitrate processing: a nitrate loss rate coefficient, and a denitrification rate coefficient. In both cases, we are dealing with a single compound (ion) and not a mixture of different forms. Therefore, there

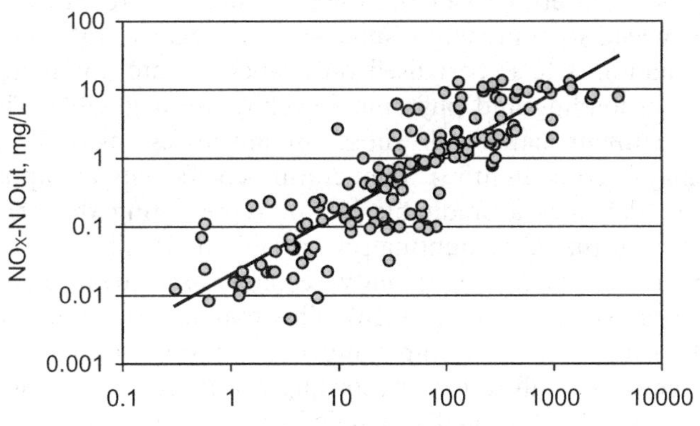

FIGURE 9.24 Load – concentration plot for nitrate nitrogen. Inlet nitrate loading from source water is used. Each point represents the POR for one wetland. The log-linear regression line has an R^2 = 0.76 (N = 167).

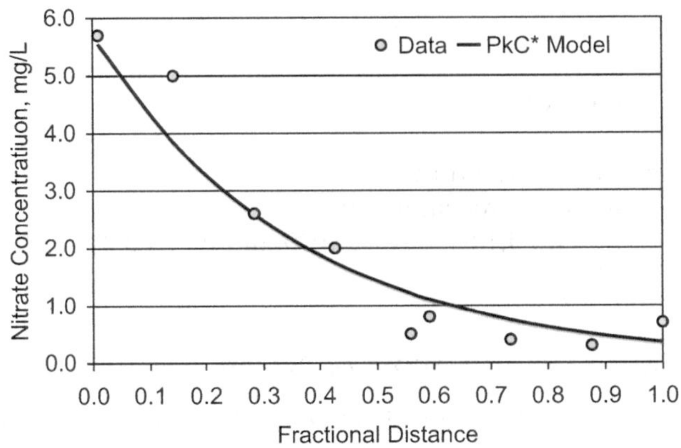

FIGURE 9.25 Decrease in nitrate nitrogen along the flow direction for the Moscow, Idaho treatment marsh in September 2010. The model parameters for this curve are P = 6, k = 32 m/yr and C* = 0, at a temperature of 14.8°C. For a theta factor of 1.1, k_{20} = 53 m/yr. Data from Beutel (2012).

is not a weathering phenomenon, and the value of P is the hydraulic flow pattern parameter N.

The loss rate concept is that nitrate flows through the wetland, and is reduced by an unspecified set of internal processes. Under the PkC* presumption, with C* = 0, as set forth in Chapter 7:

$$\frac{C_{NNout}}{C_{NNin}} = (1 + k'_{NN}/Pq)^{-P} \tag{9.46}$$

where: C_{NNin} = wetland ammonia nitrogen concentration entering, mg/L
C_{NNout} = wetland ammonia nitrogen concentration leaving, mg/L
k'_{NN} = first order areal removal rate coefficient, m/yr
P = apparent number of TIS
q = hydraulic loading rate, m/yr

The k'_{NN} value may be calculated from the HLR and the inlet and outlet concentrations for a given internal flow pattern (P value). However, internal processes of plant uptake and ammonification easily lead to confusing results, such as negative rate coefficients (k'_{NN}). On a practical level, Equation 9.46 is best used only when nitrate dominates the inflow nitrogen, and when internal loading and phytomass cycling are negligible. Those conditions are sometimes met for treatment marshes. In those circumstances, microbial processes are dominant, and carbon may become limiting. Temperature coefficients are applicable.

This first order model does a creditable job of representing the behavior of treatment marshes. It provides a means of accounting for variable hydraulic loads as well as changing temperatures. Nonetheless, random variability remains, as evidenced by the model fit for the Imperial, California wetland (Figure 9.26). This marsh was a long, narrow system, divided into four cells, some of which employing banded vegetation (Kadlec et al., 2010a). This wetland had significant infiltration. Accordingly, a limiting form of Equation 9.46 was used, for P = N → ∞, which additionally included an infiltration term (see Chapter 7).

The PkC* model also can represent internal transect profiles of nitrate concentration. The applicable form is:

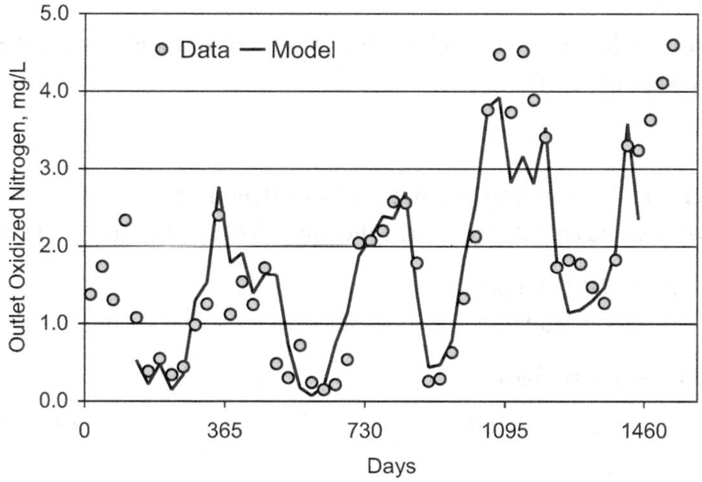

FIGURE 9.26 Model fit (line) to the monthly oxidized nitrogen data (circles) from the Imperial, California wetland. Both temperature and flow variations are accounted.

$$\frac{C_{NNout}}{C_{NNin}} = (1 + k'_{NN}/Pq)^{-Py} \tag{9.47}$$

where: y = fractional distance through the wetland, –

An example of the data fit of Equation 9.47 for an internal transect is shown in Figure 9.25.

Annual nitrate loss rate coefficients over the POR for a set of treatment marshes are shown in Table 9.16. The mean value for 167 treatment marshes is 35 m/yr.

Wetlands with modest inlet BOD do not have enough carbon in the source water to drive full denitrification of the incoming nitrate load. The requirement is not precisely defined, but a range of 5–9 g BOD per g nitrate – nitrogen has been suggested (Crites et al., 2006, based upon the U.S. EPA 1993 Nitrogen Control Manual). Nitrified municipal effluents, and agricultural and urban runoff, are typically very low in BOD, and the required carbon is not present. Accordingly, traditional denitrification requires the carbon generated in wetland sediments and litter, and those are also the locations of the highest numbers of denitrifying bacteria. Nitrate must diffuse to those locations for reaction to occur.

Denitrification Rate Coefficients

Denitrification rate coefficients focus upon the ultimate nitrogen removal flux to the atmosphere, and include the internal loading that arises from nitrification of ammonia.

In WWTP system design, reduction of nitrate is often modeled with a multi-factor Monod formulation (U.S. EPA, 1993b; Crites and Tchobanoglous, 1998):

$$J_{denitrification} = \frac{k' C_{NN}}{K_{NN} + C_{NN}} \frac{C_C}{K_C + C_C} \frac{K_{DO}}{K_{DO} + C_{DO}} \tag{9.48}$$

where C_{NN} = nitrate nitrogen concentration, mg/L

TABLE 9.16

Annual nitrate loss rate coefficients for treatment marshes. Some P = N values were measured.

Stipulations

1. Period of record averages are used in calculations.
2. For k-value calculations, the following PkC* parameters are selected:
   a. C* = 0.0 mg/L
   b. P = variable, 5.3 ± 3.33 (mean ± sd)
3. Ranges of variables:

	HLR	NO_X-N In	NO_X-N Out
	cm/d	mg/L	mg/L
mean	11.1	4.7	2.4
median	6.9	2.9	0.9
max	110.0	19.8	13.5
min	0.4	0.0	0.0

Results (N = 167 wetlands)

	Rate Coefficient, m/yr
Percentile	N = 167
0.1	4.0
0.2	7.9
0.3	13.0
0.4	20.7
0.5	25.7
0.6	29.1
0.7	37.6
0.8	49.9
0.9	78.1

C_C = carbon concentration, mg/L
C_{DO} = dissolved oxygen concentration, mg/L
$J_{denitrification}$ = nitrate nitrogen removal flux, $gN/m^2{\cdot}yr$
k' = Monod rate coefficient, m/yr
K_{NN} = half-saturation constant for nitrate, mg/L
K_C = half-saturation constant for carbon, mg/L
K_{DO} = half-saturation constant for dissolved oxygen, mg/L
The Monod rate coefficient is further modified by a temperature theta factor of 1.09.

In treatment marshes, the same factors are operative, but with much different effects. Typically, there is adequate carbon available, negating the need for the carbon factor in Equation 9.48. However, there remains an effect due to the need for nitrate transport to the solid carbon sources, which decreases the rate in comparison WWTP technologies that supply carbon in the water phase. Denitrification proceeds in biofilms, either periphytic or benthic, which have a much lower availability of DO than the bulk water.

The Monod formulation has been explored for treatment marsh performance (Karpuzcu and Stringfellow, 2012), who determined a value of K_{NN} = 43.8 mg/L. This high value means that in effect, the Monod formulation reduces to first order for treatment marshes. Other investigators have determined lower estimates. Messer et al. (2017b) determined K_{NN} of 5.6 and 6.0 mg/L for organic-based and mineral based mesocosms respectively. These lower values also indicate that treatment marshes would operate in the first order section of the Monod expression.

Some wetland literature presumes that nitrate reduction is zero order (Horne, 1995; White and Reddy, 2003; Mitsch et al., 2005), which offers the simplistic concept of an implied constant removal rate. However, treatment wetlands are most often best described by first order denitrification kinetics (with respect to nitrate), even at very high nitrate concentrations (Gale et al., 1993; Phipps and Crumpton, 1994; Spieles and Mitsch, 2000a; Hume et al., 2002a; Kadlec, 2005c). Messer et al. (2017b) examined both zero order and first order fits to bulrush mesocosm performance, as well as Monod kinetics, and found first order to be the best fit. Full-scale wetlands also show first order, exponential declines in nitrate, such as that observed in the Moscow, Idaho treatment marsh (Figure 9.25).

Therefore, an the area-based first order expression is the preferred model for marsh denitrification:

$$J_{denitrification} = k_{NN} C_{NN} \tag{9.49}$$

where: C_{NN} = wetland nitrate nitrogen concentration, mg/L
$J_{denitrification}$ = removal rate of nitrate nitrogen, gN/m^2·yr
k_{NN} = pseudo-first order denitrification rate coefficient, m/yr

The total denitrification flux can be computed from the mass balance web described above, easily on an annual basis, and with difficulty and approximations on a monthly scale.

The nitrate concentration varies along the flow path, presumptively according to a TIS model. The appropriate wetland average nitrate concentration may then be calculated from:

$$C_{NNavg} = \frac{C_{NNin} - C_{NNout}}{P\left(\left(C_{NNin}/C_{NNout}\right)^{1/P} - 1\right)} \tag{9.50}$$

where: C_{NNin} = wetland nitrate nitrogen concentration entering, mg/L
C_{NNout} = wetland nitrate nitrogen concentration leaving, mg/L
P = apparent number of TIS

Table 9.17 summarizes denitrification rate coefficients determined from annual mass balances and concentrations. The median k-value for denitrification is 43 m/yr. The 10th to 90th percentile ranges is 9–243 m/yr. As might be anticipated, the denitrification k-values are somewhat higher than the nitrate loss rate coefficients (Table 9.16), because of the internal loading of oxidized nitrogen. The median biomass requirement to supply the needed carbon for denitrification is 444 gdw/m2·yr, which is well within the range of expected marsh necromass generation.

TABLE 9.17

Annual denitrification rate coefficients for treatment marshes.

Stipulations

1. Period of record averages are used in calculations.
2. For k-value calculations, a TIS average NO_X-N concentration is used.
3. The biomass requirement is calculated as 1.29 gC/gN, and 20% useable carbon in biomass.
4. Ranges of variables:

	HLR cm/d	NO_X-N In mg/L	NO_X-N Out mg/L
mean	11.9	3.6	1.9
median	6.1	3.1	0.9
max	110.0	17.6	15.0
min	0.9	0.0	0.0

Results (N = 88 wetlands)

Percentile	Rate Coefficient m/yr	Percentile	Biomass Needed $gdw/m^2 \cdot yr$
0.1	9	0.1	74
0.2	23	0.2	193
0.3	31	0.3	301
0.4	35	0.4	406
0.5	43	0.5	444
0.6	61	0.6	554
0.7	87	0.7	713
0.8	116	0.8	1,118
0.9	243	0.9	1,484

Process Modifiers

The phenomena underlying the factors in Equation 9.48 are operative despite the choice of the simplistic first order approximation. As well, plant uptake competes with denitrification for available nitrate. And, treatment marshes can receive source waters in a variety of time sequences that must be accounted.

Plant Uptake

Plant uptake can be a significant component of the nitrate loss in a treatment marsh. Earlier in this chapter, it was noted that in pulse-dosed bulrush mesocosms, plant uptake accounted for 25–48% of the incoming nitrate load (Messer et al., 2017a). James et al. (2008) studied an 11 ha SAV marsh, of 60 cm depth, a flow-through backwater of the upper Mississippi River on the Iowa-Wisconsin border. There was little ammonia in the incoming water, and thus the plants, including their epiphytes, were implicated in the loss of nitrate that could not be attributed to denitrification. This short detention (1.7 days nominal) system removed 41% of the incoming nitrate. Plant uptake was 73% of removal, while denitrification was 27%. The source water (Mississippi River) provided essentially zero ammonia. Both of these wetlands received very low nitrogen loadings, for which plant uptake is important. Nitrate loss rate coefficients include plant uptake, while the mass-balance-derived denitrification estimates plant uptake and deducts that amount.

In Use

Care must be taken in the analysis of nitrate loss in event driven wetlands. Agricultural and urban stormwater systems do not receive steady flows, and in northern climates, do not operate in winter. Such systems require analysis using either event mean flows and concentrations, or in-use flows and concentrations (see Chapter 7). Rate coefficients determined from event mean concentrations and flows have been shown to be comparable to those for steady flow systems (see for instance Kadlec, 2010). Drake et al. (2018) present high frequency nitrate monitoring data from a 9.9 ha sparsely vegetated marsh that averaged 42 cm in depth. They found that there were not large differences between performance calculated using high versus low frequency sampling methods for the variably-loaded wetland.

Carbon Requirements

Carbon for denitrification may be available in the incoming water as dissolved organic carbon, or the supply may originate within the ecosystem. Treatment wetlands used for removal of oxidized nitrogen often receive incoming water from either of two sources: nitrified effluents from wastewater treatment plants, and runoff from fields or pastures. In the later case, the water may originate from some combination of tile drainage and overland runoff flow. These sources are usually devoid of carbonaceous materials, as measured by carbonaceous biochemical oxygen demand (CBOD). In the case of nitrified effluents, CBOD is removed upstream of the nitrification process, because CBOD inhibits nitrification, and is not required for the (autotrophic) process of ammonia oxidation. Drainage from crops is unlikely to contain more than trace amounts of carbonaceous substances, because organics are not used in fertilization, and crop residues are rarely left on the fields to contribute to carbon generation. Thus, the two most prevalent sources of water for treatment are carbon-poor in addition to being nitrate-rich. In a treatment wetland environment, the carbon supply is likely to be derived in major part from the decomposition of necromass (i.e., vegetation) in the wetland.

Suppose the annual production of biomass is 5,000 gdw/(m^2·yr) of dry material, of which 40% is carbon. On the order of 8% of this carbon has been estimated to be useable in denitrification (Hume et al., 2002a, 2002b). The useable carbon generation then totals 160 gC/(m^2·yr), which could theoretically support a like amount of nitrogen loss via denitrification, about 160 gN/m^2·yr (0.44 gN/m^2·d). Using this same concept, Baker (1998) argued that useable carbon from highly productive wetlands could be much higher, in the range of 1,700–3,000 gC/(m^2·yr), if most of the biomass carbon was available for denitrification, rather than 8%. The mesocosm studies of Lin et al. (2008) elucidated carbon limitation via the strategy of increasing the hydraulic loading. Water of 20 mg/L of NO_X-N was fed at various rates. Rate coefficients were determined both from input-output data, and from transect data. Above about 4 cm/d hydraulic loading, k-values dropped from about 35 m/yr to less than 10 m/yr, and profiles showed only gradual declines in concentration. Lin et al. (2008) attributed the effect to a carbon limitation.

Carbon-Starved Wetlands

Denitrification is sometimes observed to occur under conditions of inadequate carbon supply. Examples include the unvegetated (post) Tres Rios wetlands H1 and H2 in Phoenix,

Arizona (Kadlec, 2008a), the Imperial, California wetland (Kadlec et al., 2010a) and the Irvine Ranch Water District wetlands in San Joaquin, California (Fleming-Singer and Horne, 2006), which were essentially unvegetated ponds receiving nitrate but little or no carbon. For the first two of these, $20 < k_{NN20} < 30$ m/yr, while for Irvine Ranch, $22 < k_{NN} < 34$ m/yr. In none of these cases was vegetation-generated carbon estimated to be adequate to support the observed nitrate losses.

The possibility of algal activity contributing to nitrate removal was examined by Fleming-Singer and Horne (2006), and estimated to be only about 20% of the loss of oxidized nitrogen. Algal productivities at San Joaquin were inadequate to provide the carbon necessary for denitrification.

As discussed earlier in this chapter, possibilities exist whereby the traditional concept of carbon-fueled denitrification may be a secondary route of nitrate loss. Sulfur-driven autotrophic denitrification is possible if there is enough sulfur in the waters to provide sulfide in anoxic sediments. That was the case for Tres Rios (approximately 200 mg/L sulfate sulfur in the water), and for Imperial (approximately 600 mg/L SO_4-S), which led Wass (2003) to suggest this route for nitrate loss. The second alternative route is anammox, or equivalent close-coupled nitrification-denitrification mechanisms.

Oxygen Inhibition

Dissolved oxygen has an inhibitory effect on denitrification. For suspended growth cultures, Metcalf and Eddy (1991) suggest that the denitrification rate constant is reduced in the presence of DO, and is shut off at DO > 1.0 mg/L. In the activated sludge model of denitrification, oxygen inhibition is accounted for by an inverse proportionality relation with the denitrification rate constant, as shown in Equation 9.48. The activated sludge modeling recommended value is $K_{DO} = 0.2$ gO/m^3, which in effect makes denitrification inversely proportional to DO in the oxic range. The value of $K_{DO} = 0.5$ gO/m^3 was been used in subsurface flow (SSF) wetland modeling (McBride and Tanner, 2000), which reduces the denitrification rate by half at DO = 0.5 gO/m^3. Wynn and Liehr (2001) used $K_{DO} = 1.0$ gO/m^3 in SSF wetland modeling, and Howell et al. (2005) used $K_{DO} = 2.0$ gO/m^3 in their FWS model.

Field data have shown conclusively that treatment marshes do not behave as suspended growth WWTP processes. Denitrification has been observed in numerous treatment marsh systems that have considerable DO in their surface waters (Phipps and Crumpton, 1994; Van Oostrom, 1995). This apparent anomaly is due to the complicated spatial zonation in a wetland. Oxygen gradients occur between surface waters and bottom sediments in wetlands, allowing both aerobic and anoxic reactions to proceed in close vertical proximity (millimeters) near the sediment-water interface (Figure 9.5). Thus, nitrate in surface waters may diffuse into top anoxic soil layers where it is effectively denitrified (Reddy and Patrick, 1984). The same is presumably true for biofilms on solid surfaces within the water column.

Dissolved oxygen within the bulk water is therefore not likely to be a controlling factor. Nitrate (and possibly carbon) are known to diffuse to these surfaces, at which location the DO concentration may be much lower than in bulk solution. For instance, Fleming-Singer and Horne (2006) measured strong decreasing nitrate gradients from bulk solution into both litter and sediments. Crumpton and Phipps (1992) used microelectrodes to measure oxygen gradients from bulk solution down into the sediments, and showed decreases from 8–10 gO/m^3 in the bulk water of a treatment wetland to near zero only 2 mm deep in the sediment (Figure 9.5). James et al. (2008) used peepers to establish that nitrate disappeared in the top 1.0 cm of the sediment in a SAV marsh. Microsensor studies of biofilm communities in systems with absent or low flow have shown that oxygen respiration can occur in the outer layers while denitrification is present in deeper, anaerobic layers (Eriksson, 2001).

Therefore, it seems that in the wetland environment, it is not necessary to have low bulk water DO for denitrification to occur.

Temperature

Denitrification rate coefficients are well known to be temperature sensitive. Because nitrate loss rate coefficients represent processes that are often dominated by denitrification, these too display strong temperature dependence. As noted in Chapter 7, the modified Arrhenius equation is most often used to describe that temperature dependence:

$$k = k_{20}\theta^{(T-20)} \tag{7.84}$$

where k = nitrate loss rate coefficient, m/yr
k_{20} = nitrate loss rate coefficient at 20°C, m/yr
θ = temperature coefficient, –
T = water temperature, °C

Metcalf and Eddy (1991)typical θ = 1.14–1.16 for suspended growth denitrification in wastewater treatment. Although treatment wetlands share some features with wastewater treatment processes, temperature coefficients do not necessarily transfer from one technology to the other (Kadlec and Reddy, 2001). Accordingly, wetland-derived theta values should be used.

Data from 53 treatment marshes have a mean θ-value of 1.092 ± 0.008 (mean ± standard error) (Figure 9.27). The corresponding k_{20} values are 39 ± 6 m/yr (mean ± standard error). The average annual temperatures for the 53 wetlands range from 6.8 to 23.9 °C. These wetlands treat urban and agricultural runoff, river water and nitrified WWTP effluents. Microcosm and mesocosm data are not included, but values derived from small-scale studies are in basic agreement. For instance, the microcosm data of Wood et al. (1999) indicated θ = 1.088 ± 0.007 (mean ± standard error); and Westermann and Ahring (1987) data showed θ = 1.113. Bastviken et al. (2009) used θ = 1.088 to correct for temperature in 18 pilot scale microbial marshes in a two year study near Halmstad, Sweden The PkC* model was used, with P = 4.5. Emergent systems had k_{NN20} = 22 m/yr, submersed vegetation systems had k_{NN20} = 15 m/yr, and algal systems had k_{NN20} = 13 m/yr.

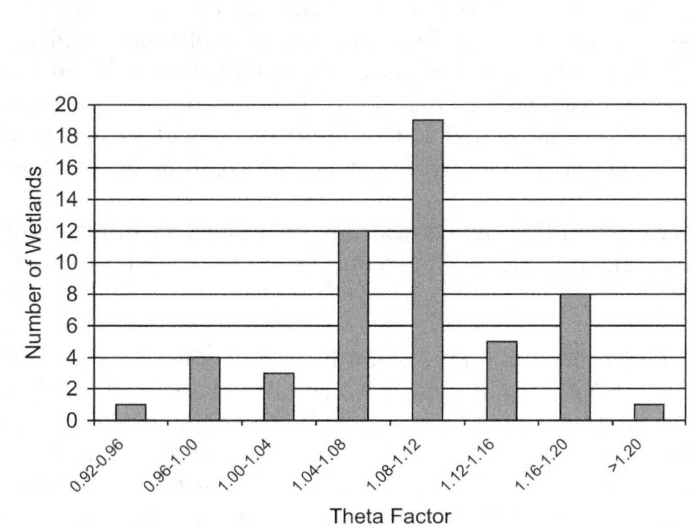

FIGURE 9.27 Frequency distribution of theta factors for treatment marshes.

The field-scale theta of 1.092 means that the rate of nitrate removal changes by a factor of two for an 8 °C temperature increase. Wetlands in warm climates have larger annual nitrate loss rates than those in cooler climates. Seasonality has a similar strong effect in temperate climates, with cold winters and warm summers. The cold temperature slowdown means that the required detention time is increased.

Nitrate Removal via Infiltration

Treatment wetlands receiving strong source waters are often lined to prevent the infiltration of contaminated waters to ground water. The category of treatment marshes receive comparatively benign waters, and most often do not require liners. If the infiltrating water is rich in organic and ammonia nitrogen, then the downward passage through unsaturated mineral soils promotes ammonification and nitrification, and thus contributes to nitrate contamination of groundwaters.

Nitrate treatment wetlands represent an entirely different situation, in which carbonaceous wetland sediments and soils promote the denitrification of infiltrating nitrate. Seepage that occurs through an organic layer on the wetland bottom, and through a carbonaceous action zone below, has the potential to promote nitrate reduction. That situation is comparable to the insertion of a "carbon wall" into the downward flow, which is a known technique for nitrate reduction (Schipper et al., 2010; Christianson et al., 2012; Hua et al., 2016; Wang and Chu, 2016).

If the constructed wetland does not leak to groundwater, then diffusive mass transfer alone must convey the nitrate from the water column to the soils and biofilms. But, if there is downward percolation through the wetland sediments and soils, then nitrate is carried by vertical advection to the microbes and the anoxic environment of the action zone. Denitrification in this zone is very rapid, because it contains the usual wetland organic sedimentary material. For instance, Whitmire and Hamilton (2005) found that nitrate injected into wetland surface soil horizons disappeared with an average half-life of 2.4 hr (N=5 wetlands). Rapid nitrate disappearance was also found beneath an infiltrating treatment wetland in Arizona (Nahar, 1998; Nahar et al., 1998).

Vertical nitrate attenuation in the action zone has been studied in treatment wetlands. A leaky wetland at the Tres Rios project in Phoenix, AZ infiltrated more nitrate than was denitrified (Kadlec, 2007). A subsurface study (Nahar, 1998) found that nitrate was efficiently removed during the first foot of vertical infiltration under the constructed marsh. Kovacic et al. (2000), and Larson et al. (2000), found 10–20% of the incoming nitrate was exported to seepage, but the fate of that nitrate was not quantitatively determined. The leaky wetland at Imperial, CA infiltrated about a quarter of the incoming nitrate (see Figure 9.11) (Kadlec, 2010), but no information on the fate of that nitrate was collected.

Other studies have quantified the vertical flow of water and nitrate, and concluded that those vertical flows were very effective in reducing nitrate concentrations in the infiltrating water.

- Brauer et al. (2015) measured infiltration flows and concentrations located adjacent to the San Joaquin River, California. The main flow path was partially vegetated with *Polygonum lapathifolium* (curlytop knotweed). The wetland was flow-through during the irrigation season, at a nominal detention time of 0.9 days. Nitrate entered at 13 mg/L and exited at 8 mg/L Approximately 72% of the water infiltrated. Surface water NO_X-N was reduced by 86%, and infiltrating NO_X-N was reduced by 88%, to less than 1.0 mg/L below the flowpath. Vertical profiles of NO_X-N are shown in Figure 9.28.

- Wetland Solutions, Inc. (2012) (Principal author: Keller, C.H.) reported results from two treatment marsh systems designed to infiltrate reclaimed wastewater, one at the Kanapaha Water Reclamation Facility (KWRF), and the other at Kanapaha Middle School, both in the Gainesville, Florida area. These were planted with *Pontederia cordata, Sagittaria lancifolia, Eleocharis cellulosa,* and *Schoenoplectus validus.* Inflow hydraulic loadings to each of the three KWRF cells were on the order of 5–10 cm/d, and nominal detention times were 7–15 days. Virtually all of the inflow water infiltrated, as designed. Surface water NO_X-N entered at 2–4 mg/L, and decreased with distance from the inflow, by 50–95%. Surface water nitrate loss rate coefficients varied from 12–155 m/yr, which do not include infiltration losses. About 30% of the NO_X-N was estimated to infiltrate, of which only trace amounts were found in groundwater samples at the 30 cm depth.

The conclusion from these studies is that marsh soils have the capacity to remove large nitrate loads from vertically percolating water with low risk to groundwater. Brauer et al. (2015) therefore concluded that:

> In contrast to the notion that seepage water from wetlands may be considered as a source of groundwater nitrate contamination, this study shows that under the conditions present in our wetland, seepage through wetland soils can actually prevent some nitrate contamination of groundwater.

Since it is not economically practical to line treatment marshes in agricultural and urban settings, it appears this potential advantage will generally be available, perhaps unintentionally.

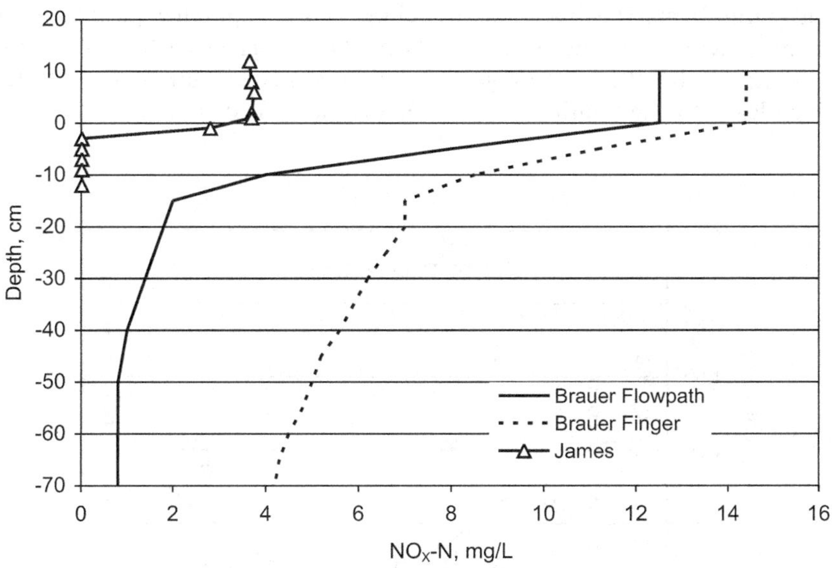

FIGURE 9.28 Vertical profiles of oxidized nitrogen in treatment marshes. The James et al. (2008b) data is from peepers in a Mississippi River marsh in June, presumably under conditions of zero infiltration. The Brauer et al. (2015) system was a marsh receiving agricultural runoff in the San Joaquin valley in California. Curves are spline fits of piezometer samples taken in the main flowpath and in dead-end side finger areas averaged over the summer.

Performance Variability

In common with other forms of nitrogen, oxidized nitrogen is susceptible to random variation around annual trends. Those trends are reasonably well described by a monthly or sinusoidal annual trend, or by the PkC* model, but data scatter is nontrivial (Figure 9.29). The 95th percentile of that monthly scatter is contained below a line of about double the deterministic annual tend (Table 9.18). This stochastic variation is superimposed on the seasonal variability in the nitrate trends.

Greenhouse Gases: Nitrous Oxide

The current intense concern about greenhouse gases (GHG) has sparked much investigation of the emission of nitrous oxide in treatment wetlands (e.g., Hernandez and Mitsch, 2006; Søvik et al., 2006; Wu et al., 2009; VanderZaag et al., 2010a; Jia et al., 2011; Yang et al., 2012; Huang et al., 2013). All constructed marshes produce greenhouse gases, predominantly carbon dioxide and methane, but also including nitrous oxide. The wetland nitrogen cycle creates tiny amounts of nitrous oxide, which also has a very greatly magnified greenhouse effect compared to carbon dioxide (multiplier of about 296). A review by Mander et al. (2014) found that FWS systems emitted 0.11% of their inlet nitrogen loading as N_2O.

N_2O is a product of incomplete nitrification and denitrification, and hence both are considered to be sources of N_2O. Nitrification makes a substantial contribution to the N_2O emission under aerobic conditions (Huang et al., 2013). N_2O may be produced during denitrification, especially under mildly oxic conditions. Under low DO conditions, N_2O will be replaced by N_2 as the end product of denitrification (Huang et al., 2013).

The median of N_2O emissions from 25 marsh studies, in all climate regions, was approximately 0.82 $gN/(m^2 \cdot yr)$ (Mander et al., 2014). For the multiplier of 296, this leads to a carbon equivalent emission of 243 $gC_e/m^2 \cdot yr$.

The data on emission rates and storage processes lead to the conclusion that most constructed (and natural) marshes in temperate climates are slight net producers of 100-year global warming potential, on the order of 800 $gC_e/m^2 \cdot yr$. Their carbon storage function is typically out-weighed by their GHG generation in terms of cumulative effect. The time element, which leads to the multipliers used for the gases, is of importance. The multipliers

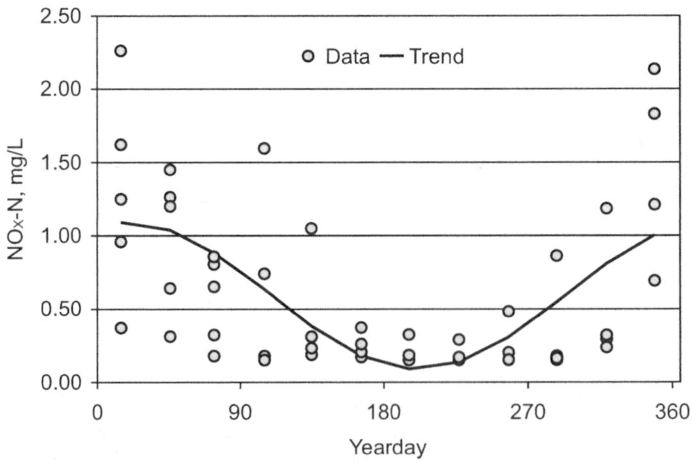

FIGURE 9.29 Monthly variability in NO_X-N concentrations for the Brawley, California wetland.

TABLE 9.18
Trend multipliers for variability of outlet oxidized nitrogen concentrations for treatment marshes with respect to the annual cyclic central tendency trends.

	Data Years	Percentile			
		50%	80%	90%	95%
Lakeland 1	4	0.79	1.26	1.47	1.65
Imperial	4	0.96	1.38	1.71	2.27
Brawley	4	1.12	1.68	2.09	2.33
Musselwhite	6	1.01	1.30	1.45	1.66
Brighton	10	0.64	1.57	2.12	2.86
Shannon	4	1.27	1.83	1.90	1.94
Tarrant	3	1.38	1.98	1.99	1.99
ENRP	4	1.04	1.08	1.11	1.13
Tres Rios Full	4	0.99	1.20	1.30	1.33
Tres Rios C1	5	0.98	1.38	1.44	1.46
Tres Rios H2	5	0.78	1.32	1.53	1.62
Tres Rios H1	5	1.02	1.58	2.67	3.19
Median		**0.99**	**1.38**	**1.53**	**1.66**
Mean		**1.00**	**1.46**	**1.73**	**1.95**

for N_2O and CH_4 decrease with the cumulation period. Generally, wetlands may be regarded as a source for GHG if evaluated on a short time scale (decades), but as a sink for GHG if evaluated over longer time scales (>100 years) (Roulet, 2000; Brix et al., 2001).

The relative importance of GHG to the utilization of nitrate reduction wetlands appears to be very small. Thiere et al. (2009) concluded that the nationally planned wetland creation (12,000 ha) in Sweden could produce 27% of the desired nitrate reduction, at an environmental risk equaling 0.04% of the national GHG emission. From another viewpoint, GHG emissions from one hectare of wetland are roughly equivalent to those from a few pigs (Vanotti et al., 2008).

Wetlands created for water quality improvement or habitat restoration have many ecological values, which far outweigh any negative global warming potential. Wetland services apart from nitrate removal create a large positive value (Prato and Hey, 2006). Therefore, GHG production is not here considered to be an important determinant of treatment marsh design.

9.10 PERFORMANCE FOR TOTAL NITROGEN

Total nitrogen (TN), which is defined as the combination of organic, ammonia and oxidized nitrogen, is subject to consideration as a group of compounds that are reduced in wetlands. This grouping is known to possess sequential conversions, including primarily ammonification followed by nitrification, followed by denitrification, all proceeding at varying rates. Total nitrogen in the water is augmented by releases from decaying vegetation and microbial biomass. As a consequence, the rate of decline of TN along the flow path is expected to be determined in part by the speciation of the incoming nitrogen. TN removal rates for nitrified influents are anticipated to be highest, because the precursor conversions of organic and ammonia nitrogen have already occurred in pretreatment.

There are zero background concentrations for ammonia and oxidized nitrogen, but all treatment wetlands display a background concentration of organic nitrogen of about one mg/L. Consequently, background TN is essentially the same as background organic nitrogen.

Loading Considerations

It is again informative to represent annual wetland performance as the effluent concentration produced (C_o) by a given inlet loading rate (LRI = HLR × C_i) and concentration (C_i). In the broad context, multiple datasets are represented by a trend that shows increasing C_o with increasing LRI, with different groupings associated with each inlet concentration (Figure 9.30). The overall slope of the inter-system data on the log-log coordinates for 72 treatment marshes is 0.31. As for other nitrogen species, inlet loading is an insufficient design specification, because hydraulic load and inlet concentration are not interchangeable factors in the load representation.

Profiles of Total Nitrogen

Treatment wetlands data display decreases in total nitrogen with contact time, or equivalently with longitudinal distance, which are consistent with first order reduction kinetics; but show a nonzero background concentration for long detention. The annual average TN profiles vary from year to year (Figure 9.31). The PkC* fit is P = 4, k_{20} = 33 m/yr, θ = 1.192, and C* = 1.1 mg/L.

There are typically clear downward trends in concentration along the flow path, as total nitrogen is removed from the water during travel through the wetland. Summer rates of decline are greater than in winter, which would be occasioned by either plant uptake or by temperature-enhance microbial processes.

Loss Rate Coefficients

Reductions in total nitrogen in treatment wetlands systems can be represented by loading analysis and the PkC* model. Results across systems for the assumed value P = 3 are summarized in Table 9.19. The value C* = 1.0 mg/L is selected based upon organic nitrogen considerations. The remaining model parameter is the k-value, selected to fit the model.

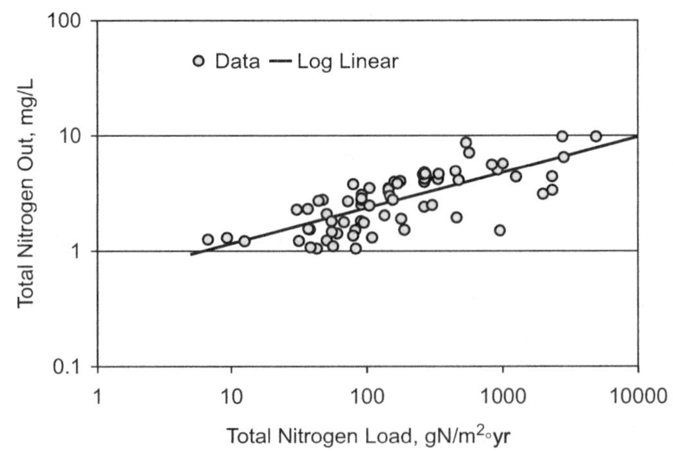

FIGURE 9.30 Load – concentration plot for total nitrogen in treatment marshes.

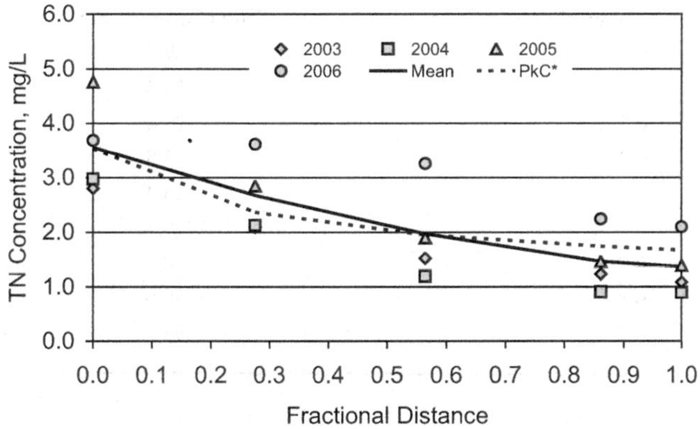

FIGURE 9.31 Total nitrogen longitudinal profiles in the Shannon wetlands near Corsicana, Texas. Points are the average of biweekly measurements.

The median annual rate constant was k_{TN} = 26.4 m/yr for 72 treatment marshes (Table 9.19). The 10th to 90th percentile range is 9.8–95.6 m/yr. On an average annual basis, temperature or season may be an important determinant of the rate constant, and is thus responsible for the some of the intersystem variability in annual k-values. Accordingly, it is necessary to examine intra-annual effects.

Seasonal and Temperature Effects

Microbially Dominated Wetlands

If the source water flows and concentrations of NH_4-N and NO_X-N combine to provide more nitrogen that is required for plant growth, then microbial processes will be present. A temperature effect is present in these microbially dominated systems, and a temperature

TABLE 9.19

Annual reduction of total nitrogen in treatment marshes.

Stipulations

1. Period of record averages are used in calculations.
2. For k-value calculations, the following kPC* parameters are selected:
   a. C* = 1.0 mg/L
   b. P = 3 TIS
3. Ranges of variables:

	HLR cm/d	TN In mg/L	TN Out mg/L
mean	21.8	5.7	3.2
median	8.2	6.1	2.8
max	181.0	16.2	9.8
min	0.5	1.3	1.0

Results (N = 72 wetlands)	
Percentile	Rate Constant, m/yr
0.1	9.8
0.2	16.4
0.3	20.4
0.4	22.8
0.5	26.4
0.6	29.1
0.7	38.1
0.8	50.9
0.9	95.6

coefficient (θ), is capable of accounting for these effects (refer to Chapter 7). For this situation, loss rate coefficients are in synchrony with wetland water temperature (Figure 9.32). The Ekeby total nitrogen loading was high, and the rate constants are seasonally synchronized with water temperature.

Results of calibration of θ-values for representative wetlands are summarized in Table 9.20. Monthly averages were used to avoid synoptic error (transit time offset). Calibrations were performed for best estimates of the internal hydraulics for each wetland. Therefore, P-values range from 3 to near plug flow conditions, based upon system geometry. The value $C^* = 1.5$ was used when data did not allow determination of a calibrated value. However, calibrated values were with a narrow range, 0.8–2.2 mg/L. The median k_{TN20}-value for total nitrogen is 42.0 m/yr, but the range is wide.

Temperature coefficients had a median value of 1.080, indicating a relatively strong thermal effect on the suite of microbial processes that contribute to TN reduction.

The wetlands in Table 9.20 include five event driven systems with high TN loadings, four in Sweden and the North Carolina State system (Merriman et al., 2017). Four stormwater wetlands in Sweden were used to model total nitrogen disappearance (Arheimer and

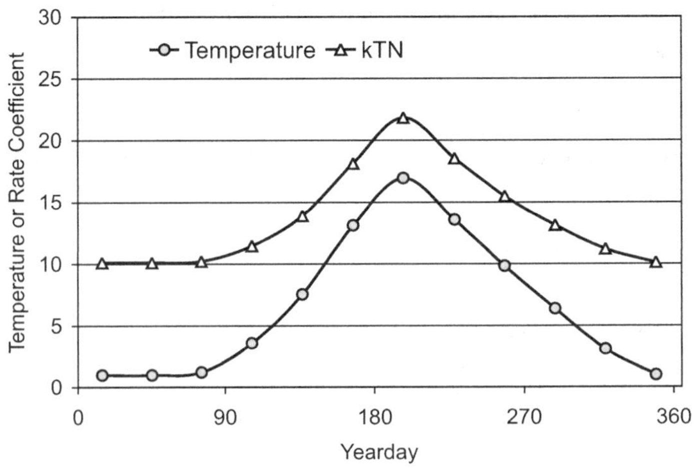

FIGURE 9.32 Total nitrogen loss rate coefficient and temperature for the Ekeby wetland at Eskilstuna, Sweden. Three years' data are represented.

TABLE 9.20

Dependence of total nitrogen loss rate coefficients on temperature for microbially dominated wetlands.

Site	Location	Wetland	T °C	P (TIS)	C* mg/L	k_{20} m/yr	θ
NCSU	NC	All	16.0	3	0.75	53	1.108
Texel	Netherlands	Pilot	10.0	∞	2.2	115	1.050
Tres Rios	AZ	H1 pre	18.9	6	0.8	49	1.061
Tres Rios	AZ	H2 pre	18.9	6	0.8	43	1.084
Tres Rios	AZ	H1 post	19.9	6	0.8	13	1.130
Tres Rios	AZ	H2 post	20.7	6	0.8	11	1.103
Imperial	CA	All	20.7	∞	1.25	76	1.048
Brawley	CA	All	20.7	∞	1.25	52	1.080
Lakeland	FL	Cell 1	22.9	3	1.0	18	1.079
Eskilstuna	Sweden	Ekeby	6.5	6	1.5	25	1.049
Linköping	Sweden	All	6.7	3	1.5	42	1.081
Soderkopingsan	Sweden	4 Wetlands	7.0			16.2	1.081
Kingman	AZ	All	11.7	20	1.5	25	1.050
Median			**18.9**		**1.1**	**42.0**	**1.080**

Wittgren, 1994). Nitrate comprised about half of the total, which was in the range $5 \leq C_{TN} \leq 20$ mg/L. The calibration spanned two calendar years.

Another study of eight Swedish agricultural runoff treatment marshes was conducted by Arheimer and Wittgren (2002). For these microbial wetlands, the mean $k_{TN20} = 15.3$ m/yr, and $\theta = 1.050$.

Agronomic Wetlands (Lightly Loaded Systems)

When the total nitrogen loading to the wetland is of the same order of magnitude as the growth requirements of the plants and algae, the removal of total nitrogen is very likely to be mediated by the growth and decay of biomass. As a rough guideline stated earlier in this Chapter, this situation occurs for total nitrogen loadings less than approximately 120 gN/m^2·yr. It is important to note that low inlet total nitrogen load very often means very low inlet total nitrogen concentration, close to background; and consequently, there is a lessened ability to obtain meaningful calibrations of total nitrogen rate constants.

Uptake presumably occurs for both the ammonia and nitrate components of total nitrogen, and release may be considered to add to the organic component. Plant uptake rates do not correspond to the annual cycle of water temperatures, and hence total nitrogen removal in agronomic wetlands cannot be characterized by modified Arrhenius θ-factors. Agronomic rate coefficients are in synchrony with the growth rate of the biomass. The monthly pattern of agronomic total nitrogen rate coefficients is illustrated for the Brighton, Ontario marsh in Figure 9.33. There is a high uptake in the spring growth period, which occurs at moderately cool temperatures, and no correspondence between monthly rate constants and temperature is present.

If a regression of rate coefficient and temperature is carried out for agronomic systems, the result will have a very low R^2 value, and the θ-value will be close to unity. For example, Merriman et al. (2017) report theta values for eight lightly loaded stormwater marshes in

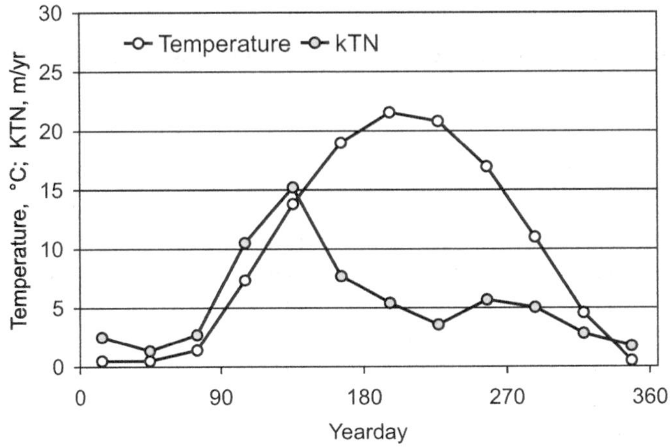

FIGURE 9.33 Total nitrogen loss rate coefficient and temperature for the Brighton, Ontario wetland. Ten years' data are represented.

North Carolina. The results show $\theta = 0.994 \pm 0.031$, and corresponding $k_{20} = 43 \pm 19$ m/yr.

Thus, for agronomic systems, the appropriate rate coefficient representation is a monthly series of values, not a theta factor.

Intrasystem Variability

The PkC* model, with appropriate seasonality, represents only the seasonal trend of effluent total nitrogen concentrations, leaving a considerable amount of probabilistic scatter in performance (refer to Chapter 7). Therefore, it is useful to examine the variability of monthly average outlet concentrations, here represented by a fractional addition to the trend value (Ψ). To account for stochastic variability, the multiplier on the trend value is $(1+\Psi)$. The set of monthly averages for a wetland over some period of record will yield a distribution of Ψ-values. The upper percentile points of that distribution are useful in design. For instance, the 90th percentile represents the fractional addition to the trend that may be expected to occur one time out of ten during the period of record.

Table 9.21 lists percentiles of the monthly Ψ-distributions for some representative treatment marshes. These data indicate that the median of the 95th percentile is an additional 57% above the trend, and that the median of the 90th percentile is an additional 47%.

9.11 MULTI-SPECIES NITROGEN MODELING

Because treatment marshes typically operate with low incoming nitrogen loadings, phytomass processes are nearly always important. The role of plants and mesophyton in the cycling of nitrogen through the course of the year cannot be ignored. It is therefore necessary to include phytomass storage and release. Therein lies a difficulty, because phytomass processes are rarely investigated as part of the wetland quantification. Yet, there is sufficient information from which to make reasonable estimates of phytomass nitrogen content through the course of the year.

The nitrogen processing network exemplified by Figures 9.10, 9.11, and 9.12 is assumed to consist of interconversions of nitrogen in the water accompanied by exchanges with the

TABLE 9.21
Trend multipliers for variability of outlet total nitrogen concentrations for treatment marshes with respect to the monthly central tendency trends.

		Percentile			
	Data Years	50%	80%	90%	95%
Brawley, CA	4	0.95	1.48	1.79	2.03
Imperial, CA	4	1.04	1.27	1.36	1.44
Shannon, TX	4	1.11	1.34	1.47	1.57
Brighton, ONT	10	0.99	1.24	1.32	1.49
Lakeland 1, FL	7	0.83	1.24	1.49	1.60
Linköping, Sweden	3	0.98	1.21	1.28	1.35
Listowel 3, ONT	4	0.86	1.29	1.65	1.72
Listowel 4, ONT	4	0.94	1.25	1.62	1.89
OEW, FL	9	0.99	1.20	1.36	1.46
Tres Rios FS, AZ	4	0.98	1.15	1.39	1.52
Tres Rios H1, AZ	6	0.96	1.42	1.65	1.70
Median		**0.98**	**1.25**	**1.47**	**1.57**
Mean		**0.97**	**1.28**	**1.49**	**1.61**

sediments and biomass and the atmosphere. Over the past two decades, this network has become recognized as the fundamental basis for simulating nitrogen flows and conversions in treatment marshes. There are, however, significant differences in literature reports, based upon different hydraulic assumptions, and upon differences in the functional form of transfer rates. As a basis for understanding, a simplified version is presented here.

Earlier in this Chapter, a speculative assessment was made of the monthly fluxes of nitrogen in The Brighton treatment marsh (see Figure 9.13). That example was done in a data-fitting mode, meaning it was based upon observed performance, including flows and concentrations of all nitrogen constituents, coupled with estimated phytomass functioning. Here, a predictive calculation is explored, in which design parameters are utilized.

Sequential Nitrogen Models: An Illustration

A highly simplified illustrative nitrogen reaction network is provided by Equations 9.33–9.36, applied on a monthly basis. Accumulation in the water column is assumed negligible, but phytomass takes up ammonium and nitrate, and releases organic nitrogen. A small residual of necromass decomposition is accreted to new sediments and soils. Organic nitrogen loss is ascribed to ammonification. Ammonia is nitrified, thus adding to the amount of nitrate in the system. Nitrate nitrogen (oxidized nitrogen) from all sources is presumed to undergo denitrification.

A simplified version of the nitrogen network restricts nitrogen biomass uptake to ammonium (J_{UA}). This is accounted by an uptake flux, where this and other fluxes are in gN/m^2·d. Biomass decomposition can also release organic nitrogen into the water via decomposition (J_{rO}). The residual of necromass nitrogen is accreted or buried, at the rate (J_b). Incoming organic nitrogen loss is ascribed to ammonification, ($J_{ammonification}$). Ammonia is nitrified at rate ($J_{nitrification}$), thus adding to the amount of nitrate in the system. Nitrate nitrogen (oxidized nitrogen) from all sources is presumed to undergo denitrification, at rate

($J_{denitrification}$). Ammonification, nitrification, and denitrification are presumed to follow first order kinetics. This simplified version omits factors such as atmospheric gains and losses, infiltration, volatilization, settling and sorption.

The hydraulic assumption for this illustration is taken to be steady, nonaugmented flow, with a flow pattern of tanks-in-series (TIS). Under more complicated assumptions, the water budgets for the TIS are required. This model has been described for a single removal reaction (see Chapter 7), but here it is necessary to extend the ideas to the network of conversions. The resulting model for the progress of concentrations as water moves through the wetland is tractable, but necessarily more complicated than for a single species.

The rates for water-borne species are specified as first order, as described in preceding sections of this chapter, according to Equations 9.37, 9.42 and 9.49. These rate coefficients are assumed to be temperature dependent, thus requiring the monthly time series of water temperatures. Further for this illustration, the uptake, decomposition and burial rates are specified as seasonally dependent. The monthly time series of phytomass nitrogen content is specified (estimated). The estimated phytomass monthly time series and monthly temperature series for this illustration are shown in Figure 9.34. The ratio of burial to return from vegetation is set, and the burial assumed to be constant throughout the year. The assumed rate constants and parameters are given in Table 9.22. Results are shown in Figure 9.35.

Despite the over-simplicity of this illustration, several important features are manifest. This example is set in the range of agronomic control, with an inlet ammonia loading of 110 $gN/m^2 \cdot yr$. Firstly, the ammonia outlet concentrations display a mid-summer minimum, and a winter maximum. However, the component nitrogen transfers are not synchronized. Plant uptake is an important contributor to the annual pattern, with a peak in the spring at maximum growth. Nitrification and denitrification peak much later, and are of the same order of magnitude as plant uptake (and release). Secondly, the choice of a high denitrification rate constant forces the nitrate levels to remain low throughout the year. Thirdly, there is a considerable return flux of organic nitrogen from decomposition (peaking at 60 $gN/m^2 \cdot yr$ during the month of October), but the average organic nitrogen content of the water remains low (average = 1.0 mg/L).

At the present time, there are no other models in the literature that contain all the processes necessary to describe nitrogen species in agronomic marshes under flow through, spatially variable conditions. As a result, the design of treatment marshes relies upon the

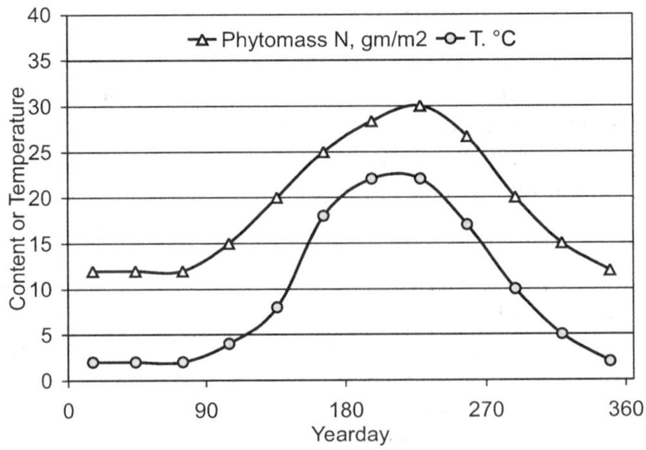

FIGURE 9.34 Time series of phytomass nitrogen and temperature for a nitrogen model illustration.

TABLE 9.22

Assumed rate constants and parameters for the nitrogen species monthly model calculations. The temperature and phytomass time series are shown in Figure 9.34.

	Value	Units
End-of-season phytomass	1,500	gdw/m^2
	30	gN/m^2@2%
Winter phytomass	600	gdw/m^2
	12	gN/m^2@2%
Phytomass burial	3	$gN/m^2{\cdot}yr$
Phytomass decomposition	27	$gN/m^2{\cdot}yr$
Ammonification k_{20}	30	m/yr
Ammonification ⊠	1.02	
Nitrification k_{20}	20	m/yr
Nitrification ⊠	1.06	
Denitrification k_{20}	30	m/yr
Denitrification ⊠	1.10	
Number of TIS	3	
System HLR	10	cm/d
Inlet Organic N	1	mg/L
Inlet Ammonia N	3	mg/L
Inlet Nitrate N	2	mg/L

ingenuity of the designer to construct spreadsheet tools similar to the illustration above. This is a relatively easy task for annual average calculations, but becomes complex when considering intra-annual behavior. Assumptions about the processing of nitrogen by plants and mesophytes are required.

9.12 SUMMARY

Organic, ammonia, and oxidized nitrogen forms interconvert in the wetland environment, with the net effect of reductions in virtually all cases. In situations of low nitrogen loading, the biogeochemical cycling of nitrogen is a critical part of the processing network for continuous flow wetlands. Plant and algal nitrogen cycling is important for a major fraction of existing operating wetlands, which leads to seasonal behavior rather than temperature driven behavior common for microbial processing. Therefore, a distinction is drawn between agronomic wetlands and microbial wetlands, although there are obviously systems with dual control. A benchmark instantaneous growing season rate is suggested to be 120 $gN/m^2{\cdot}yr$, below which agronomic control should be considered. In more heavily loaded wetlands, there is preponderance of microbial processes, and removals are strongly temperature modulated. Plants utilize nitrate and ammonium for growth, but the majority of the assimilated nitrogen is subsequently released during death and decay. A small amount is permanently stored as new soil and sediment, on the order of 2–10 $gN/m^2{\cdot}yr$.

Phytomass utilizes both ammonia and oxidized nitrogen for growth, which is maximal in springtime, well ahead of the annual maximum in water temperature. Biomass senesces and dies in autumn in temperate climates, leading to the production of a standing crop of

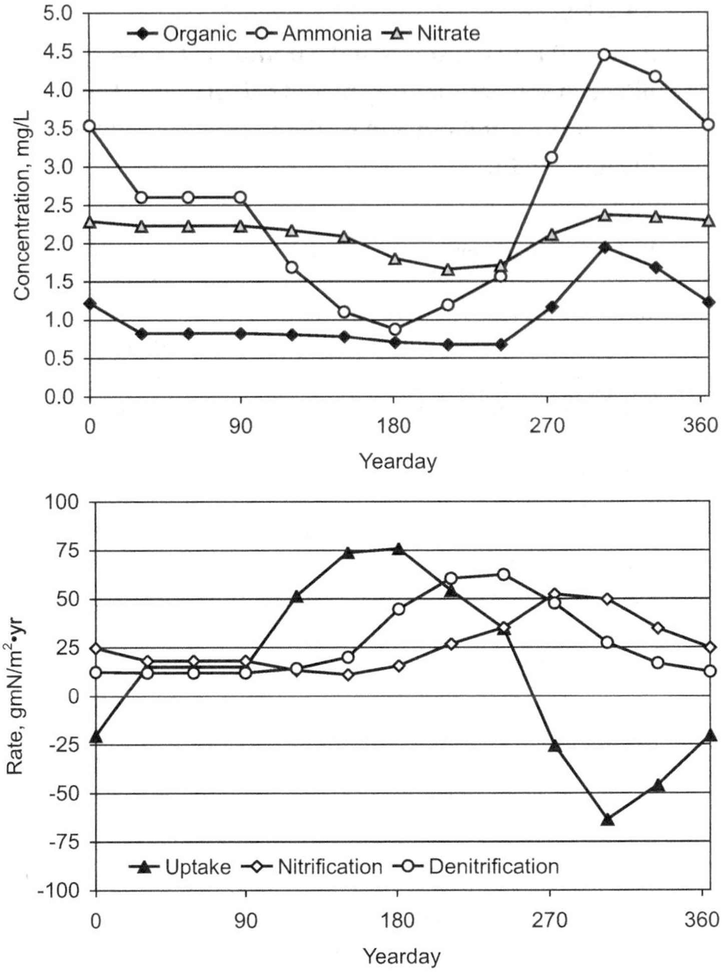

FIGURE 9.35 Illustration of the monthly time series for a 3TIS multi-species nitrogen model. Upper panel: concentrations; lower panel: monthly fluxes. The incoming water contains 1, 3 and 2 mg/L of organic, ammonia, and nitrate nitrogen, respectively.

necromass nitrogen. That necromass subsequently falls, and partially decomposes, releasing organic nitrogen. Small residual amounts of nitrogen remain in the recalcitrant accreting sediments.

The descriptions of microbial processes in conventional activated sludge and attached growth wastewater treatment systems partially carry over to processing in wetland systems under microbial control. However, there are wetland phenomena that are outside those patterns. Traditional nitrification-denitrification may not apply, because of alternative microbial routes, such as anammox or autotrophic denitrification. Conditions of oxygen supply and carbon supply are sometimes outside the bounds of conventional WWTP ideas.

Several features of nitrogen processing are known with some degree of certainty:

The profiles of TN are typically monotonic decreasing with detention time, when the input to the wetland is above background values. The background level of organic nitrogen is approximately 1.0 mg/L in treatment marshes, and hence that is the background for TN as well.

Ammonia calculations must acknowledge the ammonification process, as well as nitrification and plant uptake. The background level of ammonium nitrogen is approximately zero.

For treatment wetlands in general, nitrate rate constants divide in two groups, one that follows a slow course dictated by sequential nitrification and denitrification, and one that displays almost instantaneous removal of nitrate, with the later attributable to availability and spatial proximity of a carbon supply. Treatment marshes typically fall in the former category. The background level of nitrate nitrogen is approximately zero.

First-order, area-based nitrogen loss models provide a suitable method for design of wetland treatment systems in most circumstances. These have the advantage of correctly describing internal phenomena in flow-through wetlands, and include the effects of the principal variables of inlet concentrations and hydraulic loading rates (or the equivalent detention times). The parent mass balance equation for water movement may be adjusted to fit extreme environmental conditions of precipitation or evapotranspiration. The rate equations account for return fluxes from the wetland biomass, and thus can fit the entire range of hydraulic loadings. In parameter estimation, the sequential nature of the nitrogen transformations cannot be ignored.

The accompanying set of phytomass cycling processes interacts with the microbial transformations, to produce a dual water-phase and vegetation-phase network of interconversions. Mass balances for nitrogen species must account for both, and can readily be established for the annual average behavior of the wetland. Intra-annual effects require estimates of growth dynamics, as well as the monthly distribution of return fluxes.

Treatment marsh systems consistently reduce nitrogen concentrations for many types of source waters. The magnitude of these reductions depends on many factors including inflow concentrations, chemical form of the nitrogen, water temperature, season, organic carbon, and dissolved oxygen. Although these factors can be incorporated with some success into design of wetland treatment systems, variability is inevitable. Regardless of the operative mechanisms, there now exist large databases that allow for identification of process rates and rate constants. Both annual and seasonal rates have been measured for dozens of wetlands, and found to cluster, but with considerable inter-system variability. Designers may now choose the level of risk in sizing a new wetland. Because the major nitrogen transformation mechanisms vary seasonally, conservative design must be based on the specific situation, with different assumptions used for design to meet concentration or load limits. Data indicate that if design is based on deterministic, seasonal equations, target outflow concentrations should be divided by a sizeable factor, of about two, as compensation for the usual maximum monthly limit versus model equations.

The information in this chapter forms the basis for design sizing procedures discussed in Chapter 16. It is clear from the considerations in this chapter that nitrogen processing is too complex to admit simple rules-of-thumb for design, such as load removal specification.

10 Phosphorus

Treatment marshes remove phosphorus (P) from water and store it soils, sediments and vegetation. There is now a significantly long and successful history for numerous systems of many sizes. The size of P removal marshes is typically quite large, thus disappointing those who would hope for large removals in small footprints. The ecological value of such marshes counter-balances the size disadvantage.

When a wetland, either natural or constructed, is given a new supply of water and phosphorus, it responds by readjusting storages, pathways and structure. If those new supplies are variable and within the stochastic band of historic inputs, a mature ecosystem will not change in character or function. But, treatment of water for phosphorus removal in the wetland implies that additions will significantly exceed the historic stochastic band of the natural wetland, and a newly constructed wetland will require a successional period to adapt to the intended inputs. In both cases, a period of adaptation and change is to be expected. Thereafter, the wetland functions in a long-term sustained mode, which is tuned to the inputs, but displays probabilistic variation as well.

A large number of internal P transfer and alteration processes occur, but only soil-building provides a net long-term storage of phosphorus. Sediment and soil accretion provides phosphorus storage that can alternate between deposition and erosion on a short-term basis. The wetland environment provides appropriate conditions for net long-term buildups, because inundation slows oxidative processes. Historical accumulations in natural wetlands are the genesis of peatlands. Natural accretions are on the order of a few millimeters per year (Mitsch and Gosselink, 2015).

Marshes experience naturally occurring inputs of phosphorus are from surface inflows, and atmospheric deposition, which consists of both wet deposition and dryfall. Outputs may be in the form of surface outflows or infiltration to groundwater. Animal migration, ranging from insect movement to fish and bird travel, has been identified as a potential contribution to the phosphorus budget; but to date no quantification of this process exists.

For the purpose of understanding phosphorus cycling, two components of the ecosystem can be considered separately (Noe and Childers, 2007). The macrophyte cycle involves live and standing dead plants, and their litter (Figure 10.1). This cycle operates on an annual time scale. Plants take up P from their root system, which is supplied from overlying water by vertical downward hydraulic and transpiration flows. Senescence and death provides input to the standing dead, which in turn contributes to the litter on the wetland floor. Root and plant litter and floc decompose, and recharge the store of porewater P.

The mesophyte cycle is less well understood, and often overlooked. As discussed in Chapter 5, there are several recognized types of microbial/algal assemblages, including epiphytes, epipelon, plankton and mats. These cycle P on a daily or weekly time scale, taking P directly from the surface water for growth, and returning particulate P as the product of death and decomposition (Figure 10.2). Incoming settled and filtered particulate P joins the sloughed detritus from the mesophytes and contributes to the floc layer in the wetland.

The two cycles are physically interspersed in the wetland, but draw their P requirements from different locations: macrophytes from below ground in the action zone, and mesophytes from the surface water. As a consequence, the mesophytes are the first and fastest responders to the waterborne incoming P, while macrophytes respond more slowly. However, the macrophytes have a much greater storage capacity than the mesophytes. Residuals

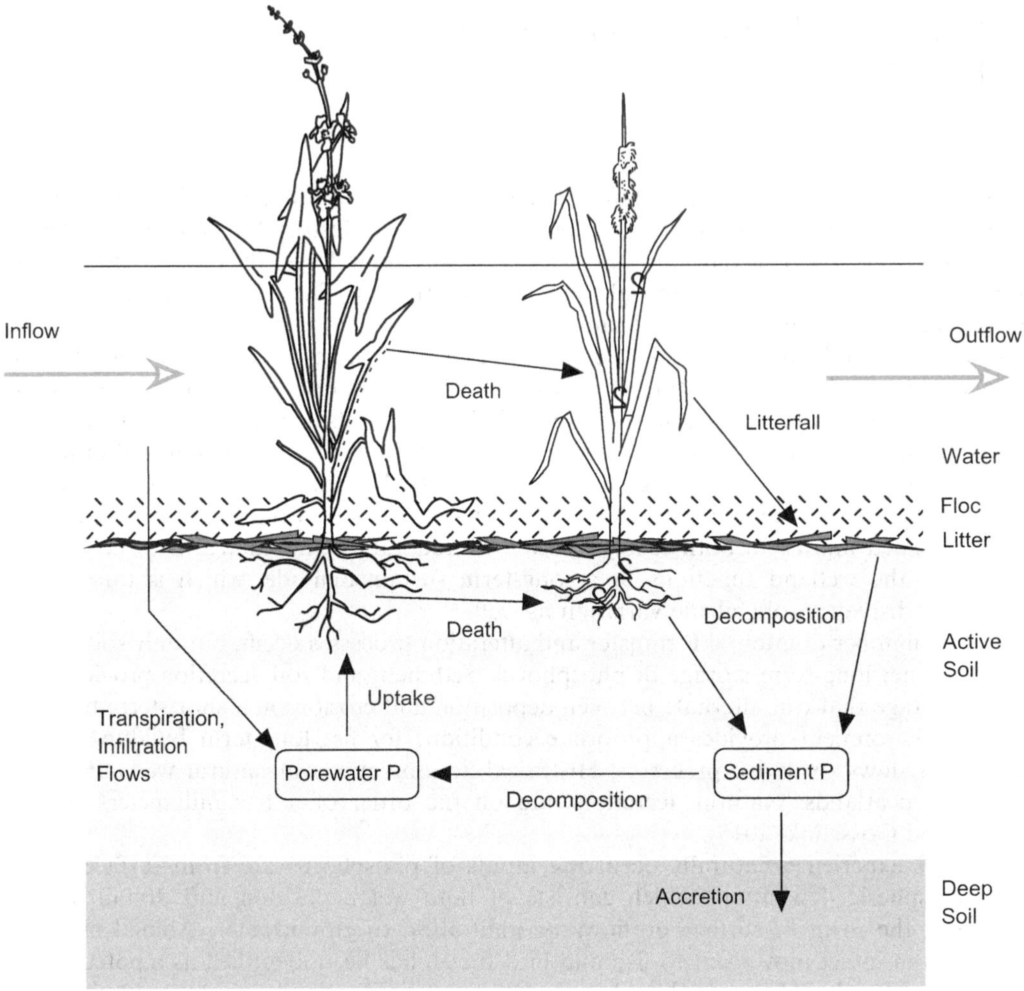

FIGURE 10.1 Phosphorus storages and transfers for the macrophytic cycle.

from both the macrophyte and mesophyte cycles accrete in the system, providing the long-term sustainable removal mechanism.

The action zone (top 10–20 cm of the soil, see Chapter 5) interacts with emergent plant parts, overlying water, and soils beneath. Overlying water carries dissolved constituents, and is subject to flow patterns determined by bathymetry, stem density and wind. Indeed, most of the data acquired from P treatment wetlands has to do with flow rates and water chemistry. Emergent plant parts provide functions of transpiration, shading and shelter from wind; and contribute to P uptake. Deep soils may be of importance in wetland treatment where there are sizable infiltration or exfiltration flows.

Phosphorus moves vertically within the action zone, driven by plant uptake, transpiration water flows, and head-driven vertical flow. This P movement is directly linked to and influenced by growth of plants, their density, and the vigor of their growth. Floc phosphorus also moves horizontally, via solid spiraling mechanisms. This horizontal P flux is dependent upon hydraulic conditions. So, in both cases, circumstances outside the action zone will modulate action zone processes.

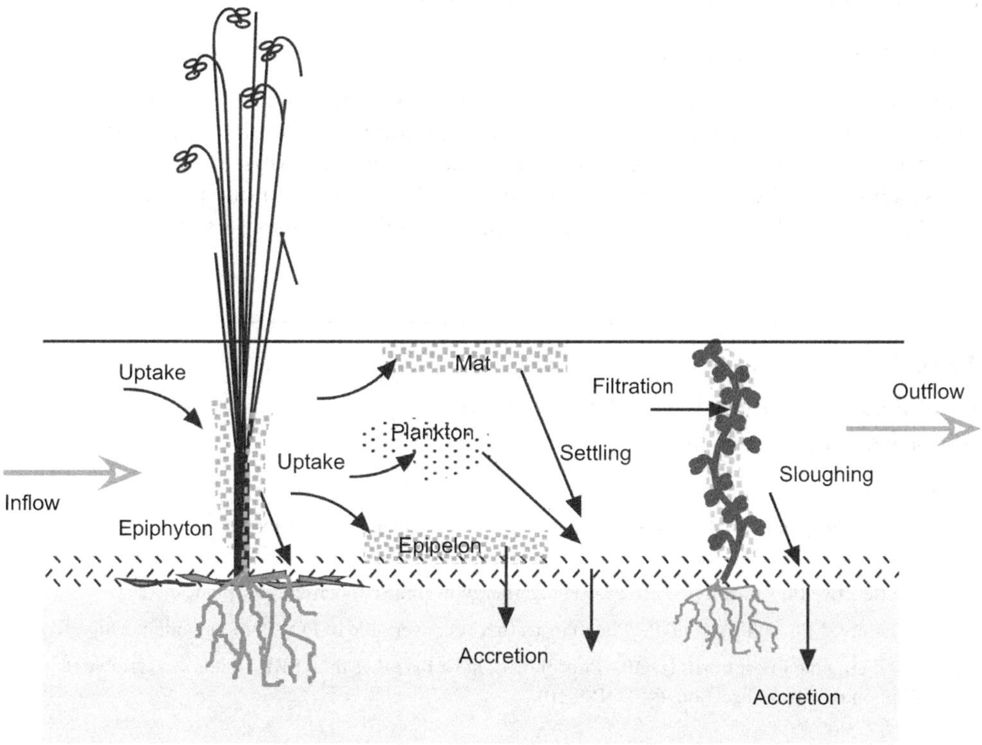

FIGURE 10.2 Phosphorus storages and transfers for the mesophytic cycle.

Plant roots are contained in the action zone, usually in the soil, but sometimes in the floc layer as adventitious roots. Although not viewable, roots undergo the same life cycle as the emergent and submergent plant parts that they support: they grow, die, and their litter decomposes. Importantly, roots extract nutrient and other chemicals from porewater, and may pass transpiration water flows to emergent plant parts.

The voluminous published information on phosphorus treatment wetlands has had a variety of focal points, including overall ecosystem performance (e.g., Reinhardt et al., 2005; O'Neill et al., 2011; Kadlec, 2016), water movement (e.g., Dierberg et al., 2005; Min and Wise, 2009), vegetation (e.g., Tanner, 1996; Toet et al., 2005), and soils and sediments (e.g., Lai and Lam, 2009; Johannesson et al., 2011). Less emphasis is found for topics such as particulate transport or soil-building processes, both associated with the uppermost sediment and soil layers, but there has been considerable emphasis on accretion in coastal wetlands (e.g., Chmura et al., 1992; Rybczyk et al., 1998; Rogers et al., 2006; DeLaune et al., 2013a), and to a much lesser extent in inland freshwater wetlands receiving excess phosphorus (Craft and Richardson, 1993; Braskerud, 2002b; Kadlec, 2009b).

The phosphorus in the action zone dominates the total wetland storage. It is useful to consider the quantities of phosphorus contained in the water, and contained in the action zone for a hypothetical P treatment marsh situation (Figure 10.3). Of course, these storage numbers will vary from system to system; and the reader might wish to alter the selection of numbers in Figure 10.3. But any estimates within reason will show that the large majority of P in the treatment marsh resides in the action zone, typically many times greater than storage in either the vegetation or in the water. A fraction of the solid-phase P is recalcitrant, and may be regarded as permanently stored, but another fraction is available to stimulate biological processes.

10.1 PHOSPHORUS FORMS AND STORAGES

Phosphorus Forms in Marsh Waters

Wetland science has evolved to focus on categories of phosphorus compounds that are defined by methods of analysis (Table 10.1). In every case, the analytical result is reported as the elemental phosphorus content of the category. The most reactive forms are the dissolved phosphates, which change hydration in response to pH. The most common species are mono- and dibasic phosphates, which dominate at all typical wetland pH values ($4 < pH < 9$) (Morel and

TABLE 10.1

Forms of phosphorus in the wetland environment.

Dissolved forms (filtered (0.45µm) samples):

- Orthophosphate (PO_4-P).
- Condensed phosphates. These consist primarily of pyro-phosphate, meta-phosphate, and poly-phosphates.
- Soluble Reactive Phosphorus (SRP). PO_4-P, together with some condensed phosphates.
- Total Dissolved Phosphorus (TDP). Phosphorus that is convertable to PO_4-P upon oxidative digestion.
- Dissolved Organic Phosphorus (DOP). Phosphorus, in forms other than SRP, that is convertable to PO_4-P upon oxidative digestion. (= TDP-SRP)

Dissolved plus associated with suspended solids. The procedures listed here performed on unfiltered samples yield, by analogy:

- Total Reactive Phosphorus (TRP).
- Total Acid Hyrolyzable Phosphorus (TAHP).
- Total Phosphorus (TP).
- Total Organic Phosphorus (TOP). (= TP-TAHP)
- Particulate Phosphorus (PP). (=TP-TDP)

Sorbed to the surface of soil particles:

- Sorbed phosphorus is removed using extractants such as water, or solutions of KCl or bicarbonate.

Contained in the structure of biomass:

- Total phosphorus may be found by analyzing for PO_4-P in digests of biomass samples. Digestion may involve dry or wet ashing, followed by re-dissolution.

Contained in the structure of soil particles:

- Structural, internal forms of phosphorus in the solid are removed (solubilized) using harsh extracts of wet soil samples. Typical extractants include:
    - Sodium Hydroxide (0.1 M). The SRP in the extract is representative of iron and aluminum bound phosphorus. The balance of the TP in the extract (TP – SRP) is representative of organic phosphorus associated with humic and fulvic acids.
    - Hydrochloric Acid (0.5 M). The SRP in the extract is representative of calcium bound phosphorus.
- Total soil phosphorus may be found by analyzing for PO_4-P in digests of soil samples. Digestion may involve dry or wet ashing, followed by re-dissolution.

Hering, 1993). The generic term used for these inorganic phosphate ions is orthophosphate, (PO_4-P). The molybdate analytical test nominally finds this form of phosphorus, but has been shown to also detect exchangeable phosphorus and colloidally bound phosphorus in eutrophic wetlands (Hens and Merckx, 2002).

Phosphorus readily combines with, and may be part of, dissolved organic materials, and in that form has the designation of Dissolved Organic Phosphorus (DOP). In fact, DOP has been characterized in great detail for treatment wetland situations, and found to consist of several kinds of organics (Turner and Newman, 2005). Some of them are readily hydrolyzed by soil enzymes, and together with PO_4-P are called Soluble Reactive Phosphorus, or SRP. The organic components of SRP can move readily in soils and sediments (Anderson and Magdoff, 2005). Phosphorus also may be associated with suspended particles, and is called Particulate Phosphorus, or PP.

Wetlands provide an environment for the interconversion of all these forms of phosphorus, with the eventual sink being one or more of the wetland solid compartments. A variety of cations can precipitate phosphate under certain conditions. The important potential mineral precipitates in the wetland environment include apatite and hydroxylapatite (Reddy and D'Angelo, 1994). In addition to direct chemical reaction, phosphorus can co-precipitate with other minerals, such as ferric oxyhydroxide and the carbonate minerals, such as calcite (calcium carbonate) (Reddy and D'Angelo, 1994). The overall P-mineral chemistry is very complex; consequently, quantitative calculations of solubilities are generally not possible. Trends for phosphorus movement in wetland soils are as follows (Reddy and D'Angelo, 1994):

- In acid soils, phosphorus may be fixed by aluminum and iron, if available.
- In alkaline soils, phosphorus may be fixed by calcium and magnesium, if available.
- Reducing conditions lead to solubilization of iron minerals and release of phosphorus co-precipitates.

Marsh Phosphorus Storages

Phosphorus compounds are a significant fraction of the dry weight of wetland plants, detritus, microbes, wildlife, and soils; although they are about ten times less than nitrogen compounds. The mass of these phosphorus storages varies in different wetland types, and with the season of the year. A general idea of the sizes of these different storage compartments is necessary to understand the P fluxes discussed below (Figure 10.3). The dominant fraction of phosphorus is contained in the wetland soils and sediments. Plants and litter comprise most of the remainder, with very little mass contained in mesophyton and water.

Phytomass Phosphorus

Plants utilize phosphorus, and decomposition processes release P back to the water (Figure 10.1). There are two direct effects of vegetation on phosphorus processing and removal in treatment wetlands:

1. The plant growth cycle seasonally stores and releases P, thus providing a "flywheel" effect for a P removal time series.
2. The creation of new, stable residuals that accrete in the wetland. These residuals contain phosphorus as part of their structure, and hence accretion represents a burial process for P.

The P content of biomass in marsh wetlands varies with species, with plant parts, with water nutrient status, and among wetland sites. Example ranges of dry weight phosphorus

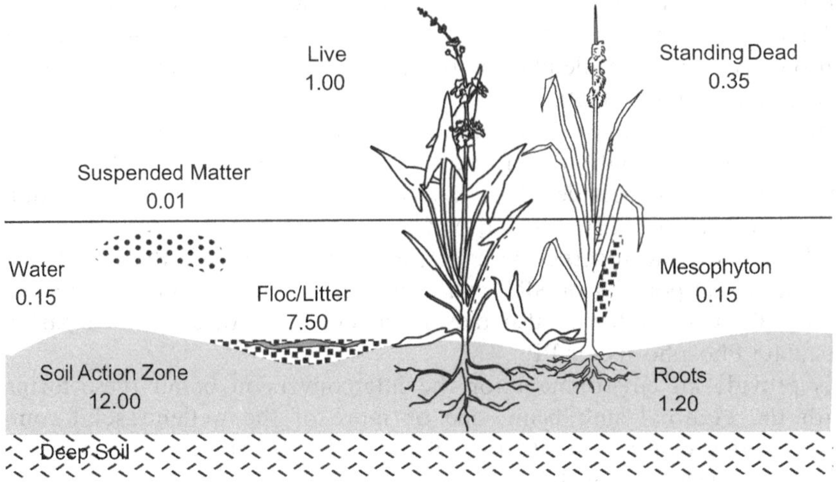

FIGURE 10.3 Hypothetical order-of-magnitude of P storages in a lightly loaded treatment marsh (gP/m^2). The action zone is comprised of the floc/litter and root zone soil (10 cm). The phosphorus proportions are: above water: within water: action zone = 1.35: 0.31: 20.7.

percentages in natural wetlands are: 0.10–0.30% for emergent plants; 0.14–0.40% for floating leaved plants; and 0.12–0.27% for submersed plants (Boyd, 1978). Bedford et al. (1999) report a range of 0.1 to 0.64% for live plant tissue in 41 marshes.

Treatment marshes are usually nutrient-enriched, and display higher values of tissue nutrient concentrations than natural wetlands. For instance, live cattail leaves in the discharge area of the Houghton Lake wetland averaged 0.33%P; those in nutrient-poor control areas averaged 0.10%P (Table 10.2). Standing dead leaves have lesser P concentrations than their live counterparts. Litter may have slightly greater or slightly lesser concentrations. In addition to the increased P content with water nutrients, plant biomass also increases. Thus, the standing crop of phosphorus in vegetation is a very strong function of water nutrient content. All plant parts respond, and the standing crops at high nutrient levels are three to ten times higher than in unimpacted wetland areas (Table 10.2). The Houghton Lake data is not an isolated instance. Inter-system data display the same strong dependence on water nutrient status. Aboveground macrophyte biomass phosphorus standing crops vary from 0.4 gP/m^2 for an oligotrophic *Cladium* marsh (Noe and Childers, 2007), to 1.6 gP/m^2 for a lightly loaded *Typha* marsh (Noe and Childers, 2007), to 4.0 gP/m^2 for a moderately loaded *Typha* marsh (Table 10.2), to 13.0 gP/m^2 for a heavily loaded *Phragmites* marsh (Hocking, 1989b).

The biomass of either live or dead mesophyton is difficult to measure, but it is small compared to the biomass of live or dead macrophytes, or macrodetritus (litter). The phosphorus content of periphyton and plankton is higher than for the macrophytes, especially in P rich waters. Vymazal (1995) reports values typically ranging from 0.2–2.0%P for a dozen different species. Noe and Childers (2007) estimated the P standing crops in the treatment zone of WCA2A in Florida. They reported a 4:1 ratio of macrophyte P to mesophyte P at a water concentration of 42 μgP/L, and a 9:1 ratio at 76 μgP/L. Most of this difference was attributable to a doubling of the P concentration in floc, mesophytes, and macrophytes.

Aboveground biomass collected at the end of the growing season displays much lower phosphorus content than in spring (Figure 10.4). Klopatek (1978) has shown trends of the same magnitude for cattail shoots, growing under lower nutrient availability. It is therefore

TABLE 10.2

Concentrations and standing crops of phosphorus in plant parts at the Houghton Lake treatment marsh (mean±sd). Four years' end-of-season data, except one year for rhizomes.

	Control TP %dw	Discharge TP %dw	Control TP gP/m^2	Discharge TP gP/m^2
Live	0.100±0.018	0.33±0.09	0.50±0.28	3.91±1.21
Standing Dead	0.050±0.024	0.26±0.06	0.21±0.20	2.80±1.77
Litter	0.122±0.018	0.36±0.09	0.10±0.05	6.71±2.19
Rhizomes	0.065	0.30	2.73	6.99

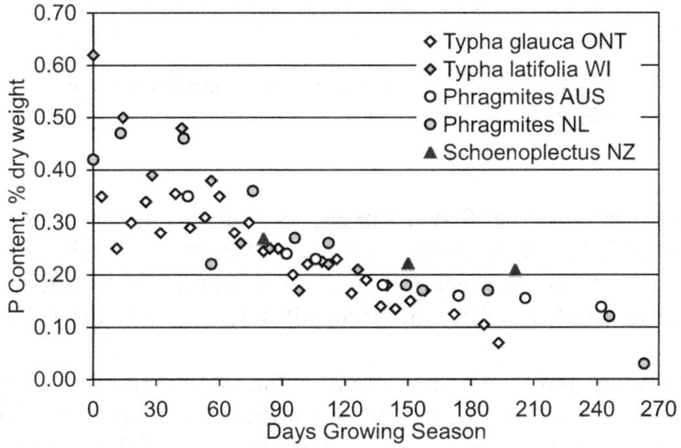

FIGURE 10.4 The seasonal decline of phosphorus content in above ground tissues of wetland plants. Data sources: *Typha glauca* Ontario: Bayly and O'Neill (1972); *Typha latifolia* Wisconsin: Prentki et al., 1978; *Phragmites* Australia: Hocking (1989b); *Phragmites* Netherlands: Mueleman et al. (2002); *Schoenoplectus* New Zealand: Tanner (2001a).

apparent that the timing of vegetation samples can greatly affect subsequent calculations of phosphorus storage and biomass translocation to below ground rhizomes.

The proportions of stored P in various plant parts changes as the seasons progress, with more pronounced patterns in temperate climates. Those growth patterns are discussed in Chapter 5 (see Figure 5.21). The time-varying P concentrations may be combined with the time-varying biomass for each compartment, which in total represent the phosphorus storage on a seasonal basis. Typically that storage follows the pattern of a growing-season increase to a maximum, followed by a senescence-season decrease to a minimum, with the cycle repeated each year (Figure 10.5). After the plants are fully mature, there is, on average, no net annual increase in plant storage. However, on a seasonal basis, plant uptake can be a very large part of P removal or release. This phosphorus is transferred to the necromass storage pool during autumn senescence.

To add to the complexity, tissue P concentrations also vary with the type of tissue sampled within the vertical structure of the plant. Hocking (1989b) reported leaf P greater than stem P, and upper strata (tip) P greater than lower strata (base) P, both by about 50%.

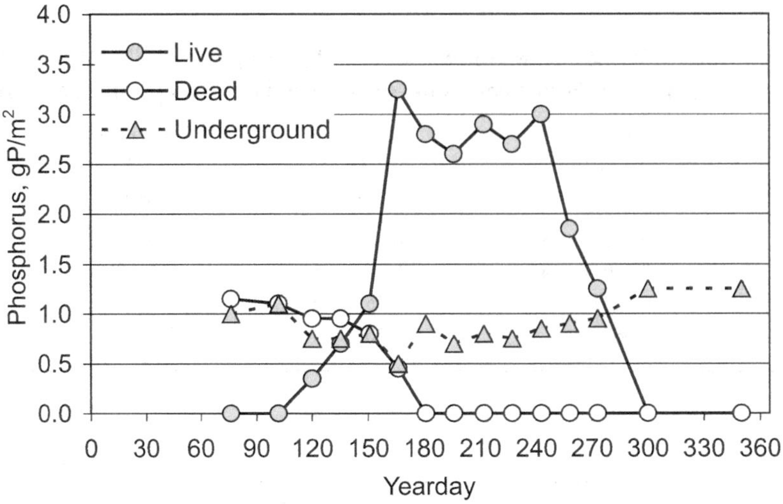

FIGURE 10.5 Seasonal plant storage of phosphorus in a lakeshore *Typha* marsh near Madison, Wisconsin. (Data from Prentki et al., 1978.)

Standing dead represents the first recipient of senesce P. Some of that phosphorus leaches and is thus added back to surface waters, but another fraction becomes plant litter as the standing dead falls to the wetland floor. The timing of litterfall is highly variable, because it is related to wind, rainfall, and snowfall, but generally litterfall occurs during the winter months.

In many cases, a major part of the P content of the litter is released during the next warm season, thus giving rise to a P return flux to the surface water that coincides with the warm temperatures that drive the decomposer organisms. However, at the same time, warm temperatures foster the growth of biofilms located on the litter. So, litterbag experiments can often display an increase in the P storage associated with the decomposing litter, even though the amount of the original litter is decreasing (Davis and van der Valk, 1978, 1983; Kominkova et al., 2000; Kao et al., 2003; Su et al., 2007). Experimentally, it is difficult if not impossible to separate the original disappearing litter necromass from the living and growing mesophytic biomass attached to it. In some cases, the warm season return of P from decomposing necromass can exceed the uptake rates during the same time period, giving rise to P release from the marsh (Lu et al., 2009; Beutel et al., 2014; Dunne et al., 2015).

Soils and Sediments

Most of the phosphorus in the soil column is structural P, both organic and inorganic. Very small fractions are found in porewater or as sorbed P (Figure 10.3). Most wetland treatment systems will build an organic sediment and soil layer with time. The first step in soil building is often the formation of an unconsolidated material of low density, called floc. This material partially decomposes, and consolidates with time, forming new soil layers in the marsh. The initial soils in a constructed treatment wetland are the choice of the wetland designer. One choice is to use on-site mineral soils; another is to use imported organic materials, such as peats or compost. Typically, about 30 cm of soil is placed, and becomes involved as the root zone of the wetland plants. These selected soils may be deficient in phosphorus, if infertile mineral soils are used; or may contain a surplus of P, if fertile, agricultural soils are employed. Here we examine the P content of soils that have adapted to treatment conditions over a period of several years.

Floc

It is now well recognized that a loose and moveable layer of unconsolidated materials exists in treatment marshes. Terminology varies: Kadlec and Bevis (1990) referred to it as "a grey gelatinous material (ooze)"; Nolte and Associates (1998) as "Layer A"; DeBusk et al. (2001) as "muck or unconsolidated peat"; and SFWMD (2005–2018) "floc." Floc is easily disturbed, and may then move with water until it resettles (see Figure 8.7). Floc is harvested from tube corers by pouring or vacuuming it out. The P content is 0.1–0.4%dw (Table 10.3), but the bulk density is low, on the order of 0.02–0.10 gdw/cm^3. Nevertheless, thicknesses range up to 20–30 cm, and can contain 4–40 gP/m^2. Movement rates have not been measured but are speculatively much lower than slow-moving wetland water speeds. Because most of the phosphorus is structural rather than sorbed in the floc, these solids provide for the slow transport of P even in waters of extremely low dissolved P (Gaiser et al., 2005). Decomposition can release a portion of the floc P, while another fraction is resistant to decomposition.

The thickness of the floc layer, and hence its storage capacity, vary along the flow direction as P is removed from the surface water (Figure 10.6). Further, the chemical composition and quantity of floc are different between EAV and SAV marshes. In the calcium-rich water environment of the Florida STAs, an SAV community fosters epiphytic growth that removes phosphorus to calcitic residuals. In contrast, an EAV plant community fosters the removal of P to organic residuals (Table 10.3). In both types, total and residual floc P decrease along the treatment gradient (Figure 10.6). Yet, the characteristics of the flocs in the community types are distinctly different, with the calcitic floc being denser and thicker, and having much greater total and residual P.

Base Soils

The P content of organic soils that have experienced only low water-phase P (ca. 50 μg/L) are generally in the range 300–500 mgP/kgdw. However, if the ecosystem has

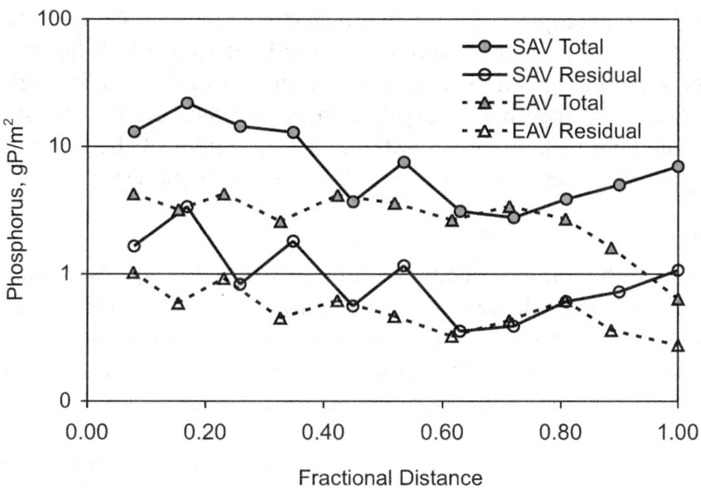

FIGURE 10.6 Floc phosphorus storage along the gradient in STA2, for EAV and SAV. Data from UF/IFAS (2017).

TABLE 10.3

Properties and phosphorus content of flocs formed in SAV and EAV plant communities.

		EAV	SAV
Thickness	cm	4.7	7.6
Bulk Density	gdw/cm^3	0.043	0.151
Total Carbon	%	40.5	18.8
Total Calcium	g/kg	76	292
Mass	gdw/m^2	1,763	11,659
Total P	mgP/kg	1,704	683
Total P	gP/m^2	3.01	8.65
Residual P	gP/m^2	0.553	1.137

Source: UF/IFAS (University of Florida Institute of Food and Agricultural Sciences) (2017).

been long exposed to higher P concentrations (e.g., > 3 mg/L), then organic soil TP is 1,000–2,000 mg/kg. These values pertain to the upper-most layer, which may still be experiencing decomposition, which is sometimes termed recently accreted soil (RAS).

Because of the competition among plant uptake, detrital decomposition and the transpiration flux in the root zone, there may exist a vertically decreasing profile of soil TP and porewater P in the soil column. Examples are given in Figure 10.7 for soil P storage and Figure 10.8 for porewater P, for various locations in the Florida STAs. The topsoil layer (0–20 cm) typically contains the most roots (see Chapter 5), which remove P from this zone of high concentration porewater. Those roots also undergo a cycle of growth, death and decomposition, and hence the return flux of available P is greatest in this top zone. Lower layers (10–20 and 20–30 cm) become progressively depleted by plant uptake, and because of lesser decomposing root biomass (Figure 10.7). The top layer is also the site of newest depositions of detrital and particulate P.

It is not common for the new sediments and soils in a treatment marsh to be inorganic in character. However, systems treating runoff may receive considerable quantities of inorganic solids from soil erosion in the watershed, which then combine with organic materials generated within the wetland. An example is Chiricahueto marsh in Mexico (Soto-Jimenez et al., 2003). Agricultural runoff brought water at about 10.2 mg/L of total phosphorus to the marsh for over 50 years. The soil column is now mostly inorganic, with less than 5% carbon. Both carbon and phosphorus decreased together as depth increased, indicating that most of the soil phosphorus was associated with the organic content. The P content of the upper 10 cm at Chiricahueto was 1,200 μgP/gdw, decreasing to 300 μgP/gdw at 60–70cm depth.

Soil P Speciation

Some of the accreted phosphorus occurs as minerals that are not particularly susceptible to remobilization, but new organic substances are subject to degradation if conditions change within the wetland. For instance, dryout can cause oxidation of some of the stored organics, which may be completely stable under submerged conditions. The associated organic-bound phosphorus may then be mineralized, and redissolve on rewetting (Olila et al., 1997; Pant and Reddy, 2001a).

The forms of soil P vary depending upon the chemistry of the wetland that formed them. For instance, two quite different speciations, one dominated by iron and aluminum compounds, and the other dominated by calcium and magnesium compounds are to be found in marshes (Kadlec and Wallace, 2009). The ferruginous system is likely to respond

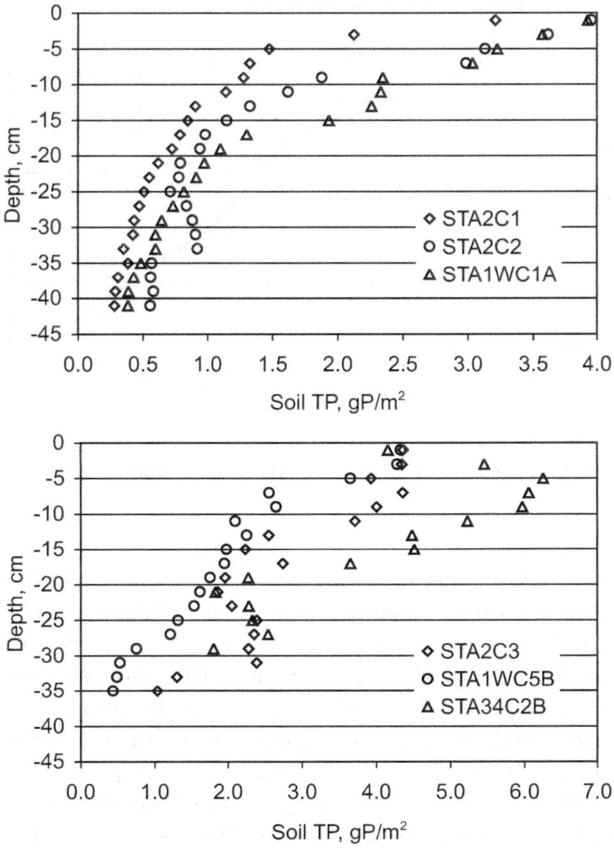

FIGURE 10.7 Depth profiles for soil phosphorus in emergent (upper panel) and submergent (lower panel) treatment wetlands. (Data from Bhomia et al., 2011, 2015.)

to lowering redox by releasing P, but the calcitic system will not. Phosphorus that is bound to ferric iron compounds under oxic conditions may be released if reducing conditions occur and convert the iron to soluble ferrous forms. Soto-Jimenez et al. (2003) speculate that under reduced conditions decomposition processes of organic matter use Mn oxides and Fe-oxyhydroxides as electron acceptors, releasing PO_4 ions to porewater solution. These are transported from the 0–8 cm sediment layer to the water column, where they are transformed into suspended particulate oxides and trapped in the surface layer. This upward migration and reincorporation into the sediment favored the enrichment of P in the upper sediments. Sorbed orthophosphate may partially desorb if the water concentrations become lower. However, under continuous inundation and stable water chemistry and redox conditions, stored P is not likely to be released.

In any case, a portion of soil P is resistant to decomposition, approximately 10–20% of the total active soil P, and forms the stable storage for removed P.

10.2 WETLAND PHOSPHORUS PROCESSING

There are three principal categories of phosphorus removal processes in wetlands: sorption, utilization to build a bigger biomass compartment, and storage as newly acquired

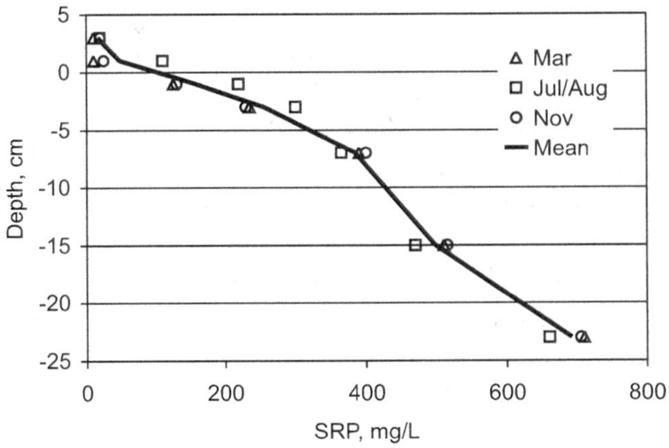

FIGURE 10.8 Depth profiles for porewater phosphorus in a submergent treatment wetland, STA2C3. (Data from DBE, 2017.)

or created, refractory residuals (burial). The first two mechanisms (sorption and biomass storage) have finite phosphorus retention capacities, whereas the third, accretion, is a sustainable process. Sorption and biomass building are sometimes termed saturable mechanisms, because they reach that limited capacity, and then can remove no more phosphorus. Particulate settling can rapidly remove large amounts of phosphorus from runoff waters that carry high amounts of suspended sediment. Residuals from the dual phytomass cycles form new stable P-containing solids, a process termed bioaccretion. Accretion of solids continues, with no capacity limit. In addition to these, secondary processes such as movement among storage compartments also exist. There may also be rearrangements of the phosphorus storages within the wetland that affect its availability and mobility.

Sorption

In some of the treatment wetland literature, there exists a seemingly indestructible misconception that wetlands provide phosphorus removal only through sorption processes on existing soils. The concept that sorption is the predominant removal mechanism is apparently a carry-over from studies of terrestrial ecosystems, which do not possess the burial mechanisms (sedimentation and bioaccretion) found in wetlands. Early wetland researchers therefore concluded (wrongly) that treatment marshes become phosphorus-saturated in a few years, with a subsequent export of excessive quantities of phosphate (Richardson, 1985). As more information became available, these same researchers (Richardson and Craft, 1993) recognized that "The efficiency of any wetland to store P on a long-term basis is thus determined by the peat or soil accretion rate …"

It is therefore difficult to understand how, in 2017, a U.S. Army Corps of Engineers' review (Currie et al., 2017) of using wetlands for P reduction in the Great Lakes region of the United States states: "… once all soil sorption sites become saturated with P, the wetland cannot retain additional P." For that same Great Lakes region, Mitsch (2017) concludes that wetlands could sustainably remove P and protect Lake Erie water quality. That conclusion was based in part upon studies that showed a retention rate of phosphorus in created wetlands receiving moderately polluted river water was 2.40 ± 0.23 $gP/m^2 \cdot yr$ over a fifteen-year period (Mitsch et al., 2014b).

Wetlands don't "saturate" with phosphorus. Saturation is a widely used term in the sorption literature to describe the situation where the soil or sediment has equilibrated with the water, and neither takes up nor releases any further quantities of phosphorus. It has unfortunately also been used to describe the response of an entire ecosystem.

Soils of all types have sorptive capacity for phosphorus, but this storage is soon saturated under any increase in phosphorus loading. When water of a given concentration is added to a substrate that has sorption capacity, sorption occurs until the entire soil of the wetland is loaded to the solid phase concentration corresponding to equilibrium with that water concentration. This time period is short for solids of low capacity, but can be longer for high capacity. Furthermore, the initial amount of P on the soil is also a determinant of the time to sorption saturation. Times are longer for soils with less of their capacity used up prior to initiation of exposure. Heavily loaded soils may actually desorb P during a startup with new P-laden water. The median times to saturation were reported at about one year (Kadlec and Wallace, 2009). It should be noted that the wetland is not "worn out" at this point; rather, one of several removal mechanisms has been exhausted.

A number of different sorption isotherms have been proposed, with the Langmuir formulation prominent among them:

$$\frac{S}{S_{\max}} = \frac{C}{C + 1/k} \tag{10.1}$$

where C = P concentration in water, mg/L

1/k = half saturation concentration, mgP/L

S = sorbed P concentration, mgP/kg

S_{max} = maximum sorbed P concentration, mgP/kg

Typical values of the Langmuir sorption parameters are shown in Table 10.4. In general, mineral soils have a larger sorption capacity than organic soils (higher S_{max}), but when the water P concentration is in the range typically encountered in marsh porewater, the solid-phase concentration of sorbed P is comparable among soil types, and in the range 12–238 mgP/kg. Estimated amounts of available (exchangeable) sorbed P in the top 20 cm of the wetland soil column are in the range 3–12 gP/m^2. This means that the underground P storages available to plants for growth uptake are many times greater than the dissolved P in porewater. Sorbed P can be a significant fraction of the total P in the action zone (see Figure 10.3). Thus, the primary effect of P sorption in marshes is to expand and increase the amount of available SRP in the soil.

The maximum sorbed concentration (S_{max}) of soils for binding phosphorus have been found to depend strongly upon the amount of iron and aluminum in the wetland soil (Reddy and Delaune, 2008):

$$S_{\max} = a + b(C_{Fe} + C_{AL}) \tag{10.2}$$

where a = intercept constant, mmolP/kg

b = proportionality constant, mmol/mmol

C_{Fe} = oxalate-extractable iron concentration in soil, mmol/kg

C_{Al} = oxalate-extractable aluminum concentration in soil, mmol/kg

S_{max} = maximum sorbed P concentration, mmolP/kg

The reported constants were a = 1.74, and b = 0.17 (Reddy and Delaune, 2008).

It is sometimes the case that a soil sample contains sorbed phosphorus when it is retrieved from the field. The amount of pre-sorbed initial phosphorus (S_i) corresponds to a water concentration termed the initial equilibrium phosphorus concentration (at Si, C_{eq} = EPC_o), and

TABLE 10.4

Phosphorus Langmuir parameters and storages for soils.

Number	Type	Source	Location	1/k mgP/L	S_{max} mgP/kg	S@1 mg/L mgP/kg	0.2 m P Storage gP/m^2	Ref.
Mineral								
13	Sands	Terrestrial	Denmark	8.1	79	12	4.1	1
5	Clays	Wetland	Korea	3.5	231	70	11.6	2
3	Soils	Wetland	Florida	6.8	271	32	3.2	3
Organic								
16	Sediments	Lakes	Florida	6.9	280	178	4	4
3	Sediments	Lakes	China	6.4	601	93	7.4	5
8	Sediments	Lakes	China	4.3	611	125	10	6
4	Soils	Marshes	Florida	46	1,145	29	1.2	7
1	Soils	Marshes	Michigan	31	2,968	93	3.7	7
1	Soils	Marshes	Michigan	6.6	1,809	238	9.5	8
3	Soils	Agricultural	Florida	81	3,033	43	6.9	9

References
1 Arias et al. (2001)
2 Yoo et al. (2006)
3 Reddy et al. (1995)
4 Belmont et al. (2009)
5 Jin et al. (2005)
6 Wang et al. (2006b)
7 Richardson and Vaithiyanathan (1995)
8 Hammer and Kadlec (1980)
9 Janardhanan and Daroub (2010)

these are related by Equation 10.1. If water at concentration EPC_o is added to the soil, there is neither adsorption nor desorption.

When water of a given concentration is added to a substrate that has sorption capacity, sorption occurs until the entire soil of the wetland is loaded to the solid phase concentration corresponding to that water concentration. The time period to saturation can be short for solids with low sorption capacity but can be long for soils with high sorption capacity. Furthermore, the initial amount of sorbed phosphorus on the soil affects the time to sorption saturation. Times are longer for soils with less of their capacity used up prior to initiation of exposure. Heavily loaded soils may actually release phosphorus if abruptly inundated with water that contains a phosphorus concentration that is lower than historical phosphorus concentrations (this is sometimes observed during startup of FWS wetlands).

For initially barren, 20 cm thick substrates exposed to 3 gP/m^2·yr of phosphorus loading, the times to are about one to three years. It should be noted that the wetland is not "worn out" at this point; rather, one of the phosphorus removal mechanisms has simply been exhausted.

Biomass Storage and Cycling

Wetland biota undergo a cycle of growth, death, and partial decomposition. This pertains to mesophytic components of the ecosystem as well as the macrophytes (see Figures 10.1

and 10.2). Of course, the mesophytic components are present in much lesser quantities, and cycle much more rapidly, than the macrophytes.

Although there are quantifications of portions of the biomass cycles of phosphorus, there have been no comprehensive measurements of all components of biomass cycling in treatment wetlands. However, a very inclusive set of studies was carried out, over many years, in Manitoba, Canada. The Marsh Ecology Research Program (MERP) was a long-term interdisciplinary study on the ecology of prairie wetlands. Ten study wetland cells, of about 5 ha each, were embedded in the Delta Marsh, a 50,000ha wetland located on the southern shore of Lake Manitoba in Manitoba, Canada. Exterior levees isolated the cells for purposes of manipulation. The study ran for ten years, and involved the efforts of over three hundred field personnel (Murkin et al., 2000). Vegetation was a mosaic of *Scolochloa festucacea, Typha glauca, Scirpus lacustris,* and *Phragmites australis,* interspersed with variable zones of submersed macrophytes, algae and open water. Depths ranged from 10 to 80 cm, and lake water was pumped to the marsh cells during the warm months, at typical annual hydraulic loadings of 0.3 cm/d. Pumping was conducted primarily to replace evaporative and seepage losses, and thus maintain the selected hydrologic regime (Kadlec, 1983, 1986). Phosphorus concentrations were on the order of 150 μgP/L.

Enormous effort went into measuring the pools of biomass and the accompanying nitrogen and phosphorus, and the transfers between eleven ecosystem compartments. Results are available by year in Murkin et al. (2000).

Mesophytic Phosphorus

The mesophyte phosphorus cycle is conceptually illustrated in Figure 10.2. Epiphyton, plankton, epipelon, and mats all take up P and ultimately deposit particulate P residuals in the wetland floc. The algal and microbial communities of wetland biofilms and soils are important both for the decomposition of organic material, and cycling of nutrients, and thus play a central role in the metabolism of the entire ecosystem (Prenger and Reddy, 2004). The relative importance of this mesophytic cycling is reported to diminish as the water nutrient levels increase (Noe and Childers, 2007).

Scinto and Reddy (2003) found extremely high P uptake rates for epiphyton, epipelon, and metaphyton for an unimpacted Florida marsh. Half-lives of added P were less than an hour. Pietro et al. (2006), based upon Pietro (1998), also found extremely high rates of uptake of orthophosphate in submerged aquatic vegetation (SAV), and associated algae, in a treatment marsh. Half-lives of spike phosphorus doses were found to be in the range of 22–61 minutes. This high rate of uptake obviously cannot be sustained over a long period of time, but does indicate that the SAV community, likely via its epiphytes, was poised to quickly claim and convert available phosphorus. Similarly, Havens et al. (1999) found that various forms of algae (epiphyton, metaphyton, epipelon, surface mat) readily took up phosphorus doses in mesocosms. A total dose of 1,100 μgP/L was reduced to 50 μgP/L in 28 days, with most of this phosphorus storage in the benthos (metaphyton and epipelon communities). Wolfe and Lind (2008) measured P uptake half-lives of 1–5 hours for epipelon on rock substrates.

Similarly, ^{32}P studies have shown that orthophosphate is very rapidly converted to particulate forms, within five minutes, probably due to microbial processes (Noe et al., 2003). Richardson and Marshall (1986) determined that small sediments, microorganisms and litter were the recipients of most ^{32}P in a *Carex* marsh.

Mesophytic organism biomass also responds to P additions, but at a somewhat slower rate. Pei et al. (2015) measured grow-in that maxed out at about 30 days in eutrophic lake waters. The Delta marsh studies followed the biomass dynamics of epiphyton, metaphyton, epipelon, and plankton over a period of five years for different hydrologic regimes

(Robinson et al., 2000). Table 10.5 shows that plankton turned over the fastest, while metaphyton (algal-cyanobacterial mats) grew more slowly.

Thus it is apparent that the life cycle of these small organisms is short, so turnover is quick and it is likely that most of the uptake is returned as DOP and PP, leaving only a small fraction as permanently buried.

Studies such as these indicate very strongly that the first lines of interaction with added phosphorus in a wetland are the microorganisms and not the macrophytes. Although the biomass of these organisms is small, they nonetheless can store significant amounts of phosphorus and, importantly, place the intercepted phosphorus in the sediment-soil systems, where other processes can operate to convert it to recalcitrant forms. It is apparent that the life cycle of these small organisms is short, so turnover is quick and it is likely that most of the uptake is returned as DOP and PP, leaving only a small fraction as permanently buried.

Macrophytes

Emergent plants (EAV) utilize phosphorus for growth, mostly from the underground action zone (Figure 10.1). Upon senescence and death, standing dead necromass is created, which may leach some P back to the surface water. Most of the standing dead becomes pronate litter, which adds to the top of the action zone and its P content. Roots and rhizomes also follow a cycle of growth and death and P release upon decomposition, which is added directly to the action zone. These two parts of the macrophyte cycle have been discussed in Chapter 4.

The biomass P cycle is pronounced for aboveground plant parts, but much less so for underground plant parts (Figure 10.9) (see also Figure 5.21A). Biomass growth begins later at the more northerly latitude and ends more abruptly.

There are two direct effects of vegetation on phosphorus processing and removal in treatment wetlands:

- Storage and release. The plant growth cycle seasonally stores and releases phosphorus, thus providing a "flywheel" effect for a phosphorus removal time series.
- Sediment accretion. The creation of new, stable residuals that accrete in the wetland. These residuals contain phosphorus as part of their structure, and hence accretion represents a burial process for phosphorus.

Macrophyte turnover in cold-climate environments is usually in the range of 1.0–2.0 reciprocal years; which means the live aboveground biomass is replaced between one and two times per year. However, aboveground growth occupies only about one-third of the year in temperate climatic zones. Belowground growth, however, continues into late autumn (Prentki et al., 1978).

Turnover may be much higher in warm climates, because growth continues over a longer growing season. Macrophyte turnover in warm climates is rapid, with as many as five turnovers of the standing aboveground crop per year (Davis, 1994). Thus, the speed of the cycle depicted in Figure 10.1 is roughly the same during the growing season, regardless of geographic location. But the cycle is much reduced during winter in high latitudes.

Leaf litter decomposes to a stable residual material in a time span of 12 to 24 warm-weather months; litter from plankton and periphyton decomposes much faster. The residual is a small fraction of the parent biomass, so there is only a small annual buildup of these residual solids. However, the accretion of biomass residuals and minerals is the only sustainable removal mechanism for phosphorus in treatment marshes.

Submersed and floating plants differ from emergents because they can draw some phosphorus from surface water (SAV), or all of their P requirement from surface water (FAV).

TABLE 10.5

Turnover times (days) for four types of mesophyton, measured over a five-year period.

	Flooding Regime		
	Low	Medium	High
Phytoplankton	3.1	3.7	6.1
Epipelon	12.2	14.5	50.4
Epiphyton	8.5	11.4	17.2
Metaphyton	25.0	32.8	40.2

Source: Robinson et al. (2000).

Seasonal biomass dynamics may be essentially absent in warm climates, although there is some evidence of seasonality (Eid and Shaltout, 2017).

Macrophyte Growth and Cycling

The removal of phosphorus from water by wetland plants has been the subject of many studies (e.g. Reddy and DeBusk, 1985; Busnardo et al., 1992; Tanner, 1996). Many such studies have been characterized by measurements of gross phosphorus uptake, with no deduction for subsequent losses due to plant death and decomposition, with the attendant leaching and resolubilization of phosphorus. The processes of growth, death, litterfall and decomposition operate year-round, and with different speed and seasonality depending on climatic conditions and genotypical habit. Even in cold climates, the total annual growth is slightly larger than the end-of-season standing crop, by about 20% (Whigham et al., 1978). In warm climates, measurements show 3.5 to 10 turnovers of the live aboveground standing crop in the course of a year (Davis, 1994). Decay and translocation processes release most of the phosphorus uptake, with the residual accreting as new sediments and soils.

From the standpoint of phosphorus removal from wetland water, it is the net effect of the macroflora on water phase concentrations that is of interest. The seasonal patterns of vegetation growth and phosphorus storage embody complex patterns of biomass allocation among plant parts, as well as the phosphorus content of those various portions of living and dead material. However, from the point of view of the annual ecosystem removal of phosphorus, uptake and return from the combination of biomass and necromass are the principal features of concern. On an annual average basis, the only concern is net removal to permanent storage. However, during the course of the year, uptake and return may occur at different times, thus influencing removals differently in different seasons. For these reasons, it is necessary to examine the transfers to and from the collective parts of the macrophytes, which is here defined as phytomass. During the course of the year, especially in temperate climates, phytomass, and its phosphorus content, increases during the growing season, and shrinks during senescence.

Annual Storages and Transfers

The several transfers of phosphorus among the phytomass compartments was the focus of the Delta Marsh studies described above. The results serve to illustrate the importance of storages and transfers on an annual basis. For simplicity, the eleven compartments in that study are simplified by lumping into seven, five phytomass compartments plus surface water and porewater (Figure 10.10). In general, all annual P transfers are significant, although uptake from porewater and return to litter are dominant. Outflow from this wetland was

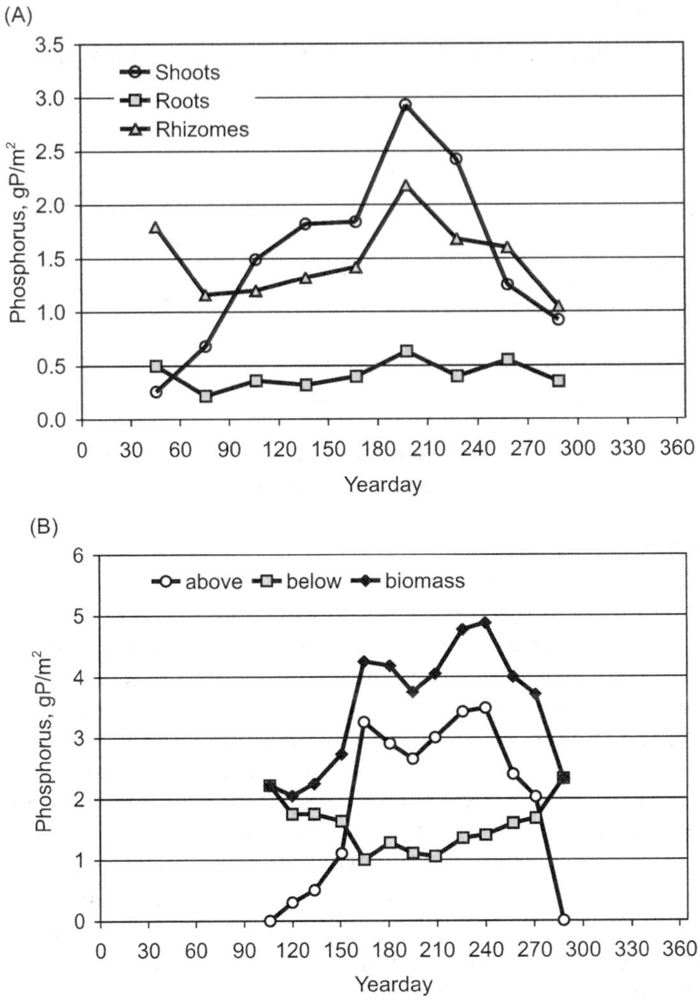

FIGURE 10.9 Seasonal patterns of above- and belowground biomass P content for marshes. (A) *Typha domingensis* at the Lake Burullus, Egypt wetland, 35.1°N. Data from Eid et al. (2012a). (B) *Typha latifolia* at the Lake Mendota, Wisconsin wetland, 43.1°N. Data from Smith et al. (1988).

mostly to evapotranspiration and seepage. It is noteworthy that surface water and its transfers are relatively small, and somewhat isolated from the major cycle of macrophyte uptake and return. As a consequence, litter forms an intermediary storage, which accumulates and decomposes on the time scale of years (Christensen et al., 2009; and see Chapter 4).

The uptake of phosphorus is via the root system, which is usually below ground. Phosphorus must therefore be transported into the rhizosphere, by processes of diffusion (minor) and vertical movement driven by transpiration flux (major) (Reddy et al., 2005). Some of the new plant growth nutrient requirement is supplied by translocation from stores in the rhizomes, and some from uptake from porewater. It is possible that the presence of phosphorus-rich porewaters causes less withdrawal from rhizomes, and results in lesser storage in below ground tissues (Tanner, 2001a).

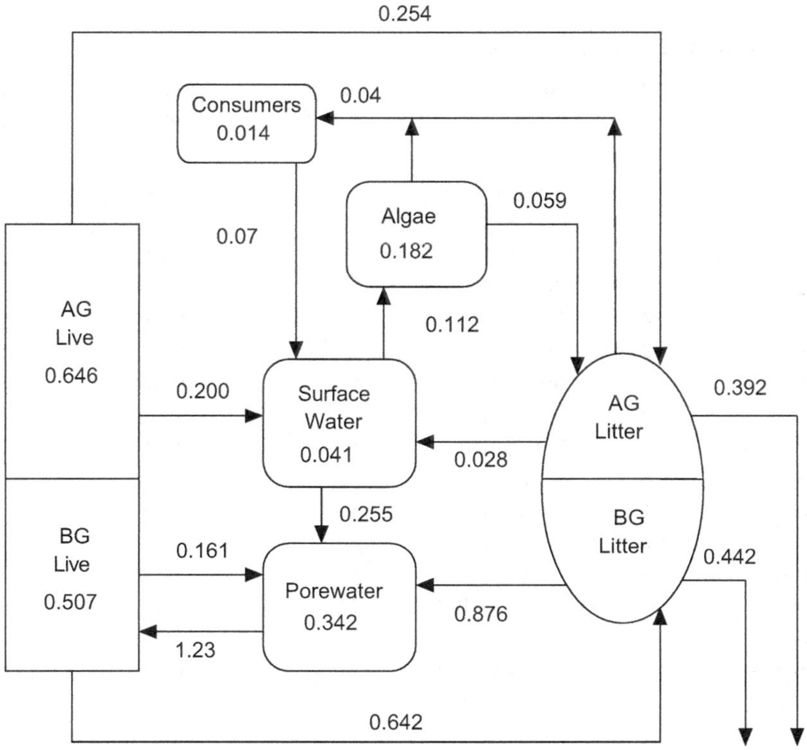

FIGURE 10.10 Example of storages and transfers of phosphorus for medium flooding at the Delta, Manitoba marsh site, averaged over five years. Transfers are $gP/m^2 \cdot yr$, and storages (boxes) are gP/m^2. Water inflow and outflows are not represented. Data from Kadlec et al. (2000).

Phosphorus is returned to surface waters and porewaters via the mechanisms of leaching and decomposition. It is likely that the majority of phosphorus in the necromass is returned, with lesser amounts transferred to permanent burial in the form of new soils and sediment. Over the course of a full calendar year, for a repetitively stable ecosystem, there is no change in the total phytomass, and the net change in phosphorus storage is zero ($\Delta P = 0$). For that annual period, plant uptake is either returned (more) or buried (less). But, as can be seen from Figure 10.9, the total phytomass phosphorus grows in spring and early summer, and recedes in autumn. This annual cycle is more pronounced in cold climates, in response to the more pronounced seasonal conditions.

At this point in the development of knowledge about plant phosphorus cycling in treatment marshes, there is insufficient data to support a compartmental description of P cycling.

Sustainable Phosphorus Removal: Bioaccretion

The phosphorus content of accreting solids forms the basis for permanent P removal. Accretion of marsh sediments and soils has been discussed in Chapter 4, and is from two principal sources: (1) the sedimentation of incoming TSS entering with water inflows, and (2) production of solids within the wetland due to biological processes, termed bioaccretion.

The burial of plant and algal detrital residuals characterizes bioaccretion. The majority of the assimilated phosphorus is subsequently released during death and decay, but 10–20%

TABLE 10.6

Phosphorus accretion rates in marshes.

Location	Ref.	Method	Water Nutrient	Accretion, cm/yr	Phosphorus Burial, $gP/m^2 \cdot yr$
Bioaccretion					
Louisiana	1	Feldspar	L	0.14	0.36
Louisiana	1	Feldspar	H	1.14	3.03
Sacramento, CA	2	Visual	H	1.50	0.51
Houghton Lake, MI	3	Resurvey	H	1.33	1.50
Isolated, China	4	137Cs & 210Pb	L	0.49	1.11
Rodman Slough, CA	5	210Pb	L	0.46	1.24
STA1W-EAV, FL	6	Breakpoints	L	0.98	1.40
STA2C1-EAV, FL	6	Breakpoints	L	1.00	1.00
STA34C12A-EAV, FL	6	Breakpoints	L	1.70	2.55
STA1WC4-SAV, FL	6	Breakpoints	L	0.70	0.70
STA1WC5-SAV, FL	6	Breakpoints	L	1.00	2.00
STA2C3-SAV, FL	6	Breakpoints	L	1.20	1.20
STA34C12B-SAV, FL	6	Breakpoints	L	1.75	4.65
Klamath Hanks, CA	7	137Cs & 210Pb	M	0.28	0.27
Klamath Upper, CA	7	137Cs & 210Pb	M	0.55	0.57
Klamath Squaw, CA	7	137Cs & 210Pb	M	0.54	0.60
Median				**0.99**	**1.16**
Sedimentation					
Ogechee, GA	8	137Cs	M	4.80	0.68
Altamaha, GA	8	137Cs	M	4.70	0.70
Saltilla, GA	8	137Cs	M	4.90	0.67
Olentangy 1, OH	9	Visual	M	0.86	3.30
Olentangy 2, OH	9	Visual	M	1.02	3.50
Riverine, China	4	137Cs & 210Pb	M	0.32	0.95
Berg, Norway	10	Traps	M	2.30	40.00
Kinn, Norway	10	Traps	M	4.00	57.00
Böl, Sweden	11	Plates	M	0.47	8.40
Wig, Sweden	11	Plates	M	0.09	1.10
Gen, Sweden	11	Plates	M	0.72	17.50
Ber, Sweden	11	Plates	M	0.41	9.10
Ski, Sweden	11	Plates	M	0.13	2.50
Eks, Sweden	11	Plates	M	0.35	5.20
Lin, Sweden	11	Plates	M	0.23	2.90
Median				**0.72**	**3.30**

References
1 Rybczyk et al. (2002b)
2 Nolte (1998b)
3 Kadlec (2009a)
4 Wang et al. (2004) 5 Kim (2003)
6 Bhomia et al. (2015)
7 Graham et al. (2005)
8 Loomis and Craft (2010)
9 Anderson and Mitsch (2006)
10 Braskerud (2002b, 2003)
11 Johannesson et al. (2011)

is permanently stored as the residual from the decomposition process. The amount of such accretion has been quantified in a number of marshes, including treatment marshes Representative accretion rates are given in Table 10.6, determined using methods described in Chapter 4. Bioaccretion for example systems is 1.0 cm/yr (range 0.14–1.75), as determined by the various methods. The corresponding median P accretion rate is 1.16 gP/m^2·yr (range 0.36–4.65).

Mass balances may also be used to calculate P removal from the water passing through the marsh. Ideally, the amount of phosphorus removed from the water should equal the amount of P in accreted sediments. Such comparisons may only be made for time periods over which accretion is determined, typically many years. The cumulative errors in both the mass balances and in the accretion determinations preclude precise comparisons, but order-of-magnitude correspondence is to be expected. In some very few cases, rough equivalence has been determined.

Because the magnitude of the vegetative cycles that produce accreting residuals are strongly dependent upon the nutrient status of the surface water, treatment wetlands display gradients in accretion from inlet to outlet. These accretion gradients have been measured in only a few instances, notably for the treatment zone of WCA2A in Florida (Craft and Richardson, 1993; Reddy et al., 1993) (Figure 10.11). This gradient information helps explain the variability among wetlands in Table 10.6, for which surface water TP concentrations are generally not available over the period of record for accretion.

SUSTAINABLE PHOSPHORUS REMOVAL: PARTICULATE SEDIMENTATION

If large concentrations of suspended material enter a quiescent wetland environment, there can be high amounts of sediment trapping compared to generation and resuspension, and buildup typically occurs preferentially in the inlet section of the wetland. Therefore, a "delta" of accreted sediments may build in the inlet region of the wetland (Brueske and Barrett, 1994; Walker, 2001; Braskerud, 2003). Such inlet deltas are the repository for incoming particulate phosphorus.

Sedimentation basins may be used as the first process component for the purpose of providing storage for accumulated sediments. However, small runoff control wetlands are usually not

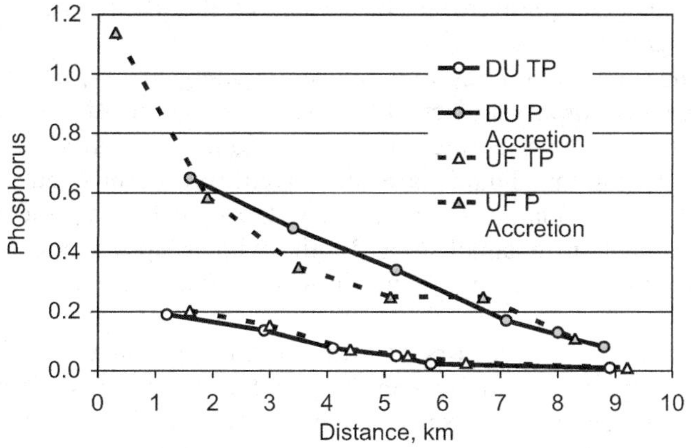

FIGURE 10.11 Phosphorus water concentration (mgP/L) and accretion rates (gP/m^2·yr) for phosphorus along the flow direction in the WCA2A treatment marsh. Duke University data (DU) from Craft and Richardson (1993), and University of Florida data (UF) from Reddy et al. (1993).

so equipped, and accretion occurs with the marsh. Representative phosphorus sedimentation rates are given in Table 10.6. Sedimentation for example systems is 0.7 cm/yr (range 0.09–4.9), as determined by the various methods. The corresponding median P accretion rate is 3.3 gP/m2·yr (range 0.67–57). The higher density of mineral sediments means that lower thicknesses can carry more phosphorus.

Settling of incoming particulate matter may be a significant process, especially in connection with crop runoff control. For example, studies on five constructed wetlands treating agricultural runoff in Norway, phosphorus retention was measured to be 26–71 $gP/m^2{\cdot}yr$ (Braskerud, 2002b). Mineral particulate phosphorus formed the large majority of these removals. Particulate phosphorus removal for incoming organic matter has been observed to be considerably less effective. The character of the particulate matter is a strong function of the hydrology and soils of the watershed that produces the runoff (Stuck et al., 2001). For peat-based farmlands in south Florida, the majority of runoff particulate phosphorus has been found to originate in the drainage canals as a result of biological growth. Conversely, the mineral soils of the Imperial Valley of California, and those of Norway, produce dense mineral particulates during rainstorms.

One might expect that sedimentation would preferentially remove particulate P, as opposed to dissolved P. But data indicate this is not necessarily the case. Kadlec and Wallace (2009) report that for 13 wetlands, median removal of particulate P was 44%, while median removal of total P was 45%.

Herbivory and Phosphorus Movement

Vegetation consumers occur at several size scales, ranging from microinvertebrates to large animals. Consumption for the phosphorus mass balances at the Delta Marsh wetlands (Figure 10.10) was considered due primarily to amphipods, gastropods, chironomids and odonates (Robinson et al., 2000), although there was also muskrat herbivory (Clark, 2000). Snails (*Lymnaea* spp.) were found to be the most abundant grazer in a Carex-Equisetum marsh near Fairbanks, Alaska (Rober et al., 2011). Snails also impacted the phosphorus removal in STA1E in Florida (Chimney, 2015). Cell 4S in STA1E experienced an enormous population increase of the island apple snail (*Pomacea maculata*) between July and August 2013. Island apple snail feeding defoliated most of the SAV, which led to the cell exporting TP from the cell. Within the treatment zone of WCA2A, slough crayfish (*Procambarus fallax*) are the dominant herbivores in the vegetated areas (*Typha*), occurring at densities of 19 animals per square meter (Hagerthey et al., 2014). They comprise a standing stock of about 70 mgP/m^2.

Fish that influence marsh ecology and nutrient cycling are typically cyprinids, such as common carp (*Cyprinus carpio*) and chichlids, such as tilapia (*Oreochromis aureus*). Most cyprinids feed mainly on invertebrates and vegetation. Some are specialized herbivores; others eat algae and biofilms. Tilapia are almost entirely herbivorous, feeding mainly on phytoplankton and aquatic plants. Unfortunately, tilapia eat algae and plants much faster than they can grow back in a small area. In the wild, tilapia schools graze over several miles. Both carp and tilapia have been found in warm climate treatment marshes, sometimes at very high densities.

Larger animals can serve as vectors of nutrient transport in wetlands, both by moving nutrients between ecosystems, and by recycling within ecosystems (Frederick and Powell, 1994). Even insect emergence has been considered (Neame, 1976) (12 mg $P/m^2{\cdot}yr$). However, the greatest potential for influence comes from animals (muskrats, nutria), fish (carp, tilapia) and birds (ducks, geese, swans). Muskrats conduct destructive herbivory in temperate marshes (Kadlec et al., 2006). The removal of standing crop by muskrats (*Ondatra zibethicus*) during an unfrozen season in southern Manitoba, Canada was estimated to be about 82 g dry biomass per kg of body mass per day, or per animal, since an adult muskrat weighs on the order of one kg (Clark, 2000). As a consequence, Clark (2000) concluded that muskrats contributed to observed vegetation declines at nontreatment wetlands at Delta, Manitoba, but were not the major factor in nutrient movement. However, it then becomes clear that densities of twenty or more muskrats per hectare may utilize the majority of the macrophyte standing crop in a given year. At such an exacerbated scale, muskrat herbivory may be termed eatout, and will be evidenced by the removal of essentially all emergent plant parts. The attendant movement of phosphorus at 0.2% dry weight would be 164 mg per animal per day, or 60 g phosphorus per animal per year. At a population density of 20 per ha, phosphorus movement would be 0.12 $gP/m^2 \cdot yr$. Nutria (*Myocastor coypus*) have been similarly implicated in the destruction of the vegetative cycle in treatment wetlands (Mokry, 2010; Shafer et al., 2015).

Birds also have the ability to move phosphorus, especially when large colonies or migratory flocks are involved. Phosphorus loading rates in historical colonies of wading birds in the Florida Everglades have been estimated at over 100 $gP/m^2 \cdot yr$ (Frederick and Powell, 1994). More importantly, migratory waterfowl have been observed to have profound effects on Phragmites stands in the Netherlands (van Den Wyngaert et al., 2003; van Den Wyngaert and Bobbink, 2009). Goose (*Anser anser*) densities of 6–10 per hectare lowered the standing crop of leaves and rhizomes by about 50% across a 3,600 ha wetland, but did not destroy the stands. The amount of phosphorus moved was approximately 0.8 gP/m^2 during the summer. Submerged macrophytes, such as *Potamogeton pectinatus*, are attractive sources of food to migratory waterfowl, such as swans (*Cygnus* spp.) (Jonzen et al., 2002). However, resident waterfowl populations on open water or SAV systems have a year-round effect in warm temperate climates. Scherer et al. (1995) found loadings of about 0.155 $gP/m^2 \cdot yr$ due primarily to coots (*Fulica americana*) and mallard ducks (*Anas platyrhynchos*) in a lake in the State of Washington.

The potential impacts of a water bird population may be roughly estimated from their numbers and the production of fecal matter. Ducks produce 17–27 g dry weight per day, coots 13, and geese 82 (Scherer et al., 1995). The phosphorus content is approximately 2% dry weight. It may easily be shown that large flocks of birds may influence the ability of treatment wetlands to achieve ultra-low phosphorus concentrations, but will not affect treatment at higher concentrations.

The effects of herbivory are currently under study in the Stormwater Treatment Areas in Florida. Preliminary results show very high densities of vertebrate consumers in the treatment marshes.

Plant Harvest to Remove Phosphorus

Harvesting is often labor intensive and costly, which is antithetical to the desired passive character treatment marshes. It is disruptive to the wetland wildlife, and to the cycling mechanisms that operate in the system. Nonetheless, the prospect of deriving added values from the wetland, and the potential for enhancing P removal, mean that much attention has been directed to harvesting as a management option. Vegetation of all types (EAV, SAV, FAV) are able to withstand cropping, and regrow readily. The exception is that EAV stubble cannot be totally inundated, because plants will then drown.

The literature contains conflicting "recommendations" concerning the potential for phosphorus removal by harvesting. Crites and Tchobanoglous (1998) conclude that: "Harvesting for nutrient removal is not practical and is not recommended." Wang and Mitsch (2000) conclude the opposite: "Harvesting macrophytes has the potential to enhance phosphorus removal in wetlands ... On average, harvesting would remove the equivalence of three quarters of the phosphorus inflow ..." Vymazal (2004) strikes the middle ground, and concludes that: "... the amount of phosphorus removed via harvesting is usually low. However, for tertiary systems with inflow < 20 g $P/m^2 \cdot yr$ this route could be significant." Vymazal and Kropfelova (2008) go on to say that the growth patterns in tropical and subtropical climates could provide expanded opportunities for harvest, compared to operations in temperate climates with marked seasonality. Zhou et al. (2017) speculated on the harvest potential of eight aquatic macrophytes found in central Florida, and concluded that cattail, water lettuce and Hydrilla had the largest potential for harvest removal, on the order of 0.8–1.0 $gP/m^2 \cdot yr$. These are speculations, because there has been no full-scale project that uses harvest for P removal. However, there are full-scale projects that target biomass harvest, and some of those report the P content of harvested biomass (Grosshans et al., 2014; Wichmann, 2017; Carson et al., 2018).

There is a theoretical potential for plant harvesting to perform part of the phosphorus removal associated with the stormwater treatment areas (STAs). However, there are very serious issues with the practicality, efficacy, and economics of such a procedure. These considerations depend strongly on the wetland vegetation type, which may be broadly separated into categories of emergent aquatic vegetation (EAV) (rooted plants), submerged aquatic vegetation (SAV), and floating aquatic vegetation (FAV).

Harvest involves cutting and removing only the aboveground plant parts. For EAV, this may be done while the plants are in their growth phase, meaning that green biomass is removed; or alternatively may be done when the plants are senesced, meaning that brown standing dead necromass is removed. For FAV and SAV, removal is of biomass. In the absence of harvest, only a small fraction of the aboveground tissue phosphorus is sequestered as buried residual in new soils and sediments. Thus harvest has the potential to provide much higher rates of P reduction than bioaccretion.

The problem is then: what to do with the harvested plant material? Several options are available, including composting, use as biomass fuel, biogas generation, and use as roof thatching. Composting requires transportation costs, and dedicated land area. Biogas generation has a sludge disposal problem. The capital cost of harvesting and gas generation is often prohibitively expensive (Chynoweth, 1987). Biomass or necromass may be dried and used as fuel. Harvesting, transportation and drying are again relatively expensive, but have been demonstrated to be practical at large scale for Typha (Grosshans et al., 2014). Phragmites is useable as roof thatching, which is a higher value product, albeit with a limited and regional market (Wichmann, 2017).

Yet another possibility is the use of biomass as a soil amendment, either within or adjacent to the treatment wetland. This has been suggested for *Typha* in the Great Lakes region of the USA (Carson et al., 2018). It has been implemented at full scale for hyacinth treatment wetlands in Florida, at several sites northeast and southwest of Lake Okeechobee. The approach is known as Floating Aquatic Vegetation with Tilling (FAVT)(Catts, 2009). FAVT systems operate as follows:

1. The FAV wetland is operated for an initial growing season, during which time the FAV assimilate nutrients and grow to a high density;
2. The wetland is drained during the dry season, thereby stranding the FAV on the soil;
3. After a natural drying process, the plant material is tilled into the soil;
4. The wetland is reflooded; and
5. Aquatic plants that remain in the deepest zones during tilling repopulate the wetland for the subsequent growth period.

Thereby, the steps of harvest, transportation, processing and marketing or disposal are avoided. These are replaced by the standard agricultural practice of tilling, or soil inversion. Data are not yet published from these FAVT projects.

EAV Harvest

Example phosphorus removals for EAV systems are shown in Table 10.7. Removals have often been relatively small, even in long-detention wetlands, with a median value of 1.52 $gP/m^2 \cdot yr$.

TABLE 10.7

Phosphorus removed by harvesting.

System	Location	Vegetation	Harvested $gP/m^2 \cdot yr$	Reference
Wellington	Florida	FAV	4.48	CH2M Hill, Inc., 2003a
De Meije	Netherlands	FAV	1.41	Mueleman et al., 2002
Montevideo	Uruguay	FAV	5.97	Rodríguez-Gallego et al., 2004
Median			**4.48**	
Listowel 1	Ontario	EAV-CT	1.38	Herskowitz, 1986
Listowel 2	Ontario	EAV-CT	1.69	Herskowitz, 1986
Listowel 3	Ontario	EAV-CT	0.18	Herskowitz, 1986
Listowel 4	Ontario	EAV-CT	0.37	Herskowitz, 1986
Listowe 51	Ontario	EAV-CT	0.78	Herskowitz, 1986
Malham	UK	EAV-Phrag	10.97	Hurry and Bellinger, 1990
Lauwersoog	Netherlands	EAV-Phrag	0.60	Mueleman et al., 2002
Texel	Netherlands	EAV-Phrag	1.66	Toet et al., 2005
Texel	Netherlands	EAV-Phrag	2.32	Toet et al., 2005
Texel	Netherlands	EAV-Phrag	1.11	Toet et al., 2005
Texel	Netherlands	EAV-Phrag	1.29	Toet et al., 2005
Texel	Netherlands	EAV-CT	4.66	Toet et al., 2005
Texel	Netherlands	EAV-CT	3.89	Toet et al., 2005
Texel	Netherlands	EAV-CT	3.92	Toet et al., 2005
Texel	Netherlands	EAV-CT	2.61	Toet et al., 2005
Dianchi	China	EAV	14.38	Lu et al., 2009, 2010
Dianchi	China	EAV	6.89	Lu et al., 2009, 2010
Pelly's Lake	Manitoba	EAV-CT	0.26	Grosshans et al., 2014
Pelly's Lake	Manitoba	EAV-CT	0.37	Grosshans et al., 2014
Weihe	China	EAV	11.22	Zheng et al., 2015
Mecklenburg	Germany	EAV-Phrag	0.45	Wichmann, 2017
Mecklenburg	Germany	EAV-Phrag	0.72	Wichmann, 2017
Median			**1.52**	
Bemmel 1X	Netherlands	SAV	0.79	Verhofstad et al., 2017
Bemmel 2X	Netherlands	SAV	1.02	Verhofstad et al., 2017
Bemmel 3X	Netherlands	SAV	0.89	Verhofstad et al., 2017
Bemmel 5X	Netherlands	SAV	0.74	Verhofstad et al., 2017
St. Paul	Minnesota	SAV	0.33	Bartodziej et al., 2017
Median			**0.79**	

Toet et al. (2005) harvested *Phragmites australis* and *Typha latifolia* in four wetlands of each species, located at Texel, Netherlands, and each receiving different P loadings. These were done on clip plots at eight locations in the wetlands. The autumn harvests were relatively uniform across loadings, and removed 1.60 ± 0.54 gP/(m^2·yr) in the *Phragmites* systems, and 3.78 ± 0.54 gP/(m^2·yr) in the *Typha* systems. This represented more than 80% of the seasonal maximum total aboveground standing stock. The once per year harvests did not affect the vitality of the vegetation.

Similar results were reported by Herskowitz (1986), who reported a four-year average removal of 3.3% of the total P loading to FWS cattail wetlands at Listowel, Ontario by single annual harvest of above-ground plant parts. In absolute terms, the P harvested was 0.75 ± 0.57 gP/m^2·yr, compared to the incoming phosphorus loading of 22 ± 11 57 gP/m^2·yr. There was also an adjacent comparison wetland that was not harvested. Although it received the same water, loadings were different and the wetland geometry was different. But the four-year average P settling rate coefficients nearly identical, 11.6 m/yr for the harvested system and 11.1 m/yr for the unharvested system. In contrast to the Texel study, attempts were made at Listowel to mechanically harvest the entire 1.3 ha wetland. The channels were 4 m wide, and a modified agricultural grain harvester was used. Considerable difficulties were experienced.

There presently are several types of mechanical harvesters available for use in wetlands (Wichmann, 2017). Standard agricultural mowers and balers have been successfully used in Manitoba (Grosshans et al., 2014), but these require dry conditions. There are also tracked vehicle harvesters that can operate at up to 60 cm water depth, such as the Softrak amphibious tractor (Loglogic, Devon, UK) (Carson et al., 2018).

In subtropical or tropical climates, growth is more luxurious, and larger crops are harvestable. Lu et al. (2009, 2010) applied the same techniques to small (2,800 m^2) constructed wetlands containing *Phragmites australis* and *Zizania caduciflora* (Manchurian wild rice), and receiving agricultural runoff at 8.6 cm/d and TP = 0.87 mg/L. About half the annual loading of 27 gP/m^2·yr was found to be in the standing crop of about 4,000 gdw/m^2, and 36% of the annual load was removed by harvesting. The method of vegetation removal was local farmers hired to cut and carry away the vegetation.

FAV Harvest

Floating plants have been intensively investigated as a means removing nutrients via harvest (DeBusk and Reddy, 1987). Thirty to forty years ago, there a number of floating plant systems implemented for wastewater treatment. The Water Environment Federation listed 18 facilities, most dating to the 1970s, (Water Environment Federation, 1990); but most of those are since abandoned. Five hyacinth facilities operated in Florida in the late 1980s (Stewart et al., 1987). Floating aquatic plants are somewhat easier to harvest than rooted emergents such as cattails. Phosphorus removal is documented for such systems. Over 20% of the total P removal was achieved by water hyacinths (*Eichhornia crassipes*) (Fisher and Reddy, 1987). Better results were obtained by Rodriguez-Gallego et al. (2004), with 39–78% of the water column P removed by water hyacinths (*Eichhornia crassipes*), water lettuce (*Pistia stratiotes*), duckweed (*Spirodela intermedia*).

Example phosphorus removals for FAV systems are shown in Table 10.7. Removals were substantial, with a median value of 4.48 gP/m^2·yr.

Harvested hyacinth systems have been specifically tested in the context of Everglades water quality improvement. The Village of Wellington, near to STA1E, conducted a pilot project in which a hyacinth system was used as a precursor to a periphyton wetland. The contractor, CH2M Hill (2003a) found that *Eichhornia* was effective at reducing phosphorus in the 0.1–0.6 mg/L range, with an average removal of 62% ascribable to wetland processes. However, only 8.6% of the inlet load was contained in harvest on a monthly schedule, which amounted to 15% of the removed load. Harvest garnered approximately 4 gP/m^2·yr.

The added complexities of biomass utilization are nontrivial. A brief description of the procedures used at the San Diego facility is informative (Crites and Tchobanoglous, 1998). The hyacinths are removed using a long-armed crane and a clamshell bucket. The channels are narrow enough that the crane reaches the entire width of the system. The hyacinths are then trucked to a grinder, preparatory to drying. The chopped biomass is spread in a thin layer, and dried to about 60%, which takes five days. The biomass is then windrowed into long piles for composting. In an additional 15 days, the remaining water is lost, at compost temperatures up to 160°F (70°C). The final destination of the compost is unspecified.

SAV Harvest

The SAV vegetation type is the most susceptible to harvest damage. Typically, only the top portions of the vertical structure are removed, permitting regrow from the remaining bottom parts. This procedure removes biomass and accompanying phosphorus, but also impairs the nutrient cycling function of the plants. Kuiper et al. (2017) suggest that this may be "battling the good guys."

Example phosphorus removals for SAV systems are shown in Table 10.7. Removals are substantial, with a median value of 0.79 gP/m^2·yr.

SAV harvest requires floating harvesters (boats with cutting and loading equipment). These are available because of the desires of lake property owners to rid their lake of submergents for aesthetic reasons. As an example, Casey Lake near St. Paul, Minnesota used a paddlewheel harvester with a cutting swath of 1.5 m and the harvesting depth set at 0.3 m was used at Casey Lake (the boat draft was 0.5 m with a full load). The mean depth of this "lake" was 60 cm. Harvested plant material was hauled off site to a local public works yard for composting.

Verhofstad et al. (2017) used two compartmentalized basins to measure P removal for *Myriophyllum spicatum* at different harvesting frequencies, from one to five times during the growing season at Bemmel, Netherlands. Manual removal was employed. Interestingly, the loading to the ponds was 0.48 gP/m^2·yr, while removals by harvest were 0.74–1.02 gP/m^2·yr. This indicates that harvesting used more than the added P load, and depleted the soil phosphorus in the system. This possibility has led to speculation that harvest can be used to remove stored phosphorus, rather than just current loads to the wetland (Carson et al., 2018).

Vertical Phosphorus Movement

Macrophyte Mining

Rooted macrophytes extract nutrients from the soil (porewater), primarily the top 20 cm, but adventitious roots sometimes utilize nutrients taken from the bottom of the water column, in the floc layer. The aboveground plant parts seasonally die and create litter, which ultimately decomposes to release some of the structural phosphorus back to the top layer of sediments and to the water column (see Figures 10.1 and 10.10). The overall effect is the removal of phosphorus from underground storages, redeposition of phosphorus on the top layer of sediments. This process has been termed phosphorus mining (White et al., 2006). This upward transfer is counteracted by at least three very important processes: downward advective flow or seepage, the transpiration phosphorus flux, and translocation of phosphorus from senescing leaves to the rhizomes.

The Transpiration Flux

Examination of Figure 10.8 reveals the driving force of vertical phosphorus transport potential: the dissolved phosphorus gradient is typically decreasing upward in the soil porewater. Thus we would expect phosphorus exports from the soil to the overlying water if diffusion were the only operative factor – which is contrary to the overwhelmingly prevalent case of phosphorus removal from the water column. One obvious mechanism that can overcome this upward diffusional driving force is vertical downflow of water into the root zone.

As noted in Chapter 4, vertical flows of water in the upper soil horizon are driven both by gravity and by plant uptake to support transpiration. In an aquatic system, without emergent transpiring plant parts, vertical downflow will be driven solely by gravity, if there is infiltration out of the wetland system. These flows move phosphorus from the overlying water downward into the root zone.

In aquatic and wetland systems with fully saturated soils or free surface water, the meteorological energy budget requires the vaporization of an amount of water sufficient to balance solar radiation and convective losses (refer to Chapter 6). Some of this vaporization is from the water surface (evaporation); some is from the emergent plants (transpiration). Emergent plants "pump" water from the root zone to the leaves, from which water evaporates through stomata, which constitutes the transpiration loss. In a densely vegetated wetland, transpiration dominates the combined process, termed evapotranspiration (ET). Water for transpiration must move through the soil to the roots. That movement is vertically downward from overlying waters in most wetland situations. In temperate climates, ET ranges from 60 to 200 cm/yr, but is concentrated in the part of the year with greatest solar radiation.

Thus, transpiration has the potential to move on the order of one meter per year of water vertically downward to the root zone. That water carries with it the phosphorus concentration associated with the bottom layer of overlying water. Porewater phosphorus concentrations near the soil surface are of the order of 200 μgP/L for systems treating agricultural runoff, for example (see Figure 4.12). The vertical transpiration flux of phosphorus due to one meter per year of transpiration is therefore 1.0 m/yr × 0.2 g/m^3 = 0.2 gP/m^2·yr. This supply contributes to plant uptake, and/or to storage in porewater and sorption. Usually, this transpiration P flux is greater than estimated diffusion fluxes (Kadlec, 1999b).

Detailed flow models, describing the vertical transpiration-driven flow of porewater, have been developed for terrestrial agricultural crops (Schoups and Hopmans, 2002; Zhu et al., 2010b). These utilize porous media flow equations with spatially dependent withdrawals of water to satisfy transpiration. Withdrawals are proportioned to the root density at a particular vertical position. Typically, a downward-decreasing root density is presumed to result in a downward-decreasing water uptake to supply transpiration. The transpiration-driven flow is presumed to carry phosphorus from porewater into the plants via the roots. Phosphorus uptake may be most simply represented by:

$$RPU = \delta \cdot RWU \cdot C_P \tag{10.3}$$

where: C_P = porewater phosphorus concentration, gP/m^3

δ = RPU factor, –

RPU = root phosphorus uptake, gP/m^3·d

RWU = root water uptake, m^3/m^3·d

The linearity of P uptake with transpiration (Equation 10.3) has been demonstrated for nitrogen and phosphorus for a variety of agricultural plants such as corn (*Zea maize*) and wheat (*Triticum aestivum*) (Schoups and Hopmans, 2002; Novak and Vidovic, 2003; Shi et al., 2007; Zhu et al., 2010b), as well as for ammonia for wetland plants (*Juncus effusus*, Wiessner et al., 2013). The RPU factor, also known as the transpiration stream concentration factor for phosphorus, has not been determined for any wetland plants. Based upon results for terrestrial crops, the value of δ for phosphorus is likely higher than unity, perhaps by a considerable margin (Russell and Shorrocks, 1959). The principal role of the transpiration flow is likely to bring P to the root surface, where plant processes extract the amount needed for growth.

Subsequent to arrival at the root surface, nutrients are subject to enhanced transport, because of transport facilitation in the roots. It is unlikely, based upon results for terrestrial crops, that the transpiration stream concentration factor for phosphorus is a constant; or that Equation 10.1 is sufficient to describe P uptake. Accordingly, early research efforts found that a Michaelis-Menten formulation was better, and is still in use today for some purposes (Silberbush and Barber, 1983; Hopkins and Bristow, 2002). This formulation results from consideration of active, facilitated transport into the root, which responds to porewater concentration in a nonlinear way, but does not depend upon the transpiration water flow:

$$RPU = RPU_{\max} \cdot \frac{C_P}{K_m + C_P} \tag{10.4}$$

where: K_m = half saturation P concentration, gP/m^3

RPU_{max} = maximum root phosphorus uptake, $gP/m^3 \cdot d$

This formulation has not been calibrated for wetland plants.

Recent wetland models have combined passive and active transport (Equations 10.3 and 10.4), with partitioning coefficients for the two mechanisms (Šimůnek and Hopmans, 2009), but again, there are no data for wetland plants.

Diffusion in the Soil Matrix

The presence of the soil matrix prevents convection currents, and therefore the diffusive process is restricted to molecular diffusion in the porespace. This process is described, for stagnant water, by Fick's Law:

$$J_D = -\phi D \frac{dC}{dz} \tag{10.5}$$

where C = P concentration, gP/m^3

D = diffusion coefficient, m^2/d

J_D = vertical diffusion flux downward, $gP/m^2 \cdot d$

z = vertical distance downward, m

ϕ = matrix blockage coefficient, –

When the P concentrations increase downward, the concentration gradient $dC/dz > 0$, and the diffusion flow is upward.

The value of the diffusion coefficient in pure water is of the order of 7.3×10^{-5} m^2/d at 25°C, for H2PO4-, which is the dominant inorganic form of phosphate at low to circumneutral pH (Lide, 1992). Values in the soil porewater are likely to be lower because of a tortuosity effect, which has been estimated to be about four times lower (Fisher and Reddy, 2001); and because of the matrix blockage coefficient, which is on the order of 0.8. In general, concentration gradients are variable in both magnitude and direction.

Equation 10.5 has been used extensively to estimate vertical P movement in wetland soils (e.g., Fisher and Reddy, 2001; Reddy and Delaune, 2008; Palmer-Felgate et al., 2011). However, it is correct only in the absence of any vertical flow, and the majority of treatment marshes will have either gravity-driven flow (seepage) or transpiration-driven flow.

For the more general case, the vertical flux of phosphorus is written (Bird et al., 1960):

$$J = uC + J_D = uC - \phi D \frac{dC}{dz} \tag{10.6}$$

where u = superficial velocity downward, m/d

Thus, a downward advection of water and its P content opposes the upward diffusion of P.

Some idea of the expected magnitudes of vertical P diffusion and P advection may be gained by examining the situation of mildly eutrophic surface waters overlying a fully saturated peatland. In Figure 4.12, for the STA2C3 curve, the porewater P gradient is about 4.0 (gP/m^3)/m in the top 25 cm. Under these circumstances, the diffusion flux predicted by equation (10.5) is:

$$J_D = -\frac{(0.8)(7.3 \times 10^{-5})}{4}(4) = 0.058\, mgP/m^2.d \tag{10.7}$$

The P concentration at the top of the soil column was 120 $\mu gP/m^3$. If the vertical water flow is 1.0 mm/d, then the vertical advective flux is uC = 0.12 $mgP/m^2{\cdot}d$. Thus a very small downward water flow forms a hydraulic barrier to P diffusion, and there is no diffusion loss from porewater to wetland surface water.

At a steady state condition, the mass balance requires dJ/dz = 0. Solution for the concentration profile, using Equation 10.6, produces the vertical P concentration profile, which is slightly curved as opposed to straight for the case of no advective flow.

It is concluded that the vertical movement of P is likely to be downward into the root zone, driven by vertical flow from the water column into the root zone, for marshes with seepage or rooted vegetation. Diffusional transport in the soil will likely be small in comparison.

Translocation

During the late summer to early autumn period, phosphorus is in part translocated to belowground plant rhizomes. For instance, Tanner (2001a) found 36% of the end of season standing crop loss of phosphorus in *Schoenoplectus* (*Scirpus*) *tabernaemontani* translocated to rhizomes at light loading (0.20 g $P/m^2{\cdot}d$), but only 13% at higher loading (0.90 g $P/m^2{\cdot}d$). This is in contradiction to Mueleman et al. (2002), who found 58% translocated in a *Phragmites* wastewater wetland, compared to 25% in a natural *Phragmites* wetland. Hocking (1989a) found 45% translocation in a nutrient enriched *Phragmites* wetland. Smith et al. (1988) found 39% translocation in natural Typha latifolia. Mixed stands (*Phragmites, Schoenoplectus, Typha* and *Scolochloa*) exhibited 29–36% translocation during a five-year study of ten low nutrient status marshes in Manitoba (Murkin et al., 2000). The central tendency of these results is the translocation of about one-third of the phosphorus standing aboveground crop to belowground storage each autumn in temperate climates.

During the spring growth spurt, phosphorus may be withdrawn from storage in the belowground biomass. The amount has been estimated to be up to 100% of the early growth needs of the plant (Bernard, 1999). It seems probable that the total return cannot exceed the excess stored in autumn, else the plant system would be unstable.

Soil Phosphorus Release

Many papers in the literature address what is called Phosphorus Release Potential for wetland soils (Fisher and Reddy, 2001; Pant and Reddy, 2001b; Malecki et al., 2004). The general procedure is to expose soil cores to a layer of phosphorus-free water, and then to measure either or both of the soil porewater phosphorus gradient and the time rate of change of the phosphorus content of the overlying water. These are typically abiotic laboratory experiments, in which plants are absent. Therefore, there is no vertical water flow and the experimental durations are relatively short (on the order of weeks).

Such experiments do place a limit on how much labile phosphorus is in a particular soil, and therefore how much might be "rinsed out" by exposure to low phosphorus floodwaters. However, it is dangerous to extrapolate the results to full-scale field situations. One need look no further than what may be termed the "removal paradox." For instance, the lab results of Fisher and Reddy (2001) indicate phosphorus release along virtually the entire gradient of a de facto treatment wetland, WCA2A of the Florida Everglades, ranging from 0.47 to 1.83 $gP/m^2 \cdot yr$ during ten days exposure to 25-μgP/L floodwater. However, as water flows through this wetland, phosphorus is stripped from the water. Entering concentrations are about 80 μgP/L, and these decrease to about 15 μg/L at downstream locations. Removals range from about 1.2 $gP/m^2 \cdot yr$ near the inlet, to about 0.2 g P/m2·yr at downstream locations. So the apparent paradox is simply that the ecosystem is taking up phosphorus (and has for over 30 years), but the core experiments show release back to overlying water.

Dryout and Re-Wetting

Continuous-flow FWS treatment wetlands maintain standing water over the entire year, which allows the maximal rates of accretion (Table 10.6). This distinguishes them from natural wetlands and stormwater wetlands, which often have hydroperiods of less than 100%. Treatment marshes treating runoff often do not receive a continuous inflow, and can be dry during significant fractions of the year. A period of water loss can be a drawdown, in which overlying water is drained, but moist soil conditions continue to prevail; or it may be a dryout, in which the soils become desiccated. Oxygen reaches the soil in either drawdown or dryout, thus providing opportunities for rearrangement of the forms of phosphorus in the soil. Dryout causes sediment oxidation and subsequent mobilization and export of sediment-bound pollutants. In some climatological regions, extended dry periods, coupled with seepage, can drop water levels in the permanent pool, and jeopardize the integrity of the wetland ecosystem. Oxidation may also cause a decrease in the sorption potential for phosphorus (Baldwin, 1996). There may be decreased phosphorus affinity due to the aging of the oxy-hydroxides of iron (Watts, 2000).

Olila et al. (1997) performed laboratory core experiments on sediments from the Lake Apopka, Florida treatment wetland, and showed that newly accreted floc materials underwent rapid mineralization during drawdown. The resulting inorganic phosphorus was released to floodwaters upon rewetting of the cores. Similar experiments in WCA2A of the Everglades showed that reflooding caused phosphorus releases intermediate between anoxic and aerobic conditions (Fisher and Reddy, 2001). The pulse of phosphorus is more rapid if the reflood occurs via underground flows (Corstanje and Reddy, 2004). When detrital material from STA1W C1, Florida was subjected to dryout and reflooding, only a few percent of the stored phosphorus was mineralized and mobilized (Pant and Reddy, 2001a).

The presence of an algal community in the desiccated material fosters a much more rapid release and recovery. Thomas et al. (2006) found that rewetting dry periphyton mats led to an initial release, for a period of about one day. However, regrowth of the biological communities led to re-absorption of that phosphorus in the ensuing few days.

There is ample evidence of the release after dryout in the performance records of the large Florida STA treatment wetlands, and in attendant mesocosm studies (White et al., 2006; Bostic and White, 2007; Moustafa et al., 2011; Dierberg et al., 2012). Figure 10.12 shows the fifth year of operation for STA6, one of the south Florida P-removal treatment wetlands. The very first waters to enter the system after the dry season produce a surge of phosphorus in the water, which declines as the wetland refills, prior to any outflow. This batch-filling mode produces treatment, and the mobilized phosphorus is essentially removed before outflow resumes about a month later.

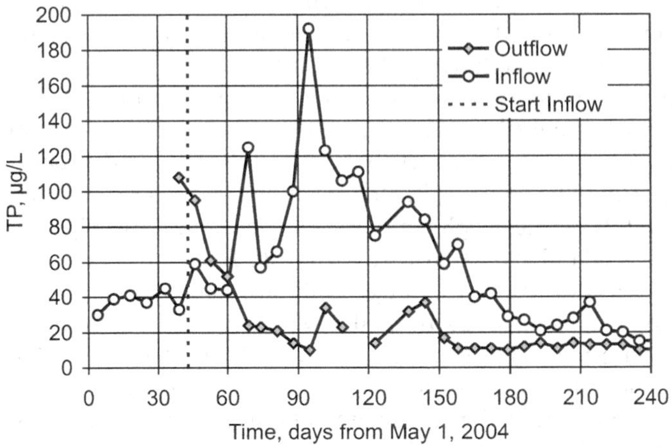

FIGURE 10.12 The behavior of STA6, Florida over a seven-month period. There is little inflow and no outflow in the spring dry season, but inflows commence in June and rehydration is complete and outflows begin in July. The flow-weighted mean outflow concentration was 16.5 µgP/L during the year, and the flow-weighted mean inflow concentration was 53.4 µgP/L. Note that the re-wetting water peaked at 110 µgP/L, but rapidly subsided over a six-week period to below 20 µgP/L. Data source: South Florida Water Management District unpublished data.

Of course, the ultimate in oxidation of phosphorus storage in wetlands is the dryout and burning of the soils and sediments. Such burns may be surficial, affecting only the top few centimeters of material; or deep burns, also called peat fires. Both types of fire have the effect of converting organic forms of phosphorus to mineral forms, in large proportions (Smith et al., 2001). The mineral forms are conducive to resolubilization on rewetting, and potential export if there is flow through the wetland. The WCA2A treatment marsh in south Florida has been subjected to burning in an attempt to change the vegetation from cattails back to vegetation characteristic of unimpacted Everglades (Miao et al. 2009, 2010; Xu et al., 2011). These burns were of surficial plant material, and not of the underlying peat and litter. Nonetheless, sudden increases in water column phosphorus resulted, but were temporary.

Accretion in some natural wetlands leads to the elevation of the wetland soil surface, which may in some circumstances reduce the hydroperiod of the wetland. If unchecked, such accretion eventually can reach a stage in which the oxidation processes during the dry periods remove the accretions formed when the system is wet. This situation is one of hydrologic burnout, in which no further net phosphorus accretion occurs, and the system stores no more phosphorus. The wetland may then alternate between storage and release during the wet and dry seasons of the year. This phenomenon, present in mature natural wetlands, has created the impression that wetlands are sometimes phosphorus sinks and sometimes phosphorus sources. However, continuously wet treatment wetlands should not be included in such a generalization.

Atmospheric Exchanges

There are atmospheric processes that involve phosphorus transfers in wetlands, but they are not normally of sufficient magnitude to warrant consideration in design. Such processes are

TABLE 10.8
Atmospheric deposition of phosphorus (mgP/m^2·yr).

	No. sites TP	PO4-P	TDP	TP
North America	38	7	22	32
South & Central America	5	2	26	36
Europe	54	13	23	22
Oceania	6	24	19	32
Africa	10	68		69
Asia	7	5	30	17
All	120	14	19	27

Source: Tipping et al. (2014).

briefly discussed here to alert practitioners to potential areas of concern in exceptional situations.

Atmospheric Deposition

Atmospheric deposition of phosphorus contributes very small but measurable quantities to receiving land areas. All forms are involved: particulate and dissolved, inorganic and organic, wetfall and dryfall. Annual total atmospheric phosphorus loadings are 17–69 mgP/m^2·yr (Table 10.8). Consequently, atmospheric sources are almost always a negligible contribution to the wetland phosphorus budget for all but ombrotrophic, nontreatment wetlands.

Phosphorus Volatilization

Gaseous forms of phosphorus, phosphine (PH_3), and diphosphine (P_2H_4), are known to exist. These are in some sense analogous to hydrogen sulfide, in that they form in low redox conditions, are volatile, and are extremely toxic. Phosphine is soluble in water, but has a high vapor pressure. It may be emitted from regions of extremely low redox potential, together with methane. It has been reported that the methane-phosphine mixture emitted from marshes and bogs can auto-ignite, forming the flickering lights known as "Will-o'-the-wisp". In the early 20th century, there were reports of failure to find these compounds, which are now thought to be the result of flawed analytical procedures. Burford and Bremner (1972) found that formation of phosphine could not be excluded on thermodynamic grounds.

Spurred by the continuing controversy over the existence and/or importance of phosphine in the biogeochemical cycle of phosphorus, Devai and Delaune (1995) measured 16 pgPH_3/kg·d wet sediment for Everglades soils at 73 mgP/kg wet soil. However, when the Everglades, Florida soil was amended to 2,200 mgP/kg wet soil, the PH_3 emission rate increased to 380 pgPH_3/kg·d wet sediment. For the top 30 cm of the Everglades peat, this is a loss of 38 mgP/m^2·yr – which is about double the median rate of atmospheric deposition. Han et al. (2000) also found that most PH_3 in rice paddy fields was matrix bound, and that emission rates averaged 16 μgP/m^2·yr.

To date, wetland phosphorus mass balance studies have ignored the possibility of a significant phosphine production or loss. In the context of treatment marshes, redox potentials are not especially low, and the contribution of phosphine to treatment marsh P budgets is very likely negligible.

10.3 SPATIAL AND TEMPORAL PHENOMENA

This section is devoted to examination of internal gradients of phosphorus concentrations in various portions of the wetland, and to the transient effects that accompany the changing storages of phosphorus within a treatment marsh.

Phosphorus Gradients

As for other pollutants that are reduced in treatment wetlands, there typically exists a decreasing profile of water concentrations from inlet to outlet. However, one must be very careful about interpretation of measurements along the direction of flow. It is well known that at any specific distance from the inflow point, there is a distribution of flows and concentrations across the width of the system (perpendicular to the flow direction). For example, five parallel P-transects were measured in STA2C3 on twenty separate occasions. The cross-flow variation, measured at five positions along the flow direction, had a coefficient of variation of 25%. Further, it is difficult to know just what depth ought to be sampled, and internal wetland sampling is an art. Multiple sampling points across the wetland width are much better than a single point, but there is no guarantee that flow proportional samples will result. Such cross-gradient protocols provide "through the wall" sampling averages instead. Those spatial averages are quite different from flow-weighted concentrations, by as much as a factor of two (Kadlec, 1999b). Dierberg et al. (Dierberg et al., 2005) sampled both slow and fast paths through a wetland, and found greatly different phosphorus gradients in the flow direction.

The true flow-weighted mean gradient is obtained if all the water is collected at various points along the flow path, and a so-called "mixing cup" concentration is obtained. This is achieved if the wetland is compartmentalized into several cells in the flow direction. The resulting cell-to-cell gradients represent flow-weighted P concentrations. An example of such a compartmentalized sampling is typified by the Shannon treatment marsh in Texas, which is divided into four cells. There is a steady but gradual reduction in TP as water moves through the cells (Figure 10.13).

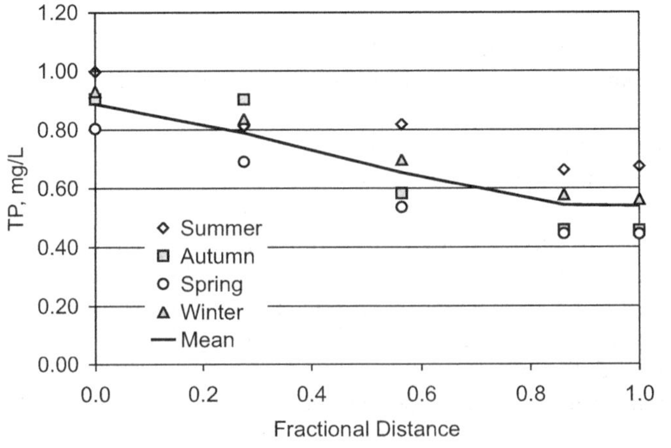

FIGURE 10.13 Profile of total phosphorus along the flow direction for the Shannon treatment marsh. Points are the mean of 11–23 cellwise transects measured over a four-year period.

At very low phosphorus concentrations, there is evidence that TP cannot be reduced below a wetland background concentration, C* (Figure 10.14). Internal cycling plus atmospheric gains and losses of water and phosphorus combine to create such a background concentration. The profiles in Figure 10.14 result from transect sampling, and do not reflect flow-weighted averages.

The existence of phosphorus gradients within treatment wetlands means that it is not appropriate to use models that are configured as one well-mixed unit. The existence of a low background concentration means it is not appropriate to use first order models that carry the predicted concentrations down to zero.

Marsh Startup

Observed wetland phosphorus removal performance may be influenced by transition from past history. Constructed treatment wetlands start from any one of several initial conditions, characterized by soil type, as well as vegetation type and density. These newly created wetlands are then exposed to a hydroperiod that is typically 100%, created by some form of wastewater. Under such conditions, it is obvious that some period of adaptation will ensue, during which soils and vegetation undergo changes dictated by ambient conditions of water depth and nutrient availability. For instance, FWS wetlands built on agricultural soils may have to deal with large stores of soil phosphorus that may temporarily result in high concentrations in floodwaters. For example, the adaptation period for the STA-1W FWS wetland in Florida extended over about one and a half years, during which time it was held in a no-discharge mode to accommodate the transition (Figure 10.15). This wetland was built on a former agricultural field, which apparently contained mobile legacy phosphorus.

In contrast, the Brawley, California treatment wetland was built on the floodplain of the New River, on land that was not previously used for agriculture. Water from the New River was pumped to the wetland, and contained about 1.4 mg/L of total phosphorus. The wetland was built in the last half of the year 2000, with the planting occurring in mid August. The transplants were healthy and robust by January 2001, but vegetative fill-in and

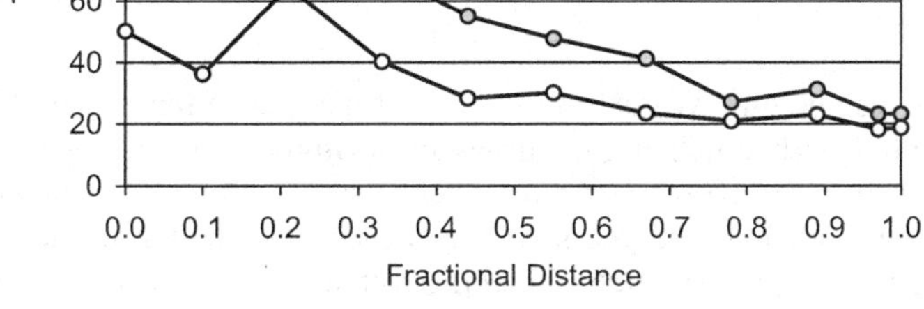

FIGURE 10.14 Profile of total phosphorus along the flow direction for a SAV treatment marsh, STA2C3. Points are the mean of five transects measured over a five-year period. The data labeled "high" represent high flow rates and high inlet concentrations. The data labeled "low" represent low flows and concentrations. (Unpublished data from DBE laboratories.)

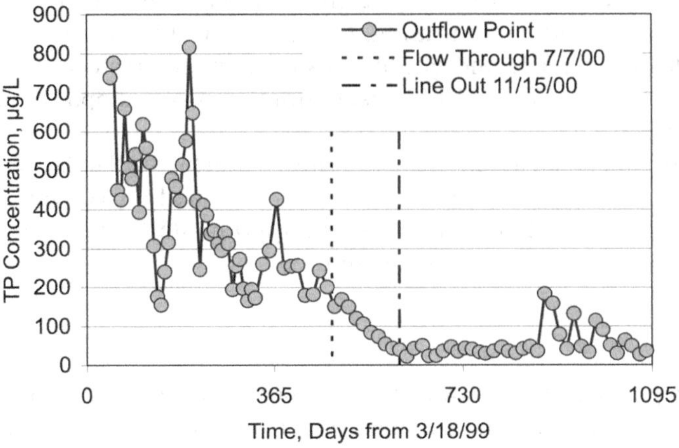

FIGURE 10.15 Floodwater phosphorus during the startup of STA-1W Cell 5 of the Everglades Protect project. Antecedent land use was agriculture. Data source: South Florida Water Management District unpublished data.

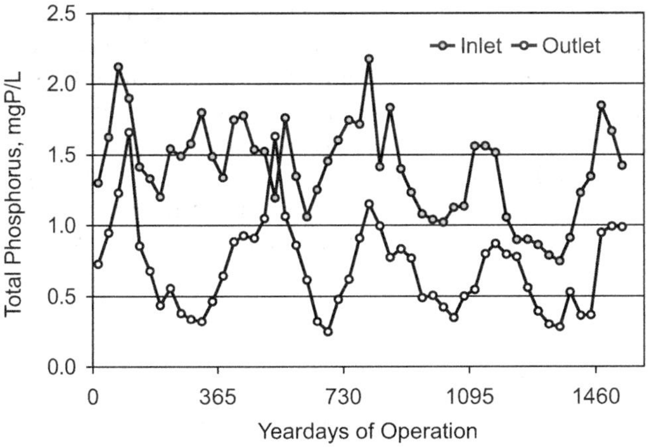

FIGURE 10.16 Time series of inlet and outlet TP concentrations for the Brawley, CA treatment marsh. There is a cyclic trend in outlet TP that is driven by an annual cycle in the flow to the wetland, but no linear trend up or down. Planting occurred about 120 days prior to the data record, which begins on January 1, 2001.

establishment had just begun. Water quality monitoring commenced in January 2001. The wetland was operated with much higher flows in springtime, resulting in the cyclic pattern of removals seen in Figures 10.16. However, there was no trend in removal percentage with time on average over the first five years of the project life (linear model has $R^2 = 0$). The loading to this wetland (approximately 50 $gP/m^2{\cdot}yr$) was far beyond that which could have been influenced by macrophyte establishment. Speculatively, the soils were already loaded with phosphorus matching the river water content (estimated $EPC_o = 1.4$ mg/L), and no further startup sorption was likely. This wetland therefore had no particular startup transient.

Antecedent conditions also often influence current performance for existing, natural wetlands receiving new phosphorus inputs. In such a case, the wetland may initially display a larger removal capacity than after stabilization. The Houghton Lake, Michigan natural treatment wetland began receiving treatment lagoon effluent in July 1978, at relatively low phosphorus loading rates, of the order of 4 $gP/m^2 \cdot yr$. The adaptation of the wetland proceeded from a fully established, low-nutrient community to a nutrient-enriched replacement community. New, very high standing crops of cattail (*Typha* spp.) incorporated more phosphorus, in an area that expanded over an initial eight-year period (Kadlec, 1997a). During this period, phosphorus removal was enhanced by vegetative expansion.

Thus it is seen that the initial period of performance of a FWS treatment wetland may exhibit lesser or greater phosphorus removal than in the long-term. The deciding factors are the antecedent phosphorus loading of the soils and the importance of the vegetative cycle in comparison to the applied phosphorus loading.

Phosphorus Movement Fronts

Frontal Development for New Source Water P Loadings

Almost all treatment marshes are flow-through systems in which phosphorus is preferentially removed in the upstream sections or compartments. It has been shown above that there exist several mechanisms and storages that act to remove phosphorus, and that some of these have finite capacity. These effects combine to produce transitory, moving "fronts" along the flow direction, and these move at different speeds. Nevertheless, frontal movements do not last forever, because there exist mechanisms that cause their eventual cessation. It is important to understand something about the approximate time scales of these processes. We will here presume that the wetland is newly constructed, and does not bear any legacy phosphorus.

Water moves through treatment wetlands in a matter of days. During that time frame, removal processes will be relatively stable, and variations will typically be due to random effects within the ecosystem. At any particular moment, there is almost certain to be a decreasing profile of water-phase phosphorus along the flow direction. This is the water column phosphorus front.

The front end of the system sees the most phosphorus. If phosphorus is the growth-limiting nutrient, a plant size gradient will likely occur. This and other growth-related phenomena are likely to last for only a year or two. During that same period, the sorption sites on the pre-existing soils will become fully occupied, and no further phosphorus removal will occur due to either sorption or plant establishment. However, the system will continue to form floc and new sediments, and phosphorus removal will continue at a reduced rate compared to the plant establishment phase.

The top layers of soils and sediments therefore build new storages of phosphorus, which are also allocated preferentially to the inlet region of the wetland. The system is building floc and soil phosphorus fronts, which are not coincidental with the water column phosphorus front. The establishment of a soil phosphorus front usually takes a matter of years, during which the vertical front of soil phosphorus profiles creeps upward due to soil building. Plant root zones can move to keep pace with this upward movement. Floc forms much more readily, and can move at speeds intermediate to the water speed and the soil phosphorus front. The rooted macrophytes respond to soil phosphorus, typically by the establishment of species that can better utilize the increased nutrient supply – unless the system was established with those more nutrient tolerant varieties. Species replacement lags the soil phosphorus front, and may take many years to develop and stabilize, although colonization may occur more rapidly. This results in a vegetation replacement front.

Supposing a stable and continuing input of water and phosphorus, the whole set of changes stops when a new set of vegetation has developed, and soil building is removing a fixed amount of phosphorus per unit time on average. The balance of phosphorus will then be exported, as a combination of continual outflows in the water, plus episodic (perhaps) outflows of floc, TSS, or detritus.

However, even then the system will continue to evolve on the time scale of decades, because accretion (approximately one centimeter per year) will eventually change the flow patterns and depths in the FWS wetland.

Although only a few projects have had the longevity to track these various fronts in full-scale situations, they have confirmed the concepts outlined here (DeBusk et al., 2001; Rutchey et al., 2008, Kadlec, 2009a, 2009b, 2009c).

Apart from accretion, wetland solids form a large pool of phosphorus, some of which is available for exchange with surface waters and porewaters. As noted above, sorption and cation exchange are active processes in the wetland environment. These solid storages will stabilize under continuous operation of a treatment wetland, but are nonetheless active, and exchange compounds with their surroundings. Thus the image of phosphorus compounds traveling with the flowing water is incorrect; phosphorus follows more of a "park and go" trajectory through the wetland.

Reduction of Source P Loading

Virtually all of the reports of phosphorus removals, fronts and adaptations are concerned with new sources of phosphorus entering a marsh, and the gradients that appear due to longitudinal processing. However, there now exists a summary report on the results of reducing the P additions to marshes that have been the recipient of elevated P source waters; which pertains to the reduction of source water P to Water Conservation Area 2A (WCA-2A) (Juston et al., 2015). WCA2A was subject to about six decades of elevated surface water P inflows. Studies conducted in 1989–1990 identified that most of the excess P load

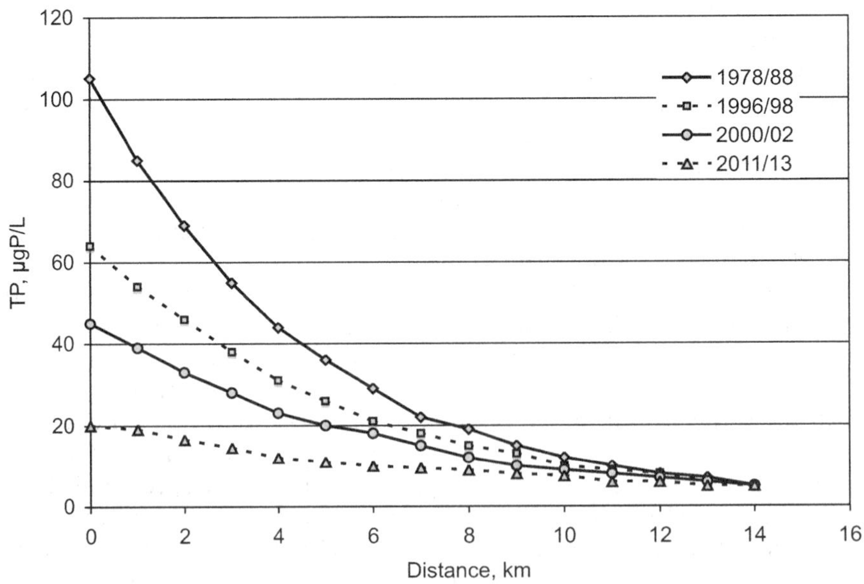

FIGURE 10.17 Surface water TP along the flow direction for WCA2A during different periods after reductions in source water TP. (Redrawn from Juston et al., 2015.)

to WCA-2A was removed and stored along a several kilometer gradient of newly accrued floc, sediments and soil. Because WCA2A is regarded as part of the Everglades ecosystem requiring protection, new P removal marshes were constructed upstream of WCA2, and have functioned to reduce the P load entering WCA2A for the past thirteen years. As a result, water column TP gradients have been greatly reduced (Figure 10.17).

The wide expanse of flocculent unconsolidated sediment (floc), containing a large pool of labile soil P, was largely supplanted by a consolidated, and presumably more stable, soil layer since load reduction measures were implemented. P enrichment in consolidated soils was persistent 13-years after load reductions, with no significant change in TP concentrations in the upper 0–2 cm along the entire 14-km study gradient, or in the 0–10 cm depth interval in the front 0–4 km range of the impacted zone. There was no strong evidence of either expansion or contraction of the extent of enriched soil P concentrations and storage since load reduction measures went into effect. Despite P-enriched soil, the impacted region in WCA2A continues to effectively remove low-level P inflows, now a fraction of pre-STA levels, to background levels over about 8-km flow distance, as it has done for over 40 years. While there is clear evidence that water P impacts are receding, it is a slow recession, and elevated values due to internal loading are likely to persist for decades.

10.4 PHOSPHORUS REMOVAL

The performance of a treatment wetland for phosphorus removal is partly deterministic, describable by central tendency equations; and partly stochastic, describable by the distributions of random effects (Kadlec, 1997b). Both pieces of the description are important, and both may be well-quantified. Alternate approaches for the deterministic part involve loading graphs and rate constants, and both contribute to design in different ways. Phosphorus processing is also constrained by lower limits to the achievable TP concentration.

It is important to recognize that there are two basic types of treatment marshes used to remove phosphorus: bioaccretion wetlands and sedimentation wetlands. These exploit different mechanisms, and therefore display different behavior in response to system variables. In general, bioaccretion marshes have long detention times, remove low amounts of phosphorus per unit area, store P in new organic soils, and have numerous ecological values in addition to P removal. Bioaccretion systems tend to be large, and receive flows of treated wastewater or contaminated river or lake water. Sedimentation wetlands have shorter detention times, remove larger amounts of P per unit area, store P as accumulated mineral sediments, and have fewer ecological values. Sedimentation wetlands tend to be small, and treat urban or agricultural field runoff. In larger sedimentation systems, a settling basin is a separate, first component, but in smaller wetlands, an internal settling zone accomplishes that purpose. There are also differences in the stability of the accreted residuals.

Background Concentrations of Total Phosphorus

Along the flow path, there is typically an exponential decrease in concentration, from the inlet value to a lower nonzero exit value (Kadlec and Knight, 1996). If the wetland is large compared to the added water flow and phosphorus load, then the concentration displays a plateau towards the outlet of the system, along which further reductions do not take place. The concentration along the plateau is the treatment wetland background concentration, C^*. The portion of the wetland that fully contains the gradient is the zone of total containment (ZTC). The gradient-less region is unimpacted by the discharge in question, but may exhibit the effects of other current or previous impacts.

Background concentrations also exist for pristine wetlands that do not incur anthropogenic water and chemical inputs, although these are arguably rare in light of global atmospheric transport. Phosphorus cycling in the natural environment leads to low concentrations in surface waters, and it cannot be expected that treatment wetlands will reduce phosphorus below those ambient levels. The phosphorus concentrations in such unimpacted wetlands are also considered to be background levels. The median flow-weighted total phosphorus concentration in 85 relatively undeveloped basins of the United States is 0.022 mgP/L (Clark et al., 2000). Areas of the Rocky Mountains and central plains generally had the highest background, up to 0.10 mgP/L. The northeastern and southeastern regions typically have lower ambient background phosphorus concentrations, about 0.01 mgP/L. Levels are very low in the Florida Everglades, often in the range of 0.006–0.010 mgP/L.

Mechanisms Creating Background Concentrations

In the treatment marsh situation, there are several possible reasons for the existence of a real or apparent nonzero background concentration for phosphorus. First, there may be some portion of the incoming phosphorus that is resistant to storage or conversion in the wetland environment. Some portion of the organic phosphorus may be highly resistant to uptake by the biogeochemical cycle. An extreme example would be an organophosphate pesticide, which is not readily degraded in wetlands. However, more benign sources may contain a biologically unavailable fraction, by virtue of the size and character of the molecules embodying the phosphorus (Proctor et al., 1999). Such phosphorus fractions may pass through the system untouched.

A second reason for a nonzero background concentration for TP is the association with particulates. Non-Reactive Particulate Phosphorus (NRPP) is part of the total water column phosphorus, and is an important part of the phosphorus fractionation of wetland sediments (Reddy et al., 1993). Because phosphorus is associated with suspended particulate matter, a nonzero background level of TSS entails a nonzero background level of phosphorus. For instance, at the Des Plaines, Illinois wetlands, the export of 8 mg/L of TSS carried 16 μgP/L of phosphorus, which is due to a phosphorus content of 0.2% in the exported TSS. Although TSS is notoriously difficult to measure inside wetlands, background levels of 5–10 mg/L are commonly found in densely vegetated systems (Kadlec and Knight, 1996).

In FWS wetlands with more open water surface, the presence of phytoplankton in the water column is enhanced because of high light levels. The presence of these organisms implies an accompanying suspended particulate phosphorus concentration. Classic examples of planktonic phosphorus for wetlands are Muskego Lake in Wisconsin (R.A. Smith & Associates, 1995), and the Lake Apopka Treatment Marsh in Florida (Dunne et al., 2015). Muskego Lake is about one meter deep, with a central area of dense submerged aquatic vegetation and a broad cattail fringe. Phosphorus loadings come from a mostly agricultural watershed, and internal loadings from storages created by discontinued sewage discharges. Central portions reach 600 μg/L TP, and a typical export concentration is 100 μg/L TP, which is nearly 100% particulate (planktonic). Lake Apopka contains about 175 μgP/L TP, 95% of which is planktonic particulate. This water is passed through a treatment marsh, which accretes particulate phosphorus, but converts a small portion to dissolved phosphorus. The marsh background is approximately 30–44 μgP/L (Dunne et al., 2015).

A third reason for a nonzero background concentration is a set of wetland processes that provide inputs distributed across the entire areal extent of the system. Groundwater discharge and rainfall may bring phosphorus into all portions of a wetland. Phosphorus may be utilized in the biogeochemical cycle, which is also distributed across the entire wetland area. That same cycle can produce return of the substance to the water column, usually by the processes of decomposition and leaching (Kadlec, 1999a). Two examples of rainfall-driven systems are the Houghton Lake natural wetland, with a background of 40 μgP/L (Kadlec, 1997a), and the Loxahatchee National Wildlife Refuge in south Florida, with a background of 10 μgP/L (McCormick et al., 1997a). Groundwater effects are illustrated in the study of Reid's Wetland in northeastern Victoria, Australia (Raisin et al., 1999). Ninety percent of the water input was groundwater discharge, at about 80 μg/L. The resulting phosphorus concentration in the surface discharge was 63 μg/L.

A fourth factor influencing concentration gradients, and the possibility of plateaus, is hydraulic bypass of the reactive wetland environment. Bypassed water carries with it the inlet substances, which may re-blend with treated water at downstream wetland locations (Kadlec, 2000). This process cannot create a true plateau or background, but may easily lead to an inferred background concentration, derived from extrapolation of gradients in the upstream portion of the system. For some chemicals, notably phosphorus, very few treatment wetlands extend beyond the zone of total containment, and such extrapolation, via curve fitting, is the norm rather than the exception. As shown in Kadlec (2000), the nature of internal flow patterns leads to a data-fitted background concentration, which varies strongly with the hydraulic loading rate to the wetland. Higher loading rates lead to higher background concentrations for flow-through wetlands.

Loading Considerations

Phosphorus removal data from full-scale treatment marshes span five orders of magnitude in size, from about 0.01 to 2,000 ha (see Table 10.9 for examples). They are hydraulically loaded from less than one cm/d to as much as a meter per day. Inlet phosphorus concentrations range from less than 50 μgP/L to about 6 mgP/L. Because of the variability present in any of these datasets, it is necessary to utilize comparisons in terms of the primary forcing variables and model parameters.

The outlet phosphorus concentration depends upon three primary variables (area, water flow and inlet concentration), as well as numerous secondary variables (vegetation type, internal hydraulics, depth, event patterns and others). It is presumed that the area effect may be combined with flow as the hydraulic loading rate (HLR) since two side-by-side wetlands with double the flow should produce the same result as one wetland. Therefore, two primary variables are often considered: HLR and inlet concentration (C_i). Both mass removal models (e.g., the PkC* model) and performance regressions are based upon these two variables (Kadlec and Knight, 1996). An equivalent approach is to rearrange the primary variables, without loss of generality, by using phosphorus inlet loading rate (PLI = HLR × C_i) and concentration (C_i). Thus, it might be expected that the effluent phosphorus concentration (C_o) will depend upon PLI and C_i. A graphical display has often been adopted in the literature (Kadlec and Knight, 1996; U.S. EPA, 2000a; Wallace and Knight, 2006; Kadlec and Wallace, 2009).

Significant differences would be expected between bioaccretion systems and sedimentation systems, because of the difference between biological cycling and sediment deposition. There is no sharp line of demarcation, and some treatment marshes embody both mechanisms in similar proportions.

Bioaccretion Systems

A number of bioaccretion intersystem performances provide the basis for comparison. The sample set used here includes information from 105 wetland cells of various sizes, but not including mesocosms. These represent many phosphorus removal applications, at widely varying concentrations, in a variety of climatic zones in North America and several other countries.

In the broad context, this inter-system dataset is represented by a trend that shows decreasing C_o with decreasing PLI (Figure 10.18). The right hand asymptote at very high

TABLE 10.9

Examples of phosphorus treatment wetlands. Sizes span 3 orders of magnitude, while loadings span 2 orders of magnitude.

Name & Source	Ref.	Years	Area	HLR	TP In	TP Out	PLI	PLR	CR	k
				cm/d	gP/m³		gP/m²·yr		%	m/yr
Patterson Creek Urban Runoff	1	2	1.5	12.1	0.081	0.014	3.6	3.0	82	130
Hidden River Urban Runoff	2	2	1.2	0.6	0.100	0.045	0.2	0.1	55	2.1
Apopka B1 Lake	3	14	73	10.2	0.109	0.088	4.0	1.4	31	15.8
Yakima N Ag Runoff	4	7	0.8	12.2	0.117	0.074	5.3	1.9	37	22.4
Råbytorp Ag Runoff	5	10	0.8	37.5	0.120	0.012	16.4	14.8	90	612
Olentangy W1 River	6	11	1.0	9.9	0.143	0.098	5.2	1.7	31	15.4
STA56 FW1 Ag Runoff	7	15	975	8.1	0.147	0.081	5.6	0.7	45	25.7
Titusville WWTP	8	7	101	0.8	0.281	0.048	0.8	0.7	83	6.9
Commerce WWTP	9	5	2.0	17.1	0.296	0.260	18.4	2.2	12	8.4
Brighton Lagoon	10	10	6.2	7.0	0.373	0.255	6.8	2.3	32	8.4
Shannon River	11	6	171	5.5	0.888	0.539	17.7	6.9	39	10.3
Imperial Ag Runoff	12	4	4.7	19.3	1.338	0.826	94.3	36.1	38	37.1
Brawley River	12	4	1.8	9.2	1.416	0.707	47.8	23.9	50	26.6
Hebert Lagoon	13	2	1.9	0.9	3.210	0.550	10.4	8.6	83	7.3
Tres Rios H1 WWTP	14	2	1.3	10.5	6.344	3.360	244.0	114.7	47	26.5

References

1 Reinelt and Horner (1995)
2 Carr and Rushton (1995)
3 Dunne et al. (2015)
4 Beutel et al. (2014)
5 Braskerud et al. (2005a)
6 Mitsch et al. (2012)
7 SFWMD Unpublished Data
8 City of Titusville Unpublished Data
9 Kadlec (2011)
10 Kadlec et al. (2012b)
11 Kadlec et al. (2011)
12 Kadlec et al. (2010a)
13 Hanson (2002)
14 City of Phoenix Unpublished Data

phosphorus loading is an outlet concentration equal to the inlet concentration – or in other words, no removal.

The greater the external P loading, the greater the outlet P concentration (Figure 10.18). The data display considerable scatter, but can be represented by a log-linear regression:

$$C_{TPout} = A(PLI)^m \tag{10.8}$$

where: A = fitting parameter
m = fitting exponent
C_{TPout} = outlet total phosphorus concentration, mgP/L
PLI = incoming total phosphorus load, gP/m^2·yr

The values for the entire dataset in Figure 10.18 are A = 0.029 and m = 0.925, with R^2 = 0.83, N = 102 systems for their POR.

Data suggest that inlet concentration is a second important determinant of the loading relation. Two (arbitrary) groups are shown, 0–0.5 mgP/L and 0.5–6.4 mgP/L. For any specific inlet concentration, or a narrow inlet concentration range, the slope of the data cloud is lower. The values for the low P dataset in Figure 10.18 are A = 0.026 and m = 0.694, with R^2 = 0.52, N = 55 systems for their POR. The values for the high P dataset in Figure 10.18 are A = 0.123 and m = 0.587, with R^2 = 0.68, N = 50 systems for their POR.

If the entire set of points in Figure 10.18 is considered, thus ignoring the effect of inlet concentration, the general trend line has a higher slope. However, such a single variable plot is nonunique, and may be misleading. For instance, use of a small intersystem dataset might result in use of left data points for high C_i, as well as right data points for low C_i, thus exaggerating the slope. Consequently, the global C_i – PLI graph advocated in some literature (Richardson et al., 1997; U.S. EPA, 2000a) is inadequate. The PkC* model has the ability to span the entire cloud of results when exercised for various choices of C_i, k and C* (Kadlec, 1999c), and is consequently the preferred alternative.

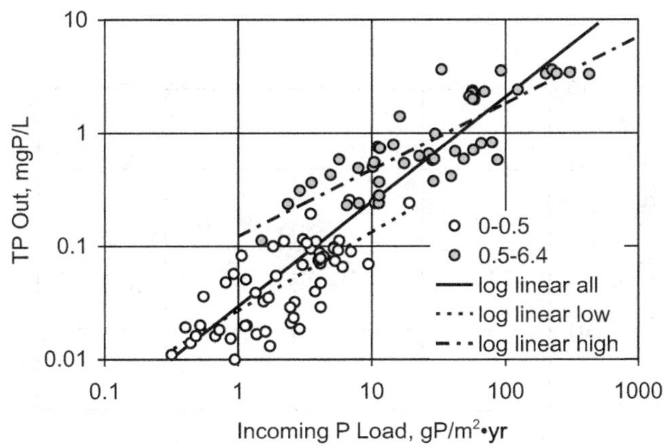

FIGURE 10.18 Intersystem phosphorus responses for bioaccretion treatment marshes. Each point represents the entire period of record for one wetland cell. Data span four orders of magnitude in both inlet and outlet concentrations and phosphorus loading, for the entire period of record for 105 individual systems. Data groups are coded in two ranges to illustrate the effect of inlet concentration on the trends.

Sedimentation Systems

The removal of particulate P depends upon settling and filtration processes, which have been seen to result in background levels of TSS (see Chapter 8), and will consequently result in background levels of the associated phosphorus. Therefore, outlet TP is not expected to be sensitive to PLI, and this is indeed the case (Figure 10.19). The best fit line is horizontal, at about 100 μgP/L, with no trend (R^2 = 0.00). Sedimentation marshes are not affected by inlet phosphorus loading.

Quantification of Rates

Despite the apparent complexity of the several removal mechanisms, data analysis shows that relatively simple equations can describe sustainable phosphorus removal processes. Profiles of phosphorus concentrations within wetlands typically show a decreasing trend that approaches the background concentration asymptote if the wetland is large enough. Global first order removal rates characterize such removal profiles, but do not incorporate any biotic features. Alternatives include regression equations for the wetland effluent TP concentration. It is useful to segregate treatment marsh performances based upon the schedule of water deliveries (event driven *vs.* continuous flow), and based upon data collection (sparse *vs.* extensive). Urban and agricultural stormwater wetlands usually are event driven and are not as intensively monitored as large wetlands treating river, lake or WWTP waters. Often, the data from small stormwater systems is insufficient to support a rate coefficient analysis, and hence empirical descriptions have been developed.

The PkC* Model

The PkC* model (Kadlec and Wallace, 2009) embodies a first order, areal net uptake of phosphorus, coupled to a zero order return flux from the static compartments of the wetland (soils, litter, biota), and is constrained to be applied to long-term average wetland performance. It is a three-parameter modified version of the first order, areal model, containing a lower limit on achievable phosphorus concentrations (C*), proposed by Kadlec and Knight (1996). As suggested by Kadlec and Wallace (2009), it also contains

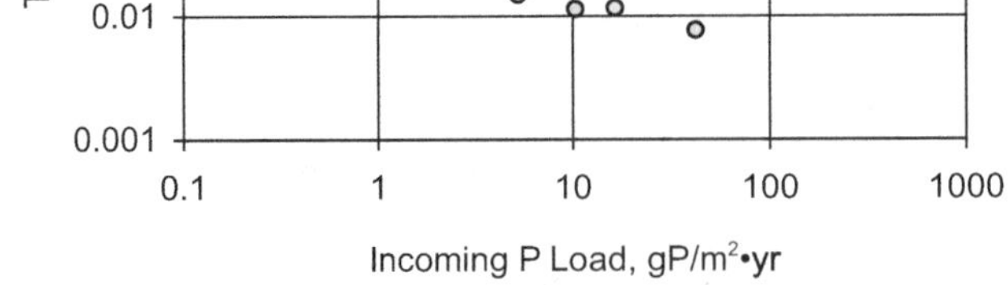

FIGURE 10.19 Intersystem phosphorus responses for sedimentation treatment marshes. Each point represents the entire period of record for one wetland cell. Data span four orders of magnitude in both inlet and outlet concentrations and phosphorus loading, for the entire period of record for 51 individual systems. No trend is apparent.

a hydraulic mixing parameter, P, related to the number of tanks in series that describe mixing (see Chapter 7). For an extreme simple case, the long-term average phosphorus removal, locally variable, is calculated from:

$$\frac{C - C*}{C_i - C*} = \frac{1}{(1 + k/Pq)^P} \tag{10.9}$$

where: P = mixing and speciation parameter
k = reaction rate coefficient, m/yr
q = hydraulic loading rate, m/yr

This allows C* as a calibration parameter, with typical values in the 4–40 μgP/L range for FWS wetlands. P is estimated, perhaps based upon a hydraulic tracer study, for which P ≤ NTIS. This rate requires the appropriate water mass balance to compute the performance of the entire wetland, as previously discussed in Chapter 7:

More complex models have been formulated to provide performance forecasting. However, most of these have proven too cumbersome to calibrate or exercise in support of design explorations. Examples are discussed later in this Chapter.

Annual Rate Coefficients

Most of the available calibration information comes from bioaccretion systems, which are considered first. Numerous individual studies have determined rate coefficients, and there are comprehensive surveys available also (see, for instance, Kadlec, 2016). Example results across continuous flow systems are given here, for measured P = NTIS values, or for P = 4 when not measured. The value C* = 0.006 mgP/L is used, and the remaining model parameter is the k-value, selected to fit Equation 10.7. The median annual rate constant for 105 full-scale systems was k = 12.9 m/yr (Table 10.10). The 10th to 90th percentile range is 2.7–42.9 m/yr. There is sometimes a significant seasonal dependence of phosphorus k-values. Even on an average annual basis, temperature or climate may be an important determinant of the rate constant, and these factors are thus responsible for the some of the intersystem variability in annual k-values. Accordingly, it is necessary, in a subsequent section, to examine such intra-annual effects.

There is less information from event driven systems, treating agricultural and urban runoff. Event mean concentrations over the course of an entire operating season are

TABLE 10.10

Annual reduction of total phosphorus in bioaccretion treatment marshes.

Stipulations

1. Period of record averages are used in calculations.
2. For k-value calculations, the following P-k-C* parameters are selected.
   a. C^* = 0.006 mg/L
   b. If not measured, P = 4. Otherwise, the measured value
3. Ranges of variables:

	HLR	TP In	TP Out
	cm/d	mg/L	mg/L
mean	6.8	1.1	0.6
median	4.5	0.3	0.2
max	40.4	6.3	3.7
min	0.3	0.02	0.01

Results (N = 105 wetlands)			
Percentile	P Load Removed (PLR) $gP/m^2 \cdot yr$	Percentile	Rate Coefficient (k) m/yr
0.1	0.4	0.1	2.7
0.2	0.7	0.2	4.3
0.3	1.0	0.3	6.8
0.4	1.4	0.4	9.5
0.5	2.0	0.5	12.9
0.6	2.9	0.6	15.8
0.7	5.5	0.7	19.7
0.8	8.2	0.8	30.6
0.9	15.2	0.9	42.9

required, which are not typically available. Example results across systems are given here, for measured P = NTIS values, or for P = 4 when not measured. The value C* = 0.006 mgP/L is used, and the remaining model parameter is the k-value, selected to fit Equation 10.7. The median annual rate constant for 51 full-scale systems was k = 18.4 m/yr (Table 10.11). The 10th to 90th percentile range is -8.6–272 m/yr. It is notable that the range of rate coefficients is large, and encompasses negative values. Those negative values are the result of data indicating more phosphorus leaving the marsh than entering. Negative removals are linked to phosphorus speciation, as discussed later herein.

First order models are suggested to be used to describe event driven systems, including urban runoff wetlands (Konyha et al., 1995; Wong and Geiger, 1997; Carleton et al., 2001; Wong et al., 2006; Merriman et al., 2017). These need to be used in the flow-weighted, or in use, sense, in which annual concentrations and flows are presumed to be treated according to the rate coefficients derived for continuous systems (Wong and Geiger, 1997). Since urban stormwater HLRs are close to the HLRs for continuous-source wetlands, it is reasonable to expect that stormwater wetlands would perform somewhere near the average for the continuous flow marsh database set. When the two source types are compared, this is indeed the case (Table 10.11, compare to Table 10.10).

TABLE 10.11

Annual reduction of total phosphorus in agricultural and urban runoff treatment marshes.

Stipulations

1. Period of record averages are used in calculations.
2. For k-value calculations, the following P-k-C* parameters are selected.
   a. $C^* = 0.006$ mg/L
   b. If not measured, $P = 4$. Otherwise, the measured value
3. Ranges of variables:

	HLR cm/d	TP In mg/L	TP Out mg/L
mean	31.0	0.3	0.2
median	13.2	0.2	0.1
max	251.2	1.5	1.3
min	0.3	0.02	0.01

Results (N = 51 wetlands)			
Percentile	P Load Removed (PLR) $gP/m^2{\cdot}yr$	Percentile	Rate Coefficient (k) m/yr
0.1	-2.7	0.1	-8.6
0.2	-0.5	0.2	-0.4
0.3	0.1	0.3	1.8
0.4	0.6	0.4	5.6
0.5	2.2	0.5	18.4
0.6	3.7	0.6	32.0
0.7	5.5	0.7	45.2
0.8	11.7	0.8	135.6
0.9	24.1	0.9	272.1

EFFECTS OF WETLAND VARIABLES

Performance of marshes for P removal depends upon several factors, prominently including mode of water delivery, season of the year, vegetation, water depth and P speciation.

Seasonal Effects

Two factors are likely candidates for modifying phosphorus uptake over the course of a year: temperature and vegetation growth patterns. Water temperature would be expected to modify microbial processes, which are involved in phosphorus release during the decomposition of detritus from bacteria and algae, and from above and below ground plant parts, but less in phosphorus uptake. Plant and algal growth requires substantial amounts of phosphorus, which is withdrawn from porewaters during the annual growth period. Warm climate systems differ markedly from temperate wetlands, because the latter have a period of plant dormancy occasioned by winter conditions. Warm climate systems may also have periods of relative low vegetative activity, but they do not experience the complete growth stoppage seen in temperate systems.

It is necessary to utilize data from several replicated annual periods to establish trends that are valid against the background of other stochastic variations. Only minimal temperature trends have been previously reported for phosphorus removal for emergent marshes (Kadlec and Reddy, 2001; Kadlec and Wallace, 2009). It is concluded that temperature is not a major factor in P uptake in marshes. However, seasonal effects are often quite pronounced. As a consequence, the modified Arrhenius temperature dependence of an uptake rate coefficient is not a good model. Kadlec and Wallace (2009) report a median θ = 1.006, and only a small amount of the seasonal variance explained (median R^2 = 0.07). Monthly values of k provide a better description.

Return Flux Temperature Effect

Cold climate bioaccretion systems display two peaks of uptake, in spring and again in autumn (Figure 10.20).

The spring peak in phosphorus uptake is readily visualized as plant uptake during the peak growing season. The secondary peak in autumn is not as intuitive, but likely has its explanation in terms of a summer maximum of a return flux due to decomposition, coupled with belowground phenomena in autumn. The visual early autumn senescence of above ground plant parts can be misleading, because phosphorus accumulation in shoot bases appears to continue into December (Prentki et al., 1978). Further, roots and rhizomes accumulate phosphorus during autumn, which provides the storage needed to partially support the spring aboveground growth spurt. Translocation of phosphorus from aboveground sources to rhizomes can occur in treatment wetlands, ranging from zero (Tanner, 2001a) to 58% of the standing phosphorus crop (Mueleman et al., 2002). Additionally, studies have shown that both above ground (Puriveth, 1980; Gessner, 2000) and belowground (van der Valk and Attiwil, 1983; Wrubleski et al., 1997) litter may increase in phosphorus mass during the cold season. In short, microbial and plant soil processing of phosphorus continues well after the senescence of above ground plant parts is complete.

The return flux of phosphorus from decomposition processes may also contribute to the pattern of uptake observed in Figure 10.20. The rate constant is a parameter in a lumped model, which describes a net transfer, which is the difference between uptake processes and release processes to and from the active compartments in the ecosystem. Thus, in mid-summer under warmest temperatures, microbial uptake may be operating rapidly, but so are leaching and decomposition processes. High returns of phosphorus from the static compartments can depress net rates of uptake and help create the mid-summer low in the overall rate constant.

Although uptake rate coefficients are seasonal and not temperature dependent, it is likely that the P return flux is temperature dependent. Return fluxes, exemplified by C*, have typically been considered to be constant throughout the year. That may not be the case for cold climate wetlands. Arheimer and Pers (2017) used a first-order uptake and a zero-order, temperature dependent return flux:

$$J = kC - k_{ret}(1.2)^{(T-20)}C_{in} \qquad (10.10)$$

where: C = TP concentration, gP/m^3

C_{in} = Incoming TP concentration, gP/m^3

J = net P uptake rate coefficient, $gP/m^2 \cdot yr$

k = uptake rate coefficient, m/yr

k_{ret} = return rate coefficient, m/yr

Arheimer and Pers (2017) calibrated Equation 10.10 to data from nine Swedish agricultural runoff wetlands, and determined k = 33 m/yr, and k_{ret} = 18 m/yr. The calibration wetlands are estimated to have had 65% concentration reduction, which causes the temperature for zero uptake to be about 18°C. Above 18°C, net P release would be observed.

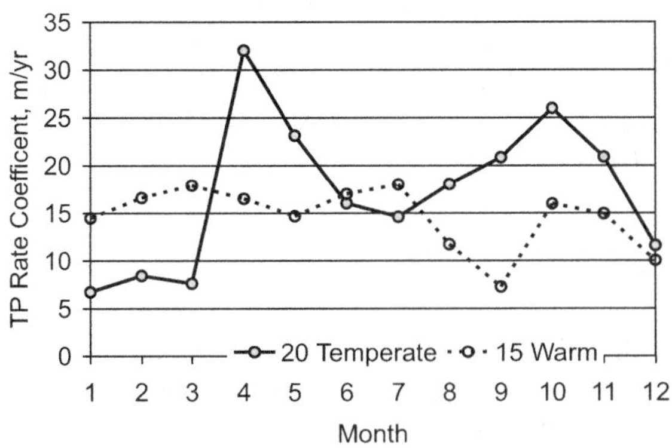

FIGURE 10.20 Mean monthly phosphorus removal rate coefficients for northern temperate marshes, by month (N = 20 wetlands). The average annual value is 17 m/yr. Also shown is the annual pattern for warm climate marshes, by month (N = 15 wetlands). The average annual value is 15 m/yr.

Warm Climate Wetlands

The examination of wetland phosphorus removal data for warm climates also shows seasonal effects (Figure 10.20), but again the modified Arrhenius model is not a good representation. For instance, the model is calibrated by $\theta = 1.006$ for the Orlando Easterly Wetland in Florida, but accounts for none of the variance in rate constants ($R^2 = 0.001$).

Warm climates often experience dry and wet seasons, which may nearly coincide with annual temperature extremes. For example, in southern Florida, the wet season commences in June and runs through October, which is also the period of highest temperatures. Because water flow influences marsh P removal, there is difficulty in ascertaining whether the causative factor is flow or temperature. The fact that continuous flow marshes in warm climates do not demonstrate marked seasonality is a strong indication that flow is the dominant causative factor.

Vegetative Cover

It is difficult to find information on the phosphorus removal potential for unvegetated wetlands. While open-water unit process wetlands are in existence and advocated for nitrate, pathogens and trace organics (Bear et al., 2017), there are not reports on their use for P reduction. The Commerce, Michigan, wetland was essentially devoid of emergent and submerged macrophytes, thanks to extensive herbivory by waterfowl (Kadlec, 2011). That system had an average annual first-order P removal rate coefficient of 7.5 m/yr

Presumably, shallow facultative lagoons represent this end of the vegetation spectrum, but data for such systems is almost entirely absent (Water Environment Federation, 2001; Abis, 2002). In the design of lagoon systems, phosphorus removal is usually presumed to be minimal, and chemical addition is normally used when a phosphorus standard is to be met (Surampalli et al., 1995). Data for lakes shows lower removal rate coefficients than marshes, with $k = 5.1 \pm 0.6$ m/yr (Brett and Benjamin, 2008). Walker and Havens (2003) calibrated Lake Istokpoga, Florida with $k = 3.5$ m/yr. Walker (2005) determined with $k = 2.7 \pm 1.8$ m/yr for 17 systems in Florida. Given the mechanisms discussed earlier, it is reasonable to presume that plants, either submerged or emergent, and accompanying mesophyton, promote a major fraction of observed phosphorus removal in wetlands. As a consequence, vegetated systems are to be preferred to unvegetated lakes and ponds.

Initial establishment of plants in a constructed wetland requires more phosphorus than that needed to sustain the mature annual cycle. This establishment requirement occurs at rates that increase as the new plant density increases. For instance, Lakshman (1982) found phosphorus uptake strongly correlated with new stem density ($R^2 \approx 0.94$). A similar effect was present during the startup of the Orlando Easterly Wetland (Kadlec and Wallace, 2009).

Water Depth

Water depth may also influence phosphorus removal by affecting the amount of submersed biota, and by affecting the hydraulics of flow. Hydraulics are more likely to be a factor in large wetlands that have considerable topographic irregularity, allowing "islands" and deeper zones, which may cause channelization or short-circuiting. In smaller wetlands with more uniform inundation, the amount of submerged "biology" is a more likely cause of short-circuiting.

Wetland data strongly support the idea of an areal rate model for small wetlands, as opposed to a volumetric rate coefficient (see Chapter 7). If processes are areally distributed, then the volumetric rate coefficient will depend inversely upon depth. For instance, the

Listowel, Ontario, Canada, systems 4 and 5 received the same water over a four-year period, but with variable depths. An inverse depth dependence was present for volumetric rate coefficients with $R^2 = 0.65$ (Figure 10.21).

Water Flow

Areal rate coefficients are likely to depend upon the water flow rate as well as or instead of depth. Performance data suggest that flow dependence is operative in many instances. There is considerable evidence that the rate of uptake of contaminants in general, and phosphorus in particular, may be in large part governed by the transfer from the bulk flowing water to the biofilms where the actual uptake takes place (Kadlec and Wallace, 2009). The mass transfer coefficients, or the equivalent local dispersion coefficients, depend upon velocity, as discussed in Chapter 7. Thus, if the rate limiting step is transfer of phosphorus from the bulk water to mesophyton, a strong dependence of the overall rate coefficient upon flow rate is expected. This situation is expected to prevail in bioaccretion systems. Sedimentation systems display HLR dependence, but for entirely different reasons. The rate coefficient for particulate P sedimentation closely resembles the TSS particle settling rate, and thus tends to be very large. At the same time, sedimentation wetlands tend to be quite small and therefore have very high HLRs. In this situation, removal is unrelated to mass transfer to biota.

Side-by-side testing has been done at the Tres Rios, Arizona, project. Twelve test cells, each with area 1,200 m^2, were subjected to different loadings of the same source water. Phosphorus removal was not large (13% concentration reduction) for incoming TP = 4.2 mgP/L. These were arguably bioaccretion wetlands, because the incoming water was essentially devoid of TSS. Weekly measurements of TP were supported by daily measurements of flow. Hydraulic tests indicated that the cells had indistinguishable differences in internal mixing patterns (NTIS = 4), and other system water quality characteristics were also virtually identical (Kadlec, 2007). Rate coefficients were directly proportional to HLR (Figure 10.22).

Large inter-system data sets may also be examined for a hydraulic loading effect upon k_{TP}. A treatment marsh dataset of 181 systems, with incoming $12 < TP < 700$ μgP/L was

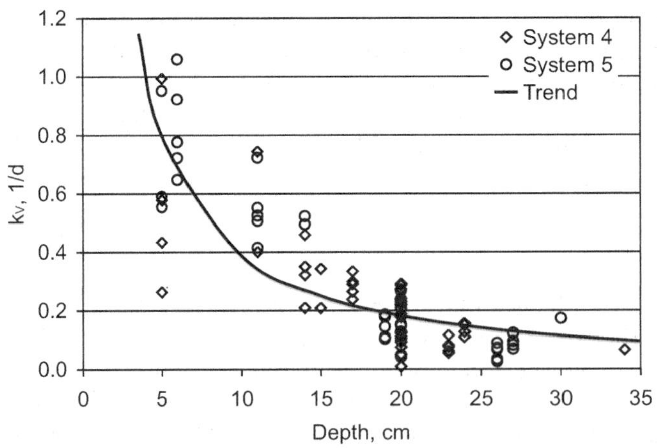

FIGURE 10.21 Dependence of the volumetric phosphorus rate constant on depth for Listowel, Ontario, FWS Systems 4 and 5. The values selected were P = 3TIS and C* = 0.02 mgP/L. Data span four years at monthly frequency. The line is an inverse proportion between k_V and depth. Data source: Herskowitz (1986).

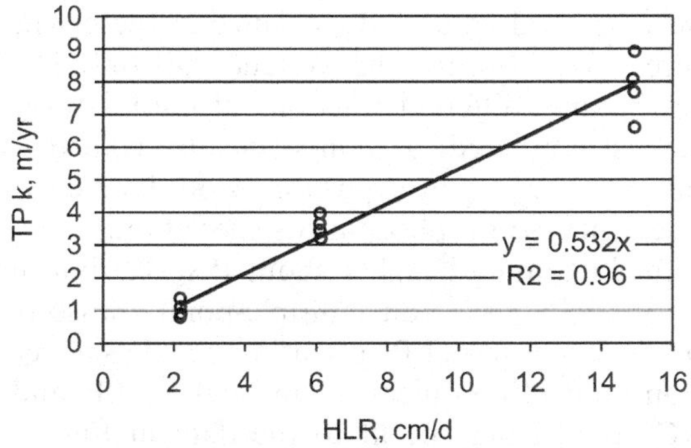

FIGURE 10.22 Dependence of the phosphorus rate coefficient on hydraulic loading rate for the Tres Rios, Arizona, test cells. Each of three HLRs were utilized in quadruplicate for a year-long study. Source: City of Phoenix unpublished data.

used to compute POR TP k-values, using P = 4 and C* = 4 μgP/L. Hydraulic loading rates spanned over two orders of magnitude (0.6–181 cm/d). On average, these marshes removed about 50% of the incoming phosphorus, and had a median k_{TP} = 22.7 m/yr. There is a surprisingly strong dependence of rate coefficients upon HLR, despite the wide range of other system differences (Figure 10.23).

Speciation

Total phosphorus is a lumped constituent in water entering a treatment wetland, comprised of several types of compounds. Total Phosphorus (TP) is comprised of Particulate Phosphorus (PP), Dissolved Organic Phosphorus (DOP), and Soluble Reactive Phosphorus (SRP). Each of these analytical categories is in turn a mixture. Particulate phosphorus is associated with a spectrum of particle sizes and types. DOP is associated with a suite of organic molecules of varying types and sizes. Even SRP contains components, such as

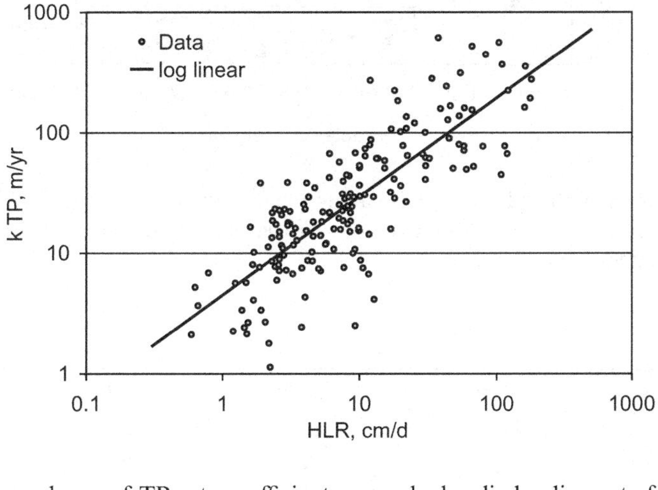

FIGURE 10.23 Dependence of TP rate coefficients upon hydraulic loading rate for 181 systems. The log linear regression is $k_{TP} = 4.5(HLR)^{0.82}$ with $R^2 = 0.65$.

mono- and di-basic phosphates, and polyphosphates. Each component of the overall TP mixture will have its own removal rate, and presumably these will be different. As water containing such a mixture passes through the wetland, not only is TP reduced, but its species composition changes because different fractions of the mixture are reduced at different rates. The mixture becomes weathered, a term originally coined to describe the selective stripping of light volatile materials upon exposure to outdoor environments. To date, only minimal attention has been focused on phosphorus speciation.

Transect data are useful in gaining insights about P speciation in treatment cells. Juston and DeBusk (2011) measured P species on multiple-point transects over multiple years in STA2 Cell 3 (Figure 10.24). The internal P speciation profile showed that SRP was reduced to the analytical detection limit more quickly than DOP or PP, and that SRP was reduced to lower levels. The PkC* model may be fit to the data in Figure 10.24, with the results shown. The fact of the low values of C* is of less importance than the marked differences in the rate coefficients k. SRP is removed at a rate roughly double the rate for DOP, and triple the rate for PP. So, after the SRP is gone, the remaining STA acres are less effective in removing TP, and the design acres needed to complete the removal to some low TP value will be large.

There is a general decreasing trend of all species with distance in most cases, but that does not preclude the formation of SRP, DOP and PP along the flow path. SRP may be formed from decomposition of dead biota, sediments and PP; and enzymatic decay of DOP. PP and DOP may be formed from decomposing bacteria, algae and plant litter; and PP may be resuspended from sediments or generated by mesophyton.

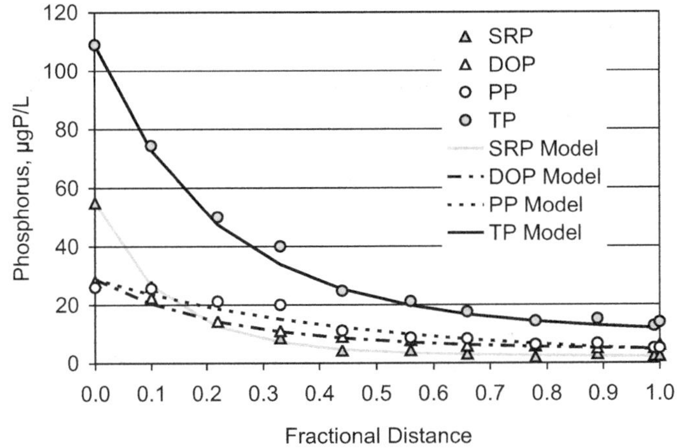

FIGURE 10.24 Profiles of phosphorus species along the flow direction in STA2C3 in Florida during 26 sampling events spanning six years. Adapted from Juston and DeBusk (2011). Model parameters are P = 6 = NTIS (measured), and:

	SRP	DOP	PP	TP
k, m/yr	153	82	38	88
C*, µgP/L	2	4	0	9
R^2	0.98	0.95	0.73	0.98

Seasonality may differ between the various P species, leading to complex seasonality for total phosphorus. The behavior of the Florida Apopka marshes exemplifies such complex patterns (Dunne et al., 2015). Data from four parallel marshes spanning nine years shows that total phosphorus was removed effectively during the first six months of the year (January–June), but not during the second six months (July – December) (Figure 10.25A). These marshes differ in source water from other Floridian systems. The STAs receive agricultural runoff, with some organic particulates; while the OEW, OCESA, and Titusville wetlands receive highly treated WWTP effluent, with essentially no particulate P. The Apopka system receives lake water dominated by algal particulate P, on a steady flow basis. The speciation of the waters helps explain the TP behavior. Particulate phosphorus is reduced very effectively during the first six-month period, and about half as effectively during the second six-month period (Figure 10.25B). However, dissolve phosphorus increases through the marshes in all months of the year, and in larger amounts during the second half of the year.

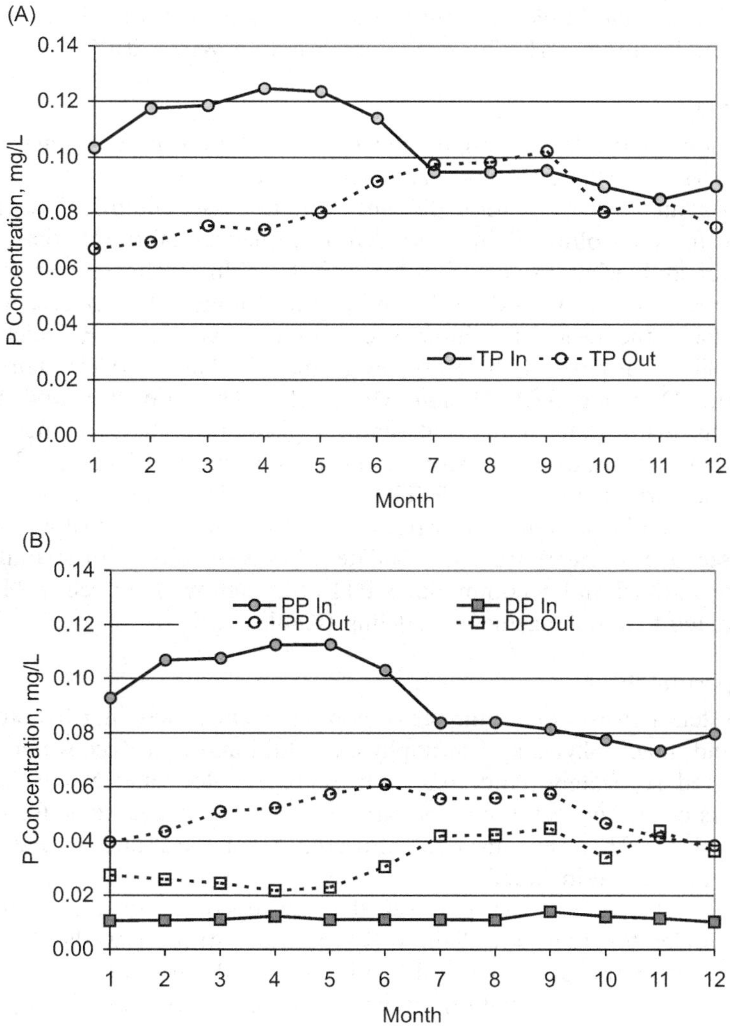

FIGURE 10.25 The annual patterns of P concentrations in the Apopka marshes. Panel A: total phosphorus. Panel B: dissolved and particulate phosphorus.
Source: Unpublished project data.

More Detailed Models

Process analysis in this book is founded on a one-dimensional spatial framework, with internal mixing and first-order removal kinetics. The reason is a desire for simplicity while preserving known longitudinal variability. Other choices may be, and have been, made, by adding detail with respect to dimensionality, ecosystem compartmentalization, and P speciation.

Two Spatial Dimensions

Several modeling studies have opted for using two planar dimensions instead of one. These require a two-dimensional description of water depth and movement, which can be keyed to the type and extent of vegetative cover across the length and width of the wetland. This detailed hydraulic description replaces the tanks-in-series approach, by utilizing the 2D flow pattern and one or more dispersion coefficients (Min and Wise, 2010; Paudel and Jawitz, 2012). This approach may also include an interconnected network of dead zones and flow zones (Martinez and Wise, 2003; Paudel et al., 2010). There exist commercial computer simulation packages that implement these ideas, such as the MIKE 21 package of the Danish Hydraulic Institute (DHI, 2017a, 2017b; Min and Wise, 2010).

Two Compartments

A simple extension of the PkC* design model is the addition of P storage in the biota of the wetland ecosystem. This extension has been applied to treatment wetlands in Michigan (Kadlec, 1997a), but that application did not include pulsed inflows. Treatment wetlands are now in use for controlling P in pulse driven sequences of runoff that often occur in both agricultural and urban basins. During periods of high phosphorus availability, that storage will increase, and in periods of P-famine, the storage will decrease. At all times, the ecosystem produces the residual sediments containing unavailable P, which characterized the WCA2A calibration data of the STA design model (Walker, 1995). This extension has been named the Dynamic STA Design Model (DMSTA) (Walker and Kadlec, 2005). DMSTA is an unsteady state model that removes phosphorus to permanent burial in proportion to the amount of labile P in storage. There is an implied labile pool of phosphorus, in addition to the permanently buried P. That labile pool is presumed to be drawn down by a return flux, or bleed-back. Temporal variations in the water budget (flows, rainfall, evapotranspiration, seepage, storage) are also simulated. DMSTA allows for spatial variability via cells in series or parallel and accommodates PTIS for each wetland cell. DMSTA is part of a broader approach to P management modeling (Kadlec, 2011).

Multiple Compartments

A marsh ecosystem consists of a number of compartments, including live and dead, above and belowground, macrophytes and mesophytes. Additionally, surface water, and porewater can be considered separately. And, one may segregate floc, active soils and deep soils. Clearly, transfers occur between these compartments, which in aggregate form the biological cycles in the wetland. There is considerable discretion for the modeler to select those compartments which are to be included.

For example, Wang and Mitsch (2000) developed a dynamic one-tank multi-compartment model for the Des Plaines River demonstration wetlands (Figure 10.26). There are eight compartments included, and only total phosphorus is considered. The model was calibrated to information drawn from four wetlands, for operational years 1989–1991. During that time period, these wetlands were tracer-tested and found to behave as approximately four tanks in series (Kadlec, 1994), and subsequently modeled in that hydraulic framework according to a first order areal removal (Kadlec, 2001a). This

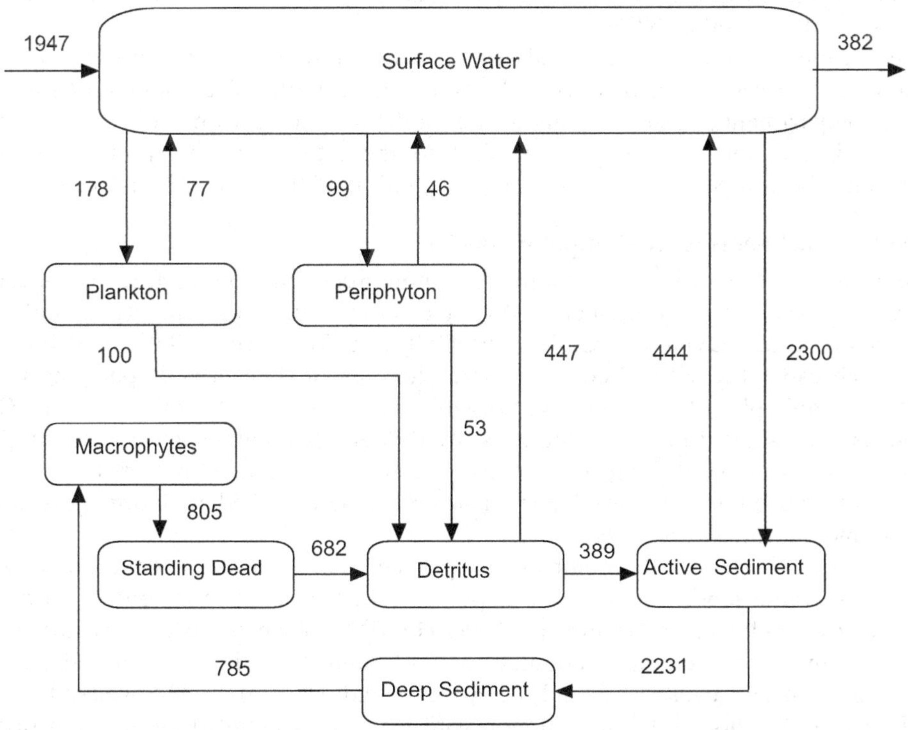

FIGURE 10.26 Phosphorus fluxes in mgP/m^2·yr for Des Plaines River wetland EW3 (modified from Wang and Mitsch, 2000). The mean loading was 1.95 gP/m^2·yr. By difference, the P load buried was 1.45 gP/m^2·yr. The period of observation was two years, during which the system was "not quite in steady state."

framework continues to be used (Marois and Mitsch, 2016). Importantly, these Mitsch models are of zero dimensionality, meaning they are developed for one stirred tank (P = 1).

A similar choice was made by Jawitz et al. (2008), however they decided to omit standing dead and floc/detritus layer, and split soils into mineral and organic categories. Additionally, they three water compartments: surface water, porewater and ground water. Two forms of water-phase phosphorus were considered: SRP and PP. All solid compartments were assumed to have fixed P concentrations. The model was dubbed transport and reaction simulation engine (TARSE). For a simulation of a treatment marsh, STA1W Cell4, TARSE required 15 initial conditions, and 10 calibration parameters.

This model solves the advection-dispersion equation on an unstructured triangular mesh and incorporates a wide range of user-selectable mechanisms for phosphorus uptake and release parameters (Jawitz et al., 2008). In general, the phosphorus model contains transfers between stores; examples of stores that can be included are soil, water column (solutes), pore water, macrophytes, suspended solids (plankton), and biofilm. Examples of transfers are growth, senescence, settling, and diffusion, described with first-order, second-order, and Monod types of transformations. Local water depths and velocities are determined from an existing two-dimensional, overland-flow hydrologic model, the South Florida Water Management District Regional Simulation Model. The model was applied to one cell of STA1W, (Cell 4). Inflow total phosphorus concentrations and flow rates for the period between 1995 and 2000 were used to simulate Cell 4 phosphorus removal. Flows to this cell were relatively steady during this period, and thus the pulse response of the model was not severely stressed as in other STAs.

There are nine state variables, including particulate and soluble P in groundwater, porewater, and surface water; together with plants and microbiota. There about 17 transfers of P among compartments, each requiring a rate model and at least one adjustable parameter. Calibration is therefore difficult, and subject to the estimating ability of the user. This model has not been used in design, or in the evaluation of the other 32 STA cells.

Multiple Compartments with Multiple P Species

The literature is rich with models of aquatic communities that include tracking phosphorus in the various ecosystem components. As an example, the water quality module of the Environmental Fluid Dynamics Code (EFDC) (Tetra Tech, Inc., 2007; USEPA, 2016b) includes such capability. EFDC considers refractory particulate organic phosphorus, labile particulate organic phosphorus, dissolved organic phosphorus, and total phosphate. Organic phosphorus is considered in three states: dissolved, labile particulate, and refractory particulate. Only a single mineral form, total phosphate, is considered. However, the model considers equilibrium among dissolved phosphate, phosphate sorbed to inorganic solids, and phosphate incorporated in algal cells.

Adaptations of this code have subsequently been developed for wetlands. For example, Jin et al. (2007) first modified EFDC to apply to Lake Okeechobee in south Florida, creating the Lake Okeechobee Environment Model (LOEM). Then the code was further modified for wetlands, by including modules for SAV and EAV in the wetland specifically implementing nutrient cycles (Jin and Ji, 2015; Ji and Jin, 2016). The wetland version is called LOEM-CW. This model, in common with other very detailed models, has dozens of state variables, and a still greater number of calibration parameters. It must be run on a fine spatial grid, and with very short time steps. As a result, the model is very cumbersome. Further, the goodness of fit, as measured by various standards including Nash-Sutcliffe efficiency, is not particularly good.

The utility of detailed models is questionable, as detailed by Paudel and Jawitz (2012). Their results revealed that the most complex model structure may not necessarily be the most effective in simulating the dynamics of total phosphorus (TP) concentrations in a wetland. The most complex model reproduced the field observations best, but only marginally.

Complex models are challenged by data availability. Comprehensive data sets are needed to constrain each transfer pathway, and are rarely available, especially for the nonaqueous model compartments and transfers. Consequently, it is crucial to identify the appropriate level of complexity for given data and questions of interest.

Stochastic Behavior

The framework for modeling treatment wetland performance, in general, is comprised of a deterministic component, characterized by appropriate equations or mean intra-annual trends, together with a stochastic component, characterized by one or more probability distributions (see Chapter 7). Treatment wetland performance is typified by large coefficients of variation, and therefore both pieces of the description are necessary:

$$C = C_{trend} + E \tag{10.11}$$

where: C_{trend} = annual trend P concentration, gP/m^3
E = stochastic component of outlet concentration, gP/m^3

The trend concentration may be determined from the PkC* model, or from a sinusoidal variation through the year, or from folded intra-annual data. The random variation, E, can be captured through the use of a multiplier on the trend value, C_{trend}:

$$\Psi = \frac{E}{C_{trend}} \tag{10.12}$$

Data from the Brighton wetland in Ontario provides an idea of the variability to be expected with respect to the mean long-term performance for phosphorus removal. This wetland received an average inlet concentration of 0.373 mgP/L for 2001–2010, at an average hydraulic loading rate of 5.02 cm/day (nominal detention time ≈ 7 days), and produced a mean outlet concentration of 0.255 mgP/L. Figure 10.27 shows a considerable probability band around the mean cyclic trend performance for the monthly means of daily TP measurements. The maximum monthly value was 0.79 mgP/L at the outlet, about triple the annual average outlet concentration.

Anticipated Excursion Frequencies

Probabilistic effects are important in the utilization of design models for predicting phosphorus removal performance of treatment wetlands. Regulatory requirements vary from location to location but may be a standard other than a long-term average. There may be a maximum monthly concentration not to be exceeded, or a specified concentration not be exceeded for more than a certain percentage of samples. To meet this type of regulatory requirement, the designer needs information on the variability of wetland outlet concentrations. Models represent only the seasonal trends of effluent TP concentrations, leaving a considerable amount of probabilistic scatter in performance. Therefore, it is useful to examine the variability of monthly average outlet concentrations, here represented by a fractional addition to the trend value (Ψ), or the multiplier on the trend value (1+ Ψ).

Because designers are interested in preventing or controlling exceedances, the upper percentile points of that distribution are useful in design. For instance, the 90th percentile represents the fractional addition to the trend that may be expected to occur one time out of

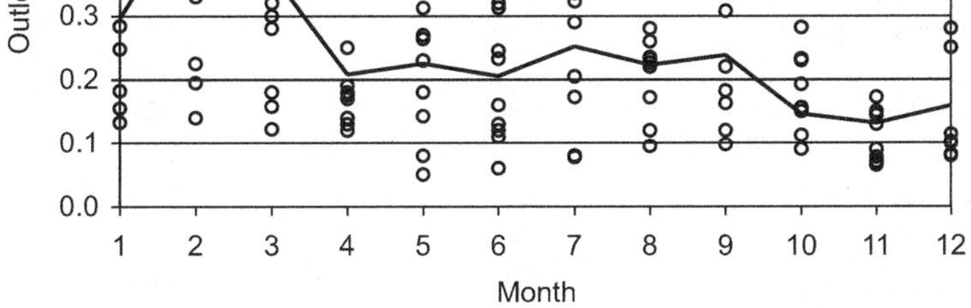

FIGURE 10.27 Variability of outlet TP concentrations compared to a deterministic first order model for the Brighton, Ontario, treatment marsh.

ten during the period of record. Note that these Ψ-values are not quite compliance percentiles, because of the finite number of months involved in a calibration set.

Table 10.12 lists percentiles of the monthly Ψ-distributions for phosphorus for some representative marshes. These data indicate that the median of the 95th percentile is an additional 139% above the trend, and that the median of the 90th percentile is an additional 90%. To incorporate this variability into design, the wetland designer must oversize the wetland. Suppose, for example, it is necessary to meet a monthly limit of 0.3 mg/L of total phosphorus at least nine times out of ten (90th percentile). The designer should proceed in three steps:

1. Use seasonal variations (either temperature-driven or month-driven) to establish the bottleneck period, which may be either winter or late summer. The month with the highest expected outlet trend concentration is selected as the controlling month.
2. Account for the random part of the effluent concentration distribution during the controlling month. For instance, this could be accomplished by choosing Ψ to be the median value of 90% (0.90) from Tables 10.10–10.13.
3. Increase the wetland size so that the calculated seasonally variable effluent concentration (C_{trend}) does not exceed $0.3/(1 + 0.90) = 0.158$ mg/L during the bottleneck period.

The reader should note that this correction is not a safety factor. It is known from the nature of existing datasets that excursions will occur. It becomes necessary to plan for those excursions, so that potential exceedances occur at less than the predetermined frequency. Wetland data may sometimes contain sufficient intrasystem variability to place the 100th

TABLE 10.12

Trend multipliers (1+Ψ) for effluent total phosphorus concentrations in treatment marshes. Data are analyzed on a monthly basis. The trend value is computed from the PkC* model.

Excursion Frequency Years		0.5	0.8	0.9	0.95
Lakeland 1	7	1.19	1.47	1.75	1.81
Linköping	3	1.08	1.40	1.97	2.15
New Hanover	2	1.00	1.61	1.72	2.04
Musselwhite	4	0.97	1.42	1.79	2.17
Imperial	4	1.44	2.87	3.80	4.94
Brawley	4	1.02	1.23	1.38	1.46
Ekeby	3	0.89	1.47	1.90	2.40
OEW	10	1.04	2.63	3.29	4.30
Commerce Twp	5	0.98	1.13	1.23	1.40
Estevan	6	0.81	1.27	3.56	6.11
Boney Marsh	9	0.88	1.22	1.53	1.91
Listowel 3	4	0.99	1.87	2.76	3.00
Listowel 4	4	0.95	1.86	2.63	3.20
Listowel 5	4	0.98	1.72	2.94	3.41
Brighton	10	1.04	1.51	1.87	2.39
Median		**0.99**	**1.47**	**1.90**	**2.39**
Mean		**1.02**	**1.64**	**2.27**	**2.84**

percentile above the median inlet concentration. Thus it may not be possible to design a wetland to totally avoid the possibility of monthly exceedances.

Longevity of P Removal Marshes

There has been confusion about the life expectancy of FWS treatment wetlands for phosphorus removal. As discussed in this Chapter, soil sorption and biomass expansion are processes that have finite phosphorus removal capacities. However, soil accretion is a sustainable mechanism that requires only a long hydroperiod to prevent oxidative release. Two lines of information allow confirmation of the long-term sustainability of phosphorus removal in treatment marshes: the life histories of wetlands that have had continuous monitoring, and the current performance of systems that have been receiving wastewater with phosphorus for extended (unmonitored) periods.

Monitored Systems

At this point in the history of treatment wetland technology, there now exist some monitoring histories of modestly long duration. Time trends allow confirmation of the long-term sustainability of P removal in these marshes.

As an illustration, data was analyzed for eight steady flow treatment wetlands, including two large natural wetlands. Linear regression was used to establish the time trend slope of the k-values, expressed as percent change per year. Time trends for k-values for these steady flow systems are shown in the top part of Table 10.13. Only slight trends were

TABLE 10.13

Time trends in annual rate coefficients (k) for large P wetlands. Two natural wetlands are included because of their long history.

Site	System	Years Trend	k slope %/yr	R^2
Steady Flow				
Boney	West	11	2.4	0.094
OEW	Stratum 4	12	4.7	0.061
Titusville		7	4.5	0.014
Lakeland	Cells 1–4	12	2.7	0.041
Shannon	Cells 1–4	6	7.5	0.347
Bayou Marcus	Natural	14	0.1	0.000
Houghton Lake	Natural	26	-0.2	0.004
Apopka	All	9	1.3	0.031
Median		**11.5**	**2.6**	**0.036**
Event Flow				
STA1W	East	14	-2.4	0.041
STA1W	West	12	0.8	0.012
STA1W	North	16	6.1	0.323
STA2	FW1	12	4.8	0.095
STA2	FW2	13	1.2	0.030
STA2	FW3	13	-0.7	0.010
STA56	FW1	16	1.9	0.014
STA56	FW2	16	-0.1	0.000
Median		**13.5**	**1.0**	**0.022**

present in performance, as quantified by linear regression, with a slight increase in median k-values over time of 2.6%/yr. These trends account for only a small amount of the variability, with $R^2 = 0.036$. In other words, the interannual data scatters, and displays almost no trends.

The south Florida stormwater treatment areas form a well-studied subset of wetlands designed to remove phosphorus. Data was also analyzed for eight of these event-flow treatment marshes. The systems receive pulses of flow on an intermittent basis. Linear regression was also used to establish the time trend slope of the in k-values. Again, only slight trends were present in performance. There was a slight increase in median k-values over time of 1.0%/yr. The linear trend accounts for only a small fraction of the annual variability, with a median $R^2 = 0.022$. In other words, as for the continuous flow examples, the interannual data scatters, and displays almost no trends for CR and k.

Long-Term Track Records

Some of the earliest treatment wetland projects have continued under continuous monitoring for up to 30 years. Two are reviewed here, to illustrate the long-term behavior of FWS wetlands for phosphorus removal.

The Houghton Lake, Michigan, system is a natural wetland project, which is described in detail in an entire issue of Ecological Engineering (Vol. 35, No. 9). The wetland was originally primarily a sedge meadow of modest hydroperiod (around 50%). Mesocosm studies (1973–1974) and a 360 m^3/d pilot project (1975–1977) indicated positive results for phosphorus removal. A 9,000 m^3/d full-scale project started in 1978 and continues to present. Because the impact zone is less than the full extent of the wetland, analysis focuses on a 100-ha subsystem around the discharge, which is defined by established sampling locations, and fully containing the visually impacted zone of the wetland (80+ ha). The incoming lagoon water averaged 3.21 mgP/L, at a hydraulic loading of 0.2 cm/d for the first 28 years of operation. Removal was 2.08 gP/m^2·yr, or 92% of the incoming mass, with no allowance for background concentration. The time sequence of events (Figures 10.28) shows a slight decreasing trend, to 88% mass reduction at Year 30.

Perhaps more interesting is the appearance of increased variability in the percent removal during the last ten years of operation. This scatter can be traced to variations in system hydraulics and hydrology, more so than to variations in inlet phosphorus concentrations. The outgoing load, and hence the mass removal, reflects the water loss to evapotranspiration, which has been high and variable in recent years at the site.

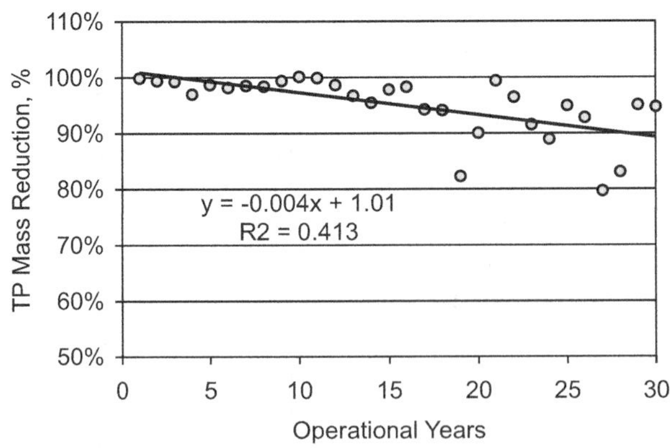

FIGURE 10.28 The time series of percent mass removals of phosphorus for the Houghton Lake, Michigan, treatment marsh.

A second long-term track record is associated with the Orlando Easterly Wetland (OEW). The City of Orlando, Florida, employs a 490-ha FWS constructed wetland to polish tertiary wastewater. The wetland began operation in 1987, and over the next thirty years received water at about 55,000 m^3/d, with TP concentration of 267 μg/L. The facility is comprised of 17 cells in series and parallel, 16 of which are mixed community marshes, plus one deep water cell (Wang et al., 2006a; Wang and Jawitz, 2006). TP concentrations were reduced to 64 μgP/L over the period 1988–2017, at a hydraulic loading rate of 1.15 cm/d. The corresponding phosphorus loading rate was 1.16 g P/m^2·yr. Over the first ten years of its history, incoming TP concentrations dropped from about 800 to 200 μgP/L, and then fluctuated between 200 and 400 (Figure 10.29). During the entire history, there was no trend in concentration reduction, which was 73 ± 9%.

Although the annual average performance continued to be relatively stable, an amplified seasonal pattern began to emerge in year ten. This created winter monthly performances that were unacceptable to the City. Studies determined that sediment accretion in the front-end cells was causing hydraulic problems. Accordingly, sediment removal was conducted in the front-end cells during 2002–2003 (Year 15 of operation) (Wang et al., 2006). The front-end performance in the first year after the rehabilitation was much improved for hydraulics, and for TP removal (5 cells: P = 12, k = 30 m/yr, C* = 40 μgP/L).

The lesson from this long-term project is that new sediment accretion may become a problem after a decade or two of operation. The accretions are not likely to be spatially uniform, with both gradient effects in the flow direction, and patchy development. The hydraulics of the system may be impaired long before the freeboard of the containment levees is jeopardized. Design features such as multiple controllable inlet points and adjustable water level controls are beneficial in the long-term control of performance.

Historical Unplanned Projects

In the 20th century, it was not uncommon for wastewater treatment plants to discharge to existing marshes. Phosphorus removal was not part of pretreatment, and hence considerable phosphorus loadings reached those marshes. Two such wetland discharges are reviewed here to illustrate the continuing ability of wetlands to remove phosphorus.

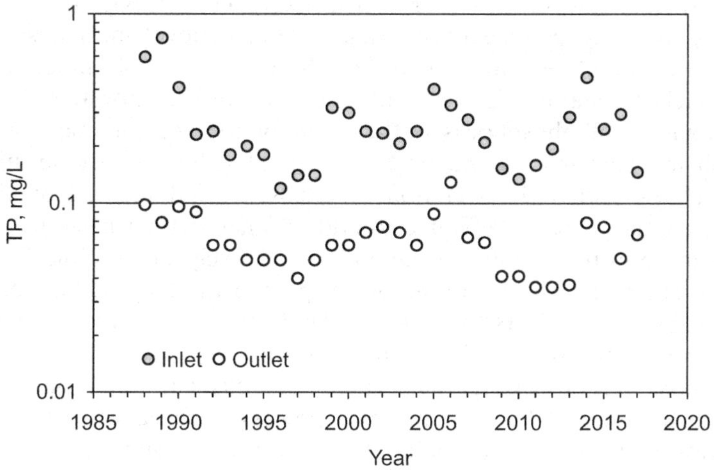

FIGURE 10.29 Time series of inlet and outlet TP concentrations for the Orlando, Florida, Easterly Wetland. Flows to this system were nearly uniform over the period of record. Data source: City of Orlando unpublished data.

The Brillion Marsh in central Wisconsin is a 156-ha natural wetland impacted by incoming wastewater beginning in 1923. The exact discharge quantities and changes over time are not known, but an intensive study was conducted over the two-year period 1974–1975 (Spangler et al., 1976; Fetter et al., 1977; Fetter et al., 1978). The marsh was a uniform stand of *Typha latifolia* and *Typha angustifolia*, with a channel input dominated by treated wastewater. Outflow was also confined to a channel, and there were small tributary streams to the marsh conveying agricultural runoff. During the study, 16,300 m^3/d of influent carried 3.3 ± 2.2 mgP/L into the marsh. The hydraulic loading rate of one centimeter per day carried a load of 12.6 gP/m^2·yr. The outgoing TP was 2.6 mgP/L, for a concentration reduction of 22%. The load removed was about 2.8 gP/m^2·yr. These studies indicate a continuing ability of this marsh to remove phosphorus.

The Great Meadows National Wildlife Refuge in Concord, Massachusetts, consists of diked wetland areas constructed around 1930. Sewage treatment plants had discharged into the constructed wetlands, indirectly from 1930–1962, and directly 1962–1978, with 1978 being the year of a hydrologic and water quality assessment (Yonika et al., 1979). Vegetation was *Phalaris arundinaceae* and *Typha latifolia* in shallower areas, and *Nuphar variegatum* and *Nelumbo lutea* in deeper zones. The 22-ha system received 2090 m^3/d of secondary effluent, for a hydraulic loading rate of about one centimeter per day. The inlet TP was 2.00 ± 0.88 mgP/L, producing a phosphorus loading of 6.9 gP/m^2·yr. The outlet TP was 0.51 ± 0.40 mgP/L, for a 75% concentration reduction. Phosphorus removal was seasonal, with an early spring maximum rate of 12 gP/m2·yr, and a mid-winter release of 5 gP/m^2·yr, and an average annual removal of 3.7 gP/m^2·yr. This treatment wetland was functional for phosphorus removal even after several decades of operation.

10.5 SUMMARY

Treatment marshes are capable of absorbing new phosphorus loadings. Phosphorus interacts strongly with wetland soils and biota, which provide both short term and sustainable long-term storage of this nutrient. Soil sorption may provide initial removal, but this partly reversible storage eventually equilibrates with incoming source water phosphorus. Uptake by biota, including mesophytes as well as macrophytes, forms an initial removal mechanism. Cycling through growth, death, and decomposition returns most of the biotic uptake, but an important residual contributes to long-term accretion in newly formed sediments and soils. Despite the apparent complexity of these several removal mechanisms, data analysis shows that relatively simple equations can describe the sustainable processes. A first-order removal rate model can characterize removal, but does not incorporate any biotic features.

Sustainable removal of phosphorus is the result of the accumulation of new sediments and soils and their associated P content. New wetland sediments accrue due to settling of incoming particulates, and by bioaccretion, which is the end product of biotic cycling. If sedimentation is too large, it is preferable to add an inlet settling basin instead of allowing sediments to build up in the marsh. Bioaccretion is relatively small, typically in the range of 0.5–3.0 gP/m^2·yr. Herbivory can be an important part of the P cycle, but generally does not result in sustainable removal. Harvest of wetland vegetation is a potential sustainable removal, but is constrained by technical difficulties.

Marsh phosphorus removal is spatially and temporally variable. Longitudinal gradients in P concentrations are normally present. Some processes are rapid, such as particulate settling and filtration and mesophytic uptake. Others are much slower, such as the establishment of the vegetative biomass and cycle, and the possible alteration of the biological communities in response to new P loadings. Both transient and sustainable removal processes occur, with periods of initial adaptation of up to several years. Similarly, a wetland may exhibit a slow response (one to two years) to changing inlet conditions.

There exist return fluxes of phosphorus from biological solids (floc and detritus), and consequently net P removal is a balance between uptake and return. Simple, highly aggregated descriptions, such as the PkC* model, can explain a good deal of both intra-system and inter-system variations. More detailed models have been developed, but are often too cumbersome to be of value in design or operations management.

Wetland water temperature appears to be only a minor modifying factor on phosphorus sequestration, and in many cases, seasonality is dominant. In some instances, spring and autumn performance is higher compared to warmer or colder seasons, for reasons that are not fully understood. The absence of vegetation lessens removal capability. Further, the community type (EAV, SAV) has significant effects, but not in all concentration ranges. However, species composition of a given community type appears to be a relatively minor factor. Deterministic models are not capable of describing the purely stochastic variations in phosphorus removals. That probabilistic component of behavior can be adequately described using statistical descriptors.

Treatment marshes now have a well-understood track record spanning several decades. They do not "wear out," or "become saturated." Therefore, they can fulfill a useful role in phosphorus reduction in many situations. Improvements in water quality for tertiary effluents are possible, but there is perhaps an even greater role in controlling nutrients in urban and agricultural runoff. This positive connotation must, however, be tempered by the large land requirement that usually accompanies the use of a passive wetland to immobilize phosphorus.

11 Event-Driven Systems

The treatment of stormwaters of various origins is of growing concern, as attempts to rectify point source pollution reach maturity. Runoff from urban, agricultural and industrial sources comprises a sizeable fraction of the total pollution load to receiving waters in many locations. Often, these sources are relatively dilute compared to primary or secondary domestic sewage, but nevertheless they may negatively impact receiving waters.

Urban stormwater wetlands were first surveyed by Strecker et al. (1992), who documented the performance of 25 natural and constructed wetlands treating runoff. The implementation of the technology and the knowledge base continued to build, resulting in a compilation of data from 76 wetlands worldwide by Wong et al. (1999), and from 49 wetlands in the United States by Carleton et al. (2001). There is now an international database that contains information on 107 urban stormwater wetland systems, covering over 2200 events (Leisenring et al., 2014; Clary et al., 2017a).

These are virtually all pond-marsh systems. A typical configuration consists of a sedimentation basin as a forebay, followed by some combination of marshes and deeper pools. Design guidelines have now been promulgated by a number of governmental agencies (for example: North Carolina – Burchell et al., 2010; Hunt et al., 2010; Minnesota – Minnesota Steering Committee, 2018; Saskatoon, Saskatchewan – CH2MHill, 2014; Melbourne, Australia – Melbourne Water, 2013; Washington D.C. – Hoffman et al., 2013).

Agricultural stormwater occurs as runoff from crops and pastures. Non Point Source (NPS) pollution from agriculture may occur when nutrients are applied at rates greater than crops can utilize or when timing of nutrient applications occurs in close proximity to heavy rains (Stone et al., 2003). A fraction of the fertilizers applied to fields is unavoidably lost to runoff and shallow groundwater. Tile drainage systems collect subsurface waters, and vent them to receiving streams, where their nitrogen content, together with phosphorus, may cause problems. Wetlands are being used at various points in the agricultural landscape, corresponding to drainage areas ranging from individual fields (U.S. Department of Agriculture, 1991; Tanner et al., 2003), to small order streams (Stone et al., 2003), to large regional landscape units of thousands of hectares (Reilly et al., 2000). Some of these receive pumped water at relatively constant rates, and these have been included under the discussion of continuous (but possibly variable) flow wetlands. Those that receive water as a result of meteorological events are considered here.

Early work on constructed wetlands for row crop runoff control was centered in the state of Maine (Higgins et al., 1993) (Wengrzynek and Terrell, 1990; U.S. Department of Agriculture, 1991). Runoff from sugar cane is being treated in the huge FWS wetlands (> 24,000 ha) called stormwater treatment areas (STAs) of south Florida (Goforth, 2001). Water quality problems in the Gulf of Mexico, stemming from excessive nitrate in watershed runoff, have stimulated numerous wetland projects in the agricultural Midwestern United states (Crumpton et al., 2012). In Sweden, the focus has also been on excessive nitrogen loads from agriculture, impacting receiving waters including the Baltic Sea. Over 1500 small wetlands were built to alleviate water quality problems from 1996–2006 (Arheimer and Pers, 2017). Although numerous and fairly effective, the aggregate wetland area is not sufficient to affect much of the pollution load.

Whatever the source, wetlands that receive pulsed flows utilize the same suite of wetland removal mechanisms as their continuous-flow counterparts, but in different ways, dictated by the transient hydraulics. Data acquisition is by no means commensurate with the numbers of event-driven wetlands. Both urban and agricultural systems are often not monitored extensively enough to quantify their performance. The available information comes from isolated study wetland systems, and occasionally from selected groups of small wetlands. Design is often technology-based, meaning that some agency has decided what a certain size wetland can do, and committed that opinion to rules and regulations.

The focus here is upon sediments and nutrients, but most metals and pesticides are also reduced in these systems. However, there is, in general, not a sufficient database or mechanistic understanding to develop quantitative, process-based descriptions. These contaminants are discussed a bit further in Chapter 12.

11.1 SOURCE CHARACTERIZATION

A large part of the design of event-driven treatment wetlands is the determination of the flows and loads to be treated. Compared to continuous or pumped treatment marshes, event-driven systems do not have easily characterized source waters. The anticipated flows and their temporal distribution will not have record upon which to base design, thus these require estimation. There is no track record of the source water quality to be treated, requiring estimates to be made. Fortunately, extensive databases of runoff water quality have been compiled, especially for urban runoff. Predictive methods are available, ranging from simple equations and graphs to sophisticated computer software. Several models are available that can forecast flows, and in some instances concentrations of contaminants, generated by the contributing watershed. Such source watershed models take in account land use types and patterns, as well as the characteristics of rainfall, which forms the primary driver of runoff.

Incoming Flows

The amount of water to be expected from a given watershed is variable, and is keyed to rainfall in the contributing basin. The land uses and soils in the contributing basin are an important modifier of the runoff and infiltration, as is the mode of water collection, whether by surface runoff or tile drainage. The antecedent dryness in the basin is also a contributing factor. An example of a tile drainage flow from a pasture into a wetland is shown in Figure 11.1

The focus here is the treatment wetland, and it will be presumed that the incoming flows will be characterized to the appropriate level of detail.

All predictive methods require knowledge of the area of the contributing watershed (catchment). Performance predictions of all kinds are dependent upon the area of the treatment marsh. It is therefore logical to characterize the combination by the relative sizes of wetland and watershed. Therefore, the wetland–watershed area ratio (WWAR) is an often-used descriptor of event-driven treatment marshes. The effect WWAR on performance is not well quantified or predictable, but can provide a high level understanding of an approximate size of wetland.

Approximate Methods – The Rational Method

For purposes of rough estimation, the rational formula may be used to estimate peak flow, Q_P (ASCE, 2006):

$$Q_P = C_R \cdot I \cdot A_b \tag{11.1}$$

where: A_C = catchment watershed area, m^2

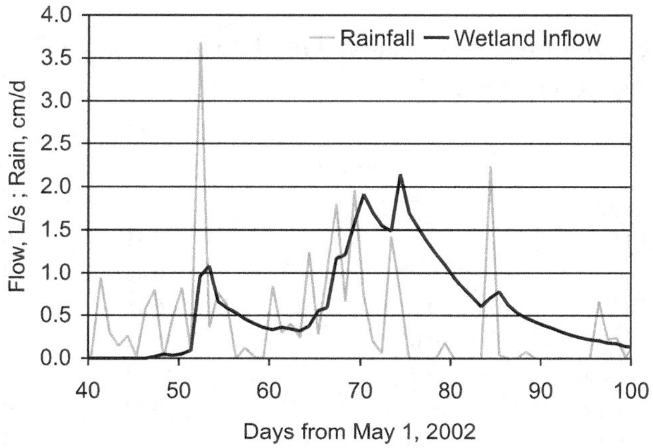

FIGURE 11.1 Time series of flows entering a treatment wetland from an improved pasture in Toenepi, New Zealand. Several rain events occurred during this winter wet season, as indicated by the repeated spikes in inflow. The runoff coefficients were 0.23–0.30, and were the result of tile drains. (Adapted from Tanner et al., 2005b)

C_R = runoff coefficient, –
Q_P = peak runoff flow, m^3/hr
I = average rainfall intensity, m/hr

The rainfall intensity, I, is selected from rainfall intensity-duration-frequency tables using a recurrence interval and a duration equal to an appropriate rainfall averaging time. The watershed time of concentration, t_C, is most commonly used as the rainfall averaging time. The time of concentration is defined as the travel time for water to flow from the most hydrologically remote point in the watershed to the watershed outlet. It depends primarily on land slope and travel distance in the catchment; for example:

$$t_C = a \cdot L^m \cdot S^{-n} \tag{11.2}$$

where: a,m,n = regression constants
L = travel distance, km
S = catchment land slope, m/m
t_C = time of concentration, hr

Typically, the exponent m = 0.5–0.7, and the exponent n = 0.33. The coefficient a varies; one example is a = 0.128 used for watersheds in Kansas (Young et al., 2009).

The values of C_R are a function of land use and storm intensity, as well as the climatological region. An example of the range of C_R for an arid region is shown in Table 11.1. To illustrate the regional effect, consider that agricultural land in Arizona is rated at C_R = 0.1–0.2 for high return frequency events, while the value for agricultural land in south Florida is $C_R \approx 0.5$. Water is used consumptively in arid regions, while the high water table and rainfall in a wet region leads to more runoff. Modifications to this simplest formula have been presented on a site-specific basis (see, for example, Brezonik and Stadelmann, 2002; ASCE, 2006).

Approximate Methods – The SCS Method

It is apparent that the rational method is very simplistic, and requires table lookup to estimate values for a particular type of watershed and rainfall return frequency. Another

TABLE 11.1

Runoff coefficient (C_R) for use in the rational method in Maricopa County, Arizona. Values are mid-range ((max+min)/2).

	Return Frequency			
	2–10 yr	25 yr	50 yr	100 yr
Very Low Density Residential	0.38	0.43	0.50	0.55
Low Density Residential	0.45	0.51	0.57	0.62
Medium Density Residential	0.57	0.63	0.68	0.70
Multiple Family Residential	0.70	0.78	0.84	0.88
Light Industrial	0.65	0.72	0.78	0.82
General and Heavy Industrial	0.75	0.83	0.90	0.92
Light Neighborhood Commercial	0.60	0.67	0.72	0.75
Central, Office Commercial	0.80	0.89	0.93	0.95
Pavement and Rooftops	0.84	0.92	0.95	0.95
Gravel Roadways & Shoulders	0.68	0.75	0.80	0.88
Agricultural	0.16	0.17	0.19	0.25
Undeveloped Desert Rangeland	0.37	0.40	0.43	0.50

Source: Gerlach et al. (2013).

historical and frequently used procedure is the SCS (U.S. Soil Conservation Service of the U.S. Department of Agriculture) method (McCuen, 1982; Novotny, 1995), which has been repeatedly updated (SCS, 1999, 2004). Infiltration and surface storage in the catchment are combined under a variety of soil and cover conditions. There are two steps. First, the general character of the catchment land use is identified and categorized by soil type. The result is a "curve number" that characterizes the runoff by event magnitude. These are shown in Figure 11.2. Table 11.2 gives examples of how catchment characteristics govern the choice of which curve to use. Basically, the SCS procedure adds the event magnitude to watershed features in the determination of a runoff coefficient.

Computer Methods

For modeling purposes, the time series of water flows entering the treatment wetland inlet hydrograph is needed. This time series is driven by the pattern of rainfall associated with a particular event (represented by a hyetograph), together with the collection and transfer characteristics of the basin. In some instances, such transfer is solely by gravity, but in some cases, pumps may be used. Usually, the inlet hydrograph contains both a rising and falling limb, although the rising limb is normally quite steep. The required specificity can be achieved with the use of computer models of watershed flows. Novotny (1995) discusses eleven such computer models, and there are currently about two dozen commercially available software packages, with manuals. Examples of these are:

- SWAT (Soil and Water Assessment Tool) (Nietsch et al., 2011). This program was developed to predict the impact of land management practices on water and chemical yields in large complex watersheds. SWAT is physically based, requiring information on weather, soils, topography, vegetation, and land management practices. It is a continuous time model. It has been implemented for description of the flows and loads reaching the Iowa CREP (Conservation Reserve Enhancement Program) treatment marshes (Ikenberry et al., 2017a, 2017b).

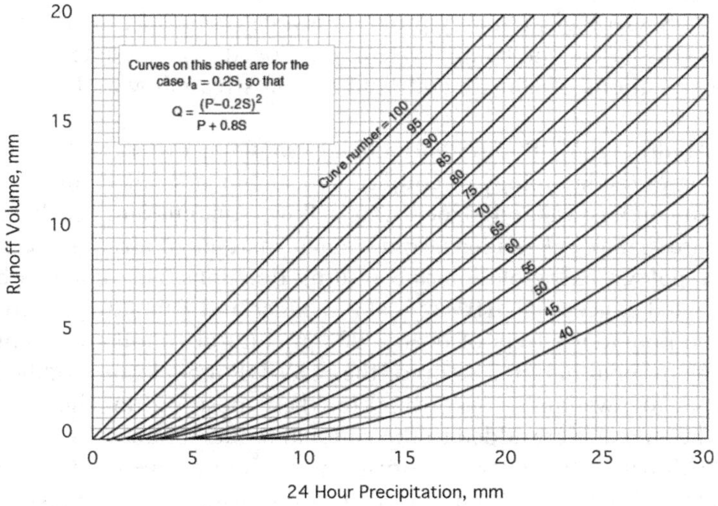

FIGURE 11.2 Relation between rainfall and runoff for the SCS procedure. Modified to metric units from SCS (1999).

TABLE 11.2

Example SCS curve numbers for different soil types and catchment land uses.

		Sand or sandy loam	Silt loam	Sandy clay loam	Clay or clay loam
Agricultural					
Fallow	Bare	77	86	91	94
Fallow	Crop Residue (CR)	74	83	88	90
Row Crops	Contoured (C)	65	75	82	86
Row Crops	C+CR	64	74	81	85
Row Crops	C+Terraced	62	71	78	81
Row Crops	C+CR+Terraced	61	70	77	80
Row Crops	Straight	67	78	85	89
Row Crops	Straight+CR	64	75	82	85
Urban					
Open	Grass 50–75%	49	69	79	84
Paved	Parking lots, etc.	98	98	98	98
Streets & Roads	Paved	83	89	92	93
Streets & Roads	Gravel	76	85	89	91
Urban districts	Commercial	89	92	94	95
Urban districts	Industrial	81	88	91	93
Residential	1/8 acre	77	85	90	92
Residential	1 acre	51	68	79	84

Source: SCS (1999).

- PRMS-IV, the precipitation-runoff modeling system, version 4. (Markstrom et al., 2015) This U.S. Geological Survey program is a deterministic, distributed-parameter, physical-process-based modeling system developed to evaluate the response of various combinations of climate and land use on streamflow and general watershed hydrology.
- HBV-NP (Hydrologiska Byråns Vattenbalansavdelning) (Pers, 2007). The HBV-NP model simulates the hydrology of coupled subcatchments called subbasins. HBV-NP. The latter integrates water balance, water flow, soil nutrient leaching, additional nutrient sources, nutrient residence and transformations in ground and surface water as well as typology, extrapolation and routing functions The model is based on the semi-distributed conceptual HBV model Arheimer and Pers (2017) used HBV-NP to illustrate the potential effects of over 1500 agricultural runoff treatment marshes in Sweden. The original national setup of the HBV-NP model was modified to simulate several types of wetlands, according to their position in the landscape. The constructed wetlands were added to the national model set up.
- MUSIC (Model for Urban Stormwater Improvement Conceptualization) (Department of Planning and Local Government, 2010). This software was constructed to accompany Water Sensitive Urban Design (WSUD) for the Greater Adelaide Region. WSUD is an approach to urban planning and design that integrates the management of the total water cycle into the urban development process (Castonguay et al., 2018). MUSIC is designed to simulate urban runoff systems operating at a range of temporal and spatial scales and provides a user friendly interface to allow complex stormwater management scenarios to be quickly and efficiently created, with results viewed using a range of graphical and tabular formats. MUSIC provides the ability to simulate both quantity and quality of runoff from catchments and the effect of treatment facilities on these components (Montaseri et al., 2015).

Several or such computer models also have a capability to forecast the effectiveness of the treatment marsh that receives the runoff. This is accomplished through built-in models of wetland pollutant removal.

In many instances, rain events are separated by inter-event periods of no inflow to the treatment wetland. These periods are important, because the wetland will act as a batch reactor during much of these no-inflow durations. A typical sequence is (1) wetland filling with no outflow, (2) flow-through with both inflow and outflow, (3) draining with no inflow, and (4) finally a batch holding mode with neither inflow nor outflow. The rates and duration of these inflows and outflows are in part controlled by structures. The volume of water in the wetland is also in part controlled by structures, but evapotranspiration may be an important component during inter-event periods. The durations of no-flow periods are very much a function of the climatological region in which the system is located. For instance, in Florida, during the rainy season, the most frequent dry periods are measured in hours (4–20 hr, Wanielista and Yousef, 1991). But the dry season of central north island New Zealand has periods of months with no rain, and inter event periods can be several months (Tanner et al., 2005b).

Many stormwater wetlands are fed by pumps. These may range in size from small sump pumps for small localized urban systems, to the huge pumps that send water to the Florida stormwater treatment areas (STAs). Those Florida inflow and outflow pumps are among the largest in the world, up to three stories tall, with capacities up to 10,000,000 m^3/d (Figure 11.3). A key feature of pumped systems is intermittent feed to the treatment wetland, usually at fixed but incremental rates, corresponding to the number of pumps that are operated, for periods of time dictated by conditions in the contributing watershed.

FIGURE 11.3 The outflow pump station from STA1E (Stormwater Treatment Area 1 East) of the Everglades Protection Project. The capacity of this pump station is 9,740,000 m^3/d, and it serves to drain a 2,700-ha wetland.

There is typically a hydraulic limitation to the event size that may be treated in a given wetland. It is generally not feasible to size a wetland to treat the 100-year return frequency storm, because of cost and footprint considerations, and because of the need for maintenance water for the (large) wetland under normal conditions. Consequently, part of the design decision process is the determination of the maximum design storm to be treated. Even if the system is sized to treat a modest sized storm flow, there remains the possibility that dry conditions might persist in some years, to the point that the integrity of the wetland ecosystem is jeopardized. If such detrimental dryout conditions are expected, then a source of irrigation water for ecosystem maintenance should be identified.

Incoming Concentrations and Loads

Urban Stormwater

Concentrations of most parameters in stormwater are time-dependent, as are the flows. Stormwater concentrations and loads are episodic due to periods of dryfall and deposition, followed by the first flush of runoff after rain, followed by exponential decreases in runoff constituent concentrations as storages rinse from the landscape, and finally, dry conditions and deposition until the next storm event. Event mean concentrations (EMCs) are used to quantify observations (see Chapter 7, Equation 7.91).

The National Stormwater Quality Database (NSQD) is a U.S. EPA effort to compile stormwater quality data of the USA (Pitt and Maestre, 2015). It includes the monitoring results of NPDES Communities, the National Urban Runoff Program (NURP), the BMP Database, and special projects that have been added since 2001. It includes Event Mean Concentration (EMC) records, information about sampling methods, climatological information, and land use. The NSQD is being maintained as a separate stand-alone database, serving as an important resource for municipal stormwater managers and researchers who are seeking urban runoff characterization data. It has recently moved to a new long-term home as a companion project to the International Stormwater BMP Database (Leisenring et al., 2014; Clary et al., 2017a; www.bmpdatabase.org/). Concentration distributions for various stormwater pollutants are summarized in Table 11.3.

Pollutant concentrations generally range from low levels from undeveloped and park lands, to low density residential and commercial, to agricultural, to higher density residential and commercial, and finally to high density commercial, industrial, and residential land uses. Suspended solids are significant, with a median of 64 mg/L across 7637 monitored events (Table 11.3). Concentrations of organics (BOD_5) are fairly low, with a median of about 9 mg/L. Oxidized and ammonia nitrogen are at about 0.6 and 0.4 mgN/L respectively. A ubiquitous organic nitrogen component is present, causing total nitrogen to be at about 2.2 mgN/L. Total phosphorus has a median of 0.25 mgP/L, which is high enough to cause eutrophication problems in many receiving waters. Trace metals are also present, such as iron and lead (Table 11.3).

In some watersheds, the chemograph is not synchronized with the hydrograph, but instead provides higher concentrations early in the inflow event. This phenomenon is termed "first flush" behavior, referring to the surge of pollutants contained in the first water to leave the contributing basin. For instance, Wanielista and Yousef (1991) report that in Florida, the first 25 mm of runoff from urban systems typically carries 90% of the pollution.

The mass loading rates represent normalized pollutant loads that are somewhat independent of local rainfall amounts. Because pollutant loads per area per time are relatively constant between similar land use areas, variable local rainfall washes these loads off the land in a few large events or over many smaller events. Urban pollutant loads increase with the imperviousness of the watershed. Although 20 to 40% of the material on street surfaces is organic, it does not biodegrade easily because it comes from leaf and wood litter, rubber, and road surface material (Novotny, 1992). The high metal content of highway solids

TABLE 11.3

Concentrations of common pollutants in urban stormwater.

	TSS mg/L	BOD5 mg/L	TN mg/L	NH3N mg/L	NOXN mg/L	TP mg/L	Iron µg/L	Lead µg/L
Observations	7637	5152	1211	2946	5675	7943	608	4960
Percentile								
0.10	13	3.3	0.70	0.12	0.17	0.083	190	2.0
0.20	22	4.8	1.17	0.19	0.30	0.120	263	5.0
0.25	28	5.1	1.30	0.22	0.35	0.140	306	5.8
0.30	34	6.0	1.47	0.26	0.40	0.160	349	7.0
0.40	47	7.1	1.79	0.33	0.50	0.200	470	10.0
0.50	64	8.8	2.16	0.43	0.60	0.250	605	14.0
0.60	87	11.0	2.60	0.56	0.72	0.310	876	19.0
0.70	120	14.0	3.19	0.72	0.89	0.390	1244	27.0
0.75	141	16.0	3.47	0.83	0.99	0.443	1498	32.1
0.80	178	19.0	3.85	1.00	1.10	0.520	1890	41.6
0.90	313	29.0	5.50	1.70	1.60	0.770	4000	86.0

Source: National Stormwater Quality Database, Pitt and Maestre (2015).

comes from vehicle emission. Novotny (1992) reported that the average total nitrogen load from urban lands is 5 kg/ha·yr (1 to 38.5 kg/ha·yr), and the total phosphorus load averages 1 kg/ha·yr (0.5 to 6.25 kg/ha·yr).

Agricultural Runoff

Nitrogen from field runoff is prevalent problem that has been recognized in the USA and Scandinavia as a threat to marine environments. Agricultural activity in the Midwestern USA, primarily fertilizer use and soil loss, has been previously identified to be the dominant source of N to the Mississippi River. The Illinois River is an important tributary to the upper Mississippi River, and NO_3-N concentrations in the Illinois River are among the highest of the large tributaries feeding the Mississippi River. Although land use in the Illinois River watershed is dominated by row crop agriculture and most of the NO_3-N load to the river is from agricultural sources (ca. 80%), discharge of treated wastewater from the Chicago region accounts for a considerable amount (Kovacic et al., 2000; Panno et al., 2008).

Similar results have been found for phosphorus. For example, Foy and Lennox (2000) estimated about 80% of the P load to the River Main in Ireland was attributable to agriculture, mainly animal husbandry. They suggested that the imbalance between P inputs and outputs is a standard feature of livestock farming in Europe and the USA, and that these regional surpluses provide a long term threat to water quality.

Although agricultural runoff systems are superficially similar to urban stormwater systems, there are very large differences in the timing and quality of the waters to be treated. The rates of surface runoff from agricultural landscapes also depend upon rainfall within the catchment, and patterns resemble those of urban watersheds. However, many agricultural watersheds collect and remove water via a tile drainage system. Tile systems are also rain-driven, but serve to provide a measure of flow equalization, as subsurface storage continues well after the cessation of rainfall.

The target contaminant in many agricultural watersheds in nitrogen, either as nitrate or as total nitrogen. In other situations, phosphorus and/or solids may be the focus, such as in the Everglades protection projects in south Florida. Field runoff quality is strongly influenced by fertilizer applications. As a consequence, nutrient concentrations and loads from row crops depend on fertilization practices and crop management practices. Reductions in field nitrate can be accomplished by using conservation tillage, or by using no-till management (Table 11.4). Cover crops, planted primarily to manage soil erosion, also help suppress nitrate levels in drainage waters.

The large agricultural watershed of south Florida illustrates the levels of phosphorus that may be expected in field runoff. Sugarcane covers a major portion of the landscape, and runoff is typically about 170–210 μgP/L (Figure 11.4). Row crop vegetables exhibit somewhat higher TP concentrations. Heightened awareness of downstream water quality problems cause the implementation of on-farm practices to reduce the P loads from the fields, which have been successful in providing outflow P loads reduced by about 50% since 1994. Flows and loads from the farms are documented yearly (South Florida Environmental Reports). Nonetheless, 30,000 ha of treatment marshes have been built to cope with the remaining phosphorus.

Tile drained watersheds also behave differently from urban watersheds with respect to the first flush phenomenon. High frequency (15 minute) monitoring of the flows and concentrations emanating from the Slough Creek, Iowa, and the behavior of the receiving treatment marsh, were monitored by Drake et al. (2018). Instead of high concentrations being rinsed off watershed surfaces early in an event, the inverse was observed (Figure 11.5). It appears that rainfall diluted the nitrate-rich source waters, creating low concentrations early in the response period.

TABLE 11.4

Nitrate nitrogen in tile drains leaving corn and soybean fields in Iowa.

	Count	1st Quartile	Median	3rd Quartile
Tillage Practice				
		Concentration, mg/L		
Conventional	11	9.3	13.4	19.6
Conservation	22	11.2	12.6	19.1
No Till	16	8.8	11.3	12.5
		Load, kg/ha·yr		
Conventional	11	20.4	28.0	45.0
Conservation	17	17.0	18.4	29.7
No Till	14	14.2	24.1	38.0
Cover Crop				
		Concentration, mg/L		
Absent	17	9.9	12.0	15.3
Present	12	7.9	8.7	9.3
		Load, kg/ha·yr		
Absent	16	12.7	23.2	39.8
Present	14	16.1	23.5	27.2

Source: Clary et al. (2017a).

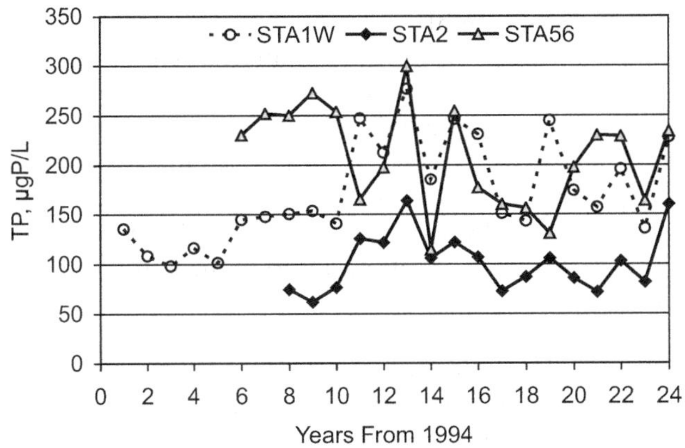

FIGURE 11.4 Example inlet TP concentrations to the Florida STAs over their POR. Means are STA1W = 172, STA2 = 102, STA56 = 209 µgP/L. Unpublished data from South Florida Water Management District.

11.2 HYDRAULICS OF PULSED AND SEASONAL SYSTEMS

Wetland Dynamics

Stormwater wetlands experience event flows and concentrations, followed by periods of batch operation. It is then necessary to account for the dynamics of water storage within the wetland. Event-driven wetlands are dynamic in all respects, and the principal underlying hydraulics exhibit variable water depths and flows. The behavior is strongly conditioned by the nature of inflow and outflow structures, which may be designed to improve detention and treatment (U.S. EPA, 2009). The most general urban situation may have a staged set of outflow control structures, but agricultural treatment marshes usually have only single structure. Here a simple and common case is explored for purposes of illustration. It will be presumed that the wetland is relatively small, and consequently behaves as a level-pool system, with no gradients in stage.

A simplified version of wetland water mass balance (see Equation 2.15) may be used to illustrate event dynamics. Bankloss, infiltration and snowmelt are not included, leading to the dynamic, level-pool water mass balance for the wetland for unsteady state inflows and meteorology:

$$\frac{d(Ah)}{dt} = Q_{in} - Q_{out} + A \cdot (P - ET - I) \tag{11.3}$$

where: A = wetland wetted area, m^2

ET = evapotranspiration, m/d

h = water depth, m

I = infiltration, m/d

P = rainfall, m/d

Q = instantaneous flow, m^3/hr

t = time, d

The stage is controlled by an outlet structure, here supposed to be a rectangular weir. Thus the outflow–stage relation would be given by the Francis formula (French, 1985):

$$Q_o = \beta w\, h_{ow}^{\alpha} \tag{11.4}$$

where h_{ow} = height over the weir, = H-H_w, m

H = wetland stage, m

H_w = elevation of weir, m

Q_o = outflow rate, m^3/d

α = calibration exponent, –

β = calibration constant, $(m^3/d)/m^{\alpha}$

It is noted that the wetted area is not constant, but changes with stage according to the bathymetry of the wetland, represented by the stage-area-volume relationship (refer to Chapter 2). Further, the wetland catches rainfall over an area typically greater than the wetted area, but loses water to ET from just the wetted area. Infiltration further contributes to water losses, and may be a significant fraction of the output from the wetland, as in the cases of the Hidden River urban wetland in Tampa, Florida (Carr and Rushton, 1995) and the Embarrass River agricultural marshes in Illinois, described by Kovacic et al. (2000).

An example of this analysis is illustrated in Figure 11.6. For illustration, a wetland of 2,500 m^2 is subjected to a 5 cm rain event, and the associated runoff from the contributing basin. The presumed conditions and wetland design are shown in Table 11.5. The wetland does not have a level bottom, rather, there is a presumed quadratic stage-volume relation, which implies that the wetted area increases linearly with stage. The hypothetical sequence of events is as follows:

- The wetland starting condition is partially full, with the water level 10 cm below the outlet weir.
- A steady rain begins at time zero, and persists for one day.

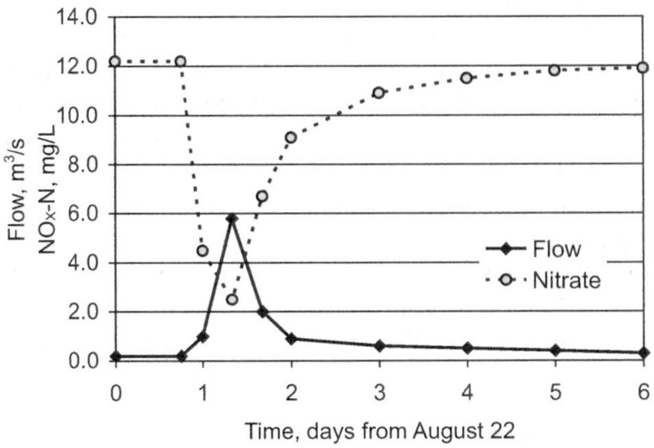

FIGURE 11.5 The time series of flows and nitrate concentrations reach the Slough Creek treatment marsh in Iowa in 2016. (Based on data from Drake et al., 2018)

- Runoff from the watershed begins after t = 0.5 days, and persists for 1.5 days.
- The wetland loses some water to ET (1.0 cm/d), and some to infiltration (1.0 cm/d).
- For the first half day (t = 0–0.5 days), the wetland fills due to direct collection of rain.
- At t = 0.65 days, the wetland has filled to the top of the outflow weir, and outflow commences.
- For the next half day (t = 0.5–1.0 days), the wetland fills because of direct rain and incoming runoff.
- For the next day (t = 1.0–2.0 days), the wetland fills because of incoming runoff.
- During t = 2.0–3.1 days, there is no inflow, and the wetland is draining over the outflow weir. At t = 3.1 days, the level has decreased to the top of the weir, and outflow stops.
- After time t = 3.1 days, the wetland loses water to ET and infiltration, and the level continues a slow drop.

Because of the assumed bathymetry, the wetted area first grows, and then shrinks through this course of events.

This simple model has been found to fit event wetland data quite well (Hey et al., 1994; Kadlec, 2001b). However, there are several different inlet and outlet structures that may be used, and the level pool model must be modified accordingly for other situations.

Flow and Capture

Event-driven wetlands operate with periods of flow-through and periods of water holding, or batch processing. Any water that does not escape the wetland during a particular event will be held until the next event, and possibly longer. Therefore, it is subject to the water quality improvement functions of the wetland for not only the event duration, but also the inter-event period. Conversely, water that enters and leaves the wetland during the event is subject to treatment only during the (possibly brief) period of detention during the event. Therefore, at the first level of consideration, it would seem desirable to contain the entire volume of water associated with an event within the wetland. This requirement is often referred back to the contributing watershed, in terms of the rain amount and runoff coefficient that generated the storm

TABLE 11.5

Parameters for a hypothetical stormwater wetland event scenario.

Basin			
	Area =	120000	m^2
	Runoff Coefficient =	0.5	
Wetland			
	WWAR =	2	%
	Nominal Maximum Depth =	1.00	m
	Wetland Area Full =	2500	m^2
	Wetland Volume Full =	1250	m^3
	Stage-Area: A = aH, a =	2500	m^2/m
	Weir Height =	0.90	m
	Weir Coefficient =	31623	$(m^3/d)/(m$^1.5)
	ET Rate =	1.00	cm/d
	Infiltration Rate =	1.00	cm/d
	Rain Catchment =	3000	m^2
Rainfall			
	Period =	1.00	d
	Rain Rate =	5.00	cm/d
	Basin Catch =	6000	m^3
	Inflow Start =	0.50	d
	Inflow End =	2.00	d
	Inflow Rate =	2000	m^3/d
	Runoff =	3000	m^3
	Storm/Wetland Volume Ratio =	2.40	

volume. The wetland may be sized to nominally contain the runoff volume associated with a rain event of specified amount or return frequency.

One limit on performance is the expulsion of water that resides in the wetland at the time of event initiation. This water is at a background concentration if the events are widely spaced. In the example of Figure 11.6, the length of time for the wetland water surcharge to dissipate, and thus for stage to reach the weir elevation, was of the order of two or three days. If the events are closely spaced, then the resident antecedent water will not be at background, but rather will exhibit a concentration representing only partial treatment of the previous event. A second limit on performance is the ultimate removal that would be associated with a continuous water input at the event average flow rate. This is a relatively low percentage reduction when high flow events are under consideration.

A complicating factor in analysis is that wetlands are not hydraulically simple, and do not operate on a basis of plug-flow displacement. Therefore, the fraction of incoming water that remains in the system for a particular event is not determined merely from displacement. Depending on the configuration of the wetland, some of the very first water that enters the wetland may find its way to the exit far in advance of nominal retention time during the event. This is partly due to preferential flow paths, and partly due to the positioning of water level control structures.

If it is assumed that the wetland behaves hydraulically as several mixed tanks in series, it is reasonable to use the known detention time distribution to compute how much of the incoming water is still in the wetland after a period of flow of known magnitude. Figure 11.7 shows the results as a function of the nominal number of wetland volumes of water

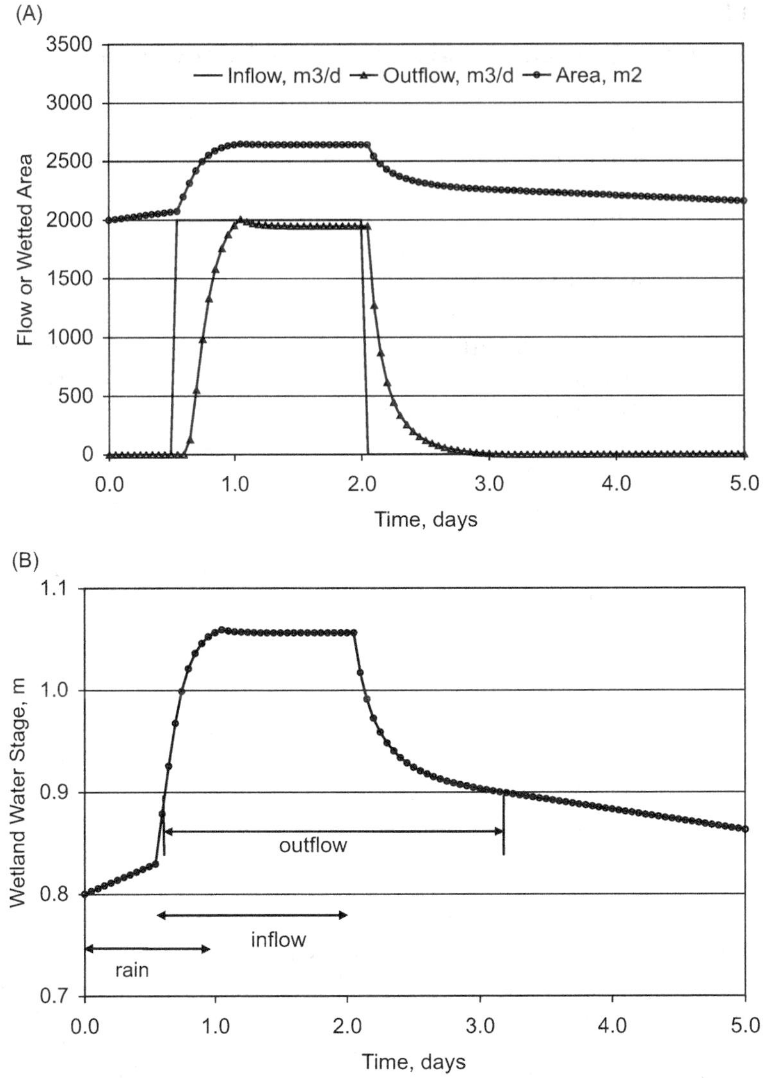

FIGURE 11.6 Response of a hypothetical stormwater wetland to a one-day steady rain. (A) Flows and wetted area, (B) stage. Runoff into the wetland begins after half a day. The wetland fills, and outflow persists for just over three days. Infiltration and ET deplete the water after the event. See Table 11.5 for parameters.

that have passed through the system. If the flow were plug flow, then the retention is 100% until the nominal detention time is reached, after which the new event water begins to exit. For less ideal flow conditions, the amount of the new event water that is held is less.

The presumption of Figure 11.7 is that all of the wetland water is involved in flow. That can be far from true because of wetland design factors. For instance, if there is a point inlet to the system, and a point outlet aligned directly opposite, then the incoming pulse of water can travel directly to the outlet, never reaching zones to the side of the most direct flow path. That renders the corner areas essentially out of the flow path, and reduces the volumetric efficiency quite markedly. Walker (1998) investigated this situation, using computational fluid mechanics (simulations) of unvegetated FWS basins. He determined that aspect

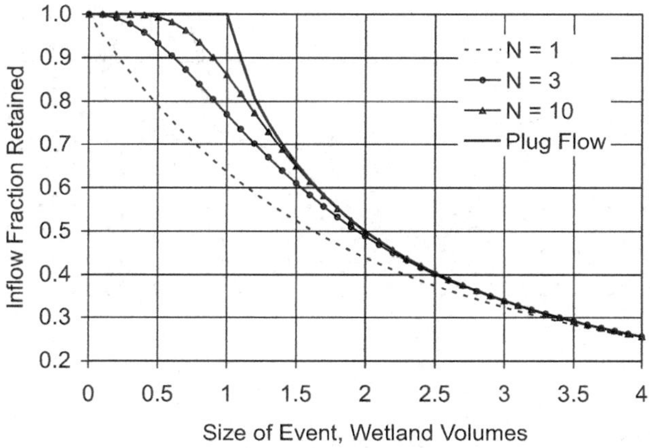

FIGURE 11.7 Fraction of the inflow retained for a range of numbers of tanks in series for event volumes of different nominal displacements.

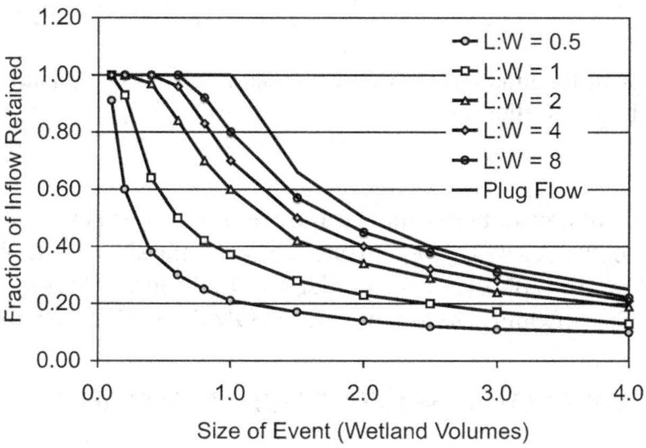

FIGURE 11.8 Fraction of an inflow event contained in a wetland as a function of the event volume and the aspect ratio (L:W). (Replotted from Walker, 1998)

ratio (L: W) was the major determinant of such short-circuiting, and calculated a retention-displacement chart (Figure 11.8).

For individual, separated inflow events, the concept of retained volume provides a strong correlating factor for performance. If the event is small, it is wholly contained and held until the next displacing flow. Treatment proceeds in the batch mode for the inter-event duration. If the event is large, then it is processed by flow-through at the detention time of the event flow. Since such flows are typically large, the wetland will have a short detention time, which results in decreased treatment performance. Therefore, mass reductions decrease (exponentially) as the number of nominal displacements caused by the event increases.

Ultimately, the computed description of performance will either be expressed as long-term mass removal or by dynamic modeling that produces time series of outflow and effluent concentrations in response to the inlet time series. Most of the stormwater wetland literature reports mass removals over some period of record. However, the past ten years have

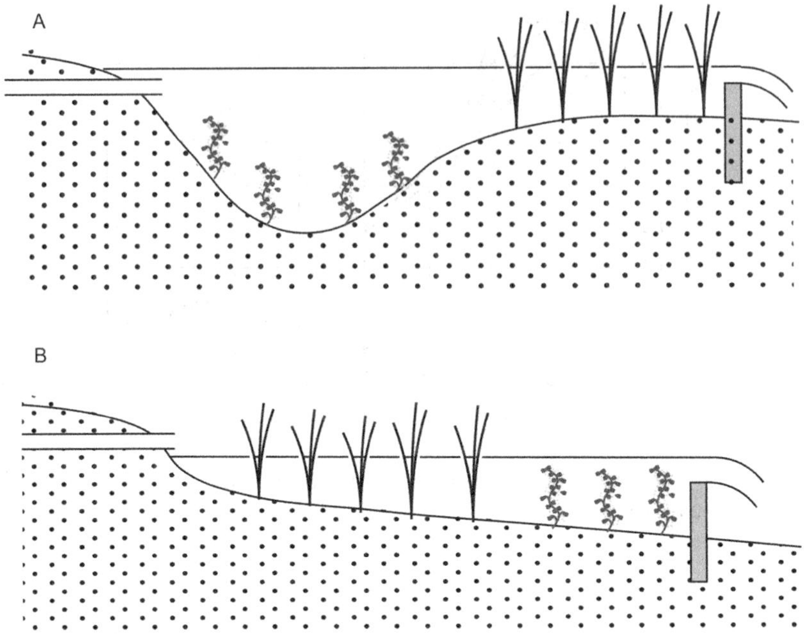

FIGURE 11.9 Conceptual longitudinal sections of agricultural runoff wetlands. Panel A: Sedimentation wetland. Panel B: Denitrification wetland.

seen the emergence of dynamic performance models for a number of situations The SWAT model is in use for characterizing the nitrate control marshes in Iowa (Ikenberry et al., 2017a, 2017b). A flow and phosphorus model for the south Florida STAs is in use, the Dynamic Model for Stormwater Treatment Areas (DMSTA) (Walker and Kadlec, 2005).

11.3 TECHNOLOGY STATUS OF EVENT-DRIVEN MARSHES

Agricultural Stormwater

Two categories of agricultural treatment wetlands may be distinguished: those treating field and pasture runoff, and those treating wastewaters from animal husbandry. Treatment wetlands for animal wastewaters, cattle, swine, and poultry, receive very strong effluents, with very high concentrations of TSS, BOD_5, and ammonia nitrogen. These are here excluded from consideration, but the interested reader can find extensive literature on the subject (e.g. Reaves et al., 1994; Knight et al., 2000; Poach et al., 2004; USDA NRCS, 2009).

The last fifteen years have seen a rapid growth in understanding of runoff wetlands, founded in numerous individual project studies across the globe (e.g., California – Díaz et al., 2012; Denmark – Hoffmann et al., 2012; New Zealand – Wilcock et al., 2012; Sweden – Kynkäänniemi et al., 2013; Illinois – Groh et al., 2015). Such studies have documented a significant removal of N and P in wetlands receiving agricultural stormwater.

Many agricultural treatment marshes, especially small systems, are built to a technology-based specification, usually a range of WWARs. For example, in Norway the trend has been to design small wetlands at 0.06–0.4% of the catchment area. These are focused on sedimentation processes, with an inlet deep pond for sedimentation followed by a shallow vegetation filter. Larger runoff control wetlands are built to performance-based specifications, such as the Florida STAs. Although not so designed, small wetlands are often monitored and analyzed according to performance models.

Configuration

The preponderance of runoff wetlands are of free-form outline, as dictated by the topography of the land at the end of the field. They are typically embedded within a single farm footprint. The small runoff wetlands are therefore highly variable in the layout of the flow path. Two different concepts are in use for determination of the longitudinal depth profiles in small runoff wetlands. The first employs a sedimentation basin in the inlet region of the system, to collect incoming sediments and the associated nutrients. The second utilizes a depth profile that increases along the flow direction, which is common for nitrate control wetlands with minimal sediment loads (Figure 11.9). Both are based on simplicity, meaning uncomplicated in form, and design; without embellishments for human or wildlife use. Earthmoving is held to a minimum, and structures are usually simple weirs. Basins are often located in depressions or in a small stream floodplain. The resulting bottom contours are then either dish-shaped, or fringing vegetated with a deeper central flow path. Examples of these low-tech end-of field wetlands are shown in Figures 11.10 and 11.11.

Exceptions to the rule of simplicity are the large P reduction wetlands of south Florida, which are generally rectangular in plan, corresponding to the boundaries of the lands purchased for the wetlands. These have inlet and outlet collector ditches, but no sedimentation basin.

Motivation

Runoff wetlands are intended to reduce the pollution loads emanating from agricultural activities. As a consequence of environmental impacts of diffuse pollution from agriculture on waterways, various national and international initiatives have been developed to drive the implementation of management strategies to help safeguard waterbodies (Collins and McGonigle, 2008). Constructed wetlands, almost entirely marsh systems, are among the remedies.

The problem of excessive nitrogen runoff has been identified as detrimental to the Baltic Sea. One such international initiative is the EU Water Framework Directive (EU WFD)

FIGURE 11.10 A field runoff wetland in Berg, Norway. (Photo from Braskerud, 2001c)

FIGURE 11.11 A CREP wetland in Iowa. (Photo from Iowa Department of Agriculture and Land Stewardship)

(2000/60/EC) requires the protection of water resources, promote their sustainable use, and prevent further contamination. The principal targets were perceived to be nitrogen and phosphorus. Member states responded differently to the directive with respect to treatment marshes for field runoff. For example, Sweden responded with the implementation of over a thousand small end-of-field wetlands (Arheimer and Pers, 2017), but with little or guidance on how to size or situate them. But, Belgium has only begun testing three systems, situated to treat runoff after it has entered small watercourses (Donoso et al., 2017). The Belgian concepts are: (a) shallow wetland within the watercourse; (b) wetland plants along the river banks; (c) wetlands next to the watercourse functioning as a partial bypass. Thus, despite the EU Directive, there is not yet a systematic, science-based methodology for design. This deficiency led Voulvoulis et al. (2017) to conclude, "Unless current implementation efforts are reviewed or revised, allowing the Directive to deliver its systemic intent in order to reach its full potential, the fading aspirations of the initial great expectations could disappear for good."

The impetus for implementation of runoff wetlands is governmental, in the USA and in Sweden and elsewhere. However, the treatment marshes are located on agricultural lands in private ownership, and so there must exist coordination between the landowners and government. The taking of private lands, either through willing sellers or through forced government actions, is not the best path. For instance, governmental land acquisition for the Florida STAs created an adversarial situation that has continued for three decades. Land costs were about 17,000 \$/ha. That situation is the exception, and most small on-farm wetlands are based upon a mutually agreeable financial compensation package. Economics will be discussed in a later Chapter, but the general concepts are provided here.

The nitrogen content of the streams and rivers of the Midwestern United States is of particular importance at this point in history, because of hypoxia in the Gulf of Mexico, together with the associated ecological and economic consequences (Diaz and Solow, 1999). Both point and nonpoint sources contribute to the nitrogen content of waters within the

Mississippi River drainage basin. About 60% of the water-borne total nitrogen is in the form of nitrate (Goolsby and Battaglin, 2000).

The source of the nitrogen is about two-thirds from agriculture, and one-third from other sources, including urban runoff, atmospheric deposition, and point sources. The result is a surface water nitrate concentration, in the upper portions of the basin, of about 4 mgN/L.

Federal and State officials have agreed on an action plan that includes a component intended to promote restoration and enhancement of natural systems for nitrogen retention and denitrification (U.S. EPA, 2001a). The premiere natural system that has the capability to effectively remove nitrate from surface water is the free water surface wetland, based simply on its ability to place contaminated water in intimate contact with the biogeochemical cycle that removes nitrogen. More than half of the pre-settlement wetlands in the upper Mississippi River basin have been lost to drainage (Dahl, 1990). It is therefore synergistic to restore wetlands that are positioned to effectively function for nitrate reduction. Considerable attention has therefore been devoted to constructed and restored wetlands for nitrogen control in the Midwestern United States (Kovacic et al., 2000; Hey et al., 2005; Mitsch et al., 2005a; Kadlec, 2005b; Crumpton et al., 2012b). Similarly, the Chesapeake Bay in the eastern United States is impacted by agricultural runoff from farms in Maryland (Jordan et al., 1999). Studies on seven constructed (restored) wetlands demonstrated that nutrients could be removed from field runoff.

The leading example is the Iowa arrangement. The Conservation Reserve Enhancement Program (CREP) is implemented by the Iowa Department of Agriculture and Land Stewardship in cooperation with the U.S. Department of Agriculture Farm Services Agency and is available in 37 counties in the tile-drained region of north central Iowa. Financial incentives are provided to private landowners to develop and restore wetlands that intercept tile drainage from agricultural watersheds (IDALS, 2017). Landowners receive annual land payments for up to 15 years (ca 110 \$/ha·yr) and reimbursement for costs of wetland and buffer establishment (ca 53,000 \$/ha). The landowner also receives payment for an easement to maintain the wetlands and buffers are required for a minimum of 30 years (ca 32,000 \$/ha), and permanent easements can also be developed. The demand for the program greatly exceeds available funding and the waiting list extends at least 5 years.

Sweden provides a second example. Farmers receive agri-environmental payments, intended to compensate farmers for changed land use (Andersson, 2012). The payments are calculated based on lost revenue and reduced income. Cropland that has been transformed to wetlands can receive a payment of 440 \$/ha·yr. The Rural Development Programme (RDP) offers an investment support for costs related to wetlands construction (up to 25,740 \$/ha) and maintenance support (193–644 \$/ha·yr).

Sizing

The scientific knowledge base for agricultural wetlands is not lagging that for other treatment marsh applications. Concepts set forth in previous chapters, including hydrology and hydraulics and reaction rate representations, are all in use for the purposes of monitoring and quantifying wetland performances. Several comprehensive simulation packages that embody those concepts are also in use for agricultural runoff wetlands performance description. Yet, these methods are not in use for design. Rather, very simple presumptive performance rules are utilized – or perhaps only best professional judgment. For example, the CREP systems must meet these requirements:

- The drainage area above wetland pool must be greater than 500 acres.
- The wetland must be between 0.5 and 2.0% of the drainage area.
- 75% of the wetland pool must be less than 0.91 meters deep.

- Wetland construction shall not negatively impact drainage rights of nearby upstream and downstream landowners.

About 100 constructed wetlands have been installed in Iowa for water quality improvement under CREP, treating. CREP is a major state/federal initiative of strategically located wetlands using advanced computer technology and designed to remove nitrate from tile-drainage water from cropland areas. These systems typically include a 2 to 5 ha wetland, surrounded by a 10 to 15 ha grass buffer. The typical system handles water from roughly 400 ha. One CREP wetland provides roughly the same amount of nitrate removal from the 400 ha as converting 200 ha to permanent grass. The demand for the program greatly exceeds available funding and the waiting list extends at least 5 years. Despite the effectiveness of these treatment marshes, the enormity of the drainage area flows that require treatment means that thousands of such small marshes would be needed to significantly abate the nitrate emissions. The situation is similar in Sweden (Arheimer and Pers, 2017). Other states in the USA have very much lesser numbers and areas of runoff wetlands. For instance, Minnesota had one CREP wetland as of 2016 (Lenhart et al., 2016), and Ohio is just beginning to consider wetlands for control of agricultural runoff (Schumacher et al., 2015).

The sizes of the Iowa and Swedish runoff wetlands are summarized in the distribution data of 11.12A, and the WWARs in Figure 11.12B. The Iowa wetlands are larger, and have a lower WWAR.

The Florida STAs, which treat agricultural runoff for P reduction, were and are being designed with the PkC* method described in Chapter 7. They have a regulatory design target of 13 μgP/L, and thus differ from most other systems that target an implied P load reduction on the order of 50%. The large size required to meet the low outlet concentration means the required WWAR is 9.7%.

Urban Stormwater

Runoff from roofs, lawns, parking lots and other urban landscape features often contains several classes of pollutants (see Table 11.3). There will be small amounts of nitrogen and phosphorus and BOD, but possibly relatively greater amounts of TSS, metals and perhaps pesticides. An important subset of urban runoff control is focused on highway runoff (Knight et al., 1998; Shutes et al., 1999, 2004; Bulc and Sajn Slak, 2003; Pontier et al., 2003; Terzakis et al., 2008; Winston et al., 2011; Gill et al., 2017; Qualls and Heyvaert, 2017). However, the greater number of systems are used to treat rainwater runoff from more generalized urban areas.

Such systems are often quite small, serving small catchments (Figure 11.13). However, if space permits, much larger systems may be employed. An example is the Shepard Wetland, also in Calgary (Figure 11.14). At 156 hectares, it is the largest constructed stormwater treatment wetland in Canada. It was constructed as part of the stormwater quality retrofit program to help solve a stormwater management issue for the east side of Calgary. This treatment wetland system can store over 6 million cubic meters of water. The facility is designated as the Ralph Klein Park, which has an Environmental Education Centre featuring an indoor classroom, resource library, art studio, interpretive signage, and meeting space. (Information from City of Calgary, available at: www.calgary.ca/CSPS/Parks/Pages/Planning-and-Operations/Conserving-our-wetlands.aspx accessed 8/7/18).

There is large variability of water quality of urban runoff, which is related to the land uses in the watershed. Examples of pollutant concentrations from a variety of uses are shown in Table 11.6. In general, variability spans one to two orders of magnitude. Lawns produce much higher levels of nutrients and TSS than roofs and parking lots, but streets

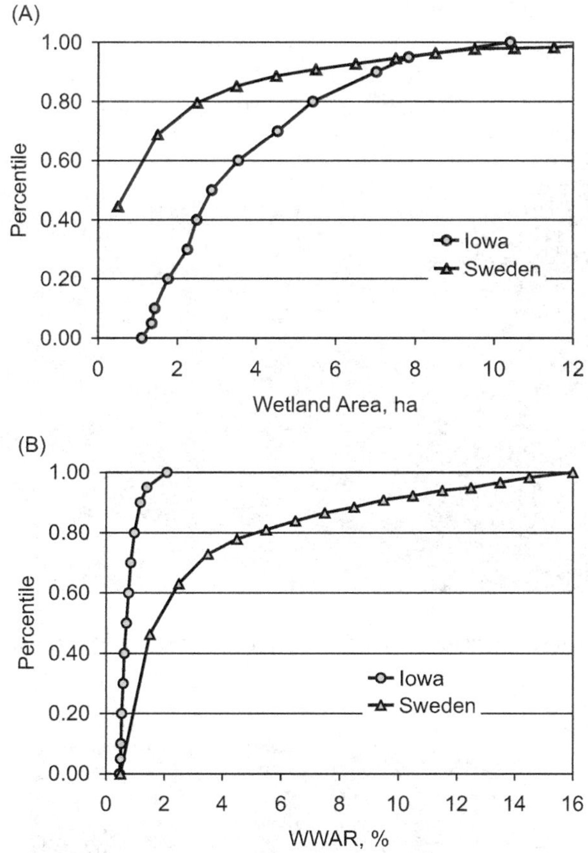

FIGURE 11.12 Sizes of agricultural runoff wetlands in Iowa and in southern Sweden. Panel A: Wetland area, ha. N = 83 wetlands for Iowa, 1156 wetlands for Sweden. Panel B: Wetland Watershed Area Ratio, %. N = 83 wetlands for Iowa, 417 wetlands for Sweden. Data for Iowa from IDALS (2017), and from Arheimer and Pers (2017) for Sweden.

and highways have higher concentrations of metals than lawn runoff. It is virtually impossible to estimate runoff concentrations accurately. As discussed previously, the volume of runoff is subject to reasonable estimating procedures with reasonable accuracy. Therefore, sizing rules have evolved to rely upon volume considerations.

Table 11.6 does not include snowmelt, which is a runoff source for northern urban treatment wetlands. Northern communities clear snow from streets, and pile it. The piles, which are typically very dirty and contain street trash, may not melt entirely until mid June in the far north. The meltwater is of generally slightly better quality than rain runoff for dissolved constituents (Bennett et al., 1981). Snowmelt water is also treated in constructed marshes (Figure 11.15).

Configuration

The preponderance of urban wetlands are of free-form outline, as dictated by the shape of the available land. While there are many variations, the general sequence of elements is a forebay followed by a marsh followed by a pool. Nomenclature is also variable. For instance, the New York manual recognizes four variants: shallow wetland, extended

FIGURE 11.13 An urban stormwater wetland in the city of Calgary, Manitoba.

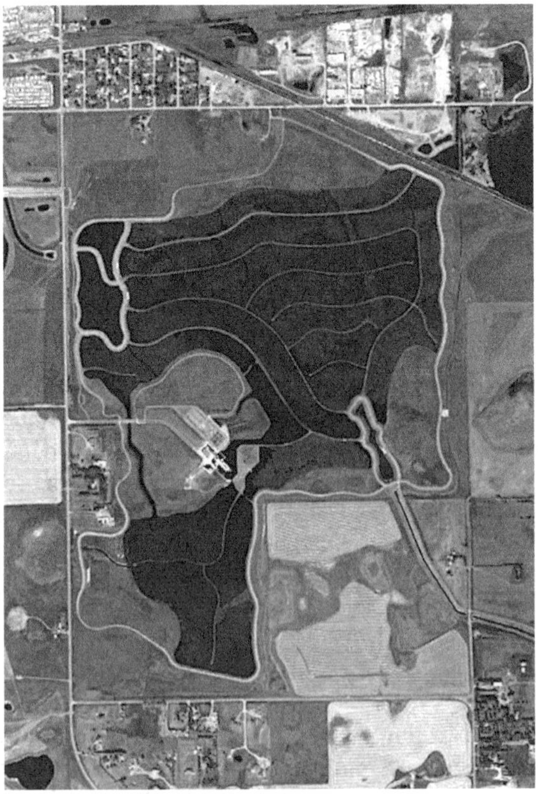

FIGURE 11.14 The Shepard urban wetland in Calgary, Manitoba.

FIGURE 11.15 An urban runoff treatment wetland in Anchorage, Alaska. This system treats both rainfall runoff and snowmelt during the warm months.

detention shallow wetland, pond/wetland, and pocket wetland. Example layouts of two of these are shown in Figure 11.16, and an example hydraulic profile in Figure 11.17. Intentional irregularity is the dominant theme, driven by concepts of detention, sedimentation and ecological values. Clearly, a good deal of earth moving is required to establish the desired bathymetry.

The hydraulic profiles are controlled by outlet structure(s) that typically define four or five internal levels. Of course, these urban systems are small and generally function as level pool systems. The lowest flowing water stage is termed the permanent pool, and is mostly synonymous with the water quality volume (WQV). The next highest stage is set to provide storage that allows control of erosion in the receiving water channel, and hence is termed the channel protection stage. Very large storms can overwhelm these low stage structures, and threaten the integrity of confining embankments, and hence an emergency spillway is usually provided. Finally, an outlet pool drain is provided, typically to used to assist in maintenance activities.

Motivation

Runoff wetlands are intended to reduce the pollution loads emanating from urban activities, and ensure that better water quality reaches the receiving water bodies. Starting about 25 years ago, problem recognition led to the first uses of wetlands as a solution, and to the first guidelines for sizing (Schueler, 1992; Strecker et al., 1992). At about the same time, governmental agencies began to promulgate rules for how urban stormwater wetlands were to be built. At present, in the USA, there are numerous such regulations, set forth by individual states, and nationally.

At the national level, USEPA administers the National Pollutant Discharge Elimination System (NPDES). The NPDES stormwater program regulates some stormwater discharges from three potential sources: municipal separate storm sewer systems, construction activities, and industrial activities. Operators of these sources might be required to obtain an NPDES

TABLE 11.6

Pollutant concentrations for source areas for stormwaters.

Constituent	TSS mg/l	TP mg/l	TN mg/l	Cu µg/l	Pb µg/l	Zn µg/l
Reference	1	2	3	1	1	1
Residential Roof	19	0.11	1.5	20	21	312
Commercial Roof	9	0.14	2.1	7	17	256
Industrial Roof	17	–	–	62	43	1,390
Business Parking	27	0.15	1.9	51	28	139
Industrial Parking	228	–	–	34	85	224
Residential Street	172	0.55	1.4	25	51	173
Comm. Street	468	–	–	73	170	450
Rural Highway	51	–	22	22	80	80
Urban Highway	142	0.32	3	54	400	329
Lawns	602	2.1	9.1	17	17	50
Landscaping	37	–	–	94	29	263
Driveway	173	0.56	2.1	17	–	107

References:
1 Claytor and Schueler (1996).
2 Average of Steuer et al. (1997), Bannerman et al. (1993), and Waschbusch (2000).
3 Steuer et al. (1997).

permit before they can discharge stormwater. This permitting mechanism is designed to prevent stormwater runoff from washing harmful pollutants into local surface waters.

Each of the individual states has its own set of standards and requirements, and they differ greatly. A full summary of these has been assembled by USEPA (U.S. EPA, 2016c). Each state's post construction program is described in a consistent format. Additional information on special criteria, offset or mitigation options, inspection and maintenance requirements and definitions are included. Links to the sources of information including regulations, design manuals, or other information published by each program are included as of July 2016. Overall, 28 states, the District of Columbia, and Puerto Rico have some type of retention standard. Ten states have a treatment only standard (with no retention requirements) and 12 states only have a narrative standard.

Similar initiatives have developed and matured in the UK and Australia. In the UK, Sustainable Urban Drainage Systems (SUDS) have established guidelines (Illman and Wilson, 2017). In Australia, Water Sensitive Urban Design (WSUD) has an extensive and detailed wetland design manual (Melbourne Water, 2017). In general, WSUD are based on the idea that the management, protection and conservation of the urban water cycle must take place to ensure that the urban water environment is such that the ecological services of urban waterways are maintained. There are different WSUD elements are in practice for stormwater nutrient attenuation, and constructed wetlands is one of them.

Sizing

The scientific knowledge base for urban wetlands is not lagging that for other treatment marsh applications. As for agricultural runoff wetlands, concepts set forth in previous chapters, including hydrology and hydraulics and reaction rate representations, are all in use for the purposes of monitoring and quantifying urban wetland performances. Comprehensive simulation packages that can employ those concepts are also in use for urban runoff

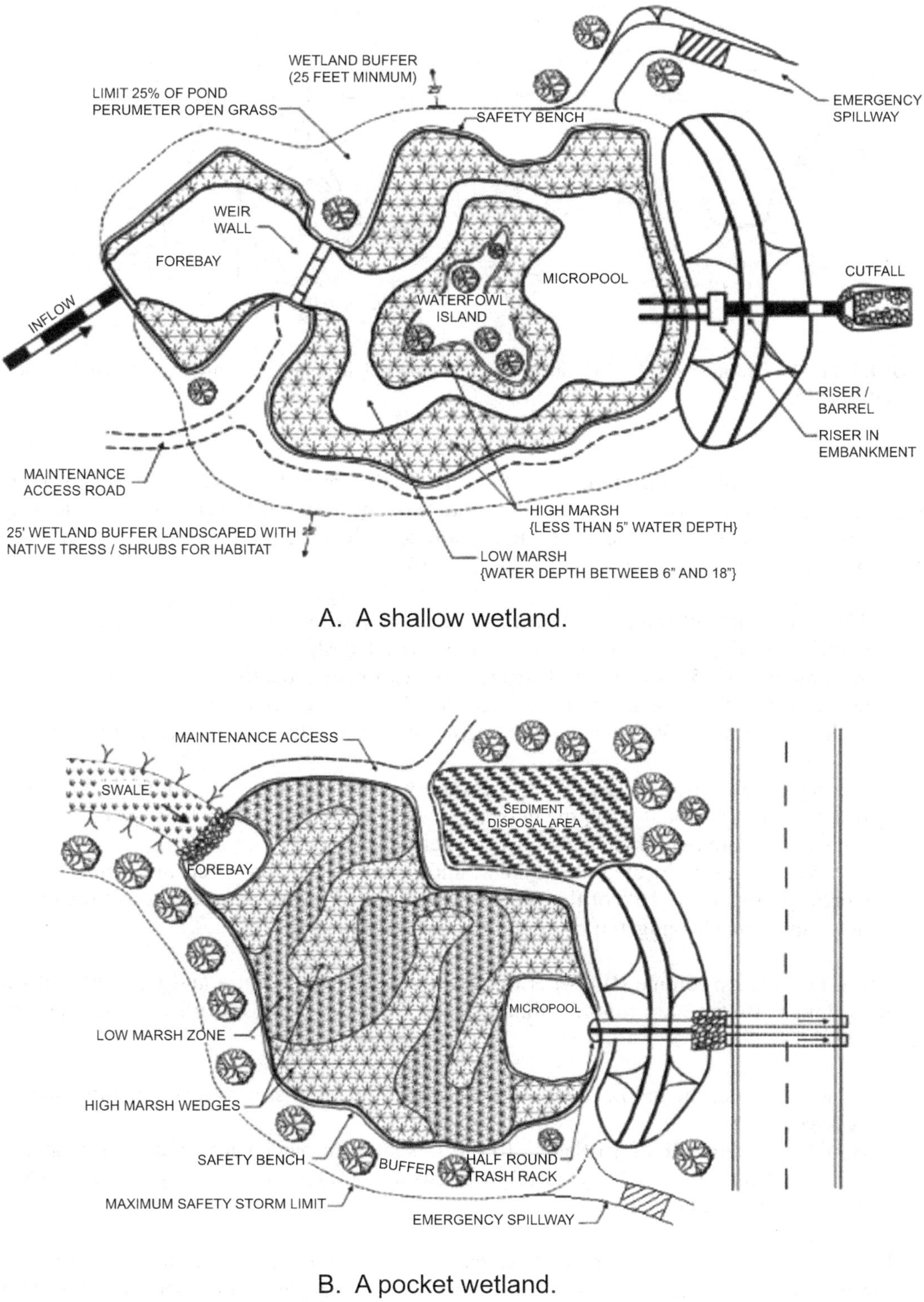

A. A shallow wetland.

B. A pocket wetland.

FIGURE 11.16 Two variants of urban stormwater wetland layouts. (Source: NYDEC, 2015)

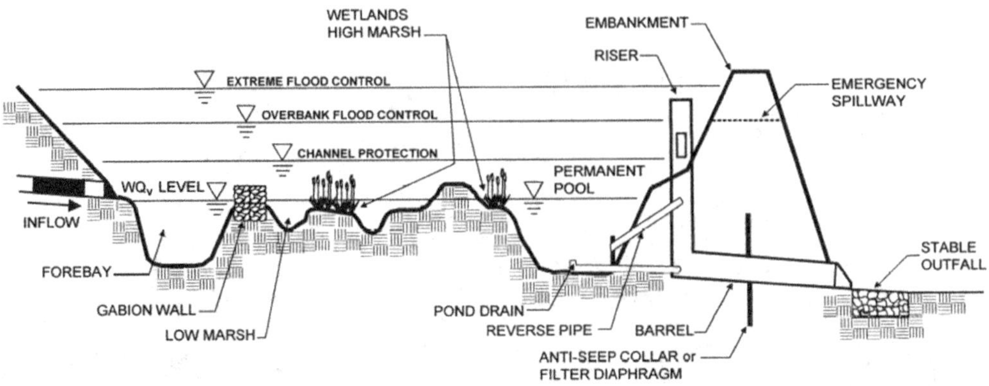

FIGURE 11.17 Example of the hydraulic profiles expected in a shallow urban wetland. (Source: NYDEC, 2015)

wetlands performance description. Yet, these methods are not in use for design. Rather, volumetric rules are most often utilized – and these are not necessarily simple, and they vary from state to state in the USA. For example, urban wetland systems in New York State must meet these requirements (NYDEC, 2015):

- The surface area of the entire stormwater wetland shall be at least one percent of the contributing drainage area (1.5% for shallow marsh design).
- A minimum of 35% of the total surface area can have a depth of six inches or less, and at least 65% of the total surface area shall be shallower than 18 inches.
- At least 25% of the WQv shall be in deepwater zones with a depth greater than four feet.
- If extended detention is used in a stormwater wetland, provide a minimum of 50% of the WQv in permanent pool; the maximum water surface elevation of WQv-ED shall not extend more than three feet above the permanent pool.
- A forebay shall be located at the inlet, and a four to six foot deep micropool that stores approximately 10% of the WQv shall be located at the outlet to protect the low flow pipe from clogging and prevent sediment resuspension.

The water quality volume (WQV) is the storage needed to capture and treat 90% of the average annual stormwater runoff volume. New York also imposes other design requirements, such as:

- Flowpaths from the inflow points to the outflow points of stormwater wetlands shall be maximized.
- A minimum flowpath of 2:1 (length to width) shall be provided across the stormwater wetland. This path may be achieved by constructing internal berms (e.g., high marsh wedges or rock filter cells).

In other words, a sinuous flowpath is required. There are also additional landscaping requirements. Some receiving water bodies may be deemed phosphorus-sensitive, and then New York invokes additional goals/requirements for enhance phosphorus removal:

- The WQv is sized for the one-year 24 hr event.

- Achieve less than 15% bypass, by running a continuous simulation model.
- The median effluent concentration of particulate phosphorus shall be below 0.100 mgP/L. This is regarded as 80% removal of PP.
- The median effluent concentration of dissolved phosphorus shall be below 0.060 mgP/L. This is regarded as 60% removal of TDP.

The reader will recognize that these P removal requirements are not necessarily in harmony. The first P requirement implies a detention time, yet there is no guarantee that it will be sufficient to meet the 80% PP and 60% TDP removals. Further, the New York requirements are not in accord with the database (Table 11.7). Additionally, the state of P removal technology does not yet allow forecasting of removal rates of individual P species.

Specifications such as the New York prescriptions imply considerable effort be expended in compliance. These "city cousins" to the rural runoff wetlands are far more complicated to design and build than the agricultural runoff systems. However, as we shall see in subsequent sections, urban system performance is comparable to agricultural system.

Floating Raft Add-Ons

There is, at present, some enthusiasm for the use of floating rafts on the pond elements of urban stormwater wetland systems (Headley and Tanner, 2012; Borne, 2014; Borne et al., 2015; Lynch et al., 2015). One of the main advantages of floating rafts (also termed floating treatment wetlands, FTW) over conventional sediment-rooted wetlands is their ability to cope with the variable water depths that are typical of event-driven stormwater systems. Because they float on the water surface, the plants in a floating raft are not affected by fluctuations in water levels that would submerse and adversely stress bottom-rooted emergent plants in stormwater systems. Owing to the youth of floating raft systems as a water quality improvement technology there are limited published performance data available at present, very little of which is actually from stormwater systems (Wang and Sample, 2014, Wang et al., 2015). Nonetheless, the level of interest is high, and the knowledge base is growing quickly.

Guidelines are also in the early emergence stage. The first such effort was initiated by the Chesapeake Bay Program in Maryland (Schueler et al., 2016), which proposed design criteria and performance credits.

The recommendation for a FTW application within an existing pond must adhere to the following recipe:

- Achieve a minimum pond surface coverage of 10% and a maximum cover of no more than 50%
- Have an initial planting density of 2 plugs per square foot and attain a 80% plant coverage on the raft by the end of the growing season
- The raft should be placed perpendicular to the stormwater flow path and be at least 3.5 feet above the bottom of the pond
- Be adequately anchored or tethered in the pond to protect its flood control function during major storms and enable retrieval for periodic maintenance, yet anchoring should not be too taut to inundate the surface and flood the raft
- Not be infested with invasive plants and should be initially protected with netting from geese and turtles during plant establishment
- Use native wetland plant species that meet the five criteria laid out in Wang and Sample (2014), which are appropriate for the ecoregion.

Schueler et al. (2016) determined that the incremental effect of FTWs in boosting pond pollutant removal was modest (5 to 20%), but was consistently positive in the monitoring studies where it was evaluated. In general, TP removal was consistently higher than TN removal in FTW pond retrofits. The proposed system of performance credits was based on Wang and Sample (2013), who developed and calibrated a first order removal rate model, which is a version of the PkC* model described herein (Table 11.7). Parameters of the model include water reaction rate of the water body, and apparent uptake velocity of the FTW; which is a function of the vegetation and the raft materials. Additional parameters include the area of the raft and the volume of the water body.

FTW add-ons are very expensive, with costs of the order of $300 per square meter ($3,000,000 per ha). Thus the enthusiasm generated by novelty and the prospect of retrofitting must be somewhat dampened. Economics will be discussed in more detail in Chapter 15.

11.4 PERFORMANCE ANALYSIS

METHODS

There is a large discrepancy between the regulatory design sizing procedures and requirements, and the analyses of event-driven treatment marshes. Sizing is most often determined by a WWAR, or a capture volume, not a target removal of load or concentration reduction. In turn, these specifications derive from (limited) databases of percent removals. An example of such a database is the International Stormwater BMP Database, a publicly accessible repository for BMP performance, design, and cost information for urban systems (Clary et al., 2017b) (Table 11.8). The overall purpose of the project is to provide scientifically sound information to improve the design, selection and performance of BMPs. The report provides a summary of BMP performance for total suspended solids, nutrients, metals, and bacteria for the most commonly monitored and reported BMP types. BMPs include grass strips, bioretention, bioswales, composite/treatment train BMPs, extended detention basins, media filters (mostly sand filters), porous pavement, retention ponds (wet ponds), wetland basins, and wetland channels. Data summaries include basic summary statistics for influent and effluent concentrations.

In contrast, performance analyses have, for the past twenty years, utilized some version of the PkC* model described herein. Historically, Wong and Geiger (1997), Carleton et al. (2001) and Wong et al. (2006) presented rate coefficients that are intended to be used in such a model. Their definition of rate coefficients involved the use of long-term average hydraulic loading rates, not event mean loading rates. The median of such areal rate coefficients reported by Carleton et al. (2001) for FWS constructed gravity-fed systems (N = 9) were:

TABLE 11.7

Additional pollutant removal credited for a FTW retrofit.

	Raft Coverage in Pond, %				
	10	20	30	40	50
Total Nitrogen, %	0.80	1.70	2.50	3.30	4.10
Total Phosphorus, %	1.60	3.30	4.90	6.50	8.00
Total Suspended Solids, %	2.30	4.70	7.00	9.20	11.50

Source: Schueler et al. (2016).

TABLE 11.8
Pollutant concentrations entering and leaving urban treatment wetlands.

		Systems	Events	%tile	Inlet	Outlet
TSS	mg/L	22	492	25th	13.1	4.7
				50th	31.0	14.1
				75th	75.9	31.0
TKN	mg/L	8	96	25th	0.61	0.62
				50th	0.99	0.94
				75th	1.33	1.24
NO_X-N	mg/L	15	466	25th	0.23	0.05
				50th	0.43	0.33
				75th	0.69	0.60
TP	mg/L	20	595	25th	0.10	0.07
				50th	0.16	0.12
				75th	0.27	0.22
TDP	mg/L	7	311	25th	0.03	0.03
				50th	0.05	0.05
				75th	0.10	0.08
Iron	µg/L	18	404	25th	269	165
				50th	692	278
				75th	2090	480
Lead	µg/L	8	145	25th	1.14	0.76
				50th	2.37	1.30
				75th	7.72	3.60

Source: Clary et al. (2017b).

- Total Phosphorus: 8.3 m/yr
- Ammonia: 5.0 m/yr
- Nitrate: 6.7 m/yr

These are very low compared to the k-values reported for continuous flow systems in the preceding Chapters. This is due to the fact that long-term loading rates include periods of no flow. Individual events are afforded much better treatment.

At the present time, there are numerous analyses of event-driven wetlands based upon the PkC* model or modifications thereof. These include agricultural runoff systems (e.g., Ikenberry et al., 2017a) and urban systems (e.g., Wong et al., 2006; Merriman et al., 2017), as well as floating rafts on urban systems (e.g., Wang and Sample, 2013). The PkC* methodology has also been added to, or embedded in, computer simulation packages, such as MIKE SHE (Crumpton et al., 2012b) and SWAT (Ikenberry et al., 2017a).

Characterizing Events

There are some specific peculiarities of event-driven systems that should be emphasized as precursors to data analysis. These systems are dynamic, and two choices can be made for analyses, averaging methods and time series analysis. In this section, averaging methods are outlined. In special circumstances, event time series analyses can be applied, but the data requirements are quite onerous. Event dynamics are discussed in the next section of this Chapter. Other special features are related to mixing and water temperature.

Event Mean Concentrations

As discussed in Chapter 7 the PkC* model should be applied to in use periods, and should not include times when the system is not in use. In other words, it is not reasonable to average over periods that include down time.

For an event-driven system, the selected period(s) would be the times when the system is receiving and discharging water, and the concentrations would be the event mean concentrations. The definitions detailed by McCarthy et al. (2018) are first reiterated, with the notation defined in Chapter 7. These authors suggest use of Site Mean Concentrations (SMCs), which were the volume-weighted average of the EMCs for each event.

$$EMC = \frac{\sum_1^n C_i V_i}{\sum_1^n V_i} \tag{7.90}$$

$$SMC = \frac{\sum_1^p EMC_j V_j}{\sum_1^p V_j} \tag{7.91}$$

The flows involved can be similarly defined.

$$EMQ = \frac{\sum_1^n V_i}{t_i} \tag{7.92}$$

$$SMQ = \frac{\sum_1^p EMQ_j V_j}{\sum_1^p V_j} \tag{7.93}$$

The use of these deducts the inter-event periods from the event and system averages.

Internal Mixing and Flow Patterns

The number of tanks in series accounts for internal mixing, and is most important for situations in which a high percentage removal of a pollutant is contemplated. That is normally not the case for runoff wetlands, which often target load reductions or concentration reductions in the 10–50% range. Figure 11.18 shows an idealized case of first order removal with $C^* = 0$ and $k_V = 0.4\ d^{-1}$. There is relatively little difference in fraction remaining due to varying P values in the 10–50% range. In other words, event-driven systems are typically not sensitive to mixing, and the P-value is not critical.

The more important feature of internal water movement is the volumetric efficiency of the wetland. As discussed in Chapter 7, factors involved include aspect ratio, vegetation patterns, and inlet/outlet configurations. Small event-driven wetlands unfortunately contain the least effective options in all categories. They therefore are expected to have low volumetric efficiency. Persson et al. (1999) and Jenkins and Greenway (2005) determined that wide short wetlands with opposed point inlets and outlets had low efficiency. Further, they found that fringing vegetation along an unvegetated channel had very low volumetric efficiency compared to cross-flow banded vegetation.

Most often, there are point inlets and point outlets for both urban and agricultural systems. The layouts shown in Figure 11.16 show this to be the design expectation for urban systems. It is also a design expectation that there be a central (sinuous) channel with fringing vegetation. Aspect ratios vary, but urban wetlands sites often constrain the length to width ration to low values. Fringing vegetation is commonly employed, presumably in an effort to maximize wildlife and aesthetic values otherwise lacking in many urban environments.

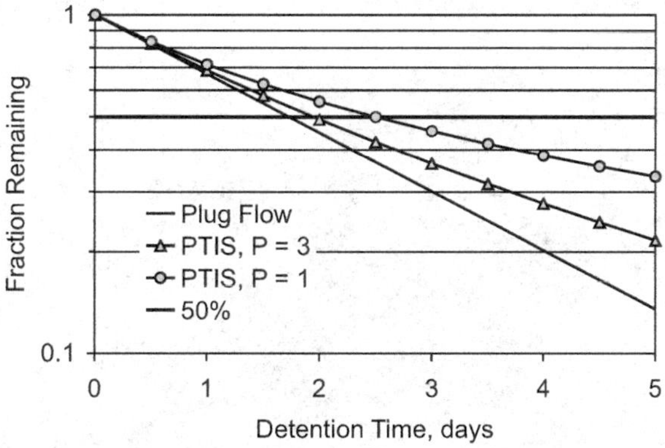

FIGURE 11.18 Comparison of fractions remaining along the flowpath (detention time) for three levels of internal mixing.

Field runoff wetlands take advantage of the local topography, by damming a swale or stream channel. The result is a preferential channel from inlet to outlet (Figure 11.19), as well as point inlets and outlets. The resulting bathymetry produces fringing vegetation patterns.

The internal flow patterns were measured on 30 tests of 5 CREP wetlands (Green, 2016). Analysis proceeded with relations following those outlined in Werner and Kadlec (1999b, 2000), for event-driven systems. He found a mean NTIS = 2.6, but with a large coefficient of variation (42%). The volumetric efficiencies averaged 77%, but also with a large coefficient of variation (50%). These values are not very different from those for uncompartmentalized continuous flow wetlands (see Chapter 7).

The low volumetric efficiencies are usually absorbed into the uptake rate coefficient, (Chapter 7), thus causing them to be at the low end of the distributions shown in earlier chapters.

Water Temperatures

Urban source waters usually travel on the land surface, and tend to reach a balance temperature before arriving at the wetland. That is also the case for some agricultural runoff wetlands, but a significant proportion receive tile drainage water. The subsurface tiles are situated in a cooler soil environment, and deliver water that is below the balance temperature (Drake et al., 2018). The wetland warms the water, which is a benefit to the removal of nitrate, due to a strong temperature dependence of the nitrate loss coefficient. The amount of warming is very significant in early summer (>7 °C), but tapers during late summer and early autumn to essentially no differential (Drake et al., 2018). The temperature coefficient for nitrate loss rate coefficients is $\psi = 1.092$ (Chapter 9), and the 7 °C temperature rise then about doubles the rate coefficient. Energy balance calculations (Chapter 6) do a creditable job of forecasting the wetland balance temperatures. For instance for the Iowa CREP Slough wetland, the RMS error is about 1.0 °C. Predicted accommodation times are less than one day, and thus water reaches the balance temperature by the time it reaches the wetland outlet, with the possible exception of very high runoff flows. It is notable that empirical water temperature forecast relations, such as that employed in the SWAT model ($T_{water} = 5.0+0.75T_{air}$), are not acceptably accurate.

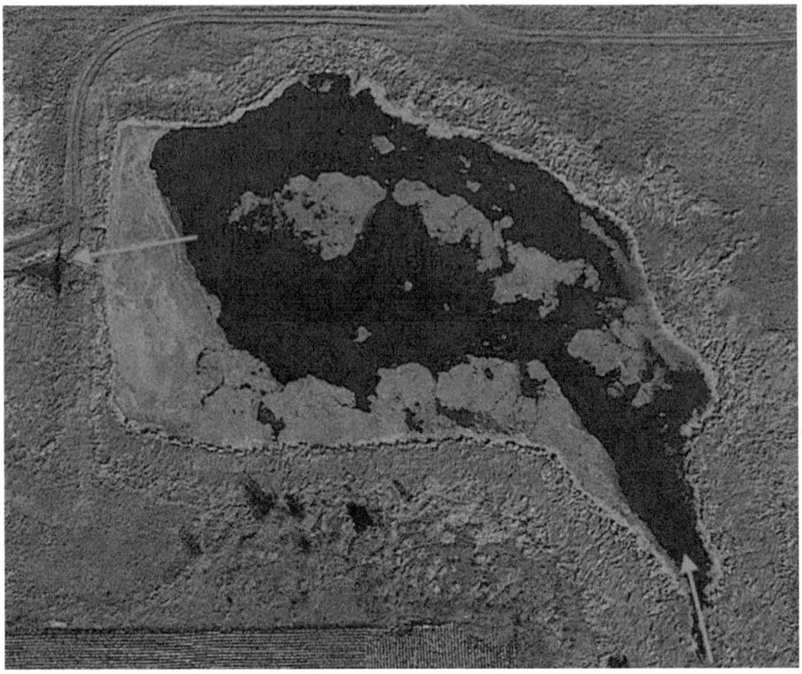

FIGURE 11.19 Layout of the KS CREP wetland in Iowa.

TSS in Event-Driven Marshes

Event-driven treatment wetlands utilize the same suite of TSS processes as continuous flow systems, but there are important differences in the mode of operation. In contrast to continuous flow wetlands, there is a greater need to account for the flow and chemical dynamics. Often, the source water TSS is unknown. Performance data acquisition is difficult because flows occur during storm conditions, and close-interval sampling is needed to define the course of a single event. Internal to the wetland, processes operate on a flow-and-stop time sequence.

Inter-System Performance: Agricultural Systems

Agricultural runoff wetlands that target TSS reduction come in all imaginable sizes and shapes, and receive widely variable concentrations and loads (Table 11.9). Their performance defies any attempt at simple inter-system analysis. Incoming concentrations and hydraulic loads both span two orders of magnitude, and reflect WWARs that span three orders of magnitude. Percent removals are inadequate, because these reflect the relation between incoming and background TSS, as discussed in Chapter 8. Basically, if source waters are at or near background, there will be very low percent removals. Load stored/removed is likewise inadequate, because low incoming loads must result in low load removals. The wetland layout is often opportunistic, and results in very poor hydraulic efficiency, including short-circuiting and dead zones. Lack of knowledge of the extent of nonideality of hydraulics defeats attempts to utilize the PkC* methodology. About the most that can be said is that incoming mineral particulates settle rapidly, and are retained. Despite these problems, percent removal remains a commonly used measure (e.g., Koch et al., 2014).

Inter-System Performance: Urban Systems

As for continuous flow systems, there is a sizeable database of urban stormwater wetland performance. In the United States, this database originated with the work of Strecker et al. (1992) and Schueler (1992), and grew through continued efforts of Godrej et al. (1999), and Carleton et al. (2001). Clary et al. (2017b) are continuing to add to the available information.

In contrast to continuous flow system data, event systems have much greater variability in design. Loadings are extremely variable, as are the geometries and arrangements of deep pools, shallow marsh zones and interconnecting channels. Therefore, the intra-system database gives an impression of what has been accomplished, rather than what might be accomplished by "optimal" designs. Percent mass removals are high in many cases, with a median of about 70% for United States data. Table 11.10 shows examples of performances of urban stormwater systems for TSS reduction.

Duncan (1998) (as referenced and reported by Wong et al., 1999) suggested the use of regression equations for purposes of sizing. Relations are based on analysis of subsets of 76 stormwater wetland systems, presumably including pumped systems, including those in the USA, and complemented by systems in Australia. The proposed relation for TSS, rewritten for concentrations, was:

TABLE 11.9

Samples of performance of agricultural runoff wetlands.

Site	Wetland	WWAR %	Area m^2	HLR cm/d	TSS In mg/L	TSS Out mg/L	Load Removed $kg/m^2 \cdot yr$	Ref.
California	W1	0.278	45,000	40.0	238	103	19,710	1
California	W2	0.71	23,000	83.3	242	20	67,525	1
California	W3	0.77	25,000	46.9	242	15	38,838	1
California	W4	7.5	1,500,000	7.5	583	38	14,919	1
Sweden	Nyb	0.23	1,000	32.9	137	131	685	2
Sweden	Ber	0.31	800	19.2	108	65	3,024	2
Sweden	Ski	0.36	800	16.4	143	19	7,436	2
Maryland	B1	9.29	13,000	0.59	165	135	65	3
Maryland	B2	13.00	26,000	0.42	110	30	123	3
Maryland	Braun	19.17	23,000	0.30	70	55	16	3
Maryland	Foster	11.40	4,900	0.48	150	35	202	3
Finland	Hovi	5.00	6,000	1.90	530	297	1,617	4
Finland	Alastaro	0.50	4,800	12.80	56	47	419	4
Finland	Flytträsk	3.00	600,000	3.80	29	24	72	4
UK	6	0.03	125				112	5
UK	7	0.11	320				31	5
UK	8	0.03	50				320	5
UK	9	0.05	8				600	5

References:
1 Díaz et al. (2012)
2 Gerenmayeh et al. (2018)
3 Jordan et al. (1999)
4 Koskiaho et al. (2003) 5 Ockenden et al. (2012, 2014)

TABLE 11.10

Sample TSS performances for urban runoff wetlands.

Location	Wetland	Area m^2	WWAR %	HLR cm/d	TSS In mg/L	TSS Out mg/L	Ref.
North Carolina	JSC2	14,500	0.23%	30	50	4	1
North Carolina	JSC1	16,700	0.27%	10	31	3	1
North Carolina	RB	1,400	0.30%	13	8	8	1
North Carolina	Bass	10,000	0.42%	20	20	14	1
North Carolina	EB	2,000	0.55%	15	24	22	1
North Carolina	NCSU	300	0.71%	46	155	82	1
North Carolina	UNCA	700	1.71%	100	352	31	1
North Carolina	BES	1,300	2.03%	10	68	21	1
North Carolina	CMS	2,000	2.08%	15	20	38	1
North Carolina	DB	3,300	2.64%	10	79	12	1
Florida	Freedom Park	46,000	0.15%	78	4	4	2
Sweden	Växjö	68,000	2.13%	4	153	22	3
North Carolina	Mooresville	3,300	2.64%		77	9	4
Washington	B31	20,000	1.09%	18	58	34	5
Washington	PC12	15,000	1.76%	12	4	2	5
California	Tahoe	6,366	2.77%	4.7	34	9	6
Median			**1.09%**	**15**	**42**	**14**	

References:
1. Merriman et al., 2017
2. Mitsch, 2017
3. Al-Rubaei et al., 2016
4. Hathaway and Hunt, 2010
5. Reinelt and Horner, 1995 6. Heyvaert et al., 2006

$$C_o = 0.78q^{0.33}C_i^{0.51} \tag{11.5}$$

$R^2 = 0.80$
Range of HLR: 3–200 m/yr (0.8–55 cm/d)
Range of C_i: 10–300 mg/L
where: C_i = inlet TSS concentration, mg/L
C_o = outlet TSS concentration, mg/L
q = hydraulic loading rate, m/yr

This regression is entirely consistent with the inter-system continuous flow treatment marsh data.

The removed TSS accretes in the treatment system, preferentially in the inlet end (Braskerud, 2001a). Eventually, the deposits require removal, as was done in the Lake Tahoe system (Heyvaert et al., 2006; Qualls and Heyvaert, 2017) (see Table 11.10). After 16 years of operation the system was drained, and the sediments analyzed. Qualls and Heyvaert (2017) measured an average accretion of 3.2 cm/yr. Based upon sediment chemistry, they estimated that 59% of accreted material was due to bioaccretion (autocthonous), with the remainder due to sedimentation (allochthonous). This is a cogent reminder that biogeochemical cycling is an important feature of all treatment marshes, and urban stormwater wetlands are no exception.

The ability of urban wetlands to reduce TSS is conceptually sustainable, and results support that concept. Al-Rubaei et al. (2016) analyzed performance of a system at Växjö, Sweden three times during its 19-year lifetime (Table 11.10). In 1997, it reduced TSS from 187 to 7 mg/L; in 2013–2014, it reduced TSS from 207 to 16 mg/L. During that period, sediment from the forebay was dredged four times, but no other maintenance was conducted. Thus, despite the lack of maintenance, the wetland performed more efficiently and stably for most pollutants after 19 years, compared to newly constructed.

Dynamic Responses

Both water and solids dynamics influence the performance of event-driven wetlands. If the event period is long compared to the detention time of the wetland under the event flow, then the first part of the event completely flushes the antecedent water, and the outflow concentration ramps up to the level dictated by the sustained event flow After the cessation of event flow, the wetland outflows and outlet concentrations ramp back down to nonflow conditions. The majority of the treatment occurs during the period of high flow, and the expulsion of the highly treated antecedent water contributes little to the TSS export of the event.

Data acquisition for individual events involves close-interval sampling, with numerous samples required to define an individual response. Consequently, there are very few time series studies. It is easier and less costly to use flow-proportional autosampling, but the result is then restricted to the EMC for the event.

Phosphorus in Event-Driven Marshes

Event-driven treatment marshes systems have demonstrated ability to reduce phosphorus concentrations in both agricultural and urban runoff. They depend upon both sedimentation and bioaccretion as mechanisms. Small systems are typically controlled by sedimentation, but larger wetlands, such as the Florida STAs, rely upon bioaccretion. Additionally, the retrofitting of floating rafts on urban systems affects the overall P removal performance.

A fraction of the fertilizers applied to fields is unavoidably lost to runoff and shallow groundwater. Tile drainage systems collect subsurface waters, and vent them to receiving streams, where their phosphorus content may cause problems. Wetlands are being used at various points in the agricultural landscape, corresponding to drainage areas ranging from individual fields (Braskerud, 2002b; Tanner et al., 2003), to large regional landscape units of thousands of hectares (SFWMD, 2018). The later are typified by the stormwater treatment areas in south Florida, which strip phosphorus from agricultural and other runoff to protect the Everglades. Systems that receive pumped water at relatively constant rates have been included under the discussion of continuous (but possibly variable) flow wetlands. Those wetlands that receive water as a result of meteorological events are considered here.

Accordingly, wetland-watershed area ratios (WWARs) vary greatly, from less than 0.1% to 20%. Runoff total phosphorus varies from 0.07–1.15 mg/L on average. It is difficult to draw generalities, but First order models may be used to describe phosphorus removal in agricultural runoff wetlands over averaged periods of operation (e.g., Tanner et al., 2005b; Beutel et al., 2014).

Small Agricultural Systems

The design concepts for agricultural runoff wetlands vary from very high loadings primarily intended for sediment trapping, to very low loadings intended to provide for nutrient removal via microbiological processes and nutrient cycling. It is clear that very high loading rates do not allow time for interaction with the wetland biogeochemical cycle. It is also evident that wetlands produce or pass through only very low background levels of phosphorus, and therefore may be expected to reduce TP to very low levels if sufficient area is provided.

End-of field treatment marshes target phosphorus removal as a means of protecting the receiving water bodies. As previously noted, there are no rules and regulations governing the sizes and construction details of these systems. Sizes typically range between 0.01–10 ha, and WWAR range between 0.1 to 20% (Table 11.11).

The example results in Table 11.11 fall in two categories, well performing and poorly performing systems. Percent removals in the former group are about 60%, compared to about 20% in the later group. Rate coefficients have a median of 65 m/yr for well-performing systems, and are essentially nil for poorly performing wetlands. The reasons for the differences are not well understood, but it appears that the Scandinavian systems function by sedimentation of P-bearing clay particles.

The phosphorus performance of agricultural wetlands in Sweden (there are more than a thousand) has been analyzed utilizing a modified first-order model (Arheimer and Pers, 2017). The small systems (median size = 2 ha) are considered to have P =1, or one well-mixed tank. Both an uptake and a return flux are considered, with the following forms:

$$J_{uptake} = kC \tag{11.6}$$

$$J_{return} = k_{R20}\theta^{(T-20)} C_{inlet} \tag{11.7}$$

where: C = outlet concentration, gP/m^3

C_{inlet} = inlet concentration, gP/m^3

J_{uptake} = uptake flux, $gP/m^2 \cdot yr$

J_{return} = return flux, $gP/m^2 \cdot yr$

k = uptake flux rate coefficient rate, m/yr

k_{R20} = return flux rate coefficient rate, m/yr

T = temperature, °C

ψ = temperature coefficient, –

The selected median uptake rate coefficient was k_{uptake} = 33 m/yr, and $_{kR20}$ = 18 m/yr. The temperature coefficient was ψ = 1.200.

The Florida STAs

Much of the research on agricultural phosphorus removal in treatment wetlands has originated from the Everglades protection projects of south Florida. The receiving Everglades ecosystem is conditioned by very low total phosphorus, in the range of 6–12 μg/L. Even modest amounts of field runoff phosphorus pose a threat to that ecosystem (Davis, 1994). Tests were performed on dozens of mesocosm tanks, thirty 0.2-ha test cells, and four 2-ha field scale wetlands. In parallel, a 1,500 ha, five-cell wetland, the Everglades Nutrient Removal Project (ENRP) was operated as a research device. These FWS systems variously contained emergent vegetation, submerged plants, or algae, and testing was conducted over the period 1994–2005. Details of ENRP results were presented in a special issue of Ecological Engineering (Vol. 27, Issue 4, 2006). Although these studies were directed toward the design of pumped event-driven systems, the trials were in general conducted at conditions of steady flow. Plans to study event operation were abandoned, because the full-scale wetlands were placed in operation, and became the study platforms.

There are presently seven Stormwater Treatment Areas (STAs) in operation, aggregating over 29,000 ha and treating an annual average flow of 4.6×10^6 m^3/d. These FWS wetlands have been in operation for periods of 13 to 24 years, and receive deliveries of pumped inflows, which arrive mostly during the wet season.

TABLE 11.11

Example performances of small agricultural wetlands for reduction of total phosphorus. Well and poor performing groups are represented. Rate coefficients are calculated in the reference, or by setting P = 4 and C* = 0.006 mgP/L.

Wetland	Location	WWAR %	HLR cm/d	TP in mg/L	TP out mg/L	Removal %	k m/yr	Ref.
Well Performing								
Titoki	NZ	2.0	19	0.075	0.015	80%	184	6
Crookes	Australia	0.1	21.8	0.33	0.241	27%	27	8
W1	California	0.28	40	0.33	0.25	24%	43	9
Genarp	Sweden	0.33	21.9	0.12	0.034	72%	136	10
Leirvollbekk	Norway	1.0	34	1.53	1.26	18%	25	10
Lier	Norway	0.15	66	0.6	0.112	81%	521	10
Råbytorp	Sweden	0.2	37.5	0.12	0.012	90%	612	10
Sonnhof	Switzerland	1.15	9.3	0.14	0.032	77%	68	10
Böl	Sweden	0.09	177	0.293	0.221	24%	192	11
Lin	Sweden	0.84	6	0.199	0.046	77%	43	11
Wig	Sweden	0.04	110	0.346	0.155	55%	368	11
Yakima S	Washington		14	0.117	0.066	44%	39	12
Median						**58%**	64.8	
Poorly Performing								
B1	Maryland	9.29	0.6	1.150	0.650	43%	1.3	1
B2	Maryland	13.00	0.4	0.970	0.400	59%	1.5	1
Braun	Maryland	19.17	0.3	1.150	0.630	45%	0.7	1
Foster	Maryland	11.40	0.5	0.200	0.250	–25%	–0.4	1
A	Illinois	4.00	2.2	0.210	0.170	21%	1.8	2
B	Illinois	5.88	1.5	0.120	0.080	35%	2.4	2
D	Illinois	3.13	2.8	0.110	0.130	–18%	–1.8	2
Hovi	Finland	5.00	1.9	0.510	0.190	62%	7.8	3
Flytträsk	Finland	3.00	3.8	0.070	0.060	15%	2.5	3
G	Illinois	0.16	16.5	0.060	0.060	–1%	0.0	4
Toenepi	New Zealand	1.00	6.9	0.015	0.070	–76%	–39.0	5
W4	California	7.50	7.5	0.310	0.410	–32%	–7.5	9
Median						**18%**	1.0	

References:
1 Jordan et al. (1999)
2 Kovacic et al. (2000)
3 Koskiaho et al. (2003)
4 Miller et al. (2002)
5 Tanner et al. (2005b)
6 Tanner et al. (2003)
7 Ballaron (1998)
8 Raisin et al. (1999)
9 Díaz et al. (2012)
10 Braskerud et al. (2005b)
11 Johannesson et al. (2011)
12 Beutel et al. (2014)

The performance of these wetlands is described in Chapter 10, and in greater detail in the series of South Florida Environmental Reports issued yearly by the South Florida Water Management District (e.g., SFWMD, 2018). However, the intermittent aspect of performance analysis is clearly illustrated by the STAs. Table 11.12 shows the effect of annualizing the flows in contrast to using the System Mean Volumes and Concentrations (SMCs and SMQs). In effect, the wetlands process water about half the time. Thus the instantaneous hydraulic loadings are about double the annual averages. This carries over to removal rate coefficients, which are also about double those based on annual average flows.

Inter-System Performance: Urban Systems

There are great numbers of urban stormwater wetlands. For example, Murray et al. (2013) reports that in suburban Melbourne, Australia, 117 stormwater retention systems have been built to capture nutrients and hold water in order to control flooding and to provide public amenity and environmental benefits. Nonetheless, there are relatively few performance analyses that tie together the effects of inlet concentrations, flow rates and system characteristics to model outlet P concentrations.

As noted previously, over three decades of research directed at stormwater treatment have produced much information on constructed wetland systems to improve runoff quality. Most of the results from these compilations are in the form of percent removals, which can give a sense of the general range of performances, but cannot tie the removals to inlet flows, concentrations and system characteristics. Statistics on the inlet and outlet concentrations are also available (see Table 11.8), but these leave the question of HLR effects unanswered.

In an effort to bring both hydraulic loading and inlet concentration into process descriptions, Wong et al. (1999) suggested the use of a regression equation for purposes of sizing, similar to that for TSS. The proposed relation for TP, rewritten for concentrations, was:

$$C_o = 0.12 q^{0.44} C_i \tag{11.8}$$

$R^2 = 0.71$
Range of HLR: 2–140 m/yr (0.5–38 cm/d)
where: C_i = inlet TP concentration, mg/L

TABLE 11.12
Comparison of annualized and instantaneous rate coefficients for the Florida STAs. The PkC* model parameters are P = 6 and C* = 6 μgP/L. A nine-year POR is considered.

		HLR		k-value	
	Days in Use	In Use cm/d	Annualized cm/d	In Use m/yr	Annualized m/yr
STA2FW1	261	4.32	3.05	82.7	57.5
STA2FW2	177	8.55	4.05	69.7	32.0
STA2FW3	221	5.90	3.47	53.6	31.6
STA34FW1	203	4.06	2.14	41.8	22.8
STA34FW2	229	3.46	2.10	24.5	15.0
STA34FW3	242	5.49	3.47	47.4	30.0
STA56FW1	123	4.43	1.40	31.5	10.5
STA56FW2	137	4.15	1.41	31.8	11.8
Mean	**199**	**5.05**	**2.64**	**47.9**	**26.4**

C_o = outlet TP concentration, mg/L
q = hydraulic loading rate, m/yr
This regression is not much different from that which is obtained from a broader set of treatment marshes. For 316 systems:

$$C_o = 0.32q^{0.33}C_i^{1.10} \quad (11.9)$$

$R^2 = 0.80$
Range of HLR: 0.6–917 m/yr (0.2–251 cm/d)
Because of the empirical nature of this regression, it cannot be used outside the stated range of HLRs.

Since urban stormwater HLRs are close to the HLRs for point-source-driven wetlands, it is reasonable to expect that stormwater wetlands would perform somewhere near the average for the continuous flow FWS database set. These regressions highlight the necessity for bringing the HLR into the description of P removal. Equation 11.9 indicates a positive correlation of phosphorus remaining with increases in hydraulic loading rate. This trend is intuitively correct, but the confidence in such correlations cannot be high, because of the numerous factors that contribute to variability.

Sample results for reductions phosphorus from constructed, gravity-fed systems treating urban runoff waters are shown in Table 11.13. It is seen that reductions have a median of

TABLE 11.13

Example performances of urban stormwater wetlands for reduction of total phosphorus. Rate coefficients for reference 1 systems are from reference 1, the others are calculated in by setting P = 3 and C* = 0.006 mgP/L.

Wetland	Site	WWAR %	HLR cm/d	TP In mg/L	TP Out mg/L	Removal %	k m/yr	Ref.
BES	N. Carolina	2.03%	10.0	0.320	0.170	47%	14	1
EB	N. Carolina	0.55%	15.0	0.190	0.090	53%	4	1
NCSU	N. Carolina	0.71%	46.0	0.300	0.200	33%	58	1
CMS	N. Carolina	2.08%	15.0	0.170	0.180	–6%	37	1
DB	N. Carolina	2.64%	10.0	0.260	0.110	58%	37	1
RB	N. Carolina	0.30%	12.5	0.170	0.200	–18%	18	1
JSC1	N. Carolina	0.27%	10.0	0.210	0.050	76%	36	1
Naples	Florida	0.260%	4.5	0.196	0.032	84%	46	2
Naples	Florida	0.260%	7.3	0.110	0.050	55%	27	3
B31	Washington	1.070%	18.1	0.172	0.137	20%	16	4
PC12	Washington	1.724%	12.1	0.028	0.025	11%	7	4
Växjö	Sweden	2.13%	4.4	0.330	0.040	88%	54	5
Tahoe	California	2.77%	4.7	0.279	0.094	66%	24	6
Median						**53%**	**27**	

References:
1 Merriman et al. (2017)
2 Bishop et al. (2012)
3 Griffiths and Mitsch (2017)
4 Reinelt and Horner (1995)
5 Al-Rubaei et al. (2016)
6 Heyvaert et al. (2006)

53%. WWAR is seen to not be a very good predictor of removals for this dataset across multiple sites and conditions, which supports an observation of Schueler (1992).

First order models are suggested to be used to describe event-driven systems, including urban runoff wetlands (Konyha et al., 1995; Wong and Geiger, 1997; Carleton et al., 2001; Wong et al., 2006; Merriman et al., 2017). These may be used in the time-average sense, in which annual concentrations and flows are presumed to be treated according to the rate coefficients derived for continuous systems (Wong and Geiger, 1997). Table 11.13 lists rate coefficients for example urban wetland systems. The median k = 27 m/yr for these examples, which is at about the 75th percentile for bioaccretion wetlands (see Table 10.10).

Dynamic Responses

The dynamics of tracer and phosphorus removal and export were studied at the Des Plaines River wetland demonstration site. Wetland pumping events were monitored for all hydrologic variables, including pumping, rain, storage change, and outflow. Tracer input and response were monitored at inflow, outflow and interior stations. Phosphorus was measured at high frequency (every six hours) for the outflows, and at lower frequency for inflows and interior stations. The events were isolated in time, with sufficient inter-event spacing to allow complete equilibration before the subsequent event. A calibrated dynamic water mass balance was developed as the framework for interpreting results (Kadlec, 2001b). Internal hydraulics were characterized by tanks in series (TIS) models calibrated to tracer studies. The residence time distribution was describable by five TIS for wetland C1, for instance. Dynamic mass balances were used, in conjunction with a first order areal removal model, to describe the time sequence of phosphorus concentrations and flows. Parameter estimation produced instantaneous rate coefficients.

A typical response time series and the model representation are shown in Figure 11.20. This two-day pumping event nominally flushed the wetland more than seven times, and the hydraulic loading rate was 87 cm/d. As a result, the outflow concentration during pumping was 213 μg/L just below that of the incoming water to the wetland, which was 258 μg/L. During the course of pumping, this corresponds to a high removal rate of 14 $gP/m^2 \cdot yr$. After the pumping stops, outflow quickly falls to zero, and the phosphorus concentrations in the wetland waters relaxes down to a background level of about 54 μg/L. Removal is sustained at high levels, with k = 101 m/yr. Prior studies at this site indicated that this high rate of removal is not sustained during continuous flow operation. The results of this study suggest that great caution must be exercised in transferring steady flow results to event-driven wetlands. For instance, the calculation approach of Wong and Geiger (Wong and Geiger, 1997), using transferred steady flow rate constants and C* values, may not be the best descriptor of pulsed treatment wetlands.

The performance of stormwater wetlands for isolated events is contingent on the flow rate and duration of the event, and the capacity of the wetland. The relevant variable is the number of nominal displacements of the antecedent wetland water that occur, as discussed in the preceding section on capture. When these concepts are applied to the Des Plaines data set, Figure 11.21 results. Phosphorus mass removal decreases as the number of displacements increases. At very low displacement fraction, only a small amount of (treated) antecedent water is "bumped" out of the system. As the number of displacements becomes large, the system approaches continuous-flow performance at the event flow rate.

Floating Rafts

Many recent publications have been intrigued by the possibility of improving P removal in the forebays of urban wetlands, and other aquatic treatment features. At this stage of the development of a knowledge base, virtually all information is from mesocosm and small

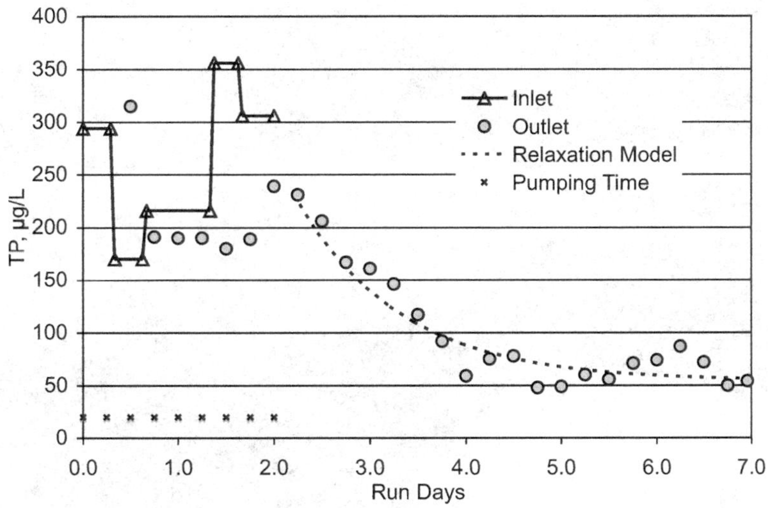

FIGURE 11.20 Response of wetland C1 at Des Plaines, Illinois to a two-day pumped pulse addition. The two-day pumping nominally flushed the wetland 7.2 times. The concentration dropped after pumping to 54 µg/L. The model line has parameters P = 5, k = 101 m/yr and C* = 54 µg/L. The incoming river water averaged 258 µg/L TP, and during pumping the average outflow was at 213 µg/L.

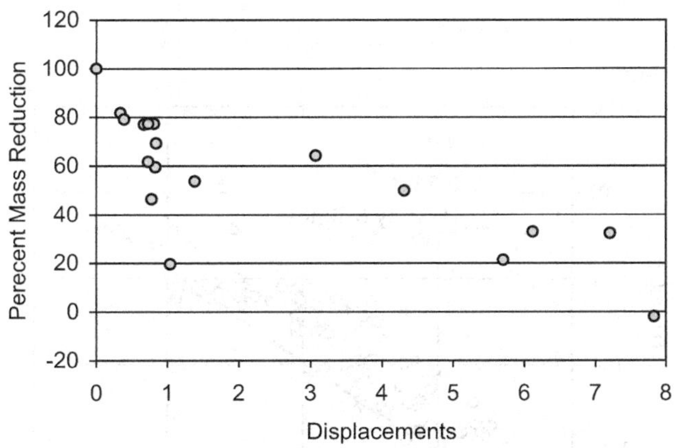

FIGURE 11.21 The effect of varying displacement on the phosphorus removal efficiency of event-driven wetlands.

pilot studies, although larger systems are becoming more common (Figure 11.22). Many of those have been concerned with treatment of strong wastewaters.

There seems to be only minor differences between the performance capabilities of floating raft wetlands and conventional vegetated wetlands. Headley and Tanner (2012) provide a survey of 20 FTW research project results, analyzed using methods outlined in earlier chapters. These have a median P rate coefficient of k = 30 m/yr, for P = 3 and C* =0.002 mgP/L. This is at the high end of the spectrum for bioaccretion wetlands (Table 10.10) and near the median for agricultural and urban wetlands (Table 10.11).

FIGURE 11.22 A floating wetland in New Zealand. This system is used to enhance treatment in a facultative pond. (Photo courtesy of C. Tanner).

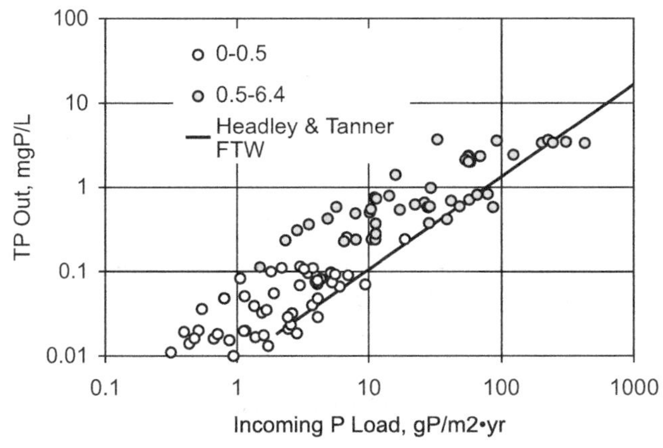

FIGURE 11.23 Intersystem phosphorus responses for FTW systems compared to bioaccretion treatment marshes. Each point represents the entire period of record for one wetland cell (see Figure 10.18). The FTW line is that representing 20 FTW mesocosms ($R^2 = 0.86$), Based on data from Headley and Tanner (2012).

The Headley and Tanner (2012) results are displayed on a loading graph in Figure 11.23, which compares the FTW information to that for other wetland systems given in Chapter 10, Figure 10.18. The FTW performance is essentially the lower bound for other marshes.

These results are similar to those of Wang and Sample (2013), for 14 FTW mesocosm platforms. They measured a mean k = 46 m/yr for C* = 0.002 mgP/L in the batch mode (P = ∞). In another survey of multiple FTW systems, Pavlineri et al. (2017) reported data for 23 mesocosms studies. The resulting mean k = 31 m/yr (median 15 m/yr), evaluated for P = 3 and C* = 0.006 mgP/L.

These reported performances are quite respectable, but must be placed in the context of land availability and costs. In Chapter 15, costs and benefits will be explored in more detail, but a this juncture, it is noted that FTWs are very expensive, and would not likely be selected for new construction if space permits a slightly larger treatment marsh.

Nitrogen in Event-Driven Marshes

Nitrogen compounds are among the principal pollutants of concern in fresh and marine waters because of their role in eutrophication, their effect on the oxygen content of receiving waters, and their potential toxicity to aquatic invertebrate and vertebrate species. Because treatment marshes deal with low ammonia loadings, toxicity and DO problems are not a principal focus. In almost all urban stormwater cases, nitrogen loadings are low enough to place them in the category of agronomic control. However, in most agricultural systems, nitrate is the dominant species, and microbial processing is of considerable importance. Often, ammonia concentrations are very low in all types of runoff wetlands. There is an omnipresent background level of organic nitrogen, both entering and leaving event-driven wetlands, which is on the order of 0.6–1.5 mgN/L.

As detailed in Chapter 9, treatment marshes possess a complex set of processes that are the basis for interconversions of nitrogen species, and the cycling of nitrogen in the biosphere. Microbial processes dominate the relations among organic, ammoniacal and oxidized nitrogen, and the vegetation cycles impart a seasonality to uptake and release processes. In a broad sense, some wetlands are dominated by microbial processes, and others by the phytomass cycle (agronomic wetlands).

The urban and agricultural treatment marsh literature is essentially devoid of any analyses of the contributing wetland processes. The reasons for this egregious omission are not apparent, but may in part reside with the regulatory rules and expectations. Design recipes do not require any knowledge of nitrogen conversion routes; rather, they are mostly based upon volumetric and ecological considerations. Those design recipes rely upon percent removals as a justification, and often those involve only total nitrogen.

There are, however, a number of significant investigations of the loss rate coefficients for various nitrogen species, and for total nitrogen.

Inter-System Performance: Agricultural Systems

The design concepts for agricultural runoff wetlands vary from very high loadings primarily intended for sediment trapping, to very low loadings intended to provide for nutrient removal via microbiological processes and nutrient cycling. Accordingly, wetland-watershed area ratios (WWARs) vary greatly, from less than 0.1% to 20%. Runoff total nitrogen varies from 1–16 mg/L on average, and is dominated by nitrate in most crop runoff cases (Table 11.14). Ammonia nitrogen is typically less than one mg/L entering the wetlands. However, pasture runoff may contain episodically high organic nitrogen. It is difficult to draw generalities from datasets such as Table 11.14, but it is clear that very high loading rates do not allow time for interaction with the wetland biogeochemical cycle, or for microbial processes. The Norwegian and California data in Table 11.14 exemplify this lack of nitrogen removal

TABLE 11.14

Examples of nitrogen species entering and leaving agricultural treatment marshes. Arranged in order of increasing inlet TN.

			NH_4-N		NO_3-N		TN		
Location	**Wetland**	**HLR cm/d**	**In mgN/L**	**Out mgN/L**	**In mgN/L**	**Out mgN/L**	**In mgN/L**	**Out mgN/L**	**Ref.**
California	W5	4.3	0.05	0.11	0.23	0.05	0.98	2.34	1
California	W6	4.3	0.05	0.04	0.23	0.001	0.98	2.31	1
California	W7	4.3	0.05	0.07	0.23	0.21	0.98	1.56	1
Norway	F	163.0	0.07	0.07	0.75	0.73	1.61	1.50	3
Denmark	Egeskov 2	0.96			1.47	0.12	2.03	0.47	4
Maryland	Braun	0.30	0.17	0.14	0.04	0.04	2.17	1.76	5
Finland	Flytträsk	3.80	0.09	0.09	2.40	2.33	3.10	3.10	7
Maryland	B1	0.59	0.36	0.31	0.25	0.07	3.13	3.48	5
Norway	A	170.0	0.16	0.17	2.28	2.29	3.20	3.10	3
Maryland	Foster	0.48	0.27	0.14	0.33	0.11	3.31	1.54	5
Norway	C	181.0	0.37	0.36	2.16	2.18	3.52	3.35	3
Maryland	B2	0.42	0.39	0.17	0.78	0.01	3.53	1.85	5
Norway	G1	123.0	0.68	0.65	2.77	2.50	5.15	4.39	3
Norway	G2	67.0	0.68	0.65	2.77	2.57	5.15	4.38	3
California	W1	40.0	0.14	0.13	4.72	4.21	5.70	5.57	1
New Zealand	Bog Burn	22.43	0.11	0.16	4.22	2.49	5.76	4.10	2
Denmark	Egeskov 1	1.69			7.56	2.84	8.26	3.26	4
Finland	Alastaro	12.80	0.04	0.04	6.80	6.23	8.40	8.40	7
Illinois	D	2.80	0.20	0.10	8.50	5.90	8.70	6.30	6
Finland	Hovi	1.90	0.11	0.10	7.90	7.13	9.80	9.80	7
Illinois	B	1.45	0.10	0.10	10.30	7.40	10.40	7.70	6
New Zealand	Titoki	14.54	2.03	1.67	8.10	3.69	10.68	7.10	2
California	W4	7.5	0.26	0.52	9.67	0.78	10.98	2.49	1
Denmark	Stor A	1.62			11.87	7.28	12.54	7.83	4
New Zealand	Toenepi	10.96	0.08	0.37	11.14	7.66	13.39	8.66	2
Illinois	A	2.19	1.00	0.80	13.00	11.20	14.00	12.30	6
California	W2	83.3	0.81	0.35	12.87	7.87	16.17	9.76	1
California	W3	46.9	0.81	0.66	12.87	7.27	16.17	9.77	1
Median		**4.3**	**0.17**	**0.16**	**3.50**	**2.50**	**5.43**	**3.79**	

References:
1 Díaz et al. (2012)
2 Tanner et al. (2003)
3 Braskerud (2002a)
4 Hoffmann et al. (2012)
5 Jordan et al. (1999)
6 Kovacic et al. (2000)
7 Koskiaho et al. (2003)

performance high loading. It is also evident that wetlands produce or pass through a background level of about 0.6–1.5 mg/L of organic nitrogen, and therefore should not be expected to reduce TKN or TN to below about 2 mg/L, even at low loadings (high WWAR).

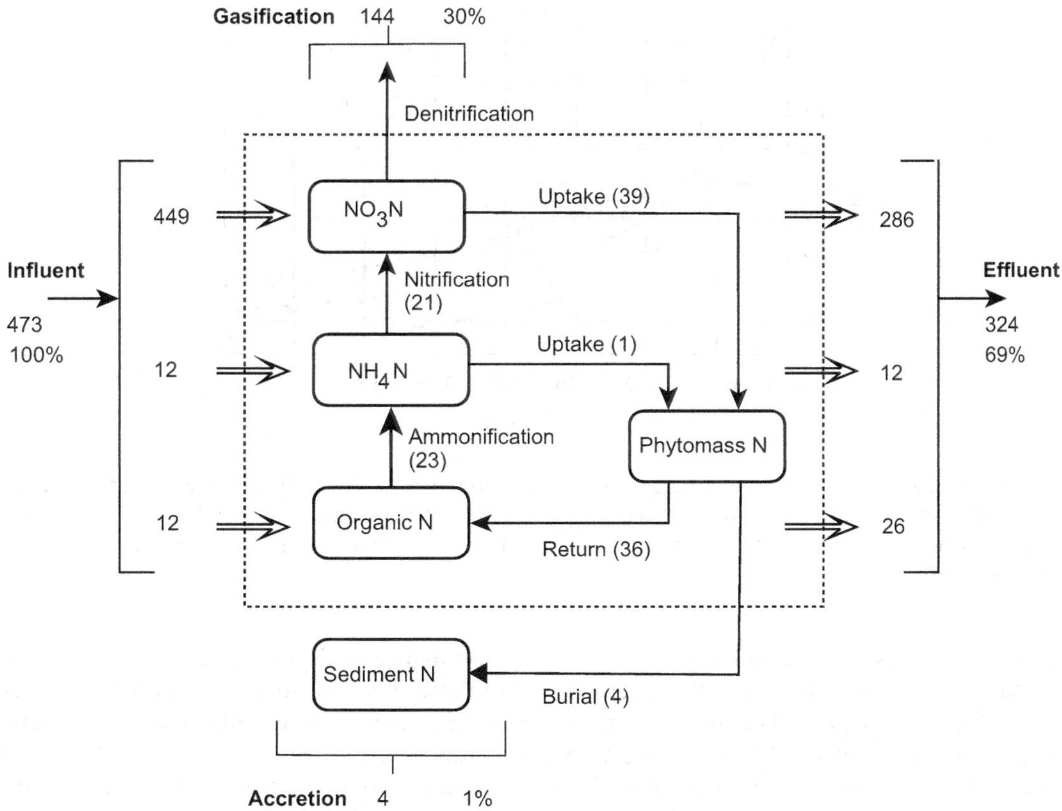

FIGURE 11.24 Estimated annual nitrogen fluxes in the Toenepi, New Zealand treatment wetland ($gN/m^2 \cdot yr$). Nitrate processing dominates this microbial system, and fuels the vegetation cycle.

Agricultural Runoff Mass Balances

Annual mass balances for nitrogen have been explained and illustrated in Chapter 9. As noted there, multiple interconversions and internal loadings due to vegetation cycling indicate that inlet-outlet percent removals for individual nitrogen species are virtually meaningless. Mass balances to form an integrated concept of nitrogen fluxes in event-driven treatment wetlands as well as for continuous flow systems. Mass balances provide a diagnostic tool for interpreting and understanding marsh performance. It is worth repeating the rationale of Chapter 9: this interpretive step is very important, because it:

1. Identifies the true rates of ammonification, ammonia oxidation and denitrification;
2. Places the role of the vegetative nitrogen cycle in the context of the microbial processes; and
3. Allocates the fate of added nitrogen to storage, leakage and gasification.

An example of process partitioning via an annual mass balance is shown in Figure 11.24. The wetland is a pasture runoff control system located at Toenepi, New Zealand (Tanner et al., 2003, 2005b). During year 5 of operation, entering water averaged 9 mgN/L NO_3-N, and less than 0.5 mgN/L of ammonia and organic nitrogen, and the average HLR = 13 cm/d. In this example, as for virtually all such examples, the phytomass cycle was estimated, with a total annual turnover of 2000 g/m^2 (above and below ground macrophytes,

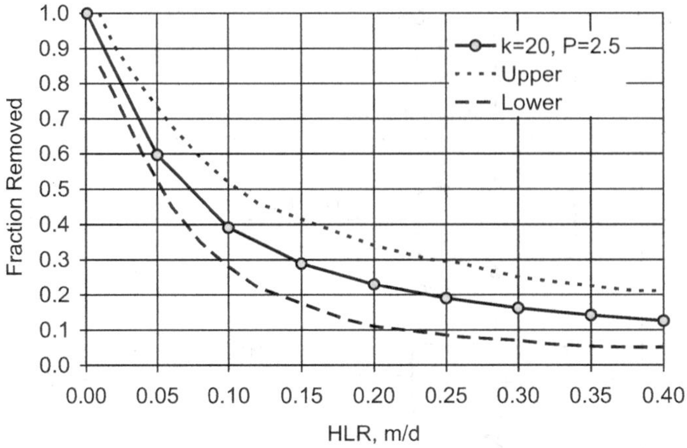

FIGURE 11.25 The PkC* representation of CREP wetland data. The upper and lower dashed lines represent data boundaries, visually set to contain about 90% of the data from 83 wetlands. The line with circles is a model calculation for C* = 0, P = 2.5 and k = 20 m/yr. Based on data from Crumpton and Stenback (2017).

plus mesophytes). Microbial nitrate processing dominates, with secondary importance of phytomass cycling. However, the necromass produced is vital to denitrification, and is ample for that purpose. The annual denitrification rate coefficient, calculated as the denitrification flux divided by the mean nitrate concentration, was k = 20 m/yr. Loss rate coefficients, based on individual entering and leaving concentrations, for organic and ammonia nitrogen are negative or zero. However, ammonification of the decomposing necromass requires an ammonification rate coefficient of 62 m/yr. The subsequent nitrification requires a nitrification rate coefficient of 88 m/yr.

The Florida STAs represent a unique case of nitrogen-starved wetlands. In common with other agricultural wetlands, these have low levels of incoming ammonia, and the ubiquitous background of about 2 mg/L of organic nitrogen (see Table 9.10). But in contrast to other agricultural wetlands, the STAs also have low levels of incoming oxidized nitrogen, averaging 0.43 mg/L. This feature may be attributable to the highly organic character of the agricultural soils, which would promote on-field denitrification.

Agricultural Loss Rate Coefficients

First order models may be used to describe agricultural runoff wetlands over averaged periods of operation (e.g., Stone et al., 2003; Tanner et al., 2005b; Ikenberry et al., 2017a). These and almost all others target nitrate reduction. Because agricultural runoff flows are often dominated by nitrate, there are only small differences between loss rate coefficients, and denitrification rate coefficients.

An important set of nitrate reduction runoff wetlands, the CREP wetlands in Iowa, have been analyzed using first order analysis. Each of the nearly one hundred wetlands has been calibrated, using C* = 0, for both k_{20} and θ in the PkC* model, implemented using the SWAT model structure. SWAT does not contain provisions for spatial profiles or detention time distributions, and thus the implied value is P = 1. However, the level of mixing has been measured at about P = 3 (Green, 2016).

Typical calibrated values are k_{20} = 17 m/yr and θ = 1.08 for the KS wetland, and k_{20} = 40 m/yr and θ = 1.08 for the AL wetland (Ikenberry et al., 2017a). Summary results are provided each year, in the form of a scatter plot of percent nitrate remaining versus

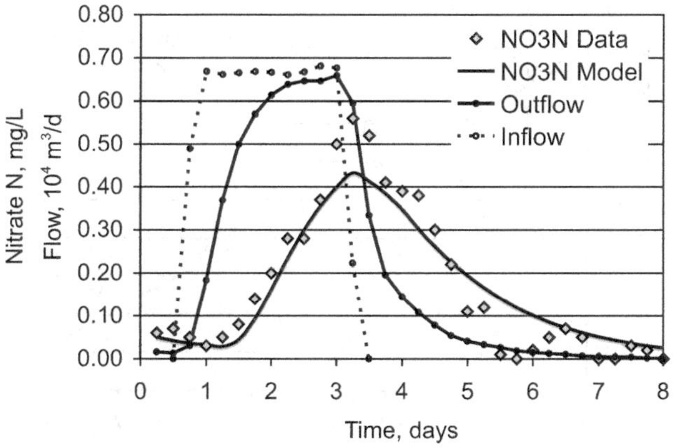

FIGURE 11.26 Response of wetland EW3 to a 60-hour flow pulse containing nitrate at 1.35 mg/L at Des Plaines, Illinois. The model parameters were P = 3, C* = 0, and k = 125 m/yr (R^2 = 0.88)

hydraulic loading rate (Crumpton and Stenback, 2017). The historical data from the period 2004–2016 are shown in Figure 11.25, along with a central tendency PkC* fit.

Nitrate Dynamics

The dynamics of tracer and nitrate removal and export were studied at the Des Plaines River wetland demonstration site (Kadlec, 2010). Twenty-eight wetland pumping events were monitored for seven wetlands, for all hydrologic variables, including pumping, rain, storage change, and outflow. Tracer input and response were monitored at inflow, outflow and interior stations. Nitrate was measured at high frequency (every six hours) for the outflows, and at lower frequency for inflows and interior stations. The events were isolated in time, with sufficient interevent spacing to allow complete equilibration before the subsequent event. A calibrated dynamic water mass balance was developed as the framework for interpreting results (Kadlec, 2001b). Internal hydraulics were characterized by tanks in series (TIS) models calibrated to tracer studies. The residence time distribution was describable by three TIS for wetland EW3, for instance. Dynamic nitrate mass balances were used, in conjunction with a first-order areal removal model, to describe the time sequence of nitrate concentrations and flows. Parameter estimation produced instantaneous rate coefficients (Werner and Kadlec, 2000).

A typical response time series and the model representation are shown in Figure 11.26. For the first day of this two-day pumping event, the outflow builds to equal the inflow, while the outflow concentration is essentially that of the antecedent water in the wetland, about 0.05 mg/L. Both data and the model calibration show that the wetland exports about 0.42 mg/L nitrate, reduced from 1.35 mg/L in the inflow, at the full-flow steady state. However, the early flush of clean antecedent water gives a lesser event mean outflow concentration of 0.29 mg/L.

The dynamic removal rate coefficient was 125 m/yr, but the mass removal is a modest 42 $g/m^2 \cdot yr$ (116 $mg/m^2 \cdot d$) over the duration of the event. For the set of pulse experiments, the dynamic k = 101 ± 14 m/yr (median ± se). The results were also analyzed via event mean concentrations and flows. For the set of pulse experiments, the EMC k = 64 ± 29 m/yr (median ± se).

It is noteworthy that the wetland response is very fast, both hydraulically and in terms of concentration reduction. Storage in the wetland biomass and sediments apparently are

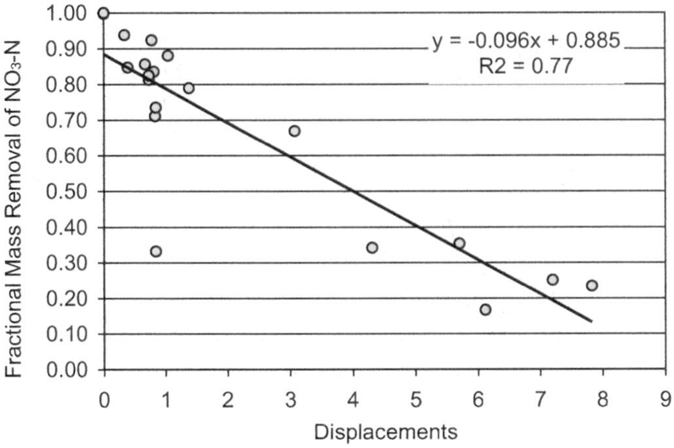

FIGURE 11.27 The effect of varying wetland water displacement on the nitrate mass removal efficiency of event-driven wetlands at Des Plaines, Illinois.

not critical to the response. This is convenient in calculations, because only water-phase storages and concentrations suffice to characterize the behavior. However, biomass provides the locus for denitrifiers and their carbon supply.

The performance of stormwater wetlands for nitrate reduction is contingent on the flow rate and duration of the event, and the capacity of the wetland. The relevant variable is the number of nominal displacements of the antecedent wetland water that occur, as discussed in the preceding section on capture. When these concepts are applied to the Des Plaines data set, Figure 11.27 results. Nitrate mass removal decreases as the number of displacements increases. At very low displacement fraction, only a small amount of (treated) antecedent water is "bumped" out of the system. As the number of displacements becomes large, the system approaches its continuous flow performance at the event flow rate.

Studies have also been conducted in which wetlands receive pulses of nitrate by virtue of a concentration spike at constant flow. In general, the response of the wetland is to reduce such new additions without delay. For example, Trujillo (1997) utilized simultaneous bromide and nitrate impulse additions to large FWS mesocosms to determine the hydraulic characteristics and the intrinsic rate of nitrate removal. The short detention times (6–18 hr) implied that plant uptake was not important, and nitrate dose removal was attributed to denitrification. Intrinsic denitrification rate coefficients ranged from 16 to 113 m/yr.

In rain-driven systems, flow and nitrate concentration pulses occur simultaneously. For tile-drain systems, concentration is inversely related to flow. An example of the response of a full-scale wetland to runoff event flows is shown in Figure 11.28. Drake et al. (2018) used high frequency measurements (every 15 minutes) to establish the time series for flows and concentrations. Because inlet concentrations are lower during events, the wetland response is to a negative impulse of nitrate nitrogen. A salient feature of the time series is the fact that outlet concentrations "track" inlet concentrations, but at a lower level as the result of removal in the system. There is essentially no time lag. Drake et al. (2018) did not model performance, and thus there are not dynamic k-values. However, the Slough Creek wetland is one of the Iowa CREP systems, whose behavior is included on Figure 11.25. Three years of summary data produce k_{20} = 40 m/yr and θ = 1.104, based upon P =2.5 TIS (Drake et al., 2018).

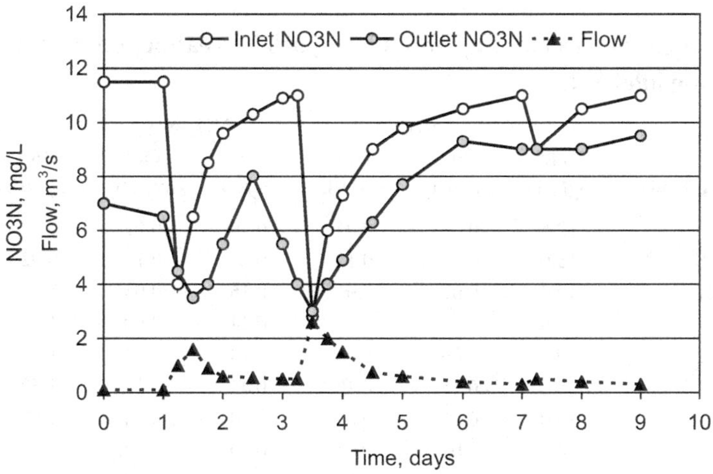

FIGURE 11.28 Response of the Slough Creek wetland in Iowa to runoff events. Inlet nitrate decreases as flow increases, and outlet nitrate follows the same time pattern. Based on data from Drake et al. (2018).

Inter-System Performance: Urban Systems

The same information base that was used for phosphorus is in major part also a source of information for nitrogen species. The reader is referred to the phosphorus section of this chapter for a discussion of those sources. Historically, urban stormwater data compilations report percent removals (e.g., Carleton et al., 2001; Kadlec and Wallace, 2009). In general, nitrate reduction has the highest median percentages (45%), compared to ammonia (31%) and TN (30%) (Kadlec and Wallace, 2009). It must be borne in mind that percent removals, even if based upon load or event-mean concentrations, may reflect a low level of an incoming constituent, an additionally may reflect background concentration limitations.

Information from published databases is presented in Tables 11.3 and 11.8, but these do not contain information on hydraulic loadings. They simply confirm that only small amounts of DIN are involved, and total nitrogen is dominated by the organic component.

Sample results for inlet and outlet concentrations of nitrogen species from constructed, gravity-fed urban systems are shown in Table 11.15, which reports information from the summary of Merriman et al. (2017) as well as other sources. Reductions for dissolved inorganic nitrogen (DIN) species (nitrate and ammonia) have medians in the 60–70% range. However, the ever-present organic nitrogen in source waters and treatment marsh effluents, 1.05 mg/L for the median influent and 0.75 mg/L for the median effluent, limits percent removals of total nitrogen. If these are irreducible background concentrations, then the TN reduction cannot exceed 30% for the example systems median.

There is an obvious need to consider factors beyond I/O data and percent removals. Hydraulic loading has been suggested as an appropriate correlator of TN reduction, as given in Equation 11.10 (Wong et al., 1999). The proposed relation, rewritten for concentrations, was:

$$C_o = 0.14q^{0.43}C_i \qquad (11.10)$$

$R^2 = 0.78$
Range of HLR: 3–100 m/yr (0.8–27 cm/d)

TABLE 11.15

Examples of nitrogen species entering and leaving urban treatment marshes. Arranged in order of increasing inlet TN.

			NH_4-N		NO_3-N		TN		
		HLR	In	Out	In	Out	In	Out	
Location	Wetland	mgN/L	mgN/L	mgN/L	mgN/L	mgN/L	mgN/L	mgN/L	Ref.
N. Carolina	RB	12.5	0.06	0.04	0.10	0.03	0.79	0.77	1
N. Carolina	CMS	15.0	0.12	0.18	0.22	0.15	0.92	1.13	1
N. Carolina	JSC1	10.0	0.07	0.06	0.16	0.05	1.17	0.63	1
Florida	Naples	7.0			0.14	0.00	1.17	0.86	4
N. Carolina	JSC2	30.0	0.09	0.03	0.14	0.04	1.25	0.56	1
N. Carolina	EB	15.0	0.15	0.10	0.47	0.14	1.52	1.05	1
N. Carolina	DB	10.0	0.21	0.02	0.40	0.14	1.57	0.86	1
California	Tahoe	4.7	0.02	0.01	0.59	0.10	1.60	0.81	2
N. Carolina	Mooresville		0.28	0.03	0.44	0.07	1.74	0.72	3
Sweden	Växjö	4.4					2.12	0.94	5
N. Carolina	BES	10.0	0.24	0.10	0.69	0.50	2.18	1.36	1
N. Carolina	UNCA	100.0	0.22	0.08	0.34	0.15	2.30	0.97	1
N. Carolina	Bass	20.0	0.50	0.40	0.50	0.10	2.46	1.90	1
N. Carolina	NCSU	46.0	0.56	0.33	0.51	0.31	2.72	1.94	1
Crete	FS12	38.2			0.60	0.40	3.00	1.50	6
Crete	FS24	19.1			0.60	0.40	3.00	1.50	6
Median		**15**	**0.18**	**0.07**	**0.44**	**0.14**	**1.67**	**0.96**	

References:
1 Merriman et al. (2017)
2 Heyvaert et al. (2006)
3 Hathaway and Hunt (2010)
4 Griffiths and Mitsch (2017)
5 Al-Rubaei et al. (2016)
6 Terzakis et al. (2008)

where: C_i = inlet TN concentration, mg/L
C_o = outlet TN concentration, mg/L
q = hydraulic loading rate, m/yr

Because of the empirical nature of this regression, it cannot be used outside the stated range of HLRs. This regression does not account for the background concentrations of TN.

The response outlet TN of urban wetlands to TN loading is distinctly different from that of the broader marsh database (Figure 11.29). Outlet TN is in general much lower, because virtually all DIN is consumed and because the return flux of organic nitrogen is lower, creating lower background concentrations.

Urban Runoff Mass Balances

As for agricultural runoff wetlands, mass balances provide a diagnostic tool for interpreting and understanding marsh performance.

An example of process partitioning via an annual mass balance is shown in Figure 11.30. The Bass wetland is an urban runoff control system located in North Carolina (Merriman et al., 2017). Entering water averaged 0.5 mgN/L NO_3-N, and 0.5 mgN/L of ammonia. Organic nitrogen averaged 0.5 mgN/L, and the average HLR = 20 cm/d (Table 11.15).

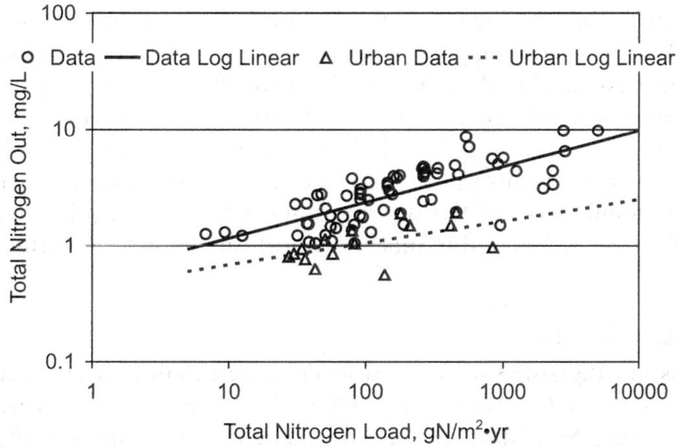

FIGURE 11.29 Load–concentration plot for total nitrogen in treatment marshes. The circles represent the broad treatment marsh database (see Figure 9.30), and the triangles represent example urban runoff marshes shown in Table 11.15.

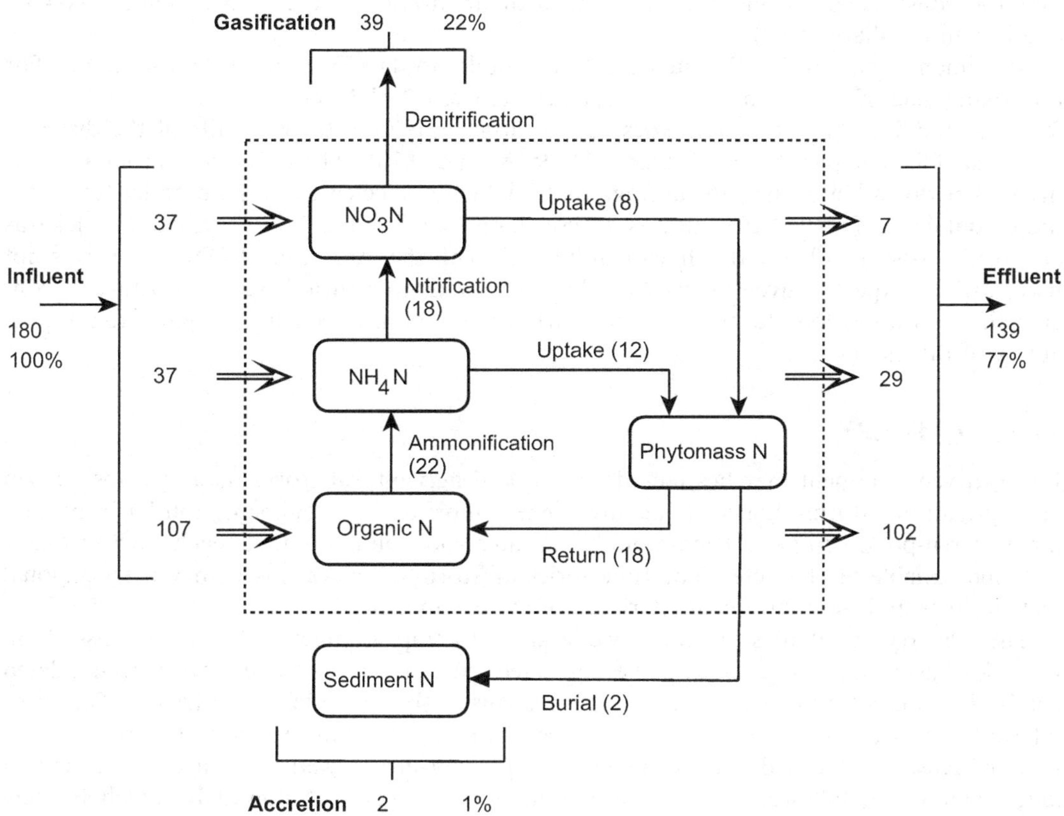

FIGURE 11.30 Estimated annual nitrogen fluxes in the Bass treatment wetland in Edenton, North Carolina, wetland ($gN/m^2 \cdot yr$). There are very low levels of DIN entering and leaving, and hence influxes and effluxes are dominated by organic nitrogen. Based on data in Merriman et al., 2017.

In this example, as for virtually all such examples, the phytomass cycle was estimated, with a total annual turnover of 1000 gdw/m^2 (above- and belowground macrophytes, plus mesophytes). Microbial nitrate processing and phytomass cycling are of about equal importance. However, the necromass produced is vital to denitrification, and is ample for that purpose. The annual denitrification rate coefficient, calculated as the denitrification flux divided by the mean nitrate concentration, was k = 130 m/yr. Loss rate coefficients, based on individual entering and leaving concentrations, for organic and ammonia nitrogen are negative or zero. However, ammonification of the decomposing necromass requires an ammonification rate coefficient of 15 m/yr. The subsequent nitrification requires a nitrification rate coefficient of 40 m/yr.

Urban Loss Rate Coefficients

First order models are suggested to be used to describe urban event-driven systems, including urban runoff wetlands (Konyha et al., 1995; Wong and Geiger, 1997; Carleton et al., 2001; Wong et al., 2006; Merriman et al., 2017). These may be used in the time-average or EMC sense, in which annual concentrations and flows are presumed to be treated according to the rate coefficients derived for continuous systems (Wong and Geiger, 1997). Since urban stormwater HLRs are close to the HLRs for point-source-driven marshes, it is reasonable to expect that stormwater wetlands would perform somewhere near the average for the continuous flow FWS database set. For instance, the median reduction for nitrate in 72 continuous-flow FWS wetlands was 53% at an average HLR = 7.1 cm/d; the median reduction for constructed stormwater marshes, with an average HLR = 5.4 cm/d, was 45% (Kadlec and Wallace, 2009).

Merriman et al. (2017) calibrated the PkC* model to data from ten urban wetlands. For a median value of P = 3, the loss rate coefficients were: NH_4N k_{20} = 34 m/yr; NO_3N k_{20} = 39 m/yr; and TN k_{20} = 44 m/yr. These are at about the 75–85th percentile of the distributions listed in Chapter 9 (see Tables 9.11, 9.13, and 9.16). These North Carolina urban marshes perform better than the average that might be expected from the broader treatment marsh database. This led the authors to conclude that "... the P-k-C* kinetic model was able to describe the observed behavior quite well, with the exception of ON, which was not necessarily unexpected given internal residuals and wetland return fluxes." They further concluded, "... (more) long-term event monitoring datasets are needed to expand the applicability of the model."

11.5 SUMMARY

Event-driven treatment marshes include urban and agricultural stormwater systems. Urban and agricultural stormwater wetlands are often comprised of a sedimentation basin preceding a marsh-pond complex. The amounts of water and pollutants that reach such wetlands are time-variable in character. The time series of stormwater systems is driven by regional meteorology, and the runoff characteristics of the catchment.

The primary intent of stormwater wetlands is to trap as much sediment as possible in the inlet basin, so as not to clog the vegetated portion of the system with accumulated solids. The secondary focus of stormwater systems is the reduction of nutrients. The same internal processes as for continuous flow wetlands reduce phosphorus and nitrogen compounds. However, the hydraulic sequence for an event-driven wetland is a displacement of antecedent water, followed possibly by a flow-through period, followed by batch holding during the inter-event period.

Attempts at simplistic performance correlations have been largely unsatisfactory, because of the high degree of variability in system designs and flow characteristics. Percentage removals help to give a broad perspective on system capabilities, but are nor sufficient for system interpretation or sizing. While it is intuitively obvious that a larger wetland to watershed area ratio is desirable, this effect has not been tightly quantified. The use of continuous-flow rate constants holds some promise, but more calibrations for event-driven wetlands are needed. Dynamic modeling has been demonstrated to be feasible, but is only affordable for large projects. Until more knowledge is gained, technology-based designs are likely to continue to be the rule.

The source waters entering rain-driven systems vary dramatically. Principal types are field runoff of TSS-laden waters, field runoff of nitrate-rich tile drainage waters, pumped phosphorus rich runoff, and nutrient-poor urban drainage waters. The flows are usually determined from some level of watershed modeling. Compendia of source water contaminant concentrations are increasingly available, and thus estimates may be made of watershed flows and loads.

Sizing is currently imposed by widely variable regulatory prescriptions, involving volumetric considerations. For urban systems, the volume of a wetland is typically specified in terms of the runoff volume associated with a rain of a specific return frequency. Both agricultural and urban systems may be required to have specific wetland to watershed area ratios.

In contrast, performance analyses are increasingly proceeding via rate coefficient models. The rate coefficient distributions from continuous flow marshes are inclusive of the values for event-driven wetlands. Speciated nitrogen mass balances also demonstrate the similarity between continuous and event wetlands.

12 Ancillary Substances

This chapter is concerned with some of the pollutants and ancillary chemicals found in treatment marshes that are not principal targets for design. Some have the potential to modify wetland processes but are not themselves design targets. Modifiers include pH, dissolved oxygen, and sulfur. Others are secondary targets for treatment, meaning their removal is of considerable interest, but not a direct factor in design. BOD and pathogens fall in this category. Subsequent chapters will deal with the larger categories of trace metals (Chapter 13) and trace organics (Chapter 14).

12.1 MAJOR IONS

In addition to the pollutants discussed in earlier chapters, source waters typically contain many other substances. These additional materials include salts, acids, bases, macronutrients, micronutrients, and heavy metals, and may be can be categorized in a number of ways. A number of substances are considered measures of water quality, - but are seldom of concern as pollutants to be treated in constructed wetlands. These include common metals (such as sodium, potassium, calcium, and magnesium) as well as halogens (such as fluorine, chlorine, bromine, and iodine). These often dominate the total ion content of natural waters and wastewaters. In total, they form the major part of electrical conductivity (EC) and total dissolved solids (TDS). In some treatment wetlands, several of these collective water quality parameters have become important in their own right

Halogens

Chloride and bromide are widely regarded as being "conservative" in wetland environments, meaning that they interact with the ecosystem to a very limited extent. Therefore, they can be used as tracers of water movement in the wetland. Usually, chloride is present at concentrations that preclude its use as an injected tracer, but it sometimes serves as a means of confirming the wetland water budget. Fluoride is usually a very minor trace constituent in aquatic systems. Bromide is often present at low background concentrations, and injected bromide may then be used to trace internal water movements. Very little is known about the fate and transport of iodine in freshwater systems.

U.S. EPA (2018a) sets a Criteria Maximum Concentration (CMC), which is an estimate of the highest concentration of a material in surface water to which an aquatic community can be exposed briefly without resulting in an unacceptable effect. The value for chloride is 860 mg/L, and for chlorine is 19 μg/L. U.S. EPA (2018a) also sets a Criterion Continuous Concentration (CCC), which is an estimate of the highest concentration of a material in surface water to which an aquatic community can be exposed indefinitely without resulting in an unacceptable effect. The CCC value for chloride is 230 mg/L, and for chlorine is 11 μg/L.

The wastewater treatment plant at Genoa-Oceola, Michigan received very high chlorides (approximately 400–550 mg/L) because of the widespread use of water softeners in the region. The treated water was discharged onto rapid infiltration basins on a hilltop, and eventually reached an adjacent wetland. The hydrologic gradient moved the water slowly toward the wetland, and the high chloride arrived in wells at the wetland edge after about three years. After about six years, the wetland surface waters reflected the chloride of the wastewater, with some dilution from the other flows in the aquifer.

Chlorine is not present in runoff waters but may be present in disinfected tertiary effluents that reach treatment marshes. Studies of the loss of TRC were conducted in the Tres Rios, Arizona FWS wetlands (Wass, Gerke and Associates, 2004). Disappearance was approximated by first-order behavior. The rate constants were calculated to be 0.86 d^{-1} for the Cobble C1 and Hayfield H1 wetlands, based on transect data. Thus, there would be more than 90% reduction in TRC for a three-day detention time.

Alkali Metals

Sodium, potassium, calcium, and magnesium are rarely the object of regulatory concern, because under most circumstances they do not pose any toxicity threat. Nevertheless, each of these has a role in wetland functioning, and can yield valuable information about pollutant processing. Examples of measured values entering and leaving marsh systems are shown in Table 12.1.

Because most freshwater wetland species have low sodium requirements, the dissolved sodium content of wastewater passing through wetlands changes little (Table 12.1). Thus, sodium concentrations can be used as a conservative tracer for calculating dilution and concentration and for tracking groundwater discharges from wetlands. For instance, the concentration of sodium in arid land treatment wetlands is likely to increase during the summer season due to evaporative concentration.

Sodium is useful as a marker for added salt (NaCl), which may enter wastewater treatment systems because of its use in water softening and road deicing. For example, the Cumberland County, Pennsylvania, wetland (Table 12.1) showed a large increase in sodium during a spring flushing event, which was also accompanied by a large pulse of chloride (data not shown: C_i = 14 mg/L and C_o = 140 mg/L). The origins of the sodium may have been the accumulation of road deicing salt, contributed by the highway runoff the wetland

TABLE 12.1

Examples of major cations entering and leaving treatment marshes (mg/L).

Wetland	Location	Sodium		Potassium		Calcium		Ref.
		In	Out	In	Out	In	Out	
Estevan	Saskatchewan	353	364	25.7	23.5	87	90	1
ENRP	Florida	110	119	–	–	84	69	1
Imperial	California	371	348	11.2	8.9	182	160	1
Brawley	California	698	774	18.6	18.7	181	183	1
Boney Marsh	Florida	10	10	1.2	1.3	14	14	1
Musselwhite	Ontario	55	71	19.0	22.5	94	113	2
Monroe County	New York	410	430	269.0	196.0	180	70	3
West Lafayette	Indiana	110	85	2.5	2.7	95	54	4
Cumberland	Pennsylvania	28	63	6.5	7.9	17	16	5

References:
1 Unpublished data
2 Bishay and Kadlec (2005)
3 Eckhardt et al. (1999)
4 Harbor et al. (2000)
5 Nelson et al. (2002)

was designed to treat. At the Genoa-Oceola, Michigan, site, discussed in the preceding chloride section, the water softener salt also had elevated sodium (150 mg/L), compared to the wetland background of about 5 mg/L. The underground plume reached the wetland after about six years, at which time the wetland surface water sodium had increased to 95 mg/L.

Potassium concentrations in plants are generally in the range of 1.0–4.0 %dw. Potassium regulates the opening and closing of stomata on plant leaves. Stomates are the valves that allow gases inside the plant to be exchanged with the atmosphere. Potassium also is used as an enzyme activator in protein synthesis in most cells. Potassium typically comprises about 2.6% of the dry weight of wetland plants. Potassium concentrations in water of surface flow treatment wetlands are typically between 1.0 and 40 mg/L (Table 12.1), with an average world river concentration of about 3.4 mg/L (Hutchinson, 1975). Potassium has not been the target of treatment wetland design. In general, there is not much change in potassium from wetland inlet to outlet (Table 12.1).

Calcium is biologically active because it is used as a nutrient by invertebrates and vertebrates, and because of its role in the carbonate cycle. Calcium is required by, and present in sizeable amounts, in angiosperm plants. The median concentrations in a variety of wetland plants is 0.77% dry mass, and in similar in freshwater planktonic algae (Vymazal, 1995). However, levels in floating plants and filamentous green algae range upward to 5–7% dry mass (Vymazal, 1995). The photosynthetic organs of plants and algae may develop calcium carbonate (calcite) encrustations in hard water environments. Because there is generally an excess of calcium in surface water and wastewater, calcium concentration does not change appreciably in many wetland treatment systems (see Table 12.1).

In some treatment wetlands, there is an iron deficiency, and calcium biogeochemistry is dominant. When this occurs, the wetland sediments contain a high proportion of calcium carbonate, which is referred to as calcitic mud or marl. The southern Everglades contain extensive areas of these calcitic muds, which form under conditions of shorter hydroperiod, as a result of calcium carbonate precipitation mediated by periphyton. These materials are very dense, and low in organic content, and are typically low in phosphorus content. Calcareous periphyton in the south Florida environment contribute to high soil calcium, with concentrations ranging from 3 to 4% in peats, to 20–40% in calcitic wetland sediments (DeBusk et al., 2004).

Magnesium is an essential micronutrient because of its role in phosphate energy transfer and because it is a structural component in the chlorophyll molecule (Wetzel, 1983). Because magnesium concentration of surface water almost always exceeds the requirements for plant growth, elevated magnesium concentrations are not affected when wastewater travels through wetland treatment systems. Magnesium is more soluble than calcium, and precipitate formation does not occur. Plant tissue concentrations are of the order of 0.25% dry weight, but may be higher for floating plants and filamentous algae (Vymazal, 1995).

Collective Parameters

Two chemical parameters are commonly used to indicate the collective concentrations of dissolved substances: total dissolved solids and specific conductance. These parameters do not specify the distribution of contributing ions and organic compounds that contribute, but they are helpful in support of the wetland water budget. Further, the total dissolved solids content of water is sometimes a regulated parameter, especially in arid regions, where salt buildup is a water quality concern.

Total Dissolved Solids (TDS) is used to quantify the degree of salinity, which is of interest in agricultural situations. The total quantity of dissolved solids in a water sample is measured by filtration followed by sample evaporation. This quantity contains both inorganic ions and organic compounds. TDS is nearly as conservative in wetlands as specific conductance and chloride. Because total dissolved solids concentrations are high in many source waters and the individual components of these solids greatly exceed the biological requirements for growth, wetlands generally have a negligible effect on this parameter.

The electrical conductivity (EC), also called specific conductance, of an aqueous solution is the reciprocal of the resistance between two platinum electrodes, one cm apart and with a surface area of one square cm. The reciprocal of electrical conductivity is equal to resistance and is a function of the total quantity of ionized materials in a water sample. Specific conductance usually is reported at a temperature of 25°C and in units of μS/cm, or μmhos/cm. Specific conductance is nearly proportional to the total dissolved solids in many surface waters and is a convenient measure of the salt content.

Total ionic salts in wetlands, as measured by specific conductance, may be somewhat altered by biological conditions in wetlands, but physical processes of dilution and evaporation represent the major influences. Therefore, EC is a relatively accurate indicator of dilution and concentration effects by rainfall and runoff and evapotranspiration in wetland treatment systems.

Treatment marshes are usually, but not always, dominated by the introduced flows. Rainfall and evapotranspiration are minor in comparison, except possibly for the duration of extreme events. Therefore, in the long term, EC of the inlet and outlet waters are close to the same. For the 11 example systems of Table 12.2, the outlet EC averages 99% of the inlet EC. This represents long-term mean performance, over an average of five years. Those wetlands receive an average annual source water hydraulic loading that is far greater than precipitation or evapotranspiration.

There are circumstances in which conductivity, chloride, and TDS change from inlet to outlet, and data then present a challenge for finding the cause. An illustration is the Incline Village, Nevada wetland. This arid-climate system does not receive water in the summer, because it is used to irrigate fodder crops. The design of the wetland was for disposal,

TABLE 12.2

Examples of conservative materials entering and leaving treatment marshes.

Wetland	Location	EC		Chloride		TDS	
		In μS/cm	Out μS/cm	In mg/L	Out mg/L	In mg/L	Out mg/L
Tres Rios H1	Arizona	1,559	1,497	259	257	902	906
Tres Rios H2	Arizona	1,558	1,511	254	253	913	924
Tres Rios C1	Arizona	1,500	1,494	264	272	934	940
Tres Rios C2	Arizona	1,500	1,488	263	264	934	949
Brawley	California	4,503	4,713	–	–	–	–
Imperial	California	2,718	2,642	–	–	–	–
OEW	Florida	550	522	66	65	–	–
Estevan	Saskatchewan	2,332	2,457	198	201	1,644	1,722
Isanti-Chisago	Minnesota	1,367	1,147	79	76	770	760
ENRP	Florida	1,033	1,037	160	167	–	–
Pensacola	Florida	2,125	1,931	–	–	1,458	1,385
Average		**1,886**	**1,858**	**193**	**194**	**1,079**	**1,084**

primarily (≈ 90%) to evaporative losses, and secondarily (≈ 10%) to infiltration. The hydraulic loading is very small, and large concentration increases in chloride and TDS are produced as the water moves through the sequence of cells (Kadlec et al., 1990).

Conductivity has also been used as a diagnostic tool for internal processes in treatment wetlands, in a number of ways. For example, EC is often much higher in the porewaters of FWS wetland soils and sediments than it is in overlying waters. DBE (2003) measured the electrical conductivity of surface waters and porewaters in Cell 1 of the ENRP in 2001. The porewater EC was higher than that of the overlying surface waters. Among the potential reasons is that rooted plants extract their transpiration requirement from porewater but reject some or all of the associated salts. The result is an upward positive gradient in EC in the top soil layer. That gradient may continue into the overlying water and produce stratification of the EC of the water column. The pattern of these results indicates an internal recycle loop, in which dissolved substances are drawn down into the root zone by transpiration or other flows, only to be rejected by the plants to avoid buildup of TDS within their tissues. This creates an upward gradient, which causes diffusive solute movements back into the water column from the porewater.

Conductivity is also a tool to understand the phenomenon of density stratification in constructed and other wetlands. Wetlands are sometimes too shallow to stratify due to thermal gradients, but the same is not true for density segregation, which may exist due to the character of incoming wastewater, rainfall, or added tracers. Two examples will serve to illustrate the potential for vertical stratification.

Various salts, such as sodium bromide and lithium chloride, are convenient tracers for water movement. Considerable quantities are needed, and consequently it is tempting to add concentrated solutions, in order to deal with manageable tracer solution volumes for addition. However, when dense solutions are introduced into the bottom of wetlands, there may be a strong energy barrier to vertical mixing, resulting in the dense material remaining on the bottom of the wetland. For instance, tracer feed concentrations can be as much as 7.5% LiCl (density = 1.04 g/cm^3) (Söhnel and Novotny, 1985). Concentrations of sodium bromide in a Tres Rios, Arizona (Whitmer, 1998; Whitmer et al., 2000) tracer study were about 20% NaBr (density = 1.16 g/cm^3) (Söhnel and Novotny, 1985). It is noteworthy that demonstration of stratification in limnology laboratory courses utilizes densities of 1.105 and 1.05 g/cm^3 (Wetzel and Likens, 1991). More directly, the work of Schmid et al. (2003, 2004) shows that both tracer injections would lead to stable, unmixed layers of tracer on the wetland bottom. Both tracer studies showed very poor tracer recovery, suggesting that the tracer "got stuck" in the wetland sediments.

Further evidence of vertical stratification was reported by Chimney et al. (2006), for a constructed wetland in the Florida Everglades (Figure 12.1). In this case, a likely cause is the back-diffusion of salts from the concentrated porewater.

12.2 HYDROGEN ION

Healthy aquatic systems can function only within a limited pH range. As a consequence, surface water discharge permits frequently require $6.5 < pH < 9.0$. Wetland water chemistry and biology are likewise affected by pH. Many treatment bacteria are not able to exist outside the range $4.0 < pH < 9.5$ (Metcalf and Eddy Inc, 1991). In addition to controlling various microbial processes, pH is also a determinant of several important chemical reactions. Ammonium changes to free ammonia at pH above neutral and at higher temperatures (refer to Chapter 9). The protonation of phosphorus changes with pH (refer to Chapter 10), and the hydroxide and oxyhydroxide precipitates of iron, manganese and aluminum are pH-sensitive. The pH value profoundly influences hydroxide, carbonate, sulfide, phosphate, and silicate equilibria in submerged soils. These equilibria regulate the

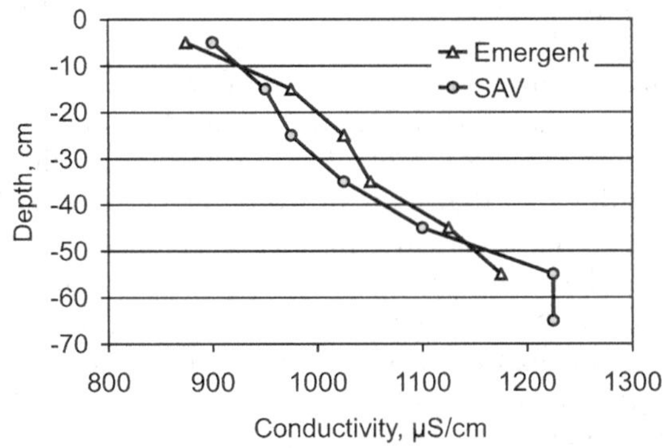

FIGURE 12.1 Vertical stratification of electrical conductivity in the ENRP, Florida, treatment marsh. Points represent the averages for 141 dates during May 1995 to October 1997. Data source: Chimney et al. (2006).

precipitation and dissolution of solids, carbon equilibria, the sorption and desorption of ions, and the concentrations of nutritionally significant ions or substrates (Ponnamperuma, 1972).

Natural wetlands exhibit pH values ranging from slightly basic in alkaline fens (pH = 7–8) to quite acidic in sphagnum bogs (pH = 3–4) (Mitsch and Gosselink, 2000). Natural freshwater marsh pH values are generally slightly acidic, (pH = 6–7) (Mitsch and Gosselink, 2000). The organic substances generated within a wetland via growth, death and decomposition cycles are the source of natural acidity. The resulting humic substances are large complex molecules with multiple carboxylate and phenolate groups. The protonated forms have a tendency to be less soluble in water, and precipitate under acidic conditions. As a consequence, wetland soil/water systems are buffered against incoming basic substances. They may be less well buffered against incoming acidic substances, since the water column contains a limited amount of soluble humics.

Treatment marsh source water hydrogen ion concentrations are typically circumneutral, and so are the marsh effluent waters. The notable exceptions are those wetlands receiving acid mine drainage, which reflect the low pH of the incoming waters. This special type of treatment wetland is not considered here; the reader is referred to Wieder (1989) and Younger et al. (2002). Furthermore, there is an important distinction between FWS and SSF systems in the ability of algae to conduct photosynthetic modulation of pH.

In aquatic systems, algal photosynthetic processes peak during the daytime hours, creating a diurnal cycle in pH. Photosynthesis utilizes carbon dioxide and produces oxygen, thereby shifting the carbonate-bicarbonate-carbon dioxide equilibria to higher pH. During nighttime hours, photosynthesis is absent, and algal respiration dominates, producing carbon dioxide and using oxygen. Open water zones within wetlands can develop high levels of algal activity, which in turn creates a high pH environment. Open water areas in wetlands also exhibit these phenomena. Diurnal pH fluctuations are not evident in areas with dense emergent vegetation. Data collected at the Sacramento, California wetlands illustrate these phenomena (Figure 12.2). In a densely vegetated zone near the outlet, there is no diurnal cycle in pH. However, there is a large diurnal cycle in the outlet deep zone, in which the detention time is about one day. Large exports of TSS occurred episodically, indicating high algal activity, which is in turn consistent with the large pH swing.

Vegetated FWS wetlands produce effluent waters with pH just above neutrality. This occurs whether the incoming water is acidic or basic. Examples are shown in Table 12.3.

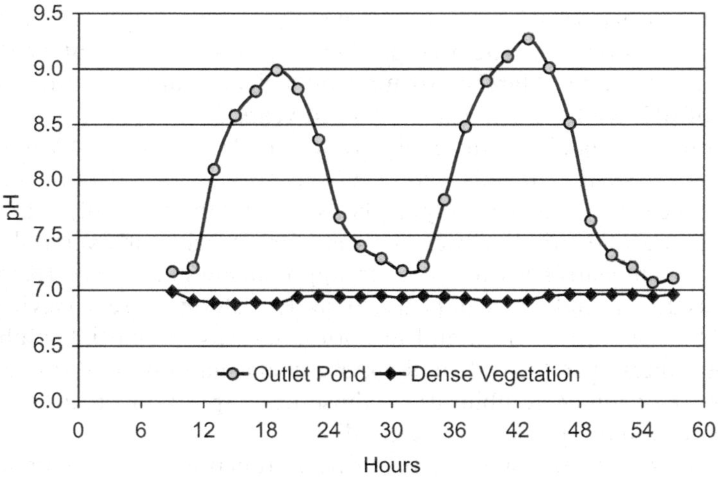

FIGURE 12.2 Diurnal variation in pH a near the exit of Cell 7 at Sacramento, California. In dense vegetation just prior to the outlet deep zone, pH does not vary. In the outlet deep zone, there is a large diurnal swing in pH, presumably driven by algal activity in the open water. Data source: Nolte and Associates (1997).

TABLE 12.3

Examples of pH of entering and leaving waters for treatment marshes.

Wetland	Location	Inlet pH	Outlet pH
OEW	Florida	7.11	6.91
Tres Rios H1	Arizona	6.94	7.14
Tres Rios H2	Arizona	6.95	7.22
Tres Rios C1	Arizona	6.92	7.36
Tres Rios C2	Arizona	6.92	7.23
Listowel 3	Ontario	7.59	7.06
Listowel 4	Ontario	7.28	7.02
Onaway	Michigan	7.39	7.28
Estevan	Saskatchewan	8.68	8.15
Brighton	Ontario	7.71	7.62
Imperial	California	7.61	7.81
Brawley	California	7.35	7.82
Shannon	Texas	8.02	7.95
ENRP	Florida	7.38	7.39
San Joaquin W1	California	8.00	8.40
San Joaquin W3	California	8.60	8.70
Sacramento W4	California	8.00	8.50
Median		**7.39**	**7.39**
s.e.		0.13	0.13

The annual trends in marsh effluent pH are typically quite weak. Because of these weak annual trends, marsh pH behavior can be adequately described by an annual mean and the associated standard deviation. The pH produced in FWS treatment wetlands is within a surprisingly narrow band. Marshes treating municipal effluents produce an intersystem annual average of pH = 7.18 ± 0.35 (N = 20, total years data = 56).

Natural treatment wetlands produce slightly lower pH, by about 0.5 units. This is possibly due to the antecedent peat soils that occupied most of these. Continued application of circumneutral wastewater to a naturally acidic wetland can eventually alter the pH of the surface waters in the wetland. This was the case for an acid sphagnum-black spruce bog, which received circumneutral wastewater for approximately 25 years (Kadlec and Bevis, 1990), as well as for a slightly acid peatland at Houghton Lake receiving slightly basic lagoon water. The effect on the peatland in both cases was the partial solubilization of the solid humic substances that formed under more acidic natural conditions. In addition to the chemical effect of humic solubilization, those decomposition processes that were acid-inhibited can resume under the less acidic conditions.

Treatment marshes may therefore be regarded as functioning at circumneutral pH, with the caveat that open water near the outlet may foster algal growth and attendant diurnal cycles of pH that include high values.

12.3 DISSOLVED OXYGEN

Dissolved oxygen (DO) is of interest in treatment wetlands for two principal reasons: it is an important participant in some pollutant removal mechanisms, and it is a regulatory parameter for discharges to surface waters. In the first instance, DO is the driver for nitrification and for aerobic decomposition of CBOD and soil organic matter. In the second instance, DO is critical for the survival of fish and other aquatic organisms, and for the general health of receiving water bodies. In many permits in the United States, a minimum DO of 5 mg/L is specified.

Water entering the treatment wetland has carbonaceous and nitrogenous oxygen demand (NOD). After entering the wetland, several competing processes affect the concentrations of oxygen, biochemical oxygen demand (BOD), and nitrogen species. Dissolved oxygen is depleted to meet wetland oxygen requirements in four major categories: sediment/litter oxygen demand, respiration requirements, dissolved carbonaceous BOD, and dissolved NOD. The sediment oxygen demand is the result of decomposing detritus generated by carbon fixation in the wetland, as well as decomposition of accumulated organic solids that entered with the water and accreted organic soils. The NOD is exerted primarily by ammonium nitrogen; but ammonium may be supplemented by the mineralization of dissolved organic nitrogen. Decomposition processes in the wetland also contribute to NOD and BOD. Microorganisms, primarily attached to solid, immersed surfaces, mediate the reactions between DO and the oxygen consuming chemicals. Plants and animals within the wetland require oxygen for respiration. In the aquatic environment, this effect is seen as the nighttime disappearance of dissolved oxygen. Oxygen transfers from air, and generation within the wetland, supplements any residual DO that may have been present in the incoming water. Three routes have been documented for transfer from air: direct mass transfer to the water surface, convective transport down dead stems and leaves, and convective transport down live stems and leaves. The latter two combine to form the plant aeration flux, (PAF). These transfers are largely balanced by root respiration but may contribute to other oxidative processes in the root zone.

Despite this complexity, wetlands are not particularly efficient at obtaining oxygen in sufficient quantities to deal with heavy pollutant loads. But in the general sphere of operation defined for treatment marshes, there is not a heavy load of BOD and ammonia.

Atmospheric aeration is usually sufficient to provide all the oxygen necessary for the various marsh demands.

Physical Oxygen Transfers

The transfer of oxygen from air to the wetland water has been discussed in Chapter 9, where it is shown that a mass transfer coefficient (K_L) model is appropriate. Atmospheric aeration drives the wetland water toward its DO solubility limit. The concentration of dissolved O_2 (DO) in water varies with temperature, dissolved salts and biological activity. The effect of temperature on the equilibrium solubility of oxygen in pure water exposed to air has been widely studied, and can be calculated from regression presented in Equation 5–8 (Elmore and Hayes, 1960):

$$C_{DO}^{sat} = 14.652 - 0.41022T + 0.007991T^2 - 0.00007777T^3 \qquad (12.1)$$

where C_{DO}^{sat} = equilibrium DO concentration at 1.0 atmosphere, mg/L
T = water temperature, °C

This relation shows that at 25°C, the equilibrium DO = 8.2 mg/L, while at 5°C, the equilibrium DO = 12.8 mg/L. However, this is not an absolute limit for marshes, because internal processes can, and do, drive DO to levels well above saturation.

Open Water Zones

Treatment wetlands are sometimes configured with open water zones, either as deep zones or as areas of submerged aquatic vegetation. These offer enhanced opportunity for atmospheric oxygen transfer. Despite the considerable uncertainty in the mass transfer coefficient, calculations show that physical reaeration is faster in the absence of emergent or floating vegetation.

An open water, unvegetated wetland was monitored for DO in Commerce Township, Michigan for a period of three years. Ammonia and BOD were very low in this "polishing" wetland, typically less than 0.2 mg/L for ammonia and less than 2.0 mg/L for $CBOD_5$. Inlet DO averaged 83% of saturation, and outlet DO was 91% of saturation after 3.3 days detention. The corresponding mean K_L value was 0.42 m/d (Kadlec, unpublished data).

A study of the effect of deep zones on marsh water quality parameters was conducted in a set of twelve side-by-side pilot marsh cells at the Phoenix, Arizona, Tres Rios site (Kadlec, 2007). The percentage open water was varied in triplicates of 12.5, 20, 28, and 35% open. The outlet DO averaged 3.6±1.3, 3.4±1.0, 3.9±1.1 and 3.6±0.7 mg/L respectively. Although there was no effect of fraction open water, higher hydraulic loadings created lower effluent DO. The HLR regression was DO Out = 4.7–14.6·HLR (mg/L, cm/d) with an $R^2 = 0.77$.

Biochemical Production of Oxygen

Oxygen is the by-product of photosynthesis (Equation 12.1). When photosynthesis takes place below the water surface, as in the case of epiphyton and plankton, oxygen is added to the water internally. A large algal bloom can raise oxygen levels to 15–20 mg/L, more than double the saturation solubility, as a result of wastewater addition. This process requires sunlight, and algal photosynthesis is suppressed in wetlands with dense covers of emergent macrophytes.

$$6CO_2 + 12H_2O + \text{light} \rightarrow C_6H_{12}O_6 + 6O_2 + 6H_2O \qquad (12.2)$$

Nonshaded aquatic micro-environments within the wetland therefore display a large diurnal swing in dissolved oxygen due to the photosynthesis-respiration cycle. Nutrients stimulate the algal community and increase the DO mean and amplitude. When large amounts of nutrients are added to the wetland, and water depths are shallow enough for emergent rooted plants, other components of the carbon cycle are increased, such as photosynthesis by macrophytes. It is then possible for other wetland processes to become dominant in the control of dissolved oxygen. The effect is typically a depression of average DO, and a decrease in the amplitude of the diurnal cycle (Figure 12.3). This suppression of the diurnal DO cycle is a characteristic of all treatment wetlands receiving moderate to high loads of carbonaceous and nitrogenous oxygen demand.

Plant Oxygen Transfer

Early in the history of treatment wetlands, it was supposed that plants acted as oxygen pumps, serving to inject large quantities of oxygen into the root zone. There is now considerable direct measurement evidence that strongly suggests that emergent plants do not contribute "extra" oxygen transfer to any appreciable degree, although they do send oxygen to the root zone to protect themselves and conduct respiration. However, communities of both floating plants (FAP) and submerged vegetation (SAV) potentially contribute oxygen directly into surface water. Great care must be exercised in the interpretation of the literature concerning oxygen transfer by plants in wetlands. While it is certain that plant oxygen transfer does occur at modest rates, the amount that is transferred in excess of plant respiration requirements is much less certain. Further, methods of measurement have been variable, and some are purely presumptive.

Emergent Plant Oxygen Transfer

One group of estimates relies upon measurements for individual plants or roots, commonly in hydroponic environments, and extrapolation via root dimensions and numbers. For

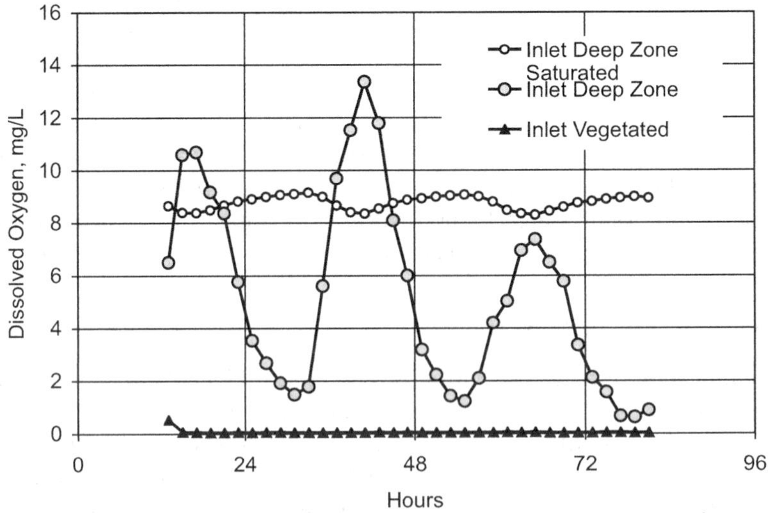

FIGURE 12.3 Diurnal cycles in dissolved oxygen in Cell 7 of the Sacramento, California treatment marsh project, May 28–31, 1996. The inlet deep zone exceeds saturation in late afternoon. Just 46 meters downstream, in a dense community of cattails and bulrush, there is essentially no dissolved oxygen. Data source: Nolte and Associates (1998a).

example, Armstrong et al. (1990) calculated 5–12 $gO/m^2 \cdot d$, and Gries et al. (1990) calculated 1–2 $gO/m^2 \cdot d$. Nonetheless, such hydroponic experiments serve to elucidate the effects of variables. For example, Wu et al. (2000) used hydroponic experiments to estimate 0.04 $gO/m^2 \cdot d$ supplied by *Typha latifolia*, versus 0.60 $gO/m^2 \cdot d$ supplied by *Spartina pectinata*.

A second group of estimates relies upon the disappearance of CBOD and ammonia to infer an oxygen supply. Differences between side-by-side systems are then used to infer the amount of the inferred supply that came from plants. This procedure also has considerable uncertainty, because it is founded on the presumption of oxygen consumption being due to oxidative processes for ammonia and CBOD, and to specific stoichiometric relations. That presumed chemistry is in question, because of alternative loss and gain mechanisms for both ammonia and CBOD. Consequently, such determinations are here termed "implied oxygen supply" rates. Because treatment marshes do not usually have large concentrations of ammonia or CBOD, this method is not accurate in marshes.

Because treatment marshes do not rely upon high oxygen transfer rates, the transfer potential of EAV is essentially moot.

Floating Plants

Open water zones, in the presence of elevated nutrient supplies, may be colonized by floating plants, such as *Pistia* spp., *Eichhornia crassipes, Lemna* spp., *Hydrocotyle umbellata*, and *Azolla* spp. These form a physical cover that is a barrier to oxygen transfer. Additionally, wind can cause the formation of very thick mats by drifting and compression. Root oxygen release rates from a number of free-floating plants in batch hydroponic laboratory studies was calculated in the range of 0.26–0.96 gO/m2·d (Moorhead and Reddy, 1988; Perdomo et al., 1996; Soda et al., 2007).

Submerged Plant Oxygen Transfer

Submerged aquatic vegetation (SAV), including associated epiphyton, photosynthesizes within the water column, and therefore contribute oxygen directly to the water. This activity is driven by sunlight, leading to very strong diurnal cycles in the resultant DO content of the water column. The magnitude of DO enhancement can be large, especially in treatment marshes. Root oxygen release rates from a number of submerged plants in natural environments are reported to be in the range of 0.5 to 5.2 $gO/m^2 \cdot d$ (Sand-Jensen et al., 1982; Kemp and Murray, 1986; Caffrey and Kemp, 1991). More recent work by Laskov et al. (2006) shows a calculated range of 0.15–0.60 $gO/m^2 \cdot d$ based on 200 plants per square meter.

DB Environmental (DBE, 2002) operated SAV mesocosms and 0.2 ha SAV wetlands during 1999–2002. Dissolved oxygen was found to be at or above saturation during the day in the surface water layer but was very much lower at night and in bottom water layers.

Vertical Stratification

Vertical dissolved oxygen profiles have not been extensively studied in treatment wetlands. Pond studies have shown some variable but strong vertical gradients over the top 25 cm of the water column (Abis, 2002). Because concentrations often exceed saturation in the top pond water layer, algal photosynthetic reaeration is present. The high values of DO at the water surface are caused by the preferential interception of photosynthetically active radiation (PAR) in the upper water layers.

Given that physical transfer occurs from the atmosphere, and biochemical generation can occur within the water column, vertical profiles of DO are anticipated in FWS wetlands, and in fact are found in the field. Measurements were made in the lightly loaded treatment wetlands of the Everglades Nutrient Removal Project in Florida (Chimney et al., 2006)

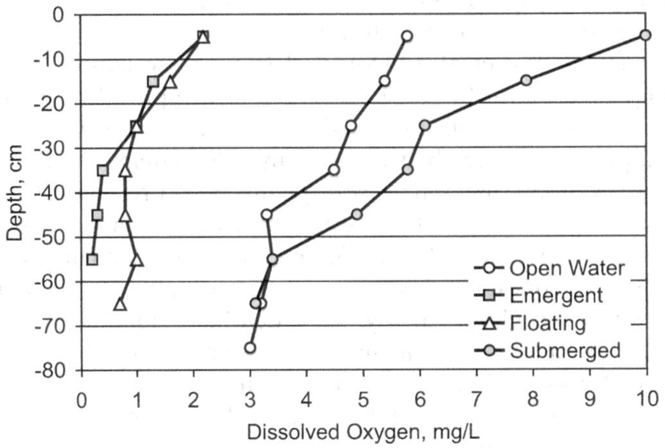

FIGURE 12.4 Vertical profiles of dissolved oxygen in the various vegetation types in the Everglades Nutrient Removal Project FWS wetlands, Florida. Data are from 141 profiles collected over a 2.5-year period. Data source: Chimney et al. (2006).

(Figure 12.4). The highest DO values were found in the open water and submerged vegetation zones, with a strong decreasing gradient with depth. In contrast, DO values in areas of floating plants and emergent vegetation were low, only 1–2 mg/L on average. As in algal ponds, the upper water zones are preferentially active.

Measurement

Evaluating the dissolved oxygen status of a marsh is not a simple matter. Measurement devices are capable of accurate measurements, whether hand-held or automated, but it matters a great deal when and where the measurement point is located. Data from marsh samples is strongly influenced by the time of day of the sample (Figure 12.5; also see Figure 12.3), because of the strong diel cycle present in many situations. Cycles display a minimum at about 8:00 AM (800 hours), and a maximum at about 6:00 PM (1800 hours). Amplitudes decrease along the gradient, while daily mean values increase. In the case of WCA2A, there is a diel swing of nearly a factor of two. Thus, if grab samples of DO are taken in the morning, they will give low readings; and if taken in the afternoon, they will produce high values.

Diel cycles of DO can be accurately described by sinusoidal trends. For instance, the cycles in Figure 12.5 are fitted with $R^2 = 0.93$. In recognition of this, marsh water quality standards can be set that depend upon the hour of the day at which measurements are taken. Such a system is in effect in the state of Florida, where the expected marsh DO is set according to a (double) sine function, which is a bit more accurate for those marshes (Weaver, 2000).

The "where" is important in two respects, both illustrated in Figure 12.4. Most treatment marshes contain a mosaic of community types, such as the four listed on Figure 12.4, which are all in the same system. If samples are taken in open water or SAV, they will register DO many times higher than in adjacent EAV or FAP areas. Additionally, samples taken at the water surface will read about double those taken at 40 cm depth.

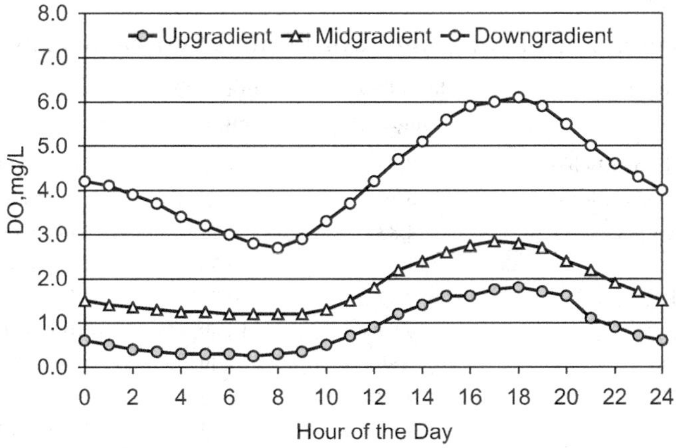

FIGURE 12.5 Diel profiles of dissolved oxygen along the gradient in the WCA2A treatment marsh in Florida.

Samples taken at inlet, outlet, and intercell transfer structures account for the spatial variability by mixing waters passing through all community types at all vertical positions and register a flow-weighted mean DO. The temporal diel cycles are still present, however.

Input/Output Behavior

Treatment marshes do not exert a strong effect upon the DO of the water passing through (Table 12.4). Source waters are not generally anoxic but do vary according to their origin. Advanced treatment plant effluents (potential sources for marshes) tend to be in the mid range of saturation (e.g., Tres Rios, Orlando Easterly). Agricultural runoff, river waters, and groundwaters are more highly oxygenated. Urban runoff is also highly oxygenated (see Table 12.6).

Some systems show minor DO depletion upon passage, while others show minor DO increases. Treatment marshes do not possess anoxic surface waters.

Longitudinal Gradients

When wastewaters with BOD and ammonia nitrogen are discharged to rivers and streams, an oxygen sag analysis is often applied (Metcalf and Eddy Inc, 1991). This Streeter and Phelps (1925) analysis is predicated on the assumption that oxygen is increased in the flow direction by mass transfer from the air above, and by photosynthesis occurring within the water column; and decreased by consumption of BOD and ammonium nitrogen oxidation, and decreased by consumption of Sediment Oxygen Demand (SOD) and respiration. In the wetland environment, both sediments and litter consume oxygen during decomposition. Decomposition processes also release carbon and nitrogen compounds to the overlying water, which can exert an oxygen demand. It is therefore apropos to designate the sum as Decomposition Oxygen Demand (DOD). The resulting downstream DO profiles therefore show an initial decrease as BDO and ammonia are oxidized, followed by an increase as atmospheric reaeration proceeds.

Treatment wetlands with appreciable incoming BOD and ammonia follow similar patterns. Here, the concern is for treatment marshes, which do not have appreciable incoming BOD and ammonia. It is therefore not surprising that oxygen sag profiles are not observed.

TABLE 12.4

Examples of DO of entering and leaving waters for treatment marshes.

Wetland System	Inlet DO mg/L	Outlet DO mg/L	
Commerce Twp., Michigan	8.32	9.86	Unpublished data
Orlando Easterly, Florida	6.10	2.68	Unpublished data
Augusta, Georgia	4.83	7.21	Eidson et al. (2005)
Yakima N	11.10	6.30	Beutel et al. (2013)
Olentangy 1	9.87	10.87	Mitsch et al. (2005c)
Olentangy 2	9.87	10.49	Mitsch et al. (2005c)
Hillsdale WT	9.46	7.49	Unpublished data
Hillsdale ET	7.42	8.56	Unpublished data
Imperial	9.81	7.14	Unpublished data
Brawley	11.13	7.98	Unpublished data
Tres Rios H1	3.39	3.81	Unpublished data
Tres Rios H2	3.39	3.28	Unpublished data
Shannon	9.03	6.86	Unpublished data
Median	**9.03**	**7.21**	

Example flow direction profiles of dissolved oxygen are shown in Figure 12.6 for four marshes situated in warm climates (OEW, Florida; Tres Rios, Arizona; Shannon, Texas; Imperial, California). There are neither increases in DO (due to reaeration), nor large decreases (due to WOD) along the flow path. It is concluded that the Streeter-Phelps analysis is not suitable for treatment marshes, due to lack of the ability to accurately quantify wetland oxygen demand (WOD), or atmospheric reaeration.

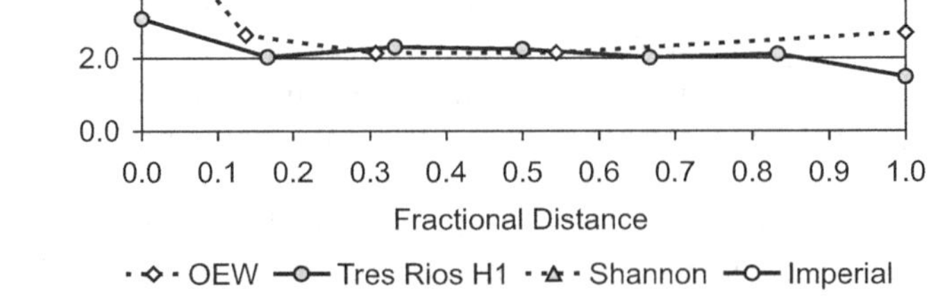

FIGURE 12.6 Dissolved oxygen profiles along the flow path for example marshes. These are averages of monthly data collected over four or more years. The sampling points were located in deep zones or at inter-cell structures.

Trends and Variability

The annual temperature cycle in FWS systems creates a similar cycle in the saturation concentrations of dissolved oxygen, with greater solubility in the colder months. Consequently, the driving force for physical reaeration is maximum in cold months. The photosynthetic production of oxygen in the water column, by algae and/or submerged macrophytes, is driven by a seasonal cycle in solar radiation (PAR). That internal production cycle would be expected to maximize in midsummer, under conditions of maximum solar radiation. If the solubility effect is dominant, marsh water dissolved oxygen will follow a seasonal cycle with larger values in cold months but muted by the summer productivity effect. This is indeed the case the example systems shown in Table 12.5.

The annual trend in effluent DO can be represented by a sine function:

$$DO = DO_{avg}[1 + A \cdot \cos(\omega(t - t_{\max}))] + E \tag{12.3}$$

where A = amplitude factor, –

DO = dissolved oxygen concentration, mg/L

DO_{avg} = annual average dissolved oxygen concentration, mg/L

E = stochastic error, mg/L

t = time, Julian day

t_{max} = time of maximum DO, Julian day

ω = annual frequency, 1/d

For the systems in Table 12.5, maximum DO occurs at the start of February (median max day = 32). The median amplitude of the annual cycle is 0.65, meaning that the summer minimum DO is only one-third of the annual average. The stochastic portion of the description is dominant, representing about two-thirds of the variability.

The percentile points of the DO scatter around the annual cosine trends are given in Table 12.5, as trend multipliers that have been defined previously. It is seen that with some frequency, the excursions from the trend DO values are lower by a considerable margin. For instance, 5% of the time, the median DO is only 25% of the trend value. This means that none of the example FWS systems in Table 5.3 satisfy the United States DO requirement for discharge to receiving waters at the 95% level of confidence (greater than 5 mg/L 19 times out of 20). This means that extra design features (such as cascade aeration) must be implemented to meet the DO requirement for surface discharges.

12.4 CARBON

Carbon compounds interact strongly with marsh ecosystems. The carbon cycle in wetlands is vigorous and typically provides carbon exports from the wetland to receiving ecosystems. Many internal wetland processes are fueled by carbon imports and by the carbon formed from decomposition processes. Treatment marshes as defined herein do not receive large external supplies of carbon from added source waters.

Degradable carbon compounds are rapidly utilized in wetland carbon processes. At the same time, a variety of wetland decomposition processes produce available carbon. The balance between uptake and production provides the carbon exports. The growth of wetland plants requires carbon dioxide (CO_2) for photosynthesis. A variety of organisms release CO_2 as a product of respiration. Many pathways lead to the microbial production of CO_2, as well as methane (CH_4). Both gases dissolve in water to a limited extent, so there are active transfers of carbon to and from the atmosphere.

In terms of treatment, it is therefore not surprising to find a confusing array of carbon reductions or productions in treatment marshes, accompanied by nonzero background

TABLE 12.5

Annual cycle parameters for dissolved oxygen and the associated trend multipliers for specified exceedances percentiles.

Wetland System	Mean	Amplitude	Max Day	R^2	Percentiles			
					0.05	**0.1**	**0.2**	**0.5**
OEW, Florida	2.64	0.41	21	0.21	0.44	0.53	0.74	1.04
Hillsdale, Michigan EA	6.92	0.65	32	0.37	0.08	0.17	0.35	1.00
Hillsdale, Michigan ET	6.94	0.65	32	0.37	0.08	0.17	0.35	1.00
Hillsdale, Michigan WT	8.65	0.78	43	0.51	1.08	1.17	1.35	2.00
Hillsdale, Michigan WA	8.70	0.61	44	0.39	0.05	0.11	0.59	0.97
Tres Rios, Arizona H1	2.54	0.91	10	0.35	0.21	0.29	0.40	0.79
Tres Rios, Arizona H2	2.29	0.72	4	0.36	0.28	0.35	0.46	0.81
Tres Rios, Arizona C1	3.29	0.84	364	0.28	0.17	0.20	0.31	0.65
Tres Rios, Arizona C2	2.77	0.79	11	0.28	0.16	0.23	0.31	0.65
Listowel, Ontario 3	3.51	0.63	360	0.29	0.06	0.13	0.26	0.89
Musselwhite, Ontario	5.33	0.65	42	0.35	0.50	0.59	0.71	0.93
Titusville, Florida	2.55	0.38	33	0.44	0.55	0.68	0.78	1.02
ENRP, Florida	3.70	0.44	41	0.42	0.34	0.52	0.72	0.99
Commerce, Michigan	10.23	0.20	61	0.60	0.83	0.87	0.92	0.98
Median		**0.65**	**32**	**0.36**	**0.25**	**0.32**	**0.52**	**0.98**

levels of various carbon compounds and the related BOD. Treatment marshes are not normally designed for BOD removal, and as a consequence, relatively little attention has been paid to carbon processing performance.

Wetland Carbon Speciation and Processing

A wide spectrum of carbon compounds exists in either dissolved or particulate forms in aquatic systems. The usual dividing line is a 0.45-μm filter. The following distinctions are made as a result of analytical methods:

- TC = Total Carbon (includes all dissolved and suspended forms)
- PC = Particulate Carbon (includes organic and inorganic forms)
- DC = Dissolved Carbon (includes organic and inorganic forms)
- IC = Inorganic Carbon (includes all dissolved and suspended forms)
- DIC = Dissolved Inorganic Carbon (usually comprised of CO2, carbonate and bicarbonate)
- TOC = Total Organic Carbon (includes all dissolved and suspended forms)
- DOC = Dissolved Organic Carbon
- POC = Particulate Organic Carbon
- NDOC = Non-Dissolved Organic Carbon
- VOC = Volatile Organic Carbon (Compounds)

In soils or biomass, samples are subjected to combustion and dissolution, followed by analysis for total carbon.

The most commonly measured parameters are TOC and DOC. Total Organic Carbon (TOC) is measured by chemical oxidation followed by analysis for carbon dioxide. DOC is measured the same way, but on filtered samples.

BOD and COD

In addition to measures of carbon, two other analytical techniques are used to measure the amount of organic material in the water. Biochemical Oxygen Demand (BOD) is a measure of the oxygen consumption of microorganisms in the oxidation of organic matter. It is measured as the oxygen consumption in an air-tight incubation of the sample. This test normally runs for five days, and the result is then more properly designated as BOD_5. Some oxygen may be used in nitrification if the necessary organisms are present in the sample. If this potential nitrogenous oxygen demand is inhibited chemically during the test, the result is Carbonaceous Biochemical Oxygen Demand, $CBOD_5$.

Chemical Oxygen Demand (COD) is the amount of a chemical oxidant, usually potassium dichromate, required to oxidize the organic matter. This measure is larger than BOD, because the strong oxidant attacks a larger group of compounds. However, nitrogenous compounds such as ammonia are not oxidized by the COD test. Oxygen or oxidant consumption may be measured before or after filtration, leading to measures of total and soluble BOD and COD. In the wetland environment, the presence of humic materials leads to COD values that are larger than BOD values. In a northern peatland, the ratio was approximately 0.05 (BOD_5 = 5 mg/L: COD = 100 mg/L) (unpublished data from the Houghton Lake, Michigan, peatland). At Tres Rios, Arizona, wetlands treating nitrified secondary effluent, four wetlands gave ratios of 0.055 ± 0.004, averaged over seven years.

In a northern peatland, the ratio $CBOD_5$: TOC was approximately 0.2 ($CBOD_5$ = 5 mg/L: TOC = 25 mg/L) (unpublished data from the Houghton Lake peatland), and was 0.28 at Estevan, Saskatchewan. At the Tres Rios, wetlands treating nitrified secondary effluent, four wetlands gave ratios of $CBOD_5$:TOC = 0.25 ± 0.08, averaged over seven years. The interpretation of these ratios is that natural marshes cycle at low levels of biologically usable carbon compounds. Many municipal wastewaters are much richer in usable carbon compounds, but these fall outside the boundaries of treatment marshes.

TABLE 12.6

Measures of carbon for marshes treating various source waters (mg/L).

		$CBOD_5$		TOC		DOC		COD	
Wetland	**Source**	**In**	**Out**	**In**	**Out**	**In**	**Out**	**In**	**Out**
Database	Urban	8.8		8.6				53.0	
Imperial	Agricultural	6.2	5.2	8.0	7.6				
ENRP	Agricultural			35.2	34.8	35.5	34.3		
Brawley	River	12.0	6.6	8.5	8.8				
OEW	Municipal	1.8	1.7	9.0	10.9				
Tres Rios H1	Municipal	2.1	2.4	8.8	15.7			33.5	35.6
Tres Rios H2	Municipal	2.1	1.8	8.9	8.7			36.5	34.0
Tres Rios C1	Municipal	2.4	2.9	8.8	8.6			41.3	40.2
Tres Rios C2	Municipal	2.5	2.4	8.8	16.9			41.2	44.2
Estevan	Lagoon	12.0	5.9	19.4	20.6	18.0	19.6		

Source: Urban, Clary et al. (2017b).

Examples of these various measures are shown in Table 12.6. TOC and DOC are not very different, but COD tends to be considerable higher than $CBOD_5$. Advanced treated municipal wastewaters are very low in $CBOD_5$, while other treatment marsh source water types are somewhat higher. None of these source waters contains levels of carbonaceous materials that would require removal.

Wetland Chemistry of Dissolved Inorganic Carbon

Dissolved inorganic carbon consists primarily of carbon dioxide, carbonate and bicarbonate.

In pure water solution, the principal carbonate species are related to atmospheric CO_2 by the temperature and pH dependent dissolution and dissociation reactions:

Henry's Law:

$$H_2CO_3^* \Leftrightarrow H_2O + CO_2\,(gas) \tag{12.4}$$

Hydration

$$H_2CO_3 \Leftrightarrow H_2O + CO_2 \tag{12.5}$$

First Dissociation

$$H_2CO_3 \Leftrightarrow HCO_3^- + H^+ \tag{12.6}$$

Second Dissociation

$$HCO_3^- \Leftrightarrow CO_3^= + H^+ \tag{12.7}$$

The various concentrations are all pH and temperature dependent. These forms are distributed in water at 25°C according to pH (Pankow, 1991). In the pH range of 6–10 seen in treatment marshes, bicarbonate is the dominant species. However, it must be noted that wetland waters are more complex than the pure water system and therefore will not follow such idealized chemistry precisely. Production and consumption of carbon dioxide in the wetland may significantly alter the chemical balance in the water.

An important feature of the carbonate system is its influence on pH under mediation by algae. Algal consumption of CO_2 drives pH upward, and may give rise to $9 < pH < 10$ in unshaded wetland environments or ponds.

A variety of cations can precipitate carbonates under certain conditions. The most important is calcium carbonate, $CaCO_3$. A major process in periphyton-dominated wetlands is chemical precipitation of $CaCO_3$ under conditions of high pH created by the algae (Gleason, 1972). Similarly, in beds of submerged aquatic vegetation, carbon dioxide and bicarbonate are consumed during photosynthesis, thereby raising the water column pH and promoting $CaCO_3$ precipitation (Dierberg et al., 2002). Calcium carbonate saturation indices may be calculated in a number of ways (APHA, 1992). However, overall carbon mineral chemistry is very complex; consequently, accurate calculations of solubilities are generally not possible in wetland environments.

Phytomass Chemistry of Organic Carbon

The wetland cycle of growth, death, and partial decomposition uses atmospheric carbon, and produces gases, dissolved organics, and solids (Figure 12.7). Decomposition involves the sugars, starches and low molecular weight celluloses in the dead plant material. Gaseous products include methane and regenerated carbon dioxide. A spectrum of soluble large

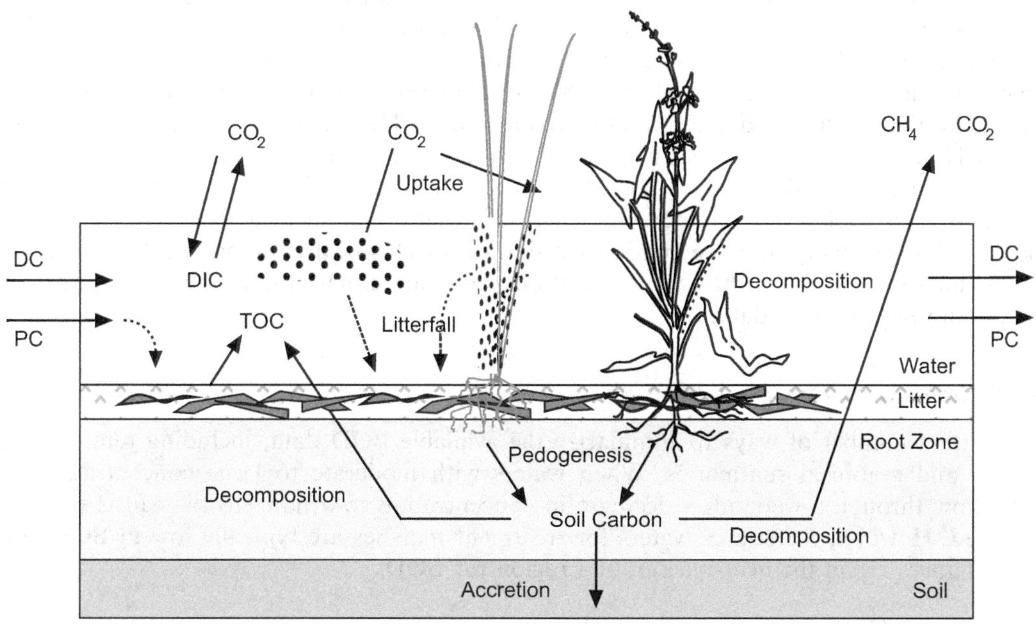

FIGURE 12.7 Carbon storages and transfers in the wetland environment. DC = Dissolved Carbon, PC = Particulate Carbon, DIC = Dissolved Inorganic Carbon, TOC = Total Organic Carbon, CH_4 = Methane, CO_2 = Carbon Dioxide. Biomass carbon consists of living and dead biomass and organic decomposition products.

organic molecules, collectively termed humic substances, are released into the water. The solid residual of plant decomposition is peat or organic sediment, which originated as celluloses and lignins in the plants. These wetland soil organics are broadly classified as fulvic material, humic material, and humin, based upon whether they are acid soluble, base soluble, or insoluble (NRCC, 1979).

The sediments, soils and biomass in a wetland contain major proportions of carbon, and the internal wetland carbon cycle is large. A general idea of the magnitudes of the various carbon transfers in a northern treatment marsh may be gained from considering the annual growth and decomposition patterns (refer to Chapter 5). A eutrophic treatment marsh grows about 3,000 gdw/m^2 of aboveground biomass each year, with a carbon content of about 43%. This translates to an annual average requirement for 35 kg/ha·d of carbon. In northern climates, this requirement is utilized during a growing season of approximately four months. In the case of emergent macrophytes, some of this carbon may be withdrawn from the atmosphere. However, submerged vegetation draws carbon from the aquatic carbonate system.

Decomposition of the resultant litter returns a significant portion of that carbon to the atmosphere and to wetland waters, but in treatment wetlands, a small fraction, on the order of 15–20%, is stored in accreted soil and sediments. That storage (burial) fraction therefore amounts to about 5 kg/ha·d as an annual average for the eutrophic marsh example. The balance, about 30 kg/ha·d, is processed via one or more mechanisms involving a variety of electron acceptors (oxidants), or via anaerobic digestion that generates methane.

Some of the submerged sediments and litter are processed into soluble organic compounds that contribute to CBOD in the water, thus creating a nonzero background CBOD in a wetland environment.

As discussed in Chapter 4, marsh necromass contributes to a buildup of material in the action zone, which is usually about 20–30 cm thick. There, it undergoes a variety of aerobic, facultative anaerobic, and obligate anaerobic processes at different depths in the action zone. Typical soil-based chemical reactions that involve organic carbon are respiration, nitrate, iron and sulfate reduction, and methanogenesis. These have been described in Equations 4.12–4.18. They occur in sequence as depth increases (see Figure 4.10).

It is apparent that the wetland provides a spectrum of potential pathways for the utilization of organic carbon compounds. Sufficient information is not available to quantify both the complex chemistry and the spatial distribution of chemical compounds. Therefore, the interactions must be described via correlations and rate equations, which are supportable by wetland performance data.

BOD Removal in Treatment Marshes

There are a number of ways to summarize the available BOD data, including removal rate models and graphical summaries. When waters with moderate to large concentrations of BOD flow through a wetland, a decrease in concentration to a nonzero plateau is typically observed. However, the source waters for treatment marshes are typically low in BOD, and performances are in the near-background region for BOD.

Annual Input-Output Concentration Relations

The concentration of carbonaceous compounds is reduced in wetlands for incoming concentrations above background. If, however, incoming BOD is below background, then concentrations may increase upon passage through the system. As inlet concentrations increase, outlet concentrations increase, in a log-linear progression (Figure 12.8). There is considerable intersystem variability, but treatment marsh data is approximated by a log-linear regression:

$$C_{out} = A(C_{in})^m \tag{12.8}$$

where: A = fitting parameter
m = fitting exponent
C_{in} = inlet $CBOD_5$ concentration, mg/L
C_{out} = outlet $CBOD_5$ concentration, mg/L

The values for the entire dataset in Figure 12.8 are A = 1.63 and m = 0.524, with R^2 = 0.60, N = 62 systems for their POR.

First-Order Modeling

The result of the operation of treatment marshes in the near-background concentration region is a loss of calibration efficiency for the PkC* model. Depending on hydraulic conditions, and the character of the incoming BOD, individual wetlands will typically exhibit different C* values as PkC* model calibration parameters. Datasets are very often poorly conditioned to produce good estimates of both k and C*.

Transects show an early exponential decline (governed by k), together with the late plateau (governed by C*). If inlet concentrations are well above background, transect data will provide a good estimate of k, but a very poor estimate of C*. Input/output data for these sites may nonetheless be fitted to the model. Conversely, for near-background inlet concentrations, there is insufficient data in the exponential region to get a good estimate of k, but plenty of data to define C*. It is tempting to arbitrarily pick some low concentration to represent C*, but there is not an existing method to make such an estimate with confidence.

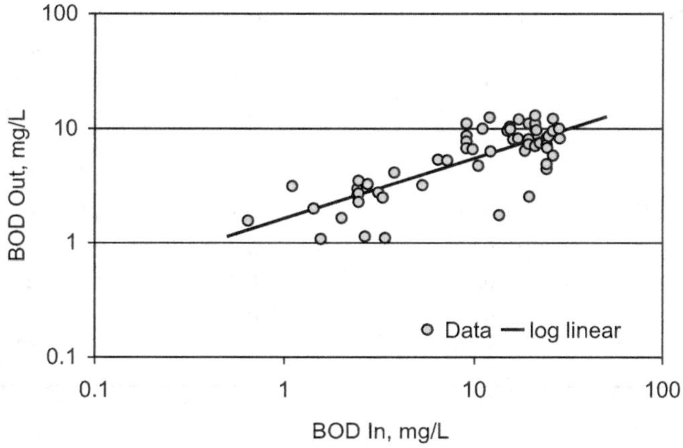

FIGURE 12.8 Input-output concentration for BOD in treatment marshes. Each point represents a POR average for one wetland. There are 173 wetland·years of data, for 62 wetlands. The trend line is y = 1.63x0.52 (R^2 = 0.60 logarithmic).

It is instructive to examine multiple I/O datasets, which provide a distribution of k-values. In general, treatment marshes operate in a concentration region where the dominant processes are aerobic removal coupled with a return flux from decomposing necromass. In this region, internal BOD loading is more important than external BOD loading. In contrast, for wetlands operating at high external BOD loadings, aerobic processes become less of a probable factor, and are replaced by anoxic processes.

Annual average k-values, determined from input/output data, are given in Table 12.7. The median k = 14.2 m/yr is quite low compared to treatment wetlands in general (Kadlec and Wallace, 2009). The spread of the distribution is quite large, implying that the characteristics of individual wetlands, or individual years in the period of record, can have strong influences on performance. Only very small amounts of BOD are removed.

TABLE 12.7

Annual reduction of BOD in treatment marshes.

Stipulations

1. Period of record averages are used in calculations.
2. For k-value calculations, the following P-k-C* parameters are selected.
   a. *C** = 1.0 mg/L
   b. *P* = 3.

3. Ranges of variables:

	HLR	BOD In	BOD Out
	cm/d	mg/L	mg/L
mean	5.5	17.1	7.6
median	4.6	19.4	7.9
max	17.7	28.3	13.0
min	0.04	3.1	1.1

Results (N = 50 wetlands)

Percentile	BOD Load Removed $g/m^2 \cdot yr$	Percentile	Rate Coefficient (k) m/yr
0.1	−0.03	0.1	0.7
0.2	0.01	0.2	3.8
0.3	0.02	0.3	7.7
0.4	0.07	0.4	10.8
0.5	0.21	0.5	14.2
0.6	0.32	0.6	16.9
0.7	0.48	0.7	25.2
0.8	0.81	0.8	33.7
0.9	1.17	0.9	52.6

Temperature and Season

For treatment wetlands in general, the temperature coefficients for BOD removal rates have been reported to be near unity (Kadlec and Wallace, 2009). Treatment marshes operate in a different processing environment, in which rate coefficients display seasonal dependence rather than temperature dependence. Further, seasonal behavior is contingent upon the water source and its associated BOD level, as well as the climate zone in which the wetland is located (Table 12.8).

Seasonal behavior patterns in this low BOD range are currently outside the capabilities of quantitative modeling. The complex set of factors described earlier in this section interact in ways we cannot yet forecast or model.

12.5 SULFUR

Sources of sulfur include geochemical weathering of minerals, wind-blown sea salt and emissions from fossil fuel combustion (Wetzel, 1983). Large quantities of sulfur enter the atmosphere from natural industrial sources and return to earth as acid precipitation containing sulfates (ca. 10 mgS/L). Treatment marshes receive these atmospheric inputs as well as sulfur compounds that may be included in the chemicals in the water to be treated. Treated municipal wastewater contains sulfur compounds, originating from the potable water supply, and augmented by waste products. Water that is high in sulfate can be found throughout the western United States. Water can be high in sulfates in both surface and ground water sources, and ultimately produces significant sulfate in both agricultural irrigation runoff and treated municipal effluents. For example, the Tres Rios wetlands received nitrified effluent at about 180 mgS/L.

Sulfur is an essential plant nutrient required by all crops. Plants take up and use S in the sulfate (SO_4-S) form. Fertilizer application rates range up to 25–30 kgS/ha·yr. The result is elevated levels of sulfate in runoff. Sulfate, like nitrate, is very mobile in the soil and is prone to leaching in wet soil conditions, particularly in sandy soils. As an illustration, the agricultural runoff reaching the ENRP in Florida was 50–60 mgS/L. Urban runoff contains sulfates, in the range of 10–50 mgS/L (Zhang et al., 2013c).

Forms of Sulfur in Marshes

Hydrogen Sulfide

Hydrogen sulfide exists in water solution as un-ionized (H_2S) or singly or doubly ionized (bisulfide, HS^- or sulfide, S^{-2}), depending on water temperature and pH. The two dissociation reactions are:

TABLE 12.8

Seasonal BOD performance for example marsh systems. A is a cold climate environment, and display low rate coefficients in mid-summer. B and C are warm climate systems, and display low rate coefficients in winter.

A.	Brighton	Ontario	Lagoon Effluent		N=3, C*=2.0
Month	HLR	Temp	BOD In	BOD Out	BOD k
	cm/day	°C	mg/l	mg/l	m/yr
1	5.16	0.500	7.260	3.885	23.0
2	5.03	0.500	8.697	5.303	14.6
3	5.78	1.401	8.756	4.822	21.4
4	6.12	7.353	7.974	4.361	24.3
5	5.45	13.790	6.366	2.942	39.8
6	4.85	18.989	3.350	2.633	15.2
7	4.53	21.559	2.861	2.718	3.1
8	4.37	20.811	4.528	2.600	29.4
9	4.46	16.946	2.868	3.234	-5.4
10	4.82	10.998	4.122	2.329	45.5
11	4.90	4.560	3.372	2.322	33.3
12	5.63	0.500	4.224	2.257	65.0
Mean		9.8			25.8

B.	Brawley	California	River		N=4, C*=2.0
Month	HLR	Temp	BOD In	BOD Out	BOD k
	cm/day	°C	mg/l	mg/l	m/yr
1	7.2	12.6	5.4	11.5	-23.7
2	8.1	14.1	7.3	13.2	-20.1
3	9.9	19.2	9.4	9.3	0.3
4	11.7	21.5	18.4	4.8	95.3
5	12.9	24.3	11.9	5.2	60.3
6	12.1	27.3	10.5	5.2	48.8
7	11.4	29.7	14.8	5.4	65.3
8	5.6	29.9	14.7	5.1	35.0
9	5.5	27.2	17.9	4.1	52.6
10	7.6	22.7	16.2	4.3	63.4
11	7.2	16.8	10.6	5.0	31.7
12	7.2	12.7	7.0	5.7	8.1
Mean		21.5			34.7

C.	Imperial	California	Agricultural Runoff		N=4, C*=2.0
Month	HLR	Temp	BOD In	BOD Out	BOD k
	cm/day	°C	mg/l	mg/l	m/yr
1	18.59	12.00	4.42	5.75	–28.1
2	18.31	13.66	6.17	6.68	–7.7
3	18.33	18.50	6.95	8.96	–21.9
4	19.00	20.44	11.68	9.37	19.6
5	19.96	23.57	15.96	6.93	86.7
6	19.41	26.50	8.67	4.77	69.5

(Continued)

TABLE 12.8 (Cont.)

C.	Imperial	California	Agricultural Runoff		N=4, C*=2.0
Month	HLR	Temp	BOD In	BOD Out	BOD k
	cm/day	°C	mg/l	mg/l	m/yr
7	19.08	29.17	6.50	3.61	81.4
8	19.14	29.41	8.15	4.54	69.2
9	19.81	26.75	5.94	3.94	56.1
10	20.02	22.29	3.22	2.64	50.5
11	20.04	16.43	4.06	2.67	95.7
12	20.12	12.27	4.62	2.93	86.4
Mean		20.9			46.5

$$H_2S \leftrightarrow H^+ + HS^- \tag{12.9}$$

$$HS^- \leftrightarrow H^+ + S^{-2} \tag{12.10}$$

At equilibrium, the unionized form (H_2S) is predominant at low pH (<6.0), and bisulfide (HS^-) is dominant at high pH (>8.5) in aqueous systems. However, equilibrium is not necessarily attained in wetland systems, because of continual influxes of sulfate, and a large number of microbially mediated processes that may occur. There may be large fractions of unionized H_2S, which is volatile and may be lost to the atmosphere.

Volatilization of hydrogen sulfide requires mass transport to the air-water interface, followed by transfer into the air, and follows rules analogous to those discussed for ammonia volatilization (see Chapter 9). The Henry's law constant for H_2S at 25°C is 3.49 mg/L·atm, which indicates large volatility. The volatility of H_2S means that losses in the wetland environment are controlled by mass transport processes in the water phase.

Organic Sulfur

Organic sulfur compounds account for a good share of the sulfur found in wetland sediments. For instance, 84–88% of the total sulfur in a New Jersey peat was organic sulfur (Spratt et al., 1987), and over 90% in a West Virginia peat (Wieder and Lang, 1988). In treatment wetlands, the storage of sulfur is also in major part associated with humic materials.

Additionally, there are several low molecular weight organic sulfur compounds that may be found in wastewater. Methanethiol (CH_3SH) and dimethyl sulfide (DMS, or $(CH_3)_2S$) are perhaps the most common, and are quite volatile (Faulkner and Richardson, 1989; Lomans et al., 2002). Both are extremely odiferous. There are several mechanisms that can produce and destroy these volatile organic sulfur compounds (see Lomans et al., 2002). Kiene and Hines (1995) found both were formed in natural fen peat at the same rate of 40 nmol/L·d (256 μg/m^2·d in the top 20 cm of soil).

Chemical Processes

The processing of sulfur in wetland ecosystems is represented by interconversions of several sulfur compounds in the different micro-regions of the ecosystem (Figure 12.9). Oxidized forms, such as sulfite, sulfate and thiosulfate, are found in the oxygenated portion of the water column. Reduced forms, including sulfide, bisulfide, and elemental sulfur, are found in the soils and sediments under conditions of low redox potential. Ionic and molecular

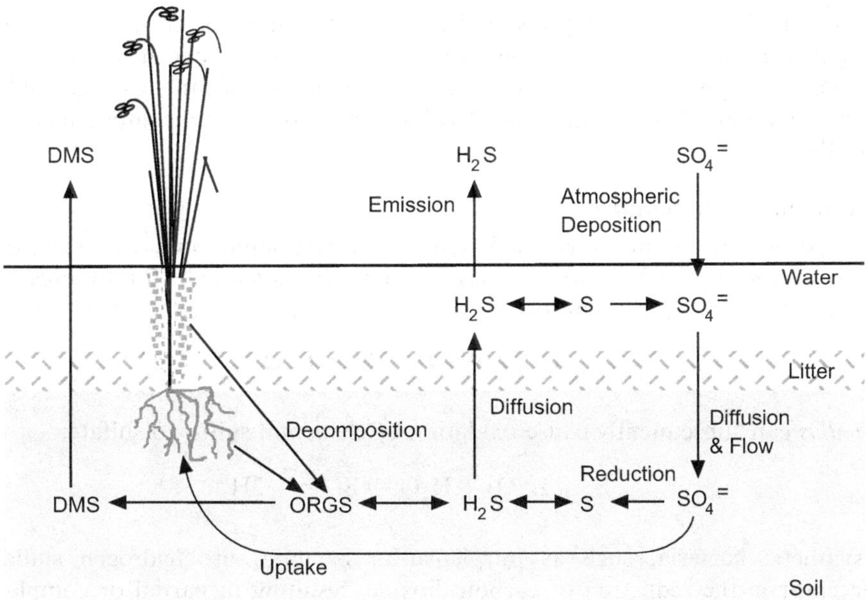

FIGURE 12.9 Simplified schematic of forms and pathways of sulfur in treatment marshes.

forms are prevalent. Hydrogen sulfide, and methylated sulfur compounds, are volatile, and may be lost from the wetland to the atmosphere. Sulfate is an essential nutrient because its reduced, sulfhydryl (-SH) form is used in the formation of amino acids. Because there is usually enough sulfate in surface waters to meet the sulfur requirement, sulfate rarely limits overall productivity in wetland systems.

Although seldom a water quality target in its own right, sulfur is an important part of the chemical processing in wetlands. From a treatment perspective, sulfur has a critical role in the formation and storage of metal sulfides. In this section, the principal reactions of sulfur in the environment are explored, together with the treatment and storage potential.

Dissimilatory Sulfate Reduction

Aerobic organisms excrete sulfur as sulfate. However, upon death and sedimentation, heterotrophic bacteria release the sulfur in the reduced state, which can result in the accumulation of high levels of hydrogen sulfide in wetland sediments. A second process that transforms sulfate and other oxidized sulfur forms (sulfite, thiosulfate, and elemental sulfur) to hydrogen sulfide in anaerobic sediments, dissimilatory sulfate reduction, is mediated by anaerobic, heterotrophic bacteria such as *Desulfovibrio* and *Desulfotomaculum*, which use sulfate as a hydrogen acceptor (Castro et al., 2002; Lloyd et al., 2004). The presence of decaying organic matter in the wetland sediments and soils depletes oxygen and creates acid porewaters. Organic matter fuels sulfate reduction. For example:

$$SO_4^{-2} + 2CH_2O + 2H^+ \rightarrow H_2S + 2H_2O + 2CO_2 \quad (12.11)$$

$$SO_4^{-2} + 2CH_2O \rightarrow HS^- + H^+ + 2HCO_3^- \quad (12.12)$$

Where CH_2O represents the organic substrate required by the microbes.

Equation 12.11 is favored at low pH, while Equation 12.12 dominates at higher pH. Since ferrous sulfide (FeS) is highly insoluble, hydrogen sulfide does not tend to accumulate until the reduced iron is removed from solution. When iron concentrations are low, or when sulfate and organic matter concentrations are high, significant hydrogen sulfide concentrations can occur. Several other metal sulfides are also very insoluble, including ZnS, CdS, and others.

Oxidation of Sulfur and Sulfides

When it is exposed to air or oxygenated water, hydrogen sulfide may be oxidized back to sulfate. This may occur via by sulfur bacteria such as *Beggiatoa*, which promotes the oxidation of hydrogen sulfide to elemental sulfur:

$$H_2S + 0.5O_2 \rightarrow S^o + H_2O \tag{12.13}$$

Thiobacillus can subsequently cause oxidation of elemental sulfur to sulfate:

$$S^o + 1.5O_2 + H_2O \rightarrow SO_4^{-2} + 2H^+ \tag{12.14}$$

Photosynthetic bacteria, such as purple sulfur bacteria, use hydrogen sulfide as an oxygen acceptor in the reduction of carbon dioxide, resulting in partial or complete oxidation back to sulfate:

$$H_2S + CO_2 \rightarrow S^o + CH_2O \tag{12.15}$$

Where CH_2O represents organic matter.

Under some circumstances, treatment wetlands have been observed to turn purple, as happened in an FWS treating high-strength potato processing wastewater (P. Burgoon, personal communication).

In any case, these microbially mediated reactions suggest that elemental sulfur may be found in treatment wetlands. Anecdotal reports of elemental sulfur have been made for the Houghton Lake, Michigan, system, the Tres Rios, Arizona, wetlands, and the Brighton, Ontario, system.

Phytotoxicity

Lamers et al. (1998) document that sulfate has negative effects on the growth rate of *Carex nigra, Juncus acutiflorus* and *Gallium palustre*, at concentrations of 64 and 128 mgS/L. Koch and Mendelssohn (1989) report that 32 mgS/L of sulfide produced negative effects in *Panicum hemitomon* and *Spartina alterniflora*. The presence of sulfide is coupled with anaerobic conditions in the root zone, but the effects of sulfide go beyond mere anoxia (Koch et al., 1990). Hydrogen sulfide apparently inhibits the activity of alcohol dehydrogenase, thereby limiting the ability of plants to avail themselves of alternative anoxic energy pathways. This effect was confirmed by measuring a reduced 15N uptake rate in the presence of sulfide. However, the availability of free sulfide is strongly mediated by the presence of iron, because of the formation of iron sulfides.

Phytotoxicity was found to be very serious at the 45 mgS/L level in *Phragmites australis* (Armstrong et al., 1996). These authors found that aeration pathways became blocked, interfering with the diffusive connection to the atmosphere, and thus reducing the plant's ability to oxygenate the rhizosphere. Smolders and Roelofs (1996) found, for *Stratiotes aloides*, an aquatic macrophyte characteristic for mesotrophic freshwater marshes, that levels of 320 mgS/L were toxic to the roots. Lamers et al. (2002) found root parts, growing in 1.7–3.4 mgS/L of sulfate into the peaty sediment, clearly showed sulfide toxicity by becoming black, slimy, and unfit for nutrient uptake from the sediment. Free sulfide could not be detected in the surface water. They concluded that only roots in the surface water would survive. *Nuphar lutea* did not propagate in the sulfate-treated enclosures. However, the sensitivity of a wetland plant species to free sulfide not only depends on the actual sulfide levels in the rhizosphere, but also on detoxification mechanisms like radial oxygen loss. As noted earlier, high sulfide concentrations in freshwater sediments may also lead to higher fluxes of volatile organic sulfur compounds to the atmosphere due to the microbial methylation of hydrogen sulfide.

Performance of Wetlands for Sulfur Removal

Since sulfate inputs in surface flow wetland treatment systems frequently exceed the biological requirements of wetland biota, wetlands generally are not as effective for removal of sulfur as for other contaminants (Wieder, 1989). Although microbial routes provide for gaseous losses of H_2S and DMS, these require the very low redox potentials usually found only in deeper wetland sediments. Metal sulfide precipitation often blocks much of the gaseous loss by immobilizing sulfides in the sediments. Plant storage is minimal, for instance plant uptake was estimated at 1.5 $gS/m^2{\cdot}yr$ in a natural bog (Hemond, 1980). The majority of sulfur removal will generally be to organic, elemental, and metal sulfide forms in the wetland sediments.

As a consequence, the median observed concentration reduction is only 4% for 14 treatment marshes (Table 12.9).

Sulfate removal should not be viewed as a process that is independent of other wetland chemistry and processes. For instance, studies by Wiessner et al. (2005) determined that sulfate reduction was strongly dependent on wastewater BOD, presumably acting as a carbon

TABLE 12.9

Example performances of treatment wetlands for sulfate reduction.

		Years	Inlet mg/L	Outlet mg/L	Reduction %
Agricultural Runoff					
STA5, FL	All	1	10	5	50%
STA6, FL	All	1	27	24	9%
ENRP, FL	All	6	70	54	22%
STA1W, FL	All	1	90	88	2%
STA2, FL	All	1	103	77	24%
Brawley, CA	All	4	675	731	-8%
Imperial, CA	All	4	722	676	6%
Municpal Wastewater					
OEW, FL	WP1-MM7	10	54	47	13%
Lakeland, FL	All	3	138	111	20%
Tres Rios, AZ	C1	8	172	181	-5%
Tres Rios, AZ	C2	8	173	175	-1%
Tres Rios, AZ	H1	8	177	181	-2%
Tres Rios, AZ	H2	8	181	185	-2%
Estevan, SAS	All	10	636	633	0%
Median					4%

source, in a manner analogous to denitrification. BOD of 200 mg/L led to 100% removal of sulfate. Wiessner et al. (2005) conclude that the importance of the sulfur transformation processes inside the rhizosphere of constructed wetlands, even in the case of treatment of domestic wastewater, has been underestimated. Extreme variations of removal processes in large-scale treatment wetlands may reflect this fact.

Sulfur-Induced Eutrophication

Interactions also may exist between sulfur processing and phosphorus removal in treatment wetlands. In some systems, but not all, a fraction of the accreted phosphorus is bound by iron containing substances. Other fractions are bound in calcium rich materials, or in organic components of soils and sediments. Two effects have been reported: (1) a reduction of phosphorus uptake due to sulfide toxicity, and (2) sulfide binds iron and interferes with that component of phosphorus storage that relies upon the iron-phosphorus link (Lamers et al., 1998).

The response of wetlands to new sources of sulfate, and ultimately sulfide in the sediments, therefore differs considerably depending upon the sources and quantities of iron in the wetland (Lamers et al., 2002). For example, a *Typha-Carex* wetland at Tienhoven near Utrecht, the Netherlands, had relatively high iron, and a continuing supply from groundwater. Addition of sulfate caused oxidation of iron in response to sulfate reduction, and considerable quantities of phosphorus were released. However, a second wetland, dominated by *Nuphar* surrounded by *Phragmites*, in Weerribben National Park, the Netherlands, did not have a groundwater supply of iron. Sulfate additions to that wetland caused large quantities of sulfide formation, but no phosphorus was released.

12.6 PATHOGENS

Natural treatment technologies have the potential to reduce populations of enteric pathogens because of natural die-off rates and hostile environmental conditions. Wetlands have been found to reduce pathogen populations with varying but significant degrees of effectiveness. This chapter reviews the pathogens typically found in treatment marsh source waters and describes the effect those marshes have on the pathogen populations passing through them.

Pathogens are disease-causing microorganisms. They are broadly classified in four groups: Bacteria, viruses, and helminth and protozoan parasites.

Bacterial pathogens are living organisms, not chemicals; nor are they plants or animals. They belong to the kingdom of *Eubacteria*. Some are not capable of movement, others have limited motility. Pathogenic bacteria move with water or attached to solids that may move with water. Accordingly, they are considered water quality parameters.

Pathogen reduction is not a design goal for treatment marshes, but bacteria, viruses and parasites may enter with source waters of various types and may be contributed within the marsh by the animal populations. Historically, concern has focused upon pathogens that cause diseases in humans (Table 12.10).

Distinct from considerations of discharges from human wastewater treatment, animal wastes from agriculture are important potential sources of pathogens to treatment marshes and other receiving waters. To further complicate matters, treatment marshes are often home to large populations of warm-blooded wildlife, which contribute pathogens directly to the marsh. Zoonoses are infectious diseases that can be naturally transmitted between animals and humans. The most significant zoonotic pathogens causing food borne diseases are *Escherichia coli, Campylobacter*, and *Salmonella*. Cattle are an important reservoir of cryptosporidiosis.

TABLE 12.10
Some human pathogens typical of domestic wastewater.

	Pathogen	Illness
Bacteria	*Campylobacter jejuni*	Diarrhea
	Escherichia coli	Diarrhea
	Legionella pneumophila	Fever, respiratory tract infections
	Leptospira (150 spp.)	Leptospirosis
	Salmonella typhi	Typhoid Fever
	Salmonella (~ 1700 spp.)	Salmonellosis
	Shigella (4 spp.)	Diarrhea, dysentery
	Vibrio spp.	Cholera, diarrhea
	Yersinia spp.	Yersiniosis
Viruses	Adenovirus (31 types)	Respiratory disease
	Enteroviruses (67 types)	Diarrhea, respiratory disease, polio
	Hepatitis A	Infectious hepatitis
	Norwalk Agent	Gastroenteritis
	Rotavirus	Diarrhea
	Reovirus	Gastroenteritis
	HIV	Aids

Sources: Shiaris (1985); Krishnan and Smith (1987); Cabelli (1977); Leclerc et al. (1977); Metcalf and Eddy (1991); Crites and Tchobanoglous (1998).

Marsh-related pathogens are not restricted to warm-blooded animals. All reptiles carry a range of germs including bacteria, viruses, parasites and worms (CDC, 2018). Healthy reptiles and amphibians can carry *Salmonella* and other germs that make people sick. Salmonella can spread by either direct or indirect contact with amphibians (e.g., frogs), reptiles (e.g., turtles, lizards or snakes) or their droppings. *Salmonella* infections can result from having contact with reptile or amphibian environments.

Pathogens are measured as colony-forming units (cfu), usually for a sample size of 100 milliliters, hence the reporting unit is cfu/100mL. It is a measure of the number of viable bacteria in a sample, where viable is defined as the ability to multiply via binary fission under the controlled conditions. It is not an enumeration of single organisms, because colonies may originate from clumps of cells.

Indicator Organisms

Measurement of pathogenic organisms in natural and waste waters is expensive and technically challenging. Consequently, it has been customary to first look for indicator organisms that are easy to monitor and correlate with populations of pathogenic organisms. The indicator concept rests upon the presence of groups of bacteria that reside in the feces of warm-blooded animals. Indicators include total coliforms (TC), fecal coliforms (FC), *Escherichia coli* (EC), and fecal enterococci (streptococci)(FS).

Examples of the magnitudes of fecal contributions of pathogenic indicator organisms of these various types are shown in Table 12.11.

In the United States, total coliforms were the first approved indicator organisms. Total coliforms include bacterial species that are rod-shaped, stain gram-negative, do not form spores, are facultatively anaerobic, and ferment lactose with gas production in 48 hours at

TABLE 12.11

Examples of the magnitudes of fecal contributions of pathogenic indicator organisms.

Indicator & Animal	Indicator Density cfu/gram x 10^{-6}	Reference
Fecal coliforms		
Human	13	Maier et al. (2000).
Cow	0.23	Maier et al. (2000).
Pig	3.3	Maier et al. (2000).
Chicken	1.3	Maier et al. (2000).
Cat	7.9	Maier et al. (2000).
Turtle	1.8	Habersack et al. (2011)
Fecal streptococci		
Human	3	Maier et al. (2000).
Cow	1.3	Maier et al. (2000).
Pig	84	Maier et al. (2000).
Chicken	3.4	Maier et al. (2000).
Cat	27	Maier et al. (2000).
Swan	2.2	Moriarty et al. (2011)
Turtle	0.53	Habersack et al. (2011)
Escherichia coli		
Human	3.56	Bailey et al. (2010).
Cow	1.8	Horton et al. (2011)
Pig	7.5	Horton et al. (2011)
Chicken	3.5	Horton et al. (2011)
Coots	100	Meerburg et al. (2011)
Gulls	120	Meerburg et al. (2011)
Swan	4	Moriarty et al. (2011)
Turtle	0.61	Habersack et al. (2011)

a temperature of 35°C. Because many members of the total coliform group are not limited to fecal sources, methods have been developed to differentiate organisms of fecal origin. Fecal coliforms (FC) are separated from total coliforms by their ability to ferment lactose with gas production in 24 hours at a temperature of 44.5°C.

An even narrower group, *E. coli*, is being used more frequently as an indicator organism, because it may readily be separated from the rest of the fecal group, and because several strains are capable of causing severe human health problems. The coliform groups in general suffer an indicator disadvantage in that they may regrow in aquatic environments.

Another indicator, enterococci, is largely synonymous subgroup of fecal streptococci, and here the acronym FS is used for both. The use of enterococci as indicators of human fecal pollution is not conclusive, because enterococci are also found in animal feces. The possible contributors of enterococci in environmental waters include human sources, direct and indirect agricultural contributions, and wildlife. The particulate matter in storm water contributes to the transport of enterococci in receiving waters and eventual deposition into sediments. Because fecal streptococci bacteria seem to survive longer in receiving waters than fecal coliforms, they are used as an indicator of fecal contamination.

It is generally recognized that fecal coliforms are not suitable indicators of viral contamination in surface water because viruses respond differently to chlorination and environmental deactivation than bacteria (Kraus, 1977; Gersberg et al., 1987; Gerba, 2000). Bacteriophages (viruses that infect bacteria, such as coliphage MS-2) have been used as viral indicators, in wetland treatment systems (Gersberg et al., 1987; Schuerman et al., 1989). Enumeration of bacteriophages is technically simpler and more rapid than enumeration of the target pathogenic viruses. Also, MS-2 is nearly the same size as enteroviruses and is more resistant to ultraviolet light, heat, and disinfection than most enteric viruses. In a FWS study at San Jacinto, California, Chendorain et al. (1998) added cultured MS-2 virus to the wetland influent wastewater to determine die-off rates.

Regulatory Standards

Pathogenic organisms are required to be absent from drinking waters. And, there are regulatory standards for indicator organisms discharged from wastewater treatment facilities, which vary from state to state within the United States, and from country to country. Typically, secondary limits are about 100–200 cfu/100mL. For example, Ontario requires *E. coli* of less than 200 cfu/100mL. For the tertiary source waters relevant to treatment marshes, disinfection is normally required.

There are not aquatic life or sediment criteria set for pathogens, as there are for other water quality parameters. However, U.S.EPA has set standards for pathogens in the form of recreational water quality criteria (U.S.EPA, 2012). They recommend using one of two fecal indicator bacteria: *Enterococcus* spp. (formerly classified as *Streptococcus*) (95% of human gut content is *Enterococcus faecalis*) and *Escherichia coli* (*E. coli*) as indicators of fecal contamination for fresh water. The criteria is expressed using two components: the geometric mean (GM) and a statistical threshold value (STV). The STV approximates the 90th percentile of the water quality distribution and is intended to be a value that should not be exceeded by more than 10% of the samples. For *Enterococci,* the criteria are GM = 35 cfu/100mL and STV = 130 cfu/100mL. For *E. coli,* the criteria are GM = 126 cfu/100mL and STV = 410 cfu/100mL.

Sources of Pathogens

Municipal Source Waters

Municipal effluents can contain significant populations of pathogens, but treatment marshes are typically the recipients of disinfected effluents. The efficiencies of conventional treatment technologies that reduce pathogens to noninfective levels have been studied thoroughly, and wastewater treatment plants regularly add processes to accomplish necessary removals (Metcalf and Eddy Inc, 1991; Crites and Tchobanoglous, 1998). The most common add-on disinfection processes are chlorination, ozonation, and ultraviolet irradiation. As a consequence, treated municipal source waters to treatment marshes are often essentially devoid of bacterial pathogens.

Over 100 virus types are known to occur in human feces, with minimum infective doses as low as one organism for some species. Human parasites in wastewater include protozoa and helminths. Two common protozoan parasites are *Entamoeba histolytica* and *Giardia lamblia* which both cause diarrhea in infected humans. The phylum Aschelminthes (cavity worms) includes all parasitic worms incapable of adult life without a host organism. A number of helminths, including tapeworms and flukes, are found in human feces from infected individuals and can be spread through wastewater pathways. For a summary discussion of environmentally transmitted pathogens, refer to Maier et al. (2000).

Lagoon treatment of municipal wastewater often results in low numbers of indicator organisms, because of long retention times and exposure to solar radiation (Davies-Colley, 2005). Lagoon effluents may or may not be disinfected prior to entering a follow-on treatment marsh.

Agricultural Source Waters

Agricultural runoff sources are of two principal kinds: those emanating from pasture runoff and those from crop runoff. Moderately high levels of pathogens would be expected from animal yards and pastures, and data confirms this to be the case (Harmel et al., 2010; Forbes et al., 2011).

It might be presumed that crop runoff would not be subject to much contamination by pathogens, but that is not the case. The primary cause of pathogens in crop runoff appears to be the use of contaminated irrigation waters. At the low end, Beutel et al. (2013) found a seven-year average of 139 cfu/100mL of fecal coliforms in field runoff from irrigated crops in the Yakima Valley of the state of Washington. Harmel et al. (2010) measured 1277 cfu/100mL of *E. coli* in irrigated field runoff in Texas. Diaz et al. (2010) measured 317 cfu/100mL of *E. coli* and 3113 cfu/100mL of enterococcus in field runoff in the San Joaquin Valley of California. Tile drains contain lesser amounts of pathogens, such as the 12.6 cfu/100mL of *E. coli* measured for corn and barley fields in Nova Scotia, Canada (VanderZaag et al., 2010b). At the high end, the field runoff in the Imperial Valley of California originates from contaminated river water. Kadlec et al. (2010a) reported a four-year field runoff average of 37,117 cfu/100mL of total coliforms, 5157 cfu/100mL of fecal coliforms, and 1123 cfu/100mL of *E. coli*.

Urban Runoff Sources

Urban runoff contains pathogens that originate from animals in the watershed, both pets (cats, dogs) and wild animals (geese, raccoons). Table 12.12 lists the percentile occurrences

TABLE 12.12

Event mean concentrations of pathogen indicators in urban runoff (cfu/100mL). FC = fecal coliforms, FS = fecal streptococcus, TC = total coliforms.

	FC	FS	E. coli	TC
Observations	2168	1317	139	282
Percentile				
0.10	200	1,220	113	825
0.20	680	3,000	400	2,800
0.25	981	5,000	542	3,950
0.30	1,470	7,000	705	6,300
0.40	2,400	11,000	1,000	12,000
0.50	4,800	18,800	1,556	24,000
0.60	9,500	26,680	2,300	50,000
0.70	18,000	39,940	3,845	110,000
0.75	25,000	50,000	4,178	170,000
0.80	36,000	70,000	5,700	240,000
0.90	90,000	170,000	17,500	900,000

Source: The National Stormwater Quality Database, Pitt and Maestre (2015).

of the principal indicator organisms for urban runoff. In general, the medians are far above the recreational water quality criteria.

Animal Contributions

Treatment marshes contain many kinds of animals, some as permanent residents and others as migratory visitors. Densities can be quite high. Animals are also a source of pathogenic organisms. Farm animals produce feces with high numbers of bacteria. Cows, sheep, pigs and poultry produce 10^5–10^7 fecal coliforms per gram, and 10^6–10^8 fecal streptococci per gram. Cats, dogs, mice, and chipmunks produce similar concentrations (Maier et al., 2000). Waterfowl and other wetland birds also contribute coliforms to the wetland environment and may do so in great quantities during migratory concentrations. Beavers, muskrats and other warm-blooded wetland animals are also producers of enteric pathogens. Beavers have been implicated in transmission of *Giardia*. About half the animals in Colorado were found to be infected and shedding 10^8 cysts per animal per day. Ninety-five percent of muskrats in this study were found to be infected with *Giardia* (Erlandsen, 1995).

Pathogen Processes

Bacterial and protozoan pathogens are living organisms, and hence we are dealing with population dynamics in addition to physical and chemical effects. They reproduce via binary fission, a form of asexual reproduction. Binary fission results in the reproduction of a living cell by dividing the cell into two parts, each with the potential to grow to the size of the original. Pathogens are of concern as enteric organisms, which survive and multiply in warm, dark conditions with ample sources of carbon to support their metabolism. Similar conditions can prevail in marshes, and hence pathogens are able to survive there for varying periods of time.

The observed net removal of pathogenic organisms in treatment marshes is the result of a variety of processes; some which remove pathogens; some which introduce pathogens. Generally speaking, removal mechanisms are much more prevalent than reintroduction mechanisms, and most constructed wetlands demonstrate large net removals of pathogenic organisms down to a nonzero background concentration (C^*). The following is a brief summary of potential processes in the marsh environment.

Disinfection

Municipal wastewaters are sometimes subject to disinfection to prevent pathogens from reaching receiving waters. The abilities of microorganisms to survive disinfection vary both with the organism and with the disinfection method. The most common disinfection strategy in the United States is the utilization of chlorine (or chlorine derivatives such as chloramines), followed by UV light irradiation. and ozonation.

Because of its low cost and proven effectiveness, chlorination has been the disinfection method of choice for many years. However, negative side effects of effluent chlorination have become apparent in the past two decades. Residual free chlorine harms a variety of aquatic organisms and causes chronic and acute toxicity to microorganisms and fish. Also, when it comes in contact with organic compounds in wastewater or receiving waters, free chlorine forms trihalomethanes and other organochlorine compounds known to be carcinogenic. These findings have resulted in the increased use of dechlorination techniques, and in the development of the ozonation and ultraviolet disinfection technologies.

Ultraviolet radiation is a potent agent for killing bacteria. An exponential relation is commonly used to describe the inactivation in disperse batch cultures:

$$N = N_o e^{(-kIt)} \tag{12.16}$$

where I = intensity of UV light in solution, $J/m^2 \cdot d$ ==
k = inactivation rate coefficient, m^2/J
N_o = initial number of dispersed organisms, –
N = surviving number of dispersed organisms, –
t = time, d

Most organisms are present in association with wastewater particulates, rather than in a dispersed phase (Emerick et al., 2000). Consequently, UV inactivation rates are lowered, due to shielding of the organisms by the particle structure. Emerick et al. (2000) found little variation in rates for fecal coliforms across eleven treatment technologies, with $0.022 < k < 0.054$ m^2/J for activated sludge, trickling filter, aerated and facultative pond effluents. The ultraviolet wavelengths that are absorbed by microorganisms, and hence are effective in disinfection, comprise the range 240–280 nanometers (Crites and Tchobanoglous, 1998). Lamps used for in-line disinfection of wastewaters are designed to provide radiation in this range.

Starvation

The transition from an animal gastrointestinal tract to oligotrophic environmental waters exposes enterococci to nutrient starvation, and endangers their survival (Byappanahalli et al., 2012). For instance, studies have indicated that *E. faecalis* can survive for extended periods in sterilized, carbon-containing wastewater, but decline rapidly in relatively sterile lake water, indicating that oligotrophic conditions are deleterious to the survival of enterococci. Marshes, however, contain ample carbon supplies to support the organisms.

Solar Disinfection

In ponds and open-water wetlands, sunlight provides a source of UV radiation. However, the fraction of the incoming solar radiation that is in the UV range is small, typically less than 1 %. Therefore, solar inactivation rates based upon total solar radiation are much smaller than those determined for UV lamp sources, and these are designated k_S. For example, Davies-Colley et al. (1999) present pond-water data for E. coli for which $k_S \approx 0.7$ m^2/MJ, which is about 100,000 times lower than for a UV lamp. Salih (2003) determined a similar value for E. coli, $k_S \approx 1.5$ m^2/MJ. Davies-Colley et al. (1999) also demonstrated that dissolved oxygen and pH had important influences on rates of inactivation.

The overall solar inactivation rate for a shallow water body (pond) is further reduced by the lack of penetration of light (Mayo, 1995). The top few centimeters of a stabilization pond see the required radiation, but deeper sections do not. Accordingly, the solar inactivation rate constant is approximately inversely proportional to pond depth:

$$k_s = \frac{k'_s}{K_L h} \tag{12.17}$$

where h = water depth, m ==
k_S = overall solar inactivation rate coefficient, m^2/J
k'_S = intrinsic solar inactivation rate coefficient, m^2/J
K_L = light attenuation coefficient, m^{-1}

Mayo (1995) reported values $0.012 < hk_S < 0.016$ m^3/MJ for five stabilization pond studies. Additionally, high rate ponds (0.2–0.3 m depth and a few days detention) have been shown to have $k_S = 0.0675$ m^2/MJ (Davies-Colley et al., 2003); and $k_S = 0.083$ m^2/MJ (Craggs et al., 2004).

The past decade has seen much more detailed understanding of sunlight destruction of pathogens in treatment marshes, notably those without EAV or SAV that interferes with irradiation (Nguyen et al., 2014; 2015; Silverman et al., 2015; Silverman and Nelson, 2016; Bear et al., 2017; Nelson et al., 2018; Silverman et al., 2019). Bacteria and viruses are susceptible to the three sunlight disinfection mechanisms: direct and indirect endogenous mechanisms, and an exogenous mechanism. The direct mechanism involves absorption of UV light directly by cell DNA/RNA to cause damage, while the indirect mechanism relies upon cell internal sensitizers to produce reactive oxygen species, which in turn cause damage. Exogenous mechanisms rely upon sensitizers in the surrounding water to produce reactive oxygen species, which in turn cause damage. Rate coefficients for these processes have been correlated with dissolved organic matter and solar intensity for a few pathogens (Silverman et al., 2019). Thus far, this more detailed approach has only been tested for the special case of no vegetation, but the concepts are surely applicable to open water sections of more typical marshes.

Solar disinfection depends upon sunlight reaching and penetrating into the water column. Dense vegetation intercepts sunlight in the wetland environment and diminishes the potential for solar disinfection. This negative effect is putatively present for emergent and floating plant communities. It may or may not be present in submersed algal or submersed macrophyte communities, in which subsurface oxygenation may occur. The presence of enhanced amounts of DO fosters the photo-oxidation route of disinfection, and may compensate for reduced penetration of UV radiation (Davies-Colley et al., 1999; Sun et al., 2003).

However, when pond performance is compared to vegetated wetland performance for fecal coliform reduction across large numbers of both types of systems receiving varying numbers of pathogens, little difference can be seen in the general level of reduction (Kadlec, 2005d). Other mechanisms of organism mortality and removal, beyond just solar disinfection, become operative in the wetland environment.

Predation

Grazing by protozoa, bacteriophage infection, and predation by some bacteria are among the biotic effects that control the abundance of prokaryotic organisms in the environment. Bacteriophage infection affects a much wider range of bacteria, and viral infection was suggested to be a mechanism responsible for the elimination of up to 50% of autochthonous bacteria from aquatic habitats (Byappanahalli et al., 2012).

Most pathogens are food for nematodes, rotifers and protozoa (Decamp and Warren, 1998). Among these, rotifers and flagellated and ciliated protozoa have been implicated as important contributors to the reduction of bacteria in treatment wetlands (Panswad and Chavalprit, 1997; Laybourn-Parry et al., 1999; Proakis, 2003; Stott and Tanner, 2005), as well as in natural aquatic systems (Menon et al., 2003). While pathogenic organisms span a wide size range (0.2–100 μm), so do the associated predator/grazing communities.

Proakis (2003) determined that rotifers consumed enterococci at the rate of about six hundred per hour when the bacterial density was approximately 10^6 per 100ml. From this rate, it was determined that rotifers at a density of ten per mL could suitably disinfect stormwater in a 1.2-hour detention in a marsh. The question then becomes what concentrations of predators can be expected in the wetland waters.

Heterotrophic nanoflagellates (HNAN) were found to dominate the protozoan community in both lagoons and grass filters treating sewage (Laybourn-Parry et al., 1999). Ciliates contributed only about 10 % of the grazer populations. The grass filters harbored larger numbers (approximately 10^8 per 100ml) of these bacterial predators than the lagoons (approximately 10^7 per 100ml).

The overall effect of grazing on bacterial numbers in aquatic systems was quantified by Menon et al. (2003). Grazing by protozooplankton was responsible for more than 90% of the overall mortality rate of both fecal and autochthonous bacteria in the river Seine. For example, the overall first-order volumetric mortality coefficient was 34×10^{-3} hr^{-1}, while it was 2×10^{-3} hr^{-1} in the absence of protozoa.

The studies described earlier in this chapter indicate a strong potential for pathogen removal due to predation in constructed wetlands. They do not, unfortunately, allow quantification of this removal mechanism. While predation/grazing is an important removal mechanism in marshes, it is not the only removal mechanism. Settling and filtration also play important roles in pathogen reduction, and these removal mechanisms have a cumulative effect in treatment wetland systems.

Settling and Filtration

A measurable proportion of wastewater microorganisms are found either associated with particulates, or as aggregates of many organisms. Bacteria may clump together to form aggregates. These are highly amorphous and porous, but the internal voids are apparently filled with exopolymeric material (Li and Yuan, 2002). Sedimentation of these particles contributes to removal from the water column, with removals of 25–75% due to gravitational sedimentation in conventional treatment plants (Metcalf and Eddy Inc, 1991). Tests of algal settling ponds produced 40% removal of E. coli by sedimentation (Davies-Colley et al., 2003).

In the wetland environment, submersed plant parts and their associated biofilms form "sticky traps" for particles, including all sizes of microorganisms (refer to Chapter 8). These biofilms are capable of trapping considerable numbers of organisms (Flood and Ashbolt, 2000; Stott and Tanner, 2005). Such biofilms are enhanced in quantity by larger quantities of submersed surfaces, and by exposure to light. There may be an optimal plant density that allows light and provides the necessary surfaces for biofilm growth.

Mortality

Bacteria, protozoa, helminths, and viruses typically do not survive longer than about 30 days in freshwater environments, and about 50 days in soil environments (Crites and Tchobanoglous, 1998). Similar survivals might therefore be predicted for wetlands, but there are many site-specific factors and processes that may materially increase or decrease survival.

A part of the overall removal of pathogens consists of the death or inactivation of organisms for reasons other than radiation damage or predation. This is also called the dark death rate (Khatiwada and Polprasert, 1999b). Mortality has been found to be approximately 25% of the overall removal rate in ponds (Craggs et al., 2004), and a similar fraction in a *Typha* marsh in Thailand (Khatiwada and Polprasert, 1999b). However, care must be taken to separate sedimentation and filtration from mortality in this dark death rate. For instance, Khatiwada and Polprasert (1999b) estimated that only 6.5% of the overall fecal coliform removal rate was due to temperature-modulated death of organisms.

There is no definitive study of the temperature effect on mortality of bacteria in wetlands. Estimates of temperature coefficients (θ) for ponds range from 1.000 (no effect) to 1.070 (strong effect).

Very low temperatures may result in freezing conditions. Ice formation, followed by a thaw, drastically reduces the populations of pathogens in water. For instance, Torrella et al. (2003) found that only 1.2% of fecal coliforms survived four days of freezing, but more coliphages survived. However, care must be taken in enumeration, because organisms may be viable after freezing, but unculturable in routine tests (Parker and Martel, 2002). Further, many bacteria survive the snow-making process. Less than one log loss was measured for fecal streptococcus due to snow formation and melting (Parker et al., 2000).

Regrowth

To complicate matters even more, some indicator organisms are capable of regrowing and multiplying in an aquatic environment. All members of the coliform group have been observed to regrow in natural surface water (Gerba, 2000). Such regrowth is fostered by high concentrations of organic matter and by elevated temperatures. This phenomenon is recognizable when disinfected wastewater is introduced into treatment wetlands, and bacterial populations increase along the direction of flow (Figure 12.10).

Reintroduction

Indicator organisms can be produced by sources other than incoming source waters. This is particularly true for the coliform group, which may originate from many different warm-blooded animals that frequent wetlands. As an example, consider the performance of the Orlando Easterly Wetlands (Figure 12.11). Incoming fecal coliforms and streptococcus (FC and FS) were very low, but the animal (bird) populations are high, especially near the outlet. Background geometrical annual averages range from 50 to 142 cfu/100ml. Monthly values show an exceedance frequency of 40% of the permit limit of 100 cfu/100ml. The five-year median FC was 64 cfu/100ml, and the fecal streptococcus (FS) was 106 cfu/100ml. The ratio FC/FS = 0.6, and thus the presumption may be made that the source of contamination is animal rather than human (criterion = FC/FS < 0.7). In this case, the treatment wetland has been measured to have very large bird populations, which also supports the concept of re-introduction of pathogens from avian populations (U.S. EPA, 1993a, 1993d).

These naturally-caused bacterial populations are generally low, but they may be variable and seasonally high because of wildlife activity patterns. Total and coliform bacteria have been measured in natural wetlands that receive no wastewater. For example, Fox et al. (1984) measured between 109 and 456 cfu/100ml of fecal coliforms in cypress wetlands in Florida that were not receiving any external water inputs.

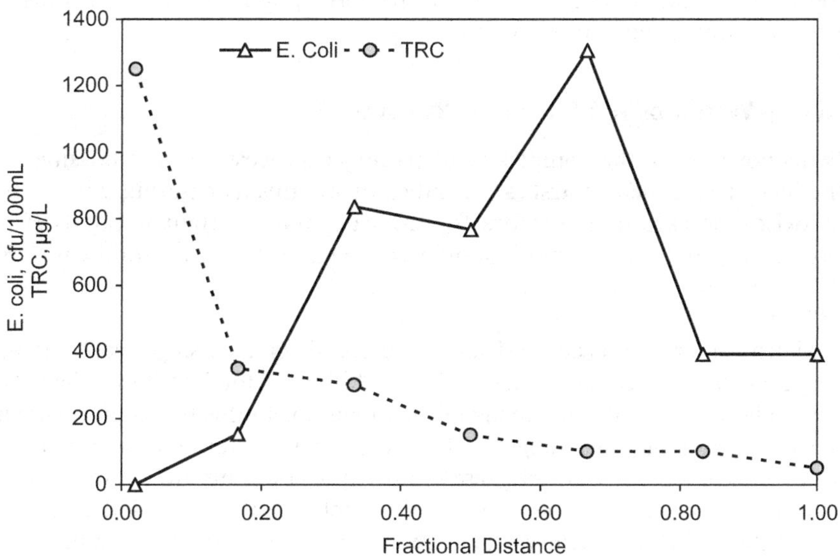

FIGURE 12.10 Longitudinal profiles of total residual chlorine (TRC) and *E. coli* through Tres Rios, Arizona, Hayfield wetland H1.

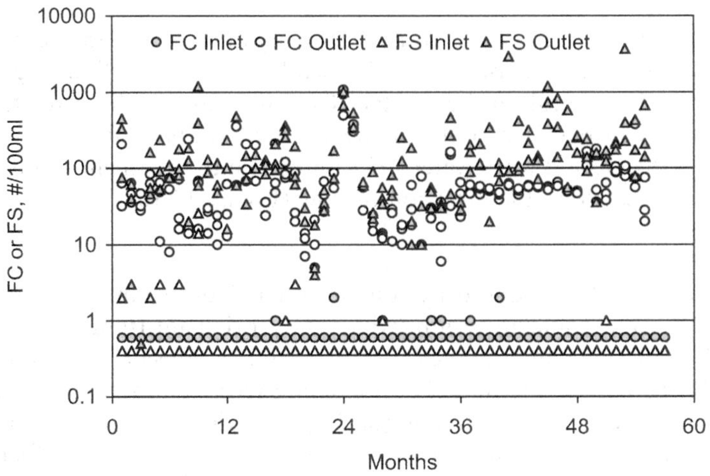

FIGURE 12.11 Fecal coliforms and streptococcus entering and leaving the Orlando, Florida, Easterly Wetland during 1993–1997. The incoming wastewater is disinfected, with only a few detections of the indicator organisms. The balance of the nondetect samples are plotted as half the detection limit of 1.0 per 100ml.

Hsu et al. (2017) attempted to trace the origins of indicator bacteria in small wetlands treating river water, by use of microbial source tracking markers. They were able to identify the sources of pathogen increases along the flow direction as due to human, avian (geese) and animal (deer) origins. Human-sourced pathogens decreased along the flow direction, while avian-sourced pathogens increased. They suggested that wildlife management was needed to reduce potential health risks. They pointed out that goose management techniques could include pyrotechnics, lasers, trained dogs and egg oiling.

Because natural sources of coliforms and fecal streptococcus bacteria are found in all wetlands open to wildlife, outflow indicator bacteria populations in treatment wetlands cannot be consistently reduced to near zero.

Performance of Wetlands for Pathogen Removal

While fully acknowledging the complexity of pathogen processing, it is possible to consider the overall effect of treatment marshes on indicator organisms entering with source waters. A database exists for various indicators for the category of urban wetlands (Clary et al., 2017b) (Table 12.13), but there is not a similar compilation for other source waters.

Total Coliforms

The total coliform group has been used as the standard for assessing fecal contamination of waters throughout the 20th century (Gerba, 2000). Although total coliforms have been widely used in older studies and monitoring, many of the organisms in this broad group are not limited to fecal sources. Accordingly, the subgroup fecal coliforms became the preferred indicator for wastewater treatment operations. At the present time, fecal coliforms are not the preferred indicator for most human contact situations, having been replaced by E. coli and enterococcus (FS). Here it is noted that FWS treatment wetlands reduce high numbers of incoming total coliforms but possess relatively large natural populations. Wetland biota reintroduce total coliforms.

TABLE 12.13

Pathogens reduction in urban wetlands.

	Systems	Events	Inlet (cfu/100mL)	Outlet (cfu/100mL)	Reduction
Enterococcus					
Wetland	4	53	840	330	61%
Wetland/Pond	6	86	780	170	78%
Escherichia coli					
Pond	4	69	2,000	80	96%
Wetland	6	77	2,800	100	96%
Wetland/Pond	10	146	2,300	450	80%
Fecal coliform					
Pond	10	121	12,000	900	93%
Wetland	5	42	5,000	1,200	76%
Wetland/Pond	15	163	6,000	4,000	33%

Source: Clary et al. (2017b).

Fecal Coliforms

Historically, the most common indicator organism group is fecal coliforms, and consequently the wetland database is largest for this group. FWS treatment wetlands reduce high numbers of incoming fecal coliforms (Kadlec and Wallace, 2009), but by definition, treatment marshes receive low numbers. Treatment marsh data has been acquired for a number of systems, with examples listed in Table 12.14. The median for these examples shows no reduction.

Input-output annual data for 106 wetland-years is displayed in Figure 12.12. The scatter in these data is quite high, but there is an absence of any trend ($R^2 = 0.000$).

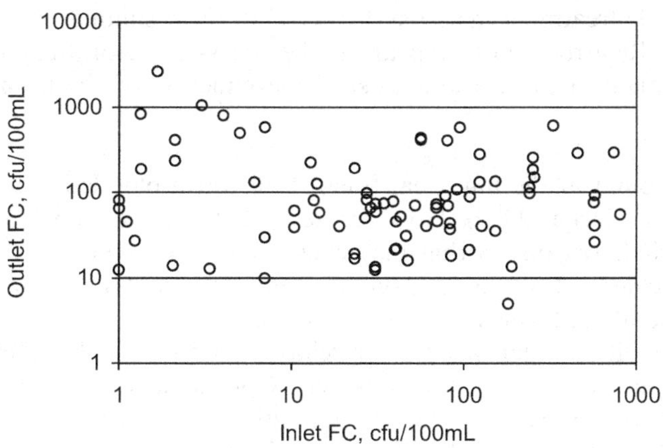

FIGURE 12.12 Outlet fecal coliforms versus inlet fecal coliforms for treatment marshes receiving less than 1000 cfu/100mL. There are 106 annual data points for 36 different wetlands.

TABLE 12.14

Examples of fecal coliforms entering and leaving treatment marshes (cfu/100mL). TWE = treated wastewater effluent, LAG = lagoon effluent, STM = urban stormwater, AGR = agricultural runoff, RIV = river.

Site	Source Water	Inlet	Outlet	$\log_{10}$ Reduction	Reference
Titusville, FL	TWE	0	48	–2.32	Unpublished data
Orlando Easterly, FL	TWE	1	24	–1.53	NADB Database (1998)
Hillsboro, OR	TWE	1	189	–2.15	NADB Database (1998)
Orange County Eastern, FL	TWE	256	150	0.23	NADB Database (1998)
Estevan, SAS	LAG	47	16	0.47	Unpublished data
Cannon Beach, OR	LAG	46	31	0.18	NADB Database (1998)
West Jackson County, MS	LAG	95	578	–0.79	NADB Database (1998)
Haines Junction, Yukon	LAG	3	62	–1.32	Unpublished data
Lenoir, NC	STM	1,857	1,029	0.26	Hathaway et al. (2011)
Lenoir, NC	STM	7,980	1,931	0.62	Hathaway et al. (2011)
Yakima, WA	AGR	139	188	–0.13	Beutel et al. (2013)
Yakima, WA	AGR	139	142	–0.01	Beutel et al. (2013)
Imperial, CA	AGR	102,366	229	2.65	Unpublished data
Brawley, CA	RIV	293,943	239	3.09	Unpublished data
Montreal, Quebec	RIV	78	91	–0.07	Vincent (1992)
Tarrant, TX	RIV	23	193	–0.92	Unpublished data
Median				**–0.04**	

Fecal Streptococcus/Enterococcus

The fecal streptococci belong to the genera *Enterococcus* and *Streptococcus.* These are considered to have advantages over fecal coliforms as an indicator because: (1) they rarely multiply in water, (2) they are more resistant to environmental stress, and (3) they generally persist longer in the environment (Gerba, 2000). A high ratio of fecal streptococci (FS) to fecal coliforms (FC) is taken as strong evidence that the source is of human origin.

Inflow and outflow densities of fecal streptococcus bacteria have been monitored at a number of full-scale treatment marsh systems. Table 12.15 summarizes these data for various source waters. Reintroduction appears to be a lesser factor than for other indicator organisms. The median for these examples shows a reduction of about one $\log_{10}$.

Escherichia Coli

Escherichia coli is found in the intestinal tract of all warm-blooded animals, and is usually considered harmless (Gerba, 2000). However, several strains are capable of causing gastroenteritis, which include organisms that may cause hemorrhagic colitis, a dangerous disease, especially for the elderly. For this reason, *E. coli* has found favor in some instances as a replacement for fecal coliforms as an indicator organism.

E. coli has the ability to regrow or to be reintroduced in marshes. Table 12.16 summarizes input-output data for various source waters. Considerable numbers were found in the wetland effluent at Tres Rios, Arizona, for a disinfected influent. Conversely, large reductions may occur when the wetland influent has large numbers. Such reductions may be quite large for highly compartmentalized wetlands of high aspect ratio, for the Brawley, California, wetlands, for instance.

TABLE 12.15

Examples of fecal streptococcus and enterococcus entering and leaving treatment marshes (cfu/100mL). TWE = treated wastewater effluent, LAG = lagoon effluent, STM = urban stormwater, AGR = agricultural runoff, RIV = river.

Site	Source Water	Inlet	Outlet	$\log_{10}$ Reduction	Reference
Fecal streptococcus					
OEW, FL	TWE	1	75	–1.87	Unpublished data
Titusville, FL	TWE	1	118	–2.07	Unpublished data
Arcata, CA	LAG	545	254	0.33	Unpublished data
Enterococcus					
Hillsboro, OR	TWE	5	75	–1.18	U.S.EPA (1993a)
Discovery Bay, CA	TWE	851	67	1.11	Nguyen et al. (2015)
San Joaquin, CA	AGR 1	2,157	283	0.88	Diaz et al. (2010)
San Joaquin, CA	AGR 2	2,681	155	1.24	Diaz et al. (2010)
San Joaquin, CA	AGR 3	2,681	121	1.35	Diaz et al. (2010)
San Joaquin, CA	AGR 4	1,388	52	1.43	Diaz et al. (2010)
Riverside, CA	RIV	4,850	10	2.69	Ibekwe et al. (2007)
Elsah, IL	STM	39,811	2,818	1.15	Waller and Bruland (2016)
Wilmington, NC	STM 1	1,018	316	0.51	Hathaway and Hunt (2012)
Wilmington, NC	STM 2	866	510	0.23	Hathaway and Hunt (2012)
Median				**0.88**	

TABLE 12.16

Examples of *E. coli* entering and leaving treatment marshes (cfu/100mL). TWE = treated wastewater effluent, LAG = lagoon effluent, STM = urban stormwater, AGR = agricultural runoff, RIV = river.

Site	Source Water	Inlet	Outlet	$\log_{10}$ Reduction	Reference
Tres Rios, AZ H1	TWE	2	468	–2.28	Unpublished data
Tres Rios, AZ H2	TWE	2	616	–2.40	Unpublished data
Tres Rios, AZ C1	TWE	1	998	–2.90	Unpublished data
Tres Rios, AZ C2	TWE	1	2,840	–3.35	Unpublished data
Brighton, ONT	LAG	153	129	0.07	Unpublished data
Imperial, Ca	AGR	1,035	70	1.17	Unpublished data
Brawley, CA	RIV	47,092	44	3.03	Unpublished data
Melbourne, Aus	STM Base	260	36	0.86	Meng et al. (2018)
Melbourne, Aus	STM Rain	1,000	130	0.89	Meng et al. (2018)
Wilmington, NC	STM 1	824	836	–0.01	Hathaway and Hunt (2012)
Wilmington, NC	STM 2	425	503	–0.07	Hathaway and Hunt (2012)
San Joaquin, CA	AGR 1	227	40	0.75	Diaz et al. (2010)
San Joaquin, CA	AGR 4	166	8	1.32	Diaz et al. (2010)
Jeongeup, Korea	STM	423	60	0.85	Cheng et al. (2013)
Philadelphia, PA	STM	500	100	0.70	Wadzuk et al. (2010)
Columbus, OH	RIV	596	568	0.02	Hsu et al. (2017)
Median				**0.39**	

Seasonal Effects

The large stochastic scatter of FC data for treatment marshes makes it difficult to discern seasonal trends. In a few cases, there are long periods of record, which allow pattern recognition. Ten years' data from Estevan, Saskatchewan, showed an increase in fecal coliforms in the wetland (0.24 log_{10} increase). The seasonal trends for both inlet and outlet values display higher summer values (Figure 12.13). Cosine time trends are quite weak, with R^2 = 0.13 for the inlet and R^2 = 0.49 for the outlet. FC increases, as indicated by the separation of the two trend lines, are not synchronous with the water temperature, with lowest values in early spring, then steadily increasing throughout summer and early autumn. Increases in FC are not unexpected, because of a large and varied animal population in the marsh. Throughout the spring and fall migration periods, large numbers of waterfowl were observed, as well as smaller numbers throughout the entire season. The system also supports a large population of leopard frogs, mice, voles and muskrats (Duncan, 1998).

A second example of seasonality is the behavior of the Imperial, California system (Figure 12.14). Four years' data from showed a decrease in *E. coli* in the wetland (1.17 log_{10} decrease, or from 1123 to 75 cfu/100mL). The seasonal trends for the inlet values showed a maximum at the end of May, while the outlet peaked at the end of August (Figure 12.14). *E. coli* decreases, as indicated by the separation of the two trend lines, are not synchronous with the water temperature, with highest values in April, and lowest values in October.

A third example of seasonality is the behavior of the Brighton, Ontario, system (Figure 12.15). Ten years' data from showed a slight decrease in the annual average *E. coli* in and out of the wetland, from 153 to 131 cfu/100mL, or a 0.07 log_{10} reduction. Decreases occurred during half the year, December through May. During the other half year, June through November, increases occurred on passage through the wetland. The lagoon source water had high *E. coli* in winter. This marsh had extensive wildlife usage during the summer months, when it was rated the best bird watching spot in the region. In contrast to

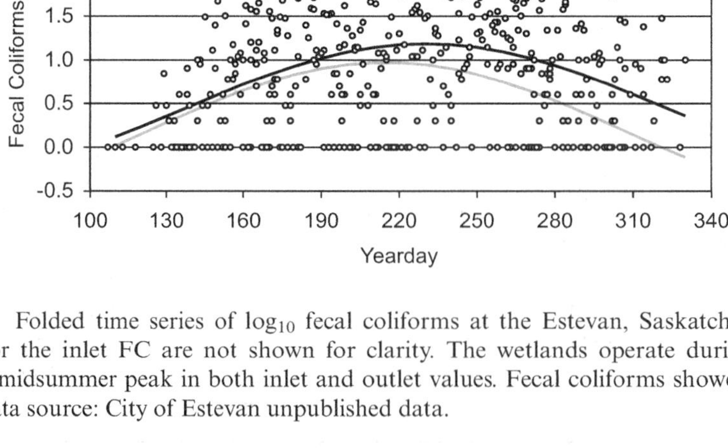

FIGURE 12.13 Folded time series of log_{10} fecal coliforms at the Estevan, Saskatchewan, treatment marsh. Data for the inlet FC are not shown for clarity. The wetlands operate during the unfrozen season. Note a midsummer peak in both inlet and outlet values. Fecal coliforms showed an increase in log_{10} = 0.24. Data source: City of Estevan unpublished data.

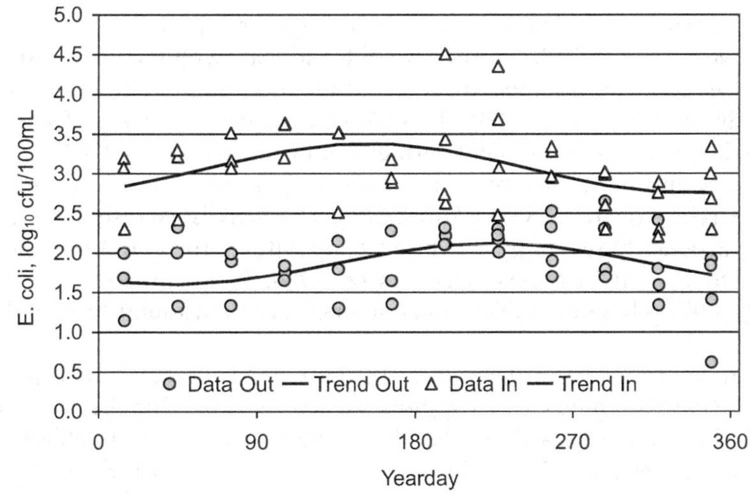

FIGURE 12.14 Folded time series of $\log_{10}$ *E. coli* at the Imperial, California treatment marsh. E. coli showed an annual decrease in $\log_{10} = 1.17$. Data source: Imperial Water District unpublished data.

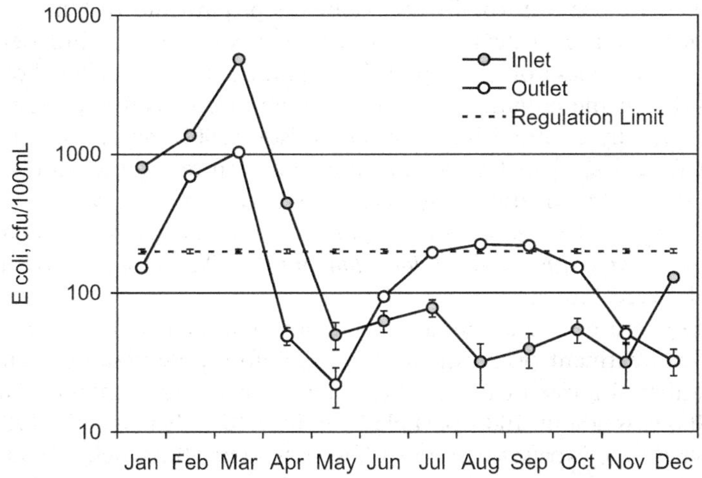

FIGURE 12.15 Monthly progression of *Escheria coli* entering and leaving the Brighton, Ontario, treatment marsh. Data are ten years, folded.

Estevan, the Brighton marsh functioned in all seasons, which in the relatively cold climate meant an ice cover winter. The combined lagoon marsh system failed to meet the regulatory guideline of 200 cfu/100mL about half the time; in winter because of high lagoon concentrations, and in summer because of presumed wildlife activity.

Clearly, there is no general rule concerning seasonality for indicator organisms. Treatment marshes operate in concentration ranges that are dominated by a complex suite of processes that are highly variable, and currently unpredictable.

Other Pathogenic Organisms

Upon some few occasions, individual pathogenic organisms have been studied. These have included Salmonella spp., which causes typhoid and paratyphoid fevers, and is found in a large variety of both warm-blooded and cold-blooded animals (Gearheart et al., 1983; Herskowitz, 1986; Karim et al., 2004). *Yersinia enterocolitica* and *Pseudomonas aeruginosa* were found in large numbers in the Listowel, Ontario lagoons, a bacterium that causes gastroenteritis. Because these survive well at cold temperatures, they were monitored in the Listowel project (Herskowitz, 1986). *Clostridium perfringens* is exclusively of fecal origin. It is regarded as particularly resistant to pH and temperature stress, and was therefore chosen for monitoring at several locations (Herskowitz, 1986; Hill and Sobsey, 2001; Stenstrom and Carlander, 2001). These organisms, plus others, have been found to be reduced in treatment marshes.

Parasites include protozoa and helminths. In the United States, the two protozoa of special interest are *Giardia lamblia* and *Cryptosporidium parvum*. The former is the most frequently identified intestinal parasite in the United States (Maier et al., 2000). The later was implicated in an outbreak in 1993 in Milwaukee, Wisconsin, in which over 400,000 people were infected. *Giardia* is of especial interest in connection with wetlands, because beavers and muskrats are believed to be reservoirs of infection. Worldwide, *Ascaris lumbricoides* is the most prevalent parasitic infection, with an estimated 22% of the world population affected (Maier et al., 2000). Adult *Ascaris* are roundworms that inhabit the intestine, but they are ingested as eggs. Two other prevalent infectious worms are *Trichuris trichiuria* and *Taenia saginata*, and these are of especial concern in warm climates.

Treatment marshes are not normally the recipients of parasites of human origins, because only a fraction of the source waters come from human sources. The broader group of FWS wetlands reduce the numbers of these parasitic organisms by reducing the number of eggs or cysts that survive in the aquatic environment. One mechanism is presumably sedimentation of eggs, followed by eventual inactivation. Nelson (2003) reports egg numbers in pond sludges, ranging from less than ten per dry gram in France, to thousands of *Ascaris* eggs per dry gram in Brazil. Inactivation may take more than a year. However, Stott et al. (2001) determined that oocysts of *Cryptosporidium parvum* were effectively ingested by protozoans, notably *Paramecium caudatum* and *Stylonychia mytilus*. Therefore, predation may also be a factor in the observed reductions.

Similarly, viruses are not generally a concern for treatment marshes. However, they can survive wastewater treatment processes, and are therefore potentially of concern for treated effluent source waters for treatment marshes. The average viral content of domestic sewage in the United States is about 7000 particles per liter (Gersberg et al., 1989). Because the potential infective dose of viruses can be so low (less than 10 particles), and because viruses are generally hardier in natural environments than bacterial pathogens, there has been considerable concern about their fate in wetland and conventional treatment systems. However, because viruses are more costly and difficult to monitor analytically, they have not received as much study as bacterial indicator populations. Existing research indicates that wetlands are generally hostile environments for viruses (Kadlec and Wallace, 2009).

12.7 SUMMARY

Ancillary chemicals are those substances found in treatment marshes that are not principal targets for design. This chapter summarizes the current knowledge about the fate and effects of major ions, pH, DO, BOD, sulfur, and pathogens in marshes. Wetlands do not provide removal of common salts such as alkali metal chlorides. This limitation exists because the typical surface water concentrations of most salts exceed growth requirements and because salts are soluble and tend to leach from sediments and tissues. There are no microbial removal mechanisms. Collectively, salts form the basis for the electrical conductivity and total dissolved solids content of wetland waters, and hence these parameters are not much affected by passage through wetlands.

The hydrogen ion content of the wetland surface water is an important regulator of several chemical and biological processes. Treatment marshes operate at circumneutral pH. This is the result of photosynthesis and the associated carbonate-bicarbonate-carbon dioxide equilibria. There large diel swings in pH.

Dissolved oxygen is vital participant in wetland microbial processes including nitrification and for aerobic decomposition of CBOD and soil organic matter. The principal source is mass transfer from the atmosphere, which, together with plant oxygen transfer and production, are is adequate for support of biological processes for the range of conditions found in treatment marshes. DO exhibits large seasonal variation, as well as large diel cycles. There are often vertical profiles in DO.

The carbon cycle in wetlands is vigorous, and many internal wetland processes are fueled by both carbon imports (usually small for treatment marshes) and by the carbon formed from decomposition processes. CBOD is both produced and consumed by internal processes, as part of the growth/death/decomposition cycles in the marsh. The overall effect is to produce a background concentration of CBOD that is independent of source water imports.

The processing of sulfur in wetlands is represented by interconversions of several sulfur compounds in the different ecosystem compartments. Atmospheric inputs as well as source water compounds bring sizeable amounts of sulfur into a treatment marsh. A complex set of redox reactions creates varied speciation ranging from sulfate to elemental sulfur, some of which are volatile. Reduced forms (sulfides) are important in binding of trace metals.

Pathogenic organisms enter marshes with most source waters, with the exception of disinfected effluents from wastewater treatment facilities. Because of the generally low density of pathogens and the difficulty associated with their enumeration, indicator organisms such as fecal coliforms, *E. coli*, and enterococcus are used to monitor the probable pathogenic activity in wetland waters and receiving waters.

Both humans and wildlife can contribute to pathogen inputs to marshes, via source waters and via direct inputs from marsh animal residents. Because treatment marshes are not intended as primary disinfection devices, the populations of pathogens are low and highly variable.

A growing body of knowledge demonstrates that a significant die-off of indicator organisms and pathogenic species occurs in marsh treatment systems. But, it is also evident that marshes contain numerous animals that contribute base populations of these same indicator organisms. These opposing effects produce highly variable wetland populations of pathogens. There are no reported human or wildlife health problems associated with the operation of treatment marshes.

13 Trace Metals

This chapter examines considerations relevant to the reduction of selected trace metals in treatment marshes. For many of the trace metals, information is very scarce or absent. That is because they have heretofore not been the subjects of regulatory concern in treatment marshes, although these are sometimes the concern for treatment wetlands dealing with industrial or mine source waters. The spectrum of commonly measured trace metals is represented by the databases for urban runoff systems (Table 13.1). Those urban wetlands remove about half the total metal concentrations, excepting nickel; but are not as effective at reducing dissolve metal concentrations.

Two metals that are of great concern for marshes in general, not just treatment marshes, are selenium and mercury. These are included here because of that intense interest.

Those that are not here included are: manganese, aluminum, antimony, tin, molybdenum, beryllium, thallium, silver, barium, and boron. Nonetheless, there are instances where these may be of either regulatory or treatment interest.

13.1 GENERAL CONSIDERATIONS

A number of trace metals are essential micronutrients at low concentrations, but some trace metals may occur in source waters at concentrations that are toxic to sensitive organisms. The probability of exceeding sensitive toxicity levels in mine water, leachates and some industrial waters is much higher, but these are not here under consideration. For most trace elements, biochemical transformations and chemical characteristics can lead to biomagnification, a phenomenon in which increasing concentrations occur in consumers along a food chain. Although most trace metals are more concentrated in biological tissues and soils than they are in surface water, hazardous situations do not always occur.

Surface flow wetland treatment systems are open to biota that may be exposed to potentially dangerous levels of metals, primarily in the wetland sediments. The removal of metals can result in storage in sediments that is inimical to the subset of wetland organisms that live or feed in those sediments. To prevent this problem from occurring, wetland treatment system designers and regulators should consider pretreatment to reduce influent metal concentrations. As a design measure, deep water systems with floating plants or rafts send sediments to depths that are out of reach of top feeders.

In the United States, treatment marshes may be considered waters of the United States and would therefore be required to meet water quality guidelines internally, as well as for the waters they are designed to protect. The levels of metals that may be tolerated by sensitive organisms have been promulgated in the form of guidelines for the protection of receiving waters and associated sediments. Examples of such guidelines are presented in Tables 13.2 and 13.3. These may or may not be considered applicable to treatment wetlands, which are not necessarily themselves protected by U.S. water quality guidelines.

13.2 METALS PROCESSING AND STORAGE

Abiotic Metal Partitioning

Depositing sediments are capable of adsorbing significant quantities of trace metals directly, or indirectly through the accumulation of coatings such as organic matter, iron, and

TABLE 13.1

Metals reduction in urban wetlands.

	Systems	Events	Inlet (μg/L)	Outlet (μg/L)	Removal
Total					
Iron	18	404	692	278	60%
Lead	47	763	7.07	2.25	68%
Nickel	10	113	4.39	2.20	50%
Zinc	61	1,075	49.0	21.0	57%
Copper	51	975	8.24	4.00	51%
Cadmium	32	508	0.34	0.20	41%
Chromium	13	170	4.18	1.50	64%
Arsenic	4	25	1.20	1.00	17%
Dissolved					
Iron	5	115	52	64	-23%
Lead	15	199	0.98	0.82	16%
Nickel	4	17	1.80	2.30	-28%
Zinc	24	466	22.6	12.0	47%
Copper	22	469	4.51	3.20	29%
Cadmium	7	90	0.20	0.12	40%
Chromium	4	41	1.00	1.00	0%

Source: Clary et al. (2017b).

manganese oxyhydroxides, which will in turn act as trace element collectors. Organic matter, which may exist as a surface coating or as a particulate, may play an important role in metal speciation and bioavailability. Microbial decomposition of organic matter typically results in sediments that are anoxic under a thin oxic surface layer. Under these anoxic conditions, divalent cationic metals such as cadmium, copper, lead, nickel, and zinc readily form metal sulfides. As long as amorphous sulfide concentrations are in excess of the total trace-metal concentration (on a molar basis), these metals will occur predominantly as insoluble metal sulfides. In anoxic sediments, metals in excess of sulfide may complex with organic matter. This buffers organisms against metal toxicity (Doig and Liber, 2006).

Sorption Relations

The Langmuir isotherm and fits metal sorption data for a number of wetland substrates:

$$C_S = \frac{KC_L}{1 + aC_L} \tag{13.1}$$

where a = Langmuir parameter, L/mg
C_L = water concentration, mg/L
C_S = solid concentration, mg/gdw
K = Langmuir parameter, L/gdw

The maximum capacity for metal is C_{smax} =K/a, which is only achieved at high water concentrations. The half-saturation water concentration, where C_S = 0.5C_{Smax}, is equal to

TABLE 13.2

U.S. EPA guidelines for metal concentrations in water. For those that are hardness dependent, 100 mg/L is used. The Criteria Maximum Concentration (CMC) is an estimate of the highest concentration of a material in surface water to which an aquatic community can be exposed briefly without resulting in an unacceptable effect. The Criterion Continuous Concentration (CCC) is an estimate of the highest concentration of a material in surface water to which an aquatic community can be exposed indefinitely without resulting in an unacceptable effect.

Metal	Symbol	Hardness mg/L	Freshwater CMC µg/L	Freshwater CCC µg/L	Human Health µg/L
Silver	Ag	100	3.2		
Aluminum	Al		750	87	
Arsenic	As		340	150	0.018
Cadmium	Cd	100	1.8	0.72	
Chromium	Cr(III)	100	570	74	
Chromium	Cr(VI)		16	11	
Iron	Fe			1,000	300
Manganese	Mn				50
Mercury	Hg		1.4	0.77	
Nickel	Ni	100	470	52	610
Lead	Pb	100	65	2.5	
Zinc	Zn	100	120	120	7,400

Source: U.S. EPA (2018a, 2018b).

1/a. At low water concentrations, $C_L \ll 1/a$, the Langmuir relation reduces to a linear partition equation:

$$C_S = KC_L \tag{13.2}$$

Freundlich isotherms are also often used to fit partitioning data:

$$C_S = K_F C_L{}^n \tag{13.3}$$

where n,K_F = Freundlich parameters

Lesage et al. (2006) reported that removal of Co, Ni, Cu and Zn by gravel and straw could be well described by Langmuir isotherm with $R^2 \geq 0.97$ for gravel and $R^2 \geq 0.93$ for straw. El-Gendy (2006) reported that sorption of heavy metals from the landfill leachate by roots of water hyacinth (*Eichhornia crassipes*) could be described by both Langmuir and Freundlich isotherms with respective correlation coefficients R^2 of 0.94 and 0.93. Organic sediments contain polar functional groups such as acids and phenolics that are responsible for cation exchange capacity, and can be involved in chemical binding (Ho and McKay, 2000). The metal-sediment reaction may be represented as:

TABLE 13.3

Guidelines for metal concentrations in sediments (μg/g). Wisconsin (WDNR, 2003) levels are a Threshold Effect Concentration (TEC) and a Probable Effect Concentration (PEC). Background levels for Ontario are those of the Great Lakes precolonial sediment horizon.

		----- Ontario -----			-- Wisconsin --	
Metal	Symbol	Background Level	Lowest Effect Level	Severe Effect Level	TEC	PEC
Silver	Ag				1.6	2.2
Arsenic	As	4.2	6	33	9.8	33
Cadmium	Cd	1.1	0.6	10	0.99	5
Chromium	Cr	31	26	110	43	110
Copper	Cu	25	16	110	32	150
Iron	Fe	31,200	20,000	40,000	20,000	40,000
Manganese	Mn	400	1,100		460	1,100
Mercury	Hg	0.1	0.2	2	0.18	1.1
Nickel	Ni	31	16	75	23	49
Lead	Pb	23			36	130
Zinc	Zn	65	120	820	120	460

Source: Ontario guidelines, Persaud et al. (1993).

$$2HP + M^{+2}P_2 + 2H^+ \tag{13.4}$$

where "P" represents the organic material (peat). If Equation 13.4, cation exchange, is presumed to represent data, then the amount of metal bound is represented by (Kadlec and Keoelian, 1986):

$$K_{eq} = \frac{[MP_2][H^+]^2}{[HP]^2[M^{+2}]} \tag{13.5}$$

where the bracket notation denotes molar concentration. This suggests that the partition coefficient ($[MP_2]/[M^{+2}] = C_S/C_L = K$) should go down markedly with decreasing pH, which is in fact the observation of many investigators, as shown in Kadlec and Keoelian (1986).

Equilibrium Metal Chemistry Calculations

There are several computer codes that have been developed to compute the theoretical thermodynamic equilibria in water solutions. For instance, MINTEQA2 is an equilibrium speciation model that can be used to calculate the equilibrium composition of dilute aqueous solutions in the laboratory or in natural aqueous systems (U.S. EPA, 1991). The model calculates the equilibrium mass distribution among dissolved species, adsorbed species, and multiple solid phases under a variety of conditions. A comprehensive database is included that is adequate for solving a broad range of problems without need for additional user-supplied equilibrium constants. The model employs a pre-defined set of components that includes free ions and neutral and charged complexes. The database of reactions is written in terms of these components as reactants. An ancillary program, PRODEFA2, serves as an interactive pre-processor to help produce the required MINTEQA2 input files. Code to achieve similar results, PHREEQE, is available from the U.S. Geological Survey (Parkhurst et al., 1980). These equilibrium calculations may be used to prepare solubility diagrams for metals in the presence of a variety of anions.

There are, however, substantive discrepancies that occur between predictions and observed wetland water chemistry. Solubility calculations for a mesocosm of homogenized sediment indicated supersaturation with respect to the sulfides of iron, copper, nickel, and zinc, yet measurements demonstrated a substantial supply of both trace metals and sulfide from the solid phase to the pore waters (Naylor et al., 2005). Ratios of metals measured in pore waters were consistent with their release from iron and manganese oxides, indicating that supply, as much as removal processes, determines the pseudo-steady state concentrations in the pore waters. The Naylor et al. (2005) observations suggest that trace metals are not immediately bound in an insoluble, inert form when they are in contact with sulfide. As a result of the complex wetland situation, forecasts from computer programs such as MINTEQA2 are not accurate representations of wetland situations. For example, Frandsen and Gammons (2000) found predictions of zinc remaining in solution to be underestimated by several factors of ten at low water concentrations, and to be strongly dependent upon the assumed form of the solid sulfide.

The reasons for such large discrepancies are presumably related to the complexities of the treatment wetland environment, compared to abiotic laboratory experiments with controlled water chemistry. The calculations are for thermodynamic equilibrium conditions, which may not be satisfied for the dynamic conditions of flow-through wetlands. Unfortunately, it would appear that equilibrium calculations are of little or no value in predicting treatment wetland performance for metal removal.

Storage in Plants and Sediments

Sediments are the primary location for metal storage, and plant storages are less important. Furthermore, most of the metal found in plants is located in the roots and rhizomes (Table 13.4). On a per unit area basis, the active sediment zone has mass that is many times greater than phytomass. As a consequence, harvest of aboveground plant parts is not an effective means of removing metals from the wetland.

There are two ways to consider the buildup of stored metals in a treatment wetland. The first presumes that metal storage occurs throughout the root zone of the FWS wetland, with transfers to sediments and roots driven by processes such as sorption and transpiration flows. The second presumes deposition by layering.

Additionally, wetland sediments contain a variety of organisms that ingest sediments over a wide range of depths, and redeposit their gut contents primarily at the sediment surface (Robbins, 1986). By such means, metals may reach the entire root zone layer, and continue to build up in concentration. As for other contaminants in wetlands, the major storage will be in the roots, soils and sediments, rather than in aboveground tissues. Thus, this approach considers the root zone to be a well-mixed region, with ever increasing solid concentrations. The metal load removed contributes to the increase, in simple form as:

$$\rho_b H \frac{dC_S}{dt} = J_{metal} \qquad (13.6)$$

This integrates to yield:

TABLE 13.4

Metal content in marsh plant tissues and sediments (μg/g). N is the number of literature reports.

	Above ground		Below ground		Sediments	
	N	Median	N	Median	N	Median
Selenium	22	6.6	12	3.9	24	5.3
Arsenic	14	2.0	14	3.9	7	9.7
Cadmium	14	0.8	14	4.0	14	6.4
Nickel	9	1.9	5	7.1	2	32.5
Copper	18	2.0	19	5.0	6	35.5
Chromium	15	12.0	15	50.0	6	21.5
Lead	15	19.0	14	30.0	5	150
Zinc	18	41.2	18	31.7	11	75
Iron	11	199	11	5,399	17	9,400

$$C_S = C_{S0} + \frac{J_{metal}t}{\rho_b H} \tag{13.7}$$

where C_{S0} = initial solid-phase metal concentration, mg/gdw
H = root zone thickness, m
r_b = solids bulk density, gdw/m^3
t = time, yr

This method was applied by Nolte and Associates (1998a), to calculate the time required to reach toxicity sediment standards. For example, in the Sacramento, California treatment marsh, copper was 27 mg/kg in 1996, and the applicable threshold limit was taken to be 35.7 mg/kg. The measured accretion of copper was 132 mg/m^2·yr. The top 25 cm of the wetland contained 50 kg/m^2 of dry solids (at ρ_b = 200,000 g/m^3), and hence the rate of concentration increase was 2.65 mg/m2·yr. Therefore, the increase from 27 to 35.7 mg/kg would take 3.3 years. Storage lifetimes for other metals ranged from zero to 120 years for lead.

This calculation presumes vertical uniformity in the root zone, which is not typically observed. Further, it does not include the effects of new sediment deposition, which is ubiquitous in treatment marshes. Accretion, precipitation of metals, and downward diffusion with sorption all tend to create layering in the upper soil-sediment horizon.

The "Layer Cake" Assumption

The second way to interpret metal storage is to presume that all of the current storage is contained in a newly formed top sediment layer, which does not mix with previous layers. This is a limiting concept, because there is very likely to be some sediment metal mixing as a result of bioturbation and vertical chromatographic flows.

Vertical stratification of wetland sediments/soils usually includes a top-most floc layer, which has been described as "a slurry of dark, decomposing, loosely-structured detrital material that pours out when the sampler is tipped" and termed the A layer (Nolte and Associates, 1997). This layer has been observed in a range of treatment wetland types, ranging from those receiving secondary effluent in Sacramento, California, to those receiving agricultural runoff in south Florida. The thicknesses are on the order of 5–30 cm; for instance, 11.3 ± 2.7 cm along the flow direction at Sacramento Cell 7 in 1996. Below this floc, there may be a second layer dominated by litter and sediment, termed the B layer by Nolte and Associates (1997). This B layer averaged 4.6 ± 1.0 cm for Sacramento Cell 7 in 1996. Below this, there existed a C layer of gleyed clay soil at Sacramento, in turn atop the base clays of the wetland basins. The metals content of these various layers were found to be markedly different, with 2–13 fold higher concentrations in the A layer than the C layer. It is probable that the layering pattern in concentrations is driven by processes of precipitation and sorption.

Metal storage in the top layer creates sustained sediment concentrations that reflect the dilution of new metal deposits by the accretion of new wetland solids. Such accretions result from microbial, algal and macrophyte detritus production, and lead to replacement of floc layers, which are depleted by consolidation. In a sense, this concept amounts to plug flow of metals and biomass downward into the top sediments, or "conveyor-belt" deposition. Under this assumption, a mass balance on the top layer of deposited material determines the concentration in the surficial solid layer:

$$C_S = \frac{J_{Metal}}{J_{Sed}} \tag{13.8}$$

where C_S = top layer solid concentration, mg/gdw
J_{Metal} = metal accumulation rate, mg/m^2·yr
J_{Sed} = sediment accumulation rate, gdw/m^2·yr

For example, consider J_{sed} = 200 gdw/m^2·yr of new sediments, originating from 1,000 dry g/m^2·yr of biomass production. Suppose the metal removal rate is J_{metal} = 10 mg/m^2·yr. When this metal stream is diluted by the new sediment stream, a new, top sediment concentration of 50 μg/g (0.05 mg/g) is forecast. This may then be compared to the appropriate sediment quality criterion for the metal in question. This example may be continued to estimate the corresponding water concentration, under a first order removal assumption:

$$C = \frac{J_{Metal}}{k} \tag{13.9}$$

For the example under consideration, if the value of k = 50 m/yr, then C = 0.2 μg/L. From the magnitudes of the numbers in this hypothetical example, it is easily seen (Table 12.11) that if the wetland is effective, and there is elevated metal in the incoming water, it will be difficult to meet stringent sediment quality standards.

Design Equations for Metal Removal

The literature does not currently contain a clear indication of what first-level design calculation should be used for treatment marsh forecasts for metal removal. The two simplest choices are a fixed load removal or a first-order areal calculation. The former is advocated in the mine water treatment literature (Hedin et al., 1994; Younger, 2000; Younger et al., 2002), and is termed the area-adjusted contaminant removal rate. This is essentially a zero-order model, which fixes the removal rate per unit area of the wetland:

$$R_A = \frac{Q}{A}(C_{in} - C_{out}) \tag{13.10}$$

where A = wetland area, m^2

C_{in} = inlet concentration, mg/L
C_{out} = outlet concentration, mg/L
Q = flow rate, m^3/d
R_A = area-adjusted contaminant removal rate, $g/m^2 \cdot d$

The subsequent sections of this chapter show that metal removal is not constant for most wetlands, but strongly correlated to the metal loading to the wetland.

The second choice, a first-order model, presumes that more detention time (lower hydraulic loading) will result in greater metal removal, and that higher inlet concentrations will result in more removal. It is consistent with the concept of mass transfer of the metal to the reactive sediment layer, which is driven by the concentration difference between water and sediment porewater metals. Younger states that "However, as the design of passive systems advances, it is likely that volume-based or retention-time based measures of performance will prove to be more appropriate in many circumstances." Tarutis et al. (1999) concluded that evidence showed that the first-order model was a better choice, and advocated the plug-flow form:

$$\frac{C_{out} - C^*}{C_{in} - C^*} = \exp\left(-\frac{kA}{Q}\right) \tag{13.11}$$

where k = areal rate constant, m/d
C^* = background concentration, μg/L

Goulet et al. (2001) examined the efficacy of the first-order model for iron, manganese, copper and zinc in a stormwater wetland in Kanata, Ontario over a two-year period. These authors found that it was acceptable for some metals (e.g., zinc) but not for others (e.g., iron and manganese). They observed that such a model could not account for releases and might also be affected by the low concentrations that existed in the incoming water. Crites et al. (1997) found exponentially decreasing profiles for zinc in water along the flow direction at the Sacramento wetlands (Figure 13.1), as indicated by Equation 13.11. There appears to be a background C^* = 6.9 μg/L. The value of R^2 = 0.99.

One of the premier uses of mesocosm experimentation is to control external factors, in an effort to elucidate processes. The efforts of Manyin et al. (1997) assist in the understanding of which model may be more realistic. Side-by-side mesocosms were fed iron solutions of varying concentrations and at varying flow rates. Data fit to Equation 13.11, across three

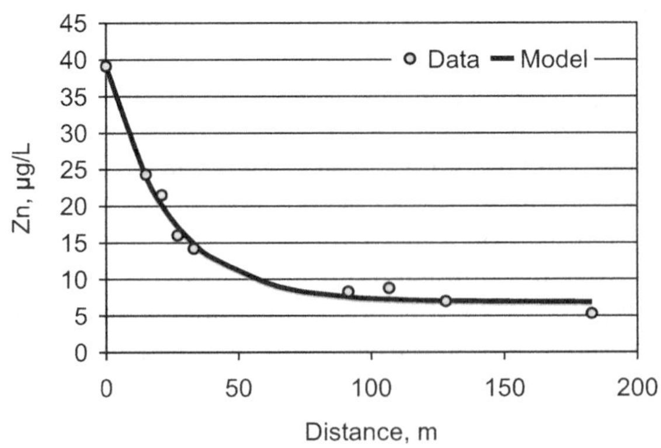

FIGURE 13.1 Profile of water-phase zinc along the flow direction in Sacramento, California Cell 7B in May 1995. The corresponding k = 45 m/yr. Data sources: Crites et al. (1995, 1997).

inlet concentrations and four hydraulic loadings produced a mean k = 53 m/yr, with R^2 = 0.83. Thus, when there are not artifacts of variable pH, temperature, flow or ancillary chemistry, the first-order model is quite effective in describing data. As a corollary, the use of an area-adjusted contaminant removal rate was entirely inappropriate for these controlled conditions.

Another approach to investigation of removal models relies upon the accumulation of metals in new wetland sediments. If for the sake of simplicity, the plug flow model is used as a basis for interpretation, it may be shown that exponential reductions in water-borne metals along the flow path result in the appearance of exponentially-distributed top sediment accretions of those metals. The buildup of those new top sediments may be considered the result of the wetland carbon cycle, which produces a stable annual accretion for a fully developed wetland. When combined with Equation 13.11, the concentration in the new sediments can be computed at each location. The rate of deposition at some fractional distance y along the wetland flow path is:

$$J_{metal} = kC(y) = kC_{in}\exp\left(-\frac{kAy}{Q}\right) \tag{13.12}$$

Then a mass balance on the top layer of deposited material determines the concentration in the surficial solid layer:

$$C_S = \frac{J_{metal}}{J_{Sed}} = \frac{kC_{in}}{J_{Sed}}\exp\left(-\frac{kAy}{Q}\right) \tag{13.13}$$

where C_S = top layer solid-phase metal concentration, mg/gdw
C_{in} = inlet concentration, μg/L = mg/m^3
J_{metal} = metal accumulation rate, $mg/m^2{\cdot}yr$
J_{Sed} = sediment accumulation rate, $gdw/m^2{\cdot}yr$
y = fractional distance along flow path, –

A benefit of this analysis is that the temporal variability of water concentrations is averaged by the sampling of many months or years of accretion. The Sacramento, California project sampled longitudinal profiles after 3.5 years of operation (Nolte and Associates, 1998b), and designated the new top sediments as the A layer. Interpretation of the Sacramento data was via the first-order areal model (Dombeck et al., 1998b). Results indicated that exponential declines explained a considerable amount of the variability in A layer solids concentrations (Figure 13.2). Rate constants fall in the range 22–71 m/yr. Thus it appears that the first-order model is applicable to field situations on a long-term average basis.

Thus, it appears that the PkC* model is appropriate for describing metals reduction in treatment marshes. However, there is a distinct lack of information concerning potential background concentrations that may prevail.

In the next sections, examples of overall marsh treatment are presented, with the intent of placing expectations in perspective. In each case, there is an extensive literature on marsh projects intended to reduce that metal concentration, which should be consulted as a prerequisite to any new application of marsh technology.

Because high metal concentrations are not the target for treatment marshes, the reader must consult other sources of information for wetland treatment of various mine waters and strong industrial waters. Nonetheless, the removal of trace amounts of heavy metals is the target of urban stormwater treatment (see Table 13.1), and the metals of concern in that application are discussed further here. Two metals, copper and mercury, are of interest

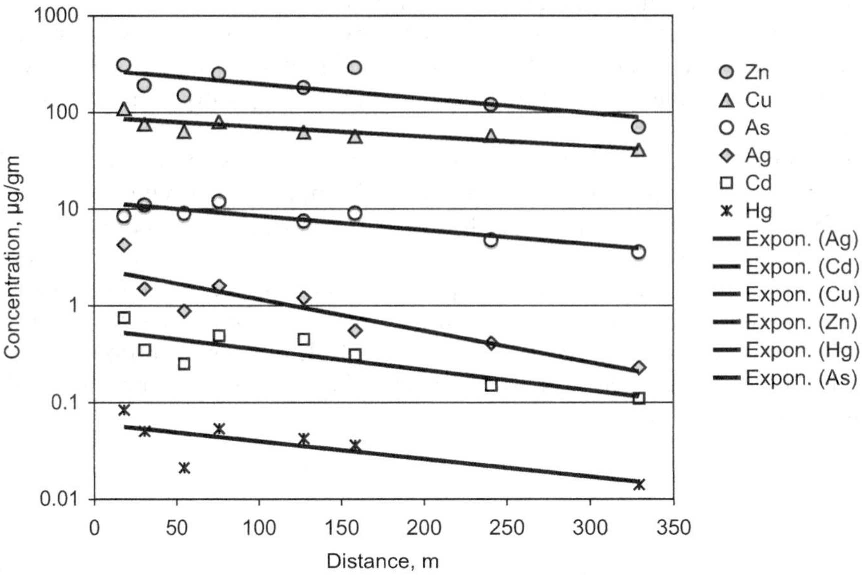

FIGURE 13.2 Profiles of metals in sediments along the flow direction for Sacramento, California Cell 7 in 1997. Data source: Nolte and Associates (1998a). The corresponding areal k-values are:

	R^2	k, m/yr
Ag	0.808	71
Cd	0.709	47
Cu	0.755	22
Zn	0.582	33
Hg	0.565	40
As	0.795	32

because treatment marshes are in use to reduce them. Similarly, two metalloids, arsenic and selenium, are also being treated at very low levels in marshes designed for that purpose.

13.3 IRON

Wetlands interact strongly with iron in a number of ways, and thus are capable of significant metal removal. Three major mechanisms are operative:

1. Binding to soils, sediments, particulates, and soluble organics;
2. Precipitation as insoluble salts, principally sulfides and oxyhydroxides;
3. Uptake by plants, including algae (however, after the plant senescence or algal die back most iron is leached out).

Iron is found at moderately high concentrations in wetland plants, algae and sediments (Table 13.4).

Information on the effects of wetlands on water iron concentrations has been reported at a low level of detail, with emphasis on removal percentages. Example performances are shown in Table 13.5.

The iron content of pretreated municipal wastewater is likely to be less than 1 mg/L in many regions, although higher in regions with high occurrence of iron minerals. Removals may or may not occur for very low iron content waters (Table 13.5).

It is likely that wetland removal performance in marsh systems is area-specific (Younger et al., 2002), rather than volume-specific. Three simple predictive approaches have been suggested:

- Constant Percent Removal. Specification of a supposedly constant percentage removal.
- Zero-Order Removal. Specification of a supposedly constant areal removal rate, in $gFe/m^2 \cdot d$. This corresponds to zero order removal kinetics, for which removal is independent of the iron concentration.
- First-Order Removal. The reduction rate is proportional to the iron concentration (Tarutis et al., 1999).

Some of the contributing processes are first-order, such as oxidation of ferrous iron and bacterial sulfate reduction (Younger et al., 2002), as is the process of particulate settling of precipitates. However, global removal does not necessarily follow such a model.

Data from 35 FWS natural wetlands receiving inputs of ferruginous acid mine drainage in western Pennsylvania served as a basis for evaluating the merits of these three approaches. Stark and Williams (1995) found that percentage removal was a better index of treatment than areal removal. Tarutis et al. (1999) further analyzed the 35-wetland dataset, and concluded that percentage removal should not be used, and that areal removal did not separate the effects of inlet concentration and flow rate. They recommended the first-order

TABLE 13.5

Example removals of iron in treatment marshes.

	Inlet mg/L	Outlet mg/L	Reduction %	Reference
Leachates & Industrial				
Champion, FL (6 systems)	0.33	0.53	-61	NADB Database (1998)
Kanata, Ontario	0.36	0.27	25	Goulet et al. (2001)
Santo Tome, Argentina	14.23	0.16	99	Di Luca et al. (2011)
Aiken, SC	0.03	0.34	-1,017	Knox et al. (2006)
Urban Runoff				
Hidden River, FL	0.28	0.34	-21	Carr and Rushton (1995)
Naples, FL	0.17	0.05	69	Bishop et al. (2012)
Municipal				
Estevan, SAS	0.30	0.20	33	Unpublished data
OEW, FL	0.02	0.07	-294	NADB Database (1998)
Brookhaven, NY	7.40	2.10	72	Hendrey et al. (1979)
Arcata, CA (12 systems)	0.60	0.55	8.3	NADB Database (1998)
Tres Rios, AZ	0.20	0.23	-17	Wass Gerke and Associates (2002)
Christchurch, NZ	0.34	0.27	21	Chague-Goff (2005)

model (Equation 13.11), and found a median rate constant of 0.18 m/d (66 m/yr) (N = 35) for calibrations of a plug flow version of the model.

Manyin et al. (1997) performed a factorial experiment varying flow rate and inlet concentrations to wetland mesocosms. They found that outlet iron concentrations increased with increasing inlet iron loadings. Analysis of their data produces a rate constant of 42 m/yr according to Equation 13.11.

In the low loading region, behavior probably reflects localized recycling of iron to and from precipitates and adsorbed forms, in response to localized variations in pH and mediated by diffusion to and from the water column. Background levels may also result from suspended particulate forms of iron. Uptake into aboveground macrophyte plant parts is low, but larger amounts of iron are found in roots, so that overall plant cycling can be of importance. Batty and Younger (2002) found that where dissolved iron concentrations in wetland waters were at or below one mg/L, direct uptake of iron by plants could account for 100% of iron removal. Batty and Younger (2002) also found 1 mg/L iron was also an optimum for healthy growth of *Phragmites australis*: at greater concentrations, the plants were not as productive, while at lower concentrations were healthy.

13.4 LEAD

Lead occurs in surface waters most often in the divalent, Pb(II) form, which forms insoluble precipitates with sulfides, carbonates, sulfates, and chlorides. Lead combines with organic ligands to form soluble and colloidal complexes. Lead solubility is likely to be less than about 1 μg/L at a pH above 8.5, with increasing solubility at lower pH values and lower alkalinities. Lead is largely unavailable in most natural conditions because of its formation of insoluble salts and partitioning to dissolved organics and suspended particulates. Lead forms alkyl compounds, such as tetra ethyl lead, a former anti-knock additive in gasoline. Methylation can occur microbially in the environment (Odum et al., 2000).

The average concentration of lead in water reported for the world's rivers is 0.2 μg/L. For example, the Sacramento River in California had 0.77 μg/L dissolved, and 1.16 μg/L total lead in a 1991–1992 study (Jones & Stokes Associates, 1993).

Lead is a physiological and neurological toxin to humans, and therefore human health standards have been set, typically at 50 μg/L. Lead poisoning of waterfowl from direct ingestion of mine tailings water has been reported at Coeur d'Alene, Idaho (DeVolder et al., 2003). U.S. EPA (2018a, 2018b) has determined that lead is a toxic priority pollutant, and sets water quality criteria for freshwater concentrations of lead according to hardness (Table 13.2). At hardness of 100 mg/L, the maximum concentration allowable is 65 μg/L, and the continuous concentration allowable is 2.5 μg/L, as dissolved lead. U.S. EPA (2018a, 2018b) provides formulae for the limits at other hardnesses, with soft waters having lower criteria.

Sediment standards for lead have been set by the State of Wisconsin, and adopted in some of the United States (Table13.2). These set a threshold effect level of 36 μg/gdw, and a probable effects level of 130 μg/gdw. For more details on the cycling of lead in aquatic ecosystems, refer to Nriagu (1978).

Lead is found in wetland plants, algae, and sediments. Concentrations in tissues are relatively low when the plants or algae grow in waters with only trace amounts of lead, as in treatment marshes. In uncontaminated waters, macrophyte tissue concentrations of 1–20 μg/gdw are common, and the same range exists for algae in clean waters (Vymazal, 1995).

Wetland sediments are typically the major repository of stored lead in wetlands. Any of several compounds may precipitate, and there is strong partitioning to wetland organic sediments. Therefore, the lead removed by the wetland builds up in the sediments, primarily in the top layers. For instance, O'Sullivan et al. (2004) found lead in the top 15 cm of sediments in a treatment wetland accumulated from an initial 31 μg/gdw to 3,300–3,900 μg/gdw after 4.58 years of exposure to variable water-phase lead. DeBusk et al. (1996) found that the top two centimeters of sediments in cattail mesocosms increased from about 10 μg/gdw to 295 μg/gdw upon exposure to 396 μgPb/L for 13 months.

Lead is removed in wetlands when the incoming amounts are greater than background (see Table 13.6). For 18 FWS systems, with inlet lead ranging from 1–320 μgPb/L, the median concentration reduction is 63%.

Lead fate in wetlands is that removed lead is found in sediments and to a lesser extent in belowground plant parts. Lead is removed from marsh surface waters by particulate settling, chemical precipitation, and sorption. Lead forms very insoluble compounds with sulfide, sulfate and carbonate, among others:

$$S^{-2} + Pb^{+2} \rightarrow PbS \tag{13.14}$$

$$SO_4^{-2} + Pb^{+2} \rightarrow PbSO_4 \tag{13.15}$$

$$CO_3^{-2} + Pb^{+2} \rightarrow PbCO_3 \tag{13.16}$$

DeBusk et al. (1996) hypothesized that the lead was bound as sulfides, based upon acid extractions with analyses for lead and sulfides. However, other lead precipitates were identified by DeVolder et al. (2003), namely carbonates (up to 66%) and sulfates (up to 20%). Carbonates are ubiquitous in wetland environments.

TABLE 13.6

Example removals of lead in treatment marshes.

	Inlet μg/L	Outlet μg/L	Reduction %	Reference
Leachates & Industrial				
Albright, PA	2	1	50	Hoover et al. (1998)
Tom's Gully, NT, AUS	9	1	94	Noller et al. (1994)
Woodcutters, NT, AUS	12	0	98	Noller et al. (1994)
Isanti-Chisago, MN	1	1	0	Loer et al. (1999)
New Hanover Co., NC	3	3	0	Unpublished data
Monroe Co., NY	13	5	62	Eckhardt et al. (1999)
Aiken, SC	1.2	0.4	67	Knox et al. (2006)
Urban Runoff				
Hidden River, FL	2	1	40	Carr and Rushton (1995)
Greenwood, FL	63	5	92	McCann and Olson (1994)
Crestwood, VA	12	3	74	Godrej et al. (1999)
Växjö, Sweden	318.2	30.6	90	Al-Rubaei et al. (2016)
San Joaquin, CA	19.84	14.52	27	Hafeznezami et al. (2012)
Municipal				
Estevan, SAS	33	13	61	Unpublished data
Sacramento, CA	35.82	6.74	81	Nolte and Associates (1998a, 1998b)
Christchurch, NZ	17.7	4.5	75	Chague-Goff (2005)
Bloomington, IL	64	23	64	Jin et al. (2003)
Bloomington, IL	64	52	18	Jin et al. (2003)
Lakeland, FL	8.2	6.7	18	NADB Database (1998)

Sulfide is formed by sulfate reducing bacteria in the anaerobic zones of treatment wetlands, while incoming sulfate persists in aerobic zones. If a source of sulfate is present in the incoming water, the wetland can be configured to provide a sustainable supply of sulfide (Sobolewski, 1999).

Lead may partition to organic constituents in the wetland sediments, or to inorganic substrates, such as oxy-hydroxides of iron and manganese (Koelmans and Radovanovic, 1998). In most non-minewater situations, a simple linear partition coefficient suffices. Koelmans and Radovanovic (1998) found a value of K = 45.6 L/g for peat, but was much higher for mineral sediments dominated by manganese, K = 894 L/g.

The strong propensity for partitioning to sediments means that if ionic lead is present in the water, there will be associated lead in large amounts in the sediments at equilibrium. However, in the presence of dissolved organic matter (DOM) typical of wetlands, the large majority of dissolved lead occurs as complexes with organic ligands and not as ionic lead. For example, Koelmans and Radovanovic (1998) suggest over 97% is soluble bound to DOM in lakes in the Netherlands.

13.5 NICKEL

The oxidation state for nickel in surface waters is divalent Ni(II). Under aerobic conditions, divalent nickel may be complexed with hydroxides, sulfates, or carbonates. Under anaerobic conditions, nickel forms sulfide precipitates. Much of the nickel in surface waters is associated with soluble organic ligands (Xue et al., 2001). Depending upon the nature of the complexing ligand, the soluble complex may not be bioavailable. Hutchinson (1975) indicated that the average nickel concentration in surface waters is 5 μg/L. Xue et al. (2001) reported total nickel concentration in lakes and rivers in Switzerland as 0.2–1.8 μg/L.

Nickel is sometimes an essential micronutrient for the growth, but it also is toxic at modest concentrations. Nickel is not biomagnified in the food chain. U.S. EPA (2018a, 2018b) sets criteria for freshwater concentrations of nickel according to hardness. At a hardness concentration of 100 mg/L as $CaCO_3$, the maximum allowable nickel concentration is 470 μg/L, and the continuous concentration allowable is 52 μg/L, as dissolved nickel (Table 13.2). There are formulae for the limits at other hardnesses, with soft waters having lower criteria. The maximum level for human ingestion is 0.61 mg/L. Sediment standards for nickel have been set by the Province of Ontario, and adopted in some of the United States. These set a lowest effect level of 16 μg/gdw, and a severe effects level of 75 μg/gdw (Table 13.3). Interestingly, Persaud et al. (1993) has also identified the level for pre-settlement deposits as 31 μg/gdw in Great Lakes sediments.

Wetland plants have trace amounts of nickel in their tissues, which are the lowest of those for the heavy metals (Table 13.4). At nonmine water sites in Minnesota the average nickel concentration in cattails ranged from 4.9 mg/kg in the leaves to 18.0 mg/kg in the roots. Wetland sediments have the potential to contain large amounts of nickel, in response to chemical precipitate formation and cation exchange with nickel in the water. In most situations, that capacity is not realized, because of the low concentrations in the wetland waters. For instance, background waters at White Cedar Bog in Minnesota were at 3–8 μg/L, and the associated peats had 26–39 μg/gdw (Eger et al., 1980).

Nickel forms very insoluble sulfides (Sobolewski, 1999):

$$S^{-2} + Ni^{+2} \rightarrow NiS \tag{13.17}$$

Sulfide and bisulfide are formed by sulfate reducing bacteria in the anaerobic zones of treatment wetlands.

Organic sediments contain polar functional groups such as acids and phenolics that are responsible for its cation exchange capacity, and can be involved in chemical binding (Ho and McKay, 2000). As a consequence, organic solids have the ability to store very large quantities of nickel. There are a finite number of exchange sites, and hence the uptake of nickel has a maximum.

Nickel is removed in wetlands when the incoming amounts are greater than background (Table 13.7). For 16 systems, with inlet nickel ranging from 2–65 µg/L, the median concentration reduction is only 8%. It is not clear that treatment marshes are effective at nickel removal at low concentrations.

13.6 ZINC

Zinc is an essential element for both plants and animals. It activates some enzymatic reactions important in respiration and serves as a cofactor in plant photosynthesis and DNA synthesis. It is present in stormwaters due to runoff from galvanized roofs (Table 13.1), but it is typically present only at very low concentrations in municipal wastewaters. Usually, zinc is present in surface waters in particulate forms, with very little ionic Zn(II). The ratio of free ionic to total zinc is often less than 1%.

U.S. EPA (2018a) sets criteria for freshwater concentrations of zinc according to hardness (Table 13.2). At hardness of 100 mg/L, the maximum concentration allowable, and the continuous concentration allowable, are both 120 µg/L, as dissolved zinc. U.S. EPA (2018a) provides formulae for the limits at other hardness values, with soft waters having lower criteria. The maximum level for human ingestion is 7.4 mg/L.

TABLE 13.7

Example removals of nickel in treatment marshes.

	Inlet µg/L	Outlet µg/L	Reduction %	Reference
Leachates & Industrial				
Musselwhite, ONT	7.0	1.3	82	Bishay and Kadlec (2005)
Santo Tome, Argentina	30.7	12.3	60	Di Luca et al. (2011)
New Hanover Co., NC	19.0	4.0	79	Unpublished data
Isanti-Chisago, MN	28.0	28.0	0	Loer et al. (1999)
Monroe Co., NY	65.0	60.0	8	Eckhardt et al. (1999)
Urban Runoff				
Orlando, FL	4.0	3.0	25	Schiffer (1989)
Municipal				
Hayward, CA	7.4	16.0	-116	NADB Database (1998)
Lakeland, FL	14.8	12.0	19	NADB Database (1998)
Bloomington, IL	11.0	4.0	64	Jin et al. (2003)
Bloomington, IL	11.0	10.0	9	Jin et al. (2003)
Tres Rios, AZ	4.6	4.8	-4	Wass Gerke and Associates (2002)
OEW, FL	11.9	10.2	14	Simon (2003)
Sacramento, CA	2.7	2.8	-5	Dombeck et al. (1998b)
Sacramento, CA	7.3	6.8	6	Dombeck et al. (1998b)
Sacramento, CA	6.9	8.4	-23	Dombeck et al. (1998b)
Sacramento, CA	5.6	6.0	-7	Dombeck et al. (1998b)

Sediment standards for zinc have been set by the Province of Ontario and adopted in some of the United States (Table 13.3). These set a lowest effect level of 120 μg/gdw, and a severe effects level of 820 μg/gdw (Persaud et al., 1993). The level for presettlement deposits was identified as 65 μg/gdw in Great Lakes sediments. Meeting rigorous sediment standards on a sustainable basis is not easily achieved for any but trace concentrations, such as those occurring in stormwater and municipal wastewater.

Wetland plants have trace amounts of zinc in their tissues, typically with higher values for roots and rhizomes than for aboveground plant parts (Table 13.4). Plant shoots contain 10–40 μg/g dry weight for nonminewater situations. Wetland sediments have the potential to contain large amounts of zinc, in response to chemical precipitate formation and cation exchange with zinc in the water. In most situations, that capacity is not realized, because of the low concentrations in the wetland waters. For instance, secondary incoming waters at Sacramento, California were at 27–49 μg/L, and the wetland sediments had 12–38 μg/gdw (Nolte and Associates, 1998b).

Zinc is removed from marsh surface waters by particulate settling and trapping, chemical precipitation and co-precipitation, partitioning to sediments, and to the least extent by plant uptake. Zinc forms very insoluble compounds with sulfide and carbonate, among others:

$$S^{-2} + Zn^{+2} \rightarrow ZnS \tag{13.18}$$

$$CO_3^{-2} + Zn^{+2} \rightarrow ZnCO_3 \tag{13.19}$$

Sulfide and bisulfide are formed by sulfate reducing bacteria in the anaerobic zones of treatment wetlands. Carbonates may be available from atmospheric aeration, and from carbonate minerals in the wetland substrate.

Partition coefficients for zinc to aquatic sediments vary considerably, and depend upon sediment pH and Eh (Rinklebe et al., 2016). Wepener et al. (2000) found a range of KD from 134 to 542 L/g at river sites in South Africa. Koelmans and Radovanovic (1998) determined that most zinc partitioned to iron and manganese oxyhydroxides in lakes in the Netherlands, and measured K_D of about 54 L/g. Ringqvist and Öborn (2002) determined K_D = 10.8 L/g for two peats.

Zinc is removed in wetlands when the incoming amounts are greater than background (Table 13.8). For 15 marsh systems, with inlet zinc ranging from 18–320 μgZn/L, the median concentration reduction is 60%.

Transect data suggest that the PkC* model is operative for zinc (Figure 13.1). The exponential profile in the water phase is reflected in an exponential profile in the sediments (Figure 13.2), from which an average k = 33 m/yr for the Sacramento data. Similar declining sediment zinc profiles have been reported at the Aiken, SC treatment marsh (Knox et al., 2010). Preliminary indications are that the background concentration is of the order of 7 μg/L (Figure 13.1). The first-order areal removal rate constant for zinc is 28 m/yr for the data of Sinicrope et al. (1992). This is comparable to the results of Delgado et al. (1993) for hyacinths, which give k = 20 m/yr. The data of Gillespie et al. (2000) produce 55 and 121 m/yr for two different loadings.

13.7 CADMIUM

Cadmium is currently used primarily in the production of Ni-Cd batteries (75%), whereas it was formerly used in electroplating (75% prior to 1940). However, it is estimated that less than 2% of cadmium entering the environment is from such sources, because most comes from waste incineration and the metallurgical industry (Laws, 1993; Plachy, 2000). It is

TABLE 13.8
Example removals of zinc in treatment marshes.

	Inlet μg/L	Outlet μg/L	Reduction %	Reference
Leachates & Industrial				
Albright, PA	83	37	55	Hoover et al. (1998)
Springdale, PA	89	24	73	Hoover et al. (1998)
Santo Tome, Argentina	41.7	16.7	60	Di Luca et al. (2011)
Aiken, SC	18.4	8.8	52	Knox et al. (2006)
Urban Runoff				
Hidden River, FL	64	19	70	Carr and Rushton (1995)
Greenwood, FL	147	72	51	McCann and Olson (1994)
Crestwood, VA	131	96	27	Godrej et al. (1999)
Naples, FL	10.8	4.8	56	Bishop et al. (2012)
Växjö, Sweden	318	31	90	Al-Rubaei et al. (2016)
San Joaquin, CA	19.84	14.52	27	Hafeznezami et al. (2012)
Municipal				
Estevan, SAS	33	13	61	Unpublished data
Sacramento, CA	35.8	6.7	81	Nolte and Associates (1998a, 1998b)
Christchurch, NZ	17.7	4.5	75	Chague-Goff (2005)
Bloomington, IL	64	23	64	Jin et al. (2003)
Bloomington, IL	64	52	18	Jin et al. (2003)

present in urban stormwaters at very low concentrations (Table 13.1). There are not treatment wetlands that have been designed for the primary purpose of removing cadmium from contaminated waters. However, cadmium is a constituent of leachates from coal ash piles, and hence has been the co-target of treatment wetlands treating those leachates.

Cadmium is a naturally occurring heavy metal with no known nutritional requirement for biota. In surface waters, cadmium typically occurs as Cd(II) and is most soluble at low pH in waters with low hardness. In solution, cadmium is present as the ion and in a number of complexed, soluble compounds with carbonate, sulfate, chloride, hydroxides, and humates. These cadmium complexes are highly adsorbable on organic particulates.

U.S. EPA (2018a) sets criteria for freshwater concentrations of cadmium according to hardness (Table 13.2). At hardness of 100 mg/L, the maximum concentration allowable is 1.8 μg/L, and the continuous concentration allowable is 0.72 μg/L, as dissolved zinc. Soft waters have lower criteria. Sediment standards for cadmium have been set by the province of Ontario, and adopted in some of the United States (Table 13.3). These set a lowest effect level of 0.6 μg/gdw, and a severe effects level of 10 μg/gdw (Persaud et al., 1993). They identified the level for pre-settlement deposits as 1.1 μg/gdw in Great Lakes sediments. The State of Wisconsin has a guideline for a probable effects sediment concentration of 5 μg/gdw.

Wetland plants have trace amounts of cadmium in their tissues, typically with higher values for roots and rhizomes than for aboveground plant parts (Table 13.4). Plant shoots contain less than 4 μg/gdw for nonminewater situations

Cadmium was transferred primarily to soils (42% of removal) and fine roots (35%) in the Sinicrope et al. (1992) study. DeBusk et al. (1996) found that the belowground storages were 7.0% in plants (nearly all in roots), 93.0% in sediments for Typha domingensis; and 9% in plants, 91% in sediments for Lemna minor, in a Florida mesocosm study. Meeting rigorous sediment standards on a sustainable basis is not easily achieved for any but trace concentrations, such as those occurring in stormwater and municipal wastewater (Table 13.1).

Cadmium is removed by particulate settling and trapping, chemical precipitation and co-precipitation, partitioning to sediments, and to the least extent by plant uptake. Cadmium forms very insoluble compounds with sulfide and carbonate, among others:

$$S^{-2} + Cd^{+2} \rightarrow CdS \tag{13.20}$$

$$CO_3^{-2} + Cd^{+2} \rightarrow CdCO_3 \tag{13.21}$$

Precipitation of the carbonate can be microbially mediated, for example, by *Alcaligenes denitrificans* (Remacle et al., 1992). The removal of cadmium can also take place through co-precipitation with iron, manganese and aluminum oxyhydroxides. Organic substrates have been found to have considerable cadmium sorption capacity, as do soluble complexing ligands.

Cadmium is removed in marshes at low levels of inlet concentrations (Table 13.9). For 8 marsh systems, with inlet cadmium ranging from 0.01–3 μgCd/L, the median concentration reduction is 61%.

Duckweed (Lemna spp.) can accumulate 7–14 kgCd/ha·yr (2–4 mgCd/m^2·d) (DeBusk et al., 1996). Accordingly, harvest could remove that amount, and such a procedure could be used if determined to be economical. The Sinicrope et al. (1992) data for SSF wetlands corresponds to removal rate constants in the range $25 < k < 35$ m/yr. The DeBusk et al. (1996) data suggest $k = 37$ m/yr. Maine et al. (2001) measured cadmium uptake by four floating plants: *Pistia stratiotes, Hydromistia stolonifera, Eichhornia crassipes*, and *Salvinia herzogii*. There was little difference among species, and the approximate first-order volumetric rate constant for removal in batch mesocosms was 0.5 d^{-1}. Removal was primarily (90%) to the plant roots.

TABLE 13.9

Example removals of cadmium in treatment marshes.

	Inlet μg/L	Outlet μg/L	Reduction %	Reference
Leachates & Industrial				
Palmerton, PA	1.06	0.02	98	Haffner (1993)
Biala River, Poland	0.01	0.00	60	Wojcik and Wojcik (2000)
Urban Runoff				
Hidden River, FL	2.00	0.00	100	Carr and Rushton (1995)
Greenwood, FL	0.40	0.20	50	McCann and Olson (1994)
Crestwood, VA	3.04	2.17	29	Godrej et al. (1999)
Växjö, Sweden	0.22	0.02	91	Al-Rubaei et al. (2016)
San Joaquin, CA	0.39	0.28	28	Hafeznezami et al. (2012)
Municipal				
Sacramento, CA	0.08	0.03	63	Nolte and Associates (1998a, 1998b)

13.8 CHROMIUM

Chromium's principal uses and entry into the environment include electroplating, production of paints and pigments, tanning, wood preservation, chromium chemicals production, metal smelting, and pulp and paper production. In nature, trivalent chromium sorbs to various ligands and forms insoluble entities becoming unavailable for uptake by biota. Hexavalent chromium forms many soluble salts that can enter body membranes and induce a toxic response.

Chromium typically occurs in the trivalent [Cr(III)] or hexavalent [Cr(VI)] forms in surface waters. Major Cr(III) dissolved species are $Cr(OH)_2^+$ and $Cr(OH)_3$, as well as several organic and a few inorganic complexes, whereas $HCrO_4^-$ and CrO_4^{-2} are the main forms of Cr(VI) in natural waters (Richard and Bourg, 1991). The Cr(VI) form is the most toxic and is generally associated with the presence of industrial wastewaters. Cr(VI) is relatively unstable under most environmental conditions, and converts to the less toxic trivalent form in surface waters, especially when organic matter is present. Trivalent chromium hydroxides and chlorides are relatively insoluble and their formation significantly reduces chromium availability to biota.

U.S. EPA (2018a) criteria for freshwater concentrations of Cr(III) are hardness dependent (Table 13.2). At hardness of 100 mg/L the maximum concentration allowable is 570 μg/L, and the continuous concentration allowable is 74 μg/L, as dissolved chromium. The criteria for Cr(VI) are not hardness dependent. The maximum concentration allowable for hexavalent chromium is 16 μg/L, and the continuous concentration allowable is 11 μg/L, as dissolved chromium.

Sediment standards for chromium have been set by the Province of Ontario, and adopted in some of the United States (Table 13.3). These set a lowest effect level of 26 μg/gdw, and a severe effects level of 110 μg/gdw (Persaud et al., 1993). They also identified the level for pre-settlement deposits as 31 μg/gdw in Great Lakes sediments. The state of Wisconsin has a guideline for a probable effects sediment concentration of 110 μg/gdw (WDNR, 2003).

Wetland plants have trace amounts of chromium in their tissues, typically with higher values for roots and rhizomes than for aboveground plant parts (Table 13.4). Plant shoots contain less than 12 μg/gdw for nonindustrial situations, but roots and sediments show higher concentrations. Chromium is transferred primarily to soils and sediments. Concentrations may be somewhat higher than in plant tissues, but the primary cause is the very high amount of sediment in a FWS wetland, compared to plant biomass. For instance, the sediment concentration at Sacramento was about 36 μg/g, compared to 1–4 μg/gdw for plant tissues. But the mass balance indicated 97% of the stored chromium was in the sediment (Nolte and Associates, 1998b). Meeting rigorous sediment standards on a sustainable basis is not easily achieved for any but trace concentrations, such as those occurring in stormwater and municipal wastewater (Table 13.1).

Chromium is removed by chemical co-precipitation, partitioning to sediments, and to the least extent by plant uptake. In general, wetlands are effective at removing Cr(VI) and converting it to Cr(III). Xu and Jaffe (2006) found that Cr(III) was the dominant form of chromium in both sediments and plants, and precipitation of Cr(III) in the sediment was the major pathway responsible for the disappearance of aqueous Cr(VI) from the pore water. Lytle et al. (1998) showed that *Eichhornia crassipes* (water hyacinth), supplied with Cr(VI) in nutrient culture, accumulated nontoxic Cr(III) in root and shoot tissues.

Treatment wetlands have been implemented for the purpose of removing chromium from contaminated waters from leather tanning (Sultana et al., 2014), electroplating (Ranieri, 2004), and metallurgical processing (Maine et al., 2006). However, chromium is also a trace contaminant in landfill leachates and urban runoff, and hence has been the co-target of treatment wetlands built for treatment of those waters.

TABLE 13.10

Example removals of chromium in treatment marshes.

	Inlet µg/L	Outlet µg/L	Reduction %	Reference
Leachates and Industrial				
Santo Tome, Argentina	172.3	18.2	89	Di Luca et al. (2011)
Clemson, SC	48.3	14.3	70	Dorman et al. (2009)
Urban Runoff				
Växjö, Sweden	0.22	0.02	91	Al-Rubaei et al. (2016)
Municipal				
Tres Rios H1	5.00	7.00	-40	Wass, Gerke and Associates (2002)
Tres Rios C2	1.60	0.50	69	Wass, Gerke and Associates (2002)
Brookhaven, NY	160	20.00	88	Hendrey et al. (1979)
San Diego, CA	100	16	84	Sinicrope et al. (1992)
San Diego, CA	100	32	68	Sinicrope et al. (1992)

Chromium is removed in treatment marshes at low levels of inlet concentrations (Table 13.10). For eight marsh systems, with inlet lead ranging from 0.2–172 µgCr/L, the median concentration reduction is 70%.

Experiments with dosed mesocosms yielded an areal k-value of 25 to 35 m/yr in a FWS/SSF system (Sinicrope et al., 1992). FWS microcosm data indicate values of $20 < k < 80$ m/yr (Kadlec and Srinivasan, 1995) for Cr(VI).

A wetland was constructed at a metallurgical plant in Santo Tomé, Argentina, for the purpose of treating a combination of metallurgical and domestic wastewaters (Maine et al., 2004a; Di Luca et al., 2011). The 2000 m^2 FWS system received about 100 m^3/d. The wetland contained primarily *Eichhornia crassipes, Pontederia cordata* and *Typha domingensis*, and operated at depths of 0.5–0.8 meters and nominal detention of 7–12 days. Chromium was reduced by 86% from inlet to outlet, from 22 to 3.3 µg/L over an initial 16-month study period, but was less in later years (Table 13.10). The authors speculated that the observed shift in vegetation from *Eichhornia crassipes* to *Typha domingensis* was beneficial, which would lead to increased sediment storage.

13.9 COPPER

Copper is an essential micronutrient for plants and animals because it is used for protein synthesis and in blood pigments. Plants and animals require minimal amounts of this element, and deficiencies in nature are rare (Wetzel, 1983). The average copper concentration in the world's lakes and rivers is about 10 µg/L. Typically, copper is present in surface waters as chelated compounds of Cu(II). The ratio of free ionic to total dissolved copper is often less than 1% and decreases with increasing organic loads and pH above neutrality (Morel and Hering, 1993). When chelated with organic compounds, copper may remain relatively soluble. Depending upon the nature of the complexing ligand, the soluble complex may not be bioavailable.

Copper is a biocide that is commonly used to control algae and other organisms, such as schistosomes (swimmer's itch), which require snails and waterfowl as intermediate hosts. Copper has relatively low toxicity to benthic macroinvertebrates and to fish, which may be unaffected to concentrations as high as 500 μg/L. The sensitivity of algae and the tolerance of fish and benthos to copper has resulted in the widespread use of copper sulfate as an algaecide and molluscicide in lakes.

Copper has special, detailed criteria for the protection of freshwater aquatic life (U.S. EPA, 2007). Aquatic life should be protected if average concentrations do not exceed the acute and chronic criteria concentrations calculated by the Biotic Ligand Model (BLM), a metal bioavailability model. The numerical value of the CMC is determined by hardness and the amount of dissolved organic carbon in the water (Figure 13.3). The CCC is taken to be 3.22 times lower than the CMC.

Sediment standards for copper have been set by the province of Ontario, and have been adopted in some of the United States. These set a lowest effect level of 16 μg/gdw, and a severe effects level of 110 μg/gdw (Persaud et al., 1993). They also identified the level for presettlement deposits as 25 μg/gdw in Great Lakes sediments.

The storage of copper in wetlands entails a risk of creating potentially impaired sediments. It is possible to estimate the sediment concentrations created by sustainable removals of copper, in terms of the dilution of the copper storage by the accretion of new wetland solids. Meeting rigorous sediment standards on a sustainable basis is not easily achieved for any but trace concentrations, but has been accomplished for systems like Savannah River, South Carolina.

Copper is found in wetland plants, algae and sediments. Copper concentrations in plants are quite low, approximately 1–20 μg/gdw (Table 13.4). Aboveground plant parts are of lowest concentration, and belowground organs are somewhat higher. Wetland sediments have the highest levels and are therefore the major repository of stored copper.

Copper Removal Processes

Marshes have the potential to trap and remove copper contained in source waters. Long-term removal is expected to occur by accumulation and burial in the plant detritus in a manner similar to the removal of phosphorus.

Copper forms very insoluble compounds with sulfur, including both cupric (Cu2+) and cuprous (Cu^+) sulfides (Sobolewski, 1999):

$$S^{-2} + Cu^{+2} \rightarrow CuS \tag{13.22}$$

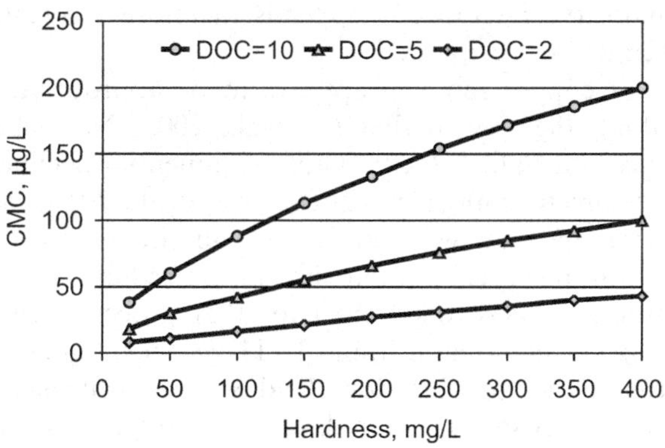

FIGURE 13.3 Maximum concentration criteria for copper as a function of hardness and dissolved organic carbon (Source: U.S.EPA, 2007).

$$S^{-2} + 2Cu^{+} \rightarrow Cu_2S \tag{13.23}$$

Sulfide and bisulfide are formed by sulfate reducing bacteria in the anaerobic zones of treatment marshes.

Organic substrates contain polar functional groups such as acids and phenolics that are responsible for its cation exchange capacity, and can be involved in chemical binding (Ho and McKay, 2000). Thus, the peat-copper reaction may be represented in two ways:

$$2P^{-} + Cu^{+2} \leftrightarrow CuP_2 \tag{13.24}$$

$$2HP + Cu^{+2} \leftrightarrow CuP_2 + 2H^{+} \tag{13.25}$$

where P represents the organic material (peat).

If Equation 12.40, cation exchange, is presumed to represent system performance, then the amount of copper bound is represented by (Kadlec and Keoleian, 1986):

$$K_{eq} = \frac{[CuP_2][H^{+}]^2}{[HP]^2[Cu^{+2}]} \tag{13.26}$$

where the bracket notation denotes molar concentration. This suggests that the partition coefficient ($[CuP_2]/[Cu^{+2}] = C_S/C_L$) should go down markedly with decreasing pH, which is in fact the observation of many investigators, as shown in Kadlec and Keoleian (1986). Other models of copper binding are also used, such as Langmuir or Freundlich isotherms. Whichever models are used, peats have the ability to store very large quantities of copper. By extrapolation, organic substrates in general have similarly large capacities. For instance, peats can store 5–10 mg/gdw for water concentrations of 1–10 mgCu/L and circumneutral pH

Performance of Wetlands for Copper Removal

Copper is effectively removed in treatment marshes (Table 13.11). For 21 systems, with inlet copper ranging from 1–57 μg/L, the median concentration reduction is 57%. Plants are a minor repository for removed copper. The Murray-Gulde et al. (2005) Savannah River, South Carolina, study determined that *Scirpus californicus* took up 6.76% of copper entering the system. The majority of removed copper is therefore associated with wetland sediments, in sorbed or chemically precipitated forms.

A first-order model of copper removal appears to be appropriate, because of observed exponential declines along the flow path (Jin et al., 2003; Nelson et al., 2006; Murray-Gulde et al., 2008; Knox and Paller, 2018). Those exponential declines are also reflected in similar declines in the sediment copper content (Knox et al., 2010) (see also Figure 12.11). It is probable that for overland flow wetlands, the rate-limiting step is the transfer of copper from the water to sorption and precipitation sites in the sediments. Traditional mass transfer modeling would utilize a first-order model for that process (refer to Chapter 7). It is also likely that percentage reductions in Table 13.11 would be greater if the hydraulic efficiencies were higher, e.g., closer to plug flow. Kadlec and Srinivasan (1995) report copper uptake in FWS cattail microcosms of $25 < k_{PF} < 120$ m/yr. The Savannah River, South Carolina, results give a first order areal plug flow removal rate constant $k_{PF} \approx 75$ m/yr (Murray-Gulde et al., 2008).

TABLE 13.11

Example removals of copper in treatment marshes.

	Inlet μg/L	Outlet μg/L	Reduction %	Reference
Leachates and Industrial				
Albright, PA	1.0	2.0	-100	Hoover et al. (1998)
Woodcutters, NT, AUS	1.4	0.9	36	Noller et al. (1994)
Springdale, PA	12.0	13.0	-8	Hoover et al. (1998)
Tom's Gully, NT, AUS	13.0	0.5	96	Noller et al. (1994)
Musselwhite, ONT	14.0	3.0	79	Bishay and Kadlec (2005)
New Hanover Co., NC	5.0	3.0	40	Unpublished data
Isanti-Chisago, MN	10.0	10.0	0	Loer et al. (1999)
Monroe Co., NY	30.0	10.0	67	Eckhardt et al., 1999
Aiken, SC	25.8	4.4	83	Knox et al. (2006)
Urban Runoff				
Hidden River, FL	4.0	3.0	25	Carr and Rushton (1995)
Greenwood, FL	14.0	6.0	57	McCann and Olson (1994)
Norco, LA	22.4	15.0	33	Hawkins et al. (1997)
Crestwood, VA	32.9	11.7	64	Godrej et al. (1999)
Naples, FL	2.2	1.7	21	Bishop et al. (2012)
Växjö, Sweden	41.3	3.8	91	Al-Rubaei et al. (2016)
San Joaquin, CA	7.8	5.6	28	Hafeznezami et al. (2012)
Municipal				
Estevan, SAS	5.0	3.0	40	Unpublished data
Sacramento, CA	8.6	3.6	59	Nolte and Associates (1998a, 1998b)
Christchurch, NZ	11.5	2.2	81	Chague-Goff (2005)
Bloomington, IL	56.6	7.9	86	Jin et al. (2003)
Bloomington, IL	56.6	11.0	81	Jin et al. (2003)

An Example Treatment Marsh for Copper Removal

Stormwater and process water discharges from the Technical Center of the U.S. Department of Energy contain unacceptably high levels of copper, typically in the range of 30–50 μg/L. Conventional treatment systems for metal removal (e.g., ion exchange, chemical precipitation, etc.) were estimated to be very expensive for the volume of water that needed to be treated (3,700 m^3/d average, 10,500 m^3/d peak; and the low concentrations that were to be achieved in the water before it was released to the stream (22 μg/L) (Nelson et al., 2002). Mesocosm testing indicated that peat-based wetlands could easily achieve the desired goals (Gladden et al., 2001). Construction of a 3.24-ha, four-cell FWS facility was carried out during 2000 (Nelson et al., 2002). Removals were acceptable (82%, outlet copper 3–8 μg/L) over the first four years of operation (Nelson et al., 2004), and have continued (Nelson et al., 2006; Murray-Gulde et al., 2008; Knox and Paller, 2018). Murray-Gulde et al. (2005) determined that shoots and roots of *Scirpus californicus* took up 0.88% and 5.88%, respectively, of copper entering the Savannah River, South Carolina system. The balance of the removal (82% (total removal) – 7% (plants) = 75% of the input) can be attributed to storage in the wetland sediments. Knox et al. (2004) determined that the majority (80%) of the stored copper in sediments was not readily remobilized.

13.10 MERCURY

Mercury differs from other trace metals in relation to treatment marshes because:

- It is rarely a target of wetland design.
- It has severe toxicological effects.
- Atmospheric deposition and storage dominate the wetland mass balance in almost all cases.
- Volatilization is a not insignificant loss pathway.

Mercury is released from a variety of anthropogenic sources such as solid and medical waste incineration, the combustion of fossil fuels for electric power production, and historically from gold mining processes. For instance, mercury amalgamation was used in the milling of gold and silver ore from the Comstock Mining District in and around Virginia City in west-central Nevada. Estimated losses were of the order of 7,500 tons of elemental mercury were lost during these operations. Mercury is now widely distributed in the lower Carson River basin (Tuttle et al., 2001). Atmospheric mercury is ubiquitous, and deposition is a large factor in wetlands. The methylated form biomagnifies, and can cause serious effects in top-level consumers, including humans. Wetlands provide an environment that may sometimes trap mercury, and in many cases convert a small fraction to methymercury. Treatment wetlands are not in use for the reduction of mercury in any source waters, but environmental effects are so large that mercury in wetlands has gathered considerable attention and study.

Forms and Storages of Mercury

Mercury may exist in three main forms in aquatic and wetland systems: elemental mercury, divalent ions, and methylated. Mercury entering marshes is subject to transformation through a variety of complex biogeochemical processes (Figure 13.4). End points of these processes can result in mercury sequestration in sediments, reduction and emission back to the atmosphere, or production of methylmercury (MeHg). MeHg is produced from inorganic mercury in anaerobic environments by sulfate-reducing bacteria.

Although inorganic forms of mercury dominate the environmental cycle of mercury, some few percent can be transformed into methylmercury. Methylmercury is primarily produced by sulfate-reducing bacteria naturally present in the sediment where oxygen is absent, and a sulfur compound (sulfate) is present. These bacteria take up inorganic mercury and manufacture methylmercury as a byproduct of the normal life processes. Studies in freshwater and marine systems have demonstrated that sulfate-reducing bacteria (SRB) are the major biological contributors of methylmercury formation in wetlands (e.g., Corrales et al., 2011; Gilmour et al., 2011). Factors that facilitate methylation include the presence of organic matter, low-oxygen sediment, high microbial activity, and water-level fluctuations. In particular, marsh dryout and rewetting causes increased methylation (Feng et al., 2014). Methylation rates also depend on pH, temperature, and oxygen levels. The presence of high levels of sulfide causes the precipitation of insoluble mercury sulfide (Figure 13.4). This process effectively prevents methylation. These competing processes lead to the interesting situation in which either too little or too much sulfate inhibits methylation, but at intermediate concentrations, the sulfate promotes methylation. However, other mechanisms can cause methylation in the absence of sulfur or sulfate reducing bacteria (Loseto et al., 2004).

Methylmercury is readily absorbed, much faster than it is excreted, by microorganisms living in the sediments or overlying water. As these are fed on by other organisms,

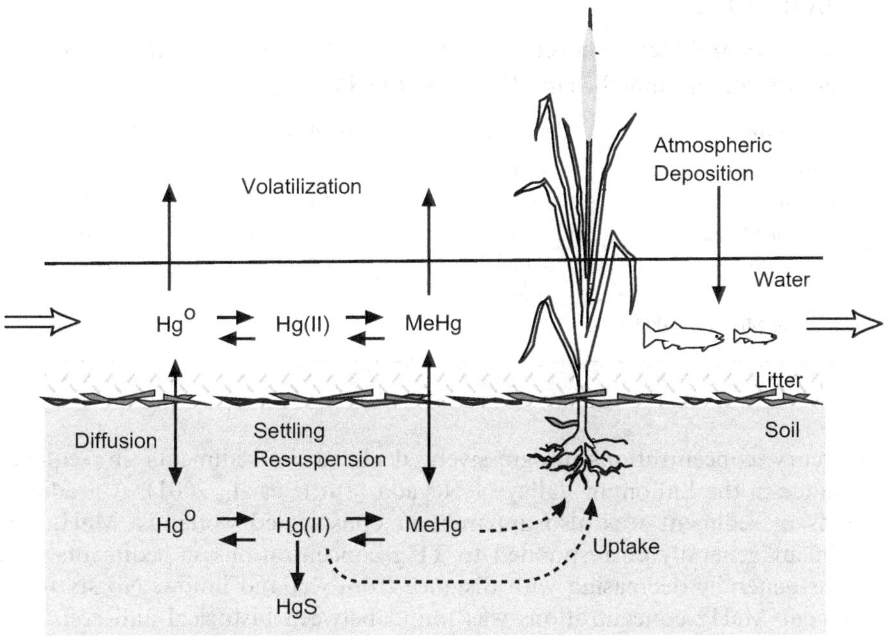

FIGURE 13.4 Simplified conceptual model for mercury processing in wetlands.

methylmercury becomes progressively more concentrated at each higher level of the aquatic food web. This process is termed bioaccumulation.

Elemental mercury and methyl mercury are volatile and may be lost from wetlands as a vapor. Both evaporation and transpiration are candidate processes, but in sum these losses are usually minor. Vaporization of mercury is also known as evasion or ebullition.

Plant uptake of mercury also occurs but is not a major route for processing in a wetland (Wyndham-Myers et al., 2014).

Atmospheric deposition of mercury varies from region to region but is everywhere large enough to affect mass balances in marshes. Deposition is tracked for North America by the National Atmospheric Deposition Program (MDN, 2016). Although there are slight downward trends, the median wet deposition concentration for North America was 9.1 ng/L over the twenty-year period 1996–2016, representing over 90,000 measurements at more than 100 locations. It is higher in south Florida, at 12.9 ng/L, and lower in California, at 6.5 ng/L. The Hg load that reaches the marsh surface also depends on rainfall. The North American median is 8.7 $\mu g/m^2 \cdot yr$, but Florida receives 17.8 $\mu g/m^2 \cdot yr$, while California gets only 4.7. To place these mass loads in perspective, the McCarron's, MN urban wetland received 1.11 $\mu g/m^2 \cdot yr$ and discharged 0.55 $\mu g/m^2 \cdot yr$ (Monson, 2007). The atmospheric deposition in Minnesota is 7.9 $\mu g/m^2 \cdot yr$ (MDN, 2016).

Except for industrial source waters, mercury in wetland surface waters is typically very low, in the range of 1–4 nanograms per liter (ng/L). As seen in Table 13.12, this is the case for agricultural, urban and natural wetlands. Methylmercury comprises only a minority of the water phase total mercury, 5–25%. The dominant storage location for total mercury is in wetland sediments, and it is there that methylation may take place (Wyndham-Myers et al., 2014).

TABLE 13.12

Mercury and methylmercury in the surface waters of treatment and non-treatment marshes in Minnesota (μg/L).

Type	Number	Total Hg	Methyl Hg
Agricultural	9	2.00	0.41
Urban	6	2.18	0.24
Designed Urban	10	2.48	0.40
Reference	4	1.96	0.09

Source: Monson (2007).

Total mercury concentrations progressively declined in sediments in sequential pre-existing wetlands in the Lahontan Valley of Nevada (Tuttle et al., 2001). A gradient of THg concentrations in sediment was also apparent in constructed wetlands. MeHg concentrations in sediment generally corresponded to THg concentrations in sediment, with MeHg concentrations generally decreasing with distance from wetland inflow. No statistical difference in sediment MeHg concentrations was found between historical and constructed wetlands. Sediment THg concentrations typically ranged from 0.2 to 1.0 mg/kg, while methylmercury was very much lower, in the range of 1 to 2 μg/kgdw.

U.S. EPA (2018a) sets criteria for freshwater concentrations of mercury (Table 13.2). The maximum concentration allowable is 1.4 μg/L, and the continuous concentration allowable is 0.77 μg/L. These standards are high compared to wetland Hg concentrations, indicating that treatment marshes are generally not a threat to aquatic organisms. The greater threat is to the higher trophic levels, involving both wildlife (fish eaters) and human consumers. The U.S. EPA human health criterion is based upon methyl mercury in fish consumed, at 0.3 μg/gdw.

Sediment standards for mercury have been set by the province of Ontario, and have been adopted in some of the United States (Table 13.3). These set a lowest effect level of 0.2 μg/g, and a severe effects level of 2.0 μg/g (Persaud et al., 1993). They also identified the level for pre-settlement deposits as 0.1 μg/g in Great Lakes sediments.

Performance of Wetlands for Mercury Removal

Total mercury is effectively removed in treatment marshes (Table 13.13). For 13 systems, with inlet THg ranging from 1–130 ng/L, the median concentration reduction is 71%. Additionally, it must be borne in mind that these removals are from surface water inflows to surface water outflows, and do not include atmospheric deposition. However, because of the propensity for marshes to methylate mercury, the 9 systems in Table 13.13 show effectively no median removal of MeHG (-4%), with inlet MeHg ranging from 0.18–10.6 ng/L. Further evidence of variable methylation is stated by Bachand et al. (2014b), who found MeHg removals ranging from -193% to 99% in seven California wetlands.

A better understanding of marsh mercury processing is gained by considering mass balances. Table 13.14 shows that surface inflows and outflows are actually the minority for low concentration source waters. Most of the input is atmospheric, and most of the output is storage.

Overall, there are trade-offs between beneficial functions of a marsh and environmental risks of methyl mercury production. Stamenkovic et al. (2005), in an evaluation of a potential treatment marsh in Nevada, concluded that:

TABLE 13.13

Example removals of mercury and methylmercury in treatment marshes.

Source and Location	Inlet ng/L	Outlet ng/L	Reduction %	Reference
------ Total Mercury ------				
Leachates and Industrial				
Clemson,SC	15.1	1.04	93	Dorman et al. (2009)
Aiken, SC	44	7	84	Nelson et al. (2006)
Sherrills Ford, NC	130	12	91	Mooney and Murray-Gulde (2008)
St. Marys, KS	50	1	98	Talley (2012)
Runoff				
ENRP, FL	2.28	0.67	71	He (2007)
STA1W, FL	4.20	0.84	80	Unpublished data
STA2, FL	0.95	0.75	21	Zheng et al. (2013)
STA5, FL	1.92	1.49	22	Unpublished data
STA6FW7, FL	1.61	1.70	-6	Fink (2004)
Reno, NV	100	25	75	Stamenkovic et al. (2005)
Twitchell Island, CA	2.30	2.60	-13	Ackerman et al. (2015)
Municipal				
Sacramento, CA	10.61	4.71	56	Crites et al. (1997)
Reno, NV	5.0	25.0	-400	Stamenkovic et al. (2005)
------ Methyl Mercury ------				
Runoff				
ENRP, FL	2.28	0.67	71	He (2007)
STA1W, FL	0.18	0.10	44	Unpublished data
STA2, FL	0.59	0.13	22	Zheng et al. (2013)
STA5, FL	0.27	0.28	-4	Unpublished data
STA6FW7, FL	0.31	0.64	-106	Fink (2004)
Reno, NV	0.5	0.68	-36	Stamenkovic et al. (2005)
Twitchell Island, CA	0.38	0.85	-124	Ackerman et al. (2015)
Municipal				
Sacramento, CA	10.61	4.71	56	Crites et al. (1997)
Reno, NV	0.26	2.66	-923	Stamenkovic et al. (2005)

Based on the developed data, the environmental risk of the proposed large-scale wetland would be an increase of methyl mercury concentration in creek water that reaches the Truckee River by as much as 20 to 75%. However, the wetland would also be a significant sink for nutrients, suspended solids, and total mercury, decreasing the amount of mercury available for methylation downstream.

In general, the same is true for marshes targeting other contaminants, but receiving mercury. The Aiken, SC, full-scale marsh project (Nelson et al., 2002) was designed to reduce copper, but also received mercury. The treatment system consists four wetland cells (3.24 hectares total). The cells are vegetated with *Scirpus californicus*, and water retention time in the system is approximately 48 hours. Operation and maintenance of the system is minimal,

TABLE 13.14
Mass balances for total mercury for Florida marshes (μg/m²·yr).

	Surface Input	Atmospheric Deposition	Evasion	Surface Output	Stored
WCA1	0.48	19.92	3.60	0.33	16.47
WCA2	1.67	20.04	3.60	1.67	16.43
WCA3	1.25	19.96	3.60	1.06	16.54
STA2	9.72	20.00	3.60	8.06	18.06

Source: Gu et al. (2012), Zheng et al. (2013).

and mainly consists of checking for growth of the vegetation and free flow of the water through the system. This has provided effective copper treatment of large volumes of permitted water discharges from an industrial area.

Mercury levels after passage through the first cell are typically reduced to 40% of the influent value and reduced by an additional 30 to 40% after passage through the second cell in the series. The final concentration steadily went down as the wetland system matured. During the June 2001 sampling event, the total mercury level fell below the permit limit (13 ng/L), and has remained below the permit limit since (Table 13.13). Performance was better than the anticipated mercury reductions based upon mesocosm testing. Methylmercury production in the system occurred but was a small component of the total mercury in the effluent. Methylmercury has increased during the warm summer months but has never exceeded 10% of the total mercury at the discharge point.

The Everglades Stormwater Treatment Areas, Florida, comprise six individual units totaling 29,000 ha. Each of the six is compartmentalized, with from two to ten cells in one to three flow paths. On an annual average basis, these event-driven wetlands have design detention time of about three weeks, but under peak flow conditions, it is less than four days. Vegetation was allowed to volunteer into these large wetlands, but for some cells, submerged aquatic vegetation was fostered by maintaining deeper water. The result is a mixture of emergent, submergent and floating plants. Although these treatment wetlands were designed for phosphorus removal, mercury performance has also been monitored. That is because the downstream ecosystem is the Florida Everglades, a wetland complex that is notoriously plagued by high levels of mercury and methylmercury.

Operating data for mercury shows that there is typically removal total mercury in the surface flows (Table 13.13). Methylmercury is sometimes higher in the STA outflows than it is in the inflows; and sometimes lower. There is variable methylation occurring in the STAs.

Free water surface constructed wetlands were investigated at Sacramento, California, for the reduction of mercury in secondary municipal effluent, although that was not their primary target (Crites et al., 1997). The 8.9-ha system consisted of ten cells, vegetated with bulrushes (Scirpus acutus), and operated at about one-week detention time. Inlet mercury of 10.61 ng/L was reduced on average to 4.71 ng/L. Mercury in soils and vegetation was below detection limits.

13.11 ARSENIC

The term metalloid refers to a chemical element that has properties between those of metals and nonmetals. Two metalloids, arsenic and selenium, have acquired some interest in connection with treatment marshes. They are both present in coal and other mining leachates, and in flue gas scrubber waters. They are both toxic to aquatic organisms and humans.

A large portion of industrial arsenic in the United States is currently used as a wood preservative, but arsenic is also used in paints, dyes, metals, drugs, soaps and semiconductors. Arsenic was a widely used pesticide prior to the introduction of dichlorodiphenyl-trichloroethane (DDT) (Knox et al., 1999). Chromate-copper-arsenate (CCA) is a common wood preservative, used in bridges and in boardwalks through wetlands. Industry activities such as copper smelting, mining and coal burning also contribute to arsenic in our environment. Arsenic is usually associated with other metallic contaminants, such as lead and zinc, because it is a component of some mined minerals.

Arsenic can be the primary target of treatment wetland systems dealing with mine effluents (e.g., Duncan et al., 2004), but such applications are outside the realm of treatment marshes. However, not all high arsenic waters are the result of industrial activities. For instance, groundwaters in Bangladesh are naturally very high in Arsenic, and yet are a large fraction of the drinking water for the country (Kinniburgh and Smedley, 2001). Accordingly, a treatment marsh study was conducted at Clemson University for the purpose of improving such unacceptable drinking waters (Schwindaman et al., 2014).

Chemistry of Arsenic in Marshes

Arsenic is a naturally occurring element in surface waters, occurring as As(III) and As(V). The arsenate form predominates in aerobic waters (Eh > 0), existing in anionic forms of either $H_2AsO_4^-$, or $HAsO_4^{-2}$, at circumneutral pH (Masscheleyn et al., 1991a, 1991b; Lizama et al., 2011). The chemistry of As(V) resembles that of phosphorus. Arsenite occurs under reducing conditions (Eh < 0, pH ≈ 6–8), as arsenious acid (H_3AsO_3). At higher soil redox levels (+200 to +500 mV) As(V) is the predominant species. The reduction of As(V) to As(III) occurs at redox levels corresponding within the nitrate reducing zone of soils characterized by a soil redox level of approximately +200 to +300 mV. Arsenic is found in organometallic forms (Frank, 2005), such as methylarsonic acid ($H_2AsO_3CH_3$) and dimethylarsinic acid [$(CH_3)_2AsO_2H$], which are active ingredients in pesticides.

Arsenic is of concern in aqueous environments because it is a known human carcinogen and because it is acutely and chronically toxic to aquatic organisms. U.S. EPA has set the arsenic standard for drinking water at 10 µg/L to protect consumers served by public water systems from the effects of long-term, chronic exposure to arsenic. There are geographic "hot spots" in the United States that have higher levels of arsenic in well water, resulting from natural underground sources. In the author's home state of Michigan, there are eight counties that have reported arsenic >50 µg/L in well water, and there are three such reports within 5 km of my house.

U.S. EPA (2018a) sets criteria for freshwater concentrations of arsenic (Table 13.2). The maximum concentration allowable is 340 µg/L, and the continuous concentration allowable is 150 µg/L. These standards are high compared to wetland arsenic concentrations, indicating that treatment marshes are generally not a threat to aquatic organisms.

Sediment standards for arsenic have been set by the province of Ontario and have been adopted in some of the United States (Table 13.3). These set a lowest effect level of 6 µg/gdw, and a severe effects level of 33 µg/gdw (Persaud et al., 1993). They also identified the level for pre-settlement deposits as 4.2 µg/gdw in Great Lakes sediments.

Wetland Processing and Storage of Arsenic

Arsenic is found in wetland plants, algae, and sediments. Arsenic concentrations in plants are quite low, approximately 1–5 µg/g dry mass (Table 13.4). Aboveground plant parts are of lowest concentration, and belowground organs are somewhat higher. Wetland sediments have the highest levels, and are therefore the major repository of stored

As. The arsenic content of peat from Finnish bogs was found to be in the range 0.1–8.0 μg/g (Frank, 2005). This bog arsenic is mostly of atmospheric origin, because there are no other sources.

The adsorption of As(V) and As(III) on various Fe(III) and Al hydroxides has been investigated extensively employing both macroscopic and spectroscopic experiments. Iron oxides are a good sorbent for As(V) (Katsoyiannis and Zouboulis, 2004). Arsenate adsorbs via a ligand exchange reaction replacing surface OH^- groups (Williams et al., 2003), for instance:

$$FeOH + AsO_4^{-3} + 2H^+ \leftrightarrow FeHAsO_4^- + H_2O \quad (13.27)$$

Smith et al. (2002) measured linear partition coefficients for As(V) for a specific soil type (an Alfisol), with low amounts of iron and aluminum, to be 1.7–3.5 L/kg. In contrast, for two other soil types (a Vertisol and an Oxisol), with high iron and aluminum (7–12 g/kg), the values were 23–62 L/kg. Arsenate anion exchange processes are analogous to those of phosphate, with competition favoring phosphorus compounds (Mkandawire and Dudel, 2005). Arsenate can also be adsorbed by clays, particularly by kaolinite and montmorillonite at low pH (Knox et al., 1999).

Performance of Wetlands for Arsenic Removal

The observed removals of arsenic in constructed wetlands are variable (Table 13.15). The higher concentrations in industrial, mining and leachate waters allows for measurable and high reductions. However, urban stormwaters, treated municipal wastewaters, and many industrial source waters do not contain appreciable arsenic, and removals are low, and in some cases are influenced by method detection limits. The median concentration reduction in 19 treatment marshes is 19%.

13.12 SELENIUM

Selenium is a nutritionally essential element for animals in small amounts, but toxic at higher concentrations. It has created severe problems for both fish and waterfowl. Selenium bioaccumulates in the aquatic food chain and chronic exposure in fish and aquatic invertebrates can cause reproductive impairments (e.g., larval deformity or mortality). Selenium can also adversely affect juvenile growth and mortality. In high concentrations, selenium acts as an environmental contaminant.

Selenium concentrations are typically quite low (< 1 μg/L) in surface waters. However, sources of pollution include waste materials from certain mining, agricultural, petrochemical, and industrial manufacturing operations. In semi-arid areas of the western United States, excess water is applied to fields to flush out salts leached onto the surface soils. This return flow water either infiltrates into the soil or runs off into nearby wetlands, ponds, or streams. When irrigation water is applied to soils containing selenium, Se can be leached from the soil. There are an estimated 11,000 square kilometers in the western United States identified as being susceptible to irrigation-induced selenium contamination (Seiler et al., 1999).

Oil sands mining in western Canada produces large volumes of process water containing selenium and other constituents that limit discharge into receiving systems. Treatment marshes can be designed to treat process water, and to serve as passive (reclaimed) wetlands following a mine's lifespan. Such marshes must be designed to mitigate ecological risks for aquatic organisms.

Coal mining and coal burning also are potential sources of selenium pollution. Mining creates leachates with elevated SE and burning produces flue gas scrubbing water with

TABLE 13.15
Example removals of arsenic in treatment marshes.

	Inlet µg/L	Outlet µg/L	Reduction %	Reference
Leachates & Industrial				
Albright, PA	3.0	2.0	33	Hoover et al. (1998)
Champion Paper, FL	9.0	9.3	-3	NADB Database (1998)
Woodcutters, NT, Australia	5.0	2.0	60	Noller et al. (1994)
Isanti-Chisago, MN	6.9	4.2	40	Loer et al. (1999)
Clemson, SC	145.0	50.4	65	Schwindaman et al. (2014)
Clemson, SC	145.0	17.0	88	Schwindaman et al. (2014)
Runoff				
ENRP, FL	2.8	1.5	46	He (2007)
Naples, FL	8.8	5.0	43	Bishop et al. (2012)
Tarrant County, TX 1	3.7	3.0	19	NADB Database (1998)
Tarrant County, TX 2	3.7	2.3	38	NADB Database (1998)
Tarrant County, TX 3	3.7	2.7	27	NADB Database (1998)
Imperial, Ca (2009–2011)	3.8	3.55	7	Bucher (2012)
Brawley, Ca (2009–2011)	5.2	4.3	17	Bucher (2012)
Municipal				
Estevan, SAS	4.7	5.1	-8	Unpublished data
Sacramento, CA	2.4	2.8	-18	Nolte & Associates (1998b)
Hayward, CA	1.9	3.8	-100	NADB Database (1998)
Oxnard, CA	2.5	2.5	0	NADB Database (1998)
Trés Rios, AZ	3.5	3.5	0	NADB Database (1998)
Trés Rios, AZ	4.5	3.8	16	Wass Gerke & Associates (2002)

elevated levels of Se. Oil refineries produce waters that have elevated levels of selenium. Treatment marshes have been used to reduce Se contamination in these situations.

Chemistry of Selenium in Marshes

Selenium is a metalloid, with reactivity and chemistry similar to sulfur. Four oxidation states exist: selenide (-II), elemental selenium (O), selenite (IV), and selenate (VI). Selenium soil solution chemistry is controlled by pH and redox conditions; however, organic carbon and microbial transformation also influence the availability of Se for plant uptake (Zhang and Moore, 1997) Under most conditions, inorganic Se in solution is found in one of two oxidation states, as selenite (SeO_3^{-2}) and as the more mobile selenate (SeO_4^{-2}). Selenate is the dominant form at high redox and high pH, is more readily taken up by plants and considered to be the most bioaccumulative inorganic ion; however, organic Se has been shown to be more mobile and bioaccumulative than selenate. Selenium readily forms organic complexes, such as dimethyl selenide and selenomethionine.

U.S. EPA (2016a) has set a selenium monthly average exposure criterion of 1.5 µg/L to protect aquatic life in still waters, including wetlands. This criterion is based on limits for Se in fish of 15.1 mg/kg in eggs/ovaries, and 8.5 mg/kg for the whole body.

Marsh Processing and Storage of Selenium

The selenium concentrations in vegetation in treatment wetlands exposed to selenium are typically in the range of 1–10 μg/gdw for both aboveground and belowground parts (Table 13.4). However, the fallen litter at Corcoran, California contained 10–20 μg/gdw. Treatment wetland sediments have 1–10 μg/gdw. Although these values are low, there is still a possibility that bioaccumulation can reach unacceptable levels under some circumstances. However, the results of the Richmond and Corcoran, California, studies have been encouraging, because toxic effects have not been identified.

Constructed wetlands remove selenium by reduction to insoluble forms that are deposited in the sediments, by sorption, by accumulation into plant tissues, and by volatilization to the atmosphere, via biological volatilization through plants.

Selenium Reduction

Water column removal processes include reduction of selenate and selenite to elemental selenium, and subsequent sequestration in plants and sediments. The sequence is:

$$SeO_4^{-2} \rightarrow SeO_3^{-2} \rightarrow Se \rightarrow Se^{-2} \qquad (13.28)$$

Lin and Terry (2003) determined that surface sediments at the Corcoran wetlands, receiving water containing Se(VI), contained Se(VI):Se(IV):Se(O):Se(-II) = 0:13:41:46. In *Polypogon monspeliensis* aboveground tissues, the proportions were 32:17:27:24.

Sediments provide a reducing environment and can produce and contain selenide and elemental selenium. Detention times of 10–16 days were found to give 89–92% reduction of Se(VI) to Se(O) in mesocosm trains (Zhang and Frankenberger, 2003).

Dissimilatory selenate reduction is carried out by anaerobic bacteria that use selenate or selenite in the oxidation of organic matter:

$$2CH_2O + SeO_4^{-2} \rightarrow 2HCO_3^{-} + H_2Se \qquad (13.29)$$

where CH_2O represents organic matter.

Volatilization

There is evidence that significant losses of selenium, through biologically mediated methylation and volatilization, occurs in wetlands. Upon release from organisms, selenium may be converted to the volatile dimethyl selenide form (Masscheleyn et al., 1991a; Masscheleyn and Patrick, 1993). This removal route involves conversion to organic forms such as dimethylselenide, which can volatilize from the wetlands, mediated by plants (Azaizeh et al., 2003):

$$Organic + SeO_4^{-2} \rightarrow (CH_3)_2Se \qquad (13.30)$$

The field measurements of Hansen et al. (1998) indicated volatilization rates of 60–190 $\mu g/m^2 \cdot d$. Plexiglas chambers that covered whole plants were being used to monitor selenium volatilization rates. However, Lin and Terry (2003) measured lower rates, 5–30 $\mu g/m^2 \cdot d$, at the Corcoran, California, site, also using flux chambers. This represented only about 1–5% of the applied selenium (Gao et al., 2003a, 2003b).

Sorption

Selenite has a stronger affinity for sorption sites than Se(VI) and thus is readily adsorbed to sediments and soil constituents such as Fe-, Mn-, and Al-oxyhydroxides, which are the major adsorbents for removing soluble selenium from water.

Plant Uptake

During microbial metabolism, selenium is organically bound as selenide compounds (seleno-amino acids; methyl selenides; and methyl selenones. Upon release from organisms, selenium may be converted to the volatile dimethyl selenide form (Masscheleyn and Patrick, 1993). Carvalho and Martin (2001) studied the hydroponic uptake of selenite by cattails (*Typha domingensis*), duckweed (*Lemna obscura*), *Hydrilla verticillata*, and swamp lilies (*Crinum americanum*). In one week, all four were able to remove 50–100% of selenium solutions of 1–200 mg/L. Tissue concentrations in *Lemna* and *Hydrilla* were increased to 0.1–1.3 μg/gdw, and to 30–80 μg/gdw in cattails and swamp lilies. However, mass balances on field-scale operating systems over several years indicate that very little of the applied selenium load is stored in plant tissues (Gao et al., 2003a, 2003b).

Performance of Wetlands for Selenium Removal

The observed removals of selenium in constructed wetlands are variable (Table 13.16). The higher concentrations in industrial, mining and leachate waters allows for measurable and

TABLE 13.16

Example removals of selenium in treatment marshes. The entries for Corcoran, CA, represent averages for 10 wetlands.

	Inlet μg/L	Outlet μg/L	Reduction %	Reference
Leachates and Industrial				
Albright, PA	4.0	2.0	50	Hoover et al. (1998)
Springdale, PA	2.0	2.0	0	Hoover et al. (1998)
Richmond, CA	25.0	8.0	68	Hansen et al. (1998)
Sherrills Ford, NC	170.0	143.0	16	Mooney and Murray-Gulde (2008)
St. Marys, KS	45	9	80	Talley (2012)
Clemson, SC FGS	1540	575	63	Eggert et al. (2008)
Clemson, SC Ash basin	229.0	170.0	26	Dorman et al. (2009)
Clemson, SC Oil well	52	5.2	90	Spacil et al. (2011b)
Clemson, SC Refinery	47.3	2.0	96	Spacil et al. (2011a)
Clemson, SC Oil sands	5.8	3.7	36	McQueen et al. (2017b)
Runoff				
Imperial, Ca (2001–2004)	7.1	5.9	17	Unpublished data
Brawley, Ca (2001–2004)	10.1	9.9	2	Unpublished data
Imperial, Ca (2009–2011)	4.3	2.78	35	Bucher (2012)
Brawley, Ca (2009–2011)	2.28	1.66	27	Bucher (2012)
Corcoran, CA (1997)	22.0	17.5	20	Lin and Terry (2003)
Corcoran, CA (1998)	21.0	14.0	33	Lin and Terry (2003)
Corcoran, CA (1999)	16.0	9.0	44	Lin and Terry (2003)
Municipal				
Estevan, SAS	1.7	1.8	-6	Unpublished data
Sacramento, CA	0.6	0.5	17	Nolte and Associates (1998b)
Oxnard, CA	12.0	9.0	25	CH2M Hill (2005)
Oxnard, CA	12.0	6.0	50	CH2M Hill (2005)
Tres Rios, AZ	5.2	5.0	4	Wass, Gerke and Associates (2002)

high reductions. However, runoff waters, treated municipal wastewaters, and many industrial source waters do not contain appreciable arsenic, and removals are low, and in some cases are influenced by method detection limits. The median concentration reduction in 22 treatment marshes is 30%.

The mass balance for the ten-cell Corcoran project defines the fate of removed selenium over the history of that project (Gao et al., 2003a) (Table 13.17). Approximately 39% of the added selenium passed through the wetlands, mostly in surface flow. Of the remaining 61%, most was founding surface sediments (new deposits 18%) and soils (0–20 cm layer 33%). Only 2% was found in plant litter, and less than 1% in plants. Flux measurements indicated that about 2% was lost to volatilization.

The Stimulus for Studies

Much of the acquired knowledge of Se processing in wetlands was motivated by the severe ecological consequences of aquatic pollution that emerged in the period 1960–1990. Young et al. (2010) document a dozen examples, some of them tragic. The Kesterson project in the San Joaquin Valley in California is a prime example of wetland soil and surface water contamination derived from agricultural practices. Seleniferous soils leached by agricultural drainage waters, at a selenium concentration of 300 μg/L, were directed by tile drains and transported to the Kesterson National Wildlife refuge, a series of man-made marshes that were originally constructed between 1969 to 1975. The impoundment of selenium-rich waters led to soil concentrations 100 times higher than those in nearby control sites. Elevated selenium concentrations in soil solution led to high concentrations of selenium in plants, invertebrates, reptiles, fish, mammals, and birds (Ohlendorf et al., 1986). The most pronounced effects were seen in waterfowl and shore birds, with common occurrences of embryonic mortality and deformity, and mortality in adult birds (Ohlendorf et al., 1990).

TABLE 13.17

Selenium mass balance on the combined ten cells at Corcoran, CA, May 1997 through September 2000 (percent).

Total In		100
Outflows		
Surface Outflow	35	
Seepage	4	
Total Out with Water		39
Volatilization		2
Internal Storage		
Surface Soils 0–20 cm	33	
Surface Sediments	18	
Litter	2	
Plants	<1	
Water	<1	
Total Storage		53
Unaccounted		6
Total Out		100

Source: Gao et al. (2003a, 2003b).

The "Kesterson Effect" of selenium bioconcentration in arid land wetlands receiving agricultural drain waters was a wakeup call.

Studies at existing constructed wetlands were undertaken, and new pilot systems were built for the express purpose of understanding Se processing in wetlands. These studies have allowed the forecasting of potential effects of other constructed wetland projects, which proved to be acceptable for the Irvine Ranch Water District project (Horne, 2004; Fleming-Singer and Horne, 2006).

The Chevron oil company conducted a study of a 36-hectare FWS constructed wetland located adjacent to the San Francisco Bay, California (Duda, 1992; Hansen et al., 1998). Flows of about 6,500 m3/d created nominal detention of about 7–10 days. Analysis of the wetland inlet and outlet waters showed that the constructed wetland was successful in removing 89% of the selenium from the wastewater passing through it. Inflow selenium concentrations of 20–30 μg/L decreased to < 5 μg/L in the outflow (Hansen et al., 1998).

A flow-through experimental wetland system was studied from 1996 through 1999, which was designed to remove selenium from agricultural drainage water in the Tulare Lake Drainage District of California (Lin and Terry, 2003; Gao et al., 2003a, 2003b). The Corcoran, CA, system consisted of ten FWS cells, 15 m × 76 m, operated at 7–21 days nominal detention time. In 1999, the wetland cells reduced total selenium by 32 to 65% in concentration and 43 to 89% in mass. The outflow contained an average of 47% Se(VI), 32% Se(IV), and 21% org-Se, indicating reduction processes. The surface sediment was the largest sink for selenium.

The successes of these early projects have created optimism for marsh projects for other selenium source waters. Two full-scale constructed wetland systems are in operation for control of Se in flue gas scrubber waters:

- The Sherrills Ford, North Carolina, system started operation in 2006 (Mooney and Murray-Gulde, 2008). It is a 6.1 ha cattail- bulrush marsh that receives 4,700 m^3/d of flue gas scrubber water containing about 170 μg/L of Se.
- The St. Marys, Kansas, is a 0.82 ha hybrid wetland system, consisting of horizontal and vertical subsurface flow units as well as marshes (Talley, 2012). The system receives about 120 m^3/d of flue gas scrubber water containing about 45 μg/L of Se.

Additionally, a number of mesocosm projects have targeted marsh Se removal form a variety of source waters (Table 13.16): flue gas scrubber water (Eggert et al., 2008), ash basin water (Dorman et al., 2009), refinery process water (Spacil et al., 2011a), produced water from oil wells (Spacil et al., 2011b), and oil sands process water (McQueen et al., 2017b).

13.13 SUMMARY

Treatment marshes are being considered as an option for treating a variety of trace metals. Urban runoff contains several metals at concentrations of concern for receiving waters, and wastewater treatment plants also typically pass through heavy metals, Design of wetland systems to treat these elements is currently limited by a small but growing knowledge of the factors that control removal rates and of the biological effects of concentrating many of these elements in wetlands. However, for copper there is now a large body of knowledge and alternative design procedures and recommendations. Copper is used in agriculture, and hence is of concern in agricultural runoff.

Wetlands are effective at retaining significant loads of several trace metals. There are no microbial removal mechanisms, and removal is to storage in the wetland sediments. Uptake by vegetation is often a minor removal mechanism. Storage capacity included in and sorbed upon pre-existing wetland sediments and biota will eventually be exceeded, and sooner at high influent loadings. Thereafter, there remains only a lesser removal potential to the newly forming sediments. Burial of metals by accreting sediments is a sustainable mechanism, but it is desirable that metal levels be reduced to levels that will not result in chronically toxic conditions to biota. In treatment marshes, this is not always practical. Limits on metal loadings to soils are based on accumulated concentrations that are kept below toxicity thresholds of sensitive biota.

Mercury is of special concern because it poses a health risk to humans in its methylated form. The principal source of mercury is atmospheric deposition, which is unique among metals in wetlands. Marshes reduce mercury concentrations, but methylate some of it in the process.

The metalloids arsenic and selenium are of concern because of their severe toxicity. There are historical examples of serious environmental problems, which have generated concerns about marshes that receive seleniferous waters.

14 Trace Organics

By definition, treatment marshes do not treat source waters that are heavily contaminated with organic compounds. Yet, an amazing array of trace organics exists in the various source waters that are in the realm of treatment marshes. Virtually all trace organics do not occur in nature, and possess no biogeochemical cycle. They are strictly the result of human chemical production.

Pesticides and herbicides are important contaminants in agricultural runoff, but also occur in urban stormwaters. There is an ever-increasing variety of pharmaceuticals and personal care products that result primarily from human use, but also from animal husbandry. Some trace organics are termed endocrine disruptors (ED) because they can induce physiological changes in biota, often at very low concentrations. Municipal WWTPs often do not treat these trace organics, and treatment of animal wastes may not occur.

Trace amounts of petroleum hydrocarbons may well exist in contaminated groundwaters, as well as in various petroleum processing waters. The same is true of chlorinated hydrocarbons, such as the chlorinated ethenes and benzene. Finally, industrial activities have resulted in cyanide contamination of groundwaters. Treatment wetlands, and especially treatment marshes have emerged as a candidate technology for remediation of all these trace organics, with a large increase in studies and projects occurring over the past decade.

Laboratory analytical techniques and devices have also evolved at a fast pace. It is now possible to determine trace organics of many kinds that were difficult or impossible to quantify in the recent past. Procedures often require sophisticated instrumentation that is both expensive and not widely available. Yet, there are hundreds of trace organics that have been identified as "contaminants of concern." The result is that the voluminous wetland research does not exhaustively cover the entire pharmacopoeia, or the over 350,000 current and historic pesticide products registered in the United States. This section attempts to set forth examples that characterize the findings in general.

Many trace organics have the potential for causing deleterious effects in receiving waters, and consequently the capabilities of treatment marshes to remove these substances is of considerable interest. But an ambiguity exists: treatment marshes are also home to the same suite of organisms that exist in other receiving water bodies. It is therefore necessary to consider chronic and acute effects in the treatment marsh as well as in the receiving water body. Therefore, aquatic life criteria serve as both a target for treatment and a constraint upon treatment marsh conditions.

14.1 GENERALITIES

NOMENCLATURE

Because of the large number of compounds of interest, nomenclature presents a first challenge. There are four basic ways to identify a specific trace organic:

1. The chemical formula, which specifies the number of atoms of each element contained in the compound.
2. The chemical structure, which specifies the interconnectivity of the constituent elements and groups of elements.
3. The CAS Number, which is a unique numerical identifier assigned by the Chemical Abstracts Service (CAS) to every chemical substance described in the open scientific

literature including all substances described from 1957 through the present (>143 million).
4. The common name for the principal ingredient, or trade name(s) under which the chemical is marketed.

Current literature on trace organics, including both use and environmental effects, utilizes common or trade names (Item 4). There remains ambiguity, because multiple manufacturers use different trade names for the same active ingredient compound. For instance, glyphosate, N-(phosphonomethyl) glycine, is one of the most widely used herbicides. Glyphosate is the active ingredient in products such as Roundup, Rodeo, Remuda, Ranger, Eraser, Pronto, AquaNeat, and Weed Impede (all trade names). Glyphosate is a broad-spectrum herbicide that targets broadleaf weeds, grasses, and woody plants. It was brought to market in 1974 under the trade name Roundup. Examples of pesticide nomenclature are shown on Figure 14.1.

Source Waters and Timing

There are important distinctions among source waters, both for the types of trace organics and the timing of their delivery to a treatment marsh.

Contaminated groundwaters would typically be retrieved by pumping, and therefore deliveries are at the discretion of the design. Source concentrations would be stable. For this application, steady flows and concentrations exist. Further, the number of contaminants is likely to be few.

Treated municipal wastewaters also are delivered at relatively constant flows. Most pharmaceuticals and personal care products are used throughout the year, on a more or less continuous basis. For this application also, steady flows and concentrations exist. However, pesticides are not used uniformly throughout the year in an urban environment, and hence concentrations in treated municipal wastewaters vary throughout the year.

Diazinon
O,O-Diethyl O-[4-methyl-6-(propan-2-yl) pyrimidin-2-yl] phosphorothioate
$C_{12}H_{21}N_2O_3PS$
CAS Number 333-41-5

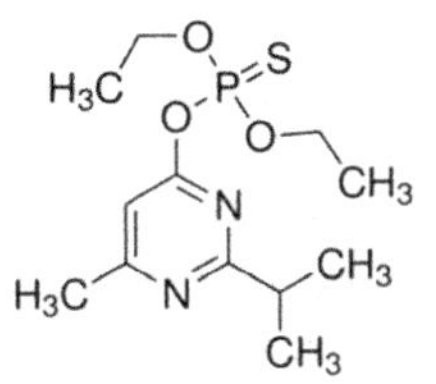

Chlorpyrifos
O,O-Diethyl O-3,5,6-trichloropyridin-2-yl phosphorothioate
$C_9H_{11}Cl_3NO_3PS$
CAS Number 2921-88-2

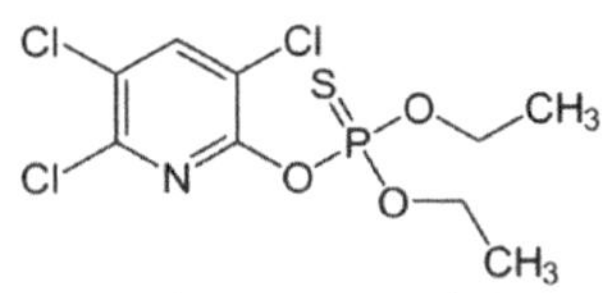

Atrazine
1-Chloro-3-ethylamino-5-isopropylamino-2,4,6-triazine
$C_8H_{14}ClN_5$
CAS Number 1912-24-9

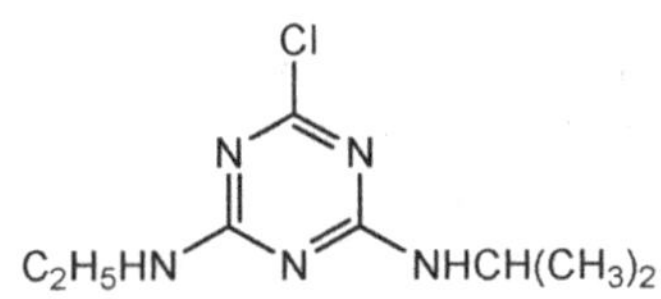

FIGURE 14.1 Examples of pesticide structures and naming.

Urban stormwaters are a doubly stochastic source of pesticides and pharmaceuticals and personal care products. Firstly, the flows are episodic; and secondly, the concentrations are episodic, depending upon use patterns in the watershed. For instance, DeLorenzo et al. (2012) studied a stormwater pond in South Carolina and found widely variable percentages of five pesticides over a five-year period. They also found widely variable numbers of "hits" for antibiotics in 540 samples, ranging from 10% without a hit, to 54% for one hit, to 15% with five or more hits.

Agricultural runoff waters are a source of pesticides. Agricultural chemicals are used on a seasonal basis, which may or may not coincide with the runoff flow patterns. In temperate climates, there is neither application to fields or runoff in the winter months. Braskerud and Haarstad (2003) found nearly coincidental flows and concentrations for thirteen agricultural chemicals delivered to a small treatment marsh, but Budd et al. (2009) found separate patterns for five insecticides in agricultural source waters that did not coincide with regional rainfall (runoff).

Wetland Performance Evaluation Methods

As a result of the variability of source patterns, several different methods of marsh performance assessment have been implemented. These may be categorized as follows:

A. For steady source flows and concentrations, methods used for nutrients can be used. See for example Barber et al. (2006)
B. Batch dose response. A known amount of the trace organic is added (uniformly) to the system, usually in a manner approximating an impulse addition. This gets increasingly difficult as the size scale moves from mesocosms to larger pilot and field scale wetlands. Batch reaction analysis is applied, and spatial variability is not a factor. See for example Rose et al. (2008).
C. Dose and flush response. In this procedure, a known amount of trace organic is added to the wetland inlet, followed by a prescribed water flow regime. Analysis proceeds via event mean flows and concentrations, which may be applied at several distances in the flow direction. See for example Moore et al. (2006).
D. Dynamic modeling of the time series of incoming flows and concentrations. See for example Alvord and Kadlec (1996).

From these tests, it is possible to derive several measures of performance. The easiest, but perhaps least useful, is simply the input-output-retention values. When collected in tabular form, these give an overall appreciation for whether wetlands are or are not effective in removing/retaining trace organics. As documented in the following sections, the overall result is that marshes are indeed quite effective. Vymazal and Březinová (2015) show positive removals for virtually all of the 87 pesticides that were included in their review. Their results sort to some degree with the chemical nature of the pesticide, that is, organochlorines, pyrethroids, triazines, and so on. However, they found only very weak bivariate correlations with pesticide feature such as solubility, sorption coefficients, and reported soil half-lives.

Of course, sizing and performance predictions require some quantification of the rate of removal. At the present time, the limit of our understanding usually allows only the application of presumed (pseudo) first order kinetics, parameterized as half-lives or removal rate coefficients. However, first steps are being taken to include more detailed parameterization of the collection of processes that comprise removal.

There are several processes involved in pesticide dissipation in marshes (Figure 14.2). Both physical transfers and chemical transformations can occur. The proportions of these vary considerably among the numerous trace chemicals of potential interest.

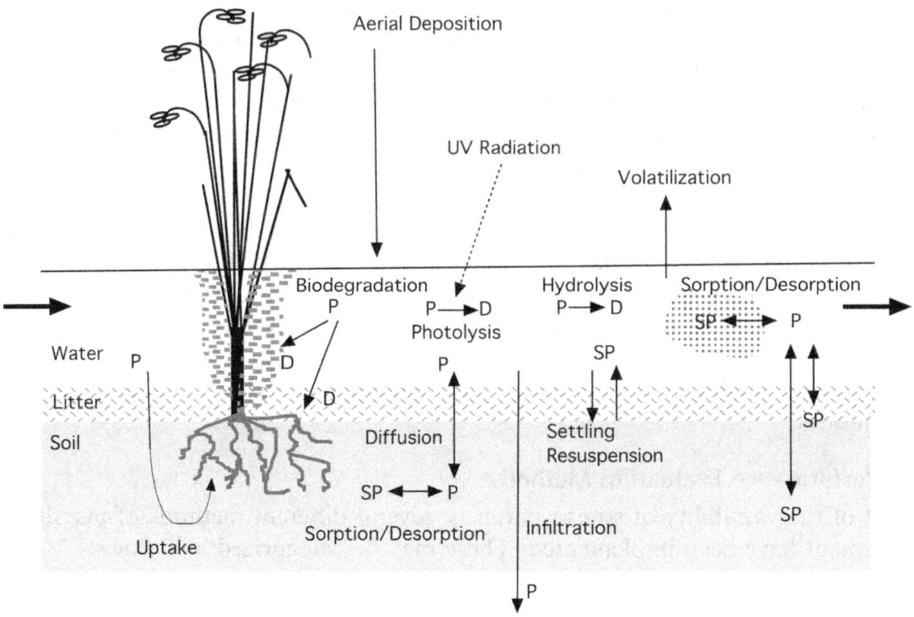

FIGURE 14.2 Simplified schematic of forms and pathways of pesticides in treatment marshes. P = water phase pesticide, SP = solid phase pesticide, D = degradation products.

14.2 PHYSICAL MECHANISMS OF REMOVAL

VOLATILIZATION

Some pesticides have low but measurable volatility, which suggests that transfer to the air may be a minor contributing mechanism. Herbicide volatilization occurs in two steps – evaporation from soil and plant material, followed by dispersion into the atmosphere by diffusion and turbulent mixing. For instance, Gish et al. (2011) used vapor sampling to measure 5% and 18% volatilization of applied atrazine and metalochlor, respectively, in an eight-year study. This is a surprising result, given that the vapor pressures at 20°C of atrazine and metalochlor, respectively, are 0.04 mPa and 1.7 mPa. These are six to seven orders of magnitude less than water (2.34 kPa).

ADSORPTION/DESORPTION

Pesticides are organic molecules that display an affinity for sorption to soils, particularly organic soils. Commonly, equilibrium between water-phase compounds and solid-phase compounds is rapidly achieved, either with soils in the action zone of the wetland or with TSS in the water column.

Partition coefficients for organic sediments, such as are found in marshes, are defined in two ways, on a total mass basis and an organic carbon basis:

$$C_S = K_d C_w = K_{OC} f_{OC} C_w \tag{14.1}$$

where C_W = water concentration, mg/L

C_S = solid concentration, mg/kg

f_{OC} = fraction organic carbon in solid, –

K_{OC} = organic carbon partition coefficient, L/kgC
K_d = partition coefficient, L/kg

The total concentration in a wetland soil is the sum of water-phase and solid-phase contributions:

$$C_T = \theta C_w + \rho_b C_S = \theta C_w + \rho_b K_d C_w \quad (14.2)$$

where C_T = total concentration, mg/L
θ = volume fraction water, –
ρ_b = soil bulk density, kg/L = g/cm^3

Many studies have been conducted to determine wetland pesticide partition coefficients (e.g., Passeport et al., 2011; Gebremariam et al., 2012). The reader is referred to the tabulation of Vymazal and Březinová (2015), who list single values of K_{OC} for 75 pesticides for wetlands. Such lists need to be interpreted with great care. Many pesticides have been the subject of several studies individually, such the chlorpyrifos results for aquatic sediments listed by Gebremariam et al. (2012), not included in the Vymazal and Březinová (2015) list. They found a median K_d = 355 L/kg (range 40–767), and a median K_{OC} = 15,500 L/kg (range 3,000–25,600). As might be expected, there is variability among the various components of marsh substrates. Alvord and Kadlec (1995) determined atrazine partition coefficients for 14 soils (peats), sediments and litter for three different wetlands. They found a median K_d = 202 L/kg (range 38–363), and a median K_{OC} = 712 L/kg (range 321–1055). This median K_{OC} is four times higher than that listed by Vymazal and Březinová (2015). All types of marsh solids displayed strong binding.

Budd et al. (2011) found that In the runoff water passing through two treatment marshes in California, pyrethroids were associated overwhelmingly with suspended solids <2 µm, and the sorbed fraction accounted for 38–100% of the total concentrations. The derived K_d values for the suspended solids were in the order of 10^4–10^5 L/kg, substantially greater than those reported for bulk soils and sediments.

The proportion of the pesticide in the solid phase is $\rho_b K_d/\theta$ he proportion of the pesticide in the solid phase is lk soils and sediments.d Ktion accounted for 38–100% of, and 40% carbon. The median for K_{OC} values given by Vymazal and Březinová (2015) is 364 L/kgC, which for 40% carbon gives K_d = 146 L/kg. The fraction of pesticide in the solid phase is then 97%.

The larger the K_d value, the more strongly the pesticide adsorbs to soil, which then determines how a pesticide moves in the runoff. A pesticide with a low K_d and high solubility will move in the dissolved form, while pesticides with high K_d will move with suspended sediment particles. For some compounds, sorption is mostly reversible (Passeport et al., 2011), and hence binding is not permanent in the likely event of flushing. For others, the binding is stable, and the pesticide remains associated with the solid. Strong sorption to sediment may limit biochemical availability and thereby inhibit microbial degradation (Budd et al., 2011). For weakly bound substances, if degradation does not occur, then sorption simply contributes to a chromatographic effect, in which an incoming pulse of pesticide is retained for a period, but then later released.

Strongly bound pesticides, which are the majority, are subject to accumulation in and accumulating marsh sediment layer. Two approaches to a mass balance description of this dual accretion process were presented in the previous section (see Equations 13.6–13.9). The result of such stable accretion is a decreasing gradient of stored pesticide along the flow direction, which has been measured by Budd et al. (2009) for several pesticides.

Sedimentation/Resuspension

Because pesticides partition to solids, they are subject to removal from the water column be settling and filtration, and to return to the water by resuspension. Boithias et al. (2014)

TABLE 14.1

Mass and concentration reductions for diazinon in two studies.

		Concentration Reduction	Water Loss	Mass Reduction
Tres Rios, AZ	H1	60%	26%	70%
Tres Rios, AZ	H2	29%	28%	49%
Tres Rios, AZ	C1	22%	74%	80%
Tres Rios, AZ	C2	12%	18%	28%
San Joaquin River, CA	Wetland 1	22%	87%	92%
San Joaquin River, CA	Wetland 2	82%	68%	94%

Sources: Tres Rios, unpublished data; San Joaquin, Budd et al. (2009).

discuss the connection between TSS and pesticide partitioning, vis-à-vis transport in flowing waters. However, pesticides may be preferentially partitioned to lighter or finer suspended solids, and are not easily removed in sedimentation basins (Budd et al., 2011). The enhanced TSS removals in vegetated wetlands may be required for this mechanism to be effective, but there is no guarantee that will happen (Budd et al., 2011).

Infiltration

Because treatment marshes do not require lined basins, there is a possibility of infiltration of incoming source waters. Infiltration removes water-borne contaminants from marsh surface waters and their subsequent discharge to receiving waters. It is a subjective decision as to whether such infiltration constitutes removal/retention, or is simply the redirection of inflows to a different receiving water body (groundwater). It is virtually certain that passage of infiltration waters through the wetland action zone results in sorption.

Infiltration losses are manifested in the mass reductions achieved in the marsh, as compared to the concentration reductions. These differences can be substantial (Table 14.1).

14.3 CHEMICAL MECHANISMS OF REMOVAL

Photolysis

Many pesticides break down under the influence of solar radiation. This is a second-order process, depending upon both the concentration of the pesticide and the intensity of radiation. But, in the event of a constant radiation source (likely) it becomes pseudo-first order. As a result, batch photodegradation experiments produce log-linear decrease rates. Zeng and Arnold (2012) found such first order behavior with $R^2 > 0.98$ for 16 different pesticides.

Studies have shown that photochemical reactions contribute to the transformation of a wide array of pesticides via direct and indirect pathways. Direct photolysis occurs when a contaminant directly absorbs light energy and undergoes a chemical reaction. Direct photolysis involves the ultraviolet portion of the sunlight spectrum (wavelength ca. 200–300 nm), which is a very small fraction of incoming solar radiation. Hence direct photolysis is rarely an important pathway for pesticides.

Indirect photolysis occurs when a photosensitizer chemical species becomes electronically excited due to light absorption, and produces reactive intermediates that can degrade the contaminant. Treatment marsh waters contain several photosensitizers, including dissolved organic matter (DOM), oxidized nitrogen (NO_XN), and carbonate and bicarbonate, which are capable of generating a suite of reactive intermediate free radicals. Concentrations of such radicals are sufficient to promote the photodegradation of aquatic contaminants (Zeng and Arnold, 2012; Jasper and Sedlak, 2013).

Rate coefficients for pesticide photolysis are generally quite high, resulting in short half-lives in irradiated water. Zeng and Arnold (2012) used waters from an agriculturally impacted wetland in North Dakota to measure rates of direct and indirect photolysis (Table 14.2). Indirect mechanisms involved hydroxyl, oxygen, carbonate and DOM radicals. Half-lives are generally observed to be a few hours.

These results encouraged investigators to recommend open water systems for reduction of trace organics (Jasper et al., 2013). Fast photolytic removal of trace organics is dependent upon sunlight penetration, and hence vegetation must be virtually absent. Macrophytes, including EAV, SAV, FAV, create shading and cannot be present. Most mesophytic components must also be absent, including plankton and metaphyton, because they too block incoming radiation. The only component to be encouraged is epipelon (benthic biofilm mats), which has the potential to compatibly contribute biotransformation to augment photolysis.

There are, therefore, strong management considerations implied, to maintain the open water character. In addition to vegetation removal, bioturbation must be avoided, because turbid waters are inimical to photolysis. Consequently, animal activity must be prevented. Aquatic rodents, waterfowl and fish all have the potential to create turbid conditions. Thus, an open water system should probably not be characterized as a treatment marsh, because it needs to be essentially devoid of flora and fauna.

TABLE 14.2

Rate coefficients and half-lives for pesticide reduction in waters from a North Dakota wetland.

	Direct k	**Direct Half-life**	**Overall k**	**Indirect Half-life**
Pesticide	**1/day**	**hr**	**1/day**	**hr**
Atrazine	1.80	10.7	6.38	2.6
Cyanazine	1.39	13.8	3.75	4.4
Acetochlor	2.92	6.6	6.29	2.6
Alachlor	2.71	7.1	6.57	2.5
Metolachlor	2.59	7.4	6.62	2.5
Mesotrione	0.48	39.8	4.54	3.7
Diuron	43.20	0.4	30.24	0.6
Isoproturon	16.90	1.1	65.32	0.3
Metoxuron	30.00	0.6	79.83	0.2
Bentazon	24.90	0.8	47.78	0.3
Clopyralid	0.10	190.6	0.94	17.7
Chlorpyrifos	1.98	9.7	3.54	4.7
Propiconazole	0.02	848.0	3.42	4.9
Ethalfluralin	460.00	0.0	397.44	0.0
Pendimethalin	27.20	0.7	23.24	0.7
Trifluralin	381.00	0.1	328.32	0.1

Source: Zeng and Arnold (2012).

Nevertheless, open water portions of vegetated systems contribute to the removal of trace organics via photolysis.

Hydrolysis

Hydrolysis of a pesticide is basically a reaction with a water molecule involving specific catalysis by proton or hydroxide, and sometimes inorganic ions such as phosphate ion, present in the aquatic environment that play a role in general acid-base catalysis. Hydrolysis transformation rates are dependent on pH and temperature (Katagi, 2002).

Dannenberg and Pehkonen (1998) studied the hydrolysis of four organophosphorus pesticides (demeton S, diazinon, disulfoton, and thiometon) in the presence or absence of metal oxides, which can act as catalysts for hydrolysis. Half-lives were found to be quite long in natural waters, ranging from 26 to 149 days (Table 14.3).

Microbial Degradation

Biodegradation is an important route of pesticide degradation. Whereas plants, animals, and fungi (Eukaryota) typically transform pesticides for detoxification or through fortuitous metabolism by broad-spectrum enzymes, bacteria (Prokaryota) more commonly metabolize them for assimilation as essential nutrients and energy. Treatment marshes provide a broad suite of both types of organisms. Algal and microbial degraders are found preferentially in biofilms characterized by epipelon, epiphyton, plankton and metaphyton (see Chapter 5).

It is to be expected that the specific degrading organisms associated with each pesticide will differ. Only a few organism identification studies for wetlands have been conducted. Lee et al. (2004) found that the transformation of pyrethroids (bifenthrin and permethrin) was relatively rapid in solid-free media (half lives less than 100 hr), but that degradation slowed down by at least 10-fold in the presence of sediment. They suggested that strong sorption to the sediment phase made the pyrethroids less available to microorganisms and prolonged their persistence. Bifenthrin was degraded by *Stenotrophomonas acidaminiphila*. Permethrin was degraded by *Aeromonas sobria, Erwinia carotovora*, and *Yersinia frederiksenii*. These studies were done in microcosms, which are not necessarily accurate representations of field results (Jasper et al., 2014).

Runes et al. (2001) evaluated the microbial degradation of atrazine in sediments taken from a constructed wetland treating irrigation runoff from a container nursery near Portland, Oregon. To stimulate atrazine – degrading ability in wetland sediments dried shoot material from cattail were incorporated and sediment soil was bioaugmented with an atrazine spill-site soil known to contain an atrazine-mineralizing microbial populations. The population density

TABLE 14.3

Example parameters for homogeneous hydrolysis of pesticides in water.

	pH 5.7	pH 8.5	pH 5.7	pH 8.5	pH 5.7	pH 8.5
	k	k	t 1/2	t 1/2	θ	θ
	1/d	1/d	days	days		
Diazinon	0.017	0.005	40	149	1.120	1.184
Disulfoton	0.027	0.018	26	38	1.100	1.146
Demetron	0.011	0.020	62	35	1.140	1.167
Thiometon	0.012	0.015	57	47	1.120	1.133

Source: Dannenberg and Pehkonen (1998).

of atrazine-degrading microorganisms in unbioaugmented sediment was increased from about 10^2/g to 10^4/g by bioaugmentation (1:100 w/w) and increased by another 60-fold (6×10^5/g) after incubation with 10 μg/g of atrazine. A high population of atrazine degraders (~10^6/g) and enhanced rates of atrazine mineralization also developed in bioaugmented sediment after incubation in flooded mesocosms planted with *Typha latifolia*.

Anderson et al. (2002) studied the mineralization of atrazine in sediments from a constructed wetland in central Ohio. Atrazine mineralization potential was measured by $14CO_2$ evolution from [U-ring-14C]-atrazine in biometers. The constructed wetland showed 70 to 80% mineralization of atrazine within one month. The highest levels of mineralization were localized to the top 5 cm of the wetland sediment.

Plant Uptake

The basic mechanisms involved in plant uptake of trace organics are not yet well understood, but all evidence points to the existence of plant uptake in wetland environments. Zhang et al. (2014) note that in the cell membranes of plants, there are no specific transporters for trace organic compounds like pesticides and pharmaceuticals. Consequently, the uptake of such micropollutants into the plant is considered to be driven by diffusion, which depends on the chemical properties of a specific compound, especially its hydrophobicity.

The review of Carvalho et al. (2014) indicates that organic compounds can be transported across plant membranes. Low molecular weight compounds can be removed from the soil or water and released through leaves via evapotranspiration processes (phytovolatilization). Other compounds can be degraded via enzymatic modification and sequestration in the plant (phytodegradation, phytoextraction). However, Carvalho et al. (2014) conclude that despite the occurrence of plant uptake, it usually represents a low percentage of the overall removal.

Trace organics can have toxic effects on wetland vegetation. Indeed, there exists a suite of herbicides that target eradication of aquatic plants. Apart from these, pharmaceuticals, insecticides and terrestrial herbicides are normally present in treatment marsh source waters at low concentrations. Thus it is not expected that phytotoxic effects will occur in treatment marshes.

There is a building set of substance-specific studies that examine the role of plant uptake in more detail. Zarate et al. (2012) report the accumulation of the pharmaceuticals triclosan (TCS) and triclocarban (TCC) in plant tissues and sediments. Three wetland macrophytes: *Typha latifolia, Pontederia cordata*, and *Sagittaria graminea* were sampled from a constructed wetland in Denton, Texas, USA. Their results show that TCS and TCC readily accumulate in the root tissues of free-living wetland plants and show species specific differences.

Lv et al. (2016) conducted hydroponic studies on imazalil and tebuconazole, which are used as pesticides to control fungi in various crops. Four macrophytes were used, *Typha latifolia, Phragmites australis, Iris pseudacorus* and *Juncus effusus*. A large fraction of the pesticides disappeared from the hydroponic solution (14% – 90%). However, the accumulation by wetland plants was relatively low (2.8–14.4% of the added pesticides). Overall loss rate coefficients for imazalil were 0.14, 0.25, 0.09 and 0.10 d^{-1} for *Typha, Phragmites, Iris* and *Juncus*, respectively; corresponding to half-lives of 5.1, 2.8, 7.7 and 7.0 days. Overall loss rate coefficients for tebuconazole were 0.10, 0.09 and 0.12 d^{-1} for *Typha, Phragmites*, and *Juncus*, respectively; corresponding to half-lives of 6.9, 7.7 and 6.0 days. However, the two competing processes of microbial degradation in solution and metabolism following plant uptake could not be distinguished.

Synthesis of Mechanisms

It is possible to combine the separate mechanisms described above into a detailed comprehensive model of pesticide removal. One class of such combinations is receiving waterbody models, the purpose of which is to assess the concentration levels and persistence in receiving waterbodies, which are systems without vegetation. These are relevant, because the plant uptake route is expected to be secondary even in treatment marshes. Xie (2016) reviewed three of these, including the Variable Volume Water Model (VVWM) developed by the U.S. Environmental Protection Agency Office of Pesticide Programs (U.S. EPA/OPP). VVWM is described in detail by Young (2014), and includes photolysis, hydrolysis, volatilization, sorbed metabolic degradation, and aqueous-phase metabolic degradation; in addition to mass transfer between the water and benthic region, and sorption to benthic and suspended sediments and dissolved organic matter. As is the case with all detailed models, the demand for input parameters is onerous, with dozens of rate coefficients and other parameters required. The European equivalent is the TOXSWA software, an acronym for toxic substances in surface waters (Beltman et al., 2014).

Similar models are available that purport to be specific to treatment marshes. Sharif et al. (2014) developed a detailed model using AQUASIM, programmed by EAWAG (Swiss Federal Institute of Aquatic Science and Technology) (Reichert, 1998). The basic fate-transport equation includes some of the same removal mechanisms as waterbody models; namely photolysis, hydrolysis, biodegradation and sorption. However, the wetland context presents a few new, unaccounted variables, such as the fraction open water that controls photolysis. The demand for input parameters is very large, with many parameters required, which are not available. Further, the effects of nonideal internal hydraulics are not accounted.

For these reasons, it is not yet feasible to use detailed models in the design forecasting mode. Instead, reliance must be placed in overall pesticide loss rates for wetlands, in accordance with the PkC* simplified model framework. This procedure has been suggested by Silverman et al. (2019). They suggest $C^* = 0$ for trace organics, since the wetland does not manufacture those substances. The parameter P is suggested to estimated from design considerations. Importantly, they also suggest that the loss rate coefficient for trace organics is the sum of the rate coefficients for the component processes, including photolysis, sorption and bio transformation. The photolysis rate coefficient is itself dependent upon light intensity and other water chemistry, for which Silverman et al. (2019) provide estimation procedures based on latitude (solar radiation), pH and concentration of dissolved organic matter. This expanded PkC* approach shows promise.

14.4 PESTICIDES AND HERBICIDES

The list of pesticides is very long, and therefore wetland fate and transport studies lag far behind their introduction and use. A distinction may be drawn between the persistent chemicals used prior to the 1950s and the more degradable substances used since that time. At the present time, there are many banned pesticides, 64 in the USA as of 2011, and 109 in the EU as of 2008. These include well-known names such as atrazine, aldrin, DDT, dieldrin, diazinon, parathion, and mirex. Many of these "old" pesticides, such as DDT, are very persistent in the environment. Modern pesticides degrade more readily, and wetlands have been found to generally reduce levels of many of these compounds

Toxic Effects

These trace organics potentially have deleterious effects on aquatic ecosystems, including both treatment marshes and the receiving waters they might seek to protect. Therefore, design targets are defined by environmental effects, and treatment marshes are constrained to avoid those same effects. In the USA, there exists a dual system of criteria concerning aquatic effects, one administered by EPA's Office of Pesticide Programs (OPP) and a second by EPA's Office of Water (OW). Both the OW and OPP methods assess and characterize effects for protecting aquatic communities.

OW uses aquatic toxicity data to develop ambient water quality criteria that can be adopted by states and tribes to establish water quality standards under the Clean Water Act. National recommended water quality criteria are available for 21 pesticides (U.S. EPA, 2018a) (Table 14.4). Procedures for deriving the criteria are described in OW's Guidelines for Deriving Numerical National Water Quality Criteria for Protection of Aquatic Organisms and require, for acute criteria, that data be available for at least eight families.

OPP uses aquatic toxicity data in its ecological risk assessments for a large number of pesticides and rely on data for the most sensitive tested effects concentration for each taxon. Aquatic life benchmarks, developed for baseline risk assessments, are estimates of the concentrations below which pesticides are not expected to harm aquatic life. Aquatic life benchmarks are available for 584 pesticides (U.S. EPA, 2017c). Examples of the aquatic life benchmarks for freshwater species and ambient water quality criteria for registered pesticides in the United States are provided in Table 14.5.

Such criteria may be interpreted in a risk reduction context for treatment marshes (Stehle et al., 2011). Threshold levels are transformed to toxic units as a complementary measure to concentrations:

TABLE 14.4

U.S. EPA National Recommended Aquatic Life Criteria table for pesticides in freshwater (μg/L). CMC: Criterion Maximum Concentration; CCC: Criterion Continuous Concentration; P = Priority Pollutant.

Pollutant	CAS Number	CMC (acute)	CCC (chronic)
Aldrin (P)	309002	3	–
alpha-Endosulfan (P)	959988	0.22	0.056
Atrazine	1912249		
beta-Endosulfan (P)	33213659	0.22	0.056
Carbaryl	63252	2.1	2.1
Chlordane (P)	57749	2.4	0.0043
Chlorpyrifos	2921882	0.083	0.041
Demeton	8065483	–	0.1
Diazinon	333415	0.17	0.17
Dieldrin (P)	60571	0.24	0.056
Endrin (P)	72208	0.086	0.036
gamma-BHC (Lindane) (P)	58899	0.95	–
Guthion	86500	–	0.01
Heptachlor (P)	76448	0.52	0.0038
Heptachlor Epoxide (P)	1024573	0.52	0.0038
Malathion	121755	–	0.1
Methoxychlor	72435	–	0.03
Mirex	2385855	–	0.001
Parathion	56382	0.065	0.013
Toxaphene (P)	8001352	0.73	0.0002
4,4′-DDT (P)	50293	1.1	0.001

TABLE 14.5

Examples of aquatic life benchmarks (μg/L) listed by the Office of Pesticide Programs (OPP). Compounds in bold are banned in the European Union.

Pesticide	Fish		Invertebrates		Nonvascular Plants	Vascular Plants
	Acute	Chronic	Acute	Chronic	Acute	Acute
Acephate	416000	5760	550	150	>50000	
Alachlor	900	187	1250	110	1.64	2.3
Aldicarb	26	78	10	20	>50000	>88700
Ametryn	1800	700	14000	240	3.67	13
Atrazine	2650	5	360	60	<1	4.6
Bifenthrin	0.075	0.04	0.8	0.0013		
Chlorpyrifos	0.9	0.57	0.05	0.04	140	
Cyfluthrin	0.034	0.01	0.0125	0.0074	>181	
Diazinon	45	<0.55	0.105	0.17	3700	
Endosulfan	0.05	0.023	0.3	0.01	428	
Glyphosate	21500	25700	26600	49900	12100	11900
Lindane	0.85	2.9	0.5	54		
Malathion	2.05	8.6	0.049	0.06	2040	24000
Methoxychlor	7.5		0.7			
Metolachlor	1900	30	550	1	80	21
Propachlor	85		395		13.5	
Pyrethrin	2.55	1.9	5.8	0.86		
Simazine	3200	960	500	40	2.24	140

Source: U.S. EPA (2017c).

$$TU = \ln\left(\frac{C}{C_C}\right) \tag{14.3}$$

where C = pesticide concentration, μg/L
C_C = pesticide criterion concentration, μg/L
TU = pesticide toxic unit, –

Stehle et al. (2011) surveyed 31 systems (22 marshes and 9 ditches), and classified the results according to pesticide type. These had very short detention times, with a median of 0.5 days. Using criteria of the European Commission, herbicides showed a mean TU reduction of 0.64, insecticides 1.26, and fungicides 0.71. The overall mean reduction was 0.90, from TU = –0.39 to TU = –1.29. Negative values indicate that concentrations are below the criterion. The median concentration reduction for the entire 188 systems inventoried was 74%.

Source Waters for Pesticides

Reductions depend upon the compound in question, which varies among source waters. The U.S. Geological Survey (1999) characterized common substances in urban and agricultural use. Pesticides – primarily herbicides – are found frequently in agricultural streams and shallow ground water. Herbicide use in agricultural areas, accounts for about 70% of total national use of pesticide. The highest rates of detection were for atrazine, metolachlor, alachlor, and cyanazine. Insecticides were frequently detected in streams draining watersheds with high insecticide use.

Urban areas traditionally have not been recognized as important contributors to pesticide contamination via runoff. However, findings showed a widespread occurrence of some insecticides commonly used around homes and gardens and in commercial and public areas. In fact, these insecticides occurred at higher frequencies, and usually at higher concentrations, in urban streams than in agricultural streams. Most common were diazinon, carbaryl, chlorpyrifos, and malathion. Some herbicides–including atrazine, simazine, and prometon, which are used to control weeds in lawns and golf courses, and along roads and rights-of-way–also occurred frequently in samples collected from streams and shallow ground water in urban areas.

Household pesticides find their way to municipal wastewater treatment plants, as do some via storm sewers. Wastewater treatment facilities may provide only partial treatment. As a consequence, treated wastewater discharges represent a source to receiving waters.

Pesticide Reductions in Marshes

Marshes provide pesticide reduction for all pesticides tested, and in most cases, significant reductions. As previously noted, percentage reductions are inadequate for either interpretation of performance results or for forecasting. First order, or pseudo first order behavior is found in the majority of instances. This is of course the preferred model for batch systems. For flow-through systems, the concept needs to expand to include the nonideal flow patterns that are very often present in field-scale systems. In the absence of hydraulic efficiency information, the default approximation is a plug flow condition. Thus both batch and default approaches presume exponential decreases with either time (batch) or detention time (flow). Detention time is also represented by the pesticide half-life:

$$t_{1/2} = \frac{0.693}{k} \tag{14.4}$$

where k = first order loss rate coefficient, 1/d

$t_{1/2}$ = pesticide half-life, days

Common pesticides have been the focus of multiple wetland studies (Table 14.6). Loss rate coefficients vary considerably from study to study, but the corresponding half-lives are measured in days. Brief descriptions of these are as follows (see also Figure 14.1):

- Atrazine, a triazine herbicide, is used to prevent pre- and post-emergence broadleaf weeds in crops such as maize (corn) and sugarcane and on turf, such as golf courses and residential lawns. It is one of the most widely used herbicides in US and Australian agriculture. It was banned in the European Union in 2003. In the United States as of 2014, atrazine was the second-most widely used herbicide after glyphosate. Studies suggest that atrazine is an endocrine disruptor.
- Diazinon is an insecticide that belongs to a group of chemicals known as organophosphates. Diazinon is used in agriculture to control insects on fruit, vegetable, nut and field crops. Diazinon was used extensively in the United States starting in the 1950s. Before the cancellation of residential uses in 2004, diazinon was used for household insects, lawn and garden insect control, and to control insects on pets. It is banned in the European Union.
- Chlorpyrifos is an organophosphate insecticide used to control many different kinds of pests, including termites, mosquitoes, and roundworms. Chlorpyrifos is very toxic to many bird species such as grackles and pigeons. Chlorpyrifos is also very toxic to fish and aquatic invertebrates. Chlorpyrifos is very toxic to bees.

TABLE 14.6

Example concentration reductions for pesticides with multiple studies.

Size	HRT days	Reduction %	k 1/d	Reference
Atrazine				
Mesocosm	10	50	0.069	Moore et al. (2017)
Mesocosm	2.1	33	0.191	Moore et al. (2013)
Field	2.75	64	0.372	Passeport et al. (2013)
Field EW3	6.8	33	0.063	Alvord and Kadlec (1996)
Field EW4	51.2	64	0.024	Alvord and Kadlec (1996)
Field EW5	8.4	26	0.038	Alvord and Kadlec (1996)
Field ENRP	16.5	72.5	0.078	He (2007)
Field STA1W	12.2	-6	-0.005	He (2007)
Field STA2	9.36	50.2	0.074	He (2007)
Field STA5	12	-108	-0.061	He (2007)
Field STA6	8.7	62.1	0.112	He (2007)
Diazinon				
Field C1	3.2	22	0.078	Unpublished data
Field C2	6.3	12	0.020	Unpublished data
Field H1	6.9	60	0.133	Unpublished data
Field H2	6.8	28	0.048	Unpublished data
Field	13	50	0.053	Moore et al. (2007)
Mesocosm	4.5	50	0.154	Moore et al. (2017)
Field	7	68	0.163	Moore et al. (2007)
Field	4.11	42	0.133	Krone-Davis et al. (2013)
Chlorpyrifos				
Field	22.5	95	0.133	Darby (1995)
Field	1.7	100	4.063	Schulz and Peall (2000)
Mesocosm	2.5	40	0.204	Karen et al. (1998)
Mesocosm	3	98	1.304	Sherrard et al. (2004)
Pilot Field	4.6	50	0.151	Moore et al. (2002b)
Pilot Field	9.4	50	0.074	Moore et al. (2002b)
Pilot Field	13	50	0.053	Moore et al. (2002b)
Pilot Field	9.1	50	0.076	Moore et al. (2002b)
Field	1.1	61	0.856	Budd et al. (2011)
Field	5.4	52	0.136	Budd et al. (2011)
Permethrin				
Field	2.1	4	0.019	Lopez-Flores et al. (2003)
Mesocosm	2.1	80	0.766	Moore et al. (2013)
Field	1.1	90	2.093	Budd et al. (2011)
Field	5.4	94	0.521	Budd et al. (2011)

- Permethrin is a pyrethroid insecticide in the family. It is used in a number of ways to control insects. Products containing permethrin may be used in public health mosquito control programs. They may be used on food and feed crops, on ornamental lawns, on livestock and pets, in structures and buildings, and on clothing. Permethrin may also be used in places where food is handled, such as restaurants.

Other less common substances have produced only one or two studies (Table 14.7). Many of these are for dosed wetlands. The wetland type across this sampling is significantly variable, and thus the cross section includes two very important controlling factors, wetland type and pesticide. Note that specification of HRT and percent removal allows estimation of the loss rate coefficient:

$$k = \left(\frac{1}{HRT}\right) \ln\left(\frac{1 - \%\text{Removal}}{100}\right) \tag{14.5}$$

Another viewpoint on performance is the behavior of a single treatment marsh for a suite of pesticides. This removes the wetland type variation. There are three distinctly

TABLE 14.7

Concentration reductions for treatment marshes for a cross-section of pesticide types.

Pesticide	Size	Location	HRT days	Reduction %	Reference
Aldrin	Field	Spain	2.10	86	Lopez-Flores et al. (2003)
Azinphos-methyl	Field	South Africa	1.70	100	Schulz and Peall (2000)
Bifenthrin	Field	CA	0.92	84	Budd et al. (2011)
Chlorothalonil	Mesocosm	SC	0.50	98	Sherrard et al. (2004)
Cinosulfuron	Field	Italy	21	81	Ferrero et al. (2001)
Cyfluthrin	Field	MS	7.00	100	Moore et al. (2006)
Cyhalothrin	Field	CA	0.92	90	Budd et al. (2011)
Cyhalothrin	Field	MS	7.00	100	Moore et al. (2006)
DDT	Field	Spain	2.10	100	Lopez-Flores et al. (2003)
Dicamba	Pilot	Norway	0.21	78	Elsaesser et al. (2011)
Dimethoate	Pilot	Norway	0.21	100	Elsaesser et al. (2011)
Dimethomorph	Pilot	France	0.45	76	Maillard et al. (2011)
Diuron	Pilot	NSW, AUS	5.00	50	Rose et al. (2006)
Endosulfan	Field	South Africa	1.70	100	Schulz and Peall (2000)
Endosulfan	Pilot	NSW, AUS	3.00	50	Rose et al. (2006)
Esfenvalerate	Field	CA	0.92	77	Budd et al. (2011)
Fluometron	Pilot	NSW, AUS	10.50	50	Rose et al. (2006)
Glyphosate	Pilot	France	0.45	79	Maillard et al. (2011)
Lindane	Field	Spain	2.10	-19	Lopez-Flores et al. (2003)
Metalaxyl	Pilot	France	0.45	70	Maillard et al. (2011)
Metamitron	Pilot	Norway	0.21	96	Elsaesser et al. (2011)
Methomyl	Pilot	CA	4.51	12	Krone-Davis et al. (2013)
Methyl Parathion	Mesocosm	MS	4.00	100	Schulz et al. (2003)
Methyl parathion	Pilot Field	MS	10.00	100	Moore et al. (2006)
Simazine	Field	Spain	2.10	100	Lopez-Flores et al. (2003)
S-metalochlor	Field	France	2.75	80	Passeport et al. (2013)
S-metalochlor	Mesocosm	MS	6.00	50	Moore et al. (2017)
Tebuconazole	Pilot	Norway	0.21	88	Elsaesser et al. (2011)
Trifloxystrobin	Pilot	Norway	0.21	96	Elsaesser et al. (2011)
Median			**2.10**	**86**	

different ways that pesticides reach treatment marshes: tile drainage, overland flow runoff, and spray over-drift. Tile drainage has passed through the soils of the contributing catchment, and hence been subjected to some processing, such as sorption. Runoff carries particulates that may have sorbed pesticide attached, as well as dissolved pesticide. Spray drift deposits the pesticide over the wetland surface, as opposed to bringing it into the wetland at a point.

One such study, which covered three years of operation, was the treatment marsh at Bray, France (Passeport et al., 2013). This system treated tile drainage from an agricultural catchment, where cereal crops are grown. This 1280 m^2 marsh was vegetated with *Glyceria maxima* (53%), *Festuca arundinacea* (12%), *Phragmites australis* (10%) and *Phalaris arundinacea* (9%). The depth was approximately 50 cm, and the detention time 5.3 days. The measured half-lives for the sixteen pesticides are quite short, on the time scale of hours (Table 14.8).

Comprehensive studies of runoff wetlands have been conducted in Norway (Braskerud and Haarstad, 2003; Haarstad and Braskerud, 2005; Blankenberg et al., 2007). Runoff was collected in small sedimentation basins, and then passed into treatment marshes. The Lier system, located 40 km south of Oslo, has a catchment area is 0.8 km^2 in which agriculture is dominant (Christmas tree, vegetable, and cereal production). Eight parallel, 120-m^2 test wetlands were monitored continuously for two years (Blankenberg et al., 2007). Vegetation was mainly *Phalaris arundinacea, Glyceria fluitans, Typha latifolia, Sparganium erectum, and Phragmites australis.* Approximately 96% of the applied pesticides disappeared within the watershed, leaving only trace amounts entering the wetland. The measured half-lives for six pesticides were quite short, mostly in the range of 0.3–3 days (Table 14.9).

Degenhardt et al. (2011) determined dissipation rates of commonly used herbicides in the water column of two prairie wetlands: an ephemeral wetland and a semi-permanent wetland. Spray application was utilized, and herbicides were not applied to the catchments in 2008 when

TABLE 14.8

Pesticide removals in the treatment marsh at Bray, France.

Pesticide	Samples	Inlet µg/L	Outlet µg/L	Reduction %	Half-life days
Isoproturon	83	5.26	2.89	45	0.22
Chlorotoluron	72	9.88	1.58	84	0.07
Atrazine	26	0.03	0.01	64	0.13
Chlorothalonil	34	0.25	0.05	79	0.08
Prosulfocarb	22	0.32	0.02	93	0.05
Fenpropidin	46	0.03	0.02	50	0.19
Ethofumesate	23	0.15	0.01	96	0.04
S-metolachlor	25	0.03	0.01	80	0.08
Metazachlor	72	1.00	0.30	70	0.11
Napropamide	56	0.04	0.01	65	0.12
Cyproconazole	27	0.25	0.14	44	0.22
Aclonifen	10	0.18	0.04	80	0.08
Diflufenican	74	0.06	0.02	75	0.09
Tebuconazole	48	0.10	0.01	86	0.07
Mefenpyr-dietyl	36	0.09	0.02	80	0.08
Epoxiconazole	65	0.45	0.13	71	0.10

Source: Passeport et al. (2013).

TABLE 14.9

Pesticide reductions in the Lier, Norway standard test marsh (0.5 m deep).

Pesticide	Inlet µg/L	Outlet µg/L	Reduction %	k 1/d	Half-life d
2003					
Metalaxyl	0.14	0.06	57	1.30	0.53
Metamitron	0.39	0.10	74	2.10	0.33
Metribuzin	0.09	0.06	33	0.62	1.11
Propachlor	0.13	0.11	15	0.26	2.69
Linuron	0.29	0.17	41	0.82	0.84
Fenpropimorph	0.07	0.02	71	1.93	0.36
2004					
Metalaxyl	0.16	0.15	3	0.06	10.81
Metamitron	0.13	0.11	15	0.33	2.11
Metribuzin	0.26	0.19	28	0.64	1.07
Propachlor	0.62	0.29	54	1.55	0.45
Linuron	0.04	0.02	49	1.33	0.52
Fenpropimorph	0.06	0.03	50	1.37	0.50

Source: Blankenberg et al. (2007).

the study took place. Results showed that the dissipation of all six herbicides could be described by first-order reaction kinetics. Infiltration through sediment with concomitant sorption to the sediment was an important dissipation route. The measured half-lives for six herbicides were long compared to other pesticides, mostly in the range of 2–20 days (Table 14.10).

In some instances, treatment wetlands have been added to the conventional wastewater treatment plants for purposes of further reducing TSS, BOD, nutrients and pathogens. For example, treatment wetlands are the final element of the treatment train at Columbia, MO. This 37 ha marsh was vegetated with *Typha* at 70% coverage. The depth was approximately 30 cm, and the detention time 2.42 days (Kadlec et al., 2010b). Petty et al. (2004) used integrative samplers to measure pesticides entering and leaving the marsh, which sampled a 28-day period of operation. They found mostly the old, banned pesticides. The Columbia list of detected pesticides had no relation to or overlap with the list from agricultural sources such as that at Bray, France. Yet, reductions at Columbia were quite high, with half-lives in the range 0.5–2.5 days (Table 14.11).

Lagoon treatment systems also can pass pesticides into receiving treatment systems. At Grand Marais, Manitoba, Canada, wastewater is treated passively in a sewage lagoon prior to passage through a treatment wetland and subsequent release into surface waters (Anderson et al., 2013). Six pharmaceuticals and pesticides were detected in samples from the lagoon-wetland system: herbicides 2,4-D and atrazine, pharmaceuticals carbamazepine and gemfibrozil, and antibiotics sulfamethoxazole and sulfapyridine. They were only found episodically, at nanogram per liter concentrations, and thus did not present an opportunity for performance analysis.

PkC* Interpretations

A few studies of pesticides in flow-through marshes have gone beyond the assumptions of plug flow and zero background concentration. While there are but few published

TABLE 14.10

Herbicide reductions in the Brandon, Manitoba prairie wetlands.

	C_0 (μg/l)	k 1/d	R^2	Half-life (d)
Small Ephemeral Wetland				
Dicamba	93.6	0.07	0.85	9.8
Bromoxynil	61.6	0.30	0.99	2.3
MCPA	125	0.06	0.89	11.0
2,4-D	98.7	0.06	0.93	12.0
Mecoprop-P	178	0.04	0.62	18.0
Dichlorprop	105	0.02	0.43	31.0
Large Semi-permanent Wetland				
Dicamba	126	0.05	0.79	14.0
Bromoxynil	61.6	0.18	0.90	3.9
MCPA	161	0.07	0.96	9.6
2,4-D	98.7	0.05	0.96	13.0
Mecoprop-P	231	0.05	0.92	14.0
Dichlorprop	105	0.03	0.83	23.0

Source: Degenhardt et al. (2011).

TABLE 14.11

Pesticide removals in the treatment marsh at Columbia, MO, USA. This system treats municipal effluent.

Pesticide	Inlet μg/L	Outlet μg/L	Reduction %	k 1/d	Half-life days
Lindane	3300	100	97	1.44	0.48
Dacthal	240	120	50	0.29	2.42
Oxychlordane	160	18	89	0.90	0.77
Heptachlor epoxide	350	54	85	0.77	0.90
trans-Chlordane	1400	310	78	0.62	1.11
trans-Nonachlor	510	120	76	0.60	1.16
p,p0-DDE	140	25	82	0.71	0.97
cis-Chlordane	1500	590	61	0.39	1.80
Dieldrin	630	290	54	0.32	2.16
cis-Nonachlor	170	31	82	0.70	0.99
p,p0-DDD	210	9	96	1.30	0.53
p,p0-DDT	11	2.8	75	0.57	1.23
Methoxychlor	890	99	89	0.91	0.76

Source: Petty et al. (2004).

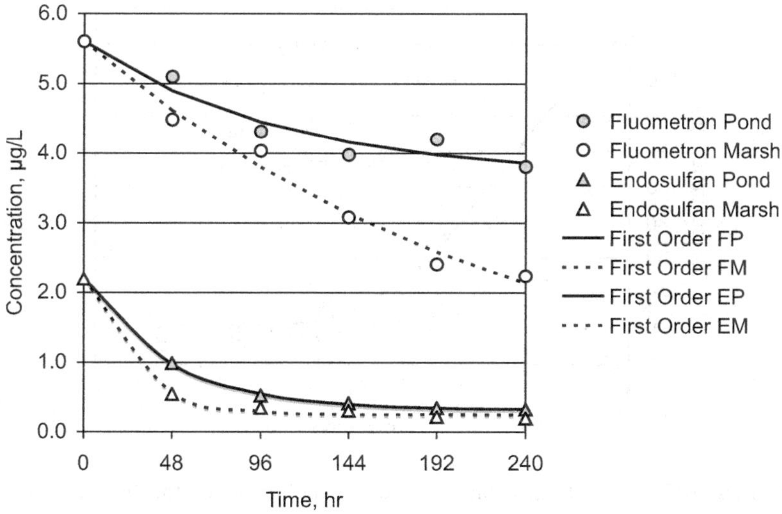

FIGURE 14.3 First order model with background concentration fit for two pesticides in a pond and a marsh. Data from Rose et al. (2008).

investigations relevant to possible marsh background concentrations, data seems to indicate the possibility of nonzero background concentrations. Although the marsh does not manufacture a pesticide, there can be return fluxes. Mechanisms include desorption, resuspension and decomposition of phytomass containing the pesticide. For batch studies, there is presumably not an issue of wetland hydraulic efficiency, and the question becomes whether there is a plateau in concentration at long times.

Two examples suggest that such plateaus do exist, at least on the time scale of an exposure event. Figures 14.3 and 14.4 display plateaus for four different pesticides. An exponential decrease in concentration to a nonzero background describes the time series of concentrations very well, with very high R^2 values, ranging from 0.94–1.00. These correlation coefficients are much higher than those for the case of zero background. Further, the plateau concentrations are a nontrivial fraction of the starting concentrations, in the range of 1–15% of initial values. It appears that the nearly universal assumption of zero background concentrations for pesticides may need to be revisited.

Parameters:

		k	**C***	**R^2**
		1/d	**µg/L**	
Fluometuron	Pond	0.0094	3.66	0.941
Fluometron	Marsh	0.0041	0.08	0.986
Endosulfan	Pond	0.0219	0.32	1.000
Endosulfan	Marsh	0.0376	0.25	0.997

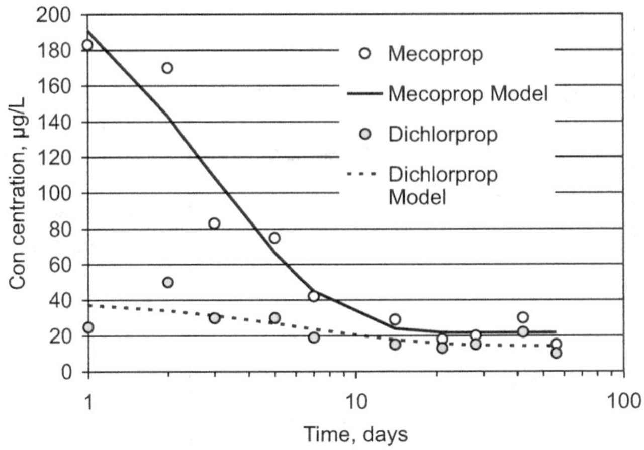

FIGURE 14.4 First order model with background concentration fit for two pesticides in a prairie marsh. Data from Degenhardt et al. (2011).

Fitting Parameters:

	k 1/d	C* µg/L	R^2
Mecoprop	0.33	21.6	0.952
MPCA	0.31	11.8	0.941

The hydraulic efficiency of a flow-through wetland is undoubtedly a major factor in pesticide removal in the majority of such systems. Exceptions do exist; for example the pesticide removal modeling at the Moleras marsh in the Salinas Valley of California. That marsh had an aspect ratio of 43:1, and model fitting determined a value of 17 tanks in series (Krone-Davis et al., 2013).

Some studies have utilized the technique of using the ratio of the pesticide to an inert tracer, which allows first order disappearance to be coupled to the actual detention time distribution. Lizzote et al. (2012) processed information on a semi-batch system in Tunica County, Mississippi. Simulated, dosed events were used, in which the wetland was not a flow through system, that is, all water stayed within the wetland. Chloride was used s = as the tracer to determine the position of "packets" of water as they progressed into the marsh.

Hydraulic efficiency was embodied in the modeling of atrazine removal in the Des Plaines, Illinois wetlands (Alvord and Kadlec, 1995, 1996). A three unit finite stage model was used, in which three tanks in series conducted the flow, while exchanging water and pesticide with three "side tanks" (see Chapter 7). This hydraulic model was separately determined in tracer studies. Dynamic water budgets were measured for the wetlands, including inflows, outflows, seepage, rain and ET. Atrazine sorption was studied separately in accompanying laboratory work. Sorption was effective for soils and sediments, but the more organic materials, such as litter, showed a stronger affinity for atrazine than the mineral base soils of the wetland cells at Des Plaines, Illinois (Alvord and Kadlec, 1995, 1996).

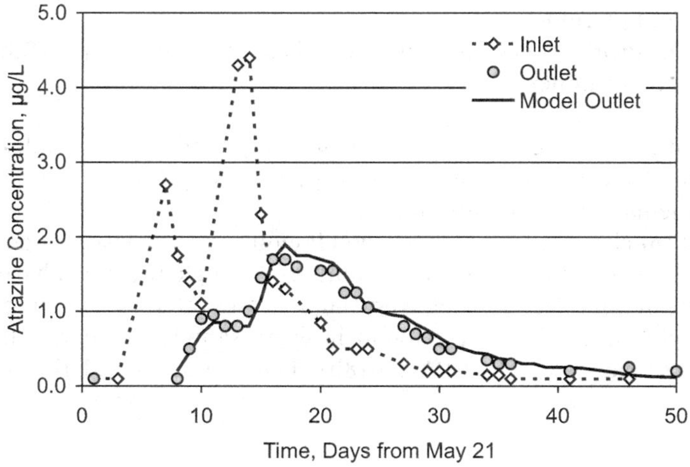

FIGURE 14.5 Des Plaines wetland EW3 atrazine inflow and outflow concentrations as a function of time. The model is a dynamic simulation that fits the hourly changes in flows of water and atrazine. Data source: Alvord and Kadlec (1996).

Atrazine was found to degrade on those sediments according to a first order rate law, with a half-life of 40 to 90 days. However, degradation was faster on cattail litter, precluding a rate measurement. The estimated half-life on cattail detritus was on the order of 5 days. Thus, a very large amount of supporting information was utilized in addition to flows and concentrations, to implement dynamic modeling.

Outflows from the Des Plaines wetland cells contained reduced amounts of atrazine compared to the river water inputs (Figure 14.5). Atrazine peaked in the river due to two rain events. Outflowing atrazine peaked a few days later, indicating a water transport delay. Runes et al. (2003) also observed such a transport delay. Event mean analysis showed that EW3 removed 33% of the incoming atrazine, but 64% was removed in wetland cell EW4, perhaps due to longer detention time. Either time averaged or dynamic models, both embodying a nonideal mixing model, fit the data for wetland EW3. A loss rate coefficient (k) of approximately 14 m/yr calibrated the models to the data. The nominal detention time for EW3 was 6.8 days, resulting in an EMC loss rate coefficient of 11.5 m/yr for three tanks in series.

14.5 PHARMACEUTICALS AND PERSONAL CARE PRODUCTS

The term "contaminant of emerging concern" (CEC) is used to identify chemicals that have no regulatory standard, and have received relatively recent attention because new analytical techniques have allowed their identification in natural waters at very low concentrations. They may potentially cause deleterious effects in aquatic life at identified existing concentrations. They are pollutants not currently included in routine monitoring programs and may be candidates for future regulation. CECs are not necessarily new chemicals. They include pollutants that have often been present in the environment, but whose presence and significance are only now being evaluated. CECs include two related types of chemicals that are the focus of this section:

- Pharmaceuticals and personal care products (PPCPs), including a wide suite of prescribed human drugs (e.g., antidepressants, blood pressure medicines),

over-the-counter medications (e.g., ibuprofen), bactericides (e.g., triclosan), sunscreens, and perfumes.
- Veterinary medicines such as antimicrobials, antibiotics, anti-fungals, growth promoters and hormones.

In common with the pesticides, many of these compounds have complex chemical structure (Figure 14.6). They are differentiated from pesticides by their low concentrations, most of which are in the nanogram per liter range.

U.S. EPA (2018b) cites their survey of WWTP effluents as a basis for preliminary indications of environmental hazards. Concentrations of 56 high priority pharmaceuticals were measured in effluent samples from 50 very large wastewater treatment plants across the USA. Although pharmaceuticals were found in every sample, the concentrations were low and in-line with predictions. U.S. EPA (2018b) therefore suggested that risks to humans,

Metropolol
(*RS*)-1-[4-(2-Methoxyethyl)phenoxy]-3-[(propan-2-yl)amino]propan-2-ol
$C_{15}H_{25}NO_3$
CAS Number 51384-51-1

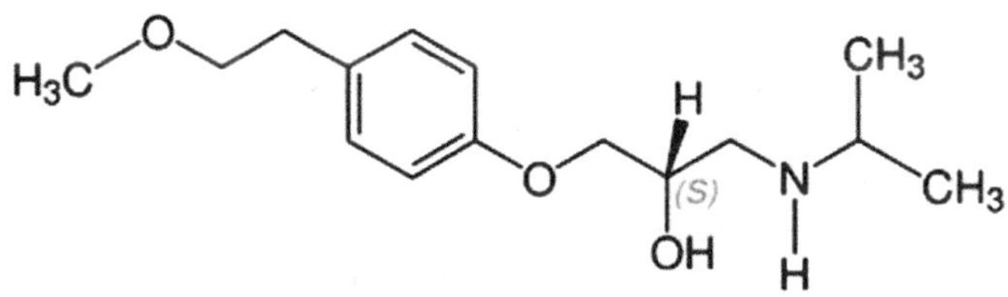

Ibuprofen
2-(4-Isobutylphenyl)propanoic acid
$C_{13}H_{18}O_2$
CAS Number 15687-27-1

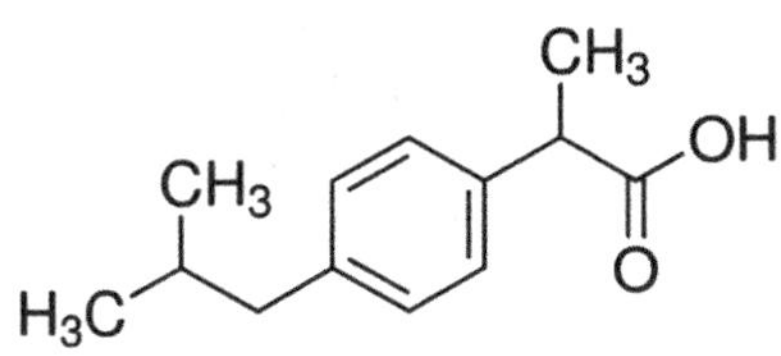

Atenolol
(*RS*)-2-{4-[2-Hydroxy-3-(propan-2-ylamino)propoxy]phenyl}acetamide
$C_{14}H_{22}N_2O_3$
CAS Number 29122-68-7

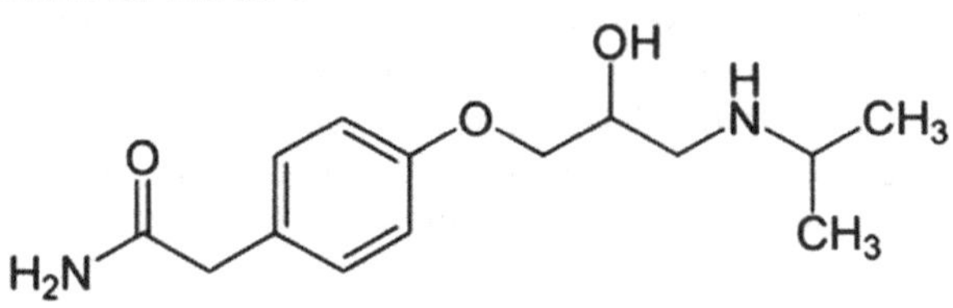

FIGURE 14.6 Examples of pharmaceutical chemical structures and naming.

from aquatic exposure (including drinking water) to single drugs or mixtures, are low (Kostich and Lazorchak, 2008). It also suggests that risks to aquatic life from most pharmaceuticals are low. However, for a few pharmaceuticals, risks to aquatic life are more difficult to dismiss due to uncertainties about both exposure and physiological responses, suggesting the need for further research.

More recently, Kostich and Länge (2015) concluded that the limited ecotoxicity data suggest that typical pharmaceutical concentrations are well below effect concentrations for all pharmaceuticals, including antibiotics, estrogens, analgesics, antidepressants, blood pressure medicines. They also conclude that there is little direct evidence that suggests important effects on ecological health. Others have continued to express concerns. For instance, Ebele et al. (2017) expressed that "An increasing number of studies has confirmed the presence of various PPCPs in different environmental compartments, which raises concerns about the potential adverse effects to humans and wildlife."

The potential for environmental effects, together with a desire to advance treatment wetland technology, has stimulated a quite large body of investigations of the ability of treatment wetlands of all kinds, and treatment marshes in particular, to reduce or eliminate PPCPs. In fact, the last four years have seen the publication of no less than seven review articles on the use of wetlands to reduce PPCPs (Carvalho et al., 2014; Li et al., 2014; Verlicchi and Zambello, 2014; Zhang et al., 2014; Avila and Garcia, 2015; Gorito et al., 2017; Guan et al., 2017).

Sources

There are two main categories of PPCPs, those originating from human use, and those originating from animal husbandry. Human wastewater treatment systems, both lagoons and conventional wastewater treatment plants for larger flows, remove some but not all of the PPCPs sent to them. Urban stormwater runoff is not generally considered to be a significant source of pharmaceuticals, yet they are present (DeLorenzo et al., 2012).

Conventional wastewater treatment plants (WWTPs) are not designed to remove emerging contaminants. Both high-tech and low-tech wastewater treatment systems do reduce the amounts of PPCPs entering them, but incompletely (Avila and Garcia, 2015). Wang and Wang (2016) concluded,

> The conventional wastewater treatment process, especially activated sludge system, cannot effectively remove the PPCPs, which are frequently detected in the effluents. Thus, activated sludge system as the most widely used treatment process should be improved to meet the demand of dealing with the PPCPs-containing wastewater.

Li et al. (2013) report good treatment of PPCPs in lagoon systems, but concluded that that the effluent discharge from the lagoon system may increase the occurrence of PPCPs and elevate their concentrations in surrounding watersheds.

Potential use of wetlands for further reducing PPCPs is principally considered for domestic and municipal effluents.

Animal wastes are subject less treatment before discharge to receiving waters. Animal pharmaceuticals are primarily used to treat or prevent diseases or infections. A broad range of categories, including anti-parasitic drugs, anti-inflammatory medications, anesthetics, pain medications, antibiotics, and specialized products for managing reproductive, cardiovascular, or metabolic conditions. The U.S. Food and Drug Administration (FDA) publishes a compilation of approved animal drug products, known as the "Green Book", which presently totals 1573 products (U.S. FDA, 2018a). Products used are specific, and yet numerous, to the particular animal husbandry activity.

Treatment wetlands, often treatment marshes, have been investigated for:

- Poultry operations (e.g., Hussain, 2011; Hussain et al., 2012)
- Swine operations (e.g., Carvalho, 2012b; Carvalho et al., 2013; Liu et al., 2013, 2014)
- Shrimp farming (e.g., Bôto et al., 2016)

Mechanisms

The mechanisms of removal of PPCPs in marshes are those that have been discussed for pesticides: volatilization, adsorption/desorption, photolysis, sedimentation/resuspension, infiltration, photolysis, hydrolysis, microbial degradation, and plant uptake. Most PPCPs are nonvolatile, and do not hydrolyze. Therefore, the leading candidates for observed reductions are adsorption/desorption, biotransformation and plant uptake.

PPCPs are organic molecules that display an affinity for sorption to soils, particularly organic soils. Commonly, equilibrium is quickly established in the marsh. Sharif et al. (2014) measured partition coefficients for 17β-estradiol K_d = 320 L/kg, testosterone 160, atrazine 24, carbamazepine 25; which are relatively high. In some instances, sorption may be dominant, For instance, Cardinal et al. (2014) found the fate of fluoxetine and carbamazepine to be dominated by sorption in a mesocosm study.

Organic matter content has a strong influence on the sorption capacity of marsh soils and sediments. Hussain and Prasher (2011) found the sorption occurring at various pHs to be predominantly controlled by soil organic carbon and sedimentation for three poultry antibiotics. Values of K_d ranged from 18 to 98 L/kg, which are not particularly high. However, the soils in question were mostly mineral, with 5–10% organic matter: thus K values were 10–20 times higher (Equation 12.46).

Photolysis is a major factor in marshes that contain significant open water areas. Prasse et al. (2015) reported phototransformation for five antiviral drugs found in a treatment wetland located at Discovery Bay, CA. First order rate coefficients for abacavir, zidovudine, emtricitabine, acyclovir and lamivudine were 12.5, 2.2, 0.72, 0.29, and 0.26 $days^{-1}$, respectively. Indirect photolysis was stimulated by hydroxyl, oxygen, and carbonate. Jasper and Sedlak (2013) measured photolysis for four trace organic pharmaceuticals, also at the treatment wetland at Discovery Bay, CA. First order rate coefficients for atenolol, propranolol, sulfamethoxazole, and carbamazepine were 0.25, 2.7, 1.1, and 0.3 $days^{-1}$, respectively. Indirect photolysis was stimulated by hydroxyl, oxygen, DOM, and carbonate.

Biotransformation is also a potentially important route for PPCP reduction in marshes (Table 14.12). Jasper et al. (2014) report biotransformation rates for pesticides measured in bench-scale biomat (epipelon) microcosms, which were extracted from a field scale treatment open water system. Biotransformation accounted for approximately 50% of the average removal rates measured in the pilot-scale wetland for atenolol, metoprolol, and trimethoprim. Biotransformation accounted for approximately 40% of the removal of propranolol and only 10% of the removal of sulfamethoxazole. Prasse et al. (2015) continued this line of investigation for antiviral drugs. They found that phototransformation processes were dominant for abacavir, zidovudine and emtricitabine, while for acyclovir and lamivudine biotransformation was similar or more important than photolysis.

Plant uptake is not likely to be a major factor. Carvalho et al. (2013) surveyed information on uptake, and concluded that despite the occurrence of plant uptake, uptake usually represents a low percentage of the mass depleted from the systems. A similar conclusion was reached by Cardinal et al. (2014), who used mesocosm studies of PPCPs, and found no significant enhancement in removal of pharmaceuticals in mesocosms with extensive aquatic plant communities. A study of *Phragmites australis* found other mechanisms predominated (Carvalho et al., 2012b), as was also found for *Scirpus validus* (Zhang et al., 2013a).

TABLE 14.12

Biotransformation rates for trace organics in open water plus benthic biomat microcosms.

	Rate coefficient 1/d	Half-life d
General Pharmaceuticals		
Atenolol	0.80	0.87
Metoprolol	0.22	3.15
Propranolol	0.11	6.30
Trimethoprim	0.29	2.39
Sulfamethoxazole	0.08	8.66
Carbamazepine	0.02	43.31
Antivirals		
Abacavir	0.19	3.61
Zidovudine	0.23	3.04
Acyclovir	0.04	19.25
Emtricitabine	0.14	5.07
Lamivudine	0.13	5.25

Sources: General pharmaceutical data, Jasper et al. (2014); antiviral data, Prasse et al. (2015).

As described in connection with pesticides above, there are initial attempts to synthesize these mechanisms into comprehensive model of treatment marsh performance for PPCP reduction (Sharif et al., 2014; Westerhof et al., 2014). As previously noted, parameterization remains an obstacle for use of such models, and loss rate coefficients are presently the best available approach.

Marsh Performance for PPCP Reduction

There are several distinguishing features of this class of trace organics that need to be recognized to place published results in perspective. First, pharmaceuticals mainly occur in continuous flow systems, originating from some form of treatment for other constituents. Thus investigators are spared the difficulties of dealing with event-driven systems. There may be seasonality in some cases, for instance, sunscreens are used in summer, but cold medicines are more prevalent in winter. Second, full-scale system monitoring for concentrations is difficult and costly, and consequently continuous time series of measurements are not feasible. This leads to studies based on "campaigns," in which measurements at inflows and outflows are monitored for a period of only a few days. The sources, although usually continuous, are subject to variability over time. This leads to questions of the effects of transport delays and the associated potential for synoptic error. This potential is high, because full-scale systems typically have detention times that are of the same order of magnitude as PPCP half-lives.

Much of the research on pharmaceuticals in treatment wetlands has been conducted in Europe, where full-scale free water systems, that is, marshes, are in short supply. As a result, mesocosms are often used to test for the treatability of PPCPs. There is also a predilection in Europe for subsurface flow wetlands, and hence such mesocosm studies include SSF tanks (see, for example, Hijosa-Valsero et al., 2011; Reyes-Contreras et al., 2012). The small scale of such studies allows for flow and concentration equalization for the input waters. Drawbacks of marsh mesocosm systems, as prototypes for full-scale systems, include uncharacteristic soils and sediments, uniform plant densities, unrealistic meteorological conditions, and a lack of influence of large organisms. Small-scale testing has been done for pharmaceuticals, with good success (Berglund et al., 2014). And, there are full-scale projects that show excellent removals (Matamoros et al., 2008).

Here, two examples of PPCP reductions in full-scale systems are presented to illustrate the general aspects of performance.

Four Marshes in Sweden

Breitholtz et al. (2012) determined concentrations of 92 pharmaceuticals in inlet and outlet waters in four Swedish treatment wetlands, during winter conditions. All receive approximately secondary municipal effluent. The Ekeby wetland in Eskilstuna covers 28 ha, and has a detention time of 6–7 days. It is about 20% EAV (*Glyceria maxima, Typha* spp., *Schoenoplectus lacustris*), with the balance SAV (*Ceratophyllum demersum, Myriophyllum spicatum, Elodea* spp.). The Nynäshamn wetland Alhagen also covers 28 ha, and has a detention time of 12 days. The plant community is dominated by emergent species such as *Phragmites australis, Typha* and *Carex*, with areas of SAV (*Elodea Canadensis, Ceratophyllum demersum*). The Oxelösund wetland covers 23 ha dominated by emergent macrophytes, mainly Typha *latifolia* (Figure 14.7). Deeper sections contain large stands of submerged plants (*Elodea canadensis, Potamogeton* sp.) The detention time is 4–6 days. The wetland at Trosa is 6 ha in size, with a detention time of 6 days. Vegetation is principally *Typha, Phragmites*, and *Scirpus*.

Synoptic testing was conducted over brief periods of a few days, and thus the results rest upon the assumption of steady flows and concentrations. Of the 92 pharmaceuticals checked, 27 did not occur. Some of the compounds were below the limit of analytical quantitation, either at inlet or outlet, leaving 26 for which reductions could be computed (Table 14.13). Loss rate coefficients, based on ideal flow, were extremely variable, and the resulting half lives spanned 1–60 days.

FIGURE 14.7 The treatment wetland at Oxelösund, Sweden.

TABLE 14.13

Pharmaceutical removals in four treatment marshes in Sweden. E = Eskilstuna, N = Nynäshamn, T = Trosa, O = Oxelösund. Negative removals not included in medians. Mid range concentrations are the average of the lowest and highest observed for the four inlets and outlets.

	E	N	O	T	Median	Median	Mid range
	Concentration Reduction				k	Half-life	concentration
Pharmaceutical	(%)	(%)	(%)	(%)	1/day	days	ng/L
Atenolol	27	53	53	53	0.062	9	1265
Bisoprolol	26	22	29	36	0.046	14	118
Carbamazepine	12	11	21	−19	0.020	35	645
Citaprolam	45	84	63	97	0.150	4	115
Codeine	29	75	74	47	0.113	7	385
Diclofenac	31	24	36	30	0.057	14	455
Diltiazem	30	74	68	88	0.110	5	35
Dipyridamol	80	100	94	98	0.358	1	3721
Eprosartan	75	39	48	24	0.109	12	700
Fexofenadine	28	−26	7.8	9.1	0.032	51	170
Fluconazole	−8.3	36	43	−40	0.065	13	112
Ibuprofen	38	80	88	5	0.131	7	790
Ibersartan	58	8	3	27	0.007	60	900
Ketoprofen	56	2.7	32	19	0.064	19	1315
Memantin	5	29	5	32	0.009	53	17
Metoprolol	−3	30	18	27	0.031	21	1050
Naloxone	84	−116	−28	3.7	0.282	75	5
Naproxen	34	46	75	50	0.064	9	178
Ranitidine	−39	92	88	56	0.280	3	71
Rosuvastatin	60	58	53	−16	0.126	6	122
Sotalol	15	−21	17	−18	0.028	25	800
Telmisartan	73	51	14	87	0.058	8	200
Terbutaline	14	83	33	−11	0.067	10	4
Tramadol	5	10	26	−12	0.009	81	575
Trimetoprim	25	51	69	86	0.058	8	160
Venlavafaxine	40	17	18	65	0.033	15	230

Source: Breitholtz et al. (2012).

The Empuriabrava Wetland in Catalonia

This treatment marsh, (or surface flow constructed wetland) has three parallel 0.8 ha cells with an average water depth, 0.5 m, and a large polishing shallow wetland in series (4.5 ha; average depth, 0.2 m) (Matamoros and Salvadó, 2012). The HRT is about 8.5 d. The system is planted with patches of EAV (*Phragmites australis* and *Typha latifolia*) (Figure 14.8). Campaigns were done quarterly, with daily sampling for a one week period. A group of 27 emerging contaminants was selected on the basis of their concentration and detection frequency in a previous screening campaign. Table 14.14 lists the reductions observed for a subset of 12 PPCPs. Half-lives range narrowly between 6 and 11 days. Because synoptic testing was conducted over brief periods of a few days, the results again rest upon the

FIGURE 14.8 The Empuriabrava treatment wetland in the northeast of Spain.

TABLE 14.14

Pharmaceutical removals in the Empuriabrava wetlands. Values are annual averages of quarterly synoptic study results.

	Concentration Reduction	k	Half-life
	%	1/d	d
Ketoprofen	90.5	0.11	6.2
Diclofenac	85.8	0.11	6.5
Galaxolide	92.0	0.11	6.0
Tonalide	81.3	0.10	6.9
Ibupfrofen	90.3	0.11	6.2
Cashmeran	83.7	0.10	6.7
Naproxen	79.0	0.10	7.1
Triclosan	66.0	0.08	8.6
Oxybenzone	63.5	0.08	8.9
Terbutrin	69.5	0.09	8.1
Benzothiazole	58.5	0.07	9.7
Caffiene	52.8	0.06	10.8

Source: Matamoros and Salvadó (2012).

assumption of steady flows and concentrations. Matamoros and Salvadó (2012) speculated that elimination of these compounds was accounted by the simultaneous occurrence of biodegradation, sorption and photodegradation processes.

14.6 ENDOCRINE DISRUPTORS

General

In the groups pesticides and pharmaceuticals, there are substances that are termed endocrine disruptors (ED), or endocrine disrupting chemicals (EDC), because they can induce physiological changes in biota, sometimes at very low concentrations. It has been suggested that humans and domestic and wildlife species have suffered adverse health consequences resulting from exposure to chemicals in the environment that interact with the endocrine system (Colborn et al., 1997).

Feminized male fish of many species have been discovered in lakes and rivers throughout North America and Europe, and estrogen-like chemicals are the suspected culprits. The presence of female eggs in male fish testes indicates some kind of hormonal confusion. Frogs in suburban lakes tend to be mostly female, and suggests that urbanization and estrogenic wastes are likely turning male frogs female. The development of such "intersex" traits have been linked to endocrine-disrupting chemicals.

There are several groups of endocrine disruptors, including natural steroid hormones, synthetic steroid pharmaceuticals, and some of the older pesticides (DDT, PCBs, dioxins) (Lintelman et al., 2003). Estrogens, androgens and progestins are produced by both males and females, but in different amounts. Common estrogens are 17βestradiol, estrone and estriol; common androgens are testosterone and androsterone Figure 14.9). Both human and animal sources can be important. In addition to animal sources, there are plant sources

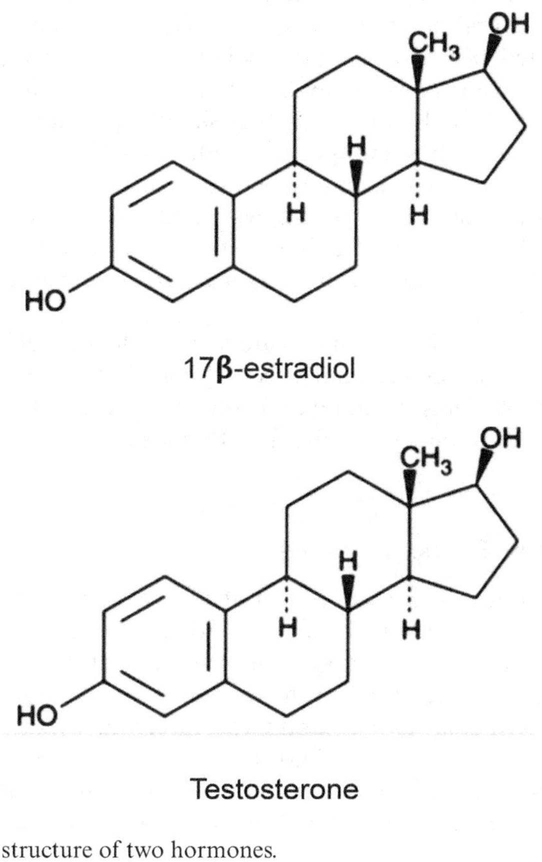

FIGURE 14.9 Chemical structure of two hormones.

of endocrine disruptors. These include molds (zearanol, cattle growth promoter) and plants (βsitosterol, medicinal lipid lowerer) (Lintelman et al., 2003). Modern pesticides have been implicated as potential endocrine disruptors, including metalaxyl, dimethoate, linuron, folpet and others. Also, several common industrial chemicals are potential endocrine disruptors, including PCBs, PAHs, alkylphenol ethoxylate byproducts (nonylphenol), bisphenol A (plastics), and organotin compounds.

Wastewater treatment plant effluents have now been well studied, and found to typically contain 10–50 ng/L of common estrogens (Lintelman et al., 2003). However, advanced treatment, including nitrification, denitrification and phosphorus removal, leads to lower numbers, typically 1–2 ng/L (Spengler et al., 2001). All classes of endocrine disruptors were found to be reduced along the treatment trains of several wastewater treatment plants in Australia (Tan, 2006).

Some pharmaceuticals are endocrine disruptors, consisting mainly of oral contraceptives (ovulation-inhibiting hormones) as well as steroids used for substitution therapy during menopause. The most common estrogen-like synthetic steroids are ethinylestradiol and mestranol. A further important group of pharmaceutical hormones are steroids used in agriculture during cattle farming. In the mid-20th century, diethylstilbestrol (DES), a synthetic steroid, was used for fattening of cattle as well as in humans to prevent miscarriages, which led to devastating health consequences (Lintelman et al., 2003).

Problematic Pesticides

Regulatory concern in the USA came into full play about ten years ago, with U.S. EPA's initiation of the Endocrine Disruptor Screening Program (EDSP) to screen chemicals rapidly for bioactivity in several endocrine pathways. At the present time, 52 suspect chemicals have been evaluated per "Tier 1" protocols (U.S. EPA, 2015). Of the 52 chemicals evaluated, there was no evidence for potential interaction with any of the endocrine pathways for 20 chemicals, and for 14 chemicals that showed potential interaction with one or more pathways, EPA already has enough information to conclude that they do not pose risks. Of the remaining 18 chemicals, all 18 showed potential interaction with the thyroid pathway, 17 of them with the androgen pathway, and 14 also potentially interacted with the estrogen pathway. These 18 require further testing to ascertain their endocrine pathway effects (Table 14.15). The reader will recognize several as being monitored in the various studies discussed above.

Many pesticides (34) have thus been "cleared" of endocrine effects, but that does not mean they do not have other detrimental effects. Many of the cleared are definitely implicated in environmental problems, including atrazine, diazinon, malathion and permethrin – all banned in the EU. Additionally, glyphosate (Roundup) has recently been implicated in the decline of honeybees.

Endocrine Disruptors in Treatment Marshes

Because endocrine disruptors have the potential for causing deleterious effects in receiving waters, the capabilities of treatment marshes to remove these substances is of considerable interest. At the same time, treatment marshes are also home to the same suite of organisms that exist in other receiving water bodies. It is therefore necessary to consider effects in the treatment marsh as well as in the receiving water body. This situation is complicated by the (current) absence of hormone criteria for ecosystem health. The criteria for pesticides that are also candidate endocrine disruptors do exist, and have been discussed in the preceding section. Therefore, the focus here is on the processing of hormones.

TABLE 14.15

Pesticides requiring Tier 2 analysis for endocrine pathway effects.

Carbaryl	Linuron
Chlorothalonil	Metalaxyl
Cypermethrin	Metribuzin
DCPA	Myclobutanil
Dichlobenil	O-phenylphenol
Dimethoate	PCNB
Flutolanil	Propargite
Folpet	Propiconazole
Iprodione	Tebuconazole

In common with other contaminants, wetlands generally do not provide complete removal of endocrine disruptors, due to wetland hydraulic and chemical cycling constraints. The literature on marsh treatment effectiveness for ED chemicals is building at the present time, with several studies completed.

Mechanisms for attenuation of estrogenic activity are speculatively biodegradation, photodegradation, volatilization, sorption and plant uptake and destruction. The common property of most of the natural estrogenic compounds is that they are easily biodegradable. But the synthetics have poor biodegradability. Volatilization is probably not an important route, because of the low vapor pressure of the substances. Plant uptake and destruction may account for a small amount of removal (Schröder et al., 2008). Sorption to wetland organic substrates probably accounts for a significant delay in the transport of EDs, and renders them susceptible to degradation by surface-bound microorganisms (Gray and Sedlak, 2004; Dai et al., 2016). Photodegradation is very rapid (Mazellier et al., 2008), and is effected by wavelengths in the ultraviolet.

There are several mesocosm studies of the reduction of EDs by plant-soil-water systems (e.g., Kumar et al., 2011; Jasper and Sedlak, 2013; Dai et al., 2016). Although useful for comparison purposes, mesocosm studies contain artifacts related to their size. Accordingly, a better representation of expectations is found in studies of full-scale or large pilot systems. Available results are characterized by the examples below, but further full-scale studies have been reported by Shappell et al. (2001), Hemming et al. (2001), Hsieh et al. (2015), and Chen et al. (2014).

Hawkesbury Agricultural Station

Norris and Burgin (2011) studied mosquito fish (*Gambusia holbrooki*) two treatment marsh trains of three units each located near Richmond, New South Wales, Australia. One treatment system received urban stormwater, and the other tertiary municipal effluent. They found morphological responses that were consistent with endocrine disruption in both trains, with a decline along the flow direction from cell to cell. Importantly, their results implicate both tertiary treated effluent and stormwater as sources of EDs. Thus, the treatment marshes reduced EDS, but their piscine inhabitants showed ED effects in the inlet zones.

Prado Constructed Treatment Wetlands, Orange County Water District

Approximately half of the Santa Ana River in Riverside County in California is diverted to treatment wetlands (Reilly et al., 2000). The estrogenic hormones 17β-estradiol (E_2)

17-ethinyl estradiol (EE_2) have been detected in municipal wastewater effluent and surface waters at concentrations sufficient to cause feminization of male fish. To evaluate the fate of steroid hormones in an engineered treatment wetland, lithium chloride, E_2, and EE_2 were added to a treatment wetland test cell (Gray and Sedlak, 2004). Comparison of hormone and tracer data indicated that 36% of the E_2 and 41% of the EE_2 were removed during the cell's 84-h hydraulic retention time (HRT). The observed attenuation was most likely the result of sorption to hydrophobic surfaces in the wetland coupled with biotransformation. Sorption was indicated by the retardation of the hormones relative to the conservative tracer. Biotransformation was indicated by elevated concentrations of the E_2 metabolite, estrone. These results were confirmed by Xie et al. (2004) for *in vitro* (yeast estrogen screening) assays. For water entering and exiting Prado Wetland YES EEQ levels were 2.57 and <1 ng/L, respectively. However, Xie et al. (2004) went further, and used *in vivo* methodology (rainbow trout vitellogenin (VTG) expression) to test waters entering and leaving the wetlands. For trout were exposed to entering water, VTG concentrations were increased above those of controls and were not significantly different from fish exposed to water exiting the wetland.

Tucson Sweetwater Wetland

Bjolseth (2004) and Bjolseth et al. (2003) examined the Sweetwater Wetlands for endocrine disruptors (Vidales-Contreras et al., 2006). The Sweetwater Wetlands are a part of the Sweetwater Recharge Facilities, owned and operated by the City of Tucson, Arizona. The wetland component is comprised of 0.7 ha of settling basins, a stream system and 15 acres (6.1 ha) of wetland cells. The dominant vegetation in the settling basins and wetland cells includes several different species of bulrush and cattail. The Sweetwater Wetlands receive a mixture of secondary effluent and backwash water from the City of Tucson Reclaimed Water Plant. Detention in the wetland is a few weeks. Samples were analyzed using the competition binding and reporter gene assays. A Yeast Estrogen Screen (YES) assay was used. Removals in the wetland were 98%, with inlet concentrations of approximately 60 ng/L.

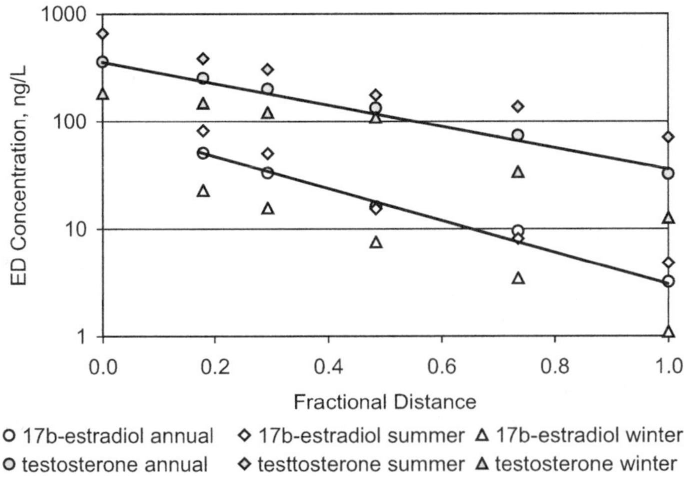

FIGURE 14.10 Reductions of endocrine disruptors along the flow path in the Greenmount wetlands. (Based on data in Cai et al., 2012)

Greenmount Agricultural Station

Cai et al. (2012) followed the time and distance progression of 17b-estradiol and testosterone in a series of EAV cells located at Muckamore near Antrim in County Antrim, Northern Ireland. Further information on this wetland system is found in Forbes et al. (2011). Source waters emanated from a herd of 140 dairy cattle, and contained significant quantities of ED hormones. The treatment train consisted of a sedimentation basin, followed by five EAV cells in series. The first two cells were vegetated by *Phragmites australis* and *Typha latifolia*, and the three downstream cells contained *Carex riparia, Sparganium erectum* and *Iris pseudacorus.* The area of 1.25 ha was operated at a depth of about 0.3 m. The system had very long detention, of approximately 75 days. Analyses were performed monthly for all cells for 17β-estradiol equivalent (EEQ) and testosterone equivalent (TEQ). The declines along the flow direction were well described as first order (Figure 14.10). The loss rate coefficients were k_V = 0.036 1/d for 17β-estradiol (R^2 = 0.990) and k_V = 0.031 1/d for testosterone (R^2 = 0.991). Declines were about the same, winter or summer, although concentrations were lower in winter. The reductions of 95 ± 4% are encouraging, for a relatively small wetland on a per cow basis.

14.7 PETROLEUM HYDROCARBONS

In contrast to pesticides and pharmaceuticals, which originate from human and animal sources, trace hydrocarbons originate from industrial activities. Small amounts find their way into and through wastewater treatment plants, and small amounts occur in urban and agricultural runoff. But these are not the principal sources for treatment marshes. Treatment marshes sometimes are used to polish pre-treated wastewaters from industry, but they are also the recipients of contaminated groundwaters. That contamination can be the result of industrial spillage, or of leachates from landfills. Chemical spillage was common at industrial sites in the 20th century, resulting in contaminated groundwater beneath them. Old petroleum and petrochemical facilities were mainly responsible, but smaller sites also existed. Remediation generally involves pumping the contaminated water and treating it. The production of petroleum from underground sources often involves the use of water, which becomes contaminated with organics in the process. Such produced waters emanate from oil wells, and from oil shale and oil sands operations (Akob et al., 2015).

Because underground contamination often proceeds toward and vents to surface waters, regulatory criteria are often couched in terms of values to be achieved at the groundwater–surface water interface (GSI), or in other words at the point where the contaminated water enters surface waters. Criteria are set for many of the potential pollutants. For example the State of Michigan lists GSI criteria for 287 substances, including many hydrocarbons and chlorinated hydrocarbons (MDEQ, 2018) (Table 14.16). Some of these hydrocarbons have considerable water solubility, but polycyclic aromatic hydrocarbons (PAHs) generally do not.

There is considerable information on the use of treatment wetlands in the petroleum industry. Some of the general principles and available data have been summarized in a 1998 industry report (Knight et al., 1997, 1999). However, that compilation did not focus upon specific hydrocarbon classes, such as BTEX and its constituent components (benzene, toluene, ethyl benzene and xylenes). The alkanes are generally not considered as a separate class in environmental studies. The low molecular weight alkanes are extremely volatile, and with the exception of methane, do not appear in wetlands. Two other hydrocarbon designations of interest include Gasoline Range Organics (GRO) and Diesel Range Organics (DRO). There is some overlap and ambiguity in these designations. Typically, GRO consists of hydrocarbons with 6 to 9 carbon atoms, while DRO contains 10 to 40 carbon atoms (Chapple et al., 2002). Total Petroleum Hydrocarbons (TPH) is a measure of the sum of paraffinic and aromatic constituents.

TABLE 14.16

Examples of GIS criteria set by the State of Michigan for trace hydrocarbons (μg/L).

	Drinking Water Criteria	GSI Criteria	Water Solubility
Hydrocarbons			
Benzene	5	200	1.75E+06
Ethylbenzene	74	18	1.69E+05
Toluene	790	270	5.26E+05
Xylenes	280	49	1.86E+05
Styrene	100	80	3.10E+05
Naphthalene	520	11	3.10E+04
Phenol	4,400	450	8.28E+07
Fluorene	880	12	1.98E+03
Fluoranthene	210	1.6	2.06E+02
Anthracene	43		4.34E+01
Diethyl phthalate	5,500	110	1.08E+06
Pyrene	140		1.35E+02
Chlorinated Hydrocarbons			
Trichloroethylene	5	200	1.10E+06
1,1-Dichloroethylene	7	130	2.25E+06
Vinyl chloride	2	13	2.76E+06
Methylene chloride	5	1,500	1.70E+07
Chlorobenzene	100	25	4.72E+05
1,2,4-Trichlorobenzene	70	99	3.00E+05
Hexachlorobenzene (C-66)	1	0.2	6.20E+03

Source: MDEQ (2018).

BTEX Mechanisms

Biodegradation

The susceptibility of hydrocarbons to microbial degradation can be generally ranked as follows: linear alkanes > branched alkanes > small aromatics > cyclic alkanes. Some compounds, such as the high molecular weight polycyclic aromatic hydrocarbons (PAHs), may not be degraded at all (Das and Chandran, 2011). Degradation of many trace organic pollutants occurs under aerobic conditions. Several bacteria are known to feed exclusively on hydrocarbons. Reports list 22–25 genera of hydrocarbon degrading bacteria, and 25–31 genera of hydrocarbon degrading fungi (Das and Chandran, 2011). Biodegradation of BTEX in the wetland environment is partly regulated biofilms on submerged plant parts, plant litter, and sediments. Although small in terms of mass per unit volume, these biofilms are very active in biodegradation, and consequently serve as important sinks for organics.

Chang et al. (2001) established that all BTEX constituents degraded rapidly, (half-lives of one to two days) when inoculated with a microbial consortium conditioned with toluene. Their study showed that benzene, toluene and ethylbenzene were directly consumed as carbon sources, while xylene was removed by co-metabolism.

Moore et al. (2002a) measured both aerobic and anaerobic biodegradation of BTEX in peat. Aerobic degradation was tracked via oxygen consumption and carbon dioxide production, and averaged 56 mg/kg·d. Anaerobic degradation was inferred from the consumption of other electron acceptors, including nitrate, sulfate, iron, manganese and methane. About one-third of the BTEX loss could have been attributed to anaerobic degradation.

Volatilization

As discussed in Chapter 7, the rate of dissipation of volatile substances is most often controlled by the rate at which they make their way the water surface, or in other words, that rate is waterside mass transfer controlled.

BTEX constituents are volatile, and therefore may be easily lost from water, especially shallow water bodies, such as marshes. Lee et al. (2004) have reported that first-order loss rate constants for BTEX constituents exhibit a fixed ratio to the loss rate constant for benzene, independent of the water mixing regime, thus suggesting volatilization. Further, the presence of more than one BTEX compound, or of surfactants, had only minor effects on this ratio.

Of direct interest are the estimates, based on wetland data, of Keefe et al. (2004b) for wetland H1 at the Tres Rios site in Arizona. They reported a synoptic first order plug flow rate coefficient k = 100 m/yr for toluene. This is a relatively high rate, compared to other pollutants treated in FWS wetlands.

Plant Uptake

Plants are capable of producing hydrocarbons, mostly volatile organic compounds (VOCs), including terpenoids and green leaf volatiles (GLVs). GLVs are best known as the smell that is produced by freshly mown grass; this group of volatiles mainly consists of C_6-aldehydes, C_6-alcohols, and their acetates. These releases are minor. Additionally, necromass decomposition produces organics, consisting of very high molecular weight substances, including long-chain alkanes and alkenes, waxes, bitumen and humic and fulvic acids. In the BTEX category, only toluene has been identified as a possible decomposition product (Richards and Sandau, 2018).

The principal direct plant effect is uptake. Trace hydrocarbons are xenobiotics, and thus there are no transporters for these compounds in plant membranes. Hydrocarbons therefore move into and within plant tissues driven by diffusion and thus linked to their chemical properties. Organics with $0.5 < \log_{10} K_{OW} < 3$ are hydrophobic enough to move through the plant's lipid membranes, and still water soluble enough to travel with intracellular water (Pilon-Smits, 2005). Only the short chain alkanes (C_1 – C_4) and benzene and toluene meet this criterion. If organics are too hydrophilic ($\log_{10} K_{OW} < 0.5$) they cannot pass membranes and get into the plant; if they are too hydrophobic ($\log_{10} K_{OW} > 3$) they remain in membranes and cell walls and do not move with intracellular water. Plants are the locus of microbes that conduct degradation, primarily in the root zone (rhizodegradation), and secondarily in the above ground tissues (phytodegradation). They may also assist in the volatilization process. Because hydrocarbons are chiefly associated with roots and rhizomes, it is unlikely that storage in plant tissues and its removal by harvest are, or will be, a significant factor in the performance of marshes for trace hydrocarbon reduction.

Sorption

Sorption onto wetland sediments is a very important mechanism for removal of trace organics. Following mass transfer to the sediment zone, most of the trace organics will partition to the sediments. There may be some degradation of the sorbed organic, but higher molecular weight materials are merely held in place. Therefore, it may be necessary to determine the level of sediment contamination that results. Because there is continual buildup of sediments in treatment mashes, sorbed materials are continually diluted by the addition of new sediments. As discussed in Chapter 13 for metals, there are two ways to consider the buildup of stored metals in a treatment wetland. The first presumes that metal storage occurs throughout the root zone of the marsh, with transfers to sediments and roots driven by processes such as sorption and transpiration flows. The second presumes deposition by layering. The mass balance on solids and organics determines the levels that may remain in the marsh solids.

BTEX components are expected to partition strongly to organic wetland substrates, although a definitive wetland studies are lacking. Wetland sediments typically have high organic content, and therefore sorption may be an important first step in overall removal. Benzene sorption onto peat was found to be strong, with a linear sorption coefficient of K_{OC} = 5.6 L/kg (Moore et al., 2002a), while binding to clay was much weaker, with K_{OC} = 0.12 L/kg.

Data for other BTEX components are presented in Table 14.17, which must be regarded as estimates because the specificity of various marsh sediment types.

BTEX Wetland System Studies

In all existing petrochemical plant applications, wetlands have been accompanied by pretreatment. However, other applications, such as those involving remediation, do not involve pretreatment. Hydrocarbon contamination of groundwater includes closed and operating sanitary landfills, army ammunition plants, and former oil refinery sites. These facilities were often the recipient of solvents and other organics, which in the course of time have contaminated nearby waters. Concentrations of petrochemicals in landfill leachate are typically lower than those associated with refineries and terminals. Some examples of the effectiveness of treatment marshes for BTEX follow.

Gulf Strachan, Rocky Mountain House, Alberta

Moore et al. (2002a) reported on the natural attenuation of BTEX in a natural peatland, which received both aqueous and nonaqueous phase hydrocarbons for more than fifteen

TABLE 14.17

Examples of the solubility, octanol-water partition coefficient and organic carbon partition coefficient for BTEX and alkylbenzenes.

Compound	Aqueous Solubility (mg/L)	Log_{10} K_{OW} g/g	Log_{10} K_{OC} L/kg
Benzene	1780	2.13	1.90
Toluene	515	2.65	2.12
Ethylbenzene	152	3.15	2.36
o-Xylene	220	3.07	2.15
m-Xylene	160	3.18	2.29
p-Xylene	215	3.16	2.47
n-Propylbenzene	61.6	3.69	2.96
n-Butylbenzene	13.9	4.31	3.64
Pentylbenzene	3.6	4.90	4.30
n-Hexylbenzene	0.96	5.52	4.99
1,2,3-Trimethylbenzene	67.4	3.60	2.84
Styrene	272	2.97	2.86

Sources: van Noort (2009); Cornelis et al. (2017a).

years. No BTEX was detected leaving the peatland. Companion studies elucidated some of the candidate removal mechanisms, including sorption, aerobic degradation, volatilization, and anaerobic degradation.

Isanti-Chisago, Minnesota Leachate Treatment System

The Isanti-Chisago Sanitary Landfill, an unlined municipal solid waste facility located near Cambridge, Minnesota, was closed in 1992 (Loer et al., 1999; Kadlec, 2003c). Leaching of soluble wastes had contaminated the surficial and increasingly deeper aquifers with toxic organic compounds and heavy metals. Extraction wells permit pumping of leachate to the top surface of the landfill mound, about eight meters above the surrounding landscape. Volatile organic compounds (VOCs), including BTEX, were largely removed by cascading the water down the side of the landfill, into a sedimentation basin, which serves to settle and store iron precipitates. The next component of the treatment train is a 0.6 ha FWS wetland. During five seasons of treatment, operating results indicate the system efficiency ranged from 85–100% for VOCs.

Only low levels of BTEX enter the system: (benzene 3.7 μg/L; toluene 0.6 μg/L; ethylbenzene 1.1 μg/L, xylene 1.8 μg/L). The cascade and settling basin do essentially all the BTEX removal, although the FWS wetland does polish out any remaining traces of BTEX. Benzene, toluene, ethyl benzene and xylenes are all individually below detection (0.1 μg/L) in the system outflow.

Saginaw Township, Michigan, Leachate Treatment System

Saginaw Charter Township's Center Road Landfill was closed in the early 1980s (Kadlec and Zmarthie, 2010). Finger drains were installed in the seepage zones, which connect to a collection pipe. Leachate is then pumped to an aerator, which provides some ammonia and BTEX stripping. The water is then held in a sedimentation basin, to promote removal of solids. Water is discharged periodically to one of two intermittent sand/gravel filters, which provide filtration and nitrification. Underdrains then convey the water to two parallel free water surface wetlands. BTEX and its constituents were monitored for ten years. A mean total BTEX of 54 μg/L was reduced to 6. However, only benzene and xylenes were detected in the outflow, each registering only 2% of hits in the samples taken (Table 14.18).

Phoenix, Arizona Wastewater Polishing

A demonstration wetland project was studied for VOC removal via a synoptic study (Keefe et al., 2004a). The Tres Rios H1 marsh was a 1.4-ha system of depth about 46 cm, and nominal detention of 3.9 days. An 80% reduction of toluene from inlet to outlet was measured. Keefe et al. (2004b) attributed the reduction to volatilization, but concluded that theoretical predictions were only good for order-of-magnitude estimation.

Alcoa, Tennessee Groundwater Remediation

A pilot wetland project was initiated and operated for a period of over one year, in a moderate north temperate continental climate. DRO and GRO were among the targeted substances (Gessner et al., 2005). GRO entered at monthly average values of 0.04 to 0.37 mg/L, and never exceeded the detection limit of 0.01 mg/L at the outlet of the FWS wetland. DRO entered at monthly average values of 0.29 to 1.08 mg/L, and exited at 0.11 to 0.44 mg/L. GRO removal rate constants could not have been less than 100 m/yr. In contrast, the DRO removal rate constants were 12 m/yr annual average.

Polycyclic Aromatic Hydrocarbons

Polycyclic aromatic hydrocarbons (PAHs) are fused ring aromatic compounds formed during the incomplete combustion of almost any organic material, and are ubiquitous

TABLE 14.18

Trace hydrocarbon parameters for the Saginaw treatment system, 1999–2008. Hits are values above the detection limit. Nineteen other monitored substances were not detected, and not subject to limits. ND= not detected. Units are μg/L. Frequencies were every two months.

	----- Concentration -----				– Hits/Samples –	
	NPDES Permit	GSI Criteria	Median Inflow	Median Outflow	Inflow	Outflow
Hydrocarbons						
Benzene		12	1.8	2	44%	2%
Toluene		140	10	ND	33%	0%
Ethylbenzene		18	12	ND	59%	0%
Xylenes – Total		35	30	4	89%	2%
Total BTEX	20	205	54	6		
Styrene		80	ND	ND	0%	0%
Fluorene	5	12	ND	ND	0%	0%
Phenanthrene	5	5	ND	ND	0%	0%
Butyl Benzyl Phthalate		6.9	ND	ND	0%	0%
Bis(2EthylHexyl) Phthalate		32	19	30	45%	29%
Naphthalene	5	13	ND	ND	0%	0%
Acenaphthene		19	ND	ND	0%	0%
Fluoranthene		1.6	ND	ND	0%	0%
Chlorinated Organics						
Chloroethane	5		11	ND	56%	0%
Chlorobenzene		47	4	ND	100%	0%
1,4-Dichlorobenzene	5	13	5	ND	9%	0%
1,1-Dichloroethane			1	ND	9%	0%
1,1-Dichloroethene	5	9.1	ND	ND	0%	0%
Methylene Chloride		47	ND	ND	0%	0%
Total PCB's	0.00002	0.2	0.25		31%	0%
Aroclor 1242 (PCB)			0.54		36%	0%
1, 1 – Dichloroethene	5	13	5		9%	0%

Source: Kadlec and Zmarthie (2010).

in the environment (Figure 14.11). Some of them are considered as dangerous substances as a function of their toxic and mutagenic or carcinogenic potentialities. The presence of PAHs in contaminated soils and sediments may pose a risk to the environment and human health. PAHs are hydrophobic compounds, and the persistence of the four- and five- ring PAHs within ecosystems is chiefly due to their low aqueous solubility. The two- and three-ring PAHs are somewhat water soluble (Table 14.19). These materials are not volatile, and no loss to the atmosphere is anticipated under wetland conditions.

Creosote consists of a mixture of organic compounds, dominated by PAHs, many of which are individually designated as hazardous wastes. Wood may be treated with creosote to enhance its resistance to decomposition. For instance, railroad ties are typically treated with 128 kg/m^3. Many of the industrial sites that are or were used to treat ties are now classified as hazardous waste sites, due to spillage and the associated soil

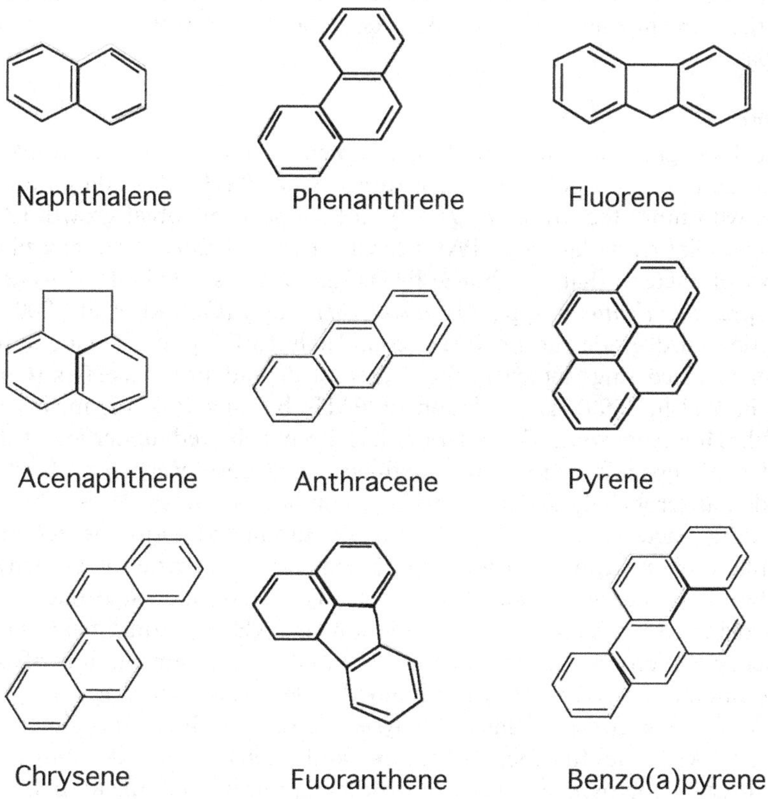

FIGURE 14.11 Example polynuclear aromatic hydrocarbons.

TABLE 14.19

Examples of the solubility, octanol-water partition coefficient and organic carbon partition coefficient for PAHs.

Compound	Aqueous Solubility (mg/L)	Log_{10} K_{OW} g/g	Log_{10} K_{OC} L/kg
Naphthalene	30.7	3.36	3.17
Acenaphthene	4.08	3.92	3.21
Biphenyl	6.89	3.96	3.26
Fluorene	1.98	4.18	4.39
Phenanthrene	1.11	4.55	3.91
Anthracene	0.0576	4.45	4.59
Pyrene	0.135	5.18	4.88
Fluoranthene	0.265	5.33	5.21
Chrysene	0.00185	6.64	5.72
Benzo[a]pyrene	0.003	6.35	5.91
Fenanthene	1.6	4.61	3.98
Acenaphthylene	16.1	4.07	3.79

Sources: van Noort (2009); Cornelis et al. (2017b).

contamination. Conversely, there is little or no risk to aquatic and wetland environments from ties as they are normally placed (Brooks, 2000b), nor from bridge or dock pilings (Brooks, 2000a).

Biodegradation

PAHs undergo biodegradation in natural environments (Haritash and Kaushik, 2009). Aromatics follow a pattern, with polyaromatic hydrocarbons (PAHs) degrading more slowly than benzene; those with more than three rings may not support microbial growth (Zander, 1980). The aerobic microbial degradation of PAHs having two and three rings is well documented. A large number of bacteria that metabolize PAHs have been isolated (*Alcaligenes denitrificans, Rhodococcus* spp., *Pseudomonas* spp., *Mycobacterium* spp.) (Giraud et al., 2001). Some varieties of bacteria can degrade certain PAHs completely to CO_2 and metabolic intermediates. As the number of fused rings increases the degree of degradation decreases (Cookson, 1995, as referenced in Walsh, 1999). Degradation of PAHs by anaerobic organisms has not been very successful. However, some degradation has been achieved under denitrifying, sulfate reducing, and methanogenic conditions. Napthalene and anthracene have been found to be slightly degraded anaerobically under denitrifying conditions (Walsh, 1999).

PAH-degrading bacteria were isolated from contaminated estuarine sediment and salt marsh rhizosphere by enrichment using naphthalene, phenanthrene, or biphenyl as the sole source of carbon and energy (Daane et al., 2001). This study indicated that the rhizosphere of salt marsh plants contains a diverse population of PAH-degrading bacteria, and the use of plant-associated microorganisms has the potential for bioremediation of contaminated sediments. Contaminated sediment was obtained from Newtown Creek in the New York Harbor, Brooklyn, New York. Chemical analysis showed the dredged sediment to contain 2 to 7 ppm of the PAHs naphthalene, anthracene, and phenanthrene. An induction period of 15 days was observed for pyrene, after which degradation was complete in 10 to 15 days. It should be noted that both the pure culture and microbial slurry experiments were performed under highly oxygenated conditions and that the limited diffusion of oxygen into organic-rich sediments has been found to restrict PAH biodegradation in the natural environment (DeLaune et al., 1981).

Sorption

Peat soils adsorb PAH compounds quite effectively, as indicated by the high values of the organic carbon partition coefficients in Table 14.19. These organics partition very strongly to carbonaceous soils (Pardue and Shin, 2000), and are not readily desorbed (Pardue and Shin, 2000; Shin and Pardue, 2002). It is therefore expected that PAHs would be sorbed and stored in wetland peats. Partition coefficients in the range of 10^3–10^4 L/kg were found for phenanthrene (Shin and Pardue, 2002), which are in line with values reported for other sediments (Table 14.19).

Plant Uptake

Polycyclic aromatic hydrocarbons are not taken up by wetland plants to any significant extent. Studies at Duluth, Minnesota have shown very low concentrations in above ground plant parts, and only trace amounts associated with roots (IT Corporation, 1997). PAHs were found in dogwood (*Cornus stolonifera*) and cattail (*Typha* spp.) shoots at the Duluth site. Willows (*Salix* spp.) and bulrushes (*Scirpus* spp.) have been reported to access tightly bound phenanthrene (Gomez-Hermosillo et al., 2000; Gomez and Pardue, 2002). Most of the PAH uptake was to the roots, and was associated with sorption. Because PAHs are not required for plant metabolism and growth, uptake is dependent on concentrations and supplies to the porewater in the root zone. However, the high K_{OC} values mean that PAHs cannot move through plant membranes.

Wetland System Studies

Testing of wetlands for PAH removal has presented mixed results. Some tests show minimal removal, while others show excellent reductions. Anecdotal monitoring of a cypress swamp (forested peatland) receiving landfill leachate in Florida indicated no PAH removal after about 15 years (Schwartz et al., 1999). However, other landfill leachate constructed wetland studies have shown no detectable PAHs in the effluent, including naphthalene at a Minnesota site (Kadlec, 2003c). Boving (2002) found virtually no retention or removal of ten frequently detected PAHs in a stormwater system comprised of ponds and wetlands. Boving showed that heavier PAHs were present in urban freeway runoff at concentrations in excess of U.S. EPA benchmarks for chronic toxicity. In particular, benzo(a)pyrene was present at over 1.0 μg/L.

Salmon et al. (1998) evaluated removal of total hydrocarbons in marsh mesocosms in France. The bed was planted with *Typha latifolia* and during the experiment, natural development of *Lemna minor* occurred. Constructed wetlands removed 90% of total hydrocarbons.

Naphthenic Acids

Exploitation of the Canadian oil sands for petroleum production has generated interest in the interactions between hydrocarbons and marshes. Oil sand deposits lie under 141,000 square kilometers of boreal forest and muskeg (peat bogs) in the province of Alberta, and contain enormous reserves of petroleum. Production of the petroleum products involves considerable quantities of water, most of which is recycled. However, large volumes of tailings are a byproduct of bitumen extraction from the oil sands and managing these tailings is one of the most difficult environmental challenges facing the oil sands industry. In 2013 that tailings ponds in the Alberta oil sands covered an area of about 77 square kilometers. However, regulations require that no process-affected water from tailings ponds may be discharged into any watercourse. Many studies have shown that naphthenic acids have acute and chronic toxicity to fish and other organisms, including wetland plants (phytotoxicity) (Armstrong et al., 2010).

Contaminants in oil sands process-affected water include metals and naphthenic acids. Numerous treatment processes have been tested and employed (Allen, 2008; Brown and Ulrich, 2015). Naphthenic acid is a generic term used to refer to all of the carboxylic acids present in petroleum. In general, they contain one or more hydrocarbon rings and an aliphatic side chain (Figure 14.12). They are the most significant contaminants issuing from oil sands petroleum extraction, because they are relatively water-soluble (ca. 50 mg L). When separated from the oil sands solids, they become a significant pollutant in the tailings pond water (oil sands pond water, OSPW).

Wetlands have been tested as a means of OSPW treatment (Nix and Bishay, 1997). However, they are not a likely candidate to treat OSPW, due to high concentrations and slow reaction rates (Brown and Ulrich, 2015). However, the mined landscapes and tailings ponds require remediation after closure, for which the reestablishment of wetlands is a strong

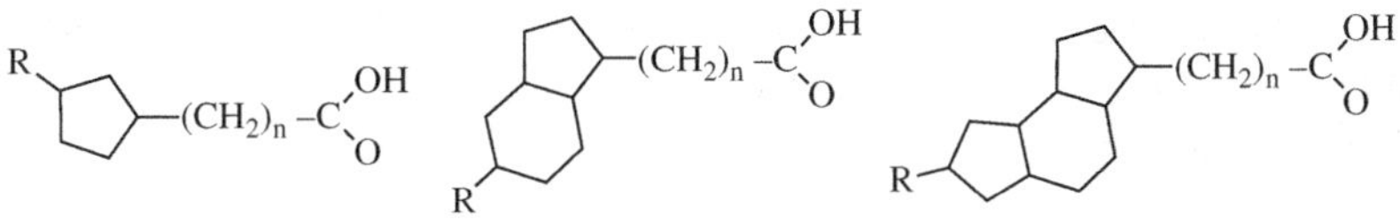

FIGURE 14.12 Examples of naphthenic acid structures.

option. Those reclaimed landscapes will use both lakes and wetlands, which will play a role in removal of the remaining traces of naphthenic acids (Quagraine et al., 2005; Toor et al., 2013). Flow-through marshes offer a strategy for the reclamation of OSPW wetland environments different from the traditional detention ponds (without an inlet and outlet) used in previous field studies. (Toor et al., 2013).

14.8 CHLORINATED HYDROCARBONS

Most organohalogens are produced and used by humans. But, there are numerous instances of the natural production and occurrence of a large number such compounds. Gribble (1992) began compiling identified natural sources, and has continued to add to the list. Of interest in the marsh context are the reports of production of organochlorine compounds by freshwater wetland organisms, including bacteria, fungi, algae and plants. Microbial producers include *Actinomyces* sp., *Pseudomonas* sp., and *Streptomyces* sp. Bluegreen algae and cyanobacteria include *Ananena* sp., *Nostoc* sp., and *Oscillatoria* sp. Higher plants are also involved, such as *Juncus roemerianus* which has been reported to contain substantial amounts of tetrachlorobenzene (Gribble, 1992). Halogenated hydrocarbons are also found in the fulvic component of peats and marsh sediments. In the context of marsh performance modeling, this means that $C^*$is not necessarily zero for chlorinated hydrocarbons.

The principal source of chlorinated hydrocarbons is contaminated groundwater. Accordingly, there are often regulations for the allowable concentrations that may be discharged to surface waters (Table 14.16). The chlorinated ethenes have large water solubility, and low octanol-water partition coefficients, which allows plant uptake (Table 14.20). Most chlorinated benzenes are not water soluble, and have high K_{OW} that block plant uptake.

TABLE 14.20

Examples of the solubility, octanol-water partition coefficient and organic carbon partition coefficient for chlorinated hydrocarbons.

Compound	Aqueous Solubility (mg/L)	Log_{10} K_{OW} g/g	Log_{10} K_{OC} L/kg
Dichloromethane	19900	1.48	1.36
Tetrachloromethane	800	2.70	2.22
Tetrachloroethylene	150	2.74	2.42
Trichloroethylene	1400	2.40	1.94
cis-1,2-Dichloroethylene	800	1.14	1.67
trans-1,2-Dichloroethylene	600	1.41	1.68
Vinylchloride	1116	1.08	
Monochlorobenzene	500	2.83	2.24
1,4-Dichlorobenzene	0.39	3.46	2.67
Trichlorobenzene	19	4.03	3.19
Tetrachlorobenzene	3.5	4.60	
Pentachlorobenzene	0.56	5.19	
Hexachlorobenzene	0.0113	5.69	4.69
2-Chlorophenol	18600	2.13	
Pentachlorophenol	17.2	4.81	

Source: Cornelis et al. (2017c).

Chlorinated Benzenes

Chlorobenzenes were and are used in the production of phenol, aniline and DDT. Mono-, di- and trichlorobenzenes are used as solvents, for instance in degreasing (Grayson, 1985). Various benzene hexachloride isomers are used as broad-spectrum insecticides, including Lindane™. Because of these varied uses, chlorobenzenes may appear in any of several different treatment marsh source waters.

Mono-, di- and trichlorobenzenes are resistant to photo-oxidation, and to both aerobic and anaerobic degradation in purely aquatic environments, with estimated half-lives of up to 6 to 24 months (Howard et al., 1991). However, wetland environments are quite different, because of the close interactions with plants and soils. The major routes for removal of chlorobenzenes from wetland waters are: biological (microbial) degradation, sorption, volatilization, and plant uptake.

Biodegradation

In general, chlorinated hydrocarbons may be dechlorinated under anaerobic conditions, and the responsible microbial consortia have been identified (van Eekert and Schraa, 2001; Lorah et al., 2014). Genuses *Dehalococcoides, Dehalobacter, Desulfitobacterium,* and *Geobacter* are known to contain species capable of reductive dechlorination. Reductive dechlorination of chlorobenzenes occurs via an anaerobic sequential pathway in wetland soils and sediments. Jackson and Pardue (2000) established that the dichlorobenzene (DCB) formed monochlorobenzene (MCB), which subsequently mineralized:

$$\text{DCB} \rightarrow \text{MCB} \rightarrow CO_2 + H_2O + Cl^- \tag{14.6}$$

Their study showed that MCB produced reached half of the initial DCB concentration in 30 days, and was accompanied by the formation of methane. In separate experiments, Jackson and Pardue (2000) recovered 13% of 14C-MCB as $14CO_2$ in a surface-sediment modulated decomposition.

Sorption

Chlorobenzenes sorb strongly to both organic and inorganic wetland substrates, with K_{OC} in the range of 10^2–10^5 L/kg (Table 14.20) (Pardue et al., 1993). Shin and Pardue (2002) showed that there are both reversible and irreversible portions of the overall sorption. For several sediment and soil samples, the reversible part ranged 1.88 ≤ log10KOC ≤ 2.88, while the irreversible part ranged $3.75 \leq \log_{10}K_{OC} \leq 5.55$. Thus, partitioning to organics is very strong, and dominated by irreversible binding.

Suspended particulate matter (TSS) forms a mobile substrate for partitioned chlorobenzenes (Shugui et al., 1994). The marsh biogeochemical cycle creates and processes very large quantities of TSS, and thus can play an important role in organic chemical cycling and removal in these systems. The wetland environment is further complicated by the presence of large molecules of humic substances, which comprise a good share of dissolved organic carbon. Organic solutes can partition to these dissolved humic substances, thus creating two soluble forms with different chemical characteristics.

Plant Uptake and Volatilization

Plants are capable of taking up organic chemicals (Trapp and Matthies, 1995), and processing them in a number of ways. They may be carried through the plant into the atmosphere via the transpiration flux, metabolized, or accumulated in plant tissues. For example, Leppich et al. (2000) found up to 100 mg/kg of various chlorobenzenes in black willow (*Salix nigra*) bark, and up to 25 mg/kgdw in leaves. The willows were growing in a contaminated swamp site, with large concentrations of di-, tri-, penta- and hexa-chlorobenzenes. Leppich et al. (2000) concluded that partitioning of these organics to the plants formed an important part of the site model.

The chlorobenzenes are volatile, and therefore may be easily lost from water, especially shallow water bodies, such as free water surface wetlands. Part of the vapor loss may be attributed to flow through plants (phytovolatilization). San Miguel et al. (2013) studied uptake of 14C-labeled 1,4-dichlorobenzene (DCB); 1,2,4-trichlorobenzene (TCB) and γ-hexachlorocyclohexane (gHCH) by *Phragmites australis*. They identified the uptake as associated with a water flow driven by the transpiration process of the plant. The transpiration stream concentration factors (see Chapter 7) were measured as 0.343 for DCB, 0.267 for TCB, and 0.095 for γHCH. Most of the absorbed chlorinated compounds were accumulated in underground plant parts such as roots and rhizomes, comprising 81±5%, 88±6%,88±2% of the total plant uptake, respectively for DCB, TCB, γHCH.

Reduction in Treatment Marshes

Despite the investigations detailed above, there is no reported treatment wetland that has specifically been designed to target chlorobenzenes. Monitoring data were acquired at the Saginaw, Michigan leachate treatment marsh (Kadlec and Zmarthie, 2010) (Table 14.18). Chlorobenzene appeared in 100% of the inlet samples, at a median concentration of 4 μg/L, and was not detected in any outlet sample in ten years of monitoring. Similarly, 1,4-dichlorobenzene appeared in 9% of the inlet samples, at a median concentration of 5 μg/L, but was not detected in any outlet sample.

Keefe et al. (2004b) conducted a synoptic sampling for chlorinated organics along the flow direction at the Tres Rios Hayfield 1 marsh. The 1.4 ha marsh was operating at a depth of 46 cm, and a detention time of 3.1 days. Their calculated plug flow removal rate coefficients, for zero background concentrations, were 123 m/yr for chlorobenzene, and 59 m/yr for 1,4 dichlorobenzene. The inlet concentrations were quite low, 30 ng/L for chlorobenzene and 740 ng/L for 1,4 dichlorobenzene. These results correspond to a 66% reduction of 1,4 dichlorobenzene from inlet to outlet. Keefe et al. (2004a) attributed the reduction to volatilization, but concluded that theoretical predictions were only good for order-of-magnitude estimation.

The Tres Rios data yield a slightly different result if the flow pattern is recognized to be described by 4 tanks in series, which was the measured value for that marsh. The removal rate coefficient, for zero background concentrations, is then 69 m/yr for 1,4 dichlorobenzene (Figure 14.13). If a nonzero background is allowed, the rate coefficient is higher.

Chlorinated Ethenes

Tetrachloroethylene, also know as perchloroethylene (PCE), and trichloroethylene (TCE), are solvents that saw extensive use for metal cleaning, clothes dry cleaning, and other applications in previous decades. At many locations, these materials found their way into groundwater, creating hazardous waste conditions, because of concern due to its carcinogenic properties. The chlorinated ethenes are water soluble to a fair extent (Table 14.20), and they are volatile, as are their decomposition products (dichloroethenes, DCE and vinyl chloride, VC). Wetlands provide environments for several mechanisms of removal of chlorinated aliphatic organics, including sorption, volatilization, reductive dechlorination, direct biological oxidation, cometabolism and plant uptake and metabolism (Pardue et al., 1993; Parkin, 1999; Pardue and Shin, 2000; Powell and Agrawal, 2011; Powell et al., 2011, 2014).

Lorah et al. (1997, 2002) observed complete removal of TCE and daughter products as contaminated water moved upward through peat to the surface. Reducing conditions were present in the peat, and both methanogenic and iron-reducing zones were identified. This

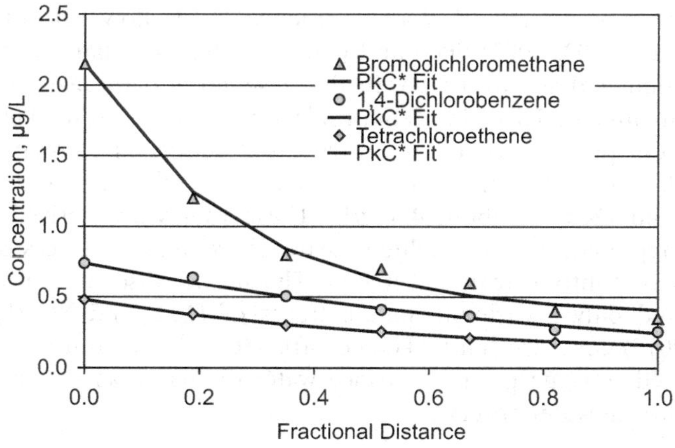

FIGURE 14.13 Concentration profiles along the flow direction for chlorinated organics at the Tres Rios H1 treatment marsh. (Data from Keefe et al., 2004a).

Fitting parameters:	**P**	**k** m/yr	**C*** µg/L	R^2
Bromodichloromethane	4	70	0.000	0.987
Dichlorobenzene	4	344	0.367	0.993
Tetrachloroethene	4	114	0.086	0.998

important study provided the impetus to examine the future role of natural and constructed wetlands in the remediation of TCE.

Sorption delays the transport of TCE but does not constitute the main removal pathway (Smets and Rittman, 1990; Mouvet et al., 1993; Deitsch et al., 2000). Partition coefficients are high (Table 14.20). Plant uptake occurs, because of favorable octanol-water partition coefficients. Bankston et al. (2002) measured taken up by cattails in microcosms, but 47% volatilized. Major degradation takes place in the marsh root zone (Powell et al., 2011, 2014).

Treatment Marsh at New Brighton, MN

A former TCE recycling plant is now the site of a city park in New Brighton, Minnesota. A TCE plume in a surficial aquifer discharges into the wetland of an adjacent lake (Bankston et al., 2002). Transect studies showed TCE in monitoring wells just upgradient from the wetland, and to a lesser degree in the fringe of the wetland. The anaerobic degradation product of TCE, cis-1,2-DCE, was detected in the aquifer and the wetland. It was hypothesized that the indigenous cattail (*Typha latifolia*) assisted in phytoremediation. Accordingly, microcosm studies were performed to determine the fate of the removed TCE. The recovery of 14C totaled 94.1%, of which 46.8% was volatilized, most likely as [14C] TCE because it was added to the microcosms by surface application. Plant tissue contained 38.2% of the 14C, 5.3% was present as [14C] CO2, and 3.7% was recovered from the soil and water. This data suggested that natural attenuation is a potential bioremediative strategy for TCE-contaminated wetlands.

Given that natural wetlands contribute to reduction of TCE, the next logical step is the reconstruction of former wetlands that are positioned in the flow path of a contamination plume. Richard et al. (2002, 2003) filled and planted a dredged channel that was conveying TCE from a contaminated site to a lake. In this case, an aquatic feature was converted to wetland to aid in treatment. This Minnesota project was completed in 2000.

The New Brighton project has now reported on eleven years of monitoring of chlorinated ethenes (Bankston et al., 2013). Entering TCE was variable, with a mean of approximately 14 mg/L, and DCE of about 4 mg/L. These values are derived from monitoring wells, and do not represent a flow-weighted mean. There are also traces of VC. Contaminated groundwaters vent into a treatment marsh. The surface discharge from the marsh contained no TCE, and only occasional minute traces of DCE and VC (less than 1 μg/L). Bankston et al. (2013) conclude that: "The constructed wetland continues to be effective in removing chlorinated solvents prior to surface water discharge as shown by the continued compliance with surface water criteria."

Treatment Marsh near Hillsdale, Michigan

In this project, a terrestrial landform was converted to constructed treatment wetlands to provide reductions in TCE. The Schilling Farm Project at Hillsdale, Michigan intercepts a TCE plume for treatment in a constructed marsh, built in a former cornfield (Kadlec et al., 2012a). The wetland system is made up of four treatment cells in parallel (refer to Figure 6.19). The two large central cells were sited to intercept the plume, and the two small flanking cells were added to ensure the full capture of that plume. Groundwater moves down a slight incline into a 4-meter deep capture trench, designed to intercept approximately the top two meters of the underground flow. This trench is filled with coarse rock to eliminate safety hazards and control rodent problems. The water flows upward and out across the FWS wetland cells, carrying TCE and DCE into the wetlands. During passage through the wetland, TCE undergoes reductive dechlorination to DCE and then to vinyl chloride (VC) in anoxic zones, and these are further degraded to carbon dioxide and water in aerobic zones.

$$\begin{array}{ccccc} & & 1,1-DCE & & \\ TCE & \rightarrow & -1,2-DCE & \rightarrow & VC \\ & & trans-1,2-DCE & & \end{array} \tag{14.7}$$

There is also volatilization of VC, and to a lesser extent DCE and TCE. Water is collected in rock-filled trenches, approximately one meter deep, at the downstream end of all four cells. These four outflows are metered, and merged to form a single project outflow for compliance monitoring. Additionally, another independent rock-filled trench is positioned across the entire downstream end of the four cells, which is drained via perforated pipe into the compliance outflow. The purpose of this trench is to capture waters that may pass totally underneath the treatment wetland.

Control of this wetland system is solely by means of setting water levels in each of the four cells, by use of the weir settings in the outlet structures. It is totally passive, with no pumps, and runs year-round in a cold climate. Therefore, water level control must compensate for ice formation.

The first four years of operation saw several internal system modifications. Herbivory and short-circuiting were created by muskrats (*Ondatra zibethicus*), necessitating removal of both the animals, and their habitat, by filling trenches with large rock, and by fencing the wetland (Kadlec et al., 2006, 2012a). Tracer testing indicated some degree of short-circuiting, which prompted the installation of finger berms to create a sinuous flow path. An interior rock-filled connection between flowpaths was installed to balance water depths. The next six years saw no substantial further modifications, and exhibited stable performance. Table 14.21 shows the performance results for the optimization and stable periods, as well as the combined period of record.

The monthly limits, applicable to the final discharge, were 150 µg/L for TCE, 45 µg/L for 1,1-DCE, and 13 µg/L for VC. During the stable operating years, there were no monthly exceedances for TCE and DCE, and only three for VC. Exceedances for VC tend to occur in late summer, when wetland surface flow contributes little to the outflow.

Figure 14.14 shows monthly performance for the stable operating period. Winter conditions provide lesser treatment. Winter is a time of elevated DO in unfrozen sections of the wetland, which may affect degradation rates. At this site, ice cover is sporadic throughout the winter, which may lower the volatilization rates. There is a second period of lesser treatment in late summer and autumn, which is occasioned by very low surface flows and dominance of underground flows.

Performance was fit to a first order, areal removal model for TCE, and a sequential version of the same for DCE (Kadlec et al., 2012a). The local removals are:

$$J_{TCE} = -k_{TCE}C_{TCE} \tag{14.8}$$

$$J_{DCE} = \alpha k_{TCE}C_{TCE} - k_{DCE}C_{DCE} \tag{14.9}$$

$$J_{VC} = \beta k_{DCE}C_{DCE} - k_{VC}C_{VC} \tag{14.10}$$

where C = concentration, mg/L
J = removal rate, g/m^2·yr
k = removal rate coefficient, m/yr

TABLE 14.21

Performance of the Schilling Farm treatment marsh for chlorinated ethenes. The optimization period of four years involved several system modifications. The ensuing six years had no system changes.

			Optimization	Stable	POR
HLR	cm/d		1.04	1.08	1.07
Depth	cm		30	26	27
HRT	days		28	24	25
TCE	µg/L	In	581	496	530
	µg/L	Out	82	28	50
	µg/L	% C Reduction	86%	94%	91%
Cis-1,2-DCE	µg/L	In	368	251	298
	µg/L	Out	73	27	45
	µg/L	% C Reduction	80%	89%	86%
Trans-1,2-DCE	µg/L	In	4.3	4.3	4.3
	µg/L	Out	0.9	0.1	0.4
	µg/L	% C Reduction	79%	98%	90%
1,1-DCE	µg/L	In	2.4	2.2	2.3
	µg/L	Out	2.1	0.6	1.2
	µg/L	% C Reduction	12%	71%	47%
VC	µg/L	In	5.2	3.4	4.1
	µg/L	Out	7.9	3.3	5.1
	µg/L	% C Reduction	–54%	4%	–19%

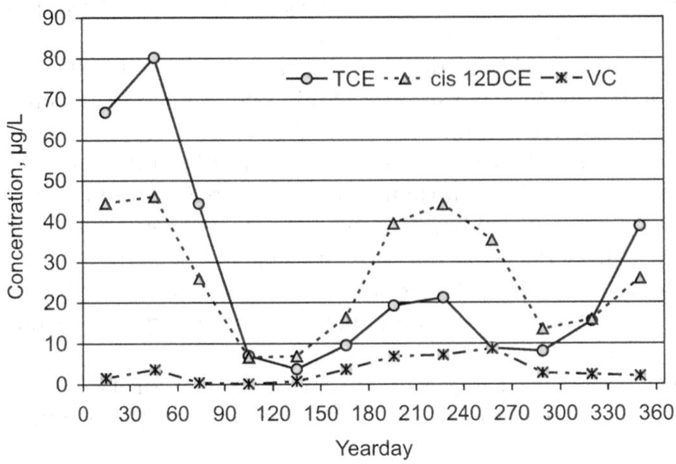

FIGURE 14.14 Annual patterns of TCE, DCE and VC in the Schilling Farm wetland system effluent. Data represent six years of stable operating conditions, folded.

α = fraction of TCE dechlorinated, –

β = fraction of DCE dechlorinated, –

This removal rate is combined with the hydraulics and hydrology of the wetland to calculate concentrations. For modeling purposes, the two treatment cells were viewed in combination, with a hydraulic pattern described by five tanks in series. A zero background concentration was assumed for all species. The long nominal detention times, coupled with the importance of rainfall as a component of the water budget, indicate the need to consider the system on a monthly, dynamic basis. The tank-to-tank calculations are described in more detail in Chapter 7 (see Equation 7.79).

The resulting monthly rate coefficients for TCE were maximum in springtime at k_{TCE} = 40 m/yr. The base values late summer through winter were approximately k_{TCE} = 10 m/yr. The rate coefficients for DCE and VC followed similar seasonal patterns. There is some variation in the numerical values, depending upon the amount of assumed dechlorination compared to degradation and volatilization. The springtime maximum for DCE was k_{DCE} = 55 m/yr, assuming α = 50% dechlorination. The minimum values 10–40 occurred in late summer through winter, and were approximately k_{DCE} = 7 m/yr.

Vinyl chloride, as the end dechlorination product, was most sensitive to assumptions about dechlorination fraction. For a = 50% of TCE producing DCE, the springtime maximum for VC was k_{VC} = 1370 m/yr, assuming b = 50% dechlorination of DCE to VC. The minimum values in late summer through winter were approximately k_{VC} = 50 m/yr.

14.9 MISCELLANEOUS

A number of other hydrocarbon chemicals have been examined for potential remediation by treatment marshes. A few examples are briefly discussed here.

EXPLOSIVES

It has been estimated that approximately 100 military bases and explosives manufacturing facilities have soil and/or groundwater contaminated with munitions Medina et al. (2000).

Studies have shown that plants are capable of transforming 2,4,6-trinitrotoluene (TNT) without microbial contribution, but very little accumulation of TNT has been found in plant material. Therefore, plant-enhanced degradation, or phytoremediation, of TNT by aquatic macrophytes has been proposed as a promising groundwater treatment process.

TNT is a reactive molecule that biotransforms readily under both aerobic and anaerobic conditions to give aminodinitrotoluenes (Hawari et al., 2000; Esteve-Nunez et al., 2001; Xiang, 2001). The resulting amines biotransform to give several other products, including azo, azoxy, acetyl, and phenolic derivatives, leaving the aromatic ring intact. Little or no mineralization is encountered during bacterial or wetland treatment.

Medina et al. (2000) performed a series of batch-scale experiments to assess the engineering kinetics of phytodegradation of TNT under a variety of operational conditions. Parrotfeather (*Myriophyllum aquaticum*) was hydroponically grown in laboratory microcosms. TNT was degraded according to a near-first order model in vegetated microcosms, but not in unvegetated systems.

Zoh and Horne (2000) also performed a series of batch-scale experiments. Utilizing straw, cattail, and bulrush litter in water, they found no degradation without litter, and first-order behavior in the presence of litter. Sorption and 14C studies indicated that removal was due to initial sorption followed by degradation to aminonitrotoluenes. Volumetric rate coefficients (kV) translate to areal rates (kA) on the order of 25 m/yr for TNT, regardless of litter type. A modified Arrhenius temperature coefficient (q) of 1.17 fit the observed difference in rates at 10 and 20°C.

Submerged aquatic vegetation was studied in mesocosms by Best et al. (1997a, 1999a, 1999b), showed promising results. Subsequently, cattails (*Typha angustifolia*) in marsh mesocosms were used for treatment of mono-, di-, and tri-nitrotoluene mixtures at the Volunteer Army Ammunition Plant, Tennessee (Best et al., 2000, 2001). Rate coefficients ranged from 16 to 45 m/yr.

The U.S. Army Corps of Engineers constructed treatment wetlands with the purpose of containing and treating residual TNT (2,4,6-trinitrotoluene) and RDX (hexahydro-1,3,5-trinitro-1,3,5-triazine) contamination at the Iowa Army Ammunition Plant (Thompson et al., 2003). Demonstration studies were then conducted to establish vegetative uptake, fate, and toxicity of TNT and RDX. Results from two years of testing indicated that TNT concentrations were undetectable in surface water or plant tissues. Concentrations of RDX in wetland surface waters were below the U.S. EPA human health advisory level of 2 mg/L and were undetectable in plant tissues during the summer months.

Despite these early successes of small scale tests, there has been no large scale treatment marsh project targeting explosives removal from contaminate waters, nonetheless, small-scale studies continue, as does optimism regarding marsh potential for remediation (Singh and Misra, 2014).

Phenols

Phenol is a compound of considerable industrial importance and a frequent pollutant in industrial effluents. It also has been studied for over four decades in the wetland treatment context. Early work at the Max Planck Institute in Krefeld, Germany, focused on SSF microcosms with bulrushes (*Schoenoplectus lacustris*) and resulted in a series of publications on removals and effects on plant physiology (for example, Seidel, 1966). That data showed degradation rates of 5 to 20 $g/m^3 \cdot d$, and in some instances, an induction period of one to three days before reduction commenced. The Seidel data may be modeled by a zero-order reaction. The zero-order rate constant for the Krefeld data was $k_0 = 81 \pm 18$ mg/d in summer, and 34 ± 10 mg/d in winter. The corresponding modified Arrhenius temperature coefficient was $\theta = 1.083$.

Microcosm research on marsh systems with cattails produced similar results (Kadlec and Srinivasan, 1995). The average value of the zero order rate constant was 76 ± 33 mg/day (loss = 0.55 $g/m^2 \cdot d$) for vegetated microcosms, and 108 ± 26 mg/d (loss = 0.78 $g/m^2 \cdot d$) for unvegetated microcosms. The sizes of these microcosms were comparable to those at Krefeld. There was little selective loss of phenol from microcosms containing only water and the phenol dose.

Polprasert et al. (1994, 1996) operated both FWS microcosms and a FWS pilot wetland in Bangkok, Thailand, for phenol reduction at high strength. Nearly complete removal of phenol was measured for loading rates up to 300 kg/ha·d. Some phenol was volatilized; some was found in the roots of the cattails. Removal followed a first order model, with rate constants on the order of 100 m/yr. About one-third of the removal was reported due to volatilization, and the remaining two-thirds due to biodegradation, plant uptake and sorption.

Dong and Lin (1994b) described a full-scale facility, built in 1990, consisting of 50 ha of wetlands and ponds treating 100,000 m^3/day. Average inflow and outflow phenol concentrations were 27 mg/L and 10 mg/L, respectively, with a mean removal efficiency of 63%. The full-scale facility had a phenol rate constant removal rate coefficient of 100 m/yr.

The treatment system at Amoco Mandan, North Dakota, received approximately 5,700 m^3/d of process water directed to an API separator for primary treatment, and then passed through the oxidation lagoon for secondary treatment. Process wastewater and stormwater was then directed to a series of ponds/wetlands (16.6 ha) before eventual discharge to the river (Litchfield and Schatz, 1989; Litchfield, 1990, 1993). Removal of phenol amounted to 94% with an average outflow concentration of 5 mg/L.

Surfactants

Linear alkylbenzene sulfonate (LAS) is a widely used synthetic surfactant for domestic detergents. This group of substances is amenable to phytoremediation using both terrestrial and wetland plants (Schwitzguebel et al., 2001).

Adsorption and degradation were found to both be operative in the wetland and in companion laboratory research (Inaba, 1992). LAS was effectively reduced in a *Typha – Phragmites* marsh. The 474-m2 system treated gray water from a 100-inhabitant community. The calculated plug flow volumetric removal coefficients (k_V) for the removal of these surfactants were in the range of 30 to 80 m/yr, depending upon chain length and isomeric form, with greater removal rates for the longer chain lengths. Seasonal dependence was strong, with a temperature factor of θ for the removal of these surfactants were in the range of 30 to 80 m/yr, depending upon chain length and isomeric form, with greater removal rates for the longer ch at an average hydraulic loading of approximately 5 cm/d. The surfactant partitioned strongly to wetland TSS, with a partition coefficient of 30 L/kg for chain length 11, isomer 5. Values for other forms of the surfactant were higher, ranging up to 520 L/kg for chain length 13, isomer 2. Sorption was not diminished at cold temperatures. Therefore, the sorption capacity and TSS trapping efficiency of the wetland partially offset the diminished degradation during winter.

Barber et al. (2001) performed synoptic sampling of several existing treatment wetlands, including analysis for LAS. Results were mixed, with removal occurring in some, but not all, systems.

Cyanide

Cyanide has an active cycle in nature, where it is present in green plants as part of the methionine cycle (ethylene production), and is therefore prevalent in autumn foliage and ripening fruit (Neuhauser and Swallow, 2005). If not handled properly, black and bing cherries (*Prunus serotina* and *Prunus avium*), and cassava can cause toxicity in humans. There are clear risks associated with cyanide release into the environment because of toxicity. The most toxic species are the free cyanides (HCN and CN^-), which are infrequent in uncontaminated soils and groundwater. Iron cyanide is less toxic, and often predominates in environmental samples. The U.S. Environmental Protection Agency and various state agencies regulate cyanide in groundwater and surface water differently. Agencies either choose to regulate the total or the free cyanide form. Because of the potential for photolytic production of free cyanide from iron cyanide, regulatory agencies may consider total cyanide to be indirectly toxic. U.S. EPA criteria for free cyanide include a drinking water standard of 200 μg/L (U.S.EPA, 2018b), and a freshwater acute criterion of 22 μg/L, and a freshwater chronic criterion of 5.2 μg/L (U.S.EPA, 2018a).

Cyanide compounds are classified into those that are readily converted into hydrogen cyanide under acidic conditions, or into CN^- under basic conditions, and those that are strongly complexed. This weak acid-dissociable cyanide (WAD) is distinguished from strong complexes formed with iron and cadmium, as examples. Total cyanide is the sum of both forms (Kavanaugh et al., 2003).

Cyanide is a pollutant of concern in several situations. Gold mine wastewaters may contain cyanide as the result of the extraction process (Kuyucak and Akcil, 2013). Manufactured gas plants historically left residues rich in cyanide and other constituents, which may subsequently leach to natural water bodies. Aluminum processing produces spent pot liners, which, when landfilled, cause similar leaching of cyanide.

Both aerobic and anaerobic microorganisms are able to destroy cyanide (Goncalves et al., 1998). Further, plants such as willows are able to ingest cyanide, and metabolically degrade it (Larsen et al., 2004, 2005). As a consequence, a natural treatment system that provides maximum opportunity for contact between cyanide-contaminated water and plants and microbes would have strong promise for removal and degradation of cyanide. Treatment wetlands are leading candidates to accomplish these goals.

The processing of cyanide in wetlands is known to involve microbial degradation, volatilization, plant uptake, and complexation with iron (Young, 2003). Additionally, iron cyanide decomposes photolytically to free cyanide (Kim et al., 1998). The aerobic and nutrient rich environment promotes the growth of the microbial population capable of uptake, conversion, sorption, and/or precipitation of thiocyanate, cyanide, ammonia, nitrate, sulfate, and metals. The microbial removal involves production of ammonia (Goncalves et al., 1998; Akcil, 2003):

$$M(CN)_2 + 4H_2O + O_2 \rightarrow M^{+2} + 2HCO_3^- + 2NH_3 \tag{14.11}$$

$$SCN^- + 3H_2O + 2O_2 \rightarrow SO_4^{-2} + NH_4^+ + HCO_3^- + H^+ \tag{14.12}$$

where M represents a metal species.

Various Pseudomonas species are responsible for oxidation of cyanide and thiocyanate. Some of the organisms known to oxidize cyanide include species of the genera *Actinomyces, Alcaligenes, Arthrobacter, Bacillus, Micrococcus, Neisseria, Paracoccus, Pseudomonas*, and *Thiobacillus* (Akcil and Mudder, 2003). Plant uptake is also potentially active (Ebbs et al., 2003; Larsen et al., 2004, 2005).

These encouraging process studies have led to several treatment wetland projects targeting cyanide reduction. A full-scale treatment marsh has been in operation at the Campbell gold mine in northwestern Ontario wetland since 2002, to polish wastewater effluents prior to discharge (Martin et al., 2015). The 16 ha marsh receives flows of ~10,000 m^3/day, with a mean HRT of 6–7 days. Vegetation is predominantly cattails (*Typha* spp.) with substrates comprised of inert mining waste-rock overlaid by organic topsoil. The system operates during the ice-free period (May through October), and serves as the final point of discharge for the mine. Over the period 2011–2013, cyanide reduction was 78%, with average influent and effluent concentrations of 0.14 mg/L and 0.07 mg/L, respectively. The mine water management system has sufficient storage to allow all mine waters to be discharged through the wetland during the ice-free period.

A large-scale trial was conducted at the Rustler's Roost mine near Darwin, Australia. A 3.2-ha site was configured as ponds and wetlands, covered with peat, and planted with *Typha orientalis* (Halford, 1999). A light hydraulic loading (<1 cm/d) was applied during the dry season of 1996 (July–September). The incoming water had total cyanide of about 2,000 μg/L, and the system discharge was 40–400 μg/L. Weak acid dissociable cyanide was reduced from about 900 to 200 μg/L.

The treatment marsh at the Musselwhite, Ontario, gold mine operated at a very high hydraulic loading (50 cm/d) (Bishay and Kadlec, 2005). As a result, total cyanide and weak acid dissociable cyanide were only slightly reduced by passage through the wetland, from 26 to 22 μg/L (15%) and from 11 to 8 μg/L (26%) respectively, as a four-year average.

Wetland remediation of groundwaters contaminated with cyanide from landfills associated with the aluminum industry has been shown to be effective (Gessner et al., 2005). A pilot project at Alcoa, Tennessee, consisted of two lined marsh cells operated in series, vegetated with *Schoenoplectus tabernaemontani*. Total cyanide inflow concentrations typically ranged from 30 mg/L to 50 μg/L, while all outflow concentrations except one were below the lowest reporting level of 10 μg/L. A second marsh pilot project was operated during 2004–2006, also in Alcoa, Tennessee (Figure 14.15). This 2,300-m2 system received about 300 μg/L of total cyanide at 130 m^3/d, resulting in a nominal detention time of about three days. This system achieved well over 90% reduction in total cyanide (Figure 14.16). The 4 TIS rate constant for total cyanide was 100 m/yr, and for WAD cyanide was 27 m/yr.

FIGURE 14.15 The Duck Spring, Tennessee, cyanide treatment marsh.

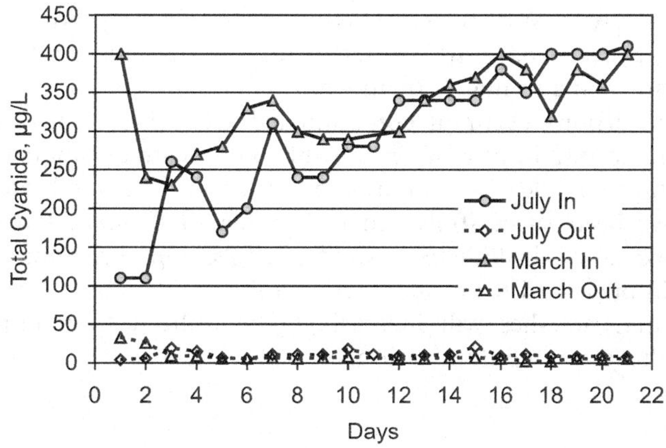

FIGURE 14.16 Input-output data for two intensive sampling periods for total cyanide at the Duck Spring treatment marsh wetland at Alcoa, Tennessee. Data source: unpublished data from Alcoa Aluminum.

14.10 SUMMARY

Source waters containing a wide variety of trace hydrocarbon constituents are presently being treated in marsh systems, and performance has been generally good. There is some removal of nearly every organic compound tested in treatment marsh systems, whether full scale, pilot scale, or microcosm. This includes organics from very important generators: pesticides, pharmaceuticals, petroleum, and petrochemicals. In each of these categories, there are multiple studies that have initiated a design database. Current knowledge shows great promise for marsh technology for organics control, especially in situations where passive systems can be fit into the landscape and accumulated organics will not pose a threat to wetlands biota. Mechanisms of removal and reduction include volatilization, sorption, photolysis, hydrolysis, microbial degradation and plant uptake. The proportions of these mechanisms vary with the chemical character of the hydrocarbon under consideration. Removal rate coefficients are typically high, especially for volatile materials.

Marshes provide significant reductions for all pesticides tested. The most important source is agricultural runoff, but urban sources also contain these pollutants. First order kinetics are often applicable, with suitable adjustment for the hydraulic efficiency of the marsh.

Contaminants of emerging concern include pharmaceuticals and personal care products, some of which are endocrine disruptors. They may potentially cause deleterious effects in aquatic life and are not completely removed in conventional wastewater treatment plants. They may also enter surface waters due use in animal husbandry. Marshes reduce these substances, often with half-lives on the order of a week. Endocrine disruptors have the potential for causing deleterious effects in receiving waters, which include marshes used to remove these substances. It is therefore necessary to consider effects in the treatment marsh as well as in the receiving water body.

Trace petroleum hydrocarbons trace hydrocarbons originate primarily from industrial activities, but small amounts find their way into and through wastewater treatment plants, and small amounts occur in urban and agricultural runoff. Treatment marshes sometimes are used to polish pretreated wastewaters from industry, and are also used to remediate contaminated groundwaters. BTEX (benzene, toluene, ethyl benzene and xylenes) represents one class of marsh-treatable compounds; polycyclic aromatic hydrocarbons (PAHs) are a second group of interest. Waters associated with petroleum production are of growing importance, as more hydrocarbons are extracted from oil shales and oil sands. These heavier substances are less amenable to marsh treatment.

Chlorinated hydrocarbons occur in contaminated groundwater, usually due to spillage. Chlorobenzenes may appear in any of several different treatment marsh source waters but have not been the target of designs. Chlorinated ethenes are groundwater contaminants of importance and have been successfully reduced in marsh systems. Other potential candidates for marsh remediation include explosives, phenols, and surfactants. Several treatment marsh systems are in operation for cyanide reduction.

Indications are that marshes will increasingly be utilized to minimize environmental impacts of trace hydrocarbon contamination.

15 Ecological Perspective

Because they treat relatively benign source waters, most treatment marshes may also be regarded as wetland restoration projects. Indeed, the large phosphorus removal marshes in south Florida are a major part of the "Restoration Strategies" initiative designed to improve water quality reaching the remnant Everglades. When wetlands are used to reduce pollutant concentrations and peak stormwater flows, ecological benefits can be achieved. This chapter summarizes those benefits of wetlands used for pollution control and recommends design features to optimize these benefits. The primary objective of most wetland pollution control projects is water quality enhancement, through assimilation and transformation of sediments, nutrients, and toxic chemicals. Secondary benefits that can be incorporated in wetland treatment designs include (1) vegetative biodiversity, (2) protection and production of fauna, and (3) aesthetic, recreational, commercial, and educational human uses. Full realization of these benefits usually requires extra land and extra expenditures.

Guiding principles for planning, siting, design, construction, operation, maintenance, and monitoring of constructed treatment marshes were set forth in U.S. EPA (2000c). These address the degree of wildlife habitat provided by constructed treatment wetlands, or sections of these wetlands, and varies across a spectrum. At one end of the spectrum are those wetlands that are intended only to provide treatment for an effluent or other water source, in order to meet the requirements of wastewater treatment objectives, and that provide little to no ecological value. At the other end are those systems that are intended to provide water reuse, wildlife habitat, and public use, while also providing a treatment function for a pretreated effluent, runoff or riverine source. The U.S. EPA (2000c) guidance document primarily addresses the latter type of wetland, which is defined herein as treatment marshes.

15.1 DISAPPEARING DICHOTOMIES

Restoration and Treatment

Moreno-Mateos and Comín (2010) suggest that conflicts and compatibilities between treatment and restoration objectives should guide strategy. At landscape scale, strategy may be either creating different uni-purpose wetlands or multipurpose wetlands, or combinations of them. They suggest that wetlands should have more ecological functions. Restored and created wetlands could then provide an array of integrated environmental services adapted to local ecological and social needs.

U.S.Da (2008) defines wetland restoration as the rehabilitation of a degraded wetland or the reestablishment of a wetland so that soils, hydrology, vegetative community, and habitat are a close approximation of the original natural condition that existed prior to modification to the extent practicable. Wetland enhancement is defined as the rehabilitation or reestablishment of a degraded wetland, and/or the modification of an existing wetland, which augments specific site conditions for specific species or purposes. Wetland creation is defined as the creation of a wetland on a site that was historically nonwetland.

U.S. EPA (2000c) recognized the overlap between treatment and restoration, and the existence of opportunities to use pretreated effluent, or other similarly benign source waters, to restore degraded wetland systems. However, this implies that treatment will occur in an existing wetland. U.S. EPA (2000c) thus states that should only be considered if (1) the source water meets all applicable water quality standards and criteria, (2) its use would result in a net environmental benefit to the receiving (degraded) wetland's natural functions and values, and (3) it would help restore the system to its historic, natural condition. Prime candidates for restoration may include wetlands that were degraded or destroyed through the diversion of water supplies, due to agricultural activities.

Yet, the concept of restoration going hand-in-hand with water quality improvement continues to be plagued by misconceptions, with lack of acceptance by the tradition-bound disciplines of both engineering and ecology. Those barriers are beginning to disappear, as evidenced by "restoration" projects that have water quality goals at the forefront, and vice versa (Greenway, 2010; Moreno-Mateos and Comín, 2010; Rodriguez and Lougheed, 2010; Hefting et al., 2013; Strand and Weisner, 2013; Darwiche-Criado et al., 2017). There is growing understanding of the dichotomy between wastewater treatment wetlands of many types, and treatment marsh polishing systems.

There exist numerous manuals that detail the procedures to be followed in wetland restoration projects (e.g., U.S. ACE, 2000; U.S. DA 2008). However, for present purposes, it is assumed that design and construction of treatment marsh projects will be largely determined by treatment requirements, although in most instances there are options for including other restoration goals,

Engineering and Ecology

Those associated with treatment and restoration wetlands have been frustrated by the lack of understanding, by both ecologists and engineers, of the motivations and toolboxes possessed the other. The question is posed: "When will ecologists learn engineering and engineers learn ecology?" (Mitsch, 2014).

Aronson et al. (2016) observed:

> Ecologists may only know a little engineering, but unless they are licensed to sign plans and permits, their engineering input is advisory. Therefore, it may not be important for an ecologist to learn engineering above a modest level. On the contrary, if engineers are to prepare ecologically viable and relevant plans, they need to be well versed in ecological principles, applied ecology, and at least comfortable with the kinds of questions ecologists are asking about ecosystems and biological communities at a restoration site.

Mitsch (2012) has labeled projects that combine water quality objectives with ecological goals as the bailiwick of "ecological engineering." There is little doubt that ecosystem restoration should be considered a part of ecological engineering. Further, Mitsch (2012) contends (correctly) that nature contributes to the final design of ecosystems as much or more than does the human designer.

This dichotomy is pervasive and will likely continue for years to come. There is a slow diminishing of the disciplinary divide, caused primarily by growing experience bases for the two fields. However, standard university curricula are strongly bound by tradition and accreditation requirements, with little cross-over permitted. Ecological majors are required to take courses in basic chemistry, statistics, biological and wildlife science, taxonomy, plant and mammalian ecology. Engineers are required to take math through differential equations, physics, advanced chemistry, and transport phenomena.

It seems logical that a new engineering field is needed – ecological engineering. As yet, only a few such university curricula exist. The first accredited ecological engineering Bachelor's degree program in the United States was developed at Oregon State University. The core courses are:

- General Chemistry I
- General Chemistry II
- Statics
- Strength of Materials
- Differential and Integral Calculus
- Vector Calculus
- Applied Differential Equations
- Matrix & Power Series
- Statistics for Engineers
- General Physics

- Principles of Biology
- Ecology
- Soil Science
- Public Speaking
- English Composition

- Ecological Engineering I – Principles
- Ecological Engineering II – Applications
- Ecological Engineering Fundamentals
- Ecological Engineering Computation

- Ecological Fluid Mechanics
- Ecohydraulics

- Ecohydrology
- Biosystems Analysis and Modeling
- Thermodynamics and Transfer Processes
- Ecological Engineering Laboratory
- Nonpoint Source Pollution Assessment and Control
- Bioremediation
- Ecological Engineering Design I
- Ecological Engineering Design II

Degree programs such this remain a rarity.

15.2 MEASURES OF ECOLOGICAL VALUE

The mosaic of vegetation and the animals that inhabit it determines the ecological values of a treatment marsh. Quantification of those values begins with a census of what occupies the wetland, including plants, animals, birds, and reptiles. This involves enumeration of the species present, and their relative abundance. The results vary both seasonally and spatially, making quantification difficult. Additionally, hydrology and water quality also have strong impacts on results.

Research and manuals on wetland evaluation techniques (Greeson et al., 1978; Adamus et al., 1992; Novitzki et al., 1997; Adamus, 2002, 2011) may be used to evaluate the ecological benefits of treatment marshes. However, these techniques are broader than needed for treatment marshes, because they consider functions such as flood protection that are not applicable to the treatment marsh situation.

Of primary concern are the floral and faunal biodiversity, and human use of the marsh. Biodiversity has moved into prominence in the past few years, as evidenced by the occurrence of biodiversity as a publication key word, as the 14th most common in 2008, compared to 45th in 1991–1996 (Zhang et al., 2010). Water quality ranked second to only "wetland" in all years surveyed.

The perceived ecological value of a system rests upon both objective and subjective considerations. The numbers and kinds of biota can be measured, and quantified, which comprises the objective part. A biodiverse system will contain many species, in varying proportions, and many metrics exist to attach numbers to the spread of the populations. But the components of such populations carry varying degrees of perceived desirability. Roughly, there are four value categories of organisms in the eyes of the human observer:

A. Rare or endangered biota native to the region and associated with "pristine" conditions. For example, snail kites (*Rostrhamus sociabilis*) are highly valued in south Florida.

B. Biota native to the region and associated with "pristine" conditions. For instance, in south Florida, the native sawgrass (*Cladium jamaicense*) is emblematic of relatively unimpacted Everglades marshes.
C. Biota native to the region and associated with "degraded" conditions. For instance, in south Florida, the native cattail (*Typha domingensis*) is emblematic of impacted Everglades marshes.
D. Biota not native to the region, labeled as exotic species. For instance, in south Florida, the exotic fern (*Lygodium japonicum*) is regarded as a threat Everglades marshes.

Ecological value is the superposition of value categories upon descriptive metrics. In other words, a system may be found to be quantitatively diverse, but include several undesirable species.

The science literature provides dozens of alternative indices that may be used to describe wetland status (Washington, 1984); here only a few of the more commonly used are summarized.

Marsh Water Quality Indices

It is generally recognized that what lives in the marsh is strongly dependent on dissolved oxygen, pH, salinity, and nutrient status. Subjective appraisals typically assign low qualitative value to low DO and high nutrient concentrations. Quantitative evaluation involves metrics derived from the concentrations of key constituents. An example is the ecological index developed to evaluate the wetland quality in the Great Lakes region of the United States.

That water quality index (WQI) was devised by Chow-Fraser (2006), and is calculated as a linear combination of the logarithms of twelve water quality variables (Seilheimer et al., 2009). The WQI measures the degree of water quality degradation as a result of nutrient enrichment and road runoff. It is supposed to be widely applicable to wetlands, and to produce measurements of wetland condition, and to measure the degree of water quality degradation as a result of nutrient enrichment and road runoff. The WQI is calculated as:

$$\begin{aligned} WQI = {} & 10.024 - 0.726 ln\,(TURB) - 0.842 ln\,(TSS) - 0.818 ln\,(ISS) - 0.866 ln(TP) - \\ & 0.432 ln\,(SRP) - 0.169 ln\,(TAN) - 0.464 ln\,(TNN) - 0.524 ln\,(TN) - \\ & 1.315 ln\,(COND) - 2.684 ln\,(TEMP) - 10.030 ln\,(pH) - 0.527 ln\,(CHLA) \end{aligned} \tag{15.1}$$

where CHLA = chlorophyll A, μg/L
COND = electrical conductivity, μS/cm
ISS = inorganic suspended solids, mg/L
pH = hydrogen ion content, s.u.
SRP = soluble reactive phosphorus, μg/L
TAN = total ammonia nitrogen, μg/L
TEMP = temperature, °C
TN = total nitrogen, μg/L
TNN = total nitrate nitrogen, μg/L
TP = total phosphorus, μg/L

TSS = total suspended solids, mg/L
TURB = turbidity, NTU

McGauley (2008) calculated the WQI for the Brighton treatment marsh, and for other wetlands in the vicinity. The average WQI for the treatment marsh was 1.19 ± 0.17 (mean ± s.e.) (range = 0.82–1.76, N = 12 sampling places). The average for three comparison wetlands was 1.39 ±0.14 (range = 1.12–1.55, N = 6 sampling places). These are all in the qualitative category labeled "very good" by Chow-Fraser (2006). Another neighboring wetland, the Presq'ile Park wetland, had WQI = 0.77 ± 0.12 (range 0.6–1.0, N = 3 years) (Chow-Fraser, 2006). That is in the qualitative category "good."

The highest score (interpreted as being the most pristine) for natural wetlands in the region, the Lake Ontario/St. Lawrence watershed south of 45° north latitude, was WQI = 1.3 (Chow-Fraser, 2006). The Brighton treatment wetlands were at the 95th percentile in this group, which comprised 87 wetland-years. Therefore, the water discharge in the Brighton wetland is of a very high quality relative to reference wetlands.

Biotic Indices and Diversity

The discussion of ecological diversity in the following sections requires some understanding of the indices of biodiversity that currently in use for treatment marshes. This topic has been the subject of numerous investigations, with many (confusing) nuances; and only a brief précis is given here. Summaries of terminology are available for further explanation (e.g., Washington, 1984; Tuomisto, 2010). The reader is cautioned that the literature terminology is often ambiguous, and care must be exercised in interpreting numerical meaning.

Suppose that a given sample is comprised of S species. The proportional abundance of species "i" is quantified as $p_i = m_i/m$, where m_i is the number of individuals classified into species "i" and m is the total number of individuals. Diversity can then defined by any of several combinations of these values.

1. The *species richness* is the value of S.
2. *Simpson's diversity* (D) (also called the Simpson concentration) is:

$$D = \sum_{i=1}^{S} {p_i}^2 \tag{15.2}$$

Index D = 0 represents infinite diversity and 1, no diversity. Alternatively, for finite sample size:

$$D = \sum_{i=1}^{S} \frac{p_i(p_i - 1/S)}{(1 - 1/S)} \tag{15.3}$$

Index D = 0 represents infinite diversity and 1, no diversity.

3. *Simpson's index of diversity* is (1-D).The greater the value, the greater the sample diversity, ranges from 0 to 1.0.
4. *Simpson's reciprocal index* is (1/D). Starts with 1 as the lowest possible and the higher the value, the greater the diversity.
5. *Shannon diversity index* (H)

$$H' = -\sum_{i=1}^{S} p_i \ln p_i \tag{15.4}$$

The Shannon diversity is zero for a monospecific population, and ranges upward, with a *de facto* limit of about 5.

The foregoing metrics apply to samples taken from a single more or less uniform system. Treatment marshes are not uniform but display patchiness and gradients in the direction of flow. Both require an added dimension to the concept of a diversity metric. Thus, three related kinds of diversity have been set forth: alpha, beta and gamma diversities. Gamma (or total) diversity describes the total species diversity in a landscape, and is determined by two different things, the mean species diversity at the habitat (or specific area) level (α) and the differentiation among habitats (β). In terms of the preceding definitions of diversity indices, multiplicative rules hold (Jost, 2007; Tuomisto, 2010; Anderson et al., 2011):

$$D_\gamma = D_\alpha \times D_\beta \tag{15.5}$$

$$\exp(H_\gamma) = \exp(H_\alpha) \times \exp(H_\beta) \tag{15.6}$$

Beta diversity describes both patchiness (nondirectional) and gradient (directional) variability in a marsh (Figure 15.1). However, it is far from clear how numerical values of beta diversity relate to the strength of observed gradients, or the variability caused by patchiness.

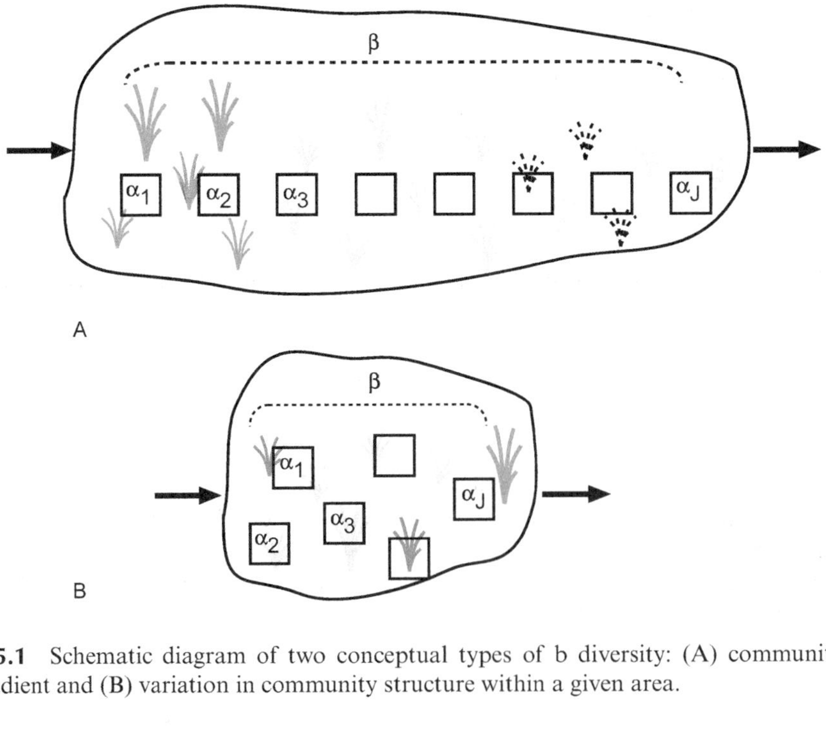

FIGURE 15.1 Schematic diagram of two conceptual types of b diversity: (A) community structure along a gradient and (B) variation in community structure within a given area.

Nonetheless, these metrics are in common use and are being applied to treatment marshes (e.g., Hefting et al., 2013).

6. *Pielou's evenness index (J)*

$$J = \frac{H'}{\ln(S)} \quad (15.7)$$

The maximum value of H' = ln(S), which occurs if all species occur at the same frequency. Index J = 1 represents uniform diversity and 1, no diversity.

Indices are usually used in a comparative sense, with the aim of quantifying the degree of difference or similarity of two different systems.

Indices of Biological Integrity

Diversity indices may be concatenated with measures of the character of the habitat associated with each species, to calculate indices of biological integrity (IBIs) (Karr, 1991). Integrity is used in the sense of the state of being sound or undamaged. Each species (or genera or other grouping) is assigned a numeric rank according to its perceived environment of occurrence. For instance, the Hilsenhoff index for macroinvertebrates is based upon assignment of an integer number to each species, between 0 (very "good" organisms, pollution intolerant) and 10 (very "bad" organisms, pollution tolerant). The index is then formed by weighting these tolerance values according to the number of individuals of each species in the sample.

This concept originated from efforts to describe the ecological status of streams (Karr, 1991). However, the floral and faunal assemblages of wetlands are quite different from those in streams. Therefore, wetland IBIs must be based upon wetland-specific information. But, in the same manner as for streams, each metric is given a rating according to whether it deviates somewhat from, or deviates strongly from values measured in least-disturbed ecosystems of a particular type within a region (Adamus, 2002).

Treatment marshes cannot achieve the highest regional IBI, because by definition, they are influenced by anthropogenic influences from added source waters, regardless of the water quality of the source. But they will not achieve the lowest regional IBI either, because also by definition they exclude high levels of contaminants. As for other indices, IBIs are most useful in a comparative context.

Water Quality Spectrum

Treatment wetlands have historically been considered to treat source waters ranging from relatively strong effluents, such as secondary WWTP waters, to lesser strength waters, such as tertiary WWTP effluents. In the context of this book, attention is directed to low strength treated wastewaters, and agricultural and urban runoff. Biodiversity is very much dependent upon nutrient content in particular, with lesser diversity associated with strong source waters. However, because there are gradients of water quality in treatment wetlands, the biological conditions at the outlet ends of wetlands treating stronger waters are relevant to the focus of the present discussion. Therefore, the following sections are not entirely restricted to treatment marshes.

15.3 VEGETATION

Two classes of vegetation are candidates for assessing the biological condition of a marsh, algae and macrophytes.

Algal Indicators

Marshes contain rich assemblages of algae, which vary with water quality, vegetative cover, and season of the year. The character of these populations is therefore a good candidate for assessing the ecological condition of the wetland. However, problems arise because of the difficulties of identification of the extremely numerous organisms, and their typically microscopic size. Species-level assessments require the great taxonomic expertise, and genera-level measurements present some of the same obstacles. Nevertheless, algal metrics have sometimes been used to quantify ecological character (Stevenson et al., 2002).

A prime example is the treatment marsh that is imbedded in the north end of WCA2a of south Florida. The marsh originally was among the most oligotrophic wetlands to be found. Agricultural runoff was added to the inlet end for many decades, creating a spatial gradient of water quality that fostered a change in the algal communities found along that treatment gradient (McCormick and O'Dell, 1996; McCormick et al., 1996). Some of those diversity changes were monotonic, and others nonmonotonic (showing both positive and negative changes variable increases along the gradient).

Table 15.1 shows the results of the McCormick et al. (1996) survey for the wet season, when most of the flow occurs. Inlet sites were near the point of water entry, and outlet sites were at the extreme downstream end of the water quality gradient. Algae were partitioned according to their growth form, as epipelon, epiphyton, and metaphyton (see Chapter 5). Epiphyton were found attached to sawgrass (*Cladium jamaicense*), eleocharis (*Eleocharis cellulosa*), and cattails *(Typha domingensis).* There was no epipelon at the inlet, or any sawgrass or eleocharis, thus the zero values for those components of the algal population. There was no identified commonality of species between the two sites, except that which may have been included in the "other" category of unspeciated metaphyton. Additionally, the biomass of algae was about two orders of magnitude larger at

TABLE 15.1

Biovolume percentages of algal constituents found at the inlet (eutrophic site) and outlet (oligotrophic site) in the treatment section of Water Conservation Area 2A in Florida.

	Epipelon		Epiphyton						Metaphyton	
			Cladium		Eleocharis		Typha			
	In	Out	In	Out	In	Out	In	Out	In	Out
Other	0	27	0	31	0	30	11	0	16	25
Synedra delicatissme	0	3	0	7	0	14	0	0	0	12
Synedra radians	0	0	0	1	0	1	0	0	0	0
Mastogloia smithii	0	7	0	22	0	21	0	0	0	5
Amphora lineolata	0	7	0	5	0	10	0	0	0	12
Scytonema hofmannii	0	16	0	14	0	14	0	0	0	16
Chroococcus turgidus	0	40	0	20	0	10	0	0	0	30
Rhopalodia gibba	0	0	0	0	0	0	2	0	2	0
Lyngbya spp.	0	0	0	0	0	0	4	0	6	0
Oscillatoria spp.	0	0	0	0	0	0	21	0	7	0
Oscillatoria amphibia	0	0	0	0	0	0	8	0	7	0
Oscillatoria princeps	0	0	0	0	0	0	55	0	23	0
Spirogyra spp.	0	0	0	0	0	0	0	0	39	0

Source: McCormick et al. (1996).

the downstream site. This strong disparity between assemblages makes the algae a primary indicator of water quality impacts on the natural marsh in which the treatment section is located.

Macrophyte Diversity

Constructed *FWS treatment wetlands* are typically dominated by emergent, floating, or submerged aquatic plant communities. Emergent species are frequently intermingled and co-dominant with populations of small floating aquatic plants such as duckweed (*Lemna* spp.). Many FWS treatment wetlands in the United States are dominated by cattails (*Typha* spp.) or bulrushes (*Scirpus* spp.), and in Europe by common reed (*Phragmites australis*) or grasses (*Phalaris arundinacea* or *Glyceria maxima*). However, many *treatment marshes* are dominated by other species or by a complex admixture of plant species that includes cattails, bulrush, and grasses. Historically, cattails, bulrushes and reeds were deemed desirable for treatment functions, because they are robust, and can survive in a wide variety of environmental conditions – in other words, they are survivors. They are inexpensive, plentiful in supply, and easy to propagate. However, they may also be viewed as nuisance species, because of their ability to crowd out other species, and assume visual dominance.

In the United States, species such as *Phalaris arundinacea, Lythrum salicaria, Hydrilla verticillata, Typha* spp. and *Phragmites australis* are considered by many to be undesirable, aggressive invaders. These may take over natural wetlands, as well as being a component of treatment wetlands. Pollutant removal processes do not require these species. They were often selected on the basis of cost and survivability. As a result, most ecologists equated FWS treatment wetlands with low biodiversity and nuisance species. This perception is slowly changing to a concept of variable biodiversity, driven by water quality and hydrology in response to a greatly increased body of knowledge. New factors include:

- A greatly expanded group of target source waters. Treatment marshes are no longer directed principally to wastewater treatment, but are used to improve the quality of urban and agricultural runoff and contaminated streams. Target constituents are no longer mainly BOD and ammonia.
- A greatly expanded set of scientific ecological studies of plant assemblages in treatment marshes. Evaluations have shifted from subjective opinions to objective, quantitative assessments.

Historical Studies

Three decades ago, U.S. EPA recognized two responsibilities with respect to treatment wetlands: (1) evaluation of the system's capability to meet water quality standards, and (2) assessment of the general habitat quality and the potential for risks to wildlife (Davis and Montgomery, 1987). The second responsibility was initiated with weeklong field studies in 1991 that measured ecological parameters at six treatment wetlands (McAllister, 1992, 1993a, 1993b). The vegetation richnesses varied from 13 to 63 species, which in each case were within the range for nontreatment wetlands of the region. However, only two of the six systems would be considered treatment marshes, and all were in warm climates. These studies called attention to the fact that treatment wetlands were not "monocultures."

In 1993, the North American Treatment Wetland Database was issued (Knight et al., 1993), and it was subsequently augmented to include ecological parameters (NADB Database, 1998; Knight et al., 2001). Although the database contains habitat data for 168 surface flow systems, very few of those qualify as treatment marshes – about a dozen. The large remaining majority are forested wetlands, and/or treat quite strong effluents. Over this broader spectrum of all FWS treatment wetlands, 693 species of EAV, 36 species of FAV and 12 species of SAV were identified. The Knight et al. (2001) report concluded that biological structure was substantial and dominated by relatively diverse assemblages of wetland plant species, typically including a few dominants and many less common species that have specific adaptations to grow in saturated soils.

Over the past decade, emphasis has refocused upon treatment marshes, which receive highly treated municipal effluents, river water, and agricultural and urban runoff. For the examples are shown in Table 15.2, the median number of macrophyte species is 25. Species richness does not address the issue of relative abundance of species, that is, it does not

TABLE 15.2

Macrophyte species richness in example treatment marshes.

Location	Wetland	Source	Area ha	Species Number	Reference
Sweden	6 EAV	Groundwater	0.004	10	1
Sweden	6 SAV	Groundwater	0.004	17	1
Sweden	6 Unplanted	Groundwater	0.004	16	1
Australia	Golden 1	Urban	0.13	10	2
Australia	Golden 2	Urban	0.13	7	2
Australia	Bridgewater	Urban	0.7	6	2
N. Carolina	20 Systems Median	Urban	0.22	7	3
Texas	Shannon Pilot T1	River	0.24	10	4
Texas	Shannon Pilot T2	River	0.24	16	4
Texas	Shannon Pilot T3	River	0.24	17	4
Texas	Shannon	River	98.4	34	5
Illinois	Des Plaines EW3	River	2.39	29	4
Illinois	Des Plaines EW4	River	2.42	48	4
Illinois	Des Plaines EW5	River	1.94	38	4
Illinois	Des Plaines EW6	River	3.38	51	4
Nevada	Incline Village	Municipal	173	24	6
Florida	Orlando Easterly	Municipal	494	63	7
Florida	Orange County	Municipal	14.2	91	4
Florida	STA1E	Agricultural	2,077	26	8
Florida	STA1W	Agricultural	2,699	105	8
Florida	STA2	Agricultural	2,565	27	8
Florida	STA34	Agricultural	6,695	40	8
Florida	STA5	Agricultural	1,663	41	8
Florida	STA6	Agricultural	352	22	8

References:
1 Weisner and Thiere (2010)
2 Greenway (2010)
3 Moore and Hunt (2012)
4 NADB (1998)
5 APAI (2008)
6 McAllister (1993b)
7 McAllister (1993a)
8 Chimney (2012)

identify the dominant species. Considerably more effort is required to measure how many of each species are present in a given location, or in a whole wetland.

Phytosociology

Phytosociology is the fundamental methodology for determination of the relative abundance of macrophytes. It deals with plant assemblages (communities) at a spatial scale of vegetation stands. Its principal goal is the characterization of vegetation types based on the details of species composition (Dengler, 2017). The method involves recording fine-scale vegetation composition. The basic unit of observation is the plot (or relevé) within which all species are recorded by vertical stratum and the abundance of each is estimated, usually using an index of cover/abundance. Related plots are combined in tabular form and groups of similar plots are defined as communities based on consistency of composition (Peet and Roberts, 2013). Sampling strategies normally attempt to cover the full variability of vegetation within the marsh, while minimizing the within-plot heterogeneity.

There are several numerical hierarchies for ranking abundance. In North America, variants of the 1–10 Domin scale are common. For instance, the macrophyte cover scale used in Kadlec and Bevis (2009) was:

Rank:	+	1	2	3	4	5	6	7	8	9	10
Percent:	solitary	<1	1	2–5	6–10	11–25	26–33	34–50	51–75	76–90	91–100

In the Houghton Lake marsh study, ninety plots, each 32 m^2, were located in three communities: (1) the original sedge and leatherleaf-bog birch zones of the wetland as controls, (2) the heavily impacted zone near the discharge pipeline, and (3) the fringe zones that contained mixtures of original and replacement vegetation. An example of the record from the control zone is shown in Table 15.3.

A comparison of the major community types may then be quantified by combining data from the appropriate plots. For instance, for Houghton Lake, the relative abundances of the major components (greater than 1% of the cover) of the vegetative canopy in each of the three communities are given in Table 15.4 for the period 1992–1994. The fringe and discharge zones are distinctive, in that these contain considerable proportions of *Typha*, and the control zones do not.

Phytosociological studies for treatment wetlands have been conducted elsewhere, but are not common (e.g., Preussler et al., 2015). There is an ongoing study in the Florida STAs, in which the relative abundance species in the SAV communities are being tracked through the years (SFWMD, 2018). Surveys were conducted following a 4-point ordinal scale to estimate SAV coverage: None = no plants observed, Low = 1 to 33% coverage, Medium = 34 to 66% coverage, and High > 66% coverage. Coverage assessments were made for each SAV taxon and for all SAV taxa taken together. Eight SAV taxa were identified during the most recent survey. These data were used to generate SAV coverage maps for the cells surveyed (Figure 15.2). There is nondirectional (▯)-diversity in each of the three cells comprising the (independent) western flow path of STA1E.

Indices of Vegetation Biodiversity

Vegetation indices are not often used to quantify vegetation in treatment wetlands or marshes. Indices of biological integrity have been discussed in the literature (Fennessy et al., 2002), with the suggestion of the Floristic Quality Assessment Index (FQAI). As for other IBIs, the FQAI assigns values to the species found in the marsh. A value of zero is assigned

TABLE 15.3

Example of the (partial) phytosociological record for four plots in the Houghton Lake unimpacted zone in 1993.

Plot number	87	88	89	90
Number of species	13	12	15	18
Species	Abundance Rank			
Carex lasiocarpa	7	7	9	5
Betula pumila	5	6	5	7
Chamaedaphne calyculata	5	8	5	6
Andromeda glaucophylla	4	4	3	6
Parmelia physodes	3	4	+	3
Sphagnum fuscum	4	8	3	3
Sphagnum girgensohii	4	5	5	4
Drepanocladus uncinatus	4	3	5	3
Campyllium stellatum	3	3	5	+
Spiraea alba			5	3
Campanula aparinoides			3	+
Muhlenbergia glomerata				3
Lycopus virginicus			3	
Viola palustris		5		

TABLE 15.4

Relative abundance of major plant species on plots in different community types at the Houghton Lake treatment marsh in 1992–1994.

		Unimpacted	**Fringe**	**Discharge**
	Species/Plot	24.6	13.6	3.7
	Number of Plots	10	31	19
Scientific Name	Common Name			
Sphagnum spp., *etc.*	Mosses and lichens	30		
Carex lasiocarpa	Woolly fruit sedge	15		
Carex aquatilus	Water sedge	36		
Salix spp.	Willow	31		
Spiraea alba	Meadow-sweet	47		
Acer rubrum	Red maple	20		
Betula pumila	Bog birch	50	69	
Chamaedaphne calyculata	Leatherleaf	88	29	
Typha latifolia	Wide leaf cattail		38	100
Lemna minor	Duckweed		38	25
Calamagrostis canadensis	Blue-joint reedgrass		13	
Galium trifidum	Small bedstraw			1
Thelypteris palustris	Fern		5	

to nonnative species and opportunistic native species, while the highest value of 10 is awarded to "high fidelity species." This vegetative IBI appears not to have been implemented in treatment marsh studies.

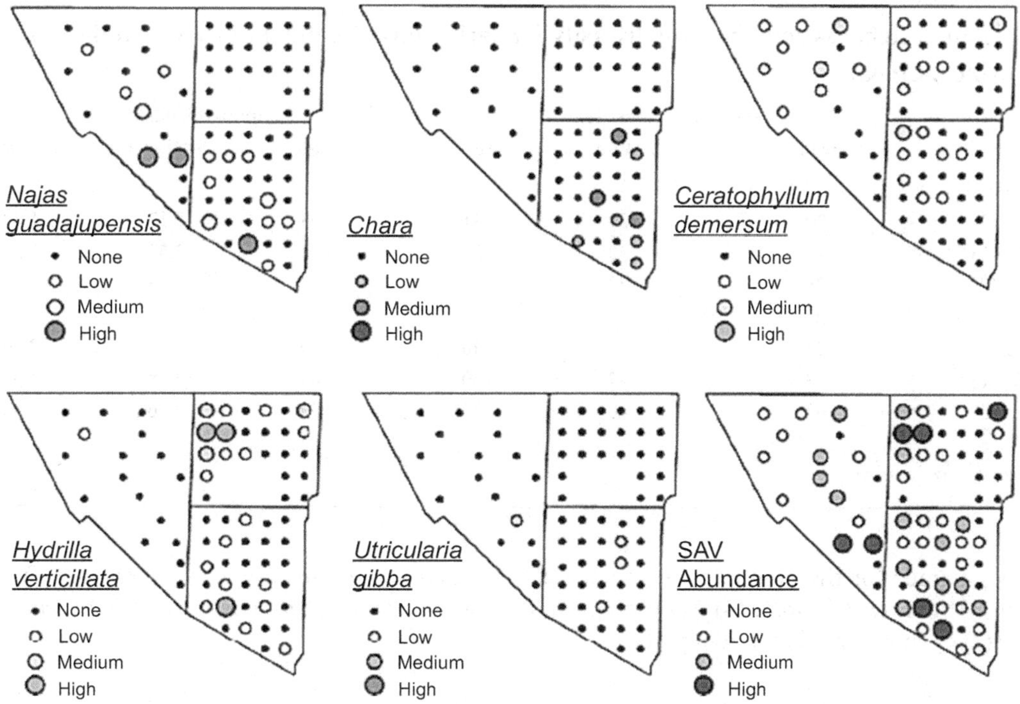

FIGURE 15.2 Relative abundance of five SAV species in the western flow path of STA1E. Source: SFWMD (2018) Draft 2019 South Florida Environmental Report.

Galbrand et al. (2007) evaluated the biological integrity of a constructed wetland receiving landfill leachate and stormwater runoff from the Burnside Industrial Park in Dartmouth, Nova Scotia. This system qualifies as a treatment marsh, with median concentrations of NO_3N = 3 mg/L, NH_4N = 0.25 mg/L, and TP = 0.01 mg/L (Galbrand et al., 2008). They found no significant difference in the plant species diversity between the constructed wetland and a reference site. Species richnesses were 46 and 41 for the treatment wetland and reference wetland, respectively, and the Shannon indices were 2.069 and 2.053, respectively. The most abundant species in the treatment wetland were *Juncus brevicaudatus, Juncus effusus*, and *Glyceria striata*. The most abundant species in the reference wetland were *Juncus effusus, Myrica gale*, and *Scirpus cyperinus.*

A second example of vegetation indices for treatment marshes is the documented by Wei (2013). Two types of wetlands receiving runoff from rice agriculture in Jiangxi Province in south central China were investigated: those interspersed with farms (Farmland-wetland) and those receiving collected runoff (Wetlands). The measured richnesses and Shannon diversities are shown in Table 15.5, at three different times over a year, and at three different sites. In general, both indices are similar to those observed by Galbrand (2007). However, the dominant species are quite different, with those in China being *Paspalum paspaloides* (knotgrass), *Nelumbo nucifera* (lotus), *Oxalis* spp. (woodsorrel), *Eleusine indica* (goosegrass), and *Alternanthera philoxeroides* (alligator weed).

Mitsch et al. (2005b) applied the Shannon relationship to plant communities within the Olentangy wetlands at Columbus, Ohio, which treat river water. They identified seven distinct

TABLE 15.5

Vegetation richness and Shannon diversity for agricultural treatment marshes in Jiangxi Province, China.

	Species Richness (S)			Shannon-Wiener Index (H′)		
	Inside	Station 2	Station 3	Inside	Station 2	Station 3
Wetland						
Sep-12	20	29	31	0.63	2.12	1.47
May-13	28	40	38	0.97	2.49	2.25
Sep-13	28	32	23	0.60	1.79	1.67
Farmland-wetland System						
Sep-12	38	37	48	1.30	2.49	2.26
May-13	43	47	60	1.84	2.28	1.37
Sep-13	43	40	37	0.99	2.26	2.00

Source: Wei (2013).

types, named for the dominant species in each: (1) *Schoenoplectus tabernaemontani*, (2) *Typha spp.*, (3) *Scirpus fluviatilis*, (4) *Nelumbo lutea*, (5) *Sparganium eurycarpum*, (6) *Spartina pectinata, and* (7) open water/submersed aquatics.

$$CDI = -\sum_{i=1}^{N} C_i \ln C_i \tag{15.8}$$

where C_i = fractional cover of community "i" in the wetland
CDI = Community Diversity Index
N = number of communities

Because the CDI has not been utilized for other marshes, its principal value is for comparison of the two one-hectare Olentangy wetlands. The development of these has been tracked for 20 years (Mitsch et al., 2014b) (Figure 15.3). Both took a few years to develop diversity, and planting appeared to have little long-term advantage.

Temporal Development of Vegetative Biodiversity

Newly constructed marshes begin their existence with a designer-imposed biodiversity. However, environmental processes and conditions immediately combine to alter the design to "what mother nature wants." It is not feasible to plant new wetlands at the densities they are capable of sustaining, and hence there is opportunity for insertion of other species from outside sources. The addition of propagules of other species occurs at speeds that are in part determined by the source water for the system. Seeds and floating aquatic plants can enter if present in the incoming water. That is typically the case with urban and agricultural runoff, and river waters. But, tertiary treatment plant effluents and groundwaters carry essentially no propagules. Propagules also arrive via aerial transport, that is, carried in on the wind. Birds transfer seeds, either in their plumage or in their feces. These mechanisms all require proximity to wetlands that generate the propagules.

Additionally, some planted species may find conditions unsuitable for their survival, due to hydrologic or water quality considerations; or they may simply be out-competed by other

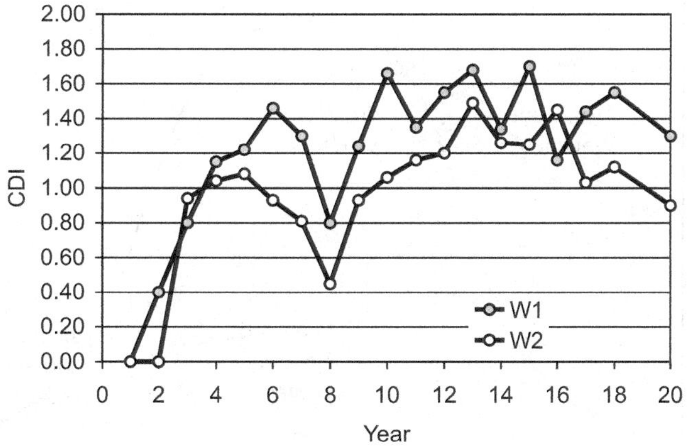

FIGURE 15.3 Development of community diversity in two treatment marshes. W1 was initially planted, W2 was not planted. Redrawn from Mitsch et al., 2014b.

species. Treatment marsh establishment experience suggests that a period of few years is typical for constructed systems to reach a condition of relative stability (see Figure 15.3).

The situation is very different for treatment systems established in existing (perhaps degraded) wetlands. The changes in vegetation under such circumstances proceed mostly via competition, which has been demonstrated to be a slow process. Prime examples are the vegetation changes that occurred in WCA2A in south Florida (Rutchey et al., 2008), and in the Houghton Lake marsh in Michigan (Kadlec and Bevis, 2009). Both experienced changes lasting many years.

Interpretation of Diversity Indices

The above discussions of macrophyte and macroinvertebrate diversity illustrate the spectrum of numerical values that have been obtained for treatment marshes and reference systems. Indices do not crisply separate natural systems from constructed systems, or treatment marshes from other marshes of a region. Some few FWS treatment wetlands deal with strong effluents and display low diversity. And some (regrettably) few pristine wetlands display high diversity, because they are not subject to human-altered source waters. But the large majority of marshes are in an intermediate category, which displays moderate diversity, because of intentional or unintentional additions of mildly polluted source waters (Figure 15.4).

The interpretation of Indices of biological integrity differs greatly from that for diversity indices, because the perceived value of a species is embedded within the numerical ranking. In fact, one may call into question the use of the word "integrity" in such indices. The commonly accepted meaning of integrity is the state of being complete or undivided, or of being sound or undamaged. All those considerations are met by treatment marsh biological assemblages, and yet not meet desirability expectations. It would be better to rename IBIs as indices of biological "pristinity" (admittedly a neologism). The commonly accepted meaning of pristine is: not yet altered by human encroachment, or belonging to an original state or condition.

There are no commonly accepted IBIs for marshes, but they have been developed regionally. For instance, Fennessy et al. (2002) details the Floristic Quality Assessment Index

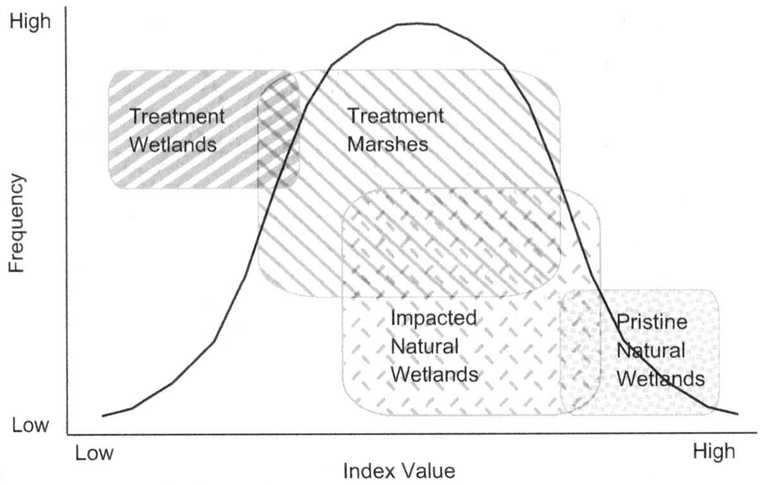

FIGURE 15.4 Index values and frequency spectrum of wetlands.

(FQAI) for wetland vegetation, and Helgen (2002) describes a depression wetland IBI for Minnesota. The former utilizes species value numbers ranging from zero to 10, the later uses a range from 1 to 5, with highest numbers given to the most desirable/pristine species. These IBIs have not been used for treatment marshes.

15.4 MACROINVERTEBRATES

Macroinvertebrates are those invertebrates that can be readily seen with the unaided eye. These organisms are commonly quantified in aquatic ecosystems, including wetlands. The macroinvertebrate groups most commonly studied are annelids (worms), arachnids (spiders and mites), crustaceans (amphipods, isopods, copepods, water fleas, and crayfish), insects, and molluscs. Individual species are so numerous that investigations are forced to consider populations at the class, order, or family level, rather than at the genus or species level. Organisms may also be classified according to their functions in the ecosystem, with the benthic invertebrate community partitioned into trophic guilds or functional feeding groups: predators, collectors, shredders, and scrapers. To make matters worse, there are varied methods of sampling, including sweep netting for emergent organisms and sediment sampling.

Nonetheless, we are left with an extremely varied array of taxa, as exemplified in Table 15.6. Information was compiled by taxon for 20 constructed stormwater wetlands in North Carolina (Moore and Hunt, 2012). Thirty taxa were listed, with some appearing in nearly all wetlands, and others in a small percentage of wetlands sampled.

Diversity Indices

Indices are commonly used to describe these varied macroinvertebrate populations. Historical data analysis is provided in the North American Treatment Wetland Database (NADB, 1998). The average Shannon diversity (H') of benthic macroinvertebrate collections was 1.36 units for constructed treatment wetlands and 2.29 units for natural treatment wetlands. Benthic macroinvertebrate evenness averaged 0.56 for constructed wetlands and 0.68 for

TABLE 15.6

Macroinvertebrate families and proportion of wetland sites in which they were collected (n = 20).

Class/order	Family	Guild	Proportion
Gastropoda	Physidae	Scraper	0.95
	Planorbidae	Scraper	0.23
Hirudinea	Hirundinidae	Collector	0.33
Coleoptera	Dysticidae	Predator	0.45
	Haliplidae	Shredder	0.59
	Hydraenidae	Collector	0.23
	Hydrophilidae	Collector	0.23
	Noteridae	Predator	0.05
Diptera	Chaoboridae	Predator	0.09
	Chironomidae	Collector	0.90
	Culicidae	Collector	0.36
	Tipulidae	Shredder	0.23
Ephemoptera	Baetidae	Collector	0.32
Hemiptera	Belostomatidae	Predator	0.52
	Corixidae	Collector	0.05
	Naucoridae	Predator	0.14
	Nepidae	Predator	0.18
	Notonectidae	Predator	0.32
	Veliidae	Predator	0.82
Lepidoptera	Pyralidae	Predator	0.14
Odonata	Aeshnidae	Predator	0.24
	Coenagrionidae	Predator	0.91
	Libellulidae	Predator	0.86
Oligochaaeta		Collector	0.95
Pelecypoda	Sphaeriidae	Collector	0.27

Source: Moore and Hunt (2012).

natural treatment wetlands. However, this analysis covers treatment wetlands for strong effluents as well as for those that are included in the concept of treatment marshes.

For present purposes, the NADB information has been screened to include just low-strength source waters, and updated to include examples from the last two decades (Table 15.7).

Indices of Biological Integrity

The Hilsenhoff (1987, 1988) macroinvertebrate IBI has been used for treatment marshes, but ironically, it was not intended for that purpose. It was designed to be an index of organic pollution of streams, with a value of zero indicating no organic pollution, and an upper limiting value of 10 for severe organic pollution. Streams with excellent water quality were supposed to fall in the range 0–3.75.

The Hilsenhoff index was evaluated for treatment marshes at Brighton, Ontario (McGauley, 2008), and in central Ohio (Spieles and Mitsch, 2000b) (Table 15.8). Hilsenhoff (1988) listed the meaning of high index ranges as follows:

TABLE 15.7

Examples of macroinvertebrate richness, diversity, and evenness in treatment marshes. OCESA (Orange County Eastern Serve Area) and Tres Rios treat tertiary municipal water; Des Plaines and Olentangy treat river water; Brighton treats lagoon effluent, WCA2A treats agricultural runoff; and Tucush treats mine waste rock leachate.

Location	Wetland	Richness S	Shannon Diversity H′	Pielou Evenness J	Reference
OCESA, FL	1	34	3.03	0.65	NADB Database (1998)
Tres Rios, AZ	C1	25	0.69	0.50	NADB Database (1998)
Tres Rios, AZ	C2	23	0.64	0.47	NADB Database (1998)
Tres Rios, AZ	H1	19	0.89	0.74	NADB Database (1998)
Tres Rios, AZ	H2	16	0.62	0.53	NADB Database (1998)
Des Plaines, IL	EW3	37	1.31	0.68	NADB Database (1998)
Des Plaines, IL	EW4	42	1.44	0.58	NADB Database (1998)
Des Plaines, IL	EW5	44	1.45	0.79	NADB Database (1998)
Des Plaines, IL	EW6	41	1.19	0.63	NADB Database (1998)
Olentangy, OH	W1	19	1.94	0.67	Gamble et al. (2004)
Olentangy, OH	W2	17	1.90	0.68	Gamble et al. (2004)
North Carolina	20 Urban	13	1.66	0.65	Moore and Hunt (2012)
Brighton, Ontario	Treatment	9	1.16	0.53	McGauley (2008)
Brighton, Ontario	3 Natural	18	2.35	0.81	McGauley (2008)
WCA2A	Inlet	21	2.30	0.76	McCormick et al. (2004)
WCA2A	Mid	23	2.50	0.80	McCormick et al. (2004)
WCA2A	Outlet	23	2.30	0.73	McCormick et al. (2004)
Tucush, Peru	Inlet	14	0.83	0.31	CNEH (2010)
Tucush, Peru	Outlet	16	1.90	0.69	CNEH (2010)
Median		**21**	**1.45**	**0.67**	

- 5.76–6.50 Fairly poor Substantial pollution likely
- 6.51–7.25 Poor Very substantial pollution likely
- 7.26–10.00 Very poor Severe organic pollution likely

Table 15.8 highlights the inappropriateness of this index for marshes. Marshes are almost all founded upon a large and active carbon cycle, which fosters carbon substrate loving organisms. It does not discriminate between natural and treatment systems, nor does it contemplate gradients in treatment systems. In other words, numerical rankings set for babbling brooks are not applicable to marshes.

Indices are set according to human standards and perceptions, but wetland fauna have very different standards and perceptions. Birds, mammals, reptiles, and fish respond to habitat and food abundance, which are not necessarily represented by pristinity.

Macroinvertebrate Problems

Some macroinvertebrates can exert very strong influences on treatment marshes, by influencing the biogeochemical cycles and by posing threats to human populations. For instance, cattail caterpillars (Henry's marsh moth, *Simyra insularis*) invaded a treatment wetland near

TABLE 15.8
Hilsenhoff indices for treatment and natural marshes.

	Wetland Type	Replicates	Location	Hilsenhoff Index
McGauley (2008)				
Brighton, ONT	Treatment	3	Open Water	6.05
	Treatment	3	Vegetated	6.88
	Natural	3	Reference	6.49
Spieles and Mitsch (2000b)				
Licking County, OH	Treatment	3	in	9.05
		3	mid	8.22
		3	out	8.19
Olentangy, OH	Treatment W1	3	in	7.72
		3	mid	7.66
		3	out	8.02
Olentangy, OH	Treatment W2	3	in	8.69
		3	mid	8.25
		3	out	7.60

Stevenson, Alabama, and damaged 60–80% of the plants (Snoddy et al., 1989). Caterpillars also episodically damaged the vegetation in the ENRP project. And, some marsh caterpillars can sting humans; such stings hospitalized the operator of the Houghton Lake wetland in its early history. Later researchers learned to avoid them.

Island apple snails (*Pomacea maculata*) are becoming increasingly abundant in the southeastern United States, with detrimental effects on wetland vegetation and potentially water quality. Cell 4S in STA-1E experienced an enormous population increase of the island applesnail between July and August 2013. Island applesnail feeding defoliated most of the cell's SAV, which led to the cell exporting TP from the cell.

Perhaps the most important wetland macroinvertebrate problem is mosquitoes. Undesirable as it may be, all wetlands produce mosquitoes. In areas with large expanses of existing wetlands, additional mosquitoes will not significantly add to the nuisance issue. But in urban areas that are accustomed to few mosquitoes, introduction of new treatment marshes can create social conflict (Knight et al., 2003b). Mosquito-borne illnesses are caused by bacteria, viruses or parasites. Diseases transmitted by mosquitoes include: malaria, dengue, West Nile virus, chikungunya, yellow fever, Zika virus, and several encephalitises: West Nile, Saint Louis, Western equine, Eastern equine, Venezuelan equine, and La Crosse. These are caused by a variety of species within the order Diptera, and the family Culicidae, including *Anopheles, Coquillettidia, Culiseta, Culex, Mansonia, Uranotaenia Aedes*, and *Ochlerotatus*. All mosquitoes need water to complete their life cycle, and they are ubiquitous in wetlands. Treatment marshes are no exception.

Within the marsh, natural controls are present in the form of predators, which can be roughly grouped as microbes (viruses, protozoans, fungi and bacteria), macroinvertebrates (nematodes, copepods, aquatic insects), and a few species of fish (*Gambusia affinis*). These do not usually reduce populations to acceptable levels, and hence mosquito control in treatment marshes depends upon addition of larvicides and adulticides, and upon marsh design – neither of which are totally effective. Applications of control agents are required several times per year, at a typical cost of several hundred dollars per hectare.

Wetland design can provide some measure of control. Firstly, mosquito production has been directly related to loading rates of organic matter and nutrients (Walton, 2012;

Duguma and Walton, 2013). Source water variables for free-water surface constructed treatment wetlands, including municipal, urban and agricultural waters, are organic loading (i.e., BOD, TSS), nutrients (nitrogen and phosphorus). By definition, treatment marshes are restricted to conditions that provide source control via water quality. The selection and configuration of emergent vegetation can have strong effects on mosquito production. Dense stands of emergent vegetation and the associated mats of decaying litter impede penetration to the water surface of mosquito control agents, such as bacterial larvicides.

Some level of mosquito population will inevitably accompany a treatment marsh, and the risks associated with them may be part of the design process.

15.5 BIRDS

Birds use wetlands for feeding, breeding and shelter, and do so in great numbers. About one-third of North American bird species use wetlands for one or more of these purposes (Kroodsma, 1979). Several hundred different bird species are either partially or wholly dependent on wetlands for some part of their life history in North America. Wetland plants, vertebrates, and invertebrates provide food for birds. Wetland vegetation provides shelter from predators and from the weather. Stewart (1996) gives a list of 155 North American birds that need wetland access during the breeding season, which are termed wetland-dependent. Seventeen of these are threatened or endangered. Wetland losses and degradation have caused loss of bird habitat. Constructed treatment marshes offer a means of substantially increasing wetland habitat for birds. The conclusion of Strand and Weisner (2013) would appear to be generally the case everywhere:

> The effects of constructed wetlands in Sweden on the species numbers and population sizes of wetland birds and amphibians were large enough to positively affect the occurrence of species in the national red list, that is on a national population level. Species numbers were high also for nutrient removal wetlands showing that it is possible to get multiple ecosystem services out of the constructed wetlands.

Species Richness and Densities

At an overview level, these may be grouped in a number of different ways, so that simple summary descriptions can be formed. One such grouping is by family, which is illustrated by the list of 21 avian families in Kadlec and Knight (1996). Another is the phylogenic tree, in which birds are sorted based upon similarities and differences in their physical or genetic characteristics (Cracraft et al., 2004). Chimney and Gawlik (2007) present a checklist for birds in the Florida STAs. Example bird groupings for treatment marshes and wetlands are shown in Table 15.9. Passerines and land birds are most common, followed closely by waterfowl and wading birds.

The numbers of bird species observed in treatment marshes increases with marsh size (Figure 15.5). Data there represent small agricultural systems in Sweden (Strand and Weisner, 2013), as well as marshes of all sizes and source waters in the USA. The log linear fit ($S= 8.4A^{0.48}$) has $R^2 = 0.64$ for 41 wetlands up to 1,000 ha. Larger sizes do not further increase the number of species. It is noteworthy that a similar relation was found by Chimney and Gawlik (2007), for the broader category of treatment wetlands, some with stronger source waters. Their log linear fit was ($S= 44A^{0.15}$) has $R^2 = 0.59$ for 20 wetlands up to 100,000 ha. This indicates lesser numbers are to be found in small wetlands with stronger source waters.

TABLE 15.9

Numbers of bird species in groups found in wetlands.

Two Florida STAs		
Passeriformes	Swallows, wrens, blackbirds	39
Charadriiformes	Stilts, sandpipers, gulls	31
Ciconiiformes	Egrets, herons	17
Anseriformes	Ducks, geese	16
Falconiformes	Hawks, kites, eagles	13
Gruiformes	Coots, rails	7
Columbiformes	Pigeons, doves	5
Pelecaniformes	Pelicans,cormorants	4
Podicipediformes	Grebes	1
Phoenicopteriformes	Flamingoes	1
Cuculiformes	Cuckoos, anis	1
Strigiformes	Owls	1
Caprimulgiformes	Nighthawks	1
Coraciiformes	Kingfishers	1
Piciformes	Woodpeckers	1
Total		139
27 Treatment Wetlands		
Land and Passerine	Swallows, wrens, blackbirds	235
Water Fowl	Ducks, geese	51
Shorebirds	Stilts, sandpipers	45
Raptors and Scavengers	Hawks, kites, eagles	29
Terns and Gulls	Gulls	24
Wading Bird	Egrets, herons	23
Fowl-like		7
Total		414

Sources: Upper panel, Chimney and Gawlik (2007); Lower panel, NADB Database (1998).

Larger treatment marshes are typified by the Florida STAs (the light dots on Figure 15.5). Chimney and Gawlik (2007) and Beck et al. (2013) have studied avian communities in these habitats

Avian densities vary seasonally in the STAs, with densities was greatest in the winter as a result of migratory waterfowl. The STAs are estimated to provide wintering habitat to a significant portion of the North American waterfowl population, including as much as 8% of the breeding population of American Coots (*Fulica americana*). Avian densities were 38 times greater in STAs than in the adjacent natural marshes, and species richness in the STAs was about four times greater than in the adjacent natural marshes (Beck et al., 2013).

In addition to numerous species, birds can episodically appear in great numbers, reflected by bird densities. Migratory birds appear in great numbers seasonally, especially in marsh systems that are situated on established flyways. Densities of over 100 birds per hectare have been measured at Hayward, CA (NADB Database, 1998), and in south Florida (Beck et al., 2013). Densities of resident and breeding birds are lower, in the range of 1–20 birds per hectare. There can also be very high densities associated with a particular species and a particular wetland and its attractants. For instance, very large numbers of black necked

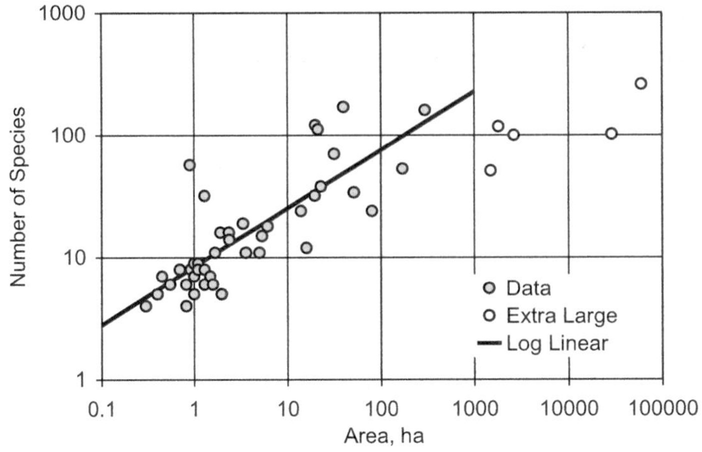

FIGURE 15.5 Bird species richness (numbers) as a function of marsh size.

stilts (*Himantopus mexicanus*) and avocets (*Recurvirostra americana)* occur on the mud flats of the Incline Village marsh in Nevada, which are the result of seasonal deliveries of source waters. A second illustration is the high densities of yellow-headed blackbirds (*Xanthocephalus xanthocephalus*) at the Tres Rios, AZ, marshes, with 150 birds per hectare, and at the Sweetwater, AZ, marsh (Figure 15.6).

Urban stormwater wetlands are typically quite small, and therefore have small numbers of bird species. Murray et al. (2013) surveyed 28 stormwater systems near Melbourne, Australia, which averaged 0.5 ha in size. They found an average of five species per wetland, and an

FIGURE 15.6 Thousands of blackbirds move in and out of the Sweetwater constructed wetlands in Tucson, AZ.

abundance of 15 birds per wetland (23 birds/ha). They also found that differences in community composition of waterbirds were related to physical variables, and not water chemistry. Larger urban systems are used by more birds. Bird usage of the 16 ha DUST (Demonstration Urban Stormwater Treatment) marsh in Coyote Hills Regional Park near Fremont, California, was studied by Duffield (1986). Weekly or biweekly bird censuses were conducted in the DUST marsh and a nearby 18.3 ha control from mid-January 1984 through mid-June 1985. Mean abundance of wetland birds in the marsh areas ranged from 100 to 300 (6.25 to 18.75 birds/ha) in the DUST marsh and 90 to 420 (4.91 to 23.0 birds/ha) in the control marsh. Mean number of species per census ranged from 14 to 23 in the DUST marsh and 10 to 18 in the control area. Highest mean species counts were observed in both marshes during the winter and spring seasons.

The CREP agricultural nitrate control wetlands in Iowa have been surveyed for bird usage, primarily focused upon waterfowl and grassland birds (Otis et al., 2010). The studies covered 27 sites containing 102 ha of wetlands (average size 3.8 ha), and 377 ha of associated grassland buffers. Species richness was 3 to 4 per wetland. Duck pair estimates ranged from 2.9 to 8.3 pairs per wetland, while estimated goose pairs ranged from 0.6 to 4.0 pairs per wetland. Total birds per site showed a strong positive correlation with wetland size.

Threatened and Endangered Birds

Strand and Weisner (2013) observed 66 wetland bird species either breeding during summer (31 wetland species), or resting during migration in autumn and spring, in 58 wetlands in Sweden. The surveys showed that a total of 22 red listed species used the wetlands and were observed in 34 of the 58 wetlands.

Chimney and Gawlik (2007) report 16 species that are state or federally listed as endangered, threatened or a species of special concern. Six frequent species were both Federally listed and breed in the region: osprey (*Pandion haliaetus*), roseate spoonbill (*Platalea ajaja*), snowy egret (*Egretta thula*), tricolored heron (*Egretta tricolor*), white ibis (*Eudocimus albus*), and wood stork (*Mycteria americana*).

It is thus apparent that treatment marshes are serving to increase the numbers of, and the habitat for, threatened and endangered bird species.

Problems Created by Marsh Birds

As opined by Post et al. (1998), dense aggregations of waterfowl can result in the destruction of wetland vegetation and agricultural crops, increase the risk of infectious wildlife disease outbreaks, and decrease water quality.

Internal and External Plant and Animal Damage

Treatment marshes, particularly those located on a major migratory flyway, can attract flocks of waterfowl, such as Canada geese (*Branta canadensis*), which can obliterate marsh planting stocks if host populations are high enough (Figure 15.7). Bird control via deterrents, such as gas cannons, is only partially effective, and exclosures are restricted to local areas.

Herbivory is also present for SAV in treatment marshes. For instance, Weisner et al. (1997) conducted exclosure experiments to evaluate the effects of waterfowl grazing on macrophytes, and found *Potamogeton pectinatus* was decreased but *Myriophyllum spicatum* was not affected.

Chaichana et al. (2011) measured herbivory in a small wetland (Brown Moss) near Whitchurch, Shropshire, UK. The 33-ha system was less than one meter deep, and about 60% vegetated with *Alisma plantago-aquatica, Alopecurus aequalis, Carex pseudocyperus, Juncus effusus, Oenanthe aquatica*, and *Typha latifolia.* The wetland suffers eutrophication

FIGURE 15.7 The Hayward, California wetland experienced great problems with waterfowl grazing. The only plantings that remain in this photo were in exclosures built to prevent such eat-out.

from agriculture and septic tank effluent (TN = 4±2.4 mg/L, TP = 0.39±25 mg/L). The density of birds was 110/ha in winter and 6/ha in summer. Ducks, mainly mallard (*Anas platyrhynchos*), and teal (*Anas crecca*), damaged plants by direct consumption, whereas larger birds, such as mute swan (*Cygnus olor*), were able to remove *Typha latifolia.*

A separate but serious problem is the foraging of marsh resident birds upon adjacent agricultural operations. Migratory birds spend time in marshes embedded in agricultural landscapes, and forage on the adjacent farms. Damage is well documented for fruit crops, grains, forage crops, and aquaculture operations.

Nutrient Cycle Effects

Food consumption by wetland birds can represent either an import or an export of nutrients; or simply a recycle of nutrients internal to the wetland. In urban environments, waterfowl may feed and rest in wetlands, but also feed and excrete on adjacent landscapes. In agricultural settings, daily feeding bouts by geese can move large quantities of nutrients from farm fields where they feed, to wetlands where they roost (Post et al., 1998). Water birds excrete nutrients in quantities corresponding to their body weights (Table 15.10), and the loadings to a marsh are proportional to their numbers.

TABLE 15.10

Estimated nitrogen and phosphorus excreted by marsh birds.

	Body Weight kg	Nitrogen mg/day	Phosphorus mg/day
Geese	2.56	1,570	490
Dabbling ducks	1.18	562	178
Coots	0.54	251	79
Smallducks	0.47	200	69
Small gulls	0.26	122	38
Grebes	0.17	71	22

Source: Chaichana et al. (2010).

Hahn et al. (2008) postulated a simple daily routine of the birds, with one foraging expedition to terrestrial areas, and the remaining time spent in a wetland or aquatic system. In this Netherlands study, nutrient imports in regions of high bird aggregation were estimated to be 43 $gN/m^2 \cdot yr$ and 0.75 $gP/m^2 \cdot yr$, which occurred during six winter months.

In the arid flyways of the USA, similar estimates have been put forth. Post et al. (1998) estimated annual averages of 1.8 $gN/m^2 \cdot yr$ and 0.22 $gP/m^2 \cdot yr$ for a 500 ha wetland, at an average annual bird density of 40/ha. Peak loadings at times of high migration were about four times higher. This study showed that about 50% of the bird excreta were deposited in the wetlands and in the forage lands. Open water areas with high bird densities displayed nitrogen and phosphorus concentrations several time higher than those with low densities (Kitchell et al., 1999).

Ducks and geese consume wetland vegetation, and return fecal matter to the system. In this internal wetland cycle, they provide a more rapid return of nutrients within the wetland than the growth-death-decomposition route; in other words, a fast cycle. Chaichana et al. (2010) measured this nutrient transfer route for the small wetland (Brown Moss) near Whitchurch, Shropshire, UK, described above. Bird loadings to the wetland resting pond were estimated at 0.74 $gN/m^2 \cdot yr$ and 0.24 $gP/m^2 \cdot yr$.

Other water quality parameters may also be affected by birds. Huge flocks (thousands) of yellow-headed blackbirds (*Xanthocephalus xanthocephalus*) have been implicated in recontamination of water by pathogens in Arizona (Orosz-Coghlan et al., 2006) (Figure 15.6). Waterfowl have been implicated in turbidity increases at the Columbia, Missouri constructed wetlands (Knowlton et al., 2002).

Conflicts with Water Quality Operations

Bird use and water quality treatment are not always totally compatible uses of constructed marshes. There may be conflicts between operations and the life styles of sentinel, threatened or endangered species. For example, the nesting habits of stilts and burrowing owls lead them to prefer the levee areas of drained or dewatered wetland cells. This has interfered with system operations in the stormwater treatment constructed wetlands in Florida. A compromise solution was found in the form of an avian protection plan (APP), which focuses on black-necked stilts (*Himantopus mexicanus*) and Florida burrowing owls (*Athene cunicularia floridana*) as sentinel species. The APP one of the first attempts to coordinate and manage operation of treatment marshes for the benefit of birds. During the 2018 nesting season (April through June), 147 black-necked stilt nests were observed within the STA cells, lower than the 15 year average of 282 nests (SFWMD, 2018a). Burrowing owls were of concern early in the STA history, but no nests have been found in recent years.

Everglade snail kites (*Rostrhamus sociabilis*) have been confirmed nesting in the STAs since 2010. They are federally- and state-listed endangered avian species. Since that time, there have been 346 snail kite nests observed with 43% of these nests successfully fledging one or more chicks.

Adjustments to STA operations are made in accordance with the APP to reduce impacts to ground-nesting birds within the STAs. Where possible, flow is routed to other cells that do not have nests. In 2018, water level restrictions were in place for 22 STA cells during April, May and June. These restrictions were not severe in terms of treatment capacity, because they occurred in the dry season.

Problems for Birds in Treatment Marshes

Chemical Toxins

Source waters delivered to treatment marshes may contain chemicals that are harmful to birds. The example of the disastrous effects selenium has been discussed in Chapter 13.

Legacy agricultural chemicals have also caused severe effects, as illustrated by the events at a restoration/treatment marsh at Lake Apopka near Orlando, Florida (Conrow et al., 2011). In 1998, wetlands were restored on former farms north of the lake, to reduce storm water discharges to Lake Apopka and related nutrient loadings. The project came with unforeseen problems, as an estimated 676 birds died during late 1998 and early 1999. Most were American white pelicans, wood storks and great blue herons. Organochlorine pesticides (OCPs: DDT, dieldrin and toxaphene) residues remaining from agricultural practices were found to be the primary cause of bird deaths. Birds accumulated OCPs by consuming contaminated fish. High levels of these now-banned OCPs were found in organic surface soils of the marsh, which resulted from years of farm pesticide applications. The bird deaths brought attention to the potential risks of restoring agricultural lands back to wetlands, particularly on organic soils in the South where multiple crop rotations each year have left a legacy of high residual pesticide loads.

Corrective actions, mainly soil inversion to bury the contaminated soils below the reach of biota, proved successful. In a dramatic reversal, new treatment marshes have been in operation for over 15 years, with significant water quality improvement and no bird problems (Dunne et al., 2012, 2013, 2015).

Conrow et al. (2011) concluded that ecological restoration is not without risks, and that at Lake Apopka pre-existing science and approaches for risk assessment underestimated the problems posed by organochlorine pesticides. They also concluded that, given the scale of habitat loss nationally and globally, it is essential that restoration efforts proceed.

Diseases

Avian botulism is caused by *Clostridium botulinum*, a strict anaerobe bacteria, which is present in the sediments of wetlands and in the digestive tracts of water birds and fishes. It produces neurotoxins that result in paralysis of the muscles and the death of the affected birds. Fertilization of a marsh with poorly treated municipal wastewater or run-off from agricultural activities can stimulate plant or invertebrate animal population growth, and results in increased necromass. Nutrients resulting litter necromass are then available to support growth of *C. botulinum*, and the accompanying toxin production (Locke and Friend, 1989). Outbreaks involve a bird-maggot cycle, in which toxins are produced in dead bird carcasses, then concentrated by maggots, then in turn the maggots are ingested by healthy birds that become infected and die.

Anza et al. (2014) monitored three wetlands in south central Spain, two of which were recipients of secondary wastewater (Veguilla and Navaseca), and the third (Tablas de Daimiel National Park) received a much less contaminated source water. This region has experienced outbreaks in the three studied wetlands, with the outbreaks being more frequent in Veguilla and Navaseca.

Migratory waterfowl concentrated in wetlands are particularly susceptible to avian cholera, a highly infectious disease caused by the bacterium *Pasteurella multocida* (Friend, 1985, 1989; Knight et al., 2003b). Transmission occurs by bird-to-bird contact, and ingestion of contaminated food or water. The spread of avian cholera through waterfowl and other migratory bird populations is enhanced by the gregarious nature of most waterfowl species and by dense concentrations of birds that result from habitat limitations.

Avian cholera has been reported at one treatment wetland, the Hayward Marsh on the east shore of San Francisco Bay, south of Oakland, California. Annual episodes of avian cholera have been noted at this site. The average number of infected waterfowl collected during a 6-year period was 127 per year (15 to 340 birds per year). This wetland supports very high waterfowl populations during the fall months, with peak numbers above 30,000 birds per day. Avian cholera is encountered in nearly all wetlands in and around San

Francisco Bay and there appears to be no relationship between the avian cholera observed and water quality.

15.6 FISH

The number of fish families that typically occur in wetland environments in North America is small compared to pond, lake and stream environments, yet the wetland species lists are surprisingly long and varied. Generalities are not meaningful, because the types and numbers of fish in wetlands depend very much on the type of wetland, and upon the region in which it is located. Lake margin wetlands are much more accessible to fish than inland marshes of short hydroperiods, and the species list for cold climates differs markedly from those for subtropical climates. Table 15.11 illustrates this aspect for lake margin marshes of the Great Lakes region of the USA and the slowly flowing marshes of south Florida. Those two regions share eleven families, out of 15 for the Great Lakes and 23 for south Florida. However, the species within each family vary with the region, as do the numbers of species. Selheimer and Chow-Fraser (2006) report 43 species for the Great Lakes region, and

TABLE 15.11

Fish families that occur in wetlands in the Great Lakes and Floridian regions of the USA. Families in common are in bold.

Great Lakes Family	Example Species	South Florida Family	Example Spcies
Amiidae	Bowfin	**Amiidae**	Bowfin
Atherinopsidae	Brook silverside	**Atherinopsidae**	Brook silverside
Catostomidae	White sucker	**Catostomidae**	Lake chubsucker
Centrarchidae	Smallmouth bass	**Centrarchidae**	Sunfish
Clupeidae	Alewife	**Clupeidae**	Gizzard shad
Cyprinidae	Fathead minnow	**Cyprinidae**	Golden shiner
Esocidae	Northern pike	**Esocidae**	Chain pickerel
Fundulidae	Banded killifish	**Fundulidae**	Marsh killifish
Ictaluridae	Black bullhead	**Ictaluridae**	Channel catfish
Percidae	Yellow perch	**Percidae**	Darter
Lepisoptiedae	Longnose gar	**Lepisosteidae**	Florida gar
Gasterosteidae	Brook stickleback	Anguillidae	American Eel
Moronidae	White bass	Anostomidae	Black-banded leporinus
Sciaenidae	Freshwater drum	Aphredoderidae	Pirate perch
Umbridae	Central mudminnow	Callichthyidae	Brown hoplo
		Cichlidae	Tilapia
		Clariidae	Walking catfish
		Cyprinodontidae	Sheepshead minnow
		Elassomatidae	Pigmy sunfish
		Gobiidae	Goby
		Poeciliidae	Mosquito fish
		Loricariidae	Sailfin catfish
		Poeciliidae	Gambusia

Sources: Great Lakes, Selheimer and Chow-Fraser (2006); Florida, Chimney and Jordan (2008).

Chimney and Jordan (2008) report 61 species for the south Florida region. Complexity increases when one considers wetlands and fish at the global scale, that is, the species native to New Zealand are not those found in North America.

Fish in Treatment Marshes and Wetlands

Seventy-seven fish species are reported from 13 treatment wetland sites in NADBv2 (NADB Database, 1998); 70 species from constructed treatment wetlands and 15 species from natural treatment wetlands. However, only four of the NADBv2 systems can be considered treatment marshes.

Fish enter treatment marshes from inlet and outlet streams, as well as from other sources, such as raptor drops. For example, smallmouth bass (*Micropterus dolomieui*) and sunfish (*Lepomis macrochirus*) swam up the outlet structures at the Hillsdale, MI treatment marsh, and took up residence in all deep zones. Drawdown of the wetlands, for purposes of maintenance and retrofitting, killed the fish. They were effectively precluded after the deep zones were filled for muskrat control. Bass and sunfish also are found in the Incline Village, Nevada constructed wetlands, which are abutted by natural wetlands with deep water habitat. Tilapia (*Oreochromis* spp.) found their way into the Tres Rios, Arizona wetlands, although there appeared to be no route open to them from either downstream with entering water (WWTP) or upstream from receiving water (drop boxes and cascades). Drawdown of the wetlands, for research purposes, killed the fish. Tilapia also are prolific in the Lakeland, Florida constructed wetlands, again with no obvious route for them to have emigrated.

A 21-ha constructed wetland has been used for final polishing of an industrial effluent since 1999 in Victoria, Texas. Fish populations in the 8 wetland cells have been used as an indicator of environmental condition at or near the top of the wetland aquatic foodchain (Reitberger et al., 2000). This constructed treatment wetland includes three different types of cells arranged in parallel and series. Treated effluent from a treatment facility is distributed into Stage 1 wetland cells that are colonized primarily by cattails (*Typha* spp.) and contain transverse deep water zones that are colonized by submerged and floating aquatic plants. Water from these cells flows into Stage 2 cells containing a diverse marsh plant community and transverse deep water zones. Water flows finally into a habitat cell with diverse water depths (up to 3 m). Six fish species have been collected: mosquitofish (*Gambusia affinis*) have the highest density, and populations of other species are variable from year-to-year, presumably in response to climatic conditions and water quality conditions. Sunfish (*Lepomis macrochirus* and *Lepomis gulosus*), bluefin killifish (*Lucania goodie*), silversides (*Menidia beryllina*) and gizzard shad (*Dorosoma cepedianum*) populations are sporadic in the habitat cell.

Chimney and Jordan (2008) collected more than 29,000 fishes representing 28 species in 16 families from the Everglades Nutrient Removal Project ENRP during 1997. Regionally, 61 species and 23 families are found, thus this treatment marsh harbored a considerable fraction of the potential fishes. Research in progress at the present time shows that the successor stormwater treatment areas are similarly populated. Throw traps collected 13 species in numbers up to 288 individuals per meter squared in STA2 in 2017. Throw traps yield mostly small fish, including mosquito fish (*Gambusia holbrooki*) sailfin mollys (*Poecilla latipinna*), and killifish (*Heterandia formosa* and *Lucania goodei*). Electrofishing is being used to capture larger fish, such as largemouth bass (*Micropterus salmoides*), gar (*Lepisosteus platyrhincus*), and tilapia (*Oreochromis aureus*). STA2 supports eleven of these larger fish species (Evans and Trexler, 2017).

Other treatment marshes have reported lesser numbers of species. Smith and Mitsch (2004) report six fish species in marshes treating river water in Ohio. Five species were reported for the Houghton Lake, MI marsh. The inlet constructed marsh at the Orlando

Eastern Service Area site in Florida had four species (NADB Database, 1998). And, the OEW marsh is reported to have 15 species of fish (Langston and Kent, 1997).

Fish Conflicts with Treatment

Fish may be problematic in treatment wetlands, particularly those that feed or nest on the bottom. For instance, at the Des Plaines constructed wetlands in Illinois, carp (*Cyprinus carpio*) were unintentionally introduced into the wetland from the pumped source water (Des Plaines River), presumably as very small individuals. Over the course of two years, the fish grew in size to about 15 cm. These larger fish were observed to stir bottom sediments and impair suspended solids removal (Figure 15.8). The carp were removed from the wetland through a winter drawdown and freeze (Figure 15.9).

In the Midwestern USA and elsewhere, carp pose a severe and widespread threat to habitat quality and function of marsh ecosystems. A wetland restoration project in central Illinois was intended to alleviate nitrate pollution in the Illinois River, as well as providing ecological benefits. The 508 ha Hennepin-Hopper project consists of open water/SAV communities along with EAV areas, located in the floodplain of the Illinois River. Carp invaded this system, with severe effects. Biomass of carp climbed to 300 kg/ha, while vegetative cover dropped from 90% to 20% (Bajer et al., 2009). Waterfowl usage dropped, both in abundance and species richness. Because authorities believed that carp entered from river water, they prohibited the restored wetland from being connected to the river. Thus, the treatment function was not realized.

A similar phenomenon occurred at the Tarrant County constructed wetlands in Texas, where floodwaters from the source river brought carp (*Cyprinus carpio*), shad (*Dorosoma cepedianum*) and bullheads (*Ictalurus* spp.) into the system. The activity of the fish caused severe bioturbation of the sediments during feeding and spawning activities (APAI, 1994). The cells were dewatered, enabling egrets and herons to effectively remove the stranded fish.

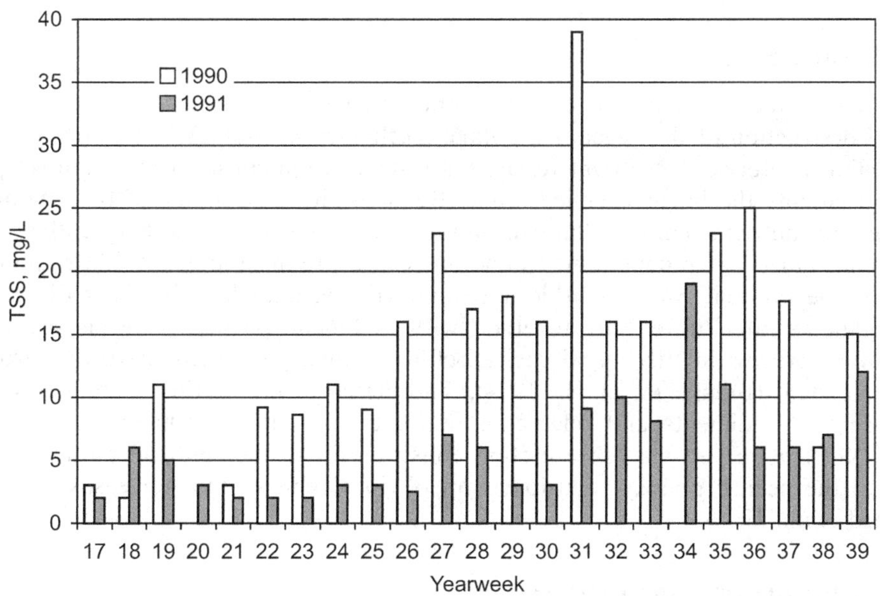

FIGURE 15.8 Effect of carp on the TSS level in the Des Plaines, Illinois treatment wetland EW4. Carp were present in 1990, but had been removed in 1991.

FIGURE 15.9 Carp frozen out of wetland EW4 at Des Plaines, Illinois.

The use of mosquito fish (*Gambusia holbrooki*) has been advocated as an effective control on nuisance mosquitoes in the USA (Knight et al., 2003b; Walton, 2012). However, *Gambusia* are regarded as exotic invaders in other parts of globe. In Australia, the species is known as the "plague minnow." There it is regarded as threat to native fauna, in particular threatened frogs. New Zealand shares this opinion, because *Gambusia* are aggressive towards native freshwater fish, with endangered galaxiids and mudfish especially vulnerable. The New South Wales National Parks and Wildlife Service declared a threatening situation in 1999, and published a Threat Abatement Plan in 2003.

Water Quality Effects

Primary water quality effects include sediment resuspension, nutrient additions via excretion, and destruction of the vegetative uptake mechanisms. Chapter 8 contains a discussion of the sediment processes. Nutrient release via fish excretion constitutes an indirect pathway that short-circuits the biogeochemical cycle. But it is the destruction of the "biomachine" that facilitates nutrient removal that can be more serious. Carp and other benthivorous fish (e.g., tilapia) graze on epipelon and epiphyton, which are an important first-response component of the nutrient cycle. Benthic and epiphytic feeding disturbs the sediments when food organisms are sucked in, together with sediment particles, to be separated by a winnowing process in the buccal cavity before removing sediment particles through the gill raker system (Breukelaar et al., 1994). The direct effects of this feeding behavior are resuspension of sediments and release of nutrients from these sediments into the water column. Nutrient release via fish excretion constitutes a further indirect pathway from the benthic to the pelagic phase. Exclusion is an obvious remedy, but not easily implemented (Lougheed et al., 1998).

15.7 AMPHIBIANS AND REPTILES

Amphibians and reptiles (known jointly as herptiles) are part of the top consumer structure of wetlands, including treatment marshes. The four most important groups are frogs (anurans), turtles, snakes, and alligators. Constructed treatment marshes are a means for increasing local amphibian species richness and abundance. Treatment marshes have been shown to support diverse populations of amphibians and can be viable replacement habitat. However, designs do not typically consider faunal use.

Anurans

The order Anura of he class Amphibia includes the frogs, toads, and tree frogs. Twenty-one anuran species are reported from 6 constructed and 3 natural treatment wetlands in the NADB Database (1998), ten from constructed treatment wetlands. This is almost certainly an under-representation, because very few studies of anurans exist. Brown et al. (2012) note that the literature on amphibian use of created and restored wetlands is limited to a small number of species, primarily in North America. Species that have been studied in the wetland context include American toads (*Bufo* [*Anaxyrus*] *americanus*), bullfrogs (*Rana* [*Lithobates*] *catesbeiana*), green frogs (*Rana* [*Lithobates*] *clamitans*), wood frogs (*Rana* [*Lithobates*] *sylvatica*), and spotted salamanders (*Ambystoma maculatum*) (Brown et al., 2012).

If the focus is narrowed to treatment marshes in North America, few anuran species are reported. Two species were found at the Olentangy, Ohio treatment marshes, bullfrogs and leopard frogs *(Rana pipiens)*(Fink and Mitsch, 2005). This result may be compared to the nine species found in a survey of 42 replacement wetlands in central Ohio, with a frequencies of occurrence between 7 and 76% (Porej and Hetherington, 2005).

At the Houghton Lake marsh, three species were found, northern green frogs (*Rana clamitans melanota*), wood frogs (*Rana sylvatica*), and northern leopard frogs. Eleven species of frogs are found in Michigan.

However, a more comprehensive study of 30 stormwater treatment sites, containing 50 wetland sites and 31 pond sites, was conducted by Hamer et al. (2012) in the Melbourne, Australia region. The wetlands had a mean size of 2.0 ha, an average of 37% vegetative cover, a mean depth of 21 cm, and a mean age of 5.7 years. Nine species of frogs were found, out of the Melbourne regional total of 14 species. The number of species at individual sites ranged from 2 to 6 (mean = 3.5), indicating that greater frog species richness is to be expected when multiple sites are surveyed. These Australian genera and species are all different from those found in the Midwestern USA.

Marsh design and operational conditions are important factors in anuran use. Most anurans prefer systems with buffer zones, which are zones of uplands surrounding the wetland. Marshes with long hydroperiods, that is, continuous flow systems, increase predation potential via fish colonization), and may promote invasion by nonnative amphibians. For example, the exotic cane toad (*Rhinella marina)* in Florida preys on native frogs, lizards, snakes, and small mammals. Gallagher et al. (2014), in a survey of 68 stormwater ponds in Maryland, found that wood frogs (*Rana sylvatica*) were more likely to breed in stormwater systems with longer hydroperiods, while toads (*Bufo americanus*) were more likely to use longer hydroperiod systems.

Anurans are sensitive to water chemistry, and consequently treatment marshes may pose health problems for them. Gallagher et al. (2014) found survival of wood frogs was related to metals and salt, with zero survival when chloride levels were above 260 mg/L. However, survival of American toads was not affected. Hersikorn and Smits (2011) found metamorphosis was delayed or never occurred in tadpoles raised in young oil sands tailings wetlands; while those exposed to older tailings developed similarly to those in reference wetlands. However, when Smits et al. (2012) tested leopard frogs for effects of naphthenic acids (principal organic oil sands contaminant), few negative physiological or toxicological effects on the frogs were found.

Glyphosate-based herbicides (e.g., Roundup®) are currently the most commonly used herbicides in the world and have been shown to affect survival and development of tadpoles under laboratory and mesocosm conditions (Lanctôt et al., 2013). They field tested several wetlands, and found that exposure to environmentally realistic concentrations of glyphosate are unlikely to cause significant deleterious effects to larval amphibian development.

Thus, questions remain concerning the effects of pollutants that may be treated in marshes, or found in source waters, upon anuran populations. But, no alarming environmental events have been reported for anurans in treatment marshes.

Turtles

Turtles are found primarily in marsh treatment systems that have areas of deeper, open water. In North America, species of turtles frequently found in these habitats include snapping turtles (*Chelydra serpentina*), mud turtles (*Kinosternon subrubrum*), and painted turtles (*Chrysemys picta*).

Snakes

A wide variety of snakes are associated with wetlands and wetland treatment systems and play an important ecological role by feeding on fish, invertebrates, birds, and small mammals. In North American treatment wetlands, the snakes that receive the most interest are the venomous species including the cottonmouth or water moccasin (*Agkistrodon piscivorus*), the copperhead (*Agkistrodon contortrix*), and timber rattlesnakes (*Crotalus horridus*) that occasionally inhabit wetlands. In fact, most snakes in wetlands are nonpoisonous species of water snakes (Genus *Natrix*), swamp snakes (*Liodytes* and *Seminatrix*), rat snakes (*Elaphe*), and king snakes (*Lampropeltis*).

To this list, it is now necessary to add an exotic, the Burmese python (and hybrids) (*Python bivittatus*). Burmese pythons escaped from a breeding facility in Florida that was destroyed during Hurricane Andrew in 1992, and there have also been pet escapes and releases. Burmese pythons can grow up to 23 feet long and weigh 200 pounds. They prey on mammals of all sizes (including deer), wading birds, and occasionally an alligator. Their habitat is marshes. It is projected that the Burmese python's range is limited to southern Florida and extreme south Texas.

Crocodilians

The order crocodylia consists of three families: alligatoridae (includes genus *Alligator* and genus *Caiman*), crocodylidae (includes genus *Crocodylus*) and Gaviaidae (genus *Gavialis*). They are found in the USA and South America, Africa, India, China, the Philippines, and Australia; but not in Canada, Europe or the former Soviet countries. Most prefer aquatic habitats, including wetlands, and in turn including treatment marshes. In North America, alligators (*Alligator mississippiensis*) have been observed at relatively high densities in wetland treatment systems with open water areas and high fish populations (Figure 15.10). Densities in the Florida stormwater treatment constructed wetlands have reached proportions sufficient to allow hunting.

15.8 MAMMALS

A wide variety of mammals reside in or visit treatment wetlands. Some of the more common for North America are listed in Table 15.12. Shrews (family Soricidae) are found

FIGURE 15.10 Alligators are at the top of the food chain in wetland systems across the American south. Photo courtesy J. Bays.

along the edges of wetlands and moist fields and feed on insects. But by far the largest and most important group of mammals associated with wetlands are rodents. Small rodents are also found in marsh systems, such as mice and voles, most of which are herbivorous species that graze on plants and seeds and are prey to fish, wading birds, and raptors. However, it is the larger rodents that have proven problematic in many treatment wetlands.

TABLE 15.12
Mammals in treatment marshes in North America.

Rodents	
Shrews	*Sorex spp.*
Mice	*Peromyscus spp.*
Voles	*Microtus spp.*
Muskrat	*Ondatra zibethica*
Beaver	*Castor canadensis*
Nutria	*Myocastor coypus*
Capybaras	*Hydrochoerus hydrochaeris*
Herbivores	
Rabbits	*Sylvilagus spp.*
Deer	*Odocoileus virginianus*
Elk	*Cervus canadensis*
Carnivores	
Raccoon	*Procyon lotor*
Opossum	*Didelphus marsupialis*
Skunks	*Mephitis spp.*
Mink	*Mustela vison*
Otter	*Lutra canadensis*
Bobcat	*Lynx rufus*
Coyote	*Canis latrans*

Muskrats

The muskrat (*Ondatra zibethicus*) is the only species in genus Ondatra. The muskrat is native to Canada and the United States, but they are not found in Florida, which may be the result of alligator predation. Muskrats are an invasive species in northwestern Europe, and a prohibited species in New Zealand. They are prolific breeders, producing two or three litters a year of six to eight young each.

Muskrats cut large numbers of emergent herbaceous plants, primarily cattails (Latchum, 1996), and build feeding platforms and nests (mounds). This grazing can change treatment wetland areas from densely vegetated to a patchwork of open and emergent areas (Kadlec et al., 2006). Muskrats consume a portion of the annual net primary productivity, primarily rhizomes, but their mounds represent a greater share of this production. Densities of twenty or more animals per ha have been found, which can destroy the majority of the macrophyte standing crop in a given year. At such an exacerbated scale, muskrat herbivory may be termed as an "eatout," and is evidenced by the removal of essentially all emergent plant parts (Figure 15.11). There have been many other reports of muskrat eatouts. For example, these occurred at the Olentangy, Ohio wetlands (Higgins and Mitsch, 2001, 2002), at the Oberlin, Ohio wetlands (Petersen et al., 2015), at Brighton, Ontario (Kadlec et al., 2012b), and at the Yakima, Washington wetlands (Beutel et al., 2014).

Destruction of the wetland vegetative infrastructure may create an attendant loss of some water quality functions, but may not harm others. No designed research studies have been conducted to quantify the effects of muskrat eatout on wetland water quality performance, but anecdotal reports are frequent. Beutel et al. (2013) report possible negative effects upon fecal coliform reduction, and Beutel et al. (2014) report negative effects for phosphorus removal.

Other potential impacts may be speculatively identified. Two water quality measures are very liable to change due to denudation: dissolved oxygen (DO) and total suspended solids (TSS). Increased potential for atmospheric reaeration is present in the plant-free wetland, and plankton can generate oxygen within the water column because there is sufficient light. A two-hectare wetland in Commerce Township, Michigan provided data to support these intuitive concepts. In an early, fully vegetated phase, this wetland had an annual average outflow of 6 mg/L TSS and 9.8 mg/L DO. After a 100% muskrat eatout, TSS rose to 12 mg/L and DO rose to 10.4 mg/L.

There were also muskrat impacts at the Olentangy, Ohio wetlands. An eatout occurred in the year 2000, five years after system startup (Higgins and Mitsch, 2001; Higgins, 2002). Thereafter, turbidity increased upon passage, in contrast to decreases that occurred prior to eatout. A similar response for total phosphorus was observed. However, such results are not conclusive, because of other factors in system operation.

Most, if not all, muskrat eatouts occur in small systems, that is, a few hectares or less.

The integrity of berms may be threatened by burrowing. Impacts on wetland hydraulics are also possible. In all cases, loss of the emergent vegetation has been viewed with dismay by owners, regulators and the general public. The damage caused by muskrats, primarily by burrowing in containment and separation berms, is not a matter of conjecture. There are several examples of compromised parallel cells, in which divider berms have been breached by burrows Estevan, Saskatchewan (Duncan et al., 1999); Corcoran, California (Gao et al., 2003a); Sacramento, California (Nolte and Associates, 1998b). At the Manitoba Interlake Site 1, it was found that muskrat burrows were extensive and threatening to breach the dikes at several locations. Roads have been damaged by burrow collapse at Saginaw, Michigan (unpublished), and at Sacramento.

When muskrats eat all the emergent plant parts, there is a large visual impact. Regardless of physical or water quality damage. A common perception is that wetlands should be lush

(A)

(B)

FIGURE 15.11 Hillsdale, Michigan wetland before (A) (July 18, 1999) and after (B) (June 29, 2001) eat-out by muskrats.

and green to be effective in treatment. This is an intuitive carry-over from agriculture, where lush green fields are a strong indicator of a good crop. Consequently, owners, regulators and the general public equate good emergent stands of plants with the highest treatment capability. This may or may not be the case, but the urge to "fix it" is irrepressible in many cases. From the point of view of treatment efficiency, an eatout calls for careful scrutiny of performance, to ascertain if the damage has in fact impaired pollutant removal. If not, then muskrat control becomes an expense associated with the ancillary benefit of a green emergent appearance. At Brighton, Ontario, where water quality was relatively unaffected by extensive muskrat herbivory, the aesthetic aspect became a concern (Kadlec et al., 2012b). Residents of the community viewed the partially vegetated wetland as ineffective and expressed concern that there might be environmental consequences.

Beavers

There are two beaver species: North American beaver (*Castor canadensis*), native to North America; and Eurasian beaver (*Castor fiber*) native to Eurasia. *Castor canadensis* was introduced into Finland, France, Poland and Russia the, as well as Argentina and Chile, where they are now considered a serious invasive species, due to their massive destruction of forest trees. *Castor canadensis* are found in wetland treatment systems from Texas to the Canadian provinces. Beavers feed on bark, twigs, and leaves from a variety of trees, but they will also eat cattails, water lilies and other aquatic vegetation, and consequently may cause undesirable herbivory damage in and around wetland treatment systems. For instance, the Roblin, Manitoba constructed marsh was integrated into the adjoining terrestrial landscape with poplar tree (*Populus* spp.) plantings (PFRA, 2002). With an abundance of their favored aquatic habitat, and their favorite food (poplar), the beavers built lodges and decimated the tree plantings.

However, beaver damage to constructed wetlands does not stop with herbivory. When beavers find flowing water, such as in a treatment wetland structure, they construct a dam to contain that water, and create a deep water habitat that will protect their nesting lodge from predators. The animals are large, up to 40 kg body weight, and can move stones of several kg, small logs, sticks and mud to create stoppages in flows (Figure 15.12). Plugging of structures is clearly inimical to the proper performance of the wetland. Problems with beaver plugging structures have been experienced at many treatment marsh locations, including Fort Worth, Texas; Tres Rios, Arizona; and Des Plaines River, Illinois. Removal of the obstructions is not effective, because the animals can rebuild a dam literally overnight. Removal of the animals is the only recourse.

Beavers construct travel trails, to facilitate movement with minimum risk from predators. These may be water-filled ditches leading to and from deeper water areas in upland margins, or deepened areas in marginal shallows. In treatment marshes, these travel trails may connect deep zones, thus providing increased opportunity for hydraulic short-circuiting (Figure 15.13).

Beavers burrow into banks as an alternative to lodges made of branches and small logs, or as an adjunct to such lodges. These burrows may create significant leakage from the wetland, or compromise the integrity of containment berms. For instance, beaver burrows caused failure of the containment berm of a constructed infiltration bed/wetland at Houghton Lake, Michigan, causing over one thousand cubic meters of soils to wash out in one event.

These several effects of the presence of *Castor canadensis* are of considerable concern, and often warrant removal of the animals.

Nutria

Nutria (*Myocastor coypus*), also known as coypu, are native to subtropical and temperate South America. In North America, they were imported in the late 19th century as part of the fur trade in Louisiana. Intentional releases were made to increase the wild fur trade. They have since been introduced Europe, Asia, and Africa. They are large rodents, with mature animals averaging 5.4 kg in weight, but reaching as much as 10 kg. Nutria are found most commonly in freshwater marshes. Females have anywhere from five to seven offspring in a litter, and they have up to three litters per year. Mature nutria are large and have fewer natural predators. Nutria are strictly herbivorous and feed on a broad range of

(A)

(B)

FIGURE 15.12 Beaver activity caused complete closure of outlet Parshall flume structures at Des Plaines River, Illinois, wetland EW3. (A) Unobstructed flume. (B) After a dam of soil, rocks and sticks was constructed. Note the staff gage in panel (B) is incorporated into the dam, along with rocks and mud.

FIGURE 15.13 Beaver travel channel at the Tres Rios Hayfield treatment wetland near Phoenix, Arizona. This short-circuit became evident during this water draw down event (panel A). The panel B shows that such channels connect the two interior deep zones (each with three islands), and the inlet and outlet deep zones.

plants in treatment wetlands including cattails, grasses, water hyacinth, duckweed, and young tree seedlings. Nutria cut vegetation in a manner similar to muskrats and build feeding platforms and nest mounds.

Nutria consume large amounts of marsh vegetation, about 25% of its body weight daily, and feeding year-round. They eat the entire plant, including the roots, which means it's less likely to grow back. Their creation of "eat-outs" areas disrupts the habitat for other animals and humans dependent on marshes.

Nutria now range across the southern, eastern and western United States. Because they are prolific, they commonly achieve damaging population densities unless they are controlled by trapping or shooting. Nutria have occurred in problematic numbers in many treatment marshes, including Halsey, Oregon (Kuehn and Moore, 1995); Victoria, Texas, West Jackson County, Mississippi, Hillsborough County, Florida (Malone, 2009), and Hammond, Louisiana (Shaffer et al., 2015).

An example of the scale and rapidity of nutria invasion on treatment marshes is the East Fork system at Seagoville, Texas (Mokry, 2010). Operation of the 745-ha river water wetland started in January 2009, and one year later, huge nutria populations had developed. It was apparent that they were causing extensive damage to the emergent vegetation as a result of their foraging and other activities. Operations staff were trained and armed to implement controlled hunting efforts. From January through December 2010, almost 13,000 nutria were eliminated from the East Fork Wetland. Additional contract services were secured, including both trapping and hunting.

Carter et al. (1999) developed a model of nutria populations and their relation to the amount of remaining vegetation on a 20 ha hypothetical site. Starting with 40 nutria, the model showed complete eatout in three years, accompanied by a complete loss of the nutria population.

Capybaras

The capybara (*Hydrochoerus hydrochaeris*), a rodent native to South America, is the largest living rodent in the world. Adult capybaras typically weigh 35 to 66 kg. These water-loving herbivores are commonly found in groups of around 10–20 individuals. An adult can eat 2.7 to 3.6 kilograms per day, and wetland plants are included, along with terrestrial plants,

Maine et al. (2013) monitored a treatment wetland in Santo Tome, Argentina, dominated by the emergent macrophyte *Typha domingensis* at a cover of roughly 80% for 5 years. Then, a population of approximately 30 capybaras caused the complete loss of the aerial parts of macrophytes. However, plant roots and rhizomes were not damaged, and *T. domingensis* showed an excellent regrowth response. The period without aerial parts lasted a few months, and water quality was only minimally impacted. Thus, this mature CW was capable of maintaining its efficiency and recovering its vegetation in the event of capybara defoliation.

There are currently an estimated 50 capybaras in northern Florida. In 1995, five animals escaped from a wildlife facility near Gainesville, and they are probably at least part of the source of those Floridian capybaras. There is a mild concern that these large invasive rodents could pose problems comparable to those caused by nutria.

Predators

The richness of the invertebrate, avian and small mammal populations in treatment marshes provides good hunting for a number of predators. Mink (*Mustela vison*) prefer muskrats as food, and will harvest a sizeable fraction of the resident population in the constructed wetland. Coyotes (*Canis latrans*) will prey upon waterfowl nestlings, provided the water is not too deep. As a consequence, wetlands that are subject to seasonal drawdown and dryout during the nesting season are vulnerable to coyote predation. This situation exists for the Incline Village, Nevada wetlands, because water is diverted to crop irrigation starting in the nesting season. Raccoons (*Procyon lotor*) and bobcats (*Lynx rufus*) are frequent visitors to wetland systems, from Arizona to Michigan in the United States. Alligators are extremely numerous in the south Florida stormwater treatment areas.

Incidental Visitors

There are numerous animals that may from time to time visit treatment wetlands, and in some instances may create problems. The Florida stormwater wetlands are very large, and have very large inlet structures, through which an occasional manatee (*Trichechus manatus*) may pass into the system. These large gentle aquatic mammals are not a threat to the treatment system, but the habitat is wrong, and efforts are made to move them back to their natural habitat. The Tarrant County wetlands in central Texas were periodically visited by wild pigs (*Sus scrofa*), an abundant introduced species (more than one million animals estimated for Texas) with pernicious effects on native ecosystems (Pimentel et al., 1999). The tubers of wetland plants were rooted up and eaten in the pilot wetland cells. On a worldwide basis, the list of potential troublemakers grows to include animals such as monkeys, which ruined a pilot treatment wetland vegetated with *Cyperus immensus* in Kenya (Abira et al., 2005).

15.9 HUMAN USE

Humans appreciate wetlands for their commercial values: plant harvesting, livestock grazing, hunting, trapping, and food production and aquaculture (Williams, 1991). Products include timber, wildlife, blueberries, cranberries, and wild rice. The USA harvest of muskrat pelts is worth over $70 million annually. Approximately 82 million acres of forested wetlands are harvested nationally for commercial timber operations. Waterfowl hunters in the United States spend over $600 million annually to harvest wetland-dependent birds. Harvest activities are virtually absent in treatment marshes, but hunting and fishing are frequent human uses.

Nonconsumptive uses of marshes constructed primarily for water quality treatment include bird watching, hiking, research and education (Reimold and Hardisky, 1978; Smardon, 1988). The observation and photography of marsh birds entices an estimated 50 million people to spend nearly $10 billion each year. Increasingly, designs of treatment marshes have incorporated access for visitors, and attractive and informative park-like areas.

Public Use Attributes and Facilities

Large Publicized Wetlands

Ghermandi and Fichtman (2015) compiled human use data for 129 treatment wetlands located in 18 different countries, with half being in the USA (Table 15.13). These include some systems treating source waters from WWTPs that are outside the realm of treatment marshes, but about two-thirds are driven by surface waters, stormwaters, and highly treated municipal effluents. Systems with reported usage were included, but not those without such reports. That effectively limits their results to large wetlands. Five activity classes were used; passive

TABLE 15.13

Human activities in treatment wetlands according to source water and activity type.

Source Water	Storm Water	Mixed, Other	Surface Water	Treated Municipal	Totals
Systems	41	16	12	60	129
Bird Watching	15	3	5	29	52
Passive Recreation	31	9	9	41	90
Contact Recreation	4	0	1	1	6
Hunting	4	0	0	8	12
Education	21	11	7	34	73
Total Activities	75	23	22	113	233

Source: Ghermandi and Fichtman (2015).

recreation (hiking, jogging), contact recreation (fishing, boating), bird watching, hunting and education (class tours). Passive recreation bird watching and education dominate.

Ghermandi (2016) established visitation frequencies using annual photo-user-days from geotagged social media data (Panoramio and Flickr). An average number of photo-user-days per year between 2004 and 2015 was calculated for each site. FWS systems were found more likely associated with geotagged photos than SSF wetlands. Public use, measured by social media data, was more likely associated with large sites and systems treating stormwater runoff or secondary treated municipal effluents. The distribution of annual visits for a subset of 45 FWS wetlands, 34 in the USA, is shown in Figure 15.14. The median site size was 62 ha, and the median visitation was 2,460 per year. The maximum was 850,000, for the Green Cay wetland in Florida (see Figure 1.17), which possesses a visitor center on a par with those at National Parks.

Little Known Wetlands

The majority of treatment wetlands, including marshes, do not have documented human use. Urban stormwater wetlands, numbering in the thousands, are enjoyed by the local communities in which they are located. They are neighborhood features that are used mainly by inhabitants for nonconsumptive purposes. Access is almost always available via perimeter trails.

Small agricultural runoff wetlands, such as the Iowa CREP marshes and the Scandinavian systems also number in the thousands. These receive no public notoriety and are situated in rural areas with few inhabitants other than the land owners. These generally do not have public access but are nonetheless used by owners for both consumptive and nonconsumptive purposes.

There are also many treatment marshes that have open public access, via perimeter roads and internal boardwalks, but are unpublicized and thus receive only small numbers of visitors. For instance, the Te Maunga marshes, near Tauranga on New Zealand's North Island, polish wastewater from domestic, commercial and industrial communities from the Mount Maunganui and Papamoa catchments. There is minimal signage, but a boardwalk provides internal access to the system (Figure 15.15).

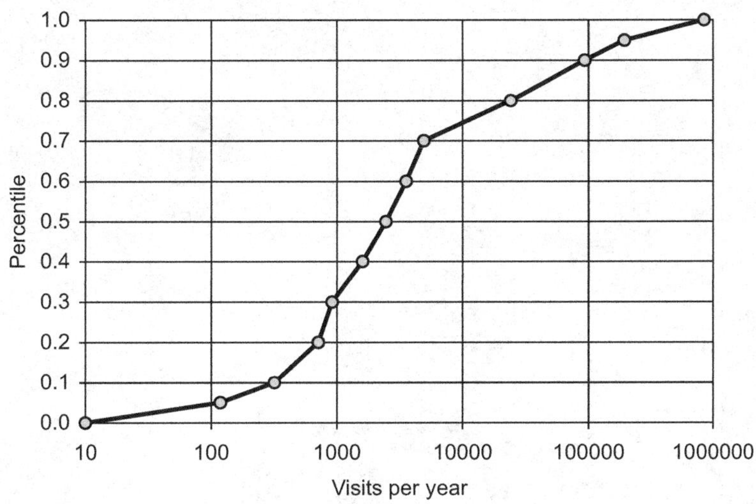

FIGURE 15.14 Distribution of visits per year for 45 treatment wetlands. (Based on data from Ghermandi, 2016).

FIGURE 15.15 The Te Maunga marsh, near Tauranga on New Zealand's North Island.

Another example is the Houghton Lake, MI, marsh. Public access is allowed, but there is no signage. There is a small parking lot, and a long boardwalk that serves as the support for the distribution piping, but is also a walkable route to the system interior (Figure 15.16). This wetland is used, to a limited extent, for deer and duck hunting and education.

Facilities

Uni-functional wastewater treatment wetlands are usually anonymous, especially SSF systems. FWS wastewater treatment wetlands typically are named for the city or county that owns

FIGURE 15.16 The boardwalk at the Houghton Lake, MI, treatment marsh.

them, for example, Columbia, MO; Lake Nebagamon, WI; Roblin, Manitoba. But when emphasis is placed upon human use, via establishment of signage, walking trails, boardwalks, and visitor centers, they become named facilities (Table 15.14). Naming facilitates publicity, in the form of websites describing public uses, driving instructions and hours of operation. It also allows easy reference on organization websites, such as that of the Audubon Society.

The infrastructure that supports human use is comprised of parking space, signage, trails and boardwalks. Early wetland human use facilities underestimated the need for parking. For instance, the Wakodahatchee wetland began with parking for about ten vehicles, which was wholly inadequate – the site now has about 800 visitors per day. A much larger parking lot was subsequently added. Signage is a necessity so that users know where to go, and which activities are allowed and which are prohibited (Figure 15.17).

Boardwalks provide interior access, and trails provide perimeter access on confining levees (see Figures 15.16 and 15.19). The former are very expensive, but the later add little cost to the project. Trails on upland islands or perimeters afford the possibility of observing wildlife that frequents treatment marshes, but also spend time on adjacent upland (Figure 15.18).

Consumptive Activities

Plant harvest can potentially provide an economic profit from treatment marshes, while assisting in nutrient removal. Principal uses of the biomass are for biogas production and combustion (Grosshans et al., 2014; Wichmann, 2017; Carson et al., 2018). Both cattails and reeds have high heating values (17–20 MJ/kg). In Europe, *Phragmites* is used for roof thatching, and provides a very high quality structure. At the present time, harvest of plants in the USA from treatment marshes is not a common practice.

Consumptive uses of treatment marshes for hunting is are more easily quantified than nonconsumptive functions. For example, 11.7 million ducks were harvested in the USA in 2016–2017. Constructed treatment marshes at Orlando, Florida, the stormwater treatment areas of Florida, and the Incline Village, Nevada system, are used for waterfowl hunting.

TABLE 15.14

Examples of named treatment wetlands.

Location	Owner/Operator	Name
Christmas, FL	City of Orlando, FL	Orlando Wetlands Park aka Orlando Easterly Wetland aka Ironbridge Wetlands
Calgary, Alberta	City of Calgary	Ralph Klein Wetlands Park aka Shepard Wetlands
Corsicana, TX	Tarrant Regional Water District	George W. Shannon
Renton, WA	City of Renton	Waterworks Gardens
Titusville, FL	City of Titusville	Blue Heron Wetlands
Tucson, AZ	City of Tucson	Sweetwater Wetlands
Arcata, CA	City of Arcata	Allen Marsh
Clayton County, GA	Clayton County	E.L. Huie Jr. Wetlands
Putrajaya, Malaysia	City of Putrajaya	Taman Wetlands
Seagoville, TX	N. Texas Municipal Water District	East Fork Wetland
Naples, FL	Collier County	Freedom Park
Boynton Bech, FL	Palm Beach County	Green Cay
Boone, NC	City of Boone	Clawson-Burnley Park

FIGURE 15.17 Access sign for the A1 wetland in south Florida. By a quirk of political labeling, this treatment marsh is called a flow equalization basin. The lower signs indicate that wildlife viewing, fishing, nonmotorized boating, hiking, bicycling and hunting are allowed.

An ancillary benefit of the STAs is that the vegetation and the shallow water have attracted many ducks, such as teal (*Anas discors*), pintails (*Anas acuta*) and widgeon (*Anas Americana*), ringnecks (*Aythya collaris*) and canvasbacks (*Aythya valsineria*). The South Florida Water Management District (SFWMD) and the Florida Fish and Wildlife Conservation Commission have opened the STAs to limited duck hunting by permit, starting in 2001. Motorized access is prohibited, and hunters canoe or kayak to spots near designated parking spots. The hunting has generally been excellent, with of the STAs producing about four ducks per hunter (Table 15.15), which is unheard of for a public waterfowl area. As a result, hunters have come from all over Florida, as well as other states, to enjoy the STAs. United Waterfowlers Florida President Newton Cook stated, "STA 5 is the #1 public waterfowling marsh in the United States and is nationally known."

STA-5 was opened for the first year for alligator hunting in 2006. Now, all the STAs, including the A1 wetland, have been opened for alligator hunting. Over 8,200 alligators were taken in 2017.

The U.S. Federal Conservation Reserve Program (CREP) has created numerous wetlands on agricultural lands, and treatment marshes (hundreds) are among them. Landowners retain ownership; and they can hunt it or lease it for hunting, thus helping to diversify income sources for farmers. They are particularly popular with duck and pheasant hunting enthusiasts and are widely used for these activities. Many farmers have opened up the land they have enrolled in CREP to public access for hunting and fishing, thus improving the relationship between landowners, state fish and wildlife agencies and sportsmen. These marshes can also be used for hiking and camping activities, provided they do not harm ecological significance or value.

FIGURE 15.18 Walking trail at the Te Maunga treatment marsh in New Zealand.

TABLE 15.15
Duck hunter success in the Florida STAs in the 2016–2017 season.

Location	Total Duck Harvest	Total Hunters	Ducks per Hunter
STA 5	5,259	1,696	3.1
STA 3/4	12,926	3,006	4.3
STA 1W	11,810	5,001	2.36
STA 2	3,081	1,497	2.06

In contrast, urban stormwater wetlands are little used for hunting. The urban context, and the perceptions of local citizens, usually prohibit hunting.

Aesthetics and Wildlife Observation

Wildlife observation is a principal human activity in treatment marshes. Because of the safety they find in those marshes, birds and animals have lost much of their fear of humans in those locations. This allows the casual observer to view and photograph them at close range without special equipment (Figure 15.19). Such close-ups are unheard of in other locations.

Bird observation for nature enjoyment is here considered to be part of the passive recreation activity. Many marsh visitors are not "professional" birders; rather, they just enjoy seeing marsh birds up close and personal.

FIGURE 15.19 Photos of (A) turtles, and (B) a double-crested cormorant (*Phalacrocorax auritus*) taken from the Wakodahatchee boardwalk with a small pocket camera.

The treatment wetlands at Arcata, California, gave an early indication of the potential for passive recreation. They are close to the urban area, and are extensively used for hiking, jogging and picnicking (Gearheart and Higley, 1993). An estimated 76,000 visits per year produced a significant maintenance activity, to clean up the trash left behind.

Such responses of the public to treatment marsh access have now been repeated many times. Their popularity for public use is very high, particularly when user-friendly facilities are provided. An example gives the general sense of this human use.

The Wakodahatchee Wetlands

The Wakodahatchee Wetlands were designed to demonstrate the following ancillary benefits of reclaimed water reuse:

- Creation of significant wildlife habitat accessible to the public in a park-like setting
- Maintenance of biological diversity and open space in a highly developed landscape
- Recharge of local water supplies through infiltration to the surficial aquifer
- Provide limited public use for passive recreational activities.

The system was completed in 1997 and included public use facilities (Figure 15.20). The wetland area totals 16 ha on the 23 ha site, and the marsh area is designed to operate at an average depth of 15 cm, but may be operated normally at depths to 45 cm. Deep zones, consisting of ponds 1.5 m or more in depth and variable in width and oriented transverse to the direction of flow, are interspersed throughout each wetland. A total of 28 deep zones and 8 habitat islands are included. The Southern Region Water Reclamation Facility pumps nearly two million gallons of highly treated wastewater into the Wakodahatchee Wetlands daily.

Marshes comprise about 70% of the wetland area, vegetated by native emergent, forested, and transitional wetland species designed to emulate native south Florida wetland plant communities. Emergent marsh zones are composed of bulrush (*Scirpus validus, S. californicus*), duck-potato (*Sagittaria lancifolia*), arrowhead (*Sagittaria latifolia*), spikerush (*Eleocharis cellulosa*), fireflag (*Thalia geniculata*), and pickerelweed (*Pontederia cordata*). The wetlands are maintained to be free of nonnative and nuisance plant species, and the surrounding berms and entry areas are landscaped to maintain an aesthetic and comfortable

a
b

FIGURE 15.20 The Wakodahatchee wetlands in Palm Beach County, Florida, has a very attractive entrance sign (A), and user-friendly boardwalk (B).

experience for wetland visitors. The combined efforts to manage the vegetation, maintain the boardwalk facility, monitor water quality, and maintain the grounds costs on the order of $150,000 per year.

Snags and perching posts were added for wildlife use. Numbers of bird species observed within the wetlands totaled 119 in 1997, 142 in 1998, and in 2018 up to 178. More than a dozen are considered threatened or endangered by state and federal agencies. Reptile diversities were high and increased over the first two years.

The public makes extensive use of the Wakodahatchee Wetlands. Parking facilities are available, and there is a boardwalk through the wetlands, with associated gazeboes, which provide an attractive and comfortable setting for walking, bird watching, nature photography, environmental education, and related passive activities. Lots of things are not permitted, including pets, food, alcohol, jogging or speed walking, biking, skating, smoking or fishing. The Audubon

Society of the Everglades has identified the Wakodahatchee as one of the birding hot spots within Palm Beach County. Ghermandi (2016) inferred an annual total visits at 304,000.

Birding

Birding is a hobby in which individuals enjoy the challenge of bird study, listing, or other general activities involving bird life. Birder is the term used to describe the person who seriously pursues the hobby of birding, and they may be professional or amateur. In the extreme, some birders are termed "twitchers," meaning those who travel long distances to see a rare bird that can then be ticked or counted on a list. Equipment commonly used for birding includes binoculars, a spotting scope with tripod, a smartphone, a notepad, and one or more field guides.

The diversity and abundance of birds in and around large treatment marshes attracts many birders who observe that their species lists are longer and their counts higher when they include wetlands in their counting areas. Where these water bodies are enriched by nutrients and organic matter from wastewater and stormwater discharges, bird watchers usually find even better success in their sport.

For example, bird populations at the Victoria, Texas, constructed treatment wetland have been quantified, by designed studies and by birder counts (Reitberger et al., 2000). A total of 188 bird species had been observed using the wetland as of June 2000. Annual species numbers range from 106 and 144, with average numbers of species per count from 34 to 53. Average annual bird densities ranged from 19 to 26 birds per hectare. Bird population numbers were dominated by grebes, cormorants, ibis, waterfowl, vultures, coots, rails, stilts, sandpipers, gulls, doves, swallows, wrens, warblers, and blackbirds. This site is open to school groups and the public and birding continues to be one of the most important ancillary benefits of the wetland.

The Brighton treatment wetlands were not designed for biodiversity or wildlife habitat, but for polishing of the lagoon discharge. The high level of bird use has taken the naturalist community somewhat by surprise. The treatment wetland proved, over the years, to be an important staging area for migratory species, which was anticipated. But it is also an important nesting area for Virginia rails (*Rallus limicola*), sora rails (*Porzana carolina*), and common moorhens (*Gallinula chloropus*). These three faired better in the Brighton wetland than in neighboring Presqu'ile Park (Campeau et al., 2009). In total, 110 species of birds have been seen in the wetlands. Some of these are rare in the region: the black-necked stilt (*Himantopus mexicanus*), the glossy ibis (*Plegadis falcinellus*), the red-necked phalarope (*Phalaropus lobatus*), the trumpeter swan (*Cygnus buccinator*) and the yellow-headed blackbird (*Xanthocephalus xanthocephalus*). A pair of ospreys (*Pandion haliaetus*) have successfully nested on a constructed platform in the treatment wetland each of the past two years (Morin, 2009). The Brighton treatment wetland has consequently become a popular birding spot, with about a thousand visitors in 2009. This level of tourism stretched the ability of the Municipality to provide guided supervision, and a permit system was implemented beginning in 2010.

The Brighton site has attracted twitchers. In 2016, a Ruff (*Calidris pugnax*), an uncommon visitor in North America, was spotted there. An internet blog then reported a trip to Brighton from Toronto for the sole purpose of viewing the bird. "Previously I had only viewed one Ruff, last winter in Morocco. Hopefully one day I can make it up to Russia or Scandinavia to watch Ruffs ..."

Educational Programs

Treatment marshes with human access and visitor centers are used to conduct educational programs (Figure 15.21). These are mostly directed at lower school classes, but can extend

FIGURE 15.21 The Environmental Education Centre at the Shepard wetlands in Calgary, Alberta. The Shepard Wetlands act as a management, filtration and cleaning system for the city's storm water. The centre sits on piles within the wetland itself. The two-story building includes classrooms, exhibition space, administrative and support facilities, and provides a location for interactive public education on wetlands.

to university students and continuing education groups. Stormwater treatment wetlands in urban settings such as Greenwood Park in Orlando, Florida, the Tollgate Wetlands in Lansing, Michigan, and Coyote Hills east of San Francisco Bay, California, are used frequently for field trips and other educational purposes. However, it is the larger systems that are more amenable to educational activities. The Sweetwater Wetlands support a learning event tied to the third grade curriculum in most Tucson area districts. Trained facilitators deliver interactive lessons that enable students to explore concepts through models and hands on experiences. Students have fun and have the unique experience of being in a wetland environment – in the middle of the Sonoran Desert.

An educational program has been in operation for over twenty years at the Invista Wetland near Victoria, Texas (formerly known as the DuPont Wetland) (Figure 15.22). The constructed wetland was built to treat wastewaters from a petrochemical plant that manufactures precursors for nylon and polyethylene. The 21.5 ha wetland receives about 12,000 m^3/d of water that has been pretreated in a biological plant. Education programs are provided to students from Grade 4 through college, via the Wetland Environmental Science Education Encounter program (WE SEE) the Victoria Independent School District. The program encompasses the treatment benefits of the wetland along with the inherent natural benefits of a wetland. A full-time on-site educator is employed, the first such person located at an industrial wetland. As of August 2016, over 65,000 students had participated in the program. Located in the heart of the plant's constructed wetland, the program is housed at a state-of-the-art education center

Conflicts

Although most reasonable human use is acceptable in treatment wetlands, there are some potential conflicts. These are generally in two categories: destructive human activities, and liabilities of the project owner, both real and perceived.

FIGURE 15.22 The Invista wetland at Victoria, Texas. The education center can be seen on the central peninsula.

Humans can be destructive, either maliciously or unintentionally. An illustration of the unintended damage issue is the case of a small treatment wetland in Dixboro, Michigan, which treats runoff from the Matthaei Botanical Garden. Elementary school children visit the facility frequently, and one of their activities was catching frogs and dragonflies in the wetland. But about 30 children running around in the 850 m^2 wetland caused severe damage to sediments and vegetation, and the activity had to be cancelled.

More problematic is malicious damage. Signage has been defaced with spray paint (Show Low, Arizona), distribution pipes cut with an axe (Houghton Lake, Michigan) stolen autos driven into the wetland (Howell, Michigan).

Issues of public safety combine with issues of facility protection to lead to limitation of access to treatment wetlands. Insurance against public injury has led to exclusion fencing at many treatment wetlands. This is more important for wetlands treating waters that may be hazardous to human health, which include some remediation sites as well as wetlands treating primary effluents. Nevertheless, there are few instances of human injury in a treatment wetland, but your author is one of them. I sustained a nasty leg injury when a rotten board in a boardwalk gave way. That boardwalk has since been replaced.

15.10 SUMMARY

Treatment marshes provide ecological benefits in addition to water quality benefits, and both are of importance. They treat relatively benign source waters, and therefore may also be regarded as wetland restoration projects. The realization of both sets of values involves the efforts of two disciplines, engineering and ecology. The engineer is tasked with achieving the water quality goals, and with getting the system built. The ecologist is tasked with preserving and enhancing the flora and fauna that inhabit the system.

Treatment marsh projects have caused the two disciplines to merge to some extent. While water quality can be easily measured, ecological values are more difficult to quantify. The goal of ecological diversity can be quantified via the use of biotic indices, which assess the numbers and kinds of biota that are present in the marsh. These measure the richness, diversity and evenness of marsh biota. They may be applied to the several components of the biological structure, ranging from macro invertebrates and algae, up to vegetation, animals and birds. These are used to evaluate the biological character of the system. Ecologists have also become accustomed to using marsh biology as a metric for the water quality therein. Indices of biological integrity are used to infer the water quality of the marsh, but these are of little use when the water quality is known in advance, and is part of marsh design.

The major easily observable features of a marsh are vegetation, birds, mammals, fish and reptiles. These are fostered in large numbers, and contribute to the human perception of the ecosystem. The result is often a considerable human use of the wetland, for recreation and consumptive uses. The type and extent of use is dependent upon the source water and location, with differences between agricultural, urban, and municipal, but in all cases use is large if opportunities are provided.

With thoughtful design, constructed wetlands can provide benefits beyond effective water treatment, such as wildlife enhancement and recreational opportunities. In fact, with the decline of the total area of natural wetlands, constructed treatment wetlands are a viable and cost-efficient way to compensate for the loss of productive wetland habitat.

16 Footprints and Performance

Design of treatment marshes may be roughly divided into two categories: sizing calculations and physical specifications. Here the various sizing procedures are discussed, and a comprehensive performance-based sizing strategy is presented. Chapter 17 will deal with physical considerations, including the number of cells, layout, depth, bathymetry, soils and plants, and structures.

The first step in design is to acquire a fundamental understanding of the size of the proposed marsh. Alternatively, the performance of a pre-determined marsh footprint can be the focal point of analysis. In either case, the tools set forth in preceding chapters are used to forecast marsh effluent water quality. Design sizing follows one of four tracks:

1. Percentage of the contributing watershed (urban and agricultural sources)
2. Design storm detention (urban sources)
3. Annual average performance (all sources)
4. Detailed dynamic modeling (all sources)

In the case of the first two procedures, performance is only implied by the experience base. Nonetheless, performance forecasting is a common accompanying or retrospective add-on, especially for larger urban systems and agricultural runoff marshes.

16.1 HISTORICAL SETTING

Much of treatment wetland sizing evolved along very simplistic lines, mostly adapted from civil and sanitary engineering. However, it has become increasingly apparent over the last 30 years that treatment wetlands are much more biologically complex than treatment processes used in the sanitary engineering field, and simplifying assumptions applied to activated sludge, trickling filter, and pond systems do not adequately describe the internal biogeochemical processes that govern treatment performance in wetlands. During the developmental years, attention was entirely focused upon secondary wastewater treatment, with the principal contaminants being BOD, TSS, ammonia and coliforms. These are not the principal contaminants targeted by treatment marshes, which are nitrate, phosphorus, and trace metals and organics.

Three principal themes have been prevalent in the history of constructed treatment wetland design: first-order removal models, regression equations, and pollutant loadings. Historically, detailed mechanistic wetland models have been of little utility, because of calibration difficulties. General rules of thumb for area requirements are used to good avail early in conceptual design. Soil accretion and the sustainable storage and dilution of irreducible constituents are other aspects not covered by existing design methods. There is also great danger of using empirical relations outside the envelope for which they were developed, and of unintentionally exceeding the capabilities of the fundamental wetland processes. Proper use of this imperfect set of design tools is a current design challenge.

First-Order Modeling

In the mid-1980s, simple first order models of pollutant removal in wetlands made their appearance in the treatment wetland literature (U.S. EPA, 1983b; Boon, 1985; Kadlec, 1985; Reed et al., 1988). This was in part a natural extension of models that had earlier been developed for waste stabilization ponds (Marais and Shaw, 1961; Marais, 1974; Thirumurthi, 1974). Some of these early models were founded on shaky ground, because they borrowed extensively from nonwetland results, and were not well-calibrated to actual treatment wetland data. Over the past three decades, considerable progress in understanding the behavior of treatment wetlands has allowed major improvements in the way that first-order models are constructed and implemented. It is the premise of this book that first-order modeling remains one of the most effective wetland design tools, but that other information can and should also be brought to bear on design.

Several variants of first order models are now in use. Many of these fail to address significant aspects of wetland performance, including effects of hydraulics, meteorology, interactions of biota and treatment, supply-side constraints and seasonal and stochastic variability.

Regression Equations

The principal variables that determine the outlet concentration (C_o) are the hydraulic loading (q) (or the equivalent detention time, ($\tau = h/q$) and the inlet concentration (C_i). Other influences include temperature, solar radiation and pH. There is therefore the possibility of simply postulating a functional relation, such as a product-power law, and regressing intersystem wetland input-output data to fit that equation. In so doing, some of the power of intra-wetland data is lost, because such regression formulae may not represent the observed longitudinal profiles for a specific wetland. However, there can be no doubt that an individual dataset can be fit with a reasonable guess at a fitting formula, because there are typically three or more fitting parameters. For example, Kadlec and Knight (1996) set forth linear regressions and power law regressions for different common pollutants. Examples are:

1. Gumbricht (1993), total nitrogen leaving a single FWS SAV system ($R^2 = 0.92$):

$$C_o = 1.63 + 0.78C_i - 1.7\ln(\tau_{nom}) \tag{16.1}$$

where: C_i = inlet concentration, mg/L
C_o = outletlet concentration, mg/L
τ_{nom} = nominal detention time, d

2. Kadlec and Knight (1996), nitrate leaving several FWS systems of different types ($R^2 = 0.35$):

$$C_o = 0.093C_i^{0.474}q^{0.745} \tag{16.2}$$

where: q = hydraulic loading rate, cm/d

Regression equations for treatment marsh performance continue to appear in the literature (e.g., Wong et al., 1999), but have not gained acceptance as a design sizing tool.

Loading Specifications

An early source of guidance utilized by wetland analysts were the methods used by wastewater stabilization pond designers. A common method was the prescription of a specified areal loading rate for BOD, on the order of 40 kg/ha·d to achieve an outflow concentration of 30 mg/L (Reed et al., 1988). BOD loading is still the recommended design method for ponds, albeit with modifications that acknowledge temperature effects (Shilton, 2005; Shilton and Mara, 2005). This concept has to some extent carried over to treatment wetland design and analysis.

Load reduction specifications, as pollutant kilograms per wetland hectare per year, in general cannot be used for design sizing. Yet, the concept of a marsh as a device that takes a fixed area-specific amount of pollutant out of the water has been frequently advanced in the literature. A prime example is the removal of nitrate. A large portion of the literature on denitrification and nitrate loss in aquatic systems uses a single-number representation of nitrate loss, namely the load removed, usually in units of $gN/m^2 \cdot d$ or equivalent. The reason is easy to understand: single number characterization can be performed without any mathematical anguish. Moreover, a single number can be used to forecast wetland size for a specified load reduction task without resorting to a desktop computer. As noted by Xue et al. (1999):

> Denitrification rate, in units of $mgN\ m^{-2}h^{-1}$, is directly useful for wetland design. By multiplying the rate with wetland area and retention time of water in the wetland, one can calculate the amount of NO_3 that may be removed from the tile drainage water by the wetland. Given NO_3 loading in a tile, one can calculate the size of the wetland …

Unfortunately, this simplistic approach doesn't work in the varied milieu of treatment marshes, for a number of reasons. For instance, as will be demonstrated below, doubling the wetland size does not double the load removed. Load removed is obviously constrained to be less than load applied, which is not reflected in a single number approach. Because removal depends on temperature, the seasonality of flows has a big effect on sizing. And, the single number for removal depends on the particulars of hydraulics. The results given by Kadlec (2012) illustrate the dilemma. Annual performances for 234 wetlands showed a median annual reduction of about 0.1 $gN/m^2 \cdot d$, but there are wetlands that do 100 times better, and some that do 100 times worse. A performance probability distribution that spans four orders of magnitude is not useful in design.

Others have observed this breadth of performances for load reduction. For instance, Mitsch et al. (2001) report wetland load removals ranging from 0 to 0.47 $gN/m^2 \cdot d$, and choose to estimate requirements based on 0.03–0.05 $gN/m^2 \cdot d$. The Kadlec (2012) database indicates that about two-thirds of the constructed marshes do better than that on an annual basis. Mitsch et al. (2005a) revised the projected "retention" rate upward to 0.08 $gN/m^2 \cdot d$, based on more data from the Olentangy, OH site, and the exclusion of some other wetlands.

The conclusion of Arheimer and Wittgren (1994) seems justified: there is no simple and general relationship between retention and load, as has been suggested in the literature. The single removal number design and forecasting approach must be rejected, because it does not deal with either primary variables (inlet flows and concentrations), or with secondary variables (temperature, hydraulic efficiency).

16.2 TREATMENT REQUIREMENTS AND GOALS

Requirements

The domain of treatment marshes contains two quite different water quality regulatory drivers: water quality based effluent standards (WQBELs) and structural best management practices (BMPs). The former are found in connection with marshes that polish municipal effluents, and the later are applied to urban and agricultural runoff wetlands.

Point Sources

Water quality goals for a receiving waterbody are defined by state water quality standards. In the United States, the Clean Water Act and its implementing regulations require development of water quality-based effluent limitations (WQBELs). They involve a site-specific evaluation of the discharge and its effect on the receiving water. A WQBEL is designed to protect the quality of the receiving water by ensuring that State water quality standards are met. Receiving water quality standards are composed of three parts:

- Use classifications
- Numeric and/or narrative water quality criteria
- Antidegradation policy.

A second aspect of water quality regulatory requirements is the implementation of a total maximum daily load (TMDL). A TMDL identifies the maximum amount of a pollutant that a body of water can receive from human activities (i.e. diffuse nonpoint sources and point discharges), while still meeting water quality standards. There are three components to the load determination:

$$TMDL = WLA + LA + MOS \tag{16.3}$$

where TMDL = allowable daily pollutant load to the water body, kg/d
WLA = waste load allocation to point sources, kg/d
LA = waste load allocation to nonpoint sources, kg/d
MOS = margin of safety, kg/d

To implement TMDLs for point sources, wasteload allocations are incorporated into discharge permits for these sources, generally NPDES permits. Nonpoint source discharges (e.g. urban, agriculture) are voluntary.

Compliance with WQBELs requires monitoring for concentrations, for numerical concentration limits, and flows and concentrations, for numerical load limitations.

The implementation vehicle for point source discharges is the national pollutant discharge elimination system (NPDES) permit, issued to the owner/operator. The responsibility for meeting the conditions of the permit rests with facility owners, which means they assume the entire risk associated with the design and facility operation. If the permit is not met (facility out of compliance), owners are faced with fines and consent orders for modifications.

Traditionally, permits have been developed to control both flows and loads of pollutants. There are typically annual averages, and monthly and weekly maxima; perhaps adjusted seasonally. The relation between averages and maximum allowable concentrations may have been determined from other technologies, and may be inappropriate for a wetland system. Similar procedures are in place for other countries.

Nonpoint Sources

Structural Best Management Practices are defined as a permit condition used in place of effluent limitations, to limit the discharge of pollutants. Best management practices (BMPs) is a term used in the United States, Canada and Australia to describe the suite of water pollution control devices. In stormwater management (both urban and rural) and wetland management, BMPs refer to a principal control or treatment technique. To distinguish these from source controls in the contributing catchment, stormwater BMPs are classified as "structural" (i.e., devices installed or constructed on a site.

Implementation of the wetland BMP occurs via prescribed designs and operational rules, usually issued by a municipal or State government, rather than at the national scale. It is the owner's responsibility to build the facility according to the prescribed requirements, but they do not bear responsibility for the realized water quality improvements or lack thereof. The risks associated with performance expectations are borne entirely by the prescribing agency.

Nonpoint source load reduction actions are implemented through a wide variety of programs at the state, local and federal level. These programs may be regulatory, nonregulatory or incentive-based, for example, a cost-share program. In addition, waterbody restoration can be assisted by voluntary actions. In the United States, federal grant programs fund projects aimed at reducing the nonpoint source pollution.

Load Reduction versus Concentration Reduction

A key feature of treatment wetlands is the ability to design or manage the system for either concentration reduction or for mass removal, but only one at the expense of the other (Trepel and Palmeri, 2002). Wetland performance, as discussed in earlier chapters, often follows the rule of mass action: the removal rate of a pollutant is greater at higher pollutant concentrations in the water. The first pseudo-order model assumes a (nearly) direct proportionality: doubling the concentration doubles the removal rate. As a result of this observed behavior, removal rates decrease as water passes through the treatment wetland, and pollutant concentrations are reduced (Figure 16.1). However, the actual mass of pollutant that is removed increases with increasing hydraulic loading. Thus increasing hydraulic loads result in more kilograms removed, but at the expense of higher effluent concentrations. This trade-off between removal efficiency and load reduction is a key feature of wetland design for nutrient control, and may be quantified via the first order model. A simple version for $C^* = 0$ and no water loss is:

$$CR = \frac{C_i - C_o}{C_i} = 1 - \left(1 + \frac{k}{Nq}\right)^{-N} \tag{16.4}$$

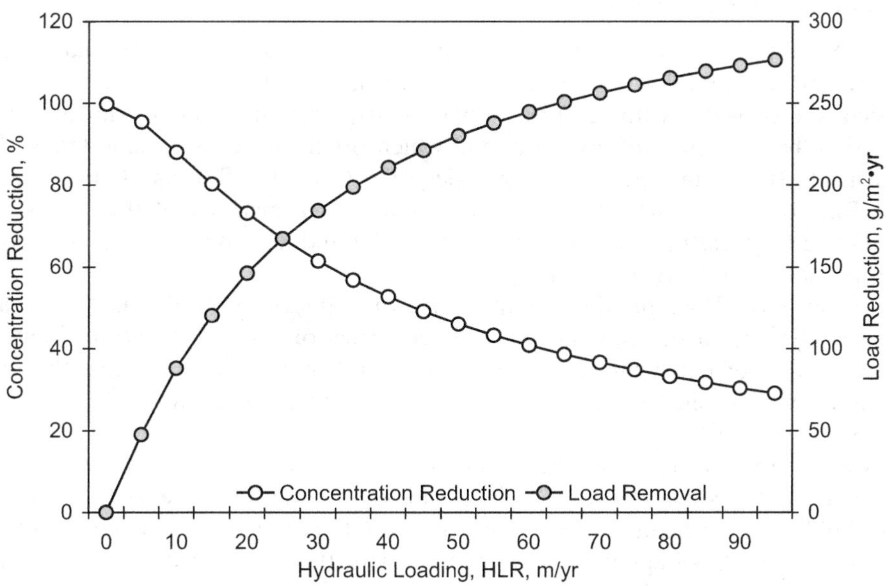

FIGURE 16.1 Concentration reduction and load reduction as a function of hydraulic load for a hypothetical treatment wetland. Parameters: $C_i = 10$ mg/L, $k = 35$ m/yr, $C^* = 0$, $P = 4$.

$$MR = MLI \cdot CR = qC_i \cdot (1 - (1 + \frac{k}{Nq})^{-N}) \tag{16.5}$$

where CR = fractional concentration reduction
C_i = inlet concentration, mg/L
C_o = outlet concentration, mg/L
MR = mass reduction, g/m2·yr
MLI = inlet mass loading, g/m2·yr
k = loss rate coefficient, m/yr
N = number of tanks in series, –
q = hydraulic loading, m/yr

The maximum mass of pollutant that can be removed in a given footprint area results from a hydraulic load so high that little or no concentration reduction is achieved. Under those conditions, the pollutant concentration is at a maximum everywhere in the wetland, thus maximizing the mass removal rate.

If the output load of a pollutant is to be held below some regulatory limit, then that has the same general effect as specifying an outlet concentration. However, because the load of a pollutant in the wetland outflow is usually taken to be that in the surface water discharge, any reduction in flow through the wetland assists in providing lower output loads, as well as higher load reductions.

Both potential design goals (concentration out, load reduction) may be sensible, depending on the pollutant in question. For instance, a low required outlet concentration can be of use in preventing toxicity in receiving waters. But if there is a load allocation, as might be the case for a discharge contributing to the TMDL) for an impaired water body, then the mass of the contaminant is of more direct interest.

Because it is easy to confuse these potentially conflicting goals, it is recommended that the designer clearly identify and state the purpose of the design.

16.3 LIMITED SPACE SYSTEMS

Treatment marshes are a land-extensive technology. Site conditions – primarily property boundaries and topography – can limit the potential size of a treatment wetland for a particular source water volume. This is particularly true of urban stormwater wetlands, which need to be sited in built-up areas, and which often utilize high-value lands. Siting of wetlands inside the boundaries of major cities, like Orlando, Florida (Palmer and Hunt, 1989) or Toronto, Ontario (Helfield and Diamond, 2004), means that the size of the wetland is dictated by existing streets, highways and buildings, not by the size needed to achieve a particular performance goal.

Topography may also limit the potential size of a treatment wetland. The presence of steep slopes adjacent to the site can preclude construction beyond a certain limit, defined by the practicality of earth moving (Figure 16.2). When the site demands, the treatment wetland may be established in terraces, with elevation drops occurring between the successive cells of the system (Inman et al., 2003).

Land ownership can also constrain opportunities for wetland construction. Above and beyond questions of acquisition costs, there is the issue of the willingness of the owner to sell property. Building a treatment wetland rarely falls into the category of eminent domain acquisition, although that has happened in connection with the phosphorus removal wetlands of south Florida. For large wetlands, suitable parcels are often already in agricultural use. Wetlands are frequently viewed as valuable landforms across the regional landscape, but aquatic and terrestrial landforms are also valuable. The construction of a treatment wetland implies

FIGURE 16.2 The treatment wetlands in the Tucush valley of the high Andes mountains of Peru are constrained to a fixed area by extremely steep slopes.(Photo courtesy Margaret McBrien)

the removal of other types of plant, animal, and human communities. Therefore, competing uses may block the construction of a wetland on a particular plot of land.

Perhaps the most serious potential constraint of pre-existing landform is the presence of naturally-occurring wetlands on property under consideration for a treatment wetland. In the United States, it is generally not allowed to build any project that destroys existing wetlands. But what if the project is a constructed wetland? That situation is obviously confusing and unclear, and it is therefore not surprising that a variety of rules and regulations pertain in the various states.

For these area-constrained situations, the design methods described herein are not used to select wetland area, but rather are used to forecast performance of the available wetland area. This predictive mode is readily accommodated in a rate coefficient approach.

16.4 SIZING BY AREA OR VOLUME SPECIFICATION

The first two sizing procedures listed above are technology-based, because they provide prescriptions that imply a general level of treatment, rather than specific numerical concentration or loading goals. They form the basis for sizing of structural BMPs. Prescriptive volume or area specifications are attractively simple, because they do not require any information about the incoming water chemistry. They suffer from an inability to be adjusted for removal targets other than the original presumptions. These methods require knowledge of the runoff volume characteristics of the contributing watershed. Agricultural stormwater systems differ somewhat from urban stormwater systems, because the target may be sediment control or nutrient control, and the target may be load or concentration reduction. Chapter 11 provides more discussion of these procedures.

Associated with event-driven wetlands is the concept of bypassing. Bypassing is necessitated when site constraints prevent the treatment of all imaginable flows. The constraint may occur because of limited land availability, which creates a hydraulic conveyance constraint on how much water can be passed through the system. For instance, the intense rains of subtropical regions can occasionally create runoff flows that simply cannot be routed through a reasonably sized wetland. The second constraint arises due to the availability of maintenance water. If the wetland is sized large to accommodate major events, then the demand for maintenance water during inter-event periods may be too large.

Percentage of the Contributing Watershed

The amount of water and the amounts of pollutants that reach stormwater treatment wetlands from contributing watersheds are not typically well known in advance. The number and duration of the events that produce input, together with the inter-event spacing, are thought to affect the efficiency of the stormwater wetland. Because of this variability of actual flow rates, detention time and hydraulic loading are difficult to define, and other single-number sizing rules have evolved. One such method of presumptive sizing is a specification of the wetland to watershed area ratio (WWAR). This rule of thumb states that the wetland size should be a specified fraction of the contributing watershed, usually in the range of 1.0 to 5.0%. A little arithmetic shows that this produces hydraulic loadings within the range of for continuous flow treatment wetlands. A simplified annualized example illustrates the principles. In a moderate climate:

annual rainfall = 60 cm
average annual rain rate = 60/365 = 0.164 cm/day
watershed runoff coefficient = 0.75
runoff = 0.75 × 0.164 = 0.123 cm/day
watershed/wetland area ratio:
WWAR = 0.025 (2.5% of the watershed)
average annual wetland HLR = 0.123/0.025 = 4.9 cm/day

Since this average annual HLR is close to the median of the distribution of HLRs for FWS wetlands (e.g., 4.0 cm/d for TSS systems and 4.8 cm/d for TP systems) (Kadlec and Wallace, 2009), it is reasonable to expect that stormwater wetlands designed in this way would perform somewhere near the average for the emergent marsh database set. For instance, the mean reduction for phosphorus in 284 FWS wetlands was 38% at a median HLR = 4.8 cm/day; the mean reduction for several constructed stormwater marshes, with an average WWAR = 4.3%, was 57 % (Strecker et al., 1992).

Ratios of 2% for marshes alone, and 1% for pond-marsh combinations, have been advanced (Schueler, 1992), but some authorities require more. Ontario and Alberta target 5%. The Iowa CREP wetlands specify between 0.5 and 2.0%. However, such generalizations are dangerous, because they do not extend to other event driven wetland applications. For example, Braskerud (2001a) found 50–60% TSS removal in four wetlands, over a six-year period, with WWAR between 0.03 and 0.07%.

Often, the WWAR methodology is accompanied by computer modeling of the runoff expected from the watershed. For example, the SWAT (Soil and Water Assessment Tool) (Nietsch et al., 2011) program is applied in connection with the Iowa CREP wetlands, and HBV-NP (Hydrologiska Byråns Vattenbalansavdelning) (Pers, 2007) is employed in Sweden.

Urban Wetland BMP Specifications

Chapter 11 has provided the experience base for design of urban and agricultural event-driven wetlands. Here the design sizing procedures are presented in general terms. However, the reader is cautioned that regional requirements are often extremely detailed. For instance, requirements for wetland BMPs in the Washington, DC area occupy 17 pages (Center for Watershed Protection, 2001), those for the State of Minnesota occupy 23 pages (Minnesota Stormwater Steering Committee, 2018) and those for Melbourne, Australia occupy 33 pages (Melbourne Water, 2017). Some of these individual municipal prescriptions require the use of sophisticated computer modeling tools to establish the runoff from a particular watershed. For instance, the MUSIC (Model for Urban Stormwater

Improvement Conceptualisation) (Melbourne Water (2017) is required for projects in the Melbourne, Australia region.

Urban stormwater wetlands are often sized to contain a specific volume of water, usually the volume associated with a rain event of a specified return frequency, or probability of occurrence. For instance, Schueler (1992) suggests that the wetland have sufficient volume to fully contain any rain event up to the 90th percentile of the rainstorm quantity distribution. Again, this may be shown to match the loading and detention design ranges used for point-source wetland systems.

A simplified annualized example illustrates the principles. In the vicinity of Washington D.C., there are 104 cm annual rainfall, and the 90th percentile storm is 3.18 cm.

watershed area = 40 ha = 400,000 m^2
watershed runoff coefficient = 0.75
design storm runoff volume = 0.75 × 0.0318·400,000 = 9,540 m^3
wetland area @ 0.3 m depth = 9,540/.3 = 31,800 m2 = 3.18 ha
WWAR = 100 × (3.18/40) = 8%
annual flow = 0.75 × 1.04 × 400,000 = 312,000 m^3
average annual detention time = 9,540/312,000 = 0.03 yr = 11 days
average annual wetland HLR = 312,000/31,800 = 9.8 m/yr = 2.7 cm/d

This single-number design technique has the advantage of allowing a variable percentage of watershed, depending upon the annual rainfall pattern and annual rainfall total. As in the case of WWAR design, the loading and detention times correspond to the median values for point-source treatment wetlands. It is therefore not surprising that Schueler (1992) lists pollutant reductions that are in the mid range for other treatment wetlands. For instance, the total phosphorus removal is projected to be 45 %, compared to the 38% mean for FWS systems in general.

The selection of volume or area also corresponds to a projected level of TSS removal. For example, Schueler (1992) suggests that 75% TSS reduction can be reliably achieved by stormwater wetlands that capture the 90th number percentile of rain events. Although confirmation of that assumption remains elusive, the criterion has been adopted in New York (Center for Watershed Protection, 2001). Other governments have adopted similar criteria, such as Wisconsin, which recommends a design rain of 3.8 cm, and suggests that this corresponds to 80% reduction (Prey et al., 1995).

16.5 RATE COEFFICIENT BASED CALCULATIONS

The PkC* model described in Chapter 7 may be used in either the design mode, with wetland area as an exploratory variable, or in the performance forecasting mode, in which the area is known. Many literature sources provide single first-order equations, into which anyone may insert numbers to compute a wetland area. While such an option is very convenient, it is not realistic except for a few highly specialized cases. Accordingly, the procedure given here explores the effect of changing wetland area (or equivalently, detention time) on the forecasted effluent concentrations of contaminants. It is presumed that the designer will conduct the design calculations via a computer spreadsheet, so that design changes may be easily explored. A tank-to-tank calculation procedure is necessitated, as depicted in Figure 16.3.

Wetland hydrology is first determined, as a necessary precursor to area calculations. Annualized calculations are considered first; the effects of season will be considered later in the Chapter.

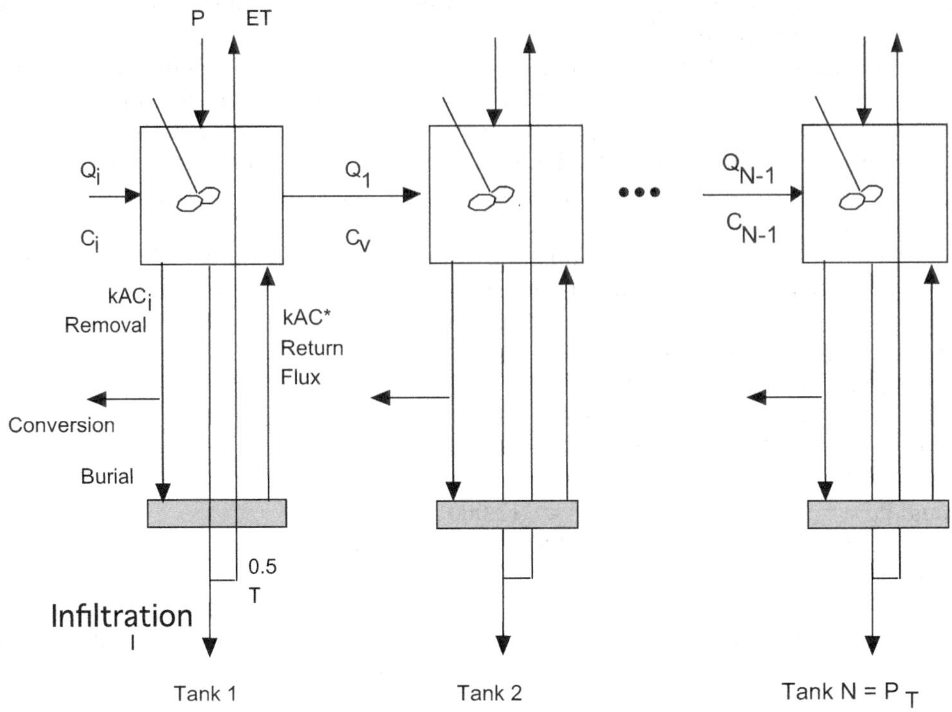

FIGURE 16.3 Conceptual TIS model for pollutant reduction.

Water Budget

The annual water budget forms the basis for a first approach to understanding pollutant reductions and the area required. Under the assumption of a constant water level, the flows out of the wetland may be computed from the inflow and the meteorological data. Components of the water mass balance have been discussed in Chapter 7, Equations 7.74 and 7.75. Some infiltration is assumed here, if leakage is present in the wetland. Groundwater discharge and non-source water inflows are assumed to be absent. The relevant steady-state equations are:

$$Q_{out} = Q_{in} + A \cdot (P - ET - I) \tag{16.6}$$

or

$$q_{out} = q_{in} + (P - ET - I) \tag{16.7}$$

where A = wetland area, m^2
ET = evapotranspiration rate, m/d
I = infiltration rate, m/d
P = precipitation rate, m/d
q = hydraulic loading rate, m/d
Q = flow rate, m^3/d

By virtue of the steady state assumption, changes in wetland water storage are not included. On an annual basis, the inlet hydraulic loading will be increased by rainfall (ca. 0.5–1.5 m/yr), and decreased by evapotranspiration (ca. 0.5–1.5 m/yr). However, there may be seasonal imbalances. These amounts are important if the wetland is to have a very low hydraulic loading, or correspondingly, a long detention time. The ratio a = (P-ET)/q_i is the atmospheric augmentation.

Evapotranspiration (rain) has two effects: lengthening (shortening) of detention time, and concentration (dilution) of dissolved constituents. The use of an average flow rate compensates for altered detention time, but not for dilution or concentration. The fractional error in a first order model prediction of concentration, due to flow averaging, is approximately equal to α, for $\alpha > -0.5$. Thus, if 25% of the inflow evaporates, use of a first order model with average flow predicts concentrations 25% lower than required by the mass balance. If rain adds 25% to the flow, use of a first order model predicts concentrations 25% higher.

A possibly important feature of ET is the transpiration carries water into the root zone (action zone) (see Chapter 4). Therefore, if the pollutant mass balances are done on surface water, then the transpiration component is the same as infiltration: it carries materials into the soil. For a fully vegetated wetland, somewhat more than half of ET is attributable to transpiration. Transpiration is typically about one-half to two-thirds of ET (Kadlec, 2006):

$$T = ET - E = \alpha ET \tag{16.8}$$

where: E = evaporation, m/d
T = transpiration, m/d
α = transpiration fraction of ET, –

The pollutant mass balances will be conducted on a cells-in-series basis (see Figure 17.1). Results of the overall water mass balance are apportioned to the cells according the chosen number of TIS. For the first unit in the series:

$$Q_1 = Q_{in} + A_1(P - ET - I) \tag{16.9}$$

where: A_1 = area of tank number 1, m^2
Q_1 = flow rate out of tank 1, m^3/d

Flows are thus computed sequentially, from inlet to outlet, for the number of tanks chosen (P_T).[1]

The input data requirements for the water mass balances are:

1. Inflow (Q_{in});
2. Rain (P);
3. Evapotranspiration (ET);
4. Infiltration (I);
5. Area (A);
6. Number of tanks (P_T).

An example is detailed in Table 16.1. This hypothetical example is configured to include small rainfall, considerable evapotranspiration, and some infiltration; in other words, a leaky arid region system. The net loss of incoming water is 41%, of which 44% is ET and 56% is infiltration.

Pollutant Mass Balance

Details of the pollutant mass balance are given in Chapter 7, Equations 7.79 and 7.80.

A first order areal model with rate constant k is selected, with a wetland background concentration C* as necessary. A volumetric first order model may also be chosen, for

TABLE 16.1

Example water budget for a treatment marsh. Sequential tank-to-tank calculations are based on Equation 16.9. This example has a combined detention time of 19.2 days, whereas averaging the flows leads to 17.2 days. The required input data is shown in bold.

Rain =	**0.05**	cm/d
ET =	**0.40**	cm/d
Infiltration =	**0.50**	
PTIS =	**3**	
Area =	**24**	ha
Area/Tank =	8	ha
Depth =	0.3	m
Porosity =	0.95	
Volume/Tank =	22800	m3

		Inflow	**Tank 1**	**Tank 2**	**Tank 3 = Outflow**	**Total**
Flow Rate	m3/d	**5000**	4320	3640	2960	
Rain	m3/d		40	40	40	120
ET	m3/d		320	320	320	960
Infiltration	m3/d		400	400	400	1200
Nominal Detention	days	13.7	5.3	6.3	7.7	19.2
HLR	cm/d	2.08	5.40	4.55	3.70	

which $k = \varepsilon h k_V$. The steady state pollutant mass balance for the first of the wetland segments, designated by subscript "1", for steady state, nonuniform flow is:

$$Q_1 C_1 = Q_{in} C_{in} - I \cdot A_1 C_1 - \alpha ETA_1 C_1 - k \cdot A_1 \cdot (C_1 - C*) \qquad (16.10)$$

By virtue of the steady state assumption, changes in wetland pollutant storage are not included. In this simple version, rainfall has been assumed to have a zero pollutant concentration, but it is easy to add an atmospheric input of the pollutant if it exists. Infiltration is assumed to occur at the outlet concentration. Transpiration flow of the contaminant has been included. Combining (16.9) with (16.10) gives the concentration exiting hypothetical segment number one:

$$C_1 = \frac{Q_{in} C_{in} + k \cdot A_1 \cdot C*}{Q_1 + (\alpha ET) \cdot A_1 + I \cdot A_1 + k \cdot A_1} \qquad (16.11)$$

or

$$C_1 = \frac{q_{in} C_{in} + k \cdot C*}{q_1 + (\alpha ET) + I + k} \qquad (16.12)$$

The TIS model is then carried forward via a sequential calculation of pollutant concentrations for each "tank" in the chosen hydraulic model (Figure 16.3). Note that the hydraulic loading rates in (16,12) are the tank loading rates, not the system loadings. This computation is then repeated sequentially for the remaining segments, in each case using the outlet concentrations and flows from the preceding unit. The wetland outlet concentration is that exiting from the final hypothetical segment.

Outgoing pollutant loads are calculated as the product of the volumetric outflow (m^3/d) and the outflow concentration (kg/m^3).

The additional input data requirements for the pollutant mass balances are:

7. Input concentration (C_{in});
8. Background concentration (C*);
9. Rate coefficient (k);
10. Transpiration fraction (α);

The hypothetical example presented for the water budget is extended to illustrate these calculations, continuing the choice of P_T = 3 TIS. An arbitrary pollutant of interest enters the wetland at 2.00 mg/L. The value of C* is chosen to be low, 0.01 mg/L. The rate constant is selected to be k = 10 m/yr. Because the transpiration flux is presumed to draw contaminant into the root zone, the fraction of ET that is transpiration must be selected, and in the example is picked as α = 0.5.

The computed concentration in the surface outflow water is C_O = 0.62 mg/L, or a concentration reduction of 69% (Table 16.2). However, the existence of ET losses and infiltration create a different result for mass removal: 82% of the contaminant entering does not leave with surface water. If infiltration water departs vertically downward without further treatment, then 15% of the contaminant mass removal is due to infiltration. Thus it is seen that load reduction and concentration reduction are two different goals, which may lead to different designs. Both of these types of design sizing require the mass balance computations for water, and rate calculations for pollutants.

Modes of Implementation

Sequential calculations have three goals: concentration forecasting, area determination, and rate coefficient calibration. The first of these is illustrated by the example above, as summarized in Tables 16.1 and 16.2. The other two uses require goal-seeking procedures, which are implemented on spreadsheets.

TABLE 16.2

Example contaminant budget for a treatment marsh. This example is based upon the water budget shown in Table 16.1. Sequential tank-to-tank calculations are based on Equation 16.10. The required input data is shown in bold. The shaded cells represent potential design targets.

Ci =	**2.00**	mg/L
C* =	**0.01**	mg/L
k =	**10**	m/yr
k =	0.0274	m/d
Area/Tank =	80000	m^2
T factor (α) =	**0.5**	

		Inflow	Tank 1	Tank 2	Tank 3 = Outflow	Removed	Reduction
Concentration	mg/L	2.00	1.42	0.96	0.62		69%
Load In Surface Water	kg/yr	3650			666	2984	82%
Load Infiltrated	kg/yr		207	140	90	437	15%
Load Stored	kg/yr					2547	70%

In the design mode, the wetland tanks areas in Equations 16.9 and 16.11 are unknowns, whose sum is the desired wetland area. With all input variables and parameters set, a (guessed) wetland area will produce the corresponding outlet concentration. In general terms, it is a trial-and-error procedure to repeat area guesses until the desired target outlet concentration is achieved. Fortunately, automated routines are available on spreadsheets that manage the trial and error goal seeking. For instance, Microsoft Excel™ contains the Solver routine, which iterates on area until the specified outlet concentration is reached.

In the calibration mode, the value of the rate coefficient, and perhaps other parameters as well, is varied until the wetland effluent concentration matches the observed data. The degree of data match may be selected in a number of ways, the most common being the sum of the squared errors (SSQE) between a time series outlet concentration data set and the prediction result set. The "best" calibration parameter values are those that minimize the SSQE. Again, spreadsheet technology saves us from a manual trial-and-error. For instance, the Microsoft Excel™ Solver routine iterates on the set of parameters until the SSQE is minimized.

Explicit Approximations

Under some conditions, explicit equations may be developed to achieve the three types of calculations described above. The set of conditions that enable this are not often met by real marsh systems, but may be assumed in order to gain a quick estimate of the result. Those conditions are:

A. No net atmospheric water loss or gain, meaning P = ET.
B. No infiltration, meaning I = 0.
C. No transpiration flow, meaning $\alpha = 0$.

The input variable are reduced to just Q and C_i, and the parameters are P (P_T), k and C*. Concentration forecasting is accomplished by the parent form of the PkC* model:

$$C_o = C^* + (C_i - C^*)\left(1 + \frac{k}{Pq}\right)^{-P} \tag{16.13}$$

In the design mode, Equation 16.13 is rearranged to be explicit in area:

$$A = \frac{PQ}{k}\left[\left(\frac{(C_o - C^*)}{(C_i - C^*)}\right)^{-\frac{1}{P}} - 1\right] \tag{16.14}$$

In the calibration mode, Equation 16.13 is rearranged to be explicit in rate coefficient:

$$k = Pq\left[\left(\frac{(C_o - C^*)}{(C_i - C^*)}\right)^{-\frac{1}{P}} - 1\right] \tag{16.15}$$

where: A = wetland area, m^2
C* = background concentration, g/m^3
C_i = background concentration, g/m^3
C_o = background concentration, g/m^3
k = rate coefficient, m/d
P = tanks in series parameter, –
Q = flow rate, m^3/d
q = hydraulic loading rate, m/d

Transect Representation

A growing interest in the internal concentration profiles in treatment marshes has resulted in the need to describe concentrations along the flow direction. In this situation, area calculations are not needed, but there remains a need to calculate concentrations as function of distance, and to determine rate parameters that best fit concentration profile data. Internal hydraulics (mixing) become more complicated, because results differ depending upon whether the internal flows occur as noninteracting parallel stream "tubes" (segregated flow), or as cross-mixed stream tubes (mixed flow), or some combination. Some treatment wetland literature presumes segregated flow (for example, see Kadlec, 2000; Carleton, 2002). Subsequent analyses have explored intermediate situations (e.g., Carleton and Montas, 2007, 2010). There are some indications that marshes are better represented by the mixed flow scenario (see Chapter 7). In the simplified case, concentrations along a flow-direction transect are:

$$C(y) = C^* + (C_i - C^*)\left(1 + \frac{k}{Nq}\right)^{-yN} \tag{16.16}$$

where y = fractional distance along the flow path, –

Goal-seeking spreadsheets may be used for calibrations. Equation 16.16 is used to predict the concentrations at discrete points along the flow direction, which are then used to form a SSQE for spatial, and perhaps also temporal, set of measured concentrations.

When the simplifying assumptions are not met (e.g., water losses/gains are not negligible), a more complicated situation occurs. Equation 16.11 allows calculation of the concentrations exiting each tank, which establishes predicted values for internal points along the gradient at a number of locations equal to P (P_T) for the mixed flow case. But those may not correspond to the internal points selected for measurement, and thus interpolation of either measurements or predictions must be made to establish paired values at internal points.

Time Period for Mass Balances

Water and pollutant mass balances are applied for a time period that is at the discretion of the analyst. Common choices are:

- Annual. For continuous sources, flows and concentrations are averaged for a year. For event-driven systems, event mean flows and concentrations are averaged for the year, providing the site mean flows and concentrations for the year (see Chapter 7). For seasonal or discretionary flow systems, the in-use flows and concentrations are used.
- Seasonal. Marshes may behave differently in warm and cold seasons, during growth and senescence periods, and during wet and dry seasons. Typically, these are partitioned into either three- or six-month time periods, although cold climates may cause partitioning into other lengths.
- Monthly. This time period is likely to be the smallest that can be used and still have a reasonable number of measurements to average. It is small enough to accommodate the annual temperature cycle that regulates some marsh processes.
- Period of Record. This time period encompasses several years, either of operating conditions for the calibration mode, or for presumed flow and concentration scenarios for the forecasting or design modes. Data and parameters are averaged by folding, and represented by any of the preceding calculation periods.

At these time scales, changes in water and pollutant storage (inventory changes) are not significant. As the time scale is decreased to weekly, daily or hourly, that assumption is no longer valid, and calculations enter the realm of dynamic modeling.

A second constraint on the time period is caused by the possibility of synoptic error. Transport delay occurs in the wetland, meaning that water that exits the system entered at an earlier time, with the nominal difference of one detention time. However, the existence of detention time distributions means that transport delay may be of the order of three detention times for a significant fraction of the departing water. Avoiding a transport delay artifact (synoptic error) requires averaging over more than those three nominal detention times. Since treatment marsh detention times are often of the order of a week, time periods for mass balances are constrained to be monthly or longer.

16.6 INPUT VARIABLES AND PARAMETERS

In common with most forecast models, the rate-based calculation procedure requires input information of three basic types: inflows and targets, hydrologic factors, and removal model parameters.

Flows and Targets

Source Water Quantity (Q_i)

The amount and timing of the water to be treated is the first and foremost item of the design basis. This information should include the possible seasonality of flows, and the anticipated progression of flows over the life of the design. This is important for treatment marsh design because of the perceived longevity of such systems. Wetlands clearly can continue to function for long periods and have been demonstrated to function for several decades. For example, the Orlando Easterly marsh has been in operation for over thirty years, and The Houghton Lake, MI marsh has been operated for over forty years, both with continuing success. Projecting flow estimates far into the future is risky, so it is necessary to be explicit about flow capacity at the time of design.

In the design forecasting procedure, source flows are established by choice of the pumps or other inflow works that are selected for most continuous sources and some runoff and river sources. Runoff flow estimation methods are presented in Chapter 11. These range from approximate techniques involving rain and runoff coefficients, to complex computer methods that take into account watershed land uses as well as rain and infiltration.

Seepage (I)

Treatment marshes generally do not require either with clay or synthetic liners, because of the relatively benign character of the water. Any water that leaks vertically downward, or horizontally outward under and through containment berms, represents a water loss that slows flow and contributes to pollutant loss from the wetland water body. Infiltrated water moves into and with the regional groundwater, and may reemerge at nearby locations as groundwater discharge. During passage through the soils under and around the wetland, these seepage waters may receive some additional treatment, by microbial degradation or adsorption for example. However, not all potential contaminants are subject to soil and aquifer treatment.

The design procedures described herein pertain to the wetland water body and the planned outflows that occur as discharges from structures. This is not to say that infiltration flows can be ignored; rather, the projects with significant seepage are to be viewed as dual-discharge projects. The seepage should meet groundwater requirements, and the structure outflows should meet the appropriate goals.

Examples of systems in which seepage is a nontrivial factor include the stormwater treatment areas of south Florida (Nungesser and Chimney, 2006), which lose and/or gain up to about 20% of the flow from groundwater. The New River wetlands in southern California infiltrated 47% of the inflow at the Brawley site, and 39% at the Imperial site, over a four-year period of operation (Kadlec et al., 2010a). And, the Kanapaha system near Gainesville, FL, is designed to infiltrate all source water (Wetland Solutions, Inc., 2012).

Three of the four wetlands at the Tres Rios, Arizona, site were built on agricultural soils, and had minimal infiltration (12–14%), but the fourth was built on very coarse river gravel bed (cobble) materials, and infiltrated 74% (Wass, Gerke and Associates, 2001). The majority of the water leaked vertically downward. Of interest in this project are the trends in the infiltration rates over the first few years of the project history. The three agricultural soil sites maintained a relatively constant infiltration rate, but the cobble site lost 40% of its infiltration (25 cm/d down to 14 cm/d) over a three-year period. Based on these limited results, only minimal self-sealing should be anticipated in treatment wetlands.

Source Water Quality (C_i)

The concentrations of the pollutants in the water to be treated are critical to the sizing process, and to the prediction of the wetland performance in the face of unknown future variations. A clear definition of the incoming water quality is essential, including the anticipated temporal distribution of concentrations. There are often seasonal fluctuations for point sources, as well as diurnal fluctuations. Incoming patterns of chemical composition propagate through the wetland and undergo modification, resulting in a spectrum of output compositions. Some of this output variability may be predicted by the design models, namely those variations which represent responses to moderately slow input changes – those which occur on monthly or greater time scales. Faster events involve ecosystem processes that are not included in the design models available at the present time, and therefore will give the appearance of generating stochastic variations.

Municipal wastewater treatment plants remove at least 70% of the biochemical oxygen demand and total suspended solids during primary and secondary treatment. Treatment requirements have generally increased over the past decades, and many treatment plants now include at least partial nitrification, perhaps denitrification, and phosphorus removal. These blur the terminology, because they range from "advanced secondary" to "tertiary" and beyond. The follow-on treatment marsh may therefore be termed "quaternary". For design purposes, it is generally the case that the specific source water quality records exist. If not, WWTP water quality is often defined by the selected treatment components.

Surface water sources, rivers and lakes, will usually have water quality records on which to base design calculations.

Concentrations of most parameters in urban stormwater are time-dependent. Stormwater concentrations and loads are cyclic due to periods of dry fall and deposition, followed by the first flush of runoff after rain, followed by exponential decreases in runoff constituent concentrations as storages rinse from the landscape, and finally dry conditions and deposition until the next storm event. Pollutant concentrations generally range from low levels from undeveloped and park lands, to low density residential and commercial, to agricultural, to higher density residential and commercial, and reach highest levels for high density commercial, and industrial land uses. Urban pollutant loads increase with the imperviousness of the watershed. Chapter 11 provides examples and summaries of the water quality reaching urban systems of the expected flows and concentrations for such event-driven systems (see Tables 11.3, 11.6, and 11.8).

Agricultural runoff concentration data are available, primarily for TSS, phosphorus and nitrogen forms (see Tables 11.9 and 11.4). Individual projects may have water quality records on which to base design calculations.

Target Outlet Water Quality (C_o)

Regulatory concentration requirements may exist for municipal water sources and occasionally for stormwater systems, but in most instances, treatment marshes will not have numerical concentration requirements. Where concentration (or load) requirements exist, design mode calculations are used. For the other situations, calculations are in the modes of performance prediction and calibration.

Climatological and Hydrologic Factors

Wetland hydrology is affected not only by the volume of applied wastewater but also by the net balance between rainfall and evapotranspiration. Wetland designers in areas with high net precipitation must take this incremental added water volume into account when deciding on wetland volume, flow control structures, and free board of berms. In areas of high evapotranspiration, wetland design calculations need to consider water loss to the atmosphere. Water temperature regulates the rate of several wetland processes, and it is dependent upon several meteorological variables as detailed in Chapter 6, including air temperature, humidity, and solar radiation.

Climate

Climate cannot be controlled when selecting a specific site for a constructed or natural wetland project. However, climate is important during project planning and alternative selection because it affects the type and size of wetland that will be used. Latitude is the most critical determinant of climate because it determines seasonal temperature ranges. Other climatic factors that are important during project planning include rainfall, insolation, wind direction and velocities, and evapotranspiration. In the United States, climatic information can be obtained from the National Oceanic and Atmospheric Administration (NOAA) at the following web address:

National Centers for Environmental Information (NCEI)
National Oceanic and Atmospheric Administration (NOAA)
Asheville, North Carolina
www.ncdc.noaa.gov/data-access/land-based-station-data/land-based-datasets/climate-normals/1981-2010-normals-data

Of primary interest for contaminant rate calculations are rainfall and evapotranspiration. Rain data are available in climatological data records, but not ET data.

Another source of information is The Weather Underground, Inc., a commercial weather service providing weather information via the Internet. Their stated goal is "to make quality weather information available to every person on this planet." Weather Underground provides weather records for most major cities across the world on its website: www.wunderground.com/history/. Its information comes from the National Weather Service (NWS), and over 250,000 personal weather stations. Many other countries provide similar services. For instance, Canadian climate data is available at: http://climate.weather.gc.ca/climate_normals/index_e.html.

Rainfall (P)

Figure 2.9 provides a map of the annual average precipitation for the United States. More detailed information concerning the average monthly net precipitation values for a specific location should be used to estimate the typical monthly water balance for a continuous flow wetland under extreme seasonal conditions. Historical values of monthly mean precipitation are available in the United States at the NCEI website given earlier. As discussed in Chapter 11, rainfall is the primary driving force for stormwater wetlands, and yet more detailed data is therefore desirable. Site specific information on the intensity and duration of rains, and the interevent spacing, is needed to perform design calculations.

Example precipitation data for a number of U.S. locations is shown in Table 16.3. In cold climates, winter precipitation is in the form of snow. Note that arid regions have an annual rainfall gain of 74 mm/yr, while wet, coastal regions have 2,235 mm/yr.

Rain affects constructed marsh hydrology in two ways, direct and indirect. Rain impinging on the wetland water surface is a direct input. In addition, the levees and buffers involved in the project intercept rainfall, and provide runoff into the marsh. Runoff from this "mini-catchment" follows the same quantification rules as for urban source waters; most simply implemented via the size of the catchment and a runoff coefficient.

$$Q_R = PA + PA_C C_R \tag{16.17}$$

where: A = wetted area, m2
A_C = catchment area, m2
C_R = runoff coefficient, –
P = rain rate, m/d
QR = rain input, m3/d

TABLE 16.3

Quarterly precipitation data for example U.S. locations.

Zone	Winter D,J,F mm/mo	Spring M,A,M mm/mo	Summer J,J,A mm/mo	Autumn S,O,N mm/mo	Total mm/yr
Imperial, CA	12	3	4	6	74
Phoenix, AZ	23	18	12	16	204
Ulysses, KS	12	66	39	30	438
Bemidji, MN	18	98	50	59	676
Houghton Lake, MI	40	83	62	117	781
Asheville, NC	75	86	82	70	940
Wellsville, NY	56	104	77	83	963
Belvidere, IL	43	122	80	79	973
Corsicana, TX	80	67	98	91	1010
Bangor, ME	75	86	90	103	1065
Columbia, MO	61	120	111	100	1176
Everglades, FL	44	243	66	117	1407
Columbus, MS	136	111	117	106	1412
Tillamook, OR	309	53	182	201	2235

Data from: 1981–2010 Normals, NCEI, www.ncdc.noaa.gov/cdo-web/datatools/normals

Calculations should include this amplification of the direct rain input. A_C is small relative to A for large wetlands but increases in importance as size is decreased.

Evapotranspiration (ET)

The mechanisms by which a wetland loses water to the atmosphere have been detailed in Chapter 6. Methods of estimation and data sources have also been detailed there, and include lake evaporation, pan evaporation, and energy balance methods. Energy balance values may be available from weather services, especially for agricultural regions.

As an example of the kind of information that wetland designers have available, Figure 16.4 shows the various climate zones of the State of California. Table 16.4 lists the monthly ET for each zone. Note that arid regions have an annual ET loss of 1,819 mm/yr, while wet, coastal regions have only 838 mm/yr. The peak ET rates are in mid-summer, and range up to 9 mm/d for the warm arid zones. For states that do not have irrigation information services, the designer can obtain pan evaporation data for locations in the United States from the National Climatic Data Center (NCDC).

The transpiration component of ET moves downward into the root zone of a marsh, and thus represents a withdrawal of water and pollutants from the surface water body. In

FIGURE 16.4 Reference evapotranspiration for the state of California. See Table 16.4 for the annual pattern for each zone. Adapted from: California Irrigation Management Information System www.cimis.water.ca.gov/cimis/info.jsp.

TABLE 16.4

Quarterly reference evapotranspiration by zone for the State of California. See Figure 16.4 for the locations of the zones.

Zone	Winter D,J,F mm/mo	Spring M,A,M mm/mo	Summer J,J,A mm/mo	Autumn S,O,N mm/mo	Total mm/yr
1	25	83	111	59	836
2	35	99	125	72	991
3	50	117	140	85	1176
4	50	112	146	87	1184
5	30	107	158	77	1115
6	50	117	161	92	1262
7	23	99	171	74	1101
8	33	122	176	88	1255
9	58	127	177	105	1401
10	30	114	190	82	1248
11	45	114	192	97	1347
12	35	130	195	92	1356
13	35	122	208	95	1379
14	45	132	205	100	1449
15	40	143	208	100	1472
16	47	173	223	110	1664
17	55	158	234	116	1688
18	68	176	236	126	1819

Adapted from: California Irrigation Management Information System www.cimis.water.ca.gov/cimis/info.jsp.

that regard, the transpiration flux resembles infiltration, and is to be treated as such in marsh design. The partitioning of ET is via a fraction α:

$$ET_W = F_V(E_V + T) + (1 - F_V)E \tag{16.18}$$

$$ET_W = T + E = \alpha ET_W + (1 - \alpha)ET_W \tag{16.19}$$

where ET_W = entire wetland evapotranspiration, mm/d
E = evaporation from open water areas, mm/d
E_V = evaporation from vegetated areas, mm/d
F_V = fraction vegetated, –
T = transpiration, mm/d
α = fraction of ET_W that is T, –

ET affects both the flow rate (detention time) and the concentrations of pollutants (evaporative concentration). Therefore, it is inaccurate to merely average the inflows and outflows to account for effects, because that misses the evaporative concentration effect. The flow is slowed, and pollutants become more concentrated. The reverse is true for rainfall: velocities are increased, and pollutants are diluted. The mass balances used herein for design correctly account for both velocity and dilution effects.

Extreme examples of systems in which ET is a dominant factor include lightly loaded wetlands in arid climates. The Incline Village, Nevada system evaporates nearly all the incoming water, and

ET was the key factor in project design (Kadlec et al., 1990). The Saginaw, Michigan project operates only in the summer, during which there is an excess of ET over rainfall even for the moderately wet climate (Kadlec and Zmarthie, 2010). On the order of half the water entering the Roblin, Manitoba FWS wetland system is, by design, lost to net evapotranspiration (PFRA, 2003).

Rate Parameters

Tanks-in-series modeling calculations require three principal parameters: the apparent number of TIS (P_T), the background concentration (C*) and the loss rate coefficient (k). The P_T parameter is determined by internal marsh hydraulics, and the other two are pollutant specific.

Apparent Number of Tanks (P_T)

The performance of the wetland depends upon the number of tanks (TIS) selected, and very strongly if the design is to approach wetland background concentrations or high degrees of removal (see Chapter 7). If a high degree of removal is required, that will necessitate a very large wetland with poor hydraulics, or a smaller wetland with good hydraulics. Figure 16.5 illustrates this high degree of sensitivity in the region of large reductions. This chart quantifies the fact that a tiny fraction of the water following a fast short-circuit will carry enough unreacted material to the outlet to make 99+% reduction almost impossible to achieve. Low P_T numbers represent flow patterns that, by virtue of either fast forward mixing or velocity profiles with high-speed elements, carry fractions of unreacted material directly to the outlet. However, in the region of low removals, there is a much lesser effect of low P_T numbers. Furthermore, once P_T reaches about 6, further increases have no significant effect on performance.

The P_T-value is somewhat at the discretion of the designer. More cells, and greater length to width ratios can increase the P_T value. As an illustration of the design decision to be made, consider a hypothetical case relative to Figure 16.5. Suppose the wetland is to achieve 90% reduction. It is possible to consider a one-cell wetland, with a presumptive P = 3, which needs 84 $m^2/(m^3/d)$. Or, the designer can opt for two cells in series, each with a presumptive P = 3, which needs 68 $m^2/(m^3/d)$. The question might be resolved based on the

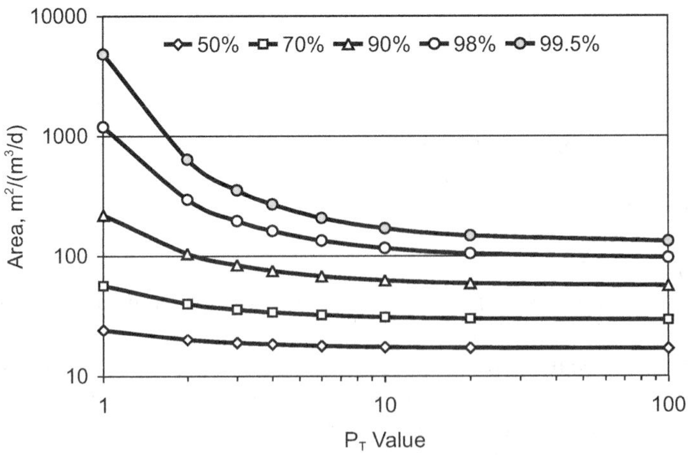

FIGURE 16.5 The effect of the number of tanks on the area required for different degrees of removal. The calculations are for k = 15 m/yr and C* = 0.

economics, that is, does the cost savings of 20% area reduction outweigh the added cost of the divider berm and structures? This illustration carried a zero background concentration, but the concepts are applicable to any pollutant provided the reduction fraction to background is used instead of percent removal.

It should be noted that the designer can control the N-value for the wetland (inert tracer tanks in series), but cannot entirely control the P-value. Pollutants that are mixtures, which may undergo weathering in the wetland, act to reduce the applicable P-value. Table 7.5 provides some guidance on the apparent P-values to be selected, relative to tracer N-values.

In the performance forecasting mode for existing systems, there may be tracer tests that determine the N-values for the marsh.

In the calibration mode, it is possible to include the P_T value as a calibration parameter.

The Limit of Large P_T

When the presumed or measured number of tanks exceeds the nominal value of six, the plug flow ($P_T = \infty$) approximation is often acceptable. The implication is that longitudinal dispersion is not an important factor. The dimensionless parameter that characterizes longitudinal mixing is the Peclet number, or its inverse, the wetland dispersion number:

$$\mathcal{D} = \frac{D}{uL} = \frac{1}{Pe} \quad (16.20)$$

where D = wetland dispersion number, dimensionless
u = velocity, m/d
D = dispersion constant, m2/d
L = distance from inlet to outlet, m
Pe = Peclet number, dimensionless
The number of tanks N is related to dispersion in the following (Kadlec and Wallace, 2009):

$$\frac{1}{N} = 2\mathcal{D} - 2\mathcal{D}^2(1 - e^{-1/\mathcal{D}}) \quad (16.21)$$

FWS wetland systems are characterized by a large amount of apparent dispersion, with $0.07 \leq D \leq 0.35$ (Kadlec, 1994). Therefore, the plug flow assumption is only applicable to the low end of the dispersion range, unless compartmentalization and/or high length to width ratios are employed.

If conditions warrant, the model then assumes a plug flow kC* form. The development of that model is given by Kadlec and Knight (1996), and for a one dimensional system results in the following:

$$C = C_a + (C_i - C_a)\left(\frac{q_i + ay}{q_i}\right)^{-(1 + (k+I)a)} \quad (16.22)$$

where

$$a = P - ET - I \quad (16.23)$$

and

$$C_a = \frac{[PC_p + kC^*]}{(P - ET + k)} \quad (16.24)$$

and

$$q_i = \frac{Q_i}{A_T} \tag{16.25}$$

and

$$y = \frac{x}{L} \tag{16.26}$$

This formulation also includes atmospheric deposition of the contaminant, with a rainfall-based concentration C_P. The long-distance plateau in concentration is not C*, but rather C_a, which reflects hydrology and atmospheric deposition as well as return flux, represented by C*. Although concentration profiles are a function of flow ratio in Equation 16.20, the internal profile decreases are well represented as exponential, with $R^2 > 0.99$.

Background Concentrations (C*)

The biomass cycles in marshes result in an internal release of particulate and dissolved biomass to the water column, which is measured as nonzero levels of BOD, TSS, TN, and TP. Additionally, wetland wildlife contribute background levels of pathogens, resulting in nonzero background for indicators such as fecal coliforms. These wetland background concentrations are here denoted by the term C*.

Consequently, many pollutants are not reduced to zero in treatment wetlands, including BOD, TSS, organic nitrogen, phosphorus and pathogenic indicators. However, it is important to distinguish between artifacts of data fitting and real wetland processes. Short-circuiting can lead to apparent high values of data-fit C* for some systems, but those high apparent C* can be dealt with by improving the hydraulics. Independent of hydraulics, the wetland can manufacture water phase organics and solids, and cycle nutrients into and out of the water body. Some individual exotic chemicals are foreign to typical wetland environments, and are not expected to exhibit background concentrations. Examples include halogenated hydrocarbons and pesticides.

Previous Chapters contain estimates of the C* values for the common pollutants. A summary is given in Table 16.5.

TABLE 16.5

Summary of approximate background concentrations for treatment marshes.

Contaminant	Units	C*
BOD_5	mg/L	1–10
TSS	mg/L	2–10
Organic N	mg/L	0.5–2.0
Ammonia N	mg/L	<0.05
Oxidized N	mg/L	<0.05
Total Phosphorus	μg/L	10–40
Fecal Coliforms	cfu/100ml	10–50
Trace Metals	μg/L	0
Trace Organics	μg/L	0
Pesticides	μg/L	0

Loss Rate Coefficients (k)

In an annualized analysis, the appropriate rate coefficients are those produced in earlier Chapters as the result of the fitting of annual data from existing wetlands. In general, values of C* and PT are used in the estimates. For some constituents, loss rate coefficients are variable across wetlands and years, thus producing frequency distributions of the fitted k-values. For others, data is more limited, and individual case studies form the database. Annual rate coefficient values are to be found in the text of preceding Chapters and Tables therein, as follows:

- TSS Chapter 8, Table 8.5
- Organic N Chapter 9, Table 9.11
- Ammonia N Chapter 9, Table 9.13
- Oxidized N Chapter 9, Table 9.16
- Total P Chapter 10, Table 10.10 (Bioaccretion systems)
- Total P Chapter 10, Table 10.11 (Runoff systems)
- BOD_5 Chapter 12, Table 12.7
- Trace Metals Chapter 13, Figure 13.2
- Pesticides Chapter 14, Table 14.7
- Pharmaceuticals Chapter 14, Table 14.13
- Trace Organics Chapter 14

Treatment marshes typically operate at near-background conditions for some contaminants, importantly including TSS, BOD_5 and pathogens. Indeed, this feature is part of the defining conditions for treatment marshes: secondary treatment of municipal effluents is not in their scope. For these substances, the PkC* model is of limited utility, and presumptive criteria are often applied.

System data for TSS system data are contained in Chapter 8, and analyzed for rate coefficients. However, incoming TSS is often reduced rather quickly (see Figure 8.9), and wetland effluent TSS results from a balance of generation and resuspension in the marsh. The removal rate coefficients that pertain to the inlet region of the wetland are often quite high, ≈ 10 m/d (3,650 m/yr). Colloidal materials are an exception to this generality. The design recommendation suggested here is the use of a high rate coefficient for TSS (perhaps 200 m/yr), unless the design-limiting bottleneck happens to be TSS. In that event, it is strongly suggested that settling tests be conducted on the candidate wetland influent waters.

Data analyses for many trace contaminants assume plug flow conditions in the test system, or the tests are for batch operation, also described by an exponential model. The results are reported as percent reduction for a given hydraulic loading rate, or as half-lives or plug flow rate coefficients. It is then necessary to disentangle the rate coefficients from the presumed (and possibly incorrect) hydraulic assumptions.

The appropriate tools are Equation 16.15 for the PkC* values and for the plug flow k-values:

$$k = q \ln\left[\frac{C_i - C^*}{C_o - C^*}\right] \tag{16.27}$$

If the test wetland platform has hydraulic inefficiency represented by P_T tanks in series, then the plug flow analysis will return an erroneously low k-value (Table 16.6). The amount of the correction is large at low P_T and high pollutant reduction (e.g., P_T =2, C_o/C_i = 0.1), but is negligible at moderate P_T and low reduction (e.g., P_T =6, C_o/C_i = 0.75).

TABLE 16.6

Corrected k-values (m/yr) for actual tanks in series for a test platform. The values of C* = 0 and q = 5.5 cm/d are used.

	C_o/C_i			
	0.10	**0.25**	**0.50**	**0.75**
	k_{PF}			
P_T Test	46.1	27.7	13.9	5.8
2	86.5	40.0	16.6	6.2
3	69.3	35.2	15.6	6.0
4	62.3	33.1	15.1	6.0
6	56.1	31.2	14.7	5.9
8	53.4	30.3	14.5	5.9
10	51.8	29.7	14.4	5.8
50	47.1	28.1	14.0	5.8

When using a specific study for estimating rate coefficients for trace contaminants, the designer must assess the likely degree of mixing in the test platform. If it is a batch system, no correction is needed. If it is a flow-through system, whether mesocosm, pilot or field scale, it is unlikely that the plug flow assumption is warranted. It is also possible to use the plug flow result, with the understanding that it is likely "conservative," meaning that the true rate coefficient is likely larger. This was the philosophy employed by Kadlec and Knight (1996), who conducted and reported plug flow analyses in many cases.

Annualized Rate Coefficients

It has been suggested the design techniques from continuous flow wetland performance may be transferred to event-driven systems by using rate coefficients to forecast average performance (Wong and Geiger, 1997; Carleton et al., 2001). However, the rate coefficients are not necessarily the same as those found for continuous flow systems. For instance, Kadlec (2001b) reported k = 100 m/yr for phosphorus, and Werner and Kadlec (2000) measured 125 m/yr for nitrate, both based on pumped event dynamics.

The designs of the first few constructed stormwater treatment areas (STA wetlands) of south Florida, which target phosphorus removal, were essentially based on annualized rate coefficients. The prototype was an event-driven natural wetland, which produced k = 10 m/yr for $C^* = 2$ μg/L and $P = \infty$ (Walker, 1995). Subsequent analyses have shown that the constructed systems have annualized k-values of the same general magnitude. However, the dynamic k-values are higher (Walker and Kadlec, 2005). Further, the k-values are sensitive to the characteristics of the event sequences. It is likely that more accuracy can be gained only by dynamic simulations.

Water Temperature

There are two reasons for needing the capability to calculate wetland water temperature: they are required as strong modifiers of rate coefficients, and they can be a design target. The calculation procedures have been presented in detail in Chapter 6. Their usage is discussed here.

A rough approximation for long detention wetlands is the mean daily air temperature during the unfrozen season. The next level of approximation involves estimating longitudinal profiles of wetland water temperature via the use of accommodation coefficients. The water temperature may depend on wetland size, due to lack of time to reach the balance condition. In that case, design iteration may become necessary. The prediction of wetland water temperatures is briefly summarized here.

Balance Temperatures

A treatment marsh may contain one or two thermal regions, depending on water loading (detention time). For long detention times, there is an inlet region in which water temperatures adjust to the prevailing meteorological conditions, and an outlet region in which that adjustment is complete. The value reached is determined by the balance of energy flows, and is termed the balance temperature. For short detention times, near the wetland inlet, the adjustment may not be completed. Forecasts are made utilizing the energy balance, as described in Chapter 6.

Annual Temperature Cycles

The annual cycle of wetland water temperatures in mild to warm climates follows a sinusoidal pattern, with a summer maximum and a winter minimum. In northern climates, the onset of frozen conditions typically is accompanied by under-ice water temperatures of 1–2 °C. The sinusoidal model, truncated for frozen conditions, is:

For the unfrozen season ($t_1 < t < t_2$):

$$T = T_{avg}(1 + A \cdot \cos[\omega(t - t_{\max})]) \tag{16.28}$$

For the frozen season ($t_2 < t < t_1$):

$$T = T_o \tag{16.29}$$

where: A = fractional amplitude of the sinusoid, unitless
ω = yearly cycle frequency = $2\pi/365$ = 0.0172 d-1
t = time, Julian day
t_1 = ice-out time, Julian day
t_2 = freeze up time, Julian day
t_{max} = time of annual maximum temperature, Julian day
T = water temperature, °C
T_{avg} = annual average water temperature, °C
T_o = under-ice water temperature, °C

Table 6.5 lists the parameters for this sinusoidal model for a variety of FWS wetlands at different latitudes. The deterministic trend represents the central tendency of water temperatures, but there are also stochastic variations. For instance, the standard deviation of the monthly means for Columbia, Missouri is 1.6°C.

The Accommodation Zone

The inlet zone of a treatment wetland exhibits temperature changes, as the water approaches the balance temperature. For short detention times (typically less than three days for FWS), the adjustment may not be completed, and the balance temperature is not reached (Kadlec, 2006).

An empirical exponential model may be easily calibrated and used to describe the approach to the balance temperature:

$$T = T_b + (T_i - T_b)\exp\left(-\frac{\eta t}{\rho c h}\right) = T_b + (T_i - T_b)\exp\left(-\frac{t}{\tau_A}\right) \tag{16.30}$$

where c_p = heat capacity of water, 4.182 x 106 J/kg·°C
h = water depth, m
T = wetland water temperature, °C
T_b = wetland balance temperature, °C
T_i = inlet water temperature, °C
η = accommodation coefficient, MJ/m2·d·°C
ρ = heat capacity of water, MJ/kg·°C
t = nominal detention time to an internal point, d
τ_A = nominal detention time to an internal point, d

The quantity $\tau_A = \rho c_p h/\eta$ represents characteristic accommodation time for the wetland water on its travel through the system, during which 63.2% of the change from inlet to balance temperature has been achieved. At $3\tau_A$, 95.0% of the change has been accomplished. Data for free water surface wetlands indicates that adaptation takes on the order of 1–3 days detention (see Table 6.9).

Equation 16.30 is the plug flow version of the kC* model, with the "background" temperature equal to the balance temperature, and the volumetric rate coefficient $k_V = 1/\tau_A$.

Temperature Coefficients and Seasonal Dependence for k-Values

Temperature effects on k are usually summarized by use of the modified Arrhenius equation:

$$k_T = k_{20}\,\theta^{(T-20)} \tag{16.31}$$

where k_T is the rate constant at temperature T = T°C and k_{20} is the rate constant at 20° C. Values of the temperature correction factor (θ) have been estimated for data sets with adequate operational temperature data. It should be recognized that annual periods require the average k-values for the monthly temperatures, and not the k-value at the average annual temperature. The difference between these can be considerable, especially for cold climate marshes operated year-round (Table 16.7).

The preceding Chapters have elucidated that most loss rate coefficients in treatment marshes are not temperature modulated, but rather display seasonal dependence unrelated to temperature. The underlying marsh feature is that performance is in the agronomic range, in which the growth patterns of wetland vegetation controls rates.

At the present time, there generally do not exist the biomass cycle balances that determine seasonality. Therefore, the designer can only estimate seasonality based on site-specific examples, as given in preceding Chapters, and in summaries like Figures 16.6 and 16.7.

Theta factors, and corresponding k_{20} values are applicable for oxidized nitrogen, and consequently for total nitrogen. Coefficients may be found as follows:

- Oxidized N Chapter 9, Figure 9.27
- Total N Chapter 9, Table 9.20

These rely on source characterization, just as for continuous flow wetlands, as discussed in Chapters 14 and 16.

16.7 INTEGRATED IMPLEMENTATION

Sizing procedures may be divided into two groups, those for trace contaminants and those for nutrients. Trace contaminants do not affect the biological components of the wetland ecosystem, while nutrients strongly affect them. Thus, for trace materials, the marsh can be considered as a reaction device, whose character is not a function of the presence of trace materials. The use of single stand-alone removal equations is normal procedure.

TABLE 16.7

Example rate coefficients as a function of temperature over an annual cycle. Temperature dependence is k_{20} = 25 m/yr, and θ = 1.090.

	Brighton		Imperial		Musselwhite	
	T	k	T	k	T	k
	°C	m/yr	°C	m/yr	°C	m/yr
Jan	1.8	5.2	12.5	13.1		
Feb	0.9	4.8	14.2	15.2		
Mar	2.5	5.5	18.5	22.0		
Apr	7.5	8.5	20.8	26.8		
May	13.6	14.4	23.9	34.9	8.3	9.1
Jun	19.9	24.7	26.6	44.1	14.8	16.0
Jul	21.6	28.7	29.3	55.7	18.0	21.0
Aug	21.2	27.7	29.5	56.7	16.8	19.0
Sep	17.2	19.7	26.8	44.9	11.7	12.2
Oct	10.5	11.0	22.2	30.3	4.0	6.3
Nov	5.9	7.4	16.3	18.2		
Dec	2.3	5.4	12.0	12.6		
Average	10.4	13.6	21.0	31.2	12.3	13.9
k@Average T		10.9		27.4		12.8
Difference		-24%		-14%		-9%

Describing nutrient removals requires consideration of both reaction rates and the biological aspects of the marsh, which come together in the mass balances for nutrients in the system. This is particularly true for treatment marshes, which by definition are agronomic systems, and not microbial wastewater treatment devices. The use of single stand-alone

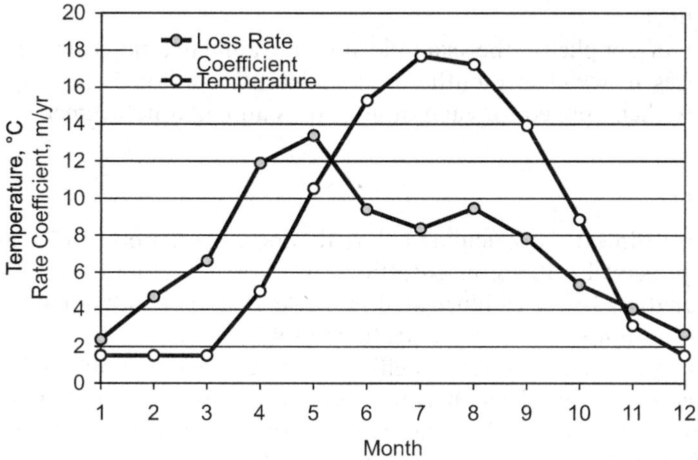

FIGURE 16.6 Apparent first order rate coefficients for ammonia removal, and temperature, in FWS treatment systems in cold climates. Model parameters are P = 3 and C* = 0. Data from eleven wetlands. The mean nitrogen loading was 130 $gN/m^2 \cdot yr$.

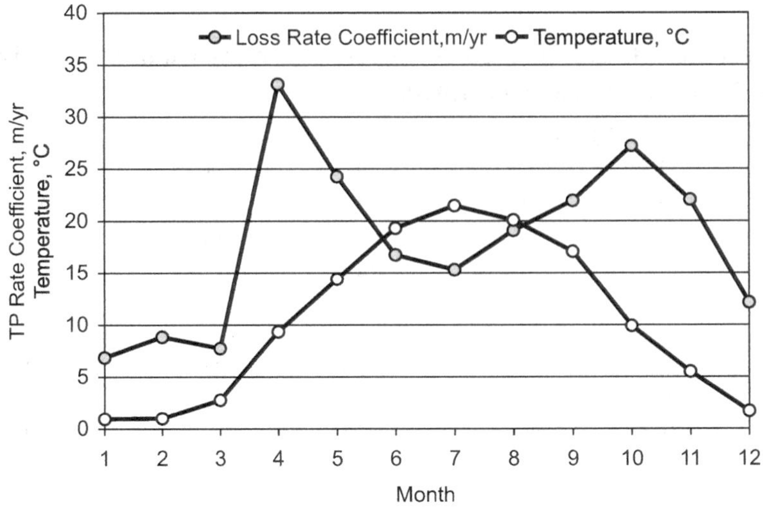

FIGURE 16.7 Apparent first order rate coefficients for total phosphorus removal, and temperature, in FWS treatment systems in cold climates. Model parameters are P = 3 and C* = 0.02 mg/L. Data from twenty wetlands.

equations for nutrient removal is therefore an exceedingly dangerous procedure, because it is all too easy to implicitly require the marsh ecosystem to do things it cannot. The author hereby makes a strong plea to avoid the "plug and chug" sizing procedure.

In this section, the quantification of the integrated description of marsh nutrient removal is presented via interactive spreadsheets for an example. There are four of these: a rate calculation sheet, a phytomass calculation sheet, a mass balance sheet, and a constraint check sheet. In the interest of simplicity, the time period for the example is annual, but it is not difficult to use analogous sheets on a monthly or other time period.

The example is in the category of bioaccretion marshes. Other situations, such as marshes dominated by sediment nutrient removal, require modifications that may be made without undue effort.

In the interest of simplicity, the example uses an annual time scale. However, there are many circumstances in which a monthly or seasonal description is needed. The procedures illustrated here are then applied for each month, thus adding a new dimension of complexity.

Rate Calculations

Marsh hydrology (flows) is calculated based upon Equation 16.7, as illustrated in Table 16.1. The projected outlet concentrations of phosphorus and the nitrogen species are calculated via Equation 16.10, as illustrated in Table 16.2. The individual nutrient fluxes are determined from those flows and concentrations (Table 16.8).

The rate coefficients are loss rate coefficients, because the individual species are considered noninteractive in this sheet. It is necessary to check that selected rte parameters are within the experiential envelope for treatment marshes, and to be cognizant of the relative frequency standing of those selections. In this example:

- k_{TP} = 10 m/yr 45th percentile (Table 10.10) HLR, TP_{in}, TP_{out} all in range
- k_{ON} = 20 m/yr 60th percentile (Table 9.11) HLR in; ON_{in}, ON_{out} out of range

TABLE 16.8

Example nutrient performance sheet. The values in bold are inputs and selected parameters.

		Hydrology				
Inflow Q, m3/d =	**1,839**			I, cm/day =	**0.50**	
Area, ha =	**8.00**			P, cm/day =	**0.40**	
Porosity =	**0.95**			ET, cm/day =	**0.36**	
Depth,m =	**0.30**			T, cm/d =	**0.18**	
HLR in, cm/d =	2.30			Downflow, cm/d =	0.86	
HLR out, cm/d =	1.84			Augmentation, m/yr =	-1.66	
HRT in, d =	12.40					
HRT, d =	13.82					
		Concentrations and Loads				
		TP	ON		NH_4-N	NO_X-N
Influent C, mg/L	Ci =	**0.100**	**1.00**		**2.00**	**10.00**
Background C, mg/L	C* =	**0.020**	**1.00**		**0.00**	**0.00**
Rate Coefficient, m/yr	k =	**10**	**20**		**20**	**35**
Effluent C, mg/L	Co =	0.028	0.75		0.19	0.45
C Reduction, %		72	25		90	96
Infiltration, mg/L		0.064	0.87		1.10	5.22
Influent Loadings, g/m2·yr		0.8	8		17	84
Effluent Loadings, g/m2·yr		0.2	5		1	3
Load Reduction, g/m2·yr		0.7	3		15	81
Load Infiltrated, g/m2·yr		0.1	2		2	10
Tank to tank						
	HLR, m/yr	Tank Concentrations				
Tank 1 Inlet	25.17	0.100	1.00		2.00	10.00
Tank 1 Outlet	23.51	0.063	0.85		0.95	3.71
Tank 2 Outlet	21.84	0.041	0.78		0.44	1.32
Tank 3 Outlet	20.18	0.028	0.75		0.19	0.45

- k_{AN} = 20 m/yr 65th percentile (Table 9.13) HLR, AN_{in}, AN_{out} all in range
- k_{NN} = 35 m/yr 67th percentile (Table 9.16) HLR, NN_{in}, NN_{out} all in range

The fact that organic nitrogen concentrations in the example are outside the range for Table 9.11 would be cause for concern if it were an important factor for the marsh under consideration. In the example, organic N concentrations enter and leave near background, and thus the exception can be tolerated.

Additional checks may also be made. For instance, the forecast P removal rate of 0.7 gP/m^2·yr is at the 20th percentile of the experience envelope of Table 10.10. And, the projected nitrate concentration reduction of 96% is quite high, and reason for questioning the selected loss rate coefficient of 35 m/yr.

Phytomass Calculations

This spreadsheet concerns the prospective wetland, and its role in nutrient processing, is brought into the picture. The rate coefficient-based analysis has provided first estimates of the quantities of materials entering and leaving the system, but there has been no allocation of removals to the various processes that comprise the entire ecosystem function. It is not feasible to go too far in breaking out processes, because the knowledge base does not support great detail. Here, the processing of carbon, nitrogen and phosphorus are estimated (see Chapter 5).

The phytomass compartments and transfers between them, and to and from air and soils, are shown in Figure 16.8. Transfers refer to carbon, nitrogen, and phosphorus, except that carbon is the only aerial transfer, representing the utilization of carbon dioxide in building plant material.

In this analysis, two points are important. First, in the design mode, the analysis is based upon estimates of what the ecosystem will be like. Such estimates cannot be precise, and consequently the analysis is "order-of-magnitude" only. Even in the performance analysis mode for an existing marsh, there is rarely quantitative information on the phytomass compartments. Second, this example analysis is for an annual period, and hence there remains the issue of seasonality.

The magnitudes of the transfer fluxes is generally determined by the storage in the generating compartment and the rate of turnover of that storage. For instance, in temperate latitudes, macrophyte biomass develops and senesces and dies in approximately a four month time period. Uptake during that period is slightly more than the peak standing crops of biomass, C, N, and P. That uptake is reversed as the standing crops are transferred to necromass, and ultimately to accretion. Discussion of standing crops and turnover times is contained in Chapter 5. Guidance is presented in Table 16.9. The magnitude of the biogeochemical cycle increases with increasing nutrient availability, up to some presumptive limit forced by availability of space and sunlight. This fertilizer response is not well quantified for treatment marshes, and therefore cannot be used directly in design. However, estimates are of value in assessing the potential importance of the biogeochemical cycle, and particularly to check that the rate-based design calculations do not imply unreasonable ecosystem function.

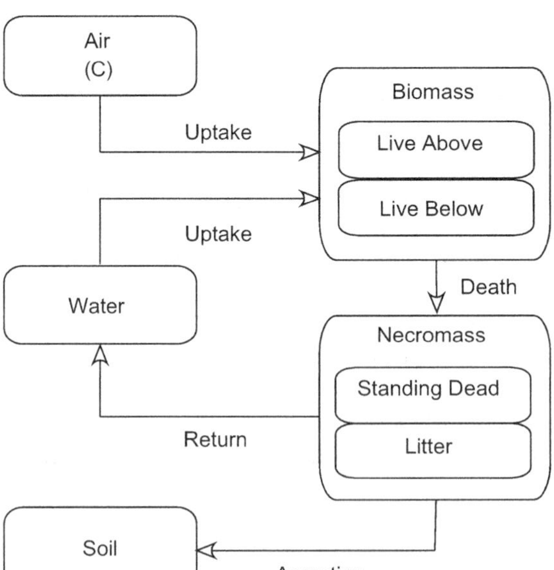

FIGURE 16.8 Conceptual annual biomass and nutrient cycles in a treatment wetland for. Biomass includes macrophytes, algae, invertebrates and microbes.

TABLE 16.9

Estimates of wetland nutrient cycling in biomass. The first two categories are based upon Davis (1994); the second two are based on Kadlec (1997a). Bacterial and algal cycling are included in these values.

		Nutrient Poor	Nutrient Moderate	Nutrient Rich	Very Nutrient Rich
Organic Matter					
Standing Crop	gdw/m^2	150	500	3000	10,000
Turnover Time	times/year	2	4	3	2
Cycling Rate (G)	gdw/m^2·yr	300	2,000	9,000	20,000
Burial Fraction (β)	%	5	10	20	25
Accretion (B)	gdw/m^2·yr	6	200	1,800	5,000
Phosphorus					
Tissue content	mgP/kg	1,000	2,000	3,000	4,000
Uptake	gP/m^2·yr	0.3	4	27	80
Return	gP/m^2·yr	0.29	3.6	21.6	60
Accretion	gP/m^2·yr	0.01	0.4	5.4	20
Nitrogen					
Tissue content	%dw	1	1.5	2	2.5
Uptake	gN/m^2·yr	3	30	180	500
Return	gN/m^2·yr	2.94	27	144	375
Accretion	gN/m^2·yr	0.06	3	36	125

The example phytomass cycling spreadsheet is shown in Table 16.10. The standing crop of phytomass is selected as the principal input variable, because that facilitates later exploration of the effect of the phytomass cycle on nutrient removals. For internal consistency, the burial of phosphorus in this system, in concert with infiltration losses, must match the rate calculation of P removal.

In this example spreadsheet, the biomass and necromass compartment sizes are presumptively set as fractions of total phytomass, one-third each of biomass, standing dead and litter. The C, N, and P contents are estimated at 40%, 2%, and 0.2% respectively for all compartments in this example. Twenty percent of production is assigned to burial, with the remaining 80% returned to the water column.

Uptakes, returns and accretions of N and P are needed in the mass balance spreadsheet, and therefore are linked to that third piece of the integrated workbook.

Mass Balance Calculations

The conceptual basis for nutrient mass balances is presented in Figure 16.9 for phosphorus, and Figure 16.10 for nitrogen. Inflows and outflows (including seepage) are available from the performance spreadsheet. Biological uptake, return and accretion are available from the phytomass spreadsheet. In the case of nitrogen, mass balances determine the internal conversion process fluxes of ammonification, nitrification and denitrification. In the case of phosphorus, the mass balance creates the need to reconcile the removal rate prediction with the phytomass accretion prediction.

Chapter 9 contains several examples of nitrogen mass balances, and a discussion of the assumptions that are needed for implementation. Those assumptions include the apportionment of nitrogen uptake from the ammonia and nitrate pools according to their respective

TABLE 16.10

Example annual biomass estimator for a marsh. Values in bold are input basis estimates. The shaded cell is the adjustable value to match accretion with performance calculations.

		Crop		Production	
Live Above	Dry Weight	296	gdw/m2	896	gdw/m2·yr
Fraction =	% Carbon	**40**			
0.333	% Nitrogen	**2**			
	% Phosphorus	**0.2**			
	Carbon Pool	118	gC/m2	358	gC/m2·yr
	Nitrogen Pool	5.9	gN/m2	17.9	gN/m2·yr
	Phosphorus Pool	0.59	gP/m2	1.79	gP/m2·yr
	Turnover Time	**0.33**	yr		
Live Below	Dry Weight	296	gdw/m2	148	gdw/m2·yr
Fraction =	% Carbon	**40**			
0.333	% Nitrogen	**2**			
	% Phosphorus	**0.2**			
	Carbon Pool	118	gC/m2	59	gC/m2·yr
	Nitrogen Pool	5.9	gN/m2	3.0	gN/m2·yr
	Phosphorus Pool	0.59	gP/m2	0.30	gP/m2·yr
	Turnover Time	**2.00**	yr		
Dead + Litter	Dry Weight	296	gdw/m2	296	gdw/m2·yr
Fraction =	% Carbon	**40**			
0.333	% Nitrogen	**2**			
	% Phosphorus	**0.2**			
	Carbon Pool	118	gC/m2	118	gC/m2·yr
	Nitrogen Pool	5.9	gN/m2	5.9	gN/m2·yr
	Phosphorus Pool	0.59	gP/m2	0.59	gP/m2·yr
	Turnover Time	**1.00**	yr		
Total	Dry Weight	**888**	gdw/m2	1339	gdw/m2·yr
	Carbon Pool	355	gC/m2	536	gC/m2·yr
	Nitrogen Pool	17.7	gN/m2	26.8	gN/m2·yr
	Phosphorus Pool	1.77	gP/m2	2.68	gP/m2·yr
Accretion	Biomass	(Fraction =	**0.2**)	268	gdw/m2·yr
	Carbon	(Fraction =	**0.2**)	107	gC/m2·yr
	Nitrogen	(Fraction =	**0.2**)	5.4	gN/m2·yr
	Phosphorus	(Fraction =	**0.2**)	0.54	gP/m2·yr
Return	Biomass			1071	gdw/m2·yr
	Carbon			429	gC/m2·yr
	Nitrogen			21.4	gN/m2·yr
	Phosphorus			2.14	gP/m2·yr

concentrations. Nitrogen return from the necromass is presumed to be organic nitrogen. The commonly accepted interconversions are ammonification, nitrification and denitrification, although anammox, nitrogen fixation, nitritation and nitrous oxide formation may be present in some cases.

For the illustrative example under consideration, particulate forms and sedimentation are not included, but may easily be added as circumstances warrant. The mass balance spreadsheet for the continuing example is given in Table 16.11. All the required input information is derived

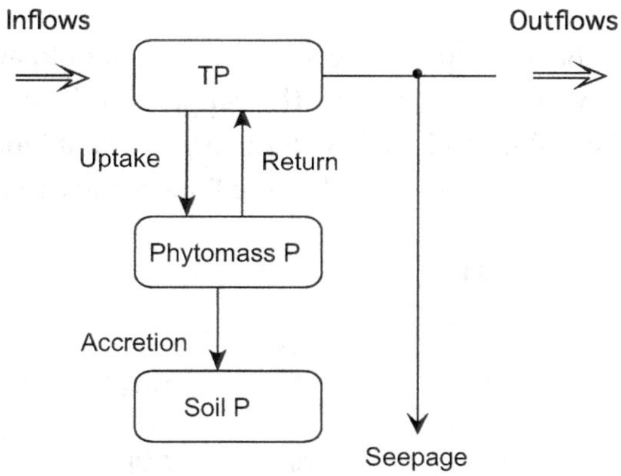

FIGURE 16.9 Conceptual annual phosphorus mass balance framework a treatment marsh.

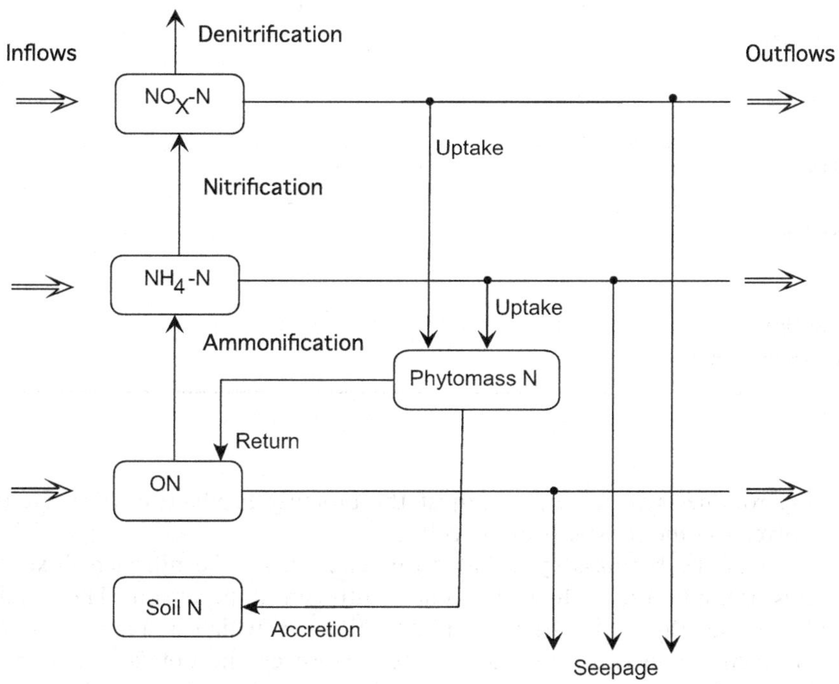

FIGURE 16.10 Conceptual annual nitrogen mass balance framework a treatment marsh.

from links to the performance and phytomass sheets. The primary output information includes the internal process fluxes of ammonification, nitrification, and denitrification; plus the ability to adjust the phytomass cycle to match the P removal flux to accretion.

The size of the biomass cycle, measured as production, is not known a priori, and is used as an adjustable parameter. The value entered on the mass balance sheet is linked to the phytomass sheet, which calculates uptakes and returns. The goal is that the phosphorus removal flux calculated on the performance sheet matches that on the phytomass sheet.

TABLE 16.11

Example nutrient mass balance spreadsheet. The values in bold are linked to the performance and phytomass spreadsheets. The supply test delta (shaded) is the difference between the inflow and outflow P fluxes, and goal biomass production (shaded) needed so that P accretion matches the P performance prediction.

Water Flows, cm/d					
HLR In	**2.30**				
HLR Out	**1.84**				
HLR Downflow	**0.86**				
	TP	ON	NH_4-N	NO_X-N	TN
Concentrations, mg/L					
C In	**0.100**	**1.00**	**2.00**	**10.00**	13.00
C Out	**0.028**	**0.75**	**0.19**	**0.45**	1.39
C Seep	0.064	0.87	1.10	5.22	7.19
Fluxes, g/m2·yr					
Water In	**0.8**	**8**	**17**	**84**	109
Water Out	**0.2**	**5**	**1**	**3**	9
Water Downflow	**0.1**	**2**	**2**	**10**	13
Biomass Uptake	**2.7**		5	22	**27**
Biomass Return	**2.1**	21			21
Accretion	**0.5**				**5**
Ammonification		23			
Nitrification			32		
Denitrification				81	
Total In	0.8				109.1
Total Out	0.8				109.1
Supply Test Delta	0.0				
Goal Biomass Production	888				

A goal-seeking routine is then used to adjust the biomass production to cause the match, such as the Solver routine in Microsoft Excel™.

Once the size of the biomass cycle has been selected, all the nitrogen fluxes are established. For this example, the cycling of biomass nitrogen is important. The required nitrogen to build the annually cycled biomass is 27 gN/m^2·yr, which is about a quarter of the amount of nitrogen supplied in the source water. However, the uptake is also provided by the return flux of nitrogen from leaching and decomposition of necromass. It should be kept in mind that these uptakes and returns involve above ground plant and below ground plant parts, as well as microbes and algae.

The ultimate fate of removed nitrogen in the example is apportioned to accretion (5%), denitrification (81%), and seepage (13%). Interestingly, this does not mean that there are only small amounts of nitrification occurring, because 30% of the incoming nitrogen load winds up being nitrified (32 out of 109 gN/m^2·yr), because of leakage and a small amount of biomass uptake. The question then arises as to whether there is a sufficient oxygen supply to support the nitrification implied by this mass balance. This, and other supply considerations are evaluated in the next section.

Additional Cross-Checks

The results of the integrated spreadsheet calculations require additional checking, to ascertain if reactant supplies are adequate, and if the process rate coefficients are within the experiential range. The example is continued, with these additional checks shown in Table 16.12. This fourth spreadsheet is linked to the other three in the workbook.

These constraints are critical to design, because marsh chemical processes cannot proceed without the necessary ancillary compounds. This is manifested in the need for oxygen to drive ammonia removal, and in the need for carbon to drive denitrification. If inadequate supplies of ancillary chemicals are forecast, rate constants and loadings must be reduced until the constraints are met. Fortunately, treatment marshes operate at loadings low enough to avoid supply constraints.

Process Rate Coefficients

Table 16.11 produces the internal fluxes of ammonification, nitrification, and denitrification, from which the rate coefficients for those processes may be calculated (see, for instance, Equation 9.39). These process rate coefficients are checked to ensure that computed values are within the experiential envelope for treatment marshes, and to be cognizant of the relative frequency standing of those selections. In this example:

- k_{Ammon} = 27 m/yr 60th percentile (Table 9.12) HLR, ON_{in}, ON_{out} all in range
- k_{Nitrif} = 29 m/yr 22nd percentile (Table 9.14) HLR, AN_{in}, AN_{out} all in range
- k_{Denit} = 16 m/yr 15th percentile (Table 9.17) HLR, NN_{in}, NN_{out} all in range

All of these are within expected ranges, and thus the workbook results can be accepted.

TABLE 16.12

Constraint checks for the example nutrient forecast workbook.

Process Rate Coefficients			Range
Ammonification	27	m/yr	5–46
Nitrification	29	m/yr	9–299
Denitrification	16	m/yr	9–243
Carbon Needed			
Denitrification	81	gN/m2·yr	
C/NO_X-N Loss Demand Ratio	**1.29**		
C Requirement	105	gC/m2·yr	
Useable Carbon Fraction	**0.3**		
Biomass Carbon	**40**	%	
Production Needed	140	gC/m2·yr	
Necromass Production	358	gC/m2·yr	
Oxygen Needed			
Nitrification	32	gN/m2·yr	
Oxygen Demand Ratio	**4.7**		
Oxygen Needed	151	gO/m2·yr	
Oxygen Needed	0.41	gO/m2·d	1.7–9.6
Accretion			
Biomass	268	gDW/m2·yr	
Bulk Density	0.10	gDW/cm3	
Depth	0.27	cm/yr	0.3–4

Carbon Supply

The carbon requirement for denitrification is discussed in Chapter 9. On an elemental basis, the requirement is 1.29 gC/gN. If the product is N_2O, then the carbon requirement is 1.07 gC/gN. The carbon produced by above-ground biomass in the example wetland (358 gC/m^2·yr) far exceeds that necessary for denitrification (140 gC/m^2·yr). Thus despite the fact that only a fraction of the biomass-produced carbon is available in the water, there is no apparent shortage of carbon for denitrification in this example.

Oxygen Supply

Oxygen is required to support nitrification, and to support CBOD removal, although CBOD is not part of the example under consideration. The oxygen requirements for the example calculated (estimated) in Table 16.12. The oxygen requirement for nitrification is discussed in Chapter 9. The traditional stoichiometric factor of 4.57 g O_2 per g NH_4-N oxidized is used here. In the example, nitrification is 32 gN/m^2·yr, which then requires 151 gO/m^2·yr, or 0.41 gO/m^2·d. This should easily be supplied by atmospheric reaeration, which has a range of 1.7–9.6 gO/m^2·d.

Accretion

The accretion process implies a rate of sediment buildup determined from the sediment bulk density. Because there is a large phytomass cycle, the accretion of solid residuals is an important factor. In the example, the calculated rate of sediment buildup is 0.27 cm/yr. This is very low for treatment wetlands, which display a range of 0.3–4 cm/yr (see Table 4.6), but well within the range of low nutrient and natural marshes, 0.15–0.60 cm/yr.

16.8 QUANTIFYING THE STOCHASTIC COMPONENT OF PERFORMANCE

All treatment technologies possess a spectrum of effluent concentrations, which is predictable only in the probabilistic sense. Therefore, in addition to the mean effluent concentration (which may vary in a deterministic way with temperature and loading) there is an associated bandwidth of effluent concentrations. The integrated calculation procedure described above provides a means of calculating the deterministic portion of performance. Observed trends in treatment marsh outlet concentrations can be obtained, and often result in sinusoidal deterministic trend functions (see, for example, Figure 9.23).

Performance targets differ among source water types and pollutant types. Infrequently, for nutrients, regulations may constrain both the mean and the maximum of the band, via specification of a limit on the maximum daily, weekly or monthly marsh effluent concentrations; together with a limit on the average annual value. This occurs if the marsh is subject to an NPDES or other permit. In design, care must be taken to accommodate the most restrictive of multiple averaging tests given by the regulation. Data availability for nutrients is sufficient to support trend analyses as well as rate-based average performance. The anticipated variability has therefore been quantified for nutrients.

For trace contaminants, there may not be permits, but rather a requirement that the marsh effluent not exceed acute or chronic receiving water standards (see, for instance, Tables 13.2 and 14.4). The deterministic portion of the marsh performance may be calculable, but data availability is generally insufficient to support trend analyses for the multitudinous target compounds. The anticipated variability has therefore generally not been quantified for trace contaminants.

It is set forth in Chapter 7, and reiterated here, that the probabilistic portion of the wetland performance model is not a "safety factor," as utilized in some published wetland design procedures (Crites et al., 2014). An estimate of the stochastic component is used to provide the required excursion containment. Tempting as it may be, it is not acceptable to ignore this "half" of wetland behavior in design.

Calculations

The applicable formulations have been presented in Chapter 7 and are summarized below.

Model based

$$C_{out} = C_{model} + E \tag{7.86}$$

Trend based

$$C_{out} = C_{trend} + E = C_{avg}[1 + A \cdot \cos(\omega(t - t_{\max}))] + E \tag{7.87}$$

where: A = fractional amplitude of the seasonal cycle, –
C_{out} = instantaneous outlet concentration, mg/L
C_{model} = model outlet concentration, mg/L
C_{trend} = model outlet concentration, mg/L
C_{avg} = average annual outlet concentration, mg/L
E = random portion of the outlet concentration, mg/L
t = time of the year, Julian day
t_{max} = time of the year for the maximum outlet concentration, Julian day

Sometimes, the annual average is used for C_{trend}.

The random variation, E, can be captured through the use of a multiplier on the trend value, C_{trend}:

$$E = C_{trend}\Psi \tag{7.88}$$

$$C_{out} = C_{trend}\,(1 + \Psi) \tag{7.89}$$

where: Ψ = stochastic portion of C/C_{trend}, –

Trend Multipliers

Information on trend multipliers for nutrients and TSS has been presented in previous Chapters:

- Total Suspended Solids Table 8.6
- Ammonia nitrogen Table 9.15
- Nitrate nitrogen Table 9.18
- Total phosphorus Table 10.12

As an interim measure to allow estimates of excursion amplitudes for trace contaminants, the average trend multipliers for these substances may be presumed to apply to other constituents. In approximate terms, the 80th percentile has 1+Ψ = 1.5, the 90th has 1+Ψ = 2.0, and the 95th has 1+Ψ = 2.5 (Table 16.13).

Implementation

Excursion containment is accomplished by lowering the target concentration in the deterministic sizing calculation, to the desired average outlet concentration divided by (1+Ψ). To illustrate, consider the representation of Figure 16.11, showing the hypothetical reduction profile curve for a marsh. According to the deterministic performance model, as wetland size increases, there

TABLE 16.13

Average trend multipliers (1+Ψ) for treatment marsh nutrients and TSS.

	Percentile			
	0.50	**0.80**	**0.90**	**0.95**
Means				
TP	1.02	1.64	2.27	2.84
NH_4-N Agronomic	0.81	1.44	2.41	3.23
NH_4-N Microbial	0.84	1.36	1.75	2.09
NO_X-N	1.00	1.46	1.73	1.95
TSS	1.00	1.72	2.25	2.83
Average	0.93	1.52	2.08	2.59
Medians				
TP	0.99	1.47	1.90	2.39
NH_4-N Agronomic	0.88	1.52	1.99	2.70
NH_4-N Microbial	0.91	1.35	1.70	1.87
NO_X-N	0.99	1.38	1.53	1.66
TSS	1.00	1.69	2.33	2.88
Average	0.95	1.48	1.89	2.30

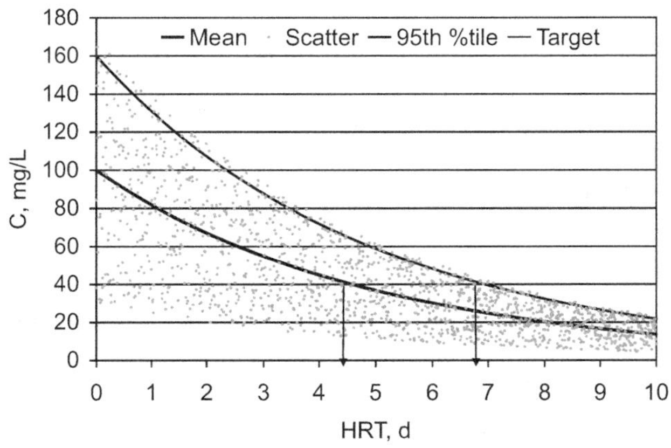

FIGURE 16.11 A hypothetical example of design for excursion containment. The trend represents P = ∞, C* = 5 mg/L, and k_V = 0.2 d^{-1}. The scatter is a uniform distribution with a ± 70% bandwidth. On average, a goal of 40 mg/L can be met with 4.6 days detention. To avoid exceedances at the 95% level (one time in twenty), the detention time should be 6.9 days. At 6.9 days, the target for the trend average is 25 mg/L.

is a downward trend in pollutant concentration. Data points from the marsh are expected to scatter about this line with an associated bandwidth of scatter in concentration values.

In this hypothetical example, that scatter is shown as a uniform distribution about the trend line, with a bandwidth proportional to the trend value. It is supposed that the regulatory limit is a concentration of 40 mg/L, as a maximum allowable. The trend model tells us what size wetland is needed to meet 40 mg/L as a long-term average, which is a detention time of 4.6 days. However, the scatter is such that half the time the measured values will be higher, up to 70 mg/L. The exceedance frequency is expected to be 50% because the design is based on the mean performance. In this illustration, this level of excursions is not unacceptable because it is desired not to exceed 40 (at least 95% of the time). It is therefore necessary to increase the size of the wetland.

At a detention time of 6.9 days (a 50% increase in wetland size), a large majority of excursions are contained (95% this hypothetical example) below the desired limit, and the system would experience exceedances only 5% of the time. In this case, the 50% increase in size is needed for excursion containment of an expected, quantified scatter. At the size that contains excursions (6.9 days detention), the trend value is 25 mg/L, or 63% of the limit value for the average design. This fraction is called the Coefficient of Reliability (COR) (Crites and Tchobanoglous, 1998).

16.9 DETAILED DYNAMIC SIMULATIONS

In some few instances, detailed design calculations have been implemented via dynamic modeling. For instance, Florida STAs have been described with dynamic phosphorus models (Min et al., 2011; Ji and Jin, 2016). And, Kadlec (2010) utilized a dynamic nitrate model for a river water treatment marsh. Dynamic models are typically complex, and difficult to calibrate. They require intricate computer procedures that are unique, meaning the software is not commercialized and user support is not available to a new design team. As a consequence, dynamic modeling is not in general use.

At present, the only generally available dynamic simulation tool developed solely for design of event driven wetlands, is for phosphorus removal in warm climates. The DMSTA (Dynamic Model for Stormwater Treatment Areas) computer routine is given at website www.wwwalker.net/dmsta/ (DMSTA; Walker and Kadlec, 2011). This model has seen use in the design of constructed wetlands in south Florida, including the most recent STAs. There are numerous other computer codes that have varying capabilities to deal with time-variable stormwater flows and concentrations, together with some form of constituent tracking. Some of these have been listed in Chapters 9 and 10. However, calibrations to wetland data are typically lacking, except for DMSTA, which has been calibrated to more than 100 wetland datasets, ranging from mesocosms to the large full-scale systems.

A key feature of dynamic simulation of event driven wetlands is the inclusion of storage of contaminants. Nutrients are stored in vegetation and sediments, and cycled on a preset time scale. Movements of pollutants into and out of storage have important effects on the long-term reductions in those pollutants.

Detailed dynamic simulations are very data-hungry. Rather extensive time series of inflows, concentrations, and meteorology are needed. The probable rainfall sequence may be constructed from historical records, but the inflows and concentrations will typically have to be estimated. As a consequence, full dynamic simulations are likely to be restricted to situations involving large and expensive wetlands.

16.10 SIZING ACCOMPANYING FEATURES

There are three potential additions to the wetted area of the marsh, thus increasing the entire project footprint. For continuous sources in cold climates, it may be necessary or desirable to store water over winter. For sediment-laden runoff sources, a front-end settling basin may be appropriate. And, wildlife and human use enhancements may require added area. The sizing of such features in not within the scope of the calculation procedures detailed in the foregoing.

Winter Storage

Urban and agricultural marshes are typically not expected to function on the winter months, but winter operation may be considered for continuous or discretionary source marshes. Most often, that means polishing marshes for lagoon effluents.

In very cold climates, winter operation of a marsh is hydraulically impossible, because ice occupies the entire water column. In North America, this situation prevails in the northern tier of states, and in central and northern Canada. The operational period is confined to a period beginning with the start of vegetative growth, and ending with the commencement of freeze-up. For instance, the Houghton Lake, MI, marsh begins operation at the beginning of May, and continues through October, a period of six months. The treated lagoon effluent is stored during the remaining six months. The area of the storage pond is 12 ha, and the treatment marsh comprises approximately 80 ha. Design of winter storage involves determination of the working depth of the storage and the anticipated seasonal flows and hydrologic inputs and losses.

Winter Storage vs. Winter Operation

In more temperate, yet cold, climates, there is an option to operate year-round, with water temperature going down near freezing, but without complete top-to-bottom ice formation. Treatment does proceed in the under-ice flows, but at slower rates. For instance, the Onaway, MI, marsh operates year-round, despite the fact that air temperatures are at or below freezing four months of the year. Water temperatures are near freezing (under ice) for approximately 100 days in winter. For nitrogen removal, winter removal process slow-down adds to the hydraulic concerns.

The two basic options for dealing with winter slow-down are:

1. Provide no winter storage, and increase the wetland size to meet desired winter performance,
2. Provide winter storage by adding storage lagoon volume, and do not attempt to operate the wetland in winter, and release water from the pond and wetland at a uniform rate throughout the warmest season.

Winter operation is hydraulically possible for lagoons, even for fairly severe cold, where ice thicknesses may range up to a meter (Heaven and Banks, 2005). Indeed, most lagoon systems receive water year-round. Winter operation is also possible for FWS wetlands, even in moderately cold conditions, because water can be managed to flow under-ice. Complete freeze-up leads to over-ice water flows and ice buildup. Cold water temperatures lead to large reductions in microbial processes, and the vegetation cycle is dormant in frozen conditions. Oxygen transfer to under-ice water is blocked by ice and snow. These individual process effects combine in complex ways to produce the overall observed slow-down of treatment.

Performance modeling may be used to resolve the pond volumes and wetland areas required for the two options. Option 2 will require the least land (smallest footprint) and the most earthmoving, while Option 1, no winter storage, will require the most land and the least earthmoving. For North American conditions, operational considerations almost invariably lead to a uniform warm season release (Option 2).

Further discussion of the winter storage option can be found in Kadlec (2003d).

Sedimentation Basins

Treatment marsh systems may include settling basins as a front-end feature, either as a stand-alone component or as an integral part of the marsh. The former are found in surface water runoff systems, and the later in urban stormwater systems. In either case, designs incorporate the concept of sediment removal from the settling pond as planned maintenance. Chapter 8 contains discussion and examples of settling basins.

Upsets in pretreatment elements of the overall treatment train can episodically pass large quantities of TSS to downstream treatment wetlands. Large depositions of solids can form a blanket that smothers the vegetation, by overpowering the plants' abilities to avoid oxygen deficiencies. Sludge overflows blanketed the inlet, killing the vegetation. Inlet deep zones may serve to protect a marsh from TSS overloads, but not if such loads are sustained (Figure 16.12). Solids overload protection is thus a necessary part of treatment wetland design.

Calculations

Empirical regressions have been developed for settling basin design for treatment wetlands (e.g., Lee et al., 2014), but these are site specific without generalized calibrations. In any case, sediment basin sizing is based on two considerations: the settling rate of the incoming TSS and the maximum accumulation depth for clean out. Chapter 8 presents the calculation methodology, in terms of the nominal detention time needed to settle a desired fraction of the incoming TSS. The first method is the use of the PkC* approach, which has the exponential form:

$$\frac{(C_o - C*)}{(C_i - C*)} = \exp\left(-\frac{wL}{uh}\right) = \exp\left(-\frac{w\tau}{h}\right) \tag{16.32}$$

where: C_i = inlet TSS concentration, g/m3
C_o = inlet TSS concentration, g/m3
h = basin depth, m
L = distance along the flow path, m
u = linear water velocity through the basin, m/d
w = TSS settling velocity, m/d
τ = nominal detention time along the flow path, d
τ_A = accommodation time (for 63% reduction), d

FIGURE 16.12 Excessive TSS filled the inlet deep zone at the Tucson, Arizona, Sweetwater wetland. Note the bird tracks that highlight the complete filling of the deep zone with relatively high-density solids.

Required inputs for the design mode are the settling rate (w), background (C*) and depth (h), and the result is the required detention time. This method has the advantage of a large literature for settling rate coefficients and background concentrations. It is also possible to measure settling rates for a new source water.

The second method is the use of an empirical formulation:

$$1 - \frac{C_o}{C_i} = \frac{N_f^a}{N_f^a + b} \tag{16.33}$$

where

$$N_{fall} = \frac{wL}{uh} \tag{16.34}$$

and where: a,b = regression parameters, –

Required inputs for the design mode are the settling rate (w), depth (h), and the empirical parameters a, b. The result is the required detention time. For values a = 1.75 and b = 1.00, the two procedures are virtually identical. There are very few published calibrations for a and b values (e.g., Fletcher et al., 2004).

Accumulation rate estimates are needed to establish the basin depth and estimated time to clean out. The rate of sediment depth buildup is calculated from removal rates and solids bulk density:

$$h_S = \frac{MR_{TSS}t}{\rho_b} = \frac{1}{\rho_b}\left(\frac{Q(TSS_{IN} - TSS_{OUT})}{A_{SB}}\right)t \tag{16.35}$$

where: A_{SB} = area of settling basin, m2
h_S = accreted solids depth, m
MR_{TSS} = mass removal rate of TSS, g/m2·yr
Q = flow rate, m3/yr
TSS_{IN} = inlet suspended solids, mg/L
TSS_{OUT} = outlet suspended solids, mg/L
t = time, yr
ρ_b = sediment bulk density, g/m3

Required inputs for the design mode are the flow rate (Q), inlet and outlet TSS concentrations, maximum accumulation depth (h_{Smax}), time until cleanout (t), and sediment bulk density (ρ_b). The result is the required settling basin area (A_{SB}).

The depth of the basin is then determined by the sum of the desired minimum working depth and the maximum accretion depth.

There is also a strong possibility that the basin operates as small number of tanks in series, perhaps as low as P = 1. Thus the use of the PkC* model may be required. Internal baffles may be added to increase the apparent number of tanks.

In the performance forecasting mode, the calculation uses the settling basin area and depth to calculate the outlet TSS concentration.

An Example

The treatment wetland system at Brawley, CA employed a 0.59 ha sedimentation basin, with a starting depth of 1.9 m (Kadlec et al., 2010a). Over a four-year period, it received 2,244 m^3/d, with inlet TSS = 216 mg/L, and outlet TSS = 35 mg/L. The mass removal rate was 16,400 g/m^2·yr. At a bulk density of 1.0 g/cm^3, the accretion rate was 1.64 cm/yr. Over a presumed 20-year cleanout interval, 17% of the initial depth would be lost.

Buffers

Treatment marshes used for urban and agricultural runoff often require added lands in the form of buffers that are intended to foster wildlife and enhance other ecological values. Such buffers are rarely required or considered for continuous source polishing or surface water treatment systems.

Urban system buffer requirements differ on slightly. As an example, the State of Minnesota recommends a buffer extending 8 meters outward from the maximum water surface elevation be provided (Minnesota Stormwater Steering Committee, 2018). However, local regulations may require a greater distance. Buildings are not allowed within the buffer. The U.S. District of Columbia has the same requirements (Hoffman et al., 2013), as does the City of Saskatoon, Saskatchewan (CH2MHill, 2014). Saskatoon goes further and recommends 16 meters when protection of adjacent wildlife habitat is a goal. In all cases, Departments of Public Health require larger (16 m) setbacks from public water supply wells.

A simple geometrical calculation determines the size of such buffers. The areal importance of buffers decreases as the marsh area increases and increases as the length-to-width ratio increases (Table 16.14). For small wetlands, the increase in total footprint ranges from about 10 to 50%.

Agricultural runoff marshes often do not have buffer requirements; for example, the Florida STAs do not. Notable exceptions are the Conservation Reserve Enhancement Program (CREP) treatment marshes in the United States. These include buffer areas of considerable extent, with sizes that vary from State to State. The State of Washington requires the total (fundable) area to include wetlands and a surrounding buffer not to exceed three times the wetland acreage. The buffer is to be planted to native and/or introduced species suited for the site. In contrast, the State of Iowa CREP wetland systems typically include a 2 to 5 ha wetland, surrounded by a 10 to 15 ha grass buffer. Otis et al. (2010) reported that 27 Iowa CREP wetland sites contained 102.4 ha of wetlands and 377.4 ha of associated grassland buffers. For these projects, wildlife values are of equal or greater importance than water quality improvement, if judged by area.

TABLE 16.14

Footprint area increases for addition of buffers for urban wetlands.

Marsh Area	Length: Width	8 m buffer	16 m buffer
ha		% Extra Area	% Extra Area
1	1	17%	35%
1	3	19%	40%
1	10	28%	58%
5	1	7%	15%
5	3	8%	17%
5	10	13%	25%
20	1	4%	7%
20	3	4%	8%
20	10	6%	13%
100	1	2%	3%
100	3	2%	4%
100	10	3%	6%

16.11 SUMMARY

This chapter has integrated the analytical elements of preceding chapters into a unified approach to sizing and evaluating a treatment marsh system.

The information-gathering phase of treatment wetland design is the first step. Design sizing of depends on the best possible knowledge of influent source water characteristics. Some general guidelines are presented for pollutant concentrations and loads from a variety of source water categories from previous chapters. Site-specific conditions, and their potential constraining effects on design, are assembled. The meteorology that the wetland will experience is determined. Importantly, the goals that need to be accomplished are to be carefully set down for use in subsequent design calculations. In the performance-based design of the wetland, all of this information plays a quantitative role.

Urban and agricultural runoff treatment marshes are often sized by prescription, rather than performance. Marsh area for these may be set as a percentage of the contributing watershed, or as the size needed to contain runoff from a design storm.

Sizing is the first step in design, for which reaction rates and biology have been integrated. First order PkC* removal equations are used first. The rate coefficients are selected for the contaminants under consideration. Hydraulic efficiency is chosen via the selection of the number of TIS. Calculations include the possibility of a nonzero background concentration, corresponding to a return flux from the ecosystem. Rate constants and background concentrations vary with season or temperature in some instances. Goals may include concentration reduction and/or load reduction, which are conflicting objectives. Excursions may need to be contained, for which exceedance frequency data is used. The model equations are solved on a spreadsheet, with a tank-to-tank calculation procedure. Nitrogen compounds further require linked species computations. In the design mode, wetland area is adjusted to meet the project objectives. In the calibration mode, rate coefficients are sought that cause calculations to match data. In the performance description mode, scenarios can be explored to ascertain the effects of process modifications on performance.

There is nothing about the rate constant approach that acknowledges the limitations of the wetland biology or ancillary chemistry. Therefore, an additional spreadsheet is invoked, which estimates the cycling of carbon, nitrogen and phosphorus in the marsh phytomass. The performance and biological cycling spreadsheets are linked via a mass balance sheet. Lastly, the carbon and oxygen supply requirements are checked to ensure that there are no unreasonable requirements for their estimated supplies. Rate coefficients may need adjustment if the presumed capacities of the ecosystem are exceeded.

For some pollutants, adjustments to the design may be required to account for seasonal or temperature driven behavior. Nitrogen reduction is the primary situation in which either seasonality can occur for agronomic systems, or temperature driven rates for microbially-driven wetlands. If the winter season creates too much of a constraint, leading to very large wetlands, then winter storage may be needed, which adds to the system area total. Other added areas may include sedimentation basins and upland buffers.

Note

1 The number of tanks modified for speciation is P_T in this context, to avoid confusion with rainfall P.

17 Interior Design and Construction

Marsh sizing calculations leave the marsh creation team with a metaphorical blank canvas: all the details remain to be filled in. Internal marsh details interact strongly with water quality functions, and have pronounced effects on ecological values. Thus, the specification of internal details requires input from both engineers and ecologists, and increasingly ecological engineers.

Ecological design includes specifications that foster the desired vegetation, and optimize the value of the system to wildlife and humans. Physical design encompasses all the choices of how the wetland basin is built, and how the water is conveyed to and through the system. With the exception of urban stormwater wetlands (Schueler, 1992), there is no single comprehensive literature source that explains options for layout and configuration of the cells that typically comprise a treatment marsh. Site civil design and construction is a broad and well-developed discipline of engineering that has been tailored to constructed wetlands through decades of experience.

17.1 FRAMEWORK FOR GUIDANCE

Metrics

From a water quality standpoint, the goal of interior design is to maximize the efficiency of the system footprint. Chapter 7 presents the two different efficiencies to be considered: volumetric and detention. The volumetric efficiency e_V describes the fraction of the marsh water body that is flushed by flow, thus excluding dead zones:

$$e_V = \frac{\tau}{\tau_n} \tag{7.43}$$

where $\tau_n = V_{total}/Q$ = nominal detention time, days

τ = tracer detention time, days

The detention time distribution efficiency e_{DTD} represents the intensity of mixing within the marsh:

$$e_{DTD} = (1 - \sigma_\theta^2) = 1 - \frac{1}{N} = \frac{t_P}{\tau} \tag{7.46}$$

where σ_θ^2 = dimensionless variance of the tracer response profile, –

N = number of tanks in series, –

t_P = time of the tracer peak, days

This efficiency is unity for plug flow, and zero for perfect mixing.

Hydraulic efficiency λ reflects the excluded volume and the mixing pattern of constructed wetlands. This measure is defined as the product of the volumetric efficiency and the DTD efficiency:

$$\lambda = e_V \cdot e_{DTD} = \frac{t_P}{\tau_n} \tag{7.47}$$

To maximize pollutant removal, both e_V and e_{DTD} should be as close to unity as practical. Bodin et al. (2013) review the data analyses that determine these efficiencies.

The literature contains other proposed indices of wetland efficiency. For instance Wahl et al. (2010) suggest a moment index, defined as the fraction of the detention time distribution that is greater than the nominal detention time. And, Wahl et al. (2012) suggest a hazard index, which combines a probability of failure of a contaminant to be removed with the DTD. Shih et al. (2013) present a correction coefficient, which is the ratio of the pollutant remaining according to a DTD approach to the pollutant remaining according to a nominal HRT calculation.

In this book, the volumetric efficiency e_V is subsumed into the loss rate coefficient:

$$k = e_V h k_V = e_V k' \tag{7.73}$$

where k' = intrinsic rate coefficient, m/yr

k = nominal rate coefficient, m/yr

Thus, volumetric inefficiency contributes to the observed spread of the loss rate coefficient distributions. The detention time distribution efficiency e_{DTD} is implemented in calculations via the number of tanks in series.

Qualifications

In the selection of internal features, efficiency considerations must be qualified according to the intended purposes of the marsh and the basis of applicable literature results. Major considerations are:

- Investigations of efficiencies have been based upon numerical simulations, laboratory physical models and field scale studies. Numerical studies can provide information on the general trends of efficiencies with ecosystem features, but should not be taken as quantitatively accurate for an actual ecosystem in the field. Laboratory physical models also provide only qualitative trend guidance, because they lack the flora and fauna and soils actually found in the field, and because they are subject to uncertainties of scale up.
- Volumetric efficiencies are so named because they are determined by tracer studies that measure detention time distributions for the whole wetland water body. However, the design calculations used herein are based upon wetland area, not volume (see Chapters 7 and 16). There is thus the implicit assumption that the wetland aerial efficiencies are equivalent to the volumetric efficiencies.
- The hydraulic efficiency λ is not directly useful when applying the PkC* modeling procedures, because the two contributing components e_V and e_{DTD} are used in separate ways in that model. As indicated by Equation 7.73, e_V is considered a modifier on the intrinsic removal rate coefficient, and is a major determinant of the spread in report k-values. It reflects the fraction of the wetland area (volume) that is actively engaged in treatment. In contrast, the DTD efficiency, or equivalently the number of TIS, represents the effect of mixing upon the uptake processes within the actively engaged portions of the wetland. It appears in quite a different way in calculations (see Equations 16.11, 16.12 and 16.13).
- While volumetric efficiency is always an important consideration, DTD efficiency or NTIS may be of little consequence for treatment marshes of some types. When the anticipated removal percentage of a contaminant is less than 50%, the effect of NTIS is practically nil (see Figure 16.5). The large majority of agricultural and urban runoff marshes display reductions of less than 50% (see Tables 11.8 through 11.15).

17.2 LOCATION AND ARRANGEMENT

The size of the wetland has been determined, at least initially, by the procedures of the preceding Chapter 16. There remain a number of design decisions on arrangement, vertical and horizontal placement, and methods of moving and controlling water.

SITING

General conditions of the potential site will have been considered during the establishment of the basis of design. In general, there will remain a number of possibilities for the location of the system within the overall site confines. Land use conflicts may exist, and require consideration in selecting the system location.

Floodplains and Floodways

A hypothetical siting situation is shown in Figure 17.1, which indicates that the regional waterways, a river and its tributary, are geographic features to be reckoned with. Constructed wetlands located in floodplains with extreme flood conditions, such as along major rivers, must have dikes that are designed to allow the passage of floods, and/or that are sized to exclude flood waters. Constructed treatment wetlands have typically been allowed

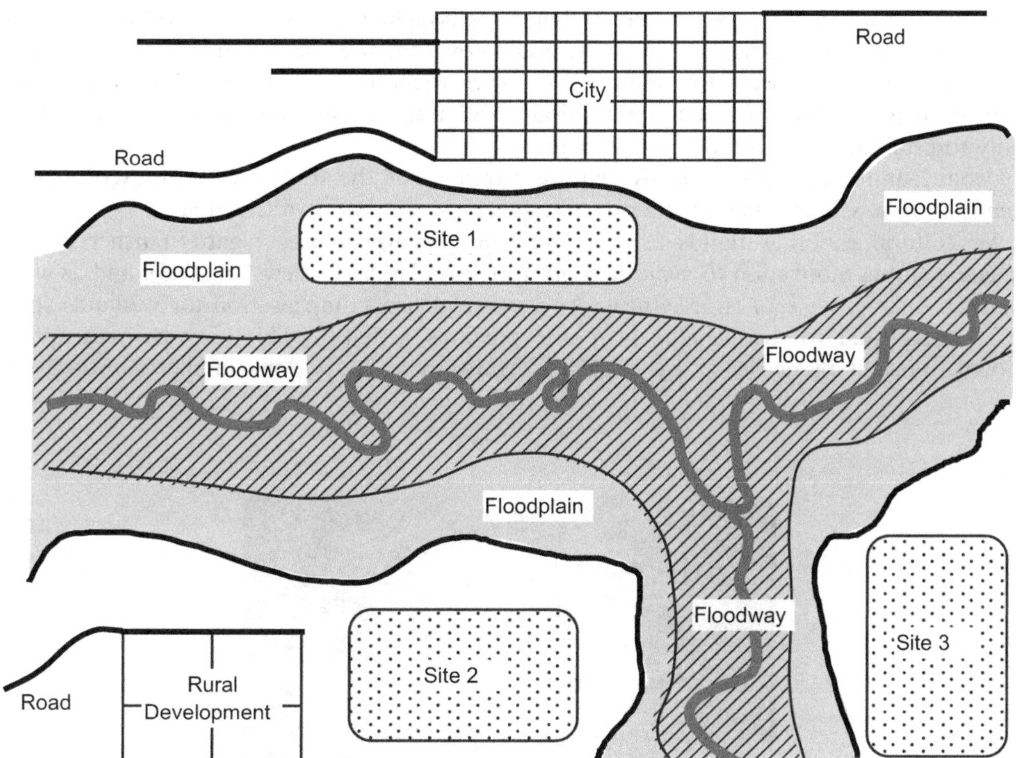

FIGURE 17.1 Siting and flood concerns for a hypothetical treatment wetlands project. Communities are adjacent to a river and its tributary. The floodway occupies the central valley corridors, and is bounded by the floodplain. Three potential constructed wetland sites are indicated, one in the floodplain.

in floodplains under constrained conditions. Designs should avoid damage due to high frequency recurrence events, and therefore would usually be built to withstand the 25-year flood event without impairment of function. Similarly, wetland facilities would usually be designed to withstand the 100-year flood event without severe damage. However, floodplains may be jurisdictional wetlands, thus creating regulatory questions. Nevertheless, many wetland systems have been constructed in floodplains, such as Columbia, Missouri; Brawley and Imperial, California, and Tres Rios, Arizona.

If a wetland is built in a floodplain, it is necessary to demonstrate that its presence will not back up floodwaters upstream of the project. In other words, the project should not block the floodway of the river. For most major rivers of the United States, there exist published maps showing the boundaries of the 100-year floodplain and the floodway.

Cultural Resources

The U.S. National Historic Preservation Act requires that sites of significance for the early peoples of America be preserved. That normally involves avoidance of construction activities that might disrupt ancient village sites and burial grounds. For example, an historic archaeological site was excluded from the Coyote Hills, California urban stormwater wetland (Figure 17.2). The Musselwhite, Ontario treatment marsh was sited with due recognition of the line-of-sight from the regional First Nations burial ground.

Land Ownership

The land needed for a constructed wetland may be in private ownership, or public ownership with use restrictions. When private lands are sought for a project, a willing seller situation must usually exist. Although various governmental units in the United States sometimes have the power of condemnation for projects deemed in the public interest, such acquisition is usually expensive, contentious, and time consuming. These conditions normally forestall any attempt at condemnation.

Urban runoff wetlands typically require purchase of the footprint land. Acquisition is usually from a willing seller, at the prevailing cost of real estate in the area.

Agricultural runoff wetlands are the most problematic in this regard. Farmer attitudes include a strong motivation to avoid loss of productive lands (Olszewska, 2005), and as well to maintain or improve economic returns. As a consequence, compensation for wetlands should be at least equal to the benefits from the previous activities. This may be accomplished through outright purchase, or by a leasing arrangement, as is done in the case of the US

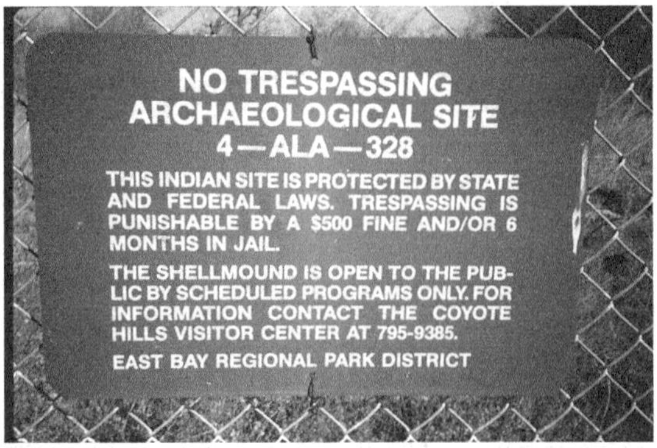

FIGURE 17.2 This archaeological site was excluded from the Coyote Hills urban stormwater wetland.

CREP wetlands. Annual land rental fees are also paid to the farmers in both the United States and Sweden (Iovanna et al., 2008; Andersson, 2012). Land productivity is an important factor, with little farmer resistance to allowing wetlands on unproductive portions of their property (Ockenden et al., 2012). The role of farmer attitude toward the CREP program has been documented (Wachenheim and Devney, 2018; Wachenheim et al., 2018a, 2018b).

Parallel Paths and Compartmentalization

Compartmentalization interacts with the design area calculation, because more cells and higher aspect ratios lead to higher NTIS numbers, and hence to higher PTIS numbers. As seen in the previous Chapters, higher PTIS does not help much for low desired removal percentages, but is a necessity if very high percentages are required. Aspect ratio is also a determinant of NTIS, with fewer for low length to width ratios. A single FWS wetland cell of modest aspect ratio ($1 < L/W < 5$) is likely to behave as three or four NTIS in tracer testing. When compounded with the weathering factor that influences most wastewater parameter reductions, it is likely that a P-value for the PkC* model will be two or three. That may be perfectly adequate if the treatment does not attempt to achieve outlet concentrations close to C*, but as the effluent target approaches C*, the value of P becomes more and more important. If the computation is sensitive to P, then it becomes desirable to use cells in series. Use of three cells in series will lead to P-values between 6 and 9, beyond which is a region of diminishing returns.

Site constraints of steeply sloped ground may mandate terraced, multi-cell design. Because it is not possible to maintain uniform, level and shallow depths on steep slopes, the change in elevation is accomplished via drop structures between cells in series. Inman et al. (2001, 2003) reported that site slopes of up to 25% called for 22 cells arranged on three separate flow paths at the Clayton County, Georgia wastewater polishing wetland facility (Figure 17.3). That project required the movement of 420,000 m^3 of earth over the 22 ha of wetlands, or close to two meters of cut and fill over the site area. There are more than 27 meters of fall from the inlet to the outlet. The Shepard urban wetland in Calgary, Alberta is another example of a terraced, multi-cell design (see Figure 11.14).

Costs increase as the number of cells increases. The number of cells required in each flow path must be determined by balancing the cost of more cells with the need for high hydraulic efficiency. The ratio of berm area to wetland surface area increases with more

FIGURE 17.3 Layout of the Inman Road treatment marsh in Clayton County Georgia.

cells. Figure 17.4 shows the design variables associated with increasing cells in series for a hypothetical fixed rectangular project footprint. Compartmentalization increases the NTIS for the footprint. The number of transfer structures increases with the number of cells, but the vegetated area becomes smaller, due to more berm area and more water distribution zone areas. If the total area is not fixed, the total footprint increases with compartmentalization.

Some wetland designs have carried compartmentalization to extremes, for unspecified reasons. The OEW system in Florida contains 34 structures for 17 cells on two flow paths. The Tucush wetland in Peru specified 30 cells on two flow paths. In neither of these two examples does the topography demand such cell numbers.

Parallel flow ways may be employed for two purposes, increasing the length-to-width ratio, and allowing for one to be shut down for maintenance. This design has been employed in the Florida STAs, which have 23 flow ways in the six STA systems.

Cell Inlets and Outlets

Both field results and computational studies have shown that point inlets and point outlets are not a good idea. The corner zones of the wetland are not easily accessed by the incoming water. This effect is obvious, and led early designers to use gated pipe spreaders to

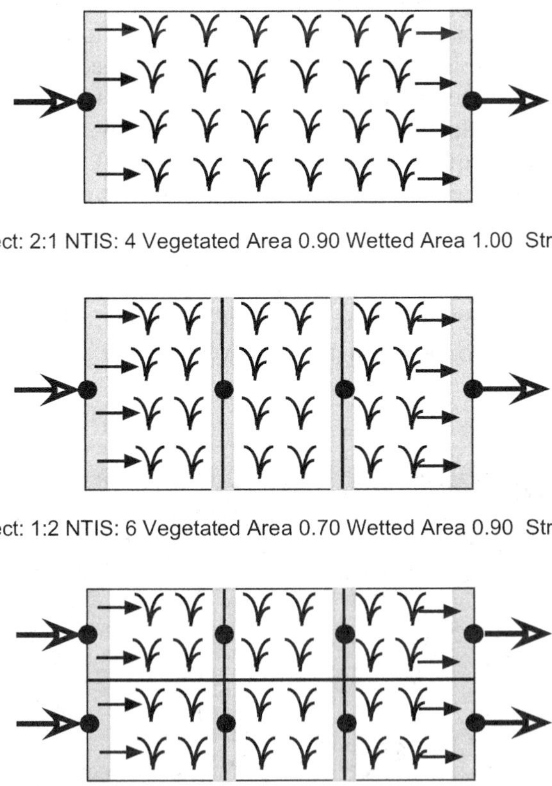

Cell aspect: 2:1 NTIS: 4 Vegetated Area 0.90 Wetted Area 1.00 Structures: 2

Cell aspect: 1:2 NTIS: 6 Vegetated Area 0.70 Wetted Area 0.90 Structures: 4

Cell aspect: 1:1 NTIS: 9 Vegetated Area 0.65 Wetted Area 0.85 Structures: 8

FIGURE 17.4 Options for cells in series and parallel. NTIS are rough estimates. Areas are examples only; actual numbers are size-dependent.

evenly distribute the water across the breadth of the system inlet (Figure 17.5). Because of expense and a propensity for clogging, newer designs replace pipe spreaders with spreader ditches or baffles.

Computational fluid mechanics suggests that if the water is distributed uniformly from side to side of the wetland inlet, rather than at a point, there will be significant improvement of the hydraulics (Persson et al., 1999). The changed distribution is forecast to increase NTIS from 1.5 to 6.7, and e_V from 79% to 89% for a basin of L:W 2:1. Wörman and Kronnäs (2005) computed that as the width of the water inlet increased relative to the wetland width, both e_V and NTIS should increase.

Stairs (1993) experimentally determined the effect of inlet cross-spreading, by performing tracer tests both before and after removal of a perforated pipe spreader. With the spreader, NTIS = 4.47 and e_V = 83.6%. Without the spreader, for a point inlet, NTIS = 3.27 and e_V = 66.5%. Thus the theoretical effect of simulation has been verified in the field. The collection of the water at the wetland outlet poses the same problems of corner dead zones. Therefore, collection across the wetland width is advantageous.

Newer numerical methods studies confirm the Persson et al. (1999) and Stairs (1993) findings. Su et al. (2009) simulated the effects of inlet and outlet flow points using a horizontal two-dimensional model, TABS-2 (Thomas and McAnally, 1985). They found e_V = 0.66 for the scenario in Figure 17.6A, and e_V = 0.72 for the scenario in Figure 17.6B. Other reports of similar effects are given by Bodin et al. (2012), and Sabokrouhiyeh et al. (2017).

Figure 17.6 illustrates some concepts for inlet and outlet spreading and collection.

Agricultural runoff marshes generally do not utilize engineered inlet and outlet features (see Figures 11.10, 11.11 and 11.19). Incoming water enters from a tile drain or overland flow. Outlets typically consist of a cross-flow berm with a V-notch or other narrow crested weir, and consequently bear some resemblance to the outlets in Figure 17.6D and 17.6E. This is done for simplicity, and the less than ideal flow patterns are accepted as necessary system feature.

FIGURE 17.5 The inlet spreader distribution gated pipe at the Houghton Lake, MI marsh.

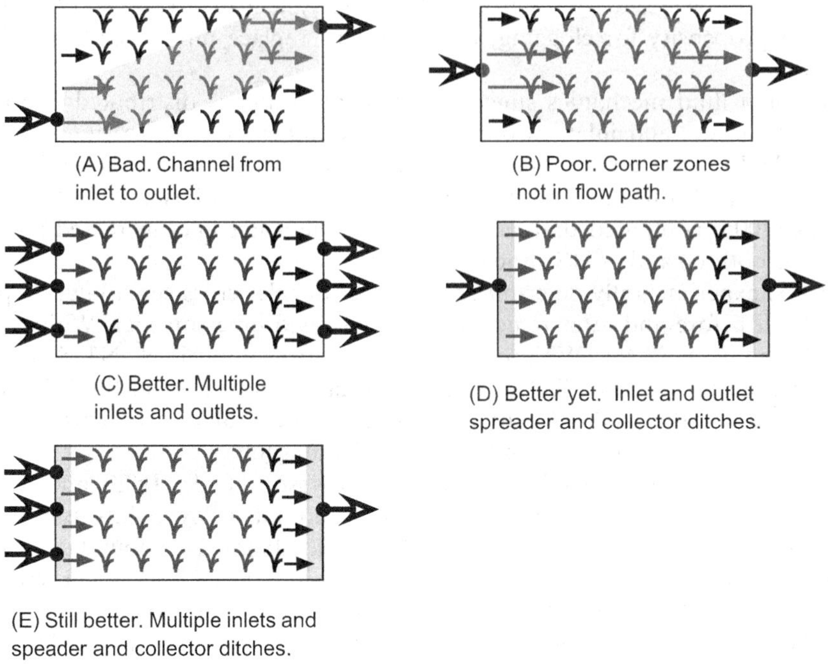

FIGURE 17.6 Concepts of cell inlet distribution and outlet collection. Dark shading indicates deep zones, light shading indicates main flow.

Urban stormwater wetlands utilize a version of Figure 11.6D, which use a point inlet to a deep zone termed the forebay, and a point outlet from a deep zone termed the micropool (see Figure 11.16). As for agricultural systems, this is done for simplicity, and the less than ideal flow patterns are accepted as necessary.

Aspect Ratio

The length-to-width (aspect) ratio is important in basin design because of its effect on flow distribution and hydraulic short-circuiting.

There is a large literature dealing with the effects of aspect ratio (length to width, L:W) on the behavior of inert tracers and on the performance of ponds and wetlands for pollutant reduction as a function of aspect ratio. Thackson et al. (1987) were among the first to present data that illustrate the importance of L:W on the effective volume (e_V) of unvegetated ponds, showing an increase with L:W. Since that time, computational capabilities have permitted theoretical exploration of the values of e_V and e_{DTD} for unvegetated basins of different aspect ratios. For instance, Persson et al. (1999) investigated the effect of L:W using the two-dimensional, depth integrated code of MIKE-21 (Warren and Bach, 1992). Jenkins and Greenway (2005) also investigated empty ponds, using the two-dimensional, depth integrated code of TDFLOW (Jenkins and Keller, 1990). Similar studies have been reported by Su et al. (2009) and Sabokrouhiyeh et al. (2017).

Volumetric Efficiency

A large aspect ratio tends to improve volumetric efficiency, quantified by the ratio of actual to theoretical detention time (e_V). In other words, it is possible to utilize the entire wetland area. The combined results of numerical studies are shown in Figure 17.7. These numerical studies utilized point inlets and point outlets. These multiple result sets are fit by an exponential regression model:

$$e_V = a(1 - \exp(-b \cdot AR)) \tag{17.1}$$

where a,b = regression parameters
AR = aspect ratio, L:W
The parameters are a = 0.885, b = 0.548, and the fit has $R^2 = 0.91$.

DTD Efficiency and Numbers of Tanks

Numerical studies also find a theoretical increase in e_{DTD} (increase in NTIS) with increasing L:W, up to ≈ 50 TIS for L:W ≈ 15 (Figure 17.8). However, these studies are based upon a selection of a dispersion coefficient, which has a marked effect upon NTIS.

Treatment wetland data show that "longer and narrower" does not necessarily mean less effect of mixing, and hence a closer approximation to plug flow. For example the tracer studies at Richmond, Australia showed NTIS = 4.6 for an open water unit of L:W = 25, and NTIS = 5.0 for a *Myriophyllum* FWS wetland of the same aspect (Bavor et al., 1988). A number of tracer tests of the wetland cells 7B and 9B, both with L:W = 12.5, at the Sacramento, California project found NTIS = 4.7 ± 0.8 and 5.0 ± 1.4 respectively (Nolte and Associates, 1998b). On the other hand, the data from the Champion, Florida project show NTIS = 11 for a FWS wetland with L:W = 10 (Knight et al., 1994). Likewise, the South Florida Water Management District periphyton treatment wetland project found NTIS = 9 for L:W = 5, and NTIS = 25 for L:W = 45 (CH2M Hill, 2003a). The effect of aspect ratio in these studies is confused by issues of wind, vegetation patterns and deep zones. Thus it is dangerous to assume that aspect ratio alone can provide the desired improvements in efficiency.

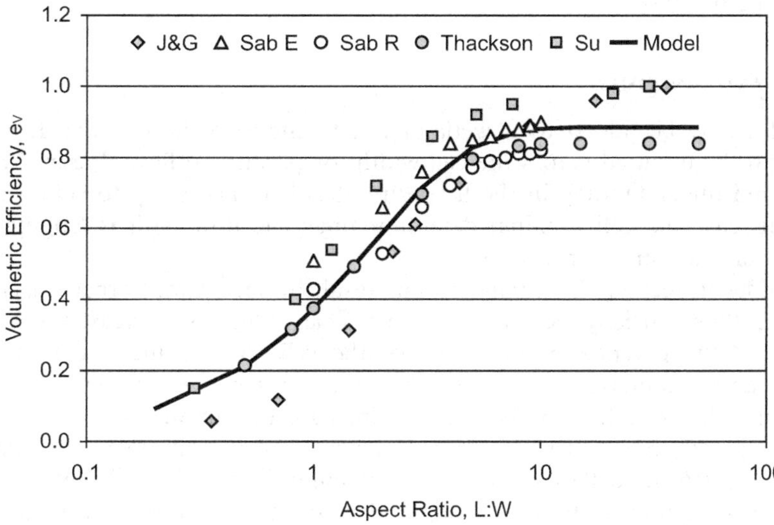

FIGURE 17.7 The effect of aspect ratio on volumetric efficiency from numerical calculations. J&G: Jenkins and Greenway (2005); Sab E: ellipses, Sabokrouhiyeh et al. (2017); Sab R: rectangles, Sabokrouhiyeh et al. (2017); Thackson: Thackson et al. (1987); Su: Su et al. (2009). See text for model.

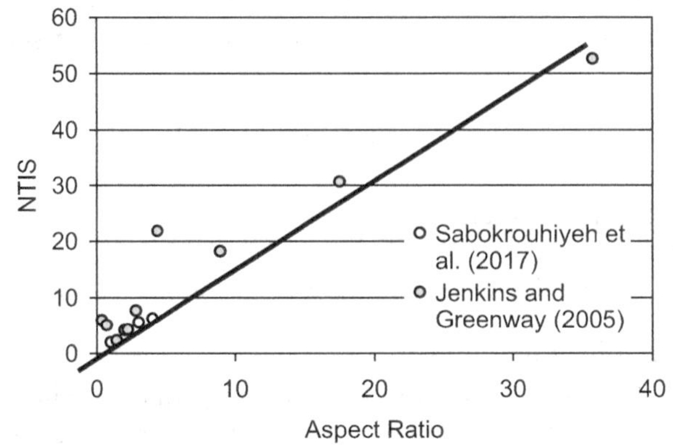

FIGURE 17.8 The effect of aspect ratio on the NTIS of unvegetated ponds with point inlets and point outlets. The regression line is y=1.60x, $R^2 = 0.88$.

There are sometimes limitations on aspect ratio due to the hydraulic profile (see Chapter 3 and subsequent sections of this Chapter). Higher aspect ratios also increase the area of external berms that must be constructed to enclose a given wetland area. Therefore, economics argue for low L:W ratios (Knight, 1987). Other methods for maintaining effective flow distribution, such as compartmentalization and deep zones, should be considered as alternatives to reduce the need for higher length-to-width ratios.

The literature contains "recommendations" for L:W ratios to be used in design. For instance Crites et al. (2014) recommend 2:1 < L:W < 4:1. U.S. EPA (2000a) comments that in general, FWS wetlands are built with L:W > 1:1. These go with the recognition that lower system aspect ratios may be used provided cells in parallel are used.

If one looks beyond efficiency improvements, data from wetland studies indicate somewhat better performance for higher length to width ratios (Herskowitz, 1986; Knight et al., 1994; CH2M Hill, 2003a), but the margin is not large when the pollutant reductions are low to moderate (0–50%).

Baffle Berms and Sinuosity

Internal baffles may be added to a wetland cell to increase the effective length to width ratio. These can be oriented across the cell width, or parallel to flow (Figure 17.9). Because of a frequent misunderstanding in the literature, it is here necessary to state that such baffles do not increase the cell nominal detention time; the flow path is lengthened and the water velocity increase in compensation.

Interior baffles in rectangular configurations result in dead zone corners at each flow direction change, thus partially defeating the beneficial effect of increased L:W. They also occupy some of the potential wetted area of the cell. As for increasing the overall cell aspect, there can be limitations due to the hydraulic profile (see Chapter 3 and subsequent sections of this Chapter). Interior baffles also increase the amount of berms that must be constructed in a given cell, and therefore have accompanying economic consequences.

Chang et al. (2016) used RMA2, a two-dimensional depth-averaged finite element hydrodynamic model to compute the flow fields and associated metrics for a cross-baffled rectangular wetland (Figure 17.9C). They found very strong increases of NTIS with increasing numbers of baffles, and only mild improvements in volumetric efficiency (Figure 17.10). Further simulation results are presented in Su et al. (2009).

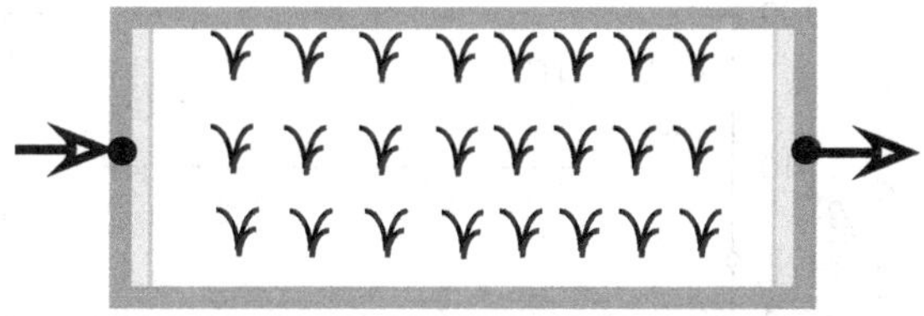

Cell aspect: 2:1 NTIS: 4 Vegetated Area 0.90 Wetted Area 1.00 Structures: 2

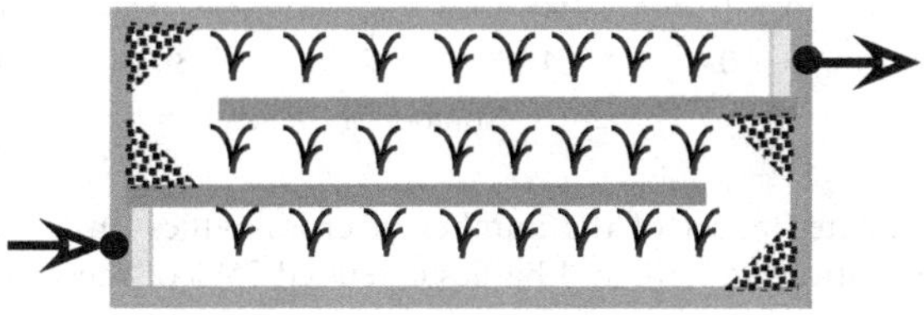

Cell aspect: 2:1 NTIS: 4 Vegetated Area 0.90 Wetted Area 1.00 Structures: 2

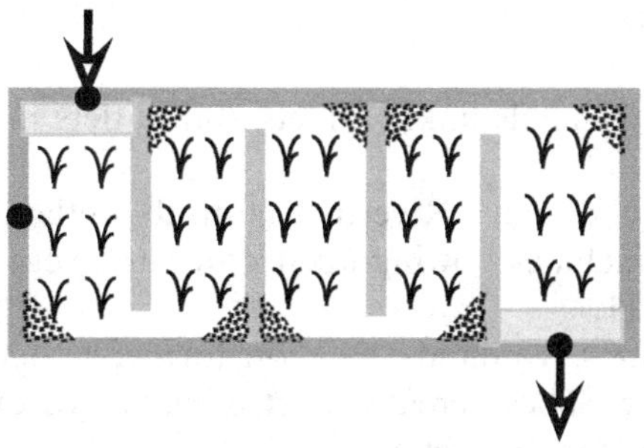

Cell aspect: 2:1 NTIS: 4 Vegetated Area 0.90 Wetted Area 1.00 Structures: 2

FIGURE 17.9 Options for interior berms creating sinuousity. Shaded areas are deep zones, stipled areas are dead zones. NTIS are rough estimates. Areas are examples only; actual numbers are size-dependent.

It is also possible to create a similar L:W effect by creation of a sinuous flow path through the cell, either by incising a sinuous deep zone, or by otherwise keeping the path free of vegetation. Numerical studies are able to avoid the deep channel, but in reality the deepening is needed as the control on vegetation. In any case, the length to width increase of the central channel is usually small, and the overall effect is short-circuiting, which is not wanted. Savickis et al. (2016) performed numerical calculations of the efficiencies of sine wave sinuous paths, which created path L:W values ranging up to 1.87. Their sinuous DTD efficiencies were consistent with Figure 17.8, NTIS = 2.90 vs. NTIS = 2.91 from the regression.

Bathymetry

Each individual wetland cell bottom may be contoured in a number of different ways. Control of the bottom elevation and vegetation density can in principle prevent poor flow distribution. But in practice, the bottom of the wetland can neither be constructed nor maintained at tolerances that promote full areal contacting. Furthermore, agricultural runoff marshes generally do not modify the existing site topography. Bottoms do not start completely level, and a number of processes contribute to the formation of local highs and lows, such as soil heaving and sediment deposition.

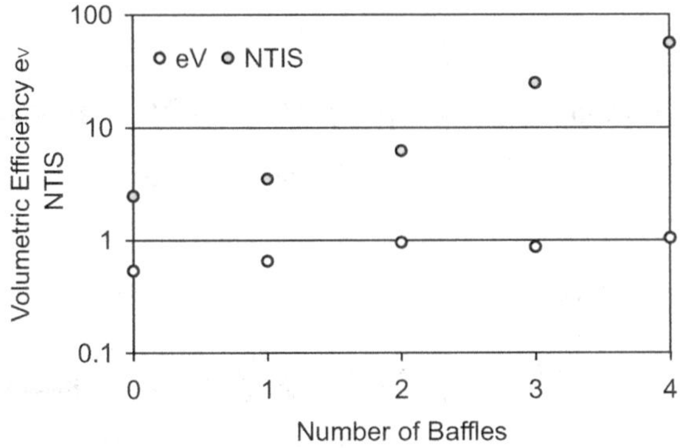

FIGURE 17.10 The calculated effect of the number of cross-baffles on the efficiencies of a treatment marsh. The overall aspect ratio was 1.88, and baffles extended 75% of the width. (Based on information in Chang et al., 2016)

Short Circuits

The several causes of deep channels linking inlets and outlets in a cell interior include:

- Intentional ecological design. Riverine natural wetlands are perceived to be a desirable form that includes flowing open water and vegetated fringes. An example is shown in Figure 17.11.
- Antecedent land features. Wetlands built on former agricultural lands can contain ditches parallel to flow. These create short circuits. An example was cell 4 of the Florida ENRP project (Dierberg et al., 2005).
- Closely opposed point inlets and outlets. For small L:W values, geometrical considerations favor direct paths, leaving the "wing" areas with little or no flow. An example is reported for Cell 7 of the OEW system in central Florida (Min and Wise, 2009) (see Figure 7.8)
- Artifacts of ill-considered construction. The construction of treatment marshes can involve external confinement levees. These require soils that may be excavated from the adjacent lands, creating borrow canals along side the levee. If interior and

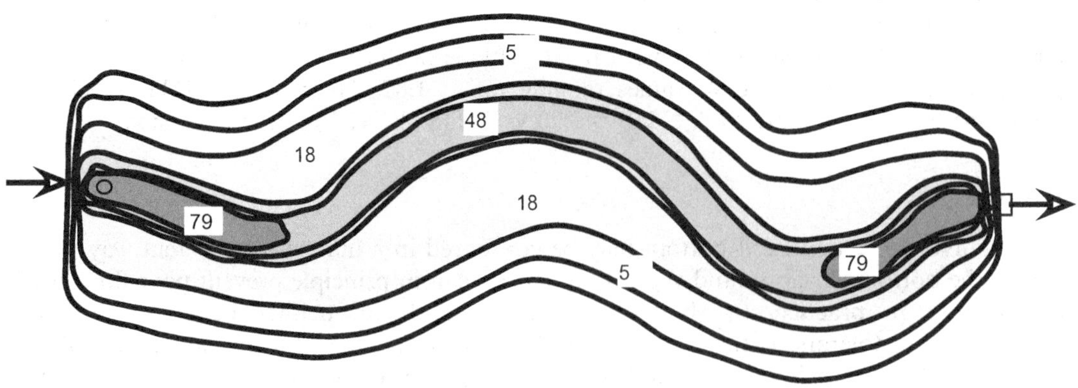

FIGURE 17.11 Bathymetry of Des Plaines wetland C1. Depths in centimeters.

adjacent to the flow direction levees, they create short circuits. An example was an early version of he Apopka, Florida treatment marshes.

- Animal trails. Beavers, muskrats and nutria travel between deep zones, following the same route repeatedly. Such trails provide for water travel and its accompanying refuge. The resulting hydraulic linkage is a short circuit. Figure 15.13 shows this phenomenon.

Design for water quality improvement unequivocally requires that all such short circuits be avoided in design and be remediated if they develop later.

Short circuits result in bimodal tracer RTD response curves, comprised of a fast portion associated with the short circuit, and a slow portion, associated with the side zones. Illustrative system data are reported by Keller and Bays (2001) and Min and Wise (2009). Computational results are given in Lightbody et al. (2009) and Musner et al. (2014). In all cases, the result of the short circuit is a drastic lowering of both volumetric and DTD efficiencies.

Deep Zones

The concept of deep zones in surface flow constructed wetlands was introduced by Knight and Iverson (1990), who supposed they would serve several purposes. These deeper areas extend below the bottom of the vegetated basin areas by at least 1 m to exclude the development of emergent macrophytes (Figure 17.12). Such unvegetated cross ditches provide a low resistance path for water to move laterally, and provide a nearly constant head across the wetland. They also provide for extra detention time, but in a deep water zone. Such ditches often become covered with duckweed *(Lemna* spp.) and can be used by wetland birds and fish as reliable habitat. These redistribution ditches change the potential for short-circuiting within the wetland, because high-speed rivulets are intercepted and mixed with quiescent water in the deep zone. However, the redistribution ditch adds a potential for wind mixing that compensates the reduced short-circuiting. But water is more effectively distributed over the wetland, improving the gross areal efficiency to some extent. Lightbody et al. (2008) investigated the Augusta, Georgia, treatment wetland cells, and found between three and six distinct flow paths in each.

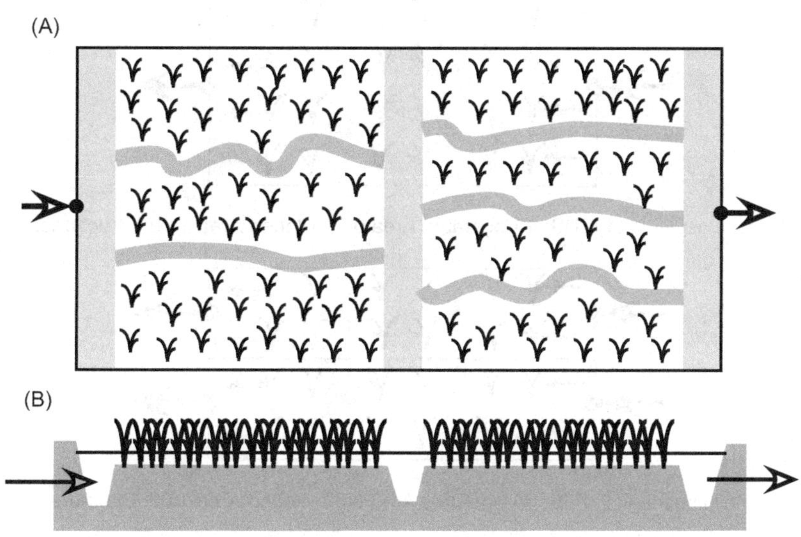

FIGURE 17.12 Schematic of a marsh cell with preferential flow paths and deep zones. (A) Plan view showing five short circuits. (B) Cross section showing bathymetry.

Lightbody et al. (2007) provided evidence from laboratory models and computations that indicated improvement due to deep zones at high degrees of removal. However, that improvement was contingent on the assumption that serious short-circuiting occurred in the vegetated sections of the system. In a follow-on numerical study, Lightbody et al. (2009) found that the effect of deep zones on pollutant reduction was essentially nil for low concentration reductions in a cell with no deep zones (i.e., <20%), but increased for greater no deep zone removals. For 35% reduction, two interior deep zones performed slightly better than none, and for 60% no deep zone removal, four deep zones provided 90% reduction.

Figure 17.13 shows options for implementing deep zones. As noted in the previous section, central deep channels parallel to flow are pathologically bad (Figure 17.13A). There is little doubt that inlet spreader canals and outlet collection canals are beneficial (Figure

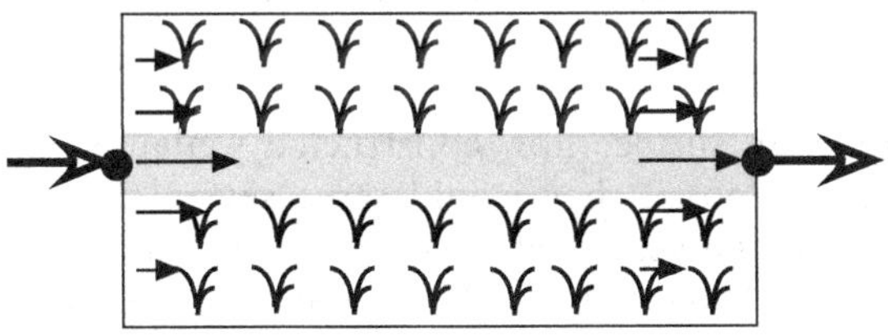

(A) Cell aspect: 2:1 NTIS: ? Vegetated Area 0.90 Wetted Area 1.00 Structures: 2

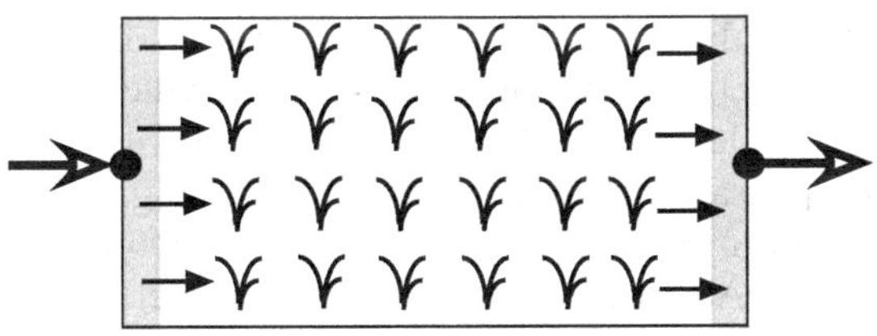

(B) Cell aspect: 2:1 NTIS: 4 Vegetated Area 0.90 Wetted Area 1.00 Structures: 2

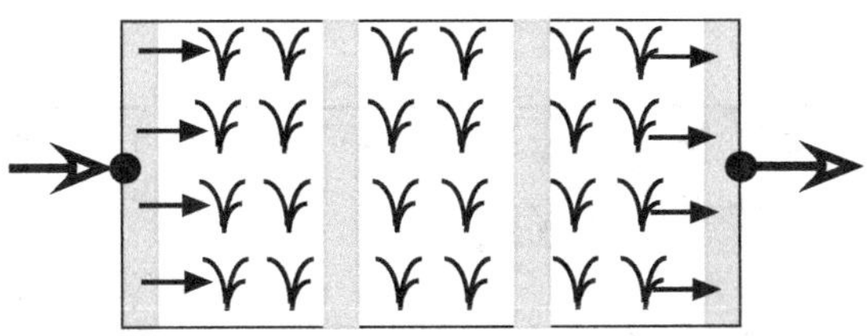

(C) Cell aspect: 2:1 NTIS: 4 Vegetated Area 0.70 Wetted Area 1.00 Structures: 2

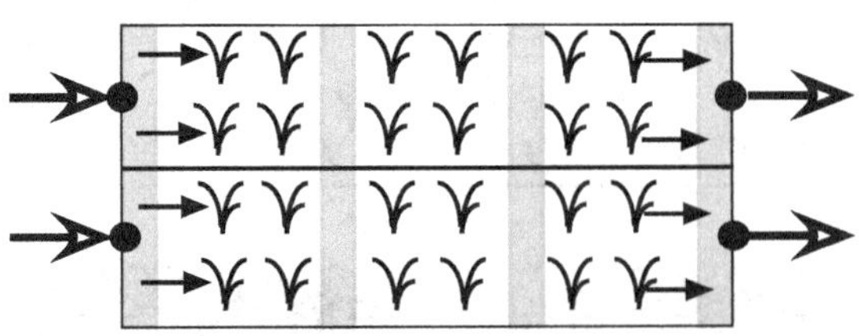

(D) Cell aspect: 4:1 NTIS: 6 Vegetated Area 0.65 Wetted Area 1.00 Structures: 4

FIGURE 17.13 Options for deep zones in series and parallel. Shaded areas are deep zones. NTIS are rough estimates. Areas are examples only; actual numbers are size-dependent.

17.13B). Interior deep zones on single or multiple flow paths are suggested above to provide advantages, according to numerical simulations. However, real system data are less conclusive.

A deep zone study of twelve research wetlands of identical L:W = 2 were operated for twenty months under varying conditions at a site west of the City of Phoenix, Arizona (Kadlec, 2007). These were constructed as a triplicated design, with zero, one, two and three internal deep zones, all containing an inlet distribution and an outlet collection deep zone, together comprising 12.5–35% of the wetland areas. There were no significant differences in the internal hydraulics with deep zone numbers, with $e_V = 0.63 \pm 0.14$, $e_{DTD} = 0.73 \pm 0.03$, and NTIS = 3.72 ± 0.39. Deep zone numbers in the wetlands did not affect nitrogen treatment performance. No differences with deep zone numbers were found for temperature, dissolved oxygen, pH, or nitrogen removals or rate constants. In this study, there was no large treatment benefit or detriment of incorporating internal deep zones.

Knight et al. (1994) operated a set of six treatment wetlands receiving treated paper pulp mill effluent for two years. Two pairs of these received the same influent water, one of each pair containing two internal deep zones comprising 25% and 45% of the area (Table 17.1). There was variability in flow (different HLRs) and aspect ratios (2.5 and 10) between the pairs. Tracer data from that project were here re-analyzed, and characterized by a tanks-in-series (TIS) detention time distribution. The effect of deep zones was to decrease the number of TIS (Table 17.1). And, for the larger cell pair (A & B), the volumetric efficiency was improved by deep zone addition. One interpretation of these results is that the net effect of better efficiency and worse NTIS results from interception of short-circuits in the wide cell, but that these are less important for the high aspect cells. Thus, wide cells profit from deep zones, but narrow cells do not.

These field studies do not demonstrate a clear advantage or a clear disadvantage for deep zones. Other wetland attributes, such as aspect ratio, vegetation community type, and compartmentalization, may be equally or more important for water quality improvement. Nevertheless, deep zones may be important for habitat value, may provide for increased reaeration, and may provide opportunity for pathogen reduction via UV radiation. The amount of open water that may be included in a cell remains a controversial issue.

Speed Bumps, Hummocks, and Islands

Compartmentalization involves emergent berms with transfer structures, but such cross-berms may also be constructed so as to be always or occasionally under water. For example,

TABLE 17.1

Performance of the Champion wetlands, July 1991 – June 1993. The DTD analyses are from this work.

Wetland	A	B	E	F
Area, m^2	4048	4048	1012	1012
Aspect	2.47	2.47	9.88	9.88
DZ #	2	2	2	2
DZ Area %	5	25	5	45
Mean Depth, m	0.27	0.38	0.27	0.58
HLR, cm/d	3.16	3.38	7.82	8.13
Nominal HRT, d	8.54	11.23	3.45	7.13
NTIS	4.6	3.6	10.7	2
e_V	0.63	0.81	0.91	0.74

Source: Hydraulic data, Knight et al. (1994).

the treatment wetland at Bulwer Island, Australia, was constructed with a repeated pattern of 0.4 → 0.2 → 1.0 meter depths along the flow path (Simi and Mitchell, 1999). The intent of this design was to break up longitudinal channeling and to promote different micro-environments. Underwater cross berms have also been placed in some of the stormwater treatment areas in south Florida. It is not known if these have positive effects on treatment. The use of cross-benches was forecast to promote good hydraulics, NTIS = 23 and e_V = 0.80, by Persson et al. (1999).

Hummocks are submersed islands within the wetted area of a treatment marsh. They create shallow planting beds situated perpendicular in the flow, and can form part of the variable water depths selected in design. Their use was suggested by Thullen et al. (2005), and implemented in wetland H1 of the Tres Rios project. The 1.2-ha wetland was reconfigured to incorporate hummocks, and tracer experiments were conducted to evaluate hydraulic performance changes (Keefe et al., 2010). Approximately 40% of Tres Rios H1 was reconfigured to include hummocks. However, the reconfiguration did not significantly influence the e_v or e_{DTD} parameters.

Chang et al. (2016) used numerical methods to test the efficacy of submerged finger baffles and compare to the emergent baffles described above. The e_V values for submerged finger berms were about 70% of those for emergent finger berms of the same dimensions, and the e_{DTD} values for submerged finger berms were about 50% of those for emergent finger berms.

Guzman et al. (2018a) utilized a bench-scale model (0.38 m^2, 3 cm deep), with no vegetation, to test the effects of islands of numerous configurations. At this micro-scale, islands were more effective than finger berms. All 34 designs effectively reduced short-circuiting metric that was present in the basin with no topography. However, the standard efficiencies (e_V, e_{DTD}, λ) were not used as performance metrics. Subsequently, Guzman et al. (2018b) have advanced a design document based upon the bench-scale work. Those design recommendations must be viewed with considerable skepticism, given the lack of flora, fauna and soils in the basis data, and the scale up needed, about a factor of 10–100,000.

Thus it appears that submerged obstructions are not particularly effective compared to emergent obstructions.

Shape

In the discussion above, nearly all systems have rectangular geometry, both field and numerical studies. However, the majority of treatment marshes are not rectangular. Site constraints and desire for simplicity mean that most urban and agricultural runoff wetlands, and many polishing and surface water treatment marshes, are of irregular shape. Figure 17.14 presents a few examples, and others may be seen in prior illustrations, such as Figures 17.3, 11.14, and 11.19.

Irregular geometries are unique, making comparisons impossible. Numerical simulation is possible, but quite difficult. It is therefore necessary to gain insights on efficiencies from intersystem values derived from field studies. As might be expected, irregular wetlands display volumetric and DTD efficiencies that are not particularly high or low. Only a few studies of efficiencies for runoff wetlands have been done. Green (2016) produced a very detailed and complete analysis (Table 17.2). The Iowa values of NTIS range from 3 to 5, which may be compared to the value of NTIS = 4.1 ± 0.4 reported by Kadlec and Wallace (2009) for a collection of 35 treatment wetlands, most of which were rectangular. However, the Iowa values of e_V range from 0.57 to 0.63, which is lower than the value of e_V = 0.91 ± 0.04 reported by Kadlec and Wallace (2009). Overall hydraulic efficiencies (λ) are low, 0.48 and 0.34, for summer and spring-autumn, respectively.

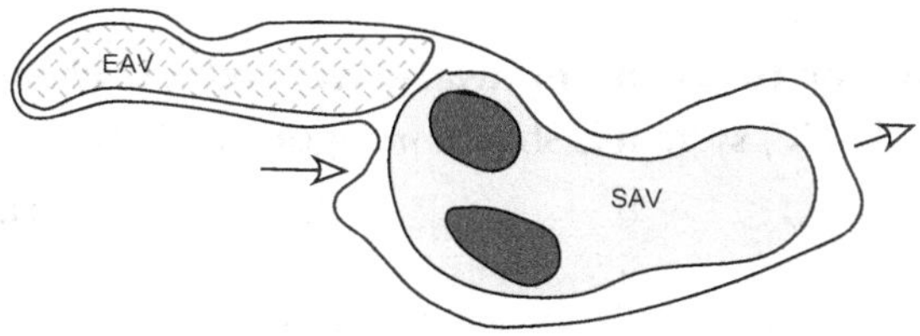

(A) Schematic of CREP Wetland W4, which receives field runoff. Shading indicates vegetation.

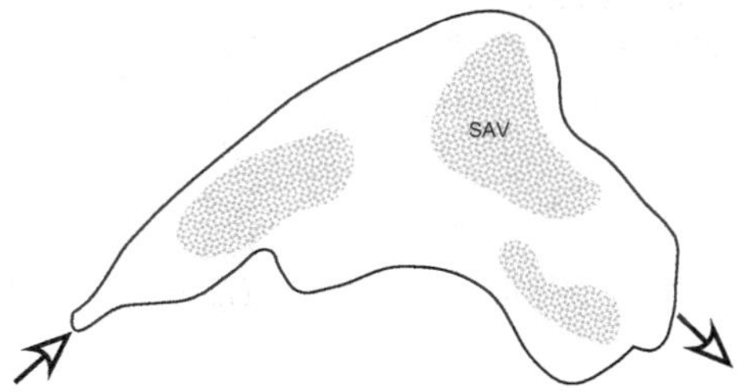

(B) Schematic of CREP weltland W5, which receives field runoff. Shading indicates vegetation.

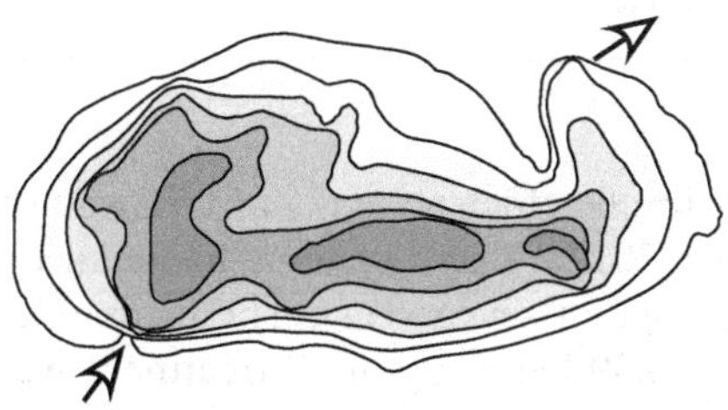

(C) Schematic of Des Plaines, IL wetland EW3, which receives river water. Shading indicates bathymetry.

FIGURE 17.14 Examples of treatment wetlands with irregular shapes. Details of e_V and e_{DTD} may be found in Green (2016) for (A) and (B), in Kadlec (1994) for (C), in Wörman and Kronnäs (2005) for (D), and in Merriman et al. (2016) for (E).

TABLE 17.2

The results of 30 tracer tests of five Iowa CREP marshes. Schematics of WL4 and WL5 are shown in Figure 17.14A and B.

—— Cover ——			—— eV ——		—— NTIS ——		—— Lambda ——	
Summer		**Spring, Autumn**	**Summer**	**Spring, Autumn**	**Summer**	**Spring, Autumn**	**Summer**	**Spring, Autumn**
WL1	92	0	0.93	0.77	6.08	1.82	0.78	0.35
WL2	93	0	0.79	0.24	3.04	1.71	0.53	0.10
WL3	40	0	0.52	0.74	3.57	2.20	0.37	0.40
WL4	64	0	0.50	0.53	10.00	5.26	0.45	0.43
WL5	99	0	0.41	0.60	2.70	3.82	0.26	0.44
Mean	**78**	**0**	**0.63**	**0.57**	**5.08**	**2.96**	**0.48**	**0.34**

Source: Green (2016).

TABLE 17.3

Data fit P-values for 10 urban stormwater wetlands. A schematic of the Jack Smith Creek, NC, system is shown in Figure 17.14E.

Wetland	P Value	Standard Deviation
Bruns Elementary	3.60	0.99
River Bend	4.22	1.89
Centennial Middle School	4.74	2.16
University of North Carolina–Asheville	3.80	1.58
North Carolina State University	3.08	0.85
Edward's Branch	2.65	1.18
Dye Branch	3.37	1.15
Jack Smith Creek 1	2.75	0.92
Jack Smith Creek A	3.48	0.76
Bass	4.00	0.95
Mean	**3.57**	**1.24**
Median	**3.54**	**1.07**

Source: Merriman et al. (2017).

Estimates of the overall hydraulic efficiencies (λ) of 25 Swedish urban stormwater wetlands were made by Al-Rubaei et al. (2017). These systems had a median size of 0.35 ha, a median WWAR of1.7%, and a median age of seven years. The median $\lambda = 0.59 \pm 0.27$, which is very close to the median $\lambda = 0.56 \pm 0.39$ based upon 28 treatment wetland tracer tests reported by Kadlec and Wallace (2009).

Merriman et al. (2017) fit the PkC* model to concentration reduction data from ten urban stormwater wetlands in North Carolina. Six constituent removals were modeled using the PkC* approach: organic nitrogen, ammonia nitrogen, oxidized nitrogen, total nitrogen, total phosphorus, and total suspended solids. Their P-values are lower estimates of NTIS (Table 17.3).

17.3 VEGETATION

Wetland systems owe much of their public acceptance to the vegetative communities they support. Plants play an extremely important role in FWS wetlands. They shade the water column, provide surface area for biofilm growth, and aid in the cycling of nutrients and organic carbon. Further discussion appears in Chapter 5.

Three plant community types are in use for water quality improvement: emergent, submergent and floating. These are largely determined by the selected wetland bathymetry, with EAV preferring water depths of 10–40 cm, SAV preferring water depths of 40–80 cm, and FAV able to exist comfortably in deeper waters.

Vegetation Type

Treatment Potential

The structure of the plant community is a more important determinant of pollutant removal performance than the species. There are numerous studies that look at the initial uptake of various nutrients and contaminants; for instance, see Zhang et al. (2011), Christiansen et al.

(2016), or Tang et al. (2017). But, to the extent that vegetation is directly involved in removals, it is the entire biogeochemical cycle that matters, not just initial uptake. Further, it is the presence of physical structures below water that determines the site densities for microorganisms and algae, which in many wetlands are a major contributing factor to removals.

Emergent wetland plants provide a wide range of treatment mechanisms in FWS wetlands, including:

1. Increased sedimentation by reducing wind-induced mixing and resuspension.
2. Additional surface area for increased biofilm growth and uptake of soluble pollutants.
3. Removal of nutrients and trace contaminants via bioaccretion.
4. Production of carbon to support denitrification.

Some mechanisms are structural in nature (1 and 2), while others are biomass-dependent (3 and 4). Consequently, selecting the plant species is not nearly as important as establishing a functional plant canopy. Thus performance results often show little difference among EAV species mixes of the same general structure. An example from a side-by-side study is shown in Figure 17.15.

Other multi-seasonal studies show much the same behavior (e.g., Liang et al., 2011; Hijosa-Valsero et al., 2012). The Liang et al. (2011) study, found that a mixed wetland community did not show significant differences in removal rates for common contaminants when compared to a monoculture wetland. The four-year mesocosm study used five EAV species: *Canna indica, Cyperus flabelliformis, Phragmites australis, Pennistum purpureum,* and *Hymenocallis littoralis.* The monoculture was *Canna indica*, and the mixed community was stable and dominated by *Phragmites* and *Cyperus.* No significant differences in removals of COD, TP, SRP, TN, NH_4-N, or NO_X-N were observed.

Submerged and floating vegetation provide much the same set of treatment mechanisms, but in different proportions. While individual studies produce variable removal rate coefficients, there is no obvious large preference for one vegetation type. As an illustration from an overall perspective, the rate coefficients for TP removal are:

- SAV N = 11 k = 49 ± 60 m/yr Knight et al. (2003a)
- EAV/SAV N = 177 k = 22 ± 17 m/yr This work

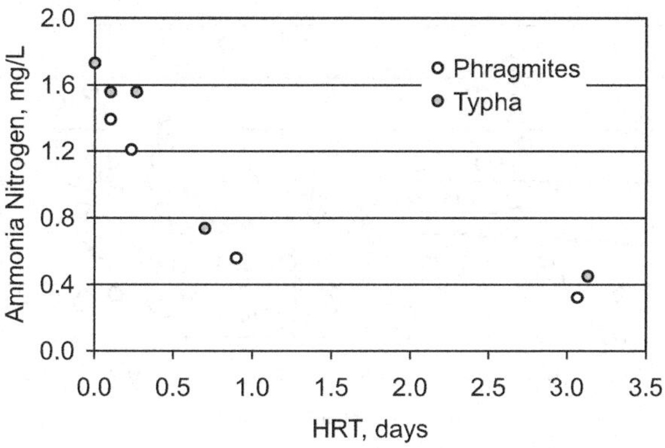

FIGURE 17.15 Reduction of ammonia nitrogen in side-by-side identical wetlands vegetated with Phragmites and Typha. Each point represents a different wetland. (Based on data from Toet, 2003.)

• EAV	N = 366	k = 20 ± 30 m/yr	This work
• FAV	N = 17	k = 32 ± 27 m/yr	Catts (2009)

The EAV/SAV designation arises from the flow ways of the Florida STAs, which consist of varying proportions of EAV and SAV. Some of the variation is due to design, but the fraction EAV has changed yearly. There is not a significant trend of P rate coefficients with EAV (Figure 17.16).

Thus, water quality performance is not a strong driver for selection of either community type or species selection and maintenance. The designer has wide latitude in the selection of plant species and mixtures thereof. Greenway (2003) studied seven wetlands in Queensland, Australia, and concluded that a large variety of aquatic plants were suitable for surface flow constructed wetlands to polish wastewater effluent. However, species such as *Typha, Phragmites, Paspalum* and *Ipomoea* were capable of displacing other species and forming dense stands. Greenway (2003) recommended a mixture of species to maximize nutrient removal and enhance the ecological value of treatment wetlands.

The mesocosm study of Bodin et al. (2012) serves to illustrate the effect of plant community type upon the various hydraulic efficiencies of a marsh. Three communities were tracer tested in triplicate, and displayed marked differences in efficiencies (Table 17.4). Emergent communities had the highest NTIS values, but submerged and mixed communities had higher volumetric efficiencies. The emergent systems also showed higher rate coefficients for nitrate removal (Bastviken et al., 2009).

Diversity

Vegetative diversity is only partially controllable in design and operation of a treatment marsh. Environmental conditions will control recruitment and development of the eventual stable community, a process of natural selection that is discussed later in this Chapter. Maintenance of a designer-selected species composition can be costly; and is almost always a losing battle. Herbiciding is usually the control mechanism, which in and of itself represents a dubious proceeding.

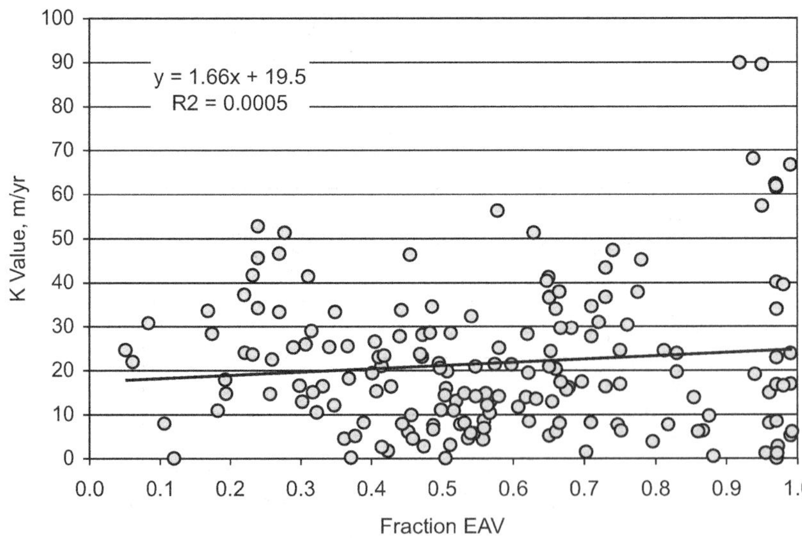

FIGURE 17.16 Rate coefficients for phosphorus removal in the flow ways of the Florida STAs. Each point represents one year for one flow way.

TABLE 17.4

Efficiencies of mesocosms with differing plant community types. EVW = emergent vegetation wetlands, SVW = submerged vegetation wetlands, FDW = free development wetlands.

	EVW	SVW	FDW
Efficiencies			
e_V	0.56	0.72	0.73
e_{DTD}	0.75	0.38	0.50
NTIS	4.00	1.60	2.00
λ	0.36	0.22	0.31
Cover Composition, %			
EAV	94	19	29
SAV	0	70	3
FAV	0	18	40
Algae	0	88	50

Source: Bodin et al. (2012).

Establishing diverse wetland vegetation is a stated goal in most wetland restorations, and to the extent practicable, should also be a goal for treatment marshes. This is desirable for a variety of reasons. A relatively large number of species means an array of environmental tolerances is represented. As the new wetland experiences fluctuations in various environmental conditions over time, such as water level, temperature, and herbivory, some plants or species will not survive, but others may thrive. If the project is dominated by a few species, there is a possibility that the project will fail with the death of only one species. Planting a variety of species increases the chances for success of at least a few species. A more diverse mix of plant species will be better able to accommodate changes in water quality and flow. The ecological benefits of diverse communities are numerous, given that the number of plant species and structural complexity of natural ecosystems generally correlate with wildlife species richness, particularly for birds. Consequently, a diversity of wetland plants within a system increases wildlife value. Various plant species association and hydrological conditions provide required habitats for different life history phases of animals, such as feeding, winter cover, and breeding.

Communities with a single dominant species are often considered to have limited ecological value. In other words, some species are regarded as weeds. Two features of plant species cause them to be deemed less than desirable: alien invaders and aggressive colonizers. In the United States, invasive species are exemplified by *Phragmites australis* and *Lythrum salicaria*. Elsewhere in the world, *Phragmites* is a valued plant, and is extensively and aggressively utilized and maintained in treatment marshes and wetlands. It should be noted that there is a native North American subspecies of Phragmites: *Phragmites australis* subsp. *americanus*. Its water quality performance is indistinguishable from the European variant (Rodríguez and Brisson, 2015). There is not agreement within the United States on what is an undesirable wetland plant: *Typha* is disliked in south Florida, *Phalaris* is considered bad in the Pacific northwest, and *Schoenoplectus* is disliked in Texas.

In some localities, the designer is presented with a list of acceptable plants, set by regulatory fiat. For instance, the Washington, DC, Stormwater Management Guidebook (Hoffman et al., 2013) states that invasive species such as cattails, *Phragmites* and purple loosestrife must not be planted. The list of acceptable plants includes those that are locally native, such as *Saggitaria latifolia, Sparganium americanum, Juncus* spp., *Scirpus* spp., *Sparganium americanum, Nuphar luteum, Pontederia cordata, Potamogeton pectinatus*, and *Elodea canadensis*.

Vegetation Coverage and Density

There are no wetlands that are devoid of all vegetation, but the proportions of macrophytes and mesophytes can vary over wide ranges. Treatment ponds (lagoons) have significant algal populations, which provide their treatment capabilities. Even open water wetlands (see Chapter 14, and Jasper et al., 2013) contain phytoplankton and epipelon. To further complicate matters, marshes can contain layered vegetation, with mesophytes and SAV comprising the understory, and EAV contributing an overstory.

The coverages and densities that develop in a marsh are not within the control of the designer. Rather, densities are dependent on water nutrient status, with higher nutrients yielding denser vegetation and more complete coverage. But in numerical fluid mechanics calculations, it is possible to vary the stem density independently, and thereby derive the efficiency parameters as a function of stem density. Sabokrouhiyeh et al. (2017) performed such computations, assuming that the stem diameter was uniform and equal to d = 5 mm, which they deemed a reasonable assumption for vegetation found in a FWS constructed wetland. The associated porosities remained high, above 97%. The resulting efficiencies are shown in Figure 17.17. The volumetric efficiency increases with stem density up to

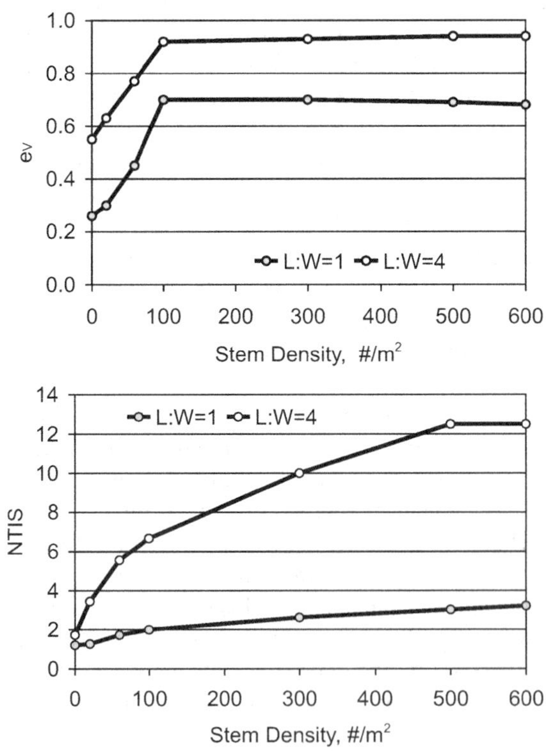

FIGURE 17.17 The calculated effect of stem density on e_V and NTIS for EAV wetlands. (Based on data from Sabokrouhiyeh et al., 2017)

a threshold at about 100 per meter squared, but NTIS continued to increase across the entire tested range of stem densities.

From the viewpoint of pollutant removal capabilities, more vegetation generally produces more capability (Kadlec, 2005d). That is the result of greater sediment removal and greater biological cycling and accretion in systems with more biomass.

The concept of adding vegetative coverage via floating vegetated rafts has gained much recent attention. As discussed in Chapter 11, such rafts do provide an improvement of a few percentage points in treatment for nutrients and TSS, compared to open water (algal) components (see Table 11.7).

Vegetation Patterns

The arrangements of internal features of a single wetland cell have very large implications for the efficiency of the cell. Both the volumetric efficiency e_V and the detention time distribution efficiency e_{DTD} are very sensitive to the vegetation patterns within the cell. It is probably these factors that give rise to the majority of the inter-system scatter in performance data.

The general categories of vegetation patterns are fringing, banded and clumped. (Figure 17.18) Many natural wetlands are riverine, meaning that the wetland contains a central

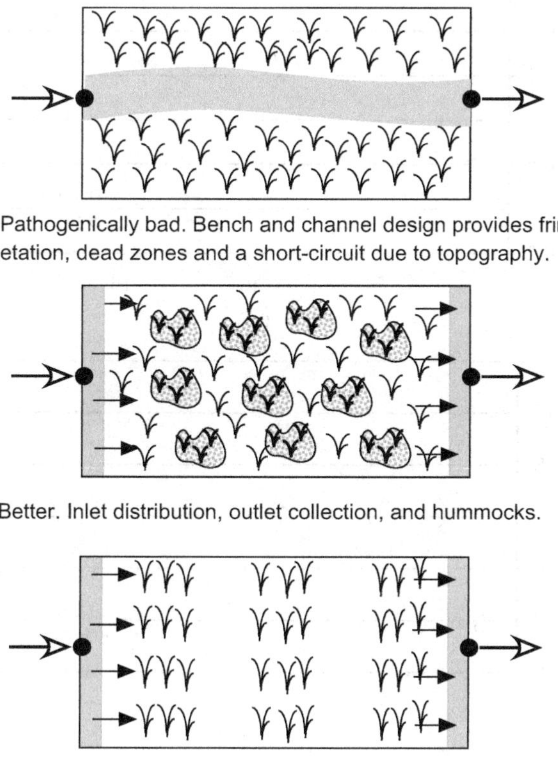

(A) Pathogenically bad. Bench and channel design provides fringing vegetation, dead zones and a short-circuit due to topography.

(B) Better. Inlet distribution, outlet collection, and hummocks.

(C) Still better. Inlet distribution, outlet collection, and banded vegetation.

FIGURE 17.18 Concepts of cell internal vegetation patterns. Fringing, patterned and banded vegetation may be used.

stream channel, which may meander through a zone of emergent vegetation. Indeed, this wetland form is so common that it is mistakenly presumed to be a good prototype for a treatment wetland. Nothing could be further from the truth. A central channel will convey most of the water, and the fringing vegetation will effectively receive flow only on rising flood conditions and discharge it on the ebbing stage. Therefore, configurations like Figure 17.18A are usually to be avoided. Simply stated, if the water needs to go through the plants, don't plant them off to the side of the flow path.

Jenkins and Greenway (2005) provide numerical calculation results for several aspect ratios and cover percentages, which show that hydraulic efficiencies are low for fringing vegetation (Figure 17.19). Their computational research also demonstrated that cross-banded vegetation is a better option for hydraulic efficiency than fringing vegetation. Therefore, any open water that may be included in the cell should be oriented across the flow direction, whether or not it is deeper than the average. However, as will be discussed in the hydraulics section later in this chapter, adequate water conveyance may be difficult to maintain if large aspect ratios are combined with significant banded or uniform vegetation. The fringing pattern results in very low computed NTIS and e_V, especially for low values of length to width ratio.

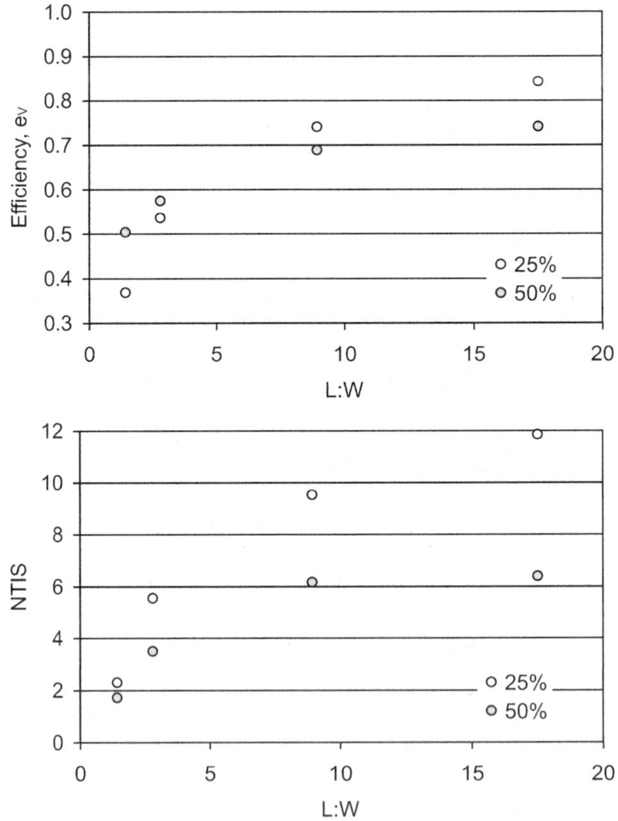

FIGURE 17.19 Efficiencies for fringing vegetation in rectangular wetlands. Upper panel: volumetric efficiency e_V; Lower panel: tanks in series. Percentages are for the fraction of the wetland area in the central unvegetated channel. Simulation data from Jenkins and Greenway (2005).

Additional numerical investigations of fringing vegetation can be found in Musner et al. (2014) and Savickis et al. (2016).

Fringing vegetation can develop in a constructed marsh, although it appears that the transition form uniform to fringing can take many years. An example is provided by the Columbia, MO system. Marsh cells had uniformly graded flat bottoms, and were planted in 1993, and placed in operation in 1994 (Kadlec et al., 2010b). Cells in unit number 1 had one point inlet and three equally spaced outlets. For the first 16 years of operation, a reasonably uniform vegetation coverage persisted, as determined from historical aerial imagery. Then, in 2010, emergent vegetation began disappearing along flow paths from inlet to all three outlets. By 2017, two-thirds of the wetland comprised a central unvegetated flow path (Figure 17.20).

It has been generally known that cross-bands of vegetation are preferable to fringing. Weisner et al. (1994) noted:

> When constructing new wetlands for nitrate removal, we suggest that deeper parts, favoring submerged vegetation (e.g., submerged species of *Potamogeton, Myriophyllum*), should alternate with shallower parts favoring emergent vegetation (e.g., *Phragmites, Typha*), along the water flow. This design should help to distribute the water flow evenly across the wetland.

These concepts were utilized in a cell of STA34 in Florida (Figure 17.21). Speed bumps were built cross-flow, with their tops under water. These were vegetated with EAV, while the balance of the area was developed and maintained as a SAV community with a large epiphytic component. The resultant efficiencies were measured to be considerably higher than for fringing vegetation.

17.4 DESIGN FOR WILDLIFE AND HUMANS

Each new treatment marsh situation brings a need to balance wildlife values and potential impairment of the structure and function of the system for water quality improvement. To some extent, wetland design can influence this balance, but management can still be an important feature of wildlife control and enhancement. Management strategies are discussed under in Chapter 18.

FIGURE 17.20 Cell 1C of the Columbia, MO wetlands. The central EAV-free zone developed in the last eight years of its 24-year life.

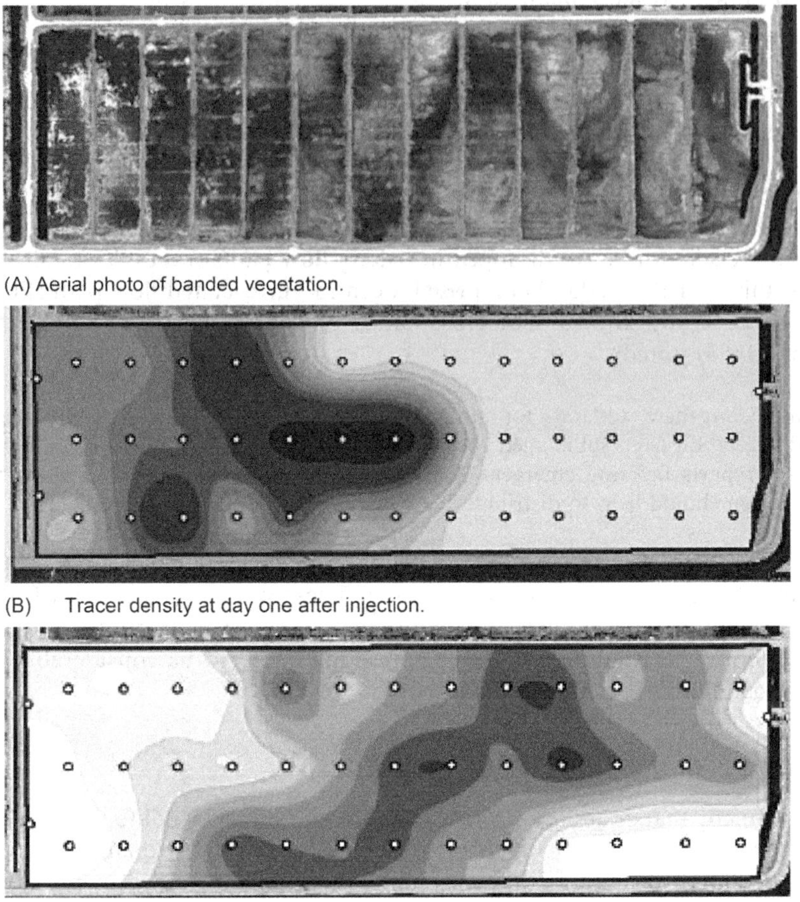

(A) Aerial photo of banded vegetation.

(B) Tracer density at day one after injection.

(C) Tracer density at day two after injection.

FIGURE 17.21 Banded vegetation in a 40-ha marsh cell (STA34 PSTA). Flow direction is left to right. Data from DBE (2009). Measured RTD parameters were: L:W = 3.38, e_V = 1.01, e_{DTD} = 0.85, NTIS = 6.5, λ = 0.85.

Design to Encourage Wildlife

The general concept of maximizing wildlife values in treatment wetlands was elucidated by Sather (1989):

> Inasmuch as plant community diversity is the key to faunal diversity, primary attention should be devoted to basic factors responsible for plant community diversity. The magnitude of ancillary benefits realized from wetlands constructed for wastewater treatment will be primarily dependent upon the degree of complexity and the amount of interspersion of the hydrophytic plant communities Engineering design, therefore, should be sensitive to the hydrologic regimes and soil characteristics required to produce these types of plant communities.

Merritt (1994) helped set the stage for the advancement of combined values. He discusses habitat requirements and the design features that promote wildlife usage. Those include islands, nesting boxes, habitat linkages, and landscape considerations. Merritt (1994) is focused upon systems in the United Kingdom. However, virtually all treatment wetlands in the United Kingdom are subsurface flow with *Phragmites*, and contain no concessions to wildlife or humans.

Worrall et al. (1997) concluded that if constructed wetlands were to support wildlife as an integral function, they should be structurally diverse, including variations in plant types and cover, water depth and substrate. Further, the wetland should be of sufficient scale and extent to support target species and populations: bigger is better.

To promote wildlife values, constructed treatment wetland designs should avoid rectangular basins and straight channels, and rather use diverse and sinuous edges and native vegetation whenever possible. It is desirable to incorporate the constructed wetland into the existing natural landscape, with margins as natural transition zones, perhaps including woody vegetated buffer areas around the site. Where appropriate, the constructed treatment wetland should provide habitat with a diversity of native species comparable to similar wetlands in the region. To the extent practicable, vegetative species diversity should be maximized, without increasing the proportion of weedy, nonindigenous, or invasive species at the expense of native species. The biological diversity of the project is likely linked to, or dependent upon, physical heterogeneity. This could include providing some areas of open water, creating nesting islands for waterfowl, and leaving some upland and buffer areas for other nesting species.

Open water sections in a cell provide better habitat value than does a uniform dense stand of emergent plants (Weller, 1978). However, habitat values must be balanced by the potential for wildlife to interfere with treatment. Excessive waterfowl use can add pathogens to the water. Muskrats, nutria, and beavers want the deep water for refuge, and can seriously impair the vegetation, hydraulics and possibly treatment (Kadlec et al., 2006).

Waterfowl populations are enhanced if open water areas are interspersed with deep emergent marsh and upland islands. An approximate 1:1 ratio of wetland area devoted to marsh and to open water will provide maximum habitat for a variety of waterfowl species (Weller, 1978). Generally, wetlands adjacent to upland habitats will have greater floral and faunal diversity in proportion to the amount of edge between adjacent ecosystems. Extending the perimeter does, however, carry added cost. Wading birds prefer a different habitat mix. These species require shallow, sparsely vegetated, littoral areas or perching substrates adjacent to open water areas. To benefit wading bird populations, constructed wetland treatment systems can be designed to provide a broad shelf of emergent marsh with water depth less than 20 to 30 cm. Deep, open water areas next to a shallow marsh provide additional foraging habitat. Open water areas and the transitional ecotones among marsh, open water, and adjacent uplands also help promote growth of herptile and fish populations, which, in turn, provide food for wading and diving birds.

Depending on regional occurrence and available habitats, other bird varieties will colonize constructed wetlands, too. For example, if living or dead trees are included in the wetland, they will serve as perching and possible nesting sites (Hair et al., 1978). Nesting boxes might attract wood ducks and owls. Upland islands surrounded by open water provide protection for ground-nesting bird species such as waterfowl.

Design to Discourage Incompatible Wildlife

Project plans may include mechanisms to control or eliminate undesirable species. Design techniques vary depending upon the perceived species of threatening animal. For instance, beavers may be prevented from plugging structures if the intakes and discharges are diffuse and noiseless, and under more than a meter of water. Inlets and outlets may be screened to block entry by larger animals, such as turtles. Chain link or coarse mesh fencing can block some intruders, such as domestic livestock.

A lack of predator habitat protects muskrats from predators such as hawks and owls. Erecting dead tree stubs or utility poles in an upright position that have sufficient height to attract raptors can serve as a perch for these aerial predators. Such structures have been very effective in the Show Low, Arizona, wetland environments. Mink, the primary muskrat predator, is ubiquitous across the range of muskrats. These small animals can and will enter treatment wetlands in search of prey and are not deterred by chain-link fencing. Attempts to allow access to larger predators, such as foxes and coyotes, proved ineffective at a remediation wetland in Hillsdale, Michigan.

Exclusion fencing is an effective, but often expensive, option for controlling access by some animals. For example, muskrats have been effectively excluded from the Ducks, Unlimited treatment wetlands at Oak Hammock, Manitoba. The exclusion fencing was erected prior to vegetating the site. Ducks Unlimited, Canada, uses 5 cm × 5 cm galvanized stucco wire mesh (hardware cloth), with about 0.30 m of the mesh vertically buried, about 0.5 m upright, and another 0.3 m set to a 45° angle sloped away from the wetland. It is important to ensure that the fencing is placed such that the muskrats cannot burrow underneath or between connected sections. Based on reports from the Ducks Unlimited domestic wastewater treatment wetland at Oak Hammock, Manitoba, similar fencing was installed at the Hillsdale, Michigan, treatment wetland, where it was also successful (see Figure 15.11) (Kadlec et al., 2006).

Riprap placement along berms and bank slopes can be very effective at preventing burrowing. This involves the placement of up to 50–150 mm riprap in a 200–300 cm layer that extends from the wetland bottom to the top of the berm. Note that while this approach reduces the potential for burrowing into the berm directly at water level, experience at Nanticoke, Ontario, showed that the muskrats burrowed down into the roadway at the topmost edge of the rip-rap, which was three meters from the waters edge along the three to one side slope. An alternative is chain-link fencing laid into the final deep zone of the wetland and extended to the top of the berm, a technique that has been used successfully at Lambton, Ontario, wetlands. There is probably a trade-off between mound building and bank burrowing, because both mounds and burrows are acceptable refugia. Thus, by denying bank access, mound building will be encouraged, and vice versa.

Design for Humans

Chapter 15 presents descriptions of the growing use of treatment marshes by humans, for both passive and consumptive activities. That expanded human use is due in major part to the incorporation of design features that promote such use. All human use is contingent upon access. For example, recreational facilities that provide public access to the STAs of south Florida include asphalt parking areas, composting toilets, landscaping, and information kiosks. In some very large treatment wetlands, boat launching facilities are provided (Figure 17.22).

Pedestrian gates, signage, and fencing are included as needed to define public access areas. Associated road improvements are sometimes needed, such as an acceleration and deceleration lane in the vicinity of the entrance to the parking lot. Some facilities include a canoe launching site, access footbridges and boardwalks. And, at selected few locations, extensive visitor centers have been built.

Although public access to a created wetland might disturb wildlife populations, disturbances can be minimized with controlled access to certain areas and with design features such as islands for roosting and nesting. Visitors typically appreciate guidance, and signage is therefore an important aspect of design for human use. Although clearly marked walking trails suffice for wetlands of limited size, boardwalks are the preferred means of protecting both the visitors and the wetland in large systems, but are not restricted to large systems.

FIGURE 17.22 Boat launch facility at STA34 in south Florida.

Boardwalks facilitate nature study and outdoor recreation, and enhance the scientific research benefits of a constructed wetland. Observation towers are valuable because they provide an opportunity for visitors to get an overview of what is generally a flat landscape. Wildlife viewing blinds can greatly enhance the ability to observe wildlife with minimal disturbance to the animals. An interpretive center allows small groups to assemble with shelter from sun and rain. Figures 15.15 through 15.22 show examples of human use features of treatment marshes.

It is not uncommon for scientists to visit treatment wetlands. This human use is typically not accompanied by any special facilities; such visitors are deemed knowledgeable enough to fend for themselves.

It is noteworthy that there are no reports of accidents involving humans and treatment wetlands.

Collective Consequences

Wetland design considerations include siting, cell size and configuration, water flow and depth control, planting, and species stocking. Because this book assumes that the treatment wetland's primary function is pollution control for wastewater, the water source itself is not an ancillary decision. However, the level of pretreatment is important to determine potential for secondary benefits, and is presumed to be high for the class of treatment marshes. Developing a wide variety of habitat type within a treatment wetland will enhance higher wildlife diversity. Controls in the design of the treatment wetland that affect habitat and plant diversity include seasonal hydroperiods, depth and flow changes, vegetative succession, and accumulation of sediments (U.S. EPA, 2000c). Habitat quality can be enhanced by providing for areas of open water, creating nesting islands for waterfowl, and leaving some upland and buffer areas for other nesting species. Because the selection of plant palette is one of the major controls on habitat quality in a treatment wetland, it is important that the following be considered in the design:

- Provide habitat with native species mixtures comparable to similar wetlands in the region (U.S. EPA, 2000c).
- Create gentle slopes to allow for plant establishment and diversity. Incorporate water level fluctuations in the operation plan.
- Maximize vegetative species diversity, without increasing use of weedy, invasive species at the expense of native species. Project plans may include means to control or eliminate undesirable species.

There are but few published papers on wetland design for wildlife and human benefits. Most focus on generalities, such as the appropriate scale for the system (Moreno-Mateos and Comín, 2010). As noted in an ancillary benefit design review by Connor and Luczak (2002), the U.S. EPA Design Manual (U.S. EPA, 2000a) "pays little attention to the conservation aspects of these treatment systems." The same may be said for Crites and Tchobanoglous (1998), Water Environment Federation (2001), and Crites et al. (2006, 2014), all of which focus entirely on treatment, and most of which focus on source waters stronger than those for treatment marshes. The book by Campbell and Ogden (1999) considers sustainable development, and how wildlife considerations may be woven into design; however Connor and Luczak (2002) note that: "the lack of detail limits its usefulness for design purposes." Only a few manuals attempt to blend the concepts of mitigation wetlands and treatment wetlands, such as Sutula and Stein (2003).

17.5 HYDRAULIC PROFILES

The primary purpose of calculating the expected hydraulic profile is to determine if the proposed wetland can pass the design flow without an excessive "pile-up" of water at the inlet. The water depths for any particular cell configuration and flow may be calculated using known procedures of varying complexity. Complicated, two-dimensional calculations may be required for complex wetland geometries. Examples of readily obtainable computer codes include the suite of Hydraulic Engineering Center (HEC) models (Brunner, 2016). However, simpler one-dimensional spreadsheet calculations often suffice to determine the approximate shape of the water surface profile through a FWS wetland cell along the flow direction. Chapter 3 presents the methods for these simpler calculations.

The water balance, friction equations and outlet structure equation combine to give the equation for depth within a single cell (see Chapter 3):

$$aW\,h^b\left(-\frac{d(h+B)}{dx}\right)^c = Q \tag{3.5}$$

where a,b,c = constants
B = elevation of the bed bottom above datum, m
h = water depth, m
H = elevation of the water surface
Q = volumetric flow rate, m^3/d
x = distance in flow direction, m

The boundary condition necessary to solve Equation 3.5 is typically a specification of the outlet water level, as determined by a weir or receiving pool:

$$H(x = L) = H_L \tag{3.6}$$

The commonly used equation for a rectangular weir is:

$$Q = C_E W_W (H_L - H_W)^{1.5} \tag{3.7}$$

where C_E = effective discharge coefficient, $(m^3/d)/(m^{2.5})$
W_W = width of weir, m
H_W = weir crest elevation, m
H_L = water surface elevation at wetland outlet, m

The results of these equations have been explored in Chapter 3, and are shown in Figures 3.8, 3.9, and 3.10. These show that small wetlands, and those with low aspect ratios, typically have flat water surfaces, and are not subject to increased water elevations at the inlet end.

Here it is assumed that the bottom is uniform across the flow direction, but possibly sloped, and a power law describes vegetative resistance to flow. The coefficients a, b, and c are selected to be representative of the type and density of the vegetation (refer to Chapter 3). The usual specification for the outlet depth is determined as a weir setting plus the height over that weir, but other devices may also result in an outlet depth specification. Because the depth is known (set) at the outlet, the integration of Equation 3.5 proceeds from the outlet, upstream to the inlet, thus resulting in the designation as a backwater calculation.

The wetland sizing procedure will have determined a hydraulic loading for the design, q = Q/LW, where Q is the design flow. The wetland width is set from the selected aspect ratio and the design area. The backwater calculation then determines the hydraulic profile, h(x).

As an illustration, consider a wetland designed for a hydraulic loading of 10 cm/day (10,000 m^3/d on 10 ha). Different aspect ratios may be chosen, in recognition of site constraints and the desire for high hydraulic efficiency provided by high aspect ratio. For this illustration, three ratios are selected: 1:1 (316 m × 316 m), 5:1 (710 m × 142 m), and 10:1 (1,000 m × 100 m). The exit weir is set for a depth of 30 cm. For the three cases, the inlet depths for dense vegetation are 33, 41 and 46 cm, respectively (Figure 17.23). These are all probably acceptable hydraulic profiles.

A field example of an unacceptable head loss situation occurred for the Imperial, California wetland. This system is comprised of four cells, with a combined aspect ratio of 30:1 (Kadlec et al., 2010a). In 2003 the hydraulic loading rate to the Imperial wetland was increased in order to determine the impact it would have on wetland performance. However, during this period the water over-topped the levees, and thus loading had to be reduced. However, the reduced hydraulic loading rates used after the over-topping event were still well above the original design flow rate at which the wetland was operated throughout 2001 and most of 2002. The levees over-topped because physical constraints

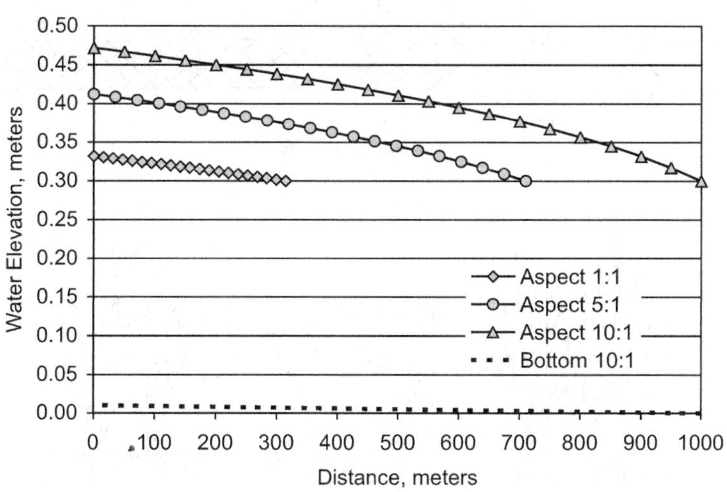

FIGURE 17.23 Wetland hydraulic profiles for an aspect ratios of L/W = 1:1, 5:1, and 10:1. The height over the weir is set at 0.3 m (left hand side). Bottom has been set as nearly level. The design hydraulic loading is 10 cm/d. Note that water surface is not a level pool, nor is it parallel to the bottom.

made it impossible for water to pass through the wetland at the rate it was being applied. This demonstrates that when considering different wetland designs, it is important to take into account the fact that the physical characteristics of the wetland may constrain the hydraulic loading rate.

17.6 DIKES, BERMS, AND LEVEES

All three of these terms are used more or less interchangeably. The earthwork that goes into the preparation of a new wetland basin is in the realm of standard civil engineering. It is not the intent here to reiterate that very large and sophisticated discipline, but rather to provide some general concepts that need to be understood by all members of the marsh design team.

Berm Design

Berms are designed based on hydraulic and geotechnical considerations. The purpose of berms is to regulate and contain water within specific flow paths. Figure 17.24 shows typical design features of constructed wetland berms for lined and unlined systems. Exterior wetland berms are kept as small as possible while still providing adequate containment and

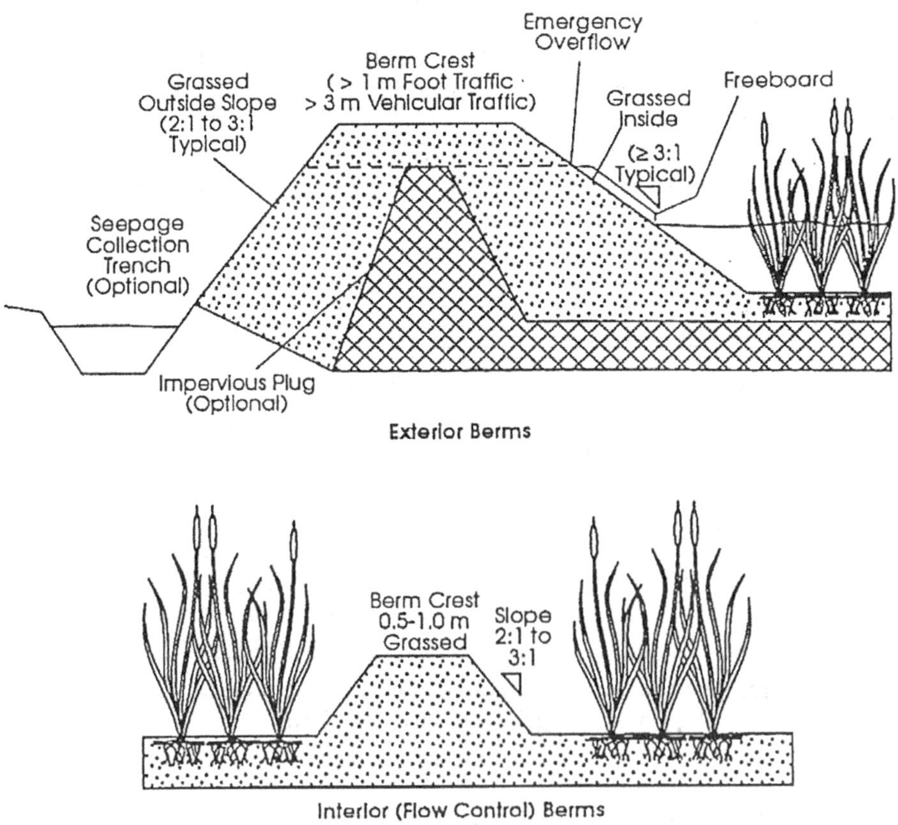

FIGURE 17.24 Design considerations for constructed wetland berms.

freeboard to prevent unauthorized flow releases. Interior berms may be used to augment flow distribution but do not have to be designed to control offsite water releases.

Exterior berm freeboard should be adequate to prevent overtopping during sudden storm events (based on a storm event frequency of 10, 25, or more years) and allow overflow of less frequent storm events through controlled and protected emergency overflow points. Berm freeboard should also consider that the wetland will gradually fill with vegetation and with mineral and organic sediments, which increase flow resistance and decrease freeboard during system life. Berm height should equal the sum of the maximum desired normal water level (for example, 45 cm), the return storm rainfall amount (for example, 20 cm for a 25-year storm event), and the lifetime loss of freeboard due to sediment and plant accumulation (up to 1 cm/yr). Thus, this example wetland requires external containment berms of 85 cm to account for the design operating depth, extreme rain events, sediment accumulation. In very large marshes, such as those in south Florida, it may also necessary to consider wind set, which increases the water level in down-wind areas at the expense of decreases in up-wind zones. Open water systems should also have freeboard adequate to contain wave action. The above example provides sufficient freeboard to deal with wind set and waves for an unobstructed fetch of about one kilometer (U.S. Army Corps of Engineers, 2000).

Marsh cells may require emergency spillways, for the purposes of berm protection in the unlikely event of water levels exceeding those planned for in design. Emergency spillways are simply armored low sections in the containment levees, with a preset elevation of their bottom. They prevent erosive washout of berms due to overtopping. A typical spillway elevation is set about 20 cm below the top of external containment berms (Figure 17.25).

Berms should be constructed on the basis of standard geotechnical considerations. The materials that are available dictate how berms will be designed and constructed. Internal clay plugs may be required to minimize berm seepage if sandy or other permeable materials

FIGURE 17.25 An emergency spillway at the Onaway, Michigan, treatment marsh. This low spot in the levee has been armored with coarse riprap.

are used for berm construction. External seepage collection channels may be necessary if external berms leak.

Berm slope is dictated by geotechnical considerations and a slope-stability analysis. Minimum berm slopes typically used in constructed wetlands are 2:1 (horizontal:vertical), and slopes up to 10:1 or 20:1 are used when a shallow littoral shelf is desired to create habitat diversity or for human safety.

Berm width should be adequate for the intended use, which is usually for access by walking or driving. A vehicle access dike needs to be more than three meters wide at the top; interior divider berms designed for pedestrian access may be as narrow as one meter. Dikes greater than about 5 m in width are less likely to be fully penetrated by muskrats. The side slopes of these are typically at a three to one slope and may be riprapped with stone to prevent erosion or rodent burrowing. The interior of a containment dike may contain a compacted core, or a layer of sealant that extends above water level. Water containment dikes are subject to local dam safety regulations.

Basin Bottom Contouring; Cut and Fill

Not all treatment marshes have contoured bottoms. Notable exceptions are agricultural runoff systems, most of which accept the antecedent landscape contours. In contrast, small wetlands may be shaped to designer specifications, as is often the case for urban wetlands.

A constructed wetland complex of cells and dikes requires earthmoving. The cost for this activity consists of two distinct components: the cost for moving soils within the site boundaries (cut and fill), and the costs for importing or exporting material from the site, from external borrow sources or to spoil disposal sites. Equal cut and fill is the preferred option, since it avoids import/export costs. Selection of the bottom elevation of the cells, together with proper positioning on the site with respect to its topography, generally allow balancing of cut and fill.

Rough estimates of the internal volume of a wetland cell may be made based upon the intended geometry. For the common rectangular geometry, the excavated volume may be calculated from:

$$V_B = \frac{h_E}{6}[LW + (L - 2sh_E)(W - 2sh_E) + 4(L - sh_E)(W - sh_E)] \tag{17.2}$$

where h_E = excavated depth, m
L = wetland length at grade, m
s = side slope, m/m
V_B = excavated volume, m^3
W = wetland width at grade, m
The volume of earth needed for the levees for one cell, neglecting the corners, is:

$$V_L \approx (2(L + W_T + 2sh_L) + 2(W + W_T + 2sh_L))(W_T h_L + sh_L^2) \tag{17.3}$$

where h_L = height of levee above grade, m
V_L = levee volume, m^3
W_T = levee top width, m
The overall depth of the created basin is:

$$h = h_L + h_E \tag{17.4}$$

This overall depth of basin is required to meet the freeboard and water depth requirements of the design. As an example, suppose that the length and width at grade are 100 and 50 m respectively, and the total depth is to be 1.5 m. The levee is set at a top width of 3 m, and side slopes of 3:1. Equations 17.2 and 17.3 balance at a cut depth of 0.45 m, with 1.05 m of levee above grade. For many projects, there is enough importance of earthmoving to warrant the use of one of the several available software packages that allow very accurate balancing of cut and fill over variable topography.

For treatment marshes, the bottom of the wetland, as well as the core of containment dikes, may be formed of compacted local soils. Clays will minimize leakage. The topsoil from the site should be stockpiled and replaced within the wetland to form a rooting medium, and for the seed bank, which it may contain. The roots and rhizomes of emergent macrophytes usually occupy the top 30–40 cm of the soil column at most. Therefore, a layer of rooting soil of that thickness should be used and needs to be considered in the cut-and-fill calculation. The original topsoil from the site may be useable; and if so, it should be stockpiled separately from the other soils during construction. This topsoil will contain seeds of the wetland plants of the region, which may assist in vegetating the wetland. If acceptable topsoil is not available at the site, it may need to be imported, or soil amendments used, to optimize plant survival and growth. Final grading to tolerances of about ±3 cm are necessary to maintain sheet flow conditions in shallow marsh areas.

A treatment marsh site may be unavoidably in the floodplain of a river. The question of protection of the wetland from flooding then arises. There are two aspects to potential flooding: maintaining the physical integrity of the system, and the effect on treatment during flood events.

Physical damage to dikes and structures is not likely to be serious if the system is not placed in the floodway of the river. The main flood currents should not impinge upon the wetlands. Several treatment wetlands have easily survived gentle inundation without any significant damage: Des Plaines, Illinois; Jackson Bottoms, Oregon; Tarrant County, Texas; and Columbia, Missouri; for example. Outlet structures should be designed to allow back-flooding (not through-flooding) of wetland cells.

Damage to the wetland vegetation is not likely for inundation up to a week or two. However, prolonged submergence, such as the midwestern U.S. floods of 1993, which lasted for many weeks, will drown the wetland plants.

The treatment efficiency of the wetland will obviously be reduced under flood conditions. Generally, the dilution effect of the flood flow is so great that reduced efficiency can easily be tolerated for the brief duration of the flood. And, some of the floodwater will receive beneficial detention in the wetland after the flood recedes.

17.7 WATER CONTROL STRUCTURES

Structures are needed to bring water into the wetland, maintain a desired water level, and to collect the discharge from the wetland. A very large number of options are available, and the choice will depend upon the magnitude of the flow to be managed. Source water may be conveyed to the marsh treatment system site by use of a pressurized forcemain, by a gravity pipeline, or by an open channel. Construction of pump stations, pipelines, and open channels are not described in this chapter and the reader should consult general references on hydraulic design and construction for more information. Once the water is brought to the wetland site, it may be distributed into the wetland via a system consisting of pipes or deep zones. Inlet control devices may be used to split the flow between parallel treatment trains. Adjustable weirs are preferred for splitter structures, as this allows the flow split to be adjusted to balance flows to the treatment trains.

Inlet Devices

For both agricultural and urban runoff marshes, point inlets are typically used. Water arrives by gravity from surface streams, tile drains or stormwater collection systems. Inlet structures are typically perceived to be unnecessary. In other situations, water flows entering the wetland complex may require flow balancing among flow paths, or complete shut-off of a cell or flow path for maintenance. Therefore, inflow structures may be needed. The complexity of inlet structures depends upon their intended use. A simple upturned pipe may suffice for small flows, or complex remote controlled gates might be needed for large flows. Examples are shown in Figure 17.26 for one flow path cell of a municipal polishing marsh; and Figure 17.27 for the single flowpath of an agricultural runoff marsh.

The inlet device initiates the flow of wastewater into the constructed wetland. Because water quality treatment is the primary objective, the inlet device initiates a sheet flow of water in the wetland by broadly and evenly distributing the influent. Multiple inlet devices may be used to distribute flows to large wetlands. The inlet device must be able to shut off during maintenance or resting of the constructed wetland. In some cases, the device is used to monitor flow and water quality. Given the strong impetus to spread the entering water across the wetland width, there remains the subject of how to accomplish that objective. The alternatives include:

- Point inlet
- Sprinklers and dribblers
- Spreader canal
- Gated pipe
- Multiple tees
- Perforated pipe
- Large rock zone
- Spreader swale
- Castellated weir

FIGURE 17.26 A simple inlet structure at the Columbia, MO, wetlands. This is the single inlet to one of the eight flow paths in this unit of the system.

FIGURE 17.27 Wetland inlet structure at the Imperial, CA, system.

A single inlet point is usually not an appropriate design based on hydraulic efficiency considerations. However, in very small systems, point inlets may be acceptable. Selection of one of these alternative inlet configurations is based on wastewater characteristics, cost, and engineer preference. In cold climates, inlet water distribution and its associated plumbing must be kept below the ice layer.

The key to successful water spreading is to provide both side-to-side introduction and head equalization from side-to-side. The latter is accomplished by use of an inlet spreader canal, or deep zone. This provides a low resistance flow path across the flow direction, and thus equalizes the water stage across the inlet. This spreader canal may be filled with coarse rock if desired for animal control.

Inflow can be distributed by gravity flow or by pressurized flow. Gravity flow is desirable because it conserves energy and reduces operation and maintenance costs. Gravity flow distribution systems may require larger pipe sizes to reduce head loss to acceptable levels. Pressurized inflows may result in high local velocities that can cause sediment erosion or physical damage to emergent wetland plants. By providing a greater number of individual outlets and greater cross-sectional area, erosion is less likely to result.

Level spreader swales and castellated weirs also provide, in theory, a gravity-fed distribution system. Outlets are hard structures that are precisely leveled during construction. These are difficult to install, because a very high degree of vertical control on positioning is required. Because these typically provide no adjustment, they are not recommended for treatment marshes.

Inlet selection is partly based on the need for fail-safe operation under the prevailing climatic conditions. Problems to be avoided include freeze-up in northern climates, and clogging by debris or animals. Freezing can be prevented by use of buried feeder mains followed by point discharge structures or submerged distribution headers. Allowance must be made for ice formation on the surface of the wetland waters. Incoming waters that originate in lagoons or other surface sources can often contain leaves, sticks and other debris. These can block small orifices in a distribution system. The collection well in the lagoon system is frequently a favorite habitat for turtles, which can block flow. Screens or other appropriate prevention measures should be considered.

If the inlet distribution system should fail, the route of the incoming water should be acceptable for continued treatment and unimpaired system integrity. This is accomplished by incorporation of an emergency overflow at the inlet (see Figure 17.26).

Outlet Devices

Wetland outlet design is important for maintaining sheet flow distribution, for controlling water level, and for monitoring flow and water quality. Many outlet designs have been used in successful constructed wetlands, ranging in complexity from a corrugated metal pipe embedded in a berm to remote controlled motor-actuated gates.

Flow distribution in a rectangular wetland basin with a single outlet point is compromised and effective hydraulic retention time is decreased. This situation can be improved by using multiple outlet weirs along the downstream berm of the wetland or by using a perpendicular deep zone to collect and route flows to a single outlet weir. If a terminal deep zone is used to accomplish this flow control goal, it should be kept as small as possible to discourage a long residence time and subsequent algal growth.

Outlet structures are designed to be the primary control on water level within the wetland. Selection of a particular device should include ease of operation, as well as ability to set the desired water level. The water level within a marsh cell is controlled, at least in part, at the downstream end of the wetland through the use of a hydraulic control structure. Having a separate control structure in each cell allows the water level in the cells to be adjusted independently. This operational flexibility is very important for managing water levels during plant establishment, allowing access for maintenance and planning for winter operations.

The hydraulic control structure should have enough vertical adjustment to accommodate the full range of water levels that will be expected to occur over the life of the facility. The outlet structure should have enough capacity to pass the peak storm events within the designed freeboard range.

For small wetlands, common outlet control structures utilize stacked stoplogs (flashboards) or adjustable weir gates. Stoplog structures made from channel iron and dimensioned lumber are not suitable for small-scale projects since their high leakage rates can interfere with maintaining a stable water level. Manufactured stoplog systems have much lower leakage rates and are better suited to wetland projects. Water level adjustment is performed through the addition or removal of stoplogs. With stoplog systems, adjustment increments are limited to the height of the stoplog. Adjustable weir gates offer the significant advantage of being continuously adjustable. A deep-water zone at the outlet provides a sump that allows for greater height adjustment, which is useful in the event that the cell must be drained.

For large applications, these weir designs can become quite sophisticated and may rely on electric power and remote control, and be combined with other flow measurement elements such as flumes (Figure 17.28).

Flows at the wetland outlet device are usually calibrated to stage and can be monitored with a staff gauge or water level recorder. For a variable outlet configuration or for a hybrid weir design, a stage-discharge curve can be prepared by using periodic volumetric flow measurements at various stage values.

An outlet control device with a fixed elevation should not be used for a marsh, as these do not provide the degree of operator control necessary for successful system operation.

Outlet weirs or pipes do not need to be sized to handle peak wastewater inflow rates because of the amplitude dampening effect of volumetric storage in the wetland cell. However, weirs should be designed to pass maximum average daily or weekly flows plus the volume of rainwater that might fall on the wetland surface from a storm event, while not compromising the design depth range. Weirs can be designed to pass these storm flows quickly (broad-crested horizontal weir) or slowly (v-notch or vertical notch weirs). Weir

FIGURE 17.28 The effluent structure at the 500 ha Lakeland, Florida, treatment wetland. The average flow is about 35,000 m^3/d. A sampling line is positioned in the center of the flow.

design is also important in quantifying wetland outflow rates. Integrating flow meters can be designed to work with almost any weir configuration.

Urban stormwater control structures differ somewhat, because of the erratic arrival of water to the wetland. It is usually not feasible to consider operation at a fixed depth. The outlet structure for the system is designed to provide a workable hydropattern for the treatment system. A fixed weir or orifice is normally not a good choice (Somes and Wong, 1997), but there are simple passive alternatives that do work well. A siphon can act to regulate the water level between two limits, with a well-modulated discharge during interim level conditions. The lower limit reflects a volume of water termed the permanent pool, while the upper limit is extended storage. A second alternative utilizes multiple outlets (Figure 17.29). Again, a permanent pool is maintained by a first (small) overflow outlet. During high flow events, that outlet is not capable of passing the full flow, and the water level rises until a second (larger) outlet begins to function. During extreme flow events, both outlets may be overwhelmed, and levels would rise to the emergency overflow spillway. A drain may also be provided, so that all water can be removed from the wetland.

Trash Racks

Outlet structures are prone to accumulation of debris. The combination of water direction and the litter production of the wetland ensure the potential for an accumulation at outlets

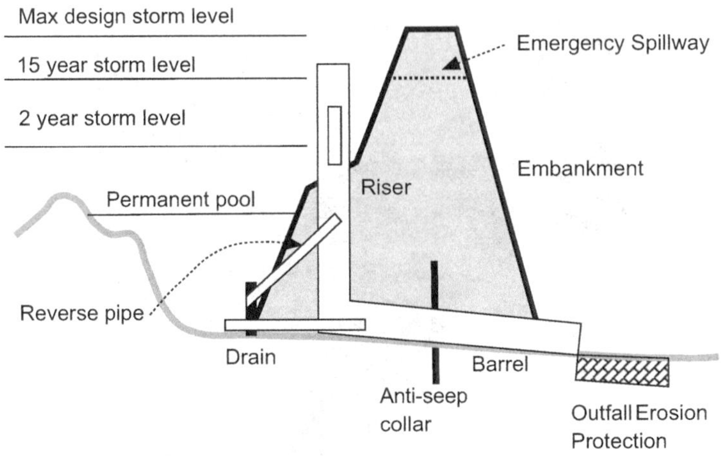

FIGURE 17.29 Example of an outflow structure for a stormwater wetland. (Based upon Hoffman et al., 2013)

(Figure 17.30). A large-mesh debris fence, placed a meter or two from the structure, can alleviate this problem, but it will become clogged and form a barrier to flow upstream of the discharge structure. The use of floating skimmer boards can keep floating debris from the outlet. Its use is coupled with water withdrawal at a subsurface location. In very large systems, automated cleaning devices have been used to good advantage.

FIGURE 17.30 Both live and dead plant material can collect at wetland outlets. This structure uses fencing to keep people out of the structure, but not vegetative debris. Photo from the Imperial, California, treatment wetland.

Pumps

Pumps are required for one of four reasons: supplying water to the wetland for treatment, removing water from the wetland to downstream uses, recycling water back to the inlet, and returning groundwater leakage to the wetland. Some systems have none of these; no North American system presently has more than three. Pumping incurs a capital cost, which is repeated on a time scale shorter than the life span of the ecosystem. Further, the operations and maintenance costs add to the capitalized total cost of the facility.

In very flat terrain, a low-head inlet pump may be needed to transfer water from the source to the wetlands at an elevation sufficient to move the water cross-country to the outlet. For the same reason, it may be necessary to utilize pumping to lower the water level at the system outlet (Figure 17.31). Redundancy is usually built into these pumping facilities, so that a portion of the capacity is normally out of service and accessible for maintenance.

If the wetland leaks an appreciable amount of water, it is a common practice to surround the wetland with a perimeter ditch to collect lateral seepage. Seepage water is then pumped back to the wetlands. This is the case in South Florida, where water conservation wetlands and stormwater treatment areas are so configured.

Recycle inevitably requires pumping. Treatment advantages that may be gained by this procedure are therefore expensive, and sometimes are more economically accomplished by increasing the size of the wetland.

17.8 CONSTRUCTION

Marsh construction is not a topic yet addressed in civil engineering textbooks, so contractors will typically need to consult the various guidance manuals that exist, particularly for urban wetlands. There is no substitute for actual experience. The purpose of this section is to provide basic information that will increase chances for successful wetland construction

FIGURE 17.31 These large pumps move treated water from the Columbia, MO, treatment wetlands to the Eagle Bluffs Conservation area.

the first time around, and to help with understanding and overcoming difficulties that may arise during construction of wetland treatment systems.

Wetland treatment system construction can be broken into two topics: site civil construction and vegetation establishment. Site civil construction is a broad discipline of engineering and has been refined through centuries of experience. This chapter does not attempt to discuss all aspects of this engineering practice, but rather describes the highlights of site civil construction that are most relevant to wetland treatment system construction. About three decades ago, there was recognition of the need for guidance for contractors concerning wetland construction (Tomljanovich and Perez, 1989). The U.S. Army Corps of Engineers responded with the production of a very detailed (705 page) *Wetlands Engineering Handbook* (U.S. Army Corps of Engineers, 2000). More specific guidance is now available in the many urban wetlands manuals (see Chapter 11), and in the many mitigation wetland manuals (e.g., Biebighauser, 2011).

The degree to which advanced construction methodology is used depends upon the size and intended purpose of the treatment marsh. Small agricultural runoff control wetlands generally utilize a minimum of construction activities. Very large systems typically do not involve interior grading. In contrast, urban wetlands often have prescriptive rules about contouring. Thus, not all of the possible construction steps are necessarily used in all projects. The following sections give a general idea of the gamut of potential construction tasks.

Site Preparation

Wetland treatment systems are usually constructed in nonwetland sites. The greatest amount of site preparation is typically required when an upland site has existing vegetation or structures. In some cases, these vegetation and/or structures must be removed to allow construction of wetland cells. Clearly, construction costs can be reduced if existing vegetation and structures are avoided during project design. Assuming that design cannot avoid these features, the first step in construction is typically site clearing. Site clearing might include demolition and removal of rubble from existing structures or alternatively involve clearing of trees, rocks, or other obstacles present on the site.

Grading and Subgrade Preparation

Following site clearing and grubbing, basic landforming can begin. Using soil grading and hauling equipment, the contractor will generally remove and stockpile topsoils from the footprint of future wetland cells to be reused for final wetland planting media. Following removal of the topsoil, the general contours of the wetland site are graded and all site features are horizontally located by standard surveying methods. The constructed wetlands site is excavated to the contours according to the detailed specifications prepared during the design process, in the form of cross-sectional and contour drawings showing the current and finished grades.

Preparation of the subgrade is an important part of the construction process. Grading tolerance specifications should be kept within ± 5 cm. Heavy construction equipment should be kept off rough grades if the subgrade is unable to support the vehicle weight without deformation. Final site grading follows construction of berms, deep zones, and islands. Final grading typically consists of leveling the wetland cell bottom to allow for a uniform thickness of rooting soil in the completed wetland. Generally, bottom elevation is constant from side-to-side within the wetland cell, and decreases slightly or is flat from the upstream to the downstream end of the wetland. Finish subgrades of the wetland bottom may have tighter tolerances, ± 3 cm, which can be accomplished by use of a laser leveling system.

Wetland cell berms are typically constructed in lifts or successive steps of placing and compacting soils. Interior berm cores and any wire meshing needed for muskrat burrow prevention should be placed in the berm during this process. Soil moisture is critical during compaction of berms and internal cell liners and construction may be delayed by excessively moist weather conditions.

As soon as berms have been shaped and dressed with any topsoils that are specified, they should be protected from erosion by use of straw and/or cover crops of grasses. Typically, a mixture of grass species is specified to allow establishment under variable climatic conditions. Hydroseeding with a mixture of grass seeds, shredded newsprint, and fertilizer has been found to be an effective method for rapid cover establishment on wetland berms. All cover crops should be irrigated as needed to insure good germination and growth of plants.

Rooting Soil Placement

Following site preparation and grading, a layer of appropriate soil is placed to serve as a medium for plant establishment and growth. Suitable substrate (at least 20 to 30 cm deep) is brought in to establish the topographic contours of the constructed wetlands to final grade. Topsoil for backfilling should consist of a soil suitable for supporting plants in that region. Rooting soil may vary from topsoil to subsoil depending on the application and the design requirements. Generally, hydric soils need not be used as fill material, because in wetlands hydrologic regimes are such that hydric parameters will develop over time. The organic content typical of wetland soils will develop as a function of both above ground and below ground plant production. However, the use of hydric soils as fill can assist in the introduction of wetland vegetative species into the system, depending on the location from where the soils were obtained. Because of the need to protect natural wetlands, it is necessary in many cases to obtain regulatory permission for the borrowing of wetland soils from off-site. The original topsoils from the site may be usable, and may be transferred from a stockpile to the wetland floor.

In those applications where the wastewater is nutrient poor (i.e., groundwater remediation or tertiary effluents), this topsoil may be amended with agricultural fertilizer at a rate typically used to cultivate a grass crop, which can be determined by soil analyses and recommendations from the local agricultural extension agent.

Piping and Structures

Wetland water conveyance systems may be fabricated offsite and then brought to the site and installed or in some cases may be totally fabricated onsite. Wastewater conveyance systems typically include piping and pump stations; water control structures such as splitter boxes, weirs, and flumes. Wastewater may be conveyed to a wetland treatment system site by use of a pressurized forcemain, by a gravity pipeline, or by an open channel. Construction of pump stations, pipelines, and open channels are not described in this chapter and the reader should consult general references on hydraulic design and construction for more information. Once the wastewater is brought to the wetland site, it is typically distributed into the wetland via some distribution system consisting of pipes, channels, or coarse rock beds. Construction related to these wetland distribution systems is briefly discussed.

Weirs and flumes should be constructed using standard site civil construction techniques. Heavy structures require adequate footings on noncompressible soils, and adequate reinforcement. If organic or high-clay soils are present in locations where footings are to be constructed, they should be removed and replaced with better geotechnical materials.

17.9 VEGETATION ESTABLISHMENT

This section deals with the process of marsh plant establishment in newly configured basins. Early literature on this subject was based upon the premise that the designer could choose the plants to be used, and that cost and robustness were the primary determinants. In the United States, that implied the use of *Typha* spp. and *Scirpus* spp. (e.g., U.S. EPA, 2000a; Crites et al., 2006). It was presumed that the wetland would be planted, and discussion focused upon where to get the plants and how to install them. A number of references were, and are, available to provide guidance with wetland plant establishment (e.g., Thurnhorst, 1993; Hammer, 1996; U.S. Army Corps of Engineers, 2000; U.S. Department of Agriculture, 2008; Romanowski, 2009). Vegetative diversity and progressive community composition changes were not considered.

Treatment marshes by definition place considerable emphasis on vegetative diversity, with the caveat that water quality goals not be compromised. Further, research has become available that shows that selected plantings almost never survive intact, and natural recruitment and competitive die-out will always affect the continually changing status of the plant communities. There are now numerous instances in which no planting is done, and natural recruitment is relied upon to create an acceptable vegetative ecology (e.g., the CREP wetlands in Iowa).

Therefore, marsh vegetation establishment is not just "select and plant." The appropriate concept is assisted natural selection (ANS), where an acronym has been awarded to emphasize the dual nature of the process.

Natural Recruitment

The previous sections of this chapter have described the creation of the requisite hydrogeomorphic features of the new wetland basin, which set the stage for vegetation development. Natural recruitment has also been called the '"Field of Dreams"' hypothesis: '"If you build it, they will come" (Hilderbrand et al., 2005). There is no doubt about that, but questions remain as to how they get there, and how many and what kind get there.

Importation Mechanisms

There are four ways in which plant propagules can reach the new wetland basin:

1. With water flows.
2. With the wind.
3. With animals.
4. With imported or antecedent soils, that is, seedbanks.

Hydrochory is defined as seed dispersal by water. Propagules entering with source waters may be expected to vary considerably. Groundwaters, industrial and treated municipal effluents will not contain seeds or other propagules. In contrast, surface waters, agricultural and urban runoff are likely bring in plant materials, which process is termed hydrochory. In most cases these will be seeds, but surface flows can also bring in whole floating aquatic plants, such as *Lemna* spp. *Eichhornia crassipes*, and *Pistia stratiotes*. Many wetland plant species use hydrochory, and seeds can travel for extremely long distances. There is a body of literature on seed water transport (see for example Nilsson et al., 2010). In the wetlands context, Soomers et al. (2013) studied and modeled the water transport of seeds of *Carex pseudocyperus* and *Phragmites australis*. Water dispersal from nonsource waters depends on aquatic connectivity in the landscape. For isolated wetlands, direct water import should be rare because water connections between wetlands are lacking.

Anemochory is defined as seed dispersal by wind. Wind driven seed dispersal is a major mechanism in natural recruitment. The import of seeds via aerial transport and deposition has been termed seed rain (Kettenring and Galatowitsch, 2011). They deployed traps to quantitatively determine the numbers of seeds falling upon wetland environments in northwestern Iowa, meaning that the seed sources were immediately adjacent. Nonetheless, their study identified the species most involved in aerial transport. Most common were invasive perennials (*Typha angustifolia/xglauca*, 864 seeds/m^2•yr; *Phalaris arundinacea*, 380 seeds/m^2•yr) and introduced annuals (*Amaranthus tuberculatus*, 3700 seeds/m^2•yr; *Setaria pumila*, 1030 seeds/m^2•yr; *Echinochloa* sp., 520 seeds/m^2•yr). Travel distances are largely unstudied, but exceed one kilometer for *Phragmites* (Soomers et al., 2013). It may be supposed that inter-wetland distance is a large factor in aerial import (O'Connell et al., 2013).

Zoochory refers to the import of plant seeds by animals. Seed dispersal via ingestion by vertebrate animals (mostly birds) is termed *endozoochory*. Animals disperse seeds of their preferred food types. Thus waterfowl disperse mainly annual seeds, whereas small mammal herbivores may disperse moderately water tolerant perennials (O'Connell et al., 2013).

Soons et al. (2016) concluded that: "… seed dispersal by dabbling ducks appears to be a major dispersal pathway for a far broader spectrum of plant species than previously considered." Their study of seven dabbling ducks (*Anas* spp.) reported 66–728 seeds found in their foreguts. Their analysis of 71 diet studies, covering a grand total of 8335 duck individuals, identified at least 445 plant species of 189 different genera and 57 different families that were ingested by dabbling ducks. The families represented by the most species are the Poaceae (grasses, 73 species identified) and Cyperaceae (most of the EAV found in marshes, 57 species). Similar numbers of seeds were found in duck fecal matter (Mueller and van der Valk, 2002), but only a small fraction was viable.

Seedbanks

Seeds of wetland plants remain viable for decades after drainage (Wienhold and van der Valk, 1989). Therefore, if a treatment marsh is established on agricultural soils that were formerly wetlands, there is an expectation that the seedbank will initiate vegetation establishment. Studies, primarily targeting the restoration aspects of newly created wetlands, have shown this is indeed the case.

van der Valk (2013) examined the seedbanks of 20 drained wetlands in Iowa. The most common taxa in wetlands of the region totaled 32; while drained wetlands in the region contained only 17 species in their seedbanks. Because restoration was the motivation of the study, it was concluded that "The results clearly show that the seed banks of drained wetlands do not contain the propagules needed to reestablish predrainage vegetation in restored wetlands." However, if the goal is to provide treatment along with some measure of diversity, those drained wetland seedbanks clearly qualify as acceptable, because they contained many of the plants that are known to function well for water quality:

Ammannia cocinnea	*Panicum* spp.
Ammanania robusta	*Polygonum* spp.
Cyperus spp.	*Rumex* spp.
Echinochloa spp.	*Leersias oryzoides*
Eleocharis spp.	*Scirpus* spp.
Juncus spp.	*Typha* spp.
Lindernia dubia	*Lemna minor*

Toth and van der Valk (2012) examined the natural reestablishment of wetlands on drained pasture lands (>30 years) in Florida. They found that obligate and facultative wetland plants quickly established, with 79% having been found in the seed bank of the drained lands. Although this seed bank contributed to the establishment of wetland communities, species composition indicated that 90% of the propagules of species that became established arrived through dispersal pathways. Again because of a restoration goal was the sole consideration, the authors were disappointed that the new plant communities did not result in predicted increase in species richness and diversity.

It is also possible to import rooting soils that contain wetland plant seeds. This approach to establishing a wetland plant community is to harvest the seed bank from a neighboring wetland (natural or constructed) that has a plant community similar to the one desired for the new constructed wetland. This seed bank is harvested by scraping the top 10 to 20 cm of topsoil or muck from the donor wetland, and then redistributing this muck in strips or over the entire surface of the new wetland. An adequate moisture regime resulting in saturated but not flooded soil conditions is absolutely essential for success of this approach. It is difficult to predict the development of vegetative cover from a transferred seed bank. Sometimes seedbank studies in greenhouses provide reasonable forecasts, but this is not always the case. The postrecruitment vegetative composition and structure from a seedbank cannot be predicted. Although many site factors, such as erosion, inundation, and drying, will affect emergence, results of a seedbank study can be used with caution to estimate the types of vegetation that will emerge on the project site.

Trusting seedbanks and dispersal does not always achieve desired results. For instance, the CREP wetlands in Iowa have had varying success in vegetation establishment (Crumpton and Stenback, 2017). Nutrient reduction has been acceptably good, but diversity and habitat values have not lived up to expectations. Accordingly, 37 of these treatment marshes were seeded with a mix containing 19 wetland plant species. That effort had no significant effect on vegetation coverage, but did increase the abundance of some shallow water species.

Trajectories

Most treatment marshes undergo a process of alteration and self-design after construction completion and the commencement of flows. The plant community that develops over time is a function of nutrients, hydrology, and climate. This adaptation process is not well understood. Hilderbrand et al. (2005) opined:

> A fundamental assumption ... is that the community and ecosystem assembly process follow a repeatable trajectory, and uncertainty is implicitly ignored. Any ecological restoration or management effort involves both explicit and implicit attempts to prescribe and predict the ecological future of a site. These efforts require extrapolating far beyond our predictive abilities.

The plants that are best adapted to the hydrologic regime and water quality of the wetland will prevail eventually, if not controlled by weeding or herbiciding. Natural recruitment begins with the sprouting in the spring season, provided that moist soil conditions prevail. The key to encouraging development of the existing seed bank is to maintain moist soil conditions, not deep- water conditions. In tropical climates, this process can be complete within one growing season, but may require two or more seasons in colder temperate climates.

It is instructive to examine the results of a long-term study of marsh development f planted and unplanted wetlands conducted in Ohio (Mitsch et al., 2012, 2014b). The time sequence of changes in plant community diversity is shown in Figure 15.3. If the frequency of occurrence is considered, it is seen that the number of species increases over time (Table 17.5), but abundant species do not (Table 17.6). Occasionals increase dramatically in number, while common plants decrease slightly. If the species that comprise the abundant category are examined, it is seen that there is no "convergence" of the dominants. Few of

TABLE 17.5

Time sequence of vegetation at the Olentangy wetlands.

	Wetland W1 Planted			Wetland W2 Unplanted		
Frequency	1996	1998	2008	1996	1998	2008
Abundant	6	9	8	10	9	7
Common	32	24	18	22	25	18
Present	30	55	66	22	45	58
Total	68	88	92	54	79	83

TABLE 17.6

Abundant plant species in the Olentangy wetlands over time.

1994 Planted	1996 Abundant		2008 Abundant	
W1	W1	W2	W1	W2
Acorus				
Cephalanthus				
		Cyperus esculentus		
		Cyperus strigosus		
		Echinochloa		
			Impatiens	*Impatiens*
Juncus				
				Leersia
	Lemna		*Lemna*	*Lemna*
			Lysimachia	*Lysimachia*
		Mimulus		
Nelumbo	*Nelumbo*			
Nymphaea				
	Polygonum	*Polygonum*		
Pontederia				
Potamogeton			*Potamogeton*	
Sagittaria				
Saururus				
		Salix exigua		
		Salix nigra		
Schoenoplectus		*Schoenoplectus*		*Schoenoplectus*
Scirpus	*Scirpus*	*Scirpus*	*Scirpus*	
Sparganium	*Sparganium*		*Sparganium*	
Spartina				
	T. angustifolia	*T. angustifolia*	*T. angustifolia*	*T. angustifolia*
			T. glauca	*T. glauca*

the plantings are abundant after 14 years, and *Typha* spp. has increased to dominate the systems.

Reinartz and Warne (1993) reported finding 142 species of vascular plants in naturally colonized created marshes of southeastern Wisconsin. They found that the diversity and richness of native wetland plants and the proportion of total plant cover that was comprised of native marsh plant species increased from one- to three-year-old wetlands. The same holds true for treatment marshes. For instance, Schwartz et al. (1994) found numerous volunteer species in a Florida constructed wetland. The system was established by planting a diverse set of 21 species. After one year, 185 species were found in the system. It is possible to start the treatment wetland with a variety of regionally prevalent plants, so that these have an opportunity to compete during the successional development of the system. This philosophy has been implemented at a number of treatment wetlands, including the 21 species at the Orange County Eastern Service Area, 20 species at Victoria, Texas (Reitberger et al., 2000), and 14 species at Wakodahatchee, Florida (Bays et al., 2000a). The Oregon Gardens treatment wetland used dozens of native wetland plants (Hill and Stoops, 2006).

The general observation may be made that the initial design does not control vegetation trajectories. Development appears to follow only a few vague and general rules:

- First colonizers are likely to be algae, presumably because their propagules are ubiquitous in many landscapes.
- The natural palette of regional species is larger than any planting list that may be developed. Development follows the hydrogeochemical features of the system, coupled with factors of climate and dispersal.
- Robust and aggressive plants tend to become the most abundant over time.
- Because treatment marsh water quality is not the most pristine, the diversity of the wetland vegetation is not as high as for pristine reference wetlands of the region.
- Water quality objectives may not reflect vegetative diversity, but rather depend upon community characteristics such as growth habit and density.
- There is no harm caused by initial planting of the system as a boost to the natural recruitment and development processes.

Planted Systems

The steps involved in providing assistance to natural selection are selection of species, selection of propagules, locating plant material sources, and installation.

Selection of Plants

This task is constrained by local rules and guidelines that contain both prescriptions and proscriptions. Within those constraints, there remains considerable latitude for the system designer to exercise best professional judgment. Factors that enter the decision process include:

- Hydrology. Natural wetlands develop a plant community that self-designs in response to the frequency and duration of saturation and/or inundation (U.S. Army Corps of Engineers, 1987). Some constructed marshes my have a simplified hydrology, if they treat a relatively constant source flow. In more variable inflows, designers should identify areas that will normally be inundated and the depth of water within these areas. It is important to evaluate the water levels at both normal and peak flow rates to ensure plant survival.
- Climate and latitude. Selected plants must be able to survive in the local climate. Since some species such as cattails (*Typha* spp.) have very wide geographic distributions, it is important to specify planting stock that is available within a reasonably close proximity to the project site and/or produced in the same climatic region.

- Cost and availability of planting stock. Plant species native to the region are preferred. Introduction of exotic plant species can result in penalties ranging from significant permitting delays to violation of state/provincial and federal laws. Designers need to understand which plants are native to the region, and among those native plants, which ones are available in reasonable quantities at a reasonable cost.
- Plant size. Wetland vegetation should be selected based upon the project budget and the size of plant that is available from local suppliers. Plants with larger root biomass will have a higher survival rate. However, larger plants cost more, so the designer must balance plant cost (and thus, plant size) against the project budget.
- Rate of colonization of the selected plant species. Plant species that spread rapidly require less planting stock, which will reduce cost and minimize the potential for weed invasion. However, aggressively spreading species will likely crowd out other plants, resulting in a less diverse plant community in the wetland.
- Project goals. Is the wetland expected to provide high diversity and high-quality wildlife habitat? If so, a more diverse planting may be appropriate. If treatment is the primary project goal, a simplified planting scheme, resulting in less diversity, may better suit project goals.
- Maintenance. Vegetation management is crucial during the first year of plant establishment, primarily via water depth control.

Establishing diverse wetland vegetation is a stated goal in most wetland mitigations, and to the extent practicable, should also be a goal for treatment marshes. This is desirable for a variety of reasons. A relatively large number of species means an array of environmental tolerances is represented. As the new wetland experiences fluctuations in various environmental conditions over time, such as water level, temperature, and herbivory, some plants or species will not survive, but others may thrive. If the project is dominated by a few species, there is a possibility that the project will fail with the death of only one species. Planting a variety of species increases the chances for success of at least a few species. A more diverse mix of plant species will be better able to accommodate changes in water quality and flow.

Plant Propagules and Sources

When considering planting the marsh as an option, designers must take steps to ensure that planting stock will be available in the quantity and quality needed for the project at a cost-effective price. Options for plant materials include:

- Commercially available planting stock. This is the most common method of obtaining planting stock for small to mid-size projects; large projects may require contract growing with a commercial nursery.
- Donor plants from another treatment wetland. This option is often chosen for small projects when the plant species involved are very easy to propagate,for example, common reed (*Phragmites australis*), reed canary grass (*Phalaris arundinacea*) or cattails (*Typha* spp.).
- Onsite plant nursery. For large projects, a plant nursery can be created at the project site to provide the required planting materials. Of course, this requires lead time in the overall schedule.
- Natural wetlands. Field harvesting materials from roadside ditches and native wetlands is labor-intensive if done manually; often plant collection permits are required.

Propagation and sale of wetland plant species has become a big business in several areas of the United States and in Great Britain and Europe. Wetland plant nurseries supply thousands of plants that are used to renovate altered landscapes. Although most of these plants are currently being propagated for use in habitat creation projects, this market has attracted suppliers who can propagate the types and quantities of plants required for constructing large wetland treatment systems. Plant propagules that are frequently used include seeds, bare-root seedlings, rhizomes, greenhouse-grown potted seedlings, and field harvested plants.

Wetland plants can be established directly from seeds, given suitable seed stock, soil moisture, light, and temperature conditions. Seed germination is highly dependent on the soil/seed contact and the moisture regime of the soil. It is preferable to establish wetland plants using a native soil rooting substrate. Some species have seeds that can be field harvested in very large numbers, while others produce few seeds. For example, a typical cattail seed head contains thousands of individual seeds, while mature bulrush culms may contain only 20 to 30 seeds each. Seed viability varies widely among plant species. If sufficient quantities can be obtained, most wetland seeds can be applied by using rotary seeders (or applied by hand) and then lightly raked into the surface soil layer. Broadcast methods have a disadvantage due to the development of areas that have poor cover, which may subsequently be dominated by undesirable species. Establishing new wetlands from seed has the advantage of lower initial cost. Seeds can generally be harvested and planted for a considerably lower cost than bare-root seedlings. However, it takes considerable time to establish a viable plant community that will provide adequate cover to allow system startup and initiation of water quality treatment. Because bare-root seedlings pass their first several months of growth in greenhouse conditions, they have a comparable time advantage over seed germination in the constructed wetland.

Herbaceous wetland plants may be provided as bare-root plants. Bare-root seedlings are typically planted in shallow, hand-dug holes. The survival rate of bare-root seedlings is significantly higher than for field germination of seeds and can generally be maintained at 80% or better with healthy plant stock and an adequate moisture regime. Initial success is largely dependent on the size of the bare-root plants; plants with small root stock typically lack the energy reserve necessary to survive the transplanting process. To produce bare-root seedlings, nurseries plant germinated seeds in flats containing potting soil. The flats may hold dozens of individual plants that are later thinned. Bare root seedlings are harvested from these flats after a few months when they have reached a height of 20 to 50 cm. Their roots are cleaned of potting soil and the plants are wrapped together with moistened paper towels for temporary storage and eventual transportation to the planting site.

Herbaceous wetland plants may also be obtained as rhizomes or tubers. Rhizomes may be shipped to the project site while still dormant; plants that have broken dormancy often have the stems clipped back to facilitate shipment and planting. Larger rhizomes have more stored energy available to support plant establishment. Sprouting is typically from nodes along the rhizome, and cuttings should have at least two undamaged nodes. They are planted in moist soil.

The energy stored in the rhizome provides the energy to get the plant started; larger tubers or rhizomes offer a much better chance of success than small stock. Rootstock in the form of tubers should be of healthy size. Plant count should be based on the number of roots or tubers, not the number of buds or shoots, since each tuber may have multiple buds or shoots.

Nursery-grown seedlings can be planted and grown in containers filled with potting soil. Potted plants have a greater advantage for initial growth over seeds and bare-root seedlings, primarily because they have larger stored energy reserves, and damage to the root structure during planting is minimal. However, the disadvantage of potted plants is the high initial cost. They are not only more expensive to grow, they are also more expensive to transport to the new constructed wetland

Field Harvested Plants

In some cases, herbaceous wetland plant stock can be field harvested and used to plant a new constructed wetland. This planting technique is used in areas with abundant natural wetlands and high regional water tables, where wetland plants such as cattails and bulrush are common in roadside ditches, man-made ponds, and along canals. A plant collection permit may be required, depending on local regulations.

Existing treatment wetlands are another potential source for planting stock. Some of the Florida STAs have had SAV transplantings done by helicopter. Source materials were extracted from one STA and air-lifted to another.

Field harvesting can occur by hand harvesting or by using a backhoe to scoop wetland plants from the ground and spread them on an open, upland area where they are separated by hand into manageable units. This method tends to be the most labor-intensive of any of the plant stock sources.

Field harvested plants have several advantages over nursery grown stock. Because herbaceous wetland plants store most of their growth reserves in their roots, rhizomes, corms, and tubers, these plant parts can produce shoots and mature plants faster than newly developing seedlings. In fact, a carefully tended rhizome generally produces numerous daughter plants in a few weeks, each of which in turn produces new rhizomes and a second generation of daughter plants. This growth results in faster development of adequate plant cover so that water quality treatment can begin earlier than otherwise possible. A second advantage of using plants that have been locally harvested is that these plants are more likely to be adapted to local climatic conditions including their phenology and genetic tolerance to temperature and moisture regimes.

Invariably, other wetland plant species are introduced into a newly constructed wetland in the rootstock and attached soil of field-harvested plants. These additional plant species come in as small seedlings or as seeds in the existing seed bank from the donor wetland. For example, at the West Jackson County, Mississippi, constructed wetland near Ocean Springs, use of field harvested cattails and the existing seed bank in the site soils resulted in the presence of 48 different plant species growing in the constructed wetlands one year after construction. This introduction of volunteer plant species is generally an advantage because it provides diversity in a constructed wetland that might otherwise be nearly monotypic.

Planting Density

Only a small fraction of the ultimate plant density is planted in a new wetland. Planting densities range from 1,000–40,000 plants per hectare, depending on the rate of spread of the selected plant species in the climatic region, and the acceptable timeframe and economics of plant establishment. As a reference, one plant per square meter, (a one-by-one meter spacing), corresponds to 10,000 plants per hectare. Through vegetative reproduction, these plants will eventually spread to much greater densities. The counting process after vegetative development usually involves stem counts rather than plant counts, with densities of 100–800 stems per square meter being common. As the first round of plants mature and die, rhizomes send up new shoots. Consequently, the rate of rhizome spread is a factor that should be considered in design, since rapidly-spreading species will eventually occupy much larger areas within the wetland than will slow-spreading species, even if the initial planting quantities were the same. For instance, *Phragmites* and *Typha* are capable of spreading 10–20 meters per year. Some species, such as *Schoenoplectus*, spread concentrically, with ever-increasing clump diameters.

The potential for plants and seeds to grow and increase their coverage on the project site depends on several factors. If hydrology, nutrients, competition, and herbivory are not limiting to plant growth, it is not uncommon in wetland projects for rhizomatous grasses, other herbs, and some shrubs planted to less than 5% coverage to reach 100% cover in less than three years by spreading vegetatively.

Planting

Most often, planting is executed by a professional planting contractor. After delivery to the job site, plant material is stored in the shade and kept cool and moist. Plant stock left out in the sun can dry out and be useless within a few hours. Most wetland plant species have extremely poor drought tolerance. Field harvested plants must be protected from excessive heat or cold, kept moist but not wet, and must be used within a few days to minimize mortality. Planting is usually done by hand on most projects. Attempts to plant field harvested rhizomes by mechanical planting have had varied success.

Plants have a growing season and are dormant for the rest of the year in temperate and subtropical climates. Annual plants die each fall or winter and must be re-established from seed the following spring or summer. Fortunately, most plants used in treatment wetlands are perennials, which lose their aboveground tissues and become dormant during the cold season and regrow from storage reserves in belowground tissues during the next growing season. In some cases, new growth is timed to take advantage of predictable rainfall and moisture conditions rather than specific day length or temperatures. Knowledge of these climatic influences is important in planting specifications and during wetland planting.

The best time to establish new plants in a constructed wetland is in the beginning or height of the growing season, which usually coincides with spring or early summer. During this period, available light is increasing daily, and competition from previously established weeds or damage from pests and pathogens is minimal. Most plants are genetically adapted to grow during this period. Also, planting at the beginning of the growing season provides ample time for full plant development and attainment of suitable plant cover before the onset of cold temperatures and declining plant growth rate. A wetland treatment system can typically begin operation when plant cover is at least 60 to 80%. If this percent cover can be obtained during the first growing season after planting, the wetland system can begin operating during the ensuing fall.

Cover Establishment

The best technique for establishing rapid plant cover is to maintain saturated soil conditions without surface flooding. The higher soil oxygen conditions resulting from the absence of floodwaters allow maximum root metabolism, effective nutrient use, and rapid development of the original plant propagules and their daughter plants within the wetland. This soil condition should optimally be maintained until the plants achieve complete cover (100%) or at least the minimum cover required for system startup (about 60 to 80%).

Once the soil is in place and the plants have been planted, water may be increased gradually to match plant growth. The best method to maintain soil saturation without excessive flooding is to start planting at the downgradient end of the wetland and continue planting upgradient while gradually raising water levels using the wetland outlet water level control weirs. When planting is complete, water levels can be dropped or raised as needed to maintain saturated soil conditions.

Incorrect control of soil moisture is the most frequent cause of failure to establish wetland plants (Figure 17.32). Inadequate soil moisture results in desiccation of roots and shoots and causes wetland species to be replaced by weedy upland plant species that may be in the seed bank of the constructed wetland soils. Too much water results in oxygen depletion in the root zone and slow growth or plant death because of insufficient oxygen for root metabolism – otherwise known as drowning. The correct amount of moisture can be maintained through adequate planning and attention during the construction period.

To maintain suitable soil moisture during plant establishment, there must be a reliable and adequate supply of water for site irrigation. This requirement can be quite large when constructing wetland systems more than a few hectares in size. When practical, the best water source for site establishment is usually the source water that will be treated in the wetland. If the wetland influent water is not available, then another irrigation supply must be provided. This irrigation source might be a potable water supply, an existing or temporary well, a pond or lake, or a nearby stream or river.

17.10 SUMMARY

The physical design of a treatment marsh requires numerous decisions concerning siting and layout. The compartmentalization and number of flowpaths are specified, as well as the aspect ratios of the cells. The internal design of a cell involves possible use of deep zones, as well as patterns of vegetation that enhance treatment. All such decisions are subject to the constraint that the wetland must pass the design flows without excessive headloss and inlet flooding. The height of the berms and depth of excavation are balanced so that material is neither exported nor imported to the site. The type and thickness of a liner are chosen, if such is necessary. Structures are selected to control the depth of water in the wetland, and to spread and collect the water efficiently.

As long as a few simple guidelines are followed and common sense is employed by contractors and field engineers, wetland project construction can help insure the success of

FIGURE 17.32 After planting, the Columbia, Missouri, wetland was allowed to dry because of disagreements about responsibility for watering. The initial planting was lost.

a good design. Site civil construction activities, pump stations, forcemains, and open-channel conveyance systems are standard experience for most wetland contractors. For wetland systems, it is important to not create conditions that will ultimately limit the wetland treatment system from accomplishing its water quality treatment or habitat creation goals.

The presence of a healthy, dense population of emergent plants is an important requirement for pollutant assimilation in wetland treatment systems. Plants provide the structural surface area occupied by the microbial flora that attack dissolved pollutants; they provide the physical filtration that reduces particulate pollutants; and they shade the water surface, preventing algal regrowth and dampening temperature fluctuations. In addition, the wetland plants provide a vast amount of fixed organic carbon important for microbial growth and denitrification, and are the basis for a complex foodweb of wetland biota.

Wetland plants can be established successfully by a number of different methods that have varying benefits and limitations. The most successful involve providing planting assistance to the natural recruitment process. Designer preference and planting cost will guide selection of the appropriate plant propagation technique. Care must be taken with the health of the plant propagules, soil and site preparation, planting, and soil moisture maintenance. The time required for vegetation establishment is often critical to meet scheduled system startup.

18 Management, Operations, and Maintenance

Most treatment marshes are designed and constructed to require infrequent operational control or maintenance. Through conservative design, and simple, low maintenance mechanical controls, wetland treatment systems will experience minimal ecological changes and will continue to meet water quality objectives for a long period of time.

The menu of activities is generally the same for all types of source waters, but there are major differences in emphasis. The list has been set forth in many guidance documents, typified by those reported by U.S. EPA (2009), Hunt et al. (2011), Walshe (2016), IDALS (2016) and Blecken et al. (2017). For present purposes, a general list is:

- Nurse the system into a fully operational state;
- Monitor and maintain water levels in the wetland;
- Maintain and control access;
- Trim and cut grass and vegetation to maintain access along pathways and embankments and in buffers around ponds;
- Monitor condition of plants, minimize need for replanting of wetland plants;
- Ensure even flow across the wetland area is being achieved;
- Keep pipe inlets/outlets to/from wetland free from plant and vegetation overgrowth;
- Repair damage to berms and structures;
- Replace worn-out infrastructure;
- Monitor condition of plants, minimize need for replanting of wetland plants;
- Remove weeds and tree saplings;
- Control/remove overly dominant plants and replant wetlands plants where required;
- Remove sediment and organic debris build up in wetland inlet (10–20 years after the establishment);
- Monitor and report.

All these potential tasks are often required for urban treatment marshes, but in contrast, very few are contemplated for agricultural runoff wetlands. For large systems, including municipal polishing marshes and surface water treatment systems, the general categories remain, but are greatly different in level of sophistication and effort.

All too often, maintenance gets lost after the system is built. Erickson et al. (2010) surveyed 19 municipalities with numerous urban wetlands in Minnesota and Wisconsin and found that 38% received maintenance less than once per year. On the other hand, emphasis should be placed on the avoidance of unnecessary operational and maintenance activities, including the management of nuisance species, when this achieves no measurable improvement.

18.1 STARTUP

It should be fairly obvious that a newly built wetland is not in a condition to perform the functions that will ultimately prevail in the system. Yet, there have been many misunderstandings resulting from startup conditions that do not meet long-term expectations. Startup transients are not necessarily brief, because it can take considerable time for the wetland biology to develop. Several processes occur during startup, importantly including:

- Plants increase in density and areal coverage.
- Senesced plant parts form a litter layer.
- Newly placed soils either release constituents if initially loaded; or absorb constituents until they are fully loaded.

Because plants form part of the "biomachine" that accretes phosphorus and other chemicals, and because plant detritus forms the sites for periphyton and bacterial assemblages, it is expected that removal will be enhanced by grow-in. One pattern of startup is typified by the rapid vegetation of a bare soil wetland after an initial planting. If that planting is sparse, the startup period will be characterized by the time for plant fill-in, plus the time for litter development. That can be relatively brief, especially in warm climates. At the Orlando Easterly Wetland in Florida, the startup duration was approximately 24 months, during which considerable change in the ecosystem took place. Figure 18.1 shows two of the principal variables: vegetation density and phosphorus rate constant. Note the close parallel between the amount of vegetation and the rate constant during the first two years. After that period, the vegetation density levels off, and the rate constant drops to its long-term average value of 13.5 m/yr. The phosphorus requirement for building the new standing crop of biomass leads to a peak uptake rate constant that is roughly double the long-term average value.

An interesting feature of the unvegetated wetlands is the absence of a carbon source to fuel the conventional heterotrophic denitrification. That process would conventionally require a supply of carbon of approximately equal mass to that of nitrogen lost (Ingersoll and Baker, 1998). A newly constructed wetland does not have the carbon source created by litter decomposition, unless the system is mulched initially with carbonaceous material.

Algae are an opportunistic first-colonizer in new wetlands, because the canopy is not yet extensive, and sunlight is available, perhaps accompanied by copious nutrient supplies (Figure 18.2). Although this initial vegetative cover develops rapidly, it utilizes only a small fraction of the nutrients in the system, and temporarily creates a pond environment.

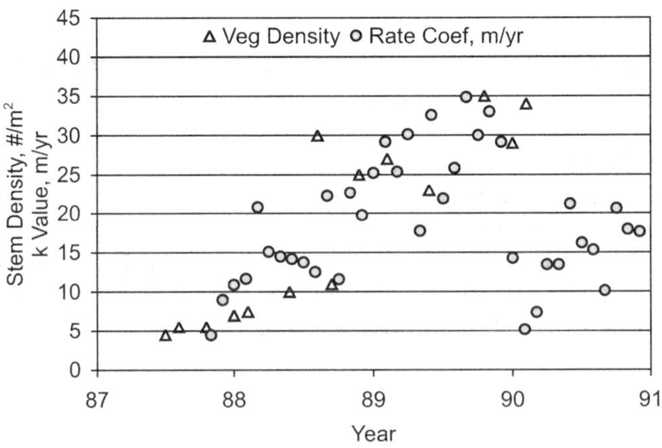

FIGURE 18.1 Vegetation density and phosphorus rate coefficient for the OEW system during startup. (City of Orlando, unpublished data.)

FIGURE 18.2 Algae were the first colonizers of this 25 ha constructed wetland cell near Carson City, Nevada.

Antecedent Conditions

The initial soils and sediments of the may contain extra nutrients or contaminants, or they may possess an initially high sorption capacity. The antecedent load of pollutants should be assessed, to determine if there is treatment liability that must be accommodated before the incoming water can be treated. The same processes that will eventually provide treatment for incoming water can also remove and immobilize such antecedent pollutants. However, if the water flow is immediately commenced, then the wetland can temporarily serve as a "source" of contaminants. If unanticipated, such releases can create regulatory problems. For instance, a wetland treatment system in Orlando, Florida, released phosphorus and nitrogen from antecedent land uses and conditions, during a startup in the flow-through mode (see Figure 10.15). The antecedent soils were laden with labile phosphorus, which was released to overlying water during startup. This initial load was absorbed and flushed over the first 200 days of operation. Despite the improvements to performance that ultimately resulted in phosphorus removal, the regulatory expectation was that removal should occur "from day one." Very large fines were levied just as the wetland came out of startup, causing closure of the facility.

The keys to avoiding such water quality startup problems are an understanding of potential treatment liability and an operations plan to avoid discharges until the wetland has "settled down" to a post- startup condition. However, it is difficult to forecast the direction or magnitude of potential startup releases. As an example, an operational strategy has been developed in south Florida for startup of the STA runoff control marshes. These systems are regulated by water quality permits. Three periods in the (early) life of a treatment marsh are identified:

- Startup phase. The permittee shall manage water depths in the treatment cells to facilitate the recruitment of marsh vegetation in accordance with an Operations Plan, which may include recirculating waters within the system. The start-up test for an individual flow-way is based on when the above samples demonstrate, over a four-week period, a net reduction in the target pollutant (phosphorus for the STAs) occurs. Discharge/flow-through operations, from an individual flow-way that has passed the start-up test may commence once documentation and all supporting data and analyses are submitted. The intent of this phase is to prevent flow-through operation until it can be demonstrated that the project is producing net benefit.

- Stabilization phase. Once flow-through discharges from a flow-way begin, the permittee shall initiate water quality monitoring for that flow-way consistent with the monitoring program. Performance during this period is not expected to be optimal, because the wetland is not yet fully developed. The stabilization test for is met when the long-term criteria for the facility are met, that is, the concentration limits are being met. If, after the two years of full flow-through operation, the facility has not met this stabilization test, the permittee is to submit a report evaluating reasons for not meeting the stabilization test.
- Routine operations phase. During the routine operations phase, the wetland is deemed in compliance if the permit effluent limitation is being achieved.

In general, the south Florida projects have achieved startup and stabilization in periods ranging from a few weeks to more than two years, which is typical of warm climates. In cold climates, a grow-in period of approximately two growing seasons may be anticipated, depending of the planting density and rate of vegetation propagation.

Vegetation Startup

Design specifications for vegetation establishment in constructed wetlands should clearly delegate responsibility for plant maintenance from the time of planting until system startup. This task may be the responsibility of the contractor or planting subcontractor, the engineer, the owner, or some combination of these parties, and should be carefully described in the planting specifications. The key requirements for healthy wetland plant propagules to succeed are water, soil, nutrients, and light. The first two components must be controlled to some extent by the engineer and the contractor, and nature generally provides the other requirements for plant growth. The establishment of vegetation is usually a task of a planting contractor, and not the facility operator. Water level control is of primary importance during the plant establishment phase. Therefore, clear lines of responsibility need to be spelled out: who is required to supply startup irrigation water, and who is responsible for watering the new plants? Failure to supply water in appropriate amounts can result in planting failure (see Figure 17.32).

18.2 WATER LEVEL AND FLOW MANAGEMENT

There are not many controls on the operation of a treatment marsh. Water depth and water flow rate are the two principal controllable features.

Flow Management

For many continuous flow wetlands, there is not a choice of the amount of water that reaches the wetland on a daily basis. Agricultural and urban stormwater systems do not typically include inflow controls. However, there are choices for some stormwater systems, for surface and ground water treatment wetlands, and for lagoon-wetland combinations. For larger continuous flow systems, there are multiple flow paths, and hence decisions on how much water to allocate to these parallel paths. Any system may incorporate recycle, which will be under the control of the operator.

Flow Balancing

In the case of multiple parallel paths, the operator has the choice of how much of the incoming flow to apportion to each flow path. Often, the adjustment of the proportion is not physically easy, but rather involves opening and closing of multiple valves or structures.

The performance of each flow path depends upon the amount of water it receives, with the optimum occurring when the hydraulic loadings to the various flow paths are equal, so that no one flow path is under worked or over worked. It is informative to examine the sensitivity of the overall system performance to the flow split, or potential imbalanced loading. To do so, consider a hypothetical example of a system with two flow paths of equal area in parallel, both of which perform according to the PkC* model.

For the conditions indicated in Figure 18.3, if the equal fractions of the flow are sent to each of the two paths, the inlet concentration is reduced by about 80%, from 100 to 19.3 mg/L. As more water is diverted to Path 2, its performance starts to decline, until the extreme when all the water goes to Path 2, doubling its hydraulic loading compared to design. At that extreme, the outlet concentration is about double the design value, at 39.3 mg/L (about 60% reduction). Of course, path 1 becomes under-loaded, and its performance increases. The combined outflow from the system shows that imbalance as well. But interestingly, a modest amount of imbalance is not particularly harmful. For the hypothetical example, a 60–40 split is forecast to raise the outlet concentration to 20.1 mg/L, about 5% higher than design.

The method of flowpath allocation varies from simple to very complex. If distribution is via a gated distribution pipe, the gates may be opened or closed at the discretion of the operator. Although conceptually easy, the effort involved in adjusting many gates is onerous, and therefore not a frequently exercised option. The same is true for multiple flowpaths with individual inlet structures. For instance, the Columbia, MO, wetland units have multiple parallel cells with dozens of manually adjustable inlets and outlets. The operator has chosen to adjust these on a weekly basis, thus requiring a large effort. At the sophisticated extreme, the Florida STAs utilize remotely operated controls on their many inflow structures. A central control facility receives water level data via telemetry and can adjust structure flows from that same control center.

The extreme of shutting down one flow path may be necessitated by a need for unusual maintenance activities, which may include repair of storm damage or protection of endangered species. To the extent that such events may be anticipated, allowance can be made in design. Water may be diverted elsewhere for treatment if such alternatives exist, or extra area and paths may be included in anticipation of path shutdowns. As an example, the Orlando Easterly system is subject to hurricanes. In 2017, in connection with hurricane

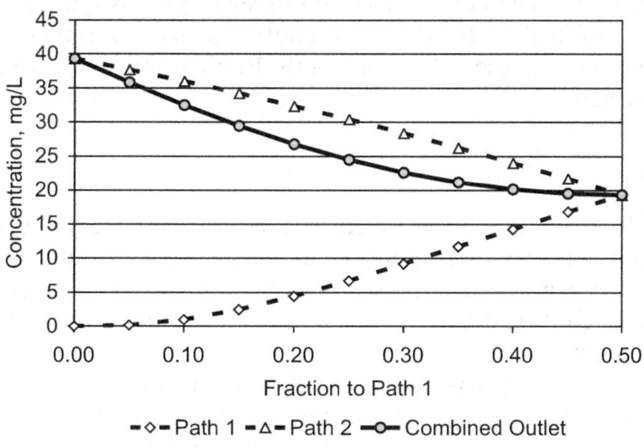

FIGURE 18.3 A hypothetical example of flow imbalance between two parallel paths in a treatment wetland, for the conditions shown in the table below. Path 1 is overloaded at the expense of Path 2.

Total Flow Q =	5000	m^3/d	C_i =	100	mg/L
Path 1 Area =	100000	m^2	C* =	0	mg/L
Path 1 Area =	100000	m^2	k =	20	m/yr
Wetland Area =	200000	m^2	P =	3	TIS

Irma, there were two months of weir stoplog adjustments at the 67 structures, totaling 100 changes (Slayton, 2018).

Systems with Storage

Flow management is an important part of those systems that involve seasonal or event storage or alternate allocation, because the stored water may be discharged according to any of several schedules. In all of these systems, the overriding goal is to bleed down the stored water in a timely fashion, to make room for the next event or for the next storage season.

In small-scale systems with winter storage, cost considerations will usually lead to storage that does not much exceed the volume of inflow over the cold season. In spring, the operator is faced with selecting the program of withdrawals from storage, to be sent to the treatment wetland. Such withdrawals may be limited by pump capacities, which would also be constrained by cost considerations, or if gravity discharge is possible, the withdrawal rates may be much less constrained. For instance, the Houghton Lake, Michigan, system has a wetland discharge season extending from May 1 to October 15 (170 days), but the transfer pump is capable of transferring the water over about 100 days. Therefore, the operator has discretion over which days there is inflow to the wetland, just so long as the storage is at its minimum level at the end of the wetland operating season. Strategies ranging from paced uniform discharge, to fast emptying followed by minimum pond level maintenance, have been used.

Bypassing

It is often not feasible to treat large event flows in stormwater wetlands, for hydraulic or biological reasons. If the hydraulics allow, treating all the water would generally result in the greater load removal, but will pass higher concentrations to receiving waters for the large events. Hydraulic integrity often requires that extreme flows be avoided, else damage to the wetland may result. Therefore, the design of such systems typically allows for bypass. This is usually accomplished by the design of control structures, rather than by overt operator action. Exceptions are the STAs of south Florida, most of which have operator-controlled bypass structures.

Depth Management

The depths of small wetlands are controlled by the outlet structures. Agricultural wetlands typically utilize weirs that may or may not be adjustable. Urban wetlands typically use passive, unadjustable multidepth outlet structures (see Figure 17.29). Larger systems use either outlet structures with controllable elevation settings, or outlet pumps. These require operator decisions and attention.

It has been shown that depth is not a crucial factor in the areal removal rates of many wastewater parameters. The extra detention time of deeper water does not foster a higher removal, and in many cases, shallower water shows better removals. It is much more important to manage water depth for the health and vigor of the plant community, and those are typically sensitive to water depth.

Depth Controllability

In very large marshes, the hydraulic resistance of the vegetation creates a dilemma for operations. The water depth is not controllable by changes in outlet structure settings, except for a small region near the outlet (refer to Chapter 3). Over most of the system, the water flow rate and the vegetation resistance will control the water depth. Therefore, it is important to distinguish between wetlands that can have depth control via outlet structures and those that cannot. In the latter case, depth control is unavoidable linked to inflow control and vegetation management, and not to any modification of structure settings.

As an illustration, consider the two hypothetical FWS wetlands, for which different level scenarios are calculated in Figures 18.4 and 18.5. Both have the same hydraulic loading rate of 5 cm/d, and both are presumed to have the same (high) vegetation resistance to flow. First consider a modest size one hectare wetland with L:W = 4. If operation is at an exit weir setting of 0.5 m, then there is essentially a zero gradient in the water surface (level pool). If the exit weir is then lowered to 0.15 m, the water level goes down everywhere, with only a few centimeters of head loss forecast (Figure 18.4) (18.9 cm down to 15.0 cm). The average depth is lowered to 17.1 cm. Next consider the long, narrow wetland, with L:W = 25. When the outlet weir elevation is 0.5 m, the inlet elevation is 3.9 cm higher, and the mean depth is 0.52 m. For this system, when the outlet elevation is lowered to 0.15 m, there is not a drawdown of the entire wetland (Figure 18.5). The inlet depth drops only to 0.48 m, and the mean depth is reduced only to 36.5 cm. Also note that the mean depth is substantially greater than the average of the inlet and outlet depths. For the specified flow, the vegetation is holding back the water, and lowering the outlet level cannot change that.

Design should be such that water depth control is at the outlet structure if possible, else the operator has no option for depth control except to lower the flow rate.

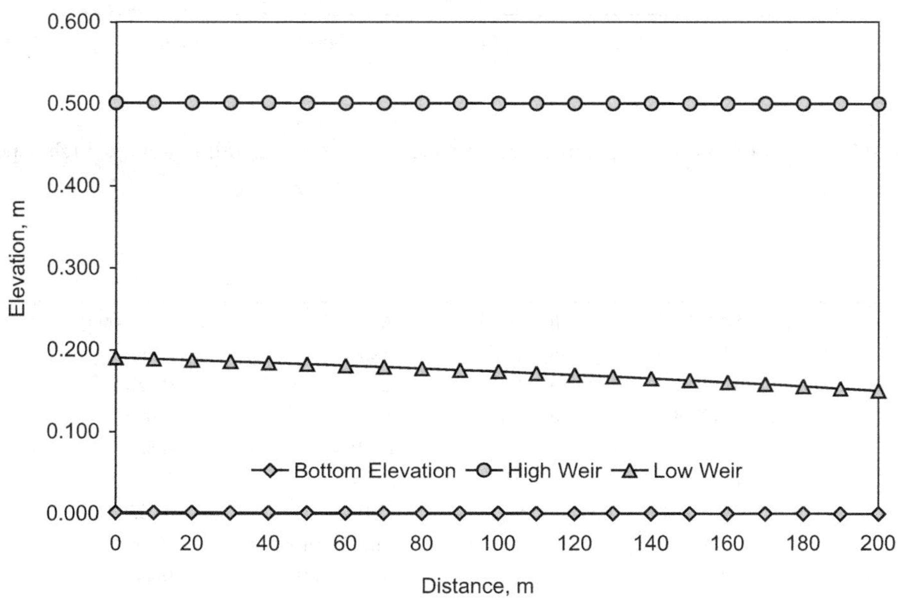

FIGURE 18.4 Effect of lowering the outlet weir setting on the water depth in a small, moderate aspect ratio wetland.

L=	200	m	A =	10,000	m^2
W =	50	M	Aspect =	4.0	
Q =	500	m^3/d	Q =	0.006	m^3/s
a =	1.00E+07	$m^{(1-b)}/d$	q =	0.050	m/d
b+1 =	3		HLR =	5.00	cm/d
c=	1		Δx =	10	m
dB/dx =	0.000010	m/m	Δh_B =	–0.002	m
High Weir =	0.500	m	Inlet Depth =	0.500	m
Low Weir =	0.150	m	Inlet Depth =	0.189	m

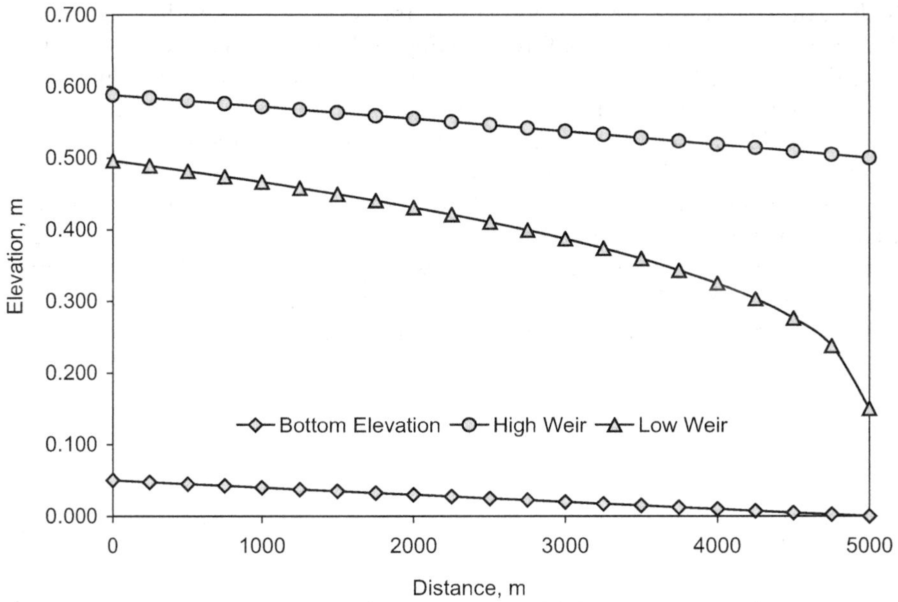

FIGURE 18.5 Effect of lowering the outlet weir setting on the water depth in a large, high aspect ratio wetland.

L=	5000	m	A =	1,000,000	m^2
W =	200	m	Aspect =	25.0	
Q =	5,000	m^3/d	Q =	0.058	m^3/s
a=	1.00E+07	$m^{(1-b)}/d$	q =	0.005	m/d
b+1 =	3		HLR =	0.50	cm/d
c=	1		Δx =	250	m
dB/dx =	0.000010	m/m	ΔhB =	-0.050	m
High Weir =	0.500	m	Inlet Depth =	0.539	m
Low Weir =	0.150	m	Inlet Depth =	0.447	m

Water Depth and Plants

Water depth is an important variable in the survival of wetland vegetation. In mitigation and habitat wetlands, it is coupled with a hydropattern to provide the desired growth conditions for any particular species. The depth that may be tolerated for continuous inundation is less than that which can be withstood for brief periods of time (Table 18.1). Some of the more robust plants, such as *Typha*, can withstand prolong periods of drought, lasting many months, and can be temporarily be completely submerged. At continuous flooding water depths of 1 m or more, emergent plants in treatment wetlands can no longer survive.

There are many sources of recommendations for water depths and associated hydroperiods for the common wetland plants (for example, Hotchkiss, 1972; Niering, 1985; Dressler et al., 1987; Boon and Groe, 1990; Thurnhorst, 1993; Eggers and Reed, 1997; Pineywoods RC&D, 2001; Shaw and Schmidt, 2003). The treatment wetland operator should understand the allowable depth and hydroperiod ranges for the plants at a particular site.

As noted in Chapter 5, floating mats are prone to form under excess depth conditions in FWS wetlands. For instance, if a good growth of cattails (*Typha* spp.) develops under shallow water conditions, it will likely have a shallow, interlocked rhizosphere. This root mat has inherent buoyancy, created by various gas-generating processes (such as methanogenesis). If water depths are then raised, the entire mat may tear loose from the rooting media and form a floating mat. If the mat covers the entire wetland cell, then it can function effectively in treatment. But if floating islands develop, they are subject to wind-driven motion, and severe disruption of the wetland. This happened in several Florida STA cells, with disastrous impairment of treatment. Removal of the plants was necessary.

Seasonal Depth Adjustments

While shallow water fosters plant growth, and is effective in promoting treatment in marshes, it can be problematic in winter, due to ice formation. In densely vegetated wetlands, the plant stems hold the ice layer in place. If water levels subsequently drop, the ice layer generally drops with it, although an air gap can form under the ice at some localities within the wetland if the ice sheet is held up by plant stems. However, this anchoring of the ice layer can create severe damage if water levels are subsequently raised. The ice layer floats and tears out embedded vegetation as it moves upward. If the ice layer becomes thick

TABLE 18.1

Water depths for common treatment wetland plants.

Common Name	Scientific Name	Best Depth cm	Flood Depth cm
Bulrush	*Scirpus californicus*	30–60	120
Soft-Stem Bulrush	*Scirpus validus*	15–30	60
Broadleaf Cattail	*Typha latifolia*	30	60
Narrowleaf Cattail	*Typha angustifolia*	30	60
Common Reed	*Phragmites australis*	30–60	90
Arrowhead	*Sagittaria* spp.	3–8	30
Yellow Flag	*Iris pseudocorus*	3–8	45
Pickerel Weed	*Pontederia cordata*	8–15	30
Soft Rush	*Juncus effusus*	8–15	25
Woolgrass	*Scirpus cyperinus*	3–15	30

Source: Pineywoods RC&D (2001).

enough, under-ice flow is impeded, which can lead to over-ice flow (Kadlec, 1987). Water flows upward through holes or cracks, emerges and flows downgradient over the ice and under the snow. For example, this mode of flow is experienced each winter at the Frank Lake, Alberta, treatment wetland.

It is possible to provide more space for flow under the ice by raising water levels prior to ice formation. However, care must be taken that such increased water depths do not entirely submerge the plants, because the oxygen transfer necessary for their survival is in major part obtained via the standing dead stalks and their airways. Indeed, an effective method of destroying cattails is to cut them and over-flood the stubble. This procedure of water level increase was used effectively at the Listowel, Ontario, wetlands (Herskowitz, 1986), and was employed at the Brighton, Ontario, wetland.

18.3 CLEANING, REPAIRS AND REPLACEMENTS

It is obvious that wetland operation involves keeping the physical features in good condition, including inflow and outflow structures, pumps, piping, berms and levees.

Structure Cleanout

Treatment wetlands export plant debris, primarily as floating detritus (see Figure 17.30). Blockage of maintenance manholes, outlet weirs and structures, outflow piping, and rodent guards by dead floating dead leaf litter from emergent vegetation, floating aquatic plants (duckweed and algae), and large turtles can play havoc when trying to maintain flow rates and water levels.

In the U.S. EPA urban stormwater wetland maintenance guidance manual (U.S. EPA, 2009) it is noted that clogged low flow weirs represent the most frequent, persistent maintenance item. However, without frequent maintenance, even openings with trash racks can become clogged. Clogging occurs when debris or sediment accumulates at riser/weir openings or outfalls, blocking the flow of water. Debris includes vegetative materials, and also other litter and trash. Vandalism and nuisance problems such as beavers contribute to clogging as well.

As an example, at Florida STA5, wind-driven floating aquatic plants, and dead material after these were herbicided, caused blockage of outflow structures (Figure 18.6). As can be seen, there is a problem of transport and disposal after removal.

For small wetlands, structure cleaning is a manual task, and there are not large amounts of detritus to dispose. But low-tech structures are also used in some large wetlands. For example, the Orlando Easterly wetland has 67 stoplog structures that are manually maintained. These are cleared several times each year (Slayton, 2018). For larger treatment wetlands, the situation can be more complex, because of the size of the structures. Automatic self-cleaning trash racks can be used to span inlets that are several meters in width (Figure 18.7). Removal of the debris then requires truck transport.

Berms and Levees

Berms and levees require inspection and maintenance. Routine activities include mowing, and pothole repair on drivable levees. Bank erosion is a common problem for small urban systems. Erickson et al. (2010) report that 11% of urban systems surveyed in Minnesota and Wisconsin required berm maintenance. A primary concern for urban systems is berm leakage. As noted in U.S. EPA (2009), leaks through embankments or wetland bottoms are often difficult to locate. They also observe that if the permanent pool becomes low during or immediately following construction, it can be a sign of poorly compacted berms.

FIGURE 18.6 Plant debris buildup at STA5 in Florida.

FIGURE 18.7 The automated trash rack at the effluent pump station for the Everglades Nutrient Removal Project. Rakes pull the clogging floating vegetation up the inclined bar screen, and deposit it on a conveyor which removes it to a stockpile. The drained material is removed by truck.

Guidance for berm maintenance for CREP agricultural treatment marshes is provided in IDALS (2016). Because the only berms used in these wetlands is an outlet dam with a weir, the outlet dam is termed the "wetland structure." It is recommended that the dam be mowed and that woody vegetation be removed. Trees and other woody plants growing on the structure are perceived to cause problems, because their roots will weaken the dam. Herbiciding is to be used if necessary

Berm damage can occur as a result of animal activity, as discussed in the following section (see Figure 18.11). Animal control may be required in addition to repairs. Vegetative damage is also likely, and nuisance species may need removal.

Erosion damage, from wave, rain or flood action, will occur if the original design and construction did not provide adequate stabilization (Figure 18.8). Extreme events are the principal culprit. For instance, substantial rainfall associated with hurricane Irma in 2017 caused washout of some berms in the OEW, which were subsequently repaired (Slayton, 2018). In some regions, hurricane force winds are to be expected with some regularity. These cause abnormal wave size and intensity that can exceed any bank protection measures.

In almost all regions, periodic flooding can be encountered. Flood events of 100- to 300-year return frequency are presently occurring at shorter return intervals than historical. The author has been involved in a dozen such events impacting treatment marsh projects.

Poor construction procedures can result in a need for repairs. Improper materials and installation invite severe damage from water even under normal operating conditions (Figure 18.9).

FIGURE 18.8 This berm at the Hayward, California, treatment wetland was need of maintenance, because of wave erosion.

FIGURE 18.9 This collection canal in the Lakeland, Florida, system was built using unstable materials from on site. Despite the attempt to reinforce the embankment with concrete matting, erosion caused the discharge structure to drop into the water. Extensive repairs were necessary.

Whatever the cause, repair and restabilization with riprap or appropriate berm plantings are required.

Pumps and Structures

Conveyance and Distribution Piping

Marshes that use surface conveyance and distribution piping are most at risk for maintenance required because of vandalism. The Houghton Lake system, which was and is open to unsupervised public access, has dealt with a number of such incidents.

- Unknown persons took an axe and chopped holes in the pipeline leading out into the marsh.
- The irrigation piping is assembled in sections, held together by clips that join two 10-meter sections. These are designed for ease of assembly by the irrigator. In the marsh, they were removed and thrown in the water.
- The water leaves through openings (gates) in the header pipe, with flow control achieved by slide valves arranged for ease of irrigator operations. It was apparently entertaining to someone to knock off several of these slide closures.
- Water is split between two gated distribution branches that span the width of the wetland. Flow to the left and right branches is controlled by multiple-turn hand crank valves at the tee. An unknown visitor decided they should both be cranked shut just before commencement of water deliveries – unbeknownst to the operator. When the transfer pump was turned on (two kilometers distant at the source pond), it was faced with a closed delivery system. The distribution piping blew apart at the tee.

Needless to say, repair activities ensued in each case.

As noted earlier, such multiple outlet surface distribution systems require operator attention, and a basic understanding of the intent of the system. These may not always occur. As an illustration, the Drummond, Wisconsin, lagoon-wetland system utilizes a pump to seasonally transfer treated water through a half-kilometer of forest to the gated distribution header in the treatment wetland. One season, an operator turned on the pump but failed to inspect the piping and distribution system for the entire season. As it happened, the transfer piping had become disconnected in the forest. That year, the entire contents of the lagoon were applied to the forest and did not reach the wetland. A subsequent operator went to an opposite extreme, and decided that only a few of the dozens of outlet gates should be open. Again, water was forced out of the transfer system joints, with almost none reaching the wetland. Because this happened just before a monitoring visit by the author, the situation was soon corrected.

These anecdotes serve as warning: operators must know how the system is to function and make frequent inspections.

Boardwalks

Whether their purpose is to support distribution piping or to facilitate visitor access, boardwalks do not last forever. The lifespan the Houghton Lake boardwalk was about 25 years, after which it became dangerous to walk on, and was replaced. It may be necessary to replace worn out boards before that time.

Pumps

In general terms, pumps and piping may last on the order of thirty years, and repair frequencies are known. Small pumps require rather infrequent inspection, and maintenance consists of lubrication, checking for wear and corrosion, checking for leaks, and appropriate replacement of seals and packing. They do not require the presence of an operator. Very large pumping facilities are referred to as pump stations and can require one or more full-time operators. In the case of the STA pump stations, there may be three shifts of operational staff. Pump life expectancies are typically about thirty years, but that is less than the life expectancy of the treatment marsh, which is 50 years or more. Consequently, pump replacement is likely to occur during the life of the system.

18.4 CONTROL OF NUISANCE ANIMALS

Experience has shown that a large variety of creatures may interfere, in some way, with the desired operational conditions treatment wetlands. The list of animals that have caused some degree of concern includes: deer, cattle, wild hogs, monkeys, groundhogs/woodchucks, prairie dogs, badgers, pocket gophers, and Richardson ground squirrels. However, most problematic are some birds, rodents, some fish, and mosquitoes.

There are over 2000 species in the order rodentia, but it is primarily the water-loving types that are of concern in treatment wetlands. There are design techniques that can help in controlling muskrats, nutria and beavers in treatment wetlands, but often these are not completely successful, and the operator should be prepared to exercise appropriate control measures.

Birds

Birds affect marsh management in several ways:

- They can cause damage to vegetation.
- Required encouragement of nesting can curtail operational and maintenance activities.
- They attract birders, thus creating human use management issues.
- Hunting of waterfowl (agricultural marshes) creates human use management issues.

Damage

Waterfowl can be problematic, because of their large size, and their large numbers during migration. Canada geese (*Branta canadensis*), ducks, and other waterfowl are known to cause significant depredation through grazing and uprooting of plant materials in wetlands. Constructed treatment marshes located in a major migratory flyway are virtually guaranteed to experience large numbers of birds, and the attendant effects on vegetation and pathogens. Municipal water sources are usually at moderate temperatures year-round, which in cold climates means that areas of open water in the marsh persist throughout the winter. These become the refuge for overwintering waterfowl, often at high densities (Figure 18.10).

Wetlands in an urban setting are very attractive to geese, which congregate in large numbers at such isolated pockets of habitat. It is herbivory that can cause considerable grief for operators. For example, the treatment wetland at Commerce Township, Michigan, was denuded of vegetation early in its life. The presence of uncontrollable deep water, and the continued presence of large numbers of waterfowl, have thwarted efforts at vegetation reestablishment (Figure 18.10).

Even if their grazing does not seriously harm the wetland vegetation, they can deposit large quantities of fecal matter on walking paths and grassy margins of the wetlands It is herbivory that can cause considerable grief for operators, especially during the vegetation establishment phase (refer to Figure 15.7). Young plants and tubers are preferred food for geese, swans, and some ducks. If the wetland planting happens to coincide with either the spring or autumn migration, the wetland can be essentially denuded in a day or two.

FIGURE 18.10 Waterfowl at the Commerce Township, Michigan, municipal wastewater polishing wetland. Residents feed them.

U.S. EPA (2009) offers only a few means of control of waterfowl in urban wetlands: using dogs to chase the birds, using proprietary products for discouraging waterfowl, and introducing (unidentified) predators. However, an exhaustive search of bird control techniques was conducted by Bishop et al. (2003). They identified auditory deterrents (3), visual techniques (15), chemical techniques (3), exclusion procedures (3), habitat modifications (6), and lethal techniques (3). In urban settings, attempts at control may be frustrated by humans who like to feed and observe the waterfowl.

Operational Considerations

Bird use creates two operations management considerations, the encouragement of breeding populations and the protection of endangered species. Agricultural systems in the United States have sponsor requirements to foster duck populations. Management must be completed prior to or after the nesting season that runs from May 15 to August 1. This is the primary nesting season for Iowa. Banning any maintenance activities during this time is used to prevent harassing and/or destroying birds as well as their nests or their young (IDALS, 2016). Between 2005 and 2011, CREP wetlands contributed to an estimated net increase of 1.5 million additional ducks per year.

There are a number of threatened and endangered birds that make use of wetlands in general, and some are attracted to treatment marshes. According to federal regulations in the United States, their presence requires that the area in question be managed on their behalf. Under the Endangered Species Act, the U.S. Fish and Wildlife Service (USFWS) is mandated to identify and list plant and animal species that are threatened or endangered by extinction, or that are considered likely to be threatened in the future. Threatened and endangered species cannot be harmed, killed, or otherwise negatively affected by human activities. The USFWS reviews potential impacts to federally listed species and can veto a discharge permit by issuing a "jeopardy" opinion. The potential for occurrence of threatened or endangered species inhabiting a project site should be considered during project planning.

Experience indicates that the presence of federally protected species does not always necessitate substantial changes to wetland treatment project siting, design, or operation. In a number of cases, constructed treatment marshes have enhanced the population of threatened or endangered, wetland-dependent species by creating additional habitat and food resources. For example, populations of bald eagles (*Haliaeetus leucocephalus*), wood storks (*Mycteria americana*), and snail kites (*Rostrhamus sociabilis*), all federally protected bird species, have increased in and around the 485-ha Orlando Easterly constructed wetland treatment system in Florida. Conversely, the presence of protected species may require modifications of operations. For instance, the STAs of south Florida are attractive to burrowing owls and stilts, which nest in or on the levees. Therefore, flooding depths that endanger their nests are precluded. The treatment wetland operator should be aware that operations may have to be modified if threatened or endangered species begin using the wetland.

Fish

Fish may be problematic in treatment wetlands, particularly those that feed or nest on the bottom. Cyprinids, which include the common carp (*Cyprinus carpio*), are omnivorous. They eat aquatic plants but prefer to scavenge the bottom for insects and crustaceans. In so doing, they disturb sediments and create turbidity. They naturally live in temperate climates, and prefer temperatures in the range 20–30 °C. However, they easily survive in ice covered water bodies, and can withstand temperatures higher than 30°C.

Also problematic are fish of the genus *Tilapi*a. They are considered invasive species in the USA and elsewhere. They consume floating aquatic plants (e.g., duckweed [*Lemna* sp.]), submerged aquatic plants, and some forms of algae. They are warm-water fish, and do not survive below temperatures of 7–17°C, depending upon species. They therefore do not pose a threat to temperate ecologies but do disrupt tropical and semi-tropical environments such as those found in Texas and Florida in the United States. They are found in vast numbers in some cells of the Florida STAs.

The American gizzard shad (*Dorosoma cepedianum*) is planktivorous in its early life, and switches to zooplankton when it becomes older. Their consumptive demand is so heavy it can cause collapses in the zooplankton community (Shepard and Mills, 1996). It ranges from Florida north to about 40°north, from the east coast to the central plains of the United States.

Management for fish control is an infrequent occurrence in treatment marshes. Examples from Des Plaines, IL, and Corsicana, TX, have been discussed in Chapter 15 (see Figures 15.8 and 15.9). Control involved drawdown of the water levels to strand them and expose the fish to predators.

Muskrats (Ondatra Zibethicus)

This species has a range stretching from the southern border of the United States to the Arctic Circle in North America, with the exception of Florida, parts of Texas and California. They are virtually guaranteed to move into treatment marshes soon after their establishment. Muskrat damage has been discussed in Chapter 15. Muskrat invasions of treatment wetlands are sometimes quite impressive in terms of numbers. Knowlton et al. (2002) note that muskrats have been problematic at the 37-ha Columbia, Missouri, project, with removal of over 1,300 muskrats in 2.5 years.

Kadlec et al. (2006) describe control measures that have been employed at several treatment marshes, which include live, lethal and colony trapping, habitat modification, encouragement of predators, noise and scent deterrents, toxicants and shooting. Further descriptions of control measures are available (Hygnstrom et al., 2014; Miller, 2018). For urban wetlands, muskrat management may be contentious, with strong feelings both for and against their presence. Agricultural wetlands either ignore muskrats, or rely upon the design of the containment berms to mitigate muskrat damage. Tolerance may be the best procedure, provided that the berms of the wetland are not compromised by muskrat burrows.

Mink (*Neovison vison*), the primary muskrat predator, is ubiquitous across the range of muskrats. These small animals can and will enter treatment wetlands in search of prey and are not deterred by chain-link fencing. But there is not any known way to establish or enhance their populations, and hence they are not a feasible management tool. Trapping is an effective method of control, and includes live, leg-hold and colony traps. Live trapping involves the capture of live muskrats, which can be released outside of the treatment wetland environment. Lethal foothold traps capture muskrats around their feet and typically the animals are killed after capture to harvest their pelts. Such traps are nonspecific, and a drawback is that they also capture muskrat predators, especially mink. Colony traps can capture multiple muskrats at once and can be used as lethal or nonlethal devices. For muskrat captures, a colony trap is commonly placed underwater to quickly drown muskrats that enter it. Shooting muskrats is another control option. Concentrated efforts should be made at dusk or in the early morning hours for effective control. It is more selective than other methods as it ensures that only target animals are killed. Commercial, licensed hunting can be very expensive; one quotation was \$100 per hour plus \$25 per animal (\$USD). Additionally, it may be difficult to control large muskrat populations with this method alone. Before this approach is used, local firearms regulations and permitting must be reviewed and abided by.

Beavers (Castor Canadensis)

American beavers are found throughout North America, except for the extreme arctic and desert areas with no flowing water. Therefore, their presence in a treatment marsh can be expected for the entire continent.

As outlined in Chapter 15, beavers are capable of inflicting considerable damage to a treatment marsh, as a result of their predilection for water level management that is at odds with that usually selected for water quality. They also cause herbivory damage, and compromise berms with their burrowing activities. Design and management strategies are required to keep the treatment marsh operational

Structure Clogging

The first, and far more serious, is their proclivity for damming sources of running water. This leads to the plugging of treatment wetland structures (see Figure 15.12). At a small wetland site, it is common for the animals to completely clog a structure in one night.

Because of a long history of interactions between beavers and man-made water structures, principally culverts, there has been a great deal of consideration given to methods of prevention. There is a long list of patented devices designed to prevent beavers from damming culverts (Goedderz, 1987; Lambert, 1987; Carroll, 1991; Thurber, 1991; Fleury, 2002). These are typically grills, grates, and small mesh devices are often installed to protect the upstream openings of culverts: however, they usually require frequent cleaning since they are easily dammed by beavers. Two of the lower maintenance alternatives are the "beaver deceiver" and the "beaver baffler" (U.S. Army Corps of Engineers, 2001).

A beaver deceiver is a fence that discourages damming due to its unnatural shape and large perimeter. Deceivers are usually trapezoidal but may be pentagonal or form long rectangles at narrow flowages. Fences are constructed of braced cedar posts and 6-gauge, wide mesh fencing. A beaver baffler is a cylindrical flow device constructed from multiple layers of wide mesh wire. This system excludes beavers from a culvert opening or dam breach while allowing free flow of water. The likelihood of beaver clogging is reduced by spreading the water intake over a large area. A baffler consists of two 5-m long, concentric cylinders of concrete reinforcing wire, including an inner cylinder that fits inside the culvert and an outer cylinder that is 30 cm larger in diameter.

Herbivory

Beavers are herbivorous but can utilize an extremely varied diet. Various parts of trees are the preferred food at times of the year, including the inner bark (cambium) and roots, shoots, leaves and buds. Favorites are poplars and aspen, willows, birches and alder. However, they will also make use of water plants, including reeds, rushes and sedges (Long, 2000). Many treatment wetlands are built with a view toward integrating the system into the surrounding landscape, by surrounding the system with appropriately sized clumps or bands of terrestrial vegetation. Trees are often an important component of these auxiliary plantings, which strive for habitat diversity. The combination of a wetland, particularly one with deep zones, and an adjacent tree plantation succeeds in providing ideal habitat for beavers. Once established, the animals go about their life style, which includes felling trees, usually within a few tens of meters of the water. North American beavers prefer the genus *Populus* and will cut numerous trees. In cold climates beavers cut down trees in autumn because they are preparing for the winter. Beavers do not hibernate, so they cache edible sticks to eat in winter.

Although agricultural wetlands are rarely in a setting with extensive tree cover, riparian and urban wetlands often promote or require tree plantings in their buffer zones. For instance, the US District of Columbia requires stormwater wetland design to integrate trees and shrubs as tree islands, peninsulas, and fringe buffer areas (Hoffman et al., 2013). They are stated to be desirable for promoting a mixed wetland community.

Burrowing

Beavers utilize both constructed lodges and bank dens as refugia, and a wintering domicile in which to raise a family. The banks along lakes and streams are places for dens, which relieve the animals of the effort necessary to build elaborate in-water lodges (Müller-Schwarze and Sun, 2003). Less than half of Eurasian beavers build lodges, but slightly more than half of North American animals do so. The berms and levees of treatment wetlands apparently qualify as acceptable burrowing sites. The animals are not small (their weight ranges upward to 25 kilograms), and bank burrows are therefore not small either. Considerable amounts of wetland water leakage or transfer may occur as a result of such burrows. More seriously, the integrity of a containment levee may be compromised, ultimately resulting in failure (Figure 18.11).

Control

Taylor et al. (2017) provides guidance on management activities to minimize beaver damage. They note it is important to identify and anticipate beaver problems in advance, because management is more difficult and costly once beaver colonies become established. A treatment wetland operator should be prepared to deal with beaver problems, despite proper preventative efforts of design and construction. In some rare instances, complete

FIGURE 18.11 Beaver excavation on the containment berm of the Tres Rios Hayfield treatment wetland near Phoenix, Arizona.

tolerance may be an option, but water quality and ecological goals are usually incompatible with beaver activities. There are various ways to manage beaver so population goals can be met. Lethal methods include shooting and trapping. Nonlethal methods include exclusion, habitat modification, repellents, and relocation.

While relocation might seem an attractive humane method, it may be ineffective for several reasons. Wetland system features that make it appealing to one beaver will also attract other beavers. The wetland will likely be reoccupied by other beavers, as young males are forced to find their own habitat areas each spring. The existence of alternate locations willing to accept relocated beavers is limited. Beaver relocation is more expensive and challenging than lethal trapping. Beaver populations can be controlled effectively by dam and lodge removal but is difficult and expensive.

Nutria (Myocastor Coypus)

As discussed in Chapter 15, nutria are a nonnative species in North America, and there are now feral populations in 22 states across the United States. Fortunately, their range exists as pockets of populations, together with a large contiguous range across the U.S. Gulf Coast, from Florida to Texas. Control methods are identical to those for muskrats and beavers (LeBlanc, 1994).

Insects

The wetland environment offers a wide variety of niches for various insects. Aquatic insects comprise a diverse assemblage of 11 orders in freshwater systems and fill a wide spectrum of ecological niches, especially in marshes. Common are the nymphal and/or adult stages of the Odonata (dragonflies and damselflies), Hemiptera (true bugs such as back swimmers, creeping water bugs, and water scorpions), Coleoptera (aquatic beetles), and Diptera (mosquitoes, horse flies, deer flies, and midges).

Some stages of certain insects can cause extensive herbivory in treatment wetlands (Snoddy et al., 1989). However, it is often observed that performance is not materially affected by such insect destruction, and a robust wetland ecosystem can recover quickly from these temporary events. Therefore, insect herbivory can generally be tolerated with no maintenance activity, despite the outward appearance of a vegetative disaster. It should be recalled that many of the treatment functions do not rely upon the green parts of plants. Further, insects typically attack one species of vegetation, and a diversity in the wetland plant community will lessen the impact of any one species being temporarily impaired. If the attack does not destroy the below ground plant parts, regrowth is probable. The general advice to the operator is: don't panic, watch the treatment performance.

Aquatic diptera are the most important macroinvertebrates in FWS wetlands in terms of number of species and population size. Mosquitoes (Family Culicidae) and midges (Families Chironomidae and Heleidae) are common wetland inhabitants. Only the larvae and pupae of these groups actually occur in the wetland water column because the adults emerge from their pupal skins and lead a terrestrial life feeding on nectar or blood. The nuisance potential of mosquitoes in wetlands and normal control methods are discussed here.

Mosquitoes

Mosquitoes present a biting nuisance, for both humans and animals. However, the bigger difficulty for treatment wetlands can be the potential for transmission of human diseases. Wetlands and other shallowly flooded habitats provide breeding habitat for mosquitoes. These mosquitoes are a nuisance and may transmit debilitating diseases to humans and livestock. Nonetheless, mosquitoes are an essential part of wetland biodiversity (Schäfer, 2004).

Mosquitoes are viewed quite differently in different climatic regions. In the far north, and in moist temperate zones plus the humid subtropics, mosquitoes are pervasive, and accepted as a part of the environment. However, in arid climates, mosquitoes are much less abundant, and viewed as a controllable nuisance. States such as California and Arizona have mosquito control agencies that operate under strict guidelines. Treatment wetlands in those states have been heavily regulated for mosquito control. In states where mosquitoes are naturally abundant, there is little or no regulation. That includes most agricultural treatment marshes. Therefore, the sections that follow are directed to treatment wetland projects that will be subject to heavy regulation. Disease transmission by mosquito vectors is virtually nonexistent in cold climates.

Some level of mosquito population will inevitably be present in a treatment marsh, and the risks associated with them may be part of the design process (see Chapter 15). However, it has been found that the residential zones of the urban environment produce more West Nile mosquitoes than urban wetlands (Johnson et al., 2012). Surveys undertaken in the state of New Jersey showed that increased bird diversity (the intermediate hosts) and increased mosquito diversity in wetlands caused lesser vector production. Further, this effect increased with wetland size.

Mosquito vector issues are much more severe in wetlands treating strong municipal effluents than in those with more benign sources that typify treatment marshes (Medlock and Vaux, 2014). Nonetheless, urban treatment marsh management should include a mosquito management strategy.

Mosquitoes and Wetlands

Mosquito species found in treatment wetlands can be classified into two groups based on their egg laying and hatching behavior. Females of some species lay their eggs directly on the water surface or on the leaves of aquatic plants. The eggs hatch usually within a few days and do not need an external hatching stimulus. These behavioral traits are characteristic of mosquitoes of the following genera (or subgenera): *Anopheles, Coquillettidia, Culiseta, Culex, Mansonia* and *Uranotaenia*. By contrast, the eggs of floodwater mosquitoes in the genera *Aedes* and *Psorophora* normally are deposited on moist soil or debris on the shore around aquatic systems and do not hatch until submerged by rising water levels

Generally speaking, adequately designed and properly operated treatment marsh systems will produce the same amount of mosquitoes as equivalent natural wetlands. Immature mosquito populations in treatment wetlands typically indicate low average densities.

It is not possible to exclude egg-laying adult mosquitoes from treatment wetlands. Mosquito production is greatest in thick, heavy, monotypic stands of vegetation. Dense stands of vegetation physically obstruct predator access to mosquitoes and hinder control efforts. Unfortunately, plant species commonly occurring in marsh treatment systems, such as cattail, bulrush, and common reed, are favorable for mosquito production (Knight et al., 2004).

Heavy accumulation of plant detritus results in areas where mosquito larvae are sheltered from predators. Open water areas with water depths greater than 30 to 40 cm rarely support mosquitoes (Merz, 2000). Deep zones provide refuge for mosquito fish and other predators, but unfortunately they are also preferred habitat for nuisance rodents.

Control Strategies

There now exist several comprehensive guidelines for wetland mosquito management, mostly targeting the urban situation (for instance, Rey et al., 2012; Webb, 2013).

The goal of system operators should be to minimize the number of larvae that can successfully survive to become adult, biting mosquitoes. This is generally achieved by either creating conditions favorable for mosquito larvae predators or through chemical/biological control of the mosquito larvae. Predators of mosquito larvae include a variety of aquatic insects (Greenway and Chapman, 2002) and fish (Knight et al., 2004; Walton, 2012). In the United States, the most commonly used predator is the mosquito fish (*Gambusia holbrooki*). Mosquito fish do not survive in cold climatic zones, and their northern analogs (the topminnows) are not nearly as effective for control of immature mosquito stages. In other parts of the world, there are niche-equivalents to *Gambusia*. In Australia, mosquito fish predators include rainbowfish (*Melanotaenia duboulayi*), blue-eye (*Pseudomugil signifier*), and gudgeons (*Hypseleotris* spp.) (Webb, 2013). While predatory fish do eat mosquito larvae, their efficacy as a control mechanism is questionable.

Bacterial control agents are effective in mosquito control in wetlands. Two bacteria species, *Bacillus thuringiensis* variety *israelensis* (Bti) and *Bacillus sphaericus*, are commonly used (Williams et al., 1996). Pellet and granular forms are available for vegetated environments but are effective only if they can be distributed into areas where the mosquito larvae are present. As a result, effective use of Bacillus toxins requires functional access points into the wetland, including:

- Drivable access around all wetland cells.
- Boat ramps, boardwalks or other structures to allow access into deep-water zones.
- Embankments and wetland areas that are free of power lines and other obstructions that might limit aerial spraying.
- Piping and valving that allows application of chemical and biological control agents directly into the flow.

Because treatment marshes are flushed by flow-through, control agents have to be reapplied periodically. This feature makes this control technique costly and time consuming.

Conventional chemical pesticides, such as the organophosphate compound temephos, are used for vector control operations in the United States because the chemicals are effective against a wide range of insects and remain economical for larviciding in large-scale abatement efforts (FCCMC, 1998). Such chemicals are not particularly toxic to mammals. Larviciding using either mosquito-specific or rapidly degrading control agents is preferred to adulticiding programs that use residual insecticides such as organophosphates (e.g., malathion, chlorpyrifos) or natural and synthetic pyrethroids (e.g., pyrethrin, resmethrin). Other mosquitocides include insect growth regulators (e.g., methoprene), neonicotenoids, and chloronicotinyls.

After adult mosquitoes emerge from developmental sites and disperse into the region surrounding a wetland, control is more difficult than are efforts focused on the developmental sites of the immature stages. However, adulticiding may be necessary if control efforts focusing on the aquatic, immature stages are ineffective.

18.5 VEGETATION MANAGEMENT

Previous chapters have elucidated the many common attributes of marshes treating different source waters. But, issues of vegetation management highlight stark differences between wetlands according to their source waters and intended purposes. Management strategies span the following range:

- Agricultural runoff wetlands are motivated by providing water quality benefits and waterfowl habitat. Very little vegetation management is utilized.
- Urban stormwater wetlands are focused upon vegetation aesthetics and diversity, with an accompanying expectation of some level of water quality improvement.
- Municipal treated wastewater polishing wetlands, and surface and ground water treatment marshes, are motivated almost entirely by water quality improvement, and strive to maintain vegetation to that purpose.

Agricultural systems require a maintenance agreement as a condition of funding, but it is typically minimal in its requirements. In contrast, urban systems often require a very detailed legalistic operations plan, which has inherent financial penalty provisions (Hoffman et al., 2013; Paul, 2013). Municipal polishing systems, although permitted and regulated for water quality, do not have vegetation requirements. Thus, vegetation management spans a range from detailed contractual botanical activities to benign neglect.

There are two perceptions that the treatment marsh operator should be aware of, and deal with appropriately: (1) the wetland only works if it is a large green expanse, and (2) the plants that were originally planted are the only ones that should be in the wetland. The first may be termed the "cosmetic concept." While it is certainly true that full vegetation promotes the biogeochemical cycles that underlie many treatment processes, it is also true that the brown component (detritus and litter) is of equal or greater importance than the green leaves. This is especially true when considering cold climate wetlands, which, despite being brown (or even white) during many months, continue to provide treatment during those months. Thus, although human values place a premium on visible green plants, the correlation to treatment is not necessarily strong.

The second idea may be termed the "cornfield concept." As a carry-over from agriculture, if the intent is to raise a particular crop, then anything else is a weed, and should be removed so as to prevent interference with the crop. However, treatment wetlands are not generally intended to provide a specific crop, but rather to promote utilization of nutrients, uptake of chemicals, and formation of microbial substrate. Therefore, it is often less important what species are present than their contributions to treatment.

General System Care

Treatment wetlands require access facilities, including walkable or drivable levees. These need to be kept in good condition. Berms may need mowing, particularly if local regulations prescribe that activity, as is often the case for treatment lagoons. A perceived problem is the growth of trees in wetland berms, such as cottonwood (*Populus deltoides*), willow (*Salix* spp.), and salt cedar (*Tamarix chinensis*). These are most easily controlled by mowing. This requires that berms be designed with flat enough side slopes for mowing equipment. Systems that are designed for public use will require that walking trails and other facilities be kept in a safe and useable condition.

Weeds

In urban wetlands in particular, some plants are considered weeds that must be removed (see Chapter 15). For example, Hunt et al. (2007) set forth the following for North Carolina urban wetlands:

> Despite being a native species, cattails are well adapted to develop monocultures that shelter mosquitoes from their predators. In short, if a stormwater wetland is to be located near a population center, such as a commercial center parking lot or a residential neighborhood, keep cattail populations under control. If cattails colonize more than 15% of a stormwater wetland located near populated areas, remove the majority – if not all – of them.

Of course, blaming cattails for mosquitoes is stretching a bit, but the message is clearly one of intense aversion to *Typha*.

Treatment wetland vegetation is intended to enhance water quality functions in the system, and not necessarily to provide habitat. Plants that are weeds from a habitat perspective may be desirable from a treatment perspective, but the two categories are not mutually exclusive. It is useful to understand which categories of plant species are considered undesirable from a habitat perspective, and what measures are recommended for their control (Hayes et al., 2000). Table 18.2 lists categories that include exotic species, and some native species, from the North American habitat perspective. If a treatment wetland is intended to serve multiple purposes, including the ancillary benefit of providing optimal habitat, then these species may be excluded from the planting list, and viewed as weeds if they subsequently volunteer into the wetland.

In general, control measures for undesirable species include burning, mowing, hand pulling, mechanical removal, and herbicide application.

Herbiciding

The use of chemicals to control marsh vegetation is questionable, because herbicides are not generally species-specific, and hence there a risk of collateral damage to desirable species. However, in some instances there may also be reasons to favor the exclusion or eradication of some species, from the point of view of enhancing treatment. A sampling of the herbicides that maybe used to control marsh vegetation is given in Table 18.3.

The Wisconsin Department of Natural Resources offers fact sheets on herbicides used on aquatic plants (e.g., WDNR, 2012). Flumioxazin controls invasive Eurasian watermilfoil (*Myriophyllum spicatum*) and curly-leaf pondweed (*Potamogeton crispus*). Triclopyr is used to treat the invasive Eurasian watermilfoil (*Myriophyllum spicatum*). Glyphosate can be used to control reed canarygrass (*Phalaris arundinacea*), cattails (*Typha* spp.), purple loosestrife (*Lythrum salicaria*), phragmites (*Phragmites australis*), water hyacinth (*Eichornia*

TABLE 18.2

Wetland species that are sometimes considered to be weeds from a North American habitat perspective and from an Australian perspective.

Scientific Name	**Common Name**
North America	
Eichhornia crassipes	Water hyacinth
Hydrilla verticillata	Hydrilla
Lythrum salicaria	Purple loosestrife
Phalaris arundinacea	Reed canary grass
Phragmites australis	Common reed
Salix spp.	Willow
Typha latifolia	Cattail
T. domingensis	Cattail
T. angustifolia	Cattail
Elaeagnus angustifolia	Russian olive
Sapium	Chinese tallow
Schinus terebinthifolius	Brazilian pepper tree
Casurina	Australian pine
Tamarisk	Salt cedar
Alternanthera philoxeroides	Alligator weed
Spiraea douglasii	Douglas spiraea
Ranunaulus repens	Creeping buttercup
Juncus effusus	Soft rush
Cytisus scoporius	Scot's broom
Lysinachia terrestris	Yellow loosestrife
Iris pseudacorus	Yellow iris
Australia	
Egeria densa	Egeria
Gymnocoronis spilanthoides	Senegal tea
Hydrocleys nymphoides	Hydrocleys
Lythrum salicaria	Purple loosestrife
Nymphoides geminata	Mashwort
Osmunda regalis	Royal fern
Sagittaria montevidensis,	Arrowhead,
Spartina alterniflora,	Spartina
Typha latifolia	Reedmace
Zizanialatifolia	Manchurian wild rice

Sources: North America, Hayes et al. (2000); Australia, de Winton et al. (2013).

TABLE 18.3
Herbicides used to control aquatic plants.

Herbicide	Vegetation Type		
Diquat	SAV	FAP	EAV
Endothall	SAV		
2,4-D	SAV	FAP	EAV
Fluridone	SAV	FAP	EAV
Glyphosate		FAP	EAV
Simazine		FAP	

crassipes), and water lettuce (*Pistia stratiotes*). Diquat and 2,4-D are effective at treating the invasive Eurasian watermilfoil (*Myriophyllum spicatum*).

Agricultural wetlands in the prairie regions of the United States have employed herbicides to protect adjacent crops for over 20 years (Linz and Homan, 2011). Sunflower crops, near wetlands provide roost for large numbers of blackbirds that cause economic damage Beginning in 1991, USDA Wildlife Services began aerially spraying cattails with glyphosate herbicide to reduce roosting habitat. The program has had some success, and the glyphosate-induced cattail reduction showed increased use by waterfowl and increased abundance of invertebrate species important to waterfowl (Linz and Homan, 2011).

Herbicide treatment is the primary means of vegetation control at the OEW in Florida. Focus is on cattails (*Typha* spp.) and water lettuce (*Pistia stratoides*). Cattails are disliked because of their aggressive character, and the perception that they do not measure up for water quality improvement. Water lettuce is disliked because of its proclivity for moving with the water and wind to structures and clogging them. Therefore, removal is perceived to "promote diverse habitat and ensure a healthy system for water quality standards" (Slayton, 2018). In 2017, over 160 hours were spent applying aquatic herbicides, primarily on just a few cells of 18 total.

The Florida STAs represent another instance of the regular use of herbicides. The purpose is partially driven by a perception of the treatment capabilities of various vegetation types, but more importantly by the perceived need to manage marsh water flows. Floating aquatic plants have a propensity for moving to and obstructing marsh outflow structures. These are also uniquely large systems, with attendant peculiarities of wind fetch. Hyacinths are moved by the wind and form dense windrows, piled above water, which clog structures (see Figure 18.6) Some invasive species are moderately effective in phosphorus reduction, such as water hyacinths (*Eichhornia crassipes*), hydrilla (*Hydrilla verticillata*), and torpedo grass (*Panicum repens*). But, torpedo grass and hydrilla form floating leaf canopies that block sunlight and shade out SAV. Accordingly, these plants were regarded as undesirable, and removed by herbiciding. Several different chemicals have been used, including 2,4-D, Glyphosate, Imazapyr, Triclopyr, Flumioxazin and Diquat (SFWMD, 2019). In 2018, 4800 ha (about 20% of the total) received 60,800 liters of liquid herbicides and 822 kg of dry herbicides.

Burning

The value of prescribed fire in managing vegetation varies according to region and vegetation type. Fire may be used in treatment wetlands for the purpose of reducing and removing excessive thatch buildups that develop in arid climate systems. However, those climatic regions may also be prone to air pollution problems, which may preclude fire as a management tool. Fire is being used on a limited basis in some constructed FWS treatment wetland systems (e.g., West Jackson County, Mississippi) to periodically reduce the density of macrophyte growth. In arid regions, fire is often regarded as a necessary management tool to reduce the thatch in SSF wetlands (Wallace and Knight, 2006).

Fires are typically set in the late fall or winter after plant senescence and wetland burns can be accomplished with or without shallow standing water in the wetland cells.

Fire mobilizes the stored nutrients and pollutants into water-soluble forms, and permits the flushing of the materials the wetland was designed to store (Smith et al., 2001). The litter and sediment layers may be destroyed, thus necessitating a new microbial startup period. If fire is used as a management tool it is essential that the treatment wetland be designed with multiple parallel flow paths to allow bypassing flows around the cells receiving controlled burns. Following a planned fire in a wetland cell, water should be held for a period of several weeks to a few months before startup and discharge to avoid the release of newly dissolved constituents.

Flores et al. (2011) studied fire effects in brackish marshes in Maryland. Results showed a decrease the accumulation of litter, and an increase the biomass and stem densities of wetland plants generally considered less desirable for wildlife.

White et al. (2002, 2008) studied the effects of fire in the Orlando Easterly treatment marsh in Florida. The accumulation of sediments had been identified as a major factor responsible for hydraulic short-circuiting, and a reduction in P removal performance. A paired cell design was implemented, with a burn in one cell, while the adjacent cell received no burn (Figure 18.12). Water quality parameters including soluble reactive phosphorus, total phosphorus, ammonium, TKN and dissolved organic carbon were measured before and up to 4 months after the burn. A pre- and post-burn quantification of aboveground biomass was conducted to determine the net reduction in biomass as a result of the burn. The prescribed burn significantly decreased the amount of senescent organic matter in the wetland. The results showed that there was phosphorus released from the burned cell immediately after the fire, but over time, the concentrations dropped below that of the no burn treatment, showing greater performance of the burned cell for treatment of water column phosphorus. White et al. (2008) concluded management strategies should include hydraulically isolating the burned area for a period after the burn, to prevent nutrient export.

FIGURE 18.12 Controlled burning is a management strategy at the Orlando Easterly Wetlands. (Photo courtesy of Mark D. Sees, Wetlands Manager)

Fire has also been employed in an attempt to facilitate the reduction of cattails in the northern portion of Water Conservation Area 2A, which for several decades served as an (unintended) phosphorus reduction marsh (Tian et al., 2010; Xu et al., 2011). Liu et al. (2010) found that phosphorus was released soon after burning. The quick release of TP suggested that prescribed burning accelerated P release from cattail and sawgrass.

Harvesting

The harvest of wetland plants for the purpose of removal of stored chemicals is often suggested, but seldom practiced. Harvest is not effective in heavily loaded treatment wetlands, because only a small fraction of the removal operates through the vegetation. But harvest can be considered for treatment marsh (lightly-loaded) systems. Harvesting of treatment marshes has yet to be attempted, although some other marshes have been harvested. Chapter 10 discusses this topic in more detail. Disposal of harvested plant material has associated costs, with few opportunities for product recovery. Harvest removes vegetation that would otherwise form litter, and serve valuable treatment functions in the system, such as insulation via snow trapping and carbon supply for denitrification. Techniques for harvesting would presumably involve system dry-down or on-ice methods.

Buffers

Both agricultural and urban systems come with buffer zones, and these also require vegetation management. Curiously, the extensive rules and regulations concerning urban wetlands generally do not address buffer maintenance. An exception is the guidelines provided by the Queensland Department of Environment and Resource Management (Deacon et al., 2011). In this region, it is required to prepare a Wetland Buffer Implementation Plan including weed and pest management and monitoring. However, no specifics are included in this document.

The CREP agricultural wetlands do have specific buffer maintenance recommendations, Because those buffers are envisioned to be grasslands (prairie) containing a mix of native grasses and forbs, they differ greatly from urban wetland buffers, which typically contain several tree and shrub species. CREP buffers are also large, ranging from four to ten times the wetland area. The stated goals are to control noxious weeds and other undesirable plants, insects, and pests. IDALS (2016) offers the following activities:

- Mowing the site early and often is the best way to ensure proper establishment of the buffer area. We recommend mowing 3–4 times per year for the first 1–2 years after seeding and spot moving 2–4 times per year for 3–4 years. The more it is cared for early on, the easier it will be to care for in the long run.
- Mowing reduces competition from early succession weed species, such as: giant ragweed, foxtail, lambs quarter, and Canada thistle; therefore, it is important to mow before these weeds have gone to seed.
- It's recommended that you use a rotary (brush hog) or shredding or flail type mower (stalk chopper) to chop up and spread out the mowed weeds. This method is preferred to a sickle type mower that cuts the weeds at the base and lays them flat which can smother a new seeding.
- Mow the buffer high (~8–10 inches) to only clip the weed growth above the emerging native plants.

Mid-contract management is required within the 15 years of the CREP contract. Specified activities include spraying, inter-seeding, disking, and burning. A prescribed burn is the most common, to control volunteer trees in the buffer and invigorate growth of native grasses and forbs. Fire management for the whole site is done over several years. For example, 1/3 of the area is burned in year 5, 1/3 in year 7, and 1/3 is burned in year 9. Buffer management must be completed prior to or after the nesting season that runs from May 15 to August 1.

18.6 HUMAN USE MANAGEMENT

Human use is featured as a major activity at a number of treatment marshes, as set forth in Chapter 15 (see Figures 1.16 and 15.21). The facilities that foster human uses, while not part of the treatment system *per se*, require staffing and maintenance. Interpretive centers require operational staff, either as employees or *pro bono* docents. Other requirements include building maintenance and utilities (Figure 18.13). The costs for this can be very substantial.

Human use facilities at most treatment marsh sites are much more modest than full-blown visitor centers. Yet, significant maintenance can be needed, which includes the following:

1. At the simplest level, public walking trails and boardwalks require upkeep. Maintenance can be as simple as mowing the grass and trimming vegetation to maintain access along pathways and embankments.
2. Signage is a critical need for human use. Visitors need to know what they may or may not do in the wetland, and where they can go. Ecological interpretive signs are

FIGURE 18.13 The Green Cay treatment wetland has very large and active human use. The services of a park ranger are required.

of great interest to most visitors and require maintenance. Regrettably, maintenance also involves repairing vandalism (Figure 18.14).

3. Upkeep of entry points requires periodic tasks. Boat ramps and parking lots need periodic attention. These are also trash magnets, because humans routinely discard travel refuse at these journey terminal points. Thus, garbage collection and disposal are on the task list.
4. Hunting activities typically are supervised, to check for appropriate permission documents and provide guidance to the assigned hunting area.
5. Passive visitor use may require wetland staff attention but may also be provided by outside agencies. For instance, the Audubon Society in partnership with the South Florida Water Management District offers escorted birding trips to Stormwater Treatment Area 5 (STA-5/6), occurring every two weeks. Trips are limited to 60 participants, and registration is required. The Friends of the Orlando Wetlands guide walking tours and free tram tours through the Orlando Easterly Wetlands. The Friends also periodically care for the entrance area, education center, and landscaped areas throughout the wetland/park.

18.7 SEDIMENT REMOVAL

Solids removal from sedimentation basins, forebays, and wetland cells is an anticipated maintenance activity, estimated to be needed about every 20 years. Dredging of sediments or soils can be accomplished, but at considerable expenditure of time, money and loss of treatment potential. In all cases, appropriate diagnostic information is a prerequisite to initiating maintenance or control steps.

Sediment bioaccretion does not typically threaten wetland vegetation, because the continual root turnover allows the plants to follow the upward progression of soil elevation on the time scale of years. In contrast, incoming sediment deposition can occur at rates that smother wetland plants. In forebays and sedimentation basins, the potential problem is the filling of the facility, and impairment of the settling times. In wetland cells, the difficulty is subtler, in the form of bathymetric alterations that impair volumetric and hydraulic efficiencies.

FIGURE 18.14 Graffiti mars this sign at the Pintail treatment marsh in northern Arizona.

Most agricultural treatment wetlands are too young to have required sediment removal, and there are no reports detailing results. However, many urban systems and some municipal polishing systems have been operating long enough to warrant attention.

The Orlando Easterly Wetland has had sediment removal in 80% of its cell areas (375/486 ha) over the past 17 years of its 30-year operational life (Wang et al., 2006; Slayton, 2018). Cells were drained and then mechanically scraped (Figure 18.15). The resulting vegetation, sediments, and organic debris were disposed of on-site as an upland soil amendment. Cell interiors were then graded, and islands constructed to lessen short-circuiting and provide wildlife habitat. The first five cells produced 130,000 m^3 of solids, or an average depth of 22 cm removed. Over 160,000 giant bulrush (*Scirpus californicus*) were planted in the dredged area. The expectation was that growin would create a dense stand and fill in the cells within a year or two. Tracer testing showed dramatic increases in the volumetric efficiency of the first cells dredged, from 0.34 ± 0.14 to 0.74 ± 0.16. The accumulations were the result of bioaccretion, because the tertiary effluent entering the system carried essentially no TSS.

Urban stormwater systems are known to foster sediment buildup, but removal is sometimes neglected. Erickson et al. (2010) survey 37 municipalities, and found a quarter of them listing sediment buildup as a problem. U.S. EPA (2009) provides guidance for sediment removal. In small wetlands, the water may be drawn down to dry residuals in place. Heavy equipment can then remove the sediment. Where dredging cannot be accomplished mechanically from the shore, sediment may be removed using hydraulic dredging methods. Sediments from ponds and wetlands are usually dewatered and then disposed of onsite or land filled.

Sediments were removed from an urban stormwater wetland in the Tahoe City, California (Qualls and Heyvaert, 2017). The wetland was designed to remove nutrients, fine sediments, and other pollutants from runoff from 23 ha comprised of commercial (24%), residential (38%), roadway (21%), and vegetated (17%). The wetland had a design surface area of 4,046 m^2, for a WWAR = 1.76%. It accreted sediment and organic matter such that it reached the end of its service life in 16 years. Core analyses showed the surface of the

FIGURE 18.15 Most of the cells of the Orlando Easterly Wetlands have been demucked once in the 30-year life of the system. (Photo courtesy of Mark D. Sees, Wetlands Manager.)

wetland solids rose in elevation by 3.2 ± 0.3 cm/yr, for a total of 51 cm over the 16-year period of accumulation. Qualls and Heyvaert (2017) determined that accumulations were about equally the result of bioaccretion and sedimentation of incoming solids.

The annual accretion rates for N and P from sediment cores corresponded reasonably well with removals measured via the water phase mass balances (Heyvaert et al., 2006). Removals in the Heyvaert et al. study were 13 gN/m^2•yr and 3.2 gP/m^2•yr, compared to 17.7 and 3.7, respectively for accretion in the cores.

18.8 MONITORING AND REPORTING

All wetland treatment systems should be monitored for at least inflow and outflow water quality, water levels, and indicators of biological condition. Following proper design and construction, monitoring is the next most important factor in successful operation of treatment marshes. The appearance of the wetland is not necessarily representative of what it is doing for water quality, because many wetland functions do not tie strongly to the above-water canopy. Consequently, it is water quality and quantity data that are the primary indicators of wetland operational status.

WATER QUALITY

Marshes with water quality permits, which are mainly municipal polishing systems, will require compliance monitoring of a specified list of chemical constituents and inflow rate. The sampling locations can vary from project to project, and usually includes some or all of the wetland inflow and outflow points. However, some projects may require internal monitoring points in the treatment wetland, and in some cases, monitoring of the receiving water body. The locations, types and frequency of compliance sampling are spelled out in the individual permit(s), as are the averaging periods and limits for the various regulated constituents. Laboratory protocols for chemical analysis are also officially prescribed, and include a variety of established methods, as defined in Standard Methods (APHA, 2017) or by the U.S. EPA, for example.

However, the methods for acquiring wetland water samples are neither well defined nor standard. There is no apparent difficulty with obtaining samples at inlet and outlet structures, yet there can be serious anomalies related to sampling methodology. For instance, either autosamplers or grab sampling may be employed. The autosampler can be set to flow-proportion, and thus (hopefully) produce flow-weighted concentrations for use in mass balances. However, not only do the sampling methods yield different results, there can be an apparent bias of the autosamplers to give higher results. Speculatively, the reason is that autosampler intake tubing may "suck mud" if the intake strays from its optimal position.

Acquiring representative internal wetland water samples in a shallow marsh environment is very difficult. Floating materials, such as *Lemna*, can often be excluded by careful sampling. Water depths may be quite shallow, and there can be large amounts of readily suspended materials in the vicinity. The inclusion of particulates in a dipped sample will not be representative of the particulates that move with water in the undisturbed environment. The local velocities created by the sampling activity resuspend an abnormally high amount of TSS. That solid material can bias the lab analyses for total amounts of contaminants such as total phosphorus or total nitrogen. The intense vertical gradients in water quality near the wetland bottom can also create anomalous results. In very shallow water, the sample can contain some or all of porewater, which usually has much different (higher) concentrations from surface waters (see Figures 4.12 and 4.13). All of these factors create a need to avoid disturbance in the act of sampling, and to avoid very shallow depths that may involve porewater. In the various Everglades sampling activities, depths less than

20 cm are not bottle-dip sampled for these reasons. However, carefully aspirated samples may be acquired at smaller depths.

Flows and Mass Balances

Compliance monitoring may not be enough to gain an understanding of wetland function, because it is focused upon measuring the variables that may affect the receiving waters and is commonly restricted to outflows. A fuller understanding of wetland performance rests upon constituent mass balances, which in turn depend upon the water mass balance for the wetland.

The water budget has been described in detail in Chapter 2. The principal inflow is usually the water to be treated, which is augmented by rainfall, evapotranspiration and seepage before forming the wetland outflow. Rainfall records may be kept at the site or procured from a nearby weather station. Evapotranspiration may be derived from weather bureau pan evaporation, or from governmental irrigation services, as discussed in Chapter 6. Therefore, the primary monitoring requirements are usually the wetland inflow rate and outflow rate. Both are required if there is significant seepage, so that the seepage may be estimated by difference.

Flow measurement can be accomplished using any one of several techniques. For pumped inflows, either pump-hours plus the pump discharge curve may be used, or a turbine meter can be installed. For small flows, a calibrated weir, possibly with continuously recording head, is a common choice (Figure 18.16). Most large inlet and outlet devices that operate with a measurable head loss can be calibrated to determine flow from headwater and tailwater elevations. For large inflows without a calibratable structure, devices such as Doppler flow meters or ultrasonic velocity meters have been employed. It is noteworthy that virtually every type of flow measurement relies upon clean conditions and periodic re-calibration. Any of the measurements are prone to being somewhat in error, with field accuracies on the order of ± 15%. For any of the devices that include sensors and data logging, frequent manual crosschecks are recommended.

In the long term, water storage changes in the wetland are almost always negligible in comparison to other flows. However, water depth is very useful parameter for other purposes than the water budget, and hence water stage in each cell is a desirable dataset. Staff gages should be installed in each cell of the wetland and surveyed for vertical control at the time of startup (Figure 18.17). When combined with a topographic survey of the wetland, stage measurements provide a quantitative tool for assessing the average, maximum, and minimum water depths in the wetland and the frequency with which these depths occur. These data are essential for interpreting tracer measurements of hydraulic residence time and for assessing any detrimental hydroperiod effects on biota. A variety of water level measuring instruments (e.g., analog chart recorders, ultrasonic, pressure-sensor, and capacitance-sensor units) are available for continuously recording water levels at various key points within the wetland.

Constituent Concentrations

The water chemistry parameters that require monitoring vary with the water quality target for a particular system. This list includes measurements of the water quality of all major inflows and outflows associated with the treatment wetland. Inflows include the source waters entering the wetland as well as natural inflow streams that may have a significant effect on the relevant water quality or the hydrologic budget of the marsh treatment system. Clearly, water quality monitoring is specific to the contaminant(s) in question, whether

FIGURE 18.16 A V-notch weir, with recorded head, is a user-friendly method of measuring small inflows and outflows. Photo from the Duck Spring wetland system near Alcoa, Tennessee.

these are nutrients (Chapters 9 and 10), trace metals (Chapter 13), or trace organics (Chapter 14).

To illustrate typical monitoring, the case of municipal wastewater polishing is generically described in Table 18.4. The parameter list is to be tested at all major inflows and outflows at least monthly includes all regulated pollutants and integrative measures such as TSS, pH, dissolved oxygen, water temperature, conductivity, NO_2-N, NO_3-N, ammonia nitrogen, TKN, total phosphorus, chloride, and possibly sulfate. The frequency of operational monitoring for system control is dictated by the size and capacity of the system, the sophistication of the owner's staff and sampling equipment, and site-specific factors related to influent quality variability and climatic factors.

Inflow and outflow stations may also be monitored less frequently for selected heavy metals or organics that might be present in the wastewater and for whole effluent acute and chronic toxicity. If water quality characteristics are highly variable for any of the inflow or outflow locations, or if there are weekly or monthly permit limits, sampling should be more frequent than monthly or quarterly.

These water quality data as well as water budget parameters should be organized and recorded in computerized spreadsheets for visual analysis of variability and trends. The constituent loadings may be compared to design criteria, to ascertain if the system is being operated according to design. The constituent mass balances are easily constructed from system data on such a spreadsheet, and may be used to determine if design conditions are being achieved, and whether the system is performing as expected.

Table 18.4 reflects the example of municipal wastewater treatment, but constructed wetland systems are used for a wide variety of other purposes, including groundwater remediation, and urban and agricultural stormwater. Each of these other applications generates a similar requirement for a strategy to promote understanding. All are based upon

FIGURE 18.17 A staff gage provides valuable information on the water depth and volume in the wetland. Photo from the Lake Nebagamon, Wisconsin, treatment wetland system, in the exit deep zone. The screen box covers the intake for the effluent siphon. The berm is stone rip-rapped to discourage muskrats.

knowledge of the water budget, but the suite of chemicals of interest may change depending upon the application.

Urban stormwater wetlands generally do not have water quality monitoring requirements. A certain level of water quality performance is assumed, based on prior studies. And, continuing interest in their performance has generated research studies that carry out water quality monitoring, often at the event time series level. Some of these are described in Chapter 11.

A unique aspect of the Iowa CREP nitrate control wetlands is that nitrate reduction is not simply assumed based on wetland acres, but is based on the measured performance of CREP wetlands. A subset of wetlands is monitored and mass balance analyses performed to document nitrate reduction. The work is performed by outside agencies (Universities). Monitoring supports continued refinement of modeling and analytical tools. Wetlands are instrumented with autosamplers and flow meters at wetland inflows and outflows. Water levels are monitored continuously at outflow structures in order to calculate changes in pool volume and discharge and wetland water temperatures are recorded continuously for modeling nitrate loss rates.

TABLE 18.4

Example of possible monitoring requirements for operation of a marsh treatment system polishing tertiary treated municipal effluent. Some of these are needed for permit reasons; others for mass balancing.

Monitored Parameters	Sample Locations	Sampling Frequency
In the field		
Flow	Inflow(s) and outflow(s)	Daily
Rainfall, Air temperature	Adjacent to wetland	Daily
Water stage	Within wetland	Daily
Water T, DO, pH, EC	Inflow(s) and outflow(s)	Weekly
Plant cover (Dominant species)	Near inflow, near outflow	Annually
In the laboratory		
NO_XN, NH_4N, ON, TP, TSS		
Chloride, Pathogens	Inflow(s) and outflow(s)	Monthly
Metals, Organics, Toxicity	Inflow(s) and outflow(s)	Quarterly

Biological Monitoring

Biological monitoring within a wetland treatment system provides information concerning the structural integrity (health) of the vegetation and fauna. Protection of this biological integrity is important from an environmental habitat perspective and because of the biota's control of wetland operational performance. It has been suggested that every component of the wetland ecosystem should be monitored, including sediments, water, vegetation, benthic invertebrates, fish, amphibians, reptiles, birds, and mammals (Ecological Services Group, 1996; Wren et al., 1997). This concept originates from the fact that some of these ecological components have been routinely found to be above pristine background for some contaminants, as expected for a treatment system.

However, there have been no ecological disasters associated with the thousands of treatment wetlands over the 40-plus years of the technology, and thus the question involves (possibly) impaired wetland ecology, not acute problems for wildlife. Therefore, biological studies fall in two categories: monitoring needed to foster the design water quality functions, and research needed to more fully define impacts of less-than-perfect water on the various organisms that may inhabit the treatment wetland. There is usually an important distinction between constructed wetlands and natural wetlands as recipients of wastewater: the former are usually not subject to intensive biological monitoring, but the latter may be in some states.

A compromise is typically put in place, with some facet of biological monitoring or research being conducted in addition to the requisite compliance and mass balance monitoring. Typical categories are bird, mammal and fish studies. Broad-scale potential for ecological problems is commonly assessed by examining or forecasting the concentrations of specific chemicals or toxins. For instance, great care is used in the arid regions of the United States because of the known acute effects of selenium that has impacted wildlife wetlands (e.g., Kesterson, California).

Wetlands designed for ancillary benefits can have nontreatment goals and requirements. Trends in use of the treatment wetlands by plant and wildlife species may indicate needed changes in system management to support those nontreatment goals. Surveys for rare or threatened species may also be conducted when appropriate.

Cover estimates and observations concerning plant health should be a routine part of operational monitoring in a constructed wetland treatment system. Because plants grow slowly and are important for maintaining the performance of wetland treatment systems used for water quality treatment, problems must be anticipated and prevented before they are serious or have progressed too far. Reestablishing a healthy plant community in a constructed wetland is a slow process when the plants have been harmed because of operator neglect.

Plant growth is monitored by estimating percent cover, species composition and possible physical attributes such as plant height. These nondestructive techniques are used to ascertain the status of plant development before and during wetland treatment system operation. Plant cover is an estimate of the percentage of the total ground area covered by stems and leaves. This parameter can be estimated by walking through or next to a plant stand and visually determining a cover category for the plants, or by aerial photography. Cover estimates can be made on a finer scale by using a frame to delineate specific areas. Cover estimates should be made at enough locations in the constructed wetland to provide reasonable statistical averages for comparison between cells and between dates.

Aerial photography is an excellent supporting tool that can provide estimates of the type and percent cover of emergent vegetative communities in treatment wetlands. However, aerial reconnaissance cannot characterize submerged vegetation with any accuracy. The various areas of different vegetation are clearly visible, but ground-truthing is required to establish which plants are growing in the zones.

Documentation

Treatment marsh projects have varying requirements for plans and reports concerning system maintenance. These can include maintenance agreements, operating plans, operations and maintenance manuals, and project performance reports.

Maintenance Agreements

Some treatment marshes, prominently including urban systems, may require maintenance that is beyond the capabilities of the local owners, such as homeowners associations. The services of contractor maintenance organizations may be needed, and the necessary inspections may be done by the regulating governmental agency. Under such circumstances, a formal agreement may have to be executed. For instance, the District of Columbia in the United States will not issue final approval until the applicant has executed a declaration of covenants binding current and subsequent owners of the land served by the wetlands and buffers to an inspection and maintenance agreement. Such agreement provides for access to the site at reasonable times, for regular inspection by the regulatory agency, and for regular or special assessments of property owners, as needed, to ensure that the wetland is maintained in proper working condition and the land covers are retained as approved (Hoffman et al., 2013).

Operation Plan

Especially for large projects, there may be a requirement for a project operation plan. For example, the stormwater treatment areas (STAs) of south Florida all have such documents. A typical outline, for STA6, (SFWMD, 2004) is:

I. Project Description
II. Structure and Canal Descriptions
III. Operations
IV. Downstream Effects

V. Additional Flows
VI. Coordination with Other Plans
VII. Permits for Operation

The operations section of the overall plan discusses various modes of operation. The design intent for normal operation is specified, as to which structures are operating and what water depths are to prevail. Designs may include operations during extreme events, such as the "standard project storm," which in south Florida is 55 cm of rain over a 24-hour period. The operating plan therefore includes information on extreme flow operations. The design calculations also indicated frequent dry periods, and the operations plan includes provisions for dealing with this lack of water. Should it become necessary to take treatment cells out of service (this has happened several times in the various flowpaths of the STAs), the documents contain the design intent for earthwork, and operations strategies, to accommodate the temporary reconfiguration. Contemplated expansions may be spelled out in the operation plan.

Operations and Maintenance Manual

It is not only the wetland itself that goes through startup. Very often the system operator also goes through a kind of startup also, during which there is a steep learning curve. Familiarity with treatment wetlands in general is a prerequisite qualification for the wetland operator, but in many circumstances that may be "book knowledge." In virtually all cases, the operator will not have been involved in the wetland design, and therefore will not be familiar with the intended function of the system. Two steps can alleviate potential difficulties: (1) a startup walk-through of the treatment system, from pretreatment through to final discharge, and (2) preparation of a comprehensive operations and maintenance manual. The walk-through includes discussion between designer and operator, of the intended function of structures and controls, and of the wetland vegetation, detritus and soils. It serves the purpose of defining what activities may be required on a seasonal basis.

The project O&M manual is the reference document that the operator may consult as needed during future operation of the facility (Table 18.5). It describes the design intent of the project. Startup and normal operations are explained. Monitoring requirements, protocols and reporting are listed. Routine O&M steps are outlined. Some of the foreseeable emergency situations, notably high rainfall events, are discussed, and emergency measures defined. For later reference, it includes the as-built drawings for the wetland, and the current regulatory permit(s).

Performance Reports

Many treatment marsh projects produce annual or other performance reports, some of which may be required by permit conditions, and others that describe operations and water quality results. Examples are:

- The Kanapaha Infiltrating Wetlands in Gainesville, Florida, are monitored by Wetlands Solutions, Inc., and they produce reports on facility performance (WSI, 2012), which are available on the internet.
- The Houghton Lake, MI, marsh project produced 30 annual reports on hydrology, water quality, soils, and vegetation; and another 30 annual reports on wetland birds and animals. These had very limited distribution and are not publicly available.
- The Orlando Easterly project has likewise produced a 30-year long set of annual reports, the most recent of which is available on the Internet (Slayton, 2018).
- The Florida STAs commenced public annual reporting in 1999 with a series of six Everglades Consolidated Reports that contained annual STA details

TABLE 18.5

Example table of contents for an operations and maintenance manual.

Introduction

Background

- User's Guide
- Treatment Wetland Processes

Unit Processes

Inflow Control and Measurement

- Wetland Flow
- Wetland Function in Shallow Zones
- Wetland Function in Deep Zones
- Outflow and Level Control

Startup Procedures

Treatment Wetland

- Planting
- Startup
- Startup Flow Diagram

System Operation

Operations

- Winter Operations
- Mosquito Control
- Sampling Procedures
- Grab Sampling
- Sampling and Monitoring
- Startup Period
- Steady State Sampling
- Laboratory Tests
- Recording Data
- Daily/Weekly Operating Logs

Monitoring

Sampling Procedures

- Grab Sampling
- Sampling and Monitoring
- Startup Period
- Steady State Sampling
- Laboratory Tests
- Recording Data
- Daily/Weekly Operating Logs
- Laboratory Records
- Monthly Reporting
- Annual Report
- Maintenance Records
- Operating Costs
- Emergency Conditions
- Trend Charts
- Laboratory Records
- Monthly Reporting

(Continued)

TABLE 18.5 (Cont.)

Annual Report
Maintenance Records
Operating Costs
Emergency Conditions
Trend Charts
Treatment Wetland Maintenance
Process Control Problems
Water Control Structures
Short-Circuiting
Deep Zones
Vegetation
Embankments
Piping
Warranty Provisions
Emergency Plans and Procedures
Hydraulic Overload
Process Failure
Safety
Appendixes
Design Criteria
Treatment Wetland Design Drawings
Sample Volumes and Storage Requirements
Operating Log

Source: J. Pries, CH2M Hill, Kitchener, Ontario; for the Nanticoke FWS wetland.

(SFWMD, 1999-2004). Starting in 2005, reporting has been in the form of South Florida Environmental Reports, focused in major part on the STAs (SFWMD, 2005-2018).

Summaries

Publication of individual project performance may occur when significant periods of record are available, and are produced by interested researchers, not the project operating and management personnel. Examples have been referenced at many places in the preceding chapters. A second type of summary consists of intersystem performance collections, also referenced in preceding chapters.

18.9 SUMMARY

Treatment marshes are sufficiently different from other treatment technologies to make learning their operation a nontrivial effort. Although they involve few sophisticated controls, they offer hydraulic and ecological challenges. Thus, although they are "low tech," they are "high bio." Given a conservatively designed marsh treatment system, successful operation is not a challenging task. There are few operational controls in most treatment wetlands, and these controls must be used wisely and with common sense to maintain top performance. Typically, the only system controls available are variation of inflow loading rates through storage or diversion of flows, system rotation between alternate discharge locations or wetlands, and control over water levels within the wetland.

Although wetlands contain only a few physical components, they do require maintenance and repair. Berms can be eroded or damaged by animals or extreme rain or winds. Structures easily become clogged with debris that requires clean out. Pumps require periodic inspection and care if they are small; and may need full-time staff if they are large.

Vegetation management may include depth management, selective herbiciding, and the controlled use of fire; and perhaps manual removal and replanting if diversity is of large importance. Animal control is increasingly understood to be an important facet of FWS wetland operation. In particular, rodents can be quite destructive, and require control in some circumstances. Mosquito control is an important issue in some wetland treatment projects, notably those in warm, arid climates. The owner and operator of these wetland systems should be prepared to deal with this possible issue. Human use creates tasks such as access control and cleanup, tour guidance, signage maintenance and road, boardwalk and path preservation.

Solids buildup has been identified as a potential performance degrading process in treatment wetlands. Forebays, settling basins and wetland cells accumulate at rates that require attention on an approximate 20-year cycle.

Monitoring is one of the most important aspects of treatment wetlands operation. Monitoring of inflow and outflow water quality provides a system-level barometer of wetland health and performance. Monitoring of the internal wetland structure provides a reference for correlating changes in water quality performance with system structure. Routine monitoring and data analysis is essential for making informed decisions concerning control of operational variables such as water depth and mass loadings. Additional monitoring beyond regulatory requirements may be performed to accomplish specific operational goals. Because of its importance in optimizing system performance and averting unacceptable biological changes, some monitoring should be conducted in all treatment marshes.

While the totality of potential management issues is large, all of the issues identified in this chapter do not occur at any one wetland. In general, treatment wetland systems are operator-friendly, but they cannot be left unattended for prolonged periods.

19 Economics

Financial considerations for treatment marshes are comprised of several factors. The actual costs incurred on operating projects consist of capital costs and operating and maintenance costs. In the process of marsh design, cost estimation is a key task. And, it often useful to consider the life costs of the project, whereby capital and operating costs are brought together in a present worth analysis. Financial benefits of a marsh project are certainly considerable, but very hard to quantify monetarily. To complicate matters still further, there are significant differences among the three major types of treatment marshes, surface water and wastewater polishing, urban runoff treatment, and agricultural runoff treatment. Because of the changing value of money over time, there is ambiguity about the value of the dollar. In this chapter, values are adjusted to 2018, unless otherwise stated or implied.

19.1 MAJOR COST ELEMENTS

Capital Cost Components

While it is not possible to offer universal cost guidelines, every system shares a similar set of construction components. Therefore, it is possible to estimate the cost of each component within a regional market. The basic direct cost components of a marsh treatment system include:

- Land
- Site investigation and system design
- Earthwork
- Pumps and conveyance
- Water control structures
- Planting
- Human use facilities and access

These costs include material, labor, overhead, and profit, and represent the contractors installed cost. Additionally, there are indirect costs associated with permitting, engineering, financing, mobilization and construction management. In general, these costs are all incurred prior to system startup. Detailed estimates are usually made after final sizing and siting. More precise economic estimating is possible after final design drawings have been prepared.

Liners

A major difference between treatment marshes and treatment wetlands is that treatment marshes are generally not lined. The reason is that marsh source waters typically do not pose an increased threat to groundwater resources. The decision to install a liner in a wetland system and which type of liner to use depends on the source water, regulatory requirements and feasibility. Some wetlands are unlined, either because the in-situ native soils are deemed to have sufficient sealing properties, or because groundwater recharge is a function of the system. Very large wetlands cannot be plastic lined because it is not feasible for systems of more than a few hectares; but clay lining has been implemented on systems up to 40 ha, such as Columbia, Missouri treatment wetland. If the wetland must be lined, there are a variety of liner materials available. The two most common liner materials used are 0.76 mm polyvinyl chloride (PVC) and 1.0 mm high-density polyethylene (HDPE). PVC liners are generally factory-seamed, one-piece liners used on small projects less than 0.1 ha in size. HDPE liners generally come in rolls and are field-seamed for larger projects.

The total installed cost of the liner includes not only the material cost, but also the labor cost associated with field seaming, seam testing, material inspection, and leak testing. If local soil conditions include sharp or angular rocks, a layer of sand or other granular material may have to be placed before the liner can be installed. If the gravel used to line the bed has sharp or angular edges, it may be necessary to cover the interior of the liner with a protective layer of geotextile fabric, since sharp rock can puncture a liner. Table 19.1 shows approximate installed costs of a variety of liner materials. Any form of lining is very expensive, and therefore not realistic, especially for large marshes.

Construction Cost Indices

Because wetland systems are constructed using local labor and local materials, it is not possible to offer precise universal cost estimates that will apply to all treatment systems. Both land and construction costs are quite time-variable. Major changes in the value of a dollar, and the cost of various marsh capital components, have occurred in the roughly four decades of treatment marsh history. There are major differences in the values of urban and agricultural lands, and these will be documented in the following sections. However, there is more uniformity in construction costs across source water types, meaning that all types require pumps, levees, and structures, the cost of which is not greatly dependent upon regional considerations within a single country.

Within the United States, Construction Cost Indices (CCI) are published by the Engineering News Record (ENR). These cost indices track inflationary changes within the construction industry over time. The ENR CCI started at 100 in the year 1913 and has

TABLE 19.1

Approximate cost of installed liners.

Material	Thickness mm	Installed Cost $1,000/ha
bentonite	10 kg/m^2	115
native clay*	30 cm	110
clay geotextile sandwich	NS	70
polyvinyl chloride	0.76	60
high-density polyethylene	1.02	70
polypropylene	1.02	85
reinforced polypropylene	1.14	100
hypalon	0.76	100
hypalon	1.52	115
XR-5	NS	160

* Assumes $25/m^3 delivered, placed and compacted.
Sources: U.S. EPA (2000a); ITRC (2003).

increased to 11,184 in November 2018. The time frame of treatment marsh construction began in approximately 1980, when the CCI = 3,237. Thus the cost of wetland construction is estimated to have more than tripled over history. Cost indexes are available as a national average, but are also published for 20 major metropolitan areas in the United States. For the purposes of this chapter, costs are based on the 2018 United States national average ENR CCI of 11,184 (U.S. DA, 2018a). The time series of the CCI is shown in Figure 19.1.

Global Variation

Economics vary geographically because of differing unit costs, and because of differences in the selection of materials and other design features. The cost of labor and materials within a particular regional market plays a large role in the cost of a wetland treatment system. Cost differentials are even greater when comparing across worldwide geographic locations and their accompanying economies. Most treatment marshes are located in the developed countries of the world, prominently including the United States, Sweden, Norway, Spain, the United Kingdom, Australia, and New Zealand. Their economies are similar, and for purposes of this book, monetary equivalence will be assumed to be represented by the currency exchange rate.

19.2 SURFACE WATER AND MUNICIPAL SYSTEMS

These systems may be continuous flow or event- or discretion-driven. However, they share major features: they are typically built on purchased agricultural lands, they usually require pumping, and ecological features are distinctly in second place to water quality features. Thus, planting is not high priority, and natural recruitment may be relied upon in many cases.

Construction Costs and Estimates

Land

Land usually needs to be purchased for the project, and this is a significant cost component. There are infrequent instances of land being freely available because of partnership with the owner, usually governmental. Over the history of treatment marshes in the United

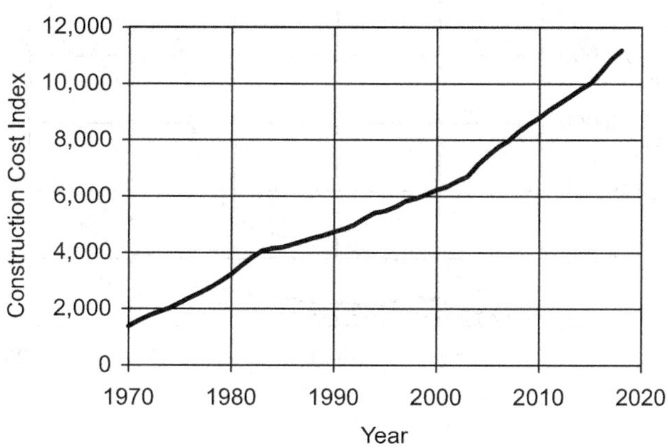

FIGURE 19.1 The historical values of the Construction Cost Index for the United States.

States, agricultural land prices have gone from \$1,820/ha in 1980 to \$7,756/ha in 2018 – a fourfold increase (Figure 19.2).

As will be discussed later in this chapter, land costs occupy a special role in the present worth analysis of a wetland system.

Site Investigation and System Design

The design and construction of a wetland requires that site characteristics be well understood, including soils, groundwater elevations, and site topography. These activities usually precede or accompany engineering design but are additional to that design.

Most marsh projects require a topographic survey. The ground elevation of the site is a critical factor in design, because it controls the cut and fill calculations that normally dictate the elevations of berms and bottoms. Associated with balancing of earth import and export is the issue of the potential for gravity flow. It is sometimes the case that proper cell elevations can eliminate the need for one or more pumps, and thus pump selection may interact with cut and fill considerations. The topographic survey will influence the directions of flow and the sequencing of cells, and the need for long or short distribution and collection canals, which for large systems can influence the project cost. The potential need to level cell bottoms for purposes of evening the hydraulic flow distribution can only be evaluated with detailed site topography. The shallow water depths in wetlands, especially large systems, creates a need for accurate as-built topography as an aid to understanding water depth and movement. The costs for such surveys are typically 50–500 \$/ha, depending upon scale and grid size requirements. For instance, the topographic survey for the Incline Village, Nevada, treatment wetland cost \$556/ha for the 175 ha wetland. However, survey work for cell 4 of STA2 of the Everglades phosphorus control project required only \$68/ha.

For medium to large-scale projects, it is often necessary to characterize the soils at the intended project site, which is accomplished through geotechnical investigations. In these situations, it is common to use shallow soil borings or backhoe pits to determine local soil characteristics. Costs for these initial site investigations vary with the size of the project. Even though treatment marsh projects rarely require liners, due to the low strength waters involved, there is a need to assess the potential for seepage from the project, and hence the need for seepage collection canals. It is critical to determine if site soils are adequate for the

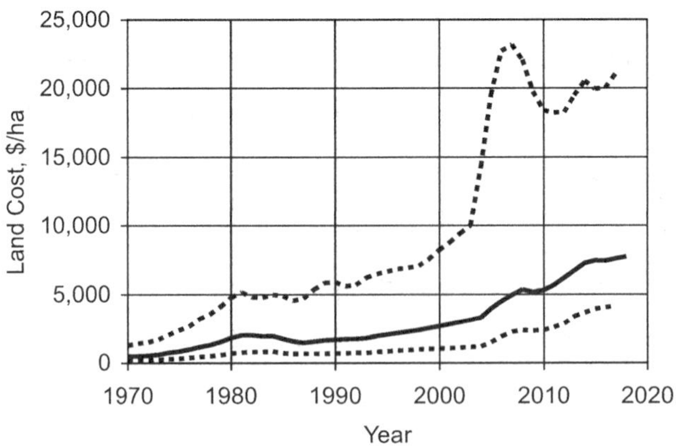

FIGURE 19.2 Historical agricultural land prices in the United States. Solid line is the national mean, dashed lines indicate the 10th and 90th percentile. (Source: U.S. DA, 2018b)

berms and levees, and if so how much compaction may be required, and how much allowance for subsidence.

Soils that may be used for rooting media may need to be assessed for contaminants of concern, including nutrients. A modest amount of chemical testing may be required, to identify any potential problems, or to form the basis for forecasting the sorption life expectancy for contaminants.

The location of the groundwater table and direction of groundwater movement can be a critical factor in wetland design. Groundwater levels and movement are determined through site-specific hydrogeological investigations. The depth to groundwater is an important consideration.

In some cases, there may be a concern for regional use of the groundwater as a potable water supply. The treatment marsh might be viewed as a possible source of contamination if partially-treated water entered into the unprotected drinking water aquifer and moved to the wells that withdraw potable water. This would usually occasion the need for a hydrogeological study, to ascertain the depths and directions of regional groundwater flow, and the consequences of even small leaks from the treatment wetland on the water quality of the aquifer. This was the case at the Columbia, Missouri treatment wetland project, for which the source water was less than secondary municipal effluent. The hydrogeological study included calibration and modeling to address this issue, at a cost of $1,127/ha for the 36 ha wetland (Brunner et al., 1992).

Earthwork

If the undeveloped site has undesirable vegetation, buildings, or other existing features that are incompatible with the wetland, these will need to be removed as part of the earthwork process. Brush and trees were removed from the treatment wetland sites at Ouray, Colorado; Sorrento, Louisiana; and West Jackson County, Mississippi, at an average cost of about $10,000 per hectare. On larger projects, roads and ditches may require degrading, and buildings may require removal.

The construction of a wetland treatment system requires excavation and grading of the site to produce level basins that are enclosed by earthen berms. For small systems (generally less than 0.05 ha) backhoes or similar types of excavation equipment are commonly used. Larger basins are generally constructed using bulldozers or construction scrapers. Project-specific earthwork costs are the product of two components. The first component is the cost to move earth, which is a volumetric (per cubic meter) cost. This cost component is a function of equipment costs, labor costs, and the source of earth supply (onsite or imported). The second component is the amount of earth that must be moved to grade the site. This areal requirement (m^3 of earth per m^2 of wetland area) is a function of the project site conditions. Earthwork costs are lowest on flat sites that require minimal grading and have suitable soils onsite (small areal grading requirement, small volumetric cost). Sloping a site requires terracing, which increases earthwork costs (due to the large areal grading requirement). If fill needs to be imported, earthwork costs will increase (due to the high volumetric cost).

In a study of two municipal treatment wetland systems (West Jackson County, Mississippi, and Gustine, California) the U.S. Environmental Protection Agency estimated a volumetric cost of $16.22/$m^3$ (U.S. EPA, 2000a). The 20.2-ha wetland system in West Jackson County, Mississippi, had an areal grading requirement of only 0.26 m^3/m^2, and the 9.7 ha system in Gustine, California, had an areal grading requirement of 0.35 m^3/m^2.

Larger treatment wetlands require far less earth moving on a per hectare basis than do small systems. For instance, Cell 4 of STA2 of the Everglades phosphorus control project required only 0.056 m^3/m^2 of earth moving based on the 816-ha wetland footprint.

However, the construction of such a large system does not involve scraping a thin layer from the entire footprint. Rather, the inlet spreader canal, outlet collection canal, and seepage return canal are the source of the fill material for the containment levees.

Even large wetlands may require significant earthmoving if built on sloping terrain. For example, the Inman Road treatment wetland in Clayton County, Georgia, has 22 wetted hectares in a terraced arrangement of 22 cells. The site presented slopes of 2–25%, and about 30 meters of vertical variation; Inman et al. (2001, 2003). Construction required moving 420,000 m^3 of earth, or 1.9 m^3/m^2.

Regional factors and site conditions can alter the cost of earth moving. Cell 4 of STA2 of the Everglades phosphorus control project moved 460,000 m^3 of material, in a nearly balanced cut and fill. The breakdown of per cubic meter costs was: \$3.92 to blast rock, \$3.20 to excavate, and \$3.14 to build levees, for a total of \$10.26/m^3.

Pumps and Conveyance

Source waters from municipal treatment facilities often require pumping for transfer to the treatment marsh site (see Figure 17.31, for example). Similarly, source waters from lakes, rivers, drainage ditches and groundwater usually require pumping. This is to be contrasted to agricultural and urban stormwater runoff marshes, many of which receive water by gravity. Exceptions are the Florida STAs, which require lift stations to get the water into the marsh, and in some instances, also require outlet pump stations.

The size range of wetland pumps spans the several orders of magnitude of the volume flow rates sent to the wetland (Figure 19.3).

There is ambiguity in the cost data literature as to whether the capital cost of pumps is attributed to the marsh project or not. Municipal polishing marshes receive water from a treatment plant, and the pump may well be located there, and its cost considered separately.

As a benchmark for perspective, the flow rate for a treatment marsh loaded at 5 cm/d is 500 (m^3/d)/ha. For a ten-hectare wetland, the pump would then have an estimated cost of \$456,000, which pro-rates to \$45,600 per hectare – considerably more than the anticipated land cost (Figure 19.3).

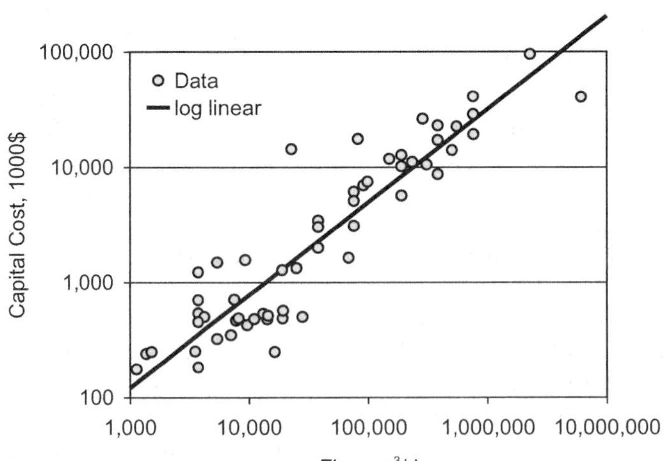

FIGURE 19.3 Capital costs for stormwater pump stations. The log linear regression is y= 0.46x^0.81, $R^2 = 0.84$.

FIGURE 19.4 The inflow conveyance canal for STA2 in south Florida

Marsh sites are not always located close to the source water point of origin. For instance, the OEW wetland is 21 km (straight line) from the Iron Bridge treatment facility that is the source of the tertiary effluent. The Incline Village wetland in Nevada is 27 km (straight line) from the source of the tertiary effluent. STA2 in Florida required 13.6 km of inflow canal, and 1.8 km of outflow canal (Figure 19.4). Pipeline costs are nearly proportional to the pipe diameter. For instance, in the state of North Carolina, C = 5700D, where D = diameter, cm and C = cost, $/km (BCC, 2016). Thus a 36 cm pipe costs $170,000 per kilometer.

It is not common for treatment wetlands to be supplied via canals; in fact, south Florida is one of the few places they are used. Canals range greatly in cost, because of size, excavation, and material disposition. Sizing depends upon the flow rate and the desired maximum headloss. Often, the canal material is used to construct a parallel berm, and thus the cost of the canal is combined into the canal-levee total. Some locations may require rock blasting. Large canals can be very expensive; for example the major canals associated with STA1WX cost about $3,000/m, or $3,000,000 per kilometer. They are over 30 m wide, and are designed to carry over a million cubic meters per day.

Water Control Structures

Hydraulic control of wetlands is generally provided through distribution piping and water level control structures. Inlet distribution, outlet collection and water level control are required in all systems.

In many marshes, inlet distribution is accomplished using an inlet deep zone across the width of the wetland, also called an inlet spreader canal. The cost for this is part of the cut and fill for the overall project. These deep zones may be filled with coarse rock to discourage rodents, thus incurring the cost of the rock.

Sometimes a perforated pipe is placed in the bottom of the deep zone, to aid in uniform introduction of the water. The cost of 10 cm diameter perforated PVC is roughly $25 per meter. In warm climates, distribution piping may be above ground and above water. Gated aluminum irrigation piping costs on the order of \$40 per meter for 30 cm diameter, with lesser costs for smaller diameters. The 30 cm size can convey up to 15,000 m^3/d without excessive headloss, and smaller diameters may be used for smaller flow rates.

The water level within a wetland cell is regulated by a downstream water level control structure. A typical outlet structure and associated cost for small marsh is shown in Figure 19.5. Larger projects may use pre-cast or cast-in-place concrete structures with a weir device (stop log, fixed plate, or adjustable weir gate). The ability to adjust the water level is an important control element and providing operational control over the water depth in the wetland is well worth the incremental cost over fixed-level structures.

Large inlet and outlet structures for FWS systems can be fairly complicated, with many cost components (see Figures 17.26, 17.27, and 17.28). These may involve remote sensing of water levels, sent to an operations center by telemetry, and remote control of opening and closing. Therefore, there is a need for electrical power, and the associated control equipment. The structure requires appropriate bedding, which usually means imported gravel. The cost breakdown for an example is given in Table 19.2. Despite the apparent complexity and cost, such structures are more economical when measured in terms of cost per unit flowrate. The structure of Table 19.2 cost approximately \$0.60 per m^3/d, whereas the Agri-Drain™ units cost about \$0.83 per m^3/d.

Planting

The plant component of capital cost varies according to the method chosen for vegetation establishment. It is presupposed that the soil media is in place. Options include seeding and planting, and natural recruitment.

The choice to plant originates with the desire to propagate a selected suite of plant species in a short period of time. Planting costs are a function of a several different factors. These include the form of plant material (plugs or rootstock), the source of material (locally harvested, onsite

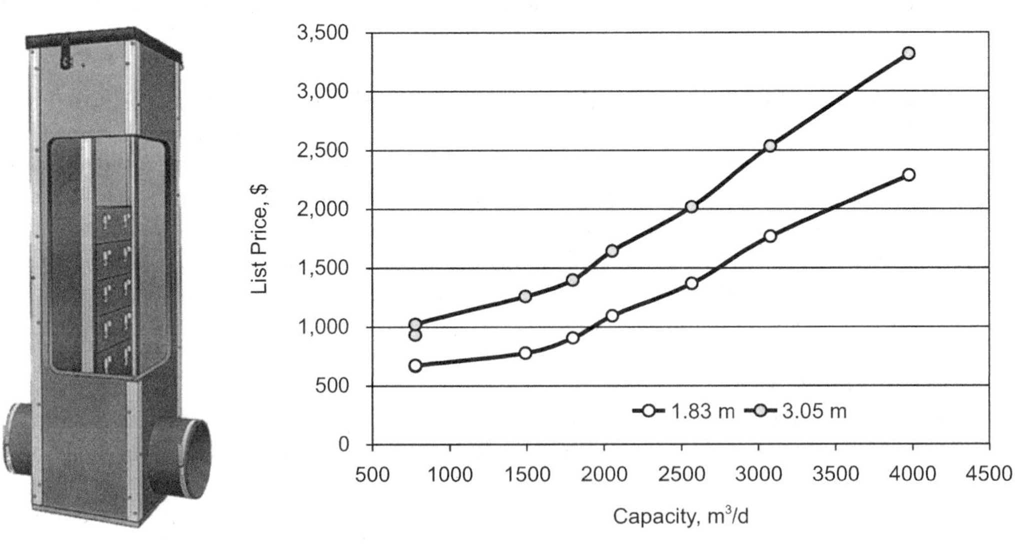

FIGURE 19.5 Example pricing of simple inline stoplog structures (\$). The height of these contained units is variable; data is shown for 1.83 m and 3.05 m total structure height. The capacity is for a height over the weir of 25 cm. Data source: Agri Drain Corporation, Adair, Iowa. www.agridrain.com/shop/c85/water-level-control-structures/p901/inline-water-level-control-structures/.

TABLE 19.2

Capital cost of one inlet structure for STA2 Cell 4. There are six such structures for the 816-ha wetland, and each has a design flow of 412,000 m³/d.

Item	Description	Cost
Piping	40 meters of 1.68-m diameter	$56,740
Structure	slide gate	$42,703
Excavate, bed, install	917 m^3 gravel	$13,123
Sensors	–	$30,653
Telemetry	–	$13,794
Stilling wells	–	$38,445
Electrical control	–	$52,111
Total		$247,569

Source: Brown and Caldwell (2005).

nursery, or commercial nursery), the method of planting (manual or mechanical, volunteer labor, contractor labor), and the overall planting density (number of plants per square meter). Because of these variables, planting costs vary widely among wetland projects. In projects that employ hand planting, plant material costs may be a small fraction of the installed cost.

Commercial nurseries sell a wide range of wetland plants as seed, individual plant plugs (small), individual plant pots (large) or bare root stock (Table 19.3). The median (2018) cost for plugs was $1.30 (N = 16), and that for pots was $3.75 (N = 16). Transportation costs must be added to these, leading to a higher delivered cost.

Seeding is also an option. Seeds may be obtained for individual species (Table 19.3), or for mixes that typify natural wetlands of a given region (Table 19.4). Seeds are less expensive than plants, but there are caveats for their use. Sprouting depends upon very careful soil moisture and burial depth control. The fraction of the purchased seed that is in viable is not 100%, despite frequent claims. And, plant growth takes a longer period of time than for rooted propagules. Seeds for wetland plants may be harvested from local sources as part of the project work, particularly for *Typha*, which produces large seed heads with extremely numerous seeds, and which are easily picked in autumn. Seeds are also available from wetland nurseries in the United States, for over a hundred wetland species. The cost is considerable, ranging from $440 to $2,300 per kilogram of live seeds for common wetland plants in the northern United States (Table 19.3). The seeding rate is typically on the order of 2–4 kg of live seed per hectare. The purchased price of the seed is consequently in the range of $100–$7,000 per hectare, with a median of about $2,100 per hectare. The cost of installation includes the seeding itself (minor) and the fostering of germination and growth, via soil moisture/water management (major). Dry seeding has been estimated at ten man-hours per hectare, and hydroseeding at $145/ha (Hayes et al., 2000). Such required activities are usually inexpensive and are less the cost of the seed itself.

Seeding techniques range from scattering in the wind (Figure 19.6), to backpack broadcasting, to the use of seed drills. The seed is then pressed into the soils using a roller, cultipacker, or by light raking. The use of foot travel has also been used effectively, at the Isanti-Chisago site in Minnesota (Loer et al., 1999). Volunteer labor has been used successfully at Brighton, Ontario, and Roblin, Manitoba, where grades 9 and 10 science students participated in seeding (PFRA, 2002). Hydroseeding has been used successfully (U.S. EPA,

TABLE 19.3
Example prices for plants and seeds.

	Seed		Plug		Pot	
	$/kg	$/ha	$/plant	$/ha	$/plant	$/ha
Iris virginica	880	2,638	1.50	15,000	3.50	35,000
Juncus effusus	528	1,583	1.30	13,000	3.50	35,000
Leersia oryzoides	704	2,110	1.30	13,000	3.75	37,500
Mimulus ringens	1,232	3,693	1.15	11,500	3.50	35,000
Panicum virgatum	35	106	1.15	11,500	3.25	32,500
Peltandra virginica	528	1,583	1.50	15,000	4.50	45,000
Pontederia cordata	528	1,583	2.00	20,000	4.75	47,500
Sagittaria latifolia	1,056	3,166	1.75	17,500	4.50	45,000
Saururus cernuus	1,584	4,748	1.75	17,500	4.25	42,500
Schoenoplectus acutus	1,056	3,166	1.30	13,000	3.75	37,500
Schoenoplectus tabernaemontani	440	1,319	1.30	13,000	3.50	35,000
Scirpus cyperinus	704	2,110	1.15	11,500	3.50	35,000
Sparganium eurycarpum	282	844	1.50	15,000	4.00	40,000
Spartina pectinata	440	1,319	1.30	13,000	3.75	37,500
Spiraea alba	2,288	6,859	2.00	20,000	4.50	45,000
Median	**704**	**2,110**	**1.30**	**13,000**	**3.75**	**37,500**

Prices from: Cardno Native Plant Nursery Walkerton, IN (www.cardnonativeplantnursery.com/docs/default-source/default-document-library/current-pricing.pdf?sfvrsn=36 Accessed 12/17/18)

2000a), but requires the addition of detergent to loosen the seed covering. In the Great Lakes region of the United States, the optimum seeding time is autumn through late spring.

It is sometimes possible to obtain plants from existing stands, such as cattails from roadside ditches. Extraction requires mechanical digging, typically with a backhoe, to a depth of about 30 cm, so as to gather the majority of the root mass. The plants are then separated from the mass by hand, with each propagule containing at least one healthy shoot and at least 20 cm of associated rhizome. The shoot is topped to a height of 30–40 cm, to reduce the amount of foliage that the transplant must maintain after the trauma of transplanting.

One of the more innovative plant transfer techniques is the helicopter air-lifting of SAV from one Florida STA to another, to foster a change to an SAV community.

On-site nurseries may also be established prior to wetland construction, for the purpose of supplying plant materials. However, the planting contractor may prefer to avoid the effort needed to extract and prepare the plants, and opt for the use of purchased plants, as was the case at Columbia, Missouri (Brunner and Kadlec, 1993). Established treatment wetlands may serve as sources of plant materials for new systems. For example, the plants for the Connell, Washington, system (Kadlec et al., 1997) were extracted from the Arcata, California, project, and successfully established in the new wetland. Marsh establishment on the Texas coast required a range from 11.3 to 29.3 man-hours per 1,000 plants to hand dig, separate, and transplant various propagule types of 11 marsh species, Dodd and Webb (1975), as referenced by Hayes et al. (2000).

The total cost of plants depends on planting density, which ranges over about 0.25–4.0 plants per square meter (spacing of 0.5–2.0 meters). If the plants are on one-meter centers,

TABLE 19.4

Example prices for wetland seed mixes seeds. Minor seed percentages not shown.

Northeast Wetland Rush/ Bulrush Mix		Northeast Wetland Hummock Mix	
$556/ha		**$1124/ha**	
37.30%	*Juncus effusus*	37.30%	*Glyceria canadensis*
34.40%	*Scirpus atrovirens*	29.30%	*Glyceria grandis*
26.90%	*Scirpus cyperinus*	7.50%	*Calamagrostis canadensis*
1.30%	*Scirpus tabernaemontanii*	7.30%	*Panicum dichotomiflorum*
Northeast Wetland Diversity Mix		4.90%	*Leersia oryzoides*
$692/ha		3.90%	*Echinochloa crusgalli*
28.82%	*Scirpus atrovirens*	3.00%	*Verbena hastata*
13.05%	*Juncus effusus*	**Northeast Wetland Grass/Forb Mix**	
12.01%	*Mimulus ringens*	**$1916/ha**	
8.35%	*Carex vulpinoidea*	37.30%	*Glyceria canadensis*
7.83%	*Penthorum sedoides*	29.30%	*Glyceria grandis*
6.68%	*Glyceria grandis*	7.50%	*Calamagrostis canadensis*
5.22%	*Scirpus cyperinus*	7.30%	*Panicum dichotomiflorum*
4.18%	*Verbena hastata*	4.90%	*Leersia oryzoides*
2.09%	*Eupatorium perfoliatum*	3.90%	*Echinochloa crusgalli*
1.57%	*Leersia oryzoides*	3.00%	*Verbena hastata*

Prices from: Southern Tier Consulting West Clarksville, NY www.southerntierconsulting.com/seedmix.htm

there are 10,000 per hectare. This value is used in Table 19.3 to establish the per hectare cost of nursery-purchased plant plugs, which ranges from $11,500–$20,000 per hectare, with a median of $13,000 per hectare.

The cost of putting the plants into the ground can be extremely variable. Most wetlands are planted by hand, with rates of up to 150 plants per person per hour for experienced planting crews. Allowing for lesser efficiency of 50 plants per person per hour, thus requiring 200 man-hours per ha at 10,000 plants per ha. This matches estimates provided by the U.S. ACE publication (by Hayes et al., 2000). For labor at $20 per hour, insertion costs would be about $2,000 per hectare at 10,000 plants per hectare, or $0.40 per plant. When combined with the per plant purchase price, the cost would be $15,000/ha installed. For small to medium size wetlands (less than 10 ha), planting costs are less than 20% of the total capital cost. For very large wetlands, planting is too expensive, and natural recruitment is relied upon.

Natural recruitment is the least costly, but the least controllable and usually of the longest duration. It is not free from cost, because moist soil conditions must be maintained, which in turn requires water management on the wetland. Virtually all of the Florida stormwater treatment wetlands (STAs) were established in this way. A newer strategy for these STAs is the use of a seeded sacrificial rice cover crop (*Oryza sativa*), which assists in soil stabilization and initial nutrient immobilization. This cover crop lasts only one growing season, does not renew itself, and gives way to other wetland vegetation.

FIGURE 19.6 Seeding of the Brighton, Ontario wetland (A), and the resulting growth (B). Photo courtesy of J. Pries.

Human Use Facilities and Access

The human use facilities that sometimes accompany a treatment marsh are not usually reckoned in the capital cost of the project. Clearly, they are not central to the missions of providing water quality improvement and ecological diversity. At the top of the scale are visitor centers, such as those shown in Figures 1.16, 15.21, and 15.22. These cost many millions of dollars and are undeniably an important feature of the system and project. At the lower end of the scale are features such as signage, walkways, boardwalks, parking lots and similar, which have been described in previous chapters. While their maintenance is generally considered part of the operating cost of the system, the capital required to build them is seldom included. As a result, reports of human use capital costs are virtually nonexistent.

Nonconstruction Costs

In addition to construction costs, there are indirect costs associated with engineering design, construction management, and administrative activities. Administrative activities include financing and legal efforts. On large projects, these tasks are usually identifiable as having separate contractual arrangements. In aggregate, these may be a significant fraction of the total project cost. For instance, Hazen and Sawyer (2011) report nonconstruction costs of 30, 32, and 38% of the project total for three large Florida systems.

Design Costs

Engineering activities include conceptual design, final sizing, preparation of plans and specifications, preparation of the O&M manual, and permitting. These costs depend upon the size, complexity and novelty of the project. Designs are usually performance-based, and require significant expertise in conceptual design and sizing. There are likely to be no well-established regulatory procedures, which means that obtaining the necessary permits can become a challenging and time-consuming process. Because of scale, the number and complexity of drawings increases. A larger project is likely to go through several design phases, and produce a conceptual design, a basis of design report, and 10%, 30%, 90%, and final design reports – each subject to scrutiny and possible revision. The project may be subjected to value engineering, a process by which outside experts examine the project plans to ascertain if further economies might be realized.

Design costs for larger projects are usually expressed as a percentage of the overall construction costs. The cost of the routine engineering design has often been estimated at approximately 10–15% of the overall construction cost (U.S. EPA, 2000a; Wallace and Knight, 2006). This amount is probably insufficient to deal with a long, drawn out permitting process.

Construction Management and Startup Services

Services provided during construction and startup typically include construction observation, inspections, testing, startup assistance and operator training. These components comprise approximately 10% of the overall construction cost (U.S. EPA, 2000a; Wallace and Knight, 2006). Construction supervision needs to involve the services of a knowledgeable treatment wetland specialist, as a complement to the usual supervision of the implementation of civil works. Startup also requires such services, for the period of vegetation establishment. Operator training should include a good foundation of the general principles of treatment wetlands, as well as the specific characteristics of the project.

Financing and Legal

Large projects in particular have money management aspects, which ensure that monies are secured and that contractors are paid on an appropriate time schedule. Construction contractors are usually required to be bonded. Unfortunately, legal issues often arise, with respect to land acquisition and contractor performance. And, in some cases, there are legal challenges to the entire project, based upon perceptions of stakeholder individuals or organizations. These components can comprise 5% or more of the overall construction cost.

Capital Cost Illustrations

There is considerable variability in the actual capital costs incurred for treatment marshes, both in terms of total cost and in the distribution of the cost across the major categories. The examples in Tables 19.5 and 19.6 illustrate this variability.

Indirect Cost Estimation

During the design phase of a project, it common practice to include estimates of the probable cost of the project. These opinions of probable cost do not include land costs, because the design engineers are not tasked with acquiring the land.

TABLE 19.5

Capital costs for the 328-ha Nubbin Slough wetland in Florida.

Item	Cost	Fraction of Total
Land Cost	\$3,211,523	13%
Construction Cost	\$14,020,940	56%
Engineering	\$5,112,215	20%
Construction Management and Administrative	\$2,715,726	11%
Total	\$25,060,404	100%

Source: Adapted from Hazen and Sawyer (2011).

TABLE 19.6

Capital costs for the 276-ha Lake Apopka wetlands in Florida.

Item	Cost	Fraction of Total
Land acquisition costs	\$4,840,502	44%
Construction	\$4,381,506	40%
Pumps	\$747,235	7%
Soil amendment	\$161,091	1%
Alum injection system	\$35,881	0%
Earthwork	\$210,822	2%
Design and engineering	\$603,609	5%
Nonland capital costs	\$6,140,145	56%
Total	\$10,980,647	100%

Source: Adapted from Dunne et al. (2015).

Nonconstruction Contractor Costs

In addition to the construction elements of site work for a wetland system, the project will incur other project costs. Contractor expense costs are incurred by the contractor but are not directly connected to the wetland construction. These typically include mobilization, bonding, insurance, traffic control (if required), and construction surveying and staking. These costs are dependent on items such as: the distance the contractor has to mobilize to the project site, the contractor's previous history with insurance carriers and bonding companies, bonding requirements in the construction contract, whether the construction work will impact existing traffic patterns, and the extent of surveying required to set reference stakes for site grading and pipe installation. Contractor overhead, or burden, is the cost of doing business that is not attached to any particular project, and it is commonly associated with the direct labor charges as a percentage. Since these items change from project to project, some variation in nonconstruction costs can be expected.

Contingency and Escalation

Escalation is an allowance for inflation. Contingency is a percentage of the base cost to cover errors in human judgment, and it is intended to be spent. The amount of money set aside or planned for unforeseen events is a function of the level of the economic estimate being formulated. A project's risk elements may include accidents, vandalism, theft, work quantity and productivity variances, unfavorable weather and bankruptcies. Contingency allotments of 10–30% are typically used.

Estimated Cost Illustrations

Capital cost estimation relies upon unit costs for the various direct cost components. Unit costs are variable with the size of the feature, and also vary regionally. Indirect costs are estimated as percentages of the direct construction cost, not including land cost. It is instructive to combine the various elements of capital cost into a summary by category. There are too many variables to do this in a general way, and consequently an illustration is presented here.

Table 19.7 shows a capital cost list for a hypothetical 10-ha treatment marsh, which might receive a hydraulic loading of 400 m^3/d (4 cm/d). It is assumed that the wetland is not planted but does require a pump to introduce the source water into the system. Land price reflects agricultural values.

The direct cost is $93,800 per hectare. It is seen that most of the direct cost is for land, earthwork, and pump. The indirect costs are a third of the direct costs.

It is informative to compare estimated and actual costs. The Nubbin Slough wetland had cost estimates prepared in 1999, which included construction costs and construction management. In 2018 dollars, the estimate was $12,116,000. The actual cost of comparable items in Table 19.5 is $16,737,000. The estimate was thus 28% lower than actual, which is at the extreme of the expected range of ±30%.

Another example of a design cost estimate is presented in Table 19.8, which contains a different suite of items. The project was the expansion of the Shannon wetland in Texas,

TABLE 19.7

Estimated capital costs for a hypothetical ten-hectare treatment marsh system.

	Unit	Quantity	Unit Cost $	Total Cost $1,000	Percent
Land Acquisition	ha	10	30,000	300	21.32
Site Evaluation	lump sum	1	2,000	2	0.14
Clear and Grub	ha	10	250	3	0.18
Earthwork	m^3	15,000	7	105	7.46
Structures	lump sum	5	3,000	15	1.07
Pump	lump sum	1	400,000	400	28.43
Conveyance	m	500	127	64	4.51
Site Work	lump sum	1	50,000	50	3.55
Total Direct Cost				**938**	**74.62**
Engineering	15%			96	7.61
Construction Observation	5%			32	2.54
Startup Services	5%			32	2.54
Nonconstruction Costs	5%			32	2.54
Contingency	20%			128	10.15
Total Indirect Cost				**319**	**25.38**
Total Cost				**1,257**	**100.00**

by the addition of 470 ha of wetland, with 19 cells in 9 trains, treating an average flow of 340,000 m^3/d. This project is a component in the Tarrant Regional Water District water supply system, which serves the Dallas-Fort Worth metroplex. The land carried no cost, because the project is a partnership with the Texas Parks and Wildlife Department, which manages the site as part of the Richland-Chambers Wildlife Management Area. The project serves to satisfy wildlife habitat enhancement. Both inlet and outlet pumps were required. Table 19.8 is for construction costs, and does not include a design cost, which might be estimated at 15%, or an additional $6,000,000.

Economy of Scale

A strong determinant in the overall cost of a wetland treatment system is the size of the system. Larger projects generally benefit from economies of scale, resulting in lower unit costs. The size of the project is usually considered to be represented by the wetland area. Cost functions relate capital expenditure to these measures, and commonly take the form of a power law, with an exponent of 0.6–0.7 often being appropriate. Cost data in the wetland literature is often ambiguous about what has been included or excluded. Land costs are often not included, and the same is true for indirect costs. Therefore, the reader is advised that the following compilations of reported costs are subject to considerable variability, and will generally be lower than the summation of the various categories enumerated earlier in this chapter.

Figure 19.7 presents the trends of costs with size for surface water and municipal systems in the United States. It is reasonably well represented by the following relationship:

$$C = 289A^{0.71} \tag{19.1}$$

$R^2 = 0.79$
$0.006 < A < 6740$
where A = wetland area, ha
C = cost, 2018 dollars in thousands

This cost function may be used to obtain an approximate cost for a prospective project but should not be used in lieu of detailed cost estimation.

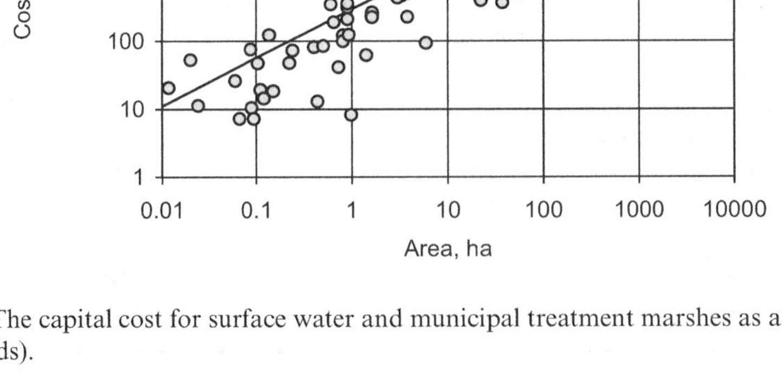

FIGURE 19.7 The capital cost for surface water and municipal treatment marshes as a function of size. (N = 119 wetlands).

TABLE 19.8

Estimated construction costs for the Shannon wetland expansion.

Category	Amount	Cost, $	Subtotal, $
Pumps and Transfer Pipe			
Trinity River Pump Station		4,720,237	
137 cm force main to sed basins	2,230m	3,210,361	
Outlet Pump Station		2,769,330	
APAI, 2008)			10,699,928
Sedimentation Basins			
Clear and Grub	14ha	117,748	
Soil Removal and Compacted Liner	168,000m^3	2,162,495	
Piping, Valves, Structures		2,501,315	
Baffles		452,151	
Roads	5.09km	337,216	
			5,570,924
Wetland Interior			
Level & Grade	416,900m^3	10,584,229	
Structures	44	4,164,159	
Emergency Spillways	16	1,539,465	
Planting and Seeding	143,100 plugs	730,775	
Roads	22.5km	1,759,115	
			18,777,744
Miscellaneous Site Work			
Road Demolition		336,422	
Electrical/Instrumentation		269,137	
Easement Acquisition		33,642	
Oil Pipeline Relocation		269,137	
Wetland Mitigation	2.4ha	121,112	
Clearing	140ha	464,262	
			1,493,712
Total Direct Costs			36,542,308
Indirect			
Mobilization/Demobilization, Overhead, Insurance, Bonds, and Profit	15%	6,083,791	
Contingency	20%	8,111,720	
			14,195,510
Total Project Construction Cost			50,737,818

Source: Based on the engineer's preliminary design report (APAI, 2008).

19.3 URBAN SYSTEMS

The urban environment gives rise to a different emphasis on the costs of treatment marshes for urban runoff. The cost of land can be extreme, but inflow is often by gravity, eliminating the need for pump stations. The set of rules and regulations is often very restrictive, which is reflected in the cost structure. Major emphasis is directed toward aesthetic and ecological values, with lesser emphasis on water quality. The public image is focused on the garden or park characteristics of the system. Most often, there are no numerical water quality goals.

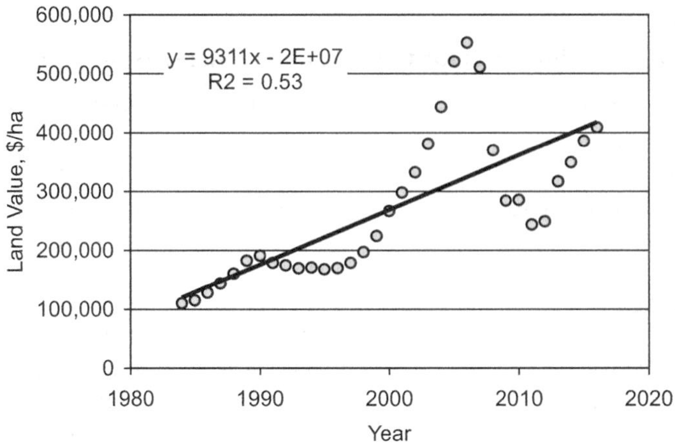

FIGURE 19.8 Time trend of large metropolitan land values. (Based upon updated version of Davis and Palumbo, 2008.)

There exist proprietary simulation software packages for urban wetlands that include cost routines. For instance, the MUSIC model, discussed in Chapter 11, has such capability. MUSIC calculates construction cost based upon a power law similar to Equation 19.1, with an exponent of 0.64 and a premultiplier that reflects Australian dollar value and cost structure (Montaseri et al., 2015).

Urban Land Costs

Land costs can be a significant fraction of the total capital cost. For example, Wossink and Hunt (2003) identify three categories of land cost for urban stormwater wetlands in North Carolina: zero cost because the project is required for community green space, $125,000 per hectare for vacant land that may be used for residential development, and $550,000 per hectare for land that may be used for commercial development. This dependence on urban use type can be separated according to labor market areas. In the United States, three major categories are defined by the Federal Government:

- Metropolitan. A metropolitan statistical area is a geographical region with a relatively high population density at its core and close economic ties throughout the area. These have a population in excess of 50,000, and in the United States, there are about 400 of these.
- Micropolitan. A micropolitan area is based on counties in the United States, and are centered on an urban cluster with a population of at least 10,000 but fewer than 50,000 people. There are about 400 of these in the United States.
- Noncore. A noncore area does not contain a substantial population nucleus, and does not have a high degree of economic and social integration, as measured through commuting ties with an associated core community.

The higher the population in a metropolitan zone, the higher the land cost (Table 19.9).

This concept has been extended to very large metro areas. Albouy et al. (2018) suggest that land values continue to rise as the urban area increases in size, according to:

TABLE 19.9

Land value of urban areas according to type. Metropolitan = >50,000 population; micropolitan = 10,000 to 50,000; non core = <10,000.

Type of Urban Area	$/ha
Large metro area of more than 1 million residents	$209,018
Small metro area of less than 1 million residents	$53,373
Micropolitan area adjacent to large metro area	$21,535
Micropolitan area adjacent to small metro area	$14,918
Micropolitan area not adjacent to a metro area	$11,195
Noncore adjacent to small metro area and contains a town of at least 2,500 residents	$11,285
Noncore adjacent to small metro area and does not contain a town of at least 2,500 residents	$8,461
Noncore not adjacent to metro or micro area and contains a town of at least 2,500 residents	$6,469
Noncore not adjacent to metro or micro area and does not contain a town of at least 2,500 residents	$5,335

Source: Larson (2015).

$$C = 70.3A^{0.35} \tag{19.2}$$

$50<A<8000$
where A = urban area, km^2
C = cost, 2018 dollars in thousands

The U.S. median for these very large cities is $1,900,000 per hectare, which clearly and obviously precludes urban stormwater wetlands in such large cities.

Land values in US cities have not followed a uniform time progression over the past three decades (Figure 19.8). It was "boom and bust" over the 2000–2010 period. Nonetheless, over the 30-year period of modern implementation of urban runoff wetlands, those land values have quadrupled.

Urban Construction Costs

Construction costs for stormwater wetlands are usually scaled to the water volume that can be retained in the system, because designs are determined on that basis. However, the water quality volume (WQV) can be converted to an areal basis, since the prescribed bathymetry usually includes about one-third deep zones (ca. 1 m deep) and two-thirds shallow zones (ca. 0.3 m deep), for an average depth of about 0.5 m. Here the areal basis is used, in order to allow ease of comparison among marsh source water types.

Weiss et al. (2007) found construction costs for 19 urban stormwater wetlands from a variety of sources (Figure 19.9). These do not include land costs, or administrative and engineering costs. A log linear regression fits the data reasonably well. According to:

$$C = 184A^{0.66} \tag{19.3}$$

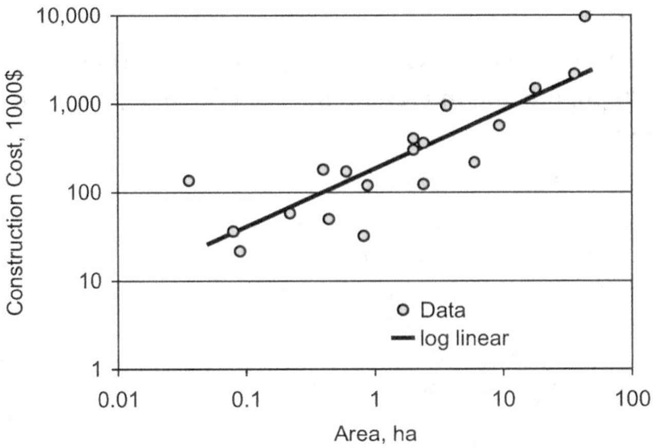

FIGURE 19.9 Reported construction costs for 19 urban wetlands.(Based on data in Weiss et al. (2007) for a mean depth of 0.5 meter and escalated.)

$R^2 = 0.71$
$0.036 < A < 44$
where A = wetland area, ha
C = cost, 2018 dollars in thousands

This cost curve may be compared to that shown in Figure 19.7, and Equation 19.1, but with vary important caveats: urban construction costs do not include engineering or land costs. Typically, engineering costs would add about 15% to the construction cost. The major factor in cost increase for the total project comes from land costs, which easily can double or triple the overall cost.

Urban Construction Cost Estimates

Construction cost estimates for stormwater wetlands are proportioned much differently than those for larger, continuous and surface water source projects. Pumps are generally not a component, but plants and seeding comprise a very large proportion of the cost. Guidelines for construction cost estimating have been set forth; for instance, the worksheet provided by the Minnesota Pollution Control Agency (Table 19.10) (MPCA, 2015). That worksheet has been cost escalated to 2018, and applied to a hypothetical project. Basin contouring, soils management and planting are the principal costs.

Design cost data for urban wetlands are generally not available. An estimate is 6.6% of the construction cost (Taylor, 2005).

Floating Raft Retrofit Costs

These systems are add-ons, and require no land or pumps. Although the literature on floating wetland performance is now fairly extensive, it contains virtually no cost information. Emphasis is entirely on water quality performance (measurably an improvement), ecological value (significant), and ecological aesthetics (greatly emphasized). Details have been discussed in Chapter 11.

TABLE 19.10

Construction cost estimate for a hypothetical one-hectare urban runoff wetland.

Description	Units	Quantity	Unit Cost	Total Estimated Price
Site Preparation				
Tree removal – up to 30 cm diameter	each	10	\$390	\$3,902
Clear and grub brush	m^2	1,000	\$2	\$2,000
Tree protection – temp. fence	m	50	\$11	\$549
Topsoil – 15 cm depth, salvage on site	m^2	5,000	\$6	\$30,003
Site Formation				
Excavation – deepwater zone – 1 m depth	m^2	3,000	\$7	\$20,002
Excavation – marsh zone – 0.3 m depth	m^2	7,000	\$1	\$9,334
Grading	m^2	5,000	\$2	\$10,001
Hauling off-site – 1.5 m depth	m^2	3,000	\$7	\$20,002
Structural Components				
Inlet structure	each		\$2,230	\$2,230
Outlet structure	each		\$3,902	\$3,902
Site Restoration				
Sod – above vegetative bench	m^2	1,000	\$6	\$6,001
Soil preparation	m^2	3,000	\$33	\$100,010
Seeding – vegetative bench	m^2	2,000	\$1	\$1,333
Planting	m^2	5,000	\$40	\$160,015
Subtotal				\$369,284
10% Contingencies				\$36,928
Total Construction Cost				\$406,213

Source: Worksheet is adapted from MPCA (2015).

Lyon et al. (2009) found no details on cost reported in the floating treatment wetland literature. Although some small projects have been conducted on a "do it yourself" basis that is rarely an option for systems of any significant size. There are many commercial raft manufacturers, but their cost information is not generally available, and pricing appears quite variable. Haynes (2014) reported that the cost breakdown consisted of the following:

- Floating wetland base material: \$43–\$344 per square meter
- Attachment hardware and anchors: \$54 per square meter
- Plants: \$54 per square meter
- Soil/planting medium: \$10 per square meter
- Range total: \$161–\$462 per square meter (\$1,610,000–\$4,620,000 per ha)

These numbers give rise to "sticker shock," which is likely the reason why they are hard to find.

19.4 AGRICULTURAL SYSTEMS

Some agricultural runoff systems have pumped inflows, usually taken from drainage ditches or canals. These have been discussed in the preceding section on surface water sources. The other large category is comprised of constructed marshes that receive runoff water by gravity, either as overland flow or as tile drainage water. Large numbers of these have been constructed in the United States and in Sweden for nitrate and sediment control. The economic setting for these is entirely different from urban or pumped systems. There are no pumps, and basin bathymetry is often simply that of the antecedent lands. Emphasis is on water quality and wildlife habitat, and minimally on vegetative biodiversity. However, there are rarely any numeric water quality standards to be met. They are event-driven, and function during the unfrozen season. The land typically remains in private ownership, with lease agreements that facilitate their use as wetlands. Descriptions and performance expectations of these are given in Chapter 11.

Conservation Reserve Enhancement Program (CREP)

The Conservation Reserve Enhancement Program (CREP) is a government funded private-land conservation program. It is administered by the Farm Service Agency of the U.S. Department of Agriculture. CREP is a conservation program involving State and federal funds. Farmers are paid an annual rental rate along with other incentives but retain ownership of the land. That means they can sell the land, but the permanent conservation easement stays on the land's title, meaning the property can never be farmed or developed. The projects require buffers to accompany the wetlands, typically several times the wetland acreage. The contract period is typically 15 years. Landowners enrolling in CREP in the State of Iowa will receive:

- Up to 15 years of annual rental payments from USDA for all enrolled (wetland plus buffer) acres paid at 150% of the average soil rental rate.
- 100% cost-share for wetland restoration and buffer establishment, comprised of design and construction costs.
- A one-time, up-front incentive payment to enter into either a 30-year or perpetual easement.

Effectively, the land cost is a rental fee, which requires the present value of a time series of payments.

Present Worth of Land Rental

The equivalent one-time cost of a series of future payments is computed via a present worth analysis. Conceptually, the present worth is the money that needs to be in the bank at a specified interest rate so that the future payments can be made. The discount rate is keyed to existing loan rates, and the current (2019) value in the United States is about 3%. For the CREP projects, payments are made over a 15-year period. Present value can be directly evaluated from:

$$P = Y\frac{(1+i)^N - 1}{i(1+i)^N} = Y \cdot F_{pw} \tag{19.4}$$

where i = discount rate, fraction per year
F_{pw} = present value factor, unitless
N = Life expectancy, yr
P = present value, \$
Y = yearly payment, \$
The present worth factor for 15 years at 3% is 11.938.

Reported CREP Wetland Costs

Wetland costs for the Iowa CREP systems are reported annually (IDALS, 2017). These include design and construction costs as incurred by each project. Land costs are incurred for both the wetland and its required associated buffer; here the choice is to apportion these on the basis of wetland area, as opposed to the contract area, which is the wetland plus buffer areas. Beyond these land rental costs, there is also a cost for obtaining a long-term easement for the land to be continued as a wetland. Table 19.11 shows the cost components that arise under this system of valuation.

Design costs are relatively low, compared to urban and surface water systems. Construction costs are relatively modest, because there is little earth movement and only one or two small structures. Land costs, consisting of leases and easements, dominate the total.

The construction cost component for the 82 wetlands has only a small trend with wetland size (Figure 19.10). Land costs, administrative and engineering costs are not included. A log linear regression fits the data reasonably well:

$$C = 101A^{0.31} \tag{19.5}$$

$R^2 = 0.14$
$1.1 < A < 10.4$
where A = wetland area, ha
C = cost, 2018 dollars in thousands

Cost Estimation for Iowa Treatment Marshes

A second approach to cost analysis is based upon the purchase of the necessary land, rather than its rental; and upon the unit prices for cost elements such as engineering, planting and structures (Christianson et al., 2013; Tyndall and Bowman, 2016). These cost estimates are lower than those reported for a rental basis.

Agricultural Runoff Wetlands in Other Regions

Agricultural treatment marshes have found extensive application in countries where land values are similar to those in the United States, such as Sweden. Their utilization is less frequent in countries where land values are quite high, such as the United Kingdom.

TABLE 19.11

Size and cost components of 82 Iowa CREP wetlands. The area basis is the wetland only.

		25%tile	50%tile	75%tile
Wetland Area	ha	2.12	2.91	5.10
Contract Area	ha	9.39	13.90	21.97
WWAR	%	0.55	0.70	0.89
Watershed Area	ha	260.7	344.8	632.0
Design	1000$/ha	5.92	11.09	16.71
Easement	1000$/ha	6.08	30.22	47.96
Construction	1000$/ha	29.93	48.36	72.94
Lease PW	1000$/ha	29.16	36.13	43.80
Total 15 Yr PW	1000$/ha	85.82	131.90	174.04

Source: IDALS (2017).

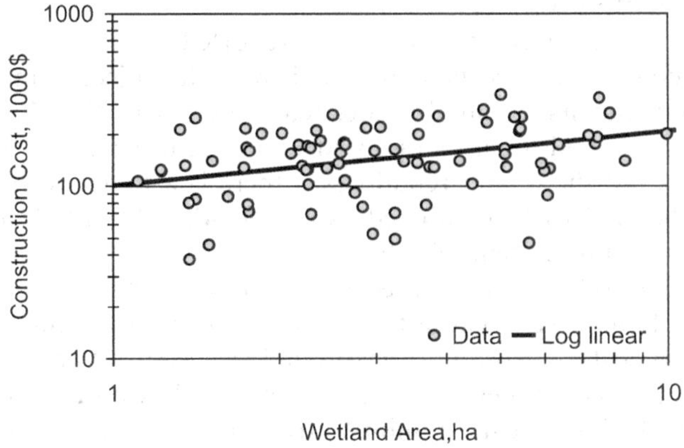

FIGURE 19.10 Reported construction costs for the Iowa agricultural runoff wetlands.

TABLE 19.12

Itemized cost estimate for an agricultural runoff wetland in Iowa.

	Cost per Hectare	Subtotal
Land		
Buffer 3.5 ha	\$60,515	
Wetland 1.0 ha	\$17,290	\$77,805
Construction		
Design	\$2,645	
Construct Basin	\$3,968	
Plants Wetland	\$1,693	
Plants Buffer	\$452	
Weir	\$15,872	
Structure	\$5,555	\$30,185
Total Capital Cost		\$107,990

Source: Tyndall and Bowman (2016).

Although the economic structure of other countries differs from that in the United States, the cost components and methods of cost estimation remain the same.

Agricultural Wetland Economics for UK Systems

Average prime arable land in the United Kingdom is priced at about \$28,000 per hectare (Eurostat, 2017), which helps explain a relative scarcity of agricultural treatment marshes. Nonetheless, several small systems have been built, primarily for sediment control (Ockenden et al., 2012). They note that establishment costs for the wetlands depend more on site specific factors, such as fencing and gateways on livestock farms, rather than on wetland size or design. They also note that these end-of-field field wetlands show promise, particularly if capital payments could be provided. Ockenden et al. (2012) report costs excavation, drainage, fencing, and vegetation. The (modest) cost for the very small requisite land may be estimated at $28,000 per hectare, and added to provide a total cost (Figure 19.11).

A log linear regression fits the data reasonably well:

$$C = 23A^{0.47} \tag{19.6}$$

$R^2 = 0.62$
$0.0005<A<0.032$
where A = wetland area, ha
C = cost, 2018 dollars in thousands

Agricultural Wetland Economics for Swedish Systems

Wetlands construction and restoration was adopted as an agrienvironmental measure in Sweden, dating to approximately year 1996. Swedish agricultural wetlands typically have an area of 0.5–1 ha, are vegetated by natural recruitment, and have source inflows of drainage water from agricultural fields (Thiere et al., 2009)

Historically, Swedish agricultural wetland capital costs were found to be directly proportional to wetland size (Byström, 1998). Since that time, over 4,000 ha of wetlands have been constructed (Arheimer and Pers, 2017). The economics of establishment are similar to those discussed earlier for Iowa. Andersson (2012) reports that the farmers can receive up to 90% of the actual capital cost, to a maximum of $29,000/ha (200,000 SEK/ha), with a 50% extra allowance if phosphorus removal is incorporated. This can cover all relevant project costs, including design. Beginning in 1996, the source of financial support for wetlands projects has been the Rural Development Programme, which is partly financed by the EU. The farmers are committed to preserving the wetland for 20 years after receiving investment support.

Farmers also receive funds for maintenance and lost revenue, termed agri-environmental payments. Payments are calculated based on lost revenue and reduced income. In 2012, these ranged from 220–580 $/ha, depending on land use. But, these are available only for a few years (5 years in 2012), and hence in most cases are deemed insufficient (Andersson, 2012; Hanson et al., 2012). Agricultural land values in Sweden averaged $8,300/ha in 2016, and the average rental rate was $200/ha·yr (Eurostat, 2017). Therefore, although the payments are generous, they do not last long enough to ensure a good economic future for the farmer.

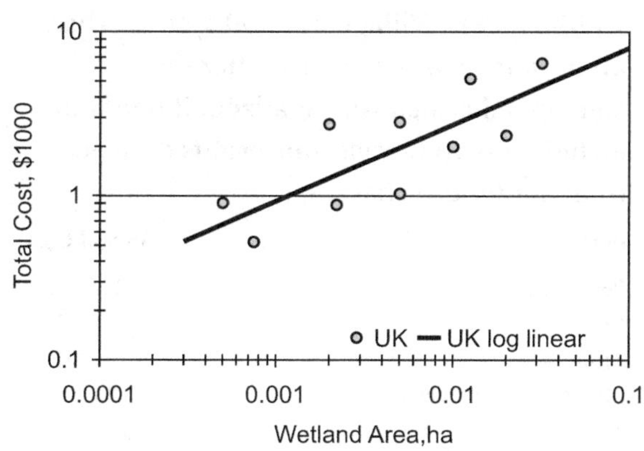

FIGURE 19.11 Reported total costs for agricultural runoff wetlands in the United Kingdom. Land cost has been estimated at $28,000 per ha. (Based on data from Ockenden et al., 2012.)

19.5 OPERATING COSTS

The operating costs of treatment marshes vary widely, according to type and source water, as detailed in Chapter 18. Agricultural marshes require very little attention, while urban systems may be intensively "gardened," primarily weeding to remove undesired plants. Large systems may incur ancillary research costs, often at the directive of the regulatory agencies; these are not considered here.

Most maintenance tasks associated with operating a wetland facility deal with servicing pumps, cleaning of debris, and repair and replacement of structures and berms and levees. Accordingly, the basic O&M cost components of a marsh treatment system include:

- Pumping energy
- Chemicals and plant materials
- Equipment repair and replacement
- Personnel
- Monitoring and reporting

The costs for the last two categories are not typically reported, especially not for small urban and agricultural marshes. However, that does not mean they are negligible; to the contrary, personnel and monitoring can dominate overall O&M (Table 19.13). These costs are high, totaling 1,742 \$/ha·yr.

Monitoring is comprised of sample collection and analysis. Sample collection is dominated by personnel costs and is accompanied by the analytical costs. For urban and agricultural systems, monitoring is usually part of ancillary research activities, and is not ascribed as part of project O&M.

Surface Water and Municipal Systems

Pumping Energy

The energy for pumping can be calculated by:

TABLE 19.13

Estimated O&M costs in 2018 dollars for the 135-ha Incline Village wetland system. This information was developed after the conceptual design was finalized. It does not include research costs, nor profits derived from hunter use charges.

Item	Annual Cost
Personnel	\$125,746
Energy	\$3,807
Monitoring	\$52,813
Maintenance materials	\$33,007
Total	\$235,178

Source: CWC, 1983.

$$P = \frac{Qg\rho\Delta H}{E} \tag{19.7}$$

where ΔH = hydraulic head, m
Q = flow rate, m^3/s
P = power required, W
E = pump efficiency, -
g = acceleration of gravity, m/s
ρ = density of water, =1,000 kg/m^3

In the United States, power may be measured in horsepower for diesel-powered pumps (1.0 kW = 1.34 hp). For example, if the marsh receives 5,000 m^3/d (0.058 m^3/s) lifted from a source with head differential 5 m, with a pump of 90% efficiency, then the required power is:

P = (0.058)(9.81)(1000)((5)/(0.9) = 3,154 W = 3.15 kW

The cost of energy for operations further depends upon the number of hours that pumping occurs. If there is continuous pumping throughout the year, then the running time is 8760 hours. For the example, the annual energy usage is 27,594 kWh. The annual energy cost then depends upon the cost of electricity, which is variable by country and by region. In the United States, the median cost among states in 2018 was $0.124/kWh (sd = $0.04/kWh). For the example, the median annual pumping energy cost for an electric pump is $3,422.

Diesel powered pumps are an alternative to electric powered pumps. Part of the decision process hinges upon the cost of fuel compared to the cost of electricity. As of 2018, in the United States, diesel fuel was priced at about $0.87 per liter, and there are 10 kWh per liter. For example, the median annual pumping energy cost for a diesel pump is $2,400.

Maintenance on a diesel motor requires more time and attention than electric motors. Depending on the delivery schedule for the treatment marsh, the pump engine may need to run for extended periods. If so, the diesel engine will constantly need refueling and the oil levels and filters will require monitoring.

Capital cost considerations will also partially dictate the choice of electric or diesel. Life cycle costs include estimates for capital cost and the present worth of O&M.

Chemicals and Plant Materials

Chemicals are sometimes used on treatment marshes, for a variety of purposes. For example, the herbicides used on the Florida STAs in 2017 to control unwanted vegetation had a market cost of approximately $300,000; which is low considering the huge size of the system (23,500 ha). The Apopka, Florida, marsh utilizes alum for phosphorus reduction, at an annual cost of about $94,000 per year for the 276 ha system (Dunne et al., 2015). Mosquito control may require larvicide application several times per year, with a chemical cost of about $15/ha for each application.

Equipment Repair and Replacement

As fully functional ecosystems, treatment wetlands may be expected to sustain their character for as long as appropriate hydrology is maintained. In general terms, pumps and piping may last on the order of 40 years, and repair frequencies can be estimated. The useful life of wastewater pump station equipment is typically limited to 20 to 30 years, with good maintenance. Berms, dikes, and levees typically have lives in excess of fifty years. Mechanical components, such as flow control structures, may need replacement at greater frequency. Hydromechanical elements such as gates and their motors have to be replaced after 30 to 50 years. Although not common, there may be deficiencies in the original construction of wetland control structures that require repair. An example is STAE in Florida, which was found to contain 46 (large) structures that were not up to specification. There was corrosion of the aluminum slide gate frames, and spliced structural members were used. The former led to corrosion of the aluminum frames. The later led to fractured gate stiffener welds. As a result, extensive repairs were required, at quite high cost.

Reported O&M Costs

The limited amount of marsh O&M data from the literature is presented in Figure 19.12, and shows a median annual cost of about $3,600 per hectare.

For rough estimating purposes, the O&M cost is proportional to the capital cost of the wetland. For the 23 wetlands that report both in Figure 19.12, the median ratio of annual O&M to capital is 3.5 ± 1.9%.

Urban and Agricultural Systems

Agricultural systems are subject to only very minimal O&M costs. These are the cost of mowing, and possibly burning, the buffer area. Mowing is considered necessary two to four times per year, at a cost of about $50 per hectare (Christianson et al., 2013; Tyndall and Bowman, 2016). The weir and flow control structures are considered to have a life of about 40 years, and thus there is a replacement cost if project life is presumed to exceed this span. If vegetation of the wetland does not meet expectations, a one-time seeding has been employed, at a cost of about $1,000 per hectare.

In contrast, urban wetlands require more diverse, frequent, and costly maintenance tasks (see Chapter 18). Erickson et al. (2010) surveyed 38 cities in the Midwestern United States to ascertain the most frequent O&M activities for urban wetlands. They found those to be associated with sediment buildup, pipe clogging, debris removal, invasive vegetation removal, and bank erosion. Weiss et al. (2007) report O&M costs as a fraction of construction cost for urban wetlands, with a median value of 5%. However, O&M cost percentage decreased with the size (capital cost) of the project.

U.S. EPA (2009) offers estimates of the unit costs for these O&M activities, which are founded in experiences in the northeastern United States. The frequency of the various activities remains as a major variable, and thus cost estimation is subjective. The excerpts

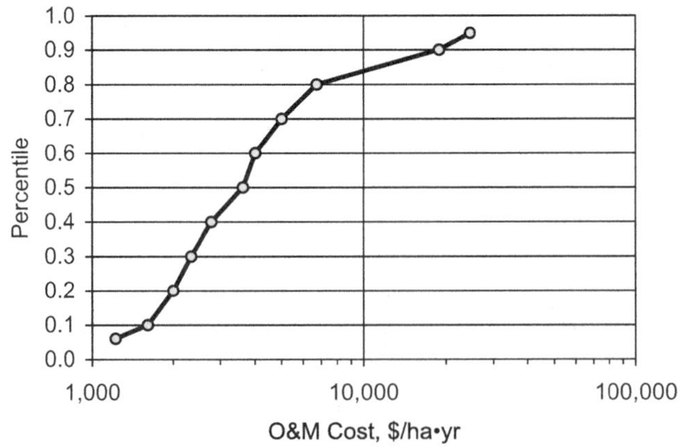

FIGURE 19.12 The O&M cost for treatment wetlands. (N = 30 wetlands).

from U.S. EPA (2009) shown in Table 19.14 do serve to highlight the more costly items. These are: invasive vegetation control, debris and sediment removal, animal control and mosquito control.

19.6 PRESENT WORTH

The capital costs for a wetland project are "up front" needs, typically expended in the first year of the project. In contrast, the operating and maintenance costs are recurring, with a need for expenditure in every year of operation. Thus, when evaluating alternatives, or planning for financing, it is necessary to combine these two types of costs to obtain a true picture of the overall project cost. Two basic choices are available. The first is to convert the future time stream of O&M costs to an amount of money that one would need to put in the bank to pay for the future O&M, and add it to the initial capital investment. This procedure produces the present worth of the project (Humphreys, 1991). The second choice is to convert the initial capital investment into a time series of payments that may be added to the time series of O&M costs, to create an annualized project cost. In essence, the capital cost is borrowed, and repaid over time. It is this later approach that is often used for financing wastewater treatment projects. In fact, in the United States, a homeowner may well be paying the two bills separately each month: one payment to pay off the debt on building the system, and a second to pay for operating costs. Unfortunately, only a very few literature sources deal with the combination of both types of project costs for treatment wetlands. For simplicity, we here consider only the present worth analysis.

Over the life of the marsh project, it is expected that some or all of the following costs will be incurred:

- Capital costs, both direct and indirect, usually in the first year of the project life.
- Annual payments for the use of the land.
- Annual O&M costs. These include monitoring and other routine activities.

TABLE 19.14
Example unit cost estimates for urban wetlands.

	Unit Cost 2018 USD	Unit	Recurrence years
Repair low spots in berm	\$334	m^3	5
Debris removal (preventative)	\$526	event	0.25
Replace structure barrel	\$4,928	m	25
Install wetland plant	\$9	ea	3
Machine remove invasive vegetation	\$4,506	event	
Spray for algae	\$8,904	ha	
Spray for cattails	\$5,194	ha	
Dredge and haul offsite	\$118	m^2	5
Sediment dump fee	\$109	tonne	
Install chainlink fence	\$128	m	
Treat pond for mosquitoes	\$3,710	ha	
Trap beavers (family of 6)	\$1,502	event	
Fill animal burrows	\$41	m^2	5
Remove graffiti	\$466	day	1
Erosion repair	\$1,652	event	2

Source: U.S. EPA (2009).

- Periodic, nonannual costs. These might include sediment cleanout, for instance.
- End of project costs and values, sometimes termed "salvage values." These may include demolition and disposal costs but also the resale value of project assets.

The preceding sections have detailed the costs in the first three categories (capital and/or rental, and O&M).

Present Worth Concepts

The total cost of a project at the time of inception is the total of capital costs, engineering services and the present worth of O&M costs over the project life. The overall project evaluation requires consideration of both capital and O&M costs, and the present worth technique is the appropriate vehicle for combining the two. The present worth of O&M costs, including equipment repairs and replacements, is the money that needs to be set aside now, at the prevailing interest rate, to pay for these future costs. It is not possible to anticipate all the lifelong, project-specific details of the economics of treatment marshes. However, there are certain common features of many projects that are considered here.

Discount Rate

The time value of money reflects the interest on borrowed sums and the inflation rate. The discount rate used in present worth calculations is the excess of the interest rate (real discount rate) over inflation; however, these are not considered separately, but combined into the nominal discount rate. In treatment marsh evaluations in the United States, a federal discount rate for water projects should probably be used Figure 19.13). Water project discount rates peaked in 1990 and have been steadily decreasing since. As will be seen, this means that present worths are now considerably higher than in past years.

Life Expectancy

Traditionally, the life expectancy of a "conventional" mechanical water quality improvement project is 20 years. A treatment marsh has a much longer life expectancy than concrete and steel equipment. There are now examples of engineered marsh systems with long periods of operation; for example the Houghton Lake, MI, system (41 years), the Vermontville, MI, system (45 years, Sutherland, 1982; U.S. EPA, 1993a), the Florida OEW system (31 years; Slayton, 2018) and Florida STA1W (SFWMD, 2019, 25 years). There are also long-lived

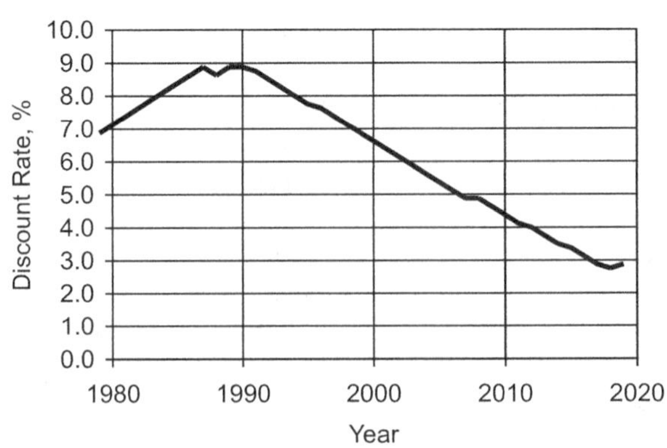

FIGURE 19.13 The U.S. federal nominal discount rate for water projects. (Source: NRCS, 2018)

natural marsh systems that have retained their effectiveness for many decades, based on ex post facto monitoring. Both the Brillion Marsh, Wisconsin (90 years, Spangler et al., 1976) and the Kinross Marsh, Michigan (60 years, Kadlec and Bevis, 1990) in later years were shown to have retained treatment efficiency. As fully functional ecosystems, treatment wetlands may be expected to sustain their character for as long as appropriate hydrology is maintained. It is a mid-life period in the history of constructed wetland facilities, so there are not many track records on life expectancies for wetland components. However, as noted earlier, pumps and piping may last on the order of 40 years, and berms, dikes and levees have lives in excess of 50 years. Therefore, it is plausible that treatment wetlands should be ascribed a basic life of 50 years or more, with mechanical components possibly requiring replacement at greater frequency.

End-of-Project Values and Decommissioning Costs

It is common practice to claim no salvage value at the end of project life in a feasibility study for mechanical water quality improvement facilities; but this does not make sense in the context of a wetland project. Typically, the entire acquisition price is charged to the project up front, and there is no "salvage" value at the end of 20 years. In contrast to the crumbling concrete and rusted steel left after the mechanical process reaches the end of its useful life, the land associated with the wetland project will have value.

It is not straightforward to forecast the value of the project after its life expectancy is reached. Firstly, there may be questions concerning possible future uses of the property. In the United States, wetlands are protected under Section 404 of the Clean Water Act. The U.S. EPA and the U.S. Army Corps of Engineers have developed draft guidance for determining whether a waterway, water body, or wetland is protected by the Clean Water Act (U.S. EPA, 2017b). Wetlands that meet the criteria are known as "jurisdictional" waters and are subject to regulatory requirements under the Clean Water Act. Aquatic areas are generally not protected by the Clean Water Act include artificially irrigated areas that would revert to upland should irrigation cease. This means that if you turn off the water supply, and a sustainable wetland remains, it may be designated as waters of the United States, and become protected under the Clean Water Act.

There are not many examples of what happens if the water is turned off, but logically systems in arid climates would revert to terrestrial conditions, while those in very wet climates might remain as wetlands. For example, when water was shut off to one of the Tres Rios, AZ, demonstration wetlands, it quickly reverted to desert ecology (Figure 19.14).

In California, the policy on the use of constructed wetlands for urban runoff states (Stromberg, 2015):

> Wetland systems constructed to treat urban runoff are intended, in part, to meet Clean Water Act requirement to reduce pollutants in urban runoff discharges. ... As treatment systems, constructed wetlands operated and maintained according to this policy will not be waters of the U.S. Constructed wetlands that are not operated and maintained according to terms of this policy and the approved management plan may forfeit the designation of "treatment system.

The jurisdictional status of agricultural wetlands is quite murky. In some instances, the owner is paid for a permanent easement that requires the wetland be preserved in perpetuity. However, other Iowa constructed wetlands are not so restricted. Crumpton et al. (2012b) report that one-third of the restored wetlands in the Galatowitsch and van der Valk (1996) study have been reconverted to cropped fields. The jurisdictional status of a wetland is determined using the Clean Water Rule (Martin, 2015). Among the categories that are excluded from federal regulation are "stormwater control features used to convey, treat or store stormwater that are created in dry land."

(A)

(B)

FIGURE 19.14 The Tres Rios, AZ, cobble cell 2 before (A) and after (B) decommissioning.

Speculatively, there are three possible scenarios for a terminated treatment marsh:

- The wetland is jurisdictional, or a permanent wetland easement is in place. In this case, the property has no marketable monetary value. Its opportunity value to the owner is zero.
- The wetland can be decommissioned, but owners and managers decide or require the wetland to remain for its human and wildlife use values. This is likely to be the case if the system is in government ownership. Government acquisition usually carries legal or implied use restrictions, which might, for instance, place it in the category of a wildlife management area or public park facility. No end-of-project monetary value would be assigned.

- The wetland is in private, or possibly public, ownership, and is to be decommissioned. The property then can be priced at the market value for vacant land.

In the last case, there may be costs for structure removal and land regrading. These were estimated at 20% of initial capital cost for pump station demolition, and 12% for regrading, in economic analyses of the Florida STAs (U.S. Army Corps of Engineers, 1996). Similar percentages may be expected for removal of urban and agricultural structures and regarding.

It is questionable to use a nominal discount rate for present worth of salvage costs, because land prices have escalated faster than the overall rate of inflation, for example, 5.6% land value increase per year vs. 2.4 % inflation over the period 1990–2018. This would suggest a negative or zero discount rate for the land salvage value. In total, the present worth of the restored land is likely to be about the same as the initial purchase price.

Present Worth Calculations

The amount of money needed for a future one-time expenditure reflects the discount rate in a simple, compound interest sense:

$$P = S\frac{1}{(1+i)^m} \tag{19.8}$$

Where i = nominal discount rate, fraction per year
m = years before need
P = present value, $
S = future amount, $

Thus, a $100,000 pump replacement 30 years from now will require $41,200 today if the discount rate is 3%.

Many O&M costs occur as a continuing stream, rather than one-time future expenditures. In analysis, it is usual to aggregate these into a yearly amount, and then to compute the sum of all such future payments. This is done sequentially using Equation 19.7, which results in the sum being directly evaluated from:

$$P = Y\frac{(1+i)^N - 1}{i(1+i)^N} = Y \cdot F_{pw} \tag{19.9}$$

where i = nominal discount rate, fraction per year
F_{pw} = present worth factor, unitless
N = life expectancy, yr
P = present value, $
Y = yearly O&M cost, $

The present worth factor for 50 years at 3% is 25.73. Thus the median marsh O&M of $3,600/ha·yr (Figure 19.12), over a 50-year lifetime, will require $92,628 today if the discount rate is 3%. Note that payment number 50 requires only $821 of today's money (Equation 19.7).

Project Present Worth

The various costs, consisting of capital plus the present worth of O&M, various replacements and salvage, are combined to yield the total present worth of the project. There are notable differences between the three major types of treatment marshes.

Surface Water and Municipal Systems

The example project in Table 19.7 (ten-hectare marsh) is extended in Table 19.15 at 3% nominal discount rate and a 50-year life. For illustration, it is presumed the pump is replaced at year 25. As suggested earlier, end-of-project costs are assumed to be 20% of initial capital cost for pump station demolition, and 12% for regrading. O&M cost is taken to be the median of $3,600 per hectare (Figure19.12). Land is assumed to have the same value at the end and beginning of the project.

The present worth of the example project is $2,129,170, nearly double the initial construction cost of $1,257,000. Although there are presumed costs for restoring the site to a condition for other uses, the land value is retrieved at the end of the project. Thus, the overall project cost is nearly independent of the land value. The per hectare cost of the example is $213,000.

Urban Systems

A hypothetical example of the present worth of an urban system is shown in Table 19.16. It is assumed to be located in a metropolitan area, with a high land cost of $200,000 per hectare. Because the system is gravity-fed, it would continue to receive water even if maintenance were abandoned. Therefore, the wetland is presumed to remain in place after a 50-year period, so that the property does not become a jurisdictional wetland. The land value is not retrieved at the end of the 50-year period. O&M costs are high, but there is not a pump that requires attention, and structure maintenance is limited. The per hectare cost of the example is $646,000, more than triple that of a comparable surface water system.

TABLE 19.15

Example present worth analysis for the example in Table 19.7. The nominal discount rate is 3%, and the life expectancy is 50 years.

Initial Costs	Cost	Present Worth	Subtotal
Land	$300,000		
Site Work	$159,500		
Structures and Conveyance	$78,500		
Pump	$400,000		
Direct Capital Cost		$938,000	
Engineering	$95,700		
Administrative and Contingency	$223,300		
Indirect Capital Cost		$319,000	
Total Capital Cost		$1,257,000	$1,257,000
O&M Costs			
Total Annual O&M	$36,000	$926,280	
Pump Replacement	$400,000	$191,042	
Total O&M			$1,117,322
End-of-Life Costs			
Demolition Pumps and Structures	$95,700	$45,707	
Regrade	$19,140	$9,141	
Land Salvage	$300,000	-$300,000	
End-of-Life			-$245,152
Project Present Worth		$2,129,170	$2,129,170

TABLE 19.16

Example present worth analysis for a hypothetical one-hectare urban runoff wetland. The nominal discount rate is 3%, and the life expectancy is 50 years.

	Cost	Present Worth	Subtotal
Initial Costs			
Land	\$200,000		
Construction Cost	\$185,000		
Direct Capital Cost		\$385,000	
Engineering	\$13,200		
Administrative and Contingency	\$10,000		
Indirect Capital Cost		\$23,200	
Total Capital Cost		\$408,200	\$408,200
O&M Costs			
Total Annual O&M @ 5% Construction	\$9,250	\$238,003	\$238,003
End-of-Life Costs	\$0	\$0	\$0
Project Present Worth		\$646,203	\$646,203

Life cycle costing may be done using commercially available software packages, such as the MUSIC program utilized for stormwater systems in Australia (Montaseri et al., 2015).

Agricultural Systems

The land required for agricultural systems is much less expensive than for urban systems. The land cost may be determined by an upfront purchase, or by a series of annual payments, as for the Iowa wetlands. In either case, the land cost is the dominant factor in the present worth analysis (Table 19.17). This example assumes a series of payments for land, costed by a present worth procedure. O&M is very low and is an insignificant economic factor. Because there is likely to be a permanent easement requiring the wetland to remain, and because the gravity driven tile drain system would not be removed, the system would remain in place after the end of the 50-year accounting period. No end-of-life costs or benefits are incurred.

The per hectare cost of the example is \$114,000, about half that of a comparable surface water system.

19.7 OTHER VALUES AND BENEFITS

The previous sections have considered what it costs to build and operate a treatment marsh. It is also possible to assess what the resulting wetland is worth to society, and this later procedure is widely used to justify existing wetland protection and govern the amount of new created wetland that putatively replaces any wetland losses. There is a huge literature on this topic, and burgeoning business activities in the United States. However, these two economic structures are separate and distinct. Detailed information is now available on some of the nonwater quality values of treatment marshes. There are also emerging details of the prospective activity of marsh plant harvest.

TABLE 19.17

Example present worth analysis for the agricultural runoff example in Table 19.12. The nominal discount rate is 3%, and the life expectancy is 50 years.

	Cost	Present Worth	Subtotal
Land Cost			
Buffer 3.5 ha		$60,515	
Wetland 1.0 ha		$17,290	
Land Present Worth			$77,805
Initial Costs			
Construction Cost	$27,540		
Design	$2,645		
Direct Capital Cost			$30,185
O&M Costs			
Total Annual O&M	$200	$5,146	
Seeding Year 5	$1,000	$863	
O&M Cost			$6,009
End-of-Life Costs	$0	$0	$0
Project Present Worth			$113,999

Categories of Values

There is a perceived distinction between created wetlands and treatment wetlands. Created wetlands are constructed on a site that never was a wetland, for the purpose of establishing a "natural" suite of (pristine) wetland functions. Treatment wetlands are constructed wetlands built specifically for the purpose of improving the quality of stormwater, agricultural runoff, or surface water, or tertiary municipal waters. In the United States, in a legal and regulatory sense, this is a "black and white" distinction.

Among the numerous lists of created wetland values that populate the literature, that of Brander et al. (2006) is one of the broadest (Table 19.18). The reader will recognize that many of these have been discussed in detail in the preceding Chapters. Brander et al. (2006) utilized a number of valuation methods to assign dollar values to wetland functions, including contingent valuation, hedonic pricing, travel cost, production functions, net factor income, total revenue estimation, opportunity cost, and replacement cost. Wetlands provide a number of ecological and physical functions that are of value to humans. The economic functions provided by wetlands (goods and services) are derived from, but separate from, their ecological and physical functions.

Adusumilli (2015) used value-function transfer analysis, based on 72 observations from 33 studies, to calculate dollar values of created wetland functions in the United States (Table 19.19). The several economic categories are significant. Water quality protection ranks highest; yet ironically, this function cannot be the *raison d'être* of the system. Importantly, these wetlands are not treatment marshes, but rather they are created wetlands designed to replace natural wetlands lost for a variety of reasons.

The aggregate values differed greatly from state to state (Figure 19.15). The variation is wide, with a median of $28,240/ha. Northeastern states had very high wetland values, with five states averaging $1,250,000/ha. In contrast, southern states had quite low values with five states averaging $5,100/ha. This disparity may in large part be attributed to

TABLE 19.18

Ecological and physical functions of wetlands, and their associated economic functions.

Ecological Function	Economic Function
Flood and flow control	Flood protection
Storm buffering	Storm protection
Sediment retention	Improved water quality
Groundwater recharge/discharge	Water supply
Water quality maintenance/nutrient retention	Improved water quality
Habitat and nursery for plant and animal species	Commercial fishing and hunting
	Recreational fishing and hunting
	Harvesting of natural materials
	Energy resources
Biological diversity	Appreciation of species existence
Micro-climate stabilization	Climate stabilization
Carbon sequestration	Reduced global warming
Natural environment	Recreational activities
	Appreciation of uniqueness to culture/heritage

Source: Brander et al. (2006).

regional societal perceptions. The metropolitan areas of the American northeast are predisposed to a puritanical ecological perception of wetlands, while the residents of the wetland-rich southern gulf coast states have a more pragmatic opinion of wetland values. This same disparity exists among the various counties across the globe, with much lower perceived values in Africa and Asia than in Europe, the Americas an Australasia (Brander et al., 2006).

Mitigation Credits

The use of constructed (created) wetlands for mitigation under the U.S Clean Water Act Section 404 is subject to approval by the U.S Army Corps of Engineers, in consultation with other Federal and State resource agencies. Decisions need to be made on a case-by-case basis, considering, among other factors, the appropriateness of the constructed wetland to fully offset the anticipated impacts from the loss of natural wetlands.

Mitigation banking is the primary vehicle for financing mitigation wetlands. A mitigation bank is a wetland area that has been restored, established, or enhanced for the purpose of providing compensation for unavoidable impacts wetland resources permitted under Section 404 or a similar state or local wetland regulation. A mitigation bank may be created when a government agency, corporation, nonprofit organization, or other entity undertakes these activities under a formal agreement with a regulatory agency.

Bank values are defined in terms of credits and debits, the terms used to designate the units of trade (i.e., currency). Credits represent the accrual or attainment of aquatic functions in a bank; debits represent the loss of functions at an impacted wetland site. Credits are debited from a bank when they are used to offset impacts (e.g. for the purpose of satisfying a Section 404 permit).

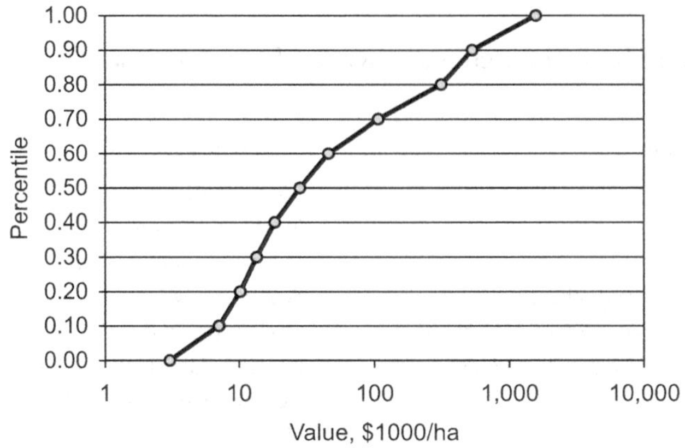

FIGURE 19.15 Distribution of wetland value estimates across the lower 48 U.S. states. Data from Adusumilli (2015).

TABLE 19.19
Economic value of wetland functions for the entire United States.

	Value, $/ha
Recreational Fishing	$6,754
Bird Hunting	$6,961
Bird Watching	$32,961
Water Quality Protection	$23,577
Water Supply Protection	$17,363
Commercial Fishing	$20,052
Flood Control	$4,257
Aggregate	$111,920

Source: Adusumilli (2015).

Because of the "black and white" distinction described earlier, treatment marshes are not eligible to receive *any* mitigation credits. U.S. EPA (2000c) states the U.S. federal regulations and policies:

> In general, wetlands constructed or restored for the primary purpose of treating wastewater will not be recognized as compensatory mitigation to offset wetland losses authorized under federal regulatory programs.

This is echoed in individual state regulations, such as that in Wisconsin (WDNR, 2013):

> No Credit for Stormwater or Wastewater Treatment Facilities. Some innovative facilities have been designed for treating stormwater and wastewater, using designs that create the physical, chemical and biological processes that occur in wetlands. These facilities have been referred to as bioretention basins, biofilters, or constructed wetlands and are considered artificial wetlands.

> While these facilities may serve an important function in alleviating impacts to natural wetlands and waterways by moderating substantially the bounce in water levels and trapping sediment loads, such single-function wetlands do not meet the intent of compensatory mitigation (Chapter NR 350.07(7), Wisconsin Administrative Code).

As a result, the guidance for treatment marshes is (ITRC, 2003):

> It should be noted that constructed wetlands should not be confused with created wetlands, which are established to mitigate impacts to natural wetlands from a construction project or to receive wetland credits in a wetland mitigation bank. Created, or "mitigation," wetlands are under considerably more regulation and are generally not used for treatment.

Despite this negative regulatory view of treatment marsh values, there is a growing number of studies that identify the partial mitigation aspects of treatment marshes. For example, Gascoigne et al. (2011) found major economic values for two ecosystem services in the CREP wetlands: carbon sequestration, and waterfowl production, with lesser values in sediment reduction. Their study found that CREP wetlands could not mitigate, hectare for hectare, the loss of native prairie potholes.

Human Use

There has been a start to monetizing the values of treatment marshes, along similar lines as for mitigation and restoration projects. Ghermandi et al. (2010) investigated whether non-water quality benefits, such as recreational activities and biological diversity, provided economic value to treatment wetlands. Relying upon meta-analysis of a database of wetland value estimates, they found the regression coefficient for natural habitat and biodiversity was positive and highly statistically significant.

Human use facilities are a distinctive feature of treatment marshes, which separates them from restoration and creation projects that highlight biological diversity. Ghermandi and Fichtman (2015) utilized opportunity cost as a metric in evaluating the recreational and educational benefits of treatment wetlands, as described by Alfranca et al. (2011). Opportunity costs are based on the wages a person may have received in lieu of the leisure activity. They found accessibility to be the main determinant of human use value. The number of visitors was strongly correlated with interpretive signs, wildlife observation points, and visitor centers. Further, wetlands that were easily reachable from large cities were more frequently visited. A North American subset of their results is shown in Table 19.20.

Ghermandi (2016, 2018)) expanded the analysis of treatment marshes and wetlands, by using a willingness to pay (WTP) metric and adding a benefit/cost calculation. The willingness to pay metric included travel expenses and time opportunity costs. Geotagged photo metadata (Flickr photographs) established the frequency of visits. Public profiles of visitors were used to establish their place of residence, and hence the travel distance to the wetland. Travel costs were calculated based upon gasoline price. The opportunity cost of time was based on one-third of the assumed wage rate. The WTP for 27 example North American wetlands had a mean of \$67, and a median of \$38 (Range: \$5 – \$200).

The yearly flows of recreational and educational benefits were calculated multiplying by the WTP by the number of visitors. The present worth of project costs was determined from reports based on the methodology described in previous sections of this Chapter. Benefit present worths were calculated by those same methods, with a lifetime of 45 years and various discount rates (4% used here). The median value for 27 North American treatment marshes was 17% (Figure 19.16). Twenty-five percent of these marshes had recreational and educational benefits equal to or greater than the project cost.

TABLE 19.20

Annual opportunity costs for recreation an education for example North American treatment marshes.

Wetland	Location	Value, \$/ha·yr
Green Cay	Florida	113,043
Wakodahatchee	Florida	78,579
Sweetwater	Arizona	26,607
Oregon Garden	Oregon	17,232
Freedom Park	Florida	10,436
Viera	Florida	4,025
Shepard	Alberta	911
Brighton	Ontario	893
Phinizy	Georgia	506
OEW	Florida	223
Blue Heron	Florida	123
E.L. Huie Jr.	Georgia	46
Incline Village	Nevada	29
Tres Rios	Arizona	25
STA5	Florida	5

Source: Ghermandi and Fichtman (2015).

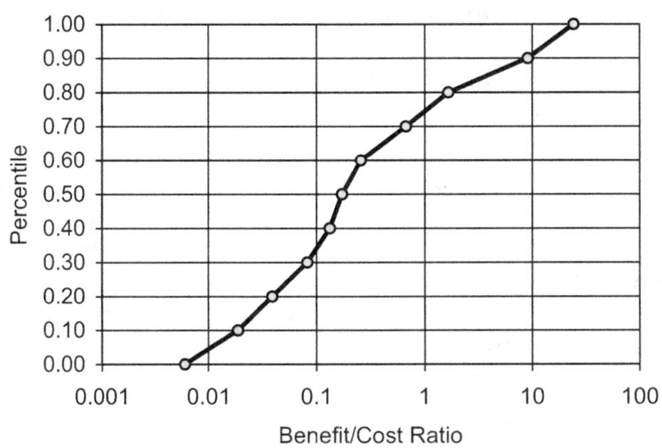

FIGURE 19.16 Recreational and educational benefit/cost ratio for 27 North American treatment marshes. (Based on information in Ghermandi, 2018.)

Other marsh values in Table 19.18 have not yet been thoroughly examined, but it is clear that treatment marshes have societal values well in excess of the dollar value of pollutant removal.

Economics of Potential Harvest

Harvest is a feature of floating aquatic plant (FAV) systems that was investigated at pilot scale in a number of historical projects. Those early projects showed poor results, primarily due to the difficulty of dealing with the biomass. Biogas generation showed poor economics, and a market for composted residuals did not develop.

Chapter 10 provides information on the potential for nutrient removal via harvest, including trials at pilot scale. Early trials of harvest of biomass from EAV treatment wetlands were conducted at pilot scale (Herskowitz, 1986) and at full scale (Johnson et al., 1987), both involving *Typha*. These efforts were hampered by the lack of equipment specifically built to work in wetlands (Figure 19.17). Many types of commercial machinery are now available, tailored to harvest of emergent wetland macrophytes.

Over the past decade, there has been greatly increased experience, derived from projects on invasive plant removal and from considerations of biofuel production. Several studies on cattail removal to promote ecological values have been conducted (e.g., Lishawa et al., 2015; Carson et al., 2018). Some of these ecological studies have focused upon the economic value of the removed biomass for fuel (e.g., Jakubowski et al., 2010; Liu et al., 2012). However, it is the reverse emphasis, on the value of plant biomass as an alternative fuel, that has generated information on costs and practical considerations. The economics generally include only the add-on costs and values; that is, they do not consider land capital and management to produce the biomass. It is implicitly assumed that the owners of cattail stands will be happy to be rid of them, and charge nothing for land use.

To date none of the donors in North America have been treatment marshes or treatment wetlands as defined herein. But, closely related features in the northern great plains of North America are retention storage sites. These are a major element of integrated flood protection in the Red River Basin, far cities such as Fargo and Grand Forks, North Dakota (Svedarsky et al., 2016). Although these sites were not originally designed for nutrient capture, they still perform this function. Many retention cells are full of cattail (Grosshans et al., 2014).

FIGURE 19.17 The modified grain binder used to harvest cattails at the Listowel wetlands.

There is not a serious impediment to treatment wetland harvest, but there are necessary design and operational considerations to facilitate harvest. The bathymetry of the system must allow access for the harvest machinery to operate. And, in most cases, water level and flow management will need to accommodate the period of machinery operation. Further, the harvest activity may be incompatible with some ecological values, including habitat; but the general perception is that ecological values will be enhanced by harvest of invasives.

Steps in Energy Production

There are three general approaches to removing harvested cattails from a wetland: cut and bale into large square or round bales, cut and blow chopped material into a hopper, and cut and gather with an amphibious machine that operates in the water. The last of these is the most difficult and has not been actively pursued.

Once harvested, the on-site material must be transported to a processing facility, and transportation is a significant cost determinant.

Processing is required because bales cannot be fed directly into an industrial boiler. Pellets or briquettes are desired end product, which are more flexible for use in a range of residential and commercial stoves. Bales are broken apart and run through a hammer mill; then sent to a pelletizer. The biomass is then in a condition to be fed to home heaters, industrial boilers, or power generation facilities.

Emerging Results

Grosshans (2016, 2017) reported that a Manitoba project had reached full-scale, with over a thousand tonnes of cattail harvested. Blended cattail/wood fuel pellets turned out to be a premium fuel pellet with high heating value and excellent burn qualities. These were used for heating at several locations in Manitoba. No dollar values have been reported.

Spinelli et al. (2017) detailed the economics of all phases of the extraction of *Phragmites* biomass for fuel in Italy. Their work was at full scale, involving over 300 large bales of biomass taken from seven fields, comprising 5.0 tonnes per hectare. Their costs were: cutting (290$/ha), windrowing (60 $/ha), baling (197 $/ha) and transport (139 $/ha). The total of 686 $/ha produced 5 dry metric tons of biomass/ha, or a cost of 137 $/tonne, which was virtually identical to the market price of the biomass fuel (132 $/tonne). The collection process used less than 5% of the energy content of dry reed fuel.

Wichmann (2017) similarly detailed the economics of several different uses of *Phragmites* in Germany. The three options were: (1) chopped biomass for biogas production, (2) round bales for direct combustion, and (3) bundles for roof thatching. For biogas, green biomass is cut in summer and windrowed. The swath is picked up and placed into a trailer. The finely chopped biomass is processed in biogas plants to generate electricity and heat. For combustion, biomass is cut and baled into round bales in a single pass. A separate vehicle picks up the bales and transports them to the edge of the field. The bales are burnt in combustion plants designed to generate heat (and in the high power range also electricity) Thatching reed is harvested in winter. Long, straight culms are mown, brushed for an initial cleaning and bound into bundles with a diameter of approximately 20 cm.

The costs considered included harvesting, transporting to the field edge and processing needed to sell the biomass. Categories were machinery costs (fuel, machine care, depreciation, insurance) and labor costs. General and administrative costs, site-specific costs, or revenues (land lease, direct payments) were excluded, as were costs of further transport and storage. The results indicated that biogas could not pay for harvesting, combustion most often would pay for harvesting, and thatching was economically well worthwhile (Table 19.21). This last conclusion is contingent upon the market for thatching, which may well be limited in Europe compared to wetland biomass availability.

TABLE 19.21

Economic analysis of *Phragmites* harvest in Germany. (tdw = tonnes dry weight, tfw = tonnes fresh weight)

	Chopped Biogas	Baled Combustion	Bundles Thatching
Yield (mode)	5 tdw/ha	8 tdw/ha	500 Bundles/ha
Price (mode)	12 \$/tfw	78 \$/tdw	2.4 \$/Bundle
Revenue (mode)	127 \$/ha·yr	558 \$/ha·yr	1,291 \$/ha·yr
Cost (mode)	476 \$/ha·yr	503 \$/ha·yr	768 \$/ha·yr
Net Benefit (median)	-340 \$/ha·yr	120 \$/ha·yr	636 \$/ha·yr

Source: Wichmann (2017).

These economic studies show that harvest of EAV is a process that should not be ignored for an appropriate subset of treatment marsh projects. The total of nutrient removal harvest values and energy production values means that harvest merits consideration as a treatment marsh add-on.

19.8 SUMMARY

Early treatment wetland literature opined that treatment wetlands were "low cost." But, in terms of common wastewater quality parameters, capital costs of treatment wetlands and marshes now have an extensive track record of being comparable to alternative technologies for accomplishing the same task.

Basic cost components of a wetland can be estimated using a line-item approach. Marsh treatment systems generally share a basic list of common components, which includes items such as: land, pumps, hydraulic control structures earthwork, and wetland plants. The cost of these basic components can be estimated using pricing and availability information for a given regional economy. However, the line-item approach makes estimating economies of scale difficult, and has limited application in areas where local pricing information is unavailable. Other components of wetland system design, such as site investigation, construction management, and engineering design fees, are also part of the project capital cost.

Historically, treatment wetlands were regarded as having very low O&M costs, but that has proven to be the case for only a subset of treatment marshes. For municipal wastewater parameters, the costs of operating a treatment wetland are typically much lower than for competing technologies. Mechanical devices are always more energy intensive, and will always be more expensive to operate than a passive wetland system.

Operating costs can be quite low, especially for agricultural systems, which require very little effort. However, O&M costs can become appreciable if it is attempted to maintain specific vegetation types. Urban wetlands are particularly susceptible to "weeding" costs, as well as mosquito and rodent control. Energy costs are typically close to zero for gravity-driven urban and agricultural wetlands. However, municipal and surface water sources often require pumping to the constructed marsh, and can be an important factor in O&M costs. Water quality monitoring is often a principal part of O&M costs for larger systems, and those with compliance monitoring requirements.

Operation and maintenance costs are combined with capital costs when developing an overall cost estimate. The life cycle cost of a wetland project, as represented by the present worth of capital and operating expenses, very often has significant contributions for both capital and O&M.

The ecological values of treatment marshes have been, and continue to be, regarded as of no consequence when evaluating wetlands in a societal and preservation context. They get no mitigation credits. In contrast, treatment marshes have human use values that can far exceed those of restoration projects. There is a good chance that in the future, the energy recovery values of treatment marsh biomass will add yet another value, which is thus far unaccounted in wetland economics or values.

References

Abira M.A., van Bruggen J.J.A., Denny P. (2005) Potential of a tropical subsurface constructed wetland to remove phenol from pre-treated pulp and papermill wastewater. *Water Science and Technology 51*(9): 173–176.

Abis K.L. (2002) *The Performance of Facultative Waste Stabilization Ponds in the United Kingdom*. PhD Dissertation, University of Leeds.

Abis K.L., Mara D. (2006) Temperature measurement and stratification in facultative waste stabilization ponds in the UK climate. *Environmental Monitoring Assessment 114*: 35–47.

Abtew W., Obeysekera J. (1995) Lysimeter study of evapotranspiration cells of cattails and comparison of three estimation models. *Transactions of the American Society of Agricultural Engineers 38*(1): 121–129.

Abtew W. (1996) Evapotranspiration measurements and modeling for three wetland systems in South Florida. *JAWRA Journal of the American Water Resources Association 32*(3): 465–473.

Abtew W. (2005) *Evapotranspiration in the Everglades; Comparison of Bowen Ratio Measurements and Model Estimations*. Technical paper EMA # 417, ASAE meeting in Tampa, Florida, July 2005.

Abtew W., Obeysekera J., Iricanin N. (2011) Pan evaporation and potential evapotranspiration trends in South Florida. *Hydrological Processes 25*(6): 958–969.

Ackerman J.T., Kraus T.E., Fleck J.A., Krabbenhoft D.P., Horwath W.R., Bachand S.M., Herzog M.P., Hartman C.A., Bachand P.A. (2015) Experimental dosing of wetlands with coagulants removes mercury from surface water and decreases mercury bioaccumulation in fish. *Environmental Science and Technology 49*(10): 6304–6311.

Acreman M.C., Harding R.J., Lloyd C.R., McNeil D.D. (2003) Evapotranspiration characteristics of wetlands: Experience from a wet grassland and a reedbed using eddy correlation measurement. *Hydrology and Earth System Sciences 7*: 11–21.

Adamus P.R. (2002) *Methods for Evaluating Wetland Condition: Biological Assessment Methods for Birds*. EPA-822-R-02-023. Office of Water, U.S. Environmental Protection Agency, Washington, DC.

Adamus P.R. (2011) Manual for the Wetland Ecosystem Services Protocol for the United States (WESPUS). Available at: www.oregonstate.edu/~adamusp/WESPUS. Acessed 10/16/18.

Adamus P.R., Clairain E.J. Jr., Smith D.R., Young R.E. (1992) *Wetland Evaluation technique (WET). Volume I. Literature Review and Evaluation Rationale*. U.S. Army Corps of Engineers, Waterways Experiment Station: Vicksburg, MS.

Adcock P.W., Ganf G.G. (1994) Growth characteristics of three macrophyte species growing in a natural and constructed wetland system. *Water Science and Technology 29*(4): 95–102.

Adusumilli N. (2015) Valuation of ecosystem services from wetlands mitigation in the United States. *Land 4*(1): 182–196.

Adyel T.M., Hipsey M.R., Oldham C.E. (2017) Temporal dynamics of stormwater nutrient attenuation of an urban constructed wetland experiencing summer low flows and macrophyte senescence. *Ecological Engineering 102*: 641–661.

Ahn C., Mitsch W.J. (2002) Scaling considerations of mesocosm wetlands in simulating large created freshwater marshes. *Ecological Engineering 18*(3): 327–342.

Akcil A. (2003) Destruction of cyanide in gold mill effluents: Biological versus chemical treatments. *Biotechnology Advances 21*: 501–511.

Akcil A., Mudder T. (2003) Microbial destruction of cyanide wastes in gold mining: Process review. *Biotechnology Letters 25*: 445–450.

Akob D.M., Cozzarelli I.M., Dunlap D.S., Rowan E.L., Lorah M.M. (2015) Organic and inorganic composition and microbiology of produced waters from Pennsylvania shale gas wells. *Applied Geochemistry 60*: 116–125.

Albouy D., Ehrlich G., Shin M. (2018) Metropolitan land values. *The Review of Economics and Statistics 100*(3): 454–466.

Alfranca O., García J., Varela H. (2011) Economic valuation of a created wetland fed with treated wastewater located in a peri-urban park in Catalonia, Spain. *Water Science and Technology 63*(5): 891–898.

Allen E.W. (2008) Process water treatment in Canada's oil sands industry: I. Target pollutants and treatment objectives. *Journal of Environmental Engineering and Science 7*(2): 123–138.

Allen R.G., Pereira L.S., Raes D., Smith M. (1998) Crop evapotranspiration-guidelines for computing crop water requirements-FAO Irrigation and drainage paper 56. *FAO, Rome 300*(9): D05109.

Allen R.G., Pereira L.S., Raes D., Smith M. (2005) FAO Irrigation and Drainage Paper No. 56; Crop Evapotranspiration. *Journal of Irrigation and Drainage Engineering 131*(1).

Allen R.G., Walkter I.A., Elliot R., Mecham B., Jensen M.E., Itenfisu D., Howell T.A., Snyder R., Brown P., Eching S., Spofford T., Hattendorf M., Cuenca R.H., Wright J.L., Martin D. (2000) Issues, requirements, and challenges in selecting and specifying a standardized ET equation. Proceedings of the 4th National Irrigation Symposium; American Society of Agricultural Engineers: Phoenix, Arizona.

Allen S.T., Edwards B.L. Reba M.L., Keim R.F. (2016) Sub-canopy evapotranspiration from floating vegetation and open water in a swamp forest. *Wetlands 36*(4): 681–688.

Al-Rubaei A.M., Engström M., Viklander M., Blecken G.T. (2016) Long-term hydraulic and treatment performance of a 19-year old constructed stormwater wetland—Finally maturated or in need of maintenance? *Ecological Engineering 95*: 73–82.

Al-Rubaei A.M., Merriman L.S., Hunt W.F. III, Viklander M., Marsalek J., Blecken G.T. (2017) Survey of the operational status of 25 Swedish municipal Stormwater management ponds. *Journal of Environmental Engineering 143*(6): 05017001.

Alvarez J.A., Becares E. (2006) Seasonal decomposition of *Typha latifolia* in a free-water surface constructed wetland. *Ecological Engineering 28*(2): 99–105.

Alvord H.H., Kadlec R.H. (1995) The interaction of atrazine with wetland sorbents. *Ecological Engineering 5*(4): 469–479.

Alvord H.H., Kadlec R.H. (1996) Atrazine fate and transport in the Des Plaines Wetlands. *Ecological Modelling 90*(1): 97–107.

Ameen A.D., Kolker A.S., Taylor C.M. (2017) Vegetation and shear strength in a Delta-splay mouth bar. *Wetlands 37*(6): 1159–1168.

Anda A., Teixeira Da Silva J.A., Soos G. (2014) Evapotranspiration and crop coefficient of common reed at the surroundings of Lake Balaton, Hungary. *Aquatic Botany 116*: 53–59.

Anderson B.H., Magdoff F.R. (2005) Relative movement and soil fixation of soluble organic and inorganic phosphorus. *Journal of Environmental Quality 34*(6): 2228–2233.

Anderson C.J., Mitsch W.J. (2006) Sediment, carbon, and nutrient accumulation at two 10-year-old created riverine marshes. *Wetlands 26*(3): 779–792.

Anderson J.C., Carlson J.C., Low J.E., Challis J.K., Wong C.S., Knapp C.W., Hanson M.L. (2013) Performance of a constructed wetland in Grand Marais, Manitoba, Canada: Removal of nutrients, pharmaceuticals, and antibiotic resistance genes from municipal wastewater. *Chemistry Central Journal 7*(1): 54.

Anderson K.L., Wheeler K.A., Robinson J.B., Tuovinen O.H. (2002) Atrazine mineralization potential in two wetlands. *Water Research* 36(19): 4785–4794.

Anderson M.J., Crist T.O., Chase J.M., Vellend M., Inouye B.D., Freestone A.L., Sanders N.J., Cornell H.V., Comita L.S., Davies K.F., Harrison S.P. (2011) Navigating the multiple meanings of β diversity: A roadmap for the practicing ecologist. *Ecology Letters 14*(1): 19–28.

Andersson K. (2012) *Multifunctional Wetlands and stakeholder engagement: Lessons from Sweden.* Stockholm Environment Institute, Stockholm, Sweden. Available at: www.weadapt.org/sites/weadapt.org/files/legacy-new/knowledge-base/files/1232/524c41a5ea855sei-wp-2012-08-sweden-wetland (Accessed 8/1/18).

Andradóttir H.Ó., Nepf H.M. (2001) Impact of exchange flows on wetland flushing. *Water Resources Research 37*(12): 3265–3273.

Angstrom A. (1924) Solar and terrestrial radiation. *Quarterly Journal of Royal Meteorological Society 50*: 121–125.

Anza I., Vidal D., Laguna C., Díaz-Sánchez S., Sánchez S., Chicote Á., Florín M., Mateo R. (2014) Risk factors for avian botulism outbreaks in wetlands receiving effluents from urban wastewater treatment plants: Eutrophication and bacterial pathogens. *Applied and Environmental Microbiology 80*(14): 4251–4259.

APAI (1994) *Pilot-Scale Constructed Wetlands Demonstration Project: Phase I Report*, Phase I Report by APAI (Alan Plummer and Associates, Inc.) to the American Water Works Association Research Foundation and the Tarrant County Water Control and Improvement District No. 1. AWWA: Forth Worth, TX.

APAI (Alan Plummer Associates, Inc.) (2008) *Tarrant Regional Water District Field-Scale Constructed Wetland, Phase 1 Operations, June 3, 2003–January 9, 2007*. Final Report to Tarrant Regional Water District, Fort Worth, TX, January 31, 2008.

Apfelbaum S., Duvall K., Nelson T., Mensing D., Bengtson H., Eppich J., Penhallegon C., Thompson R. (2013) *Wetland Water Cooling Partnership: The Use of Constructed Wetlands to Enhance Thermoelectric Power Plant Cooling and Mitigate the Demand of Surface Water Use.* Applied Ecological Services Incorporated. U.S. Department of Energy, Office of Scientific and Technical Information. Available at: www.osti.gov/biblio/1121759 (Accessed 3/25/18).

APHA. (1992) *Standard Methods for the Examination of Water and Wastewater*, Greenberg A.E. (ed.), 18th edition. American Public Heath Association (APHA), American Water Works Association (AWWA), and the Water Environment Federation (WEF): Washington, DC.

APHA. (2005) *Standard Methods for the Examination of Water and Wastewater*, Eaton A.D., Clesceri L.S., Rice E.W., Greenberg A.E., Franson M.A.H., 21st edition (eds.) American Public Health Association (APHA), American Water Works Association (AWWA), and the Water Environment Federation (WEF): Washington, DC.

APHA (2017) *Standard Methods for the Examination of Water and Wastewater*, 23rd edition, Clesceri L.S., Rice E.W., Baird R.B., Eaton A.D. (eds.) American Public Health Association (APHA), American Water Works Association (AWWA), and the Water Environment Federation (WEF): Washington, DC.

Arheimer B., Pers B.C. (2017) Lessons learned? Effects of nutrient reductions from constructing wetlands in 1996–2006 across Sweden. *Ecological Engineering 103*: 404–414.

Arheimer B., Wittgren H.B. (1994) Modelling the effects of wetlands on regional nitrogen transport. *Ambio 23*(6): 378–386.

Arheimer B., Wittgren H.B. (2002) Modelling nitrogen removal in potential wetlands at the catchment scale. *Ecological Engineering 19*(1): 63–80.

Arias C.A., Del Bubba M., Brix H. (2001) Phosphorus removal by sands for use as media in subsurface flow constructed reed beds. *Water Research 35*(5): 1159–1168.

Armanini A., Righetti M., Grisenti P. (2005) Direct measurement of vegetation resistance in prototype scale. *Journal of Hydraulic Research 43*(5): 481–487.

Armstrong J., Afreen-Zobayed F., Armstrong W. (1996) *Phragmites* dieback: Sulphide and acetic acid induced bud and root death, lignifications, and blockages within aeration and vascular systems. *New Phytologist 134*(4): 601–614.

Armstrong J., Armstrong W., Beckett P.M. (1990) Measurement and modeling of oxygen release from roots of *Phragmites australis.* In: *Constructed Wetlands in Water Pollution Control*, Cooper P.F., Findlater B.C. (eds.) Pergamon Press: Oxford, United Kingdom, p. 41–52.

Armstrong S.A., Headley J.V., Peru K.M., Mikula R.J., Germida J.J. (2010) Phytotoxicity and naphthenic acid dissipation from oil sands fine tailings treatments planted with the emergent macrophyte *Phragmites australis. Journal of Environmental Science and Health Part A 45*(8): 1008–1016.

Arndal M.F., Illeris L., Michelsen A., Albert K., Tamstorf M., Hansen B.U. (2009) Seasonal variation in gross ecosystem production, plant biomass, and carbon and nitrogen pools in five high arctic vegetation types. *Arctic, Antarctic, and Alpine Research 41*(2): 164–173.

Aronson J., Clewell A., Moreno-Mateos D. (2016) Ecological restoration and ecological engineering: Complementary or indivisible? *Ecological Engineering 91*: 392–395.

Asaeda T. Hai D.N., Manatunge J., Williams D., Roberts J. (2005) Latitudinal characteristics of below-and above-ground biomass of Typha: A modelling approach. *Annals of Botany 96*(2): 299–312.

Asaeda T., Hietz P., Tanaka N., Karunaratne S. (2002) Seasonal fluctuations in live and dead biomass of Phragmites australis as described by a growth and decomposition model: Implications of duration of aerobic conditions for litter mineralization and sedimentation. *Aquatic Botany 73*(3): 223–239.

Asaeda T., Karunaratne S. (2000) Dynamic modeling of the growth of *Phragmites australis*: Model description. *Aquatic Botany 67*(4): 301–318.

ASCE. (1975) *Sedimentation Engineering*, Vanoni V.O. (ed.) American Society of Civil Engineers (ASCE): New York.

ASCE. (1990) *Evapotranspiration and Irrigation Water Requirements*, Jensen M.E., Burman R., Allen R. (eds.) American Society of Civil Engineers: New York.

ASCE. (2006) *Standard Guidelines for the Design of Urban Stormwater Systems*, ASCE/EWRI 45-05, American Society of Civil Engineers (ASCE): Reston, VA.

ASTM (American Society for Testing and Materials) (2000) *Standard test methods for determining sediment concentration in water samples: D 3977-97*. ASTM International: West Conshohocken, PA.

Åström B.T., Pipes R.B., Advani S.G. (1992) On flow through aligned fiber beds and its application to composites processing. *Journal of Composite Materials 26*(9): 1351–1373.

Austin D.C., Lohan E., Verson E. (2003) Nitrification and denitrification in a tidal vertical flow wetland pilot. Proceedings, WEFTEC 2003 National Conference, 76th Annual Conference and Exhibition; Water Environment Federation, Alexandria, Virginia.

Ávila C., García J. (2015) Pharmaceuticals and personal care products (PPCPs) in the environment and their removal from wastewater through constructed wetlands. In: *Comprehensive Analytical Chemistry*, Zeng E. (ed.) Vol. 67. Elsevier: New York, pp. 195–244.

Azaizeh H.A., Salhani N., Sebesvari Z., Emons H. (2003) The potential of rhizosphere microbes isolated from a constructed wetland to biomethylate selenium. *Journal of Environmental Quality 32*(1): 55–62.

Bachand P.A.M., Bachand S.M., Fleck J., Anderson F., Windham-Myers L. (2014a) Differentiating transpiration from evaporation in seasonal agricultural wetlands and the link to advective fluxes in the root zone. *Science of the Total Environment 484*: 232–248.

Bachand P.A.M., Bachand S.M., Fleck J.A., Alpers C.N., Stephenson M., Windham-Myers L. (2014b) Methylmercury production in and export from agricultural wetlands in California, USA: The need to account for physical transport processes into and out of the root zone. *Science of the Total Environment 472*: 957–970.

Bachand P.A.M., Vaithiyanathan P., Richardson C.J. (2000) Phase II low intensity chemical dosing (LICD): Development of management practices. Final Report Submitted to Florida Department of Environmental Protection in Fulfillment of Contract No. WM720. December, 2000. Available at: http://aquaticcommons.org/1876/1/Phase_2_LICD_Final_Report.pdf (Accessed 4/24/18).

Bachmann R.W., Hoyer M.V., Canfield D.E. (1999) The restoration of Lake Apopka in relation to alternative stable states. *Hydrobiologia 394*: 219–232.

Badiou P.H., Goldsborough L.G. (2010) Ecological impacts of an exotic benthivorous fish in large experimental wetlands, Delta Marsh, Canada. *Wetlands 30*(4): 657–667.

Badiou P.H., Goldsborough L.G. (2015) Ecological impacts of an exotic benthivorous fish, the common carp (*Cyprinus carpio* L.), on water quality, sedimentation, and submerged macrophyte biomass in wetland mesocosms. *Hydrobiologia 755*(1): 107–121.

Bailey J.E., Ollis D.F. (1986) *Biochemical Engineering Fundamentals*. McGraw-Hill: New York.

Bailey J.K., Pinyon J.L., Anantham S., Hall R.M. (2010) Commensal Escherichia coli of healthy humans: A reservoir for antibiotic-resistance determinants. *Journal of Medical Microbiology 59*(11): 1331–1339.

Bailey L.D. (1976) Effects of temperature and root on denitrification in a soil. *Canadian Journal of Soil Science 56*: 79–87.

Bajer P.G., Sullivan G., Sorensen P.W. (2009) Effects of a rapidly increasing population of common carp on vegetative cover and waterfowl in a recently restored Midwestern shallow lake. *Hydrobiologia 632*(1): 235–245.

Baker L.A. (1998) Design consideration and applications for wetland treatment of high-nitrate waters. *Water Science and Technology 38*(1): 389–395.

Bal K., Struyf E., Vereecken H., Viaene P., De Doncker L., de Deckere E., Mostaert F., Meire P. (2011) How do macrophyte distribution patterns affect hydraulic resistances? *Ecological Engineering 37*(3): 529–533.

Baldwin D.S. (1996) Effects of exposure to air and subsequent drying on the phosphate sorption characteristics of sediments from a eutrophic river. *Limnology and Oceanography 41*(8): 1725–1732.

Ballaron P.B. (1998) *Use of a field drain and an artificial Wetland to minimize Groundwater contamination from an agricultural site*, Publication No. 197, Susquehanna River Basin Commission: Harrisburg, PA.

Balogh M. (1982) Patent: A method to prevent the eutrophication of surface waters. Hungary 8202226.

Banerjee T., Muste M., Katul G. (2015) Flume experiments on wind induced flow in static water bodies in the presence of protruding vegetation. *Advances in Water Resources 76*: 11–28.

Bankston J., Blodgett E., Karthikeyan S., Richard D. (2013) Chlorinated Solvent treatment Wetland. *Remediation Journal 23*(4): 59–69.

Bankston J.L., Komor A.T., Sola D.V., Dwyer D.F. (2002) Degradation of trichloroethylene in wetland microcosms containing broad-leaved cattail and eastern cottonwood. *Water Research 36*: 1539–1546.

Bannerman R., Owens D., Dodds R., Hornewer N. (1993) Sources of pollutants in Wisconsin stormwater. *Water Science and Technology 28*(35): 241–259.

Barber L.B., Keefe S.H., Antweiler R.C., Taylor H.E., Wass R.D. (2006) Accumulation of contaminants in fish from wastewater treatment wetlands. *Environmental Science & Technology 40*(2): 603–611.

Barber L.B., Leenheer J.A., Noyes T.I., Stiles E.A. (2001) Nature and transformation of dissolved organic matter in treatment wetlands. *Environmental Science and Technology 35*(24): 4805–4816.

Barlocher F., Graças M.A.S., Gessner M.O. (2005) *Methods to Study Litter Decomposition*. Springer: Berlin, Germany.

Bartlett M.S., Brown L.C., Hancs N.B., Nickerson N.H. (1979) Denitrification in freshwater wetland soil. *Journal of Environmental Quality 8*: 460–464.

Bartodziej W.M., Blood S.L., Pilgrim K. (2017) Aquatic plant harvesting: An economical phosphorus removal tool in an urban shallow lake. *Journal of Aquatic Plant Management 55*: 26–34.

Bastviken S.K., Eriksson P.G., Martins I., Neto J.M., Leonardson L., Tonderski K.S. (2003) Potential nitrification and denitrification on different surfaces in a constructed treatment wetland. *Journal of Environmental Quality 32*(6): 2414–2420.

Bastviken S.K., Eriksson P.G., Premrov A., Tonderski K. (2005) Potential denitrification in wetland sediments with different plant species detritus. *Ecological Engineering 25*(2): 183–190.

Bastviken S.K., Weisner S.E., Thiere G., Svensson J.M., Ehde P.M., Tonderski K.S. (2009) Effects of vegetation and hydraulic load on seasonal nitrate removal in treatment wetlands. *Ecological Engineering 35*(5): 946–952.

Batchelor B., Lawrence A.W. (1978) Autotrophic denitrification using elemental sulfur. *Journal of the Water Pollution Control Federation 55*: 1986–2001.

Batty L.C., Younger P.L. (2002) Critical role of macrophytes in achieving low iron concentrations in mine water treatment wetlands. *Environmental Science and Technology 36*: 3997–4002.

Batzer D.P., Sharitz R.R. (eds.) (2014) *Ecology of Freshwater and Estuarine Wetlands*. Univ of California Press: Oakland, CA.

Bavor H.J., Roser D., McKersie S.A., Breen P.F. (1988) *Treatment of Secondary Effluent*. Report to Sydney Water Board: Sydney, NSW, Australia.

Bavor H.J., Roser D.J., Adcock P.W. (1995) Challenges for the development of advanced constructed wetlands technology. *Water Science and Technology 32*(3): 13–20.

Bayly I.L., O'Neill T.A. (1972) Seasonal ionic fluctuations in a *Typha glauca* community. *Ecology 53*(4): 714–719.

Bays J., Dernlan G., Hadjimiry H., Vaith K., Keller C. (2000a) Treatment wetlands for multiple functions: Wakodahatchee wetlands, Palm Beach county, Florida. *Proceedings of the Water Environment Federation 2000*(9): 15–37.

Bays J.S., Dernlan G., Hadjimiry H., Vaith K., Keller C. (2000b) *Treatment wetlands with multiple functions*. Proceedings of the WEFTEC 2000 National Conference, 73rd Annual Conference and Exhibition; Water Environment Federation: Alexandria, VA.

Bays J.S., Knight R.L., Frederick D.J. (1993) *Projected Wetland Effluent TSS Concentrations*, Report to the City of Lakeland: June 16, 1993.

BCC. (2016) Comparison of water main pipe installation lengths and costs in North and South Carolina: Raleigh, Charlotte, and Spartanburg/Greenville. Report by BCC Research to American Chemistry Council. Available at: www.uni-bell.org/portals/0/E-NEWS/MediaFiles/bcc_pipe_report_carolinas.pdf (Accessed 12/21/18).

Bear S.E., Nguyen M.T., Jasper J.T., Nyguen S., Nelson K.L., Sedlak D.L. (2017) Removal of nutrients, trace organic contaminants, and bacterial indicator organisms in a demonstration-scale unit process open-water treatment wetland. *Ecological Engineering 109*: 76–83.

Beck T.J., Gawlik D.E., Pearlstine E.V. (2013) Community patterns in treatment wetlands, natural wetlands, and croplands in Florida. *Wilson Journal of Ornithology 125*: 329–341.

Bedford B.L., Walbridge M.R., Aldous A. (1999) Patterns in nutrient availability and plant diversity of temperate North American wetlands. *Ecology 80*(7): 2151–2169.

Beebe D.A., Castle J.W., Molz F.J., Rodgers J.H. (2014) Effects of evapotranspiration on treatment performance in constructed wetlands: Experimental studies and modeling. *Ecological Engineering 71*: 394–400.

Belanger T.V., Korzun E.A. (1990) Rainfall-reaeration effects. *Journal of Irrigation and Drainage Engineering 116*(4): 582–587.

Belmont M.A., White J.R., Reddy K.R. (2009) Phosphorus sorption and potential phosphorus storage in sediments of Lake Istokpoga and the upper chain of lakes, Florida, USA. *Journal of Environmental Quality 38*(3): 987–996.

Beltman W.H.J., Ter Horst M.M.S., Adriaanse P.I., de Jong A., Deneer J.W. (2014) *FOCUS_TOXSWA Manual 4.4.2: Users Guide Version 4.* WOt-technical report 14, Wettelijke Onderzoekstaken Natuur & Milieu: Wageningen, The Netherlands.

Bennett C.O., Myers J.E. (1982) *Momentum, Heat, and Mass Transfer.* McGraw-Hill: New York.

Bennett E.R., Linstedt K.D., Nilsgard V., Battaglia G.M., Pontius F.W. (1981) Urban snowmelt-characteristics and treatment. *Journal of the Water Pollution Control Federation* 53(1): 119–125.

Benoy G.A., Kalff J. (1999) Sediment accumulation and Pb burdens in submerged macrophyte beds. *Limnology and Oceanography 44*(4): 1081–1090.

Bentzen T.R., Larsen T., Rasmussen M.R. (2008) Wind effects on retention time in highway ponds. *Water Science and Technology 57*(11): 1713–1720.

Berglund B., Khan G.A., Weisner S.E., Ehde P.M., Fick J., Lindgren P.E. (2014) Efficient removal of antibiotics in surface-flow constructed wetlands, with no observed impact on antibiotic resistance genes. *Science of the Total Environment 476*: 29–37.

Bernard J.M. (1999) Seasonal growth patterns in wetland plants growing in landfill leachate. In: *Constructed Wetlands for the Treatment of Landfill Leachate*, Mulamoottil G., McBean E., Rovers F. A. (eds.) CRC Press: Boca Raton, FL, pp. 223–234.

Bernard J.M., Fiala K. (1986) Distribution and standing crop of living and dead roots in three wetland Carex species. *Bulletin of the Torrey Botanical Club 113*(1): 1–5.

Best E.P.H., Miller J.L., Larson S.L. (2000) Explosives removal from groundwater at the Volunteer Army Ammunition Plant, Tennessee, in small-scale wetland modules. In: Means J.L., Hinchee R. E. (eds.), *Wetlands & Remediation: An International Conference.* Battelle Press: Columbus, OH, pp. 365–374.

Best E.P.H., Miller J.L., Larson S.L. (2001) Tolerance toward explosives, and explosives removal from groundwater in treatment wetland mesocosms. *Water Science and Technology 44*(11–12): 515–521.

Best E.P.H., Sprecher S.L., Frederickson H.L., Zappi M.E., Larson S.L. (1997a) *Screening Aquatic and Wetland Plant Species for Phytoremediation of Explosives-Contaminated Groundwater from the Iowa Army Ammunition Plant*, Technical Report EL-97-2, U.S. Army Corps of Engineers Waterways Experiment Station: Vicksburg, MO.

Best E.P.H., Sprecher S.L., Larson S.L., Fredrickson H.L., Bader D.F. (1999a) Environmental behavior of explosives in groundwater from the Milan Army Ammunition Plant in aquatic and wetland plant treatments. Removal, mass balances and fate in groundwater of TNT and RDX. *Chemosphere 39*: 3383–3396.

Best E.P.H., Sprecher S.L., Larson S.L., Fredrickson H.L., Bader D.F. (1999b) Environmental behavior of explosives in groundwater from the Milan Army Ammunition Plant in aquatic and wetland plant treatments. Uptake and rate of TNT and RDX in plants. *Chemosphere 39*: 2057–2072.

Best M.J. (1998) A model to predict surface temperatures. *Boundary-Layer Meteorology 88*(2): 279–306.

Beutel M.W. (2012) Water quality in a surface-flow constructed treatment wetland polishing tertiary effluent from a municipal wastewater treatment plant. *Water Science and Technology 66*(9): 1977–1983.

Beutel M.W., Morgan M.R., Erlenmeyer J.J., Brouillard E.S. (2014) Phosphorus removal in a surface-flow constructed wetland treating agricultural runoff. *Journal of Environmental Quality 43* (3): 1071–1080.

Beutel M.W., Newton C.D., Brouillard E.S., Watts R.J. (2009) Nitrate removal in surface-flow constructed wetlands treating dilute agricultural runoff in the lower Yakima Basin, Washington. *Ecological Engineering 35*(10): 1538–1546.

Beutel M.W., Whritenour V., Brouillard E. (2013) Fecal coliform removal in a lightly loaded surface-flow constructed treatment wetland polishing agricultural runoff. *Water Science and Technology 68*(4): 909–915.

Bhadha J.H., Jawitz J.W., Min J-H. (2011) Phosphorus mass balance and internal load in an impacted subtropical isolated wetland. *Water Air Soil Pollution 218*: 619–632.

Bhomia R.K., Inglett P.W., Reddy K.R. (2015) Soil and phosphorus accretion rates in subtropical wetlands: Everglades stormwater treatment areas as a case example. *Science of the Total Environment 533*: 297–306.

Bhomia R.K., Inglett P.W., Wright A.L., Reddy K.R. (2011) *Soil Characterization in Selected Stormwater Treatment Areas: Soil Accretion Rates and Nutrient Storage.* Report to South Florida Water Management District: West Palm Beach, FL, January 2011.

Biebighauser T.R. (2011) *Wetland Restoration and Construction: A Technical Guide.* Upper Susquehanna Coalition: Oswego, NY.

Biebighauser T.R. (2015) *Wetland Restoration and Construction: A Technical Guide*, 2nd edition. Center for Wetland and Stream Restoration: Morehead, KY.

Biljetina R., Srivastava D.P., Chynoweth D.P., Hayes T.D. (1987) Anaerobic digestion of water hyacinth and sludge. In: *Aquatic Plants for Water Treatment and Resource Recovery*, Reddy K.R., Smith W. H. (eds.) Magnolia Press: Orlando, KL, pp. 747–753.

Billore S.K., Dass P., Hyas H. (1994) Ammonia volatilization through plant species in domestic wastewater applied agriculture field and wetland. In: Jiang C. (ed.) *Proceedings of the 4th International Conference on Wetland Systems for Water Pollution Control, 6–10 November 1994.* IWA Publishing: Guangzhou, PR China, pp. 171–179.

Bird R.B., Stewart W.E., Lightfoot E.N. (1960) *Transport Phenomena.* Wiley: New York.

Birge E.A. (1916) The work of the wind in warming a lake. *Transactions of the Wisconsin Academy of Sciences, Arts, and Letters 18*(Part 2): 341.

Bishay F., Kadlec R.H. (2005) Wetland treatment at musselwhite mine. In: *Natural and Constructed Wetlands: Nutrients, Metals, and Management*, Vymazal J. (ed.) Backhuys Publishers: Leiden, The Netherlands, pp. 176–198.

Bishop J., McKay H., Parrott D., Allan J. (2003) *Review of International Research Literature regarding the Effectiveness of Auditory Bird Scaring Techniques and Potential Alternatives.* Produced by Central Science Laboratories for the Department for Environmental Food and Rural Affairs, London, United Kingdom. Available at: www.researchgate.net/profile/Dave_Parrott/publication/242454383_Review_of_international_research_literature_regarding_the_effectiveness_of_auditory_bird_scaring_techniques_and_potential_alternatives/links/55ba176608aed621de0aae17/Review-of-international-research-literature-regarding-the-effectiveness-of-auditory-bird-scaring-techniques-and-potential-alternatives.pdf (Accessed 12/9/18).

Bishop M., Bays J., Griffin M., Gramer W. (2012) More than a pretty space: Stormwater treatment Wetlands with multiple benefits at freedom park, Naples FL. In: *Proceedings of WEFTEC 2012, New Orleans, LA.* Water Environment Federation: Alexandria, VA, pp. 8095–8115.

Bjolseth I.M. (2004) *Fate of Estrogenic Activity during Wetland Treatment of Wastewater Effluent and Stormwater Runoff.* MS Thesis. Department of Chemical and Environmental Engineering. The University of Arizona. Tucson, AZ.

Bjolseth I.M., Quanrud D.M., Karpiscak M.M., Ela W.P., Lansey K.E., Arnold R.G. (2003) Fate of estrogenic activity during wetland treatment of wastewater effluent and stormwater runoff. Proceedings, 11th Biannual Symposium on Groundwater Recharge, Phoenix, AZ, June 5–7, 2003.

Blankenberg A.G., Haarstad K., Braskerud B.C. (2007) Pesticide retention in an experimental wetland treating non-point source pollution from agriculture runoff. *Water Science and Technology 55*(3): 37–44.

Blecken G.T., Hunt III W.F., Al-Rubaei A.M., Viklander M., Lord W.G. (2017) Stormwater control measure (SCM) maintenance considerations to ensure designed functionality. *Urban Water Journal 14*(3): 278–290.

Bloesch J. (1994) A review of methods used to measure sediment resuspension. *Hydrobiologia 284*: 13–18.

Bodin H., Mietto A., Ehde P.M., Persson J., Weisner S. (2012) Tracer behaviour and analysis of hydraulics in experimental free watersurface wetlands. *Ecological Engineering 49*: 201–211.

Bodin H., Persson J., Englund J.E., Milberg P. (2013) Influence of residence time analyses on estimates of wetland hydraulics and pollutant removal. *Journal of Hydrology 501*: 1–12.

Bodine B.R., Hebler M.T. (1978) *Wave Runup and Wind Setup -Computational Model.* USACE Southwestern Division: Dallas, TX.

Bois P., Childers D.L., Corlouer T., Laurent J., Massicot A., Sanchez C.A., Wanko A. (2017) Confirming a plant-mediated "Biological Tide" in an aridland constructed treatment wetland. *Ecosphere 8* (3): e01756.

Boithias L., Sauvage S., Merlina G., Jean S., Probst J.L., Pérez J.M.S. (2014) New insight into pesticide partition coefficient Kd for modelling pesticide fluvial transport: Application to an agricultural catchment in south-western France. *Chemosphere 99*: 134–142.

Bolster C.H., Saiers J.E. (2002) Development and evaluation of a mathematical model for surface-water flow within the Shark River Slough of the Florida Everglades. *Journal of Hydrology 259*(1): 221–235.

Bonneville M.C., Strachan I.B., Humphreys E.R., Roulet N.T. (2008) Net ecosystem CO 2 exchange in a temperate cattail marsh in relation to biophysical properties. *Agricultural and Forest Meteorology 148*(1): 69–81.

Boon A.G. (1985) *Report on Visit by Members and Staff of WRc to Germany to Investigate Root Zone Method for Treatment of Wastewaters*, WRc Report 376-S/1, Stevenage, United Kingdom.

Boon B., Groe H. (1990) *Nature's Heartland Native Plant Communities of the Great Plains*. Iowa State University Press: Ames, IA.

Bormans M., Webster I.T. (1998) Dynamics of temperature stratification in lowland rivers. *Journal of Hydraulic Engineering 124*(10): 1059–1063.

Bormans M., Maier H., Burch M., Baker P. (1997) Temperature stratification in the lower River Murray, Australia: Implication for cyanobacterial bloom development. *Marine and Freshwater Research 48*(7): 647–654.

Borne K.E. (2014) Floating treatment wetland influences on the fate and removal performance of phosphorus in stormwater retention ponds. *Ecological Engineering 69*: 76–82.

Borne K.E., Fassman-Beck E.A., Winston R.J., Hunt W.F., Tanner C.C. (2015) Implementation and maintenance of floating treatment wetlands for urban stormwater management. *Journal of Environmental Engineering 141*(11): 04015030.

Bostic E.M., White J.R. (2007) Phosphorus and vegetation effects on P retention in wetland soil after a drawdown/reflood event. *Soil Science Society of America Journal 71*: 238–244.

Bothe H., Jost G., Schloter M., Ward B.B., Witzel K.-P. (2000) Molecular analysis of ammonia oxidation and denitrification in natural environments. *FEMS Microbiology Reviews 24*: 673–690.

Bôto M., Almeida C., Mucha A. (2016) Potential of constructed wetlands for removal of antibiotics from saline aquaculture effluents. *Water 8*(10): 465.

Boutwell J.E. (2001) *Floating Island Construction and Placement in Las Vegas Bay, Lake Mead, Nevada*, Technical Memorandum No. 8220-01-16, U.S. Bureau of Reclamation: Denver, CO.

Boving T.B. (2002) *Chemical Retention Capacity of a Newly Constructed Roadway Runoff Detention Pond System*. Final Report to the University of Rhode Island Transportation Center: Kingston, RI.

Boyd C.E. (1971) Further studies on productivity, nutrient, and pigment relationships in *Typha latifolia* populations. *Bulletin of the Torrey Botanical Club 98*(3): 144–150.

Boyd C.E. (1978) Chemical composition of wetland plants. In: *Freshwater Wetlands: Ecological Processes and Management Potential*, Good R.E., Whigham D.F., Simpson R.L. (eds.) Academic Press: New York, pp. 155–167.

Brander L.M., Florax R.J., Vermaat J.E. (2006) The empirics of wetland valuation: A comprehensive summary and a meta-analysis of the literature. *Environmental and Resource Economics 33*(2): 223–250.

Braskerud B.C. (2001a) Clay particle retention in small constructed wetlands. *Journal Environmental Quality 30*(4): 1447–1457.

Braskerud B.C. (2001b) The influence of vegetation on sedimentation and resuspension of soil particles in small constructed wetlands. *Journal of Environmental Quality 30*(4): 1447–1457.

Braskerud B.C. (2001c) *Sedimentation in Small Constructed Wetlands*. Doctor Scientiarum Thesis, Agricultural University of Norway, Ås, Norway.

Braskerud B.C. (2002a) Factors affecting nitrogen retention in small constructed wetlands treating agricultural non-point source pollution. *Ecological Engineering 18*(3): 351–370.

Braskerud B.C. (2002b) Factors affecting phosphorus retention in small constructed wetlands treating agricultural non-point source pollution. *Ecological Engineering 19*(1): 41–61.

Braskerud B.C. (2003) Clay particle retention in small constructed wetlands. *Water Research 37*(16): 3793–3802.

Braskerud B.C., Dunne E., Reddy K.R., Carton O. (2005a) Retention of soil particles and phosphorus in small constructed wetlands in agricultural watersheds. In: *Nutrient Management in Agricultural Watersheds: A Wetlands Solution*, Dunne E.J., Reddy K.R., Carton O.T. (eds.) Wageningen Academic Publishers: Wageningen, The Netherlands, pp. 121–131.

Braskerud B.C., Haarstad K. (2003) Screening the retention of thirteen pesticides in a small constructed wetland. *Water Science and Technology 48*(5): 267–274.

Braskerud B.C., Lundekvam H., Krogstad T. (2000) The impact of hydraulic load and aggregation on sedimentation of soil particles in small constructed wetlands. *Journal of Environmental Quality 29* (6): 2013–2020.

Braskerud B.C., Tonderski K.S., Wedding B., Bakke R., Blankenberg A.G., Ulen B., Koskiaho J. (2005b) Can constructed wetlands reduce the diffuse phosphorus loads to eutrophic water in cold temperate regions? *Journal of Environmental Quality 34*(6): 2145–2155.

Brauer N., Maynard J.J., Dahlgren R.A., O'Geen A.T. (2015) Fate of nitrate in seepage from a restored wetland receiving agricultural tailwater. *Ecological Engineering 81*: 207–217.

Breen P.F., Lawrence I. (1998) *Design Guidelines: Stormwater Pollution Control Ponds and Wetlands.* The Cooperative Research Centre for Freshwater Ecology: Canberra, Australia.

Breitholtz M., Näslund M., Stråe D., Borg H., Grabic R., Fick J. (2012) An evaluation of free water surface wetlands as tertiary sewage water treatment of micro-pollutants. *Ecotoxicology and Environmental Safety 78*: 63–71.

Bremner J.M., Shaw K. (1958) Denitrification in soil: II. Factors affecting denitrification. *Journal of Agricultural Science 51*: 40–52.

Brenner M., Kenney W.F. (2013) Dating wetland sediment cores. In: *Methods in Biogeochemistry of Wetlands,* DeLaune R.D., Reddy K.R., Richardson C.J., Megonigal J.P. Soil Science Society of America: Madison, WI, pp. 879–900.

Brett M.T., Benjamin M.M. (2008) A review and reassessment of lake phosphorus retention and the nutrient loading concept. *Freshwater Biology 53*(1): 194–211.

Breukelaar A.W., Lammens E.H.R.R., Klein-Breteler J.G.P., Tatrai I. (1994) Effects of benthivorous bream (*Abramis brama*) and carp (*Cyprinus carpio*) on sediment resuspension and concentrations of nutrients and chlorophyll*a*. *Freshwater Biology 32*(1): 113–121.

Brezonik P.L., Stadelmann T.H. (2002) Analysis and predictive models of stormwater runoff volumes, loads, and pollutant concentrations from watersheds in the Twin Cities metropolitan area, Minnesota, USA. *Water Research 36*(7): 1743–1757.

Bristow J.M. (1974) Nitrogen fixation in the rhizosphere of freshwater angiosperms. *Canadian Journal of Botany 52*(1): 217–221.

Brix H. (1990) Gas exchange through the soil-atmosphere interphase and through dead culms of *Phragmites australis* in a constructed reed bed receiving domestic sewage. *Water Research 24* (2): 259–266.

Brix H. (1993) Macrophyte-mediated oxygen transfer in wetlands: Transport mechanisms and rates. In: *Constructed Wetlands for Water Quality Improvement*, Moshiri G.A. (ed.) Lewis Publishers: Boca Raton, FL, pp. 391–398.

Brix H. (1997) Do macrophytes play a role in constructed treatment wetlands? *Water Science and Technology 35*(5): 11–17.

Brix H. (2006) *Course Notes: Onsite Treatment of Wastewater in Willow Systems.* Department of Biological Sciences, Aarhus University: Aarhus, Denmark.

Brix H., Dyhr-Jensen K., Lorenzen B. (2002b) Root-zone acidity and nitrogen source affects *Typha latifolia* L. growth and uptake kinetics of ammonium and nitrate. *Journal of Experimental Botany 53* (379): 2441–2450.

Brix H., Gregersen P. (2002) Water balance of willow dominated constructed wetlands. Mbwette T.S.A. (ed.), Proceedings of the 8th International Conference on Wetland Systems for Water Pollution Control, 16–19 September 2002. IWA Publishing and University of Dar Es Salaam: Arusha, Tanzania, pp. 669–670.

Brix H., Orr P.T. (1992) Internal pressurization and convective gas flow in some emergent freshwater macrophytes. *Limnology and Oceanography 37*(7): 1420–1433.

Brix H., Schierup H.-H. (1990) Soil oxygenation in constructed reed beds: The role of macrophyte and soil-atmosphere interface oxygen transport. In: *Constructed Wetlands in Water Pollution Control,* Cooper P.F., Findlater B.C. (eds.) Pergamon Press: Oxford, United Kingdom, pp. 53–66.

Brix H., Sorrell B.K., Lorenzen B. (2001) Are *Phragmites*-dominated wetlands a net source or net sink of greenhouse gases? *Aquatic Botany 69*: 313–324.

Broadbent F.E., Clark F. (1965) Denitrification. *Agronomy 10*: 344–359.

Brooks K.M. (2000a) *An Assessment of the Environmental Effects Associated with Bridges Constructed of Wood Preserved with Creosote, Pentachlorophenol, or Chromate-Copper-Arsenate (CCA-C).* Report to the USDA Forest Service: Madison, WI.

Brooks K.M. (2000b) Migration of polycyclic aromatic hydrocarbons (PAH) from new and used railway crossties into ballast and adjacent wetland environments. Paper presented at American Wood-Preservers' Association 96th Annual Meeting, San Francisco, CA, May 7–10, 2000.

Brown and Caldwell (2005) Opinion of probable construction cost. Basis of Design Report, Stormwater Treatment Area 2 Cell 4 Expansion Project. Report to South Florida Water Management District, West Palm Beach, FL.

Brown D.J., Street G.M., Nairn R.W., Forstner M.R. (2012) A place to call home: Amphibian use of created and restored wetlands. *International Journal of Ecology* Vol. 2012: 1–11.

Brown L.D., Ulrich A.C. (2015) Oil sands naphthenic acids: A review of properties, measurement, and treatment. *Chemosphere 127*: 276–290.

Brueske C.C., Barrett G.W. (1994) Effects of vegetation and hydrologic load on sedimentation patterns in experimental wetland ecosystems. *Ecological Engineering 3*(4): 429–448.

Brunner C.W., Kadlec R.H. (1993) Effluent restores wetland hydrology. *Water Environment Technology 5*(7): 60–63.

Brunner C.W., Smith J.W., Stucky N.P., Kempf D.W., Baskett R.K., Dakec R.H., Patterson L.B., White J.E., Zdeb J.A. (1992) *Design of constructed wetlands with effluent reuse wetlands in Columbia, Missouri.* Paper presented at the 65th Annual Water Environment Federation Conference, 20–24 September 1992, New Orleans, Louisiana; Water Environment Federation: Alexandria, VA, pp. 153–164.

Brunner G.W. (2010) *HEC-RAS River Analysis System User's Manual, Version 4.1.* Hydrologic Engineering Center, Institute For Water Resources. US Army Corps of Engineers: Davis, CA.

Brunner G.W. (2016) *HEC-RAS River Analysis System User's Manual.* US Army Corps of Engineers, Hydrologic Engineering Center: Davis, CA.

Bucher T.N. (2012) Removal of Dissolved Metals by the Imperial and Brawley Constructed Wetlands, Imperial Valley, CA. MPH Thesis, San Diego State University, San Diego, CA

Budd R., O'Geen A., Goh K.S., Bondarenko S., Gan J. (2009) Efficacy of constructed wetlands in pesticide removal from tailwaters in the Central Valley, California. *Environmental Science and Technology 43*(8): 2925–2930.

Budd R., O'Geen A., Goh K.S., Bondarenko S., Gan J. (2011) Removal mechanisms and fate of insecticides in constructed wetlands. *Chemosphere 83*(11): 1581–1587.

Bulc T., Sajn Slak A. (2003) Performance of constructed wetland for highway runoff treatment. *Water Science and Technology 48*(2): 315–322.

Burba G.G., Verma S.B., Kim J. (1999a) Surface energy fluxes of *Phragmites australis* in a prairie wetland. *Agricultural and Forest Meteorology 94*: 31–51.

Burba G.G., Verma S.B., Kim J. (1999b) A comparative study of surface energy fluxes of three communities (*Phragmites australis, Scirpus acutus*, open water) in a prairie wetland ecosystem. *Wetlands 19*(2): 451–457.

Burchell M.L., Hunt W., Bass K.L., Wright J.D. (2010) Stormwater Wetland Construction Guidance. Publication AG-588-13 NC State Extension Service, N.C. State University, Raleigh, NC. Available at: https://content.ces.ncsu.edu/stormwater-wetland-construction-guidance (Accessed 7/19/18).

Burchell M.R., Skaggs R.W., Broome S.W., Lee C.R. (2002) *Effect of Substrate Organic Matter Addition on Nitrate Removal Efficiency in Surface-Flow Constructed Wetlands.* American Society of Agricultural Engineers (ASAE) Annual International Meeting in Chicago, Illinois: St. Joseph, MI.

Buresh R.J., Patrick W.H. Jr. (1978) Nitrate reduction to ammonium in anaerobic soil. *Soil Science Society of America Journal 42*: 913–918.

Buresh R.J., Patrick W.H. Jr. (1981) Nitrate reduction to ammonium and organic nitrogen in an estuarine sediment. *Soil Biology and Biochemistry 13*: 279–283.

Burford J.R., Bremner J.M. (1972) Is phosphate reduced to phosphine in waterlogged soils? *Soil Biology and Biochemistry 4*(489): 495.

Burgin A.J., Hamilton S.K. (2007) Have we overemphasized the role of denitrification in aquatic ecosystems? A review of nitrate removal pathways. *Frontiers in Ecology and the Environment* 5(2): 89–96.

Burgin A.J., Hamilton S.K., Gardner W.S., McCarthy M.J. (2013) Nitrate reduction, denitrification, and dissimilatory nitrate reduction to ammonium in wetland sediments. Chapter 46 in: *Methods in Biogeochemistry of Wetlands*, DeLaune R.D., Reddy K.R., Richardson C.J., Megonigal J.P. (eds.) Soil Science Society of America, Madison, WI, USA, 519–537.

Burgoon P.S. (2001) Denitrification in free water surface wetlands receiving carbon supplements. *Water Science and Technology 44*(11–12): 163–170.

Burke E.N., Wadzuk B.M. (2009) The effect of field conditions on low Reynolds number flow in a wetland. *Water Research 43*(2): 508–514.

Burrows C.J. (2012) *Processes of Vegetation Change.* Springer Science & Business Media: Berlin, Germany.

Busnardo M.J., Gersberg R.M., Langis R., Sinicrope T.L., Zedler J.B. (1992) Nitrogen and phosphorus removal by wetland mesocosms subjected to different hydroperiods. *Ecological Engineering 1*(4): 287–308.

Butt M.J., Bilal M. (2011) Application of snowmelt runoff model for water resource management. *Hydrological Processes 25*(24): 3735–3747.

Byappanahalli M.N., Nevers M.B., Korajkic A., Staley Z.R., Harwood V.J. (2012) Enterococci in the environment. *Microbiology and Molecular Biology Reviews 76*(4): 685–706.

Byström O. (1998) The nitrogen abatement cost in wetlands. *Ecological Economics 26*(3): 321–331.

Cabelli V.J. (1977) Indicators of recreational water quality. In: *Bacterial Indicators/Health Hazards Associated with Water*, Hoadley A.W., Dutka B.J. (eds.) American Society for Testing and Materials (ASTM): Philadelphia, PA, pp. 222–238.

Caffrey J.M., Kemp W.M. (1991) Seasonal and spatial patterns of oxygen production, respiration, and root-rhizome release in *Potamogeton perforliatus* L. and *Zostera marina* L. *Aquatic Botany 40*: 109–128.

Cahoon D.R., Marin P.E., Black B.K., Lynch J.C. (2000) A method for measuring vertical accretion, elevation, and compaction of soft shallow-water sediments. *Journal of Sedimentary Research 70* (5): 1250–1253.

Cai K., Elliott C.T., Phillips D.H., Scippo M.L., Muller M., Connolly L. (2012) Treatment of estrogens and androgens in dairy wastewater by a constructed wetland system. *Water Research 46*(7): 2333–2343.

Caldwell P.V., Vepraskas M.J., Skaggs R.W., Gregory J.D. (2007) Simulating the water budgets of natural Carolina bay wetlands. *Wetlands 27*(4): 1112–1123.

Callaway J.C., Cahoon D.R., Lynch J.C. (2013) The surface elevation table–Marker horizon method for measuring wetland accretion and elevation dynamics. Chapter 46. In: *Methods in Biogeochemistry of Wetlands*, DeLaune, R.D., Reddy K.R., Richardson C.J., Megonigal J.P., (eds.) Soil Science Society of America: Madison, WI, pp. 901–917.

Campbell C.S., Ogden M.H. (1999) *Constructed Wetlands in the Sustainable Landscape*. John Wiley & Sons: New York.

Campeau M., Lee R.V., Lee K. (2009) Brighton Constructed Wetland. Ancillary Benefit Report to the Municipality of Brighton, October, 2009.

Cardinal P., Anderson J.C., Carlson J.C., Low J.E., Challis J.K., Beattie S.A., Bartel C.N., Elliott A.D., Montero O.F., Lokesh S., Favreau A. (2014) Macrophytes may not contribute significantly to removal of nutrients, pharmaceuticals, and antibiotic resistance in model surface constructed wetlands. *Science of the Total Environment 482*: 294–304.

Carleton J.N. (2002) Damköhler number distributions and constituent removal in treatment wetlands. *Ecological Engineering 19*(4): 233–248.

Carleton J.N., Grizzard T.J., Godrej A.N., Post H.E. (2001) Factors affecting the performance of stormwater treatment wetlands. *Water Research 35*(6): 1552–1562.

Carleton J.N., Montas H.J. (2007) A modeling approach for mixing and reaction in wetlands with continuously varying flow. *Ecological Engineering 29*(1): 33–44.

Carleton J.N., Montas H.J. (2010) An analysis of performance models for free water surface wetlands. *Water Research 44*(12): 3595–3606.

Carpenter K., Snyder D., Duff J., Triska F. Lee K. Avanzino R., Sobieszczyk S. (2009) *Hydrologic and Water-Quality Conditions during Restoration of the Wood River Wetland, Upper Klamath River Basin, Oregon, 2003–05*. US Geological Survey Scientific Investigations Report 2009-5004. US Department of the Interior, US Geological Survey: Reston, VA.

Carr D.W., Rushton B.T. (1995) *Integrating a native Herbaceous Wetland into stormwater management*. Report by the Southwest Florida Water Management District (July 1995): Boca Raton, FL.

Carroll K.T. (1991) Patent: Protective grating with pivoting sections for culvert pipe. United States US 5,037,542.

Carson B. D., Lishawa S.C., Tuchman N.C., Monks A.M., Lawrence B.A., Albert D.A. (2018) Harvesting invasive plants to reduce nutrient loads and produce bioenergy: An assessment of Great Lakes coastal wetlands. *Ecosphere 9*(6): e02320.

Carter J., Foote A.L., Johnson-Randall A. (1999) Modeling the effects of nutria (*Myocastor coypus*) on wetland loss. *Wetlands 19*(1): 209–219.

Carvalho K.M., Martin D.F. (2001) Removal of aqueous selenium by four aquatic plants. *Journal of Aquatic Plant Management 39*: 33–36.

Carvalho P.N., Araújo J.L., Mucha A.P., Basto M.C.P., Almeida C.M.R. (2013) Potential of constructed wetlands microcosms for the removal of veterinary pharmaceuticals from livestock wastewater. *Bioresource Technology 134*: 412–416.

Carvalho P.N., Basto M.C.P., Almeida C.M.R. (2012b) Potential of Phragmites australis for the removal of veterinary pharmaceuticals from aquatic media. *Bioresource Technology 116*: 497–501.

Carvalho P.N., Basto M.C.P., Almeida C.M.R., Brix H. (2014) A review of plant–Pharmaceutical interactions: From uptake and effects in crop plants to phytoremediation in constructed wetlands. *Environmental Science and Pollution Research 21*(20): 11729–11763.

Caskey W.H., Tiedje J.M. (1979) Evidence for clostridia as agents of dissimilatory reduction of nitrate to ammonium in soils. *Soil Science Society of America Journal 43*: 931–936.

Castonguay A.C., Iftekhar M.S., Urich C., Bach P.M., Deletic A. (2018) Integrated modelling of stormwater treatment systems uptake. *Water Research 142*: 301–312.

Castro H., Reddy K.R., Ogram A. (2002) Composition and function of sulfate-reducing prokaryotes in eutrophic and pristine areas of the Florida Everglades. *Applied and Environmental Microbiology 68*(12): 6129–6137.

Catts C.W. (2009) Water Hyacinth Treatment Systems in Agricultural Watersheds: Influence of Biomass Incorporation into Soil on Phosphorus Retention. MS Thesis, University of Florida, Gainesville, FL.

CDC. (2018) *Take Care with Pet Reptiles and Amphibians.* National Center for Emerging and Zoonotic Infectious Diseases, Centers for Disease Control and Prevention, Atlanta, GA. Available at: www.cdc.gov/Features/SalmonellaFrogTurtle/ (Accessed 9/30/18).

Center for Watershed Protection. (2001) *New York State Stormwater Management Design Manual.* Report by the Center for Watershed Protection (CWP) for the New York State Department of Environmental Conservation: Albany, NY.

CH2M HILL. (1998). Treatment Wetland Habitat and Wildlife Use Assessment Project. Report prepared for the U.S. Environmental Protection Agency, U.S. Bureau of Reclamation, and City of Phoenix with funding from the Environmental Technology Initiative Program. Available at: www.wetlandsolutionsinc.com/resources/ (Accessed 10/29/18).

CH2M Hill. (2003a) *Village of Wellington Aquatics Pilot Program.* Prepared for the Village of Wellington, FL.

CH2M Hill. (2003b) *PSTA Research and Demonstration Project, Phases 1, 2, and 3 Summary Report.* Prepared for the South Florida Water Management District West Palm Beach, FL.

CH2M Hill. (2005) *Additional Testing for the Membrane Concentrate Pilot Wetlands Project.* Report to the City of Oxnard Water Division, Oxnard, CA.

CH2M Hill. (2014) Wetland Design Guidelines. Report prepared for City of Saskatoon. Available at: www.saskatoon.ca/sites/default/files/documents/transportation-utilities/construction-design/new-neighbourhood-design/wetlands_design_guidelines.pdf (Accessed 7/19/18).

Chague-Goff C. (2005) Assessing the removal efficiency of Zn, Cu, Fe, and Pb in a treatment wetland using sequential chemical extraction. *Water, Air, and Soil Pollution 160*(1–4): 161–179.

Chaichana R., Leah R., Moss B. (2010) Birds as eutrophicating agents: A nutrient budget for a small lake in a protected area. *Hydrobiologia 646*(1): 111–121.

Chaichana R., Leah R., Moss B. (2011) Seasonal impact of waterfowl on communities of macrophytes in a shallow lake. *Aquatic Botany 95*(1): 39–44.

Chang J., Liu D., Cao H., Chang S.X., Wang X., Huang C., Ge Y. (2010) NO3−/NH4+ ratios affect the growth and N removal ability of Acorus calamus and Iris pseudacorus in a hydroponic system. *Aquatic Botany 93*(4): 216–220.

Chang S.W., La H.J., Lee S.J. (2001) Microbial degradation of benzene, toluene, ethylbenzene and xylene isomers (BTEX) contaminated groundwater in Korea. *Water Science and Technology 44*(7): 165–171.

Chang T.J., Chang Y.S., Lee W.T., Shih S.S. (2016) Flow uniformity and hydraulic efficiency improvement of deep-water constructed wetlands. *Ecological Engineering 92*: 28–36.

Chanton J.P. (2005) The effect of gas transport on the isotope signature of methane in wetlands. *Organic Geochemistry 36*(5): 753–768.

Chapple M., Cooper P.F., Cooper D.J., Revitt D.M. (2002) Pilot trials of a constructed wetland system for reducing the dissolved hydrocarbon in the runoff from a decommissioned oil refinery. Mbwette T.S.A. (ed.), *Proceedings of the 8th International Conference on Wetland Systems for Water Pollution Control, 16–19 September 2002*. Comprint International Limited: University of Dar Es Salaam, Tanzania, pp. 877–883.

Chapra S.C., Canale R.P. (1991) Long-term phenomenological model of phosphorus and oxygen for stratified lakes. *Water Research 25*(6): 707–715.

Chaudhry M.H., Cass D.E., Edinger J.E. (1983) Modeling of unsteady-flow water temperatures. *Journal of Hydraulic Engineering 109*(5): 657–669.

Chazarenc F., Gérard M., Gontheir Y. (2003) Hydrodynamics of horizontal subsurface flow constructed wetlands. *Ecological Engineering 21*(2–3): 165–173.

Chen B., Liu D., Han W., Fan X., Cao H., Jiang Q., Liu Y., Chang J., Ge Y. (2015) Nitrogen-removal ability and niche of Coix lacryma-jobi and Reineckia carnea in response to NO3−/NH4+ ratio. *Aquatic Botany 120*: 193–200.

Chen H., Zamorano M.F., Ivanoff D. (2010) Effect of flooding depth on growth, biomass, photosynthesis, and chlorophyll fluorescence of Typha domingensis. *Wetlands 30*(5): 957–965.

Chen Z., Cuervo D.P., Müller J.A., Wiessner A., Köser H., Vymazal J., Kästner M., Kuschk P. (2016) Hydroponic root mats for wastewater treatment—a review. *Environmental Science and Pollution Research 23*(16):15911–15928

Chen T.C., Yeh K.J.C., Kuo W.C., Chao H.R., Sheu S.C. (2014) Estrogen degradation and sorption onto colloids in a constructed wetland with different hydraulic retention times. *Journal of Hazardous Materials 277*: 62–68.

Chen Y.C., Kao S.P. (2011) Velocity distribution in open channels with submerged aquatic plant. *Hydrological Processes 25*(13): 2009–2017.

Chendorain M., Yates M., Villegas F. (1998) The fate and transport of viruses through surface water constructed wetlands. *Journal of Environmental Quality 27*(6): 1451–1458.

Cheng J., Niu S., Kim Y. (2013) Relationship between water quality parameters and the survival of indicator microorganisms–Escherichia coli–In a stormwater wetland. *Water Science and Technology 68*(7): 1650–1656.

Cheng N.S., Nguyen H.T. (2010) Hydraulic radius for evaluating resistance induced by simulated emergent vegetation in open-channel flows. *Journal of Hydraulic Engineering 137*(9): 995–1004.

Cheng X.Y., Wang M., Zhang C.F., Wang S.Q., Chen Z.H. (2014) Relationships between plant photosynthesis, radial oxygen loss and nutrient removal in constructed wetland microcosms. *Biochemical Systematics and Ecology 54*: 299–306.

Chimney M.J. (2012) The plant assemblage of constructed treatment Wetlands in South Florida used for Everglades restoration. *Florida Scientist 75*(2): 131–151.

Chimney M.J., Gawlik D.E. (2007) The avifauna of constructed treatment wetlands in south Florida used for Everglades restoration. *Florida Field Naturalist 35*(1): 9–21.

Chimney M.J., Jordan F. (2008) The fish assemblage of a constructed treatment wetland in South Florida. *Florida Scientist 71*(3): 246–264.

Chimney M.J., Nungesser M., Newman J., Pietro K., Germain G., Lynch T., Goforth G., Moustafa M. (2000) Chapter 6: Stormwater Treatment Areas – Status of research and monitoring to optimize effectiveness of nutrient removal and annual report on operational compliance. In: *SFWMD Everglades Consolidated Report*. SFWMD, pp. 6–1 through 6–127.

Chimney M.J., Pietro K.C. (2006) Decomposition of macrophyte litter in a subtropical constructed wetland in south Florida (USA). *Ecological Engineering 27*(4): 301–321.

Chimney M.J., Wenkert L., Pietro K.C. (2006) Patterns of vertical stratification in a subtropical constructed wetland in south Florida (USA). *Ecological Engineering 27*(4): 322–330.

Chimney M.J. (ed.) (2015) Performance of the Everglades Stormwater Treatment Areas. Chapter 5B. In: *2015 South Florida Environmental Report*, South Florida Water Management District: West Palm Beach, FL. Available at: www.sfwmd.gov/science-data/scientific-publications-sfer (Accessed 7/4/18).

Chin D.A. (2011) Thermodynamic consistency of potential evapotranspiration estimates in Florida. *Hydrological Processes 25*(2): 288–301.

Chmura G.L., Costanza R., Kosters E.C. (1992) Modelling coastal marsh stability in response to sea level rise: A case study in coastal Louisiana, USA. *Ecological Modeling 64*(1): 47–64.

Choi J, Harvey J.W., J.T. Newlin (2003) Measuring Wetland microtopography and modeling its role in water flow in the Central Everglades. Draft Manuscript, June 6, 2003.

Choi J., Harvey J.W. (2014) Relative significance of microtopography and vegetation as controls on surface water flow on a low-gradient floodplain. *Wetlands 34*(1): 101–115.

Chow-Fraser P. (2006) Development of the water quality index (WQI) for assess effects of basin-wide land-use alteration on coastal marshes of the Laurentian Great Lakes. In: *Coastal Wetlands of the Laurentian Great Lakes: Health, Habitat and Indicators*, Simon T.P., Stewart P.M. (eds.) Authorhouse: Bloomington, IN, pp. 137–166.

Christensen J.R., Crumpton W.G., van der Valk A.G. (2009) Estimating the breakdown and accumulation of emergent macrophyte litter: A mass-balance approach. *Wetlands 29*(1): 204–214.

Christensen N., Mitsch W.J., Jørgensen S.E. (1994) A first generation ecosystem model of the Des Plaines River experimental wetlands. *Ecological Engineering 3*(4): 495–521.

Christiansen N.H., Andersen F.Ø., Jensen H.S. (2016) Phosphate uptake kinetics for four species of submerged freshwater macrophytes measured by a 33P phosphate radioisotope technique. *Aquatic Botany 128*: 58–67.

Christianson L., Tyndall J., Helmers M. (2013) Financial comparison of seven nitrate reduction strategies for Midwestern agricultural drainage. *Water Resources and Economics 2*: 30–56.

Christianson L.E., Bhandari A., Helmers M.J. (2012) A practice-oriented review of woodchip bioreactors for subsurface agricultural drainage. *Applied Engineering in Agriculture 28*(6): 861–874.

Chu J., Zhang J., Zhou X., Liu B., Li Y. (2015) A comparison of anammox bacterial abundance and community structures in three different emerged plants-related sediments. *Current Microbiology 71*(3): 421–427.

Chynoweth D.P. (1987) Biomass conversion options. In: *Aquatic Plants for Water Treatment and Resource Recovery*, Reddy K.R., Smith W.H. (eds.) Magnolia Publishing: Orlando, FL, pp. 621–642.

Clark G.M., Mueller D.K., Mast M.A. (2000) Nutrient concentrations and yields in undeveloped stream basins of the United States. *Journal of the American Water Resources Association 36*(4): 849–860.

Clark M.W., Reddy K.R. (1998) *Analysis of Floating and Emergent Vegetation Formation in Orange Lake*. Final Report to St. Johns River Water Management District, Palatka, FL.

Clark W.R. (2000) The ecology of Muskrats in Prairie Wetlands. In: *Prairie Wetland Ecology*, Murkin H.R., van der Valk A.G., Clark W.R. (eds.) Iowa State University Press: Ames, IA.

Clary J., Jones J., Leisenring M., Hobson P., Strecker E. (2017b) *International Stormwater BMP Database 2016 Summary Statistics*. The Water Environment & Reuse Foundation, Alexandria, VA. Available at: www.bmpdatabase.org/Docs/03-SW-1COh%20BMP%20Database%202016%20Summary%20Stats.pdf (Accessed 7/18/18).

Clary J., Jones J., Leisenring M., Strecker E., Koryto K. (2017a) *Agricultural Best Management Practices Database (Agbmpdb), Version 2.0 Data Summary*. The Water Environment & Reuse Foundation, Alexandria, VA. Available at: www.bmpdatabase.org/Docs/SIWM14T16_web.pdf (Accessed 8/23/18).

Claytor R., Schueler T.R. (1996) *Design of Stormwater Filtering Systems*. Center for Watershed Protection: Ellicott City, MD.

Clement A., Somylyodi L., Koncsos L. (1998) Modeling the phosphorus retention of the Kis-Balaton Upper Reservoir. *Water Science and Technology 37*(3): 113–120.

CNEH. (2010) *Final Report of the Assessment and Monitoring of the Tucush Wetland – Antamina*. Report by Centro Neotropical de Entrenamiento en Humedales - Peru to Compañia Minera Antamina S.A., Lima, Peru.

Colborn T., Dumanoski D., Myers J.P. (1997) *Our Stolen Future: Are We Threatening Our Fertility, Intelligence, and Survival?* Plume: New York.

Cole J.A., Brown C.M. (1980) Nitrite reduction to ammonia by fermentative bacteria: A short circuit in the biological nitrogen cycle. *FEMS Microbiology Letters 7*: 65–72.

Cole T.M., Wells S.A. (2016) *CE-QUAL-W2: A Two-Dimensional, Laterally Averaged, Hydrodynamic and Water Quality Model, User Manual, Version 4.0*. Waterways Experiment Station, Hydraulics Laboratory. US Army Corps of Engineers, Vicksburg MS.

Collins A.L., McGonigle D.F. (2008) Monitoring and modelling diffuse pollution from agriculture for policy support: UK and European experience. *Environmental Science and Policy 11*(2): 97–101.

Colmer T.D., Pedersen O. (2008) Underwater photosynthesis and respiration in leaves of submerged wetland plants: Gas films improve CO2 and O2 exchange. *New Phytologist 177*(4): 918–926.

Condie S.A., Webster I.T. (2001) Estimating stratification in shallow water bodies from mean meteorological conditions. *Journal of Hydraluic Engineering 127*(4): 286–292.

Connor M.A., Luczak A. (2002) Designing wetland treatment systems that contribute to wildlife conservation. Mbwette T.S.A. (ed.), Proceedings of the 8th International Conference on Wetland Systems for Water Pollution Control, 16–19 September 2002. IWA Publishing and University of Dar Es Salaam: Arusha, Tanzania, pp. 1024–1037.

Conrow R., Lowe E.F., Coveney M.F., Rauschenberger R.H., Masson G. (2011) Restoration of Lake Apopka's north shore marsh: High hopes, tough times, and persistent progress. In: *Wildlife Ecotoxicology*, J.E. Elliott, et al. (ed.) Springer: New York, pp. 189–211.

Cooke J.G. (1994) Nutrient transformations in a natural wetland receiving sewage effluent and the implications for waste treatment. *Water Science and Technology 29*: 209–217.

Cookson J.T. Jr. (1995) *Bioremediation Engineering: Design and Application*. McGraw Hill: New York.

Cooper A.B. (1992) Coupling wetland treatment to land treatment: An innovative method for nitrogen stripping. Proceedings of the 3rd International Conference on Wetland Systems for Water Pollution Control, 30 November to 3 December 1992; Australian Water and Wastewater Association and IWA: Sydney, Australia, p. 37.1–37.9.

Cooper P.F., Job J.D., Green B., Shutes R.B.E. (1996) *Reed Beds and Constructed Wetlands for Wastewater Treatment*. WRc Publications. 184 plus data diskette: Swindon, United Kingdom.

Cornelis C., Bierkens J., Standaert A. (2017a) *S-Risk substance data sheets–Part 2: BTEX, styrene and trimethylbenzenes.* Available at: https://s-risk.be/sites/s-risk.be/files/Substance%20data%20sheet%20BTEXS.pdf (Accessed 10/8/18).

Cornelis C., Bierkens J., Standaert A. (2017b) *S-Risk Substance Data Sheets–Part 4: Polycyclic Aromatic Hydrocarbons (PAHs)*. Available at: https://s-risk.be/sites/s-risk.be/files/Substance%20data%20sheet%20PAH.pdf (Accessed 10/8/18).

Cornelis C., Bierkens J., Standaert A. (2017c) *S-Risk substance data sheets–Part 3: Chloroaliphatic Substances, Chlorobenzenes and Chlorophenols.* Available at: https://s-risk.be/sites/s-risk.be/files/Substance%20data%20sheet%20chlorinated.pdf (Accessed 10/8/18).

Corrales J., Naja G.M., Dziuba C., Rivero R.G., Orem W. (2011) Sulfate threshold target to control methylmercury levels in wetland ecosystems. *Science of the Total Environment 409*(11): 2156–2162.

Corstanje R., Grunwald S., Reddy K. R., Osborne T. Z., Newman S. (2006a) Assessment of the spatial distribution of soil properties in a Northern Everglades Marsh. *Journal of Environmental Quality 35*: 938–949.

Corstanje R., Reddy K.R. (2004) Response of biogeochemical indicators to a drawdown and subsequent reflood. *Journal of Environmental Quality 33*(6): 2357–2366.

Corstanje R., Reddy K.R., Portier K.M. (2006b) *Typha latifolia* and *Cladium jamaicense* litter decay in response to exogenous nutrient enrichment. *Aquatic Botany 84*(1): 70–78.

Coveney M.F., Stites D.L., Lowe E.F., Battoe L.E., Conrow R. (2002) Nutrient removal from eutrophic lake water by wetland filtration. *Ecological Engineering 19*(2): 141–159.

Cowardin L.M., Carter V., Golet F.C., LaRoe E.T. (1979) *Classification of Wetlands and Deepwater Habitats of the United States.* US Fish and Wildlife Service, Office of Biological Services: Washington, DC. Available at: web.archive.org/web/20140121073314/www.fws.gov/wetlands/Documents/classwet/index.html (Accessed 2/2/19).

Cox C.D., Payne W.J. (1973) Separation of soluble denitrifying enzymes and cytochromes from *Pseudamonas perfectomarinus. Canadian Journal of Microbiology 19*: 861–872.

Cracraft J., Barker F.K., Braun M., Harshman J., Dyke G.J., Feinstein J., Stanley S., Cibois A., Schikle P., Beresford P., García-Moreno J. (2004) Phylogenetic relationships among modern birds (Neornithes): Towards an avian tree of life. In: Cracraft J., Donoghue M.J. (eds.), *Assembling the Tree of Life.* Oxford University Press: New York, pp. 468–489.

Craft C.B. (2016) *Creating and Restoring Wetlands.* Elsevier: Amsterdam, The Netherlands.

Craft C.B., Richardson C.J. (1993) Peat accretion and phosphorous accumulation along a eutrophication gradient in the Northern Everglades of Florida, United States. *Biogeochemistry 22*: 133–156.

Craft C.B., Richardson C.J. (1998) Recent and long-term organic soil accretion and nutrient accumulation in the Everglades. *Soil Science Society of America Journal 62*(3): 834–843.

Craggs R.J., Zwart A., Nagels J.W., Davies-Colley R.J. (2004) Modeling sunlight disinfection in a high rate pond. *Ecological Engineering 22*(2): 113–122.

Crepeau K.L., Miller R.L. (2014) Water temperature differences by plant community and location in re-established wetlands in the Sacramento–San Joaquin Delta, California, July 2005 to February 2008. U.S. Geological Survey Data Series 882. Available at: http://dx.doi.org/10.3133/ds882 (Accessed 3/22/18).

Crites R., Dombeck G.D., Watson R.C., Williams C.R. (1997) Removal of metals and ammonia in constructed wetlands. *Water Environment Research 69*(2): 132–135.

Crites R., Tchobanoglous G. (1998) *Small and Decentralized Wastewater Management Systems.* McGraw-Hill: New York.

Crites R., Watson R.C., Williams C.R. (1995) Removal of metals in constructed wetlands. In: *Proceedings of the 68th Annual WEFTEC Conference in Miami, Florida.* Water Environment Federation: Alexandria, VA.

Crites R.W., Middlebrooks E.J., Bastian R.K. (2014). *Natural Wastewater Treatment Systems*, 2nd edition. CRC/Taylor Francis: Boca Raton, FL.

Crites R.W., Middlebrooks E.J., Reed S.C. (2006) *Natural Wastewater Treatment Systems*. CRC/Taylor Francis: Boca Raton, FL.

Crohn D.M., Ruud N.C., Decruyenaere J.G., Carlon D.B. (2005) Goodness-of-fit test for modeling tracer breakthrough curves in wetlands. *Journal of Environmental Engineering 131*(2): 242–251.

Cronk J.K., Fennessy M.S. (2001) *Wetland Plants: Biology and Ecology*. Lewis Publishers: Boca Raton, FL.

Crumpton W., Stenback G. (2017) *2016 Annual Report on Performance of Iowa CREP Wetlands: Monitoring and Evaluation of Wetland Performance*. Report to Iowa Department of Agriculture and Land Stewardship, Des Moines, IA. Available at: www.iowacrep.org/files/page/files/crep_wetland_monitoring_and_evaluation_2016_final_report.pdf (Accessed 8/14/18).

Crumpton W., van der Valk A., Hoyer W., Osterberg D. (2012a) *Wetland Restoration in Iowa*. Report of the Iowa Policy Project, Iowa City, IA. Available at: www.iowapolicyproject.org/2012docs/120510-wetlands.pdf (Accessed 1/1/19).

Crumpton W.G., Helmers M.J., Stenback G.A., Lemke D.W., Richmond S. (2012b) *Integrated Drainage-Wetland Systems for Reducing Nitrate Loads from Des Moines Lobe Watersheds*. Agricultural and Biosystems Engineering Technical Reports and White Papers 13, Iowa State University, Ames, IA. Available at: https://lib.dr.iastate.edu/abe_eng_reports/ (Accessed 7/27/18).

Crumpton W.G., Phipps R. (1992) Fate of nitrogen loads in experimental wetlands. In: *Volume III, in the Des Plaines River Wetlands Demonstration Projects*. Report by Wetlands Research, Inc., Chicago, IL.

Crusius J., Wanninkhof R. (2003) Gas transfer velocities measured at low wind speed over a lake. *Limnology and Oceanography 48*(3): 1010–1017.

Cueto A.J. (1993) Development of criteria for the design and construction of engineered aquatic treatment units in Texas. In: *Constructed Wetlands for Water Quality Improvement*, Moshiri G.A. (ed.). Lewis Publishers: Boca Raton, FL, pp. 99–105.

Currie S.J., VanZomeren C.M., Berkowitz J.F. (2017) *Utilizing Wetlands for Phosphorus Reduction in Great Lakes Watersheds: A Review of Available Literature Examining Soil Properties and Phosphorus Removal Efficiency U.S. Army Corps of Engineers Report No. ERDC/EL SR-17-4)*. ERDC-EL Vicksburg United States.

CWC. (1983) *Draft Design Memorandum, Incline Village General Improvement District Wetlands*. Culp, Wesner, & Culp (CWC): Robin Lake, Cameron Park, CA.

Daane L.L., Harjono I., Zylstra G.J., Häggblom M.M. (2001) Isolation and characterization of polycyclic aromatic hydrocarbon-degrading bacteria associated with the rhizosphere of salt marsh plants. *Applied and Environmental Microbiology 67*(6): 2683–2691.

Dahl T.E. (1990) *Wetland Losses in the United States: 1780s to 1980s*. U.S. Department of the Interior: Washington, DC.

Dai Y.N., Yang Y., Tam N.F.Y., Tai Y.P., Tang X.Y. (2016) Factors affecting behavior of phenolic endocrine disruptors, estrone and estradiol, in constructed wetlands for domestic sewage treatment. *Environmental Science and Technology 50*(21): 11844–11852.

Daigger G.T. (2014) Oxygen and carbon requirements for biological nitrogen removal processes accomplishing nitrification, nitritation, and anammox. *Water Environment Research 86*(3): 204–209.

Daly C., Widrlechner M.P., Halbleib M.D., Smith J.I., Gibson W.P. (2012) Development of a new USDA plant hardiness zone map for the United States. *Journal of Applied Meteorology and Climatology 51*(2): 242–264.

Dan T.H., Brix H. (2009) Growth responses of the perennial legume Sesbania sesban to NH4 and NO3 nutrition and effects on root nodulation. *Aquatic Botany 91*(3): 238–244.

Danckwerts P.V. (1953) Continuous flow systems: Distribution of residence times. *Chemical Engineering Science 2*(1): 1–13.

Daniels R.E., Parr T.W. (1990) *1989 Survey of Reed Growth in UK Reed Bed Treatment Systems*, WRc/NERC Contract Report T02058F1, The Institute of Terrestrial Ecology: Dorset, United Kingdom.

Danish Hydraulic Institute. (2011) *MIKE SHE User Manual, Volume 1: User Guide*. DHI Water and Environment, Danish Hydraulic Institute: Hørsholm, Denmark.

Dannenberg A., Pehkonen S.O. (1998) Investigation of the heterogeneously catalyzed hydrolysis of organophosphorus pesticides. *Journal of Agricultural and Food Chemistry 46*(1): 325–334.

Darby M.A. (1995) *Modeling the Fate of Chlorpyrifos in Constructed Wetlands*. MS Thesis, The University of Mississippi.

Darcy H. (1856) *Les Fontaines Publiques De La Ville De Dijon*. Dalmont: Paris.

Darley W.M. (1982) *Algal Biology: A Physiological Approach.* Blackwell Scientific: Oxford, United Kingdom.

Darwiche-Criado N., Comín F.A., Masip A., García M., Eismann S.G., Sorando R. (2017) Effects of wetland restoration on nitrate removal in an irrigated agricultural area: The role of in-stream and off-stream wetlands. *Ecological Engineering 103*: 426–435.

Das N., Chandran P. (2011) Microbial degradation of petroleum hydrocarbon contaminants: An overview. *Biotechnology Research International* Vol.2011: 1–13.

Davidsson T.E., Stahl M. (2000) The influence of organic carbon on nitrogen transformation in five wetland soils. *Soil Science Society of America Journal 64*(3): 1129–1136.

Davies-Colley R.J. (2005) Pond disinfection. In: *Pond Treatment Technology*, Shilton A. (ed.) IWA Publishing: London, United Kingdom, pp. 100–136.

Davies-Colley R.J., Craggs R.J., Nagels J.W. (2003) Disinfection in a pilot-scale "advanced" pond system (APS) for domestic sewage treatment in New Zealand. *Water Science and Technology 48*(2): 81–87.

Davies-Colley R.J., Donnison A.M., Speed D.J., Ross C.M., Nagels J.W. (1999) Inactivation of faecal indicator organisms in waste stabilization ponds. *Water Research 33*(5): 1220–1230.

Davis C.B., van der Valk A. (1978) The decomposition of standing and fallen litter of *Typha glauca* and. *Scirpus Fluviatilis. Canadian Journal of Botany 56*(6): 662–675.

Davis C.B., van der Valk A. (1983) Uptake and release of nutrients by living and decomposing Typha glauca Godr. tissues at Eagle Lake, Iowa. *Aquatic Botany 16*(1): 75–89.

Davis D.G., Montgomery J.C. (1987) EPA's regulatory and policy considerations on wetlands and municipal watewater treatment. In: *Aquatic Plants for Water Treatment and Recovery*, Reddy K.R., Smith W.H. (eds.) Magnolia Publishing: Orlando, FL.

Davis M.A., Palumbo M.G. (2008) The price of residential land in large US cities. *Journal of Urban Economics 63*(1): 352–384. Data located at Land and Property Values in the U.S., Lincoln Institute of Land Policy. Availble at: https://datatoolkits.lincolninst.edu/subcenters/land-values/metro-area-land-prices.asp (Accessed 12/27/18).

Davis S.M. (1989) Sawgrass and cattail production in relation to nutrient supply in the Everglades. In: Sharitz R.R., Gibbons J.W. (eds.), *U.S. Department of Energy Conference No. 8603101, Held in Charleston, South Carolina*. National Technical Information Service: Springfield, VA, pp. 325–341.

Davis S.M. (1994) Phosphorus inputs and vegetation sensitivity in the Everglades. In: *Everglades: The Ecosystem and Its Restoration*, Davis S.M., Ogden J.C. (eds.) St. Lucie Press: Delray Beach, FL, pp. 357–378.

Dawson Q., Kechavarzi C., Leeds-Harrison P.B., Burton R.G.O. (2010) Subsidence and degradation of agricultural peatlands in the Fenlands of Norfolk, UK. *Geoderma 154*(3–4): 181–187.

DBE. (1999) *A Demonstration of Submerged Aquatic Vegetation/Limerock Treatment System Technology for Removal of Phosphorus from Everglades Agricultural Area Water: Final Report*. Prepared for the South Florida Water Management District (SFWMD) and the Florida Department of Environmental Protection (FDEP). Contract No. C-E10660, DB Environmental (DBE).

DBE. (2002) *Submerged Aquatic Vegetation/Limerock Treatment System Technology: Follow-On Study.* Final Report, Prepared for the SFWMD and FDEP, Contract No. C-E10660, DB Environmental (DBE).

DBE. (2003) *Assessment of Hydraulic and Ecological Factors Influencing Phosphorus Removal in Stormwater Treatment Area 1 West*, Final report to Florida Department of Environmental Protection, Contract No. WM795, April 2003, DB Environmental (DBE).

DBE. (2009) *Tracer Study: PSTA Cell, STA-3/4*. Report to South Florida Water Management District, West Palm Beach, FL, by DB Environmental, Inc., Rockledge, FL.

DBE (DB Environmental, Inc.) (2017) *Field and Supplemental Laboratory Support for Evaluating Phosphorus Sources, Flux, and Transformation Processes in the Stormwater Treatment Areas.* Annual report to South Florida Water Management District.

de Langre E. (2008) Effects of wind on plants. *Annual Review of Fluid Mechanics 40*: 141–168.

de Winton M, Jones H, Edwards T, Özkundakci D, Wells R, McBride C, Rowe D, Hamilton D, Clayton J, Champion P, Hofstra D. (2013). *Review of Best Management Practices for Aquatic Vegetation Control in Stormwater Ponds, Wetlands, and Lakes.* Prepared by NIWA and the University of Waikato for Auckland Council. Auckland Council technical report, TR2013/026.

Deacon G., Ronan M., Saunders M., Miller G., Burcher J., Kennedy A., Moore S. (2011) *Queensland Wetland Buffer Planning Guideline.* Department of Environment and Resource Management, Queensland Wetlands Program: Brisbane, Queensland, Australia.

Dean R.G., Dalrymple R.A. (1991) *Water Wave Mechanics for Engineers and Scientists*. World Scientific: Singapore, pp. 157–158.

DeBusk T.A., Grace K.A., Dierberg F.E., Jackson S.D., Chimney M.J., Gu B. (2004) An investigation of the limits of phosphorus removal in wetlands: A mesocosm study of a shallow periphyton-dominated treatment system. *Ecological Engineering 23*(1): 1–14.

DeBusk T.A., Laughlin R.B., Schwartz L.N. (1996) Retention and compartmentalization of lead and cadmium in wetland microcosms. *Water Research 30*(11): 2707–2716.

DeBusk T.A., Reddy K.R. (1987) Wastewater treatment using floating aquatic macrophytes: Contaminant removal processes and management strategies. In: *Aquatic Plants for Water Treatment and Resource Recovery*, Reddy K.R., Smith W.H. (eds.). Magnolia Publishing: Orlando, FL, pp. 643–656.

DeBusk W.F., Newman S., Reddy K.R. (2001) Spatio-temporal patterns of soil phosphorus enrichment in Everglades Water Conservation Area 2A. *Journal of Environmental Quality 30*(4): 1438–1446.

DeBusk W.F., Reddy K.R. (1998) Turnover of detrital organic carbon in a nutrient-impacted Everglades marsh. *Soil Science Society of America Journal 62*(5): 1460–1468.

Decamp O., Warren A. (1998) Bactivory in ciliates isolated from constructed wetlands (reed beds) used for wastewater treatment. *Water Research 32*(7): 1989–1996.

DeGaetano A.T., Eggleston K.L., Knapp W.W. (1994) *Daily Evapotranspiration and Soil Moisture Estimates for the Northeastern United States*. Publication No. RR 94-1, Northeast Regional Climate Center, Cornell University, Ithaca, NY.

Degenhardt D., Cessna A.J., Raina R., Farenhorst A., Pennock D.J. (2011) Dissipation of six acid herbicides in water and sediment of two Canadian prairie wetlands. *Environmental Toxicology and Chemistry 30*(9): 1982–1989.

Deitsch J.J., Smith J.A., Culver T.B. Brown R.A., Riddle S.A. (2000) "Distributed-rate model analysis of 1,2-dichlorobenzene batch sorption and desorption rates for five natural sorbents. *Environmental Science and Technology 34*(8): 1469–1476.

DeLaune R., Patrick W.H., Casselman M.E. (1981) Effect of sediment pH and redox conditions on degradation of benzo(a)pyrene. *Marine Pollution Bulletin 12*: 251–253.

DeLaune R.D., Kongchum M., White J.R., Jugsujinda A. (2013a) Freshwater diversions as an ecosystem management tool for maintaining soil organic matter accretion in coastal marshes. *Catena 107*: 139–144.

DeLaune R.D., Reddy K.R., Richardson C.J., Megonigal J.P. (eds.) (2013b) *Methods in Biogeochemistry of Wetlands*. Soil Science Society of America: Madison, WI

Deletic A. (1999) Sediment behaviour in grass filter strips. *Water Science and Technology 39*(9): 129–136.

Delgado M., Biggeriego M., Guardiola E. (1993) Uptake of Zn, Cr, and Cd by water hyacinths. *Water Research 27*(2): 269–272.

DeLorenzo M.E., Thompson B., Cooper E., Moore J., Fulton M.H. (2012) A long-term monitoring study of chlorophyll, microbial contaminants, and pesticides in a coastal residential stormwater pond and its adjacent tidal creek. *Environmental Monitoring and Assessment 184*(1): 343–359.

Dengler J. (2017) *Phytosociology*. In: *The International Encyclopedia of Geography*, Richardson D., Castree D.N., Goodchild M.F., Kobayashi A., Liu W., Marston R.A. (eds.) Wiley-Blackwell: Chichester, United Kingdom. Available at: www.researchgate.net/publication/260986799_Phytosociology (Accessed 10/18/18).

Department of Planning and Local Government. (2010) *Modelling Process and Tools*. Chapter 15 in: Water Sensitive Urban Design Technical Manual for the Greater Adelaide Region. Government of South Australia: Adelaide. Available at: www.sa.gov.au/__data/assets/pdf_file/0020/15059/WSUD_chapter_15.pdf (Accessed 7/25/18).

Dessu S.B., Price R.M., Scinto L.J., Thomas S., Fugate D.C., Perkey D.W., Smith S.J. (2017). Settling and Entrainment Properties of of STA Particulates. Presentation to Phosphorus Flux Workshop; SFWMD, West Palm Beach, FL, Nov. 3, 2017.

Dettenmaier E. M., Doucette W. J.; Bugbee B. (2009) Chemical hydrophobicity and uptake by plant roots. *Environmental Science and Technology 43*: 324–329.

Devai I., Delaune R.D. (1995) Evidence for phosphine production and emission from Louisiana and Florida marsh soils. *Organic Geochemistry 23*(3): 277–279.

DeVolder P.S., Brown S.L., Hesterberg D., Pandya K. (2003) Metal bioavailability and speciation in a wetland tailings repository amended with biosolids compost, wood ash, and sulfate. *Journal of Environmental Quality 32*(3): 851–864.

DHI. (2017a) *MIKE SHE Volume 2: Reference Guide.* Danish Hydraulic Institute, Hørsholm, Denmark. Available at: http://manuals.mikepoweredbydhi.help/2017/Water_Resources/MIKE_SHE_Printed_V2.pdf (Accessed 5/17/18).

DHI. (2017b) *MIKE 21 & MIKE3 Flow Model, Advection-Dispersion Model, Scientific Documentation.* Danish Hydraulic Institute, Hørsholm, Denmark: Available at: http://manuals.mikepoweredbydhi.help/2017/Coast_and_Sea/M3M21AD_SCD.pdf (Accessed 7/15/18).

Dhondt K., Boeckx P., van Cleemput O., Hofman G. (2003) Quantifying nitrate retention processes in a riparian buffer zone using the natural abundance of ^{15}N in NO_3^-. *Rapid Communications in Mass Spectrometry 17*: 2597–2604.

Di Luca G.A., Maine M.A., Mufarrege M.M., Hadad H.R., Sánchez G.C., Bonetto C.A. (2011) Metal retention and distribution in the sediment of a constructed wetland for industrial wastewater treatment. *Ecological Engineering 37*(9): 1267–1275.

Díaz F.J., Anthony T.O., Dahlgren R.A. (2012) Agricultural pollutant removal by constructed wetlands: Implications for water management and design. *Agricultural Water Management 104*: 171–183.

Díaz F.J., O'Geen A.T., Dahlgren R.A. (2010) Efficacy of constructed wetlands for removal of bacterial contamination from agricultural return flows. *Agricultural Water Management 97*(11): 1813–1821.

Diaz R.J., Solow A. (1999) *Ecological and Economic Consequences of Hypoxia*, Topic 2 Report for the Integrated Assessment of Hypoxia in the Gulf of Mexico, NOAA Coastal Ocean Program: Washington, DC.

Dickerman J.A., Wetzel R.G. (1985) Clonal growth in Typha latifolia: Population dynamics and demography of the ramets. *The Journal of Ecology 73*: 535–552.

Dierberg F.E., DeBusk T.A., Henry J.L., Jackson S.D., Galloway S., Gabriel M.C. (2012) Temporal and spatial patterns of internal phosphorus recycling in a South Florida (USA) stormwater treatment area. *Journal of Environmental Quality 41*: 1661–1673.

Dierberg F.E., DeBusk T.A., Jackson S.D., Chimney M.J., Pietro K. (2002) Submerged aquatic vegetation-based treatment wetlands for removing phosphorous from agricultural runoff: Response to hydraulic and nutrient loading. *Water Research 36*(6): 1409–1422.

Dierberg F.E., Juston J., DeBusk T.A., Pietro K., Gu B. (2005) Relationship between hydraulic efficiency and phosphorus removal in a submerged aquatic vegetation dominated treatment wetland. *Ecological Engineering 25*(1): 9–23.

Dieter C.D. (1990) The importance of emergent vegetation in reducing sediment resuspension in wetlands. *Journal of Freshwater Ecology 5*(4): 467–473.

Dietrich W.E. (1982) Settling velocity of natural particles. *Water Resources Research 18*(6): 1615–1626.

Dingman S.L., 2015. *Physical Hydrology.* Waveland Press: Long Grove, IL.

Dissanayaka S.H., Hamamoto S., Komatsu T., Kawamoto K. (2013) *Thermal Properties for Peaty Materials under Variable Saturation and Their Correlation to Mass Transport Parameters in Gaseous and Aqueous Phases.* 39, Research Report of Department of Civil and Environmental Engineering, Saitama University, Saitama City, Japan, pp. 21–32.

Dodd J.D., Webb J.W. (1975) *Establishment of Vegetation for Shoreline Stabilization in Galveston Bay*, Paper 6-75, U.S. Army Coastal Engineering Center: Fort Belvoir, VA.

Doig L.E., Liber L. (2006) Nickel partitioning in formulated and natural freshwater sediments. *Chemosphere 62*: 968–979.

Dolan T.J., Hermann A.J., Bayley S.E., Zoltek J. (1984) Evapotranspiration of a Florida, USA, freshwater wetland. *Journal of Hydrology 74*(3–4): 355–371.

Dombeck G., Williams C., Crites R. (1998a) Hydraulics in constructed wetlands. Proceedings of the Annual Water Environment Federation Meeting; WEF, Alexandria, VA: 297–307.

Dombeck G.D., Perry M.W., Phinney J.T. (1998b) Mass balance on water column trace metals in a free surface flow constructed wetland in Sacramento, California. *Ecological Engineering 10*(4): 313–339.

Dong K., Lin C. (1994b) Treatment of petrochemical wastewater by the system of wetlands and oxidation ponds and engineering design of the system. Jiang C. (ed.) Proceedings of the 4th International Conference on Wetland Systems for Water Pollution Control, 6–10 November 1994; IWA, Guangzhou, PR China: 628–633.

Donoso N., Gobeyn S., Boets P., Goethals P.L., Wilde D.D., Meers E. (2017) Assessing the integration of wetlands along small European waterways to address diffuse nitrate pollution. *Water 9*(6): p.369.

Dorman L., Castle J.W., Rodgers Jr J.H. (2009) Performance of a pilot-scale constructed wetland system for treating simulated ash basin water. *Chemosphere 75*(7): 939–947.

Dotro G., Langergraber G., Molle P., Nivala J., Puigagut J., Stein O., Von Sperling M. (2017) *Treatment Wetlands.* IWA Publishing: London, United Kingdom.

Douglas E.M., Jacobs J.M., Sumner D.M., Ray R.L. (2009) A comparison of models for estimating potential evapotranspiration for Florida land cover types. *Journal of Hydrology 373*(3–4): 366–376.

Downing-Kunz M.A., Stacey M.T. (2012) Observations of mean and turbulent flow structure in a free-floating macrophyte root canopy. *Limnology and Oceanography: Fluids and Environments 2*(1): 67–79.

Drake C.W., Jones C.S., Schilling K.E., Amado A.A., Weber L.J. (2018) Estimating nitrate-nitrogen retention in a large constructed wetland using high-frequency, continuous monitoring and hydrologic modeling. *Ecological Engineering 117*: 69–83.

Drapcho C.M., Brune D.E. (2000) The partitioned aquaculture system: Impact of design and environmental parameters on algal productivity and photosynthetic oxygen production. *Aquacultural Engineering 21*(3): 151–168.

Dressler R.L., Hall D.W., Perkins K.D., Williams N.H. (1987) *Identification Manual for Wetland Plant Species of Florida.* Institute of Food and Agricultural Services, University of Florida: Gainesville, FL.

Drexler J.Z., Anderson F.E., Snyder R.L. (2008) Evapotranspiration rates and crop coefficients for a restored marsh in the Sacramento–San Joaquin Delta, California, USA. *Hydrological Processes 22*(6): 725–735.

Dronova I., Taddeo S. (2016). Canopy leaf area index in non-forested marshes of the California Delta. *Wetlands 36*(4): 705–716.

Droppo I.G. (2001) Rethinking what constitutes suspended sediment. *Hydrological Processes 15*: 1551–1564.

Duda P.J. II (1992) *Chevron's Richmond Refinery Water Enhancement Wetland.* Report to the California Regional Water Quality Control Board: Oakland, CA.

Duffield J.M. (1986) Waterbird use of urban stormwater wetland system in central California, USA. *Colonial Wetlands 9*(2): 227–235.

Duguma D., Walton W.E. (2013) Effects of nutrients on mosquitoes and an emergent macrophyte, Schoenoplectus maritimus, for use in treatment wetlands. *Journal of Vector Ecology 39*(1): 1–13.

Duncan B., Pratt K., Kerr H. (1999) *SaskPower Constructed Wetlands, 1998 Operational Report.* Report to Saskatchewan Water Corporation and Saskatchewan Environment and Resource Management, Moosejaw, Saskatchewan, Canada.

Duncan H.P. (1998) Urban stormwater quality improvement. Walker D.J., Daniell T.M. (eds.) Proceedings of Hydrastorm '98: Joint 3rd International Symposium on Stormwater Management and 6th International Conference on Hydraulics in Civil Engineering, held in Adelaide, Australia, 27–30 September 1998; 203–208.

Duncan W.F.A., Mattes A.G., Gould W.D., Goodazi F. (2004) Multi-stage biological treatment system for removal of heavy metal contaminants. In: *Waste Processing and Recycling in Mineral and Metallurgical Industries V,* Rao S.R., Harrison F.W., Kosinski J.A., Amaratunga L.M., Cheng T. C., Richards G.G. (eds.) Metallurgical Society of the Canadian Institute of Mining, Metallurgy, and Petroleum: Montreal, Quebec, Canada, pp. 469–484.

Dunne E.J., Coveney M.F., Hoge V.R., Conrow R., Naleway R., Lowe E.F., Battoe L.E., Wang Y. (2015) Phosphorus removal performance of a large-scale constructed treatment wetland receiving eutrophic lake water. *Ecological Engineering 79*: 132–142.

Dunne E.J., Coveney M.F., Marzolf E.R., Hoge V.R., Conrow R., Naleway R., Lowe E.F., Battoe L.E. (2012) Efficacy of a large-scale constructed wetland to remove phosphorus and suspended solids from Lake Apopka, Florida. *Ecological Engineering 42*: 90–100.

Dunne E.J., Coveney M.F., Marzolf E.R., Hoge V.R., Conrow R., Naleway R., Lowe E.F., Battoe L.E., Inglett P.W. (2013) Nitrogen dynamics of a large-scale constructed wetland used to remove excess nitrogen from eutrophic lake water. *Ecological Engineering 61*: 224–234.

Ebbs S.D., Bushey J., Poston S., Kosma D., Samiotakis M., Dzombak D. (2003) Transport and metabolism of free cyanide and iron cyanide complexes by willow. *Plant, Cell, and Environment 26*: 1467–1478.

Ebele A.J., Abdallah M.A.E., Harrad S. (2017) Pharmaceuticals and personal care products (PPCPs) in the freshwater aquatic environment. *Emerging Contaminants 3*(1): 1–16.

Eckhardt D.A.V., Surface J.M., Peverly J.H. (1999) A constructed wetland system for treatment of landfill leachate. In: *Constructed Wetlands for the Treatment of Landfill Leachate*, Mulamoottil G., McBean E., Rovers F.A. (eds.) CRC Press: Boca Raton, FL.

Ecological Services Group. (1996) *A Proposed Monitoring Protocol for Constructed Wetlands*, Report prepared for the Canadian Wildlife Service and Environment Canada, Ecological Services for Planning Ltd.: Guelph, ON.

Edinger J.E. (1970) Vertical temperature structure and water surface heat exchange. *Water Resources Research 6*(5): 1392–1395

Edinger J.E., Duttweiler D.W., Geyer J.C. (1968) The response of water temperatures to meteorological conditions. *Water Resources Research 4*(5): 1137–1143.

Eger P., Lapakko K., Otterson P. (1980) Trace metal uptake by peat interaction of a white cedar bog and mining stockpile leachate. Proceedings of the 6th National Peat Congress; Duluth, Minnesota, 542–547.

Eggers S.D., Reed D.M. (1997) *Wetland Plants and Plant Communities of Minnesota and Wisconsin*. Second edition, U.S. Army Corps of Engineers: St. Paul, MN.

Eggert D. A., Rodgers Jr., J. H., Huddleston G. M., Hensman C. E. (2008) Performance of pilot-scale constructed wetland treatment systems for flue gas desulfurization waters. *Environmental Geosciences 15*(3): 115–129.

Eid E.M., Shaltout K.H. (2017) Growth dynamics of water hyacinth (Eichhornia crassipes): A modeling approach. *Rendiconti Lincei 28*(1): 169–181.

Eid E.M., Shaltout K.H., Asaeda T. (2012a) Modeling growth dynamics of Typha domingensis (Pers.) Poir. ex Steud. in Lake Burullus, Egypt. *Ecological Modelling, 243*: 63–72.

Eid E.M., Shaltout K.H., El-Sheikh M.A., Asaeda T. (2012b) Seasonal courses of nutrients and heavy metals in water, sediment and above-and below-ground Typha domingensis biomass in Lake Burullus (Egypt): Perspectives for phytoremediation. *Flora-Morphology, Distribution, Functional Ecology of Plants 207*(11): 783–794.

Eidson G.W., Flite O.P., Garman A.M. (2005) *Constructed Wetlands Optimization Study*, Final Report to Augusta Utilities Department: Augusta, Georgia.

El-Gendy A.S. (2006) Modeling of heavy metals removal from municipal landfill leachate using living biomass of water hyacinth. Dias V., Vymazal J. (eds.) Proceedings of the 10th International Conference on Wetland Systems for Water Pollution Control, 23–29 September 2006; Ministério de Ambiente, do Ordenamento do Territóri e do Desenvolvimento Regional (MAOTDR) and IWA: Lisbon, Portugal, 1473–1482.

Elimelech M., Gregory J., Jia X. (2013) *Particle Deposition and Aggregation: Measurement, Modelling and Simulation*. Butterworth-Heinemann, Oxford, Uinted Kingdom.

Elmore H.L., Hayes T.W. (1960) Solubility of atmospheric oxygen in water: 29th Progress Report of the Committee on Sanitary Engineering Research. *Journal of the Sanitary Engineering Division (ASCE) 86*(SA4): 41–53.

Elsaesser D., Blankenberg A.G.B., Geist A., Mæhlum T., Schulz R. (2011) Assessing the influence of vegetation on reduction of pesticide concentration in experimental surface flow constructed wetlands: Application of the toxic units approach. *Ecological Engineering 37*(6): 955–962.

Emerick R.W., Loge F.L., Ginn T., Darby J.L. (2000) Modeling the inactivation of particle-associated coliform bacteria. *Water Environment Research 72*(4): 432–438.

Ergun S., 1952. Fluid flow through packed columns. *Chemical Engineering Progress 48*(2): 89–94.

Erickson A.J., Gulliver J.S., Kang J-H., Weiss P.T., Wilson C.B. (2010) Maintenance of stormwater treatment practices. *Journal of Contemporary Water Research and Education 146*: 75–82.

Eriksson P.G. (2001) Interaction effects of flow velocity and oxygen metabolism on nitrification and denitrification in biofilms on submersed macrophytes. *Biogeochemistry 55*: 29–44.

Eriksson P.G., Weisner S.E.B. (1997) Nitrogen removal in a wastewater reservoir: The importance of denitrification by epiphytic biofilms on submersed vegetation. *Journal of Environmental Quality 26*(3): 905–910.

Erlandsen S.L. (1995) Biotic transmission - Is giardiasis a zoonosis? In: *Giardia: From Molecules to Disease*, Thompson R., Reynoldsen J., Lymbery A. (eds.) CAB International: Wallinford, United Kingdom, pp. 83–97.

Erler D.V., Eyre B.D., Davison L. (2008) The contribution of anammox and denitrification to sediment N2 production in a surface flow constructed wetland. *Environmental Science and Technology 42*(24): 9144–9150.

Esteve-Nunez A., Caballero A., Ramos J.L. (2001) Biological degradation of 2,4,6-trinitrotoluene. *Microbiology and Molecular Biology Review 65*(3): 335–352.

Eurostat. (2017) *National and Regional Prices of Arable Land, 2016*. Report of the statistical office of the European Union: Luxembourg. Available at: http://appsso.eurostat.ec.europa.eu/nui/show.do?dataset=apri_lrnt&lang=en (Accessed 12/29/18).

Evans N., Trexler J. (2017) Quantification ot he fish and macroinvertebrate community and its effect on water column phosphorus in the STAs. Presentation to Science Plan Technical Representatives; West Palm Beach, Florida, November 2017.

Ewel K.C., Odum H.T. (1984) *Cypress Swamps*. University of Florida Press: Gainesville, FL.

Farnsworth R.K., Thompson E.S. (1982) *Mean Monthly, Seasonal, and Annual Pan Evaporation for the United States*. NOAA Technical Report NWS34, US Department of Commerce, National Oceanic and Atmospheric Administration: Washington, DC.

Faulkner S.P., Richardson C.J. (1989) Physical and chemical characteristics of freshwater wetlands soils. In: *Constructed Wetlands for Wastewater Treatment; Municipal, Industrial, and Agricultural*, Hammer D.A. (ed.) Lewis Publishers: Boca Raton, FL, pp. 41–72.

FCCMC (1998) *Florida Mosquito Control: The State of the Mission as Defined by Mosquito Controllers, Regulators, and Environmental Managers*, Florida Coordinating Council on Mosquito Control (FCCMC): Vero Beach, FL.

Feng S., Ai Z., Zheng S., Gu B., Li Y. (2014) Effects of dryout and inflow water quality on mercury methylation in a constructed wetland. *Water, Air, and Soil Pollution, 225*(4): 1929.

Fennessy M.S., Brueske C.C., Mitsch W.J. (1992) Sediment deposition patterns in restored freshwater wetlands using sediment traps. *Ecological Engineering 3*(4): 409–428.

Fennessy S., Gernes M., Wardrop D.H. (2002) *Using Vegetation to Assess Environmental Conditions in Wetlands*. EPA-822-R-02-020. Office of Water, U.S. Environmental Protection Agency: Washington, DC.

Fermor P.M., Hedges P.D., Brown P.F. (1999) Evapotranspiration of a reedbed within a created surface water fed wetland nature preserve. In: *Nutrient Cycling and Retention in Natural and Constructed Wetlands*, Vymazal J. (ed.) Backhuys Publishers: Leiden, The Netherlands, pp. 165–175.

Ferrero A., Vidotto F., Gennari M., Negre M. (2001) Behaviour of Cinosulfuron in paddy surface waters. *Journal of Environmental Quality 56*(2): 333–339.

Fetter C.W. Jr., Sloey W.E., Spangler F.L. (1977) Phosphorus accumulation - discharge cycles in marshes. *Water Resources Bulletin (Now Journal of the American Water Resources Association) 13*(6): 1191–1201.

Fetter C.W. Jr., Sloey W.E., Spangler F.L. (1978) Use of a natural marsh for wastewater polishing. *Journal of the Water Pollution Control Federation 50*(2): 290–307.

Fiala K. (1973) Growth and production of underground organs of *Typha angustifolia, Typha latifolia*, and *Phragmites communis. Polskie Archiwum Hydrobiologii, 20*: 59–66.

Finch J.W., Hall R.L., 2001. *Estimation of Open Water Evaporation: A Review of Methods*. R&D Technical Report W6-043/TR, Environment Agency: Bristol, United Kingdom Available at www.gov.uk/government/uploads/system/uploads/attachment_data/file/290579/sw6-043-hb-e-e.pdf (Accessed 3/26/18).

Fink D.F., Mitsch W.J. (2005) Fish and amphibian abundance and diversity in two created freshwater marshes. In: W.J. Mitsch, L. Zhang, and C.L. Tuttle (eds.), *Olentangy River Wetland Research Park at the Ohio State University, Annual Report 2004*. The Ohio State University, Columbus, OH, pp. 125–128.

Fink L. (2004) *STA-6 Mercury Special Studies Interim Report*. Appendix 2B-6, 2004 Evergalades Consolidated Report, South Florida Water Management District: West Palm Beach, FL

Finney B.A. (2000) *Internet Accessible Constructed Treatment Wetlands Descriptive and Performance Database*. EPA Assistance Agreement CX826884-01-0. Final Technical Report or the treatment wetland database, TWDB, http://firehole.humboldt.edu/wetland/twdb.html.Sitenolongeraccessible.

Finnigan J. J. (2000) Turbulence in plant canopies. *Annual Review of Fluid Mechanics 32*: 519–571.

Fisher M.M., Reddy K.R. (1987) Water hyacinth for improving EUTROPHIC Lake Water: Water quality and mass balance. In: *Aquatic Plants for Water Treatment and Resource Recovery*, Reddy K.R., Smith W.H. (eds.) Magnolia Publishing, Orlando, FL, pp. 969–976.

Fisher M.M., Reddy K.R. (2001) Phosphorus flux from wetland soils affected by long-term nutrient loading. *Journal of Environmental Quality 30*(1): 261–271.

Fleming-Singer M.S., Horne A.J. (2006) Balancing wildlife needs and nitrate removal in constructed wetlands: The case of the Irvine Ranch Water District's San Joaquin Wildlife Sanctuary. *Ecological Engineering 26*(2): 147–166.

Fletcher T.D., Poelsma P., Li Y., Deletic A. (2004) Wet and dry weather performance of constructed stormwater wetlands. In: International Conference on Water Sensitive Urban Design 2004; Adelaide, Australia, 21–25 November 2004.

Fleury M. (2002) Patent: Beaver control screen for culvert pipe. United States US 6,447,206.

Flood J.A., Ashbolt N.J. (2000) Virus-sized particles can be trapped and concentrated one hundred fold within wetland biofilms. *Advances in Environmental Research 3*(7): 403–411.

Flood J.A., Ashbolt N.J., Pollard P.C. (1999) Complimentary independent molecular, radioisotopic, and fluorogenic techniques to assess biofilm communities in two wastewater wetlands. *Water Science and Technology 39*(7): 65–70.

Flores C., Bounds D.L., Ruby D.E. (2011) Does prescribed fire benefit wetland vegetation? *Wetlands 31*(1): 35–44.

Focht D.D., Verstraete W. (1977) Biochemical ecology of nitrification and denitrification. *Advances in Microbial Ecology 1*: 135–214.

Fogler H.S. (1992) *Elements of Chemical Reaction Engineering*. Second edition, Prentice Hall: Englewood Cliffs, NJ.

Fogler H.S. (2016) *Elements of Chemical Reaction Engineering*. Fifth edition, Prentice Hall: Englewood Cliffs, NJ.

Fonder N., Headley T. (2013) The taxonomy of treatment wetlands: A proposed classification and nomenclature system. *Ecological Engineering 51*: 203–211.

Forbes E.G.A., Foy R.H., Mulhollan M.V., Brettell J.L. (2011) Performance of a constructed wetland for treating farm-yard dirty water. *Water Science and Technology 64*(1): 22–28.

Fox J.L., Price D.E., Allinson J. (1984) Distribution of fecal coliform bacteria in and around experimental cypress domes. In: *Cypress Swamps*, Ewel K.C., Odum H.T. (eds.) University of Florida Press: Gainesville, FL, pp. 225–226.

Foy R.H., Lennox S.D. (2000) Contributions of diffuse and point sources to the phosphorus loads in the River Main over a 22-year period. *Boreal Environment Research 5*(1): 27–37.

Frandsen A.K., Gammons C.H. (2000) Complexation of metals with aqueous sulfide in an anaerobic treatment wetland. Means J.L., Hinchee R.E. (eds.) Wetlands & Remediation: An International Conference; Battelle Press: Columbus, OH, 423–430.

Frank J. (2005) *Determination of Arsenic and Arsenic Species in Ombrotrophic Peat Bogs from Finland.* PhD Dissertation, University of Heidelberg: Heidelberg, Germany.

Frederick P.C., Powell G.V.N. (1994) Nutrient transport by wading birds in the Everglades. In: *Everglades: The Ecosystem and Its Restoration*, Davis S.M., Ogden J.C. (eds.) St. Lucie Press: Delray Beach, FL, pp. 571–584.

Freeze R.A. (1994) Henry Darcy and the fountains of Dijon. *Groundwater 32*(1): 23–30.

Freeze R.A., Cherry J.A. (1979) *Groundwater.* Prentice-Hall: Englewood Ciffs, NJ.

French R.H. (1985) *Open-Channel Hydraulics.* McGraw-Hill: New York.

Freney J.R., Leuning R., Simpson J.R., Denmead O.T., Muirhead W.A. (1985) Estimating ammonia volatilization from flooded rice fields by simplified techniques. *Soil Science Society of America Journal 49*: 1049–1054.

Friend M. (1985) Wildlife health implications of sewage disposal in wetlands. In: *Ecological Considerations in Wetlands Treatment of Municipal Wastewaters*, Godfrey P.J., Kaynor E.R., Pelczarski S., Benforado J. (eds.) Van Nostrand Reinhold: New York, pp. 262–269.

Friend M.N. (1989) 13.2.5. Avian cholera: A major new cause of waterfowl mortality. In: *Waterfowl Management Handbook*. U.S. Fish and Wildlife Service, National Wildlife Health Research Center: Madison, WI. Available at: http://digitalcommons.unl.edu/icwdmwfm/4 (Accessed 10/30/18).

Furukawa K., Rouse J.D., Imajo U., Sugino H., Fujii T. (2001) Sustainable nitrogen removal by anaerobic oxidation during soil aquifer treatment. Proceedings of the WEFTEC 2001 National Conference, 74th Annual Conference and Exhibition; Water Environment Federation: Alexandria, VA.

Gaiser E.E., Trexler J.C., Richards J.H., Childers D.L., Lee D., Edwards A.L., Scinto L.J., Jayachandran K., Noe G.B., Jones R.D. (2005) Cascading ecological effects of low-level phosphorus enrichment in the Florida Everglades. *Journal of Environmental Quality 34*(2): 717–723.

Galatowitsch S.M., van der Valk A.G. (1996) The vegetation of restored and natural prairie wetlands. *Ecological Applications 6*(1): 102–112.

Galbrand C., Lemieu I.G., Cote R., Verma M. (2008) Water quality assessment of a constructed wetland treating landfill leachate and industrial park runoff. *American Journal of Environmental Sciences 4*(2): 111–120.

Galbrand C.C., Snow A.M., Ghaly A.E., Côté R. (2007) Assessment of the biological integrity of the native vegetative community in a surface flow constructed wetland treating industrial park contaminants. *Online Journal of Biological Sciences 7*(1): 21–29.

Gale P.M., Reddy K.R., Graetz D.A. (1993) Nitrogen removal from reclaimed water applied to constructed and natural wetland mesocosms. *Water Environment Research 65*(2): 162–168.

Gallagher M.T., Snodgrass J.W., Brand A.B., Casey R.E., Lev S.M., Van Meter R.J. (2014) The role of pollutant accumulation in determining the use of stormwater ponds by amphibians. *Wetlands Ecology and Management 22*(5): 551–564.

Gamble D.L., Washburn S.W., Mitsch W.J. (2004) Macroinvertebrate abundance and diversity in two ten-year-old created wetlands. In: *2004 Annual Report, Olentangy River Wetland Research Park*, Mitsch W.J. (ed.), The Ohio State University: Columbus, OH, pp. 109–123.

Gambrell R.P., Khalid R.A., Patrick W.H. Jr. (1987) Capacity of a swamp forest to assimilate the TOC loading from a sugar refinery wastewater stream. *Journal of the Water Pollution Control Federation 59*(10): 897–904.

Gantes P., Torremorell A. (2005) Production and decomposition in floating soils of the Iberá wetlands (Argentina). *Limnetica 24*(3–4): 203–209.

Gao S., Tanji K.K., Peters D.W., Lin Z., Terry N. (2003a) Selenium removal and mass balance in a constructed flow-through wetland system. *Journal of Environmental Quality 32*(4): 1557–1570.

Gao S., Tanji K.K., Peters D.W., Lin Z., Terry N. (2003b) Selenium removal from irrigation drainage water flowing through constructed wetland cells with special attention to accumulation in sediments. *Water, Air, and Soil Pollution 144*: 263–284.

Garrigues E., Doussan C., Pierret A. (2006) Water uptake by plant roots: I–Formation and propagation of a water extraction front in mature root systems as evidenced by 2D light transmission imaging. *Plant and Soil* 283(1–2): 83–98.

Gascoigne W.R., Hoag D., Koont L., Tangen B.A., Shaffer T.L., Gleason R.A. (2011) Valuing ecosystem and economic services across land-use scenarios in the Prairie Pothole Region of the Dakotas, USA. *Ecological Economics 70*: 1715–1725.

Gaudet J.J. (1977) Natural drawdown on Lake Naivasha, Kenya, and the formation of Papyrus swamps. *Aquatic Botany 3*: 1–47.

Gavin H., Agnew C.A. (2004) Modelling actual, reference and equilibrium evaporation from a temperate wet grassland. *Hydrological Processes 18*(2): 229–246.

Gearheart R.A., Higley M. (1993) Constructed open surface wetlands: The water quality benefits and wildlife benefits. In: *Constructed Wetlands for Water Quality Improvement*, Moshiri G.A. (ed.) Lewis Publishers: Boca Raton, FL, pp. 561–567.

Gearheart R.A., Klopp F., Allen G. (1989) Constructed free surface wetlands to treat and receive wastewater: Pilot project to full scale. In: *Constructed Wetlands for Wastewater Treatment: Municipal, Industrial, and Agricultural*, Hammer D.A. (ed.) Lewis Publishers: Chelsea, MI, pp. 121–137.

Gearheart R.A., Wilbur S., Williams J., Finney B.A., Sundberg S. (1983) *City of Arcata Marsh Pilot Project, Effluent Quality Results – System Design and Management*. Final Report to the North Coast Regional Water Quality Board (Santa Rosa, California) and State Water Resources Board: Sacramento, CA.

Gebremariam S.Y., Beutel M.W., Yonge D.R., Flury M., Harsh J.B. (2012) Adsorption and desorption of chlorpyrifos to soils and sediments. In: *Reviews of Environmental Contamination and Toxicology*, Whitaker D.M. (ed.) Springer: New York, pp. 123–175).

Geranmayeh P., Johannesson K.M., Ulén B., Tonderski K.S. (2018) Particle deposition, resuspension and phosphorus accumulation in small constructed wetlands. *Ambio 47*(1): 134–145.

Gerba C.P. (2000) Indicator Microorganisms. In: *Environmental Microbiology*, Maier R.M., Pepper I.L., Gerba C.P. (eds.) Academic Press: San Diego, CA.

Gerlach R.M., Motamedi A.M., Loomis T.R. (2013) *Drainage Design Manual for Maricopa County, Arizona: Hydrology*. Maricopa County Flood Control District and City of Phoenix: AZ. Available at: www.maricopa.gov/DocumentCenter/View/2370/2015-01-Drainage-Design-Manual-for-Maricopa-County-Volume-I-Hydrology-PDF?bidId= (Accessed 7/23/18).

Gersberg R.M., Brenner R., Lyon S.R., Elkins B.V. (1987) Survival of bacteria and viruses in municipal wastewaters applied to artificial wetlands. In: *Aquatic Plants for Water Treatment and Resource Recovery*, Reddy K.R., Smith W.H. (eds.) Magnolia Publishing: Orlando, FL, pp. 237–245.

Gersberg R.M., Elkins B.V., Goldman C.R. (1983) Nitrogen removal in artificial wetlands. *Water Research 17*(9): 1009–1014.

Gersberg R.M., Elkins B.V., Goldman C.R. (1984) Use of artificial wetlands to remove nitrogen from wastewater. *Journal of the Water Pollution Control Federation 56*: 152–156.

Gersberg R.M., Gearheart R.A., Ives M. (1989) Pathogen reduction in constructed wetlands. In: *Constructed Wetlands for Wastewater Treatment; Municipal, Industrial and Agricultural*, Hammer D. A. (ed.) Lewis: Chelsea, MI, pp. 431–445.

Gessner M.O. (2000) Breakdown and nutrient dynamics of submerged *Phragmites* shoots in the littoral zone of a temperate hardwater lake. *Aquatic Botany 66*(1): 9–20.

Gessner T.P., Reaves R.P., Kadlec R.H. (2005) Wetland remediation of cyanide and hydrocarbons. *Ecological Engineering 25*(4): 457–469.

Ghermandi A. (2016) Analysis of intensity and spatial patterns of public use in natural treatment systems using geotagged photos from social media. *Water Research 105*: 297–304.

Ghermandi A. (2018) Integrating social media analysis and revealed preference methods to value the recreation services of ecologically engineered wetlands. *Ecosystem Services 31*: 351–357.

Ghermandi A., Fichtman E. (2015) Cultural ecosystem services of multifunctional constructed treatment wetlands and waste stabilization ponds: Time to enter the mainstream? *Ecological Engineering 84*: 615–623.

Ghermandi A., Van Den Bergh J.C., Brander L.M., de Groot H.L., Nunes P.A. (2010) Values of natural and human-made wetlands: A meta-analysis. *Water Resources Research* 46: 1–12.

Gill L.W., Ring P., Casey B., Higgins N.M., Johnston P.M. (2017) Long term heavy metal removal by a constructed wetland treating rainfall runoff from a motorway. *Science of the Total Environment 601*: 32–44.

Gillespie W.B. Jr., Hawkins W.B., Rodgers J.H. Jr., Cano M.L., Dorn P.B. (2000) Transfers and transformations of zinc in constructed wetlands: Mitigation of refinery effluent. *Ecological Engineering 14*(3): 279–292.

Gilmour C.C., Elias D.A., Kucken A.M., Brown S.D., Palumbo A.V., Schadt C.W., Wall J.D. (2011) The sulfate-reducing bacterium *Desulfovibrio desulfuricans* ND132 as a model for understanding bacterial mercury methylation. *Applied and Environmental Microbiology 77*(12): 3938–3951.

Giraud F., Guiraud P., Kadri M., Blake G. (2001) Biodegradation of anthracene and fluoranthene by fungi isolated from an experimental constructed wetland for wastewater treatment. *Water Research 35*(17): 4126–4136.

Gish T.J., Prueger J.H., Daughtry C.S., Kustas W.P., McKee L.G., Russ A.L., Hatfield J.L. (2011) Comparison of field-scale herbicide runoff and volatilization losses: An eight-year field investigation. *Journal of Environmental Quality 40*(5): 1432–1442.

Gladden J.B., Specht W.L., Nelson E.A. (2001) *Comparison of Constructed Wetland Mesocosms Designed for Treatment of Copper-Contaminated Wastewater*, Report WSRC-MS-2001-00128, U.S. Department of Energy: Oak Ridge, TN.

Gleason P.J. (1972) *The Origin, Sedimentation, and Stratigraphy of Calcitic Mud Located in the Southern Fresh-Water Everglades.* PhD Dissertation, The Pennsylvania State University: University Park, PA.

Gleason P.J., Stone P. (1994) Age, origin and landscape evolution of the Everglades peatland. In: *Everglades: The Ecosystem and Its Restoration*, S.M. Davis and J.C. Ogden, (eds.) St. Lucie Press: Delray Beach, FL, pp. 149–198.

Glenn E., Thompson T.L., Frye R., Riley J., Baumgartner D. (1995) Effects of salinity on growth and evapotranspiration on *Typha domingensis* Pers. *Aquatic Botany 52*: 75–91.

Gnatowski T., Szatyłowicz J., Brandyk T., Kechavarzi C. (2010) Hydraulic properties of fen peat soils in Poland. *Geoderma 154*(3–4): 188–195.

Godrej A.N., Grizzard T.J., Kenel P.P., Lampe L.K., Carleton J.N. (1999) *Evaluating the Use of Constructed Wetlands in Urban Areas*, Final Report on Project 92-NPS-1, Water Environment Research Foundation: Alexandria, VA.

Godshalk G.L., Wetzel R.G. (1978a) Decomposition in the littoral zone of lakes. In: *Freshwater Wetlands: Ecological Processes and Management Potential*, Good R.E., Whigham D.F., Simpson R.L. (eds.) Academic Press: New York, pp. 131–144.

Godshalk G.L., Wetzel R.G. (1978b) Decomposition of aquatic angiosperms. III. *Zostera marina* L. and a conceptual model of decomposition. *Aquatic Botany 5*: 329–354.

Godshalk G.L., Wetzel R.G. (1978c) Decomposition of aquatic angiosperms. II. Particulate components. *Aquatic Botany 5*: 301–327.

Goedderz S.J. Jr. (1987) Patent: Removable culvert gate. United States US 4,713,179.

Goforth G.F. (2001) Surmounting the engineering challenges of Everglades restoration. *Water Science and Technology 44*(11–12): 295–302.

Goldsborough G. (2001) Sampling Algae in Wetlands. Chapter 13 in: *Bioassessment and Management of North American Freshwater Wetlands*, Rader R.B., Batzer, D.P., Wissinger S.A. (eds.) Wiley: New York.

Gomez C., Pardue J.H. (2002) Bioavailability of non-ionic organics to wetland plants. In: *Wetlands and Remediation II*, Nehring K.W., Brauning S.E. (eds.) Battelle Press: Columbus, OH, pp. 49–56.

Gomez-Hermosillo C., Pardue J.H., Shin W.S. (2000) Bioavailability of desorption-resistant phenanthrene to wetland plants. Means J.L., Hinchee R.E. (eds.) Wetlands & Remediation: An International Conference; Battelle Press: Columbus, OH, 145–152.

Goncalves M.M.M., Pinto A.F., Granato M. (1998) Biodegradation of free cyanide, thiocyanate, and metal complexed cyanides in solutions with different compositions. *Environmental Technology 19*: 133–142.

Goolsby D.A., Battaglin W.A. (2000) Nitrogen in the Mississippi Basin - Estimating Sources and Predicting Flux to the Gulf of Mexico, USGS Fact Sheet No. 135-00, United States Geological Survey.

Gorito A.M., Ribeiro A.R., Almeida C.M.R., Silva A.M. (2017) A review on the application of constructed wetlands for the removal of priority substances and contaminants of emerging concern listed in recently launched EU legislation. *Environmental Pollution 227*: 428–443.

Goulden M. L., Litvak M., Miller S.D. (2007) Factors that control Typha marsh evapotranspiration. *Aquatic Botany 86*: 97–106.

Goulet R.R., Pick F.R., Droste R.L. (2001) Test of the first-order removal model for metal retention in a young constructed wetland. *Ecological Engineering 17*(4): 357–371.

Grace J.B. (1989) Effects of water depth on *Typha latifolia* and *Typha domingensis*. *American Journal of Botany 76*(5): 762–768.

Graham S.A., Craft C.B., McCormick P.V., Aldous A. (2005) Forms and accumulation of soil P in natural and recently restored peatlands—Upper Klamath Lake, OR, USA. *Wetlands 25*(3): 594–606.

Grant W.D., Long P.E. (1985) Environmental microbiology. In: *The Handbook of Environmental Chemistry. Volume 1, Part D: The Natural Environment and Biogeochemical Cycles*, Hutzinger O. (ed.) Springer-Verlag: Berlin, Germany, pp. 125–237.

Gray J.L., Sedlak D.L. (2004) The fate of estrogenic hormones in an engineered treatment wetland with dense macrophytes. *Water Environment Research 77*(1): 24–31.

Gray J.R., Glysson G.D., Turcios L.M., Schwartz G.E. (2000) *Comparability of Suspended Sediment Concentration and Total Suspended Solids Data*, Water Resources Investigations Report 00-4091, U.S. Geological Survey: Reston, VA.

Grayson M. (1985) *Kirk-Othmer Encyclopedia of Chemical Technology*. Grayson M. (ed.) Wiley: New York.

Green J. (2006) Effect of macrophyte spatial variability on channel resistance. *Advances in Water Resources 29*: 426–438.

Green D.I. (2016) *Environmental Factors Affecting the Residence Time Distribution Dynamics of Constructed Agricultural Wetlands*. PhD Dissertation, Iowa State University: Ames, IA.

Greenway M. (2002) Seasonal Phragmites biomass and nutrient storage in a subtropical subsurface flow wetland receiving secondary treated effluent in Brisbane, Australia. Mbwette T.S.A. (ed.) Proceedings of the 8th International Conference on Wetland Systems for Water Pollution Control, 16–19 September 2002; University of Dar es Saalam and IWA Publishing: Arusha, Tanzania, 242–253.

Greenway M. (2003) Suitability of macrophytes for nutrient removal from surface flow wetlands receiving secondary treated sewage effluent in Queensland, Australia. *Water Science and Technology 48*(2): 121–128.

Greenway M. (2010) Wetlands and ponds for stormwater treatment in subtropical Australia: Their effectiveness in enhancing biodiversity and improving water quality? *Journal of Contemporary Water Research & Education 146*(1): 22–38.

Greenway M., Chapman H. (2002) Constructed wetlands for wastewater treatment: Macrophytes, macroinvertebrates and mosquitoes. Mbwette T.S.A. (ed.) Proceedings of the 8th International Conference on Wetland Systems for Water Pollution Control, 16–19 September 2002; Comprint International Limited: University of Dar Es Salaam: Tanzania, 1009–1023.

Greeson P.E., Clark J.R., Clark J.E.E. (1978) Wetland functions and values: The state of our understanding. Proceedings of the National Symposium on Wetlands, 7–10 November 1978; American Water Resources Association: Lake Buena Vista, FL, 674 pp

Gregory C.T. (2009) *Temperature and Infiltration Characterization of a Constructed Wetland for Wastewater Treatment*. MS Thesis, Oregon State University, Corvallis, OR.

Greiner R.W., de Jong J. (1984) *The Use of Marsh Plants for the Treatment of Waste Water in Areas Designated for Recreation and Tourism*, RIJP Report No. 225, Lelystad, The Netherlands.

Gribble G.W. (1992) Naturally occurring organohalogen compounds–A survey. *Journal O Natural Products 55*(10): 1353–1395.

Gries C.L., Kappen L., Lösch R. (1990) Mechanism of flood tolerance in reed, *Phragmites australis* (Cav.) Trin. ex Strudel. *New Phytologist 144*: 589–593.

Griffiths L.N., Mitsch W.J. (2017) Removal of nutrients from urban stormwater runoff by storm-pulsed and seasonally pulsed created wetlands in the subtropics. *Ecological Engineering 108*: 414–424.

Groh T.A., Gentry L.E., David M.B. (2015) Nitrogen removal and greenhouse gas emissions from constructed wetlands receiving tile drainage water. *Journal of Environmental Quality 44*(3): 1001–1010.

Gross M.F., Hardisky M.A., Wolf P.L., Klemas V. (1993) Relationships among *Typha* biomass, pore water methane, and reflectance in a Delaware (USA) brackish marsh. *Journal of Coastal Research 9*(2): 339–355.

Grosshans R. (2016) *Cattails into Clean Energy: Where to from Here?* Report of The International Institute for Sustainable Development (IISD), Winnipeg, Manitoba, Canada. Available at: www.iisd.org/blog/cattails-clean-energy-where-here (Accessed 1/5/19).

Grosshans R. (2017) Cattail harvesting for nutrient capture and sustainable energy: The stacked benefits of water retention green infrastructure that drives economic sustainability. Paper presented at 3rd International Sustainable Wetland Plant Management Conference; North Dakota State University: Fargo, ND, May 31st & June 1st, 2017. Available at: www.ag.ndsu.edu/bioepic/3rd-international-sustainable-wetland-plant-management-conference (Accessed 1/5/109).

Grosshans R., Grieger L., Ackerman J., Gauthier S., Swystun K., Gass P., Dimple R. (2014) *Cattail Biomass in a Watershed-Based Bioeconomy: Commercial-Scale Harvesting and Processing for Nutrient Capture, Biocarbon, and High Value Bioproducts*. The International Institute for Sustainable Development: Winnipeg, Manitoba. Available at: www.iisd.org/sites/default/files/publications/cattail-biomass-watershed-based-bioeconomy-commerical-scale-harvesting.pdf (Accessed 7/5/18).

Grover S.P.P., Baldock J.A. (2013) The link between peat hydrology and decomposition: Beyond von Post. *Journal of Hydrology 479*: 130–138.

Gu B., Axelrad D.M., Lange T., et al. (2012) *Regional Mercury and Sulfur Monitoring and Environmental Assessment*. Chapter 3B in: 2012 South Florida Environmental Report, South Florida Water Management District: West Palm Beach, FL.

Guan H., Wilson J.L. (2009) A hybrid dual-source model for potential evaporation and transpiration partitioning. *Journal of Hydrology 377*(3–4): 405–416.

Guan Y., Wang B., Gao Y., Liu W., Zhao X., Huang X., Yu J. (2017) Occurrence and fate of antibiotics in the aqueous environment and their removal by constructed wetlands in China: A review. *Pedosphere 27*(1): 42–51.

Guardo M. (1999) Hydrologic balance for a subtropical treatment wetland constructed for nutrient removal. *Ecological Engineering 12*(3–4): 315–337.

Guganessharajah R.K. (2001) *Development of Computational Hydraulic and Water Quality Models for Rivers, Estuaries, Reservoirs, and Aquifers, with Particular Reference to WSPs*. PhD Dissertation, University of Surrey: United Kingdom.

Gumbricht T. (1993) Nutrient removal processes in freshwater submersed macrophyte systems. *Ecological Engineering* 2(1): 1–30.

Guntenspergen G.R., Stearns F. (1981) Ecological limitations on wetland use for wastewater treatment. Richardson B. (ed.) Selected Proceedings of the Midwest Conference on Wetland Values and Management, 17–19 June 1981; Freshwater Society: St. Paul, MN, 273–284.

Guo C., Cui Y., Dong B., Liu F. (2017) Tracer study of the hydraulic performance of constructed wetlands planted with three different aquatic plant species. *Ecological Engineering 102*: 433–442.

Gustafson T.D. (1976) *Production, Photosynthesis and Storage and Utilization of Reserves in a Natural Stand of Typha Latifolia*. PhD Thesis, University of Wisconsin, Madison, WI.

Guzman C.B., Cohen S., Xavier M., Swingle T., Qiu W., Nepf H. (2018a) Island topographies to reduce short-circuiting in stormwater detention ponds and treatment wetlands. *Ecological Engineering 117*: 182–193.

Guzman C.B., Nepf H., Berger A.M. (2018b) *Design Guidelines for Urban Stormwater Wetlands. ISSUU Document Sharing* Website. Available at: https://issuu.com/prurobinson/docs/design_guidelines_report (Accessed 11/26/18).

Haarstad K., Braskerud B. (2005). Pesticide retention in the watershed and in a small constructed wetland treating diffuse pollution. *Water Science and Technology 51*(3–4): 143–150.

Habersack M.J., Dillaha T.A., Hagedorn C. (2011) Common Snapping Turtles (*Chelydra serpentina*) as a Source of Fecal Indicator Bacteria in Freshwater Systems 1. *JAWRA Journal of the American Water Resources Association 47*(6): 1255–1260.

Hafeznezami S., Kim J.L., Redman J. (2012) Evaluating removal efficiency of heavy metals in constructed wetlands. *Journal of Environmental Engineering 138*(4): 475–482.

Haffner W.M. (1993) *Palmerton Zinc Superfund Site Constructed Wetlands, Final Report Phase II.* Report to Zinc Corporation of America, Palmerton, PA.

Hagerthey S.E., Cook M.I., Mac Kobza R., Newman S., Bellinger B.J. (2014) Aquatic faunal responses to an induced regime shift in the phosphorus-impacted E verglades. *Freshwater Biology 59*(7): 1389–1405.

Hahn S., Bauer S., Klaassen M. (2008) Quantification of allochthonous nutrient input into freshwater bodies by herbivorous waterbirds. *Freshwater Biology 53*(1): 181–193.

Haines D.A., Bryson R.A. (1961) An empirical study of wind factor in Lake Mendota. *Limnology and Oceanography 6*(3): 356–364.

Hair J.D., Hepp G.T., Luckett L.M., Reese K.P., Woodward D.K. (1978) Beaver pond ecosystems and their relationship to multi-use natural resource management. In: *Proceedings of Strategies for Protection and Management of Floodplain Wetlands and Other Riparian Ecosystems, 11–13 December 1978, Callaway Gardens, Georgia*, Johnson R.R., McCormick J.F. (eds.) U.S. Department of Agriculture: Washington, DC, pp. 80–92.

Halford K.J. (1999) Evaluation of the effectiveness of wetland treatment at Rustler's Roost gold mine in the northern territory. *Mineral Resources Engineering 8*(4): 453–469.

Hall B.R., Freeman G.E. (1994) Study of hydraulic roughness in wetland vegetation takes new look at Manning's *n*. *The Wetlands Research Program Bulletin 4*(1): 1–4.

Hamer A.J., Smith P.J., McDonnell M.J. (2012) The importance of habitat design and aquatic connectivity in amphibian use of urban stormwater retention ponds. *Urban Ecosystems 15*(2): 451–471.

Hamilton D. P., Mitchell S. F. (1996) An empirical model for sediment resuspension in shallow lakes. *Hydrobiologia 317*: 209–220.

Hamilton S.K., Sippel S.J., and Melack J.M. (1995) Oxygen depletion and carbon dioxide and methane production in waters of the Pantanal wetland of Brazil. *Biogeochemistry 30*(2): 115–141.

Hammer D.A. (1996) *Creating Freshwater Wetlands. 2nd edition*. CRC Press, Boca Raton, FL.

Hammer D.A., Bastian R.K. (1989) Wetlands ecosystems: Natural water purifiers? In: *Constructed Wetlands for Wastewater Treatment; Municipal, Industrial and Agricultural*, Hammer D.A. (ed.) Lewis Publishers: Chelsea, MI, pp. 5–20.

Hammer D.A., Knight R.L. (1994) Designing constructed wetlands for nitrogen removal. *Water Science and Technology 29*(4): 15–27.

Hammer D.E., Kadlec R.H. (1980) Ortho-phosphate adsorption on peat. Proceedings of the 6th National Peat Congress; Duluth, MN, 563–569.

Hammer D.E., Kadlec R.H. (1986) A model for wetland surface water dynamics. *Water Resources Research 22*(13): 1951–1958.

Han S.-H., Zhuang Y.-H., Liu J.-A., Glindeman D. (2000) Phosphorus cycling through phosphine in paddy fields. *Science of the Total Environment 258*(3): 195–203.

Hansen A.T., Dolph C.L., Foufoula-Georgiou E., Finlay J.C. (2018) Contribution of wetlands to nitrate removal at the watershed scale. *Nature Geoscience 11*(2): 127.

Hansen D., Duda P.J., Zayed A., Terry N. (1998) Selenium removal by constructed wetlands: Role of biological volatilization. *Environmental Science and Technology 32*(5): 591–597.

Hanson A. (2002) An overview of wastewater treatment performance and wildlife habitat use of two small constructed wetlands in Nova Scotia, Canada. In: *Treatment Wetlands for Water Quality Improvement*, Pries J.H. (ed.) Pandora Press: Ontario, Canada, pp. 73–82.

Hansson A., Pedersen E., Weisner S.E. (2012) Landowners' incentives for constructing wetlands in an agricultural area in south Sweden. *Journal of Environmental Management 113*: 271–278.

Harbaugh A.W. (2005) *MODFLOW-2005, the US Geological Survey Modular Ground-Water Model: The Ground-Water Flow Process.* Report 6-A16, US Department of the Interior, US Geological Survey: Reston, VA.

Harbaugh A.W., Banta E.R., Hill M.C., McDonald M.G. (2000) *MODFLOW-2000, the U. S. Geological Survey Modular Ground-Water Model-User Guide to Modularization Concepts and the Ground-Water Flow Process*. Open-file Report 92, US Department of the Interior, US Geological Survey: Reston, VA.

Harbor J., Tatlovich S., Turnco R., Reicher Z., Spacie A., Poole V. (2000) Using constructed wetlands to reduce nonpoint source pollution control in urban areas. Proceedings from the National Conference on Tools for Urban Water Resource Management and Protection. Report EPA/625/R-00/001; U.S. EPA: Cincinnati, OH, 303–313.

Hardej M., Ozimek T. (2002) The effect of sewage sludge flooding on growth and morphometric parameters of *Phragmites australis* (Cav.) Trin. ex Steudel. *Ecological Engineering 18*(3): 343–350.

Haritash A.K., Kaushik C.P. (2009) Biodegradation aspects of polycyclic aromatic hydrocarbons (PAHs): A review. *Journal of Hazardous Materials 169*(1–3): 1–15.

Harmel R.D., Karthikeyan R., Gentry T., Srinivasan R. (2010) Effects of agricultural management, land use, and watershed scale on E. coli concentrations in runoff and streamflow. *Transactions of the ASABE 53*(6): 1833–1841.

Harter S.K., Mitsch W.J. (2003) Patterns of short-term sedimentation in a freshwater created marsh. *Journal of Environmental Quality 32*(4): 325–334.

Harvey J.W., Krupa S.L., Gefvert C.J., Jungyill C., Mooney R.H., Giddings J.B. (2000) *Interaction between Ground Water and Surface Water in the Northern Everglades and Relation to Water Budget and Mercury Cycling; Study Methods and Appendixes (No. 2000-168)*. US Department of the Interior, US Geological Survey; Branch of Information Services, [distributor]: Reston, VA.

Harvey J.W., Krupa S.L., Krest J.M. (2004) Ground water recharge and discharge in the central Everglades. *Groundwater 42*(7): 1090–1102.

Harvey J.W., Noe G.B., Larsen L.G., Nowacki D.J., McPhillips L.E. (2011) Field flume reveals aquatic vegetation's role in sediment and particulate phosphorus transport in a shallow aquatic ecosystem. *Geomorphology 126*: 297–313.

Hathaway J.M., Hunt W.F. (2010) Indicator bacteria removal by stormwater BMPs in Coastal North Carolina. World Environmental and Water Resources Congress 2010: Challenges of Change; 3218–3225.

Hathaway J.M., Hunt W.F. (2012) Indicator bacteria performance of storm water control measures in Wilmington, North Carolina. *Journal of Irrigation and Drainage Engineering 138*(2): 185–197.

Hathaway J.M., Hunt W.F., Graves A.K., Bass K.L., Caldwell A. (2011) Exploring fecal indicator bacteria in a constructed stormwater wetland. *Water Science and Technology* 63(11): 2707–2712.

Hatvani I.G., Clement A., Kovács J., Kovács I.S., Korponai J. (2014) Assessing water-quality data: The relationship between the water quality amelioration of Lake Balaton and the construction of its mitigation wetland. *Journal of Great Lakes Research 40*(1): 115–125.

Hatvani I.G., Kovács J., Kovács I.S., Jakusch P., Korponai J. (2011) Analysis of long-term water quality changes in the Kis-Balaton Water Protection System with time series-, cluster analysis and Wilks' lambda distribution. *Ecological Engineering 37*(4): 629–635.

Hauck R.D. (1984) Atmospheric nitrogen chemistry, nitrification, denitrification, and their relationships. In: *The Handbook of Environmental Chemistry*, Hutzinger O. (ed.) Springer-Verlag: Berlin, Germany, pp. 105–127.

Havens K.E., East T.L., Rodusky A.J., Sharfstein B. (1999) Littoral periphyton responses to nitrogen and phosphorus: An experimental study in a subtropical lake. *Aquatic Botany 63*(3–4): 267–290.

Hawari J., Beaudet S., Halasz A., Thibotot S., Ampleman G. (2000) Microbial degradation of explosives: Biotransformation vs. mineralization. *Applied and Environmental Microbiology 54*(5): 605–618.

Hawkins W.B., Rodgers J.H. Jr., Gillespie W.B., Dunn A.W., Dorn P.B., Cano M.L. (1997) Design and construction of wetlands for aqueous transfers and transformations of selected metals. *Ecotoxicology and Environmental Safety 36*(3): 238–248.

Hayes D.F., Olin T.J., Fischenich J.C., Palermo M.J. (2000) *Wetlands Engineering Handbook*, ERDC/EL TP-WRP-RE21 (719), U.S. Army Corps of Engineers: Washington, DC.

Haynes A. (2014) *A Floating Wetland Handbook for San Francisco's Southeast Waterfront*. Report from Piero N. Patri Fellowship for Urban Design. Available at: www.spur.org/sites/default/files/wysiwyg/u169/2014%20Patri%20Fellow%20%E2%80%94%C2%A0Floating%20Wetlands.pdf (Accessed 12/27/18).

Hazen and Sawyer, P.C. (2011) *Compilation of Benefits and Costs of STA and Reservoir Projects in the South Florida Water Management District*. Report to World Wildlife Fund: Washington, DC.

Available at: www.fresp.org/pdfs/Compilation%20of%20STA%20and%20REZ%20Benefits%20Costs%20HandS%2011_2011.pdf (Accessed 12/17/18).

He G. (2007) *Efficiency of Stormwater Treatment Areas in Removing Contaminants*. Report to Everglades National Park: Homestead, FL.

He G., Engel V., Leonard L., Croft A., Childers D., Laas M., Deng Y., Solo-Gabriele H.M. (2010) Factors controlling surface water flow in a low-gradient subtropical wetland. *Wetlands 30*(2): 275–286.

Headley T.R., Kadlec R.H. (2007) Conducting hydraulic tracer studies of constructed wetlands: A practical guide. *Ecohydrology & Hydrobiology 7*(3–4): 269–282.

Headley T.R., Tanner C.C. (2006) *Application of Floating Wetlands for Enhanced Stormwater Treatment: A Review*. Report Prepared for Auckland Regional Council. Available at: www.aucklandcity.govt.nz/council/documents/technicalpublications/TP324%20%20Floating%20Wetland%20Review-Final.pdf (Accessed 2/18/18).

Headley T.R., Tanner C.C. (2012) Constructed wetlands with floating emergent macrophytes: An innovative stormwater treatment technology. *Critical Reviews in Environmental Science and Technology 42*(21): 2261–2310.

Heaven S., Banks C. (2005) Cold and continental climate ponds. In: *Pond Treatment Technology*, Shilton A. (ed.) IWA Publishing: London, United Kingdom, pp. 381–407.

Hedin R., Nairn R., Kleinman R.L. (1994) *Passive Treatment of Coal Mine Drainage*, Bureau of Mines Circular 9389. 35, U.S. Bureau of Mines: Pittsburgh, PA.

Hefting M.M., van Den Heuvel R.N., Verhoeven J.T. (2013) Wetlands in agricultural landscapes for nitrogen attenuation and biodiversity enhancement: Opportunities and limitations. *Ecological Engineering 56*: 5–13.

Hefty R. (2002) Floating vegetated platforms close on open water ditch in peat wetland (Wisconsin). *Ecological Restoration 20*(4): 284–285.

Helfield M., Diamond M.L. (2004) Use of constructed wetlands for urban stream restoration: A critical analysis. *Environmental Management 21*(3): 329–341.

Helgen J. (2002) *Methods for Evaluating Wetland Condition. #9 Developing an Invertebrate Index of Biological Integrity for Wetlands*. EPA-822-R-02-019. Office of Water, U.S. Environmental Protection Agency: Washington, DC.

Hemming J.M., Waller W.T., Chow M.C., Denslow N.D., Venables B. (2001) Assessment of the estrogenicity and toxicity of a domstic wastewater effluent flowing through a constructed wetland system using biomarkers in male fathead minnows (*Pimephales promelas* Rafinesque, 1820). *Environmental Toxicology and Chemistry 20*(10): 2268–2275.

Hemond H.F. (1980) Biogeochemistry of Thoreau's Bog, Concord, Massachusetts. *Ecological Monographs 50*(4): 507–526.

Hendrey G.R., Clinton J., Blumer K., Lewin K. (1979) *Lowland Recharge Project Operations, Physical, Chemical, and Biological Changes 1975–1978*, Final report to the Town of Brookhaven, Brookhaven National Laboratory, Brookhaven, New York.

Hens M., Merckx R. (2002) The role of colloidal particles in the speciation and analysis of "dissolved" phosphorus. *Water Research 36*(12): 1483–1492.

Herb W.R., Stefan H.G. (2005) Model for wind-driven vertical mixing in a shallow lake with submersed macrophytes. *Journal of Hydraulic Engineering 131*(6): 488–496.

Herbst M., Kappen L. (1999) The ratio of transpiration versus e vaporation in a reed belt as Influenced by weather conditions. *Aquatic Botany 62*(2): 113–125.

Hernandez M.E., Mitsch W.J. (2006) Influence of hydrologic pulses, flooding frequency, and vegetation on nitrous oxide emissions from created riparian marshes. *Wetlands 26*(3): 862–877.

Hersikorn B.D., Smits J.E. (2011) Compromised metamorphosis and thyroid hormone changes in wood frogs (*Lithobates sylvaticus*) raised on reclaimed wetlands on the Athabasca oil sands. *Environmental Pollution 159*(2): 596–601.

Herskowitz J. (1986) *Listowel Artificial Marsh Project Report*. Ontario Ministry of the Environment, Water Resources Branch: Toronto, Ontario, Canada.

Hey D.L. (2002) Nitrogen farming: Harvesting a different crop. *Restoration Ecology 10*(1): 1–11.

Hey D.L., Kenimer A.L., Barrett K.R. (1994) Water quality improvement by four experimental wetlands. *Ecological Engineering 3*(4): 381–398.

Hey D.L., Kostel J.A., Kadlec R.H. (2005) *Nutrient Farming and Traditional Removal: An Economic Comparison*, Final Report on Project 03-WSM-6CO, Water Environment Research Foundation: Alexandria, VA.

Heyvaert A.C., Reuter J.E., Goldman C.R. (2006) Subalpine, cold climate, stormwater treatment with a constructed surface flow wetland. *JAWRA Journal of the American Water Resources Association 42*(1): 45–54.

Higgins C.R., (2002) *Ecosystem Engineering by Muskrats (Ondatra Zibethicus) in Created Freshwater Marshes.* Doctoral dissertation, The Ohio State University.

Higgins C.R., Mitsch W.J. (2001) *The Role of Muskrats (Ondatra zibethicus) as Ecosystem Engineers in Created Freshwater Marshes.* In: *2001 Report of Olentangy River Wetland Research Park*, School of Natural Resources, The Ohio State University, Columbus, OH, 81–85. Available at: https://kb.osu.edu/bitstream/handle/1811/358/muskrat?sequence=1 (Accessed 10/15/18).

Higgins M.J., Rock C.A., Bouchard R., Wengrzynek B. (1993) Controlling agricultural runoff by use of constructed wetlands. In: *Constructed Wetlands for Water Quality Improvement*, Moshiri G.A. (ed.) Lewis Publishers: Boca Raton, FL, pp. 359–367.

Hijosa-Valsero M., Fink G., Schlüsener M.P., Sidrach-Cardona R., Martín-Villacorta J., Ternes T., Bécares E. (2011) Removal of antibiotics from urban wastewater by constructed wetland optimization. *Chemosphere 83*(5): 713–719.

Hijosa-Valsero M., Sidrach-Cardona R., Bécares E. (2012) Comparison of interannual removal variation of various constructed wetland types. *Science of the Total Environment 430*: 174–183.

Hilderbrand R.H., Watts A.C., Randle A.M. (2005) The myths of restoration ecology. *Ecology and Society 10*(1).

Hiley P.D. (1990) Wetlands treatment revival in Yorkshire. In: *Constructed Wetlands in Water Pollution Control*, Cooper P.F., Findlater B.C. (eds.) Pergamon Press: Oxford, United Kingdom, pp. 279–288.

Hill B.H. (1987) Typha productivity in a Texas pond: Implications for energy and nutrient dynamics in freshwater wetlands. *Aquatic Botany 27*(4): 385–394.

Hill C.R., Robinson J.S. (2012) Phosphorus flux from wetland ditch sediments. *Science of the Total Environment 437*: 315–322.

Hill D., Stoops R. (2006) The Oregon garden wetlands. *Water Gardeners International 1* (1). Available at: www.watergardenersinternational.org/journal/index_author.html (Accessed 12/4/18).

Hill D.T., Payne V.W.E., Rogers J.W., Kown S.R., (1997) Ammonia effects on the biomass production of five constructed wetland plant species. *Bioresource Technology 62*: 109–113.

Hill V.R., Sobsey M.D. (2001) Removal of salmonella and microbial indicators in constructed wetlands treating swine wastewater. *Water Science and Technology 44*(11–12): 215–222.

Hilsenhoff W.L. (1987) An improved biotic index of organic stream pollution. *Great Lakes Entomologist 20*: 31–39

Hilsenhoff W.L. (1988) Rapid field assessment of organic pollution with a family-level biotic index. *Journal of the North American Benthological Society 7*(1): 65–68.

Ho Y.B. (1979) Shoot development and production studies of *Phragmites australis* (Cav.) Trin. ex Steudel in Scottish lochs. *Hydrobiologia 64*(3): 215–222.

Ho Y.S., McKay G. (2000) The kinetics of sorption of divalent metal ions onto sphagnum moss peat. *Water Research 34*(3): 735–742.

Hocking P.J. (1989a) Seasonal dynamics of production, nutrient accumulation, and cycling by *Phragmites australis* (Cav.) Trin ex Stuedel in a nutrient enriched swamp in inland Australia. Part I: Whole plants. *Australian Journal of Marine and Freshwater Research 40*: 421–444.

Hocking P.J. (1989b) Seasonal dynamics of production, nutrient accumulation, and cycling by *Phragmites australis* (Cav.) Trin ex Stuedel in a nutrient enriched swamp in inland Australia. Part II: Individual Shoots. *Australian Journal of Marine and Freshwater Research 40*: 445–464.

Hoffman G., Stack R., Van Wye B. (2013) *Stormwater Management Guidebook.* Prepared for District Department of the Environment, District of Columbia, Washington, DC. Available at: https://doee.dc.gov/publication/2003-stormwater-guidebook (Accessed 7/23/18).

Hoffmann C.C., Heiberg L., Audet J., Schønfeldt B., Fuglsang A., Kronvang B., Ovesen N.B., Kjaergaard C., Hansen H.C.B., Jensen H.S. (2012) Low phosphorus release but high nitrogen removal in two restored riparian wetlands inundated with agricultural drainage water. *Ecological Engineering 46*: 75–87.

Hogg E.H., Wein R.W. (1987) Growth dynamics of floating Typha mats: Seasonal translocation and internal deposition of organic material. *Oikos 50*(2): 197–205.

Hogg E.H., Wein R.W. (1988) The contribution of *Typha* components to floating mat buoyancy. *Ecology 69*(4): 1025–1031.

Holland J.F., Martin J.F., Granata T., Bouchard V., Quigley M., Brown L. (2005) Analysis and modeling of suspended solids from high frequency monitoring in a stormwater treatment wetland. *Ecological Engineering 24*(3): 159–176.

Holland J.F., Martin J.F., Granata T., Bouchard V., Quigley M., Brown L. (2004) Effects of wetland depth and flow rate on residence time distribution characteristics. *Ecological Engineering 23*(3): 189–203.

Holloway P.E. (1980) A criterion for thermal stratification in a wind-mixed system. *Journal of Physical Oceanography 10*: 861–869.

Hondulas J.L. (1994) Patent: Treatment of polluted water using wetland plants in a floating habitat. United States US 5,337,516.

Hoover K.L., Rightnour T.A., Collins R., Herd R. (1998) Applications of passive treatment to trace metals removal. Proceedings of the American Power Conference in Chicago, Illinois; Pennwell Publishing: Tulsa, OK.

Hopkins J.W., Bristow K.L. (2002) Current capabilities and future needs of root water and nutrient uptake modeling. *Advances in Agronomy 77*: 103–183.

Horne A.J. (1995) Nitrogen removal from waste treatment pond or activated sludge plant effluents with free-surface wetlands. *Water Science and Technology 31*(12): 341–351.

Horne A.J. (2004) Predicted changes in bioaccumulation and toxicity of selenium following construction of the natural treatment systems in sand diego creek watershed relative to the current status quo, Technical memorandum in the Irvine Ranch Water District Selenium Action Plan (6 January 2004).

Horne A.J., Goldman C.R. (2001) *Limnology*. 3rd Edition. Academic Press, New York.

Horppila J., Nurminen L. (2001) The effect of of an emergent macrophyte (Typha angustifolia) on sediment resuspension in a shallow north temperate lake. *Freshwater Biology 46*(11): 1447–1455.

Horppila J., Nurminen L. (2003) Effects of submerged macrophytes on sediment resuspension and internal phosphorus loading in Lake Hiidenvesi (Southern Finland). *Water Research 37*(18): 4468–4474.

Horton R.E. (1938) The interpretation and application of runoff plat experiments with reference to soil erosion problems. *Soil Science Society of America Journal* 3: 340–349.

Horton R.A., Randall L.P., Snary E.L., Cockrem H., Lotz S., Wearing H., Duncan D., Rabie A., McLaren I., Watson E., La Ragione R.M. (2011) Fecal carriage and shedding density of CTX-M extended-spectrum β-lactamase-producing Escherichia coli in cattle, chickens, and pigs: Implications for environmental contamination and food production. *Applied and Environmental Microbiology, 77*(11): 3715–3719.

Hotchkiss N. (1972) *Common Marsh, Underwater, and Floating-Leaved Plants of the United States and Canada*. Dover Publications: New York.

Howard R.H., Boethling R.S., Jarvis W.F., Meylan W.M., Michalenko E.M. (1991) *Handbook of Environmental Degradation Rates*. Lewis Publishers: Chelsea, MI.

Howell C.J., Crohn D.M., Omary M. (2005) Simulating nutrient cycling and removal through treatment wetlands in arid/semiarid environments. *Ecological Engineering 25*(1): 25–40.

Hsieh C.Y., Liaw E.T., Fan K.M. (2015) Removal of veterinary antibiotics, alkylphenolic compounds, and estrogens from the Wuluo constructed wetland in southern Taiwan. *Journal of Environmental Science and Health, Part A 50*(2): 151–160.

Hsu T.T.D., Mitsch W.J., Martin J.F., Lee J. (2017) Towards sustainable protection of public health: The role of an urban wetland as a frontline safeguard of pathogen and antibiotic resistance spread. *Ecological Engineering 108*: 547–555.

Hua G., Salo M.W., Schmit C.G., Hay C.H. (2016) Nitrate and phosphate removal from agricultural subsurface drainage using laboratory woodchip bioreactors and recycled steel byproduct filters. *Water Research 102*: 180–189.

Huai W., Hu Y., Zeng Y., Han J. (2012) Velocity distribution for open channel flows with suspended vegetation. *Advances in Water Resources 49*: 56–61.

Huang J., Wang S.H., Yan L., Zhong Q.S. (2010) Plant photosynthesis and its influence on removal efficiencies in constructed wetlands. *Ecological Engineering 36*(8): 1037–1043.

Huang L., Gao X., Guo J., Ma X., Liu M. (2013) A review on the mechanism and affecting factors of nitrous oxide emission in constructed wetlands. *Environmental Earth Sciences 68*(8): 2171–2180.

Huang Y.H., Saiers J.E., Harvey J.W., Noe G.B., Mylon S. (2008) Advection, dispersion, and filtration of fine particles within emergent vegetation of the Florida Everglades. *Water Resources Research 44*: W04408.

Humbert S., Zopfi J., Tarnawski S.E. (2012) Abundance of anammox bacteria in different wetland soils. *Environmental Microbiology Reports 4*(5): 484–490.

Hume N.P., Fleming M.S., Horne A.J. (2002b) Denitrification potential and carbon quality of four aquatic plants in wetland microcosms. *Soil Science Society of America Journal 66*(5): 1706–1712.

Hume N.P., Fleming M.S., Horne A.J. (2002a) Plant carbohydrate limitation on nitrate reduction in wetland microcosms. *Water Research 36*: 577–584.

Humphreys K.K. (1991) *Jelen's Cost and Optimization Engineering*. Third edition, McGraw-Hill: New York.

Hunt W. F., Burchell M.R., Wright J.D., Bass K.L. (2010) *Stormwater Wetland Design Update*. Publication AG-588-12 NC State Extension Service, N.C. State University, Raleigh, NC. Available at: https://content.ces.ncsu.edu/stormwater-wetland-construction-guidance (Accessed 7/19/18).

Hunt W.F., Burchell M.R., Wright J.D., Bass K.L. (2007) *Urban Waterways–Stormwater Wetland Design Update: Zones, Vegetation, Soil, and Outlet Guidance*. AG-588-12. North Carolina Cooperative Extension Service, Raleigh, North Carolina.

Hunt W.F., Greenway M., Moore T.C., Brown R.A., Kennedy S.G., Line D.E., Lord W.G. (2011) Constructed storm-water wetland installation and maintenance: Are we getting it right? *Journal of Irrigation and Drainage Engineering 137*(8): 469–474.

Hurry R.J., Bellinger E.G. (1990) Potential yield and nutrient removal by harvesting of *Phalaris arundinacea* in a wetland treatment system. In: *Constructed Wetlands in Water Pollution Control*, Cooper P. F. and Findlater B. C. (eds.) Pergamon Press: Oxford, United Kingdom, pp. 543–546.

Hussain S.A. (2011) *Removal of Poultry Pharmaceuticals by Constructed Wetlands*. PhD Dissertation, McGill University: Montreal, Canada.

Hussain S.A., Prasher S.O. (2011) Understanding the sorption of ionophoric pharmaceuticals in a treatment wetland. *Wetlands 31*(3): 563–571.

Hussain S.A., Prasher S.O., Patel R.M. (2012) Removal of ionophoric antibiotics in free water surface constructed wetlands. *Ecological Engineering 41*: 13–21.

Hussner A., Meyer C., Busch J. (2009) The influence of water level and nutrient availability on growth and root system development of *Myriophyllum aquaticum*. *Weed Research 49*(1): 73–80.

Hutchinson G.E. (1975) *A Treatise on Limnology*. Wiley Interscience: New York.

Hygnstrom S.E., Timm R.M., Curtis P.D., Nolte D.L., Tobin M.E., VerCauteren K.C. (2014) *Prevention and Control of Wildlife Damage*. 5th edition. Staff Publications, 1782. USDA-APHIS National Wildlife Research Center: Fort Collins, CO.

Ibekwe A.M., Lyon S.R., Leddy M., Jacobson-Meyers M. (2007) Impact of plant density and microbial composition on water quality from a free water surface constructed wetland. *Journal of Applied Microbiology 102*(4): 921–936.

IDALS (Iowa Department of Agriculture and Land Stewardship) (2016) *Iowa Conservation Reserve Enhancement Program (CREP). Landowner Guide to Operation and Maintenance*. Iowa Department of Agriculture and Land Stewardship, Des Moines, IA. Available at: https://iowaagriculture.gov/sites/default/files/water_resources/CREP%20Landowner%20Maintenance%20Guide%202016.pdf (Accessed 12/6/18).

IDALS (Iowa Department of Agriculture and Land Stewardship) (2017) *Iowa Conservation Reserve Enhancement Program (CREP). 2016 Annual Performance Report*. Iowa Department of Agriculture and Land Stewardship, Des Moines, IA. Available at: www.iowaagriculture.gov/waterresources/CREP.asp (Accessed 7/29/18).

Ikenberry C.D., Crumpton W.G., Arnold J.G., Soupir M.L., Gassman P.W. (2017a) Evaluation of existing and modified wetland equations in the SWAT model. *JAWRA Journal of the American Water Resources Association 53*(6): 1267–1280.

Ikenberry C.D., Soupir M.L., Helmers M.J., Crumpton W.G., Arnold J.G., Gassman P.W. (2017b) Simulation of Daily Flow Pathways, Tile-Drain Nitrate Concentrations, and Soil-Nitrogen Dynamics Using SWAT. *JAWRA Journal of the American Water Resources Association 53*(6): 1251–1266

Illman S., Wilson S. (2017) *Guidance on the Construction of SuDS*. CIRIA (Construction Industry Research and Information Association) report C768, CIRIA, London, United Kingdom. Available at: www.ciria.org/Resources/Free_publications/Guidance_on_the_construction_of_SuDS_-_C768.aspx (Accessed 8/4/18).

Inaba I. (1992) Quantitative assessment of natural purification in wetland for linear alkylbenzene sulfonates. *Water Research 26*(7): 893–898.

Ingersoll L.R., Zobel O.J., Ingersoll A.C. (1948) *Heat Conduction*. McGraw-Hill: New York.

Ingersoll T.L., Baker L.A. (1998) Nitrate removal in wetland microcosms. *Water Research 32*(3): 677–684.

Inglett P.W., Rivera-Monroy V.H., Wozniak J.R. (2011) Biogeochemistry of nitrogen across the Everglades landscape. *Critical Reviews in Environmental Science and Technology 41*(S1): 187–216.

Ingram H.A.P. (1978) Soil layers in mires: Function and terminology. *Journal of Soil Science 29*(2): 224–227.

Inman B.L., Thomas M., Kirk J. (2003) *Construction and startup of treatment wetlands on a sloping site (Session 36 Presentation)*. Proceedings, WEFTEC 2003 National Conference, 76th Annual Conference and Exhibition; Water Environment Federation: Alexandria, VA.

Inman B.L., Thomas M., Knight R.L. (2001) Design and construction of treatment wetlands on a sloping site. Proceedings of the WEFTEC 2001 National Conference, 74th Annual Conference and Exhibition; Water Environment Federation: Alexandria, VA.

Inoue T.M., Tsuchiya T. (2008) Interspecific differences in radial oxygen loss from the roots of three Typha species. *Limnology 9*(3): 207–211.

Iovanna R., Hyberg S., Crumpton W.G. (2008) Treatment wetlands: Cost-effective practice for intercepting nitrate before it reaches and adversely impacts surface waters. *Journal of Soil and Water Conservation 63*(1): 14A–15A.

Irmak S., Kabenge I., Rudnick D., Knezevic S., Woodward D., Moravek M. (2013) Evapotranspiration crop coefficients for mixed riparian plant community and transpiration crop coefficients for Common reed, Cottonwood and Peach-leaf willow in the Platte River Basin, Nebraska-USA. *Journal of Hydrology 481*: 177–190.

Irmak S., Mutiibwa D. (2010) On the dynamics of canopy resistance: Generalized linear estimation and relationships with primary micrometeorological variables. *Water Resources Research 46*(8).

Ishikawa M., Mizuno K. (1998) Patent: Floating island with water purifying effect, *Plant Cultivation Bag and Method for Purifying Water. United States US 5,799,440.*

IT Corporation. (1997) Appendix A: Wetland delineation and evaluation, Remedial Investigation Data Report, May 1997.

ITRC (2003) *Technical & Regulatory Guidance Document for Constructed Treatment Wetlands.* Interstate Technology and Regulatory Council, Washington, DC. Available at: www.itrcweb.org/Guidance Documents/WTLND-1.pdf (Accessed 1/3/19).

Jackson J.A., Sees M. (2001) Rerating capacity of a constructed treatment wetland system. *Water Science and Technology 44*(11–12): 435–440.

Jackson W.A., Pardue J.H. (2000) *Natural attenuation case study for chlorobenzenes in a forested wetland.* Means J.L., Hinchee R.E. (eds.) Wetlands & Remediation: An International Conference; Battelle Press: Columbus, OH, 9–16.

Jacobs A.F., Van Boxel J.H. (1988) Changes of the displacement height and roughness length of maize during a growing season. *Agricultural and Forest Meteorology 42*(1): 53–62.

Jacobs J.M., Mergelsberg S.L., Lopera A.F., Myers D.A. (2002) Evapotranspiration from a wet prairie wetland under drought conditions: Paynes Prairie Preserve, Florida, USA. *Wetlands 22*(2): 374–385.

Jadhav R.S., Buchberger S.G. (1995) Effects of vegetation on flow through free water surface wetlands. *Ecological Engineering 5*(4): 481–496.

Jakubowski A.R., Casler M.D., Jackson R.D. (2010) The benefits of harvesting wetland invaders for cellulosic biofuel: An ecosystem services perspective. *Restoration Ecology 18*(6): 789–795.

Jalonen J., Järvelä J., Aberle J. (2012) Leaf area index as vegetation density measure for hydraulic analyses. *Journal of Hydraulic Engineering 139*(5): 461–469.

James A.I., Jawitz J.W., Muñoz-Carpena R. (2009). *Development and Implementation of a Transport and Reaction Simulation Engine (Tarse) Based on the Godunov-Mixed Finite Element Method.* U.S. Geological Survey Scientific Investigations Report 2009-5034, U.S. Geological Survey: Reston, VA.

James B.B., Bogart R. (1989) Wastewater treatment/disposal in a combined marsh and forest system provides for wildlife habitat and recreational use. In: *Constructed Wetlands for Wastewater Treatment: Municipal, Industrial, and Agricultural*, Hammer D.A. (ed.) Lewis Publishers: Chelsea, MI, pp. 597–605.

James C.S., Birkhead A.L. Jordanova A.A., O'sullivan J.J. (2004) Flow resistance of emergent vegetation. *Journal of Hydraulic Research 42*(4): 390–398.

James C.S., Goldbeck U.K., Patini A., Jordanova A.A. (2008) Influence of foliage on flow resistance of emergent vegetation. *Journal of Hydraulic Research 46*(4): 536–542

James W.F., Barko J.W. (1991) Estimation of phosphorus exchange between littoral and pelagic zones during nighttime convective circulation. *Limnology and Oceanography 36*(1): 179–187.

James W.F., Barko J.W., Butler M.G. (2004b) Shear stress and sediment resuspension in relation to submersed macrophyte biomass. *Hydrobiologia 515*(1–3): 181–191.

James W.F., Best E.P.H., Barko J.W. (2002) *Suspended Sediment Dynamics and Light Attenuation Characteristics in Peoria Lake, Illinois: Can Submersed Macrophyte Communities Improve Water Quality in This Shallow System?* Report ERDC/EL TR-02-2, prepared for the U.S. Army Corps of Engineers, Washington, DC.

James W.F., Richardson W.B., Soballe D.M. (2008b) Contribution of sediment fluxes and transformations to the summer nitrogen budget of an Upper Mississippi River backwater system. *Hydrobiologia 598*(1): 95–107.

Jampeetong A., Brix H. (2009) Nitrogen nutrition of Salvinia natans: Effects of inorganic nitrogen form on growth, morphology, nitrate reductase activity and uptake kinetics of ammonium and nitrate. *Aquatic Botany 90*(1): 67–73.

Jampeetong A., Konnerup D., Piwpuan N., Brix H. (2013) Interactive effects of nitrogen form and pH on growth, morphology, N uptake and mineral contents of *Coix lacryma-jobi* L. *Aquatic Botany 111*: 144–149.

Janardhanan L., Daroub S.H. (2010) Phosphorus sorption in organic soils in South Florida. *Soil Science Society of America Journal 74*(5): 1597–1606.

Jasper J.T., Jones Z.L., Sharp J.O., Sedlak D.L. (2014) Biotransformation of trace organic contaminants in open-water unit process treatment wetlands. *Environmental Science and Technology 48*(9): 5136–5144.

Jasper J.T., Nguyen M.T., Jones Z.L., Ismail N.S., Sedlak D.L., Sharp J.O., Luthy R.G., Horne A.J., Nelson K.L. (2013) Unit process wetlands for removal of trace organic contaminants and pathogens from municipal wastewater effluents. *Environmental Engineering Science 30*(8): 421–436.

Jasper J.T., Sedlak D.L. (2013) Phototransformation of wastewater-derived trace organic contaminants in open-water unit process treatment wetlands. *Environmental Science and Technology 47*(19): 10781–10790.

Jawitz J.W., Muñoz-Carpena R., Muller S., Grace K.A., James A.I. (2008) *Development, Testing, and Sensitivity and Uncertainty Analyses of a Transport and Reaction Simulation Engine (Tarse) for Spatially Distributed Modeling of Phosphorus in South Florida Peat Marsh Wetlands.* U.S. Geological Survey Scientific Investigations Report 2008-5029, U.S. Geological Survey: Reston, VA.

Jenkins G.A., Greenway M. (2005) The hydraulic efficiency of fringing versus banded vegetation in constructed wetlands. *Ecological Engineering 25*(1): 61–72.

Jenkins G.A., Keller R.J. (1990) Numerical Modelling of Flows in Natural Rivers. In: Conference on Hydraulics in Civil Engineering, 1990: Preprints of Papers. Institution of Engineers, Australia.

Jensen M.E., Burman R.D., Allen R.G. (1990) *Evapotranspiration and Irrigation Water Requirements.* ASCE Manuals and Reports on Engineering Practice No. 70, American Society of Civil Engineers, New York.

Jenter H. L., Schaffanek R.W., Smith T.J. (2003) Thermally-Driven Vertical Mixing in the Everglades. Poster presented April 2003, at the Greater Everglades Ecosystem Restoration Conference. Available at: https://sofia.usgs.gov/geer/2003/posters/thermmix/print.html (Accessed 12/17/17).

Jenter H.L., Duff M.P. (1999) *Locally-forced wind effects on shallow waters with emergent vegetation. Proceedings of the Third International Symposium on Ecohydraulics*; Salt Lake City, UT.

Jenter H.L., Schaffanek R.W., Smith T.J.I. (2003b) *Thermally-driven vertical mixing in the everglades.* Preprint from the Greater Everglades and Florida Bay Ecosystem Conference; Palm Harbor, Florida, 290–292.

Jetten M., Logemann S., Muyzer G.M., Robertson L.A., de Vries S., van Loosdrecht M.C.M., Kuenen J.G. (1997) Novel principles in the microbial conversion of nitrogen compounds. *Antonie Van Leeuwenhoek 71*: 75–93.

Ji Z.G., Jin K.R. (2016) An integrated environmental model for a surface flow constructed wetland: Water quality processes. *Ecological Engineering 86*: 247–261.

Jia W., Zhang J., Li P., Xie H., Wu J., Wang J. (2011) Nitrous oxide emissions from surface flow and subsurface flow constructed wetland microcosms: Effect of feeding strategies. *Ecological Engineering 37*(11): 1815–1821.

Jin G., Kelley T., Vargas N., Callahan M. (2003) Preliminary evaluation of metals removal in three pilot-scale constructed wetland systems. *Management of Environmental Quality 14*(3): 323–332.

Jin K.R., Ji Z.G. (2015) An integrated environment model for a constructed wetland–hydrodynamics and transport processes. *Ecological Engineering 84*: 416–426.

Jin K-R., Sun D. (2007) Sediment resuspension and hydrodynamics in Lake Okeechobee during the late summer. *Journal of Engineering Mechanics 133*(8): 899–910.

Jin X., Wang S., Pang Y., Zhao H., Zhou X. (2005) The adsorption of phosphate on different trophic lake sediments. *Colloids and Surfaces A: Physicochemical Engineering Aspects 254*: 241–248.

Jin K.R., Ji Z.G., James R.T. (2007) Three-dimensional water quality and SAV modeling of a large shallow lake. *Journal of Great Lakes Research*, 33(1): 28–45.

Jitka V., Jiří D., Stanislav S., Lenka M., Hana ⊠. (2017) Effect of hummock-forming vegetation on methane emissions from a temperate sedge-grass marsh. *Wetlands 37*(4): 675–686.

Johannesson K., Andersson J., Sundblad-Tonderski K. (2011) Efficiency of a constructed wetland for retention of sediment associated phosphorus. *Hydrobiologia 674*(1): 179–190.

Johannesson K.M., Kynkäänniemi P., Ulén B., Weisner S.E.B., Tonderski K.S. (2015) Phosphorus and particle retention in constructed wetlands—A catchment comparison. *Ecological Engineering 80*: 20–31.

Johnson B.J., Munafo K., Shappell L., Tsipoura N., Robson M., Ehrenfeld J., Sukhdeo M.V. (2012) The roles of mosquito and bird communities on the prevalence of West Nile virus in urban wetland and residential habitats. *Urban Ecosystems 15*(3): 513–531.

Johnson W.E., Erickson S.E., Pratt D.C., Dubbe D.R., Garvey E.G. (1987) *Typha (Cattail) Biomass Production at American Crystal Sugar Company.* University of Minnesota.: Crookston, MN. 1986 Progress Report to American Crystal Sugar Company, Hillsboro, Nporth Dakota.

Johnston C.A. (1993) Mechanisms of wetland-water quality interaction. In: *Constructed Wetlands for Water Quality Improvement*, Moshiri G.A. (ed.) Lewis Publishers: Boca Raton, FL, pp. 293–299.

Jones & Stokes Associates. (1993) *CWFATE: Constructed Wetland Fate and Aquatic Transport Evaluation Model [Computer Program].* Report to the City of Sacramento, CA.

Jones C.G., Lawton J.H., Shachak M. (1997) Positive and negative effects of organisms as physical ecosystem engineers. *Ecology 78*(7): 1946–1957.

Jones M.P., Hunt W.F. (2010) Effect of storm-water wetlands and wet ponds on runoff temperature in trout sensitive waters. *Journal of Irrigation and Drainage Engineering 136*(9): 656–661.

Jones N.L. (1999) *SEEP2D Primer.* GMS documentation, Environmental Modeling Research Laboratory, Brigham Young University: Provo, Utah.

Jones Z.L., Jasper J.T., Sedlak D.L., Sharp J.O. (2017) Sulfide-induced dissimilatory nitrate reduction to ammonium supports anammox in an open-water unit process wetland. *Applied and Environmental Microbiology 83*(15): 1–15 AEM-00782.

Jones Z.L., Mikkelson K.M., Nygren S., Sedlak D.L., Sharp J.O. (2018) Establishment and convergence of photosynthetic microbial biomats in shallow unit process open-water wetlands. *Water Research 133*: 132–141.

Jonzen N., Nolet B.A., Santamaria L., Svensson M.G.E. (2002) Seasonal herbivory and mortality compensation in a swamp-pondweed system. *Ecological Modelling 147*: 209–219.

Jordan T.E., Whigham D.F. (1988) The importance of standing dead shoots of the narrow leaved cattail, Typha angustifolia L. *Aquatic Botany 29*(4): 319–328.

Jordan T.E., Whigham D.F., Hofmockel K.H., Gerber N. (1999) Restored wetlands in crop fields control nutrient runoff. In: *Nutrient Cycling and Retention in Natural and Constructed Wetlands*, Vymazal J. (ed.) Backhuys Publishers: Leiden, The Netherlands.

Jost L. (2007) Partitioning diversity into independent alpha and beta components. *Ecology 88*(10): 2427–2439.

Juston J.M., DeBusk T.A. (2011) Evidence and implications of the background phosphorus concentration of submerged aquatic vegetation wetlands in stormwater treatment areas for Everglades restoration. *Water Resources Research 47*: WO1511.

Juston J.M., DeBusk T.A., Grace K.A., Jackson S.D. (2013) A model of phosphorus cycling to explore the role of biomass turnover in submerged aquatic vegetation wetlands for Everglades restoration. *Ecological Modeling 251*: 135–149.

Juston J.M., Kadlec R.H., DeBusk W.F., Jerauld M.J., DeBusk T.A. (2015) Persistence of legacy soil P and elevated background water P concentrations in water conservation area 2A, a Northern Everglades wetland. *Water Resources Research 51*(12): 9746–9762.

Kadlec J.A. (1983) Water budgets for small di9ked marshes. *Journal of the American Water Resources Association 19*(2): 223–230.

Kadlec J.A. (1986) Input-output nutrient budgets for small diked marshes. *Canadian Journal of Fisheries and Aquatic Sciences 43*(10): 2009–2016.

Kadlec J.A., Murkin H.R., van der Valk A.G. (2000) The baseline and deep flooding years. Chapter 4. In: *Prairie Wetland Ecology*, Murkin H.R., van der Valk A.G., Clark W.R. (eds.) Iowa State University Press: Ames, IA, pp. 55–74.

Kadlec R.H. (1985) Aging phenomena in wastewater wetlands. In: *Ecological Considerations in Wetland Treatment of Municipal Wastewaters*, Kaynor E.R., Godfrey P.J., Benforado J. (eds.) Van Nostrand Reinhold: New York, pp. 338–347.

Kadlec R.H. (1987) *Wetland hydrology and water pollution control functions.* Kusler J.A., Brooks G. (eds.) Proceedings of the National Wetland Symposium on Wetland Hydrology, 16–18 September 1987; Chicago, IL, 168–173.

Kadlec R.H. (1990) Overland flow in wetlands: Vegetation resistance. *Journal of Hydraulic Engineering 116*(5): 691–706.

Kadlec R.H. (1994) Detention and mixing in free water wetlands. *Ecological Engineering 3*(4): 345–380.

Kadlec R.H. (1997a) An autobiotic wetland phosphorus model. *Ecological Engineering 8*(2): 145–172.

Kadlec R.H. (1997b) Deterministic and stochastic aspects of constructed wetland performance and design. *Water Science and Technology 35*(5): 149–156.

Kadlec R.H. (1998) *Wetland Utilization for Management of Community Wastewater*, 1997 Operations Summary (March 1998). Report to Michigan Department of Natural Resources.

Kadlec R.H. (1999a) Chemical, physical and biological cycles in treatment wetlands. *Water Science and Technology 40*(3): 37–44.

Kadlec R.H. (1999b) Transport of phosphorus in wetlands. In: *Phosphorus Biogeochemistry in Subtropical Ecosystems*, Reddy K.R., O'Connor G.A., Schelske C.L. (eds.) Lewis Publishers: Boca Raton, FL, pp. 355–376.

Kadlec R.H. (1999c) The limits of phosphorus removal in wetlands. *Wetlands Ecology and Management 7*: 165–175.

Kadlec R.H. (2000) The inadequacy of first-order treatment wetland models. *Ecological Engineering 15* (1–2): 105–119.

Kadlec R.H. (2001a) Thermal environments of subsurface treatment wetlands. *Water Science and Technology 44*(11–12): 251–258.

Kadlec R.H. (2001b) Phosphorus dynamics in event driven wetlands. In: *Transformation of Nutrients in Natural and Constructed Wetlands*, Vymazal J. (ed.) Backhuys Publishers: Leiden, The Netherlands, pp. 365–392.

Kadlec R.H. (2003a) Effects of pollutant speciation in treatment wetlands design. *Ecological Engineering 20*(1): 1–16.

Kadlec R.H. (2003b) Integrated natural systems for landfill leachate treatment. In: *Wetlands - Nutrients, Metals, and Mass Cycling*, Vymazal J. (ed.) Backhuys Publishers: Leiden, The Netherlands, pp. 1–33.

Kadlec R.H. (2003c) Pond and wetland treatment. *Water Science and Technology 48*(5): 1–8.

Kadlec R.H. (2005a) Vegetation effects on ammonia reduction in treatment wetlands. In: *Natural and Constructed Wetlands: Nutrients, Metals, and Management*, Vymazal J. (ed.) Backhuys Publishers: Leiden, The Netherlands, pp. 233–260.

Kadlec R.H. (2005b) Nitrogen farming: The use of wetlands for nitrate removal. *Journal of Environmental Science and Health, Part A: Toxic/Hazardous Substances & Environmental Engineering 40*(6–7): 1307–1330.

Kadlec R.H. (2005c) Constructed wetlands to remove nitrate. In: *Nutrient Management in Agricultural Watersheds: A Wetland Solution*, Dunne E.J., Reddy K.R., Carton O.T. (eds.) Wageningen Academic Publishers: Wageningen, The Netherlands, pp. 132–143.

Kadlec R.H. (2005d) Wetland to pond treatment gradients. *Water Science and Technology 51*(9): 291–298.

Kadlec R.H. (2006) Water temperature and evapotranspiration in surface flow wetlands in hot arid climate. *Ecological Engineering 26*: 328–340.

Kadlec R.H. (2007) The effects of deep zones on wetland nitrogen processing. *Water Science and Technology 56*(3): 101–108.

Kadlec R.H. (2008a) The effects of wetland vegetation and morphology on nitrogen processing. *Ecological Engineering 33*(2): 126–141.

Kadlec R.H. (2008b) *The lagoon-wetland treatment system at Onaway, Michigan.* Proceedings of the 37th Annual WEAO Technical Symposium, May, 2008; Collingwood, Ontario.

Kadlec R.H. (2009a) Wastewater treatment at the Houghton Lake wetland: Hydrology and water quality. *Ecological Engineering 35*(9): 1287–1311.

Kadlec R.H. (2009b) Wastewater treatment at the Houghton Lake wetland: Soils and sediments. *Ecological Engineering 35*(9): 1333–1348.

Kadlec R.H. (2009c) Wastewater treatment at the Houghton Lake wetland: Temperatures and energy balance. *Ecological Engineering 35*(9): 1349–1356.

Kadlec R.H. (2010) Nitrate dynamics in event driven wetlands. *Ecological Engineering 36*(4): 503–516.

Kadlec R.H. (2011) *Commerce Township treatment wetland*. Proceedings of the 40th Annual WEAO Technical Symposium, April, 2011; Toronto, Ontario.

Kadlec R.H. (2012) Nitrate removal in treatment marshes. *Critical Reviews in Environmental Science and Technology 42*(9): 934–1005.

Kadlec R.H. (2016) Large constructed wetlands for phosphorus control: A review. *Water 8*(6): 243.

Kadlec R.H., Alvord H.H. (1989) Mechanisms of water quality improvement in wetland treatment systems. In: *Wetlands Concerns and Successes*, Fisk D.W. (ed.) American Water Resources Association: Middleburg, Virginia, pp. 489–498.

Kadlec R.H., Bastiaens W., Urban D.T. (1993) Hydrological design of free water surface treatment wetlands. In: *Constructed Wetlands for Water Quality Improvement*, Moshiri G.A. (ed.) Lewis Publishers: Boca Raton, FL, pp. 77–86.

Kadlec R.H., Bays J.S., Mokry L.E., Andrews D., Ernst M.R. (2011) Performance analysis of the Richland-Chambers treatment wetlands. *Ecological Engineering 37*(2): 176–190.

Kadlec R.H., Bevis F.B. (1990) Wetlands and wastewater: Kinross, Michigan. *Wetlands 10*(1): 77–92.

Kadlec R.H., Bevis F.B. (2009) Wastewater treatment at the Houghton Lake wetland: Vegetation response. *Ecological Engineering 35*(9): 1312–1332.

Kadlec R.H., Burgoon P.S., Henderson M.E. (1997) Integrated natural systems for treating potato processing wastewater. *Water Science and Technology 35*(5): 263–270.

Kadlec R.H., Cuvellier C., Stober T. (2010b) Performance of the Columbia, Missouri treatment wetland. *Ecological Engineering 36*(5): 672–684.

Kadlec R.H., Hammer D.E. (1988) Modeling nutrient behavior in wetlands. *Ecological Modelling 40*: 37–66.

Kadlec R.H., Hammer D.E., Girts M. (1990) A total evaporative constructed wetland treatment system. In: *Constructed Wetlands in Water Pollution Control*, Cooper P.F., Findlater B.C. (eds.) Pergamon Press: Oxford, United Kingdom, pp. 127–138.

Kadlec R.H., Hammer D.E., Nam I-S., Wilkes J.O. (1981) The hydrology of overland flow in wetlands. *Chemical Engineering Communications 9*: 331–344.

Kadlec R.H., Keoleian G.A. (1986) Metal ion exchange on peat. In: *Peat and Water*, Fuchsman C.H. (ed.) Elsevier Applied Science Publishers: New York, pp. 61–93.

Kadlec R.H., Knight R.L. (1996) *Treatment Wetlands*. First edition, CRC Press: Boca Raton, FL.

Kadlec R.H., Martin D.C., Tsao D. (2012a) Constructed marshes for control of chlorinated ethenes: An 11-year study. *Ecological Engineering 46*: 11–23.

Kadlec R.H., Pries J., Lee K. (2012b) The Brighton treatment wetland. *Ecological Engineering 47*: 56–70.

Kadlec R.H., Pries J., Mustard H. (2006) Muskrats (*Ondatra zibethicus*) in treatment wetlands. *Ecological Engineering 29*(2): 143–153.

Kadlec R.H., Reddy K.R. (2001) Temperature effects in treatment wetlands. *Water Environment Research 73*(5): 543–557.

Kadlec R.H., Robbins J.A. (1984) Sedimentation and sediment accretion in Michigan coastal wetlands (USA). *Journal of Chemical Geology 44*(1/3): 119–150.

Kadlec R.H., Roy S.B., Munson R.K., Charlton S., Brownlie W. (2010a) Water quality performance of treatment wetlands in the Imperial Valley, California. *Ecological Engineering 36*(8): 1093–1107.

Kadlec R.H., Srinivasan K. (1995) *Wetland Treatment of Oil and Gas Well Wastewaters*, Final Report DOE/MT/92010-10 (DE95000176), U.S. Department of Energy: Bartlesville, OK.

Kadlec R.H., Walker W.W. (1999) Management models to evaluate phosphorus impacts in wetlands. In: *Phosphorus Biogeochemistry in Subtropical Ecosystems*, Reddy K.R., O'Connor G.A., Schelske C.L. (eds.) Lewis Publishers: Boca Raton, FL, pp. 621–642.

Kadlec R.H., Wallace S.D. (2009) *Treatment Wetlands*. 2nd edition, CRC Press: Boca Raton, FL.

Kadlec R.H., Williams R. B., Scheffe R.D. (1987) Wetland evapotranspiration in temperate and arid climates. In: *The Ecology and Management of Wetlands*, Hook D.D. (ed.) Croom Helm: Beckenham, United Kingdom, pp. 146–160.

Kadlec R.H., Zmarthie L.A. (2010) Wetland treatment of leachate from a closed landfill. *Ecological Engineering 36*(7): 946–957.

Kania B.G., Stewart F.M., 2011. Patent application: Adjustably rigid floating island system. United States US 2011/030602 A1.

Kania B.G., Stewart F.M. (2012) *Patent: Highly Buoyant and Semi-Rigid Floating Islands*. United States US 8,132,364 B2.

Kansiime F., Nalubega M. (1999) Wastewater treatment by a natural wetland: the Nakivubo swamp, Uganda. Processes and implementations, Ph. D. Dissertation, Wageningen Agricultural University, Netherlands.

Kansiime F., van Bruggen J.J.A. (2001) Distribution and retention of faecal coliforms in the Nakivubo wetland in Kampala, Uganda. *Water Science and Technology 44*(11–12): 199–206.

Kao J.T., Titus J.E., Zhu W.X. (2003) Differential nitrogen and phosphorus retention by five wetland plant species. *Wetlands 23*(4): 979–987.

Karen D.J., Joab B.M., Wallin J.M., Johnson K.A. (1998) Partitioning of chlorpyrifos between water and an aquatic macrophyte (*Elodea densa*). *Chemosphere 37*(8): 1579–1586.

Karim M.R., Manshadi F.D., Karpiscak M.M., Gerba C.P. (2004) The persistence and removal of enteric pathogens in constructed wetlands. *Water Research 38*: 1831–1837.

Karpuzcu M.E., Stringfellow W.T. (2012) Kinetics of nitrate removal in wetlands receiving agricultural drainage. *Ecological Engineering 42*: 295–303.

Karr J.R. (1991) Biological integrity: A long-neglected aspect of water resource management. *Ecological Applications 1*: 66–84.

Karunaratne S., Asaeda T. (2000) Verification of a mathematical growth model of Phragmites australis using field data from two Scottish lochs. *Folia Geobotanica 35*(4): 419–432.

Karunaratne S., Asaeda T., Yutani K. (2003) Growth performance of *Phragmites Australis* in Japan: Influence of geographic gradient. *Environmental and Experimental Botany 50*(1): 51–66.

Katagi T. (2002) Abiotic hydrolysis of pesticides in the aquatic environment. *Reviews of Environmental Contamination and Toxicology 175*: 79–262.

Katoh S., Horiuchi J., Yoshida F. (2015) *Biochemical Engineering*. Second edition, Wiley-VCH: New York.

Katsoyiannis I.A., Zouboulis A.I. (2004) Application of biological processes for the removal of arsenic from groundwaters. *Water Research 38*(1): 17–26.

Kavanaugh M., Deeb R.A., Markowitz D., Dzombak D.A., Zheng A., Theis T.L., Young T.C., Luthy R.G. (2003) Transport and fate of cyanide species in surface waters. In: *Cyanide Formation and Fate in Complex Effluents and Its Relation to Water Quality Criteria*, Water Environment Research Foundation: Alexandria, Virginia.

Kayranli B., Scholz M., Mustafa A., Hedmark Å. (2010) Carbon storage and fluxes within freshwater wetlands: A critical review. *Wetlands 30*(1): 111–124.

Kechavarzi C., Dawson Q., Leeds-Harrison P.B. (2010) Physical properties of low-lying agricultural peat soils in England. *Geoderma 154*(3–4): 196–202.

Keddy P.A. (2010) *Wetland Ecology: Principles and Conservation. 2nd Edition*, Cambridge University Press: Cambridge, United Kingdom.

Keefe S.H., Barber L.B., Runkel R.L., Ryan J.N. (2004b) Fate of volatile organic compounds in constructed wastewater treatment wetlands. *Environmental Science and Technology 38*(7): 2209–2216.

Keefe S.H., Daniels J.S.T., Runkel R.L., Wass R.D., Stiles E.A., Barber L.B. (2010) Influence of hummocks and emergent vegetation on hydraulic performance in a surface flow wastewater treatment wetland. *Water Resources Research 46*(11): W11518.

Keefe S.H., Runkel R.L., Ryan J.N., McKnight D.M., Wass R.D. (2004a) Conservative and reactive solute transport in constructed wetlands. *Water Resources Research 40*(1): W01201.

Keeney D.R. (1973) The nitrogen cycle in sediment-water system. *Journal of Environmental Quality 2*: 15–29.

Keller C.H., Bays J.S. (2001) Tracer studies in treatment wetlands. Pries J.H. (ed.) *Treatment Wetlands for Water Quality Improvement: Quebec 2000 Conference Proceedings*; Pandora Press: Kitchener, Ontario, 173–182.

Kellner E., Pires E. C. (2002) The influence of thermal stratification on the hydraulic behavior of waste stabilization ponds. *Water Science and Technology 45*(1): 41–48.

Kemp W.M., Murray L. (1986) Oxygen release from roots of the submersed macrophyte *Potamogeton perfoliatus* L.: Regulating factors and ecological implications. *Aquatic Botany 26*: 271–283.

Kettenring K.M., Galatowitsch S.M. (2011) Seed rain of restored and natural prairie wetlands. *Wetlands 31*(2): 283–294.

Khatiwada N.R., Polprasert C. (1999a) Assessment of effective specific surface area for free water surface constructed wetlands. *Water Science and Technology 40*(3): 83–89.

Khatiwada N.R., Polprasert C. (1999b) Kinetics of fecal coliform removal in constructed wetlands. *Water Science and Technology 40*(3): 109–116.

Kiene R.P., Hines M.E. (1995) Microbial formation of dimethyl sulfide in anoxic sphagnum peat. *Applied and Environmental Microbiology 61*(7): 2716–2720.

Killham K. (1994) *Soil Ecology.* Cambridge University Press: Cambridge, United Kingdom.

Kim B.R., Podsialik D.H., Kalis E.M., Hartlund J.L., Gaines W.A. (1998) Photochemical destruction of cyanide in landfill leachate. *Journal of Environmental Engineering 124*(11): 1108–1113.

Kim H., Ogram A., Bae H.S. (2017) Nitrification, Anammox and denitrification along a nutrient gradient in the Florida Everglades. *Wetlands 37*(2): 391–399.

Kim J.G. (2003) Response of sediment chemistry and accumulation rates to recent environmental changes in the Clear Lake watershed, California, USA. *Wetlands 23*(1): 95–103.

Kinniburgh D.G., Smedley P.L. (eds.) (2001) *Arsenic Contamination of Groundwater in Bangladesh. British Geological Survey Technical ReportWC/00/19, Volume 2.* Final Report. British Geological Survey: Keyworth, Available at: www.bgs.ac.uk/downloads/start.cfm?id=2222 (Accessed 9/4/18).

Kitchell J.F., Schindler D.E., Herwig B.R., Post D.M., Olson M.H., Oldham M. (1999) Nutrient cycling at the landscape scale: The role of diel foraging migrations by geese at the Bosque del Apache National Wildlife Refuge, New Mexico. *Limnology and Oceanography 44*(3part2): 828–836.

Kjellin J., Wörman A., Johansson H., Lindahl A. (2007) Controlling factors for water residence time and flow patterns in Ekeby treatment wetland, Sweden. *Advances in Water Resources 30*(4): 838–850.

Kleeberg A., Köhler J.A.N., Sukhodolova T., Sukhodolov A. (2010) Effects of aquatic macrophytes on organic matter deposition, resuspension and phosphorus entrainment in a lowland river. *Freshwater Biology 55*(2): 326–345.

Klopatek M. (1978) Nutrient dynamics of freshwater riverine marshes and the role of emergent macrophytes. In: *Freshwater Wetlands: Ecological Processes and Management Potential*, Good R.E., Whigham D.F., Simpson R.L. (eds.) Academic Press: New York, pp. 195–216.

Knight R., Adams R., O'Brien C., Davis E. (1998) Beltway 8 wetland water quality project: Constructed wetlands for storm water polishing and wetland mitigation banking. *Transportation Research Record: Journal of the Transportation Research Board 1626*: 11–20.

Knight R.L. (1987) Effluent distribution and basin design for enhanced pollutant assimilation by freshwater wetlands. In: *Aquatic Plants for Water Treatment and Resource Recovery*, Reddy K.R., Smith W.H. (eds.) Magnolia Publishing: Orlando, FL, pp. 913–921.

Knight R.L., Clarke R.A., Bastian R.K. (2001) Surface flow (SF) treatment wetlands as a habitat for wildlife and humans. *Water Science and Technology 44*(11–12): 27–38.

Knight R.L., Gu B., Clarke R.A., Newman J.M. (2003a) Long-term phosphorus removal in Florida aquatic systems dominated by submerged aquatic vegetation. *Ecological Engineering 20*(1): 45–63.

Knight R.L., Hilleke J., Grayson S. (1994) Design and performance of the Champion pilot-constructed wetland treatment system. *TAPPI Journal 77*(5): 240–245.

Knight R.L., Iverson M.E. (1990) Design of the Fort Deposit, Alabama constructed wetlands treatment system. In: *Constructed Wetlands in Water Pollution Control*, Cooper P.F., Findlater B.C. (eds.) Pergamon Press: Oxford, United Kingdom, pp. 521–524.

Knight R.L., Kadlec R.H., Ohlendorf H.M. (1997) *The Use of Treatment Wetlands for Petroleum Industry Effluents*, Prepared for the American Petroleum Institute (API), Publication Number 4672, API Publishing Services: Washington, DC.

Knight R.L., Kadlec R.H., Ohlendorf H.M. (1999) The use of treatment wetlands for petroleum industry effluents. *Environmental Science and Technology 33*(7): 973–980.

Knight R.L., Payne V.W.E., Borer R.E., Clarke R.A., Pries J.H. (2000) Constructed wetlands for livestock wastewater management. *Ecological Engineering 15*(1–2): 41–55.

Knight R.L., Ruble R., Kadlec R.H., Reed S.C. (1993) Wetlands for wastewater treatment performance database. In: *Constructed Wetlands for Water Quality Improvement*, Moshiri G.A. (ed.) Lewis Publishers: Boca Raton, FL, pp. 35–58.

Knight R.L., Walton W.E., O'Meara G., Reisen W.K., Wass R.D. (2004) Strategies for effective mosquito control in constructed treatment wetlands. *Ecological Engineering 21*: 211–232.

Knowles G., Downing A.L., Barrett M.J. (1965) Determination of kinetic constants for nitrifying bacteria in mixed culture, with the aid of an electronic computer. *Microbiology 38*(2): 263–278.

Knowles R. (1982) Denitrification. *Microbial Reviews 46*: 43–70.

Knowlton M.F., Cuvellier C., Jones J.R. (2002) Initial performance of a high capacity surface-flow treatment wetland. *Wetlands 22*(3): 522–527.

Knox A.S., Gamerdinger A.P., Adriano D.C., Kolka R.K., Kaplan D.I. (1999) Sources and practices contributing to soil contamination. In: *Bioremediation of Contaminated Soils*, Adriano D.C., Bollag J.-M., Frankenberger W.T., Sims R.C. (eds.) American Society of Agronomy: Madison, WI, pp. 53–87.

Knox A.S., Nelson E.A., Halverson N.V., Gladden J.B. (2010) Long-term performance of a constructed wetland for metal removal. *Soil and Sediment Contamination 19*(6): 667–685.

Knox A.S., Paller M.H. (2018) Metal retention in constructed wetland sediment. Paper presented st the 12th International Symposium on Biogeochemistry of Wetlands; Coral Springs, FL.

Knox A.S., Paller M.H., Nelson E.A., Specht W.L., Halverson N.V., Gladden J.B. (2006) Metal distribution and stability in constructed wetland sediment. *Journal of Environmental Quality 35*(5): 1948–1959.

Koch B.J., Febria C.M., Gevrey M., Wainger L.A., Palmer M.A. (2014) Nitrogen removal by stormwater management structures: A data synthesis. *JAWRA Journal of the American Water Resources Association 50*(6): 1594–1607.

Koch M.S., Mendelssohn I.A. (1989) Sulphide as a soil phytotoxin: Differential responses in two marsh species. *Journal of Ecology 77*(2): 565–578.

Koch M.S., Mendelssohn I.A., McKee K.L. (1990) Mechanism for the hydrogen sulfide-induced growth limitation in wetland macrophytes. *Limnology and Oceanography 35*(2): 399–408.

Koch M.S., Rawlik P.S. (1993) Transpiration and stomotal conductance of two wetland macrophytes (*Cladium jamaicense* and *Typha domingensis*) in the subtropical everglades. *American Journal of Botany 80*: 1146–1154.

Koelmans A.A., Radovanovic H. (1998) Prediction of trace metal distribution coefficients (K_D) for aerobic sediments. *Water Science and Technology 37*(6–7): 71–78.

Koenig A., Liu L.H. (2001) Kinetic model of autotrophic denitrification in sulphur packed-bed reactors. *Water Research 35*(8): 1969–1978.

Koerselman W., Beltman B. (1988) Evapotranspiration from fens in relation to Penman's potential free water evaporation (E_o) and pan evaporation. *Aquatic Botany 31*: 307–320.

Koike I., Hattori A. (1978) Denitrification and ammonia formation in anaerobic coastal sediment. *Applied and Environmental Microbiology 35*: 278–282.

Kominkova D., Kuehn K.A., Büsing N., Steiner D., Gessner M.O. (2000) Microbial biomass, growth, and respiration associated with submerged litter of Phragmites australis decomposing in a littoral reed stand of a large lake. *Aquatic Microbial Ecology 22*(3): 271–282.

Komor A., Fox P. (2001) Nitrate removal in constructed wetlands: Batch tests versus field results. Proceedings of the WEFTEC 2001 National Conference, 74th Annual Conference and Exhibition; Water Environment Federation: Alexandria, Virginia.

Komor A., Fox P. (2002) Evaluation of denitrification rates and mechanisms in microcosms with sediments and plants. Proceedings of the 75th Annual WEFTEC Conference in Anaheim, California; Water Environment Federation: Alexandria, Virginia.

Konnerup D., Brix H. (2010) Nitrogen nutrition of Canna indica: Effects of ammonium versus nitrate on growth, biomass allocation, photosynthesis, nitrate reductase activity and N uptake rates. *Aquatic Botany 92*(2): 142–148.

Konyha K.D., Shaw D.T., Weiler K.W. (1995) Hydrologic design of a wetland: Advantages of continuous modeling. *Ecological Engineering 4*(2): 99–116.

Korlyakov K.A., Nokhrin D.Y. (2011) Specific surface area and biomass of zoophytos in different waterbodies of the southern trans-Ural region. *Inland Water Biology 4*(3): 327–331.

Koskiaho J. (2003) Flow velocity retardation and sediment retention in two constructed wetland ponds. *Ecological Engineering 19*(5): 325–337.

Koskiaho J., Ekholm P., Räty M., Riihimäki J., Puustinen M. (2003) Retaining agricultural nutrients in constructed wetlands - experiences under boreal conditions. *Ecological Engineering 20*(1): 89–104.

Kostich M.S., Länge R. (2015) Ecotoxicology, environmental risk assessment and potential impact on human health. In: *Pharmaceuticals in the Environment*, Hester R.E., Harrison R.M. (eds.) Royal Society of Chemistry: London, United Kingdom, pp. 180–215.

Kostich M.S., Lazorchak J.M. (2008) Risks to aquatic organisms posed by human pharmaceutical use. *Science of the Total Environment 389*(2–3): 329–339.

Kovacic D.A., David M.B., Gentry L.E., Starks K.M., Cooke R.A. (2000) Effectiveness of constructed wetlands in reducing nitrogen and phosphorus export from agricultural tile drainage. *Journal of Environmental Quality 29*: 1262–1274.

Kraus M.P. (1977) Bacterial indicators and potential health hazards of aquatic viruses. In: *Bacterial Indicators/Health Hazards Associated with Water*, Hoadley A.W., Dutka B.J. (eds.) American Society for Testing and Materials (ASTM): Philadelphia, Pennsylvania, pp. 196–217.

Kreissl J.F., Brown D.S. (2000) "Blue collar" workers: A new US Environmental Protection Agency manual addresses the use of constructed wetlands as a major wastewater treatment process. *Water Environment and Technology 12*(10): 36–40.

Krishnan S.B., Smith J.E. (1987) Public health issues of aquatic systems used for wastewater treatment. In: *Aquatic Plants for Water Treatment and Resource Recovery*, Reddy K.R., Smith W.H. (eds.) Magnolia Publishing: Orlando, FL, pp. 855–878.

Krom M.D., Berner R.A. (1980) The diffusion coefficients of sulfate, ammonium, and phosphate ions in anoxic marine sediments. *Limnology and Oceanography 25*(2): 327–337.

Krone-Davis P., Watson F., Los Huertos M., Starner K. (2013) Assessing pesticide reduction in constructed wetlands using a tanks-in-series model within a Bayesian framework. *Ecological Engineering 57*: 342–352.

Kronzucker H.J., Glass A.D.M., Siddiqi M.Y., Kirk J.D. (2000) Comparative kinetic analysis of ammonium and nitrate acquisition by tropical lowland rice: Implications for rice cultivation and yield potential. *New Phytologist 145*(3): 471–476.

Kroodsma D.E. (1979) Habitat values for nongame wetland birds. In: *Wetland Functions and Values the State of Our Understanding*, Greeson P.E., Clark J.R., Clark I.E. (eds.) American Water Resources Association: Minneapolis, MN, pp. 320–343.

Krusi B.O., Wein R.W. (1988) Experimental studies on the resiliency of floating *Typha* mats in a freshwater marsh. *Journal of Ecology 76*: 60–72.

Kryven I., Lazzari S., Storti G. (2014) Population balance modeling of aggregation and coalescence in colloidal systems. *Macromolecular Theory and Simulations 23*(3): 170.

Kuehn E., Moore J.A. (1995) Variability of treatment performance in constructed wetlands. *Water Science and Technology 32*(3): 241–250.

Kuehn K.A., Ohsowski B.M., Francoeur S.N., Neely R.K. (2011) Contributions of fungi to carbon flow and nutrient cycling from standing dead *Typha angustifolia* leaf litter in a temperate freshwater marsh. *Limnology and Oceanography 56*(2): 529–539.

Kuehn K.A., Suberkropp K. (1998) Decomposition of standing litter of the freshwater emergent macrophyte *Juncus effusus*. *Freshwater Biology 40*(4): 717–727.

Kühl H., Kohl J.-G. (1993) Seasonal nitrogen dynamics in reed beds (*Phragmites australis* (Cav.) Trin ex. Steudel) in relation to productivity. *Hydrobiologia 251*: 1–12.

Kuiper J.J., Verhofstad M.J., Louwers E.L., Bakker E.S., Brederveld R.J., van Gerven L.P., Janssen A. B., de Klein J.J., Mooij W.M. (2017) Mowing submerged macrophytes in shallow lakes with alternative stable states: Battling the good guys? *Environmental Management 59*(4): 619–634.

Kumar A.K., Chiranjeevi P., Mohanakrishna G., Mohan S.V. (2011) Natural attenuation of endocrine-disrupting estrogens in an ecologically engineered treatment system (EETS) designed with floating, submerged and emergent macrophytes. *Ecological Engineering 37*(10): 1555–1562.

Kustas W.P. (1990) Estimates of evapotranspiration with a one-and two-layer model of heat transfer over partial canopy cover. *Journal of Applied Meteorology 29*(8): 704–715.

Kustas W.P., Norman J.M. (1997) A two-source approach for estimating turbulent fluxes using multiple angle thermal infrared observations. *Water Resources Research 33*(6): 1495–1508.

Kustas W.P., Norman J.M. (1999) Evaluation of soil and vegetation heat flux predictions using a simple two-source model with radiometric temperatures for partial canopy cover. *Agricultural and Forest Meteorology 94*(1): 13–29.

Kustas W.P., Rango A., Uijlenhoet R. (1994) A simple energy budget algorithm for the snowmelt runoff model. *Water Resources Research 30*(5): 1515–1527.

Kuusemets V., Lohmus K., Mander Ü. (2002) Nitrogen and phosphorus assimilation and biomass production by Scirpus sylvaticus and Phragmites australis in a horizontal subsurface flow constructed wetland. Mbwette T.S.A. (ed.) Proceedings of the 8th International Conference on Wetland Systems for Water Pollution Control, 16–19 September 2002; Comprint International Limited: University of Dar Es Salaam, Tanzania, 930–937.

Kuyucak N., Akcil A. (2013) Cyanide and removal options from effluents in gold mining and metallurgical processes. *Minerals Engineering 50*: 13–29.

Kuzyakov Y. (2006) Sources of CO2 efflux from soils and review of partitioning methods. *Soil Biology & Biochemistry 38*: 425–448.

Kvet J. (1982) Production of organic matter in macrophyte stands (in Czech). Proceedings of Macrophytes in Water Management, Water Hygiene, and Fishery; Dum Techniky CSVTS: Ceské Budejovice, Czech Republic, 73–81.

Kvet J., Jeník J., Soukupová L. (2002) *Freshwater Wetlands and Their Sustainable Future*. Parthenon Publishing Group: New York.

Kynkäänniemi P., Ulén B., Torstensson G., Tonderski K.S. (2013) Phosphorus retention in a newly constructed wetland receiving agricultural tile drainage water. *Journal of Environmental Quality 42*(2): 596–605.

Laanbroek H.J. (1990) Bacterial cycling of minerals that affect plant growth in waterlogged soils: A review. *Aquatic Botany 38*: 105–125.

Lafleur P. (1988) Leaf conductance of four species growing in a subarctic marsh. *Canadian Journal of Botany 66*(7): 1367–1375.

Lafleur P.M. (1990) Evapotranspiration from sedge-dominated wetland surfaces. *Aquatic Botany 37*: 341–353.

Lagrace S., Miller C.D., Gearheart R.A. (2000) Emergent macrophyte density in a constructed wetland system for wastewater treatment. Proceedings of a Constructed Wetlands Conference held at Humboldt State University in 2000; Arcata, California.

Lai D.Y.F., Lam K.C. (2009) Phosphorus sorption by sediments in a subtropical constructed wetland receiving stormwater runoff. *Ecological Engineering 35*: 735–743.

Lai W.L., Zhang Y., Chen Z.H. (2012) Radial oxygen loss, photosynthesis, and nutrient removal of 35 wetland plants. *Ecological Engineering 39*: 24–30.

Laio F., Tamea S., Ridolfi L., D'Odorico P., Rodriguez-Iturbe I. (2009) Ecohydrology of groundwater-dependent ecosystems: 1. Stochastic water table dynamics. *Water Resources Research 45*: 5.

Lakatos G. (1998) Hungary. In: *Constructed Wetlands for Wastewater Treatment in Europe*, Vymazal J., Brix H., Cooper P.F., Green M.B., Haberl R. (eds.) Backhuys Publishers: Leiden, The Netherlands, pp. 191–206.

Lakshman G. (1982) *A Demonstration Project at Humboldt to Provide Tertiary Treatment to the Municipal Effluent Using Aquatic Plants*, SRC Publication No. E-820-11-B-82. Saskatchewan Research Council.

Lal A.M.W., Moustafa M.Z., Wilcox W.M. (2015) The use of discharge perturbations to understand in situ vegetation resistance in wetlands. *Water Resources Research 51*: 2477–2497.

Lambert W. (1987) Patent: Culvert beaver block. United States US 4,662,782.

Lamers L.P.M., Falla S.-J., Samborska E.M., van Dulken I.A.R., van Hengstum G., Roelofs J.G.M. (2002) Factors controlling the extent of eutrophication and toxicity in sulfate-polluted freshwater wetlands. *Limnology and Oceanography 47*(2): 585–593.

Lamers L.P.M., Tomassen H.B.M., Roelofs J.G.M. (1998) Sulfate-induced eutrophication and phytotoxicity in freshwater wetlands. *Environmental Science and Technology 32*(2): 199–205.

Lanctôt C., Robertson C., Navarro-Martín L., Edge C., Melvin S.D., Houlahan J., Trudeau V.L. (2013) Effects of the glyphosate-based herbicide Roundup WeatherMax® on metamorphosis of wood frogs (*Lithobates sylvaticus*) in natural wetlands. *Aquatic Toxicology 140*: 48–57.

Langston M.A., Kent D.M. (1997) Fish recruitment to a constructed wetland. *Journal of Freshwater Ecology 12*(1): 123–129.

Larsen L.G. Harvey J.W., Noe G.B., Crimaldi J.P. (2009a) Predicting organic floc transport dynamics in shallow aquatic ecosystems: Insights from the field, the laboratory, and numerical modeling. *Water Resources Research 45*: W01411.

Larsen M., Trapp S., Pirandello A. (2004) Removal of cyanide by woody plants. *Chemosphere 54*(3): 325–333.

Larsen M., Ucisik A.S., Trapp S. (2005) Uptake, metabolism, accumulation, and toxicity of cyanide in willow trees. *Environmental Science and Technology 39*: 2135–3142.

Larson A.C., Gentry L.E., David M.B., Cooke R.A., Kovacic D.A. (2000) The role of seepage in constructed wetlands receiving agricultural tile drainage. *Ecological Engineering 15*(1–2): 91–104.

Larson W. (2015) *New Estimates of Value of Land of the United States*. US Department of Commerce, Bureau of Economic Analysis: Washington, DC. Available at: www.bea.gov/research/papers/2015/new-estimates-value-land-united-states (Accessed 12/17/18).

Laskov C., Horn O., Hupfer M. (2006) Environmental factors regulating radial oxygen loss from roots of *Myriophyllum spicatum* and *Potamogeton crispus*. *Aquatic Botany 84*: 333–340.

Latchum J.A. (1996) *Ecological Engineering Factors of a Constructed Wastewater Wetland*. MS Thesis, University of Maryland: College Park, MD.

Laws E.A. (1993) *Aquatic Pollution*. Second edition, John Wiley & Sons: New York.

Laybourn-Parry J., Boyall J., Rogers P. (1999) The role of flagellated and ciliated protozoa in lagoon and grass filter treatment systems. *Water Research 33*(13): 2971–2977.

League M.T., Colbert E.P., Seliskar D.M., Gallagher J.L. (2006) Rhizome growth dynamics of native and exotic haplotypes of *Phragmites australis* (common reed). *Estuaries and Coasts 29*(2): 269–276.

LeBlanc D.J. (1994) Nutria. In: *The Handbook: Prevention and Control of Wildlife Damage*, Wildlife Damage Management Technical Series, 16. U.S. Department of Agriculture, Animal & Plant Health Inspection Service Wildlife Services, Port Allen, Louisiana. Available at: http://icwdm.org/handbook/rodents/Nutria.asp (Accessed 12/10/18).

Leclerc H., Mossel D.A., Trinel P.A., Gavini F. (1977) Microbiological monitoring - a new test for fecal contamination. In: *Bacterial Indicators/Health Hazards Associated with Water*, Hoadley A.W., Dutka B.J. (eds.) American Society for Testing and Materials (ASTM): Philadelphia, Pennsylvania, pp. 23–36.

Lee S., Gan J., Kim J.S., Kabashima J.N., Crowley D.E. (2004) Microbial transformation of pyrethroid insecticides in aqueous and sediment phases. *Environmental Toxicology and Chemistry 23*(1): 1–6.

Lee S., Klein A.G., Over T.M. (2005) A comparison of MODIS and NOHRSC snow-cover products for simulating streamflow using the Snowmelt Runoff Model. *Hydrological Processes 19*(15): 2951–2972.

Lee S., Maniquiz-Redillas M.C., Kim L.H. (2014) Settling basin design in a constructed wetland using TSS removal efficiency and hydraulic retention time. *Journal of Environmental Sciences 26*(9): 1791–1796.

Leisenring M., Clary J., Hobson P. (2014) *International Stormwater Best Management Practices (BMP) Database Pollutant Category Statistical Summary Report: Solids, Bacteria, Nutrients, and Metals.* Water Environment Research Federation, International Stormwater BMP Database. Available at: www.bmpdatabase.org. (Accessed 6/18/18).

Lenhart C., Gordon B., Gamble J., Current D., Ross N., Herring L., Nieber J., Peterson H. (2016) Design and hydrologic performance of a tile drainage treatment wetland in Minnesota, USA. *Water 8*(12): 549.

Lenters J.D., Cutrell G.J., Istanbulluoglu E., Scott D.T., Herrman K.S., Eisenhauer D.E. (2011) Seasonal energy and water balance of a *Phragmites australis*-dominated wetland in the Republican River basin of south-central Nebraska (USA). *Papers in Natural Resources 317*: http://digitalcommons.unl.edu/natrespapers/317.

Leppich J., Pardue J.H., Jackson W.A. (2000) Plant-air partitioning of chlorobenzenes in wetland vegetation at a superfund site. In: *Wetlands and Remediation*, Means J.L., Hinchee R.E. (eds.) Battelle Press: Columbus, OH, pp. 17–24.

Lesage E., Rousseau D.P.L., van de Moortel A., Tack F.M.G., De Pauw N., Verloo M.G. (2006) Effects of sorption, sulphate reduction, and Phragmites australis on the removal of heavy metals in subsurface flow constructed wetland microcosms. Dias V., Vymazal J. (eds.) Proceedings of the 10th International Conference on Wetland Systems for Water Pollution Control, 23–29 September 2006; Ministério de Ambiente, do Ordenamento do Territóri e do Desenvolvimento Regional (MAOTDR) and IWA: Lisbon, Portugal, 1687–1695.

Leto C., Tuttolomondo T., La Bella S., Leone R., Licata M. (2013) Effects of plant species in a horizontal subsurface flow constructed wetland–Phytoremediation of treated urban wastewater with *Cyperus alternifolius* L. and *Typha latifolia* L. in the West of Sicily (Italy). *Ecological Engineering 61*: 282–291.

Levenspiel O. (1995) *Chemical Reaction Engineering*. 2nd edition, John Wiley & Sons: New York.

Levenspiel O., Turner J.C.R. (1970) The interpretation of residence time experiments. *Chemical Engineering Science 25*: 1605–1609.

Levenspiel O. (1999) *Chemical Reaction Engineering*. 3rd edition, John Wiley & Sons: New York.

Lewis, III R.R. (1989) Wetland restoration/creation/enhancement terminology: Suggestions for standardization. In: *Wetland Creation and Restoration: The Status of the Science, Vol. II*, EPA 600/3/89/038B, U.S. Environmental Protection Agency: Washington, DC.

Li E.H., Li W., Liu G.H., Yuan L.Y. (2008) The effect of different submerged macrophyte species and biomass on sediment resuspension in a shallow freshwater lake. *Aquatic Botany 88*(2): 121–126.

Li X., Zheng W., Kelly W.R. (2013) Occurrence and removal of pharmaceutical and hormone contaminants in rural wastewater treatment lagoons. *Science of the Total Environment 445*: 22–28.

Li X.-Y., Yuan Y. (2002) Settling velocities and permeabilities of microbial aggregates. *Water Research 36*(15): 3110–3120.

Li Y., Fletcher T.D. (2007) Modelling wet weather sediment removal by stormwater constructed wetlands: Insights from a laboratory study. *Journal of Hydrology 338*(3–4): 285–296.

Li Y., Zhu G., Ng W.J., Tan S.K. (2014) A review on removing pharmaceutical contaminants from wastewater by constructed wetlands: Design, performance and mechanism. *Science of the Total Environment 468*: 908–932.

Liang M.Q., Zhang C.F., Peng C.L., Lai Z.L., Chen D.F., Chen Z.H. (2011) Plant growth, community structure, and nutrient removal in monoculture and mixed constructed wetlands. *Ecological Engineering 37*(2): 309–316.

Liang Z.S., Westerman P.W., Arogo J. (2002) Modelling ammonia emission from swine anaerobic lagoons. *Transactions of the American Society of Agricultural Engineers 45*(3): 787–798.

Liao X., Inglett P.W. (2012) Biological nitrogen fixation in periphyton of native and restored Everglades marl prairies. *Wetlands 32*(1): 137–148.

Lick W. (2009) *Sediment and Contaminant Transport in Surface Waters.* CRC Press: Boca Raton, FL.

Lide D.R. (ed.) (1992) *CRC Handbook of Chemistry and Physics.* 73rd edition, CRC Press: Boca Raton, FL.

Liffen T., Gurnell A.M., O'Hare M.T. (2013) Profiling the below ground biomass of an emergent macrophyte using an adapted ingrowth core method. *Aquatic Botany 110*: 97–102.

Lightbody A.F., Avener M.E., Nepf H.M. (2008) Observations of short-circuiting flow paths within a free-surface wetland in Augusta, Georgia, USA. *Limnology and Oceanography 53*(3): 1040–1053.

Lightbody A.F., Nepf H.M. (2006) Prediction of velocity profiles and longitudinal dispersion in salt marsh vegetation. *Limnology and Oceanography 51*(1): 218–228.

Lightbody A.F., Nepf H.M., Bays J.S. (2007) Mixing in deep zones within constructed treatment wetlands. *Ecological Engineering 29*(2): 209–220.

Lightbody A.F., Nepf H.M., Bays J.S. (2009) Modeling the hydraulic effect of transverse deep zones on the performance of short-circuiting constructed treatment wetlands. *Ecological Engineering 35*(5): 754–768.

Ligi T., Truu M., Oopkaup K., Nõlvak H., Mander Ü., Mitsch W.J., Truu J. (2015) The genetic potential of N2 emission via denitrification and ANAMMOX from the soils and sediments of a created riverine treatment wetland complex. *Ecological Engineering 80*: 181–190.

Lin Y.-F., Jing S.-R., Lee D.-Y., Chang Y.-F., Shih K.-C. (2008) Nitrate removal from groundwater using constructed wetlands under various hydraulic loading rates. *Bioresource Technology 99*: 7504–7513.

Lin Z.-Q., Terry N. (2003) Selenium removal by constructed wetlands: Quantitative importance of biological volatilization in the treatment of selenium-laden agricultural drainage water. *Environmental Science and Technology 37*(5): 606–615.

Linacre E.T. (1976) Swamps. In: *Vegetation and Atmosphere, Volume II: Case Studies*, Monteith J.L. (ed.) Academic Press: London, United Kingdom, pp. 329–347.

Lintelman J., Katayama A., Kurihara N., Shore L., Wenzel A. (2003) Endocrine disruptors in the environment. *Pure and Applied Chemistry 75*(5): 631–681.

Linz G.M., Homan H.J. (2011) Use of glyphosate for managing invasive cattail (*Typha*) to disperse blackbird (*Icteridae*) roosts. *Crop Protection 30*(2): 98–104.

Lishawa S.C., Lawrence B.A., Albert D.A., Tuchman N.C. (2015) Biomass harvest of invasive Typha promotes plant diversity in a Great Lakes coastal wetland. *Restoration Ecology 23*(3): 228–237.

Litchfield D.K. (1990) Constructed wetlands for wastewater treatment at Amoco Oil Company's Mandan, North Dakota Refinery. In: *Constructed Wetlands in Water Pollution Control*, Cooper P. F., Findlater B.C. (eds.) Pergamon Press: Oxford, United Kingdom, pp. 399–402.

Litchfield D.K. (1993) Constructed wetlands for wastewater treatment at Amoco Oil Company's Mandan, North Dakota Refinery. In: *Constructed Wetlands for Water Quality Improvement*, Moshiri G.A. (ed.) CRC Press: Boca Raton, FL, pp. 485–490.

Litchfield D.K., Schatz D.D. (1989) Constructed wetlands for wastewater treatment at Amoco Oil Company's Mandan, North Dakota Refinery. In: *Constructed Wetlands for Wastewater Treatment: Municipal, Industrial, and Agricultural*, Hammer D.A. (ed.) Lewis Publishers: Chelsea, MI, pp. 233–237.

Liu D., Wu X., Chang J., Gu B., Min Y., Ge Y., Shi Y., Xue H., Peng C., Wu J. (2012) Constructed wetlands as biofuel production systems. *Nature Climate Change 2*(3): 190.

Liu G., Wang J. (2012) Probing the stoichiometry of the nitrification process using the respirometric approach. *Water Research 46*(18): 5954–5962.

Liu G.D., Gu B., Miao S.L., Li Y.C., Migliaccio K.W., Qian Y. (2010) Phosphorus release from ash and remaining tissues of two wetland species after a prescribed fire. *Journal of Environmental Quality 39*(5): 1585–1593.

Liu L., Liu Y.H., Liu C.X., Wang Z., Dong J., Zhu G.F., Huang X. (2013) Potential effect and accumulation of veterinary antibiotics in *Phragmites australis* under hydroponic conditions. *Ecological Engineering 53*: 138–143.

Liu L., Liu Y.H., Wang Z., Liu C.X., Huang X., Zhu G.F. (2014) Behavior of tetracycline and sulfamethazine with corresponding resistance genes from swine wastewater in pilot-scale constructed wetlands. *Journal of Hazardous Materials 278*: 304–310.

Li Y., Jalil A., Du W., Gaob X., Wang J., Luo L., Huiyun Li H., Dai S., Hashim S., Yu Z., Achary K. (2017) Wind induced reverse flow and vertical profile characteristics in asemi-enclosed bay of large shallow Lake Taihu, China. *Ecological Engineering 102*: 224–233.

Lizama K., Fletcher T.D., Sun G. (2011) Removal processes for arsenic in constructed wetlands. *Chemosphere 84*(8): 1032–1043.

Lizotte Jr R.E., Shields Jr F.D., Murdock J.N., Kröger R., Knight S.S. (2012) Mitigating agrichemicals from an artificial runoff event using a managed riverine wetland. *Science of the Total Environment 427*: 373–381.

Lloyd J.R., Klessa D.A., Parry D.L., Buck P., Brown N.L. (2004) Simulation of microbial sulphate reduction in a constructed wetland: Microbiological and geochemical analysis. *Water Research 38*: 1822–1830.

Lloyd S. (1997) *Influence of Macrophytes on Sediment Deposition and Flow Pattern within a Stormwater Pollution Control Wetland.* M. Eng. Sci. Thesis, Monash University, Australia.

Locke L.N., Friend M.N. (1989) Avian botulism: Geographic expansion of a historic disease. *Waterfowl management handbook*. U.S. Fish and Wildlife Service, National Wildlife Health Research Center, Madison, WI. Available at: http://digitalcommons.unl.edu/icwdmwfm/3 (Accessed 10/30/18).

Loer J., Scholz-Barth K., Kadlec R.H., Wetzstein D., Julik J. (1999) An integrated natural system for leachate treatment. In: *Constructed Wetlands for the Treatment of Landfill Leachate*, McBean E., Rovers F.A. (eds.) Lewis Publishers: Boca Raton, FL, pp. 187–204.

Lomans B.P., Pol A., Op Den Camp H.J.M. (2002) Microbial cycling of volatile organic sulfur compounds in anoxic environments. *Water Science and Technology 45*(10): 55–60.

Long K. (2000) *Beavers: A Wildlife Handbook*. Johnson Books: Boulder, CO.

Loomis M.J., Craft C.B. (2010) Carbon sequestration and nutrient (nitrogen, phosphorus) accumulation in river-dominated tidal marshes, Georgia, USA. *Soil Science Society of America Journal 74*(3): 1028–1036.

Lopez-Flores R., Quintana X.D., Salvado V., Hidalgo M., Sala L., Moreno-Amich R. (2003) Comparison of nutrient and contaminant fluxes in two areas with different hydrological regimes (Emporda Wetlands, NE Spain). *Water Research 37*(12): 3034–3046.

Lorah M.M., Burris D.R., Dyer L.J. (2002) Efficiency of natural attenuation of chlorinated solvents in two freshwater wetlands. In: *Wetlands and Remediation II*, Nehring K.W., Brauning S.E. (eds.) Battelle Press: Columbus, OH, pp. 9–16.

Lorah M.M., Olsen L.D., Smith B.L., Johnson M.A., Fleck W.B. (1997) *Natural attenuation of chlorinated volatile organic compounds in a freshwater tidal wetland (Aberdeen Proving Ground, Maryland)*, Water Resources Investigations Report 97-4171, U.S. Geological Survey.

Lorah M.M., Walker C.W., Baker A.C., Teunis J.A., Majcher E.H., Brayton M.J., Raffensperger J.P., Cozzarelli I.M. (2014) *Hydrogeologic Characterization and Assessment of Bioremediation of Chlorinated Benzenes and Benzene in Wetland Areas, Standard Chlorine of Delaware, Inc. Superfund Site, New Castle County, Delaware, 2009–12*. Scientific Investigations Report 2014–5140, US Department of the Interior, US Geological Survey, Reston, VA.

Loseto L.L., Siciliano S.D., Lean D.R.S. (2004) Methylmercury production in high arctic wetlands. *Environmental Toxicology and Chemistry 23*(1): 17–23.

Lougheed V.L., Crosbie B., Chow-Fraser P. (1998) Predictions on the effect of common carp (Cyprinus carpio) exclusion on water quality, zooplankton, and submergent macrophytes in a Great Lakes wetland. *Canadian Journal of Fisheries and Aquatic Sciences 55*(5): 1189–1197.

Lowe E., FKeenan L.W. (1997) Managing phosphorus based, cultural eutrophication in wetlands: A conceptual approach. *Ecological Engineering 9*(1/2): 109–118.

Lu S.Y., Wu F.C., Lu Y.F., Xiang C.S., Zhang P.Y., Jin C.X. (2009) Phosphorus removal from agricultural runoff by constructed wetland. *Ecological Engineering 35*(3): 402–409.

Lu S.Y., Zhang P.Y., Cui W.H. (2010) Impact of plant harvesting on nitrogen and phosphorus removal in constructed wetlands treating agricultural region wastewater. *International Journal of Environment and Pollution 43*(4): 339–353.

Luhar M., Rominger J., Nepf H. (2008) Interaction between flow, transport and vegetation spatial structure. *Environmental Fluid Mechanics 8*(5): 423–439.

Lusk M.G., Toor G.S. (2016) Dissolved organic nitrogen in urban streams: Biodegradability and molecular composition studies. *Water Research 96*: 225–235.

Lv T., Zhang Y., Casas M.E., Carvalho P.N., Arias C.A., Bester K., Brix H. (2016) Phytoremediation of imazalil and tebuconazole by four emergent wetland plant species in hydroponic medium. *Chemosphere 148*: 459–466.

Lynch J., Fox L.J., Owen, Jr J.S., Sample D.J. (2015) Evaluation of commercial floating treatment wetland technologies for nutrient remediation of stormwater. *Ecological Engineering 75*: 61–69.

Lyon S., Horne A., Jordahl J., Emond H., Carlson K. (2009) *Preliminary Feasibility Assessment of Constructed Treatment Wetlands in the Vicinity of the Klamath Hydroelectric Project.* Draft Report. Prepared by CH2M HILL and Alex Horne Associates. Prepared for PacifiCorp Energy. January 2009. Available at: www.pacificorp.com/content/dam/pacificorp/doc/Energy_Sources/Hydro/Hydro_Licensing/Klamath_River/Klamath_Preliminary_Feasibility_Assessment_of_Constucted_Treatment_Wetlands.pdf (Accessed 12/27/18).

Lytle C.M., Lytle F.W., Yang N., Qian J.-H., Hansen D., Zayed A., Terry N. (1998) Reduction of Cr(VI) to Cr(III) by wetland plants: Potential for in situ heavy metal detoxification. *Environmental Science and Technology 32*(20): 3087–3093.

Mabit L., Benmansour M., Walling D.E. (2008) Comparative advantages and limitations of the fallout radionuclides 137 Cs, 210 Pb ex and 7 Be for assessing soil erosion and sedimentation. *Journal of Environmental Radioactivity 99*(12): 1799–1807.

MacIntyre S., Jonsson A., Jansson M., Aberg J., Turney D.E., Miller S.D. (2010) Buoyancy flux, turbulence, and the gas transfer coefficient in a stratified lake. *Geophysical Research Letters 37*: 1–5, L24604.

Mackay D., Leinonen P.J. (1975) Rate of evaporation of low-solubility contaminants from water bodies to atmosphere. *Environmental Science and Technology 9*: 1178–1180.

Maheu A., Poff N.L., St. Hilaire A. (2015) Classification of stream water temperature regimes in the conterminous USA. *River Research and Applications 32*(5): 896–906.

Maier R.M., Pepper I.L., Gerba C.P. (2000) *Environmental Microbiology.* Academic Press: San Diego, CA.

Maillard E., Payraudeau S., Faivre E., Grégoire C., Gangloff S., Imfeld G. (2011) Removal of pesticide mixtures in a stormwater wetland collecting runoff from a vineyard catchment. *Science of the Total Environment 409*: 2317–2324.

Maine M.A., Duarte M.V., Suñe N. (2001) Cadmium uptake by floating macrophytes. *Water Research 35*(11): 2629–2634.

Maine M.A., Hadad H., Sánchez G., Caffaratti S., Bonetto C. (2006) Removal efficiency in a constructed wetland for wastewater treatment from a tool factory. Dias V., Vymazal J. (eds.) Proceedings of the 10th International Conference on Wetland Systems for Water Pollution Control, 23–29 September 2006; Ministério de Ambiente, do Ordenamento do Territóri e do Desenvolvimento Regional (MAOTDR) and IWA: Lisbon, Portugal, 1753–1761.

Maine M.A., Hadad H.R., Sánchez G.C., Mufarrege M.M., Di Luca G.A., Caffaratti S.E., Pedro M.D. C. (2013) Sustainability of a constructed wetland faced with a depredation event. *Journal of Environmental Management 128*: 1–6.

Maine M.A., Suñe N., Hadad H., Sánchez G., Bonetto C. (2004a) *Nutrient and metal removal in a constructed wetlands for wastewater treatment from a metallurgic industry.* Liénard A., Burnett H. (eds.) Proceedings of the 9th International Conference on Wetland Systems for Water Pollution Control, 26–30 September 2004; Association Scientifique et Technique pour l'Eau et l'Environnement (ASTEE), Cemagref, and IWA: Avignon, France, 597–604.

Makoto O., Ikeda T. (1984) *Methods in Marine Zooplankton Ecology.* Wiley: New York.

Malecki L.M., White J.R., Reddy K.R. (2004) Nitrogen and phosphorus flux rates from sediment in the lower St. Johns River estuary. *Journal of Eenvironmental Quality 33*(4): 1545–1555.

Malmaeus J.M., Hakanson L. (2003) A dynamic model to predict suspended particulate matter in lakes. *Ecological Modelling 167*(3): 247–262.

Malone K., Williams H. (2010) *Growing Season Definition and Use in Wetland Delineation*. eBooks, Book 15, U.S. Army Corp of Engineers, Engineer Research and Development Center: Vicksburg, MS.

Malone S.J. (2009) *Agricultural Nonpoint Source Pollution Management: Water Quality Impacts of Balm Road Treatment Marsh, Hillsborough County, Florida*. MS Thesis, University of South Florida, Tampa, FL.

Maltby E., Barker T. (eds.) (2009) *The Wetlands Handbook, 2 Volume Set*. John Wiley & Sons.: Hoboken, NJ.

Mander Ü., Dotro G., Ebie Y., Towprayoon S., Chiemchaisri C., Nogueira S.F., Jamsranjav B., Kasak K., Truu J., Tournebize J., Mitsch W.J. (2014) Greenhouse gas emission in constructed wetlands for wastewater treatment: A review. *Ecological Engineering 66*: 19–35.

Mangelson K.A., Watters G.Z. (1972) Treatment efficiency of waste stabilization ponds. *Journal of the Sanitary Engineering Division 98*(2): 407–425.

Manning R. (1891) On the flow of waters in open channels and pipes. *Tranactions of The. Institution of Civil Engineers, Ireland 20*: 161–207.

Manrique R., Flores J., Arranda C., Gusek J.J. (2012) Antamina treatment wetland at Tucush valley–Five year performance review. Paper presented at the 3rd International Congress on Water Management in the Mining Industry, Santiago, Chile, 6–8 June, 2012.

Manyin T., Williams F.M., Stark L.R. (1997) Effects of iron concentration and flow rate on treatment of coal mine drainage in wetland mesocosms: An experimental approach to sizing of constructed wetlands. *Ecological Engineering 9*(3–4): 171–186.

Mao L.M., Bergman M.J., Tai C.C. (2002) Evapotranspiration measurement and estimation of three wetland environments in the Upper St. Johns River Basin, Florida. *JAWRA Journal of the American Water Resources Association 38*(5): 1271–1285.

MAPAQ (2013). *Water Storage and Sedimentation Basins: Concept and Sizing*. Quebec Ministry of Agriculture, Fisheries and Food (MAPAQ), Quebec, Canada. Available at: www.agrireseau.net (Accessed 4/21/18).

Marais G., Shaw V. (1961) A rational theory for the design of sewage stabilization ponds in central and South Africa. *Transactions of the South African Institution of Civil Engineers 3*: 205–227.

Marais G.v.R. (1974) Faecal bacteria kinetics in stabilization ponds. *Journal of Environmental Engineering (ASCE) 115*(1): 119–139.

Marjoribanks T.I., Hardy R.J., Lane S.N., Parsons D.R. (2014) Dynamic drag modeling of submerged aquatic vegetation canopy flows. River Flow 2014: Proceedings of the International Conference on Fluvial Hydraulics, (River Flow 2014); Lausanne, Switzerland, 3–5 September 2014, CRC Press, Taylor & Francis Group.

Markstrom S.L., Regan R.S., Hay L.E., Viger R.J., Webb R.M.T., Payn R.A., LaFontaine J.H. (2015) *PRMS-IV, the Precipitation-Runoff Modeling System, Version 4*. U.S. Geological Survey Techniques and Methods, book 6, chap. B7. https://dx.doi.org/10.3133/tm6B7.

Marois D.E., Mitsch W.J. (2016) Modeling phosphorus retention at low concentrations in Florida Everglades mesocosms. *Ecological Modelling 319*: 42–62.

Martin A., McNee C., Russell J., Gelderland D. (2015) Performance of a 16 ha engineered wetland for the treatment of neutral-pH gold mine effluents (Ontario, Canada). Paper 165 in: Agreeing on solutions for more sustainable mine water management. Proceedings of the 10th ICARD & IMWA Annual Conference; Santiago, Chile. Available at: www.imwa.info/docs/imwa_2015/IMWA2015_Martin_165.pdf (Accessed 10/12/18).

Martin C. (2015) *Is that Wetland Jurisdictional? A Practical Guide to the New Clean Water Rule*. Available at: www.williamsmullen.com/sites/default/files/files/A%20PRACTICAL%20GUIDE%20TO%20THE%20NEW%20%20CLEAN%20WATER%20RULE.pdf (Accessed 1/1/19).

Martin J.F., Reddy K.R. (1997) Interaction and spatial distribution of wetland nitrogen processes. *Ecological Modelling 105*: 1–21.

Martinez C.J., Wise W.R. (2001) *Analysis of Hydraulic Performance of the Orlando Easterly Wetland Cells: Tracer Study Results*, Report to the City of Orlando, FL.

Martinez C.J., Wise W.R. (2003) Analysis of constructed treatment wetland hydraulics with the transient storage model OTIS. *Ecological Engineering 20*(3): 211–222.

Mason C.F., Bryant R.J. (1975) Production, nutrient content and decomposition of *Phragmites communis* Trin. and *Typha angustifolia* L. *Journal of Ecology 63*: 71–95.

Masscheleyn P.H., Delaune R.D., Patrick W.H. Jr. (1991a) Arsenic and selenium chemistry as affected by sediment redox potential and pH. *Journal of Environmental Quality 20*(2): 522–527.

Masscheleyn P.H., Delaune R.D., Patrick W.H. Jr. (1991b) Effect of redox potential and pH on arsenic speciation and solubility in a contaminated soil. *Environmental Science and Technology 25*(8): 1414–1419.

Masscheleyn P.H., Patrick W.H. Jr. (1993) Biogeochemical processes affecting selenium cycling in wetlands. *Environmental Toxicology and Chemistry 12*: 2235–2243.

Massman W.J. (1992) A surface energy balance method for partitioning evapotranspiration data into plant and soil components for a surface with partial canopy cover. *Water Resources Research 28* (6): 1723–1732.

Matamoros V., García J., Bayona J.M. (2008) Organic micropollutant removal in a full-scale surface flow constructed wetland fed with secondary effluent. *Water Research 42*(3): 653–660.

Matamoros V., Salvadó V. (2012) Evaluation of the seasonal performance of a water reclamation pond-constructed wetland system for removing emerging contaminants. *Chemosphere 86*(2): 111–117.

Mayer T., Ptacek C., Zanini L. (1999) Sediments as a source of nutrients to hypereutrophic marshes of Point Pelee, Ontario, Canada. *Water Research 33*(6): 1460–1470.

Mayo A.W. (1995) Modeling coliform mortality in waste stabilization ponds. *Journal of Environmental Engineering 121*(2): 140–152.

Mazellier P., Méité L., De Laat J. (2008) Photodegradation of the steroid hormones 17β-estradiol (E2) and 17α-ethinylestradiol (EE2) in dilute aqueous solution. *Chemosphere 73*(8): 1216–1223.

McAllister L.S. (1992) *Habitat Quality Assessment of Two Wetland Treatment Systems in Mississippi - A Pilot Study*, EPA/600/R-92/229, U.S. EPA Environmental Research Laboratory: Corvallis, OR.

McAllister L.S. (1993a) *Habitat Quality Assessment of Two Wetland Treatment Systems in Florida - A Pilot Study*, EPA/600/R-93/222, U.S EPA Environmental Research Laboratory: Corvallis, OR.

McAllister L.S. (1993b) *Habitat Quality Assessment of Two Wetland Treatment Systems in the Arid West - A Pilot Study*, EPA/600/R-93/117, U.S EPA Environmental Research Laboratory.

McBride G.B., Tanner C.C. (2000) Modelling biofilm nitrogen transformations in constructed wetland mesocosms with fluctuating water levels. *Ecological Engineering 14*(1–2): 93–106.

McCann K., Olson K. (1994) *Greenwood Urban Wetland Treatment Effectiveness*, Final report to the Florida Department of Environmental Protection, Project WM427, Tallahassee, FL.

McCarter C.P., Price J.S. (2017) The transport dynamics of chloride and sodium in a ladder fen during a continuous wastewater polishing experiment. *Journal of Hydrology 549*: 558–570.

McCarthy D.T., Zhang K., Westerlund C., Viklander M., Bertrand-Krajewski J.L., Fletcher T.D., Deletic A. (2018) Assessment of sampling strategies for estimation of site mean concentrations of stormwater pollutants. *Water Research 129*: 297–304.

McCormick P., Newman S., Miao S., Sklar F., Fontaine T. (1997a) *Field Research and Monitoring Studies to Support Derivation of a Class III Water Quality Criterion for Phosphorus in the Arthur A. Marshall Loxahatchee National Wildlife Refuge: An Annual Progress Report*. South Florida Water Management District: West Palm Beach, FL.

McCormick P.V., O'Dell M.B. (1996) Quantifying periphyton responses to phosphorus in the Florida Everglades: A synoptic-experimental approach. *Journal of the North American Benthological Society 15*(4): 450–468.

McCormick P.V., Rawlik P.S., Lurding K., Smith E.P., Sklar F.H. (1996) Periphyton-water quality relationships along a nutrient gradient in the northern Florida Everglades. *Journal of the North American Benthological Society 15*(4): 433–449.

McCormick P.V., Shuford R.B., Chimney M.J. (2006) Periphyton as a potential phosphorus sink in the Everglades Nutrient Removal Project. *Ecological Engineering 27*(4): 279–289.

McCormick P.V., Shuford R.B., Rawlik P.S. (2004) Changes in macroinvertebrate community structure and function along a phosphorus gradient in the Florida Everglades. *Hydrobiologia 529*(1–3): 113–132.

McCormick P.V. Shuford, III R.B., Backus J.G., Kennedy W.C. (1997b) Spatial and seasonal patterns of periphyton biomass and productivity in the northern Everglades, Florida, USA. *Hydrobiologia 362*(1): 185–210.

McCuen R.H. (1982) *A Guide to Hydrologic Analysis Using SCS Method*. Prentice Hall, Englewood Cliffs: NJ.

McEnroe N.A., Buttle J.M., Marsalek J., Pick F.R., Xenopoulos M.A., Frost P.C. (2013) Thermal and chemical stratification of urban ponds: Are they 'completely mixed reactors'? *Urban Ecosystems 16*: 327–339.

McGauley E.K. (2008) *Evaluating the Ecological Potential of Constructed Wetlands: A Case Stdy of the Brighton Treatment Wetland.* MS Thesis, Trent University: Peterborough, Ontario.

McIntire C.D. (1973) Periphyton dynamics in laboratory streams: A simulation model and its implications. *Ecological Monographs 43*(3): 399–420.

McIntire C.D., Colby J.A. (1978) A hierarchical model of lotic ecosystems. *Ecological Monographs 48*(2): 167–190.

McJannet D. L., Webster I. T., Cook F.J. (2012) An area-dependent wind function for estimating open water evaporation using landbased meteorological data. *Environmental Modelling Software 31*: 76–83.

McKillop R., Kouwen N., Soulis E.D. (1999) Modeling the rainfall-runoff response of a headwater wetland. *Water Resources Research 35*(4): 1165–1177.

McMahon T.A., Peel M.C., Lowe L., Srikanthan R., McVicar T.R. (2013) Estimating actual, potential, reference crop and pan evaporation using standard meteorological data: A pragmatic synthesis. *Hydrology and Earth System Sciences 17*(4): 1331.

McQueen A.D., Hendrikse M., Gaspari D.P., Kinley C.M., Rodgers, Jr J.H., Castle J.W. (2017b) Performance of a hybrid pilot-scale constructed wetland system for treating oil sands process-affected water from the Athabasca oil sands. *Ecological Engineering 102*: 152–165.

MDEQ (2018) *Groundwater/Surface Water Interface Pathway Compliance Options.* Michigan Department of Environmental Quality, Lansing, MI. Available at: www.michigan.gov/documents/deq/deq-rrd-2018-05GSIPathwayComplianceOptionsResourceMaterials_624831_7.pdf (Accessed 10/6/18).

MDN (2016) *Mercury Deposition Network Results.* National Atmospheric Deposition Program, University of Illinois: Urbana Champaign, IL. Available at: http://nadp.slh.wisc.edu/data/mdn/annual.aspx (Accessed 9/3/18).

Medina V.F., Larson S.L., Bergstedt A.E., McCutcheon S.C. (2000) Phyto-removal of trinitrotoluene from water with batch kinetic studies. *Water Research 34*(10): 2713–2722.

Medlock J.M., Vaux A.G. (2014) Colonization of a newly constructed urban wetland by mosquitoes in England: Implications for nuisance and vector species. *Journal of Vector Ecology 39*(2): 249–260.

Meerburg B.G., Koene M.G., Kleijn D. (2011) Escherichia coli concentrations in feces of geese, coots, and gulls residing on recreational water in The Netherlands. *Vector-Borne and Zoonotic Diseases 11*(6): 601–603.

Mei X.Q., Yang Y., Tam N.F.Y., Wang Y.W., Li L. (2014) Roles of root porosity, radial oxygen loss, Fe plaque formation on nutrient removal and tolerance of wetland plants to domestic wastewater. *Water Research 50*: 147–159.

Melbourne Water (2013) *Water Sensitive Urban Design Guidelines.* Melbourne Water Corporation, Melbourne,Victoria, Australia. Available at: www.melbournewater.com.au/planning-and-building/developer-guides-and-resources/standards-and-specifications/constructed-0 Acessed 7/19/18.

Melbourne Water. (2017) *Wetland Design Manual.* Melbourne Water, Melbourne, Victoria, Australia. Available at: www.melbournewater.com.au/planning-and-building/developer-guides-and-resources/stan (Accessed 8/4/18).

Melzer A., Exler D. (1982) Nitrate and nitrite reductase activities in aquatic macrophytes. In: *Studies on Aquatic Vascular Plants*, Symoens J.J., Hooper S.S., Compère P. (eds.) Royal Botanical Society: Brussels, Belgium, pp. 128–135.

Meng Z., Chandrasena G., Henry R., Deletic A., Kolotelo P., McCarthy D. (2018) Stormwater constructed wetlands: A source or a sink of Campylobacter. *Water Research 131*: 218–227.

Menon P., Billen G., Servais P. (2003) Mortality rates of autochthonous and fecal bacteria in natural aquatic ecosystems. *Water Research 37*(17): 4151–4158.

Merriman L.S., Hathaway J.M., Burchell M.R., Hunt W.F. (2017) Adapting the relaxed tanks-in-series model for stormwater wetland water quality performance. *Water 9*(9): 691.

Merriman L.S., Hunt W.F., Bass K.L. (2016) Development/ripening of ecosystems services in the first two growing seasons of a regional-scale constructed stormwater wetland on the coast of North Carolina. *Ecological Engineering 94*: 393–405.

Merritt A. (1994) *Wetlands, Industry, and Wildlife.* The Wildfowl and Wetlands Trust: Slimbridge, Gloucester, United Kingdom. Available at: http://wildpro.twycrosszoo.org/S/00Ref/bookscontents/B93.htm (Accessed 11/29/18).

Merritt R.W., Lawson D.L. (1979) Leaf litter processing in floodplain and stream communities. Johnson R.R., McCormick J.F. (eds.) Proceedings of the Strategies for Protection and Management of Floodplain Wetlands and Other Riparian Ecosystems Symposium, 11–13 December 1978; U.S. Department of Agriculture: Callaway Gardens, Georgia, 93–105.

Merritts D. (2016) *Hydrosphere, Starting Point*. Retrieved Nov 21, 2017, from https://serc.carleton.edu/sp/library/examples/hydrosphere.html.

Merz S.K. (2000) *Guidelines for Using Free Water Surface Constructed Wetlands to Treat Municipal Sewage*. DNRQ00047, Queensland Department of Natural Resources: Brisbane, Queensland, Australia.

Messer T.L., Burchell M.R., Bírgand F. (2017b) Comparison of four nitrate removal kinetic models in two distinct wetland restoration mesocosm systems. *Water 9*(7): 517.

Messer T.L., Burchell M.R., Böhlke J.K., Tobias C.R. (2017a) Tracking the fate of nitrate through pulse-flow wetlands: A mesocosm scale 15N enrichment tracer study. *Ecological Engineering 106*: 597–608.

Metcalf and Eddy Inc (1991) *Wastewater Engineering, Treatment, Disposal, and Reuse*. Tchobanoglous G., Burton F.L. (eds.) Third, McGraw-Hill: New York.

Metcalf and Eddy Inc (2003) *Wastewater Engineering, Treatment, Disposal, and Reuse*. Tchobanoglous G., Burton F.L., Stensel H.D. (eds.) Fourth edition, McGraw-Hill: New York.

Meuleman A.F.M., Beltman B., Scheffer R.A. (2004) Water pollution control by aquatic vegetation of treatment wetlands. *Wetlands Ecology and Management 12*(5): 459–471.

Miao S.L., Carstenn S., Thomas C., Edelstein C., Sindhoj E., Gu B. (2009) Integrating multiple spatial controls and temporal sampling schemes to explore shortand long-term ecosystem response to fire in an Everglades wetland. In: *Real World Ecology: Large-Scale and Long-Term Case Studies and Methods*, Miao S.L. et al. (eds.) Springer: New York.

Miao S.L., Edelstein C., Carstenn S., Gu B.H. (2010) Immediate ecological impacts of a prescribed fire on a cattail-dominated wetland in Florida Everglades. *Fundamental and Applied Limnology 176*(1): 29–41.

Miao S.L., Sklar F.H. (1997) Biomass and nutrient allocation of sawgrass and cattail along a nutrient gradient in the Florida Everglades. *Wetlands Ecology and Management 5*(4): 245–264.

Mierau R., Trimble P. (1988) *Hydrologic Characteristics of the Kissimmee River Floodplain Boney Marsh Experimental Area*. Technical memorandum, South Florida Water Management District, West Palm Beach, FL (September 1988).

Miller P.S., Mitchell J.K., Cooke R.A., Engel B.A. (2002) A wetland to improve agricultural subsurface drainage quality. *Transactions of the American Society of Agricultural Engineers 45*(5): 1305–1317.

Miller R.L., Fram M., Fujii R., Wheeler G. (2008) Subsidence reversal in a re-established wetland in the Sacramento-San Joaquin Delta, California, USA. *San Francisco Estuary and Watershed Science 6*(3): 1–21.

Miller R.L., Fujii R. (2010) Plant community, primary productivity, and environmental conditions following wetland re-establishment in the Sacramento-San Joaquin Delta, California. *Wetlands Ecology and Management 18*(1): 1–16.

Miller J.E. (2018) *Muskrats*. Wildlife Damage Management Technical Series, 14. U.S. Department of Agriculture, Animal & Plant Health Inspection Service Wildlife Services, Urbandale, IA. Available at: https://digitalcommons.unl.edu/cgi/viewcontent.cgi?article=1014&context=nwrcwdmts (Accessed 12/9/18).

Min J., Paudel R., Jawitz J.W. (2011) Mechanistic biogeochemical model applications for Everglades restoration: A review of case studies and suggestions for future modeling needs. *Critical Reviews in Environmental Science and Technology 41*: 489–516.

Min J.H., Paudel R., Jawitz J.W. (2010) Spatially distributed modeling of surface water flow dynamics in the Everglades ridge and slough landscape. *Journal of Hydrology 390*(1): 1–12.

Min J.H., Wise W.R. (2009) Simulating short-circuiting flow in a constructed wetland: The implications of bathymetry and vegetation effects. *Hydrological Processes 23*(6): 830–841.

Min J.H., Wise W.R. (2010) Depth-averaged, spatially distributed flow dynamic and solute transport modelling of a large-scaled, subtropical constructed wetland. *Hydrological Processes 24*(19): 2724–2737.

Minke M., Augustin J., Hagemann U., Joosten H. (2014) Similar methane fluxes measured by transparent and opaque chambers point at belowground connectivity of Phragmites australis beyond the chamber footprint. *Aquatic Botany 113*: 63–71.

Minnesota Stormwater Steering Committee (2018) Minnesota Stormwater Manual. Minnesota Pollution Control Agency: St. Paul, MN. Available at: https://stormwater.pca.state.mn.us/index.php?title=Main_Page&oldid=38167 (Accessed 7/19/18).

Mitsch W.J. (2012) What is ecological engineering? *Ecological Engineering 45*: 5–12.

Mitsch W.J. (2014) When will ecologists learn engineering and engineers learn ecology? *Ecological Engineering 65*: 9–14.

Mitsch W.J. (2017) Solving Lake Erie's harmful algal blooms by restoring the Great Black Swamp in Ohio. *Ecological Engineering 108*: 406–413.

Mitsch W.J., Day J.W., Gilliam J.W., Groffman P.M., Hey D.L., Randall G.W., Wang N. (2001) Reducing nitrogen loading to the Gulf of Mexico from the Mississippi River Basin: Strategies to counter a persistent ecological problem: Ecotechnology—The use of natural ecosystems to solve environmental problems—Should be a part of efforts to shrink the zone of hypoxia in the Gulf of Mexico. *AIBS Bulletin 51*(5): 373–388.

Mitsch W.J., Day J.W., Zhang L., Lane R.R. (2005a) Nitrate-nitrogen retention in wetlands in the Mississippi River Basin. *Ecological Engineering 24*(4): 267–278.

Mitsch W.J., Gosselink J.G. (2000) *Wetlands*. Third edition, John Wiley & Sons: New York.

Mitsch W.J., Gosselink J.G. (2015) *Wetlands*. Fifth edition, John Wiley & Sons: New York.

Mitsch W.J., Nedrich S.M., Harter S.K. Anderson C., Nahlik A.M., Bernal B. (2014a) Sedimentation in created freshwater riverine wetlands: 15 years of succession and contrast of methods. *Ecological Engineering 72*: 25–34.

Mitsch W.J., Wang N., Zhang L., Deal R., Wu X., Zuwerink A. (2005b) Using ecological indicators in a whole-ecosystem wetland experiment. In: *Handbook of Ecological Indicators for Assessment of Ecosystem Health*, Jørgensen S.E., Costanza R., Xu F.-L. (eds.) CRC Press: Boca Raton, FL, pp. 235–260.

Mitsch W.J., Wu X., Nairn R.W., Weihe P.E., Wang N., Deal R., Boucher C.E. (1998) Creating and restoring wetlands – An ecosystem experiment in self-design. *BioScience 48*: 1019–1030.

Mitsch W.J., Zhang L., Anderson C.J., Altor A.E., Hernández M.E. (2005c) Creating riverine wetlands: Ecological succession, nutrient retention, and pulsing effects. *Ecological Engineering 25*(5): 510–527.

Mitsch W.J., Zhang L., Stefanik K.C., Nahlik A.M., Anderson C.J., Bernal B., Hernandez M., Song K. (2012) Creating wetlands: Primary succession, water quality changes, and self-design over 15 years. *Bioscience 62*(3): 237–250.

Mitsch W.J., Zhang L., Waletzko E., Bernal B. (2014b) Validation of the ecosystem services of created wetlands: Two decades of plant succession, utrient retention, and carbon sequestration in experimental riverine marshes. *Ecological Engineering 72*: 11–24.

Mkandawire M., Dudel E.G. (2005) Accumulation of arsenic in *Lemna gibba* L. (duckweed) in tailing waters of two abandoned uranium mining sites in Saxony, Germany. *Science of the Total Environment 336*: 81–89.

Mohamed Y.A., Bastiaanssen W.G.M. Savenije H.H.G., Van Den Hurk B.J.J.M., Finlayson C.M. (2012) *Evaporation from Wetland Versus Open Water: A Theoretical Explanation and An Application with Satellite Data Over The Sudd Wetland.* Report of Royal Netherlands Meteorological Institute, Utrecht, The Netherlands. projects.knmi.nl/publications/fulltexts/sudd_wetland.pdf

Mohr G.E. (2000) Roblin Engineered Wetlands and Hybrid Poplar Tertiary Wastewater Treatment Project. Western Canada Water, September 2000, 32.

Mokry L.E. (2010 April). *North texas municipal water district's east fork wetland: Initial operational issues and performance evaluation.* In: Proceedings of the Texas Water Conference, Dallas, TX, 13–16.

Mokry L.E., Andrews D., Frossard W., Plummer A.H. (2008) Tarrant Regional Water District's Field-Scale wetland Phase 1 Operations–Lessons Learned. *Proceedings of the Water Environment Federation 2008*(12): 4357–4365.

Monson B.A., (2007) *Effectiveness of Stormwater Ponds/Constructed Wetlands in the Collection of Total Mercury and Production of Methylmercury.* Final Project Report, Minnesota Pollution Control Agency, St, Paul, MN. Available at: www.pca.state.mn.us/sites/default/files/tdr-g1-05.pdf ((Accessed 9/3/18).

Montaseri M., Afshar M.H., Bozorg-Haddad O. (2015) Development of simulation-optimization model (MUSIC-GA) for urban stormwater management. *Water Resources Management 29*(13): 4649–4665.

Montgomery Watson Americas, Inc. (1999) *STA-3/4 Field Investigations and Seepage Analysis.* Report from Montgomery Watson Americas, Inc., Lake Worth, FL to South Florida water Management Dsitrict, West Palm Beach, FL..

Montieth J.L. (1981) Evaporation and surface temperature. *Quarterly Journal of the Royal Meteorological Society 107*(451): 1–27.

Mooney F.D., Murray-Gulde C. (2008) Constructed treatment wetlands for flue gas desulfurization waters: Full-scale design, construction issues, and performance. *Environmental Geosciences 15*(3): 131–141.

Moore B.J., DuPont R.R., Armstrong J.E., Headley J.V. (2002a) Study of natural attenuation processes in a hydrocarbon-impacted wetland. In *Wetlands and Remediation II*, Nehring K.W., Brauning S. E. (eds.) Battelle Press: Columbus, OH, pp. 57–64.

Moore M.T., Bennett E.R., Cooper C.M., Smith, Jr S., Farris J.L., Drouillard K.G., Schulz R. (2006) Influence of vegetation in mitigation of methyl parathion runoff. *Environmental Pollution 142*(2): 288–294.

Moore M.T., Cooper C.M., Smith S., Cullum R.F., Knight S.S., Locke M.A., Bennett E.R. (2007) Diazinon mitigation in constructed wetlands: Influence of vegetation. *Water, Air, and Soil Pollution 184*(1–4): 313–321.

Moore M.T., Locke M.A., Kröger R. (2017) Mitigation of atrazine, S-metolachlor, and diazinon using common emergent aquatic vegetation. *Journal of Environmental Sciences 56*: 114–121.

Moore M.T., Schulz R., Cooper C.M., Smith, Jr S., Rodgers H., Jr. (2002b) Mitigation of chlorpyrifos runoff using constructed wetlands. *Chemosphere 46*(6): 827–835.

Moore M.T., Tyler H.L., Locke M.A. (2013) Aqueous pesticide mitigation efficiency of *Typha latifolia* (L.), *Leersia oryzoides* (L.) Sw., and *Sparganium americanum* Nutt. *Chemosphere 92*(10): 1307–1313.

Moore T.L., Hunt W.F. (2012) Ecosystem service provision by stormwater wetlands and ponds–A means for evaluation? *Water Research 46*(20): 6811–6823.

Moorhead K.K., Reddy K.R. (1988) Oxygen transport though selected aquatic macrophytes. *Journal of Environmental Quality 17*(138): 142.

Mora M.A., Benner R., Hodson R.E. (1989) Kinetics of microbial degradation of vascular plant material in two wetland ecosystems. *Oecologia 79*(2): 158–167.

Moran M.A., Benner R., Hodson R.E. (1989) Kinetics of microbial degradation of vascular plant material in two wetland ecosystems. *Oecologia* 79(2): 158–167.

Morel F.M.M., Hering J.G. (1993) *Principles and Applications of Aquatic Chemistry.* John Wiley & Sons: New York.

Moreno-Mateos D., Comín F.A. (2010) Integrating objectives and scales for planning and implementing wetland restoration and creation in agricultural landscapes. *Journal of Environmental Management 91*(11): 2087–2095.

Moriarty E.M., Karki N., Mackenzie M., Sinton L.W., Wood D.R., Gilpin B.J. (2011) Faecal indicators and pathogens in selected New Zealand waterfowl. *New Zealand Journal of Marine and Freshwater Research 45*(4): 679–688.

Morin B. (2009) Brighton Constructed Wetland. *Ontario Birding News 14*(3): 1–2.

Moustafa M. Z., J. M. Hamrick. (2000) Calibration of the wetland hydrodynamic model to the Everglades nutrient removal project. *Water Quality and Ecosystem Modeling 1*: 141–167.

Moustafa M.Z. (1998) Long-term equilibrium phosphorus concentrations in the Everglades as predicted by a Vollenwieder-type model. *Journal of the American Water Resources Association 34* (1): 135–147.

Moustafa M.Z., White J.R., Coghlan C.C., Reddy K.R. (2011) Influence of hydropattern and vegetation type on phosphorus dynamics in flow-through wetland treatment systems. *Ecological Engineering 37*: 1369–1378.

Mouvet C., Barberi D., Bourg A.C.M. (1993) Adsorption isotherms of tri- and tetrachloroethylene by various natural solids. *Journal of Hydrology 149*: 163–182.

MPCA. (2015) Stormwater wetland cost estimate worksheet, Minnesota Stormwater Manual. Minnesota Pollution Control Agency, SDt. Paul, MN. Available at: https://stormwater.pca.state.mn.us/index.php?title=Stormwater_wetland_cost_estimate_worksheet (Accessed 12/27/18).

Mueleman A.F.M., Beekman J.P., Verhoeven J.T.A. (2002) Nutrient retention and nutrient-use efficiency in *Phragmites australis* stands after wastewater application. *Wetlands 22*(4): 712–721.

Mueller M.H., van der Valk A.G. (2002) The potential role of ducks in wetland seed dispersal. *Wetlands 22*(1): 170–178.

Müller-Schwarze D., Sun D. (2003) *The Beaver: Natural History of a Wetlands Engineer.* Comstock Publishing Associates: Ithaca, NY.

Mulling B.T., van Den Boomen R.M., Claassen T.H., van der Geest H.G., Kappelhof J.W., Admiraal W. (2013) Physical and biological changes of suspended particles in a free surface flow constructed wetland. *Ecological Engineering 60*: 10–18.

Murkin H.R., van der Valk A.G., Clark W.R. (eds.) (2000) *Prairie Wetland Ecology*. Iowa State University Press: Ames, IA.

Murray C.G., Kasel S., Loyn R.H., Hepworth G., Hamilton A.J. (2013) Waterbird use of artificial wetlands in an Australian urban landscape. *Hydrobiologia 716*(1): 131–146.

Murray-Gulde C., Huddleston G.M., Garber K.V., Rodgers J.H. (2005) Contributions of *Schoenoplectus californicus* in a constructed wetland system receiving copper contaminated water. *Water, Air, and Soil Pollution 163*(1–4): 355–378.

Murray-Gulde C.L., Bridges W.C., Rodgers Jr J.H. (2008) Evaluating performance of a constructed wetland treatment system designed to decrease bioavailable copper in a waste stream. *Environmental Geosciences 15*(1): 21–38.

Musner T., Bottacin-Busolin A., Zaramella M., Marion A. (2014) A contaminant transport model for wetlands accounting for distinct residence time bimodality. *Journal of Hydrology 515*: 237–246.

Muzathik A.M., Nik W.B.W., Ibrahim M.B., Samo K.B., Sopian K., Alghoul M.A. (2011) Daily global solar radiation estimate based on sunshine hours. *International Journal of Mechanical and Materials Engineering 6*(1): 75–80.

NADB Database. (1998) *North American Treatment Wetland Database (NADB). Version 2.0*. Compiled by CH2MHill. Gainesville, FL. Available at: www.wetlandsolutionsinc.com/resources/ (Accessed 10/22/18).

Nahar S.N., 1998. Sub-Surface Investigation of Water Quality from a Percolating Constructed Wetlands. MS thesis, Arizona State University, Tempe, AZ.

Nahar S.N., Fox P., Wass R. (1998) Subsurface investigation of water quality from a percolating constructed wetlands. Preprint from 71st National Conference of the Water Environment Federation, WEF, Alexandria, VA, pp. 325–336.

Nahar S.N., Fox P., Wass R. (2000) *Sulfur driven autotrophic denitrification in constructed wetlands*. Proceedings of the WEFTEC 2000 National Conference, 73rd Annual Conference and Exhibition; Water Environment Federation: Alexandria, VA.

Naldi M., Wheeler P.A. (2002) ^{15}N measurements of ammonia and nitrate uptake by *Ulva fenestrata* (Chlorophyta) and *Gracilaria pacifica* (Rhodophyta): Comparison of net nutrient disappearance, release of ammonium, and nitrate and ^{15}N accumulation in algal tissue. *Journal of Phycology 38* (1): 135–144.

Naylor C., Davidson W., Motelica-Heino M., van Den Berg G.A., van der Heijdt L.M. (2005) Potential kinetic availability of metals in suphidic freshwater sediments. *Science of the Total Environment 357*(1–3): 208–220.

Neame P.A. (1976) Phosphorus Flux Across the Sediment-Water Interface. In: *Interactions between Sediments and Fresh Water*, Golterman H.L. (ed.) B.V. Publishers: The Hague, The Netherlands, pp. 307–312.

Neill C. (1992) Comparison of soil coring and ingrowth methods for measuring belowground production. *Ecology 73*(5): 1918–1921.

Neitsch S.L., Arnold J.G., Kiniry J.R., Williams J.R. (2011) *Soil and Water Assessment Tool, Theoretical Documentation, Version 2009*. Texas Water Resources Institute Technical Report No. 406, Texas A&M University, College Station, TX.

Nelson E.A., Specht W.L., Knox A.S. (2006) Metal removal from water discharges by a constructed treatment wetland. *Engineering in Life Sciences 6*(1): 26–30.

Nelson E.A., Specht W.L., Bowers J.A., Gladden J.B. (2002) *Constructed Wetlands for Removal of Heavy Metals from NPDES Outfall Effluent*, Report WSRC-MS-2002-00600, Westinghouse Savannah River Company: Aiken, SC.

Nelson E.A., Specht W.L., Knox A.S. (2004) *Metal Removal from Process and Stormwater Discharges by Constructed Treatment Wetlands*, Report WSRC-MS-2004-00763 to Savannah River National Laboratory: Aiken, SC.

Nelson K.L. (2003) Concentrations and inactivation of ascaris eggs and pathogen indicator organisms in wastewater stabilization pond sludge. *Water Science and Technology 48*(2): 89–95.

Nelson K.L., Boehm A.B., Davies-Colley R.J., Dodd M.C., Kohn T., Linden K.G., Liu Y., Maraccini P. A., McNeill K., Mitch W.A., Nguyen T.H. (2018) Sunlight-mediated inactivation of health-relevant microorganisms in water: A review of mechanisms and modeling approaches. *Environmental Science: Processes & Impacts 20*(8): 1089–1122.

Nepf H.M. (1999) Drag, turbulence, and diffusion in flow through emergent vegetation. *Water Resources Research 35*(2): 479–489.

Nepf H.M. (2012a) Hydrodynamics of vegetated channels. *Journal of Hydraulic Research 50*(3): 262–279.

Nepf H.M. (2012b) Flow and transport in regions with aquatic vegetation. *Annual Review of Fluid Mechanics 44*: 123–142.

Nepf H.M., Ghisalberti M. (2008) Flow and transport in channels with submerged vegetation. *Acta Geophysica 56*(3): 753–777.

Nepf H.M., Koch E.W.K. (1994) Vertical secondary flows in submersed plant-like arrays. *Limnology and Oceanography 44*(4): 1072–1080.

Nepf H.M., Sullivan J.A., Zavistoski R.A. (1997) A model for diffusion within emergent vegetation. *Limnology and Oceanography 42*(8): 1735–1745.

Nepf H.M., Vivoni E.R. (2000) Flow structure in depth-limited, vegetated flow. *Journal of Geophysical Research: Oceans 105*(C12): 28547–28557.

Neuhauser E., Swallow P. (2005) *Cyanide in the Environment*, Report of the Aluminum Company of America (ALCOA) and Niagara Mohawk, June 2005.

Nguyen M.T., Jasper J.T., Boehm A.B., Nelson K.L. (2015) Sunlight inactivation of fecal indicator bacteria in open-water unit process treatment wetlands: Modeling endogenous and exogenous inactivation rates. *Water Research 83*: 282–292.

Nguyen M.T., Silverman A.I., Nelson K.L. (2014) Sunlight inactivation of MS2 coliphage in the absence of photosensitizers: Modeling the endogenous inactivation rate using a photoaction spectrum. *Environmental Science and Technology 48*(7): 3891–3898.

Ni B.J., Peng L., Law Y., Guo J., Yuan Z. (2014) Modeling of nitrous oxide production by autotrophic ammonia-oxidizing bacteria with multiple production pathways. *Environmental Science and Technology 48*(7): 3916–3924.

Nicholls E. (2015) Multi-Year Water Balance Dynamics of a Newly Constructed Wetland, Fort McMurray, AB. PhD Dissertation, McMaster University, Hamilton, ON.

Niering W.A. (1985) *Wetlands*. Alfred A. Knopf, Inc.: New York.

Nietsch S.L., Arnold J.G., Kiniry J.R., Williams J.R., King K.W. (2011) *Soil and Water Assessment Tool Theoretical Documentation, Version 2009*. Texas Water Resources Institute Technical Report 406, Texas A&M University System, College Station, TX.

Nikolov N.T., Zeller K.F. (1992) A solar radiation algorithm for ecosystem dynamic models. *Ecological Modelling 61*: 149–168.

Nilsson C., Brown R.L., Jansson R., Merritt D.M. (2010) The role of hydrochory in structuring riparian and wetland vegetation. *Biological Reviews 85*(4): 837–858.

Nix P.G., Bishay F.S. (1997) Application of constructed wetlands for the detoxification of oil sands wastewater: A research overview. *Canadian Technical Report of Fisheries and Aquatic Sciences 2144*: 13–23.

Nixon S.W., Lee V. (1986) *A Regional Review of Recent Research in the United States on the Role of Freshwater and Saltwater Wetlands as Sources, Sinks, and Transformers of Nitrogen, Phosphorus, and Various Heavy Metals*. Final report prepared for the Department of the Army (U.S. Army Corps of Engineers), Technical Report Y-86-2, Waterways Experiment Station, Vicksburg, MS.

Nobel P.S. (1999) *Physico-Chemical and Environmental Plant Physiology*. Academic Press: San Diego, CA.

Noe G.B., Childers D.L. (2007) Phosphorus budgets in Everglades wetland ecosystems: The effects of hydrology and nutrient enrichment. *Wetlands Ecology and Management 15*: 189–205.

Noe G.B., Harvey J.W., Schaffranek R.W., Larsen L.G. (2010) Controls of suspended sediment concentration, nutrient content, and transport in a subtropical wetland. *Wetlands 30*(1): 39–54.

Noe G.B., Scinto L.J., Taylor J., Childers D.L., Jones R.D. (2003) Phosphorus cycling and partitioning in an oligotropic Everglades wetland ecosystem: A radioisotope tracing study. *Freshwater Biology 48*: 1993–2008.

Noller B.N., Woods P.H., Ross B.J. (1994) Case studies of wetland filtration of mine wastewater in constructed and naturally occurring systems in Northern Australia. *Water Science and Technology 29*(4): 257–265.

Nolte and Associates. (1997) *Sacramento Regional Wastewater Treatment Plant Demonstration Wetlands Project*, 1996 Annual Report to Sacramento Regional County Sanitation District, Nolte and Associates: Sacramento, CA.

Nolte and Associates. (1998a) *Sacramento Regional Wastewater Treatment Plant Demonstration Wetlands Project*, 1997 Annual Report to Sacramento Regional County Sanitation District, Nolte and Associates: Sacramento, CA.

Nolte and Associates. (1998b) *Sacramento Regional Wastewater Treatment Plant Demonstration Wetlands Project: Five Year Summary Report 1994–1998*, Report to Sacramento Regional County Sanitation District, Available at: www.srcsd.com/cw.html

Norris A., Burgin S. (2011) Apparent rapid loss of endocrine disruptors from wetlands used to store either tertiary treated sewage effluent or stormwater runoff. *Water, Air, and Soil Pollution 219* (1–4): 285–295.

Novak J.M., Stone K.C., Szogi A.A. Watts D.W., Johnson M.H. (2004) Dissolved phosphorus retention and release from a coastal plain in-stream wetland. *Journal of Environmental Quality 33*(1): 394–401.

Novak J.M., Szogi A.A., Stone K.C., Watts D.W., Johnson M.H. (2007) Dissolved phosphorus export from an animal waste impacted in-stream wetland: Response to tropical storm and hurricane disturbance. *Journal of Environmental Quality 36*(3): 790–800.

Novak V., Vidovic J. (2003) Transpiration and nutrient uptake dynamics in maize (Zea mays L.). *Ecological Modeling 166*: 99–107.

Novitzki R.P., Smith R.D., Fretwell J.D. (1997) *Restoration, Creation, and Recovery of Wetlands: Wetland Functions, Values, and Assessment*. Water Supply Paper 2425, United States Geological Survey, Alexandria, VA

Novotny V. (1992) Unit pollutant loads: Their fit in abatement strategies. *Water Environment Technology 4*(1): 40–43.

Novotny V. (1995) *Nonpoint Pollution and Urban Stormwater Management*. Technomic Publishing: Lancaster, PA.

NRC (Natural Resources Canada). (1978) *Hydrological Atlas of Canada*. Fisheries and Environment Canada, Ottawa, Ontario, Canada. Available at: https://open.canada.ca/data/en/dataset/67de4f04-855d-5d23-bb4a-2a270d1488d0 (Accessed 3/9/18).

NRCC (1979) *Peat Testing Manual*, Technical Memorandum No. 125. National Research Council of Canada (NRCC): Ottawa, Ontario, Canada.

NRCS. (2018) *NRCS National Bulletin 200-17-1 ECN - Normalized Prices and Discount Rate for FY 2018*. Natural Resources Conservation Service, United States Department of Agriculture, Washington, DC, Available at: www.nrcs.usda.gov/wps/portal/nrcs/detail/national/technical/econ/prices/?cid=nrcs143_009685 (Accessed 12/17/18).

Nriagu J.O. (ed.) (1978) *The Biogeochemistry of Lead in the Environment*. Elsevier: Amsterdam, The Netherlands.

Nungesser M.K., Chimney M.J. (2006) A hydrologic assessment of a subtropical constructed wetland in south Florida (USA). *Ecological Engineering 27*(4): 331–344.

NYDEC. (2015) *New York State Stormwater Management Design Manual*. New York State Department of Environmental Conservation, Albany, NY. Available at: www.dec.ny.gov/chemical/29072.html (Accessed 8/4/18).

O'Connell J.L., Johnson L.A., Beas B.J., Smith L.M., McMurry S.T., Haukos D.A. (2013) Predicting dispersal-limitation in plants: Optimizing planting decisions for isolated wetland restoration in agricultural landscapes. *Biological Conservation 159*: 343–354.

O'Neill A., Foy R.H., Phillips D.H. (2011) Phosphorus retention in a constructed wetland system used to treat dairy wastewater. *Bioresource Technology 102*(8): 5024–5031.

O'Brien J.M., Hamilton S.K., Kinsman-Costello L.E., Lennon J.T., Ostrom N.E. (2012) Nitrogen transformations in a through-flow wetland revealed using whole-ecosystem pulsed 15N additions. *Limnology and Oceanography 57*(1): 221–234.

Ockenden M.C., Deasy C., Quinton J.N., Bailey A.P., Surridge B., Stoate C. (2012) Evaluation of field wetlands for mitigation of diffuse pollution from agriculture: Sediment retention, cost and effectiveness. *Environmental Science and Policy 24*: 110–119.

Ockenden M.C., Deasy C., Quinton J.N., Surridg B., Stoate C. (2014) Keeping agricultural soil out of rivers: Evidence of sediment and nutrient accumulation within field wetlands in the UK. *Journal of Environmental Management 135*: 54–62.

O'Connor D.J., Dobbins W.E. (1958) *Mechanism of reaeration in natural streams* (Paper No. 2934). *Transactions of the American Society of Civil Engineers*, pp. 641–684.

Odum H.T., Wojcik W., Pritchard L., Ton S., Delfino J.J., Wojcik M., Lesczynski S., Patel J.D., Doherty S.J., Stasik J. (2000) *Heavy Metals in the Environment: Using Wetlands for Their Removal*. Lewis Publishers: Boca Raton, FL.

Ohlendorf H.M., Hoffman D.J., Saiki M.K., Aldrich T.W. (1986) Embryonic mortality and abnormalities of aquatic birds: Apparent impacts of selenium from irrigation drainwater. *Science of the Total Environment 52*: 49–63.

Ohlendorf H.M., Hothem R.L., Bunck C.M., Marios K.C. (1990) Bioaccumulation of selenium in birds at Kesterson Reservoir, California. *Archives of Environmental Contamination and Toxicology 19*: 495–507.

Oldham C.E., Sturman J.J. (2001) The effect of emergent vegetation on convective flushing in shallow wetlands: Scaling and experiments. *Limnology and Oceanography 46*(6): 1486–1493.

Olila O.G., Reddy K.R., Stites D.L. (1997) Influence of draining on soil phosphorus forms and distribution in a constructed wetland. *Ecological Engineering 9*(3–4): 157–169.

Olszewska D.O. (2005) Wetland Planning in Agricultural Landscape Using Geographical Information System: A Case Study of Lake Ringsjön Basin in South Sweden. MS Thesis, Linköping University, Linköping, Sweden.

Orosz-Coghlan P.A., Rusin P.A., Karpiscak M.M. (2006) Microbial source tracking of *Escherichia coli* in a constructed wetland. *Water Environment Research 78*(3): 227–232.

Osborne T.Z., Bruland G.L., Newman S., Reddy K.R., Grunwald S. (2011) Spatial distributions and eco-partitioning of soil biogeochemical properties in the Everglades National Park. *Environmental Monitoring and Assessment 183*(1–4): 395–408.

Ostrofsky M.L., Miller C. (2017) Photosynthetically-mediated calcite and phosphorus precipitation by submersed aquatic vascular plants in Lake Pleasant, Pennsylvania. *Aquatic Botany 143*: 36–40.

O'Sullivan A.D., Moran B.M., Otte M.L. (2004) Accumulation and fate of contaminants (Zn, Pb, Fe, and S) in substrates of wetlands constructed for treating mine wastewater. *Water, Air, and Soil Pollution 157*: 345–364.

Otis D.L., Crumpton W.G., Green D.L., Loan-Wilsey A.K., McNeely R.L., Kane K.L., Johnson R., Cooper T., Vandever M. (2010). *Assessment of Environmental Services of CREP Wetlands in Iowa and the Midwestern Corn Belt.* Iowa Cooperative Fish and Wildlife Research Unit Reports, Iowa State University, Ames, IA. Available at: http://lib.dr.iastate.edu/cfwru_reports/3 (Accessed 10/30/18).

Oyegbile B., Ay P., Narra S. (2016) Flocculation kinetics and hydrodynamic interactions in natural and engineered flow systems: A review. *Environmental Engineering Research 21*(1): 1–14.

Pálfy T.G., Langergraber G. (2014) The verification of the Constructed Wetland Model No. 1 implementation in HYDRUS using column experiment data. *Ecological Engineering 68*: 105–115.

Palmer C.N., Hunt J.D. (1989) Greenwood urban wetland: A manmade stormwater treatment facility. In: *Wetlands: Concerns and Successes. Proceedings of a Symposium Held 17–22 September 1989 in Tampa, Florida*, Fisk D. (ed.), American Water Works Association: Bethesda, MD, pp. 205–214.

Palmer M.R., Nepf H.M., Pettersson T.J.R., Ackerman J.D. (2004) Observations of particle capture on a cylindrical collector: Implications for particle accumulation and removal in aquatic systems. *Limnology and Oceanography 49*(1): 76–85.

Palmer-Felgate E.J., Bowes M.J., Stratford C., Neal C., MacKenzie S. (2011) Phosphorus release from sediments in a treatment wetland: Contrast between DET and EPC0 methodologies. *Ecological Engineering 37*(6): 826–832.

Pang C.C., Wang F.F., Wu S.Q., Lai X.J. (2015) Impact of submerged herbaceous vegetation on wind-induced current in shallow water. *Ecological Engineering 81*: 387–394.

Pankow J.F. (1991) *Aquatic Chemistry Concepts.* Lewis Publishers: Boca Raton, FL.

Panno S.V., Kelly W.R., Hackley K.C., Hwang H.H., Martinsek A.T. (2008) Sources and fate of nitrate in the Illinois River Basin, Illinois. *Journal of Hydrology 359*(1–2): 174–188.

Panswad T., Chavalprit O. (1997) Water quality and occurrences of protozoa and metazoa in two constructed wetlands treating different wastewaters in Thailand. *Water Science and Technology 36* (12): 183–188.

Pant H.K., Reddy K.R. (2001a) Hydrologic influence on stability of organic phosphorus in wetland detritus. *Journal of Environmental Quality 30*(2): 668–674.

Pant H.K., Reddy K.R. (2001b) Phosphorus sorption characteristics of estuarine sediments under different redox conditions. *Journal of Environmental Quality 30*: 1474–1480.

Pardue J.H., Kassenga G., Shin W.S. (2000) Design approaches for chlorinated VOC treatment wetland. Means J.L., Hinchee R.E. (eds.), Wetlands & Remediation: An International Conference; Battelle Press: Columbus, OH, pp. 301–308.

Pardue J.H., Masscheleyn P.H., DeLaune R., Patrick W.H. Jr. (1993) Assimilation of hydrophobic chlorinated organics in freshwater wetlands: Sorption and sediment-water exchange. *Environmental Science and Technology 27*(5): 875–882.

Pardue J.H., Shin W.S. (2000) *Desorption resistance of organic compounds in wetland soils.* Means J.L., Hinchee R.E. (eds.), Wetlands & Remediation: An International Conference; Battelle Press: Columbus, OH, pp. 25–32.

Parker L.V., Martel C.J. (2002) *Long-Term Survival of Enteric Microorganisms in Frozen Wastewater,* Report ERDC/CRREL TR-02-16, Prepared for the National Science Foundation, U.S. Army Corps of Engineers: Washington, DC

Parker L.V., Yushak M.L., Martel C.J., Reynolds C.M. (2000) *Bacterial Survival in Snow Made from Wastewater,* Report ERDC/CRREL TR-00-9, Prepared for the Office of the Chief of Engineers, U.S. Army Corps of Engineers: Washington, DC

Parkhurst D.L., Thorstenson D.C., Plummer L.N. *PHREEQC* (1990) – *A Computer Program for Geochemical Calculations* [Computer Program]. U.S. Geological Survey Water Investigations Report 80–96 (195); Revised and reprinted in August 1990.

Parkhurst R.S., Winter T.C., Rosenberry D.O., Sturrock A.M. (1998) Evaporation from a small prairie wetland in the Cottonwood Lake area, North Dakota—An energy-budget study. *Wetlands 18*(2): 272–287.

Parkin G.F. (1999) Anaerobic biotransformation of chlorinated aliphatic hydrocarbons: Ugly duckling to beautiful swan. *Water Environment Research 71*(6): 1158–1164.

Passeport E., Benoit P., Bergheaud V., Coquet Y., Tournebize J. (2011) Selected pesticides adsorption and desorption in substrates from artificial wetland and forest buffer. *Environmental Toxicology and Chemistry 30*(7): 1669–1676.

Passeport E., Tournebize J., Chaumont C., Guenne A., Coquet Y. (2013) Pesticide contamination interception strategy and removal efficiency in forest buffer and artificial wetland in a tile-drained agricultural watershed. *Chemosphere 91*(9): 1289–1296.

Patrick W.H. Jr., Mikkelsen D.S., Wells B.R. (1985) Plant nutrient behavior in flooded soils. In: *Fertilizer Technology and Use,* 3rd edition. Soil Science Society of America: Madison, WI, pp. 197–228.

Paudel R., Grace K.A., Galloway S., Zamorano M., Jawitz J.W. (2013) Effects of hydraulic resistance by vegetation on stage dynamics of a stormwater treatment wetland. *Journal of Hydrology 484*: 74–85.

Paudel R., Jawitz J.W. (2012) Does increased model complexity improve description of phosphorus dynamics in a large treatment wetland? *Ecological Engineering 42*: 283–294.

Paul E.A., Clark F.E. (1996) *Soil Microbiology and Biochemistry,* 2nd edition. Academic Press: San Diego, CA.

Paudel R., Min J., Jawitz J.W. (2010) Management scenario evaluation for a large treatment wetland using a spatio-temporal phosphorus transport and cycling model. *Ecological Engineering 36*: 1627–1638.

Paul S. (ed.) (2013) *Workbook for Managing Urban Wetlands in Australia.* WET eBook, Sydney Olympic Park Authority, Sydney, Australia. Available at: www.sopa.nsw.gov.au/Resource-Centre/WET-eBook-Workbook-for-Managing-Urban-Wetlands-in-Australia (Accessed 12/10/18).

Pauwels H., Talbo H. (2004) Nitrate concentration in wetlands: Assessing the contribution of deeper groundwater from anions. *Water Research 38*(4): 1019–1025.

Pavlineri N., Skoulikidis N.T., Tsihrintzis V.A. (2017) Constructed floating wetlands: A review of research, design, operation and management aspects, and data meta-analysis. *Chemical Engineering Journal 308*: 1120–1132.

Payne E.K., Burgin A.J., Hamilton S.K. (2009) Sediment nitrate manipulation using porewater equilibrators reveals potential for N and S coupling in freshwaters. *Aquatic Microbial Ecology 54*: 233–241.

Peacock C.E., Hess T.M. (2004) Estimating evapotranspiration from a reed bed using the Bowen ratio energy balance method. *Hydrological Processes 18*(2): 247–260.

Peet R.K., Roberts D.W. (2013) Classification of natural and semi-natural vegetation. Vegetation ecology. In: *Vegetation Ecology, 2nd Edition,* van der Maarel E., Franklin J. (eds.), Wiley-Blackwell: Chichester, United Kingdom, p. 28–70.

Pei G., Wang Q., Liu G. (2015) The role of periphyton in phosphorus retention in shallow lakes with different trophic status, China. *Aquatic Botany 125*: 17–22.

Peng J., Bewtra J.K., Biswas N. (1995) Effect of turbulence on volatilization of selected organic compounds from water. *Water Environment Research 61*(1): 101–107.

Penman H.L. (1948) Natural evaporation from open water, bare soil and grass. *Proceedings of the Royal Society of London A: Mathematical, Physical and Engineering Sciences 193*(1032): 120–145.

Perdomo S., Fujita M., Furukawa K. (1996) Oxygen transport through Pistia stratiotes L.: Chapter VIII/4. Proceedings of the 5th International Conference on Wetland Systems for Water Pollution Control. Universität für Bodenkultur: Vienna, Austria.

Pers C. (2007) *HBV-NP Model Manual*. Swedish Meteorological and Hydrological Institute: Norrköping, Sweden.

Persaud D., Jaagumagi R., Hayton A. (1993) *Guidelines for the The Protection and Management of Aquatic Sediment Quality in Ontario*. Ontario Ministry of the Environment and Energy (OMOEE), Queen's Printer for Ontario: Toronto, Ontario, Canada.

Persson J., Somes N.L.G., Wong T.H.F. (1999) Hydraulics efficiency of constructed wetlands and ponds. *Water Science and Technology 40*(3): 291–300.

Petersen J.E., Brandt E.C., Grossman J.J., Allen G.A., Benzing D.H. (2015) A controlled experiment to assess relationships between plant diversity, ecosystem function and planting treatment over a - nine year period in constructed freshwater wetlands. *Ecological Engineering 82*: 531–541.

Petticrew E.L., Kalff J. (1992) Water flow and clay retention in submerged macrophyte beds. *Canadian Journal of Fisheries and Aquatic Sciences 49*(12): 2483–2489.

Petty J.D., Huckins J.N., Alvarez D.A., Brumbaugh W.G., Cranor W.L., Gale R.W., Rastall A.C., Jones-Lepp T.L., Leiker T.J., Rostad C.E., Furlong E.T. (2004) A holistic passive integrative sampling approach for assessing the presence and potential impacts of waterborne environmental contaminants. *Chemosphere 54*(6): 695–705.

Peverly J.H., Surface J.M., Steenhuis T.S., Sanford W.E. (1993) *Constructed Wetlands for Municipal Solid Waste Landfill Leachate Treatment*, Report to New York State Energy Research and Development Authority, U.S. Department of Commerce, National Technical Information Service: Springfield, VA.

PFRA. (2002) *Town of Roblin Effluent Irrigation and Constructed Wetland 2001 Annual Report*. Prairie Farm Rehabilitation Administration (PFRA): Winnipeg, Manitoba.

PFRA. (2003) *Town of Roblin Effluent Irrigation and Constructed Wetland 2003 Annual Report*. Prairie Farm Rehabilitation Administration (PFRA): Winnipeg, Manitoba.

Phipps R., Crumpton W.G. (1994) Factors affecting nitrogen loss in experimental wetlands with different hydrologic loads. *Ecological Engineering 3*(4): 399–408.

Piccone T., McBryan J., Zhao H., Yan Y. (2014) Everglades Stormwater Treatment Area average ground elevations, stage-area/stage-volume relationships and effective rreatment areas. Technical Publication ASB-WQTT-12-002, 2012 Updated, South Florida Water Management District. December 2012 (Original), May 2013 (Revision 1), August 2014 (Revision 2).

Pickett J., McKellar H., Kelley J. (1989) *Plant community composition, leaf mortality, and aboveground production in a tidal freshwater marsh*. Freshwater Wetlands and Wildlife, Department of Energy Symposium Series No. 61. Sharitz R.R., Gibbons J.W. (eds.) pp. 351–364.

Piercy C.D. (2010) Hydraulic Resistance Due to Emergent Wetland Vegetation. PhD Dissertation, Virginia Tech University, Blacksburg, VA.

Pietro K.C. (1998) Phosphorus Uptake by the Ceratophyllum/Periphyton Complex in a Southern Florida Freshwater Marsh. MS Thesis, Florida Atlantic University (Boca Raton, FL).

Pietro K.C., Chimney M.J., Steinman A.D. (2006) Phosphorus uptake by the ceratophyllum/periphyton complex in a South Florida Freshwater Marsh. *Ecological Engineering 27*(4): 290–300.

Pilon-Smits E. (2005) Phytoremediation. *Annual Review of Plant Biology 56*: 15–39.

Pimentel D., Lach L., Zuniga R., Morrison D. (1999) *Environmental and economic costs associated with non-indigenous species in the United States*. Paper presented at the 1999 Annual Meeting of the American Association for the Advancement of Science (AAAS), Anaheim, California.

Pineywoods RC&D. (2001) *Constructed Wetlands in East Texas (4 Volumes)*. Pineywoods Resource Conservation & Development Council: Nagodoches, TX.

Pitt R., Maestre A. (2015) *The National Stormwater Quality Database (NSQD, Version 4.02)*. Department of Civil and Environmental Engineering, University of Alabama: Tuscaloosa, AL. Available at: www.bmpdatabase.org/nsqd.html (Accessed 7/18/18).

Piwpuan N., Zhai X., Brix H. (2013) Nitrogen nutrition of Cyperus laevigatus and Phormium tenax: Effects of ammonium versus nitrate on growth, nitrate reductase activity and N uptake kinetics. *Aquatic Botany 106*: 42–51.

Plachy J. (2000) Cadmium. Chapter 16 in: *U.S. Geological Survey Minerals Yearbook – 2000*. U.S. Geological Survey: Reston, VA.

Poach M.E., Hunt C.L., Vanotti M.B., Stone K.C., Matheny T.A., Johnson M.H., Sadler E.J. (2003) Improved nitrogen treatment by constructed wetlands receiving partially nitrified liquid swine manure. *Ecological Engineering 20*(2): 183–197.

Poach M.E., Hunt P.G., Reddy G.B., Stone K.C., Johnson M.H., Grubbs A. (2004) Swine wastewater treatment by marsh-pond-marsh constructed wetlands under varying nitrogen loads. *Ecological Engineering 23*(3): 165–175.

Poach M.E., Hunt P.G., Sadler E.J., Matheny T.A., Johnson M.H., Stone K.C., Humenik F.J., Rice J.M. (2002) Ammonia volatilization from constructed wetlands that treat swine wastewater. *Transactions of the American Society of Agricultural Engineers 45*(3): 619–627.

Poindexter C.M., Variano E.A. (2013) Gas exchange in wetlands with emergent vegetation: The effects of wind and thermal convection at the air-water interface. *Journal of Geophysical Research: Biogeosciences 118*(3): 1297–1306.

Polatel C., Abtew W., Krupa S., Piccone T. (2014) Stormwater Treatment Area Water and Phophorus Budget Improvements, Phase I: STA-3/4 Cells 3A and 3B Water Budgets, Technical Publication WR-2014-004, South Florida Water Management District, West Palm Beach, FL.

Polprasert C., Dan N.P. (1994) Phenol removal in model constructed wetlands located in the tropics. Proceedings of the 67th Annual WEFTEC Conference. Water Environment Federation: Alexandria, VA, pp. 45–55.

Polprasert C., Dan N.P., Thayalakumaran N. (1996) Application of constructed wetlands to treat some toxic wastewaters under tropical conditions. *Water Science and Technology 34*(11): 165–171.

Polprasert C., Khatiwada N.R., Bhurtel J. (1998) Design model for COD removal in constructed wetlands based on biofilm activity. *Journal of Environmental Engineering (ASCE) 124*(9): 838–843.

Ponnamperuma F.N. (1972) The chemistry of submerged soils. *Advances in Agronomy 24*: 29–96.

Pontier H., Williams J.B., May E. (2003) Behaviour of metals associated with sediments in a wetland based system for road runoff control. *Water Science and Technology 48*(5): 291–298.

Porej D., Hetherington T.E. (2005) Designing wetlands for amphibians: The importance of predatory fish and shallow littoral zones in structuring of amphibian communities. *Wetlands Ecology and Management 13*(4): 445–455.

Post D.M., Taylor J.P., Kitchell J.F., Olson M.H., Schindler D.E., Herwig B.R. (1998) The role of migratory waterfowl as nutrient vectors in a managed wetland. *Conservation Biology 12*(4): 910–920.

Powell C.L. (2010) Biodegradation of Groundwater Pollutants (Chlorinated Hydrocarbons) in Vegetated Wetlands: Role of Aerobic Microbes Naturally Associated with Roots of Common Plants. PhD Dissertation, Wright State University, Fairborn, OH.

Powell C.L., Agrawal A. (2011) Biodegradation of trichloroethene by methane oxidizers naturally associated with wetland plant roots. *Wetlands 31*(1): 45–52.

Powell C.L., Goltz M.N., Agrawal A. (2014) Degradation kinetics of chlorinated aliphatic hydrocarbons by methane oxidizers naturally-associated with wetland plant roots. *Journal of Contaminant Hydrology 170*: 68–75.

Powell R.W. (1968) The origin of Manning's formula. *Journal of the Hydraulics Division 94*(4): 1179–1181.

Powell C.L., Nogaro G., Agrawal A. (2011) Aerobic cometabolic degradation of trichloroethene by methane and ammonia oxidizing microorganisms naturally associated with *Carex comosa* roots. *Biodegradation 22*(3): 527–538.

Prasse C., Wenk J., Jasper J.T., Ternes T.A., Sedlak D.L. (2015) Co-occurrence of photochemical and microbiological transformation processes in open-water unit process wetlands. *Environmental Science and Technology 49*(24): 14136–14145.

Prato T., Hey D.L. (2006) Economic analysis of wetland restoration along the Illinois River. *Journal of the American Water Resources Association* Paper No. 04183, 125–131.

Prenger J.P., Reddy K.R. (2004) Microbial enzyme activities in a freshwater marsh after cessation of nutrient loading. *Soil Science Society of America Journal 68*: 1796–1804.

Prentki R.T., Gustafson T.D., Adams M.S. (1978) Nutrient movements in Lakeshore Marshes. In *Freshwater Wetlands: Ecological Processes and Management Potential*, Good R.E., Whigham D.F., Simpson R.L. (eds.) Academic Press: New York, pp. 169–194.

Preussler K.H., Mahler C.F., Maranho L.T. (2015) Performance of a system of natural wetlands in leachate of a posttreatment landfill. *International Journal of Environmental Science and Technology 12*(8): 2623–2638.

Prey J., Chern L., Williamson A. (1995) *The Wisconsin Stormwater Manual*. Wisconsin Department of Natural Resources: Madison, WI.

Priban K., Jenik J., Ondok J.P., Popela P., 1992. *Analysis and Modeling of Wetland Microclimate.* Study CSAV 2-92, Czech Academy of Science Publishers, Prague, Czech Republic.

Price J.S. (1994) Evapotranspiration from a lakeshore *Typha* marsh on Lake Ontario. *Aquatic Botany 48*(3–4): 261–272.

Proakis E. (2003) *Pathogen removal in constructed wetlands focusing on biological predation and marine recreational water quality.* Proceedings, WEFTEC 2003 National Conference, 76th Annual Conference and Exhibition; Water Environment Federation: Alexandria, VA.

Proctor L.M., Shaefer G., Landing W.M. (1999) Bioavailability of dissolved organic phosphorus in the Florida Everglades: Evaluation with native bacterial and reference enzymes. Proceedings of the 6th Symposium on the Biogeochemistry of Wetlands, 11–14 July 1999: Fort Lauderdale, FL.

Pullin B.P., Hammer D.A. (1989) Comparison of plant density and growth forms related to removal efficiencies in constructed wetlands treating municipal wastewater. Proceedings of the 62nd Annual WPCF Meeting, 16–19 October 1989. Water Pollution Control Federation (WPCF).

Pumo D., Tamea S., Noto L.V., Miralles-Wilhelm F., Rodriguez-Iturbe I. (2010) Modeling belowground water table fluctuations in the Everglades. *Water Resources Research 46*: W11557.

Puriveth P. (1980) Decomposition of emergent macrophytes in a Wisconsin marsh. *Hydrobiologia 72*: 231–242.

Quagraine E.K., Peterson H.G., Headley J.V. (2005) In situ bioremediation of naphthenic acids contaminated tailing pond waters in the Athabasca oil sands region—Demonstrated field studies and plausible options: A review. *Journal of Environmental Science and Health 40*(3): 685–722.

Qualls R.G., Heyvaert A.C. (2017) Accretion of nutrients and sediment by a constructed stormwater treatment wetland in the Lake Tahoe Basin. *Journal of the American Water Resources Association 53*(6): 1495–1512.

R.A. Smith & Associates. (1995) *Wind Lake/Muskego Canal Nutrient Inactivation Project*, Report to Wind Lake Management District, 24 February 1995.

Racine C.H., Walters J.C. (1991) Groundwater-discharge Wetlands in the Tanana flats, Interior Alaska. Cold Regions Research and Engineering Lab Report 91–14, CRREL, Hanover, NH.

Raisin G., Bartley J., Croome R. (1999) Groundwater influence on the water balance and nutrient budget of a small natural wetland in Northeastern Victoria, Australia. *Ecological Engineering 12*(1–2): 133–147.

Raisin G., Mitchell D.S., Croome R. (1997) The effectiveness of a small constructed wetland in ameliorating diffuse nutrient loadings from an Australian rural catchment. *Ecological Engineering 9*(1–2): 19–35.

Ranieri E. (2004) *Chromium and nickel removal in lab and pilot scale CW plants under different clogging conditions.* Liénard A., Burnett H. (eds.) Proceedings of the 9th International Conference on Wetland Systems for Water Pollution Control, 26–30 September 2004. Association Scientifique et Technique pour l'Eau et l'Environnement (ASTEE), Cemagref, and IWA: Avignon, France, pp. 587–594.

Rasmussen T. (2008) *Methods for Evaluating Wetland Condition; #20 Wetland Hydrology.* USEPA Report EPA-822-R-08-024, Office of Water, Washington, DC.

Rasse D.P., Rumpel C., Dignac M.F. (2005) Is soil carbon mostly root carbon? Mechanisms for a specific stabilisation. *Plant and Soil 269*(1–2): 341–356.

Reaves R.P., DuBowy P.J., Jones D.D., Sutton A.L. (1994) *Design of an experimental constructed wetland for treatment of swine lagoon effluent.* Proceedings of a Workshop on Constructed Wetlands for Animal Waste Management, 4–6 April 1994, Purdue University: Lafayette, IN, pp. 82–96.

Reddy K.R., D'Angelo E.M. (1994) Soil process regulating water quality in wetlands. In: *Global wetlands: Old world and new,* Mitsch W.J. (ed.) Elsevier: Amsterdam, The Netherlands, pp. 309–324.

Reddy K.R., DeBusk W.F., DeLaune R.D., Koch M.S. (1993) Long-term nutrient accumulation rates in the Everglades. *Soil Science Society of America Journal 57*(4): 1147–1155.

Reddy K.R., DeBusk W.F., Wang Y., DeLaune R., Koch M. (1991) *Physico-Chemical Properties of Soils in the Water Conservation Area 2 of the Everglades.* Report to the South Florida Water Management District, West Palm Beach FL.

Reddy K.R., DeBusk W.R. (1985) Nutrient removal potential of selected aquatic macrophytes. *Journal of Environmental Quality 14*: 459–462.

Reddy K.R., Delaune R. (2008) *Biogeochemistry of Wetlands: Science and Applications.* CRC Lewis Publishers: Boca Raton, FL.

Reddy K.R., Diaz O.A., Scinto L.J., Agami M. (1995) Phosphorus dynamics in selected wetlands and streams of the Lake Okeechobee Basin. *Ecological Engineering 5*(2–3): 183–207.

Reddy K.R., Feijtel T.C., Patrick W.H. Jr. (1986) Effect of soil redox conditions on microbial oxidation of organic matter. In: *The Role of Organic Matter in Modern Agriculture*, Chen Y., Avnimelech Y. (eds.) Martinus Nijhoff: Dordrecht, The Netherlands, pp. 117–156.

Reddy K.R., Graetz D.A. (1988) Carbon and nitrogen dynamics in wetland soils. In *The Ecology and Management of Wetlands*, Hook D.D. (ed.) Croom Helm: London, United Kingdom, pp. 307–318.

Reddy K.R., O'Connor G.A., Schelske C.L. (eds.) (1999) *Phosphorus Biogeochemistry of Sub-Tropical Ecosystems*. CRC Press: Boca Raton, FL.

Reddy K.R., Patrick W.H. (1984) Nitrogen transformations and loss in flooded soils and sediments. *CRC Critical Reviews in Environmental Control 13*: 273–309.

Reddy K.R., Wetzel R.G., Kadlec R.H. (2005) Biogeochemistry of Phosphorus in Wetlands. In *Phosphorus: Agriculture and the Environment*, Sims J.T., Sharpley A.N. (eds.) Soil Science Society of America: Madison, WI, pp. 263–316.

Reed S.C., Crites R., Middlebrooks E.J. (1988) *Natural Systems for Waste Management and Treatment*, 1st edition. McGraw-Hill: New York.

Reed S.C., Crites R., Middlebrooks E.J. (1995) *Natural Systems for Waste Management and Treatment*, 2nd edition. McGraw-Hill: New York.

Reichert P. (1998) *AQUASIM 2.0-User Manual Computer Program for the Identification and Simulation of Aquatic Systems*. Swiss Federal Institute for Environmental Science and Technology, (EAWAG) CH - 8600 Dubendorf, Switzerland.

Reid M.C., Jaffé P.R. (2012) Gas-phase and transpiration-driven mechanisms for volatilization through wetland macrophytes. *Environmental Science and Technology 46*(10): 5344–5352.

Reilly J.F., Horne A.J., Miller C.D. (2000) Nitrate removal from a drinking water supply with large free-water surface constructed wetlands prior to groundwater recharge. *Ecological Engineering 14*: 33–47.

Reimold R.J., Hardisky M.A. (1978) Nonconsumptive use values of wetlands. In: *Wetland Functions and Values: The State of Our Understanding*, Proceedings of the National Symposium on Wetlands, 7–10 November 1978, Greeson P.E., Clark J.R., Clark J.E. (eds.) American Water Resources Association: Lake Buena Vista, FL.

Reinartz J.A., Warne E.L. (1993) Development of vegetation in small created wetlands in southeastern Wisconsin. *Wetlands 13*(3): 153–164.

Reinelt L.E., Horner R.R. (1995) Pollutant removal from stormwater runoff by palustrine wetlands based on comprehensive budgets. *Ecological Engineering 4*(2): 77–97.

Reinhardt M., Gächter R., Wehrli B., Müller B. (2005) Phosphorus retention in small constructed wetlands treating agricultural drainage water. *Journal of Environmental Quality 34*: 1251–1259.

Reitberger J.H., Mokry L.E., Knight R.L. (2000) *Achieving multiple benefits from a constructed wetland*. Reddy K.R., Kadlec R.H. (eds.) Proceedings of the 7th International Conference on Wetland Systems for Water Pollution Control, 11–16 November 2000; University of Florida and IWA: Lake Buena Vista, FL, 749–758.

Rejmankova E. (2011) The role of macrophytes in wetland ecosystems. *Journal of Ecology and Environment 34*(4): 333–345.

Rejmánková E., Komárková J. (2000) A function of cyanobacterial mats in phosphorus-limited tropical wetlands. *Hydrobiologia 431*(2): 135–153.

Remacle J., Muguruza L., Fransolet M. (1992) Cadmium removal by a strain of *Alcaligenes denitrificans* isolated from a metal-polluted pond. *Water Research 26*(7): 923–926.

Restrepo J.I., Montoya A.M., Obeysekera J. (1998) A wetland simulation module for the MODFLOW ground water model. *Groundwater 36*(5): 764–770.

Rey J.R., Walton W.E., Wolfe R.J., Connelly C.R., O'Connell S.M., Berg J., Sakolsky-Hoopes G.E., Laderman A.D. (2012) North American wetlands and mosquito control. *International Journal of Environmental Research and Public Health 9*(12): 4537–4605.

Reyes-Contreras C., Hijosa-Valsero M., Sidrach-Cardona R., Bayona J.M., Bécares E. (2012) Temporal evolution in PPCP removal from urban wastewater by constructed wetlands of different configuration: A medium-term study. *Chemosphere 88*(2): 161–167.

Richard D.E., Berns J.J., Bankston J.L., Connell D. (2003) Restored wetland remediation of a chlorinated solvent plume. Paper N-05. *Proceedings of the Seventh International in Situ and On-Site Bioremediation Symposium in Orlando, Florida, June 2003*. Battelle Press: Columbus, OH.

Richard D.E., England K.P., Connell D., Berns J.J., Hirl P.J. (2002) Assessment of a reconstructed wetland for remediation of chlorinated solvents. In: *Wetlands and Remediation II*, Nehring K.W., Brauning S.E. (eds.) Battelle Press: Columbus, OH, pp. 117–124.

Richard F.C., Bourg A.C.M. (1991) Aqueous geochemistry of chromium: A review. *Water Research 25* (7): 807–816.

Richards P.I., Sandau C.D. (2018) Forensic source attribution for toluene in environmental samples. *Environmental Toxicology and Chemistry 37*(3): 729–737.

Richardson C.J. (1985) Mechanisms controlling phosphorus retention capacity in freshwater wetlands. *Science 228*: 1424–1427.

Richardson C.J., Craft C.B. (1993) Effective phosphorus retention in wetlands: Fact or fiction. In *Constructed Wetlands for Water Quality Improvement*, Moshiri G.A. (ed.) Lewis Publishers: Boca Raton, FL, pp. 271–282.

Richardson C.J., Marshall P.E. (1986) Processes controlling movement, storage, and export of phosphorus in a fen peatland. *Ecological Monographs 56*(4): 279–302.

Richardson C.J., Qian S.S., Craft C.B., Qualls R.G. (1997) Predictive models for phosphorus retention in wetlands. *Wetlands Ecology and Management 4*(159): 175.

Richardson C.J., Vaithiyanathan P. (1995) Phosphorus sorption characteristics of Everglades soils along a eutrophication gradient. *Soil Science Society of America Journal 59*(6): 1782–1788.

Richardson C.J., Vymazal J. (2001) Sampling macrophytes in wetlands. In: *Bioassessment and Management of North American Freshwater Wetlands*, Rader R.B., Barter D.P., Wissinger S.A. (eds.) John Wiley & Sons: New York, pp. 297–337.

Richardson J.L., Vespraskas M.J. (eds.) (2001) *Wetland Soils*. Lewis Publishers: Boca Raton, FL.

Richardson C.J. (2008) *The Everglades Experiments: Lessons for Ecosystem Restoration*. Springer Science & Business Media: New York.

Riemann B., Simonsen P, Stensgaard L. (1989) The carbon and chlorophyll content of phytoplankton from various nutrient regimes. *Journal of Plankton Research 11*(5): 1037–1045.

Ringqvist L., Öborn I. (2002) Copper and zinc adsorption onto poorly humified *Sphagnum* and *Carex* peat. *Water Research 36*(9): 2233–2242.

Rinklebe J., Knox A.S., Paller M. (2016) *Trace Elements in Waterlogged Soils and Sediments*. CRC Press: Boca Raton, FL.

Ritchie J.C., McHenry J.R. (1990) Application of radioactive fallout cesium-137 for measuring soil erosion and sediment accumulation rates and patterns: A review. *Journal of Environmental Quality 19*(2): 215–233.

Ro K.S., Hunt P.G. (2006) New unified equation for wind-driven surficial oxygen transfer into stationary water bodies. *Transactions of the American Society of Agricultural and Biological Engineers 49*(5): 1615–1622.

Ro K.S., Hunt P.G., Poach M.E. (2006) Surficial oxygen transfer into treatment lagoons: Part I – Review of theory and empirical correlations. *Journal of Environmental Science and Health, Part A 41*: 1627–1638.

Ro K.S., Hunt P.G., Poach M.E. (2007) Wind-driven surficial oxygen transfer. *Critical Reviews in Environmental Science and Technology 37*(6): 539–563.

Robbins J.A. (1986) A model for particle-selective transport of tracers in sediments with conveyor belt deposit feeders. *Journal of Geophysical Research 91*(C7): 8542–8558.

Robbins J.A., Mudroch A., Oliver B.G. (1990) Transport and storage of 137Cs and 210Pb in sediments of Lake St. Clair. *Canadian Journal of Fisheries and Aquatic Sciences 47*(3): 572–587.

Robbins J.A., Reddy K.R., Holmes C.W., Moorhead N.R., Marot M., Rood R.W. (1999) *Sediment Core Dating with 137cs*. Report to South Florida Water Management District, West Palm Beach, FL

Rober A.R., Wyatt K.H., Stevenson R.J. (2011) Regulation of algal structure and function by nutrients and grazing in a boreal wetland. *Journal of the North American Benthological Society 30*(3): 787–796.

Robinson G.G.C., Gurney S.E., Goldsborough L.G. (2000) Algae in prairie wetlands. Chapter 8. In *Prairie Wetland Ecology*, Murkin H.R., van der Valk A.G., Clark W.R. (eds.) Iowa State University Press: Ames, IA, pp. 163–200.

Rodgers J.H., Dunn A. (1992) Developing design guidelines for constructed wetlands to remove pesticides from agricultural runoff. *Ecological Engineering 1*(1–2): 83–95.

Rodríguez M., Brisson J. (2015) Pollutant removal efficiency of native versus exotic common reed (Phragmites australis) in North American treatment wetlands. *Ecological Engineering 74*: 364–370.

Rodriguez R., Lougheed V.L. (2010) The potential to improve water quality in the middle Rio Grande through effective wetland restoration. *Water Science and Technology 62*(3): 501.

Rodriguez-Gallego L.R., Mazzeo N., Gorga J., Meerhoff M., Clemente J., Kruk C., Scasso F., Lacerot G., García J., Quintans F. (2004) The effects of an artificial wetland dominated by free-floating plants on the restoration of a subtropical, hypertrophic lake. *Lakes and Reservoirs: Research & Management* 9(3–4): 203–215.

Rogers K., Wilton K.M., Saintilan N. (2006) Vegetation change and surface elevation dynamics in estuarine wetlands of southeast Australia. *Esturarine, Coastal and Shelf Science 66*: 559–569.

Rogers K.H., Breen P.F., Chick A.J. (1991) Nitrogen removal in experimental wetland treatment systems: Evidence for the role of plants. *Journal of the Water Pollution Control Federation 63*(7): 934–941.

Romanowski N. (2009) *Planting Wetlands and Dams: A Practical Guide to Wetland Design, Construction and Propagation*, (2nd edition. CSIRO Publishing: Clayton VIC, Australia.

Ronkanen A.K., Kløve B. (2008) Hydraulics and flow modelling of water treatment wetlands constructed on peatlands in Northern Finland. *Water Research 42*(14): 3826–3836.

Rose M.T., Crossan A.N., Kennedy I.R. (2008) The effect of vegetation on pesticide dissipation from ponded treatment wetlands: Quantification using a simple model. *Chemosphere 72*(7): 999–1005.

Rose M.T., Sanchez-Bayo F., Crossan A.N., Kennedy I.R. (2006) Pesticide removal from cotton farm tailwater by a pilot-scale ponded wetland. *Chemosphere 63*(11): 1849–1858.

Roulet N.T. (2000) Peatlands, carbon storage, greenhouse gases, and the Kyoto protocol: Prospects and significance for Canada. *Wetlands 20*(4): 605–615.

Ruiz N., Bleiler J., Gardner K. (2013) *Framework Guidance Manual for in Situ Wetland Restoration Demonstration.* Technical Report CR-NAVFAC-EXWC-1401, Defense Technical Information Center, Fort Belvoir, VA.

Runes H.B., Jenkins J.J., Bottomley P.J. (2001) Atrazine degradation by bioaugmented sediment from constructed wetlands. *Applied Microbiology and Biotechnology 57*: 427–432.

Runes H.B., Jenkins J.J., Moore J.A., Bottomley P.J., Wilson B.D. (2003) Treatment of atrazine in nursery irrigation runoff by a constructed wetland. *Water Research 37*: 539–550.

Runkel R.L. (1998) *One-Dimensional Transport with Inflow and Storage (OTIS): A Solute Transport Model for Streams and Rivers.* U.S. Geological Survey Water-Resources Investigations Report 98-4018 Available at: https://water.usgs.gov/software/OTIS/doc/ (Accessed 4/11/2018).

Russell R.S., Shorrocks V.M. (1959) Relationship between transpiration and the absorption of inorganic ions by intact plants. *Journal of Experimental Botany 10*(2): 301–316.

Rutchey K., Schall T., Sklar F. (2008) Development of vegetation maps for assessing Everglades restoration progress. *Wetlands 28*(3): 806–816.

Rybczyk J.M., Callaway J.C., Day J.W. Jr. (1998) A relative elevation model for a subsiding coastal forested wetland receiving wastewater effluent. *Ecological Modeling 112*(1): 23–44.

Rybczyk J.M., Day J.W., Conner W.H. (2002a) The impact of wastewater effluent on accretion and decomposition in a subsiding forested wetland. *Wetlands 22*(1): 18–32.

Rybicki N.B., Reel J.T., Gammon P.T., Garrett M.K. (2002) *Vegetative Resistance to Flow in South Florida Everglades: Summary of Vegetation Sampling in Water Conservation Area 2A, September 1999.* U.S. Geological Survey Open File Report No. 02-38, USGS, Reston, VA

Rybicki N.B., Reel J.T., Ruhl H.A., Gammon P.T., Carter V., Lee J.K. (2000) *Sawgrass Density, Biomass, and Leaf Area Index: A Flume Study in Support of Research on Wind Sheltering Effects in the Florida Everglades.* Report No. 2000-172, U.S. Department of the Interior, U.S. Geological Survey, Reston, VA.

Sabokrouhiyeh N., Bottacin-Busolin A., Savickis J., Nepf H., Marion A. (2017) A numerical study of the effect of wetland shape and inlet-outlet configuration on wetland performance. *Ecological Engineering 105*: 170–179.

Saiers J.E. Harvey J.W., Mylon S.E. (2003) Surface-water transport of suspended matter through wetland vegetation of the Florida Everglades. *Geophysical Research Letters 30*(19): HLS 1–5.

Saintilan N., Rogers K., Mazumder D., Woodroffe C. (2013) Allochthonous and autochthonous contributions to carbon accumulation and carbon store in southeastern Australian coastal wetlands. *Estuarine, Coastal and Shelf Science 128*: 84–92.

Salih F.M. (2003) Formulation of a mathematical model to predict solar water disinfection. *Water Research 37*(16): 3921–3927.

Salmon C., Crabos J.L., Stambuco J.P., Bessiere J.M., Basseres A., Caumette P. (1998) Artificial wetland performances in the purification efficiency of hydrocarbon wastewater. *Water, Air, and Soil Pollution 104*: 313–329.

San Miguel A., Ravanel P., Raveton M. (2013) A comparative study on the uptake and translocation of organochlorines by Phragmites australis. *Journal of Hazardous Materials 244*: 60–69.

Sanchez C.A., Childers D.L., Turnbull L., Upham R.F., Weller N. (2016) Aridland constructed treatment wetlands II: Plant mediation of surface hydrology enhances nitrogen removal. *Ecological Engineering 97*: 658–665.

Sánchez-Carrillo S., Angeler D.G., Sánchez-Andrés R., Alvarez-Cobelas M., Garatuza-Payán J. (2004) Evapotranspiration in semi-arid wetlands: Relationships between inundation and the macrophyte-cover: Open-water ratio. *Advances in Water Resources 27*(6): 643–655.

Sand-Jensen K., Prahl C., Stokholm H. (1982) Oxygen release from roots of submerged aquatic macrophytes. *Oikos 38*: 349–354.

Sangoyomi T., Nayak A., Huang G., Xiao S. (2011) *Levee Seepage Analysis at Stormwater Treatment Area 3/4*. Report to South Florida Water Management District, West Palm Beach, FL.

Šantrůčková H., Rejmánková E., Pivničková B., Snyder J.M. (2010) Nutrient enrichment in tropical wetlands: Shifts from autotrophic to heterotrophic nitrogen fixation. *Biogeochemistry 101*: 295–310.

Sartoris J., Thullen J., Barber L.B., Salas D. (2000) Investigation of nitrogen transformations in a southern California constructed wastewater treatment wetland. *Ecological Engineering 14*(1–2): 49–65.

Sasser C.E., Gosselink J.G., Swenson E.M., Swarzenski C.M., Leibowitz N.C. (1996) Vegetation, substrate and hydrology in floating marshes in the Mississippi river delta plain wetlands, USA. *Plant Ecology 122*(2): 129–142.

Sather J.H. (1989) Ancillary benefits of wetlands constructed primarily for wastewater treatment. In: *Constructed Wetlands for Wastewater Treatment: Municipal, Industrial, and Agricultural*, Hammer D.A. (ed.) Lewis Publishers: Chelsea, MI, pp. 353–358.

Savickis J., Bottacin-Busolin A., Zaramella M., Sabokrouhiyeh N., Marion A. (2016) Effect of a meandering channel on wetland performance. *Journal of Hydrology 535*: 204–210.

Schäfer M. (2004) Mosquitoes as a part of wetland biodiversity. Acta Univesitatis Upsaliensis. *Comprehensive Summaries of Uppsala Dissertations from the Faculty of Science and Technology*, Uppsala, Sweden.

Scherer N.M., Gibbons H.L., Stoops K.B., Muller M. (1995) Phosphorus loading of an urban lake by bird droppings. *Lake and Reservoir Management: An International Journal of the North American Lake Management Society 11*(4): 317–327.

Schiffer D.M. (1989) Water-quality variability in a central Florida wetland receiving highway runoff. In: *Water: Laws and Management*, Davis F.E. (ed.) American Water Resources Association: Bethesda, MD.

Schipper L.A., Robertson W.D., Gold A.J., Jaynes D.B., Cameron S.C. (2010) Denitrifying bioreactors—An approach for reducing nitrate loads to receiving waters. *Ecological Engineering 36*(11): 1532–1543.

Schmid B.H., Hengl M.A., Stephan U. (2003) Salinity-induced density stratification in low Reynolds number free surface flows. *Publicationes Instituti Geographici Universitatis Tartuensis 94*: 186–190.

Schmid B.H., Hengl M.A., Stephan U. (2004) Salt tracer experiments in constructed wetland ponds with emergent vegetation: Laboratory study on the formation of density layers and its influence on breakthrough curve analysis. *Water Research 38*(8): 2095–2102.

Schnitzer M. (1986) Water retention by humic substances. In: *Peat and Water: Aspects of Water Retention and Dewatering in Peat*. Fuchsman C.H. (ed.), Elsevier Applied Science Publishing: London, United Kingdom, pp. 159–176.

Schoups G., Hopmans J.W. (2002) Analytical model for vadose zone solute transport with root water and solute uptake. *Vadose Zone Journal 1*(1–2): 158–171.

Schröder P., Daubner D., Maier H., Heustifter J., Debus R. (2008) Phytoremediation of organic xenobiotics – Glutathione dependent detoxification in *Phragmites* plants. *Bioresource Technology 99*(15): 7183–7191.

Schuck R.J., Schuck B., Schuck S., Hubbard R.K. (2007) Patent Application: Floating wetland raft. United States US 2007/0209277 A1.

Schueler T., Lane C., Wood D., Lane S., Sample D., Lazur A., Winston R., Streb C., Ferrier D., Linker L., Brittingham K. (2016) *Recommendations of the Expert Panel to Define Removal Rates for Floating Treatment Wetlands in Existing Wet Ponds*. Final Report to Urban Stormwater Workgroup, Chesapeake Bay Program, Annapolis, MD. Available at: www.chesapeakebay.net/documents/FINAL-FTW-EXPERT-PANEL-REPORT-072716-LONG.pdf (Accessed 8/5/18).

Schueler T.R. (1992) *Design of Stormwater Wetland Systems: Guidelines for Creating Diverse and Effective Stormwater Wetlands in the Mid-Atlantic Region*. Metropolitan Washington Council of Governments: Washington, DC

Schuerman P.R., Bitton G., Farrah S.R. (1989) Fate of microbial indicators and viruses in a forested wetland. In *Constructed Wetlands for Wastewater Treatment: Municipal, Industrial, and Agricultural*, Hammer D.A. (ed.) Lewis Publishers: Chelsea, MI, pp. 657–663.

Schulz R., Moore M.T., Bennett E.R., Farris J.L., Smith S., Cooper C.M. (2003) Methyl parathion toxicity in vegetated and nonvegetated wetland mesocosms. *Environmental Toxicology and Chemistry 22*(6): 1262–1268.

Schulz R., Peall S.K.C. (2000) Effectiveness of a constructed wetland for retention of nonpoint-source pesticide pollution in the Lourens River Catchment, South Africa. *Environmental Science and Technology 35*(2): 422–426.

Schumacher W., Harcari T., Saas E. (2015) *Nutrient Runoff Reduction via Restored Wetlands in an Agricultural* Setting – *A GIS Model*. Ohio EPA Technical Report WET/2016-1. Ohio Environmental Protection Agency, Wetland Ecology Group, Division of Surface Water, Columbus, OH.

Schwartz L.N., Wallace P.M., Gale P.M., Smith W.F., Wittig J.L., McCarty S.L. (1994) Orange County Eastern Service Area reclaimed water reuse system. *Water Science and Technology 29*(4): 273–282.

Schwartz L.N., Wiseman L.P., Melear E.L. (1999) Leachate wetland treatment system in Orange County, Florida. In: *Constructed Wetlands for the Treatment of Landfill Leachate*, Mulamoottil G., McBean E., Rovers F.A. (eds.) Lewis Publishers: Boca Raton, FL, pp. 99–140.

Schwindaman J.P., Castle J.W., Rodgers Jr J.H. (2014) Fate and distribution of arsenic in a process-designed pilot-scale constructed wetland treatment system. *Ecological Engineering 68*: 251–259.

Schwitzguebel J.-P., Aubert S., Große W., Laturnus F. (2001) Sulphonated aromatic pollutants. *Environmental Science and Pollution Research International 1*: 62–72.

Scinto L.J., Reddy K.R. (2003) Biotic and abiotic uptake of phosphorus by periphyton in a subtropical freshwater wetland. *Aquatic Botany 77*(3): 203–222.

Scott J.T. (2006) Periphyton-Nutrient Dynamics in a Gradient-Dominated Freshwater Marsh Ecosystem. PhD Thesis, Baylor University, Waco, TX.

Scott J.T., Doyle R.D., Back J.A., Dworkin S.I. (2007) The role of N2 fixation in alleviating N limitation in wetland metaphyton: Enzymatic, isotopic, and elemental evidence. *Biogeochemistry 84*: 207–218.

Scott J.T., McCarthy M.J., Gardner W.S., Doyle R.D. (2008) Denitrification, dissimilatory nitrate reduction to ammonium, and nitrogen fixation along a nitrate concentration gradient in a created freshwater wetland. *Biogeochemistry 87*(1): 99–111.

SCS. (1999) *Urban Hydrology for Small Watersheds*. Technical Release 55. Soil Conservation Service, USDA, Washington, DC Available at: www.nrcs.usda.gov/Internet/FSE_DOCUMENTS/stelprdb1044171.pdf (Accessed 7/23/18).

SCS. (2004) *National Engineering Handbook, Section 4: Hydrology*. Soil Conservation Service, USDA: Washington, DC.

Seeger E.M., Reiche N., Kuschk P., Borsdorf H., Kaestner M. (2011) Performance evaluation using a three compartment mass balance for the removal of volatile organic compounds in pilot scale constructed wetlands. *Environmental Science and Technology 45*(19): 8467–8474.

Sees M. (2005) *The Orlando Easterly Wetlands Project: 18 Years of Operation*. Paper presented at the Society of Wetland Scientists Annual Meeting, 5–10 June 2005; Charleston, South Carolina.

Seidel K. (1966) Reinigung von Gewässern durch höhere Pflanzen. *Naturwissenschaften 53*: 289–297.

Seiler R.L., Skorupa J.P., Peltz L.A. (1999) Areas susceptible to irrigation-induced selenium contamination of water and biota in the western United States. U.S. Geological Survey Circular 1180, U.S. Geological Survey. Reston, VA. Available at: www.fws.gov/mountain-prairie/contaminants/contam inants4.html (Accessed 9/5/18).

Seilheimer T.S., Chow-Fraser P. (2006) Development and use of the Wetland Fish Index to assess the quality of coastal wetlands in the Laurentian Great Lakes. *Canadian Journal of Fisheries and Aquatic Sciences 63*(2): 354–366.

Seilheimer T.S., Mahoney T.P., Chow-Fraser P. (2009) Comparative study of ecological indices for assessing human-induced disturbance in coastal wetlands of the Laurentian Great Lakes. *Ecological Indicators 9*: 81–91.

Selbig W.R., Bannerman R.T. (2011) *Characterizing the Size Distribution of Particles in Urban Stormwater by Use of Fixed-Point Sample-Collection Methods (No. 2011-1052)*. US Geological Survey: Reston, VA.

Serra T., Fernando J.S., Rodriguez R.V. (2004) Effects of emergent vegetation on lateral diffusion in wetlands. *Water Research 38*: 139–147.

SFWMD. (2003) *Chapter 4B: STA Optimization and Advanced Treatment Technologies* In: Everglades Consolidated Report 2004, South Florida Water Management District (SFWMD): West Palm Beach, FL.

SFWMD. (2004) *Everglades Consolidated Report 2003*. South Florida Water Management District (SFWMD): West Palm Beach, FL.

SFWMD. (2019) *Draft 2019 South Florida Environmental Report*, South Florida Water Management District (SFWMD), West Palm Beach, FL. Available at: https://apps.sfwmd.gov/sfwmd/SFER/2019_sfer_draft/v1/sfer_toc_v1.pdf (Accessed 1/17/19.

SFWMD (South Florida Water Management District) (1999–2004) *Everglades Consolidated Reports*. South Florida Water Management District, West Palm Beach, FL Available at: http://apps.sfwmd.gov/sfwmd/SFER/2018_sfer_final/v3/sfer_toc_v3.pdf (Accessed 12/13/18).

SFWMD (South Florida Water Management District). (2005–2018) *South Florida Environmental Reports*. SFWMD, West Palm Beach, FL. Available at: http://apps.sfwmd.gov/sfwmd/SFER/2018_sfer_final/v3/sfer_toc_v3.pdf (Accessed 8/11/18).

SFWMD (South Florida Water Management District). (2018) *2018 South Florida Environmental Report*. SFWMD, West Palm Beach, FL. Available at: www.sfwmd.gov/science-data/scientific-publications-sfer (Accessed 1/16/19).

SFWMD (South Florida Water Management District). (2017) *2017 South Florida Environmental Report*. SFWMD: West Palm Beach, FL.

Sgouridis F., Heppell C.M., Wharton G., Lansdown K., Trimmer M. (2011) Denitrification and dissimilatory nitrate reduction to ammonium (DNRA) in a temperate re-connected floodplain. *Water Research 45*(16): 4909–4922.

Sha C., Mitsch W.J., Mander Ü., Lu J., Batson J., Zhang L., He W. (2011) Methane emissions from freshwater riverine wetlands. *Ecological Engineering 37*(1): 16–24.

Shaffer G.P., Day J.W., Hunter R.G., Lane R.R., Lundberg C.J., Wood W.B., Hillmann E.R., Day J.N., Strickland E., Kandalepas D. (2015) System response, nutria herbivory, and vegetation recovery of a wetland receiving secondarily-treated effluent in coastal Louisiana. *Ecological Engineering 79*: 120–131.

Shah N., Nachabe M., Ross M. (2007) Extinction depth and evapotranspiration from ground water under selected land covers. *Groundwater 45*(3): 329–338.

Shappell N.W., Billey L.O., Forbes D., Matheny T.A., Poach M.E., Reddy G.B., Hunt P.G. (2001) Estrogenic activity and steroid hormones in swine wastewater through a lagoon constructed-wetland system. *Environmental Science and Technology 41*(2): 444–450.

Sharif F., Westerhoff P., Herckes P. (2014) Impact of hydraulic and carbon loading rates of constructed wetlands on contaminants of emerging concern (CECs) removal. *Environmental Pollution 185*: 107–115.

Shaw D., Schmidt R. (2003) *Plants for Stormwater Design - Species Selection for the Upper Midwest*. Minnesota Pollution Control Agency: St. Paul, MN.

Shepard W.C., Mills E.L. (1996) Diel feeding, daily food intake, and *Daphnia* consumption by age-0 gizzard shad in Oneida Lake, New York. *Transactions of the American Fisheries Sociey 125*: 411–421.

Shepherd H.L., Tchobanoglous G., Grismer M.E. (2001) Time-dependent retardation model for Chemical Oxygen Demand removal in a subsurface-flow constructed wetland for winery wastewater treatment. *Water Environment Research 73*(5): 597–606.

Sherrard R.M., Bearr J.S., Murray-Gulde C., Rodgers J.H. Jr., Shah Y.T. (2004) Feasibility of constructed wetlands for removing chlorthalonil and chlorpyrifos from aqueous mixtures. *Environmental Pollution 127*(3): 385–394.

Shi J., Zuo Q., Zhang R. (2007) An inverse method to estimate the source-sink term in the nitrate transport equation. *Soil Science Society of America Journal 71*(1): 26–34.

Shiaris M.P. (1985) Public health implications of sewage applications on wetlands: Microbiological aspects. In: *Ecological Considerations in Wetlands Treatment of Municipal Wastewaters*, Godfrey P.J., Kaynor E.R., Pelczarski S., Benforado J. (eds.) Van Nostrand Reinhold: New York, pp. 243–261.

Shih S.F., Rahi G.S. (1982) Seasonal variations of Manning's roughness coefficient in a subtropical marsh. *Transactions of the ASAE 25*: 116–120.

Shih S.F., Federico A.C., Milleson J.F., Rosen M. (1979) Sampling programs for evaluating upland marsh to improve water quality. *Transactions of the American Society of Agricultural Engineers 22*: 828–833.

Shih S.S., Kuo P.H., Fang W.T., LePage B.A. (2013) A correction coefficient for pollutant removal in free water surface wetlands using first-order modeling. *Ecological Engineering 61*: 200–206.

Shilton A. (ed.) (2005) *Pond Treatment Technology*. IWA Publishing: London, United Kingdom.

Shilton A., Mara D. (2005) Pond process design – A historical review. In: *Pond Treatment Technology*, Shilton A. (ed.) IWA Publishing: London, United Kingdom.

Shin W.S., Pardue J.H. (2002) Biphasic desorption of hydrophobic organic contaminants in highly organic wetland soils. In *Wetlands and Remediation II*, Nehring K.W., Brauning S.E. (eds.) Battelle Institute: Columbus, OH, pp. 41–48.

Shipin O., Koottatep T., Nguyen T.T.K., Polprasert C. (2004) Integrated natural treatment systems for developing communities: Low-tech N-removal through the fluctuating microbial pathways. In: *Proceedings of the 9th International Conference on Wetland Systems for Water Pollution Control, 26–30 September 2004*, Liénard A., Burnett H. (eds.) Association Scientifique et Technique pour l'Eau et l'Environnement (ASTEE), Cemagref, and IWA: Avignon, France, pp. 75–82.

Shugui D., Xin R., Juxian W., Qingyu L., Ruizhi Z., Rengao S. (1994) Removal of priority organic pollutants in stabilization ponds. *Water Research 28*(3): 681–685.

Shutes R., Ellis J.B., Revitt D.M., Forshaw M., Winter B. (2004) Urban and highway runoff treatment by constructed wetlands. *Wetlands Ecosystems in Asia: Function and Management 1*: 50023–50025.

Shutes R.B.E., Revitt D.M., Lagerberg I.M., Barraud V.C.E. (1999) The design of vegetative constructed wetlands for the treatment of highway runoff. *Science of the Total Environment 235*(1–3): 189–197.

Shuttleworth W.J. (1993) Evapotranspiration. In: *Handbook of Hydrology*, Maidment D.R. (ed.) McGraw-Hill: New York, pp. 4.1–4.53.

Shuttleworth W.J., Wallace J.S. (1985) Evaporation from sparse crops-an energy combination theory. *Quarterly Journal of the Royal Meteorological Society 111*: 839–855.

Silberbush M., Barber S.A. (1983) Sensitivity of simulated phosphorus uptake to parameters used by a mechanistic-mathematical model. *Plant and Soil 74*: 93–100.

Silvan N., Vasander H., Laine J. (2004) Vegetation is the main factor in nutrient retention in a constructed wetland buffer. *Plant and Soil 258*(1): 179–187.

Silverman A.I., Nelson K.L. (2016) Modeling the endogenous sunlight inactivation rates of laboratory strain and wastewater E. coli and Enterococci using biological weighting functions. *Environmental Science and Technology 50*(22): 12292–12301.

Silverman A.I., Nguyen M.T., Schilling I.E., Wenk J., Nelson K.L. (2015) Sunlight inactivation of viruses in open-water unit process treatment wetlands: Modeling endogenous and exogenous inactivation rates. *Environmental Science and Technology 49*(5): 2757–2766.

Silverman A.I., Sedlak D.L., Nelson K.L. (2019) Simplified process to determine rate constants for sunlight-mediated removal of trace organic and microbial contaminants in unit process open-water treatment wetlands. *Environmental Engineering Science 36*: 1.

Silyn-Roberts G., Lewis G. (2001) *In situ* analysis of *Nitrosomonas* in wastewater treatment wetland biofilms. *Water Research 35*(11): 2731–2739.

Simi A.L., Mitchell C.A. (1999) Design and hydraulic performance of a constructed wetland treating oil refinery wastewater. *Water Science and Technology 40*(3): 301–307.

Simon S.M. (2003) Phosphorus Retention and Release of Soils in a Constructed Wetland for Wastewater Treatment. MS Thesis, University of Florida, Gainesville, FL.

Šimůnek J., Hopmans J.W. (2009) Modeling compensated root water and nutrient uptake. *Ecological Modelling 220*(4): 505–521.

Šimůnek J., van Genuchten M.T. Šejna M. (2012) *The HYDRUS Software Package for Simulating Two- and Three Dimensional Movement of Water, Heat, and Multiple Solutes in Variably-Saturated Porous Media*. Technical Manual, Version 2.0, PC Progress, Prague, Czech Republic.

Singh S.N., Mishra S. (2014) Phytoremediation of TNT and RDX. In: *Biological Remediation of Explosive Residues*, Singh S.N. (ed.) Springer International Publishing: New York, pp. 371–392.

Singh V.P., Xu C. (1997) Evaluation and generalization of 13 mass-transfer equations for determining free water evaporation. *Hydrological Processes 11*: 311–323.

Sinicrope T.L., Langis R., Gersberg R.M., Busanardo M.J., Zelder J.B. (1992) Metal removal by wetlands mesocosms subjected to different hydroperiods. *Ecological Engineering 1*(4): 309–322.

Skaggs R.W. (1978) *A Water Management Model for Shallow Water Table Soils*. Technical Report 134, Water Resources Institute, Raleigh, NC.

Slayton K. (2018) *Orlando Easterly Wetlands Compliance and Performance Review for the City of Orlando's Easterly Wetlands Treatment System, 2017*. Annual Report. Prepared for Florida Department of Environmental Protection. Available at: www.cityoforlando.net/wetlands/wp-content/uploads/sites/59/2018/05/2017-Annual-Report.pdf (Accessed 12/12/18).

Smardon R.C. (1988) *Aesthetic, recreational, landscape values of urban wetlands.* Kusler J.A., Daly S., Brooks G. (eds.) Proceedings of the National Wetland Symposium on Urban Wetlands, 26–29 June 1988, Oakland, CA; Association of Wetland Managers: Berne, NY, 92–95.

Smesrud J.K., Boyd M.S., Cuenca R.H., Eisner S.L. (2014) A mechanistic energy balance model for predicting water temperature in surface flow wetlands. *Ecological Engineering 67*: 11–24.

Smets B.F., Rittman B.E. (1990) Sorption equilibria for trichloroethene on algae. *Water Research 24*(3): 355–360.

Smith C.S., Adams M.S., Gustafson T.D. (1988) The importance of belowground mineral element stores in cattails (*Typha latifolia* L.). *Aquatic Botany 30*(4): 343–352.

Smith E., Naidu R., Alston A.M. (2002) Chemistry of inorganic arsenic in soils: II. Effect of phosphorus, sodium, and calcium on arsenic adsorption. *Journal of Environmental Quality 31*(2): 253–260.

Smith J.L., W.J. Mitsch. (2004) Fish and amphibian abundance and diversity in two created freshwater marshes. In *Olentangy River Wetland Research Park at the Ohio State University, Annual Report 2003*, Mitsch W.J., Zhang L., Tuttle C.L. (eds.) The Ohio State University: Columbus, OH, pp. 117–125.

Smith M.P., Kalin M. (2001) Floating wetland vegetation covers for suspended solids removal. Pries J. H. (ed.), *Treatment Wetlands for Water Quality Improvement: Quebec 2000 Conference Proceedings.* Pandora Press: Kitchener, ON, 73–82.

Smith S.M., Newman S., Garrett P.B., Leeds J.A. (2001) Differential effects of surface and peat fire on soil constituents in a degraded wetland of the Northern Florida Everglades. *Journal of Environmental Quality 30*(6): 1998–2005.

Smits J.E., Hersikorn B.D., Young R.F., Fedorak P.M. (2012) Physiological effects and tissue residues from exposure of leopard frogs to commercial naphthenic acids. *Science of the Total Environment 437*: 36–41.

Smolders A.J.P., Roelofs J.G.M. (1996) The roles of internal iron hydroxide precipitation, sulphide toxicity, and oxidizing ability in the survival of *Stratiotes aloides* roots at different iron concentrations in sediment pore water. *New Phytologist 3*: 253–260.

Snoddy E.L., Brodie G.A., Hammer D.A., Tomljanovich D.A. (1989) Control of the armyworm, *Simyra henrici* (Lepidoptera: Noctuidae), on cattail plantings in acid drainage constructed wetlands at Widows Creek Steam-Electric Plant. In *Constructed Wetlands for Wastewater Treatment: Municipal, Industrial, and Agricultural*, Hammer D.A. (ed.) Lewis Publishers: Chelsea, MI, pp. 808–811.

Snyder J.M., Rejmánková E. (2015) Macrophyte root and rhizome decay: The impact of nutrient enrichment and the use of live versus dead tissue in decomposition studies. *Biogeochemistry 124*(1–3): 45–59.

Snyder R.L. (2001) *Daily Reference Evapotranspiration Calculator (Pmday.Xls)*. University of California Davis, Davis, CA.

Snyder R.L., Eching S. (2000) *Daily Reference Evapotranspiration Calculator User's Guide for PMday.xls.* University of California Davis, Davis, CA.

Soares M.I.M. (2002) Denitrification of groundwater with elemental sulfur. *Water Science and Technology 36*(5): 1392–1395.

Sobolewski A. (1999) A review of processes responsible for metal removal in wetlands treating contaminated mine drainage. *International Journal of Phytoremediation 1*(1): 19–51.

Soda S., Ike M., Ogasawara Y., Yoshinaka M., Mishima D., Fujita M. (2007) Effects of light intensity and water temperature on oxygen release from roots into water lettuce rhizosphere. *Water Research 41*(2): 487–491.

Söhnel O., Novotny P. (1985) *Densities of Aqueous Solutions of Inorganic Substances.* Elsevier: Amsterdam, The Netherlands.

Somes N.L.G., Wong T.H.F. (1997) Designing outlet characteristics for optimal wetland performance. *Water Science and Technology 36*(9): 235–240.

Somodi I., Botta-Dukat Z. (2004) Determinants of floating island vegetation and succession in a recently flooded shallow lake, Kis-Balaton (Hungary). *Aquatic Botany 79*: 357–366.

Song B., Mallin M., Long A., McIver M. (2014) Factors controlling microbial nitrogen removal efficacy in constructed stormwater wetlands. report UNC-WRRI-443, *Water Resources Research Institute of the University of North Carolina*, Chapel Hill, NC.

Song K., Xenopoulos M.A., Buttle J.M., Marsalek J., Wagner N.D., Pick F.R., Frost P.C. (2013) Thermal stratification patterns in urban ponds and their relationships with vertical nutrient gradients. *Journal of Environmental Management 127*: 317–323.

Soomers H., Karssenberg D., Soons M.B., Verweij P.A., Verhoeven J.T., Wassen M.J. (2013) Wind and water dispersal of wetland plants across fragmented landscapes. *Ecosystems 16*(3): 434–451.

Soons M.B., Brochet A.L., Kleyheeg E., Green A.J. (2016) Seed dispersal by dabbling ducks: An overlooked dispersal pathway for a broad spectrum of plant species. *Journal of Ecology 104* (2): 443–455.

Sorrell B.K., Boon P.I. (1992) Biogeochemistry of billabong sediments II. Seasonal variations in methane production. *Freshwater Biology 27*: 435–445.

Soto-Jimenez M.F., Paez-Osuma F., Bojorquez-Leyva H. (2003) Nutrient cycling at the sediment-water interface and in sediments at Chiricahueto marsh: A subtropical ecosystem associated with agricultural land uses. *Water Research 37*: 719–728.

Søvik A.K., Augustin J., Heikkinen K., Huttunen J.T., Necki J.M., Karjalainen S.M., Kløve B., Liikanen A., Mander Ü., Puustinen M., Teiter S. (2006) Emission of the greenhouse gases nitrous oxide and methane from constructed wetlands in Europe. *Journal of Environmental Quality 35*(6): 2360–2373.

Spacil M.M., Rodgers J.H., Castle J.W., Gulde C.L.M., Myers J.E. (2011a) Treatment of selenium in simulated refinery effluent using a pilot-scale constructed wetland treatment system. *Water, Air, and Soil Pollution 221*(1–4): 301–312.

Spacil M.M., Rodgers J.H. Jr., Castle J.W., Chao W.Y. (2011b) Performance of a pilot-scale constructed wetland treatment system for selenium, arsenic, and low-molecular-weight organics in simulated fresh produced water. *Environmental Geosciences 18*(3): 145–156.

Spangler F.L., Sloey W.E., Fetter C.W. Jr. (1976) *Wastewater Treatment by Natural and Artificial Marshes.* EPA 600/2-76-207. U.S. EPA Office of Water: Ada, OK.

Spengler P., Körner W., Metzger J.W. (2001) Substances with estrogenic activity in effluents of sewage treatment plants in southwestern Germany. 1. Chemical analysis. *Environmental Toxicology and Chemistry 20*(10): 2133–2141.

Spieles D.J., Mitsch W.J. (2000b) Macroinvertebrate community structure in high-and low-nutrient constructed wetlands. *Wetlands 20*(4): 716–729.

Spieles D.J., Mitsch W.J. (2000a) The effects of season and hydrologic and chemical loading on nitrate retention in constructed wetlands: A comparison of low- and high-nutrient riverine systems. *Ecological Engineering 14*: 77–91.

Spinelli R., Magagnotti N., DeFrancesco F., Aminti G., Stauder M., Pari L. (2017) Biomass recovery from invasive species management in wetlands. *Biomass and Bioenergy 105*: 259–265.

Spratt H.G. Jr., Morgan M.D., Good R.E. (1987) Sulfate reduction in peat from a New Jersey pinelands cedar swamp. *Applied and Environmental Microbiology 53*(7): 1406–1411.

Stairs D.B. (1993) *Flow Characteristics of Constructed Wetlands: Tracer Studies of the Hydraulic Regime.* MS Thesis, Oregon State University, Corvallis, OR.

Stamenkovic J., Gustin M.S., Dennett K.E. (2005) Net methyl mercury production versus water quality improvement in constructed wetlands: Trade-offs in pollution control. *Wetlands 25*(3): 748.

Stark L.R., Williams F.M. (1995) Assessing the performance indices and design parameters for H^+, Fe, and Mn retention. *Ecological Engineering 5*(4): 433–444.

Stefanson R.C. (1973) Evolution of patterns of nitrous oxide and nitrogen in sealed soil-plant systems. *Soil Biology and Biochemistry 5*: 1441–1445.

Stehle S., Elsaesser D., Gregoire C., Imfeld G., Niehaus E., Passeport E., Payrsudeau S., Schäfer R.B., Tournebize J., Schulz R. (2011) Pesticide risk mitigation by vegetated treatment systems: A meta-analysis. *Journal of Environmental Quality 40*: 1068–1080.

Stein L.Y., Roy R., Dunfield P.F. (2012) Aerobic methanotrophy and nitrification: Processes and connections. In: *Encyclopedia of Life Sciences (Els.)*, John Wiley & Sons, Chichester, United Kingdom, pp. 1–11.

Stenberg J.R., Clark M.W., Crisman T.L. (1998) *Apopka Marsh Flow-Way Restoration Project, August 1993 – March 1995*, Special Publication SJ98-SP17, Report to St. Johns River Water Management District, Palatka, FL.

Stengel E., Carduck W., Jebsen C. (1987) Evidence for denitrification in artificial wetlands. In: *Aquatic Plants for Water Treatment and Resource Recovery*, Reddy K.R., Smith W.H. (eds.) Magnolia Publishing: Orlando, FL, pp. 543–550.

Stenström T.A., Carlander A. (2001) Occurrence and die-off of indicator organisms in the sediment in two constructed wetlands. *Water Science and Technology 44*(11–12): 223–230.

Stephan U., Schmid B.H., Hengl M.A. (2004) Hydraulic restriction to the use of salt tracer experiments in wetland ponds. In: *Proceedings of the 9th International Conference on Wetland Systems for Water Pollution Control, 26–30 September 2004*, Liénard A., Burnett H. (eds.) Association Scientifique et Technique pour l'Eau et l'Environnement (ASTEE), Cemagref, and IWA: Avignon, France, pp. 169–176.

Steuer J., Selbig W., Hornewer N., Prey J. (1997) *Sources of Contamination in an Urban Basin in Marquette, Michigan and an Analysis of Concentration, Loads, and Data Quality*, Water Resources Investigation Report 97-4242, U.S. Geological Survey, Reston, VA.

Stevenson R.J., Bothwell M.L., Lowe R.L., Thorp J.H. (eds.) (1996) *Algal Ecology: Freshwater Benthic Ecosystem*. Academic press: San Diego, CA.

Stevenson R.J., McCormick P.V., Frydenborg R. (2002) *Using Algae to Assess Environmental Conditions in Wetlands*. EPA-822-R-02-023. Office of Water, U.S. Environmental Protection Agency: Washington, DC.

Stewart E.A., Haselow D.L., Wyse N.M. (1987) Review of operations and performance data on five water hyacinth based treatment systems in Florida. In: *Aquatic Plants for Water Treatment and Resource Recovery*, Reddy K.R., Smith W.H. (eds.) Magnolia Publishing: Orlando, FL, pp. 279–288.

Stewart R.E. Jr. (1996) Technical aspects of wetlands as bird habitat. In: *National Water Summary on Wetland Resources*, Water Supply Paper 2425, United States Geological Survey: Reston, VA, pp. 49–56.

Stone K.C., Hunt P.G., Novak J.M., Johnson M.H. (2003) In-stream wetland design for non-point source pollution abatement. *Applied Engineering in Agriculture 19*(2): 171–175.

Stothoff S., Mitchell-Bruker S. (2003) Modeling changes in flow induced by anisotropy of vegetation patches and topographic features. Proceedings of the Joint Conference on Greater Everglades and Florida Bay Ecosystem; Miami, FL.

Stott R., May E., Matsushita E., Warren A. (2001) Protozoan predation as a mechanism for the removal of *Cryptosporidium* oocysts from wastewaters in constructed wetlands. *Water Science and Technology 44*(11–12): 191–198.

Stott R., Tanner C.C. (2005) Influence of biofilm on removal of surrogate faecal microbes in a constructed wetland and maturation pond. *Water Science and Technology 51*(9): 315–322.

Strand J.A., Weisner S.E. (2013) Effects of wetland construction on nitrogen transport and species richness in the agricultural landscape—Experiences from Sweden. *Ecological Engineering 56*: 14–25.

Stratton F.E. (1969) Nitrogen losses from alkaline water impoundments. *Journal of Sanitary Engineering (ASCE) 95*(SA2): 223–231.

Straub L.G., Silberman E., Nelson H.C. (1958) Open-channel flow at small Reynolds numbers. *Transactions of the American Society of Civil Engineers 123*(1): 685–706.

Strecker E.W., Kersnar J.M., Driscoll E.D., Horner R.R. (1992) *The Use of Wetlands for Controlling Stormwater Pollution*. The Terrene Institute: Washington, DC.

Streeter H.W., Phelps E.B. (1925) *A Study of the Pollution and Natural Purification of the Ohio River*. Public Health Bulletin 146, U.S. Public Health Service, Treasury Department: Washington, DC.

Stringfellow W.T., Karpuzcu M.E., Spier C., Hanlon J.S., Graham J. (2012) Sizing mitigation wetlands in agricultural watersheds. *Water Science and Technology 67*(1): 40–46.

Ström L., Lamppa A., Christensen T.R. (2007) Greenhouse gas emissions from a constructed wetland in southern Sweden. *Wetlands Ecology and Management 15*(1): 43–50.

Stromberg M. (2015) *Maintenance Plan Key to Avoid Stormwater to Wetland Conversion*. Report of WRA Environmental Consultants: San Rafael, CA. Available at: https://wra-ca.com/maintenance-plan-key-avoid-stormwater-wetland-conversion/ (Accessed 1/1/19).

Stuck J.D., Izuno F.T., Campbell K.L., Botcher A.B., Rice R.W. (2001) Farm-level studies of particulate phosphorus transport in the Everglades Agricultural Area. *Transactions of the American Society of Agricultural Engineers 44*(5): 1105–1116.

Stumpner E.B., Kraus T.E.C., Liang Y.L., Bachand S.M., Horwath W.R., Bachand P.A.M. (2018) Sediment accretion and carbon storage in constructed wetlands receiving water treated with metal-based coagulants. *Ecological Engineering 111*: 176–185.

Su R., Lohner R.N., Kuehn K.A., Sinsabaugh R., Neely R.K. (2007) Microbial dynamics associated with decomposing Typha angustifolia litter in two contrasting Lake Erie coastal wetlands. *Aquatic Microbial Ecology 46*(3): 295–307.

Su T.M., Yang S.C., Shih S.S., Lee H.Y. (2009) Optimal design for hydraulic efficiency performance of free-water-surface constructed wetlands. *Ecological Engineering 35*(8): 1200–1207.

Sultana M.Y., Akratos C.S., Pavlou S., Vayenas D.V. (2014) Chromium removal in constructed wetlands: A review. *International Biodeterioration and Biodegradation 96*: 181–190.

Sun D.D., Tay J.H., Tan K.H. (2003) Photocatalytic degradation of *E. coliform* in water. *Water Research 37*(14): 3452–3462.

Sun L., Song C. (2008) Evapotranspiration from a freshwater marsh in the Sanjiang Plain, Northeast China. *Journal of Hydrology 352*(1–2): 202–210.

Sun R., Chen A., Chen L., Lü Y. (2012) Cooling effects of wetlands in an urban region: The case of Beijing. *Ecological Indicators 20*: 57–64.

Sunblad K. (1998) Sweden. In: *Constructed Wetlands for Wastewater Treatment in Europe*, Vymazal J., Brix H., Cooper P.F., Green M.B., Haberl R. (eds.) Backhuys Publishers: Leiden, The Netherlands, pp. 251–260.

Surampalli R.Y., Banerji S.K., Pycha C.J., Lopez E.R. (1995) Phosphorus removal in ponds. *Water Science and Technology 21*(12): 331–340.

Sutherland J.C. (1982) Michigan wetland wastewater tertiary treatment systems. In: *Water Reuse*, Middlebrooks E.J. (ed.), Ann Arbor Science: Ann Arbor, MI, pp. 369–378.

Sutula M., Stein E. (2003) *Habitat Value of Constructed and Natural Wetlands Used to Treat Urban Runoff: A Literature Review*. Technical Report 388, Prepared for the California State Coastal Conservancy, Westminster, CA.

Svedarsky D., Bruggman J., Ellis-Felege S., et al. (2016) *Cattail Management in the Northern Great Plains: Implications for Wetland Wildlife and Bioenergy Harvest*. Report of Northwest Research and Outreach Center, University of Minnesota, Crookston, MN. Available at: www.nwroc.umn.edu/sites/nwroc.umn.edu/files/cattail_management.pdf (Accessed 1/5/19).

Sveistrup T.E., de Melo Marcelino V., Braskerud B.C. (2008) Aggregates explain the high clay retention of small constructed wetlands: A micromorphological study. *Boreal Environment Research 13*(3): 275–284.

Swain E., Decker J. (2010) Measurement-derived heat-budget approaches for simu-lating coastal wetland temperature with a hydrodynamic model. *Wetlands 30*: 635–648.

Swain E.D., Wolfert M.A., Bales J.D., Goodwin C.R. (2004) *Two-Dimensional Hydrodynamic Simulation of Surface-Water Flow and Transport to Florida Bay through the Southern Inland and Coastal Systems (SICS)*. U.S. Geological Survey Water-Resources Investigations Report 03-4287, U.S. Geological Survey, Reston, VA.

Sweers H.E. (1976) A nomogram to estimate the heat-exchange coefficient at the air-water interface as a function of wind speed and temperature; a critical survey of some literature. *Journal of Hydrology 30*: 375–401.

Talley M.K. (2012) *Analysis of a Pilot-Scale Constructed Wetland Treatment System for Flue Gas Desulfurization Wastewater*. PhD Dissertation, Kansas State University, Manhattan, KS.

Tan B.L.L., 2006. *Chemical and Biological Analyses of Selected Endocrine Disruptors in Wastewater Treatment Plants in South East Queensland, Australia*. PhD Dissertation, Griffith University, Queensland, Australia.

Tang Y., Harpenslager S.F., van Kempen M.M., Verbaarschot E.J., Loeffen L.M., Roelofs J.G., Smolders A.J., Lamers L.P. (2017) Aquatic macrophytes can be used for wastewater polishing but not for purification in constructed wetlands. *Biogeosciences 14*(4): 755–766.

Tanino Y, Nepf H. (2008) Lateral dispersion in random cylinder arrays at high Reynolds number. *Journal of Fluid Mechanics 600*: 339–371.

Tanner C.C. (1994) Growth and nutrition of *Schoenoplectus validus* in agricultural wastewaters. *Aquatic Botany 47*(2): 131–153.

Tanner C.C. (1996) Plants for constructed wetland treatment systems – A comparison of the growth and nutrient uptake of eight emergent species. *Ecological Engineering 7*(1): 59–84.

Tanner C.C. (2001) Growth and nutrient dynamics of soft stem bulrush in constructed wetlands treating nutrient-rich wastewaters. *Wetlands Ecology and Management 9*: 49–73.

Tanner C.C., Nguyen M.L., Sukias J.P.S. (2003) Using constructed wetlands to treat subsurface drainage from intensively grazed dairy pastures in New Zealand. *Water Science and Technology 48*(5): 207–213.

Tanner C.C., Nguyen M.L., Sukias J.P.S. (2005b) Nutrient removal by a constructed wetland treating subsurface drainage from grazed dairy pasture. *Agriculture, Ecosystems and Environment 105*(1–2): 145–162.

Tanner C.C., Sukias J.P. (2011) Multiyear nutrient removal performance of three constructed wetlands intercepting tile drain flows from grazed pastures. *Journal of Environmental Quality 40*(2): 620–633.

Tanner C.C. (2001b) Plants as ecosystem engineers in subsurface-flow treatment wetlands. *Water Science and Technology 44*(11–12): 9–17.

Tao W., Wen J., Han Y., Huchzermeier M.P. (2012) Nitrogen removal in constructed wetlands using nitritation/anammox and nitrification/denitrification: Effects of influent nitrogen concentration. *Water Environment Research 84*(12): 2099–2105.

Tarutis W.J., Stark L.R., Williams F.M. (1999) Sizing and performance estimation of coal mine drainage wetlands. *Ecological Engineering 12*(4): 353–372.

Taylor A. (2005) *Structural Stormwater Quality BMP Cost–Size Relationship Information from the Literature.* Cooperative Research Centre for Catchment Hydrology, Melbourne, Australia. Available at: www.eeservices.com.au/wp-content/uploads/2016/12/Taylor-a-2005b-Structural-Stormwater-Quality-BMP-Cost.pdf (Accessed 12/28/18).

Taylor J.D., Yarrow G.K., Miller J.E. (2017) *Beavers.* Wildlife Damage Management Technical Series. USDA, APHIS, WS National Wildlife Research Center: Ft. Collins, CO. Available at: www.aphis.usda.gov/wildlife_damage/reports/Wildlife%20Damage%20Management%20Technical%20Series/Beaver-WDM-Technical-Series.pdf (Accessed 12/10/18).

Tchobanoglous G., Crites R., Gearheart R.A., Reed S.C. (2000) *A review of treatment kinetics for constructed wetlands.* Proceedings of Disinfection 2000: Disinfection of Wastes in the New Millennium, 15–18 March 2000; Water Environment Federation: New Orleans, LA.

Teasdale P.R., Batley G.E., Apte S.C., Webster I.T. (1995) Pore water sampling with sediment peepers. *Trends in Analytical Chemistry 14*(6): 250–256.

Teels B.M., Adamus P. (2002) *Methods for Evaluating Wetland Condition. #6 Developing Metrics and Indexes of Biological Integrity.* EPA-822-R-02-016. Office of Water, U.S. Environmental Protection Agency: Washington, DC.

Terzakis S., Fountoulakis M.S., Georgaki I., Albantakis D., Sabathianakis I., Karathanasis A.D., Kalogerakis N., Manios T. (2008) Constructed wetlands treating highway runoff in the central Mediterranean region. *Chemosphere 72*(2): 141–149.

Tetra Tech, Inc. (2007) *The Environmental Fluid Dynamics Code Theory and Computation, Volume 3: Water Quality Module.* Technical report, Tetra Tech, Inc., Fairfax, VA. Available at: www.epa.gov/sites/production/files/2016-01/documents/efdc_water_quality_manual.pdf (Accessed 1/18/19).

Thackson E.L., Shields F.D. Jr., Schroeder P.R. (1987) Residence time distributions of shallow basins. *Journal of Environmental Engineering (ASCE) 113*(6): 1319–1332.

Theil-Nielsen J., Søndergaard M. (1999) Production of epiphytic bacteria and bacterioplankton in three shallow lakes. *Oikos 86*: 283–292.

They N.H., Amado A.M., Cotner J.B. (2017) Redfield Ratios in inland waters: Higher biological control of c: n: p ratios in tropical semi-arid high water residence time lakes. *Frontiers in Microbiology 8*: 1505.

Thiere G., Milenkovski S., Lindgren P.-E., Sahlén G., Berglund O., Weisner S.E.B. (2009) Wetland creation in agricultural landscapes: Biodiversity benefits on local and regional scales. *Biological Conservation 142*: 964–973.

Third K.A., Paxman J., Schmid M., Strous M., Jetten M.S.M., Cord-Ruwisch R. (2005) Treatment of nitrogen-rich wastewater using partial nitrification and Anammox in the CANON process. *Water Science and Technology 52*(4): 47–54.

Thirumurthi D. (1974) Design criteria for waste stabilization ponds. *Journal of the Water Pollution Control Federation 46*(9): 2094–2106.

Tho B.T., Lambertini C., Eller F., Brix H., Sorrell B.K. (2017) Ammonium and nitrate are both suitable inorganic nitrogen forms for the highly productive wetland grass Arundo donax, a candidate species for wetland paludiculture. *Ecological Engineering 105*: 379–386.

Thom A.S. (1971) Momentum absorption by vegetation. *Quarterly Journal of the Royal Meteorological Society 97*(414): 414–428.

Thomann R.V., Mueller J.A. (1987) *Principles of Surface Water Quality Modeling and Control.* Harper and Row: New York.

Thomas C.R., Miao S.L., Sindhøj E. (2009) Environmental factors affecting temporal and spatial patterns of soil redox potential in Florida Everglades wetlands. *Wetlands 29*(4): 1133–1145.

Thomas D.N., Judd S.J., Fawcett N. (1999) Flocculation modelling: A review. *Water Research 33*(7): 1579–1592.

Thomas S., Gaiser E.E., Gantar M., Scinto L.J. (2006) Quantifying the responses of calcareous periphyton crusts to rehydration: A microcosm study (Florida Everglades). *Aquatic Botany 84*: 317–323.

Thomas W.A., McAnally W.A. Jr. (1985) *User's Manual for the Generalized Computer Program System: Open-Channel Flow and Sedimentation, TABS-2*. Final report prepared for US Army Corps of Engineers, Washington, DC.

Thompson P.L., Moses D.D., Howe K.M. (2003) Phytorestoration at the Iowa army ammunition plant. In: *Phytoremediation*, McCutcheon S.C., Schnoor J. (eds.) Wiley: New York.

Thormann M.N., Bayley S.E., Currah R.S. (2001) Comparison of decomposition of belowground and aboveground plant litters in peatlands of boreal Alberta, Canada. *Canadian Journal of Botany 79* (1): 9–22.

Thullen J., Sartoris J., Walton W.E. (2002) Effects of vegetation management in constructed wetland treatment cells on water quality and mosquito production. *Ecological Engineering 18*(4): 441–457.

Thullen J.S., Sartoris J.J., Nelson S.M. (2005) Managing vegetation in surface-flow wastewater-treatment wetlands for optimal treatment performance. *Ecological Engineering 25*(5): 583–593.

Thurber N.J. (1991) Patent: Drainage conduit which is resistant to clogging by beavers. United States US 4,998,847.

Thurnhorst G.A. (1993) *Wetland Planting Guide for the Northeastern United States: Plants for Wetland Creation Restoration, and Enhancement*. Environmental Concern, Inc.: St. Michaels, MD.

Tian H., Xu X., Miao S., Sindhoj E., Beltran B.J., Pan Z. (2010) Modeling ecosystem responses to prescribed fires in a phosphorus-enriched Everglades wetland: I. Phosphorus dynamics and cattail recovery. *Ecological Modelling 221*(9): 1252–1266.

Tilman D. (1982) *Resource Competition and Community Structure*. Princeton University Press: Princeton, NJ.

Tinker P.B., Nye P.H. (2000) *Solute Movement in the Rhizosphere*. Oxford University Press: Oxford, United Kingdom.

Tipping E., Benham S., Boyle J.F., Crow P., Davies J., Fischer U., Guyatt H., Helliwell R., Jackson-Blake L., Lawlor A.J., Monteith D.T., Rowe E.C., Toberman H. (2014) Atmospheric deposition of phosphorus to land and freshwater. *Environmental Science: Processes and Impacts 16*(7): 1608–1617.

Toet S. (2003) *A Treatment Wetland Used for Polishing Tertiary Effluent from A Sewage Treatment Plant: Performance and Processes*. PhD Dissertation, University of Utrecht, The Netherlands.

Toet S., Bouwman M., Cevaal A., Verhoeven J. T. A. (2005) Nutrient removal through autumn harvest of *Phragmites australis* and *Typha latifolia* shoots in relation to nutrient loading in a wetland system used for polishing sewage treatment plant effluent. *Journal of Environmental Science and Health, Part A-Toxic/Hazardous Substances and Environmental Engineering 40*: 1133–1156.

Tolk J.A. (1992) *Corn Aerodynamic and Canopy Surface Resistances and Their Role in Sprinkler Irrigation Efficiency*. PhD Dissertation, Agronomy, Texas Tech University, Lubbock, TX.

Tomljanovich D.A., Perez O. (1989) Constructing the wastewater treatment wetland – Some factors to consider. In: *Constructed Wetlands for Wastewater Treatment; Municipal, Industrial, and Agricultural*, Hammer D.A. (ed.) Lewis Publishers: Chelsea, MI, pp. 399–402.

Toor N.S., Franz E.D., Fedorak P.M., MacKinnon M.D., Liber K. (2013) Degradation and aquatic toxicity of naphthenic acids in oil sands process-affected waters using simulated wetlands. *Chemosphere 90*(2): 449–458.

Torrella F., Lopez J.P., Banks C.J. (2003) Survival of indicators of bacterial and viral contamination of wastewater subjected to low temperatures and freezing: Application to cold climate waste stabilization ponds. *Water Science and Technology 48*(2): 105–112.

Toth L.A., van der Valk A. (2012) Predictability of flood pulse driven assembly rules for restoration of a floodplain plant community. *Wetlands Ecology and Management 20*(1): 59–75.

Towler B.W., Cahoon J.E., Stein O.R. (2004) Evapotranspiration crop coefficients for cattail and bulrush. *Journal of Hydrologic Engineering 9*(3): 235–239.

Trapp S. (2004) Plant uptake and transport models for neutral and ionic chemicals. *Environmental Science and Pollution Research 11*: 33–39.

Trapp S., Matthies M. (1995) Generic one-compartment model for uptake of organic chemicals by foliar vegetation. *Environmental Science and Technology 29*(9): 2333–2338.

Trepel M., Palmeri L. (2002) Quantifying nitrogen retention in surface flow wetlands for environmental planning at the landscape scale. *Ecological Engineering 19*(2): 127–140.

Trujillo M.E.C. (1997) *Residence Time Distributions in Constructed Wetlands and Their Application to Modeling Nitrate Removal.* PhD Dissertation, Department of Civil and Environmental Engineering, University of Illinois, Urbana, IL.

Tsanis I.K., Prescott K.L., Shwen H. (1998) Modelling of phosphorus and suspended solids in Cootes Paradise marsh. *Ecological Modeling 114*(1): 1–17.

Tse I.C., Poindexter C.M., Variano E.A. (2016) Wind-driven water motions in wetlands with emergent vegetation. *Water Resources Research 52*(4): 2571–2581.

Tuomisto H. (2010) A consistent terminology for quantifying species diversity? Yes, it does exist. *Oecologia 164*(4): 853–860.

Tupacz E.G., Day F.P. (1990) Decomposition of roots in a seasonally flooded swamp ecosystem. *Aquatic Botany 37*: 199–214.

Turner B.L., Newman S. (2005) Phosphorus cycling in wetland soils. *Journal of Environmental Quality 34*(5): 1921–1929.

Turner B.L., Newman S., Newman J.M. (2006) Organic phosphorus sequestration in subtropical treatment wetlands. *Environmental Science and Technology 40*(3): 727–733.

Tuttle P.L., Higgins D.K., Quashnick J. (2001) Mercury characterization in Lahontan Valley Wetlands, Carson River Mercury site, Lyon and Churchill counties, Nevada, Report FF No. 1N45 to the Nevada Fish and Wildlife Service and U.S. EPA Region 9, San Francisco, CA.

Twarakavi N.K., Šimůnek J., Schaap M.G. (2009) Development of pedotransfer functions for estimation of soil hydraulic parameters using support vector machines. *Soil Science Society of America Journal 73*(5): 1443–1452.

Tylová E., Steinbachová L., Votrubová O., Lorenzen B., Brix H. (2008) Different sensitivity of Phragmites australis and Glyceria maxima to high availability of ammonium-N. *Aquatic Botany 88*(2): 93–98.

Tyndall J., Bowman T. (2016) *Iowa Nutrient Reduction Strategy Best Management Practice CostOverview Series.* Department of Ecology & Natural Resource Management, Iowa State University, Ames, IA. Available at: www.researchgate.net/profile/John_Tyndall/publication/315496577_Iowa_Nutrient_Reduction_Strategy_BMP_Cost_Decision_Tool_Overview/links/58d2b87592851c319e56ed56/Iowa-Nutrient-Reduction-Strategy-BMP-Cost-Decision-Tool-Overview.pdf (Accessed 12/28/18).

U.S. Army Corps of Engineers. (1987) *Wetlands Delineation Manual*, Wetlands Research Program Technical Report Y-87-1 (online edition), U.S. Army Corps of Engineers: Washington, DC. Available at: https://usace.contentdm.oclc.org/digital/collection/p266001coll1/id/4530 (Accessed 12/4/18).

U.S. Army Corps of Engineers. (1996) *Final Programmatic Environmental Impact Statement, Volume II, Appendix L, Desktop Evaluation of Alternative Technologies.* U.S. Army Corps of Engineers: Jacksonville, FL.

U.S. Army Corps of Engineers. (2000) *Wetlands Engineering Handbook*, Report No. ERDC/EL TR-WRP-RE-21, Hayes D.F., Olin T.J., Fischenich J.C., Palermo M.R. (eds.) U.S. Army Corps of Engineers, Army Engineer Research and Development Center: Washington, DC.

U.S. Army Corps of Engineers. (2001) *Control of Beaver Flooding at Restoration Projects*, Technical Note ERDC TN-WRAP-01-01, Wetlands Regulatory Assistance Program, U.S. Army Engineer Research and Development Center: Vicksburg, MS.

U.S. DA. (1991) *Nutrient Sediment Control System.* Environmental Quality Technical Note #4. U.S. Department of Agriculture (USDA) Soil Conservation Service: Chester, PA, 19 pp.

U.S. DA. (2008) *Part 650 Engineering Field Handbook, Chapter 13 Wetland Restoration, Enhancement, or Creation.* United States Department of Agriculture, Natural Resources Conservation Service: Washington, DC. Available at: https://directives.sc.egov.usda.gov/OpenNonWebContent.aspx?content=17765.wba (Accessed 10/16/18).

U.S. DA. (2018a) *Quick Spreadsheet of the Most Common Indexes and Discount Rates.* Natural Resources Conservation Service, United States Department of Agriculture: Washington, DC. Available at: www.nrcs.usda.gov/wps/portal/nrcs/main/national/technical/econ/prices/ (Accessed 12/17/18).

U.S. DA. (2018b) *Land Values 2018 Summary.* United States Department of Agriculture, National Agricultural Statistics Service, Washington, DC. Available at: www.usda.gov/nass/PUBS/TODAYRPT/land0817.pdf (Accessed 12/16/18).

U.S. DA NRCS. (2009) Part 637 *Environmental Engineering National Engineering Handbook. Chapter 3, Constructed Wetlands.* U.S. Department of Agriculture, Natural Resources Conservation Service: Washington, DC. Available at: www.nrcs.usda.gov/wps/portal/nrcs/detail/mi/technical/engineering/?cid=nrcs141p2_024573 (Accessed 730/18).

U.S. EPA. (1983b) *Design Principles for Wetland Treatment Systems.* EPA 600/2-83/026. Hammer D.E., Kadlec R.H. (eds.) National Technical Information Service: Springfield, VA.

U.S. EPA. (1988) *Design Manual: Constructed Wetlands and Aquatic Plant Systems for Municipal Wastewater Treatment.* EPA 625/1-88/022. U.S. EPA Office of Water: Cincinnati, OH.

U.S. EPA. (1991) *MINTEQA2/PROFEFA2, A Geochemical Assessment Model for Environmental Systems: Version 3.0 User's Manual.* EPA/600/3-91/021. Allison J.D., Brown D.S., Novo-Gradac K.J. (eds.) U.S. EPA Office of Research and Development: Athens, GA.

U.S. EPA. (1993a) *Constructed Wetlands for Wastewater Treatment and Wildlife Habitat: 17 Case Studies.* EPA 832/R-93/005. U.S. EPA Office of Research and Development: Washington, DC.

U.S. EPA. (1993b) *Design Manual: Nitrogen Control.* EPA 625/R-93/010. U.S. EPA Office of Research and Development: Washington, DC.

U.S. EPA. (1993d) *Habitat Quality Assessment of Two Wetland Treatment Systems in Florida – A Pilot Study.* EPA/600/R-93/005. U.S. EPA Environmental Research Laboratory: Corvallis, OR.

U.S. EPA. (1999) *Free Water Surface Wetlands for Wastewater Treatment: A Technology Assessment.* EPA 832/R-99/002. U.S. EPA Office of Water: Washington, DC, p. 165.

U.S. EPA. (2000a) *Constructed Wetlands Treatment of Municipal Wastewaters.* EPA 625/R-99/010. U.S. EPA Office of Research and Development: Washington, DC.

U.S. EPA. (2000c) *Guiding Principles for Constructed Treatment Wetlands: Providing Water Quality and Wildlife Habitat.* EPA 843/B-00/003. U.S. EPA Office of Wetlands, Oceans, and Watersheds: Washington, DC.

U.S. EPA. (2001a) *Action Plan for Reducing, Mitigating, and Controlling Hypoxia in the Northern Gulf of Mexico.* Report to Congress and the Gulf of Mexico Watershed Nutrient Task Force. U.S. EPA: Washington, DC.

U.S. EPA. (2002a) *Functions and Values of Wetlands.* Report EPA843-F-01-002c. Office of Wetlands, Oceans and Watersheds, U.S. Environmental Protection Agency: Washington, DC.

U.S. EPA. (2007) *Aquatic Life Ambient Freshwater Quality Criteria -Copper.* EPA 822-R-07-001. U.S. EPA Office of Water: Washington, DC.

U.S. EPA. (2009) *Stormwater Wet Pond and Wetlands Management Guidebook.* EPA 833-B-09-001, U. U. Enviromental Protection Agency: Washington, DC. Available at: www.epa.gov/sites/production/files/2015-11/documents/pondmgmtguide.pdf (Accessed 12/6/18).

U.S. EPA. (2012) *Recreational Water Quality Criteria.* Report 820-F-12-058, Office of Water, United States Environmental Protection Agency, Washington, DC. Available at: www.epa.gov/sites/production/files/2015-10/documents/rwqc2012.pdf (Accessed 9/29/18).

U.S. EPA. (2015) *Endocrine Disruptor Screening Program Tier 1 Screening Determinations and Associated Data Evaluation Records.* Available at: www.epa.gov/endocrine-disruption/endocrine-disruptor-screening-program-tier-1-screening-determinations-and (Accessed 9/25/18).

U.S. EPA. (2016a) *Aquatic Life Ambient Water Quality Criterion for Selenium – Freshwater.* EPA 822-R-16-006. U.S. EPA Office of Water: Washington, DC.

U.S. EPA. (2016b) *Exposure Assessment Models: EFDC.* U.S. EPA, Washington, DC. Available at: www.epa.gov/exposure-assessment-models/efdc (Accessed 1/21/19).

U.S. EPA. (2016c) *Summary of State Post Construction Stormwater Standards.* U.S. EPA Office of Water: Washington, DC. Available at: www.epa.gov/sites/production/files/2016-08/documents/swstdsummary_7-13-16_508.pdf (Accessed 8/4/18).

U.S. EPA. (2017a) *Mississippi River/Gulf of Mexico Watershed Nutrient Task Force, 2017 Report to Congress.* United States Environmental Protection Agency: Washington, DC. Available at: www.epa.gov/ms-htf/hypoxia-task-force-reports-congress (Accessed 6/5/18).

U.S. EPA. (2017b) *Guidance to Identify Waters Protected by the Clean Water Act.* U.S. EPA: Washington, DC. Available at: www.epa.gov/cwa-404/guidance-identify-waters-protected-clean-water-act (Accessed 1/1/19).

U.S. EPA. (2017c) *Aquatic Life Benchmarks and Ecological Risk Assessments for Registered Pesticides.* U.S. EPA: Washington, DC. Available at; www.epa.gov/pesticide-science-and-assessing-pesticide-risks/aquatic-life-benchmarks-and-ecological-risk#main-content (Accessed 9/14/18).

U.S. EPA. (2018a) *Current National Recommended Water Quality Criteria- Aquatic Life Criteria.* U.S. EPA, Washington, DC. Available at: www.epa.gov/wqc/national-recommended-water-quality-criteria-aquatic-life-criteria-table#main-content (Accessed 8/26/18).

U.S. EPA. (2018b) *National Recommended Water Quality Criteria- Human Health Criteria Table.* U.S. EPA: Washington, DC. Available at: www.epa.gov/wqc/national-recommended-water-quality-criteria-human-health-criteria-table (Accessed 8/26/18).

U.S. EPA. (2018c) *Risks to Aquatic Organisms Posed by Human Pharmaceutical Use.* U.S. EPA: Washington, DC. Available at: www.epa.gov/water-research/risks-aquatic-organisms-posed-human-pharmaceutical-use (Accessed 9/23/18).

U.S. FDA. (2018a) *Approved Animal Drug Products (Green Book).* U.S. Food and Drug Administration: Silver Spring, MD. Available at: www.fda.gov/animalveterinary/products/approvedanimal drugproducts/ (Accessed 9/23/18).

U.S. FWS. (2007) *A.R.M. Loxahatchee National Wildlife Refuge – Enhanced Monitoring and Modeling Program – 2nd Annual Report – February 2007.* LOXA06-008. U.S. Fish and Wildlife Service: Boynton Beach, FL.

U.S. Geological Survey. (1999) *The Quality of Our Nation's Waters– Nutrients and Pesticides: U.S. Geological Survey Circular 1225.* U.S. Geological Survey: Reston, VA. Available at: https://pubs.usgs.gov/circ/circ1225/pdf/ (Accessed 9/19/18).

UF/IFAS (University of Florida Institute of Food and Agricultural Sciences). (2017) *Evaluation of Soil Biogeochemical Properties Influencing Phosphorus Flux in the Everglades Stormwater Treatment Areas (Stas).* Report to South Florida Water Management District: West Palm Beach, FL.

Vaccaro L.E., Bedford B.L., Johnston C.A. (2009) Litter accumulation promotes dominance of invasive species of cattails *(Typha* spp.) in Lake Ontario wetlands. *Wetlands 29*(3): 1036–1048.

Vallentyne J.R. (1957) Principles of modern limnology. *American Scientist 45*(3): 218–244.

van Den Wyngaert I.J., Bobbink R. (2009) Ecological dynamics. II. The influences of vertebrate herbivory on ecological dynamics in wetland ecosystems. In: *The Wetlands Handbook*, Maltby E., Barker T. (eds.) John Wiley & Sons: West Sussex, United Kingdom, pp. 304–325.

van Den Wyngaert I.J.J., Wienk L.D., Sollie S., Bobbink R., Verhoeven J.T.A. (2003) Long term effects of yearly grazing by moulting greyleg geese (*Anser anser*) on reed (*Phragmites australis*) growth and nutrient dynamics. *Aquatic Botany 75*: 229–248.

van der Leeden F., Troise F.L., Todd D.K. (1990) *The Water Encyclopedia.* Second edition. Lewis Publishers: Boca Raton, FL.

Van der Peijl M.J., Verhoeven J.T.A. (1999) A model of carbon, nitrogen and phosphorus dynamics and their interactions in river marginal wetlands. *Ecological Modelling* 118(2–3): 95–130.

Van der Peijl M.J., Verhoeven J.T.A. (2000) Carbon, nitrogen and phosphorus cycling in river marginal wetlands; using a dynamic simulation model to examine the importance of landscape geochemical flows. *Biogeochemistry 50*: 45–71.

van der Valk A. (2012) *The Biology of Freshwater Wetlands.* Oxford University Press: Oxford, United Kingdom.

van der Valk A., Davis C.B. (1980) The impact of a natural drawdown on the growth of four emergent species in a prairie glacial marsh. *Aquatic Botany 9*: 301–322.

van der Valk A.G. (2013) Seed banks of drained floodplain, drained palustrine, and undrained wetlands in Iowa, USA. *Wetlands 33*(1): 183–190.

van der Valk A.G., Attiwil P.M. (1983) Above- and below-ground litter decomposition in an Australian salt marsh. *Australian Journal of Ecology 8*: 441–447.

van der Valk A.G., Davis C.B. (1978) Primary production of Prairie Glacial Marshes. In: *Freshwater Wetlands: Ecological Processes and Management Potential*, Good R.E., Whigham D.F., Simpson R.L. (eds.) Academic Press: New York, pp. 21–37.

Van der Weert R., Kamerling G.E. (1974) Evapotranspiration of water hyacinth (Eichhornia crassipes). *Journal of Hydrology 22*(3–4): 201–212.

van Dongen U.G.J.M., Jetten M.S.M., van Loosdrecht M.C.M. (2001) The SHARON® – Anammox® process for treatment of ammonium rich wastewater. *Water Science and Technology 44*(1): 153–160.

van Duzer C. (2004) *Floating Islands: A Global Bibliography.* Cantor Press: Los Altos Hills, CA.

van Eekert M.H.A., Schraa G. (2001) The potential of anaerobic bacteria to degrade chlorinated compounds. *Water Science and Technology 44*(8): 49–56.

van Genuchten M. T. (1980) A closed-form equation for predicting the hydraulic conductivity of unsaturated soils. *Soil Science Society of America Journal 44*: 892–898.

van Loosdrecht M.C.M., Jetten M.S.M. (1998) Microbiological conversions in nitrogen removal. *Water Science and Technology 38*(1): 1–7.

van Noort P.C. (2009) Estimation of amorphous organic carbon/water partition coefficients, subcooled aqueous solubilities, and n-octanol/water distribution coefficients of alkylbenzenes and polycyclic aromatic hydrocarbons. *Chemosphere 74*(8): 1018–1023.

Van Oostrom A.J. (1995) Nitrogen removal in constructed wetlands treating nitrified meat processing effluent. *Water Science and Technology 32*(3): 137–148.

Van Oostrom A.J., Russell J.M. (1994) Denitrification in constructed wastewater wetlands receiving high concentrations of nitrate. *Water Science and Technology 29*(4): 7–14.

VanderZaag A. C., Gordon R.J., Burton D.L., Jamieson R. C., Stratton G.W. (2010a) Greenhouse gas emissions from surface flow and subsurface flow constructed wetlands treating dairy wastewater. *Journal of Environmental Quality 39*(2): 460–471.

VanderZaag A.C., Campbell K.J., Jamieson R.C., Sinclair A.C., Hynes L.G. (2010b) Survival of *Escherichia coli* in agricultural soil and presence in tile drainage and shallow groundwater. *Canadian Journal of Soil Science 90*(3): 495–505.

Vanotti M.B., Szogi A.A., Vives C.A. (2008) Greenhouse gas emission reduction and environmental quality improvement from implementation of aerobic waste treatment systems in swine farms. *Waste Management 28*: 759–766.

Variano E.A., Ho D.T., Engel V.C., Schmieder P.J., Reid M.C. (2009) Flow and mixing dynamics in a patterned wetland: Kilometer-scale tracer releases in the Everglades. *Water Resources Research 45*: 8.

Veenstra S. (1998) The Netherlands. In: *Constructed Wetlands for Wastewater Treatment in Europe*, Vymazal J., Brix H., Cooper P.F., Green M.B., Haberl R. (eds.) Backhuys Publishers: Leiden, The Netherlands, pp. 289–314.

Veldhuis M.P., Berg M.P., Loreau M., Olff H. (2018) Ecological autocatalysis: A central principle in ecosystem organization? *Ecological Monographs 88*(3): 304–319.

Verhofstad M.J.J.M., Poelen M.D.M., Van Kempen M.M.L., Bakker E.S., Smolders A.J.P. (2017) Finding the harvesting frequency to maximize nutrient removal in a constructed wetland dominated by submerged aquatic plants. *Ecological Engineering 106*: 423–430.

Verlicchi P., Zambello E. (2014) How efficient are constructed wetlands in removing pharmaceuticals from untreated and treated urban wastewaters? A review. *Science of the Total Environment 470*: 1281–1306.

Verry E.S., Boelter D.H., Päivänen J., Nichols D.S., Malterer T., Gafni A. (2011) Physical properties of organic soils. In: *Peatland Biogeochemistry and Watershed Hydrology at the Marcell Experimental Forest*, Kolka R.K., Sebestyen S.D., Verry E.S., Brooks K.N. (eds.) CRC Press: New York, pp. 135–176.

Vespraskas M.J., Craft C.B. (eds.) (2016) *Wetland Soils: Genesis, Hydrology, Landscapes, and Classification*. Second edition. CRC Press: Boca Raton, FL.

Vet R., Artz R.S., Carou S., et al. (2014) A global assessment of precipitation chemistry and deposition of sulfur, nitrogen, sea salt, base cations, organic acids, acidity and pH, and phosphorus. *Atmospheric Environment 93*: 3–100.

Vidales-Contreras J.A, Gerba C.P., Karpiscak M.M., Acuna-Askar K., Chaidez-Quiroz C. (2006) Transport of coliphage PRD1 in a surface flow constructed wetland. *Water Environment Research 78* (11): 2253–2260.

Vincent G. (1992) *Use of artificial wetlands for the treatment of recreational wastewater*. Proceedings of the 3rd International Conference on Wetland Systems for Water Pollution Control, 30 November to 3 December 1992; Australian Water and Wastewater Association: Sydney, Australia, pp. 28.1–28.4.

Vollenweider R.A. (1975) Input-output models with special reference to the phosphorus loading concept in limnology. *Schweizerische Zeitschrift Fur Hydrologie (Swiss Journal of Hydrology) 37*: 53–84.

von Sperling M. (1999) A critical analysis of classical design equations for waste stabilization lagoons and other wastewater treatment systems. *Water Environment Research 71*(6): 1240–1243.

Voulvoulis N., Arpon K.D., Giakoumis T. (2017) The EU water framework directive: From great expectations to problems with implementation. *Science of the Total Environment 575*: 358–366.

Vrugt J.A., Wijk M.V., Hopmans J.W., Šimunek J. (2001) One-, two-, and three-dimensional root water uptake functions for transient modeling. *Water Resources Research 37*(10): 2457–2470.

Vymazal J. (1995) *Algae and Nutrient Cycling in Wetlands.* CRC Press/Lewis Publishers: Boca Raton, FL.

Vymazal J. (2001a) Removal of organics in Czech constructed wetlands with horizontal sub-surface flow. In: *Transformations of Nutrients in Natural and Constructed Wetlands*, Vymazal J. (ed.) Backhuys Publishers: Leiden, The Netherlands, pp. 305–328.

Vymazal J. (2004) *Removal of phosphorus via harvesting of emergent vegetation in constructed wetlands for wastewater treatment.* Liénard A., Burnett H. (eds.) Proceedings of the 9th International

Conference on Wetland Systems for Water Pollution Control, 26–30 September 2004; Association Scientifique et Technique pour l'Eau et l'Environnement (ASTEE), Cemagref, and IWA: Avignon, France; 415–422.

Vymazal J. (2006) Constructed wetlands with emergent macrophytes: From experiments to a high quality treatment technology. Dias V., Vymazal J. (eds.) Proceedings of the 10th International Conference on Wetland Systems for Water Pollution Control, 23–29 September 2006; Ministério de Ambiente, do Ordenamento do Territóri e do Desenvolvimento Regional (MAOTDR) and IWA: Lisbon, Portugal; 3–27.

Vymazal J. (2013) Emergent plants used in free water surface constructed wetlands: A review. *Ecological Engineering 61*: 582–592.

Vymazal J., Březinová T. (2015) The use of constructed wetlands for removal of pesticides from agricultural runoff and drainage: A review. *Environment International 75*: 11–20.

Vymazal J., Kropfelova L. (2008) *Wastewater Treatment in Constructed Wetlands with Horizontal Sub-Surface Flow*. Springer: New York.

Wachenheim C., Roberts D.C., Dhingra N., Lesch W., Devney J. (2018a) Conservation reserve program enrollment decisions in the Prairie Pothole region. *Journal of Soil and Water Conservation 73*(3): 337.

Wachenheim C.J., Devney J. (2018) *Landowner Conservation Attitudes and Behaviors in the Prairie Pothole Region*. Agribusiness & Applied Economics Report 267904, Department of Agribusiness and Applied Economics, North Dakota State University, Fargo, ND.

Wachenheim C.J., Roberts D.C., Addo N.S., Devney J. (2018b) Farmer preferences for a working wetlands program. *Wetlands 38*(5): 1005–1015.

Wadzuk B.M., Rea M., Woodruff G., Flynn K., Traver R.G. (2010) Water-quality performance of a constructed stormwater wetland for all flow conditions. *Journal of the American Water Resources Association 46*(2): 385–394.

Wahl M.D., Brown L.C., Soboyejo A.O., Dong B. (2012) Quantifying the hydraulic performance of treatment wetlands using reliability functions. *Ecological Engineering 47*: 120–125.

Wahl M.D., Brown L.C., Soboyejo A.O., Martin J., Dong B. (2010) Quantifying the hydraulic performance of treatment wetlands using the moment index. *Ecological Engineering 36*(12): 1691–1699.

Wake L.V., Christopher R.K., Rickard A.D., Anderson J.E., Ralph B.J. (1977) A thermodynamic assessment of possible substrate for sulfate reducing bacteria. *Australian Journal of Biological Science 30*: 155–172.

Waki M., Yasuda T., Suzuki K., Komada M., Abe K. (2015) Distribution of anammox bacteria in a free-water-surface constructed wetland with wild rice (Zizania latifolia). *Ecological Engineering 81*: 165–172.

Walker D.J. (1998) Modeling residence time in stormwater ponds. *Ecological Engineering 10*(3): 247–262.

Walker D.J. (2001) Modelling sedimentation processes in a constructed stormwater wetland. *Science of the Total Environment 266*(1–3): 61–68.

Walker W.W. (1995) Design basis for Everglades stormwater treatment areas. *Water Resources Bulletin (Now Journal of the American Water Resources Association) 31*(4): 671–685.

Walker W.W. (2005) *DMSTA Calibration & Enhancement for Reservoirs*, Report to U.S. Army Corps of Engineers and U.S. Department of the Interior, June 30, 2005. Available at: www.wwwalker.net/dmsta/reservoirs/index.htm (Accessed 7/12/18).

Walker W.W. Jr., Kadlec R.H. (2011) Modeling phosphorus dynamics in Everglades wetlands and stormwater treatment areas. *Critical Reviews in Environmental Science and Technology 41*(6): 430–446.

Walker W.W., Havens K.E. (2003) Development and application of a phosphorus balance model for Lake Istokpoga, Florida. *Lake and Reservoir Management: an International Journal of the North American Lake Management Society 19*(1): 79–91.

Walker W.W., Kadlec R.H. (2008) Dynamic model for stormwater treatment areas, model version 2, documentation update. Available at: www.wwwalker.net/dmsta/ (Accessed 7/15/18).

Walker W.W., Kadlec R.H. *Dynamic Model for Stormwater Treatment Areas, Model Version 2* (2005) Available at: www.wwwalker.net/dmsta1/ (Accessed11/18/18).

Wallace S.D., Knight R.L. (2006) *Small-Scale Constructed Wetland Treatment Systems: Feasibility, Design Criteria, and O&M Requirements*, Final Report, Project 01-CTS-5, Water Environment Research Foundation (WERF): Alexandria, VA.

Waller V.L., Bruland G.L. (2016) Fecal indicator bacteria dynamics in a surface flow constructed wetland in southwestern Illinois, USA. *Wetlands 36*(3): 539–546.

Walsh J.B. (1999) *A Feasibility Study of Bioremediation in A Highly Organic Soil.* MS Thesis, Worcester Polytechnic Institute, Worcester, PA.

Walshe K. (2016) Wastewater treatment in wetlands – An operations perspective. *Engineers Journal*, March 8, 2016. Available at: www.engineersjournal.ie/2016/03/08/wastewater-treatment-in-wetlands-an-operations-perspective/ (Accessed 1/21/19).

Walton W.E. (2012) Design and management of free water surface constructed wetlands to minimize mosquito production. *Wetlands Ecology and Management 20*(3): 173–195.

Wang C.Y., Sample D.J. (2013) Assessing floating treatment wetlands nutrient removal performance through a first order kinetics model and statistical inference. *Ecological Engineering 61*: 292–302.

Wang C.Y., Sample D.J. (2014) Assessment of the nutrient removal effectiveness of floating treatment wetlands applied to urban retention ponds. *Journal of Environmental Management 137*: 23–35.

Wang C.Y., Sample D.J., Day S.D., Grizzard T.J. (2015) Floating treatment wetland nutrient removal through vegetation harvest and observations from a field study. *Ecological Engineering 78*: 15–26.

Wang G.-P., Liu J.-S., Tang J. (2004) The long-term nutrient accumulation with respect to anthropogenic impacts in the sediments from two freshwater marshes (Xianghai Wetlands, Northeast China). *Water Research 38*: 4462–4474.

Wang H., Jawitz J.W. (2006) Hydraulic analysis of cell-network treatment wetlands. *Journal of Hydrology 330*(3–4): 721–734.

Wang H., Jawitz J.W., White J.R., Martinez C.J., Sees M.D. (2006) Rejuvenating the largest municipal treatment wetland in Florida. *Ecological Engineering 26*(2): 132–146.

Wang J., Chu L. (2016) Biological nitrate removal from water and wastewater by solid-phase denitrification process. *Biotechnology Advances 34*(6): 1103–1112.

Wang J., Wang S. (2016) Removal of pharmaceuticals and personal care products (PPCPs) from wastewater: A review. *Journal of Environmental Management 182*: 620–640.

Wang N., Mitsch W.J. (2000) A detailed ecosystem model of phosphorus dynamics in created riparian wetlands. *Ecological Modelling 126*: 101–130.

Wang P., Chen G.Q. (2017) Contaminant transport in wetland flows with bulk degradation and bed absorption. *Journal of Hydrology 552*: 674–683.

Wang S., Jin X., Pang Y., Zhao H., Zhou X., Wu F. (2006b) Phosphorus fractions and phosphate sorption characteristics in relation to the sediment compositions of shallow lakes in the middle and lower reaches of the Yangtze River region, China. *Journal of Colloid and Interface Science 289* (2005): 339–346.

Wanielista M.P., Yousef Y. (1991) Precipitation inter-event dry periods. In: *Statewide Stormwater Management Workshop: Report by the Southwest Florida Water Management District*, Rushton B.T., Cunningham J.A., Dye C.W. (eds.), Southwest Florida Water Management District, Brookside, FL.

Warren I.R., Bach H. (1992) MIKE 21: A modelling system for estuaries, coastal waters and seas. *Environmental Software 7*(4): 229–240.

Wasantha Lal A.M. (2017) Mapping vegetation-resistance parameters in wetlands using generated waves. *Journal of Hydraulic Engineering 143*(9): 04017023.

Wasantha Lal A.M., Moustafa M.Z., Wilcox W.M. (2015) The use of discharge perturbations to understand in situ vegetation resistance in wetlands. *Water Resources Research 51*(4): 2477–2497.

Wasantha Lal A.M., Van Zee R., Belnap M. (2005) Case study: Model to simulate regional flow in South Florida. *Journal of Hydraulic Engineering 131*(4): 247–258.

Waschbusch. (2000) Sources of phosphorus in stormwater and street dirt from two urban residential basins in Madison, Wisconsin, 1994–1995. Proceedings of the National Conference on Tools for Urban Water Resource Management and Protection; U.S. Environmental Protection Agency: Washington, DC; 15–55.

Washington H.G. (1984) Diversity, biotic and similarity indices: A review with special relevance to aquatic ecosystems. *Water Research 18*(6): 653–694.

Wass R.D. (1997) Tres rios demonstration constructed wetland project: 1996/1997 operation and water quality report, Report to the City of Phoenix.

Wass R.D. (2003) *Occurrence of Sulfur-Driven Autotrophic Denitrification in Demonstration Scale Treatment Wetlands.* PhD Dissertation, Arizona State University,Tempe, AZ.

Wass, Gerke and Associates. (2001) *Status Report to the 1998 Research Plan for the Tres Rios Demonstration Constructed Wetlands Project*. Prepared for the U.S. Bureau of Reclamation: Phoenix, AZ.

Wass, Gerke and Associates. (2002) *Heavy-Metal Investigation of the Tres Rios Demonstration Constructed Treatment Wetlands*, Report to the City of Phoenix.

Wass, Gerke and Associates. (2004) Residual chlorine study for the tres rios demonstration constructed treatment wetlands, Report to the U.S. Bureau of Reclamation, Phoenix, AZ.

Watanabe A., Tsutsuki K., Inoue Y., Maie N., Melling L., Jaffé R. (2014) Composition of dissolved organic nitrogen in rivers associated with wetlands. *Science of the Total Environment 493*: 220–228.

Water Environment Federation. (1990) *Natural Systems for Wastewater Treatment (WEF Manual of Practice FD-16)*. First edition. Water Environment Federation: Alexandria, VA.

Water Environment Federation. (2001) *Natural Systems for Wastewater Treatment (WEF Manual of Practice FD-16)*. Second edition. Water Environment Federation: Alexandria, VA.

Waters M., Luketina D.A. (1998) The development of temperature stratification in freshwater wetlands. Proceedings of the 13th Australasian Fluid Mechanics Conference, Melbourne, Australia; 811–814.

Waters M.T. (1997) *Effects of Vegetation on the Hydrodynamics of Freshwater Wetlands*. PhD Dissertation, University of New South Wales, Sydney, Australia.

Watters G.Z. (1972) *The Hydraulics of Waste Stabilization Ponds*. Utah Water Research Laboratory Report 18,Utah State University, Logan, UT. Available at: https://digitalcommons.usu.edu/water_rep/18 (Accessed 4/12/18).

Watts C.J. (2000) Seasonal phosphorus release from exposed, re-inundated littoral sediments of two Australian reservoirs. *Hydrobiologia 431*: 27–39.

Watts D.A., Moore G.W. (2011) Water-use dynamics of an invasive reed, Arundo donax, from leaf to stand. *Wetlands 31*(4): 725–734.

WDNR. (2003) *Consensus – Based Sediment Quality Guidelines; Recommendations for Use & Application*. Wisconsin Department of Natural Resources: Madison, WI.

WDNR. (2012). *2,4-D Chemical Fact Sheet*. Wisconsin Department of Natural Resources, Madison, WI. Available at: https://dnr.wi.gov/lakes/plants/factsheets/2,4-DFactSheet.pdf (Accessed 12/11/18).

WDNR. (2013) *Guidelines for Wetland Compensatory Mitigation in Wisconsin*. Wisconsin Department of Natural Resources: Madison, WI. Available at: https://dnr.wi.gov/topic/Wetlands/documents/mitigation/WetlandCompensatoryMitigationGuidelines.pdf (Accessed 1/3/19).

Weaver K. (2000) *Proposed Everglades Marsh Dissolved Oxygen Site Specific Alternative Criterion Technical Support Document*. Florida Department of Environmental Protection, Appendix 4-2 in 2001 Everglades Consolidated Report. Available at: https://my.sfwmd.gov/portal/page/portal/pg_grp_sfwmd_sfer/portlet_prevreport/consolidated_01/chapter%2004/chapter%204%20appendices/A04-2.pdf (Accessed 8/20/18).

Webb C. (2013) Managing mosquitoes in constructed freshwater wetlands. In: *Workbook for Managing Urban Wetlands in Australia, Section 2.6*, Paul S. (ed.) Sydney Olympic Park Authority: Sydney, Australia.

Wei X. (2013) *Biodiversity Survey and Surface Flow Constructed Wetland Design to Help Control Agricultural Non-Point Source Pollution in Poyang Lake Basin, China*. PhD Dissertation, The Ohio State University, Columbus, OH.

Weisner S.E., Miao S.L. (2004) Use of morphological variability in Cladium jamaicense and Typha domingensis to understand vegetation changes in an Everglades marsh. *Aquatic Botany 78*(4): 319–335.

Weisner S.E., Strand J.A., Sandsten H. (1997) Mechanisms regulating abundance of submerged vegetation in shallow eutrophic lakes. *Oecologia 109*(4): 592–599.

Weisner S.E., Thiere G. (2010) Effects of vegetation state on biodiversity and nitrogen retention in created wetlands: A test of the biodiversity–Ecosystem functioning hypothesis. *Freshwater Biology 55* (2): 387–396.

Weisner S.E.B. (1993) Long-term competitive displacement of Typha latifolia by Typha angustifolia in a eutrophic lake. *Oecologia 94*: 451–456.

Weisner S.E.B., Eriksson N.P., Graneli W., Leonardson L. (1994) Influence of macrophytes on nitrate removal in wetlands. *Ambio 23*(6): 363–366.

Weiss P.T., Gulliver J.S., Erickson A.J. (2007) Cost and pollutant removal of storm-water treatment practices. *Journal of Water Resources Planning and Management 133*(3): 218–229.

Weller M.W. (1978) Management of freshwater marshes for wildlife. In: *Freshwater Wetlands: Ecological Processes and Management Potential*, Good R.E., Whigham D.F., Simpson R.L. (eds.) Academic Press: New York, pp. 267–284.

Weller N.A., Childers D.L., Turnbull L., Upham R.F. (2016) Aridland constructed treatment wetlands I: Macrophyte productivity, community composition, and nitrogen uptake. *Ecological Engineering 97*: 649–657.

Welsch M., Yavitt J.B. (2003) Early stages of decay of *Lythrum salicaria* L. and *Typha latifolia* L. in a standing-dead position. *Aquatic Botany 75*(1): 45–57.

Welty J.R., Wicks C.E., Wilson R.E. (1983) *Fundamentals of Momentum, Heat, and Mass Transfer.* Third edition. John Wiley & Sons: New York.

Wen L., Recknagel F. (2002) In situ removal of dissolved phosphorus in irrigation drainage water by planted floats: Preliminary results from growth chamber experiment. *Agriculture, Ecosystems and Environment 90*: 9–15.

Wengrzynek R.J., Terrell C.R. (1990) Using constructed wetlands to control agricultural nonpoint source pollution. In: *Constructed Wetlands for Water Pollution Control*, Cooper P.F., Findlater B. C. (eds.) Pergamon Press: Cambridge, United Kingdom.

Wepener V., van Vuren J.H.J., Du Preez H.H. (2000) Application of the equilibrium partitioning method to derive copper and zinc quality criteria for water and sediment: A South African perspective. *Water SA 26*(1): 97–104.

Werner T., Kadlec R.H. (1999b) Stochastic simulation of partially-mixed, event-driven treatment wetlands. *Ecological Engineering 14*(3): 253–267.

Werner T., Kadlec R.H. (1999a) Wetland residence time distribution modeling. *Ecological Engineering 15*(1–2): 77–90.

Werner T., Kadlec R.H. (2000) An improved model for wetland tracer analysis. Proceedings of the 7th International Conference on Wetland Systems for Water Pollution Control, 11–16 November 2000, Lake Buena Vista, FL, 639–646.

Werner T.M., Kadlec R.H. (1996) Application of residence time distributions to stormwater treatment systems. *Ecological Engineering 7*(3): 213–234.

Westall J., Stumm W. (1980) The hydrosphere. In: *The Handbook of Environmental Chemistry, Volume 1: Part A. The Natural Environment and the Biogeochemical Cycles*, Hutzinger O. (ed.) Springer-Verlag: Berlin, Germany, pp. 17–49.

Westerhoff P., Sharif F., Halden R., Herckes P., Krajmalnik-Brown R. (2014) *Constructed Wetlands for Treatment of Organic and Engineered Nanomaterial Contaminants of Emerging Concerns.* Water Research Foundation: Alexandria, VA.

Westermann P., Ahring B.K. (1987) Dynamics of methane production, sulfate reduction, and denitrification in a permanently waterlogged alder swamp. *Applied and Environmental Microbiology 53*(10): 2554–2559.

Westlake D.F. (1969) Macrophytes. In: *A Manual on Methods for Measuring Primary Production in Aquatic Environments (International Biological Program Handbook No. 12)*, Vollenweider R.A. (ed.) Blackwell Science Publishers: Oxford, United Kingdom, pp. 103–107.

Wetzel R.G. (1983) *Limnology.* Second edition. Saunders College Publishing: New York.

Wetzel R.G. (2001) *Limnology. Lake and River Ecosystems.* Third edition. Academic Press: San Diego, CA.

Wetzel R.G., Likens G.E. (1991) *Limnological Analyses.* Springer-Verlag: New York.

Whigham D.F., McCormick J., Good R.E., Simpson R.L. (1978) Biomass and primary production in freshwater tidal wetlands of the Middle Atlantic Coast. In: *Freshwater Wetlands: Ecological Processes and Management Potential*, Good R.E., Whigham D.F., Simpson R.L. (eds.) Academic Press: New York, pp. 3–20.

Whigham D.F., Simpson R.L., Good R.E., Sickels F.A. (1989) Decomposition and nutrient-metal dynamics of litter in freshwater tidal wetlands. Sharitz R.R., Gibbons J.W. (eds.) *Freshwater Wetlands and Wildlife.* Proceedings of a symposium (CONF-8603101), 24–27 March 1986; U.S. Department of Energy: Charleston, SC, pp. 167–188.

White J., Prenger J., Reddy K.R. (2002) The influence of fire on P removal efficiency of a constructed wetland in Florida. Paper at the 87th Annual Meeting of the Ecological Society of America, 4–9 August 2002; Ecological Society of America: Tucson, AZ.

White J.R., Gardner L.M., Sees M., Corstanje R. (2008) The short-term effects of prescribed burning on biomass removal and the release of nitrogen and phosphorus in a treatment wetland. *Journal of Environmental Quality 37*(6): 2386–2391.

White J.R., Reddy K.R. (2003) Nitrification and denitrification rates of everglades wetland soils along a phosphorus-impacted gradient. *Journal of Environmental Quality 36*(6): 2436–2443.

White J.R., Reddy K.R., Majer-Newman J. (2006) Hydrologic and vegetation effects on water column phosphorus in wetland mesocosms. *Soil Science Society of America Journal 70*(4): 1242–1251.

White J.R., Wang H., Jawitz J.W., Sees M.D. (2004) *Rejuvenating the largest treatment wetland in Florida: Tracer moment and model analysis of wetland hydraulic performance.* Paper presented at the American Geophysical Union Fall Meeting, San Francisco, CA.

Whitmer S. (1998) *Hydraulic Characterization of Treatment Wetlands to Improve Design Modeling and Treatment Efficiency.* MS Thesis. Arizona State University, Tempe, AZ.

Whitmer S., Baker L., Wass R. (2000) Loss of bromide in a wetland tracer experiment. *Journal of Environmental Quality 29*(6): 2043–2049.

Whitmire S.L., Hamilton S.K. (2005) Rapid removal of nitrate and sulfate in freshwater wetland sediments. *Journal of Environmental Quality 34*: 2062–2071.

Wichmann S. (2017) Commercial viability of paludiculture: A comparison of harvesting reeds for biogas production, direct combustion, and thatching. *Ecological Engineering 103*: 497–505.

Wieder R.K. (1989) A survey of constructed wetlands for acid coal mine drainage treatment in the eastern United States. *Wetlands 9*: 299–315.

Wieder R.K., Lang G.E. (1988) Cycling of inorganic and organic sulfur in peat from Big Run Bog, West Virginia. *Biogeochemistry 5*(2): 221–242.

Wienhold C.E., van der Valk A. (1989) The impact of duration of drainage on the seed banks of northern prairie wetlands. *Canadian Journal of Botany 67*(6): 1878–1884.

Wießner A., Kappelmeyer U., Kaestner M., Schultze-Nobre L., Kuschk P. (2013) Response of ammonium removal to growth and transpiration of Juncus effusus during the treatment of artificial sewage in laboratory-scale wetlands. *Water Research 47*(13): 4265–4273.

Wießner A., Kappelmeyer U., Kuschk P., Kästner M. (2005) Sulphate reduction and the removal of carbon and ammonia in a laboratory scale constructed wetland. *Water Research 39*(19): 4643–4650.

Wilcock R.J., Müller K., van Assema G.B., Bellingham M.A., Ovenden R. (2012) Attenuation of nitrogen, phosphorus and E. coli inputs from pasture runoff to surface waters by a farm wetland: The importance of wetland shape and residence time. *Water, Air, and Soil Pollution 223*(2): 499–509.

Wilcox D.A., Thompson T.A., Booth R.K., Nicholas J.R. (2007) *Lake-Level Variability and Water Availability in the Great Lakes.* Circular 1311, U.S. Geological Survey, Reston, VA.

Wilhelm M., Lawry S.R., Hardy D.D. (1989) Creation and management of wetlands using municipal wastewater in northern Arizona: A status report. In: *Constructed Wetlands for Wastewater Treatment; Municipal, Industrial and Agricultural*, Hammer D.A. (ed.) Lewis Publishers: Chelsea, MI, pp. 179–185.

Williams C.R., Jones R.D., Wright S.A. (1996) *Mosquito control in constructed wetlands.* Proceedings of WEFTEC '96, 69th Annual Conference and Exhibition; Water Environment Federation: Alexandria, VA; 333–344.

Williams L.E., Barnett M.O., Kramer T.A., Melville J.G. (2003) Adsorption and transport of arsenic (V) in experimental subsurface systems. *Journal of Environmental Quality 32*(3): 841–850.

Williams M. (1991) The human use of wetlands. *Progress in Human Geography 15*(1): 1–22.

Wilson C.A.M.E. (2007) Flow resistance models for flexible submerged vegetation. *Journal of Hydrology 342*(3): 213–222.

Wilson D., Alm J., Riutta T., Laine J., Byrne K.A., Farrell E.P., Tuittila E.S. (2007) A high resolution green area index for modelling the seasonal dynamics of CO 2 exchange in peatland vascular plant communities. *Plant Ecology 190*(1): 37–51.

Windham-Myers L., Fleck J.A., Ackerman J.T., Marvin-DiPasquale M., Stricker C.A., Heim W.A., Bachand P.A., Eagles-Smith C.A., Gill G., Stephenson M., Alpers C.N. (2014) Mercury cycling in agricultural and managed wetlands: A synthesis of methylmercury production, hydrologic export, and bioaccumulation from an integrated field study. *Science of the Total Environment 484*: 221–231.

Winston R.J., Hunt W.F., Kennedy S.G., Wright J.D., Lauffer M.S. (2011) Field evaluation of storm-water control measures for highway runoff treatment. *Journal of Environmental Engineering 138*(1): 101–111.

Winter H.C., Burris R.H. (1976) Nitrogenase. *Annual Review of Biochemistry 45*: 409–430.

Winter T.C. (1981) Uncertainties in estimating the water balance of lakes. *Water Resources Bulletin (Now Journal of the American Water Resources Association) 17*: 82–115.

Wojcik W., Wojcik M. (2000) Lead and Zinc Retention in the Biala River Wetland of Poland. In: *Heavy Metals in the Environment*, Odum H.T., Wojcik W., Pritchard L., Ton S., Delfino J.J., Wojcik M., Lesczynski S., Patel J.D., Doherty S.J., Stasik J. (eds.) CRC Press: Boca Raton, FL, pp. 97–109.

Wolfe J.E., Lind O.T. (2008) Influence of suspended clay on phosphorus uptake by periphyton. *Hydrobiologia 610*(1): 211–222.

Wong T.H. (2006) Water sensitive urban design-the journey thus far. *Australasian Journal of Water Resources 10*(3): 213–222.

Wong T.H.F., Breen P.F., Somes N.L.G., Lloyd S. (1999) *Managing Urban Stormwater Using Constructed Wetlands*, Cooperative Research Centre for Catchment Hydrology, Monash University: Clayton, Victoria, Australia. Available at: www.researchgate.net/profile/Peter_Breen2/publication/237341383_Managing_Urban_Stormwater_Using_Constructed_Wetlands/links/0c960533ea34ca8cf4000000/Managing-Urban-Stormwater-Using-Constructed-Wetlands.pdf (Accessed 8/8/18).

Wong T.H.F., Fletcher T.D., Duncan H.P., Jenkins G.A. (2006) Modelling urban stormwater treatment – A unified approach. *Ecological Engineering 27*(1): 58–70.

Wong T.H.F., Geiger W.F. (1997) Adaptation of wastewater surface flow wetland formulae for application in constructed stormwater wetlands. *Ecological Engineering 9*(3–4): 187–202.

Wood S.L., Wheeler E.F., Berghage R.D., Graves R.E. (1999) Temperature effects on wastewater nitrate removal in laboratory-scale constructed wetlands. *Transactions of the ASAE 42*(1): 185.

Woolway R.I., Jones I.D., Maberly S.C., French J.R., Livingstone D.M., Monteith D.T., Simpson G.L., Thackeray S.J., Andersen M.R., Battarbee R.W., DeGasperi C.L. (2016) Diel surface temperature range scales with lake size. *PLoS One 11*(3): e0152466.

Wörman A., Kronnäs V. (2005) Effect of pond shape and vegetation heterogeneity on flow and treatment performance of constructed wetlands. *Journal of Hydrology 301*: 123–138.

Worrall F., Burt T.P., Clay G.D., Moody C.S. (2015) A 19-year long energy budget of an upland peat bog, northern England. *Journal of Hydrology 520*: 17–29.

Worrall P., Peberdy K.J., Millett M.C. (1997) Constructed wetlands and nature conservation. *Water Science and Technology 35*(5): 205–213.

Wossink A., Hunt B. (2003) *The Economics of Structural Stormwater BMPs in North Carolina*, Report UNC-WRRI-2003-344 to the North Carolina Urban Water Consortium, Water Resources Research Institute of the University of North Carolina, via North Carolina State University, Raleigh, NC.

WPCF. (1990) *Natural Systems for Wastewater Treatment. Manual of Practice FD-16*. Water Pollution Control Federation (WPCF): Washington, DC.

Wren C.D., Bishop C.A., Stewart D.L., Barrett G.C. (1997) *Wildlife and Contaminants in Constructed Wetlands and Stormwater Ponds: Current State of Knowledge and Protocols for Monitoring Contaminant Levels and Effects in Wildlife*, Technical Report Series No. 269, Canadian Wildlife Service: Burlington, ON.

WRI (1992) *The Des Plaines River Wetlands Demonstration Project*, Report to USEPA, July 1992. Seven volumes, Wetlands Research Inc. (WRI): Chicago, IL.

Wrubleski D.A., Murkin H.R., van der Valk A.G., Nelson J.W. (1997) Decomposition of emergent macrophyte roots and rhizomes in a northern prairie marsh. *Aquatic Botany 58*(2): 121–134.

WSI (Wetland Solutions, Inc.). (2012) *Kanapaha Water Reclamation Facility Infiltrating Wetlands Demonstration Project Data Report (March 2009–October 2011)*. Prepared by Wetland Solutions, Inc. for Gainesville Regional Utilities, Inc. Available at: www.wetlandsolutionsinc.com/resources/ (Accessed 12/14/18).

Wu D., Hua Z. (2014) The effect of vegetation on sediment resuspension and phosphorus release under hydrodynamic disturbance in shallow lakes. *Ecological Engineering 69*: 55–62.

Wu J., Zhang J., Jia W., Xie H., Gu R.R., Li C., Gao B. (2009) Impact of COD/N ratio on nitrous oxide emission from microcosm wetlands and their performance in removing nitrogen from wastewater. *Bioresource Technology 100*(12): 2910–2917.

Wu M.-Y., Franz E.H., Chen S. (2000) *Spartina pectinata*: A candidate species for constructed treatment wetlands. In: *Wetlands and Remediation*, Means J.L., Hinchee R.E. (eds.) Battelle Press: Columbus, OH, pp. 133–139.

Wynn T.M., Liehr S.K. (2001) Development of a constructed subsurface-flow wetland simulation model. *Ecological Engineering 16*(4): 519–536.

Xiang X. (2001) *Treatment of Trinitrotoluene (Tnt)-Contaminated Wastewater*. PhD Dissertation, Technical University of Munich, Munich, Germany.

Xie L., Sapozhnikova Y., Bawardi O., Schlenk D. (2004) Evaluation of wetland and tertiary wastewater treatments for estrogenicity using in vivo and in vitro assays. *Archives of Environmental Contamination and Toxicology 48*(1): 81–86.

Xie Y. (2016) *Developing a California-Based Receiving Waterbody Model for Pesticide Product Registration Evaluation: Review of Model Theory.* Report 293, California Department of Pesticide Regulation, Scvaramento, CA

Xu S., Jaffe P.R. (2006) Effects of plants on the removal of hexavalent chromium in wetland plants. *Journal of Environmental Quality 35*: 334–341.

Xu X., Tian H., Pan Z., Thomas C.R. (2011) Modeling ecosystem responses to prescribed fires in a phosphorus-enriched Everglades wetland: II. Phosphorus dynamics and community shift in response to hydrological and seasonal scenarios. *Ecological Modelling 222*(23–24): 3942–3956.

Xue H.B., Jansen S., Prash A., Sigg L. (2001) Nickel speciation and complexation kinetics in freshwater by ligand exchange and DPCSV. *Environmental Science and Technology 35*(3): 539–546.

Xue Y., Kovacic D.A., David M.B., Gentry L.E., Mulvaney R.L., Lindau C.W. (1999) In situ measurements of denitrification in constructed wetlands. *Journal of Environmental Quality 28*(1): 263–269.

Yang W., Choi S.U. (2010) A two-layer approach for depth-limited open-channel flows with submerged vegetation. *Journal of Hydraulic Research 48*(4): 466–475.

Yang Z., Zhao Y., Xia X. (2012) Nitrous oxide emissions from Phragmites australis-dominated zones in a shallow lake. *Environmental Pollution 166*: 116–124.

Yates C.N., Wootton B.C., Murphy S.D. (2012) Performance assessment of arctic tundra municipal wastewater treatment wetlands through an arctic summer. *Ecological Engineering 44*: 160–173.

Yi Q., Yu J., Kim Y. (2010) Removal patterns of particulate and dissolved forms of pollutants in a stormwater wetland. *Water Science and Technology 61*(8): 2083–2096.

Yonika D., Lowry D., Hollands G., Mullica W., Smith G., Bigelow S. (1979) *Feasibility Study of Wetland Disposal Wastewater Treatment Plant Effluent*, Report to Massachusetts Water Resources Commission, Project 78-04.

Yoo J.-H., Ro H.-M., Choi W.-J., Yoo S.-H., Han K.-H. (2006) Phosphorus adsorption and removal by sediments of a constructed marsh in Korea. *Ecological Engineering 27*(2): 109–117.

Young C.B., McEnroe B.M., Rome A.C. (2009) Empirical determination of rational method runoff coefficients. *Journal of Hydrologic Engineering 14*(12): 1283–1289.

Young D. (2014) *The Variable Volume Water Model.* USEPA/OPP 734F14003 U.S. Environmental Protection Agency, Washington, DC.

Young T.C. (2003) *Cyanide Formation and Fate in Complex Effluents and Its Relation to Water Quality Criteria*, Final Report, Project 98-HHE-5, Water Environment Research Foundation (WERF): Alexandria, VA

Young T.F., Finley K., Adams W.J., Besser J., Hopkins W.A., Jolley D., McNaughton E., Presser T.S., Shaw D.P., Unrine J. (2010) What you need to know about selenium? In: *Ecological Assessment of Selenium in the Aquatic Environment*, Chapman P.M., Adams W.J., Brooks M.L., Delos C.G., Luoma S.N., Maher W.A., Ohlendorf H.M., Presser T.S., Shaw D.P. (eds.) SETAC Press: Pensacola, FL.

Younger P.L. (2000) The adoption and adaptation of passive treatment technologies for mine waters in the United Kingdom. *Mine Water and the Environment 19*: 84–97.

Younger P.L., Banwart S.A., Hedin R. (2002) *Mine Water: Hydrology, Pollution, Remediation.* Kluwer Academic Publishers: London, United Kingdom.

Zahraeifard V., Deng Z. (2011) Hydraulic residence time computation for constructed wetland design. *Ecological Engineering 37*(12): 2087–2091.

Zander M. (1980) Polycyclic aromatic and heteroaromatic hydrocarbons. In: *Handbook of Environmental Chemistry*, Hutzinger O. (ed.) Springer-Verlag: New York.

Zarate F.M. Jr., Schulwitz S.E., Stevens K.J., Venables B.J. (2012) Bioconcentration of triclosan, methyl-triclosan, and triclocarban in the plants and sediments of a constructed wetland. *Chemosphere 88*(3): 323–329.

Zeng L., Wu Y.H., Chen B., Zhao Y.J., Chen G.Q., Wu Z. (2012) Effect of wind on contaminant dispersion in a wetland flow dominated by free-surface effect. *Ecological Modelling 237–238*: 101–108.

Zeng T., Arnold W.A. (2012) Pesticide photolysis in prairie potholes: Probing photosensitized processes. *Environmental Science and Technology 47*(13): 6735–6745.

Zhang C., Gao X., Wang L., Chen X. (2015) Modelling the role of epiphyton and water level for submerged macrophyte development with a modified submerged aquatic vegetation model in a shallow reservoir in China. *Ecological Engineering 81*: 123–132.

Zhang C.B., Liu W.L., Han W.J., Guan M., Wang J., Liu S.Y., Ge Y., Chang J. (2017) Responses of dissimilatory nitrate reduction to ammonium and denitrification to plant presence, plant species and species richness in simulated vertical flow constructed wetlands. *Wetlands 37*(1): 109–122.

Zhang D., Gersberg R.M., Ng W.J., Tan S.K. (2014) Removal of pharmaceuticals and personal care products in aquatic plant-based systems: A review. *Environmental Pollution 184*: 620–639.

Zhang D.Q., Gersberg R.M., Hua T., Zhu J., Goyal M.K., Ng W.J., Tan S.K. (2013a) Fate of pharmaceutical compounds in hydroponic mesocosms planted with Scirpus validus. *Environmental Pollution 181*: 98–106.

Zhang L., Jacob D.J., Knipping E.M., Kumar N., Munger J.W., Carouge C.C., Van Donkelaar A., Wang Y.X., Chen D. (2012) Nitrogen deposition to the United States: Distribution, sources, and processes. *Atmospheric Chemistry and Physics 12*(10): 4539.

Zhang L., Wang M.H., Hu J., Ho Y.S. (2010) A review of published wetland research, 1991–2008: Ecological engineering and ecosystem restoration. *Ecological Engineering 36*(8): 973–980.

Zhang M., Cao T., Ni L., Xie P., Zhu G., Zhong A., Xu J., Fu H. (2011) Light-dependent phosphate uptake of a submersed macrophyte *Myriophyllum spicatum* L. *Aquatic Botany 94*(4): 151–157.

Zhang M., Li C.W., Shen Y. (2013b) Depth-averaged modeling of free surface flows in open channels with emerged and submerged vegetation. *Applied Mathematical Modelling 37*(1–2): 540–553.

Zhang Y., Frankenberger W.T. (2003) Factors affecting removal of selenate in agricultural drainage water utilizing rice straw. *Science of the Total Environment 305*(1–3): 207–216.

Zhang Y., Moore J.N. (1997) Reduction potential of selenate in wetland sediment. *Journal of Environmental Quality 20*(3): 910–916.

Zhang Y., Sun T., Li F., Wang J., Oh K. (2013c) Effect of decing salts on ion concentrations in urban stormwater runoff. *Procedia Environmental Sciences 18*: 567–571.

Zhao D, Xie D, Zhou H, Jiang H, An S. (2012) Estimation of leaf area index and plant area index of a submerged macrophyte canopy using digital photography. *PLoS One 7*(12): e51034.

Zhao H., Piccone T. (2017) *Appendix 5C-6: Summary Report for Stormwater Treatment Area 2 Flow-ways 1, 2, and 3 Water and Total Phosphorus Budget Analyses.* 2018 South Florida Environmental Report, South Florida Water Management District, West Palm Beach, FL.

Zheng S., Gu B., Zhou Q., Li Y. (2013) Variations of mercury in the inflow and outflow of a constructed treatment wetland in south Florida, USA. *Ecological Engineering 61*: 419–425.

Zheng Y., Wang X.C., Ge Y., Dzakpasu M., Zhao Y., Xiong J. (2015) Effects of annual harvesting on plants growth and nutrients removal in surface-flow constructed wetlands in northwestern China. *Ecological Engineering 83*: 268–275.

Zhou X., He Z., Jones K.D., Li L., Stoffella P.J. (2017) Dominating aquatic macrophytes for the removal of nutrients from waterways of the Indian River Lagoon basin, South Florida, USA. *Ecological Engineering 101*: 107–119.

Zhu G., Jetten M.S., Kuschk P., Ettwig K.F., Yin C. (2010a) Potential roles of anaerobic ammonium and methane oxidation in the nitrogen cycle of wetland ecosystems. *Applied Microbiology and Biotechnology 86*(4): 1043–1055.

Zhu X., Zuo Q., Shi J. (2010b) Analyzing soil soluble phosphorus transport with root-phosphorus-uptake applying an inverse method. *Agricultural Water Management 97*(2): 291–299.

Zoh K.-D., Horne A.J. (2000) Removal of TNT using plants in constructed wetlands. In: *Wetlands and Remediation: An International Conference*, Means J.L., Hinchee R.E. (eds.) Battelle Press: Columbus, OH, pp. 357–364.

Index

F

K

L

M

O

P

S

U

V

W

X

Y

Z